PREFACE

D0103188

This book is the product of a half century of innovation in physics education. When the first edition of *University Physics* by Francis W. Sears and Mark W. Zemansky appeared in 1949, it was revolutionary in its emphasis on the fundamental principles of physics and how to apply them. This Commemorative Tenth Edition continues to emphasize principles and applications as it provides today's students with a broad, rigorous, yet accessible introduction to calculus-based physics. The success of *University Physics* with generations of students and educators in all parts of the world is a testament to the merits of this approach.

Two key objectives guided the writing of this text: helping students develop physical intuition, and helping them build strong problem-solving skills. Also reflected throughout are the results of two decades of research in physics education on the conceptual pitfalls that commonly plague beginning physics students. These pitfalls include the notions that force is required for motion, that electric current is "used up" as it goes around a circuit, and that the product of a body's mass and its acceleration is itself a force. A key focus of this edition is to discuss not only the correct way to analyze a situation or solve a problem, but also the reason why the wrong way (which may have occurred to the student first) is indeed wrong.

The prose style of the book continues to be relaxed and conversational, without being colloquial or excessively familiar. We see the student as our partner in learning, not as an audience to be lectured to from atop a platform. This style makes it much easier for us to convey to the student our own excitement and enthusiasm for the beauty, intellectual challenge, and fundamental unity of physics.

In preparing the Tenth Edition, we have relied heavily on the comments of a great many faculty and students on how best to help them meet the challenges of physics education. Based on these comments, we have designed the following features of this edition.

A GUIDE FOR THE STUDENT

Many physics students experience difficulty simply because they don't know how to make the best use of their textbook. A section entitled "How to Succeed in Physics by Really Trying," which follows this preface, serves as a "user's manual" to all the features of this book. This section, written by Professor Mark Hollabaugh (Normandale Community College), also gives a number of helpful study hints. We strongly encourage *every* student to read this section!

CHAPTER ORGANIZATION

The *Introduction* to each chapter gives specific examples of the chapter's content and connects it with what has come before. At the end of each chapter is a *Summary* of the most important principles introduced in the chapter, along with the associated *Key Equations.* The summary also includes a list of *Key Terms* that the student should have learned to use, with references to the page on which each term is first introduced.

CONTENTS

Some of the most significant content features of this edition include:

- In Chapter 2, motion diagrams help students to distinguish between position, velocity, and acceleration in one-dimensional motion (see pp. 40–41).
- Chapter 12 has been updated with new data on the supermassive black hole at the center of our Milky Way galaxy.
- We discuss the microscopic interpretation of entropy in Chapter 18.
- A qualitative introduction to the ideas behind Gauss's law is given in Chapter 23.
- Chapter 28 on magnetic fields and forces explains the attraction and repulsion of magnets and magnetic materials.
- The discussion of electromagnetic induction in Chapter 30 and of inductance in Chapter 31 has been rewritten to make these essential but challenging concepts more accessible to students.
- Every chapter now includes a selection of photographs that illustrate how physical principles manifest themselves in the natural world and in our technological society.

QUESTIONS AND PROBLEMS

At the end of each chapter is a collection of *Discussion Questions,* intended to probe and extend the student's conceptual understanding, followed by an extensive set of problems. The problems have been revised and their number increased, including many new problems drawn from astrophysics, biology, and aerodynamics. Many problems have a conceptual part in which students must discuss and explain their results. The problems are grouped into *Exercises,* which are single-concept problems keyed to specific sections of the text; *Problems,* usually requiring two or more nontrivial steps; and *Challenge Problems,* intended to challenge the strongest students. Many new questions, exercises, and problems, especially for Chapters 38, 41, 42, 43, 44, and 45, were suggested by Professor A. Lewis Ford (Texas A&M University) and Professor Tom Sandin (North Carolina A&T State University).

PROBLEM-SOLVING STRATEGIES

Problem-Solving Strategy sections, an extremely popular feature of the book, have been retained and strengthened. They have proved to be a very substantial help, especially to the many earnest but bewildered students who "understood the material but couldn't do the problems." (See, for example, pp. 110, 121, and 171.)

EXAMPLES

Each *Problem-Solving Strategy* section is followed immediately by one or more worked-out examples that illustrate the strategy. Several of these are purely qualitative, such as Examples 6–6 (Comparing kinetic energies, p. 173), 8–1 (Momentum vs. kinetic energy, p. 230), and 18–7 (Isentropic processes, p. 576). Many examples are drawn from real-life situations relevant to the student's own experience. Units and correct significant figures in examples are always carried through all stages of numerical calculations.

Example solutions always begin with a statement of the general principles to be used and, when necessary, a discussion of the reason for choosing them. We emphasize modeling in physics, showing the student how to begin with a seemingly complex situation, make simplifying assumptions, apply the appropriate physical principles, and evaluate the final result. Does it make sense? Is it what you expected? How can you check it?

"CAUTION" PARAGRAPHS

In the text of each chapter we have labeled certain paragraphs with the word **CAUTION.** These paragraphs alert the student to common misconceptions or to points of potential confusion. (See, for example, pp. 102, 140, and 167.) We think of them as being similar to the flagged paragraphs in the user's manual for a power drill or a VCR, describing potential sources of trouble when using the equipment.

ACTIVPHYSICS LINKS

An important and unique supplement to *University Physics* is the set of *ActivPhysics I* and *ActivPhysics 2* CD-ROMs and workbooks, developed by Professors Alan Van Heuvelen and Paul D'Alessandris and published by Addison Wesley Longman. By combining carefully designed interactive simulations with proven pedagogy, *ActivPhysics* helps students become adept at solving problems about dynamic physical phenomena. Icons throughout the text of *University Physics* indicate which of the 200-plus exercises in *ActivPhysics* correspond to specific topics in this book. Exercises 1.1 through 10.10 appear in *ActivPhysics I,* while Exercises 11.1 through 20.4 are in *ActivPhysics 2.* For more information about the *ActivPhysics* CD-ROMs (compatible with both Macintosh and Windows) and workbooks, see below under "Supplements."

CASE STUDIES

We have included 10 optional sections called *Case Studies,* each building on the material of its chapter. Some (Neutrinos, Black Holes, Photons) emphasize connections between classical and modern physics. Others (Automotive Power, Energy Resources, Power Distribution Systems) have an engineering flavor; still others (Baseball Trajectories, Electric Potential Maps) emphasize computer simulations and include computer exercises for the student. All case studies have corresponding end-of-chapter problems.

NOTATION AND UNITS

Students often have a hard time keeping track of which quantities are vectors and which are not. In this edition, we use boldface italic symbols with an arrow on top for vector quantities, such as $\vec{v}$, $\vec{a}$, and $\vec{F}$; unit vectors have a caret on top, such as $\hat{\imath}$. Boldface $+$, $-$, $\times$, and $=$ signs are used in vector equations to emphasize the distinction between these operations and operations with ordinary numbers.

In this edition SI units are used exclusively. English unit conversions are included where appropriate. The joule is used as the standard unit of energy of all forms, including heat.

FLEXIBILITY

The book is adaptable to a wide variety of course outlines. There is plenty of material for an intensive three-semester or five-quarter course. Most instructors will find that there is too much material for a one-year course, but it is easy to tailor the book to a variety of one-year course plans by omitting certain chapters or sections. For example, any or all of the chapters on relativity, fluid mechanics, acoustics, electromagnetic waves, optical instruments, and several other topics can be omitted without loss of continuity. Some sections that are unusually challenging or somewhat out of the mainstream have been identified with an asterisk preceding the section title; these, too, may be omitted. In any case, no one should feel constrained to work straight through the entire book. We encourage instructors to select the chapters that fit their needs, omitting material that is not appropriate for the objectives of a particular course.

STANDARD, EXTENDED, AND SPLIT VERSIONS

This edition is available in three versions. The Standard version (ISBN 0–201–60322–5) includes 39 chapters, ending with the special theory of relativity. The Extended version (ISBN 0–201–60336–5) adds seven chapters on modern physics, including the physics of atoms, molecules, condensed matter, nuclei, and elementary particles. The Split version includes all 46 chapters in three softbound volumes: Volume 1, Chapters 1–21 (ISBN 0–201–60329–2); Volume 2, Chapters 22–39 (ISBN 0–201–60335–7); and Volume 3, Chapters 40–46 (ISBN 0–201–65663–9).

SUPPLEMENTS

For the Student: The *Online Course Companion Web site* (http://www.awlonline.com/young) offers problem solving tips, interactive quizzes, key concepts for each chapter of *University Physics,* a glossary, tips for success in physics, web links to applications of physical concepts, and much more.

The *ActivPhysics* CD-ROMs and workbooks, developed by Professors Alan Van Heuvelen and Paul D'Alessandris, use interactive simulations and multiple representations to help students become better physics problem-solvers. The CD-ROMs are compatible with both Macintosh and Windows. *ActivPhysics 1* (ISBN 0–201–69482–4) covers the topics of Chapters 1–21, and *ActivPhysics 2* (ISBN 0–201–36111–6) covers the material found in Chapters 22–46. As mentioned above, icons in the text of *University Physics* show the connections between topics in the book and exercises in *ActivPhysics.*

The *Study Guide,* prepared by Professors James R. Gaines and William F. Palmer, reinforces the text's emphasis on problem-solving strategies and student misconceptions. The *Study Guide for Volume 1* (ISBN 0–201–61835–4) covers Chapters 1–21, and the *Study Guide for Volumes 2 and 3* (ISBN 0–201–61834–6) covers Chapters 22–46.

The *Student Solutions Manual,* prepared by Professor A. Lewis Ford, includes completely worked-out solutions for about two-thirds of the odd-numbered problems in *University Physics.* (Answers to all odd-numbered problems are found in this book following the Appendices.) The *Student Solutions Manual for Volume 1* (ISBN 0–201–64394–4) covers Chapters 1–21, and the *Student Solutions Manual for Volumes 2 and 3* (ISBN 0–201–64395–2) covers Chapters 22–46.

For the Instructor: The *Instructor's Solutions Manual,* prepared by Professor Mark Hollabaugh and Dr. Thomas D. Gutierrez, contains worked-out solutions to all exercises, problems, and challenge problems. The *Instructor's Solutions Manual for Volume 1* covers Chapters 1–21 (ISBN 0–201–61836–2) and the *Instructor's Solution Manual for Volumes 2 and 3* covers Chapters 22–46 (ISBN 0–201–61837–0). It is also available as a cross-platform CD-ROM (ISBN 0–201–65679–5). With the CD-ROM, you can read, edit, and print any solutions you choose, as well as post them on your secure, password-protected class web site.

The *Instructor's Guide for an Active Learning Classroom* (ISBN 0–201–65676–0) offers quick strategies for tailoring your course to include active learning techniques. This supplement is ideal for instructors who want to integrate these techniques into their course, but do not have time to create a new teaching plan.

The *Online Course Companion Web site* (http://www.awlonline.com/young) makes it easy to put your course syllabus and assignments on the web and password-protect your course information. Through the site, students can submit assignments to you or your teaching assistants.

The *Instructor's Presentation CD-ROM* contains the full-color line art figures from the text. Images may be exported into other programs, such as PowerPoint.

The *Overhead Transparencies* (ISBN 0–201–61833–8) include 200 four-color figures from the text. These are on acetate for use on an overhead projector.

The *Test Item File* (ISBN 0–201–60344–6), written by Dr. Elliot Farber and Professor Michael Browne, includes multiple-choice and short-answer problems. The accompanying TestGen software (ISBN 0–201–65662–0), compatible with both Macintosh and Windows, makes it easy to edit these test items, assemble them into an exam, and generate an answer key.

ACKNOWLEDGMENTS

In this Commemorative Tenth Edition, we would like to thank the hundreds of reviewers and colleagues who have contributed valuable comments and suggestions over the life of this textbook.

Edward Adelson (Ohio State University), Ralph Alexander (University of Missouri at Rolla), J. G. Anderson, R. S. Anderson, Alex Azima (Lansing Community College), Dilip Balamore (Nassau Community College), Harold Bale (University of North Dakota), Arun Bansil (Northeastern University), John Barach (Vanderbilt University), J. D. Barnett, H. H. Barschall, Albert Bartlett (University of Colorado), Paul Baum (CUNY, Queens College), B. Bederson, Lev I. Berger (San Diego State University), Robert Boeke (William Rainey Harper College), S. Borowitz, A. C. Braden, James Brooks (Boston University), Nicholas E. Brown (California Polytechnic State University, San Luis Obispo), Tony Buffa (California Polytechnic State University, San Luis Obispo), A. Capecelatro, Michael Cardamone (Pennsylvania State University), Duane Carmony (Purdue University), P. Catranides, Roger Clapp (University of South Florida), William M. Cloud (Eastern Illinois University), Leonard Cohen (Drexel University), W. R. Coker (University of Texas, Austin), Malcolm D. Cole (University of Missouri at Rolla), H. Conrad, David Cook (Lawrence University), Gayl Cook (University of Colorado), Hans Courant (University of Minnesota), Bruce A. Craver (University of Dayton), Larry Curtis (University of Toledo), Jai Dahiya (Southeast Missouri State University), Steve Detweiler (University of Florida), George Dixon (Oklahoma State University), Donald S. Duncan, Boyd Edwards (West Virginia University), Robert Eisenstein (Carnegie-Mellon University), William Faissler (Northeastern University), William Fasnacht (U.S. Naval Academy), Paul Feldker (St. Louis Community College), L. H. Fisher, Neil Fletcher (Florida State University), Robert Folk, Peter Fong (Emory University), A. Lewis Ford (Texas A&M University), D. Frantszog, James R. Gaines (Ohio State University), Solomon Gartenhaus (Purdue University), Ron Gautreau (New Jersey Institute of Technology), J. David Gavenda (University of Texas, Austin), Dennis Gay (University of North Florida), James Gerhart (University of Washington), N. S. Gingrich, J. L. Glathart, S. Goodwin, Walter S. Gray (University of Michigan), Howard Grotch (Pennsylvania State University), John Gruber (San Jose State University), Graham D. Gutsche (U.S. Naval Academy), Michael J. Harrison (Michigan State University), Harold Hart (Western Illinois University), Howard Hayden (University of Connecticut), Carl Helrich (Goshen College), Laurent Hodges (Iowa State University), C. D. Hodgman, Michael Hones (Villanova University), Keith Honey (West Virginia Institute of Technology), Gregory Hood (Tidewater Community College), John Hubisz (North Carolina State University), M. Iona, Alvin Jenkins (North Carolina State University), Lorella Jones (University of Illinois), John Karchek (GMI Engineering & Management Institute), Thomas Keil (Worcester Polytechnic Institute), Robert Kraemer (Carnegie-Mellon University), Jean P. Krisch (University of Michigan), Robert A. Kromhout, Robert J. Lee, Alfred Leitner (Rensselaer Polytechnic University), Gerald P. Lietz (De Paul University), Gordon Lind (Utah State University), S. Livingston, Elihu Lubkin (University of Wisconsin, Milwaukee), Robert Luke (Boise State University), Michael Lysak (San Bernardino Valley College), Jeffrey Mallow (Loyola University), Robert

Mania (Kentucky State University), Robert Marchina (University of Memphis), David Markowitz (University of Connecticut), R. J. Maurer, Oren Maxwell (Florida International University), Joseph L. McCauley (University of Houston), T. K. McCubbin, Jr. (Pennsylvania State University), Charles McFarland (University of Missouri at Rolla), Lawrence McIntyre (University of Arizona), Fredric Messing (Carnegie-Mellon University), Thomas Meyer (Texas A&M University), Andre Mirabelli (St. Peter's College, New Jersey), Herbert Muether (S.U.N.Y., Stony Brook), Jack Munsee (California State University, Long Beach), Lorenzo Narducci (Drexel University), Van E. Neie (Purdue University), David A. Nordling (U.S. Naval Academy), L. O. Olsen, Jim Pannell (DeVry Institute of Technology), W. F. Parks (University of Missouri), Jerry Peacher (University of Missouri at Rolla), Arnold Perlmutter (University of Miami), Lennart Peterson (University of Florida), R. J. Peterson (University of Colorado, Boulder), R. Pinkston, Ronald Poling (University of Minnesota), J. G. Potter, C. W. Price (Millersville University), Francis Prosser (University of Kansas), Shelden H. Radin, Michael Rapport (Anne Arundel Community College), R. Resnick, James A. Richards, Jr., John S. Risley (North Carolina State University), Francesc Roig (University of California, Santa Barbara), T. L. Rokoske, Richard Roth (Eastern Michigan University), Carl Rotter (University of West Virginia), S. Clark Rowland (Andrews University), Rajarshi Roy (Georgia Institute of Technology), Russell A. Roy (Santa Fe Community College), Melvin Schwartz (St. John's University), F. A. Scott, L. W. Seagondollar, Stan Shepherd (Pennsylvania State University), Bruce Sherwood (Carnegie-Mellon University), Hugh Siefkin (Greenville College), C. P. Slichter, Charles W. Smith, Malcolm Smith (University of Lowell), Ross Spencer (Brigham Young University), Julien Sprott (University of Wisconsin), Victor Stanionis (Iona College), James Stith (American Institute of Physics), Edward Strother (Florida Institute of Technology), Conley Stutz (Bradley University), Albert Stwertka (U.S. Merchant Marine Academy), Martin Tiersten (CUNY, City College), David Toot (Alfred University), Somdev Tyagi (Drexel University), F. Verbrugge, Helmut Vogel (Carnegie-Mellon University), Thomas Weber (Iowa State University), M. Russell Wehr, Lester V. Whitney, Thomas Wiggins (Pennsylvania State University), George Williams (University of Utah), John Williams (Auburn University), Stanley Williams (Iowa State University), Jack Willis, Suzanne Willis (Northern Illinois University), Robert Wilson (San Bernardino Valley College), L. Wolfenstein, James Wood (Palm Beach Junior College), Lowell Wood (University of Houston), R. E. Worley, D. H. Ziebell (Manatee Community College), George O. Zimmerman (Boston University)

In addition, we both have individual acknowledgments we would like to make.

I want to extend my heartfelt thanks to my colleagues at Carnegie-Mellon, especially Professors Robert Kraemer, Bruce Sherwood, Helmut Vogel, and Brian Quinn, for many stimulating discussions about physics pedagogy and for their support and encouragement during the writing of this new edition. I am equally indebted to the many generations of Carnegie-Mellon students who have helped me learn what good teaching and good writing are, by showing me what works and what doesn't. It is always a joy and a privilege to express my gratitude to my wife Alice and our children Gretchen and Rebecca for their love, support, and emotional sustenance during the writing of this new edition. May all men and women be blessed with love such as theirs. — H. D. Y.

I would like to thank my past and present colleagues at UCSB, including Francesc Roig, Elisabeth Nicol, Al Nash, and Carl Gwinn, for their wholehearted support and for many helpful discussions. I owe a special debt of gratitude to my early teachers Willa Ramsay, Peter Zimmerman, William Little, Alan Schwettman, and Dirk Walecka for showing me what clear and engaging physics teaching is all about, and to Stuart Johnson for inviting me to join this project as a co-author. I am grateful to Nathan Palmer of the Colorado School of Mines for his careful checking of the page proofs for this edition. I want to thank my parents for their continued love and support and for keeping a space open on their bookshelf for this book. Most of all, I want to express my gratitude and love to my wife Caroline, to whom I dedicate my contributions to this book. Hey, Caroline, the new edition's done at last — let's go flying! — R. A. F.

PLEASE TELL US WHAT YOU THINK!

We welcome communications from students and professors, especially concerning errors or deficiencies that you find in this edition. We have devoted a lot of time and effort to writing the best book we know how to write, and we hope it will help you to teach and learn physics. In turn, you can help us by letting us know what still needs to be improved! Please feel free to contact us either by ordinary mail or electronically. Your comments will be greatly appreciated.

September 1999

Hugh D. Young
Department of Physics
Carnegie-Mellon University
Pittsburgh, Pennsylvania 15213
hdy+@andrew.cmu.edu

Roger A. Freedman
Department of Physics
University of California, Santa Barbara
Santa Barbara, California 93106-9530
airboy@physics.ucsb.edu
http://www.physics.ucsb.edu/~airboy/

HOW TO SUCCEED IN PHYSICS BY REALLY TRYING

Mark Hollabaugh, Normandale Community College

Physics encompasses the large and the small, the old and the new. From the atom to galaxies, from electrical circuitry to aerodynamics, physics is very much a part of the world around us. You probably are taking this introductory course in calculus-based physics because it is required for subsequent courses you plan to take in preparation for a career in science or engineering. Your professor wants you to learn physics and to enjoy the experience. He or she is very interested in helping you learn this fascinating subject. That is part of the reason your professor chose this textbook for your course. That is also the reason why Drs. Young and Freedman asked me to write this introductory section. We want you to succeed!

The purpose of this section of *University Physics* is to give you some ideas that will assist your learning. Specific suggestions on how to use the textbook will follow a brief discussion of general study habits and strategies.

PREPARATION FOR THIS COURSE

If you had high school physics, you will probably learn concepts faster than those who have not because you will be familiar with the language of physics. If English is a second language for you, keep a glossary of new terms that you encounter and make sure you understand how they are used in physics. Likewise, if you are farther along in your mathematics courses, you will pick up the mathematical aspects of physics faster. Even if your mathematics is adequate, you may find a book such as Arnold D. Pickar's *Preparing for General Physics: Math Skill Drills and Other Useful Help (Calculus Version)* to be useful. Your professor may actually assign sections of this math review to assist your learning.

LEARNING TO LEARN

Each of us has a different learning style and a preferred means of learning. Understanding your own learning style will help you to focus on aspects of physics that may give you difficulty and to use those components of your course that will help you overcome the difficulty. Obviously you will want to spend more time on those aspects that give you the most trouble. If you learn by hearing, lectures will be very important. If you learn by explaining, then working with other students will be useful to you. If solving problems is difficult for you, spend more time learning how to solve problems. Also, it is important to understand and develop good study habits. Perhaps the most important thing you can do for yourself is to set aside adequate, regularly scheduled, study time in a distraction-free environment.

Answer the following questions for yourself:

- Am I able to use fundamental mathematical concepts from algebra, geometry and trigonometry? (If not, plan a program of review with help from your professor.)
- In similar courses, what activity has given me the most trouble? (Spend more time on this.) What has been the easiest for me? (Do this first; it will help to build your confidence.)
- Do I understand the material better if I read the book before or after the lecture? (You may learn best by skimming the material, going to lecture, and then undertaking an in-depth reading.)
- Do I spend adequate time in studying physics? (A rule of thumb for a class like this is to devote, on the average, 2.5 hours out of class for each hour in class. For a course

meeting 5 hours each week, that means you should spend about 10 to 15 hours per week studying physics.)

- Do I study physics every day? (Spread that 10 to 15 hours out over an entire week!) At what time of the day am I at my best for studying physics? (Pick a specific time of the day and stick to it.)
- Do I work in a quiet place where I can maintain my focus? (Distractions will break your routine and cause you to miss important points.)

WORKING WITH OTHERS

Scientists or engineers seldom work in isolation from one another but rather work cooperatively. You will learn more physics and have more fun doing it if you work with other students. Some professors may formalize the use of cooperative learning or facilitate the formation of study groups. You may wish to form your own informal study group with members of your class who live in your neighborhood or dorm. If you have access to e-mail, use it to keep in touch with one another. Your study group is an excellent resource when reviewing for exams.

LECTURES AND TAKING NOTES

An important component of any college course is the lecture. In physics this is especially important because your professor will frequently do demonstrations of physical principles, run computer simulations, or show video clips. All of these are learning activities that will help you to understand the basic principles of physics. Don't miss lectures, and if for some reason you do, ask a friend or member of your study group to provide you with notes and let you know what happened.

Take your class notes in outline form, and fill in the details later. It can be very difficult to take word for word notes, so just write down key ideas. Your professor may use a diagram from the textbook. Leave a space in your notes and just add the diagram later. After class, edit your notes, filling in any gaps or omissions and noting things you need to study further. Make references to the textbook by page, equation number, or section number.

Make sure you ask questions in class, or see your professor during office hours. Remember the only "dumb" question is the one that is not asked. Your college may also have teaching assistants or peer tutors who are available to help you with difficulties you may have.

EXAMINATIONS

Taking an examination is stressful. But if you feel adequately prepared and are well-rested, your stress will be lessened. Preparing for an exam is a continual process; it begins the moment the last exam is over. You should immediately go over the exam and understand any mistakes you made. If you worked a problem and made substantial errors, try this: Take a piece of paper and divide it down the middle with a line from top to bottom. In one column, write the proper solution to the problem. In the other column, write what you did and why, if you know, and why your solution was incorrect. If you are uncertain why you made your mistake, and how to avoid it again, talk with your professor. Physics continually builds on fundamental ideas and it is important to correct any misunderstandings immediately. Warning: While cramming at the last minute may get you through the *present* exam, you will not adequately retain the concepts for use on the *next* exam.

USING YOUR TEXTBOOK

Now let's take a look at specific features of *University Physics* that will help you understand the concepts of physics. At its heart, physics is not equations and numbers. Physics is a way of looking at the universe and understanding how the universe works and how its various parts relate to each other. And although solving quantitative problems is an important part of physics, it is equally important for you to understand concepts qualitatively. Your textbook will help you in both areas.

First of all, don't be afraid to write in your book. It is more important for you to learn the concepts of physics than to keep your book in pristine condition. Write in the margins, make cross references. Take notes in your notebook as you read. *University Physics* is your primary "reference book" for this course. Refer to it often to help you understand the concepts you hear in lecture. Become familiar with the contents of the appendices and end papers.

> **CAUTION ▶** Please note that the quantity $m\vec{a}$ is *not* a force. All that Eqs. (4–7) and (4–8) say is that the vector $m\vec{a}$ is equal in magnitude and direction to the vector sum $\Sigma\vec{F}$ of all the forces acting on the body. It's incorrect to think of acceleration as a force; rather, acceleration is a result of a nonzero net force. It's "common sense" to think that there is a "force of acceleration" that pushes you back into your seat when your car accelerates forward from rest. But *there is no such force*; instead, your inertia causes you to tend to stay at rest relative to the earth, and the car accelerates around you. The "common sense" confusion arises from trying to apply Newton's second law in a frame of reference where it isn't valid, like the non-inertial reference frame of an accelerating car. We will always examine motion relative to *inertial* frames of reference only. ◀
>
> In learning how to use Newton's second law, we will begin in this chapter with examples of straight-line motion. Then in Chapter 5 we will consider more general cases and develop more detailed problem-solving strategies for applying Newton's laws of motion.

CAUTION!

Educational research has found numerous misconceptions or misunderstandings that students frequently have when they study physics. Dr. Freedman has added *Caution!* paragraphs to warn you about these potential pitfalls. Heed them!

WORKED EXAMPLES

Your professor will work example problems in class to illustrate the application of the concepts of physics to real-world problems. You should work through all the examples in the textbook, filling in any missing steps, and making note of things you don't understand. Get help with the concepts that confuse you!

EXAMPLE 6–6

Comparing kinetic energies Two iceboats like the one in Example 5–5 (Section 5–3) hold a race on a frictionless horizontal lake (Fig. 6–9). The two iceboats have masses m and $2m$, respectively. Each iceboat has an identical sail, so the wind exerts the same constant force $\vec{F}$ on each iceboat. The two iceboats start from rest and cross the finish line a distance s away. Which iceboat crosses the finish line with greater kinetic energy?

SOLUTION If you simply use the mathematical definition of kinetic energy, $K = \frac{1}{2}mv^2$ (Eq. (6–5)), the answer to this problem isn't immediately obvious. The iceboat of mass $2m$ has greater mass, so you might guess that the larger iceboat attains a greater

kinetic energy at the finish line. But the smaller iceboat, of mass m, will cross the finish line with a greater speed, and you might guess that *this* iceboat will have the greater kinetic energy. How can we decide?

The correct way to approach this problem is to remember that *the kinetic energy of a particle is just equal to the total work done to accelerate it from rest.* Both iceboats travel the same distance s, and only the horizontal force F in the direction of motion does work on either iceboat. Hence the total work done between the starting line and the finish line is the *same* for each iceboat, $W_{tot} = Fs$. At the finish line, each iceboat has a kinetic energy equal to the work W_{tot} done on it, because each iceboat started from rest. So both iceboats have the *same* kinetic energy at the finish line!

Start   Finish   **6–9** A race between iceboats.

Problem–Solving Strategy

PROBLEMS USING MECHANICAL ENERGY

1. First decide whether the problem should be solved by energy methods, by using $\Sigma \vec{F} = m\vec{a}$ directly, or by a combination. The energy approach is particularly useful when the problem involves motion with varying forces, motion along a curved path (discussed later in this section), or both. But if the problem involves elapsed time, the energy approach is usually *not* the best choice because this approach doesn't involve time directly.

2. When using the energy approach, first decide what the initial and final states (the positions and velocities) of the system are. Use the subscript 1 for the initial state and the subscript 2 for the final state. It helps to draw sketches showing the initial and final states.

3. Define your coordinate system, particularly the level at which $y = 0$. You will use this to compute gravitational potential energies. Equation (7–2) assumes that the positive direction for y is upward; we suggest that you use this choice consistently.

4. List the initial and final kinetic and potential energies, that is, K_1, K_2, U_1, and U_2. In general, some of these will be known and some will be unknown. Use algebraic symbols for any unknown coordinates or velocities.

5. Identify all nongravitational forces that do work. A free-body diagram is always helpful. Calculate the work W_{other} done by all these forces. If some of the quantities you need are unknown, represent them by algebraic symbols.

6. Relate the kinetic and potential energies and the nongravitational work W_{other} using Eq. (7–7). If there is no nongravitational work, this becomes Eq. (7–4). It's helpful to draw bar graphs showing the initial and final values of K, U, and $E = K + U$. Then solve to find whatever unknown quantity is required.

7. Keep in mind, here and in later sections, that the work done by each force must be represented either in $U_1 - U_2 = -\Delta U$ or as W_{other}, but *never* in both places. The gravitational work is included in ΔU, so do not include it again in W_{other}.

PROBLEM-SOLVING STRATEGIES

One of the features of *University Physics* that first caught my eye as a teacher were the *Problem-Solving Strategy* boxes. This is the advice I would give to a student who came to me for help with a physics problem. Physics teachers approach a problem in a very systematic and logical manner. These boxes will help you as a beginning problem solver to do the same. Study these suggestions in great detail and implement them. In many cases these strategy boxes will tell you *how* to visualize an abstract concept.

CASE STUDIES

Physics relates to the real world, and these *Case Studies* will give you examples of applying physics to real science or engineering problems.

6-6 AUTOMOTIVE POWER

A Case Study in Energy Relations

The power requirements of a gasoline-powered automobile are an important and practical example of the concepts in this chapter. If roads were frictionless and air resistance didn't exist, there would be no need for an automobile to have an engine. All you'd need to go for a drive would be a few strong friends to give you a push to get started and a few other friends at your destination to stop you. (Steering on frictionless roads would be a problem, though.) In the real world, however, a moving car without an engine slows down because of forces that resist its motion. The engine's function is to continuously provide power to overcome this resistance. So to understand how much power is required from a car's engine, we must analyze the forces that act on the car.

Two forces oppose the motion of an automobile: rolling friction and air resistance. We described rolling friction in Section 5–4 in terms of a coefficient of rolling friction μ_r. A typical value of μ_r for properly inflated tires on hard pavement is 0.015. A Porsche 911 Carrera has a mass of 1251 kg and a weight of $(1251 \text{ kg})(9.80 \text{ m/s}^2) = 12,260 \text{ N}$, and

SUMMARY, REVIEW QUESTIONS AND PROBLEMS

The most important concepts are listed in the *Key Terms*. Keep a glossary of terms in your notebook. Your professor may indicate through the use of course objectives which terms are important for you to know. The *Summary* will give you a quick review of the chapter's main ideas and the equations that represent those ideas mathematically. Everything else can be derived from these general equations. If your professor assigns *Problems* at the end of the chapter, make sure you work them carefully with other students. If solutions are available, do not look at the answer until you have struggled with the problem and compared your answer with someone else's. If the two of you agree on the answer, then look at the solution. If you have made a mistake, go back and rework the problem. Do not simply read the problem. You will note that the *Exercises* are keyed to specific sections of the chapter and are easier. Work on these before you attempt the *Problems* or *Challenge Problems* which typically use multiple concepts.

Well, there you have it. We hope these suggestions will benefit your study of physics. Strive for understanding and excellence, and be persistent in your learning.

CONTENTS

Note: Chapters 1-21 appear in Volume 1; Chapters 22-39 appear in Volume 2; and Chapters 40-46 appear in Volume 3.

WAVES/ACOUSTICS

ELECTROMAGNETISM

ABOUT THE AUTHORS

Hugh D. Young is Professor of Physics at Carnegie-Mellon University in Pittsburgh, PA. He attended Carnegie-Mellon for both undergraduate and graduate study and earned his Ph.D. in fundamental particle theory under the direction of the late Richard Cutkosky. He joined the faculty of Carnegie-Mellon in 1956, and has also spent two years as a Visiting Professor at the University of California at Berkeley.

Prof. Young's career has centered entirely around undergraduate education. He has written several undergraduate-level textbooks, and in 1973 he became a co-author with Francis Sears and Mark Zemansky for their well-known introductory texts. With their deaths, he assumed full responsibility for new editions of these books, most recently the eighth edition of *University Physics*.

Prof. Young is an enthusiastic skier, climber, and hiker. He also served for several years as Associate Organist at St. Paul's Cathedral in Pittsburgh, and has played numerous organ recitals in the Pittsburgh area. Prof. Young and his wife Alice usually travel extensively in the summer, especially in Europe and in the desert canyon country of southern Utah.

Roger Freedman is a Lecturer in Physics at the University of California, Santa Barbara. He was an undergraduate at the University of California campuses in San Diego and Los Angeles, and did his doctoral research in nuclear theory at Stanford University under the direction of Professor J. Dirk Walecka. Dr. Freedman came to UCSB in 1981 after three years teaching and doing research at the University of Washington.

At UCSB, Dr. Freedman teaches in both the Department of Physics and the College of Creative Studies, a branch of the university intended for highly gifted and motivated undergraduates. He has published research in nuclear physics, elementary particle physics, and laser physics. In recent years he has helped to develop computer-based tools for learning introductory physics and astronomy.

Dr. Freedman holds a commercial pilot's license, and when not teaching or writing he can frequently be found flying with his wife Caroline.

T. R. Sandin is Professor of Physics at North Carolina A&T State University. He received his B.S. in physics from Santa Clara University and his M.S. and Ph.D. from Purdue University. He has received awards for excellence in teaching from both Purdue University and NC A&T. He has published research articles in low-temperature solid state physics, the Mössbauer effect, ferromagnetic anisotropy, and physics teaching, and is also the author of *Essentials of Modern Physics* (Addison-Wesley).

A. Lewis Ford is Professor of Physics at Texas A&M University. He received a B.A. from Rice University in 1968 and a Ph.D. in chemical physics from the University of Texas at Austin in 1972. After a one-year postdoc at Harvard University, he joined the Texas A&M physics faculty in 1973 and has been there ever since. Professor Ford's research area is theoretical atomic physics, with a specialization in atomic collisions. At Texas A&M he has taught a variety of undergraduate and graduate courses, but primarily introductory physics.

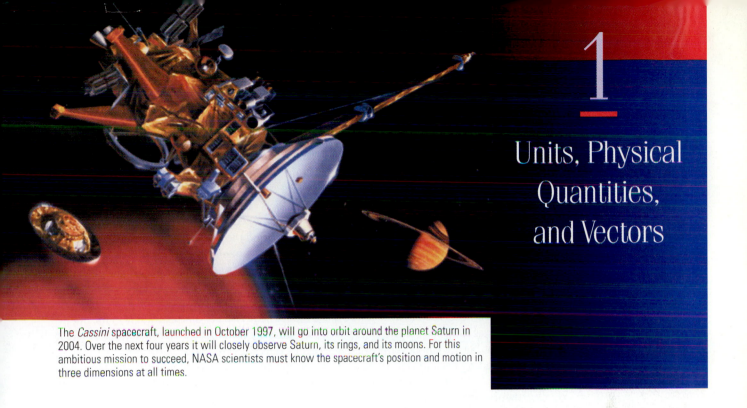

1

Units, Physical Quantities, and Vectors

The *Cassini* spacecraft, launched in October 1997, will go into orbit around the planet Saturn in 2004. Over the next four years it will closely observe Saturn, its rings, and its moons. For this ambitious mission to succeed, NASA scientists must know the spacecraft's position and motion in three dimensions at all times.

1-1 INTRODUCTION

Why study physics? For two reasons. First, physics is one of the most fundamental of the sciences. Scientists of all disciplines make use of the ideas of physics, from chemists who study the structure of molecules to paleontologists who try to reconstruct how dinosaurs walked. Physics is also the foundation of all engineering and technology. No engineer could design any kind of practical device without first understanding the basic principles involved. To design a spacecraft or a better mousetrap, you have to understand the basic laws of physics.

But there's another reason. The study of physics is an adventure. You will find it challenging, sometimes frustrating, occasionally painful, and often richly rewarding and satisfying. It will appeal to your sense of beauty as well as to your rational intelligence. Our present understanding of the physical world has been built on the foundations laid by scientific giants such as Galileo, Newton, Maxwell, and Einstein, and their influence has extended far beyond science to affect profoundly the ways in which we live and think. You can share some of the excitement of their discoveries when you learn to use physics to solve practical problems and to gain insight into everyday phenomena. If you've ever wondered why the sky is blue, how radio waves can travel through empty space, or how a satellite stays in orbit, you can find the answers by using fundamental physics. Above all, you will come to see physics as a tow-

ering achievement of the human intellect in its quest to understand our world and ourselves.

In this opening chapter, we'll go over some important preliminaries that we'll need throughout our study. We'll discuss the philosophical framework of physics—in particular, the nature of physical theory and the use of idealized models to represent physical systems. We'll introduce the systems of units used to describe physical quantities and discuss ways to describe the accuracy of a number. We'll look at examples of problems for which we can't (or don't want to) find a precise answer, but for which rough estimates can be useful and interesting. Finally, we'll study several aspects of vectors and vector algebra. Vectors will be needed throughout our study of physics to describe and analyze physical quantities, such as velocity and force, that have direction as well as magnitude.

1-2 THE NATURE OF PHYSICS

Physics is an *experimental* science. Physicists observe the phenomena of nature and try to find patterns and principles that relate these phenomena. These patterns are called physical theories or, when they are very well established and of broad use, physical laws or principles. The development of physical theory requires creativity at every stage. The physicist has to learn to ask appropriate questions, design experiments to try to answer the questions, and draw appropriate conclusions from the results. Figure 1–1 shows two famous experimental facilities.

According to legend, Galileo Galilei (1564–1642) dropped light and heavy objects from the top of the Leaning Tower of Pisa (Fig. 1–1a) to find out whether their rates of fall were the same or different. Galileo recognized that only experimental investigation could answer this question. From examining the results of his experiments (which were actually much more sophisticated than in the legend), he made the inductive leap to the principle, or theory, that the acceleration of a falling body is independent of its weight.

The development of physical theories such as Galileo's is always a two-way process that starts and ends with observations or experiments. This development often takes an indirect path, with blind alleys, wrong guesses, and the discarding of unsuccessful theories in favor of more promising ones. Physics is not simply a collection of facts and principles; it is also the *process* by which we arrive at general principles that describe how the physical universe behaves.

No theory is ever regarded as the final or ultimate truth. The possibility always exists that new observations will require that a theory be revised or discarded. It is in the nature of physical theory that we can disprove a theory by finding behavior that is inconsistent with it, but we can never prove that a theory is always correct.

Getting back to Galileo, suppose we drop a feather and a cannonball. They certainly do *not* fall at the same rate. This does not mean that Galileo was wrong; it means that his theory was incomplete. If we drop the feather and the cannonball *in a vacuum* to eliminate the effects of the air, then they do fall at the same rate. Galileo's theory has a **range of validity:** it applies only to objects for which the force exerted by the air (due to air resistance and buoyancy) is much less than the weight. Objects like feathers or parachutes are clearly outside this range.

Every physical theory has a range of validity outside of which is not applicable. Often a new development in physics extends a principle's range of validity. Galileo's analysis of falling bodies was greatly extended half a century later by Newton's laws of motion and law of gravitation.

An essential part of the interplay of theory and experiment is learning how to apply physical principles to a variety of practical problems. At various points in our study we

1–1 Two research laboratories. (a) The Leaning Tower in Pisa, Italy. According to legend, Galileo studied the motion of freely falling bodies by dropping them from the tower. He is also said to have studied pendulum motion by observing the swinging of the chandelier in the cathedral behind the tower. (b) The Hubble Space Telescope is the first major telescope to operate outside the earth's atmosphere. Its sensitive instruments measure visible, ultraviolet, and near-infrared light from astronomical objects. The Space Telescope has been used to study phenomena ranging from eruptions on the moons of Jupiter to the cores of distant galaxies. It is shown here in December 1993 after being repaired in orbit by the crew of the Space Shuttle *Endeavour.*

will discuss systematic problem-solving strategies that will help you set up and solve problems efficiently and accurately. Learning to solve problems is absolutely essential; you don't *know* physics unless you can *do* physics. This means not only learning the general principles, but also learning how to apply them in specific situations.

1–3 IDEALIZED MODELS

In everyday conversation we often use the word "model" to mean either a small-scale replica, such as a model railroad, or a person who displays articles of clothing (or the absence thereof). In physics a **model** is a simplified version of a physical system that would be too complicated to analyze in full detail.

Here's an example. Suppose we want to analyze the motion of a baseball thrown through the air. How complicated is this problem? The ball is neither perfectly spherical nor perfectly rigid; it has raised seams, and it spins as it moves through the air. Wind and air resistance influence its motion, the earth rotates beneath it, the ball's weight varies a little as its distance from the center of the earth changes, and so on. If we try to include all these things, the analysis gets hopelessly complicated. Instead, we invent a simplified version of the problem. We neglect the size and shape of the ball by representing it as a point object, or **particle.** We neglect air resistance by making the ball move in a vacuum, we forget about the earth's rotation, and we make the weight constant. Now we have a problem that is simple enough to deal with. We will analyze this model in detail in Chapter 3.

To make an idealized model of the system, we have to overlook quite a few minor effects to concentrate on the most important features of the system. Of course, we have to be careful not to neglect too much. If we ignore the effects of gravity completely, then our model predicts that when we throw the ball up, it goes in a straight line and disappears into space. We need to use some judgment and creativity to construct a model that simplifies a problem enough to make it manageable, yet keeps its essential features.

When we use a model to predict how a system will behave, the validity of our predictions is limited by the validity of the model. Going back to Galileo once more, we see that his prediction about falling bodies corresponds to an idealized model that does not include the effects of air resistance. This model works fairly well for a dropped cannonball, but not so well for a feather.

The concept of idealized models is extremely important in all physical science and technology. When we apply physical principles to complex systems, we always use idealized models, and we have to be aware of the assumptions we are making. In fact, the principles of physics themselves are stated in terms of idealized models; we speak about point masses, rigid bodies, ideal insulators, and so on. Idealized models play a crucial role throughout this book. Watch for them in discussions of physical theories and their applications to specific problems.

1–4 STANDARDS AND UNITS

Physics is an experimental science. Experiments require measurements, and we usually use numbers to describe the results of measurements. Any number that is used to describe a physical phenomenon quantitatively is called a **physical quantity.** For example, two physical quantities that describe you are your weight and your height. Some physical quantities are so fundamental that we can define them only by describing how to measure them. Such a definition is called an **operational definition.** Some examples are measuring a distance by using a ruler, and measuring a time interval by using a stopwatch. In other cases we define a physical quantity by describing how to calculate it from other quantities that we *can* measure. Thus we might define the average speed of a moving object as the distance traveled (measured with a ruler) divided by the time of travel (measured with a stopwatch).

When we measure a quantity, we always compare it with some reference standard. When we say that a Porsche 944 is 4.29 meters long, we mean that it is 4.29 times as long as a meter stick, which we define to be 1 meter long. Such a standard defines a **unit** of the quantity. The meter is a unit of distance, and the second is a unit of time. When we use a number to describe a physical quantity, we must always specify the unit that we are using; to describe a distance simply as "4.29" wouldn't mean anything.

To make accurate, reliable measurements, we need units of measurement that do not change and that can be duplicated by observers in various locations. The system of units used by scientists and engineers around the world is commonly called "the metric system," but since 1960 it has been known officially as the **International System,** or **SI** (the abbreviation for its French name, *Système International*). A list of all SI units is given in Appendix A, as are definitions of the most fundamental units.

The definitions of the basic units of the metric system have evolved over the years. When the metric system was established in 1791 by the French Academy of Sciences, the meter was defined as one ten-millionth of the distance from the North Pole to the equator (Fig. 1–2). The second was defined as the time required for a pendulum one meter long to swing from one side to the other. These definitions were cumbersome and hard to duplicate precisely, and by international agreement they have been replaced with more refined definitions.

1–2 In 1791 the distance from the North Pole to the equator was defined to be exactly 10^7 m. With the modern definition of the meter this distance is about 0.02% more than 10^7 m.

TIME

From 1889 until 1967, the unit of time was defined as a certain fraction of the mean solar day, the average time between successive arrivals of the sun at its highest point in the sky. The present standard, adopted in 1967, is much more precise. It is based on an atomic clock, which uses the energy difference between the two lowest energy states of the cesium atom. When bombarded by microwaves of precisely the proper frequency, cesium atoms undergo a transition from one of these states to the other. One **second** is defined as the time required for 9,192,631,770 cycles of this radiation.

LENGTH

In 1960 an atomic standard for the meter was also established, using the wavelength of the orange-red light emitted by atoms of krypton (^{86}Kr) in a glow discharge tube. In November 1983 the length standard was changed again, in a more radical way. The speed of light in a vacuum was *defined* to be precisely 299,792,458 m/s. The meter is defined to be consistent with this number and with the above definition of the second. Hence the new definition of the **meter** is the distance that light travels in a vacuum in 1/299,792,458 second. This provides a much more precise standard of length than the one based on a wavelength of light.

MASS

The standard of mass, the **kilogram,** is defined to be the mass of a particular cylinder of platinum-iridium alloy. That cylinder is kept at the International Bureau of Weights and Measures at Sèvres, near Paris. An atomic standard of mass would be more fundamental, but at present we cannot measure masses on an atomic scale with as much accuracy as on a macroscopic scale. The *gram* (which is not a fundamental unit) is 0.001 kilogram.

UNIT PREFIXES

Once we have defined the fundamental units, it is easy to introduce larger and smaller units for the same physical quantities. In the metric system these other units are related to the fundamental units (or, in the case of mass, to the gram) by multiples of 10 or $\frac{1}{10}$. Thus one kilometer (1 km) is 1000 meters, and one centimeter (1 cm) is $\frac{1}{100}$ meter. We usually express multiples of 10 or $\frac{1}{10}$ in exponential notation: $1000 = 10^3$, $\frac{1}{1000} = 10^{-3}$, and so on. With this notation, $1 \text{ km} = 10^3 \text{ m}$ and $1 \text{ cm} = 10^{-2} \text{ m}$.

The names of the additional units are derived by adding a **prefix** to the name of the fundamental unit. For example, the prefix "kilo-," abbreviated k, always means a unit larger by a factor of 1000; thus

$$1 \text{ kilometer} = 1 \text{ km} = 10^3 \text{ meters} = 10^3 \text{ m},$$
$$1 \text{ kilogram} = 1 \text{ kg} = 10^3 \text{ grams} = 10^3 \text{ g},$$
$$1 \text{ kilowatt} = 1 \text{ kW} = 10^3 \text{ watts} = 10^3 \text{ W}.$$

A table on the inside back cover of this book lists the standard SI prefixes, with their meanings and abbreviations.

Here are several examples of the use of multiples of 10 and their prefixes with the units of length, mass, and time.

LENGTH

$$1 \text{ nanometer} = 1 \text{ nm} = 10^{-9} \text{ m (a few times the size of the largest atom)}$$
$$1 \text{ micrometer} = 1 \text{ } \mu\text{m} = 10^{-6} \text{ m (size of some bacteria and living cells)}$$
$$1 \text{ millimeter} = 1 \text{ mm} = 10^{-3} \text{ m (diameter of the point of a ballpoint pen)}$$
$$1 \text{ centimeter} = 1 \text{ cm} = 10^{-2} \text{ m (diameter of your little finger)}$$
$$1 \text{ kilometer} = 1 \text{ km} = 10^{3} \text{ m (a 10-minute walk)}$$

MASS

$$1 \text{ microgram} = 1 \text{ } \mu\text{g} = 10^{-6} \text{ g} = 10^{-9} \text{ kg (mass of a very small dust particle)}$$
$$1 \text{ milligram} = 1 \text{ mg} = 10^{-3} \text{ g} = 10^{-6} \text{ kg (mass of a grain of salt)}$$
$$1 \text{ gram} = 1 \text{ g} = 10^{-3} \text{ kg (mass of a paper clip)}$$

TIME

$$1 \text{ nanosecond} = 1 \text{ ns} = 10^{-9} \text{ s (time for light to travel 0.3 m)}$$
$$1 \text{ microsecond} = 1 \text{ } \mu\text{s} = 10^{-6} \text{ s (time for an orbiting space shuttle to travel 8 mm)}$$
$$1 \text{ millisecond} = 1 \text{ ms} = 10^{-3} \text{ s (time for sound to travel 0.35 m)}$$

THE BRITISH SYSTEM

Finally, we mention the British system of units. These units are used only in the United States and a few other countries, and in most of these they are being replaced by SI units. British units are now officially defined in terms of SI units, as follows:

Length: 1 inch = 2.54 cm (exactly)

Force: 1 pound = 4.448221615260 newtons (exactly)

The newton, abbreviated N, is the SI unit of force. The British unit of time is the second, defined the same way as in SI. In physics, British units are used only in mechanics and thermodynamics; there is no British system of electrical units.

A mission to Mars involves various measurement scales. We measure the distance from the earth to Mars in hundreds of millions of kilometers, the landing spot is targeted to within tens of kilometers, and a mobile rover (like the *Sojourner* rover from the *Mars Pathfinder* mission, shown here) can move tens of meters in an hour. For a link to NASA's exploration of Mars, visit the *University Physics* web site.

Many everyday items make use of both SI and British units. An example is this speedometer from a U.S.-built automobile, which shows the speed in both kilometers per hour and miles per hour.

In this book we use SI units for all examples and problems, but we occasionally give approximate equivalents in British units. As you do problems using SI units, you may also wish to convert to the approximate British equivalents if they are more familiar to you. But you should try to think directly in SI as much as you can.

1–5 UNIT CONSISTENCY AND CONVERSIONS

We use equations to express relationships among physical quantities that are represented by algebraic symbols. Each algebraic symbol always denotes both a number and a unit. For example, d might represent a distance of 10 m, t a time of 5 s, and v a speed of 2 m/s.

An equation must always be **dimensionally consistent.** You can't add apples and automobiles; two terms may be added or equated only if they have the same units. For example, if a body moving with constant speed v travels a distance d in a time t, these quantities are related by the equation

$$d = vt. \tag{1–1}$$

If d is measured in meters, then the product vt must also be expressed in meters. Using the above numbers as an example, we may write

$$10 \text{ m} = \left(2\,\frac{\text{m}}{\text{s}}\right)(5\,\text{s}).$$

Because the unit 1/s on the right side of the equation cancels the unit s, the product vt has units of meters, as it must. In calculations, units are treated just like algebraic symbols with respect to multiplication and division.

CAUTION ▶ When a problem requires calculations using numbers with units, *always* write the numbers with the correct units and carry the units through the calculation as in the example above. This provides a very useful check for calculations. If at some stage in a calculation you find that an equation or an expression has inconsistent units, you know you have made an error somewhere. In this book we will always carry units through all calculations, and we strongly urge you to follow this practice when you solve problems. ◀

Problem–Solving Strategy

UNIT CONVERSIONS

1. Units are multiplied and divided just like ordinary algebraic symbols. This gives us an easy way to convert a quantity from one set of units to another. The key idea is that we can express the same physical quantity in two different units and form an equality. For example, when we say that 1 min = 60 s, we don't mean that the number 1 is equal to the number 60; we mean that 1 min represents the same physical time interval as 60 s. For this reason, the ratio (1 min)/(60 s) equals 1, as does its reciprocal (60 s)/(1 min). We can multiply a quantity by either of these factors without changing that quantity's physical meaning. To find the number of seconds in 3 min, we write

$$3 \text{ min} = (3 \text{ min})\left(\frac{60 \text{ s}}{1 \text{ min}}\right) = 180 \text{ s}.$$

2. If you do your unit conversions correctly, unwanted units will cancel, as in the example above. If instead you had multiplied 3 min by (1 min)/(60 s), your result would have been $\frac{1}{20}$ min^2/s, which is a rather odd way of measuring time. To be sure you convert units properly, you must write down the units at *all* stages of the calculation.

3. As a final check, ask yourself whether your answer is reasonable. Is the result 3 min = 180 s reasonable? The answer is yes; the second is a smaller unit than the minute, so there are more seconds than minutes in the same time interval.

EXAMPLE 1–1

Converting speed units The official world land speed record is 1228.0 km/h, set on October 15, 1997, by Andy Green in the jet engine car *Thrust SSC*. Express this speed in meters per second.

SOLUTION The prefix k means 10^3, so the speed 1228.0 km/h = 1228.0×10^3 m/h. We also know that there are 3600 s in 1 h. So we must combine the speed of 1228.0×10^3 m/h and a factor of 3600. But should we multiply or divide by this factor? If we treat the factor as a pure number without units, we're forced to guess how to proceed.

The correct approach is to carry the units with each factor. We then arrange the factor so that the hour unit cancels:

$$1228.0 \text{ km/h} = \left(1228.0 \times 10^3 \, \frac{\text{m}}{\cancel{\text{h}}}\right)\left(\frac{1 \, \cancel{\text{h}}}{3600 \text{ s}}\right) = 341.11 \text{ m/s.}$$

If you multiplied by (3600 s)/(1 h) instead of (1 h)/(3600 s), the hour unit wouldn't cancel, and you would be able to easily recognize your error. Again, the *only* way to be sure that you correctly convert units is to carry the units throughout the calculation.

EXAMPLE 1–2

Converting volume units The world's largest cut diamond is the First Star of Africa (mounted in the British Royal Sceptre and kept in the Tower of London). Its volume is 1.84 cubic inches. What is its volume in cubic centimeters? In cubic meters?

SOLUTION To convert cubic inches to cubic centimeters, we multiply by $[(2.54 \text{ cm})/(1 \text{ in.})]^3$, not just (2.54 cm)/(1 in.). We find

$$1.84 \text{ in.}^3 = (1.84 \text{ in.}^3)\left(\frac{2.54 \text{ cm}}{1 \text{ in.}}\right)^3$$

$$= (1.84)(2.54)^3 \, \frac{\cancel{\text{in.}^3} \, \text{cm}^3}{\cancel{\text{in.}^3}} = 30.2 \text{ cm}^3.$$

Also, 1 cm = 10^{-2} m, and

$$30.2 \text{ cm}^3 = (30.2 \text{ cm}^3)\left(\frac{10^{-2} \text{ m}}{1 \text{ cm}}\right)^3$$

$$= (30.2)(10^{-2})^3 \, \frac{\cancel{\text{cm}^3} \, \text{m}^3}{\cancel{\text{cm}^3}} = 30.2 \times 10^{-6} \text{ m}^3.$$

Note: While 1 centimeter is 10^{-2} of a meter (that is, 1 cm = 10^{-2} m), a cubic centimeter (1 cm^3) is *not* 10^{-2} of a cubic meter. Rather, it is the volume of a cube whose sides are 1 cm long. So 1 cm^3 = $(1 \text{ cm})^3 = (10^{-2} \text{ m})^3 = (10^{-2})^3 \text{ m}^3$, or 1 cm^3 = 10^{-6} m^3.

1–6 UNCERTAINTY AND SIGNIFICANT FIGURES

Measurements always have uncertainties. If you measure the thickness of the cover of this book using an ordinary ruler, your measurement is reliable only to the nearest millimeter, and your result will be 3 mm. It would be *wrong* to state this result as 3.00 mm; given the limitations of the measuring device, you can't tell whether the actual thickness is 3.00 mm, 2.85 mm, or 3.11 mm. But if you use a micrometer caliper, a device that measures distances reliably to the nearest 0.01 mm, the result will be 2.91 mm. The distinction between these two measurements is in their **uncertainty.** The measurement using the micrometer caliper has a smaller uncertainty; it's a more accurate measurement. The uncertainty is also called the **error,** because it indicates the maximum difference there is likely to be between the measured value and the true value. The uncertainty or error of a measured value depends on the measurement technique used.

We often indicate the **accuracy** of a measured value—that is, how close it is likely to be to the true value—by writing the number, the symbol ±, and a second number indicating the uncertainty of the measurement. If the diameter of a steel rod is given as 56.47 ± 0.02 mm, this means that the true value is unlikely to be less than 56.45 mm or greater than 56.49 mm. In a commonly used shorthand notation, the number 1.6454(21) means 1.6454 ± 0.0021. The numbers in parentheses show the uncertainty in the final digits of the main number.

We can also express accuracy in terms of the maximum likely **fractional error** or **percent error** (also called *fractional uncertainty* and *percent uncertainty*). A resistor labeled "47 ohms ± 10%" probably has a true resistance differing from 47 ohms by no more than 10% of 47 ohms, that is, about 5 ohms. The resistance is probably between

1–3 This spectacular mishap was the result of a very small percent error—traveling a few meters too far in a journey of hundreds of thousands of meters.

42 and 52 ohms. For the diameter of the steel rod given above, the fractional error is (0.02 mm)/(56.47 mm), or about 0.0004; the percent error is (0.0004) (100%), or about 0.04%. Even small percent errors can sometimes be very significant (Fig. 1–3).

In many cases the uncertainty of a number is not stated explicitly. Instead, the uncertainty is indicated by the number of meaningful digits, or **significant figures,** in the measured value. We gave the thickness of the cover of this book as 2.91 mm, which has three significant figures. By this we mean that the first two digits are known to be correct, while the third digit is uncertain. The last digit is in the hundredths place, so the uncertainty is about 0.01 mm. Two values with the *same* number of significant figures may have *different* uncertainties; a distance given as 137 km also has three significant figures, but the uncertainty is about 1 km.

When you use numbers having uncertainties to compute other numbers, the computed numbers are also uncertain. It is especially important to understand this when you compare a number obtained from measurements with a value obtained from a theoretical prediction. Suppose you want to verify the value of π, the ratio of the circumference of a circle to its diameter. The true value of this ratio to ten digits is 3.141592654. To make your own calculation, you draw a large circle and measure its diameter and circumference to the nearest millimeter, obtaining the values 135 mm and 424 mm. You punch these into your calculator and obtain the quotient 3.140740741. Does this agree with the true value or not?

First, the last seven digits in this answer are meaningless; they imply a smaller uncertainty than is possible with your measurements. When numbers are multiplied or divided, the number of significant figures in the result can be no greater than in the factor with the fewest significant figures. For example, $3.1416 \times 2.34 \times 0.58 = 4.3$. Your measurements each have three significant figures, so your measured value of π, equal to (424 mm)/(135 mm), can have only three significant figures. It should be stated simply as 3.14. Within the limit of three significant figures, your value does agree with the true value.

When we add and subtract numbers, it's the location of the decimal point that matters, not the number of significant figures. For example, $123.62 + 8.9 = 132.5$. Although 123.62 has an uncertainty of about 0.01, 8.9 has an uncertainty of about 0.1. So their sum has an uncertainty of about 0.1 and should be written as 132.5, not 132.52.

In the examples and problems in this book we usually give numerical values with three significant figures, so your answers should usually have no more than three significant figures. (Many numbers that you encounter in the real world have even less accuracy. An automobile speedometer, for example, usually gives only two significant figures.) You might do the arithmetic with a calculator having a display with ten digits. But to give a ten-digit answer is not only unnecessary, it is genuinely wrong, because it misrepresents the accuracy of the results. Always round your final answer to keep only the correct number of significant figures or, in doubtful cases, one more at most. In Example 1–1 it would have been wrong to state the answer as 341.11111 m/s. Note that when you reduce such an answer to the appropriate number of significant figures, you must *round,* not *truncate.* Your calculator will tell you that the ratio of 525 m to 311 m is 1.688102894; to three significant figures, this is 1.69, not 1.68.

When we calculate with very large or very small numbers, we can show significant figures much more easily by using **scientific notation,** sometimes called **powers-of-10 notation.** The distance from the earth to the moon is about 384,000,000 m, but writing the number in this form gives no indication of the number of significant figures. Instead, we move the decimal point eight places to the left (corresponding to dividing by 10^8) and multiply by 10^8. That is,

$$384,000,000 \text{ m} = 3.84 \times 10^8 \text{ m}.$$

In this form, it is clear that we have three significant figures. The number 4.00×10^{-7} also has three significant figures, even though two of them are zeros. Note that in scientific notation the usual practice is to express the quantity as a number between 1 and 10 multiplied by the appropriate power of 10.

When an integer or a fraction occurs in a general equation, we treat that number as having no uncertainty at all. For example, in the equation $v^2 = v_0^2 + 2a(x - x_0)$, which is Eq. (2–13) in Chapter 2, the coefficient 2 is *exactly* 2. We can consider this coefficient as having an infinite number of significant figures (2.000000...). The same is true of the exponent 2 in v^2 and v_0^2.

Finally, let's note that **precision** is not the same as *accuracy*. A cheap digital watch that says the time is 10:35:17 A.M. is very *precise* (the time is given to the second), but if the watch runs several minutes slow, then this value isn't very *accurate*. On the other hand, a grandfather clock might be very accurate (that is, display the correct time), but if the clock has no second hand, it isn't very precise. A high-quality measurement, like those used to define standards (Section 1–4), is both precise *and* accurate.

EXAMPLE 1–3

Significant figures in multiplication The rest energy E of an object with rest mass m is given by Einstein's equation

$$E = mc^2,$$

where c is the speed of light in a vacuum. Find E for an object with $m = 9.11 \times 10^{-31}$ kg (to three significant figures, the mass of an electron). The SI unit for E is the joule (J); 1 J $= 1$ kg $\cdot$ m^2/s^2.

SOLUTION From Section 1–4 the exact value of the speed of light in vacuum is $c = 299{,}792{,}458$ m/s $= 2.99792458 \times 10^8$ m/s. Substituting the values of m and c into Einstein's equation, we find

$$E = (9.11 \times 10^{-31} \text{ kg})(2.99792458 \times 10^8 \text{ m/s})^2$$
$$= (9.11)(2.99792458)^2(10^{-31})(10^8)^2 \text{ kg} \cdot \text{m}^2/\text{s}^2$$
$$= (81.87659678)(10^{[-31+(2\times8)]}) \text{ kg} \cdot \text{m}^2/\text{s}^2$$
$$= 8.187659678 \times 10^{-14} \text{ kg} \cdot \text{m}^2/\text{s}^2.$$

Since the value of m was given to only three significant figures, we must round this to

$$E = 8.19 \times 10^{-14} \text{ kg} \cdot \text{m}^2/\text{s}^2 = 8.19 \times 10^{-14} \text{ J}.$$

Most calculators use scientific notation and add exponents automatically, but you should be able to do such calculations by hand when necessary.

1–7 ESTIMATES AND ORDERS OF MAGNITUDE

We have stressed the importance of knowing the accuracy of numbers that represent physical quantities. But even a very crude estimate of a quantity often gives us useful information. Sometimes we know how to calculate a certain quantity but have to guess at the data we need for the calculation. Or the calculation might be too complicated to carry out exactly, so we make some rough approximations. In either case our result is also a guess, but such a guess can be useful even if it is uncertain by a factor of two, ten, or more. Such calculations are often called **order-of-magnitude estimates.** The great Italian-American nuclear physicist Enrico Fermi (1901–1954) called them "back-of-the-envelope calculations."

Exercises 1–16 through 1–27 at the end of this chapter are of the estimating, or "order-of-magnitude," variety. Some are silly, and most require guesswork for the needed input data. Don't try to look up a lot of data; make the best guesses you can. Even when they are off by a factor of ten, the results can be useful and interesting.

EXAMPLE 1–4

An order-of-magnitude estimate You are writing an adventure novel in which the hero escapes across the border with a billion dollars worth of gold in his suitcase. Is this possible? Would that amount of gold fit in a suitcase? Would it be too heavy to carry?

SOLUTION Gold sells for around $400 an ounce. On a particular day the price might be $200 or $600, but never mind. An ounce is about 30 grams. Actually, an ordinary (avoirdupois) ounce is 28.35 g; an ounce of gold is a troy ounce, which is 9.45% more.

Again, never mind. Ten dollars' worth of gold has a mass somewhere around 1 gram, so a billion (10^9) dollars worth of gold is a hundred million (10^8) grams, or a hundred thousand (10^5) kilograms. This corresponds to a weight in British units of around 200,000 lb, or 100 tons. Whether the precise number is closer to 50 tons or 200 tons doesn't matter. Either way, the hero is not about to carry it across the border in a suitcase.

We can also estimate the *volume* of this gold. If its density were the same as that of water (1 g/cm³), the volume would be

10^8 cm³, or 100 m³. But gold is a heavy metal; we might guess its density to be ten times that of water. Gold is actually 19.3 times as dense as water. But by guessing ten, we find a volume of 10 m³. Visualize ten cubical stacks of gold bricks, each 1 meter on a side, and ask yourself whether they would fit in a suitcase! Clearly, your novel needs rewriting.

1–8 VECTORS AND VECTOR ADDITION

Some physical quantities, such as time, temperature, mass, density, and electric charge, can be described completely by a single number with a unit. But many other important quantities have a *direction* associated with them and cannot be described by a single number. Such quantities play an essential role in many of the central topics of physics, including motion and its causes and the phenomena of electricity and magnetism. A simple example of a quantity with direction is the motion of an airplane. To describe this motion completely, we must say not only how fast the plane is moving, but also in what direction. To fly from Chicago to New York, a plane has to head east, not south. The speed of the airplane combined with its direction of motion together constitute a quantity called *velocity.* Another example is *force,* which in physics means a push or pull exerted on a body. Giving a complete description of a force means describing both how hard the force pushes or pulls on the body and the direction of the push or pull.

When a physical quantity is described by a single number, we call it a **scalar quantity.** In contrast, a **vector quantity** has both a **magnitude** (the "how much" or "how big" part) and a direction in space. Calculations with scalar quantities use the operations of ordinary arithmetic. For example, 6 kg + 3 kg = 9 kg, or 4 × 2 s = 8 s. However, combining vectors requires a different set of operations.

To understand more about vectors and how they combine, we start with the simplest vector quantity, **displacement.** Displacement is simply a change in position of a point. (The point may represent a particle or a small body.) In Fig. 1–4a we represent the change of position from point P_1 to point P_2 by a line from P_1 to P_2, with an arrowhead at P_2 to represent the direction of motion. Displacement is a vector quantity because we must state not only how far the particle moves, but also in what direction. Walking 3 km north from your front door doesn't get you to the same place as walking 3 km southeast; these two displacements have the same magnitude, but different directions.

We usually represent a vector quantity such as displacement by a single letter, such as $\vec{A}$ in Fig. 1–4a. In this book we always print vector symbols in ***boldface italic type with an arrow above them.*** We do this to remind you that vector quantities have different properties from scalar quantities; the arrow is a reminder that vectors have direction. In handwriting, vector symbols are usually underlined or written with an arrow above them (Fig. 1–4). When you write a symbol for a vector, *always* write it in one of these ways. If you don't distinguish between scalar and vector quantities in your notation, you probably won't make the distinction in your thinking either, and hopeless confusion will result.

When drawing any vector, we always draw a line with an arrowhead at its tip. The length of the line shows the vector's magnitude, and the direction of the line shows the vector's direction. Displacement is always a straight-line segment, directed from the starting point to the end point, even though the actual path of the particle may be curved. In Fig. 1–4b the particle moves along the curved path shown from P_1 to P_2, but the displacement is still the vector $\vec{A}$. Note that displacement is not related directly to the total

Handwritten notation: $\underline{A}$ or $\vec{A}$

(a)

(b)

1–4 (a) Vector $\vec{A}$ is the displacement from point P_1 to point P_2. (b) A displacement is always a straight-line segment directed from the starting point to the end point, even if the actual path is curved. When a path ends at the same place where it started, the displacement is zero.

distance traveled. If the particle were to continue on to P_3 and then return to P_1, the displacement for the entire trip would be *zero.*

If two vectors have the same direction, they are **parallel.** If they have the same magnitude *and* the same direction, they are *equal,* no matter where they are located in space. The vector $\vec{A}'$ from point P_3 to point P_4 in Fig. 1–5 has the same length and direction as the vector $\vec{A}$ from P_1 to P_2. These two displacements are equal, even though they start at different points. We write this as $\vec{A} = \vec{A}'$ in Fig. 1–5, using a boldface equals sign to emphasize that equality of two vector quantities is not the same relationship as equality of two scalar quantities. Two vector quantities are equal only when they have the same magnitude *and* the same direction.

The vector $\vec{B}$ in Fig. 1–5, however, is not equal to $\vec{A}$ because its direction is *opposite* to that of $\vec{A}$. We define the **negative of a vector** as a vector having the same magnitude as the original vector but the *opposite* direction. The negative of vector quantity $\vec{A}$ is denoted as $-\vec{A}$, and we use a boldface minus sign to emphasize the vector nature of the quantities. If $\vec{A}$ is 87 m south, then $-\vec{A}$ is 87 m north. Thus the relation between $\vec{A}$ and $\vec{B}$ of Fig. 1–5 may be written as $\vec{A} = -\vec{B}$ or $\vec{B} = -\vec{A}$. When two vectors $\vec{A}$ and $\vec{B}$ have opposite directions, whether their magnitudes are the same or not, we say that they are **antiparallel.**

We usually represent the *magnitude* of a vector quantity (its length in the case of a displacement vector) by the same letter used for the vector, but in *light italic type* with *no* arrow on top, rather than boldface italic with an arrow (which is reserved for vectors). An alternative notation is the vector symbol with vertical bars on both sides:

$$\text{(Magnitude of } \vec{A}) = A = |\vec{A}|. \tag{1–2}$$

By definition the magnitude of a vector quantity is a scalar quantity (a number) and is *always positive.* We also note that a vector can never be equal to a scalar because they are different kinds of quantities. The expression "$\vec{A} = 6$ m" is just as wrong as "2 oranges = 3 apples" or "6 lb = 7 km"!

When drawing diagrams with vectors, we'll usually use a scale similar to those used for maps, in which the distance on the diagram is proportional to the magnitude of the vector. For example, a displacement of 5 km might be represented in a diagram by a vector 1 cm long, since an actual-size diagram wouldn't be practical. When we work with vector quantities with units other than displacement, such as force or velocity, we *must* use a scale. In a diagram for force vectors we might use a scale in which a vector that is 1 cm long represents a force of magnitude 5 N. A 20-N force would then be represented by a vector 4 cm long, with the appropriate direction.

VECTOR ADDITION

Now suppose a particle undergoes a displacement $\vec{A}$, followed by a second displacement $\vec{B}$ (Fig. 1–6a). The final result is the same as if the particle had started at the same initial point and undergone a single displacement $\vec{C}$, as shown. We call displacement $\vec{C}$ the **vector sum,** or **resultant,** of displacements $\vec{A}$ and $\vec{B}$. We express this relationship symbolically as

$$\vec{C} = \vec{A} + \vec{B}. \tag{1–3}$$

The boldface plus sign emphasizes that adding two vector quantities requires a geometrical process and is not the same operation as adding two scalar quantities such as $2 + 3 = 5$. In vector addition we usually place the *tail* of the *second* vector at the *head,* or tip, of the *first* vector (Fig. 1–6a).

If we make the displacements $\vec{A}$ and $\vec{B}$ in reverse order, with $\vec{B}$ first and $\vec{A}$ second, the result is the same (Fig. 1–6b). Thus

$$\vec{C} = \vec{B} + \vec{A} \quad \text{and} \quad \vec{A} + \vec{B} = \vec{B} + \vec{A}. \tag{1–4}$$

1–5 The displacement from P_3 to P_4 is equal to that from P_1 to P_2. The displacement $\vec{B}$ from P_5 to P_6 has the same magnitude as $\vec{A}$ and $\vec{A}'$ but opposite direction; displacement $\vec{B}$ is the negative of displacement $\vec{A}$.

(a)

(b)

(c)

1–6 Vector $\vec{C}$ is the vector sum of vectors $\vec{A}$ and $\vec{B}$. The order in vector addition doesn't matter; vector addition is commutative.

1–7 The vector sum of (a) two parallel vectors and (b) two antiparallel vectors. Note that the vectors $\vec{A}$, $\vec{B}$, and $\vec{C}$ in part (a) are not the same as the vectors $\vec{A}$, $\vec{B}$, and $\vec{C}$ in part (b).

1–8 Several constructions for finding the vector sum $\vec{A} + \vec{B} + \vec{C}$.

This shows that the order of terms in a vector sum doesn't matter. In other words, vector addition obeys the commutative law.

Figure 1–6c shows an alternative representation of the vector sum. When vectors $\vec{A}$ and $\vec{B}$ are both drawn with their tails at the same point, vector $\vec{C}$ is the diagonal of a parallelogram constructed with $\vec{A}$ and $\vec{B}$ as two adjacent sides.

CAUTION ▶ A word of warning is in order about vector addition. It's a common error to conclude that if $\vec{C} = \vec{A} + \vec{B}$, the magnitude C should just equal the magnitude A plus the magnitude B. Figure 1–6 shows that in general, this conclusion is *wrong;* you can see from the figure that $C < A + B$. The magnitude of the vector sum $\vec{A} + \vec{B}$ depends on the magnitudes of $\vec{A}$ and of $\vec{B}$ *and* on the angle between $\vec{A}$ and $\vec{B}$ (see Problem 1–72). Only in the special case in which $\vec{A}$ and $\vec{B}$ are *parallel* is the magnitude of $\vec{C} = \vec{A} + \vec{B}$ equal to the sum of the magnitudes of $\vec{A}$ and $\vec{B}$ (Fig. 1–7a). By contrast, when the vectors are *antiparallel* (Fig. 1–7b) the magnitude of $\vec{C}$ equals the *difference* of the magnitudes of $\vec{A}$ and $\vec{B}$. Students who aren't careful about distinguishing between scalar and vector quantities frequently make errors about the magnitude of a vector sum. Don't let this happen to you! ◀

When we need to add more than two vectors, we may first find the vector sum of any two, add this vectorially to the third, and so on. Figure 1–8a shows three vectors $\vec{A}$, $\vec{B}$, and $\vec{C}$. In Fig. 1–8b, vectors $\vec{A}$ and $\vec{B}$ are first added, giving a vector sum $\vec{D};$ vectors $\vec{C}$ and $\vec{D}$ are then added by the same process to obtain the vector sum $\vec{R}:$

$$\vec{R} = (\vec{A} + \vec{B}) + \vec{C} = \vec{D} + \vec{C}.$$

Alternatively, we can first add $\vec{B}$ and $\vec{C}$ (Fig. 1–8c) to obtain vector $\vec{E}$, and then add $\vec{A}$ and $\vec{E}$ to obtain $\vec{R}:$

$$\vec{R} = \vec{A} + (\vec{B} + \vec{C}) = \vec{A} + \vec{E}.$$

We don't even need to draw vectors $\vec{D}$ and $\vec{E}$; all we need to do is draw the given vectors in succession, with the tail of each at the head of the one preceding it, and complete the polygon by a vector $\vec{R}$ from the tail of the first vector to the head of the last vector (Fig. 1–8d). The order makes no difference; Fig. 1–8e shows a different order, and we invite you to try others. We see that vector addition obeys the associative law.

We mentioned above that $-\vec{A}$ is a vector having the same magnitude as $\vec{A}$ but the opposite direction. This provides the basis for defining vector subtraction. We define the difference $\vec{A} - \vec{B}$ of two vectors $\vec{A}$ and $\vec{B}$ to be the vector sum of $\vec{A}$ and $-\vec{B}:$

$$\vec{A} - \vec{B} = \vec{A} + (-\vec{B}). \tag{1–5}$$

Figure 1–9 shows an example of vector subtraction. To construct the vector difference $\vec{A} - \vec{B}$, the tail of $-\vec{B}$ is placed at the head of $\vec{A}$.

A vector quantity such as a displacement can be multiplied by a scalar quantity (an ordinary number). The displacement $2\vec{A}$ is a displacement (vector quantity) in the same direction as the vector $\vec{A}$ but twice as long; this is the same as adding $\vec{A}$ to itself. In general, when a vector $\vec{A}$ is multiplied by a scalar c, the result $c\vec{A}$ has magnitude $|c|A$ (the absolute value of c multiplied by the magnitude of the vector $\vec{A}$). If c is positive, $c\vec{A}$ is in the same direction as $\vec{A}$; if c is negative, $c\vec{A}$ is in the direction opposite to $\vec{A}$. Thus $5\vec{A}$ is parallel to $\vec{A}$, while $-5\vec{A}$ is antiparallel to $\vec{A}$.

The scalar quantity used to multiply a vector may also be a physical quantity having units. For example, you may be familiar with the relationship $\vec{F} = m\vec{a};$ the net force $\vec{F}$ (a vector quantity) that acts on a body is equal to the product of the body's mass m (a positive scalar quantity) and its acceleration $\vec{a}$ (a vector quantity). The magnitude of the net force is equal to the mass (which is positive and equals its own absolute value) multiplied by the magnitude of the acceleration. The unit of the magnitude of force is the unit of mass multiplied by the unit of the magnitude of acceleration. The direction of $\vec{F}$ is the same as that of $\vec{a}$ because m is positive.

1–9 (a) Vector $\vec{A}$ and vector $\vec{B}$. (b) Vector $\vec{A}$ and vector $-\vec{B}$. (c) The vector difference $\vec{A} - \vec{B}$ is the sum of vectors $\vec{A}$ and $-\vec{B}$. The tail of $-\vec{B}$ is placed at the head of $\vec{A}$. (d) To check: $(\vec{A} - \vec{B}) + \vec{B} = \vec{A}$.

EXAMPLE 1–5

Vector addition A cross-country skier skis 1.00 km north and then 2.00 km east on a horizontal snow field. a) How far and in what direction is she from the starting point? b) What are the magnitude and direction of her resultant displacement?

SOLUTION a) Figure 1–10 is a scale diagram of the skier's displacements. By careful measurement we find that the distance from the starting point is about 2.2 km and the angle ϕ (the Greek letter "phi") is about 63°. But it is much more accurate to *calculate* the result. The vectors in the diagram form a right triangle, and we can find the length of the hypotenuse by using the Pythagorean theorem:

$$\sqrt{(1.00 \text{ km})^2 + (2.00 \text{ km})^2} = 2.24 \text{ km}.$$

The angle ϕ can be found with a little simple trigonometry. If you need a review, the trigonometric functions and identities are summarized in Appendix B, along with other useful mathematical and geometrical relations. By the definition of the tangent function,

$$\tan \phi = \frac{\text{opposite side}}{\text{adjacent side}} = \frac{2.00 \text{ km}}{1.00 \text{ km}},$$

$$\phi = 63.4°.$$

b) The magnitude of the resultant displacement is just the distance that we found in part (a), 2.24 km. We can describe the direction as 63.4° east of north or 90° − 63.4° = 26.6° north of east. Take your choice!

1–10 The vector diagram, drawn to scale, for a cross-country ski trip.

1–9 COMPONENTS OF VECTORS

In Section 1–8 we added vectors by using a scale diagram and by using properties of right triangles. Measuring a diagram offers only very limited accuracy, and calculations with right triangles work only when the two vectors are perpendicular. So we need a simple but general method for adding vectors. This is called the method of *components*.

To define what we mean by the components of a vector, we begin with a rectangular (Cartesian) coordinate system of axes (Fig. 1–11). We then draw the vector we're considering with its tail at O, the origin of the coordinate system. We can represent any vector lying in the xy-plane as the sum of a vector parallel to the x-axis and a vector parallel to the y-axis. These two vectors are labeled $\vec{A}_x$ and $\vec{A}_y$ in the figure; they are called the **component vectors** of vector $\vec{A}$, and their vector sum is equal to $\vec{A}$. In symbols,

$$\vec{A} = \vec{A}_x + \vec{A}_y. \tag{1–6}$$

By definition, each component vector lies along a coordinate-axis direction. Thus we need only a single number to describe each one. When the component vector $\vec{A}_x$ points in the positive x-direction, we define the number A_x to be equal to the magnitude of $\vec{A}_x$. When the component vector $\vec{A}_x$ points in the negative x-direction, we define the number A_x to be equal to the negative of that magnitude, keeping in mind that the magnitude of a vector quantity is itself never negative. We define the number A_y in the same way. The two numbers A_x and A_y are called the **components** of $\vec{A}$.

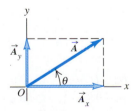

1–11 Vectors $\vec{A}_x$ and $\vec{A}_y$ are the rectangular component vectors of $\vec{A}$ in the directions of the x- and y-axes. For the vector $\vec{A}$ shown here, the components A_x and A_y are both positive.

The components A_x and A_y of a vector $\vec{A}$ are just numbers; they are *not* vectors themselves. This is why we print the symbols for components in light italic type with no arrow on top instead of the boldface italic with an arrow reserved for vectors.

We can calculate the components of the vector $\vec{A}$ if we know its magnitude A and its direction. We'll describe the direction of a vector by its angle relative to some reference direction. In Fig. 1–11 this reference direction is the positive x-axis, and the angle between vector $\vec{A}$ and the positive x-axis is θ (the Greek letter "theta"). Imagine that the vector $\vec{A}$ originally lies along the $+x$-axis and that you then rotate it to its correct direction, as indicated by the arrow in Fig. 1–11 on the angle θ. If this rotation is from the $+x$-axis toward the $+y$-axis, as shown in Fig. 1–11, then θ is *positive*; if the rotation is from the $+x$-axis toward the $-y$-axis, θ is *negative*. Thus the $+y$-axis is at an angle of 90°, the $-x$-axis at 180°, and the $-y$-axis at 270° (or $-90°$). If θ is measured in this way, then from the definition of the trigonometric functions,

$$\frac{A_x}{A} = \cos\theta \qquad \text{and} \qquad \frac{A_y}{A} = \sin\theta;$$

$$A_x = A\cos\theta \qquad \text{and} \qquad A_y = A\sin\theta. \tag{1-7}$$

(θ measured from the $+x$-axis, rotating toward the $+y$-axis)

CAUTION ▶ Equations (1–7) are correct *only* when the angle θ is measured from the positive x-axis as described above. If the angle of the vector is given from a different reference direction or using a different sense of rotation, the relationships are different. Be careful! ◀

In Fig. 1–11, A_x is positive because its direction is along the positive x-axis, and A_y is positive because its direction is along the positive y-axis. This is consistent with Eqs. (1–7); θ is in the first quadrant (between 0° and 90°), and both the cosine and the sine of an angle in this quadrant are positive. But in Fig. 1–12a the component B_x is negative; its direction is opposite to that of the positive x-axis. Again, this agrees with Eqs. (1–7); the cosine of an angle in the second quadrant is negative. The component B_y is positive ($\sin\theta$ is positive in the second quadrant). In Fig. 1–12b, both C_x and C_y are negative (both $\cos\theta$ and $\sin\theta$ are negative in the third quadrant).

(a)

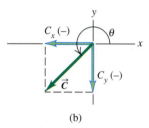

(b)

1–12 The components of a vector may be positive or negative numbers.

EXAMPLE 1–6

Finding components a) What are the x- and y-components of vector $\vec{D}$ in Fig. 1–13a? The magnitude of the vector is $D = 3.00$ m, and the angle $\alpha = 45°$. b) What are the x- and y-components of vector $\vec{E}$ in Fig. 1–13b? The magnitude of the vector is $E = 4.50$ m, and the angle $\beta = 37.0°$.

SOLUTION a) The angle between $\vec{D}$ and the positive x-axis is α (the Greek letter "alpha"), but this angle is measured toward the *negative y-axis*. So the angle we must use in Eqs. (1–7) is $\theta = -\alpha = -45°$. We find

$$D_x = D\cos\theta = (3.00\text{ m})(\cos(-45°)) = +2.1\text{ m},$$

$$D_y = D\sin\theta = (3.00\text{ m})(\sin(-45°)) = -2.1\text{ m}.$$

The vector has a positive x-component and a negative y-component, as shown in the figure. Had you been careless and substituted $+45°$ for θ in Eqs. (1–7), you would have gotten the wrong sign for D_y.

b) The x-axis isn't horizontal in Fig. 1–13b, nor is the y-axis vertical. In general, *any* orientation of the x- and y-axes is per-

missible, just so the axes are mutually perpendicular. (Later we'll use axes like these to study the motion of an object sliding on a ramp or incline; one axis will lie along the ramp, and the other will be perpendicular to the ramp.)

Here the angle β (the Greek letter "beta") is the angle between $\vec{E}$ and the positive y-axis, *not* the positive x-axis, so we

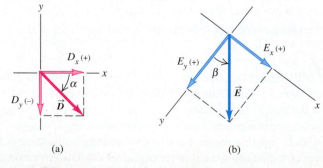

(a)   (b)

1–13 Calculating the x- and y-components of vectors.

cannot use this angle in Eqs. (1–7). Instead, note that $\vec{E}$ defines the hypotenuse of a right triangle; the other two sides of the triangle are the magnitudes of E_x and E_y, the x- and y-components of $\vec{E}$. The sine of β is the opposite side (the magnitude of E_x) divided by the hypotenuse (the magnitude E), and the cosine of β is the adjacent side (the magnitude of E_y) divided by the hypotenuse (again, the magnitude E). Both components of $\vec{E}$ are positive, so

$$E_x = E \sin \beta = (4.50 \text{ m}) (\sin 37.0°) = +2.71 \text{ m},$$

$$E_y = E \cos \beta = (4.50 \text{ m}) (\cos 37.0°) = +3.59 \text{ m}.$$

Had you used Eqs. (1–7) directly and written $E_x = E \cos 37.0°$ and $E_y = E \sin 37.0°$, your answers for E_x and E_y would have been reversed!

If you insist on using Eqs. (1–7), you must first find the angle between $\vec{E}$ and the positive x-axis, measured toward the positive y-axis; this is $\theta = 90.0° - \beta = 90.0° - 37.0° = 53.0°$. Then $E_x = E \cos \theta$ and $E_y = E \sin \theta$. You can substitute the values of E and θ into Eqs. (1–7) to show that the results for E_x and E_y are the same as given above.

Notice that the answers to part (b) have three significant figures, but the answers to part (a) have only two. Can you see why?

USING COMPONENTS

We can describe a vector completely by giving either its magnitude and direction or its x- and y-components. Equations (1–7) show how to find the components if we know the magnitude and direction. We can also reverse the process: we can find the magnitude and direction if we know the components. By applying the Pythagorean theorem to Fig. 1–11, we find that the magnitude of a vector $\vec{A}$ is

$$A = \sqrt{A_x{}^2 + A_y{}^2}, \tag{1–8}$$

where we always take the positive root. Equation (1–8) is valid for any choice of x-axis and y-axis, as long as they are mutually perpendicular. The expression for the vector direction comes from the definition of the tangent of an angle. If θ is measured from the positive x-axis, and a positive angle is measured toward the positive y-axis (as in Fig. 1–11), then

$$\tan \theta = \frac{A_y}{A_x} \quad \text{and} \quad \theta = \arctan \frac{A_y}{A_x}. \tag{1–9}$$

We will always use the notation arctan for the inverse tangent function. The notation $\tan^{-1}$ is also commonly used, and your calculator may have an INV button to be used with the TAN button. Some computer languages use atan (FORTRAN) or atn (BASIC). Pascal uses ArcTan.

CAUTION ▶ There is one slight complication in using Eqs. (1–9) to find θ. Suppose $A_x = 2$ m and $A_y = -2$ m; then $\tan \theta = -1$. But there are two angles having tangents of -1, namely, $135°$ and $315°$ (or $-45°$). In general, any two angles that differ by $180°$ have the same tangent. To decide which is correct, we have to look at the individual components. Because A_x is positive and A_y is negative, the angle must be in the fourth quadrant; thus $\theta = 315°$ (or $-45°$) is the correct value. Most pocket calculators give $\arctan(-1) = -45°$. In this case that is correct; but if instead we have $A_x = -2$ m and $A_y = 2$ m, then the correct angle is $135°$. Similarly, when A_x and A_y are both negative, the tangent is positive, but the angle is in the third quadrant. You should *always* draw a sketch to check which of the two possibilities is the correct one. ◀

Here's how we use components to calculate the vector sum (resultant) of two or more vectors. Figure 1–14 shows two vectors $\vec{A}$ and $\vec{B}$ and their vector sum $\vec{R}$, along with the x- and y-components of all three vectors. You can see from the diagram that the x-component R_x of the vector sum is simply the sum $(A_x + B_x)$ of the x-components of the vectors being added. The same is true for the y-components. In symbols,

$$R_x = A_x + B_x, \quad R_y = A_y + B_y \quad \text{(components of } \vec{R} = \vec{A} + \vec{B}\text{)}. \tag{1–10}$$

Figure 1–14 shows this result for the case in which the components A_x, A_y, B_x, and B_y are all positive. You should draw additional diagrams to verify for yourself that Eqs. (1–10) are valid for *any* signs of the components of $\vec{A}$ and $\vec{B}$.

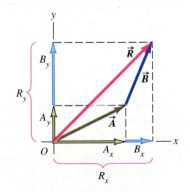

1–14 Vector $\vec{R}$ is the vector sum (resultant) of $\vec{A}$ and $\vec{B}$. Its x-component, R_x, equals the sum of the x-components of $\vec{A}$ and $\vec{B}$. The y-components are similarly related.

If we know the components of any two vectors $\vec{A}$ and $\vec{B}$, perhaps by using Eqs. (1–7), we can compute the components of the vector sum $\vec{R}$. Then if the magnitude and direction of $\vec{R}$ are needed, we can obtain them from Eqs. (1–8) and (1–9), with the A's replaced by R's.

This procedure for finding the sum of two vectors can easily be extended to any number of vectors. Let $\vec{R}$ be the vector sum of $\vec{A},\ \vec{B},\ \vec{C},\ \vec{D},\ \vec{E},\ \ldots.$ Then the components of $\vec{R}$ are

$$R_x = A_x + B_x + C_x + D_x + E_x + \cdots,$$
$$R_y = A_y + B_y + C_y + D_y + E_y + \cdots. \qquad (1\text{–}11)$$

We have talked only about vectors that lie in the xy-plane, but the component method works just as well for vectors having any direction in space. We introduce a z-axis perpendicular to the xy-plane; then in general a vector $\vec{A}$ has components A_x, A_y, and A_z in the three coordinate directions. The magnitude A is given by

$$A = \sqrt{A_x{}^2 + A_y{}^2 + A_z{}^2}. \qquad (1\text{–}12)$$

Again, we always take the positive root. Also, Eqs. (1–11) for the components of the vector sum $\vec{R}$ have an additional member:

$$R_z = A_z + B_z + C_z + D_z + E_z + \cdots.$$

Finally, while our discussion of vector addition has centered on combining *displacement* vectors, the method is applicable to all other vector quantities as well. When we study the concept of force in Chapter 4, we'll find that forces are vectors that obey the same rules of vector addition that we've used with displacement. Other vector quantities will make their appearance in later chapters.

Problem–Solving Strategy

VECTOR ADDITION

1. First draw the individual vectors being summed and the coordinate axes being used. In your drawing, place the tail of the first vector at the origin of coordinates, place the tail of the second vector at the head of the first vector, and so on. Draw the vector sum $\vec{R}$ from the tail of the first vector to the head of the last vector.

2. Find the x- and y-components of each individual vector and record your results in a table. If a vector is described by its magnitude A and its angle θ, measured from the $+x$-axis toward the $+y$-axis, then the components are given by

$$A_x = A \cos \theta, \qquad A_y = A \sin \theta.$$

Some components may be positive and some may be negative, depending on how the vector is oriented (that is, what quadrant θ lies in). You can use this sign table as a check:

Quadrant	I	II	III	IV
A_x	+	−	−	+
A_y	+	+	−	−

If the angles of the vectors are given in some other way, perhaps using a different reference direction, convert them to angles measured from the $+x$-axis as described above. Be particularly careful with signs.

3. Add the individual x-components algebraically, including signs, to find R_x, the x-component of the vector sum. Do the same for the y-components to find R_y.

4. Then the magnitude R and direction θ of the vector sum are given by

$$R = \sqrt{R_x{}^2 + R_y{}^2},$$
$$\theta = \arctan \frac{R_y}{R_x}.$$

Remember that the magnitude R is *always* positive and that θ is measured from the positive x-axis. The value of θ that you find with a calculator may be the correct one, or it may be off by 180°. You can decide by examining your drawing.

EXAMPLE 1–7

Adding vectors with components The three finalists in a contest are brought to the center of a large, flat field. Each is given a meter stick, a compass, a calculator, a shovel, and (in a different order for each contestant) the following three displacements:

72.4 m, 32.0° east of north;

57.3 m, 36.0° south of west;

17.8 m straight south.

The three displacements lead to the point where the keys to a new Porsche are buried. Two contestants start measuring immediately, but the winner first *calculates* where to go. What does she calculate?

SOLUTION The situation is shown in Fig. 1–15. We have chosen the +x-axis as east and the +y-axis as north, the usual choice for

1–15 Three successive displacements $\vec{A}$, $\vec{B}$, and $\vec{C}$ and the resultant (vector sum) displacement $\vec{R} = \vec{A} + \vec{B} + \vec{C}$.

maps. Let $\vec{A}$ be the first displacement, $\vec{B}$ the second, and $\vec{C}$ the third. We can estimate from the diagram that the vector sum $\vec{R}$ is about 10 m, 40° west of north. The angles of the vectors, measured from the +x-axis toward the +y-axis, are $(90.0° - 32.0°) = 58.0°$, $(180.0° + 36.0°) = 216.0°$, and $270.0°$. We have to find the components of each. Because of our choice of axes, we may use Eqs. (1–7), and so the components of $\vec{A}$ are

$$A_x = A \cos \theta_A = (72.4 \text{ m})(\cos 58.0°) = 38.37 \text{ m},$$

$$A_y = A \sin \theta_A = (72.4 \text{ m})(\sin 58.0°) = 61.40 \text{ m}.$$

Note that we have kept one too many significant figures in the components; we will wait until the end to round to the correct number of significant figures. The table below shows the components of all the displacements, the addition of components, and the other calculations. Always arrange your component calculations systematically like this.

Distance	Angle	x-component	y-component
A = 72.4 m	58.0°	38.37 m	61.40 m
B = 57.3 m	216.0°	−46.36 m	−33.68 m
C = 17.8 m	270.0°	0.00 m	−17.80 m
		$R_x = -7.99$ m	$R_y = 9.92$ m

$$R = \sqrt{(-7.99 \text{ m})^2 + (9.92 \text{ m})^2} = 12.7 \text{ m},$$

$$\theta = \arctan \frac{9.92 \text{ m}}{-7.99 \text{ m}} = 129° = 39° \text{ west of north.}$$

The losers try to measure three angles and three distances totaling 147.5 m, one meter at a time. The winner measured only one angle and one much shorter distance.

Notice that $\theta = -51°$, or 51° south of east, also satisfies the equation for θ. But since the winner has made a drawing of the displacement vectors (Fig. 1–15), she knows that $\theta = 129°$ is the only correct solution for the angle.

EXAMPLE 1–8

A vector in three dimensions After an airplane takes off, it travels 10.4 km west, 8.7 km north, and 2.1 km up. How far is it from the takeoff point?

SOLUTION Let the +x-axis be east, the +y-axis north, and the +z-axis up. Then $A_x = -10.4$ km, $A_y = 8.7$ km, and $A_z = 2.1$ km; Eq. (1–12) gives

$$A = \sqrt{(-10.4 \text{ km})^2 + (8.7 \text{ km})^2 + (2.1 \text{ km})^2} = 13.7 \text{ km}.$$

1–10 UNIT VECTORS

A **unit vector** is a vector that has a magnitude of 1, with no units. Its only purpose is to *point,* that is, to describe a direction in space. Unit vectors provide a convenient notation for many expressions involving components of vectors. We will always include a caret or "hat" (^) in the symbol for a unit vector to distinguish it from ordinary vectors whose magnitude may or may not be equal to 1.

In an *x-y* coordinate system we can define a unit vector $\hat{\imath}$ that points in the direction of the positive x-axis and a unit vector $\hat{\jmath}$ that points in the direction of the positive y-axis. Then we can express the relationship between component vectors and components,

1-16 Using unit vectors, we can express a vector $\vec{A}$ in terms of its components A_x and A_y as $\vec{A} = A_x\hat{\imath} + A_y\hat{\jmath}$.

described at the beginning of Section 1–9, as follows:

$$\vec{A}_x = A_x\hat{\imath},$$
$$\vec{A}_y = A_y\hat{\jmath}. \tag{1-13}$$

Similarly, we can write a vector $\vec{A}$ in terms of its components as

$$\vec{A} = A_x\hat{\imath} + A_y\hat{\jmath}. \tag{1-14}$$

Equations (1–13) and (1–14) are vector equations; each term, such as $A_x\hat{\imath}$, is a vector quantity (Fig. 1–16). The boldface equals and plus signs denote vector equality and addition.

When two vectors $\vec{A}$ and $\vec{B}$ are represented in terms of their components, we can express the vector sum $\vec{R}$ using unit vectors as follows:

$$\vec{A} = A_x\hat{\imath} + A_y\hat{\jmath},$$
$$\vec{B} = B_x\hat{\imath} + B_y\hat{\jmath},$$
$$\vec{R} = \vec{A} + \vec{B}$$
$$= (A_x\hat{\imath} + A_y\hat{\jmath}) + (B_x\hat{\imath} + B_y\hat{\jmath})$$
$$= (A_x + B_x)\hat{\imath} + (A_y + B_y)\hat{\jmath}$$
$$= R_x\hat{\imath} + R_y\hat{\jmath}. \tag{1-15}$$

Equation (1–15) restates the content of Eqs. (1–10) in the form of a single vector equation rather than two component equations.

If the vectors do not all lie in the xy-plane, then we need a third component. We introduce a third unit vector $\hat{k}$ that points in the direction of the positive z-axis. The generalized forms of Eqs. (1–14) and (1–15) are

$$\vec{A} = A_x\hat{\imath} + A_y\hat{\jmath} + A_z\hat{k},$$
$$\vec{B} = B_x\hat{\imath} + B_y\hat{\jmath} + B_z\hat{k}, \tag{1-16}$$
$$\vec{R} = (A_x + B_x)\hat{\imath} + (A_y + B_y)\hat{\jmath} + (A_z + B_z)\hat{k}$$
$$= R_x\hat{\imath} + R_y\hat{\jmath} + R_z\hat{k}. \tag{1-17}$$

EXAMPLE 1-9

Using unit vectors Given the two displacements

$$\vec{D} = (6\hat{\imath} + 3\hat{\jmath} - \hat{k})\,\text{m} \quad \text{and} \quad \vec{E} = (4\hat{\imath} - 5\hat{\jmath} + 8\hat{k})\,\text{m},$$

find the magnitude of the displacement $2\vec{D} - \vec{E}$.

SOLUTION Letting $\vec{F} = 2\vec{D} - \vec{E}$, we have

$$\vec{F} = 2(6\hat{\imath} + 3\hat{\jmath} - \hat{k})\,\text{m} - (4\hat{\imath} - 5\hat{\jmath} + 8\hat{k})\,\text{m}$$
$$= [(12 - 4)\hat{\imath} + (6 + 5)\hat{\jmath} + (-2 - 8)\hat{k}]\,\text{m}$$
$$= (8\hat{\imath} + 11\hat{\jmath} - 10\hat{k})\,\text{m}.$$

The units of the vectors $\vec{D}$, $\vec{E}$, and $\vec{F}$ are meters, so the components of these vectors are also in meters. From Eq. (1–12),

$$F = \sqrt{F_x^2 + F_y^2 + F_z^2}$$
$$= \sqrt{(8\,\text{m})^2 + (11\,\text{m})^2 + (-10\,\text{m})^2} = 17\,\text{m}.$$

1-11 PRODUCTS OF VECTORS

We have seen how addition of vectors develops naturally from the problem of combining displacements, and we will use vector addition for many other vector quantities later.

We can also express many physical relationships concisely by using *products* of vectors. Vectors are not ordinary numbers, so ordinary multiplication is not directly applicable to vectors. We will define two different kinds of products of vectors. The first, called the scalar product, yields a result that is a scalar quantity. The second, the vector product, yields another vector.

SCALAR PRODUCT

The **scalar product** of two vectors $\vec{A}$ and $\vec{B}$ is denoted by $\vec{A} \cdot \vec{B}$. Because of this notation, the scalar product is also called the **dot product.**

To define the scalar product $\vec{A} \cdot \vec{B}$ of two vectors $\vec{A}$ and $\vec{B}$, we draw the two vectors with their tails at the same point (Fig. 1–17a). The angle between their directions is ϕ as shown; the angle ϕ always lies between 0° and 180°. (As usual, we use Greek letters for angles.) Figure 1–17b shows the projection of the vector $\vec{B}$ onto the direction of $\vec{A}$; this projection is the component of $\vec{B}$ parallel to $\vec{A}$ and is equal to $B \cos \phi$. (We can take components along any direction that's convenient, not just the x- and y-axes.) We define $\vec{A} \cdot \vec{B}$ to be the magnitude of $\vec{A}$ multiplied by the component of $\vec{B}$ parallel to $\vec{A}$. Expressed as an equation,

$$\vec{A} \cdot \vec{B} = AB \cos \phi = |\vec{A}|\,|\vec{B}| \cos \phi \quad \text{(definition of the scalar (dot) product),} \quad (1\text{--}18)$$

where ϕ ranges from 0° to 180°.

Alternatively, we can define $\vec{A} \cdot \vec{B}$ to be the magnitude of $\vec{B}$ multiplied by the component of $\vec{A}$ parallel to $\vec{B}$, as in Fig. 1–17c. Hence $\vec{A} \cdot \vec{B} = B(A \cos \phi) = AB \cos \phi$, which is the same as Eq. (1–18).

The scalar product is a scalar quantity, not a vector, and it may be positive, negative, or zero. When ϕ is between 0° and 90°, the scalar product is positive. When ϕ is between 90° and 180°, it is negative. You should draw a diagram like Fig. 1–17, but with ϕ between 90° and 180°, to confirm for yourself that the component of $\vec{B}$ parallel to $\vec{A}$ is negative in this case, as is the component of $\vec{A}$ parallel to $\vec{B}$. Finally, when $\phi = 90°$, $\vec{A} \cdot \vec{B} = 0$. *The scalar product of two perpendicular vectors is always zero.*

For any two vectors $\vec{A}$ and $\vec{B}$, $AB \cos \phi = BA \cos \phi$. This means that $\vec{A} \cdot \vec{B} = \vec{B} \cdot \vec{A}$. The scalar product obeys the commutative law of multiplication; the order of the two vectors does not matter.

We will use the scalar product in Chapter 6 to describe work done by a force. When a constant force $\vec{F}$ is applied to a body that undergoes a displacement $\vec{s}$, the work W (a scalar quantity) done by the force is given by

$$W = \vec{F} \cdot \vec{s}.$$

The work done by the force is positive if the angle between $\vec{F}$ and $\vec{s}$ is between 0° and 90°, negative if this angle is between 90° and 180°, and zero if $\vec{F}$ and $\vec{s}$ are perpendicular. (This is another example of a term that has a special meaning in physics; in everyday language, "work" isn't something that can be positive or negative.) In later chapters we'll use the scalar product for a variety of purposes, from calculating electric potential to determining the effects that varying magnetic fields have on electric circuits.

We can calculate the scalar product $\vec{A} \cdot \vec{B}$ directly if we know the x-, y-, and z-components of $\vec{A}$ and $\vec{B}$. To see how this is done, let's first work out the scalar products of the unit vectors. This is easy, since $\hat{\imath}, \hat{\jmath},$ and $\hat{k}$ are all perpendicular to each other. Using Eq. (1–18), we find

$$\hat{\imath} \cdot \hat{\imath} = \hat{\jmath} \cdot \hat{\jmath} = \hat{k} \cdot \hat{k} = (1)(1) \cos 0 = 1,$$
$$\hat{\imath} \cdot \hat{\jmath} = \hat{\imath} \cdot \hat{k} = \hat{\jmath} \cdot \hat{k} = (1)(1) \cos 90° = 0. \quad (1\text{--}19)$$

Now we express $\vec{A}$ and $\vec{B}$ in terms of their components, expand the product, and use

(a)

(b)

(c)

1–17 (a) Two vectors drawn from a common starting point to define their scalar product $\vec{A} \cdot \vec{B} = AB \cos \phi$. (b) $B \cos \phi$ is the component of $\vec{B}$ in the direction of $\vec{A}$, and $\vec{A} \cdot \vec{B}$ is the product of the magnitude of $\vec{A}$ and this component. (c) $\vec{A} \cdot \vec{B}$ is also the product of the magnitude of $\vec{B}$ and the component of $\vec{A}$ in the direction of $\vec{B}$.

these products of unit vectors:

$$\vec{A} \cdot \vec{B} = (A_x \hat{i} + A_y \hat{j} + A_z \hat{k}) \cdot (B_x \hat{i} + B_y \hat{j} + B_z \hat{k})$$

$$= A_x \hat{i} \cdot B_x \hat{i} + A_x \hat{i} \cdot B_y \hat{j} + A_x \hat{i} \cdot B_z \hat{k}$$

$$+ A_y \hat{j} \cdot B_x \hat{i} + A_y \hat{j} \cdot B_y \hat{j} + A_y \hat{j} \cdot B_z \hat{k}$$

$$+ A_z \hat{k} \cdot B_x \hat{i} + A_z \hat{k} \cdot B_y \hat{j} + A_z \hat{k} \cdot B_z \hat{k}$$

$$= A_x B_x \hat{i} \cdot \hat{i} + A_x B_y \hat{i} \cdot \hat{j} + A_x B_z \hat{i} \cdot \hat{k}$$

$$+ A_y B_x \hat{j} \cdot \hat{i} + A_y B_y \hat{j} \cdot \hat{j} + A_y B_z \hat{j} \cdot \hat{k}$$

$$+ A_z B_x \hat{k} \cdot \hat{i} + A_z B_y \hat{k} \cdot \hat{j} + A_z B_z \hat{k} \cdot \hat{k}. \tag{1-20}$$

From Eqs. (1–19) we see that six of these nine terms are zero, and the three that survive give simply

$$\vec{A} \cdot \vec{B} = A_x B_x + A_y B_y + A_z B_z \qquad \text{(scalar (dot) product in terms of components)}. \tag{1-21}$$

Thus the scalar product of two vectors is the sum of the products of their respective components.

The scalar product gives a straightforward way to find the angle ϕ between any two vectors $\vec{A}$ and $\vec{B}$ whose components are known. In this case, Eq. (1–21) can be used to find the scalar product of $\vec{A}$ and $\vec{B}$. From Eq. (1–18) the scalar product is also equal to $AB \cos \phi$. The vector magnitudes A and B can be found from the vector components with Eq. (1–12), so $\cos \phi$ and hence the angle ϕ can be determined (see Example 1–11).

EXAMPLE 1–10

Calculating a scalar product Find the scalar product $\vec{A} \cdot \vec{B}$ of the two vectors in Fig. 1–18. The magnitudes of the vectors are $A = 4.00$ and $B = 5.00$.

SOLUTION There are two ways to calculate the scalar product. The first way uses the magnitudes of the vectors and the angle between them (Eq. 1–18), and the second uses the components of the two vectors (Eq. 1–21).

Using the first approach, the angle between the two vectors is $\phi = 130.0° - 53.0° = 77.0°$, so

$$\vec{A} \cdot \vec{B} = AB \cos \phi = (4.00)(5.00) \cos 77.0° = 4.50.$$

This is positive because the angle between $\vec{A}$ and $\vec{B}$ is between $0°$ and $90°$.

To use the second approach, we first need to find the components of the two vectors. Since the angles of $\vec{A}$ and $\vec{B}$ are given with respect to the $+x$-axis, and these angles are measured in the sense from the $+x$-axis to the $+y$-axis, we can use Eqs. (1–7):

$$A_x = (4.00) \cos 53.0° = 2.407,$$

$$A_y = (4.00) \sin 53.0° = 3.195,$$

$$A_z = 0,$$

$$B_x = (5.00) \cos 130.0° = -3.214,$$

$$B_y = (5.00) \sin 130.0° = 3.830,$$

$$B_z = 0.$$

The z-components are zero because both vectors lie in the xy-plane. As in Example 1–7, we are keeping one too many significant figures in the components; we'll round to the correct number at the end. From Eq. (1–21) the scalar product is

$$\vec{A} \cdot \vec{B} = A_x B_x + A_y B_y + A_z B_z$$

$$= (2.407)(-3.214) + (3.195)(3.830) + (0)(0) = 4.50.$$

We get the same result for the scalar product with both methods, as we should.

1–18 Two vectors in two dimensions.

EXAMPLE 1-11

Finding angles with the scalar product Find the angle between the two vectors

$$\vec{A} = 2\hat{\imath} + 3\hat{\jmath} + \hat{k} \quad \text{and} \quad \vec{B} = -4\hat{\imath} + 2\hat{\jmath} - \hat{k}.$$

SOLUTION The vectors are shown in Fig. 1-19. The scalar product of two vectors is given by either Eq. (1-18) or Eq. (1-21). Equating these two and rearranging, we obtain

$$\cos\phi = \frac{A_x B_x + A_y B_y + A_z B_z}{AB}.$$

This formula can be used to find the angle between any two vectors $\vec{A}$ and $\vec{B}$. For our example the components of $\vec{A}$ are $A_x = 2$, $A_y = 3$, and $A_z = 1$, and the components of $\vec{B}$ are $B_x = -4$, $B_y = 2$, and $B_z = -1$. Thus

$$\vec{A} \cdot \vec{B} = A_x B_x + A_y B_y + A_z B_z = (2)(-4) + (3)(2) + (1)(-1) = -3,$$
$$A = \sqrt{A_x^2 + A_y^2 + A_z^2} = \sqrt{2^2 + 3^2 + 1^2} = \sqrt{14},$$
$$B = \sqrt{B_x^2 + B_y^2 + B_z^2} = \sqrt{(-4)^2 + 2^2 + (-1)^2} = \sqrt{21},$$
$$\cos\phi = \frac{A_x B_x + A_y B_y + A_z B_z}{AB} = \frac{-3}{\sqrt{14}\sqrt{21}} = -0.175,$$
$$\phi = 100°.$$

As a check on this result, note that the scalar product $\vec{A} \cdot \vec{B}$ is negative. This means that ϕ is between 90° and 180°, in agreement with our answer.

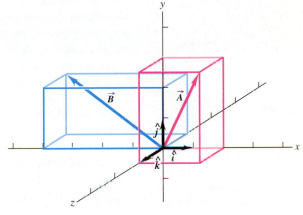

1-19 Two vectors in three dimensions.

VECTOR PRODUCT

The **vector product** of two vectors $\vec{A}$ and $\vec{B}$, also called the **cross product,** is denoted by $\vec{A} \times \vec{B}$. We will use this product in Chapter 10 to describe torque and angular momentum. Later we will also use it extensively for magnetic fields, where it will help us describe the relationships among the directions of several vector quantities.

To define the vector product $\vec{A} \times \vec{B}$ of two vectors $\vec{A}$ and $\vec{B}$, we again draw the two vectors with their tails at the same point (Fig. 1-20a). The two vectors then lie in a plane. We define the vector product to be a vector quantity with a direction perpendicular to this plane (that is, perpendicular to both $\vec{A}$ and $\vec{B}$) and a magnitude equal to $AB \sin\phi$. That is, if $\vec{C} = \vec{A} \times \vec{B}$, then

$$C = AB \sin\phi \quad \text{(magnitude of the vector (cross) product of } \vec{A} \text{ and } \vec{B} \text{).} \quad (1-22)$$

We measure the angle ϕ from $\vec{A}$ toward $\vec{B}$ and take it to be the smaller of the two possible angles, so ϕ ranges from 0° to 180°. Thus C in Eq. (1-22) is always positive, as a vector magnitude should be. Note also that when $\vec{A}$ and $\vec{B}$ are parallel or antiparallel, $\phi = 0$ or 180° and $C = 0$. That is, *the vector product of two parallel or antiparallel vectors is always zero.* In particular, *the vector product of any vector with itself is zero.* To see the contrast between the scalar product and the magnitude of the vector product, imagine that we vary the angle between $\vec{A}$ and $\vec{B}$ while keeping their magnitudes constant. When $\vec{A}$ and $\vec{B}$ are parallel, the scalar product will be maximum and the magnitude of the vector product will be zero. When $\vec{A}$ and $\vec{B}$ are perpendicular, the scalar product will be zero and the magnitude of the vector product will be maximum.

There are always *two* directions perpendicular to a given plane, one on each side of the plane. We choose which of these is the direction of $\vec{A} \times \vec{B}$ as follows. Imagine rotating vector $\vec{A}$ about the perpendicular line until it is aligned with $\vec{B}$, choosing the smaller of the two possible angles between $\vec{A}$ and $\vec{B}$. Curl the fingers of your right hand around the perpendicular line so that the fingertips point in the direction of rotation; your thumb will then point in the direction of $\vec{A} \times \vec{B}$. This **right-hand rule** is shown in Fig. 1-20a. The direction of the vector product is also the direction in which a right-hand screw advances if turned from $\vec{A}$ toward $\vec{B}$, as shown.

(a)

(b)

1-20 (a) Vectors $\vec{A}$ and $\vec{B}$ lie in a plane; the vector product $\vec{A} \times \vec{B}$ is perpendicular to this plane in a direction determined by the right-hand rule. (b) $\vec{B} \times \vec{A} = -\vec{A} \times \vec{B}$; the vector product is anticommutative.

1–21 (a) $B \sin \phi$ is the component of $\vec{B}$ perpendicular to the direction of $\vec{A}$, and the magnitude of $\vec{A} \times \vec{B}$ is the product of the magnitude of $\vec{A}$ and this component. (b) The magnitude of $\vec{A} \times \vec{B}$ is also the product of the magnitude of $\vec{B}$ and the component of $\vec{A}$ perpendicular to $\vec{B}$.

Similarly, we determine the direction of $\vec{B} \times \vec{A}$ by rotating $\vec{B}$ into $\vec{A}$ in Fig. 1–20b. The result is a vector that is *opposite* to the vector $\vec{A} \times \vec{B}$. The vector product is not commutative! In fact, for any two vectors $\vec{A}$ and $\vec{B}$,

$$\vec{A} \times \vec{B} = -\vec{B} \times \vec{A}. \tag{1–23}$$

Just as we did for the scalar product, we can give a geometrical interpretation of the magnitude of the vector product. In Fig. 1–21a, $B \sin \phi$ is the component of vector $\vec{B}$ that is *perpendicular* to the direction of vector $\vec{A}$. From Eq. (1–22) the magnitude of $\vec{A} \times \vec{B}$ equals the magnitude of $\vec{A}$ multiplied by the component of $\vec{B}$ perpendicular to $\vec{A}$. Figure 1–21b shows that the magnitude of $\vec{A} \times \vec{B}$ also equals the magnitude of $\vec{B}$ multiplied by the component of $\vec{A}$ perpendicular to $\vec{B}$. Note that Fig. 1–21 shows the case in which ϕ is between $0°$ and $90°$; you should draw a similar diagram for ϕ between $90°$ and $180°$ to show that the same geometrical interpretation of the magnitude of $\vec{A} \times \vec{B}$ still applies.

If we know the components of $\vec{A}$ and $\vec{B}$, we can calculate the components of the vector product, using a procedure similar to that for the scalar product. First we work out the multiplication table for the unit vectors $\hat{\imath}, \hat{\jmath}$, and $\hat{k}$. The vector product of any vector with itself is zero, so

$$\hat{\imath} \times \hat{\imath} = \hat{\jmath} \times \hat{\jmath} = \hat{k} \times \hat{k} = 0.$$

The boldface zero is a reminder that each product is a zero *vector*, that is, one with all components equal to zero and an undefined direction. Using Eqs. (1–22) and (1–23) and the right-hand rule, we find

$$\hat{\imath} \times \hat{\jmath} = -\hat{\jmath} \times \hat{\imath} = \hat{k},$$
$$\hat{\jmath} \times \hat{k} = -\hat{k} \times \hat{\jmath} = \hat{\imath}, \tag{1–24}$$
$$\hat{k} \times \hat{\imath} = -\hat{\imath} \times \hat{k} = \hat{\jmath}.$$

Next we express $\vec{A}$ and $\vec{B}$ in terms of their components and the corresponding unit vectors, and we expand the expression for the vector product:

$$\begin{aligned} \vec{A} \times \vec{B} &= (A_x\hat{\imath} + A_y\hat{\jmath} + A_z\hat{k}) \times (B_x\hat{\imath} + B_y\hat{\jmath} + B_z\hat{k}) \\ &= A_x\hat{\imath} \times B_x\hat{\imath} + A_x\hat{\imath} \times B_y\hat{\jmath} + A_x\hat{\imath} \times B_z\hat{k} \\ &\quad + A_y\hat{\jmath} \times B_x\hat{\imath} + A_y\hat{\jmath} \times B_y\hat{\jmath} + A_y\hat{\jmath} \times B_z\hat{k} \\ &\quad + A_z\hat{k} \times B_x\hat{\imath} + A_z\hat{k} \times B_y\hat{\jmath} + A_z\hat{k} \times B_z\hat{k}. \end{aligned} \tag{1–25}$$

We can also rewrite the individual terms as $A_x\hat{\imath} \times B_y\hat{\jmath} = (A_xB_y)\,\hat{\imath} \times \hat{\jmath}$, and so on. Evaluating these by using the multiplication table for the unit vectors and then grouping the terms, we find

$$\vec{A} \times \vec{B} = (A_yB_z - A_zB_y)\hat{\imath} + (A_zB_x - A_xB_z)\hat{\jmath} + (A_xB_y - A_yB_x)\hat{k}. \tag{1–26}$$

Thus the components of $\vec{C} = \vec{A} \times \vec{B}$ are given by

$$C_x = A_yB_z - A_zB_y, \quad C_y = A_zB_x - A_xB_z, \quad C_z = A_xB_y - A_yB_x$$
$$\text{(components of } \vec{C} = \vec{A} \times \vec{B}). \tag{1–27}$$

The vector product can also be expressed in determinant form as

$$\vec{A} \times \vec{B} = \begin{vmatrix} \hat{\imath} & \hat{\jmath} & \hat{k} \\ A_x & A_y & A_z \\ B_x & B_y & B_z \end{vmatrix}.$$

If you aren't familiar with determinants, don't worry about this form.

With the axis system of Fig. 1–22a, if we reverse the direction of the z-axis, we get the system shown in Fig. 1–22b. Then, as you may verify, the definition of the vector

1–22 (a) A right-handed coordinate system, in which $\hat{\imath} \times \hat{\jmath} = \hat{k}$, $\hat{\jmath} \times \hat{k} = \hat{\imath}$, and $\hat{k} \times \hat{\imath} = \hat{\jmath}$. (b) A left-handed coordinate system, in which $\hat{\imath} \times \hat{\jmath} = -\hat{k}$, and so on. We'll use only right-handed systems.

product gives $\hat{\imath} \times \hat{\jmath} = -\hat{k}$ instead of $\hat{\imath} \times \hat{\jmath} = \hat{k}$. In fact, all vector products of the unit vectors $\hat{\imath}, \hat{\jmath},$ and $\hat{k}$ would have signs opposite to those in Eqs. (1–24). We see that there are two kinds of coordinate systems, differing in the signs of the vector products of unit vectors. An axis system in which $\hat{\imath} \times \hat{\jmath} = \hat{k}$, as in Fig. 1–22a, is called a **right-handed system.** The usual practice is to use *only* right-handed systems, and we will follow that practice throughout this book.

EXAMPLE 1–12

Calculating a vector product Vector $\vec{A}$ has magnitude 6 units and is in the direction of the $+x$-axis. Vector $\vec{B}$ has magnitude 4 units and lies in the xy-plane, making an angle of 30° with the $+x$-axis (Fig. 1–23). Find the vector product $\vec{A} \times \vec{B}$.

SOLUTION We can find the vector product in one of two ways. The first way is to use Eq. (1–22) to determine the magnitude of $\vec{A} \times \vec{B}$ and then use the right-hand rule to find the direction of the vector product. The second way is to use the components of $\vec{A}$ and $\vec{B}$ to find the components of the vector product $\vec{C} = \vec{A} \times \vec{B}$ using Eqs. (1–27).

With the first approach, from Eq. (1–22) the magnitude of the vector product is

$$AB \sin \phi = (6)(4)(\sin 30°) = 12.$$

From the right-hand rule, the direction of $\vec{A} \times \vec{B}$ is along the $+z$-axis, so we have $\vec{A} \times \vec{B} = 12\,\hat{k}$.

To use the second approach, we first write the components of $\vec{A}$ and $\vec{B}$:

$$A_x = 6, \qquad A_y = 0, \qquad A_z = 0,$$
$$B_x = 4 \cos 30° = 2\sqrt{3}, \quad B_y = 4 \sin 30° = 2, \quad B_z = 0.$$

Defining $\vec{C} = \vec{A} \times \vec{B}$, we have from Eqs. (1–27) that

$$C_x = (0)(0) - (0)(2) \quad = 0,$$
$$C_y = (0)(2\sqrt{3}) - (6)(0) = 0,$$
$$C_z = (6)(2) - (0)(2\sqrt{3}) = 12.$$

The vector product $\vec{C}$ has only a z-component, and it lies along the $+z$-axis. The magnitude agrees with the result we obtained with the first approach, as it should.

For this example the first approach was more direct because we knew the magnitudes of each vector and the angle between them, and furthermore, both vectors lay in one of the planes of the coordinate system. But often you will need to find the vector product of two vectors that are not so conveniently oriented or for which only the components are given. In such a case the second approach, using components, is the more direct one.

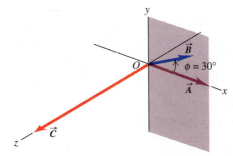

1–23 Vectors $\vec{A}$ and $\vec{B}$ and their vector product $\vec{C} = \vec{A} \times \vec{B}$. The vector $\vec{B}$ lies in the xy-plane.

SUMMARY

- The fundamental physical quantities of mechanics are mass, length, and time. The corresponding basic SI units are the kilogram, the meter, and the second. Other units for these quantities, related by powers of 10, are identified by adding prefixes to the basic unit. Derived units for other physical quantities are products or quotients of the basic units. Equations must be dimensionally consistent; two terms can be added only when they have the same units.

- The accuracy of a measurement can be indicated by the number of significant figures, or by a stated uncertainty. The result of a multiplication or division usually has no more significant figures than the input data. When only crude estimates are available for input data, we can often make useful order-of-magnitude estimates.

- Scalar quantities are numbers and are combined with the usual rules of arithmetic. Vector quantities have direction as well as magnitude and are combined according to the rules of vector addition. Graphically, two vectors $\vec{A}$ and $\vec{B}$ are added by

KEY TERMS

range of validity, 2
model, 3
particle, 3
physical quantity, 3
operational definition, 3
unit, 4
International System (SI), 4
second, 4
meter, 4
kilogram, 4
prefix, 5
dimensionally consistent, 6

placing the tail of $\vec{B}$ at the head, or tip, of $\vec{A}$. The vector sum $\vec{A} + \vec{B}$ then extends from the tail of $\vec{A}$ to the head of $\vec{B}$.

■ Vector addition can be carried out by using components of vectors. If A_x and A_y are the components of vector $\vec{A}$ and B_x and B_y are the components of vector $\vec{B}$, the components of the vector sum $\vec{R} = \vec{A} + \vec{B}$ are given by

$$R_x = A_x + B_x, \qquad R_y = A_y + B_y. \qquad (1\text{–}10)$$

■ Unit vectors describe directions in space. A unit vector has a magnitude of 1, with no units. We always write unit vectors with a caret or "hat" (^). The unit vectors $\hat{\imath}, \hat{\jmath},$ and $\hat{k}$, aligned with the x-, y-, and z-axes of a rectangular coordinate system, are especially useful.

■ The scalar product $C = \vec{A} \cdot \vec{B}$ of two vectors $\vec{A}$ and $\vec{B}$ is a scalar quantity, defined as

$$\vec{A} \cdot \vec{B} = AB\cos\phi = |\vec{A}|\,|\vec{B}|\cos\phi. \qquad (1\text{–}18)$$

The scalar product can also be expressed in terms of components:

$$\vec{A} \cdot \vec{B} = A_x B_x + A_y B_y + A_z B_z. \qquad (1\text{–}21)$$

The scalar product is commutative; for any two vectors $\vec{A}$ and $\vec{B}$, $\vec{A} \cdot \vec{B} = \vec{B} \cdot \vec{A}$. The scalar product of two perpendicular vectors is zero.

■ The vector product $\vec{C} = \vec{A} \times \vec{B}$ of two vectors $\vec{A}$ and $\vec{B}$ is another vector $\vec{C}$, with magnitude given by

$$C = AB\sin\phi. \qquad (1\text{–}22)$$

The direction of the vector product is perpendicular to the plane of the two vectors being multiplied, as given by the right-hand rule. In terms of the components of the two vectors being multiplied, the components of the vector product are

$$C_x = A_y B_z - A_z B_y, \qquad C_y = A_z B_x - A_x B_z, \qquad C_z = A_x B_y - A_y B_x. \qquad (1\text{–}27)$$

The vector product is not commutative; for any two vectors $\vec{A}$ and $\vec{B}$, $\vec{A} \times \vec{B} = -\vec{B} \times \vec{A}$. The vector product of two parallel or antiparallel vectors is zero.

DISCUSSION QUESTIONS

Q1–1 How many correct experiments do we need to disprove a theory? How many to prove a theory? Explain.

Q1–2 A guidebook describes the rate of climb of a mountain trail as 120 meters per kilometer. How can this be expressed as a number with no units?

Q1–3 Suppose you are asked to compute the tangent of 5.00 meters. Is this possible? Why or why not?

Q1–4 A highway contractor stated that in building a bridge deck he poured 250 yards of concrete. What do you think he meant?

Q1–5 What is your height in centimeters? What is your weight in newtons?

Q1–6 The U.S. National Institute of Science and Technology (NIST) maintains several accurate copies of the international standard kilogram. Even after careful cleaning, these national standard kilograms are gaining mass at an average rate of about 1 μg/y (1 y = 1 year) when compared every ten years or so to the standard international kilogram. Does this apparent change have any importance? Explain.

Q1–7 What physical phenomena (other than a pendulum or cesium clock) could be used to define a time standard?

Q1–8 Describe how you could measure the thickness of a sheet of paper with an ordinary ruler.

Q1–9 The quantity $\pi = 3.14159...$ is a number with no dimensions, since it is a ratio of two lengths. Describe two or three other geometrical or physical quantities that are dimensionless.

Q1–10 What are the units of volume? Suppose another student tells you that a cylinder of radius r and height h has volume given by $\pi r^3 h$. Explain why this cannot be right.

Q1–11 Three archers each fire four arrows at a target. Joe's four arrows hit at points 10 cm above, 10 cm below, 10 cm to the left, and 10 cm to the right of the center of the target. All four of Moe's arrows hit within 1 cm of a point 20 cm from the center, and Flo's four arrows all hit within 1 cm of the center. The contest judge says that one of the archers is precise but not accurate, another archer is accurate but not precise, and the third archer is both accurate and precise. Which description goes with which archer? Explain your reasoning.

Q1–12 A circular racetrack has radius 500 m. What is the displacement of a bicyclist when she travels around the track from the north side to the south side? When she makes one complete circle around the track? Explain your reasoning.

Q1–13 Can you find two vectors with different lengths that have a vector sum of zero? What length restrictions are required for three vectors to have a vector sum of zero? Explain your reasoning.

Q1–14 One sometimes speaks of the "direction of time," evolving from past to future. Does this mean that time is a vector quantity? Explain your reasoning.

Q1–15 Air traffic controllers give instructions to airline pilots telling them in what direction they are to fly. These instructions are called "vectors." If these are the only instructions given, is the name "vector" being used correctly? Why or why not?

Q1–16 Can you find a vector quantity that has a magnitude of zero, but components that are different from zero? Explain.

Q1–17 Let $\vec{A}$ represent any nonzero vector. Why is $\vec{A}/A$ a unit vector and what is its direction? If θ is the angle that $\vec{A}$ makes with the $+x$-axis, explain why $(\vec{A}/A) \cdot \hat{\imath}$ is called the *direction cosine* for that axis.

Q1–18 Which of the following are legitimate mathematical operations: a) $\vec{A} \cdot (\vec{B} - \vec{C})$; b) $(\vec{A} - \vec{B}) \times \vec{C}$; c) $\vec{A} \cdot (\vec{B} \times \vec{C})$; d) $\vec{A} \times (\vec{B} \times \vec{C})$; e) $\vec{A} \times (\vec{B} \cdot \vec{C})$? In each case, give the reason for your answer.

Q1–19 Consider the two repeated vector products $\vec{A} \times (\vec{B} \times \vec{C})$ and $(\vec{A} \times \vec{B}) \times \vec{C}$. Give an example that illustrates the general rule that these two vector products do not have the same magnitude or direction. Can you choose the vectors $\vec{A}$, $\vec{B}$, and $\vec{C}$ such that these two vector products *are* equal to each other? If so, give an example.

EXERCISES

SECTION 1–4 STANDARDS AND UNITS

SECTION 1–5 UNIT CONSISTENCY AND CONVERSIONS

1–1 Starting with the definition 1 in. = 2.54 cm, find the number of kilometers in 1.00 mile.

1–2 According to the label on a bottle of salad dressing, the volume of the contents is 0.473 liter (L). Using only the conversions 1 L = 1000 cm^3 and 1 in. = 2.54 cm, express this volume in cubic inches.

1–3 How many nanoseconds does it take light to travel 1.00 km in vacuum?

1–4 The density of lead is 11.3 g/cm^3. What is this value in kilograms per cubic meter?

1–5 The most powerful engine available for the classic 1963 Chevrolet Corvette Sting Ray developed 360 horsepower and had a displacement of 327 cubic inches. Express this displacement in liters (L) by using only the conversions 1 L = 1000 cm^3 and 1 in. = 2.54 cm.

1–6 Having been told he needs to set goals for himself, Billy Joe Bob decides to drink one cubic meter of his favorite beverage during the coming year. How many 16-fluid-ounce bottles should he drink each day? (Use Appendix E. A fluid ounce is a volume unit; 128 fluid ounces equals one gallon.)

1–7 The Concorde is the fastest airliner used for commercial service. It cruises at 1450 mi/h (about two times the speed of sound, or Mach 2). a) What is the cruise speed of the Concorde in km/h? b) What is it in m/s?

1–8 While driving in an exotic foreign land you see a speed limit sign on a highway that reads 180,000 furlongs per fortnight. How many miles per hour is this? (One furlong is $\frac{1}{8}$ mile, and a fortnight is 14 days. A furlong originally referred to the length of a plowed furrow.)

1–9 The gasoline consumption of a small car is advertised as 15.0 km/L (1 L = 1 liter). How many miles per gallon is this? Use the conversion factors in Appendix E.

SECTION 1–6 UNCERTAINTY AND SIGNIFICANT FIGURES

1–10 A useful and easy-to-remember approximate value for the number of seconds in a year is $\pi \times 10^7$. Determine the percent error in this approximate value. (There are 365.24 days in one year.)

1–11 Figure 1–3 shows the result of unacceptable error in the stopping position of a train. a) If a train travels 890 km from Berlin to Paris and then overshoots the end of the track by 10 m, what is the percentage error in the total distance covered? b) Would it be correct to write the total distance covered by the train as 890,010 m? Explain.

1–12 With a wooden ruler you measure the length of a rectangular piece of sheet metal to be 12 mm. You use a micrometer calipers to measure the width of the rectangle and obtain the value 5.98 mm. Give your answers to the following questions to the correct number of significant figures. a) What is the area of the rectangle? b) What is the ratio of the rectangle's width to its length? c) What is the perimeter of the rectangle? d) What is the difference between the length and width? e) What is the ratio of the length to the width?

1–13 Estimate the percent error in measuring a) a distance of about 75 cm with a meter stick; b) a mass of about 12 g with a chemical balance; c) a time interval of about 6 min with a stopwatch.

1–14 A rectangular piece of aluminum is 5.60 ± 0.01 cm long and 1.90 ± 0.01 cm wide. a) Find the area of the rectangle and the uncertainty in the area. b) Verify that the fractional uncertainty in the area is equal to the sum of the fractional uncertainties in the length and in the width. (This is a general result; see Challenge Problem 1–76.)

1–15 As you eat your way through a bag of chocolate chip cookies, you observe that each cookie is a circular disk of diameter 8.50 ± 0.02 cm and thickness 0.050 ± 0.005 cm. a) Find the average volume of a cookie and the uncertainty in the volume. b) Find the ratio of the diameter to the thickness and the uncertainty in this ratio.

SECTION 1–7 ESTIMATES AND ORDERS OF MAGNITUDE

1–16 How many gallons of gasoline are used in the United States in one day?

1–17 A box of typewriter paper has dimensions 11 in. × 17 in. × 9 in.; it is marked "10 M." Does that mean it contains 10,000 sheets or 10 million?

1–18 How many kernels of corn does it take to fill a 2-L soft drink bottle?

1–19 How many words are there in this book?

1–20 What total volume of air does a person breathe in a lifetime? How does that compare with the volume of the Houston Astrodome? (Estimate that a person breathes about 500 cm^3 of air with each breath.)

1–21 How many hairs do you have on your head?

1–22 How many times does a human heart beat during a lifetime? How many gallons of blood does it pump? (Estimate that the heart pumps 50 cm^3 of blood with each beat.)

1–23 In Wagner's opera *Ring of the Nibelung,* the goddess Freia is ransomed by a pile of gold just tall enough and wide enough to hide her from sight. Estimate the monetary value of this pile. (Refer to Example 1–4 for information on the price per ounce and density of gold.)

1–24 How many drops of water are in all the oceans on earth?

1–25 How many pizzas are consumed each academic year by students at your school?

1–26 How many dollar bills would have to be stacked to reach the moon? Would that be cheaper than building and launching a spacecraft?

1–27 How much would it cost to paper the entire United States (including Alaska and Hawaii) with dollar bills? What would be the cost to each person in the United States?

SECTION 1–8 VECTORS AND VECTOR ADDITION

1–28 Hearing rattles from a snake, you make two rapid displacements of magnitude 1.8 m and 2.4 m. In sketches (roughly to scale), show how your two displacements might add to give a resultant of magnitude a) 4.2 m; b) 0.6 m; c) 3.0 m.

1–29 A postal employee drives a delivery truck along the route shown in Fig. 1–24. Determine the magnitude and direction of

FIGURE 1–24 Exercises 1–29 and 1–34.

the resultant displacement by drawing a scale diagram. (See also Exercise 1–34 for a different approach to this same problem.)

1–30 For the vectors $\vec{A}$ and $\vec{B}$ in Fig. 1–25, use a scale drawing to find the magnitude and direction of a) the vector sum $\vec{A} + \vec{B}$; b) the vector difference $\vec{A} - \vec{B}$. From your answers to parts (a) and (b), find the magnitude and direction of c) $-\vec{A} - \vec{B}$; d) $\vec{B} - \vec{A}$. (See also Exercise 1–35 for a different approach to this problem.)

FIGURE 1–25 Exercises 1–30, 1–35, 1–40, and 1–48.

1–31 A spelunker is surveying a cave. She follows a passage 180 m straight west, then 210 m in a direction 45° east of south, then 90 m at 30° west of north. After a fourth unmeasured displacement she finds herself back where she started. Use a scale drawing to determine the magnitude and direction of the fourth displacement. (See also Problem 1–59 for a different approach to a similar problem.)

SECTION 1–9 COMPONENTS OF VECTORS

1–32 Use a scale drawing to find the *x*- and *y*-components of the following vectors. For each vector the numbers given are i) the magnitude of the vector and ii) the angle, measured in the sense from the +*x*-axis toward the +*y*-axis, that it makes with the +*x*-axis. Find a) magnitude 9.30 m, angle 60.0°; b) magnitude 22.0 km, angle 135°; c) magnitude 6.35 cm, angle 307°.

1–33 Compute the *x*- and *y*-components of the vectors $\vec{A}$, $\vec{B}$, and $\vec{C}$ in Fig. 1–26.

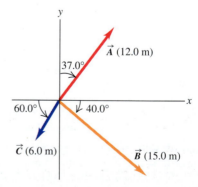

FIGURE 1–26 Exercises 1–33, 1–41, and 1–44, and Problem 1–58.

1–34 A postal employee drives a delivery truck over the route shown in Fig. 1–24. Use the method of components to determine the magnitude and direction of her resultant displacement. In a vector addition diagram (roughly to scale), show that the resultant displacement found from your diagram is in qualitative agreement with the result you obtained using the method of components.

1–35 For the vectors $\vec{A}$ and $\vec{B}$ in Fig. 1–25, use the method of components to find the magnitude and direction of a) the vector sum $\vec{A} + \vec{B}$; b) the vector sum $\vec{B} + \vec{A}$; c) the vector difference $\vec{A} - \vec{B}$; d) the vector difference $\vec{B} - \vec{A}$.

1–36 Find the magnitude and direction of the vector represented by the following pairs of components: a) $A_x = -8.60$ cm, $A_y = 5.20$ cm; b) $A_x = -9.70$ m, $A_y = -2.45$ m; c) $A_x = 7.75$ km, $A_y = -2.70$ km.

1–37 A disoriented physics professor drives 3.25 km north, then 4.75 km west, then 1.50 km south. Find the magnitude and direction of the resultant displacement, using the method of components. In a vector addition diagram (roughly to scale), show that the resultant displacement found from your diagram is in qualitative agreement with the result you obtained using the method of components.

1–38 Vector $\vec{A}$ has components $A_x = 1.30$ cm, $A_y = 2.25$ cm; vector $\vec{B}$ has components $B_x = 4.10$ cm, $B_y = -3.75$ cm. Find a) the components of the vector sum $\vec{A} + \vec{B}$; b) the magnitude and direction of $\vec{A} + \vec{B}$; c) the components of the vector difference $\vec{B} - \vec{A}$; d) the magnitude and direction of $\vec{B} - \vec{A}$.

1–39 Vector $\vec{A}$ is 2.80 cm long and is 60.0° above the x-axis in the first quadrant. Vector $\vec{B}$ is 1.90 cm long and is 60.0° below the x-axis in the fourth quadrant (Fig. 1–27). Find the magnitude and direction of a) $\vec{A} + \vec{B}$; b) $\vec{A} - \vec{B}$; c) $\vec{B} - \vec{A}$. In each case, sketch the vector addition or subtraction and show that your numerical answers are in qualitative agreement with your sketch.

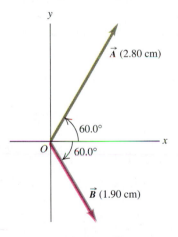

FIGURE 1–27 Exercises 1–39 and 1–50.

SECTION 1–10 UNIT VECTORS

1–40 Write each vector in Fig. 1–25 in terms of the unit vectors $\hat{\imath}$ and $\hat{\jmath}$.

1–41 Write each vector in Fig. 1–26 in terms of the unit vectors $\hat{\imath}$ and $\hat{\jmath}$.

1–42 a) Write each vector in Fig. 1–28 in terms of the unit vectors $\hat{\imath}$ and $\hat{\jmath}$. b) Use unit vectors to express the vector $\vec{C}$, where $\vec{C} = 3.00\vec{A} - 4.00\vec{B}$. c) Find the magnitude and direction of $\vec{C}$.

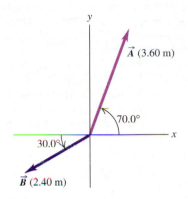

FIGURE 1–28 Exercise 1–42 and Problem 1–66.

1–43 Given two vectors $\vec{A} = 4.00\,\hat{\imath} + 3.00\,\hat{\jmath}$ and $\vec{B} = 5.00\,\hat{\imath} - 2.00\,\hat{\jmath}$, a) find the magnitude of each vector; b) write an expression for the vector difference $\vec{A} - \vec{B}$ using unit vectors; c) find the magnitude and direction of the vector difference $\vec{A} - \vec{B}$. d) In a vector diagram show $\vec{A}$, $\vec{B}$, and $\vec{A} - \vec{B}$, and also show that your diagram agrees qualitatively with your answer in part (c).

SECTION 1–11 PRODUCTS OF VECTORS

1–44 For the vectors $\vec{A}$, $\vec{B}$, and $\vec{C}$ in Fig. 1–26, find the scalar products a) $\vec{A} \cdot \vec{B}$; b) $\vec{B} \cdot \vec{C}$; c) $\vec{A} \cdot \vec{C}$.

1–45 a) Find the scalar product of the two vectors $\vec{A}$ and $\vec{B}$ given in Exercise 1–43. b) Find the angle between these two vectors.

1–46 Find the angle between each of the following pairs of vectors:
a) $\vec{A} = -2.00\,\hat{\imath} + 6.00\,\hat{\jmath}$ and $\vec{B} = 2.00\,\hat{\imath} - 3.00\,\hat{\jmath}$;
b) $\vec{A} = 3.00\,\hat{\imath} + 5.00\,\hat{\jmath}$ and $\vec{B} = 10.0\,\hat{\imath} + 6.00\,\hat{\jmath}$;
c) $\vec{A} = -4.00\,\hat{\imath} + 2.00\,\hat{\jmath}$ and $\vec{B} = 7.00\,\hat{\imath} + 14.0\,\hat{\jmath}$.

1–47 Assuming a right-handed coordinate system, find the direction of the $+z$-axis in a) Fig. 1–13a; b) Fig. 1–13b.

1–48 For the two vectors in Fig. 1–25, a) find the magnitude and direction of the vector product $\vec{A} \times \vec{B}$; b) find the magnitude and direction of $\vec{B} \times \vec{A}$.

1–49 Find the vector product $\vec{A} \times \vec{B}$ (expressed in unit vectors) of the two vectors given in Exercise 1–43. What is the magnitude of the vector product?

1–50 For the two vectors in Fig. 1–27, a) find the magnitude and direction of the vector product $\vec{A} \times \vec{B}$; b) find the magnitude and direction of $\vec{B} \times \vec{A}$.

PROBLEMS

1–51 An acre, a unit of land measurement still in wide use, has a length of one furlong ($\frac{1}{8}$ mi) and a width one-tenth of its length. a) How many acres are in a square mile? b) How many square feet are in an acre? See Appendix E. c) An acre-foot is the volume of water that would cover one acre of flat land to a depth of one foot. How many gallons are in an acre-foot?

1–52 An estate on the California coast was offered for sale for $4,950,000. The total area of the estate was 102 acres (see

Problem 1–51). a) Considering the price of the estate to be proportional to its area, what was the cost of one square meter of the estate? b) What would be the price of a portion of the estate the size of a postage stamp ($\frac{7}{8}$ in. by 1.0 in.)?

1–53 The Hydrogen Maser. The radio waves generated by a hydrogen maser can be used as a standard of frequency. The frequency of these waves is 1,420,405,751.786 hertz. (A hertz is another name for one cycle per second.) A clock controlled by a hydrogen maser is off by only 1 s in 100,000 years. For the following questions, use only three significant figures. (The large number of significant figures given for the frequency simply illustrates the remarkable accuracy to which it has been measured.) a) What is the time for one cycle of the radio wave? b) How many cycles occur in 1 h? c) How many cycles would have occurred during the age of the earth, which is estimated to be 4.6×10^9 years? d) By how many seconds would a hydrogen maser clock be off after a time interval equal to the age of the earth?

1–54 Estimate the number of atoms in your body. (*Hint:* Based on what you know about biology and chemistry, what are the most common types of atom in your body? What is the mass of each type of atom? Appendix D gives the atomic masses for different elements, measured in atomic mass units; you can find the value of an atomic mass unit, or 1 u, in Appendix F.)

1–55 a) Estimate the number of dentists in your city. You will need to consider the number of people in your city, how often they need to go to the dentist, how often they actually go, how many hours a typical dental procedure (filling, root canal, and so on) takes, and how many hours a dentist works in a week. b) Using your local telephone directory, check to see whether your estimate was roughly correct.

1–56 Physicists, mathematicians, and others often deal with large numbers. The number 10^{100} has been given the whimsical name *googol* by mathematicians. Let us compare some large numbers in physics with the googol. (*Note:* This problem requires numerical values that can be found in the appendices of the book, with which you should become familiar.) a) Approximately how many atoms make up our planet? For simplicity, take the average atomic mass of the atoms to be 14 g/mol. Avogadro's number gives the number of atoms in a mole. b) Approximately how many neutrons are in a neutron star? Neutron stars are composed almost entirely of neutrons and have approximately twice the mass of the sun. c) In the leading theory of the origin of the universe, the entire universe that we can now observe occupied at a very early time a sphere whose radius was approximately equal to the present distance of the earth to the sun. At that time the universe had a density (mass divided by volume) of 10^{15} g/cm^3. Assuming that $\frac{1}{3}$ of the particles were protons, $\frac{1}{3}$ of the particles were neutrons, and the remaining $\frac{1}{3}$ were electrons, how many particles then made up the universe?

1–57 You are to program a robotic arm on an assembly line to move in the *xy*-plane. Its first displacement is $\vec{A}$; its second displacement is $\vec{B}$, of magnitude 6.40 cm and direction 63.0° measured in the sense from the +*x*-axis toward the −*y*-axis. The resultant $\vec{C} = \vec{A} + \vec{B}$ of the two displacements should also have a magnitude of 6.40 cm, but a direction 22.0° measured in the

sense from the +*x*-axis toward the +*y*-axis. a) Draw the vector addition diagram for these vectors, roughly to scale. b) Find the components of $\vec{A}$. c) Find the magnitude and direction of $\vec{A}$.

1–58 a) Find the magnitude and direction of the vector $\vec{R}$ that is the sum of the three vectors $\vec{A}$, $\vec{B}$, and $\vec{C}$ in Fig. 1–26. In a diagram, show how $\vec{R}$ is formed from the three vectors in Fig. 1–26. b) Find the magnitude and direction of the vector $\vec{S} = \vec{C} - \vec{A} - \vec{B}$. In a diagram, show how $\vec{S}$ is formed from the three vectors in Fig. 1–26.

1–59 As noted in Exercise 1–31, a spelunker is surveying a cave. She follows a passage 180 m straight west, then 210 m in a direction 45° east of south, then 280 m at 30° east of north. After a fourth unmeasured displacement she finds herself back where she started. Use the method of components to determine the magnitude and direction of the fourth displacement. Draw the vector addition diagram and show that it is in qualitative agreement with your numerical solution.

1–60 A sailor in a small sailboat encounters shifting winds. She sails 2.00 km east, then 3.50 km southeast, then an additional distance in an unknown direction. Her final position is 5.80 km directly east of the starting point (Fig. 1–29). Find the magnitude and direction of the third leg of the journey. Draw the vector addition diagram and show that it is in qualitative agreement with your numerical solution.

FIGURE 1–29 Problem 1–60.

1–61 A cross-country skier skis 2.80 km in the direction 45.0° west of south, then 7.40 km in the direction 30.0° north of east, and finally 3.30 km in the direction 22.0° south of west. a) Show these displacements on a diagram. b) How far is the skier from the starting point?

1–62 On a training flight, a student pilot flies from Lincoln, Nebraska, to Clarinda, Iowa; then to St. Joseph, Missouri; then to Manhattan, Kansas (Fig. 1–30). The directions are shown relative to north: 0° is north, 90° is east, 180° is south, and 270° is west. Use the method of components to find a) the distance she has to fly from Manhattan to get back to Lincoln; b) the direction (relative to north) she must fly to get there. Illustrate your solution with a vector diagram.

1–63 A graphic artist is creating a new logo for her company's web site. In the graphics program she is using, each pixel in an image file has coordinates (*x, y*), where the origin (0, 0) is at the upper left corner of the image, the +*x*-axis points to the right, and the +*y*-axis points down. Distances are measured in pixels.

FIGURE 1–30 Problem 1–62.

a) The artist draws a line from the pixel location (10, 20) to the location (210, 200). She wishes to draw a second line that starts at (10, 20), is 250 pixels long, and is at angle of 30° measured clockwise from the first line. At which pixel location should this second line end? Give your answer to the nearest pixel. b) The artist now draws an arrow that connects the lower right end of the first line to the lower right end of the second line. Find the length and direction of this arrow. Draw a diagram showing all three lines.

1–64 An explorer in the dense jungles of equatorial Africa leaves his hut. He takes 40 steps northeast, then 80 steps 60° north of west, then 50 steps due south. Assume his steps all have equal length. a) Sketch, roughly to scale, the three vectors and their resultant. b) Save him from becoming hopelessly lost in the jungle by giving him the displacement, calculated using the method of components, that will return him to his hut.

1–65 Vectors $\vec{A}$ and $\vec{B}$ are drawn from a common point. Vector $\vec{A}$ has magnitude A and angle θ_A measured in the sense from the +x-axis to the +y-axis. The corresponding quantities for vector $\vec{B}$ are B and θ_B. Then $\vec{A} = A \cos \theta_A \hat{\imath} + A \sin \theta_A \hat{\jmath}$, $\vec{B} = B \cos \theta_B \hat{\imath} + B \sin \theta_B \hat{\jmath}$, and $\phi = |\theta_B - \theta_A|$ is the angle between $\vec{A}$ and $\vec{B}$. a) Derive Eq. (1–18) from Eq. (1–21). b) Derive Eq. (1–22) from Eq. (1–27).

1–66 For the two vectors $\vec{A}$ and $\vec{B}$ in Fig. 1–28, a) find the scalar product $\vec{A} \cdot \vec{B}$; b) find the magnitude and direction of the vector product $\vec{A} \times \vec{B}$.

1–67 Figure 1–6c shows a parallelogram based on the two vectors $\vec{A}$ and $\vec{B}$. a) Show that the magnitude of the cross product of these two vectors is equal to the area of the parallelogram. (*Hint:* area = base × height.) b) What is the angle between the cross product and the plane of the parallelogram?

1–68 The vector $\vec{A}$ is 3.50 cm long and is directed into this page. Vector $\vec{B}$ points from the lower right corner of this page to the upper left corner of this page. Define an appropriate right-handed coordinate system and find the three components of the vector product $\vec{A} \times \vec{B}$, measured in cm². In a diagram show your coordinate system and the vectors $\vec{A}$, $\vec{B}$, and $\vec{A} \times \vec{B}$.

1–69 Given two vectors $\vec{A} = -2.00\hat{\imath} + 3.00\hat{\jmath} + 4.00\hat{k}$ and $\vec{B} = 3.00\hat{\imath} + 1.00\hat{\jmath} - 3.00\hat{k}$, do the following. a) Find the magnitude of each vector. b) Write an expression for the vector difference $\vec{A} - \vec{B}$, using unit vectors. c) Find the magnitude of

the vector difference $\vec{A} - \vec{B}$. Is this the same as the magnitude of $\vec{B} - \vec{A}$? Explain.

1–70 Bond Angle in Methane. In the methane molecule, CH₄, each hydrogen atom is at a corner of a regular tetrahedron with the carbon atom at the center. In coordinates where one of the C—H bonds is in the direction of $\hat{\imath} + \hat{\jmath} + \hat{k}$, an adjacent C—H bond is in the $\hat{\imath} - \hat{\jmath} - \hat{k}$ direction. Calculate the angle between these two bonds.

1–71 The two vectors $\vec{A}$ and $\vec{B}$ are drawn from a common point, and $\vec{C} = \vec{A} + \vec{B}$. a) Show that if $C^2 = A^2 + B^2$, the angle between the vectors $\vec{A}$ and $\vec{B}$ is 90°. b) Show that if $C^2 < A^2 + B^2$, the angle between the vectors $\vec{A}$ and $\vec{B}$ is greater than 90°. c) Show that if $C^2 > A^2 + B^2$, the angle between the vectors $\vec{A}$ and $\vec{B}$ is between 0° and 90°.

1–72 When two vectors $\vec{A}$ and $\vec{B}$ are drawn from a common point, the angle between them is ϕ. a) Using vector techniques, show that the magnitude of their vector sum is given by

$$\sqrt{A^2 + B^2 + 2AB \cos \phi}.$$

b) If $\vec{A}$ and $\vec{B}$ have the same magnitude, for what value of ϕ will their vector sum have the same magnitude as $\vec{A}$ or $\vec{B}$? c) Derive a result analogous to that in part (a) for the magnitude of the vector difference $\vec{A} - \vec{B}$. d) If $\vec{A}$ and $\vec{B}$ have the same magnitude, for what value of ϕ will $\vec{A} - \vec{B}$ have this same magnitude?

1–73 A cube is placed so that one corner is at the origin and three edges are along the x-, y-, and z-axes of a coordinate system (Fig. 1–31). Use vectors to compute a) the angle between the edge along the z-axis (line *ab*) and the diagonal from the origin to the opposite corner (line *ad*); b) the angle between line *ac* (the diagonal of a face) and line *ad*.

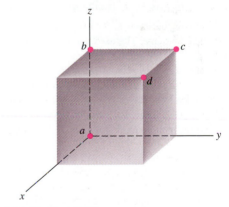

FIGURE 1–31 Problem 1–73.

1–74 Obtain a *unit vector* perpendicular to the two vectors given in Problem 1–69.

1–75 Later in our study of physics we will encounter quantities represented by $(\vec{A} \times \vec{B}) \cdot \vec{C}$. a) Prove that for any three vectors $\vec{A}$, $\vec{B}$, and $\vec{C}$, $\vec{A} \cdot (\vec{B} \times \vec{C}) = (\vec{A} \times \vec{B}) \cdot \vec{C}$. b) Calculate $(\vec{A} \times \vec{B}) \cdot \vec{C}$ for the three vectors $\vec{A}$ with magnitude $A = 5.00$ and angle $\theta_A = 26.0°$ measured in the sense from the +x-axis toward the +y-axis, $\vec{B}$ with $B = 4.00$ and $\theta_B = 63.0°$, and $\vec{C}$ with magnitude 6.00 and in the +z-direction. Vectors $\vec{A}$ and $\vec{B}$ are in the xy-plane.

CHALLENGE PROBLEMS

1–76 The length of a rectangle is given as $L \pm l$, and its width as $W \pm w$. a) Show that the uncertainty in its area A is $a = Lw + lW$. Assume that the uncertainties l and w are small, so that the product lw is very small and can be neglected. b) Show that the fractional uncertainty in the area is equal to the sum of the fractional uncertainty in length and the fractional uncertainty in width. c) A rectangular solid has dimensions $L \pm l$, $W \pm w$, and $H \pm h$. Find the fractional uncertainty in the volume, and show that it equals the sum of the fractional uncertainties in the length, width, and height.

1–77 At Enormous State University (ESU), the football team records its plays using vector displacements, with the origin taken to be the position of the ball before the play starts. In a certain pass play, the receiver starts at $+1.0\hat{\imath} - 5.0\hat{\jmath}$, where the units are yards, $\hat{\imath}$ is to the right, and $\hat{\jmath}$ is downfield. Subsequent displacements of the receiver are $+9.0\hat{\imath}$ (in motion before the snap), $+11.0\hat{\jmath}$ (breaks downfield), $-6.0\hat{\imath} + 4.0\hat{\jmath}$ (zigs), and $+12.0\hat{\imath} + 18.0\hat{\jmath}$ (zags). Meanwhile, the quarterback has dropped straight back to a position $-7.0\hat{\jmath}$. How far and in what direction must the quarterback throw the ball? (Like the coach, you will be well advised to diagram the situation before solving it numerically.)

1–78 Navigating in the Solar System. The *Mars Polar Lander* spacecraft was launched on January 3, 1999. On December 3, 1999, the day that *Mars Polar Lander* touched down on the Martian surface, the positions of the earth and Mars were given by these coordinates:

	x	y	z
Earth	0.3182 AU	0.9329 AU	0.0000 AU
Mars	1.3087 AU	−0.4423 AU	−0.0414 AU

In these coordinates, the sun is at the origin and the plane of the earth's orbit is the xy-plane. The earth passes through the $+x$-axis once a year on the autumnal equinox, the first day of autumn in the northern hemisphere (on or around September 22). One AU, or *astronomical unit,* is equal to 1.496×10^8 km, the average distance from the earth to the sun. a) In a diagram, show the positions of the sun, the earth, and Mars on December 3, 1999. b) Find the following distances in AU on December 3,

1999: i) from the sun to the earth; ii) from the sun to Mars; iii) from the earth to Mars. c) As seen from the earth, what was the angle between the direction to the sun and the direction to Mars on December 3, 1999? d) Explain whether Mars was visible from your location at midnight on December 3, 1999. (When it is midnight at your location, the sun is on the opposite side of the earth from you.)

1–79 Navigating in the Big Dipper. The stars of the Big Dipper (part of the constellation Ursa Major) may appear to all be at the same distance from the earth, but in fact they are very far from each other. Figure 1–32 shows the distances from the earth to each of these stars. The distances are given in light years (ly), the distance that light travels in one year. One light year equals 9.461×10^{15} m. a) Alkaid and Merak are 25.6° apart in the earth's sky. In a diagram, show the relative positions of Alkaid, Merak, and our sun. Find the distance in light years from Alkaid to Merak. b) To an inhabitant of a planet orbiting Merak, how many degrees apart in the sky would Alkaid and our sun be?

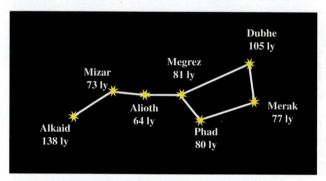

FIGURE 1–32 Challenge Problem 1–79.

1–80 The vector $\vec{r} = x\hat{\imath} + y\hat{\jmath} + z\hat{k}$, called the *position vector,* points from the origin (0, 0, 0) to an arbitrary point in space with coordinates (x, y, z). Use what you know about vectors to prove the following: All points (x, y, z) that satisfy the equation $Ax + By + Cz = 0$, where A, B, and C are constants, lie in a plane that passes through the origin and that is perpendicular to the vector $A\hat{\imath} + B\hat{\jmath} + C\hat{k}$. Sketch this vector and the plane.

Every successful space shuttle mission ends with a brief period of straight-line motion as it comes to a stop on the runway. No larger than a typical airliner, the space shuttle orbiter lands at more than 350 km/h (220 mi/h). Even with a drag parachute for braking, the fast-moving orbiter requires up to 3 km (about 2 mi) to stop.

2–1 INTRODUCTION

How do you describe the motion of a jet fighter being catapulted down the deck of an aircraft carrier? When you throw a baseball straight up in the air, how high does it go? When a glass slips from your hand, how much time do you have to catch it before it hits the floor? These are the kinds of questions you will learn to answer in this chapter. We are beginning our study of physics with *mechanics,* the study of the relationships among force, matter, and motion. The goal of this chapter and the next is to develop general methods for describing motion. The part of mechanics that deals with the description of motion is called *kinematics.* Later we will study *dynamics,* the relation of motion to its causes.

In this chapter we will study the simplest kind of motion: a single particle moving along a straight line. We will often use a particle as a model for a moving body when effects such as rotation or change of shape are not important. To describe the motion of a particle, we will introduce the physical quantities *velocity* and *acceleration.* These quantities have simple definitions in physics; however, those definitions are more precise and slightly different than the ones used in everyday language. Paying careful attention to how velocity and acceleration are defined will help you work with these and other important physical quantities.

An important part of how a physicist defines velocity and acceleration is that these quantities are *vectors.* As you learned in Chapter 1, this means that they have both mag-

nitude and direction. Our concern in this chapter is with motion along a straight line only, so we won't need the full mathematics of vectors just yet. But in Chapter 3 we'll extend our discussion to include motion in two or three dimensions, and using vectors will be essential.

An important special case of straight-line motion is when the acceleration is constant, a situation that we will encounter many times in our study of physics. An example is the motion of a freely falling body. We'll develop simple equations to describe motion with constant acceleration. We'll also consider situations in which the acceleration varies during the motion; in this case, it's necessary to use integration to describe the motion. (If you haven't studied integration yet, this section is optional.)

2–2 DISPLACEMENT, TIME, AND AVERAGE VELOCITY

Suppose a drag racer drives her AA-fuel dragster along a straight track. To study this motion, we need a coordinate system to describe the dragster's position. We choose the *x*-axis of our coordinate system to lie along the dragster's straight-line path, with the origin O at the starting line (Fig. 2–1). We will describe the dragster's position in terms of the position of a representative point, such as its front end. By doing this, we represent the entire dragster by that point and so treat the dragster as a **particle.**

2–1 Positions of an AA-fuel dragster at two times during its run.

The position of the front of the dragster—that is, the position of the particle—is given by the coordinate x, which varies with time as the dragster moves. A useful way to describe the dragster's motion is in terms of the change in x over a time interval. Suppose that 1.0 s after the start the front of the dragster is at point P_1, 19 m from the origin, and 4.0 s after the start it is at point P_2, 277 m from the origin. Then it has traveled a distance of (277 m − 19 m) = 258 m in a time interval of (4.0 s − 1.0 s) = 3.0 s. We define the dragster's **average velocity** during this time interval as a *vector* quantity whose x-component is the change in x divided by the time interval: (258 m)/(3.0 s) = 86 m/s. In general, the average velocity depends on the particular time interval chosen. For a 3.0-s time interval *before* the start of the race, the average velocity would be zero, because the dragster would be at rest at the starting line and would have zero displacement.

Let's generalize the concept of average velocity. At time t_1 the dragster is at point P_1, with coordinate x_1, and at time t_2 it is at point P_2, with coordinate x_2. The displacement of the dragster during the time interval from t_1 to t_2 is the vector from P_1 to P_2, with x-component ($x_2 − x_1$) and with y- and z-components equal to zero. The x-component of the dragster's displacement is just the change in the coordinate x, which we write in a more compact way as

$$\Delta x = x_2 - x_1. \tag{2–1}$$

The Greek capital letter Δ ("delta") represents a change *in a quantity, calculated by subtracting the initial value from the final value.* Be sure you understand that Δx is *not* the product of Δ and x: it is a single symbol that means "the change in the quantity x." We likewise write the time interval from t_1 to t_2 as $\Delta t = t_2 - t_1$. Note that Δx or Δt always means the final value minus the initial value, never the reverse.

We can now define the x-component of average velocity more precisely: it is the x-component of displacement, Δx, divided by the time interval Δt during which the displacement occurs. We represent this quantity by the letter v with a subscript "av" to signify average value:

$$v_{\text{av}} = \frac{x_2 - x_1}{t_2 - t_1} = \frac{\Delta x}{\Delta t} \qquad \text{(average velocity, straight-line motion).} \tag{2–2}$$

For the above example we had $x_1 = 19$ m, $x_2 = 277$ m, $t_1 = 1.0$ s, and $t_2 = 4.0$ s, so Eq. (2–2) gives

$$v_{\text{av}} = \frac{277 \text{ m} - 19 \text{ m}}{4.0 \text{ s} - 1.0 \text{ s}} = \frac{258 \text{ m}}{3.0 \text{ s}} = 86 \text{ m/s}.$$

The average velocity of the dragster is positive. This means that during the time interval, the coordinate x increased and the dragster moved in the positive x-direction (to the right in Fig. 2–1). If a particle moves in the *negative* x-direction during a time interval, its average velocity for that time interval is negative. For example, suppose an official's truck moves to the left along the track (Fig. 2–2). The truck is at $x_1 = 277$ m at $t_1 = 16.0$ s and is at $x_2 = 19$ m at $t_2 = 25.0$ s. Then $\Delta x = (19 \text{ m} - 277 \text{ m}) = -258$ m and $\Delta t = (25.0 \text{ s} - 16.0 \text{ s}) = 9.0$ s, and the x-component of average velocity is $v_{\text{av}} = \Delta x/\Delta t = (-258 \text{ m})/(9.0 \text{ s}) = -29$ m/s. Whenever x is positive and increasing or is negative and becoming less negative, the particle is moving in the $+x$-direction and v_{av} is positive (Fig. 2–1). Whenever x is positive and decreasing or is negative and becoming more negative, the particle is moving in the $-x$-direction and v_{av} is negative (Fig. 2–2).

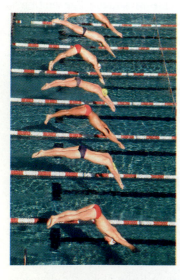

The winner of a 50-m swimming race is the swimmer whose average velocity has the greatest magnitude — that is, the swimmer who traverses a displacement Δx of 50 m in the shortest elapsed time Δt.

2–2 Positions of an official's truck at two times during its motion. The points P_1 and P_2 now refer to the motion of the truck, and so are different than in Fig. 2–1. The x-component of the truck's displacement is negative, so v_{av} is negative.

CAUTION ▶ You might be tempted to conclude that positive average velocity necessarily implies motion to the right, as in Fig. 2–1, and that negative average velocity necessarily implies motion to the left, as in Fig. 2–2. But these conclusions are correct *only* if the positive x-direction is to the right, as we have chosen it to be in Figs. 2–1 and 2–2. We could just as well have chosen the positive x-direction to be to the left, with the origin at the finish line. Then the dragster would have negative average velocity, and the official's truck would have positive average velocity. In most problems the direction of the coordinate axis will be yours to choose. Once you've made your choice, you *must* take it into account when interpreting the signs of v_{av} and other quantities that describe motion! ◀

With straight-line motion we will usually call Δx simply the displacement and v_{av} simply the average velocity. But be sure to remember that these are really the x-components of vector quantities that, in this special case, have *only* x-components. In Chapter 3, displacement, velocity, and acceleration vectors will have two or three nonzero components.

Figure 2–3 is a graph of the dragster's position as a function of time, that is, an **x-t graph**. The curve in the figure *does not* represent the dragster's path in space; as Fig. 2–1 shows, the path is a straight line. Rather, the graph is a pictorial way to represent how the dragster's position changes with time. The points on the graph labeled p_1 and p_2 correspond to the points P_1 and P_2 along the dragster's path. Line $p_1 p_2$ is the hypotenuse of a right triangle with vertical side $\Delta x = x_2 - x_1$ and horizontal side $\Delta t = t_2 - t_1$. The average velocity $v_{av} = \Delta x / \Delta t$ of the dragster is then equal to the *slope* of the line $p_1 p_2$, that is, the ratio of the vertical side Δx of the triangle to the horizontal side Δt.

Average velocity depends only on the total displacement $\Delta x = x_2 - x_1$ that occurs during the time interval $\Delta t = t_2 - t_1$, not on the details of what happens during the time interval. A second dragster might have passed through point P_1 in Fig. 2–1 at the same time t_1 as the first dragster, raced past the first dragster, then blown its engine and slowed down to pass through point P_2 at the same time t_2 as the first dragster. Both dragsters have the same displacement during the same time interval and so have the same average velocity.

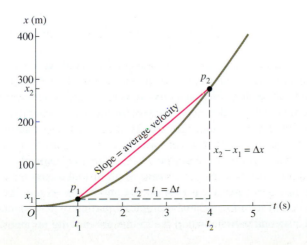

2–3 The position of an AA-fuel dragster as a function of time. The average velocity v_{av} between points P_1 and P_2 in Fig. 2–1 is the slope of the line $p_1 p_2$. This line slopes upward to the right, so the slope is positive and v_{av} is positive.

TABLE 2–1
TYPICAL VELOCITY MAGNITUDES

A snail's pace	10^{-3} m/s	Random motion of air molecules	500 m/s
A brisk walk	2 m/s	Fastest airplane	1000 m/s
Fastest human	11 m/s	Orbiting communications satellite	3000 m/s
Running cheetah	35 m/s	Electron orbiting in a hydrogen atom	2×10^{6} m/s
Fastest car	341 m/s	Light traveling in a vacuum	3×10^{8} m/s

If we express distance in meters and time in seconds, average velocity is measured in meters per second (m/s). Other common units of velocity are kilometers per hour (km/h), feet per second (ft/s), miles per hour (mi/h), and knots (1 knot = 1 nautical mile/h = 6080 ft/h). Table 2–1 shows a few typical velocity magnitudes.

2–3 INSTANTANEOUS VELOCITY

The average velocity of a particle during a time interval cannot tell us how fast, or in what direction, the particle was moving at any given time during the interval. To describe the motion in greater detail, we need to define the velocity at any specific instant of time or specific point along the path. Such a velocity is called **instantaneous velocity,** and it needs to be defined carefully.

Note that the word *instant* has a somewhat different definition in physics than in everyday language. You might use the phrase "It lasted just an instant" to refer to something that lasted for a very short time interval. But in physics an instant has *no* duration at all; it refers to a single value of time.

To find the instantaneous velocity of the dragster in Fig. 2–1 at the point P_1, we imagine moving the second point P_2 closer and closer to the first point P_1. We compute the average velocity $v_{av} = \Delta x/\Delta t$ over these shorter and shorter displacements and time intervals. Both Δx and Δt become very small, but their ratio does not necessarily become small. In the language of calculus the limit of $\Delta x/\Delta t$ as Δt approaches zero is called the **derivative** of x with respect to t and is written dx/dt. *The instantaneous velocity is the limit of the average velocity as the time interval approaches zero; it equals the instantaneous rate of change of position with time.* We use the symbol v, with no subscript, for instantaneous velocity:

$$v = \lim_{\Delta t \to 0} \frac{\Delta x}{\Delta t} = \frac{dx}{dt} \qquad \text{(instantaneous velocity, straight-line motion).} \quad (2\text{–}3)$$

We always assume that the time interval Δt is positive so that v has the same algebraic sign as Δx. If the positive x-axis points to the right, as in Fig. 2–1, a positive value of v means that x is increasing and the motion is toward the right; a negative value of v means that x is decreasing and the motion is toward the left. A body can have positive x and negative v, or the reverse; x tells us where the body is, while v tells us how it's moving.

Instantaneous velocity, like average velocity, is a vector quantity. Equation (2–3) defines its x-component, which can be positive or negative. In straight-line motion, all other components of instantaneous velocity are zero, and in this case we will often call v simply the instantaneous velocity. When we use the term "velocity," we will always mean instantaneous rather than average velocity, unless we state otherwise.

The terms "velocity" and "speed" are used interchangeably in everyday language, but they have distinct definitions in physics. We use the term **speed** to denote distance traveled divided by time, on either an average or an instantaneous basis. Instantaneous *speed* measures how fast a particle is moving; instantaneous *velocity* measures how fast *and* in what direction it's moving. For example, a particle with instantaneous velocity $v = 25$ m/s and a second particle with $v = -25$ m/s are moving in opposite directions at

Even when moving forward, this cyclist's instantaneous velocity can be negative — if he's traveling in the negative x-direction. In any problem, the choice of which direction is positive and which is negative is entirely up to you.

the same instantaneous speed of 25 m/s. Instantaneous speed is the magnitude of instantaneous velocity, and so instantaneous speed can never be negative.

Average speed, however, is *not* the magnitude of average velocity. When Alexander Popov set a world record in 1994 by swimming 100.0 m in 48.21 s, his average speed was (100.0 m)/(48.21 s) = 2.074 m/s. But because he swam two lengths in a 50-m pool, he started and ended at the same point, giving him zero total displacement and zero average *velocity* for his effort! Both average speed and instantaneous speed are scalars, not vectors, because these quantities contain no information about direction.

EXAMPLE 2–1

Average and instantaneous velocities A cheetah is crouched in ambush 20 m to the east of an observer's blind (Fig. 2–4). At time $t = 0$ the cheetah charges an antelope in a clearing 50 m east of the observer. The cheetah runs along a straight line. Later analysis of a videotape shows that during the first 2.0 s of the attack, the cheetah's coordinate x varies with time according to the equation $x = 20 \text{ m} + (5.0 \text{ m/s}^2)t^2$. (Note that the units for the numbers 20 and 5.0 *must* be as shown to make the expression dimensionally consistent.) a) Find the displacement of the cheetah during the interval between $t_1 = 1.0$ s and $t_2 = 2.0$ s. b) Find the average velocity during the same time interval. c) Find the instantaneous velocity at time $t_1 = 1.0$ s by taking $\Delta t = 0.1$ s, then $\Delta t = 0.01$ s, then $\Delta t = 0.001$ s. d) Derive a general expression for the instantaneous velocity as a function of time, and from it find v at $t = 1.0$ s and $t = 2.0$ s.

SOLUTION a) At time $t_1 = 1.0$ s the cheetah's position x_1 is

$$x_1 = 20 \text{ m} + (5.0 \text{ m/s}^2)(1.0 \text{ s})^2 = 25 \text{ m}.$$

At time $t_2 = 2.0$ s its position x_2 is

$$x_2 = 20 \text{ m} + (5.0 \text{ m/s}^2)(2.0 \text{ s})^2 = 40 \text{ m}.$$

The displacement during this interval is

$$\Delta x = x_2 - x_1 = 40 \text{ m} - 25 \text{ m} = 15 \text{ m}.$$

b) The average velocity during this time interval is

$$v_{av} = \frac{x_2 - x_1}{t_2 - t_1} = \frac{40 \text{ m} - 25 \text{ m}}{2.0 \text{ s} - 1.0 \text{ s}} = \frac{15 \text{ m}}{1.0 \text{ s}} = 15 \text{ m/s}.$$

c) With $\Delta t = 0.1$ s, the time interval is from $t_1 = 1.0$ s to $t_2 = 1.1$ s. At time t_2, the position is

$$x_2 = 20 \text{ m} + (5.0 \text{ m/s}^2)(1.1 \text{ s})^2 = 26.05 \text{ m}.$$

The average velocity during this interval is

$$v_{av} = \frac{26.05 \text{ m} - 25 \text{ m}}{1.1 \text{ s} - 1.0 \text{ s}} = 10.5 \text{ m/s}.$$

We invite you to follow this same pattern to work out the average velocities for the 0.01-s and 0.001-s intervals. The results are 10.05 m/s and 10.005 m/s. As Δt gets smaller, the average velocity gets closer to 10.0 m/s, so we conclude that the instantaneous velocity at time $t = 1.0$ s is 10.0 m/s.

d) We find the instantaneous velocity as a function of time by taking the derivative of the expression for x with respect to t. For any n the derivative of t^n is nt^{n-1}, so the derivative of t^2 is $2t$. Therefore

$$v = \frac{dx}{dt} = (5.0 \text{ m/s}^2)(2t) = (10 \text{ m/s}^2)t.$$

At time $t = 1.0$ s, $v = 10$ m/s as we found in part (c). At time $t = 2.0$ s, $v = 20$ m/s.

2–4 A cheetah attacking an antelope from ambush. The animals are not drawn to the same scale as the axis.

(a)   (b)   (c)

2–5 (a) and (b) As the average velocity is calculated over shorter and shorter time intervals, its value approaches the instantaneous velocity. (c) In the limit $\Delta t \to 0$ the slope of the line p_1p_2 approaches the slope of the tangent to the x-t curve at point p_1. The value of this slope equals the instantaneous velocity v at point P_1 in Fig. 2–1. We find the slope of the tangent by dividing any vertical interval (with distance units) along the tangent line by the corresponding horizontal interval (with time units). Here, v at P_1 equals $(160 \text{ m})/(4.0 \text{ s}) = 40$ m/s.

1.1

Analyzing Motion Using Diagrams

FINDING VELOCITY ON AN *x-t* GRAPH

The velocity of a particle can also be found from the graph of the particle's position as a function of time. Suppose we want to find the velocity of the dragster in Fig. 2–1 at point P_1. As point P_2 in Fig. 2–1 approaches point P_1, point p_2 in the x-t graph of Fig. 2–3 approaches point p_1. This is shown in Figs. 2–5a and 2–5b, in which the average velocity is calculated over shorter time intervals Δt. In the limit that $\Delta t \to 0$, shown in Fig. 2–5c, the slope of the line $p_1 p_2$ equals the slope of the line tangent to the curve at point p_1. *On a graph of position as a function of time for straight-line motion, the instantaneous velocity at any point is equal to the slope of the tangent to the curve at that point.*

If the tangent to the x-t curve slopes upward to the right, as in Fig. 2–5c, then its slope is positive, the velocity is positive, and the motion is in the positive x-direction. If the tangent slopes downward to the right, the slope and velocity are negative and the motion is in the negative x-direction. When the tangent is horizontal, the slope is zero and the velocity is zero. Figure 2–6 illustrates these three possibilities.

Note that Fig. 2–6 depicts the motion of a particle in two ways. Figure 2–6a is an x-t graph, and Fig. 2–6b is an example of a **motion diagram.** A motion diagram shows the position of the particle at various times during the motion, like frames from a movie or video of the particle's motion, as well as arrows to represent the particle's velocity at each instant. Both x-t graphs and motion diagrams are helpful aids to understanding motion, and we will use both frequently in this chapter. You will find it worth your while to draw *both* an x-t graph and a motion diagram as part of solving any problem involving motion.

2–6 (a) The x-t graph of the motion of a particular particle. The slope of the tangent at any point equals the velocity at that point. Between points A and B, x is negative but increasing. The slope of the graph is becoming more positive, so the particle is moving with increasing velocity in the positive x-direction. At B the slope and velocity are greatest. Between points B and C the particle is still moving in the positive x-direction, but the slope and velocity are decreasing. At point C the slope and velocity are zero. From C to E, x is positive but decreasing, so the velocity is negative; the most negative value of v occurs at D, where the slope is most negative. At point E the slope and velocity are approaching zero as x approaches a constant value, that is, the particle is coming to a stop. (b) A motion diagram showing the position and velocity of the particle at each of the five times labeled on the x-t graph.

(a)

(b)

2-4 AVERAGE AND INSTANTANEOUS ACCELERATION

When the velocity of a moving body changes with time, we say that the body has an *acceleration*. Just as velocity describes the rate of change of position with time, acceleration describes the rate of change of velocity with time. Like velocity, acceleration is a vector quantity. In straight-line motion its only nonzero component is along the axis along which the motion takes place.

AVERAGE ACCELERATION

Let's consider again the motion of a particle along the x-axis. Suppose that at time t_1 the particle is at point P_1 and has x-component of (instantaneous) velocity v_1, and at a later time t_2 it is at point P_2 and has x-component of velocity v_2. So the x-component of velocity changes by an amount $\Delta v = v_2 - v_1$ during the time interval $\Delta t = t_2 - t_1$.

We define the **average acceleration** a_{av} of the particle as it moves from P_1 to P_2 to be a vector quantity whose x-component is Δv, the change in the x-component of velocity, divided by the time interval Δt:

$$a_{av} = \frac{v_2 - v_1}{t_2 - t_1} = \frac{\Delta v}{\Delta t} \qquad \text{(average acceleration, straight-line motion).} \qquad (2\text{-}4)$$

For straight-line motion we will usually call a_{av} simply the average acceleration, remembering that in fact it is the x-component of the average acceleration vector.

If we express velocity in meters per second and time in seconds, then average acceleration is in meters per second per second, or (m/s)/s. This is usually written as m/s^2 and is read "meters per second squared."

CAUTION▶ Be very careful not to confuse acceleration with velocity! Velocity describes a body's speed and direction of motion at any time; acceleration describes how the speed and direction of motion *change* over time. ◀

EXAMPLE 2-2

Average acceleration An astronaut has left an orbiting space shuttle to test a new personal maneuvering unit. As she moves along a straight line, her partner on board the shuttle measures her velocity every 2.0 s, starting at time $t = 1.0$ s, as follows:

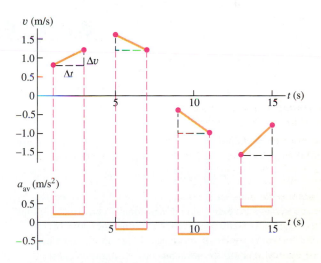

t	v	t	v
1.0 s	0.8 m/s	9.0 s	−0.4 m/s
3.0 s	1.2 m/s	11.0 s	−1.0 m/s
5.0 s	1.6 m/s	13.0 s	−1.6 m/s
7.0 s	1.2 m/s	15.0 s	−0.8 m/s

Find the average acceleration, and describe whether the speed of the astronaut increases or decreases, for each of the following time intervals: a) $t_1 = 1.0$ s to $t_2 = 3.0$ s; b) $t_1 = 5.0$ s to $t_2 = 7.0$ s; c) $t_1 = 9.0$ s to $t_2 = 11.0$ s; d) $t_1 = 13.0$ s to $t_2 = 15.0$ s.

SOLUTION The upper part of Fig. 2–7 graphs the velocity as a function of time. On this graph, the slope of the line connecting the points at the beginning and end of each interval equals the average acceleration $a_{av} = \Delta v / \Delta t$ for that interval. The values of a_{av} are graphed in the lower part of Fig. 2–7. For each time

2–7 The slope of the line connecting two points on a graph of velocity versus time (upper graph) equals the average acceleration between those two points (lower graph).

interval, we have
a) $a_{av} = (1.2 \text{ m/s} - 0.8 \text{ m/s})/(3.0 \text{ s} - 1.0 \text{ s}) = 0.2 \text{ m/s}^2$. The speed (magnitude of instantaneous velocity) increases from 0.8 m/s to 1.2 m/s.
b) $a_{av} = (1.2 \text{ m/s} - 1.6 \text{ m/s})/(7.0 \text{ s} - 5.0 \text{ s}) = -0.2 \text{ m/s}^2$. The speed decreases from 1.6 m/s to 1.2 m/s.
c) $a_{av} = [-1.0 \text{ m/s} - (-0.4 \text{ m/s})]/[11.0 \text{ s} - 9.0 \text{ s})] = -0.3 \text{ m/s}^2$. The speed increases from 0.4 m/s to 1.0 m/s.
d) $a_{av} = [-0.8 \text{ m/s} - (-1.6 \text{ m/s})]/[15.0 \text{ s} - 13.0 \text{ s}] = 0.4 \text{ m/s}^2$. The speed decreases from 1.6 m/s to 0.8 m/s.

When the acceleration has the *same* direction (same algebraic sign) as the initial velocity, as in intervals a and c, the astronaut goes faster; when it has the *opposite* direction (opposite algebraic sign), as in intervals b and d, she slows down. When she moves in the negative direction with increasing speed (interval c), her velocity is algebraically decreasing (becoming more negative), and her acceleration is negative. But when she moves in the negative direction with decreasing speed (interval d), her velocity is algebraically increasing (becoming less negative) and her acceleration is positive.

INSTANTANEOUS ACCELERATION

We can now define **instantaneous acceleration,** following the same procedure that we used to define instantaneous velocity. Consider this situation: A race car driver has just entered the final straightaway at the Grand Prix. He reaches point P_1 at time t_1, moving with velocity v_1. He passes point P_2, closer to the finish line, at time t_2 with velocity v_2 (Fig. 2–8).

To define the instantaneous acceleration at point P_1, we take the second point P_2 in Fig. 2–8 to be closer and closer to the first point P_1 so that the average acceleration is computed over shorter and shorter time intervals. *The instantaneous acceleration is the limit of the average acceleration as the time interval approaches zero.* In the language of calculus, *instantaneous acceleration equals the instantaneous rate of change of velocity with time.* Thus

$$a = \lim_{\Delta t \to 0} \frac{\Delta v}{\Delta t} = \frac{dv}{dt} \qquad \text{(instantaneous acceleration, straight-line motion).} \quad (2\text{–}5)$$

Note that Eq. (2–5) is really the definition of the x-component of the acceleration vector; in straight-line motion, all other components of this vector are zero. Instantaneous acceleration plays an essential role in the laws of mechanics. From now on, when we use the term "acceleration," we will always mean instantaneous acceleration, not average acceleration.

2–8 A Grand Prix car at two points on the straightaway.

EXAMPLE 2-3

Average and instantaneous accelerations Suppose the velocity v of the car in Fig. 2–8 at any time t is given by the equation

$$v = 60 \text{ m/s} + (0.50 \text{ m/s}^3)t^2.$$

(Note that the units for the numbers 60 and 0.50 *must* be as shown to make the expression dimensionally consistent.) a) Find the change in velocity of the car in the time interval between $t_1 = 1.0$ s and $t_2 = 3.0$ s. b) Find the average acceleration in this time interval. c) Find the instantaneous acceleration at time $t_1 = 1.0$ s by taking Δt to be first 0.1 s, then 0.01 s, then 0.001 s. d) Derive an expression for the instantaneous acceleration at any time, and use it to find the acceleration at $t = 1.0$ s and $t = 3.0$ s.

SOLUTION a) We first find the velocity at each time by substituting each value of t into the equation. At time $t_1 = 1.0$ s,

$$v_1 = 60 \text{ m/s} + (0.50 \text{ m/s}^3)(1.0 \text{ s})^2 = 60.5 \text{ m/s}.$$

At time $t_2 = 3.0$ s,

$$v_2 = 60 \text{ m/s} + (0.50 \text{ m/s}^3)(3.0 \text{ s})^2 = 64.5 \text{ m/s}.$$

The change in velocity Δv is

$$\Delta v = v_2 - v_1 = 64.5 \text{ m/s} - 60.5 \text{ m/s} = 4.0 \text{ m/s}.$$

The time interval is $\Delta t = 3.0 \text{ s} - 1.0 \text{ s} = 2.0$ s.

b) The average acceleration during this time interval is

$$a_{av} = \frac{v_2 - v_1}{t_2 - t_1} = \frac{4.0 \text{ m/s}}{2.0 \text{ s}} = 2.0 \text{ m/s}^2.$$

During the time interval from $t_1 = 1.0$ s to $t_2 = 3.0$ s, the velocity and average acceleration have the same algebraic sign (in this case, positive), and the car speeds up.

c) When $\Delta t = 0.1$ s, $t_2 = 1.1$ s and

$$v_2 = 60 \text{ m/s} + (0.50 \text{ m/s}^3)(1.1 \text{ s})^2 = 60.605 \text{ m/s},$$

$$\Delta v = 0.105 \text{ m/s},$$

$$a_{av} = \frac{\Delta v}{\Delta t} = \frac{0.105 \text{ m/s}}{0.1 \text{ s}} = 1.05 \text{ m/s}^2.$$

We invite you to repeat this pattern for $\Delta t = 0.01$ s and $\Delta t = 0.001$ s; the results are $a_{av} = 1.005$ m/s^2 and $a_{av} = 1.0005$ m/s^2 respectively. As Δt gets smaller, the average acceleration gets closer to 1.0 m/s^2. We conclude that the instantaneous accelera-

tion at $t = 1.0$ s is 1.0 m/s^2.

d) The instantaneous acceleration is $a = dv/dt$, the derivative of a constant is zero, and the derivative of t^2 is $2t$. Using these, we obtain

$$a = \frac{dv}{dt} = \frac{d}{dt}[60 \text{ m/s} + (0.50 \text{ m/s}^3)t^2]$$

$$= (0.50 \text{ m/s}^3)(2t) = (1.0 \text{ m/s}^3)t.$$

When $t = 1.0$ s,

$$a = (1.0 \text{ m/s}^3)(1.0 \text{ s}) = 1.0 \text{ m/s}^2.$$

When $t = 3.0$ s,

$$a = (1.0 \text{ m/s}^3)(3.0 \text{ s}) = 3.0 \text{ m/s}^2.$$

Note that neither of these values is equal to the average acceleration found in part (b). The instantaneous acceleration of this car varies with time. Automotive engineers sometimes call the rate of change of acceleration with time the "jerk."

FINDING ACCELERATION ON A *v-t* GRAPH OR AN *x-t* GRAPH

We interpreted average and instantaneous velocity in terms of the slope of a graph of position versus time. In the same way, we can get additional insight into the concepts of average and instantaneous acceleration from a graph with instantaneous velocity v on the vertical axis and time t on the horizontal axis, that is, a ***v-t* graph** (Fig. 2–9). The points on the graph labeled p_1 and p_2 correspond to points P_1 and P_2 in Fig. 2–8. The average acceleration $a_{av} = \Delta v / \Delta t$ during this interval is the slope of the line $p_1 p_2$. As point P_2 in Fig. 2–8 approaches point P_1, point p_2 in the v-t graph of Fig. 2–9 approaches point p_1, and the slope of the line $p_1 p_2$ approaches the slope of the line tangent to the curve at point p_1. Thus *on a graph of velocity as a function of time, the instantaneous acceleration at any point is equal to the slope of the tangent to the curve at that point*. In Fig. 2–9 the instantaneous acceleration varies with time.

Note that by itself, the algebraic sign of the acceleration doesn't tell you whether a body is speeding up or slowing down. You must compare the signs of the velocity and the acceleration. When v and a have the *same* sign, the body is speeding up. If both are positive, the body is moving in the positive direction with increasing speed. If both are negative, the body is moving in the negative direction with a velocity that is becoming more and more negative, and again the speed is increasing. When v and a have *opposite* signs, the body is slowing down. If v is positive and a is negative, the body is moving in the positive direction with decreasing speed; if v is negative and a is positive, the body is moving in the negative direction with a velocity that is becoming less negative, and again the body is slowing down. Figure 2–10 illustrates these various possibilities.

2–9 A v-t graph of the motion in Fig. 2–8. The average acceleration between t_1 and t_2 equals the slope of the line $p_1 p_2$. The instantaneous acceleration at P_1 equals the slope of the tangent at p_1.

(a)   (b)

2–10 (a) A v-t graph of the motion of a particle. The slope of the tangent at any point equals the acceleration at that point. From A to B, v is negative but increasing, and the slope and acceleration are positive. The particle slows down until time B, when it is instantaneously at rest ($v = 0$); it is still accelerating, because the slope is not zero. From B to C, v is positive and increasing, and the acceleration is positive; the particle is speeding up. At point C the velocity has its maximum value, and the acceleration is instantaneously zero. From C to D the velocity is positive but decreasing, and the acceleration is negative. The body slows down again until $v = 0$ at D; the acceleration is negative at this point. From D to E, v is negative and decreasing, and the acceleration is negative; once again the particle is speeding up. (b) A motion diagram showing the position, velocity, and acceleration of the particle at each of the times labeled on the v-t graph. The positions are consistent with the v-t graph; for instance, from t_A to t_B the velocity is negative, so at t_B the particle is at a more negative value of x than at t_A.

The term *deceleration* is sometimes used for a decrease in speed. Because this may mean positive or negative a, depending on the sign of v, we avoid this term.

We can also learn about the acceleration of a body from a graph of its *position* versus time. Because $a = dv/dt$ and $v = dx/dt$, we can write

$$a = \frac{dv}{dt} = \frac{d}{dt}\left(\frac{dx}{dt}\right) = \frac{d^2x}{dt^2}. \tag{2–6}$$

2–11 (a) The same x-t graph as shown in Fig. 2–6a. The velocity is equal to the *slope* of the graph, and the acceleration is given by the *concavity* or *curvature* of the graph. Between points A and B the graph is curved upward (concave up). The velocity is increasing, and the acceleration is positive. At B the graph is not curved either upward or downward, and the acceleration is zero. At this point the velocity is maximum. From B to D the graph is curved downward (concave down), and the acceleration is negative; the velocity is decreasing from a positive value at B to zero at C and to a negative value at D. From D to E the graph is once again curved upward (concave up). The velocity is negative and approaching zero, so the acceleration is positive. (b) A motion diagram showing the position, velocity, and acceleration of the particle at each of the times labeled on the x-t graph.

That is, a is the second derivative of x with respect to t. The second derivative of any function is directly related to the *concavity* or *curvature* of the graph of that function. At a point where the x-t graph is concave up (curved upward), the acceleration is positive and v is increasing; at a point where the x-t graph is concave down (curved downward), the acceleration is negative and v is decreasing. At a point where the x-t graph has no curvature, such as an inflection point, the acceleration is zero and the velocity is constant. All three of these possibilities are shown in Fig. 2–11. Examining the curvature of an x-t graph is an easy way to decide what the *sign* of acceleration is. This technique is less helpful for determining numerical values of acceleration, because the curvature of a graph is hard to measure accurately.

(a)

(b)

2–5 MOTION WITH CONSTANT ACCELERATION

The simplest accelerated motion is straight-line motion with *constant* acceleration. In this case the velocity changes at the same rate throughout the motion. This is a very special situation, yet one that occurs often in nature. As we will discuss in the next section, a falling body has a constant acceleration if the effects of the air are not important. The same is true for a body sliding on an incline or along a rough horizontal surface. Straight-line motion with nearly constant acceleration also occurs in technology, such as a jet fighter being catapulted from the deck of an aircraft carrier.

In this section we'll derive key equations for straight-line motion with constant acceleration. These equations will enable us to solve a wide range of problems.

Figure 2–12 is a motion diagram showing the position, velocity, and acceleration at five different times for a particle moving with constant acceleration. Figures 2–13 and 2–14 depict this same motion in the form of graphs. Since the acceleration a is constant, the **a-t graph** (graph of acceleration versus time) in Fig. 2–13 is a horizontal line. The graph of velocity versus time has a constant *slope* because the acceleration is constant, and so the v-t graph is a straight line (Fig. 2–14).

When the acceleration is constant, it's easy to derive equations for position x and velocity v as functions of time. Let's start with velocity. In Eq. (2–4) we can replace the average acceleration a_{av} by the constant (instantaneous) acceleration a. We then have

$$a = \frac{v_2 - v_1}{t_2 - t_1}. \qquad (2\text{–}7)$$

Now we let $t_1 = 0$ and let t_2 be any arbitrary later time t. We use the symbol v_0 for the velocity at the initial time $t = 0$; the velocity at the later time t is v. Then Eq. (2–7) becomes

$$a = \frac{v - v_0}{t - 0}, \qquad \text{or}$$

$$v = v_0 + at \qquad \text{(constant acceleration only).} \qquad (2\text{–}8)$$

We can interpret this equation as follows. The acceleration a is the constant rate of change of velocity, that is, the change in velocity per unit time. The term at is the product of the change in velocity per unit time, a, and the time interval t. Therefore it equals the *total* change in velocity from the initial time $t = 0$ to the later time t. The velocity v at any time t then equals the initial velocity v_0 (at $t = 0$) plus the change in velocity at. Graphically, we can consider the height v of the graph in Fig. 2–14 at any time t to be the sum of two segments: one with length v_0 equal to the initial velocity, the other with length at equal to the change in velocity during time t. The graph of velocity as a function of time is a straight line with slope a that intercepts the vertical axis (the v-axis) at v_0.

Another interpretation of Eq. (2–8) is that the change in velocity $v - v_0$ of the particle between $t = 0$ and any later time t equals the *area* under the a-t graph between those two times. In Fig. 2–13, the area under the graph of acceleration versus time is shown as a red rectangle of vertical side a and horizontal side t. The area of this rectangle is equal to at, which from Eq. (2–8) is just equal to the change in velocity $v - v_0$. In Section 2–7 we'll show that even if the acceleration is not constant, the change in velocity during a time interval is still equal to the area under the a-t curve, although in that case Eq. (2–8) does not apply.

Next we want to derive an equation for the position x of a particle moving with constant acceleration. To do this, we make use of two different expressions for the average velocity v_{av} during the interval from $t = 0$ to any later time t. The first expression comes

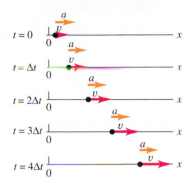

2–12 A motion diagram for a particle moving in a straight line with constant positive acceleration a. The position, velocity, and acceleration are shown at five equally spaced times. The velocity changes by equal amounts in equal time intervals because the acceleration is constant. The position changes by *different* amounts in equal time intervals because the velocity is changing.

2–13 An acceleration-time (a-t) graph for straight-line motion with constant positive acceleration a.

2–14 A velocity-time (v-t) graph for straight-line motion with constant positive acceleration a. The initial velocity v_0 is also positive.

from the definition of v_{av}, Eq. (2–2), which holds true whether or not the acceleration is constant. We call the position at time $t = 0$ the *initial position,* denoted by x_0. The position at the later time t is simply x. Thus for the time interval $\Delta t = t - 0$ and the corresponding displacement $\Delta x = x - x_0$, Eq. (2–2) gives

$$v_{av} = \frac{x - x_0}{t}. \qquad (2–9)$$

We can also get a second expression for v_{av} that is valid only when the acceleration is constant, so that the v-t graph is a straight line (as in Fig. 2–14) and the velocity changes at a constant rate. In this case the average velocity during any time interval is simply the arithmetic average of the velocities at the beginning and end of the interval. For the time interval 0 to t,

$$v_{av} = \frac{v_0 + v}{2} \qquad \text{(constant acceleration only).} \qquad (2–10)$$

(This equation is *not* true if the acceleration varies and the v-t graph is a curve, as in Fig. 2–10.) We also know that with constant acceleration the velocity v at any time t is given by Eq. (2–8). Substituting that expression for v into Eq. (2–10), we find

$$v_{av} = \frac{1}{2}(v_0 + v_0 + at)$$

$$= v_0 + \frac{1}{2}at \qquad \text{(constant acceleration only).} \qquad (2–11)$$

Finally, we equate Eqs. (2–9) and (2–11) and simplify the result:

$$v_0 + \frac{1}{2}at = \frac{x - x_0}{t}, \qquad \text{or}$$

$$x = x_0 + v_0t + \frac{1}{2}at^2 \qquad \text{(constant acceleration only).} \qquad (2–12)$$

This equation states that if at the initial time $t = 0$, a particle is at position x_0 and has velocity v_0, its new position x at any later time t is the sum of three terms: its initial position x_0, plus the distance v_0t that it would move if its velocity were constant, plus an additional distance $\frac{1}{2}at^2$ caused by the change in velocity.

Just as the change in velocity of the particle equals the area under the a-t graph, the displacement—that is, the change in position—equals the area under the v-t graph. To be specific, the displacement $x - x_0$ of the particle between $t = 0$ and any later time t equals the area under the v-t graph between those two times. In Fig. 2–14 the area under the graph is divided into a dark-colored rectangle of vertical side v_0 and horizontal side t and a light-colored right triangle of vertical side at and horizontal side t. The area of the rectangle is v_0t and the area of the triangle is $\frac{1}{2}(at)(t) = \frac{1}{2}at^2$, so the total area under the curve is

$$x - x_0 = v_0t + \frac{1}{2}at^2,$$

in agreement with Eq. (2–12).

The displacement during a time interval can always be found from the area under the v-t curve. This is true even if the acceleration is *not* constant, although in that case Eq. (2–12) does not apply. (We'll show this in Section 2–7.)

The motion of an airplane being catapulted from the deck of an aircraft carrier is an example of nearly constant acceleration. Thanks to the catapult, these airplanes attain takeoff speed in a distance of less than 100 m (see Exercise 2–18).

We can check whether Eqs. (2–8) and (2–12) are consistent with the assumption of constant acceleration by taking the derivative of Eq. (2–12). We find

$$v = \frac{dx}{dt} = v_0 + at,$$

which is Eq. (2–8). Differentiating again, we find simply

$$\frac{dv}{dt} = a,$$

as we should expect.

In many problems, it's useful to have a relationship between position, velocity, and acceleration that does not involve the time. To obtain this, we first solve Eq. (2–8) for t, then substitute the resulting expression into Eq. (2–12) and simplify:

$$t = \frac{v - v_0}{a},$$

$$x = x_0 + v_0\left(\frac{v - v_0}{a}\right) + \frac{1}{2}a\left(\frac{v - v_0}{a}\right)^2.$$

We transfer the term x_0 to the left side and multiply through by $2a$:

$$2a(x - x_0) = 2v_0v - 2v_0^2 + v^2 - 2v_0v + v_0^2.$$

Finally, simplifying gives us

$$v^2 = v_0^2 + 2a(x - x_0) \qquad \text{(constant acceleration only).} \qquad (2\text{–}13)$$

We can get one more useful relationship by equating the two expressions for v_{av}, Eqs. (2–9) and (2–10), and multiplying through by t. Doing this, we obtain

$$x - x_0 = \left(\frac{v_0 + v}{2}\right)t \qquad \text{(constant acceleration only).} \qquad (2\text{–}14)$$

Note that Eq. (2–14) does not contain the acceleration a. This is sometimes a useful equation when a is constant but its value is unknown.

Equations (2–8), (2–12), (2–13), and (2–14) are the *equations of motion with constant acceleration.* By using these equations, we can solve *any* kinematics problem involving straight-line motion of a particle with constant acceleration.

Figure 2–15 shows a graph of the coordinate x as a function of time for motion with constant acceleration. That is, it is a graph of Eq. (2–12); the x-t graph for constant acceleration is always a *parabola.* The curve intercepts the vertical axis (x-axis) at x_0, the position at $t = 0$. The slope of the tangent at $t = 0$ equals v_0, the initial velocity, and the slope of the tangent at any time t equals the velocity v at that time. The graph in Fig. 2–15 is concave up (it curves upward). The slope and velocity are continuously increasing, so the acceleration a is positive. If a is negative, the x-t graph is a parabola that is concave down (has a downward curvature).

For the particular case of motion with constant acceleration depicted in Fig. 2–12 and graphed in Figs. 2–13, 2–14, and 2–15, the values of x_0, v_0, and a are all positive. We invite you to redraw these figures for cases in which one, two, or all three of these quantities are negative.

A special case of motion with constant acceleration occurs when the acceleration is *zero.* The velocity is then constant, and the equations of motion become simply

$$v = v_0 = \text{constant},$$

$$x = x_0 + vt.$$

1.8
Seat Belts Save Lives
1.9
Screeching to a Halt
1.11
Car Starts, Then Stops
1.12
Solving Two-Vehicle Problems
1.13
Car Catches Truck
1.14
Avoiding a Rear-End Collision

2–15 A position-time (x-t) graph for straight-line motion with constant acceleration. This graph refers to the same motion as is shown in Figs. 2–12, 2–13, and 2–14. For this motion the initial position x_0, the initial velocity v_0, and the acceleration a are all positive.

Problem–Solving Strategy

MOTION WITH CONSTANT ACCELERATION

1. You *must* decide at the beginning of a problem where the origin of coordinates is and which axis direction is positive. The choices are usually a matter of convenience. It is often easiest to place the particle at the origin at time $t = 0$; then $x_0 = 0$. It is always helpful to make a motion diagram showing these choices and some later positions of the particle.

2. Remember that your choice of the positive axis direction automatically determines the positive directions for velocity and acceleration. If x is positive to the right of the origin, then v and a are also positive toward the right.

3. Restate the problem in words first, and then translate this description into symbols and equations. *When* does the particle arrive at a certain point (that is, at what value of t)? *Where* is the particle when its velocity has a specified value (that is, what is the value of x when v has the specified value)? The following example asks "Where is the motorcyclist when his velocity is 25 m/s?" Translated

into symbols, this becomes "What is the value of x when $v = 25$ m/s?"

4. Make a list of quantities such as x, x_0, v, v_0, a, and t. In general, some of these will be known and some will be unknown. Write down the values of those that are known. Be on the lookout for implicit information. For example, "A car sits at a stoplight" usually means $v_0 = 0$.

5. Once you have identified the unknowns, try to choose an equation from Eqs. (2–8), (2–12), (2–13), and (2–14) that contains only one of the unknowns. Solve this equation for the unknown, using symbols only. Then substitute the known values and compute the value of the unknown. Sometimes you will have to solve two simultaneous equations for two unknowns.

6. Take a hard look at your results to see whether they make sense. Are they within the general range of values you expected?

EXAMPLE 2-4

Constant-acceleration calculations A motorcyclist heading east through a small Iowa city accelerates after he passes the signpost marking the city limits (Fig. 2–16). His acceleration is a constant 4.0 m/s². At time $t = 0$ he is 5.0 m east of the signpost, moving east at 15 m/s. a) Find his position and velocity at time $t = 2.0$ s. b) Where is the motorcyclist when his velocity is 25 m/s?

SOLUTION We take the signpost as the origin of coordinates ($x = 0$), and choose the positive x-axis to point east (Fig. 2–16). At the initial time $t = 0$ the initial position is $x_0 = 5.0$ m, and the initial velocity is $v_0 = 15$ m/s. The constant acceleration is $a = 4.0$ m/s².

a) The unknowns are the values of the position x and the velocity v at the later time $t = 2.0$ s. We determine the first of these using Eq. (2–12), which gives position x as a function of time t:

$$x = x_0 + v_0 t + \frac{1}{2} a t^2$$

$$= 5.0 \text{ m} + (15 \text{ m/s})(2.0 \text{ s}) + \frac{1}{2}(4.0 \text{ m/s}^2)(2.0 \text{ s})^2$$

$$= 43 \text{ m}.$$

We also use Eq. (2–8), which gives velocity v as a function of time t:

$$v = v_0 + at$$

$$= 15 \text{ m/s} + (4.0 \text{ m/s}^2)(2.0 \text{ s}) = 23 \text{ m/s}.$$

Do these results make sense? The motorcyclist accelerates

2–16 A motorcyclist traveling with constant acceleration.

from 15 m/s (about 34 mi/h, or 54 km/h) to 23 m/s (about 51 mi/h, or 83 km/h) in 2.0 s while traveling a distance of 38 m (about 125 ft). This is pretty brisk acceleration, but well within the capabilities of a high-performance bike.

b) We want to know the value of x when $v = 25$ m/s. From our solution to part (a), we can see that this occurs at a time later than 2.0 s and at a point farther than 43 m from the signpost. From Eq. (2–13) we have

$$v^2 = v_0^2 + 2a(x - x_0).$$

Solving for x and substituting in the known values, we find

$$x = x_0 + \frac{v^2 - v_0^2}{2a}$$

$$= 5.0 \text{ m} + \frac{(25 \text{ m/s})^2 - (15 \text{ m/s})^2}{2(4.0 \text{ m/s}^2)}$$

$$= 55 \text{ m}.$$

Alternatively, we may use Eq. (2–8) to first find the time when $v = 25$ m/s:

$$v = v_0 + at, \quad \text{so}$$

$$t = \frac{v - v_0}{a} = \frac{25 \text{ m/s} - 15 \text{ m/s}}{4.0 \text{ m/s}^2}$$

$$= 2.5 \text{ s.}$$

Then from Eq. (2–12) we have

$$x = x_0 + v_0 t + \frac{1}{2} at^2$$

$$= 5.0 \text{ m} + (15 \text{ m/s})(2.5 \text{ s}) + \frac{1}{2}(4.0 \text{ m/s}^2)(2.5 \text{ s})^2$$

$$= 55 \text{ m.}$$

Again, this seems to be a plausible result.

EXAMPLE 2–5

Two bodies with different accelerations A motorist traveling with constant velocity of 15 m/s (about 34 mi/h) passes a school-crossing corner, where the speed limit is 10 m/s (about 22 mi/h). Just as the motorist passes, a police officer on a motorcycle stopped at the corner starts off in pursuit with constant acceleration of 3.0 m/s² (Fig. 2–17a). a) How much time elapses before the officer catches up with the motorist? b) What is the officer's speed at that point? c) What is the total distance each vehicle has traveled at that point?

SOLUTION The police officer and the motorist both move with constant acceleration (equal to zero for the motorist), so we can use the formulas we have developed. We take the origin at the corner, so $x_0 = 0$ for both, and take the positive direction to the right. Let x_P (for police) be the officer's position and x_M (for motorist) the motorist's position at any time. The initial velocities are $v_{P_0} = 0$ for the officer and $v_{M_0} = 15$ m/s for the motorist; the constant accelerations are $a_P = 3.0$ m/s² for the officer and $a_M = 0$ for the motorist. Then, applying Eq. (2–12) to each, we have

$$x = x_0 + v_0 t + \frac{1}{2} at^2,$$

$$x_M = v_{M_0} t,$$

$$x_P = \frac{1}{2} a_P t^2.$$

a) We want to find the time t when the officer catches the motorist. This is when the two vehicles are at the same position, so at this time, $x_M = x_P$. Equating the two expressions above, we have

$$v_{M_0} t = \frac{1}{2} a_P t^2,$$

$$t = 0 \quad \text{or} \quad t = \frac{2 v_{M_0}}{a_P} = \frac{2(15 \text{ m/s})}{3.0 \text{ m/s}^2} = 10 \text{ s.}$$

There are *two* times when both the vehicles have the same x-coordinate. The first, $t = 0$, is the time when the motorist passes the parked motorcycle at the corner. The second, $t = 10$ s, is the time when the officer catches up with the motorist. b) We want the magnitude of the officer's velocity v_P at the time t found in part (a). Her velocity at any time is given by Eq. (2–8):

$$v_P = v_{P_0} + a_P t = 0 + (3.0 \text{ m/s}^2)t,$$

so when $t = 10$ s, we find $v_P = 30$ m/s. When the officer overtakes the motorist, she is traveling twice as fast as the motorist is. c) In 10 s the distance the motorist travels is

$$x_M = (15 \text{ m/s})(10 \text{ s}) = 150 \text{ m,}$$

and the distance the officer travels is

$$x_P = \frac{1}{2}(3.0 \text{ m/s}^2)(10 \text{ s})^2 = 150 \text{ m.}$$

This verifies that at the time the officer catches the motorist, they have gone equal distances.

Figure 2–17b shows graphs of x versus t for each vehicle. We see again that there are two times when the two positions are the same (where the two graphs cross). At neither of these times do the two vehicles have the same velocity (i.e., where the two graphs cross, their slopes are different). At $t = 0$, the officer is at rest; at $t = 10$ s, the officer has twice the speed of the motorist.

In a real pursuit the officer would accelerate to a speed faster than that of the motorist, then slow down in order to have the same velocity as the motorist when she catches him. We haven't treated this case here because it involves a changing acceleration (but see Problem 2–56).

2–17 (a) Motion with constant acceleration overtaking motion with constant velocity. (b) Graph of x versus t for each vehicle.

(a)

(b)

2–6 FREELY FALLING BODIES

The most familiar example of motion with (nearly) constant acceleration is that of a body falling under the influence of the earth's gravitational attraction. Such motion has held the attention of philosophers and scientists since ancient times. In the 4th century B.C., Aristotle thought (erroneously) that heavy objects fall faster than light objects, in proportion to their weight. Nineteen centuries later, Galileo argued that a body should fall with a downward acceleration that is constant and independent of its weight. We mentioned in Section 1–2 that, according to legend, Galileo experimented by dropping bullets and cannonballs from the Leaning Tower of Pisa.

The motion of falling bodies has since been studied with great precision. When the effects of the air can be neglected, Galileo is right; all bodies at a particular location fall with the same downward acceleration, regardless of their size or weight. If the distance of the fall is small compared to the radius of the earth, the acceleration is constant. In the following discussion we use an idealized model in which we neglect the effects of the air, the earth's rotation, and the decrease of acceleration with increasing altitude. We call this idealized motion **free fall,** although it includes rising as well as falling motion. (In Chapter 3 we will extend the discussion of free fall to include the motion of projectiles, which move both vertically and horizontally.)

Figure 2–18 is a photograph of a falling ball taken with a stroboscopic light source that produces a series of intense flashes at equal time intervals. Each flash is so short (a few millionths of a second) that there is little blur in the image of even a rapidly moving body. As each flash occurs, the position of the ball at that instant is recorded on the film. Because of the equal time intervals between flashes, the average velocity of the ball between any two flashes is proportional to the distance between corresponding images in the photograph. The increasing distances between images show that the velocity is continuously changing; the ball is accelerating downward. Careful measurement shows that the velocity change is the same in each time interval, so the acceleration of the freely falling ball is constant.

The constant acceleration of a freely falling body is called the **acceleration due to gravity,** and we denote its magnitude with the letter g. At or near the earth's surface the value of g is approximately 9.8 m/s^2, 980 cm/s^2, or 32 ft/s^2. The exact value varies with location, so we will often give the value of g at the earth's surface to only two significant figures. Because g is the magnitude of a vector quantity, it is always a *positive* number. On the surface of the moon the acceleration due to gravity is caused by the attractive force of the moon rather than the earth, and $g = 1.6$ m/s^2. Near the surface of the sun, $g = 270$ m/s^2.

In the following examples we use the constant-acceleration equations developed in Section 2–5. We suggest that you review the problem-solving strategies discussed in that section before you study the examples below.

2–18 Multiflash photo of a freely falling ball.

1.7
Balloonist Drops Lemonade
1.10
Pole-Vaulter Lands

EXAMPLE 2–6

A 500-lira coin is dropped from the Leaning Tower of Pisa. It starts from rest and falls freely. Compute its position and velocity after 1.0, 2.0, and 3.0 s.

SOLUTION We'll take the origin O at the starting point, the coordinate axis as vertical, and the upward direction as positive (Fig. 2–19). Because the coordinate axis is vertical, let's call the coordinate y instead of x. We replace all the x's in the constant-acceleration equations by y's. The initial coordinate y_0

and the initial velocity v_0 are both zero. The acceleration is downward (in the negative y-direction), so $a = -g = -9.8$ m/s^2. (Remember that, by definition, g itself is *always* positive.)

The unknowns are the values of y and v at the three specified times. From Eqs. (2–12) and (2–8), with x replaced by y, we get

$$y = v_0 t + \frac{1}{2} a t^2 = 0 + \frac{1}{2}(-g)t^2 = (-4.9 \text{ m/s}^2)t^2,$$

$$v = v_0 + at = 0 + (-g)t = (-9.8 \text{ m/s}^2)t.$$

When $t = 1.0$ s, $y = (-4.9 \text{ m/s}^2)(1.0 \text{ s})^2 = -4.9$ m and $v = (-9.8 \text{ m/s}^2)(1.0 \text{ s}) = -9.8$ m/s; after one second, the coin is 4.9 m below the origin (y is negative) and has a downward velocity (v is negative) with magnitude 9.8 m/s.

The position and velocity at 2.0 and 3.0 s are found in the same way. The results are shown in Fig. 2–19; check the numerical values for yourself.

2–19 Position and velocity of a coin freely falling from rest.

EXAMPLE 2–7

You throw a ball vertically upward from the roof of a tall building. The ball leaves your hand at a point even with the roof railing with an upward speed of 15.0 m/s; the ball is then in free fall. On its way back down, it just misses the railing. At the location of the building, $g = 9.80 \text{ m/s}^2$. Find a) the position and velocity of the ball 1.00 s and 4.00 s after leaving your hand; b) the velocity when the ball is 5.00 m above the railing; c) the maximum height reached and the time at which it is reached; d) the acceleration of the ball when it is at its maximum height.

SOLUTION In Fig. 2–20 the downward path is displaced a little to the right of its actual position for clarity. Take the origin at the roof railing, at the point where the ball leaves your hand, and take the positive direction to be upward. First, let's collect our data. The initial position y_0 is zero. The initial velocity v_0 is +15.0 m/s, and the acceleration is $a = -g = -9.80 \text{ m/s}^2$.

a) The position y and velocity v at any time t after the ball leaves your hand are given by Eqs. (2–12) and (2–8) with x's replaced by y's, so

$$y = y_0 + v_0 t + \frac{1}{2} a t^2 = y_0 + v_0 t + \frac{1}{2}(-g)t^2$$
$$= (0) + (15.0 \text{ m/s})t + \frac{1}{2}(-9.80 \text{ m/s}^2)t^2,$$

$$v = v_0 + at = v_0 + (-g)t$$
$$= 15.0 \text{ m/s} + (-9.80 \text{ m/s}^2)t.$$

When $t = 1.00$ s, these equations give

$$y = +10.1 \text{ m}, \qquad v = +5.2 \text{ m/s}.$$

The ball is 10.1 m above the origin (y is positive), and it is moving upward (v is positive) with a speed of 5.2 m/s. This is less than the initial speed of 15.0 m/s, as expected.

When $t = 4.00$ s, the equations for y and v as functions of time t give

$$y = -18.4 \text{ m}, \qquad v = -24.2 \text{ m/s}.$$

The ball has passed its highest point and is 18.4 m *below* the origin (y is negative). It has a *downward* velocity (v is negative) with magnitude 24.2 m/s. This speed is greater than the initial speed, as we should expect for points below the ball's launching point. Note that to get these results, we don't need to find the highest point reached or the time at which it was reached. The equations of motion give the position and velocity at any time, whether the ball is on the way up or on the way down.

b) The velocity v at any position y is given by Eq. (2–13) with x's replaced by y's:

$$v^2 = v_0^2 + 2a(y - y_0) = v_0^2 + 2(-g)(y - 0)$$
$$= (15.0 \text{ m/s})^2 + 2(-9.80 \text{ m/s}^2)y.$$

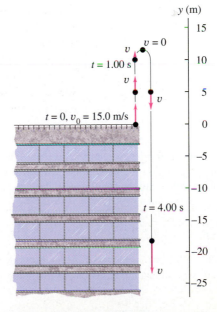

2–20 Position and velocity of a ball thrown vertically upward.

When the ball is 5.00 m above the origin, $y = +5.00$ m, so

$$v^2 = (15.0 \text{ m/s})^2 + 2(-9.80 \text{ m/s}^2)(5.00 \text{ m}) = 127 \text{ m}^2/\text{s}^2,$$

$$v = \pm 11.3 \text{ m/s}.$$

We get *two* values of v, one positive and one negative. As shown in Fig. 2–20, the ball passes this point twice, once on the way up and again on the way down. The velocity on the way up is +11.3 m/s, and on the way down it is −11.3 m/s.

c) At the highest point, the ball stops going up (positive v) and starts going down (negative v). Just at the instant when it reaches the highest point, $v = 0$. The maximum height y_1 can then be found in two ways. The first way is to use Eq. (2–13) and substitute $v = 0$, $y_0 = 0$, and $a = -g$:

$$0 = v_0^2 + 2(-g)(y_1 - 0),$$

$$y_1 = \frac{v_0^2}{2g} = \frac{(15.0 \text{ m/s})^2}{2(9.80 \text{ m/s}^2)} = +11.5 \text{ m}.$$

The second way is find the time at which $v = 0$ using Eq. (2–8), $v = v_0 + at$, and then substitute this value of t into Eq. (2–12) to find the position at this time. From Eq. (2–8), the time t_1 when the ball reaches the highest point is given by

$$v = 0 = v_0 + (-g)t_1,$$

$$t_1 = \frac{v_0}{g} = \frac{15.0 \text{ m/s}}{9.80 \text{ m/s}^2} = 1.53 \text{ s}.$$

Substituting this value of t into Eq. (2–12), we find

$$y = y_0 + v_0 t + \frac{1}{2} a t^2 = (0) + (15 \text{ m/s})(1.53 \text{ s})$$

$$+ \frac{1}{2}(-9.8 \text{ m/s}^2)(1.53 \text{ s})^2 = +11.5 \text{ m}.$$

Notice that with the first way of finding the maximum height, it's not necessary to find the time first.

d) **CAUTION ▶** It's a common misconception that at the highest point of the motion the velocity is zero *and* the acceleration is zero. If this were so, once the ball reached the highest point it would hang there in midair forever! To see why, remember that acceleration is the rate of change of velocity. If the acceleration were zero at the highest point, the ball's velocity would no longer change, and once the ball was instantaneously at rest it would remain at rest forever. ◀

The truth of the matter is that at the highest point, the acceleration is still $a = -g = -9.80$ m/s^2, the same value as when the ball is moving up and when it's moving down. The ball stops for an instant at the highest point, but its velocity is continuously changing, from positive values through zero to negative values.

Figure 2–21 shows graphs of position and velocity as functions of time for this problem. Note that the v-t graph has a constant negative slope. The acceleration is negative (downward) on the way up, at the highest point, and on the way down.

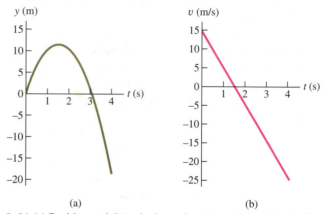

(a)   (b)

2–21 (a) Position and (b) velocity as functions of time for a ball thrown upward with an initial velocity of 15 m/s.

EXAMPLE 2–8

Find the time when the ball in Example 2–7 is 5.00 m below the roof railing.

SOLUTION We again choose the y-axis as in Fig. 2–20, so y_0, v_0, and $a = -g$ have the same values as in Example 2–7. The position y as a function of time t is again given by Eq. (2–12):

$$y = y_0 + v_0 t + \frac{1}{2} a t^2 = y_0 + v_0 t + \frac{1}{2}(-g)t^2.$$

We want to solve this for the value of t when $y = -5.00$ m. Since this equation involves t^2, it is a *quadratic* equation for t. We first rearrange our equation into the same form as the standard quadratic equation for an unknown x, $Ax^2 + Bx + C = 0$:

$$\left(\frac{1}{2} g\right) t^2 + (-v_0)t + (y - y_0) = At^2 + Bt + C = 0,$$

so $A = g/2$, $B = -v_0$, and $C = y - y_0$. Using the quadratic formula (Appendix B), we find that this equation has *two* solutions:

$$t = \frac{-B \pm \sqrt{B^2 - 4AC}}{2A}$$

$$= \frac{-(-v_0) \pm \sqrt{(-v_0)^2 - 4(g/2)(y - y_0)}}{2(g/2)},$$

$$t = \frac{v_0 \pm \sqrt{v_0^2 - 2g(y - y_0)}}{g}.$$

Substituting the values $y_0 = 0$, $v_0 = +15.0$ m/s, $g = 9.80$ m/s^2, and $y = -5.00$ m, we find

$$t = \frac{(15.0 \text{ m/s}) \pm \sqrt{(15.0 \text{ m/s})^2 - 2(9.80 \text{ m/s}^2)(-5.00 \text{ m} - 0)}}{9.80 \text{ m/s}^2},$$

$$t = +3.36 \text{ s} \quad \text{or} \quad t = -0.30 \text{ s}.$$

To decide which of these is the right answer, the key question to ask is "Are these answers reasonable?" The second answer, $t = -0.30$ s, is simply not reasonable; it refers to a time 0.30 s *before* the ball left your hand! The correct answer is $t = +3.36$ s. The ball is 5.00 m below the railing 3.36 s *after* it leaves your hand.

Where did the erroneous "solution" $t = -0.30$ s come from? Remember that we began with Eq. (2–12) with $a = -g$, that is, $y = y_0 + v_0 t + \frac{1}{2}(-g)t^2$. This has built into it the assumption that the acceleration is constant for *all* values of *t*, whether positive, negative, or zero. Taken at face value, this equation would tell us that the ball had been moving upward in free fall ever since the dawn of time; it eventually passes your hand at $y = 0$ at the special instant we chose to call $t = 0$, then continues in free fall. But anything that this equation describes happening before $t = 0$ is pure fiction, since the ball went into free fall only after leaving your hand at $t = 0$; the "solution" $t = -0.30$ s is part of this fiction.

We invite you to repeat these calculations to find the times at which the ball is 5.00 m *above* the origin ($y = +5.00$ m). The two answers are $t = +0.38$ s and $t = +2.68$ s; these are both positive

values of *t*, and both refer to the real motion of the ball after leaving your hand. The earlier time is when the ball passes through $y = +5.00$ m moving upward, and the later time is when the ball passes through this point moving downward. (Compare this with part (b) of Example 2–7.) You should also solve for the times at which $y = +15.0$ m. In this case, both solutions involve the square root of a negative number, so there are *no* real solutions. This makes sense; we found in part (c) of Example 2–7 that the ball's maximum height is only $y = +11.5$ m, so it never reaches $y = +15.0$ m. While a quadratic equation such as Eq. (2–12) always has two solutions, in some situations one or both of the solutions will not be physically reasonable.

*2–7 VELOCITY AND POSITION BY INTEGRATION

This optional section is intended for students who have already learned a little integral calculus. In Section 2–5 we analyzed the special case of straight-line motion with constant acceleration. When *a* is not constant, as is frequently the case, the equations that we derived in that section are no longer valid. But even when *a* varies with time, we can still use the relation $v = dx/dt$ to find the velocity *v* as a function of time if the position *x* is a known function of time. And we can still use $a = dv/dt$ to find the acceleration *a* as a function of time if the velocity *v* is a known function of time.

In many physical situations, however, position and velocity are not known as functions of time, while the acceleration is. How can we find the position and velocity from the acceleration function $a(t)$? This problem arises in navigating an airliner between North America and Europe (Fig. 2–22). The crew of the airliner must know their position precisely at all times, because the airspace over the North Atlantic is very congested. But over the ocean an airliner is usually out of range of both radio navigation beacons on land and air traffic controllers' radar. To determine their position, airliners carry a device called an inertial navigation system (INS), which measures the acceleration of the moving airliner. This is done in much the same way that you can sense changes in the velocity of a car in which you're riding, even when your eyes are closed. (In Chapter 4 we'll discuss how your body detects acceleration.) Given this information, along with the airliner's initial position (say, a particular gate at Kennedy International Airport) and its initial velocity (zero when parked at the gate), the INS calculates and displays for the crew the current velocity and position of the airliner at all times during the flight. Our goal in the remainder of this section is to see how these calculations are done for the simpler case of motion in a straight line with time-varying acceleration.

We first consider a graphical approach. Figure 2–23 is a graph of acceleration versus time for a body whose acceleration is not constant but increases with time. We can divide the time interval between times t_1 and t_2 into many smaller intervals, calling a typical one Δt. Let the average acceleration during Δt be a_{av}. From Eq. (2–4) the change in velocity Δv during Δt is

$$\Delta v = a_{av}\Delta t.$$

Graphically, Δv equals the area of the shaded strip with height a_{av} and width Δt, that is, the area under the curve between the left and right sides of Δt. The total velocity change during any interval (say, t_1 to t_2) is the sum of the velocity changes Δv in the small subintervals. So the total velocity change is represented graphically by the *total* area under the *a-t* curve between the vertical lines t_1 and t_2. (In Section 2–5 we showed this to be true in the special case in which the acceleration was constant.)

In the limit that all the Δt's become very small and their number very large, the value of a_{av} for the interval from any time *t* to $t + \Delta t$ approaches the instantaneous acceleration

To London

Acceleration: Known
Velocity: To be determined
Position: To be determined

N
W — E
S

From
New York

2–22 The position and velocity of an airliner crossing the Atlantic are found by integrating its acceleration with respect to time.

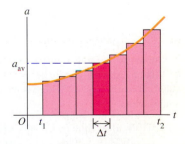

2–23 The area under an *a-t* graph between times t_1 and t_2 equals the change in velocity, $v_2 - v_1$, that occurs between those times.

a at time t. In this limit, the area under the a-t curve is the *integral* of a (which is in general a function of t) from t_1 to t_2. If v_1 is the velocity of the body at time t_1 and v_2 is the velocity at time t_2, then

$$v_2 - v_1 = \int_{v_1}^{v_2} dv = \int_{t_1}^{t_2} a \, dt. \qquad (2\text{--}15)$$

The change in velocity v is the integral of acceleration a with respect to time.

We can carry out exactly the same procedure with the curve of velocity versus time, where v is in general a function of t. If x_1 is a body's position at time t_1 and x_2 is its position at time t_2, from Eq. (2–2) the displacement Δx during a small time interval Δt is equal to $v_{av} \, \Delta t$, where v_{av} is the average velocity during Δt. The total displacement $x_2 - x_1$ during the interval $t_2 - t_1$ is given by

$$x_2 - x_1 = \int_{x_1}^{x_2} dx = \int_{t_1}^{t_2} v \, dt. \qquad (2\text{--}16)$$

The change in position x—that is, the displacement—is the time integral of velocity v. Graphically, the displacement between times t_1 and t_2 is the area under the v-t curve between those two times. (This is the same result that we obtained in Section 2–5 for the special case in which v is given by Eq. (2–8).)

If $t_1 = 0$ and t_2 is any later time t, and if x_0 and v_0 are the position and velocity, respectively, at time $t = 0$, then we can rewrite Eqs. (2–15) and (2–16) as follows:

$$v = v_0 + \int_0^t a \, dt, \qquad (2\text{--}17)$$

$$x = x_0 + \int_0^t v \, dt. \qquad (2\text{--}18)$$

Here x and v are the position and velocity at time t. If we know the acceleration a as a function of time and we know the initial velocity v_0, we can use Eq. (2–17) to find the velocity v at any time; in other words, we can find v as a function of time. Once we know this function, and given the initial position x_0, we can use Eq. (2–18) to find the position x at any time.

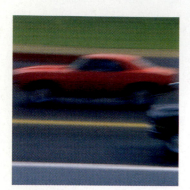

When you push your car's accelerator pedal to the floorboard, the resulting acceleration is *not* constant: the greater the car's speed, the more slowly it gains additional speed. A typical car takes twice as long to accelerate from 50 km/h to 100 km/h as it does to accelerate from 0 to 50 km/h.

EXAMPLE 2–9

Motion with changing acceleration Sally is driving along a straight highway in her classic 1965 Mustang. At time $t = 0$, when Sally is moving at 10 m/s in the positive x-direction, she passes a signpost at $x = 50$ m. Her acceleration is a function of time:

$$a = 2.0 \text{ m/s}^2 - (0.10 \text{ m/s}^3)t.$$

a) Derive expressions for her velocity and position as functions of time. b) At what time is her velocity greatest? c) What is the maximum velocity? d) Where is the car when it reaches maximum velocity?

SOLUTION a) At time $t = 0$, Sally's position is $x_0 = 50$ m, and her velocity is $v_0 = 10$ m/s. Since we are given the acceleration a as a function of time, we first use Eq. (2–17) to find the velocity v as a function of time t. The integral of t^n is $\int t^n dt = \frac{1}{n+1} t^{n+1}$ for $n \neq -1$, so

$$v = 10 \text{ m/s} + \int_0^t [2.0 \text{ m/s}^2 - (0.10 \text{ m/s}^3)t] \, dt$$

$$= 10 \text{ m/s} + (2.0 \text{ m/s}^2)t - \frac{1}{2}(0.10 \text{ m/s}^3)t^2.$$

Then we use Eq. (2–18) to find x as a function of t:

$$x = 50 \text{ m} + \int_0^t [10 \text{ m/s} + (2.0 \text{ m/s}^2)t - \frac{1}{2}(0.10 \text{ m/s}^3)t^2] \, dt$$

$$= 50 \text{ m} + (10 \text{ m/s})t + \frac{1}{2}(2.0 \text{ m/s}^2)t^2 - \frac{1}{6}(0.10 \text{ m/s}^3)t^3.$$

Figure 2–24 shows graphs of x, v, and a as functions of time. Note that for any time t the slope of the x-t graph equals the value of v, and the slope of the v-t graph equals the value of a.
b) The maximum value of v occurs when v stops increasing and begins to decrease. At this instant, $dv/dt = a = 0$. Setting the

expression for acceleration equal to zero, we obtain

$$0 = 2.0 \text{ m/s}^2 - (0.10 \text{ m/s}^3)t,$$

$$t = \frac{2.0 \text{ m/s}^2}{0.10 \text{ m/s}^3} = 20 \text{ s}.$$

As Fig. 2–24 shows, a is positive between $t = 0$ and $t = 20$ s and negative after that. It is zero at $t = 20$ s, the time at which v is maximum. The car speeds up until $t = 20$ s (because v and a have the same sign) and slows down after $t = 20$ s (because v and a have opposite signs).

c) We find the maximum velocity by substituting $t = 20$ s (when velocity is maximum) into the general velocity equation:

$$v_{max} = 10 \text{ m/s} + (2.0 \text{ m/s}^2)(20 \text{ s}) - \frac{1}{2}(0.10 \text{ m/s}^3)(20 \text{ s})^2$$

$$= 30 \text{ m/s}.$$

d) The maximum value of v occurs at time $t = 20$ s. We obtain the position of the car (that is, the value of x) at that time by substituting $t = 20$ s into the general expression for x:

$$x = 50 \text{ m} + (10 \text{ m/s})(20 \text{ s}) + \frac{1}{2}(2.0 \text{ m/s}^2)(20 \text{ s})^2$$

$$- \frac{1}{6}(0.10 \text{ m/s}^3)(20 \text{ s})^3$$

$$= 517 \text{ m}.$$

Since v is maximum at $t = 20$ s, the x-t graph has its maximum positive slope at this time. Note that the x-t graph is concave up (curved upward) from $t = 0$ to $t = 20$ s when a is positive and is concave down (curved downward) after $t = 20$ s when a is negative.

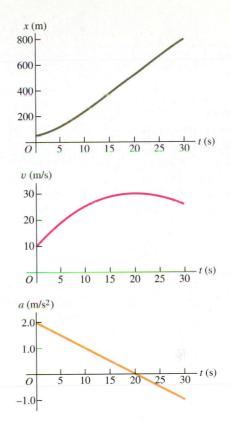

2–24 The position, velocity, and acceleration of the car in Example 2–9 as functions of time. Can you show that if this motion continues, the car will stop at $t = 44.5$ s?

EXAMPLE 2–10

Constant-acceleration formulas via integration Use Eqs. (2–17) and (2–18) to find v and x as functions of time in the case in which the acceleration is constant. Compare the results to the constant-acceleration formulas $v = v_0 + at$ (Eq. 2–8) and $x = x_0 + v_0 t + \frac{1}{2}at^2$ (Eq. 2–12), which were derived in Section 2–5 without using integration.

SOLUTION From Eq. (2–17) the velocity is given by

$$v = v_0 + \int_0^t a \, dt = v_0 + a \int_0^t dt = v_0 + at.$$

We were able to take a outside the integral because it is constant. Our result is identical to Eq. (2–8), as it should be. Substituting this expression for v into Eq. (2–18), we get

$$x = x_0 + \int_0^t v \, dt = x_0 + \int_0^t (v_0 + at) \, dt.$$

Since v_0 and a are constants, they may be taken outside the integral:

$$x = x_0 + v_0 \int_0^t dt + a \int_0^t t \, dt = x_0 + v_0 t + \frac{1}{2}at^2.$$

This result is the same as Eq. (2–12). Our expressions for v and x, Eqs. (2–17) and (2–18), which we developed to deal with cases in which acceleration depends on time, can be used just as well when the acceleration is constant.

SUMMARY

- When a particle moves along a straight line, we describe its position with respect to an origin O by means of a coordinate such as x.

■ The particle's average velocity during a time interval $\Delta t = t_2 - t_1$ is defined as

$$v_{av} = \frac{x_2 - x_1}{t_2 - t_1} = \frac{\Delta x}{\Delta t}. \qquad (2\text{--}2)$$

The instantaneous velocity at any time t is defined as

$$v = \lim_{\Delta t \to 0} \frac{\Delta x}{\Delta t} = \frac{dx}{dt}. \qquad (2\text{--}3)$$

■ The average acceleration during a time interval $\Delta t = t_2 - t_1$ is defined as

$$a_{av} = \frac{v_2 - v_1}{t_2 - t_1} = \frac{\Delta v}{\Delta t}. \qquad (2\text{--}4)$$

The instantaneous acceleration at any time t is defined as

$$a = \lim_{\Delta t \to 0} \frac{\Delta v}{\Delta t} = \frac{dv}{dt}. \qquad (2\text{--}5)$$

■ When the acceleration is constant, the position x and velocity v at any time t are related to the acceleration a, the initial position x_0, and the initial velocity v_0 (both at time $t = 0$) by the following equations:

$$v = v_0 + at \qquad \text{(constant acceleration only)}, \qquad (2\text{--}8)$$

$$x = x_0 + v_0 t + \frac{1}{2}at^2 \qquad \text{(constant acceleration only)}, \qquad (2\text{--}12)$$

$$v^2 = v_0^2 + 2a(x - x_0) \qquad \text{(constant acceleration only)}, \qquad (2\text{--}13)$$

$$x - x_0 = \left(\frac{v_0 + v}{2}\right)t \qquad \text{(constant acceleration only)}. \qquad (2\text{--}14)$$

■ Free fall is a case of motion with constant acceleration. The magnitude of the acceleration due to gravity is a positive quantity g. The acceleration of a body in free fall is always downward.

■ When the acceleration is not constant but is a known function of time, we can find the velocity and position as functions of time by integrating the acceleration function.

DISCUSSION QUESTIONS

Q2–1 Does the speedometer of a car measure speed or velocity? Explain.

Q2–2 Janet claims that a speed of 60 mi/h is the same as 88 ft/s and 27 m/s. Which relation is exact and which approximate?

Q2–3 Some European countries have highway speed limits of 110 km/h. What is the equivalent number of miles per hour?

Q2–4 Under what conditions is average velocity equal to instantaneous velocity?

Q2–5 In a given time interval, is the total displacement of a particle equal to the product of the average velocity and the time interval, even when the velocity is not constant? Explain.

Q2–6 Under what conditions does the magnitude of the average velocity equal the average speed?

Q2–7 When a Dodge Viper is at Elwood's Car Wash, a BMW Z3 is at Elm and Main. Later, when the Dodge reaches Elm and Main, the BMW reaches Elwood's Car Wash. How are the cars' average velocities between these two times related?

Q2–8 A driver in Massachusetts was sent to traffic court for speeding. The evidence against the driver was that a policewoman observed the driver's car to be alongside a second car at a certain moment, and that the policewoman had already clocked the second car as going faster than the speed limit. The driver argued that "the second car was passing me. I was not speeding." The judge ruled against the driver because, in the judge's words, "if two cars were side by side, you were both speeding." If you were a lawyer representing the accused driver, how would you argue this case?

Q2–9 Can you have a zero displacement and a nonzero average velocity? A nonzero velocity? Illustrate your answers on a *x-t* graph.

Q2–10 Can you have zero acceleration and nonzero velocity? Explain using a *v-t* graph.

Q2–11 Can you have zero velocity and nonzero average acceleration? Zero velocity and nonzero acceleration? Explain using a v-t graph.

Q2–12 An automobile is traveling west. Can it have a velocity toward the west and at the same time have an acceleration toward the east? Under what circumstances?

Q2–13 The official's truck in Fig. 2–2 is at $x_1 = 277$ m at $t_1 = 16.0$ s, and is at $x_2 = 19$ m at $t_2 = 25.0$ s. a) Sketch *two* different possible x-t graphs for the motion of the truck. b) Does the average velocity v_{av} during the time interval from t_1 to t_2 have the same value for both of your graphs? Why or why not?

Q2–14 Under constant acceleration the average velocity of a particle is half the sum of its initial and final velocities. Is this still true if the acceleration is *not* constant? Explain.

Q2–15 You throw a baseball straight up in the air so that it rises to a maximum height much greater than your height. Is the magnitude of the acceleration greater while it is being thrown or after it leaves your hand? Explain.

Q2–16 Prove these statements: i) As long as you can neglect the effects of the air, if you throw anything vertically upward, it will have the same speed when it returns to the release point as when it was released. ii) The time of flight will be twice the time it takes to get to its highest point.

Q2–17 In Example 2–7, substituting $y = -18.4$ m into Eq. (2–13) gives $v = \pm 24.2$ m/s. The negative root is the velocity at $t = 4.00$ s. Explain the meaning of the positive root.

Q2–18 If the initial position and initial velocity of a vehicle are known and a record is kept of the acceleration at each instant, can the vehicle's position after a certain time be computed from these data? If so, explain how this might be done.

EXERCISES

SECTION 2–2 DISPLACEMENT, TIME, AND AVERAGE VELOCITY

2–1 A rocket carrying a satellite is accelerating straight up from the earth's surface. At 1.15 s after liftoff, the rocket clears the top of its launch platform, 63 m above the ground. After an additional 4.75 s it is 1.00 km above the ground. Calculate the magnitude of the average velocity of the rocket for a) the 4.75-s part of its flight; b) the first 5.90 s of its flight.

2–2 In an experiment, a shearwater (a seabird) was taken from its nest, flown 5150 km away, and released. It found its way back to its nest 13.5 days after release. If we place the origin in the nest and extend the $+x$-axis to the release point, what was the bird's average velocity in m/s a) for the return flight? b) for the whole episode, from leaving the nest to returning?

2–3 You normally drive on the freeway between San Diego and Los Angeles at an average speed of 105 km/h (65 mi/h), and the trip takes 2 h and 20 min. On a Friday afternoon, however, heavy traffic slows you down and you drive the same distance at an average speed of only 70 km/h (43 mi/h). How much longer does the trip take?

2–4 **From Pillar to Post.** Starting from a pillar, you run 200 m east (the $+x$-direction) at an average speed of 5.0 m/s, then run 280 m west at an average speed of 4.0 m/s to a post. Calculate a) your average speed from pillar to post; b) your average velocity from pillar to post.

2–5 a) Your old VW van sputters along at an average speed of 8.0 m/s for 60 s, then catches hold and zips along at an average speed of 20.0 m/s for another 60 s. Calculate your average speed for the entire 120 s. b) Suppose the speed of 8.0 m/s was maintained while you traveled 240 m, followed by the average speed of 20.0 m/s for another 240 m. Calculate the van's average speed for the entire distance. c) In which case is the average speed for the entire motion the numerical average of the two speeds?

2–6 A Honda Civic travels in a straight line along a road. Its distance x from a stop sign is given as a function of time t by the equation $x(t) = \alpha t^2 - \beta t^3$, where $\alpha = 1.50$ m/s^2 and $\beta = 0.0500$ m/s^3. Calculate the average velocity of the car for the following time intervals: a) $t = 0$ to $t = 2.00$ s; b) $t = 0$ to $t = 4.00$ s; c) $t = 2.00$ s to $t = 4.00$ s.

SECTION 2–3 INSTANTANEOUS VELOCITY

2–7 A car is stopped at a traffic light. It then travels along a straight road so that its distance from the light is given by $x(t) = bt^2 - ct^3$, where $b = 2.40$ m/s^2 and $c = 0.120$ m/s^3. a) Calculate the average velocity of the car for the time interval $t = 0$ to $t = 10.0$ s. b) Calculate the instantaneous velocity of the car at i) $t = 0$; ii) $t = 5.0$ s; iii) $t = 10.0$ s. c) How long after starting from rest is the car again at rest?

2–8 A physics professor leaves her house and walks along the sidewalk toward campus. After 5 min it starts to rain and she returns home. Her distance from her house as a function of time is shown in Fig. 2–25. At which of the labeled points is her velocity a) zero? b) constant and positive? c) constant and negative? d) increasing in magnitude? e) decreasing in magnitude?

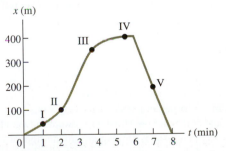

FIGURE 2–25 Exercise 2–8.

SECTION 2–4 AVERAGE AND INSTANTANEOUS ACCELERATION

2–9 A test driver at Incredible Motors, Inc., is testing a new model car having a speedometer calibrated to read m/s rather than mi/h. The following series of speedometer readings was obtained during a test run along a long, straight road:

Time (s)	0	2	4	6	8	10	12	14	16
Speed (m/s)	0	0	2	6	10	16	19	22	22

a) Compute the average acceleration during each 2-s interval. Is the acceleration constant? Is it constant during any part of the test run? b) Make a v-t graph of the data above, using scales of 1 cm = 1 s horizontally and 1 cm = 2 m/s vertically. Draw a smooth curve through the plotted points. By measuring the slope of your curve, find the instantaneous acceleration at $t = 9$ s, 13 s, and 15 s.

2–10 Figure 2–26 shows the velocity of a solar-powered car as a function of time. The driver accelerates from a stop sign, cruises for 20 s at a constant speed of 60 km/h, and then brakes to come to a stop 40 s after leaving the stop sign. Compute the average acceleration for the following time intervals: a) $t = 0$ to $t = 10$ s; b) $t = 30$ s to $t = 40$ s; c) $t = 10$ s to $t = 30$ s; d) $t = 0$ to $t = 40$ s.

FIGURE 2–26 Exercises 2–10 and 2–11.

2–11 Refer to Exercise 2–10 and Fig. 2–26. a) During what time interval does the instantaneous acceleration a have its most positive value? b) During what time interval does the instantaneous acceleration a have its most negative value? c) What is the instantaneous acceleration at $t = 20$ s? d) What is the instantaneous acceleration at $t = 35$ s? e) In a motion diagram (like Fig. 2–10b or Fig. 2–11b), show the position, velocity, and acceleration of the car at the four times $t = 5$ s, $t = 15$ s, $t = 25$ s, and $t = 35$ s.

2–12 An astronaut has left the International Space Station to test a new space scooter. Her partner measures the following velocity changes, each taking place in a 10-s interval. What are the magnitude, the algebraic sign, and the direction of the average acceleration in each interval? Assume that the positive direction is to the right. a) At the beginning of the interval the astronaut is moving toward the right along the x-axis at 15.0 m/s, and at the end of the interval she is moving toward the right at 5.0 m/s. b) At the beginning she is moving toward the left at 5.0 m/s, and at the end she is moving toward the left at 15.0 m/s. c) At the beginning she is moving toward the right at 15.0 m/s, and at the end she is moving toward the left at 15.0 m/s.

2–13 a) Based on your experiences of riding in automobiles, estimate the magnitude of a car's average acceleration when it brakes from highway speeds to a sudden stop. b) Explain why this average acceleration could be regarded as either positive or negative.

2–14 A car's velocity as a function of time is given by $v(t) = \alpha + \beta t^2$, where $\alpha = 3.00$ m/s and $\beta = 0.100$ m/s^3. a) Calculate the average acceleration for the time interval $t = 0$ to

$t = 5.00$ s. b) Calculate the instantaneous acceleration for i) $t = 0$; ii) $t = 5.00$ s. c) Draw accurate v-t and a-t graphs for the car's motion between $t = 0$ and $t = 5.00$ s.

2–15 Figure 2–27 is a graph of the coordinate of a spider crawling along the x-axis. a) Graph its velocity and acceleration as functions of time. b) In a motion diagram (like Fig. 2–10b or Fig. 2–11b), show the position, velocity, and acceleration of the spider at the five times $t = 2.5$ s, $t = 10$ s, $t = 20$ s, $t = 30$ s, and $t = 37.5$ s.

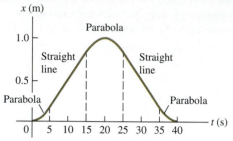

FIGURE 2–27 Exercise 2–15.

2–16 The position of the front bumper of a test car under microprocessor control is given by $x(t) = 2.17$ m $+ (4.80$ m/s$^2)t^2 - (0.100$ m/s$^6)t^6$. Find its a) position and acceleration at the instants when the car has zero velocity. b) Draw x-t, v-t, and a-t graphs for the motion of the bumper between $t = 0$ and $t = 2.00$ s.

SECTION 2–5 MOTION WITH CONSTANT ACCELERATION

2–17 An antelope moving with constant acceleration covers the distance between two points 70.0 m apart in 7.00 s. Its speed as it passes the second point is 15.0 m/s. a) What is its speed at the first point? b) What is the acceleration?

2–18 The catapult of the aircraft carrier *USS Abraham Lincoln* accelerates an F/A-18 Hornet jet fighter from rest to a takeoff speed of 173 mi/h in a distance of 307 ft. Assume constant acceleration. a) Calculate the acceleration of the fighter in m/s^2. b) Calculate the time required for the fighter to be accelerated to takeoff speed.

2–19 Automobile Airbags. The human body can survive a negative acceleration trauma incident (sudden stop) if the magnitude of the acceleration is less than 250 m/s^2 (approximately 25g). If you are in an automobile accident with an initial speed of 105 km/h (65 mi/h) and are stopped by an airbag that inflates from the dashboard, over what distance must the airbag stop you for you to survive the crash?

2–20 An airplane travels 280 m down the runway before taking off. If it starts from rest, moves with constant acceleration, and becomes airborne in 8.00 s, what is its speed, in m/s, when it takes off?

2–21 A car sits in an entrance ramp to a freeway, waiting for a break in the traffic. The driver sees a small gap between a van and an 18-wheel truck and accelerates with constant acceleration along the ramp and onto the freeway. The car starts from rest, moves in a straight line, and has speed 20 m/s (45 mi/h) when it reaches the end of the 120-m-long ramp. a) What is the acceleration of the car? b) How much time does it take the car to travel the length of the ramp? c) The traffic on the freeway is moving at a constant speed of 20 m/s. What distance does the traffic travel while the car is moving the length of the ramp?

2–22 Figures 2–12, 2–13, 2–14, and 2–15 were drawn for motion with constant acceleration with positive values of x_0, v_0, and a. Redraw these four figures for the following cases: a) $x_0 < 0$, $v_0 < 0$, $a < 0$; b) $x_0 > 0$, $v_0 < 0$, $a > 0$; c) $x_0 > 0$, $v_0 > 0$, $a < 0$.

2–23 At time $t = 0$ a Corvette is traveling on a long, straight stretch of highway in Arizona at a constant speed of 30 m/s. This motion continues for 20 s. Then the driver, worried that she is behind schedule, accelerates at a constant rate for 5 s to bring the car's speed to 40 m/s. The car travels at this new speed for 10 s. But then the driver spots a police motorcycle parked behind a large cactus up ahead and slows with a constant acceleration of magnitude 4.0 m/s² until the car's speed is reduced once again to the legal limit of 30 m/s. She then maintains this speed, and waves to the officer as she passes him 5 s later. a) For the motion of the car from $t = 0$ until when the car passes the parked motorcycle, draw accurate a-t, v-t, and x-t graphs. b) In a motion diagram (like Fig. 2–10b or 2–11b), show the position, velocity, and acceleration of the car.

2–24 At $t = 0$ a car is stopped at a traffic light. When the light turns green, the car starts to speed up, and gains speed at a constant rate until it reaches a speed of 20 m/s 8 seconds after the light turns green. The car continues at a constant speed for 60 m. Then the driver sees a red light up ahead at the next intersection, and starts slowing down at a constant rate. The car stops at the red light, 180 m from where it was at $t = 0$. a) Draw accurate x-t, v-t, and a-t graphs for the motion of the car. b) In a motion diagram (like Fig. 2–10b or 2–11b), show the position, velocity, and acceleration of the car.

2–25 The graph in Fig. 2–28 shows the velocity of a motorcycle police officer plotted as a function of time. a) Find the instantaneous acceleration at $t = 3$ s, at $t = 7$ s, and at $t = 11$ s. b) How far does the officer go in the first 5 s? The first 9 s? The first 13 s?

FIGURE 2–28 Exercise 2–25.

2–26 Figure 2–29 is a graph of the acceleration of a model railroad locomotive moving on the x-axis. Graph its velocity and coordinate as functions of time if $x = 0$ and $v = 0$ at $t = 0$.

FIGURE 2–29 Exercise 2–26.

2–27 A spaceship ferrying workers to Moon Base I takes a straight-line path from the earth to the moon, a distance of 384,000 km. Suppose it accelerates at 20.0 m/s² for the first 15.0 min of the trip, then travels at constant speed until the last 15.0 min, when it accelerates at –20.0 m/s², just coming to rest as it reaches the moon. a) What is the maximum speed attained? b) What fraction of the total distance is traveled at constant speed? c) What total time is required for the trip?

2–28 A subway train starts from rest at a station and accelerates at a rate of 1.60 m/s² for 14.0 s. It runs at constant speed for 70.0 s and slows down at a rate of 3.50 m/s² until it stops at the next station. Find the *total* distance covered.

2–29 Two cars, A and B, move along the x-axis. Figure 2–30 is a graph of the positions of A and B versus time. a) In motion diagrams (like Figure 2–10b or Figure 2–11b), show the position, velocity, and acceleration of each of the two cars at $t = 0$, $t = 1$ s, and $t = 3$ s. b) At what time(s), if any, do A and B have the same position? c) Graph velocity versus time for both A and B. d) At what time(s), if any, do A and B have the same velocity? e) At what time(s), if any, does car A pass car B? f) At what time(s), if any, does car B pass car A?

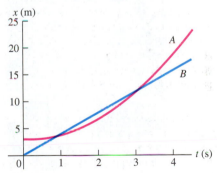

FIGURE 2–30 Exercise 2–29.

2–30 At the instant the traffic light turns green, a car that has been waiting at an intersection starts ahead with a constant acceleration of 3.20 m/s². At the same instant a truck, traveling with a constant speed of 20.0 m/s, overtakes and passes the car. a) How far beyond its starting point does the car overtake the truck? b) How fast is the car traveling when it overtakes the truck? c) Sketch an x-t graph of the motion of both vehicles. Take $x = 0$ at the intersection. d) Sketch a v-t graph of the motion of both vehicles.

2–31 As in Example 2–5, a car is traveling at a constant velocity with magnitude v_C. At the instant that the car passes a motorcycle officer, the motorcycle accelerates from rest with acceleration a_M. a) Sketch an x-t graph of the motion of both objects. Show that when the motorcycle overtakes the car the motorcycle has a speed twice that of the car, no matter what the value of a_M. b) Let d be the distance the motorcycle travels before catching up with the car. In terms of d, how far has the motorcycle traveled when its velocity equals the velocity of the car?

SECTION 2–6 FREELY FALLING BODIES

2–32 If the effects of the air acting on falling raindrops can be neglected, then we can treat raindrops as freely falling objects. a) Rain clouds are typically a few hundred meters above the ground. Estimate the speed with which raindrops would strike the ground if they were freely falling objects. Give your estimate in m/s, km/h, and mi/h. b) Estimate (from your own personal observations of rain) the speed with which raindrops actually strike the ground. c) Based on your answers to parts (a) and (b), is it a good approximation to neglect the effects of the air on falling raindrops? Explain.

2–33 a) If a flea can jump straight up to a height of 0.440 m, what is its initial speed as it leaves the ground? b) For how much time is it in the air?

2–34 Touchdown on the Moon. A lunar lander is making its descent to Moon Base I (Fig. 2–31). The lander descends slowly under the retro-thrust of its descent engine. The engine is cut off when the lander is 5.0 m above the surface and has a downward speed of 0.8 m/s. With the engine off, the lander is in free fall. What is the speed of the lander just before it touches the surface? The acceleration due to gravity on the moon is 1.6 m/s^2.

FIGURE 2–31 Exercise 2–34.

2–35 A Simple Reaction-Time Test. A meter stick is held vertically above your hand, with the lower end between your thumb and first finger. On seeing the meter stick released, you grab it with these two fingers. Your reaction time can be calculated from the distance the meter stick falls, read directly from the point where your fingers grabbed it. a) Derive a relationship for your reaction time in terms of this measured distance, d. b) If the measured distance is 17.6 cm, what is the reaction time?

2–36 A brick is dropped (zero initial speed) from the roof of a building. The brick strikes the ground in 2.50 s. Air resistance may be ignored, so the brick is in free fall. a) How tall, in

meters, is the building? b) What is the magnitude of the brick's velocity just before it reaches the ground? c) Sketch a-t, v-t, and y-t graphs for the motion of the brick.

2–37 Jillian angrily throws her engagement ring straight up from the roof of a building, 12.0 m above the ground, with an initial speed of 5.00 m/s. Air resistance may be ignored. For the motion from her hand to the ground, what are the magnitude and direction of a) the average velocity of the ring? b) the average acceleration of the ring? c) In how many seconds after being thrown does the ring strike the ground? d) What is the speed of the ring just before it strikes the ground? e) Sketch a-t, v-t, and y-t graphs for the motion.

2–38 A hot-air balloonist, rising vertically with a constant velocity of magnitude 5.00 m/s, releases a sandbag at an instant when the balloon is 40.0 m above the ground (Fig. 2–32). After it is released, the sandbag is in free fall. a) Compute the position and velocity of the sandbag at 0.250 s and 1.00 s after its release. b) How many seconds after its release will the bag strike the ground? c) With what magnitude of velocity does it strike? d) What is the greatest height above the ground that the sandbag reaches? e) Sketch a-t, v-t, and y-t graphs for the motion.

$v = 5.00$ m/s

40.0 m to ground

FIGURE 2–32 Exercise 2–38.

2–39 A student throws a water balloon vertically downward from the top of a building. The balloon leaves the thrower's hand with a speed of 6.00 m/s. Air resistance may be ignored, so the water balloon is in free fall after it leaves the thrower's hand. a) What is its speed after falling for 2.00 s? b) How far does it fall in 2.00 s? c) What is the magnitude of its velocity after falling 10.0 m? d) Sketch a-t, v-t, and y-t graphs for the motion.

2–40 An egg is thrown nearly vertically upward from a point near the cornice of a tall building. It just misses the cornice on the way down and passes a point 50.0 m below its starting point 5.00 s after it leaves the thrower's hand. Air resistance may be ignored. a) What is the initial speed of the egg? b) How high does it rise above its starting point? c) What is the magnitude of its velocity at the highest point? d) What are the magnitude and direction of its acceleration at the highest point? e) Sketch a-t, v-t, and y-t graphs for the motion of the egg.

2–41 The rocket-driven sled *Sonic Wind No. 2,* used for investigating the physiological effects of large accelerations, runs on a straight, level track 1070 m (3500 ft) long. Starting from rest, it can reach a speed of 224 m/s (500 mi/h) in 0.900 s. a) Compute the acceleration in m/s^2, assuming that it is constant. b) What is the ratio of this acceleration to that of a freely falling body (g)? c) What is the distance covered in 0.900 s? d) A magazine article states that at the end of a certain run, the speed of the sled decreased from 283 m/s (632 mi/h) to zero in 1.40 s and that during this time the magnitude of the acceleration was more than 40g. Are these figures consistent?

2–42 A large boulder is ejected vertically upward from a volcano with an initial speed of 40.0 m/s. Air resistance may be ignored. a) At what time after being ejected is the boulder moving at 20.0 m/s upward? b) At what time is it moving at 20.0 m/s downward? c) When is the displacement of the boulder from its initial position zero? d) When is the velocity of the boulder zero? e) What are the magnitude and direction of the accelera-

tion while the boulder is i) moving upward? ii) moving downward? iii) at the highest point? f) Sketch a-t, v-t, and y-t graphs for the motion.

2–43 Suppose the acceleration of gravity were only 0.98 m/s^2 instead of 9.8 m/s^2, but the initial velocities with which you can jump upward or throw a ball are unchanged. a) Calculate the height to which you could jump vertically from a standing start if you can jump to 0.75 m when $g = 9.8$ m/s^2. b) How high could you throw a baseball if you can throw it 18 m up when $g = 9.8$ m/s^2? c) Calculate the maximum height of a window from which you would care to jump to a concrete sidewalk below if for $g = 9.8$ m/s^2 you are willing to jump from a height of 2.0 m, which is the typical height of a first-story window.

*SECTION **2–7** VELOCITY AND POSITION BY INTEGRATION

*__2–44__ The acceleration of a bus is given by $a(t) = \alpha t$, where $\alpha = 1.2$ m/s^3. a) If the bus's velocity at time $t = 1.0$ s is 5.0 m/s, what is its velocity at time $t = 2.0$ s? b) If the bus's position at time $t = 1.0$ s is 6.0 m, what is its position at time $t = 2.0$ s? c) Sketch a-t, v-t, and x-t graphs for the motion.

*__2–45__ The acceleration of a motorcycle is given by $a(t) = At - Bt^2$, where $A = 1.50$ m/s^3 and $B = 0.120$ m/s^4. The motorcycle is at rest at the origin at time $t = 0$. a) Find its position and velocity as functions of time. b) Calculate the maximum velocity it attains.

PROBLEMS

2–46 On a 20-mile bike ride, you ride the first 10 miles at an average speed of 8 mi/h. What must your average speed over the next 10 miles be to have your average speed for the total 20 miles be a) 4 mi/h? b) 12 mi/h? c) Given this average speed for the first 10 miles, can you possibly attain an average speed of 16 mi/h for the total 20-mile ride? Explain.

2–47 The position of a particle between $t = 0$ and $t = 2.00$ s is given by $x(t) = (3.00 \text{ m/s}^3)t^3 - (10.0 \text{ m/s}^2)t^2 + (9.00 \text{ m/s})t$. a) Draw the x-t, v-t, and a-t graphs of this particle. b) At what time(s) between $t = 0$ and $t = 2.00$ s is the particle instantaneously at rest? Does your numerical result agree with the v-t graph in part (a)? c) At each time calculated in part (b) is the acceleration of the particle positive or negative? Show that in each case the same answer is deduced from $a(t)$ and from the v-t graph. d) At what time(s) between $t = 0$ and $t = 2.00$ s is the velocity of the particle instantaneously not changing? Locate this point on the v-t and a-t graphs of part (a). e) What is the particle's greatest distance from the origin ($x = 0$) between $t = 0$ and $t = 2.00$ s? f) At what time(s) between $t = 0$ and $t = 2.00$ s is the particle *speeding up* at the greatest rate? At what time(s) between $t = 0$ and $t = 2.00$ s is the particle *slowing down* at the greatest rate? Locate these points on the v-t and a-t graphs of part (a).

2–48 In a relay race, each contestant runs 25.0 m while carrying an egg balanced on a spoon, turns around, and comes back to the starting point. Edith runs the first 25.0 m in 20.0 s. On the return trip she is more confident and takes only 15.0 s. What is the magnitude of her average velocity for a) the first 25.0 m?

b) the return trip? c) What is her average velocity for the entire round trip? d) What is her average speed for the round trip?

2–49 Dan gets on Interstate Highway I–80 at Seward, Nebraska, and drives due west in a straight line and at an average velocity of magnitude 88 km/h. After traveling 76 km, he reaches the Aurora exit (Fig. 2–33). Realizing he has gone too far, he turns around and drives due east 34 km back to the York exit at an average velocity of magnitude 72 km/h. For his whole trip from Seward to the York exit, what is a) his average speed? b) the magnitude of his average velocity?

FIGURE 2–33 Problem 2–49.

2–50 *Freeway Traffic.* According to a *Scientific American* article (May, 1990), current freeways can sustain about 2400 vehicles per lane per hour in smooth traffic flow at 96 km/h (60 mi/h). Above that figure the traffic flow becomes "turbulent" (stop-and-go). a) If a vehicle is 4.6 m (15 ft) long on the average, what is the average spacing between vehicles at the above traffic density? b) Collision-avoidance automated control systems, which operate by bouncing radar or sonar signals off surrounding vehicles and then accelerate or brake the car when necessary, could greatly reduce the required spacing between vehicles. If the average spacing is 9.2 m (two car lengths), how many vehicles per hour can a lane of traffic carry at 96 km/h?

2–51 A world-class sprinter accelerates to his maximum speed in 4.0 s. He then maintains this speed for the remainder of a 100-m race, finishing with a total time of 9.1 s. a) What is the runner's average acceleration during the first 4.0 s? b) What is his average acceleration during the last 5.1 s? c) What is his average acceleration for the entire race? d) Explain why your answer to part (c) is not the average of the answers to parts (a) and (b).

2–52 A sled starts from rest at the top of a hill and slides down with a constant acceleration. At some later time it is 14.4 m from the top; 2.00 s after that it is 25.6 m from the top, 2.00 s later 40.0 m from the top, and 2.00 s later it is 57.6 m from the top. a) What is the magnitude of the average velocity of the sled during each of the 2.00-s intervals after passing the 14.4-m point? b) What is the acceleration of the sled? c) What is the speed of the sled when it passes the 14.4-m point? d) How much time did it take to go from the top to the 14.4-m point? e) How far did the sled go during the first second after passing the 14.4-m point?

2–53 A car 3.5 m in length and traveling at a constant speed of 20 m/s is approaching an intersection (Fig. 2–34). The width of the intersection is 20 m. The light turns yellow when the front of the car is 50 m from the beginning of the intersection. If the driver steps on the brake, the car will slow at -3.8 m/s^2. If the driver instead steps on the gas pedal, the car will accelerate at 2.3 m/s^2. The light will be yellow for 3.0 s. Ignore the reaction time of the driver. To avoid being in the intersection while the light is red, should the driver hit the brake pedal or the gas pedal?

FIGURE 2–34 Problem 2–53.

2–54 The engineer of a passenger train traveling at 25.0 m/s sights a freight train whose caboose is 200 m ahead on the same track (Fig. 2–35). The freight train is traveling at 15.0 m/s in the same direction as the passenger train. The engineer of the passenger train immediately applies the brakes, causing a constant acceleration of -0.100 m/s^2, while the freight train continues with constant speed. Take $x = 0$ at the location of the front of the passenger train when the brakes are applied. a) Will the cows

FIGURE 2–35 Problem 2–54.

nearby witness a collision? b) If so, where will it take place? c) On a single graph, sketch the positions of the front of the passenger train and the back of the freight train.

2–55 Large cockroaches can run as fast as 1.50 m/s in short bursts. Suppose you turn on the light in a cheap motel and see one scurrying directly away from you at a constant 1.50 m/s as you move toward it at 0.80 m/s. If you start 0.90 m behind it, what minimum constant acceleration would you need to catch up with it when it has traveled 1.20 m, just short of safety under a counter?

2–56 Consider the situation described in Example 2–5. The example is somewhat unrealistic, because the constantly accelerating officer continues past the car. In a real pursuit the officer would accelerate to a speed faster than that of the car, then slow down in order to have the same velocity as the car when she catches up with it. Suppose that the officer in Example 2–5 accelerates from rest at $a = 2.5$ m/s^2 until her speed is 20 m/s. She then slows down at a constant rate until she comes alongside the car at $x = 360$ m, traveling with the same speed of 15 m/s as the car. a) At what time does the officer catch up with the car? b) At what time does the officer change from speeding up to slowing down? At this time, how far is the officer from the stop sign? How far behind the car is she? c) Find the acceleration of the officer while she is slowing down. d) Draw an x-t graph for the two vehicles. e) Draw a v-t graph for the two vehicles.

2–57 An automobile and a truck start from rest at the same instant, with the automobile initially at some distance behind the truck. The truck has a constant acceleration of 2.10 m/s^2, and the automobile an acceleration of 3.40 m/s^2. The automobile overtakes the truck after the truck has moved 40.0 m. a) How much time does it take the automobile to overtake the truck? b) How far was the automobile behind the truck initially? c) What is the speed of each when they are abreast? d) On a single graph, sketch the position of each vehicle as a function of time. Take $x = 0$ at the initial location of the truck.

2–58 Two stunt drivers drive directly toward each other. At time $t = 0$ the two cars are a distance D apart, car 1 is at rest, and car 2 is moving to the left with speed v_0. Car 1 begins to move at $t = 0$, speeding up with a constant acceleration a. Car 2 continues to move with a constant velocity. a) At what time do the two cars collide? b) Find the speed of car 1 just before it collides with car 2. c) Sketch x-t and v-t graphs for car 1 and car 2. Draw the curves for each car using the same axes.

2–59 In his Mustang, Juan rounds a curve onto a straight section of country road while traveling at 20 m/s (45 mi/h), only to find a fully loaded manure spreader completely blocking the road 37 m ahead of him. Startled, he travels at a constant velocity for 0.80 s of reaction time, then hits the brakes and slows at constant acceleration so that he manages to stop just short of the spreader. With the same reaction time and acceleration, if he had roared around the corner at 25 m/s (56 mi/h) instead of 20 m/s, a) how fast would he have been going when he hit the spreader? b) how much time would he have had for his life to flash before his eyes from the time he first sees the spreader until he hits it?

2–60 A police car is traveling in a straight line with constant speed v_p. A truck traveling in the same direction with speed $\frac{3}{2} v_p$ passes the police car. The truck driver realizes that she is speeding, and immediately begins to slow down at a constant rate until she comes to a stop. This is her lucky day, however, and the police car (still moving with the same constant speed) passes the truck driver without giving her a ticket. a) Show that the truck's speed at the instant that the police car passes the truck does *not* depend on the magnitude of the truck's acceleration as it slows down, and find the value of that speed. b) Sketch the x-t graph for the two vehicles.

2–61 The driver of a car wishes to pass a truck that is traveling at a constant speed of 20.0 m/s (about 45 mi/h). Initially, the car is also traveling at 20.0 m/s and its front bumper is 24.0 m behind the truck's rear bumper. The car accelerates at a constant 0.600 m/s², then pulls back into the truck's lane when the rear of the car is 26.0 m ahead of the front of the truck. The car is 4.5 m long and the truck is 21.0 m long. a) How much time is required for the car to pass the truck? b) What distance does the car travel during this time? c) What is the final speed of the car?

*2–62 An object's velocity is measured to be $v(t) = \alpha - \beta t^2$, where $\alpha = 4.00$ m/s and $\beta = 2.00$ m/s³. At $t = 0$ the object is at $x = 0$. a) Calculate the object's position and acceleration as functions of time. b) What is the object's maximum *positive* displacement from the origin?

*2–63 The acceleration of a particle is given by $a(t) = -2.00$ m/s² + (3.00 m/s³)t. a) Find the initial velocity v_0 such that the particle will have the same x-coordinate at $t = 4.00$ s as it had at $t = 0$. b) What will be the velocity at $t = 4.00$ s?

2–64 You are on the roof of the physics building, 46.0 m above the ground (Fig. 2–36). Your physics professor, who is 1.80 m tall, is walking alongside the building at a constant speed of 1.20 m/s. If you wish to drop an egg on your professor's head, where should the professor be when you release the egg? Assume that the egg is in free fall.

2–65 A physics student with too much free time drops a watermelon from the roof of a building. He hears the sound of the watermelon going "splat" 2.50 s later. How high is the building? The speed of sound is 340 m/s. Ignore air resistance.

2–66 Estimate the maximum speed and the magnitude of acceleration of an elevator. You will need to use your observations of roughly how long it takes the elevator to go from one floor to another, the approximate vertical distance from one floor to the next, and the distance over which an elevator accelerates to its maximum speed or slows to a stop.

46.0 m

$v = 1.20$ m/s

1.80 m

FIGURE 2–36 Problem 2–64.

2–67 Visitors at an amusement park watch divers step off a platform 21.3 m (70 ft) above a pool of water. According to the announcer, the divers enter the water at a speed of 56 mi/h (25 m/s). Air resistance may be ignored. a) Is the announcer correct in this claim? b) Is it possible for a diver to leap directly upward off the board so that, missing the board on the way down, she enters the water at 25.0 m/s? If so, what initial upward speed is required? Is the required initial speed physically attainable?

2–68 A flowerpot falls off a windowsill and falls past the window below. Air resistance may be ignored. It takes 0.420 s to pass this window, which is 1.90 m high. How far is the top of the window below the windowsill from which the flowerpot fell?

2–69 A football is kicked vertically upward from the ground and a student gazing out of the window sees it moving upward past her at 5.00 m/s. The window is 12.0 m above the ground. Air resistance may be ignored. a) How high does the football go above ground? b) How much time does it take to go from the ground to its highest point?

2–70 A model rocket has a constant upward acceleration of 40.0 m/s² while its engine is running. The rocket is fired vertically, and the engine runs for 2.50 s before the fuel is used up. After the engine stops, the rocket is in free fall. The motion of the rocket is purely up and down. a) Sketch the a-t, v-t, and y-t graphs for the rocket. b) What is the maximum height that the rocket reaches? c) What will be the speed of the rocket just before it hits the ground? d) Is the total time of flight equal to twice the time the rocket takes to reach its highest point? Why or why not? (See Question Q2–16.)

2–71 Sam heaves a 16-lb shot straight upward, giving it a constant upward acceleration from rest of 45.0 m/s² for 64.0 cm. He releases it 2.20 m above the ground. Air resistance may be ignored. a) What is the speed of the shot when he releases it? b) How high above the ground does it go? c) How much time does he have to get out of its way before it returns to the height of the top of his head, 1.83 m above the ground?

2–72 Determined to test the law of gravity for himself, a student walks off a skyscraper 180 m high, stopwatch in hand, and starts his free fall (zero initial velocity). Five seconds later, Superman arrives at the scene and dives off the roof to save the student. a) Superman leaves the roof with an initial speed v_0 that he produces by pushing himself downward from the edge of the roof with his legs of steel. He then falls with the same acceleration as any freely falling body. What must be the value of v_0 so that Superman catches the student just before the ground is reached? b) On the same graph, sketch the positions of the student and of Superman as functions of time. Take Superman's initial speed to have the value calculated in part (a). c) If the height of the skyscraper is less than some minimum value, even Superman can't reach the student before he hits the ground. What is this minimum height?

2–73 Another determined student (see Problem 2–72) walks off the top of the CN Tower in Toronto, which is 553 m high, and falls freely. His initial velocity is zero. The Rocketeer arrives at the scene five seconds later and dives off the top of the tower to save the student. The Rocketeer leaves the roof with an initial downward speed v_0. In order both to catch the student and to prevent injury to him, the Rocketeer should catch the student at a sufficiently great height above ground so that the Rocketeer and the student slow down and arrive at the ground with zero velocity. The upward acceleration that accomplishes this is provided by the Rocketeer's jet pack, which he turns on just as he catches the student; before then the Rocketeer is in free fall. To prevent discomfort to the student, the magnitude of the acceleration of the Rocketeer and the student as they move downward together should be no more than five times g. a) What is the minimum height above the ground at which the Rocketeer should catch the

student? b) What must be the Rocketeer's initial downward speed so that he catches the student at the minimum height found in part (a)? c) Sketch the y-t, v-t, and a-t graphs for the student and for the Rocketeer. On each graph, draw the curves for the Rocketeer and for the student using the same axes.

2–74 A ball is thrown straight up from the ground with speed v_0. At the same instant, a second ball is dropped from rest from a height H, directly above the point where the first ball was thrown upward. There is no air resistance. a) Find the time at which the two balls collide. b) Find the value of H in terms of v_0 and g so that at the instant when the balls collide, the first ball is at the highest point of its motion.

2–75 Two cars, A and B, travel in a straight line. The distance of A from the starting point is given as a function of time by $x_A(t) = \alpha t + \beta t^2$, with $\alpha = 2.60$ m/s and $\beta = 1.20$ m/s^2. The distance of B from the starting point is $x_B(t) = \gamma t^2 - \delta t^3$, with $\gamma = 2.80$ m/s^2 and $\delta = 0.20$ m/s^3. a) Which car is ahead just after they leave the starting point? b) At what time(s) are the cars at the same point? c) At what time(s) is the distance from A to B neither increasing nor decreasing? d) At what time(s) do A and B have the same acceleration?

2–76 An apple drops from an apple tree and falls freely. The apple is originally at rest a height H above the top of the grass of a thick lawn, which is made of blades of grass of height h. When the apple enters the grass, it slows down at a constant rate so that its speed is zero when it reaches ground level. a) Find the speed of the apple just before it enters the grass. b) Find the acceleration of the apple while it is in the grass. c) Sketch the y-t, v-t, and a-t graphs for the apple's motion.

CHALLENGE PROBLEMS

2–77 A student is running at her top speed of 5.0 m/s to catch a bus, which is stopped at the bus stop. When the student is still 40.0 m from the bus, it starts to pull away, moving with a constant acceleration of 0.170 m/s^2. a) For how much time and what distance does the student have to run at 5.0 m/s before she overtakes the bus? b) When she reaches the bus, how fast is the bus traveling? c) Sketch an x-t graph for both the student and the bus. Take $x = 0$ at the initial position of the student. d) The equations you used in part (a) to find the time have a second solution, corresponding to a later time for which the student and bus are again at the same place if they continue their specified motions. Explain the significance of this second solution. How fast is the bus traveling at this point? e) If the student's top speed is 3.5 m/s, will she catch the bus? f) What is the *minimum* speed the student must have to just catch up with the bus? For what time and what distance does she have to run in that case?

2–78 In the vertical jump, an athlete starts from a crouch and jumps upward to reach as high as possible. Even the best athletes spend little more than 1.00 s in the air (their "hang time"). Treat the athlete as a particle and let y_{max} be his maximum height above the floor. To explain why he seems to hang in the air, calculate the ratio of the time he is above $y_{max}/2$ to the time it takes him to go from the floor to that height. Air resistance may be ignored.

2–79 A ball is thrown straight up from the edge of the roof of a building. A second ball is dropped from the roof 1.00 s later. Air resistance may be ignored. a) If the height of the building is 20.0 m, what must be the initial speed of the first ball if both are to hit the ground at the same time? On the same graph, sketch the position of each ball as a function of time, measured from when the first ball is thrown. Consider the same situation, but now let the initial speed v_0 of the first ball be given and treat the height h of the building as an unknown. b) What must the height of the building be for both balls to reach the ground at the same time for each of the following values of v_0: i) 6.0 m/s; ii) 9.5 m/s? c) If v_0 is greater than some value v_{max}, a value of h does not exist that allows both balls to hit the ground at the same time. Solve for v_{max}. The value v_{max} has a simple physical interpretation. What is it? d) If v_0 is less than some value v_{min}, a value of h does not exist that allows both balls to hit the ground at the same time. Solve for v_{min}. The value v_{min} also has a simple physical interpretation. What is it?

2–80 An alert hiker sees a boulder fall from the top of a distant cliff and notes that it takes 1.30 s for the boulder to fall the last third of the way to the ground. Air resistance may be ignored. a) What is the height of the cliff in meters? b) If in part (a) you get two solutions of a quadratic equation and use one for your answer, what does the other solution represent?

Navigation—the problem of determining where you are, what direction you should travel to reach your destination, and how soon you will get there—has been revolutionized by the Global Positioning System (GPS). A handheld GPS unit like this one can determine its position to within a few meters by analyzing signals from several orbiting GPS satellites.

3

Motion in Two or Three Dimensions

3-1 INTRODUCTION

When a baseball is hit by a bat, what determines where the baseball lands? How do you describe the motion of a roller coaster car along a curved track or the flight of a hawk circling over an open field? If you throw a water balloon horizontally from your window, will it take the same amount of time to hit the ground as a balloon that you simply drop?

We can't answer these kinds of questions using the techniques of Chapter 2, in which particles moved only along a straight line. Instead, we have to confront the reality that our world is three-dimensional. To understand the curved flight of a baseball, the orbital motion of a satellite, or the trajectory of a thrown projectile, we need to extend our descriptions of motion to two- and three-dimensional situations. We'll still use the vector quantities displacement, velocity, and acceleration, but now these quantities will have two or three components and will no longer lie along a single line. We'll find that several important and interesting kinds of motion take place in two dimensions only, that is, in a *plane*. These motions can be described with two coordinates and two components of velocity and acceleration.

We also need to consider how the motion of a particle is described by different observers who are moving relative to each other. The concept of *relative velocity* contains the seeds of the special theory of relativity, although we won't get to that until later in the book.

This chapter merges the vector language we learned in Chapter 1 with the kinematic language of Chapter 2. As before, we are concerned with describing motion, not with analyzing its causes. But the language you learn here will be an essential tool in later chapters when you use Newton's laws of motion to study the relation between force and motion.

3-2 POSITION AND VELOCITY VECTORS

To describe the *motion* of a particle in space, we first need to be able to describe the *position* of the particle. Consider a particle that is at a point P at a certain instant. The **position vector** $\vec{r}$ of the particle at this instant is a vector that goes from the origin of the coordinate system to the point P (Fig. 3–1). The figure also shows that the Cartesian coordinates x, y, and z of point P are the x-, y-, and z-components of vector $\vec{r}$. Using the unit vectors introduced in Section 1–10, we can write

$$\vec{r} = x\hat{\imath} + y\hat{\jmath} + z\hat{k}. \tag{3–1}$$

3–1 The position vector $\vec{r}$ from the origin to point P has components x, y, and z.

As the particle moves through space, the path that it follows is in general a curve (Fig. 3–2). During a time interval Δt the particle moves from P_1, where its position vector is $\vec{r}_1$, to P_2, where its position vector is $\vec{r}_2$. The change in position (the displacement) during this interval is $\Delta\vec{r} = \vec{r}_2 - \vec{r}_1$. We define the **average velocity** $\vec{v}_{av}$ during this interval in the same way we did in Chapter 2 for straight-line motion, as the displacement divided by the time interval:

$$\vec{v}_{av} = \frac{\vec{r}_2 - \vec{r}_1}{t_2 - t_1} = \frac{\Delta\vec{r}}{\Delta t} \qquad \text{(average velocity vector).} \qquad (3\text{–}2)$$

Note that *dividing* a vector by a scalar is really a special case of *multiplying* a vector by a scalar, described in Section 1–8; the average velocity $\vec{v}_{av}$ is equal to the displacement vector $\Delta\vec{r}$ multiplied by $1/\Delta t$, the reciprocal of the time interval.

We now define **instantaneous velocity** just as we did in Chapter 2; it is the limit of the average velocity as the time interval approaches zero, and it equals the instantaneous rate of change of position with time. The key difference is that position $\vec{r}$ and instantaneous velocity $\vec{v}$ are now both vectors:

$$\vec{v} = \lim_{\Delta t \to 0} \frac{\Delta\vec{r}}{\Delta t} = \frac{d\vec{r}}{dt} \qquad \text{(instantaneous velocity vector).} \qquad (3\text{–}3)$$

The *magnitude* of the vector $\vec{v}$ at any instant is the *speed* v of the particle at that instant. The *direction* of $\vec{v}$ at any instant is the same as the direction in which the particle is moving at that instant.

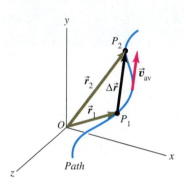

3–2 The average velocity $\vec{v}_{av}$ between points P_1 and P_2 has the same direction as the displacement $\Delta\vec{r}$.

Note that as $\Delta t \to 0$, points P_1 and P_2 in Fig. 3–2 move closer and closer together. In this limit, the vector $\Delta\vec{r}$ becomes tangent to the curve. The direction of $\Delta\vec{r}$ in the limit is also the direction of the instantaneous velocity $\vec{v}$. This leads to an important conclusion: *at every point along the path, the instantaneous velocity vector is tangent to the path at that point* (Fig. 3–3).

It's often easiest to calculate the instantaneous velocity vector using components. During any displacement $\Delta\vec{r}$, the changes Δx, Δy, and Δz in the three coordinates of the particle are the *components* of $\Delta\vec{r}$. It follows that the components v_x, v_y, and v_z of the instantaneous velocity $\vec{v}$ are simply the time derivatives of the coordinates x, y, and z. That is,

$$v_x = \frac{dx}{dt}, \qquad v_y = \frac{dy}{dt}, \qquad v_z = \frac{dz}{dt} \qquad \text{(components of instantaneous velocity).} \qquad (3\text{–}4)$$

We can also get this result by taking the derivative of Eq. (3–1). The unit vectors $\hat{i}$, $\hat{j}$, and $\hat{k}$ are constant in magnitude and direction, so their derivatives are zero, and we find

$$\vec{v} = \frac{d\vec{r}}{dt} = \frac{dx}{dt}\hat{i} + \frac{dy}{dt}\hat{j} + \frac{dz}{dt}\hat{k}. \qquad (3\text{–}5)$$

This shows again that the components of $\vec{v}$ are dx/dt, dy/dt, and dz/dt.

The magnitude of the instantaneous velocity vector $\vec{v}$—that is, the speed—is given in terms of the components v_x, v_y, and v_z by the Pythagorean relation

$$|\vec{v}| = v = \sqrt{v_x^2 + v_y^2 + v_z^2}. \qquad (3\text{–}6)$$

Figure 3–4 shows the situation when the particle moves in the xy-plane. In this case, z and v_z are zero. Then the speed (the magnitude of $\vec{v}$) is

$$v = \sqrt{v_x^2 + v_y^2},$$

and the direction of the instantaneous velocity $\vec{v}$ is given by the angle α in the figure. We see that

3–3 The instantaneous velocity $\vec{v}$ at any point is tangent to the path at that point.

$$\tan \alpha = \frac{v_y}{v_x}. \qquad (3\text{--}7)$$

(We always use Greek letters for angles. We use α for the direction of the instantaneous velocity vector to avoid confusion with the direction θ of the *position* vector of the particle.)

The instantaneous velocity vector is usually more interesting and useful than the average velocity vector. From now on, when we use the word "velocity," we will always mean the instantaneous velocity vector $\vec{v}$ (rather than the average velocity vector). Usually, we won't even bother to call $\vec{v}$ a vector; it's up to you to remember that velocity is a vector quantity with both magnitude and direction.

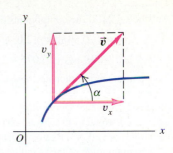

3-4 The two velocity components for motion in the xy-plane.

EXAMPLE 3-1

Calculating average and instantaneous velocity You are operating a radio-controlled model car on a vacant tennis court. Your position is the origin of coordinates, and the surface of the court lies in the xy-plane. The car, which we represent as a point, has x- and y-coordinates that vary with time according to

$$x = 2.0 \text{ m} - (0.25 \text{ m/s}^2)t^2,$$
$$y = (1.0 \text{ m/s})t + (0.025 \text{ m/s}^3)t^3.$$

a) Find the car's coordinates and its distance from you at time $t = 2.0$ s. b) Find the car's displacement and average velocity vectors during the interval from $t = 0.0$ s to $t = 2.0$ s. c) Derive a general expression for the car's instantaneous velocity vector, and find the instantaneous velocity at $t = 2.0$ s. Express the instantaneous velocity in component form and also in terms of magnitude and direction.

SOLUTION a) The car's path is shown in Fig. 3–5. At time $t = 2.0$ s the car's coordinates are

$$x = 2.0 \text{ m} - (0.25 \text{ m/s}^2)(2.0 \text{ s})^2 = 1.0 \text{ m},$$
$$y = (1.0 \text{ m/s})(2.0 \text{ s}) + (0.025 \text{ m/s}^3)(2.0 \text{ s})^3 = 2.2 \text{ m}.$$

The car's distance from the origin at this time is

$$r = \sqrt{x^2 + y^2} = \sqrt{(1.0 \text{ m})^2 + (2.2 \text{ m})^2} = 2.4 \text{ m}.$$

b) To find the displacement and average velocity, we express the position vector $\vec{r}$ as a function of time t. From Eq. (3–1), this is

$$\vec{r} = x\hat{\imath} + y\hat{\jmath}$$
$$= [2.0 \text{ m} - (0.25 \text{ m/s}^2)t^2]\hat{\imath}$$
$$+ [(1.0 \text{ m/s})t + (0.025 \text{ m/s}^3)t^3]\hat{\jmath}.$$

At time $t = 0.0$ s the position vector is

$$\vec{r}_0 = (2.0 \text{ m})\hat{\imath} + (0.0 \text{ m})\hat{\jmath}.$$

From part (a) we find that at time $t = 2.0$ s the position vector $\vec{r}_2$ is

$$\vec{r}_2 = (1.0 \text{ m})\hat{\imath} + (2.2 \text{ m})\hat{\jmath}.$$

Therefore the displacement from $t = 0.0$ s to $t = 2.0$ s is

$$\Delta\vec{r} = \vec{r}_2 - \vec{r}_0 = (1.0 \text{ m})\hat{\imath} + (2.2 \text{ m})\hat{\jmath} - (2.0 \text{ m})\hat{\imath}$$
$$= (-1.0 \text{ m})\hat{\imath} + (2.2 \text{ m})\hat{\jmath}.$$

During the time interval the car has moved 1.0 m in the negative

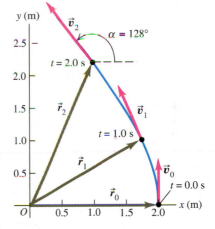

3-5 Path of a radio-controlled model car. At $t = 0$ the car has position vector $\vec{r}_0$ and instantaneous velocity vector $\vec{v}_0$. Likewise, $\vec{r}_1$ and $\vec{v}_1$ are the vectors at $t = 1.0$ s; $\vec{r}_2$ and $\vec{v}_2$ are the vectors at $t = 2.0$ s.

x-direction and 2.2 m in the positive y-direction. The average velocity during the interval from $t = 0.0$ s to $t = 2.0$ s is this displacement divided by the elapsed time (Eq. (3–2)):

$$\vec{v}_{av} = \frac{\Delta\vec{r}}{\Delta t} = \frac{(-1.0 \text{ m})\hat{\imath} + (2.2 \text{ m})\hat{\jmath}}{2.0 \text{ s} - 0.0 \text{ s}}$$
$$= (-0.50 \text{ m/s})\hat{\imath} + (1.1 \text{ m/s})\hat{\jmath}.$$

The components of this average velocity are

$$(v_{av})_x = -0.50 \text{ m/s}, \qquad (v_{av})_y = 1.1 \text{ m/s}.$$

c) From Eq. (3–4), the components of instantaneous velocity are the time derivatives of the coordinates:

$$v_x = \frac{dx}{dt} = (-0.25 \text{ m/s}^2)(2t),$$
$$v_y = \frac{dy}{dt} = 1.0 \text{ m/s} + (0.025 \text{ m/s}^3)(3t^2).$$

Then we can write the instantaneous velocity vector $\vec{v}$ as

$$\vec{v} = v_x\hat{\imath} + v_y\hat{\jmath} = (-0.50 \text{ m/s}^2)t\hat{\imath}$$
$$+ [1.0 \text{ m/s} + (0.075 \text{ m/s}^3)t^2]\hat{\jmath}.$$

At time $t = 2.0$ s, the components of instantaneous velocity are

$$v_x = (-0.50 \text{ m/s}^2)(2.0 \text{ s}) = -1.0 \text{ m/s},$$

$$v_y = 1.0 \text{ m/s} + (0.075 \text{ m/s}^3)(2.0 \text{ s})^2 = 1.3 \text{ m/s}.$$

Take a moment to compare these components of $\vec{v}$ with the components of $\vec{v}_{av}$ that we found in part (b). Just as in one dimension, the average velocity vector $\vec{v}_{av}$ over an interval is in general *not* equal to the instantaneous velocity vector $\vec{v}$ at the beginning or end of the interval (see Example 2–1).

The magnitude of the instantaneous velocity (that is, the speed) at $t = 2.0$ s is

$$v = \sqrt{v_x{}^2 + v_y{}^2} = \sqrt{(-1.0 \text{ m/s})^2 + (1.3 \text{ m/s})^2}$$

$$= 1.6 \text{ m/s}.$$

Its direction with respect to the positive x-axis is given by the

angle α, where, from Eq. (3–7),

$$\tan \alpha = \frac{v_y}{v_x} = \frac{1.3 \text{ m/s}}{-1.0 \text{ m/s}} = -1.3, \quad \text{so} \quad \alpha = 128°.$$

Your calculator will tell you that the inverse tangent of -1.3 is $-52°$. But as we learned in Section 1–9, you have to examine a sketch of a vector (Fig. 3–5) to decide on its direction. The correct answer for α is $-52° + 180° = 128°$.

We invite you to calculate the position, instantaneous velocity vector, speed, and direction of motion at $t = 0.0$ s and $t = 1.0$ s. The position vectors and instantaneous velocity vectors at $t = 0.0$ s, 1.0 s, and 2.0 s are shown in Fig. 3–5. Notice that at every point, the instantaneous velocity vector $\vec{v}$ is tangent to the path. The magnitude of $\vec{v}$ increases as the car moves, which shows that the car's speed is increasing.

3-3 THE ACCELERATION VECTOR

Now let's consider the *acceleration* of a particle moving in space. Just as for motion in a straight line, acceleration describes how the velocity of the particle changes. But now we'll generalize acceleration to describe changes in the magnitude of the velocity (that is, the speed) *and* changes in the direction of the velocity (that is, the direction in which the particle is moving in space).

In Fig. 3–6a, a particle is moving along a curved path. The vectors $\vec{v}_1$ and $\vec{v}_2$ represent the particle's instantaneous velocities at time t_1, when the particle is at point P_1, and at time t_2, when the particle is at point P_2. The two velocities may differ in both magnitude and direction. We define the **average acceleration** $\vec{a}_{av}$ of the particle as it moves from P_1 to P_2 as the *vector change in velocity,* $\vec{v}_2 - \vec{v}_1 = \Delta\vec{v}$, divided by the time interval $t_2 - t_1 = \Delta t$:

$$\vec{a}_{av} = \frac{\vec{v}_2 - \vec{v}_1}{t_2 - t_1} = \frac{\Delta\vec{v}}{\Delta t} \quad \text{(average acceleration vector)}. \qquad (3\text{–}8)$$

Average acceleration is a *vector* quantity in the same direction as the vector $\Delta\vec{v}$ (Fig. 3–6a). Note that $\vec{v}_2$ is the vector sum of the original velocity $\vec{v}_1$ and the change $\Delta\vec{v}$ (Fig. 3–6b).

As in Chapter 2, we define the **instantaneous acceleration** $\vec{a}$ at point P_1 as the limit approached by the average acceleration when point P_2 approaches point P_1 and $\Delta\vec{v}$ and Δt both approach zero; the instantaneous acceleration is also equal to the instantaneous rate of change of velocity with time. Because we are not restricted to straight-line motion, instantaneous acceleration is now a vector:

$$\vec{a} = \lim_{\Delta t \to 0} \frac{\Delta\vec{v}}{\Delta t} = \frac{d\vec{v}}{dt} \quad \text{(instantaneous acceleration vector)}. \qquad (3\text{–}9)$$

The velocity vector $\vec{v}$, as we have seen, is tangent to the path of the particle. But the construction in Fig. 3–6c shows that the instantaneous acceleration vector $\vec{a}$ of a moving particle always points toward the concave side of a curved path—that is, toward the inside of any turn that the particle is making. We can also see that when a particle is moving in a curved path, it *always* has nonzero acceleration, even when it moves with constant speed. This conclusion may seem contrary to your intuition, but it's really just contrary to the everyday use of the word "acceleration" to mean that speed is increasing. The more precise definition given in Eq. (3–9) shows that there is a nonzero acceleration whenever the velocity vector changes in *any* way, whether there is a change of speed, direction, or both.

3-6 (a) The vector $\vec{a}_{av} = \Delta\vec{v}/\Delta t$ represents the average acceleration between P_1 and P_2. (b) Construction for obtaining $\Delta\vec{v} = \vec{v}_2 - \vec{v}_1$. (c) Instantaneous acceleration $\vec{a}$ at point P_1. Vector $\vec{v}$ is tangent to the path; vector $\vec{a}$ points toward the concave side of the path.

To convince yourself that a particle has a nonzero acceleration when moving on a curved path with constant speed, think of your sensations when you ride in a car. When the car accelerates, you tend to move within the car in a direction *opposite* to the car's acceleration. (We'll discover the reason for this behavior in Chapter 4.) Thus you tend to slide toward the back of the car when it accelerates forward (speeds up) and toward the front of the car when it accelerates backward (slows down). If the car makes a turn on a level road, you tend to slide toward the outside of the turn; hence the car has an acceleration toward the inside of the turn.

We will usually be interested in the instantaneous acceleration, not the average acceleration. From now on, we will use the term "acceleration" to mean the instantaneous acceleration vector $\vec{a}$.

Each component of the acceleration vector is the derivative of the corresponding component of velocity:

$$a_x = \frac{dv_x}{dt}, \qquad a_y = \frac{dv_y}{dt}, \qquad a_z = \frac{dv_z}{dt} \qquad \text{(components of instantaneous acceleration).} \qquad (3\text{–}10)$$

In terms of unit vectors,

$$\vec{a} = \frac{dv_x}{dt}\hat{\imath} + \frac{dv_y}{dt}\hat{\jmath} + \frac{dv_z}{dt}\hat{k}. \qquad (3\text{–}11)$$

Also, because each component of velocity is the derivative of the corresponding coordinate, we can express the components a_x, a_y, and a_z of the acceleration vector $\vec{a}$ as

$$a_x = \frac{d^2x}{dt^2}, \qquad a_y = \frac{d^2y}{dt^2}, \qquad a_z = \frac{d^2z}{dt^2}, \qquad (3\text{–}12)$$

and the acceleration vector $\vec{a}$ as

$$\vec{a} = \frac{d^2x}{dt^2}\hat{\imath} + \frac{d^2y}{dt^2}\hat{\jmath} + \frac{d^2z}{dt^2}\hat{k}. \qquad (3\text{–}13)$$

4.1
Determining Acceleration Direction Graphically

4.2
Acceleration Direction of a Car

When a frog leaps, it accelerates both forward and upward. Thus its acceleration vector has both a horizontal component (a_x) and a vertical component (a_y).

EXAMPLE 3–2

Calculating average and instantaneous acceleration Let's look again at the radio-controlled model car in Example 3–1. We found that the components of instantaneous velocity at any time t are

$$v_x = \frac{dx}{dt} = (-0.25 \text{ m/s}^2)(2t),$$

$$v_y = \frac{dy}{dt} = 1.0 \text{ m/s} + (0.025 \text{ m/s}^3)(3t^2),$$

and that the velocity vector is

$$\vec{v} = v_x\hat{\imath} + v_y\hat{\jmath} = (-0.50 \text{ m/s}^2)t\hat{\imath}$$
$$+ [1.0 \text{ m/s} + (0.075 \text{ m/s}^3)t^2]\hat{\jmath}.$$

a) Find the components of the average acceleration in the interval from $t = 0.0$ s to $t = 2.0$ s. b) Find the instantaneous acceleration at $t = 2.0$ s.

SOLUTION a) From Eq. (3–8), in order to calculate the components of the average acceleration, we need the instantaneous velocity at the beginning and the end of the time interval. We find the components of instantaneous velocity at time $t = 0.0$ s

by substituting this value into the above expressions for v_x and v_y. We find that at time $t = 0.0$ s,

$$v_x = 0.0 \text{ m/s}, \qquad v_y = 1.0 \text{ m/s}.$$

We found in Example 3–1 that at $t = 2.0$ s the values of these components are

$$v_x = -1.0 \text{ m/s}, \qquad v_y = 1.3 \text{ m/s}.$$

Thus the components of average acceleration in this interval are

$$(a_{av})_x = \frac{\Delta v_x}{\Delta t} = \frac{-1.0 \text{ m/s} - 0.0 \text{ m/s}}{2.0 \text{ s} - 0.0 \text{ s}} = -0.5 \text{ m/s}^2,$$

$$(a_{av})_y = \frac{\Delta v_y}{\Delta t} = \frac{1.3 \text{ m/s} - 1.0 \text{ m/s}}{2.0 \text{ s} - 0.0 \text{ s}} = 0.15 \text{ m/s}^2.$$

b) From Eq. (3–10) the components of instantaneous acceleration are the time derivatives of the components of instantaneous velocity. We find

$$a_x = \frac{dv_x}{dt} = -0.50 \text{ m/s}^2, \qquad a_y = \frac{dv_y}{dt} = (0.075 \text{ m/s}^3)(2t).$$

We can write the instantaneous acceleration vector $\vec{a}$ as

$$\vec{a} = a_x\hat{\imath} + a_y\hat{\jmath} = (-0.50 \text{ m/s}^2)\hat{\imath} + (0.15 \text{ m/s}^3)t\hat{\jmath}.$$

At time $t = 2.0$ s, the components of instantaneous acceleration are

$$a_x = -0.50 \text{ m/s}^2, \qquad a_y = (0.15 \text{ m/s}^3)(2.0 \text{ s}) = 0.30 \text{ m/s}^2.$$

The acceleration vector at this time is

$$\vec{a} = (-0.50 \text{ m/s}^2)\hat{\imath} + (0.30 \text{ m/s}^2)\hat{\jmath}.$$

The magnitude of acceleration at this time is

$$
\begin{aligned}
a &= \sqrt{a_x{}^2 + a_y{}^2} \\
&= \sqrt{(-0.50 \text{ m/s}^2)^2 + (0.30 \text{ m/s}^2)^2} = 0.58 \text{ m/s}^2.
\end{aligned}
$$

The direction of $\vec{a}$ with respect to the positive x-axis is given by the angle β, where

$$\tan \beta = \frac{a_y}{a_x} = \frac{0.30 \text{ m/s}^2}{-0.50 \text{ m/s}^2} = -0.60,$$

$$\beta = 180° - 31° = 149°.$$

We invite you to calculate the instantaneous acceleration at $t = 0.0$ s and $t = 1.0$ s. Figure 3–7 shows the path of the car and the velocity and acceleration vectors at $t = 0.0$ s, 1.0 s, and 2.0 s. Note that the two vectors $\vec{v}$ and $\vec{a}$ are *not* in the same direction at any of these times. The velocity vector is tangent to the path at

3–7 The path of the radio-controlled car, showing the velocity and acceleration at $t = 0.0$ s ($\vec{v}_0$ and $\vec{a}_0$), $t = 1.0$ s ($\vec{v}_1$ and $\vec{a}_1$), and $t = 2.0$ s ($\vec{v}_2$ and $\vec{a}_2$).

each point, and the acceleration vector points toward the concave side of the path.

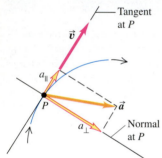

3–8 The acceleration can be resolved into a component $a_\parallel$ parallel to the path (and to the velocity) and a component $a_\perp$ perpendicular to the path (that is, along the normal to the path).

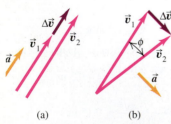

3–9 (a) When $\vec{a}$ is parallel to $\vec{v}$, the magnitude of $\vec{v}$ increases, but its direction does not change. The particle moves in a straight line with changing speed. (b) When $\vec{a}$ is perpendicular to $\vec{v}$, the direction of $\vec{v}$ changes, but its magnitude is constant. The particle moves along a curve with constant speed.

PARALLEL AND PERPENDICULAR COMPONENTS OF ACCELERATION

We can also represent the acceleration of a particle moving in a curved path in terms of components parallel and perpendicular to the velocity at each point (Fig. 3–8). In the figure these components are labeled $a_\parallel$ and $a_\perp$. To see why these components are useful, we'll consider two special cases. In Fig. 3–9a the acceleration vector is *parallel* to the velocity $\vec{v}_1$. The change in $\vec{v}$ during a small time interval Δt is a vector $\Delta\vec{v}$ having the same direction as $\vec{a}$ and hence the same direction as $\vec{v}_1$. The velocity $\vec{v}_2$ at the end of Δt, given by $\vec{v}_2 = \vec{v}_1 + \Delta\vec{v}$, is a vector having the same direction as $\vec{v}_1$ but greater magnitude. In other words, during the time interval Δt the particle in Fig. 3–9a moved in a straight line with increasing speed.

In Fig. 3–9b the acceleration $\vec{a}$ is *perpendicular* to the velocity $\vec{v}$. In a small time interval Δt, the change $\Delta\vec{v}$ is a vector very nearly perpendicular to $\vec{v}_1$, as shown. Again $\vec{v}_2 = \vec{v}_1 + \Delta\vec{v}$, but in this case $\vec{v}_1$ and $\vec{v}_2$ have different directions. As the time interval Δt approaches zero, the angle ϕ in the figure also approaches zero, $\Delta\vec{v}$ becomes perpendicular to *both* $\vec{v}_1$ and $\vec{v}_2$, and $\vec{v}_1$ and $\vec{v}_2$ have the same magnitude. In other words, the speed of the particle stays the same, but the path of the particle curves.

Thus when $\vec{a}$ is *parallel* (or antiparallel) to $\vec{v}$, its effect is to change the magnitude of $\vec{v}$ but not its direction; when $\vec{a}$ is *perpendicular* to $\vec{v}$, its effect is to change the direction of $\vec{v}$ but not its magnitude. In general, $\vec{a}$ may have components *both* parallel and perpendicular to $\vec{v}$, but the above statements are still valid for the individual components. In particular, when a particle travels along a curved path with constant speed, its acceleration is always perpendicular to $\vec{v}$ at each point.

Figure 3–10 shows a particle moving along a curved path for three different situations: constant speed, increasing speed, and decreasing speed. If the speed is constant, $\vec{a}$ is perpendicular, or *normal,* to the path and to $\vec{v}$ and points toward the concave side of the path (Fig. 3–10a). If the speed is increasing, there is still a perpendicular component of $\vec{a}$, but there is also a parallel component having the same direction as $\vec{v}$ (Fig. 3–10b). Then $\vec{a}$ points ahead of the normal to the path. (This was the case in Example 3–2.) If the speed is decreasing, the parallel component has the direction opposite to $\vec{v}$, and $\vec{a}$ points behind the normal to the path (Fig. 3–10c). We will use these ideas in Section 3–5 when we study the special case of motion in a circle.

(a)   (b)   (c)

3-10 Velocity and acceleration vectors for a particle moving through a point P on a curved path with (a) constant speed, (b) increasing speed, and (c) decreasing speed.

CAUTION ▶ Note that the two quantities

$$\frac{d|\vec{v}|}{dt} \quad \text{and} \quad \left|\frac{d\vec{v}}{dt}\right|$$

are not the same. The first is the rate of change of speed; it is zero whenever a particle moves with constant speed, even when its direction of motion changes. The second is the magnitude of the vector acceleration; it is zero only when the particle's acceleration is zero, that is, when the particle moves in a straight line with constant speed. ◀

A race car typically slows down when it enters a turn and speeds up as it exits the turn. Thus the car has a component of acceleration parallel or antiparallel to the direction of motion (which describes the change in speed) and a component perpendicular to the direction of motion (which describes the change in direction).

EXAMPLE 3-3

Calculating parallel and perpendicular components of acceleration For the radio-controlled car of Examples 3-1 and 3-2, find the parallel and perpendicular components of the acceleration at $t = 2.0$ s.

SOLUTION To find the parallel and perpendicular components, we need to know the angle between the acceleration vector $\vec{a}$ and the velocity vector $\vec{v}$. In Example 3-2 we found that at $t = 2.0$ s the particle has an acceleration of magnitude 0.58 m/s² at an angle of 149° with respect to the positive x-axis. From Example 3-1, at this same time the velocity vector is at an angle of 128° with respect to the positive x-axis. So Fig. 3-7 shows that the angle between $\vec{a}$ and $\vec{v}$ is 149° − 128° = 21° (Fig. 3-11). The parallel and perpendicular components of acceleration are then

$$a_{\parallel} = a \cos 21° = (0.58 \text{ m/s}^2) \cos 21° = 0.54 \text{ m/s}^2,$$
$$a_{\perp} = a \sin 21° = (0.58 \text{ m/s}^2) \sin 21° = 0.21 \text{ m/s}^2.$$

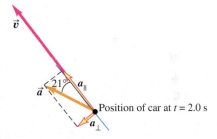

3-11 The parallel and perpendicular components of the acceleration of the radio-controlled car at $t = 2.0$ s.

The parallel component $a_{\parallel}$ is in the same direction as $\vec{v}$, which means that the speed is increasing at this instant. Because the perpendicular component $a_{\perp}$ is not zero, it follows that the car's path is curved at this instant; in other words, the car is turning.

EXAMPLE 3-4

Acceleration of a skier A skier moves along a ski-jump ramp as shown in Fig. 3-12. The ramp is straight from point A to point C and curved from point C onwards. The skier picks up speed as she moves downhill from point A to point E, where her speed is maximum. She slows down after passing point E. Draw the direction of the acceleration vector at points B, D, E, and F.

SOLUTION At point B the skier is moving in a straight line with increasing speed, so her acceleration points downhill, in the same direction as her velocity.

At point D the skier is moving along a curved path, so her acceleration has a component perpendicular to the path. There is also a component in the direction of her motion because she is

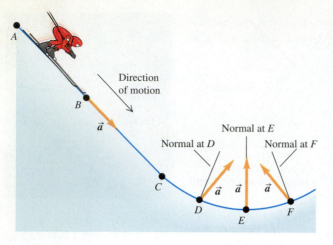

3–12 The direction of a skier's acceleration is different at different points along the path.

still speeding up at this point. So the acceleration vector points *ahead* of the normal to her path at point *D*.

The skier's speed is instantaneously not changing at point *E;* the speed is maximum at this point, so its derivative is zero. So there is no parallel component of $\vec{a}$, and the acceleration is perpendicular to her motion. The acceleration is vertical because at point *E* the skier's path is horizontal.

Finally, at point *F* the acceleration has a perpendicular component (because her path is curved at this point) and a parallel component *opposite* to the direction of her motion (because she's slowing down). So at this point, the acceleration vector points *behind* the normal to her path.

In the next section we'll examine the skier's acceleration after she flies off the ramp.

3–4 PROJECTILE MOTION

A **projectile** is any body that is given an initial velocity and then follows a path determined entirely by the effects of gravitational acceleration and air resistance. A batted baseball, a thrown football, a package dropped from an airplane, and a bullet shot from a rifle are all projectiles. The path followed by a projectile is called its **trajectory.**

To analyze this common type of motion, we'll start with an idealized model, representing the projectile as a single particle with an acceleration (due to gravity) that is constant in both magnitude and direction. We'll neglect the effects of air resistance and the curvature and rotation of the earth. Like all models, this one has limitations. Curvature of the earth has to be considered in the flight of long-range missiles, and air resistance is of crucial importance to a sky diver. Nevertheless, we can learn a lot from analysis of this simple model. For the remainder of this chapter the phrase "projectile motion" will imply that we're ignoring air resistance. In Chapter 5 we will see what happens when air resistance cannot be ignored.

3–13 All projectile motion occurs in the vertical plane containing the initial velocity vector $\vec{v}_0$.

We first notice that projectile motion is always confined to a vertical plane determined by the direction of the initial velocity (Fig. 3–13). This is because the acceleration due to gravity is purely vertical; gravity can't move the projectile sideways. Thus projectile motion is *two-dimensional*. We will call the plane of motion the *xy*-coordinate plane, with the *x*-axis horizontal and the *y*-axis vertically upward.

The key to analysis of projectile motion is that we can treat the *x*- and *y*-coordinates separately. The *x*-component of acceleration is zero, and the *y*-component is constant and equal to $-g$. (Remember that by definition, g is always positive, while with our choice of coordinate directions, a_y is negative.) So *we can analyze projectile motion as a combination of horizontal motion with constant velocity and vertical motion with constant acceleration*. Figure 3–14 shows two projectiles with different *x*-motion but identical *y*-motion; one is dropped from rest and the other is projected horizontally, but both projectiles fall the same distance in the same time.

We can then express all the vector relationships for position, velocity, and acceleration by separate equations for the horizontal and vertical components. The actual motion of the projectile is the superposition of these separate motions. The components of $\vec{a}$ are

$$a_x = 0, \qquad a_y = -g \qquad \text{(projectile motion, no air resistance).} \qquad (3\text{--}14)$$

We will usually use $g = 9.8 \text{ m/s}^2$. Since the *x*-acceleration and *y*-acceleration are both constant, we can use Eqs. (2–8), (2–12), (2–13), and (2–14) directly. For example, suppose that at time $t = 0$ our particle is at the point (x_0, y_0) and that at this time its velocity components have the initial values v_{0x} and v_{0y}. The components of acceleration are $a_x = 0$, $a_y = -g$. Considering the x-motion first, we substitute v_x for v, v_{0x} for v_0, and 0 for a in Eqs. (2–8) and (2–12). We find

$$v_x = v_{0x}, \tag{3-15}$$

$$x = x_0 + v_{0x}t. \tag{3-16}$$

For the y-motion we substitute y for x, v_y for v, v_{0y} for v_0, and $-g$ for a:

$$v_y = v_{0y} - gt, \tag{3-17}$$

$$y = y_0 + v_{0y}t - \frac{1}{2}gt^2. \tag{3-18}$$

It's usually simplest to take the initial position (at time $t = 0$) as the origin; in this case, $x_0 = y_0 = 0$. This point might be, for example, the position of a ball at the instant it leaves the thrower's hand or the position of a bullet at the instant it leaves the gun barrel.

Figure 3–15 shows the path of a projectile that starts at (or passes through) the origin at time $t = 0$. The position, velocity, velocity components, and acceleration are shown at a series of times separated by equal intervals. The x-component of acceleration is zero, so v_x is constant. The y-component of acceleration is constant and not zero, so v_y changes by equal amounts in equal times. At the highest point in the trajectory, $v_y = 0$.

We can also represent the initial velocity $\vec{v}_0$ by its magnitude v_0 (the initial speed) and its angle α_0 with the positive x-axis. In terms of these quantities, the components v_{0x} and v_{0y} of the initial velocity are

$$v_{0x} = v_0 \cos \alpha_0, \qquad v_{0y} = v_0 \sin \alpha_0. \tag{3-19}$$

Using these relations in Eqs. (3–15) through (3–18) and setting $x_0 = y_0 = 0$, we find

$$x = (v_0 \cos \alpha_0)t \qquad \text{(projectile motion)}, \tag{3-20}$$

$$y = (v_0 \sin \alpha_0)t - \frac{1}{2}gt^2 \qquad \text{(projectile motion)}, \tag{3-21}$$

$$v_x = v_0 \cos \alpha_0 \qquad \text{(projectile motion)}, \tag{3-22}$$

$$v_y = v_0 \sin \alpha_0 - gt \qquad \text{(projectile motion)}. \tag{3-23}$$

3–14 Independence of horizontal and vertical motion. The red ball is dropped from rest, and the yellow ball is simultaneously projected horizontally; successive images in this stroboscopic photograph are separated by equal time intervals. At any given time, both balls have the same y-position, y-velocity, and y-acceleration, despite having different x-positions and x-velocities.

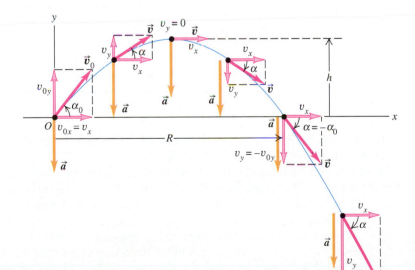

3–15 The trajectory of a body projected with an initial velocity $\vec{v}_0$ at an angle α_0 above the horizontal with negligible air resistance. The distance R is the horizontal range, and h is the maximum height.

3.7
Target Practice II

3.8
Gravitational Constant on Planet Zeus

3.9
Motorcycle Daredevil

3–16 Stroboscopic photograph of a bouncing ball, showing parabolic trajectories after each bounce. Successive images are separated by equal time intervals, as in Fig. 3–14. Each peak in the trajectory is lower than the preceding one because of energy loss during the "bounce," or collision with the horizontal surface.

Rock fragments ejected from this volcano follow nearly parabolic trajectories. This *pyroclastic* eruption is caused by the sudden release of high-pressure gases from within the erupting lava. A web site at the U.S. Geological Survey describes the physics of volcanoes (for a link, visit the *University Physics* web site)

3–17 Air resistance has a large cumulative effect on the motion of a baseball.

These equations describe the position and velocity of the projectile in Fig. 3–15 at any time t.

We can get a lot of information from these equations. For example, at any time the distance r of the projectile from the origin (the magnitude of the position vector $\vec{r}$) is given by

$$r = \sqrt{x^2 + y^2}. \tag{3-24}$$

The projectile's speed (the magnitude of its velocity) at any time is

$$v = \sqrt{v_x^2 + v_y^2}. \tag{3-25}$$

The *direction* of the velocity, in terms of the angle α it makes with the positive x-axis, is given by

$$\tan \alpha = \frac{v_y}{v_x}. \tag{3-26}$$

The velocity vector $\vec{v}$ is tangent to the trajectory at each point.

We can derive an equation for the trajectory's shape in terms of x and y by eliminating t. From Eqs. (3–20) and (3–21), which assume $x_0 = y_0 = 0$, we find $t = x/(v_0 \cos \alpha_0)$ and

$$y = (\tan \alpha_0)x - \frac{g}{2v_0^2 \cos^2 \alpha_0} x^2. \tag{3-27}$$

Don't worry about the details of this equation; the important point is its general form. The quantities v_0, $\tan \alpha_0$, $\cos \alpha_0$, and g are constants, so the equation has the form

$$y = bx - cx^2,$$

where b and c are constants. This is the equation of a *parabola*. In projectile motion, with our simple model, the trajectory is always a parabola (Fig. 3–16).

We mentioned at the beginning of this section that air resistance isn't always negligible. When it has to be included, the calculations become a lot more complicated; the effects of air resistance depend on velocity, so the acceleration is no longer constant. Figure 3–17 shows a computer simulation of the trajectory of a baseball with $v_0 = 50$ m/s and $\alpha_0 = 53.1°$, both without air resistance and with air resistance proportional to the square of the baseball's speed. We see that air resistance has a very large effect; the maximum height and range both decrease, and the trajectory is no longer a parabola. Projectile motion with air resistance is discussed in greater detail in Section 5–7.

Problem–Solving Strategy

PROJECTILE MOTION

The strategies we used in Sections 2–5 and 2–6 for straight-line, constant-acceleration problems are also useful here.

1. Define your coordinate system and make a sketch showing your axes. Usually, it is easiest to place the origin at the initial ($t = 0$) position of the projectile, with the x-axis horizontal and the y-axis upward. Then $x_0 = 0$, $y_0 = 0$, $a_x = 0$, and $a_y = -g$.

2. List the known and unknown quantities. In some problems the components (or magnitude and direction) of initial velocity will be given, and you can use Eqs. (3–20) through (3–23) to find the coordinates and velocity components at some later time. In other problems you might be given two points on the trajectory and be asked to find the initial velocity. Be sure you know which quantities are given and which are to be found.

3. It often helps to state the problem in words and then translate into symbols. For example, *when* does the particle

arrive at a certain point? (That is, at what value of t?) *Where* is the particle when its velocity has a certain value? (That is, what are the values of x and y when v_x or v_y has the specified value?)

4. At the highest point in a trajectory, $v_y = 0$. So the question "When does the projectile reach its highest point?" translates into "What is the value of t when $v_y = 0$?" Similarly, "When does the projectile return to its initial elevation?" translates into "What is the value of t when $y = y_0$?"

5. Resist the temptation to break the trajectory into segments and analyze each one separately. You don't have to start all over, with a new axis system and a new time scale, when the projectile reaches its highest point. It is usually easier to set up Eqs. (3–20) through (3–23) at the start and use the same axes and time scale throughout the problem.

EXAMPLE 3–5

Acceleration of a skier, continued Let's consider again the skier in Example 3–4. What is her acceleration at points G, H, and I in Fig. 3–18 *after* she flies off the ramp? Neglect air resistance.

SOLUTION The skier's acceleration changed from point to point while she was on the ramp. But as soon as she leaves the ramp, she becomes a projectile. So at points G, H, and I, and indeed at

all points after she leaves the ramp, the skier's acceleration points vertically downward and has magnitude g. No matter how complicated the acceleration of a particle before it becomes a projectile, its acceleration as a projectile is given by $a_x = 0$, $a_y = -g$.

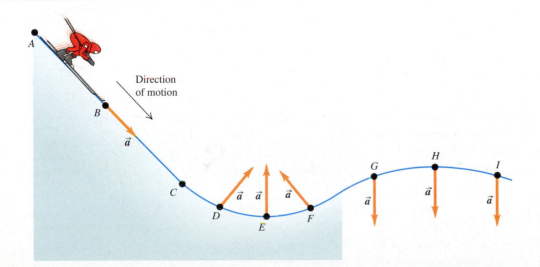

3–18 Once the skier is in projectile motion, her acceleration is constant and downward.

EXAMPLE 3-6

A body projected horizontally A motorcycle stunt rider rides off the edge of a cliff. Just at the edge his velocity is horizontal, with magnitude 9.0 m/s. Find the motorcycle's position, distance from the edge of the cliff, and velocity after 0.50 s.

SOLUTION The coordinate system is shown in Fig. 3–19. We choose the origin at the edge of the cliff, where the motorcycle first becomes a projectile, so $x_0 = 0$ and $y_0 = 0$. The initial velocity is purely horizontal (that is, $\alpha_0 = 0$), so the initial velocity components are $v_{0x} = v_0 \cos \alpha_0 = 9.0$ m/s and $v_{0y} = v_0 \sin \alpha_0 = 0$.

Where is the motorcycle at $t = 0.50$ s? The answer is contained in Eqs. (3–20) and (3–21), which give x and y as functions of time. When $t = 0.50$ s, the x- and y-coordinates are

$$x = v_{0x}t = (9.0 \text{ m/s})(0.50 \text{ s}) = 4.5 \text{ m},$$

$$y = -\frac{1}{2} gt^2 = -\frac{1}{2}(9.8 \text{ m/s}^2)(0.50 \text{ s})^2 = -1.2 \text{ m}.$$

The negative value of y shows that at this time the motorcycle is below its starting point.

What is the motorcycle's distance from the origin at this time? From Eq. (3–24),

$$r = \sqrt{x^2 + y^2} = \sqrt{(4.5 \text{ m})^2 + (-1.2 \text{ m})^2} = 4.7 \text{ m}.$$

What is the velocity at time $t = 0.50$ s? From Eqs. (3–22) and (3–23), which give v_x and v_y as functions of time, the components of velocity at time $t = 0.50$ s are

$$v_x = v_{0x} = 9.0 \text{ m/s},$$

$$v_y = -gt = (-9.8 \text{ m/s}^2)(0.50 \text{ s}) = -4.9 \text{ m/s}.$$

The motorcycle has the same horizontal velocity v_x as when it left the cliff at $t = 0$, but in addition there is a vertical velocity v_y

3–19 Trajectory of a body projected horizontally.

directed downward (in the negative y-direction). If we use unit vectors, the velocity at $t = 0.50$ s is

$$\vec{v} = v_x\hat{\imath} + v_y\hat{\jmath} = (9.0 \text{ m/s})\hat{\imath} + (-4.9 \text{ m/s})\hat{\jmath}.$$

We can also express the velocity in terms of magnitude and direction. From Eq. (3–25), the speed (magnitude of the velocity) at this time is

$$v = \sqrt{v_x^2 + v_y^2}$$
$$= \sqrt{(9.0 \text{ m/s})^2 + (-4.9 \text{ m/s})^2} = 10.2 \text{ m/s}.$$

From Eq. (3–26), the angle α of the velocity vector is

$$\alpha = \arctan \frac{v_y}{v_x}$$

$$= \arctan\left(\frac{-4.9 \text{ m/s}}{9.0 \text{ m/s}}\right) = -29°.$$

At this time the velocity is 29° below the horizontal.

EXAMPLE 3-7

Height and range of a baseball A batter hits a baseball so that it leaves the bat with an initial speed $v_0 = 37.0$ m/s at an initial angle $\alpha_0 = 53.1°$, at a location where $g = 9.80$ m/s² (Fig. 3–20). Let's see how we can predict where the ball is and how it's moving at a certain time, how we can find the ball's maximum height, and how we can find the distance from home plate to where the ball comes down. (As shown in Fig. 3–17, the effects of air resistance on the motion of a baseball aren't really negligible. We'll nonetheless ignore air resistance for this example.)

The ball is probably struck a meter or so above ground level, but we neglect this distance and assume that it starts at ground level ($y_0 = 0$). The initial velocity of the ball has components

$$v_{0x} = v_0 \cos \alpha_0 = (37.0 \text{ m/s}) \cos 53.1° = 22.2 \text{ m/s},$$

$$v_{0y} = v_0 \sin \alpha_0 = (37.0 \text{ m/s}) \sin 53.1° = 29.6 \text{ m/s}.$$

a) Find the position of the ball, and the magnitude and direction of its velocity, when $t = 2.00$ s.

Using the coordinate system shown in Fig. 3–15, we want to find x, y, v_x, and v_y at time $t = 2.00$ s. From Eqs. (3–20) through (3–23),

$$x = v_{0x}t = (22.2 \text{ m/s})(2.00 \text{ s}) = 44.4 \text{ m},$$

$$y = v_{0y}t - \frac{1}{2} gt^2$$

$$= (29.6 \text{ m/s})(2.00 \text{ s}) - \frac{1}{2}(9.80 \text{ m/s}^2)(2.00 \text{ s})^2$$

$$= 39.6 \text{ m},$$

$$v_x = v_{0x} = 22.2 \text{ m/s},$$

$$v_y = v_{0y} - gt = 29.6 \text{ m/s} - (9.80 \text{ m/s}^2)(2.00 \text{ s})$$

$$= 10.0 \text{ m/s}.$$

The y-component of velocity is positive, which means that the ball is still moving upward at this time (Fig. 3–20). The magnitude and direction of the velocity are found from Eqs. (3–25) and (3–26):

$$v = \sqrt{v_x^2 + v_y^2} = \sqrt{(22.2 \text{ m/s})^2 + (10.0 \text{ m/s})^2}$$

$$= 24.3 \text{ m/s},$$

$$\alpha = \arctan\left(\frac{10.0 \text{ m/s}}{22.2 \text{ m/s}}\right) = \arctan 0.450 = 24.2°.$$

3–20 Trajectory of a batted baseball, treated as a projectile.

The velocity is 24.2° above the horizontal.

b) Find the time when the ball reaches the highest point of its flight, and find its height h at this point.

At the highest point, the vertical velocity v_y is zero. When does this happen? Call the time t_1; then

$$v_y = 0 = v_{0y} - gt_1,$$

$$t_1 = \frac{v_{0y}}{g} = \frac{29.6 \text{ m/s}}{9.80 \text{ m/s}^2} = 3.02 \text{ s}.$$

The height h at this time is the value of y when $t = t_1 = 3.02$ s:

$$h = v_{0y}t_1 - \frac{1}{2}gt_1^2$$

$$= (29.6 \text{ m/s})(3.02 \text{ s}) - \frac{1}{2}(9.80 \text{ m/s}^2)(3.02 \text{ s})^2$$

$$= 44.7 \text{ m}.$$

Alternatively, we can apply the constant-acceleration formula Eq. (2–13) to the y-motion:

$$v_y^2 = v_{0y}^2 + 2a_y(y - y_0) = v_{0y}^2 - 2g(y - y_0).$$

At the highest point, $v_y = 0$ and $y = h$. Substituting these, along with $y_0 = 0$, we find

$$0 = v_{0y}^2 - 2gh,$$

$$h = \frac{v_{0y}^2}{2g} = \frac{(29.6 \text{ m/s})^2}{2(9.80 \text{ m/s}^2)} = 44.7 \text{ m}.$$

This is roughly half the height above the playing field of the roof of Skydome in Toronto.

c) Find the *horizontal range R,* that is, the horizontal distance from the starting point to the point at which the ball hits the ground.

First, *when* does the ball hit the ground? This occurs when $y = 0$. Call this time t_2; then

$$y = 0 = v_{0y}t_2 - \frac{1}{2}gt_2^2 = t_2\left(v_{0y} - \frac{1}{2}gt_2\right).$$

This is a quadratic equation for t_2. It has two roots,

$$t_2 = 0 \quad \text{and} \quad t_2 = \frac{2v_{0y}}{g} = \frac{2(29.6 \text{ m/s})}{9.80 \text{ m/s}^2} = 6.04 \text{ s}.$$

There are two times at which $y = 0$; $t_2 = 0$ is the time the ball *leaves* the ground, and $t_2 = 6.04$ s is the time of its return. This is exactly twice the time to reach the highest point, so the time of descent equals the time of ascent. (This is *always* true if the starting and end points are at the same elevation and air resistance can be neglected. We will prove this in Example 3–9.)

The horizontal range R is the value of x when the ball returns to the ground, that is, at $t = 6.04$ s:

$$R = v_{0x}t_2 = (22.2 \text{ m/s})(6.04 \text{ s}) = 134 \text{ m}.$$

For comparison the distance from home plate to the center field fence at Pittsburgh's Three Rivers Stadium is 125 m. Can you show that when the ball reaches this fence, it is 11.4 m above the ground? (This is more than enough to clear the fence, so this ball is a home run.)

The vertical component of velocity when the ball hits the ground is

$$v_y = v_{0y} - gt_2 = 29.6 \text{ m/s} - (9.80 \text{ m/s}^2)(6.04 \text{ s})$$

$$= -29.6 \text{ m/s}.$$

That is, v_y has the same magnitude as the initial vertical velocity v_{0y} but the opposite direction (down). Since v_x is constant, the angle $\alpha = -53.1°$ (below the horizontal) at this point is the negative of the initial angle $\alpha_0 = 53.1°$.

d) If the ball could continue to travel on *below* its original level (through an appropriately shaped hole in the ground), then negative values of y, corresponding to times greater than 6.04 s, would be possible. Can you compute the position and velocity at a time 7.55 s after the start, corresponding to the last position shown in Fig. 3–15? The results are

$$x = 168 \text{ m}, \qquad y = -55.8 \text{ m},$$

$$v_x = 22.2 \text{ m/s}, \qquad v_y = -44.4 \text{ m/s}.$$

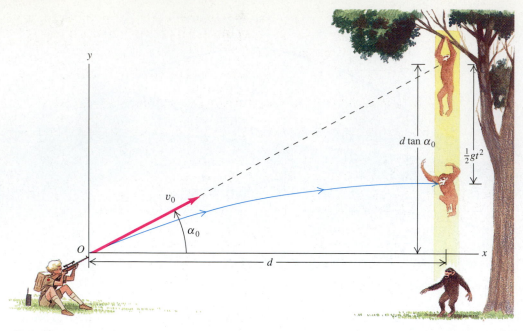

3–21 The tranquilizer dart hits the falling monkey.

EXAMPLE 3-8

The zoo keeper and the monkey A clever monkey escapes from the zoo. The zoo keeper finds him in a tree. After failing to entice the monkey down, the zoo keeper points her tranquilizer gun directly at the monkey and shoots (Fig. 3–21). The clever monkey lets go at the same instant the dart leaves the gun barrel, intending to land on the ground and escape. Show that the dart *always* hits the monkey, regardless of the dart's muzzle velocity (provided that it gets to the monkey before he hits the ground).

SOLUTION We place the origin of coordinates at the end of the barrel of the tranquilizer gun. To show that the dart hits the monkey, we have to prove that there is some time when the monkey and the dart have the same x-coordinate and the same y-coordinate. First let's ask when the x-coordinates x_{monkey} and x_{dart} are the same. The monkey drops straight down, so $x_{monkey} = d$ at *all* times. For the dart we use Eq. (3–20): $x_{dart} = (v_0 \cos \alpha_0)t$. When these x-coordinates are equal, $d = (v_0 \cos \alpha_0)t$, or

$$t = \frac{d}{v_0 \cos \alpha_0}.$$

Now we ask whether y_{monkey} and y_{dart} are also equal at this time; if they are, we have a hit. The monkey is in one-dimensional free fall; his position at any time is given by Eq. (2–12), with appropriate symbol changes. Figure 3–21 shows that the monkey's

initial height is $d \tan \alpha_0$ (the opposite side of a right triangle with angle α_0 and adjacent side d), and we find

$$y_{monkey} = d \tan \alpha_0 - \frac{1}{2} gt^2.$$

For the dart we use Eq. (3–21):

$$y_{dart} = (v_0 \sin \alpha_0)t - \frac{1}{2} gt^2.$$

So we see that if $d \tan \alpha_0 = (v_0 \sin \alpha_0)t$ at the time when the two x-coordinates are equal, then $y_{monkey} = y_{dart}$, and we have a hit. To prove that this happens, we replace t with $d/(v_0 \cos \alpha_0)$, the time when $x_{monkey} = x_{dart}$; then

$$(v_0 \sin \alpha_0)t = (v_0 \sin \alpha_0) \frac{d}{v_0 \cos \alpha_0} = d \tan \alpha_0.$$

We have proved that at the time the x-coordinates are equal, the y-coordinates are also equal; a dart aimed at the initial position of the monkey always hits it, no matter what v_0 is. This result is also independent of the value of g, the acceleration due to gravity. With no gravity ($g = 0$), the monkey would remain motionless, and the dart would travel in a straight line to hit him. With gravity, both "fall" the same distance ($\frac{1}{2}gt^2$) below their $g = 0$ positions, and the dart still hits the monkey.

EXAMPLE 3-9

Height and range of a projectile For a projectile launched with speed v_0 at initial angle α_0 (between 0 and 90°), derive general expressions for the maximum height h and horizontal range R (Fig. 3–15). For a given v_0, what value of α_0 gives maximum height? What value gives maximum horizontal range?

SOLUTION We follow the same pattern as in part (b) of Example 3–7. First, for a given α_0, when does the projectile reach its highest point? At this point, $v_y = 0$, so the time t_1 at the highest point ($y = h$) is given by Eq. (3–23):

$$v_y = v_0 \sin \alpha_0 - gt_1 = 0, \qquad t_1 = \frac{v_0 \sin \alpha_0}{g}.$$

Next, in terms of v_0 and α_0, what is the value of y at this time? From Eq. (3–21),

$$h = v_0 \sin \alpha_0 \left(\frac{v_0 \sin \alpha_0}{g} \right) - \frac{1}{2} g \left(\frac{v_0 \sin \alpha_0}{g} \right)^2$$

$$= \frac{v_0^2 \sin^2 \alpha_0}{2g}.$$

If we vary α_0, the maximum value of h occurs when $\sin \alpha_0 = 1$ and $\alpha_0 = 90°$; that is, when the projectile is launched straight up. That's what we should expect. If it is launched horizontally, as in Example 3–6, $\alpha_0 = 0$ and the maximum height is zero!

To derive a general expression for the horizontal range R, we first find an expression for the time t_2 when the projectile returns to its initial elevation. At that time, $y = 0$, and from Eq. (3–21),

$$(v_0 \sin \alpha_0)t_2 - \frac{1}{2} gt_2^2 = t_2 \left(v_0 \sin \alpha_0 - \frac{1}{2} gt_2 \right) = 0.$$

The two roots of this quadratic equation for t_2 are $t_2 = 0$ (the time of launch) and $t_2 = (2v_0 \sin \alpha_0)/g$. The horizontal range R is the value of x at the second time. From Eq. (3–20),

$$R = (v_0 \cos \alpha_0) \frac{2v_0 \sin \alpha_0}{g}.$$

We can now use the trigonometric identity $2 \sin \alpha_0 \cos \alpha_0 = \sin 2\alpha_0$ to rewrite this as

$$R = \frac{v_0^2 \sin 2\alpha_0}{g}.$$

The maximum value of $\sin 2\alpha_0$ is 1; this occurs when $2\alpha_0 = 90°$, or $\alpha_0 = 45°$. This angle gives the maximum range for a given initial speed.

Figure 3–22 is based on a composite photograph of three trajectories of a ball projected from a spring gun at angles of 30°, 45°, and 60°. The initial speed v_0 is approximately the same in all three cases. The horizontal ranges are nearly the same for the 30° and 60° angles, and the range for 45° is greater than either. Can you prove that for a given value of v_0 the range is the same for an initial angle α_0 as for an initial angle $90° - \alpha_0$?

Comparing the expressions for t_1 and t_2, we see that $t_2 = 2t_1$; that is, the total flight time is twice the time up to the highest point. It follows that the time to reach the highest point equals the time to fall from there back to the initial elevation, as we asserted in Example 3–7.

CAUTION ▶ We don't recommend memorizing the above expressions for h and R. They are applicable only in the special circumstances we have described. In particular, the expression for the range R can be used *only* when launch and landing heights are equal. There are many end-of-chapter problems to which these equations are *not* applicable. ◀

3–22 A firing angle of 45° (shown in purple) gives the maximum horizontal range. The range is less with firing angles of 30° (blue) or 60° (green).

EXAMPLE 3–10

Different initial and final heights You toss a water balloon from your window 8.0 m above the ground. When the water balloon leaves your hand, it is moving at 10.0 m/s at an angle of 20° below the horizontal. How far horizontally from your window will the balloon hit the ground? Ignore air resistance.

SOLUTION We have $v_0 = 10.0$ m/s and $\alpha_0 = -20°$; the angle is negative because the initial velocity is below the horizontal. We place the origin of coordinates at the point where the balloon leaves your hand, so that $x_0 = y_0 = 0$ and ground level is at $y = -8.0$ m. We want to find the value of x at the point where the balloon reaches the ground. The initial and final heights of the balloon are different, so we can't use the expression for the horizontal range found in Example 3–9. Instead, we first find the time t when the balloon reaches $y = -8.0$ m. We rewrite Eq. (3–21) in the standard form for a quadratic equation for t:

$$\frac{1}{2} gt^2 - (v_0 \sin \alpha_0)t + y = 0.$$

The roots of this equation are

$$t = \frac{v_0 \sin \alpha_0 \pm \sqrt{(-v_0 \sin \alpha_0)^2 - 4\left(\frac{1}{2} g\right)y}}{2\left(\frac{1}{2} g\right)}$$

$$= \frac{v_0 \sin \alpha_0 \pm \sqrt{v_0^2 \sin^2 \alpha_0 - 2gy}}{g}$$

$$= \frac{\left[\begin{array}{c}(10.0 \text{ m/s}) \sin(-20°) \\ \pm \sqrt{(10.0 \text{ m/s})^2 \sin^2(-20°) - 2(9.8 \text{ m/s}^2)(-8.0 \text{ m})}\end{array}\right]}{9.8 \text{ m/s}^2}$$

$$= -1.7 \text{ s} \quad \text{or} \quad 0.98 \text{ s}.$$

We can discard the negative root, since it refers to a time before the balloon left your hand (see Example 2–8 in Section 2–6 for a discussion of such "fictional" solutions to quadratic equations). The positive root tells us that the balloon takes 0.98 s to reach the ground. From Eq. (3–20), the balloon's x-coordinate at that time is

$$x = (v_0 \cos \alpha_0)t = (10.0 \text{ m/s})(\cos(-20°))(0.98 \text{ s})$$

$$= 9.2 \text{ m}.$$

The balloon hits the ground a horizontal distance of 9.2 m from your window.

(a)

(b)

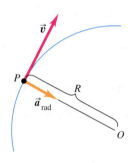

3–23 Finding the change in velocity, $\Delta\vec{v}$, of a particle moving in a circle with constant speed.

4.3

Magnitude of Centripetal Acceleration

3–5 MOTION IN A CIRCLE

When a particle moves along a curved path, the direction of its velocity changes. As we saw in Section 3–3, this means that the particle *must* have a component of acceleration perpendicular to the path, even if its speed is constant. In this section we'll calculate the acceleration for the important special case of motion in a circle.

UNIFORM CIRCULAR MOTION

When a particle moves in a circle with *constant speed,* the motion is called **uniform circular motion.** A car rounding a curve with constant radius at constant speed, a satellite moving in a circular orbit, and an ice skater skating in a circle with constant speed are all examples of uniform circular motion. There is no component of acceleration parallel (tangent) to the path; otherwise, the speed would change. The component of acceleration perpendicular (normal) to the path, which causes the direction of the velocity to change, is related in a simple way to the speed of the particle and the radius of the circle. Our next project is to derive this relation.

First we note that this is a different problem from the projectile motion considered in Section 3–4, in which the acceleration was constant in both its magnitude (g) and its direction (straight down). In uniform circular motion the acceleration is perpendicular to the velocity at each instant; as the direction of the velocity changes, the direction of the acceleration also changes. As we will see, the acceleration vector at each point in the circular path is directed toward the *center* of the circle.

Figure 3–23a shows a particle moving with constant speed in a circular path of radius R with center at O. The particle moves from P_1 to P_2 in a time Δt. The vector change in velocity $\Delta\vec{v}$ during this time is shown in Fig. 3–23b.

The angles labeled $\Delta\phi$ in Figs. 3–23a and 3–23b are the same because $\vec{v}_1$ is perpendicular to the line OP_1 and $\vec{v}_2$ is perpendicular to the line OP_2. Hence the triangles OP_1P_2 (Fig. 3–23a) and Op_1p_2 (Fig. 3–23b) are *similar.* Ratios of corresponding sides are equal, so

$$\frac{|\Delta\vec{v}|}{v_1} = \frac{\Delta s}{R}, \qquad \text{or} \qquad |\Delta\vec{v}| = \frac{v_1}{R}\Delta s.$$

The magnitude a_{av} of the average acceleration during Δt is therefore

$$a_{av} = \frac{|\Delta\vec{v}|}{\Delta t} = \frac{v_1}{R}\frac{\Delta s}{\Delta t}.$$

The magnitude a of the *instantaneous* acceleration $\vec{a}$ at point P_1 is the limit of this expression as we take point P_2 closer and closer to point P_1:

$$a = \lim_{\Delta t \to 0}\frac{v_1}{R}\frac{\Delta s}{\Delta t} = \frac{v_1}{R}\lim_{\Delta t \to 0}\frac{\Delta s}{\Delta t}.$$

But the limit of $\Delta s/\Delta t$ is the speed v_1 at point P_1. Also, P_1 can be any point on the path, so we can drop the subscript and let v represent the speed at any point. Then

$$a_{rad} = \frac{v^2}{R} \qquad \text{(uniform circular motion).} \tag{3–28}$$

We have added the subscript "rad" as a reminder that the direction of the instantaneous acceleration at each point is always along a radius of the circle, toward its center. This is consistent with our discussion in Section 3–3: the acceleration vector points toward the *concave* side of the particle's circular path, that is, toward the *inside* of the circle (never the outside). Because the speed is constant, the acceleration is always perpendicular to the instantaneous velocity. This is shown in Fig. 3–23c; compare with Fig. 3–10a.

We conclude: *In uniform circular motion, the magnitude a of the instantaneous acceleration is equal to the square of the speed v divided by the radius R of the circle. Its direction is perpendicular to $\vec{v}$ and inward along the radius.* Because the acceleration is always directed toward the center of the circle, it is sometimes called **centripetal acceleration.** The word *centripetal* is derived from two Greek words meaning "seeking the center." Figure 3–24 shows the directions of the velocity and acceleration vectors at several points for a particle moving with uniform circular motion. Compare this with the projectile motion shown in Fig. 3–15, in which the acceleration is always directed straight down and is *not* perpendicular to the path, except at one point.

We can also express the magnitude of the acceleration in uniform circular motion in terms of the **period** T of the motion, the time for one revolution (one complete trip around the circle). In a time T the particle travels a distance equal to the circumference $2\pi R$ of the circle, so its speed is

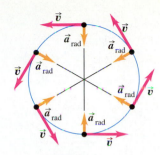

3–24 For a particle in uniform circular motion, the velocity at each point is tangent to the circle and the acceleration is directed toward the center.

$$v = \frac{2\pi R}{T}. \tag{3–29}$$

When we substitute this into Eq. (3–28), we obtain the alternative expression

$$a_{\text{rad}} = \frac{4\pi^2 R}{T^2} \quad \text{(uniform circular motion).} \tag{3–30}$$

EXAMPLE 3–11

A BMW Z3 roadster has a "lateral acceleration" of $0.87g$, which is $(0.87)(9.8 \text{ m/s}^2) = 8.5 \text{ m/s}^2$. This represents the maximum centripetal acceleration that can be attained without skidding out of the circular path. If the car is traveling at a constant 40 m/s (about 89 mi/h, or 144 km/h), what is the minimum radius of curve it can negotiate? (Assume that the curve is unbanked.)

SOLUTION We are given a_{rad} and v, so we first solve Eq. (3–28) for R:

$$R = \frac{v^2}{a_{\text{rad}}} = \frac{(40 \text{ m/s})^2}{8.5 \text{ m/s}^2} = 190 \text{ m} \quad \text{(about 620 ft).}$$

If the curve is banked, the radius can be smaller, as we will see in Chapter 5.

EXAMPLE 3–12

In a carnival ride, the passengers travel at constant speed in a circle of radius 5.0 m. They make one complete circle in 4.0 s. What is their acceleration?

SOLUTION The speed is constant, so this is uniform circular motion. We can use Eq. (3–30) to calculate the acceleration, since $R = 5.0$ m and the period $T = 4.0$ s (the time for one revolution) are both given:

$$a_{\text{rad}} = \frac{4\pi^2(5.0 \text{ m})}{(4.0 \text{ s})^2} = 12 \text{ m/s}^2.$$

We could instead calculate a_{rad} using Eq. (3–28), but this approach requires an extra step to first calculate the speed v. From Eq. (3–29), the speed is the circumference of the circle divided by the period T:

$$v = \frac{2\pi R}{T} = \frac{2\pi(5.0 \text{ m})}{4.0 \text{ s}} = 7.9 \text{ m/s.}$$

The centripetal acceleration is then

$$a_{\text{rad}} = \frac{v^2}{R} = \frac{(7.9 \text{ m/s})^2}{5.0 \text{ m}} = 12 \text{ m/s}^2.$$

As in the preceding example, the direction of $\vec{a}$ is always toward the center of the circle. The magnitude of $\vec{a}$ is greater than g, the acceleration due to gravity, so this is not a ride for the faint-hearted. (Some roller coasters subject their passengers to accelerations as great as $4g$.)

NON-UNIFORM CIRCULAR MOTION

We have assumed throughout this section that the particle's speed is constant. If the speed varies, we call the motion **non-uniform circular motion.** An example is a roller coaster car that slows down and speeds up as it moves around a vertical loop. In

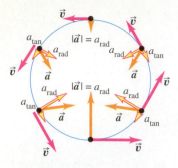

3–25 A particle moving in a vertical loop, like a roller coaster car, with a varying speed. The radial component of acceleration a_{rad} is largest where the speed is greatest (at the bottom of the loop) and smallest where the speed is least (at the top). The tangential component of acceleration a_{tan} is in the same direction as $\vec{v}$ when the particle is speeding up (going downhill) and opposite $\vec{v}$ when it is slowing down (going uphill).

non-uniform circular motion, Eq. (3–28) still gives the *radial* component of acceleration $a_{rad} = v^2/R$, which is always *perpendicular* to the instantaneous velocity and directed toward the center of the circle. But since the speed v has different values at different points in the motion, the value of a_{rad} is not constant. The radial (centripetal) acceleration is greatest at the point in the circle where the speed is greatest.

In non-uniform circular motion there is also a component of acceleration that is *parallel* to the instantaneous velocity. This is the component $a_\parallel$ that we discussed in Section 3–3; here we call this component a_{tan} to emphasize that it is *tangent* to the circle. From the discussion at the end of Section 3–3 we see that the tangential component of acceleration a_{tan} is equal to the rate of change of *speed*. Thus

$$a_{rad} = \frac{v^2}{R} \quad \text{and} \quad a_{tan} = \frac{d|\vec{v}|}{dt} \quad \text{(non-uniform circular motion).} \quad (3\text{--}31)$$

The vector acceleration of a particle moving in a circle with varying speed is the vector sum of the radial and tangential components of accelerations. The tangential component is in the same direction as the velocity if the particle is speeding up, and is in the opposite direction if the particle is slowing down (Fig. 3–25).

In *uniform* circular motion there is no tangential component of acceleration, but the radial component is the magnitude of $d\vec{v}/dt$. We have mentioned before that the two quantities $|d\vec{v}/dt|$ and $d|\vec{v}|/dt$ are in general *not* equal. In uniform circular motion the first is constant and equal to v^2/R; the second is zero.

3–6 RELATIVE VELOCITY

You've no doubt observed how a car that is moving slowly forward appears to be moving backward when you pass it. In general, when two observers measure the velocity of a moving body, they get different results if one observer is moving relative to the other. The velocity seen by a particular observer is called the velocity *relative* to that observer, or simply **relative velocity.** We'll first consider relative velocity along a straight line, then generalize to relative velocity in a plane. Recall that for straight-line (one-dimensional) motion we use the term *velocity* to mean the component of the velocity vector along the line of motion; this can be positive, negative, or zero.

RELATIVE VELOCITY IN ONE DIMENSION

A woman walks with a velocity of 1.0 m/s along the aisle of a train that is moving with a velocity of 3.0 m/s (Fig. 3–26a). What is the woman's velocity? It's a simple enough question, but it has no single answer. As seen by a passenger sitting in the train, she is moving at 1.0 m/s. A person on a bicycle standing beside the train sees the woman moving at 1.0 m/s + 3.0 m/s = 4.0 m/s. An observer in another train going in the opposite direction would give still another answer. We have to specify which observer we mean, and we speak of the velocity *relative* to a particular observer. The woman's velocity relative to the train is 1.0 m/s, her velocity relative to the cyclist is 4.0 m/s, and so on. Each observer, equipped in principle with a meter stick and a stopwatch, forms what we call a **frame of reference.** Thus a frame of reference is a coordinate system plus a time scale.

Let's call the cyclist's frame of reference (at rest with respect to the ground) A and the frame of reference of the moving train B (Fig. 3–26b). In straight-line motion the position of a point P relative to frame of reference A is given by the distance $x_{P/A}$ (the position of P with respect to A), and the position relative to frame B is given by $x_{P/B}$. The distance from the origin of A to the origin of B (position of B with respect to A) is $x_{B/A}$. We can see from the figure that

$$x_{P/A} = x_{P/B} + x_{B/A}. \quad (3\text{--}32)$$

Airshow pilots face a complicated problem involving relative velocities. They must keep track of their velocity relative to the air (to maintain enough airflow over the wings to sustain lift), their velocity relative to each other (to keep a tight formation without colliding), and their velocity relative to their audience (to remain in sight of the spectators).

(a)   (b)

3–26 (a) A woman walking in a train. (b) At the instant shown, the position of the woman (particle P) relative to frame of reference A is different from her position relative to frame of reference B.

This says that the total distance from the origin of A to point P is the distance from the origin of B to point P plus the distance from the origin of A to the origin of B.

The velocity of P relative to frame A, denoted by $v_{P/A}$, is the derivative of $x_{P/A}$ with respect to time. The other velocities are similarly obtained. So the time derivative of Eq. (3–32) gives us a relationship among the various velocities:

$$\frac{dx_{P/A}}{dt} = \frac{dx_{P/B}}{dt} + \frac{dx_{B/A}}{dt},$$

or

$$v_{P/A} = v_{P/B} + v_{B/A} \qquad \text{(relative velocity along a line).} \qquad (3\text{–}33)$$

Getting back to the woman on the train, A is the cyclist's frame of reference, B is the frame of reference of the train, and point P represents the woman. Using the above notation, we have

$$v_{P/B} = 1.0 \text{ m/s}, \qquad v_{B/A} = 3.0 \text{ m/s}.$$

From Eq. (3–33) the woman's velocity $v_{P/A}$ relative to the cyclist is

$$v_{P/A} = 1.0 \text{ m/s} + 3.0 \text{ m/s} = 4.0 \text{ m/s},$$

as we already knew.

In this example, both velocities are toward the right, and we have implicitly taken this as the positive direction. If the woman walks toward the *left* relative to the train, then $v_{P/B} = -1.0$ m/s, and her velocity relative to the cyclist is 2.0 m/s. The sum in Eq. (3–33) is always an algebraic sum, and any or all of the velocities may be negative.

When the woman looks out the window, the stationary cyclist on the ground appears to her to be moving backward; we can call the cyclist's velocity relative to her $v_{A/P}$. Clearly, this is just the negative of $v_{P/A}$. In general, if A and B are any two points or frames of reference,

$$v_{A/B} = -v_{B/A}. \qquad (3\text{–}34)$$

Problem–Solving Strategy

RELATIVE VELOCITY

Note the order of the double subscripts on the velocities above; $v_{A/B}$ always means "velocity of A relative to B." These subscripts obey an interesting kind of algebra, as Eq. (3–33) shows. If we regard each one as a fraction, then the fraction on the left side is the *product* of the fractions on the right sides: $P/A = (P/B)(B/A)$. This is a handy rule you can use when applying Eq. (3–33) to any number of frames of reference. For example, if there are three different frames of reference A, B, and C, we can write immediately

$$v_{P/A} = v_{P/C} + v_{C/B} + v_{B/A}.$$

EXAMPLE 3-13

You are driving north on a straight two-lane road at a constant 88 km/h. A truck traveling at a constant 104 km/h approaches you (in the other lane, fortunately). a) What is the truck's velocity relative to you? b) What is your velocity with respect to the truck? c) How do the relative velocities change after you and the truck have passed each other?

SOLUTION Let you be Y, the truck be T, and the earth be E, and let the positive direction be north (Fig. 3–27). Then $v_{Y/E} = +88$ km/h.
a) The truck is approaching you, so it must be moving south, giving $v_{T/E} = -104$ km/h. We want to find $v_{T/Y}$. Transcribing Eq. (3–33), we have

$$v_{T/E} = v_{T/Y} + v_{Y/E},$$

$$v_{T/Y} = v_{T/E} - v_{Y/E}$$

$$= -104 \text{ km/h} - 88 \text{ km/h} = -192 \text{ km/h}.$$

The truck is moving 192 km/h south relative to you.
b) From Eq. (3–34),

$$v_{Y/T} = -v_{T/Y} = -(-192 \text{ km/h}) = +192 \text{ km/h}.$$

You are moving 192 km/h north relative to the truck.
c) The relative velocities don't change at all after you and the truck have passed each other. The relative positions of the bodies

3–27 Reference frames for you and the truck.

don't matter. The velocity of the truck relative to you is still -192 km/h, but it is now moving away from you instead of toward you.

RELATIVE VELOCITY IN TWO OR THREE DIMENSIONS

We can extend the concept of relative velocity to include motion in a plane or in space by using vector addition to combine velocities. Suppose that the woman in Fig. 3–26a is walking not down the aisle of the railroad car but from one side of the car to the other, with a speed of 1.0 m/s (Fig. 3–28a). We can again describe the woman's position P in two different frames of reference, A for the stationary ground observer and B for the moving train. But instead of coordinates x, we use position vectors $\vec{r}$ because the problem is now two-dimensional. Then, as Fig. 3–28b shows,

$$\vec{r}_{P/A} = \vec{r}_{P/B} + \vec{r}_{B/A}. \tag{3–35}$$

Just as we did before, we take the time derivative of this equation to get a relation among the various velocities; the velocity of P relative to A is $\vec{v}_{P/A} = d\vec{r}_{P/A}/dt$, and so on for the other velocities. We get

$$\vec{v}_{P/A} = \vec{v}_{P/B} + \vec{v}_{B/A} \quad \text{(relative velocity in space).} \tag{3–36}$$

If all three of these velocities lie along the same line, then Eq. (3–36) reduces to Eq. (3–33) for the components of the velocities along that line.

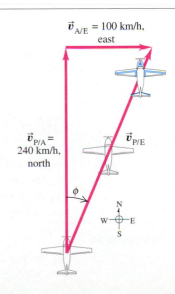

If the train's velocity relative to the ground has magnitude $v_{B/A} = 3.0$ m/s and the woman's velocity relative to the railroad car has magnitude $v_{P/B} = 1.0$ m/s, then her velocity vector $\vec{v}_{P/A}$ relative to the ground is as shown in the vector diagram of Fig. 3–28c. The Pythagorean theorem then gives us

$$v_{P/A} = \sqrt{(3.0 \text{ m/s})^2 + (1.0 \text{ m/s})^2} = \sqrt{10 \text{ m}^2/\text{s}^2} = 3.2 \text{ m/s}.$$

3–28 (a) A woman walking across a railroad car, seen from above. (b) The position vector depends on the frame of reference. (c) Vector diagram for the velocity of the woman relative to the ground.

We can also see from the diagram that the *direction* of her velocity vector relative to the ground makes an angle α with the train's velocity vector $\vec{v}_{B/A}$, where

$$\tan \alpha = \frac{v_{P/B}}{v_{B/A}} = \frac{1.0 \text{ m/s}}{3.0 \text{ m/s}}, \qquad \alpha = 18°.$$

As in the case of motion along a straight line, we have the general rule that if A and B are *any* two points or frames of reference,

$$\vec{v}_{A/B} = -\vec{v}_{B/A}. \qquad (3\text{--}37)$$

The velocity of the woman relative to the train is the negative of the velocity of the train relative to the woman, and so on.

EXAMPLE 3–14

Flying in a crosswind The compass of an airplane indicates that it is headed due north, and its airspeed indicator shows that it is moving through the air at 240 km/h. If there is a wind of 100 km/h from west to east, what is the velocity of the airplane relative to the earth?

SOLUTION Let subscript P refer to the plane and subscript A to the moving air (which now plays the role of the railroad car in Fig. 3–28). Subscript E refers to the earth. The information given is

$$\vec{v}_{P/A} = 240 \text{ km/h} \qquad \text{due north},$$

$$\vec{v}_{A/E} = 100 \text{ km/h} \qquad \text{due east}.$$

3–29 The plane is pointed north, but the wind blows east, giving the resultant velocity $\vec{v}_{P/E}$ relative to the earth.

We want to find the magnitude and direction of $\vec{v}_{P/E}$. Using Eq. (3–36), we have

$$\vec{v}_{P/E} = \vec{v}_{P/A} + \vec{v}_{A/E}.$$

The three relative velocities and their relationship are shown in Fig. 3–29; the unknowns are the speed $v_{P/E}$ and the angle ϕ. From this diagram we find

$$v_{P/E} = \sqrt{(240 \text{ km/h})^2 + (100 \text{ km/h})^2} = 260 \text{ km/h},$$

$$\phi = \arctan\left(\frac{100 \text{ km/h}}{240 \text{ km/h}}\right) = 23° \text{ E of N.}$$

The crosswind increases the speed of the airplane relative to the earth, but at the price of pushing the airplane off course.

EXAMPLE 3–15

In Example 3–14, in what direction should the pilot head to travel due north? What will then be her velocity relative to the earth? (Assume that her airspeed and the velocity of the wind are the same as in Example 3–14.)

SOLUTION Now the information given is

$$\vec{v}_{P/A} = 240 \text{ km/h} \quad \text{direction unknown,}$$

$$\vec{v}_{A/E} = 100 \text{ km/h} \quad \text{due east.}$$

Note that both this and the preceding example require us to determine two unknown quantities. In Example 3–14 these were the direction and magnitude of $\vec{v}_{P/E}$; in this example the unknowns are the direction of $\vec{v}_{P/A}$ (the direction the pilot must point the airplane) and the magnitude of $\vec{v}_{P/E}$ (her speed relative to the earth).

The three relative velocities must still satisfy the vector equation

$$\vec{v}_{P/E} = \vec{v}_{P/A} + \vec{v}_{A/E}.$$

The appropriate vector diagram is shown in Fig. 3–30. The speed $v_{P/E}$ and the angle ϕ are given by

$$v_{P/E} = \sqrt{(240 \text{ km/h})^2 - (100 \text{ km/h})^2} = 218 \text{ km/h},$$

$$\phi = \arcsin\left(\frac{100 \text{ km/h}}{240 \text{ km/h}}\right) = 25°.$$

The pilot should head 25° west of north, and her ground speed is then 218 km/h.

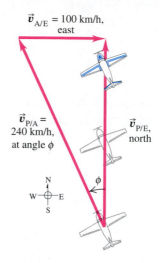

3–30 The pilot must point the plane in the direction of the vector $\vec{v}_{P/A}$ to travel due north relative to the earth.

When we derived the relative-velocity relations, we assumed that all the observers use the same time scale. This is precisely the point at which Einstein's special theory of relativity departs from the physics of Galileo and Newton. When speeds approach the speed of light, denoted by c, the velocity-addition equation has to be modified. It turns out that if the woman in Fig. 3–26a could walk down the aisle at $0.30c$ and the train could move at $0.90c$, then her speed relative to the ground would be not $1.20c$ but $0.94c$; nothing can travel faster than light! We'll return to the special theory of relativity in Chapter 39.

SUMMARY

KEY TERMS

- The position vector $\vec{r}$ of a point P in space is the vector from the origin to P. Its components are the coordinates x, y, and z.

- The average velocity vector $\vec{v}_{av}$ during the time interval Δt is the displacement $\Delta \vec{r}$ (the change in the position vector $\vec{r}$) divided by Δt:

$$\vec{v}_{av} = \frac{\vec{r}_2 - \vec{r}_1}{t_2 - t_1} = \frac{\Delta \vec{r}}{\Delta t}. \tag{3–2}$$

The instantaneous velocity vector is $\vec{v} = d\vec{r}/dt$. Its components are

$$v_x = \frac{dx}{dt}, \qquad v_y = \frac{dy}{dt}, \qquad v_z = \frac{dz}{dt}. \tag{3–4}$$

The instantaneous speed v is the magnitude of $\vec{v}$.

- The average acceleration vector $\vec{a}_{av}$ during the time interval Δt is the velocity change $\Delta \vec{v}$ divided by Δt:

$$\vec{a}_{av} = \frac{\vec{v}_2 - \vec{v}_1}{t_2 - t_1} = \frac{\Delta \vec{v}}{\Delta t}. \tag{3–8}$$

The instantaneous acceleration vector is $\vec{a} = d\vec{v}/dt$. Its components are

$$a_x = \frac{dv_x}{dt}, \qquad a_y = \frac{dv_y}{dt}, \qquad a_z = \frac{dv_z}{dt}. \tag{3–10}$$

- Acceleration can also be represented in terms of its components parallel and perpendicular to the direction of the instantaneous velocity.

- In projectile motion with no air resistance, $a_x = 0$ and $a_y = -g$. The coordinates and velocity components, as functions of time, are

$$x = (v_0 \cos \alpha_0)t, \tag{3–20}$$

$$y = (v_0 \sin \alpha_0)t - \frac{1}{2} gt^2, \tag{3–21}$$

$$v_x = v_0 \cos \alpha_0, \tag{3–22}$$

$$v_y = v_0 \sin \alpha_0 - gt. \tag{3–23}$$

The shape of the path in projectile motion with no air resistance is always a parabola.

- When a particle moves in a circular path of radius R with constant speed v, it has an acceleration with magnitude

$$a_{rad} = \frac{v^2}{R}. \tag{3–28}$$

The acceleration is always directed toward the center of the circle and perpendicular to $\vec{v}$. The period T of a circular motion is the time for one revolution. If the speed is constant, then $v = 2\pi R/T$ and

$$a_{rad} = \frac{4\pi^2 R}{T^2}. \tag{3–30}$$

When the speed is not constant, there is still a radial component of $\vec{a}$ given by Eq. (3–28) or (3–30), but there is also a component of $\vec{a}$ parallel to the path; this component is equal to the rate of change of speed, dv/dt.

- When a body P moves relative to a body (or reference frame) B, and B moves relative to A, we denote the velocity of P relative to B by $\vec{v}_{P/B}$, the velocity of P relative to A by $\vec{v}_{P/A}$, and the velocity of B relative to A by $\vec{v}_{B/A}$. If these velocities are all along the same line, their components along that line are related by

$$v_{P/A} = v_{P/B} + v_{B/A} \qquad \text{(relative velocity along a line).} \tag{3–33}$$

More generally, these velocities are related by

$$\vec{v}_{P/A} = \vec{v}_{P/B} + \vec{v}_{B/A} \qquad \text{(relative velocity in space).} \tag{3–36}$$

DISCUSSION QUESTIONS

Q3–1 A simple pendulum (a mass swinging at the end of a string) swings back and forth in a circular arc. What is the direction of its acceleration at the ends of the swing? At the midpoint? In each case, explain how you obtain your answer.

Q3–2 Redraw Fig. 3–9a if $\vec{a}$ is antiparallel to $\vec{v}_1$. Does the particle move in a straight line? What happens to the speed?

Q3–3 A projectile moves in a parabolic path without air resistance. Is there any point at which $\vec{a}$ is parallel to $\vec{v}$? Perpendicular to $\vec{v}$? Explain.

Q3–4 When a rifle is fired at a distant target, the barrel is not lined up exactly on the target. Why not? Does the angle of correction depend on the distance of the target?

Q3–5 At the same instant that a bullet is fired horizontally from a gun, you drop a bullet from the height of the barrel. If there is no air resistance, which bullet hits the ground first? Explain.

Q3–6 A package falls out of an airplane that is flying in a straight line at a constant altitude and speed. If air resistance could be neglected, what would be the path of the package as observed by the pilot? As observed by a person on the ground?

Q3–7 Sketch the six graphs of the x- and y-components of position, velocity, and acceleration versus time for projectile motion with $x_0 = y_0 = 0$ and $0 < \alpha_0 < 90°$.

Q3–8 If $y_0 = 0$ and α_0 is negative, y can never be positive for a projectile. However, the expression for h in Example 3–9 seems to give a positive maximum height for a negative α_0. Explain this apparent contradiction.

Q3–9 If a jumping frog can give itself the same initial speed regardless of the direction in which it jumps (forward or straight

up), how is the maximum vertical height to which it can jump related to its maximum horizontal range $R_{max} = v_0^2/g$?

Q3–10 Suppose the tranquilizer dart in Fig. 3–21 is fired at a relatively low initial speed v_0, so that the dart has passed the high point of its trajectory and is descending when it hits the monkey (the monkey is still in midair when hit). At the instant that the dart was at the high point of its trajectory, was the monkey's height above the ground the same, lower, or higher than that of the dart? Explain your answer with a drawing.

Q3–11 In uniform circular motion, what are the *average* velocity and *average* acceleration for one revolution? Explain.

Q3–12 In uniform circular motion, how does the acceleration change when the speed is increased by a factor of 3? When the radius is decreased by a factor of 2?

Q3–13 In uniform circular motion, the acceleration is perpendicular to the velocity at every instant, even though both change continuously in direction. Is there any other motion having this property, or is uniform circular motion unique?

Q3–14 Raindrops hitting the side windows of a car in motion often leave diagonal streaks. Why? Is the explanation the same or different for diagonal streaks on the windshield?

Q3–15 In a rainstorm with a strong wind, what determines the best position in which to hold an umbrella?

Q3–16 You are on the west bank of a river that is flowing north with a speed of 1.2 m/s. Your swimming speed relative to the water is 1.5 m/s and the river is 60 m wide. What is your path relative to earth that allows you to cross the river in the shortest time? Explain your reasoning

EXERCISES

SECTION 3–2 POSITION AND VELOCITY VECTORS

3–1 A squirrel has x- and y-coordinates (1.1 m, 3.4 m) at time $t_1 = 0$ and coordinates (5.3 m, –0.5 m) at time $t_2 = 3.0$ s. For this time interval, find a) the components of the average velocity; b) the magnitude and direction of the average velocity.

3–2 A rhinoceros is at the origin of coordinates at time $t_1 = 0$. For the time interval from $t_1 = 0$ to $t_2 = 12.0$ s, the rhino's average velocity has x-component –3.8 m/s and y-component 4.9 m/s. At time $t_2 = 12.0$ s, a) what are the x- and y-coordinates of the rhino? b) how far is the rhino from the origin?

3–3 A web page designer creates an animation in which a dot on a computer screen has a position of $\vec{r} = [4.0 \text{ cm} + (2.5 \text{ cm/s}^2)t^2]\hat{\imath} + (5.0 \text{ cm/s})t\hat{\jmath}$. a) Find the magnitude and direction of the dot's average velocity between $t = 0$ and $t = 2.0$ s. b) Find the magnitude and direction of the instantaneous velocity at $t = 0$, $t = 1.0$ s, and $t = 2.0$ s. c) Sketch the dot's trajectory from $t = 0$ to $t = 2.0$ s, and show the velocities calculated in part (b).

3–4 If $\vec{r} = bt^2\hat{\imath} + ct^3\hat{\jmath}$, where b and c are positive constants, when does the velocity vector make an angle of 45.0° with the x- and y-axes?

SECTION 3–3 THE ACCELERATION VECTOR

3–5 A jet plane is flying at a constant altitude. At time $t_1 = 0$ it has components of velocity $v_x = 90$ m/s, $v_y = 110$ m/s. At time $t_2 = 30.0$ s the components are $v_x = -170$ m/s, $v_y = 40$ m/s. a) Sketch the velocity vectors at t_1 and t_2. How do these two vectors differ? For this time interval calculate b) the components of the average acceleration; c) the magnitude and direction of the average acceleration.

3–6 A dog running in an open field has components of velocity $v_x = 2.6$ m/s and $v_y = -1.8$ m/s at $t_1 = 10.0$ s. For the time interval from $t_1 = 10.0$ s to $t_2 = 20.0$ s, the average acceleration of the dog has magnitude 0.45 m/s² and direction 31.0° measured from the +x-axis toward the +y-axis. At $t_2 = 20.0$ s, a) what are the x- and y-components of the dog's velocity? b) what are the magnitude and direction of the dog's velocity? c) Sketch the velocity vectors at t_1 and t_2. How do these two vectors differ?

3–7 The coordinates of a bird flying in the xy-plane are given by $x(t) = \alpha t$ and $y(t) = 3.0 \text{ m} - \beta t^2$, where $\alpha = 2.4$ m/s and $\beta = 1.2$ m/s². a) Sketch the path of the bird between $t = 0$ and $t = 2.0$ s. b) Calculate the velocity and acceleration vectors of the bird as functions of time. c) Calculate the magnitude and direction of the bird's velocity and acceleration at $t = 2.0$ s. d) Sketch the velocity and acceleration vectors at $t = 2.0$ s. At this instant, is the bird speeding up, is it slowing down, or is its speed instantaneously not changing? Is the bird turning? If so, in what direction?

3–8 A particle moves along a path as shown in Fig. 3–31. Between points B and D, the path is a straight line. Sketch the acceleration vectors at A, C, and E in the cases in which a) the particle moves with a constant speed; b) the particle moves with a steadily increasing speed; c) the particle moves with a steadily decreasing speed.

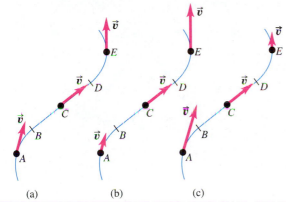

(a)   (b)   (c)

FIGURE 3–31 Exercise 3–8.

SECTION 3–4 PROJECTILE MOTION

3–9 A physics book slides off a horizontal table top with a speed of 1.10 m/s. It strikes the floor in 0.350 s. Ignore air resistance. Find a) the height of the table top above the floor; b) the horizontal distance from the edge of the table to the point where the book strikes the floor; c) the horizontal and vertical components of the book's velocity and the magnitude and direction of its velocity just before the book reaches the floor. d) Draw x-t, y-t, v_x-t, and v_y-t graphs for the motion.

3–10 A military helicopter on a training mission is flying horizontally at a speed of 60.0 m/s and accidentally drops a bomb (fortunately not armed) at an elevation of 300 m. Ignore air resistance. a) How much time is required for the bomb to reach the earth? b) How far does it travel horizontally while falling? c) Find the horizontal and vertical components of its velocity just before it strikes the earth. d) Draw x-t, y-t, v_x-t, and v_y-t graphs for the bomb's motion. e) If the velocity of the helicopter remains constant, where is the helicopter when the bomb hits the ground?

3–11 A football is thrown with an initial speed $v_0 = 15.0$ m/s at an initial angle $\alpha_0 = 45.0°$. a) Find the time T when the football is at its maximum height. b) At the three times $t_1 = T - 0.50$ s, $t_2 = T$, and $t_3 = T + 0.50$ s, find the x- and y-components of the position vector. c) At the three times t_1, t_2, and t_3, find the magnitude and direction of the velocity vector. d) At the three times t_1, t_2, and t_3, find the component of the acceleration vector that is parallel (or antiparallel) to the velocity, and find the component of the acceleration vector that is perpendicular to the velocity. e) Sketch the trajectory of the football. On your sketch, label the

position of the football at the three times t_1, t_2, and t_3. At each of these positions, draw the velocity vector and the parallel and perpendicular components of the acceleration vector. f) Discuss how the speed and the direction of motion of the football are changing at the three times t_1, t_2, and t_3, and explain how the vectors in your sketch describe these changes.

3–12 A tennis ball rolls off the edge of a table top 0.750 m above the floor and strikes the floor at a point 1.40 m horizontally from the edge of the table. Ignore air resistance. a) Find the time of flight. b) Find the magnitude of the initial velocity. c) Find the magnitude and direction of the velocity of the ball just before it strikes the floor. d) Draw x-t, y-t, v_x-t, and v_y-t graphs for the motion.

3–13 A pistol that fires a signal flare gives the flare an initial speed (muzzle speed) of 120 m/s. a) If the flare is fired at an angle of 55° above the horizontal on the level salt flats of Utah, what is its horizontal range? Ignore air resistance. b) If the flare is fired at the same angle over the flat Sea of Tranquility on the moon, where $g = 1.6$ m/s^2, what is its horizontal range?

3–14 John Elway throws a football with an initial upward velocity component of 16.0 m/s and a horizontal velocity component of 20.0 m/s. Ignore air resistance. a) How much time is required for the football to reach the highest point of the trajectory? b) How high is this point? c) How much time (after being thrown) is required for the football to return to its original level? How does this compare with the time calculated in part (a)? d) How far has it traveled horizontally during this time? e) Draw x-t, y-t, v_x-t, and v_y-t graphs for the motion.

3–15 Mark McGwire hits a baseball so that it leaves the bat at a speed of 30.0 m/s and at an angle of 36.9° above the horizontal. Ignore air resistance. a) At what *two* times is the baseball at a height of 10.0 m above the point at which it left the bat? b) Calculate the horizontal and vertical components of the baseball's velocity at each of the two times calculated in part (a). c) What is the magnitude and direction of the baseball's velocity when it returns to the level at which it left the bat?

3–16 A shot putter releases the shot some distance above the level ground with a velocity of 12.0 m/s, 51.0° above the horizontal. The shot hits the ground 2.08 s later. Ignore air resistance. a) What are the components of the shot's acceleration while in flight? b) What are the components of the shot's velocity at the beginning and at the end of its trajectory? c) How far did she throw the shot horizontally? d) Why does the expression for R in Example 3–9 *not* give the correct answer for part (c)? e) How high was the shot above the ground when she released it? f) Draw x-t, y-t, v_x-t, and v_y-t graphs for the motion.

3–17 In a carnival booth you win a stuffed giraffe if you toss a quarter into a small dish. The dish is on a shelf above the point where the quarter leaves your hand and is a horizontal distance of 2.1 m from this point (Fig. 3–32; see next page). If you toss the coin with a velocity of 6.4 m/s at an angle of 60° above the horizontal, the coin lands in the dish. Ignore air resistance. a) What is the height of the shelf above the point where the quarter leaves your hand? b) What is the vertical component of the velocity of the quarter just before it lands in the dish?

FIGURE 3–32 Exercise 3–17.

3–18 Suppose the departure angle α_0 in Fig. 3–21 is 42.0° and the distance d is 3.00 m. Where will the dart and monkey meet if the initial speed of the dart is a) 12.0 m/s? b) 8.0 m/s? c) What will happen if the initial speed of the dart is 4.0 m/s? Sketch the trajectory in each case.

3–19 A man stands on the roof of a 15.0-m-tall building and throws a rock with a velocity of magnitude 30.0 m/s at an angle of 33.0° above the horizontal. Ignore air resistance. Calculate a) the maximum height above the roof reached by the rock; b) the magnitude of the velocity of the rock just before it strikes the ground; c) the horizontal distance from the base of the building to the point where the rock strikes the ground. d) Draw x-t, y-t, v_x-t, and v_y-t graphs for the motion.

SECTION 3–5 MOTION IN A CIRCLE

3–20 On your first day at work for an appliance manufacturer, you are told to figure out what to do to the period of rotation during a washer spin cycle to triple the centripetal acceleration. You impress your boss by answering immediately. What do you tell her?

3–21 The earth has a radius of 6380 km and turns around once on its axis in 24 h. a) What is the radial acceleration of an object at the earth's equator? Give your answer in m/s^2 and as a fraction of g. b) If a_{rad} at the equator is greater than g, objects would fly off the earth's surface and into space. (We will see the reason for this in Chapter 5.) What would the period of the earth's rotation have to be for this to occur?

3–22 A model of a helicopter rotor has four blades, each 3.40 m in length from the central shaft to the blade tip. The model is rotated in a wind tunnel at 550 rev/min. a) What is the linear speed of the blade tip, in m/s? b) What is the radial acceleration of the blade tip expressed as a multiple of the acceleration of gravity, g?

3–23 In a test of a "g-suit," a volunteer is rotated in a horizontal circle of radius 7.0 m. What is the period of rotation at which the centripetal acceleration has a magnitude of a) $3.0g$? b) $10g$?

3–24 The radius of the earth's orbit around the sun (assumed to be circular) is 1.50×10^8 km, and the earth travels around this orbit in 365 days. a) What is the magnitude of the orbital velocity of the earth in m/s? b) What is the radial acceleration of the earth toward the sun in m/s^2? c) Repeat parts (a) and (b) for the motion of the planet Mercury (orbit radius = 5.79×10^7 km, orbital period = 88.0 days).

3–25 A Ferris wheel with radius 14.0 m is turning about a horizontal axis through its center (Fig. 3–33). The linear speed of a passenger on the rim is constant and equal to 7.00 m/s. What are the magnitude and direction of the passenger's acceleration as she passes through a) the lowest point in her circular motion? b) the highest point in her circular motion? c) How much time does it take the Ferris wheel to make one revolution?

FIGURE 3–33 Exercises 3–25 and 3–26.

3–26 The Ferris wheel in Fig. 3–33, which rotates counterclockwise, is just starting up. At a given instant, a passenger on the rim of the wheel and passing through the lowest point of his circular motion is moving at 3.00 m/s and is gaining speed at a rate of 0.500 m/s^2. a) Find the magnitude and the direction of the passenger's acceleration at this instant. b) Sketch the Ferris wheel and passenger showing his velocity and acceleration vectors.

3–27 A level racetrack has the shape of an ellipse. (See your math textbook, a dictionary, or an encyclopedia for an illustration of an ellipse.) A car drives around this track at a constant speed. a) Draw a picture of the track showing the car's velocity vector and acceleration vector at five or more different places around the track. b) Does the car's acceleration vector always point toward the geometrical center of the ellipse? Explain. c) At what point(s) on the ellipse does the car's acceleration have the greatest magnitude? Explain.

SECTION 3–6 RELATIVE VELOCITY

3–28 A railroad flatcar is traveling to the right at a speed of 13.0 m/s relative to an observer standing on the ground. A motor scooter is being ridden on the flatcar (Fig. 3–34). What is the velocity (magnitude and direction) of the motor scooter relative to the flatcar if its velocity relative to the observer on the ground is a) 18.0 m/s to the right? b) 3.0 m/s to the left? c) zero?

3–29 A "moving sidewalk" in an airport terminal building moves at 1.0 m/s and is 35.0 m long. If a woman steps on at one end and walks at 1.5 m/s relative to the moving sidewalk, how much time does she require to reach the opposite end if she walks a) in the same direction the sidewalk is moving? b) in the opposite direction?

3–30 Two piers A and B are located on a river: B is 1500 m downstream from A (Fig. 3–35). Two friends must make round

FIGURE 3–34 Exercise 3–28.

trips from pier A to pier B and return. One rows a boat at a constant speed of 4.00 km/h relative to the water; the other walks on the shore at a constant speed of 4.00 km/h. The velocity of the river is 2.80 km/h in the direction from A to B. How much time does it take each person to make the round trip?

FIGURE 3–35 Exercise 3–30.

PROBLEMS

3–36 A faulty model rocket moves in the xy-plane (the positive y-direction is vertically upward). The rocket's acceleration has components $a_x(t) = \alpha t^2$ and $a_y(t) = \beta - \gamma t$, where $\alpha = 2.50$ m/s^4, $\beta = 9.00$ m/s^2, and $\gamma = 1.40$ m/s^3. At $t = 0$ the rocket is at the origin and has velocity $\vec{v}_0 = v_{0x}\hat{\imath} + v_{0y}\hat{\jmath}$ with $v_{0x} = 1.00$ m/s and $v_{0y} = 7.00$ m/s. a) Calculate the velocity and position vectors as functions of time. b) What is the maximum height reached by the rocket? c) Sketch the path of the rocket. d) What is the horizontal displacement of the rocket when it returns to $y = 0$?

3–37 A student is moving in the xy-plane in a darkened room, trying to locate a lost $20 bill. The student's coordinates are given as functions of time by $x(t) = \alpha t$ and $y(t) = 15.0$ m $- \beta t^2$, where $\alpha = 1.20$ m/s and $\beta = 0.500$ m/s^2. Unknown to the student, the $20 bill is at the origin. a) At what times is the student's velocity perpendicular to his acceleration? b) At what times is the student's speed instantaneously not changing? c) At what times is the student's velocity perpendicular to his position vector? What are the locations of the student at these times? d) What is the minimum distance from the student to the $20 bill? At what time does this minimum occur? e) Sketch the path of the hapless student.

3–38 A bird flies in the xy-plane with a velocity vector given by $\vec{v} = (\alpha - \beta t^2)\hat{\imath} + \gamma t\hat{\jmath}$, with $\alpha = 2.4$ m/s, $\beta = 1.6$ m/s^3, and $\gamma = 4.0$ m/s^2. The positive y-direction is vertically upward. At $t = 0$ the bird is at the origin. a) Calculate the position and acceleration vectors of the bird as functions of time. b) What is the bird's altitude (y-coordinate) as it flies over $x = 0$ for the first time after $t = 0$?

3–31 A canoe has a velocity of 0.40 m/s southeast relative to the earth. The canoe is on a river that is flowing 0.50 m/s east relative to the earth. Find the velocity (magnitude and direction) of the canoe relative to the river.

3–32 An airplane pilot wishes to fly due west. A wind of 80.0 km/h (about 50 mi/h) is blowing toward the south. a) If the airspeed of the plane (its speed in still air) is 320.0 km/h (about 200 mi/h), in what direction should the pilot head? b) What is the speed of the plane over the ground? Illustrate with a vector diagram.

3–33 A river flows due south with a speed of 2.0 m/s. A man steers a motorboat across the river; his velocity relative to the water is 4.2 m/s due east. The river is 800 m wide. a) What is his velocity (magnitude and direction) relative to the earth? b) How much time is required to cross the river? c) How far south of his starting point will he reach the opposite bank?

3–34 a) In what direction should the motorboat in Exercise 3–33 be headed in order to reach a point on the opposite bank directly east from the starting point? (The boat's speed relative to the water remains 4.2 m/s.) b) What is the velocity of the boat relative to the earth? c) How much time is required to cross the river?

3–35 The nose of an ultralight plane is pointed south, and its air speed indicator shows 35 m/s. The plane is in a 10 m/s wind blowing toward the southwest relative to the earth. a) In a vector addition diagram, show the relation of $\vec{v}_{P/E}$ (the velocity of the plane relative to the earth) to the two given vectors. b) Letting x be east and y be north, find the components of $\vec{v}_{P/E}$. c) Find the magnitude and direction of $\vec{v}_{P/E}$.

3–39 A Piper Warrior, a small four-seat airplane, requires 300 m of runway to take off. Its liftoff speed is 88 km/h. It then climbs with a constant speed of 88 km/h along a straight-line path, just clearing a power line 15 m high at a horizontal distance of 460 m from where the Piper began its takeoff roll. a) What was the acceleration of the Piper (assumed constant) during its takeoff roll? b) After the Piper takes off, what is the angle of its flight path above the horizontal? c) What is the rate of climb (in m/s) of the Piper? d) What is the elapsed time from the beginning of the takeoff roll to when the Piper just clears the power line?

3–40 A track and field coach (who is also a physics teacher) trains an athlete to throw a javelin so that it leaves the athlete's hand at a height h above the ground, moving at a speed of $\sqrt{25gh/8}$ and at an angle of $36.9°$ above the horizontal. The javelin continues in flight until it strikes the ground. The ground around the athlete is level, and air resistance is negligible. a) Graph the javelin's horizontal velocity versus time and the javelin's vertical velocity versus time. b) Find the maximum height reached by the javelin. c) Find the horizontal distance that the javelin travels from where it leaves the athlete's hand to where it strikes the ground.

3–41 A demolition crew uses dynamite to blow an old building apart. Debris from the explosion flies off in all directions, and is later found at distances up to 50 m from the explosion. Estimate the maximum speed at which debris was blown outward by the explosion. Describe any assumptions that you make.

3–42 A movie stuntwoman drops from a helicopter that is 30.0 m above the ground and moving with a constant velocity whose components are 10.0 m/s upward and 15.0 m/s horizontal and toward the south. Ignore air resistance. a) Where on the ground (relative to the position of the helicopter when she drops) should the stuntwoman have placed the foam mats that break her fall? b) Draw x-t, y-t, v_x-t, and v_y-t graphs of the motion.

3–43 In fighting forest fires, airplanes work in support of ground crews by dropping water on the fires. A pilot is practicing by dropping a canister of red dye, hoping to hit a target on the ground below. If the plane is flying in a horizontal path 90.0 m above the ground and with a speed of 64.0 m/s (143 mi/h), at what horizontal distance from the target should the pilot release the canister? Ignore air resistance.

3–44 A girl throws a water-filled balloon at an angle of 50.0° above the horizontal with a speed of 12.0 m/s. The horizontal component of the balloon's velocity is directed toward a car that is approaching the girl at a constant speed of 8.00 m/s (Fig. 3–36). If the balloon is to hit the car at the same height at which it leaves her hand, what is the maximum distance the car can be from the girl when the balloon is thrown? Ignore air resistance.

FIGURE 3–36 Problem 3–44.

3–45   The Longest Home Run. According to the *Guinness Book of World Records,* the longest home run ever measured was hit by Roy "Dizzy" Carlyle in a minor league game. The ball traveled 188 m (618 ft) before landing on the ground outside the ballpark. a) Assuming the ball's initial velocity was 45° above the horizontal and neglecting air resistance, what did the initial speed of the ball need to be to produce such a home run if the ball was hit at a point 0.9 m (3.0 ft) above ground level? Assume that the ground was perfectly flat. b) How far would the ball be above a fence 3.0 m (10 ft) in height if the fence was 116 m (380 ft) from home plate?

3–46 On the day after graduation, you decide to toss a burning match into the top of a cylindrical wastebasket (diameter D and height $2D$) full of old homework papers. To make this event more sporting, the bottom of the wastebasket is at the same height as the point where the match leaves your hand, and the near side of the wastebasket is a horizontal distance $6D$ from the point where the match leaves your hand. You launch the match at a 45.0° angle above the horizontal. Find the *minimum* and *maxi-*

mum values of the launch speed that will make the match enter the top of the wastebasket. Ignore air resistance and give your answers in terms of g and D.

3–47 You are playing catch with a friend in the hallway of your dormitory. The distance from floor to ceiling is D, and you throw the ball with an initial speed $v_0 = \sqrt{6gD}$. What is the maximum horizontal distance (in terms of D) that the ball can travel without bouncing? (Assume that the ball is launched from the floor.)

3–48 A baseball thrown at an angle of 60.0° above the horizontal strikes a building 18.0 m away at a point 8.00 m above the point from which it is thrown. Ignore air resistance. a) Find the magnitude of the initial velocity of the baseball (the velocity with which the baseball is thrown). b) Find the magnitude and direction of the velocity of the baseball just before it strikes the building. c) Draw x-t, y-t, v_x-t, and v_y-t graphs of the motion.

3–49 A projectile is launched with speed v_0 at an angle α_0 above the horizontal. The launch point is a height h above the ground. a) Show that if air resistance is neglected, the horizontal distance that the projectile travels before striking the ground is

$$x = \frac{v_0 \cos \alpha_0}{g}\left(v_0 \sin \alpha_0 + \sqrt{v_0{}^2 \sin^2 \alpha_0 + 2gh}\right).$$

Verify that if the launch point is at ground level so that $h = 0$, this is equal to the horizontal range R found in Example 3–9. b) For the case where $v_0 = 10$ m/s and $h = 5.0$ m, graph x as a function of launch angle α_0 for values of α_0 from 0° to 90°. Your graph should show that x is zero if $\alpha_0 = 90°$, but x is nonzero if $\alpha_0 = 0$; explain why this is so. c) We saw in Example 3–9 that for a projectile that lands at the same height from which it is launched, the horizontal range is maximum for $\alpha_0 = 45°$. For the case graphed in part (b), is the angle for maximum horizontal distance equal to, less than, or greater than 45°? (This is a general result for the situation where a projectile is launched from a point higher than where it lands.)

3–50 A snowball rolls off a barn roof that slopes downward at an angle of 40° (Fig. 3–37). The edge of the roof is 14.0 m above the ground, and the snowball has a speed of 7.00 m/s as it rolls off the roof. Ignore air resistance. a) How far from the edge of the barn does the snowball strike the ground if it doesn't strike anything else while falling? b) Draw x-t, y-t, v_x-t, and v_y-t graphs for the motion in part (a). c) A man 1.9 m tall is standing 4.0 m from the edge of the barn. Will he be hit by the snowball?

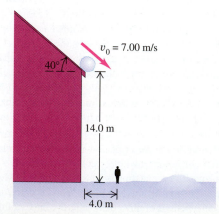

FIGURE 3–37 Problem 3–50.

3–51 a) Prove that a projectile launched at angle α_0 has the same horizontal range as one launched with the same speed at angle $(90° - \alpha_0)$. b) A frog jumps at a speed of 2.2 m/s and lands 25 cm from its starting point. At what angles above the horizontal could it have jumped?

3–52 On the Flying Trapeze. A new circus act is called the Texas Tumblers. Lovely Mary Belle swings from a trapeze, projects herself at an angle of 53°, and is supposed to be caught by Joe Bob, whose hands are 6.1 m above and 8.2 m horizontally from her launch point (Fig. 3–38). Ignore air resistance. a) What initial speed v_0 must Mary Belle have to just reach Joe Bob? b) For the initial speed calculated in part (a), what are the magnitude and direction of her velocity when Mary Belle reaches Joe Bob? c) Assuming that Mary Belle has the initial speed calculated in part (a), draw x-t, y-t, v_x-t, and v_y-t graphs showing the motion of both tumblers. Your graphs should show the motion up until the point where Mary Belle reaches Joe Bob. d) The night of their debut performance, Joe Bob misses her completely as she flies past. How far horizontally does Mary Belle travel, from her initial launch point, before landing in the safety net 8.6 m below her starting point?

FIGURE 3–38 Problem 3–52.

3–53 A physics professor did daredevil stunts in his spare time. His last stunt was an attempt to jump across a river on a motorcycle (Fig. 3–39). The takeoff ramp was inclined at 53.0°, the river was 40.0 m wide, and the far bank was 15.0 m lower than the top of the ramp. The river itself was 100 m below the ramp. Ignore air resistance. a) What should his speed have been at the top of the ramp to have just made it to the edge of the far bank? b) If his speed was only half the value found in (a), where did he land?

3–54 A rock is thrown from the roof of a building with a velocity v_0 at an angle of α_0 from the horizontal. The building has height h. Ignore air resistance. Calculate the magnitude of the velocity of the rock just before it strikes the ground, and show that this speed is independent of α_0.

3–55 A basketball player is fouled and knocked to the floor during a layup attempt. The player is awarded two free throws. The center of the basket is a horizontal distance of 4.21 m (13.8 ft) from the foul line and is a height of 3.05 m (10.0 ft) above the floor (Fig. 3–40). On the first free-throw attempt, he shoots the ball at an angle of 35° above the horizontal and with a speed of $v_0 = 4.88$ m/s (16.0 ft/s). The ball is released 1.83 m (6.0 ft) above the floor. This

FIGURE 3–39 Problem 3–53.

shot misses badly. Ignore air resistance. a) What is the maximum height reached by the ball? b) At what distance along the floor from the free-throw line does the ball land? c) For the second throw, the ball goes through the center of the basket. For this second free throw, the player again shoots the ball at 35° above the horizontal and releases it 1.83 m above the floor. What initial speed does the player give the ball on this second attempt? d) For the second throw, what is the maximum height reached by the ball? At this point, how far horizontally is the ball from the basket?

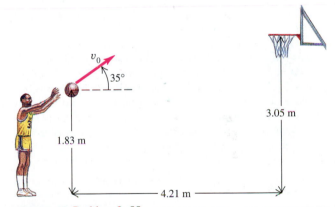

FIGURE 3–40 Problem 3–55.

3–56 Romeo tosses a pebble at Juliet's window to wake her. Unfortunately, he throws too large a pebble too fast. Just before crashing through the glass, the pebble is moving horizontally, having traveled a horizontal distance x and a vertical distance y as a projectile. Find the magnitude and direction of the pebble's velocity as it leaves Romeo's hand.

3–57 A rocket is initially at rest on the ground. When its engines are fired, the rocket flies off in a straight line at an angle of 53.1° above the horizontal with a constant acceleration of magnitude g. The engines stop at a time T after launch, after which the rocket is in projectile motion. Ignore air resistance and assume g is independent of altitude. a) Draw the trajectory of the rocket from when its engines are first fired until the rocket hits the ground. Indicate the direction of the velocity and acceleration vectors at various points along the trajectory. b) Sketch v_x-t and v_y-t graphs for the motion of the rocket from when its engines are first fired until the rocket hits the ground. c) Find the maximum altitude reached by the rocket. Your answer will be in terms of g and T. d) Find the horizontal distance from the launch point to where the rocket hits the ground (that is, the range). Your answer will be in terms of g and T.

3–58 In an action-adventure film the hero is supposed to throw a grenade from his car, which is going 90.0 km/h, to his enemy's car, which is going 110 km/h. The enemy's car is 15.8 m in front of the hero's when he lets go of the grenade. If the hero throws the grenade so its initial velocity relative to him is at an angle of 45° above the horizontal, what should be the magnitude of the initial velocity? The cars are both traveling in the same direction on a level road. Ignore air resistance. Find the magnitude of the velocity both relative to the hero and relative to the earth.

3–59 A rock tied to a rope moves in the xy-plane. Its coordinates are given as functions of time by.

$$x(t) = R \cos \omega t, \qquad y(t) = R \sin \omega t,$$

where R and ω are constants. a) Show that the rock's distance from the origin is constant and equal to R, that is, that its path is a circle of radius R. b) Show that at every point the rock's velocity is perpendicular to its position vector. c) Show that the rock's acceleration is always opposite in direction to its position vector and has magnitude $\omega^2 R$. d) Show that the magnitude of the rock's velocity is constant and equal to ωR. e) Combine the results of parts (c) and (d) to show that the rock's acceleration has constant magnitude v^2/R.

3–60 The speed of a particle moving in a plane is equal to the magnitude of its instantaneous velocity, $v = |\vec{v}| = \sqrt{v_x^2 + v_y^2}$. a) Show that the rate of change of the speed is $dv/dt = (v_x a_x + v_y a_y)/\sqrt{v_x^2 + v_y^2}$. b) Use this expression to find the value of dv/dt at time $t = 2.0$ s for the radio-controlled car in Examples 3–1, 3–2, and 3–3. Compare your answer to the components of acceleration found in Example 3–3. Explain why your answer is *not* equal to the magnitude of the acceleration found in part (b) of Example 3–2. c) Show that the rate of change of speed can be expressed as $dv/dt = \vec{v} \cdot \vec{a}/v$, and use this result to explain why dv/dt is equal to $a_\parallel$, the component of $\vec{a}$ that is parallel to $\vec{v}$.

3–61 A particle moves in the xy-plane. Its coordinates are given as functions of time by

$$x(t) = R(\omega t - \sin \omega t), \quad y(t) = R(1 - \cos \omega t),$$

where R and ω are constants. a) Sketch the trajectory of the particle. (This is the trajectory of a point on the rim of a wheel that is rolling at a constant speed on a horizontal surface. The curve traced out by such a point as it moves through space is called a *cycloid*.) b) Determine the velocity components and the acceleration components of the particle at any time t. c) At what times is the particle momentarily at rest? What are the coordinates of the particle at these times? What are the magnitude and direction of the acceleration at these times? d) Does the magnitude of the acceleration depend on time? Compare to uniform circular motion.

3–62 You are flying in a light airplane spotting traffic for a radio station. Your flight carries you due east above a highway. Landmarks below tell you that your speed is 50 m/s relative to the ground and your air speed indicator also reads 50 m/s. However, the nose of your airplane is pointed somewhat south of east and the station's weather person tells you that a 20 m/s wind is blowing. What direction is the wind blowing?

3–63 **The Carrier Pigeon Problem.** Larry is driving east at 40 km/h. His twin brother Harry is driving west at 30 km/h, toward Larry in an identical car on the same straight road. When they are 42 km apart, Larry sends out a carrier pigeon, which flies at a constant speed of 50 km/h. (All speeds are relative to the earth.) The pigeon flies to Harry, gets confused and immediately returns to Larry, gets more confused and immediately flies back to Harry. This continues until the twins meet, at which time the dazed pigeon drops to the ground in exhaustion. Ignoring turnaround time, how far did the pigeon fly?

3–64 When a train's velocity is 12.0 m/s eastward, raindrops that are falling vertically with respect to the earth make traces that are inclined 30.0° to the vertical on the windows of the train. a) What is the horizontal component of a drop's velocity with respect to the earth? With respect to the train? b) What is the magnitude of the velocity of the raindrop with respect to the earth? With respect to the train?

3–65 An airplane pilot sets a compass course due west and maintains an airspeed of 220 km/h. After flying for 0.500 h, she finds herself over a town 120 km west and 20 km south of her starting point. a) Find the wind velocity (magnitude and direction). b) If the wind velocity is 40 km/h due south, in what direction should the pilot set her course to travel due west? Take the same airspeed of 220 km/h.

3–66 In an air race, an airplane flies from a point directly over Metropolis to a point directly over Gotham City, then turns around and flies back to the starting point. The airspeed (that is, speed of the airplane relative to the air) is constant throughout the flight and equal to v. Gotham City lies a distance D due east of Metropolis. a) If there is no wind, how much time is required for the round trip? How much time is required for the round trip if a steady wind of speed w is blowing b) toward the east? c) toward the south? d) If $D = 3.00 \times 10^2$ km, $v = 4.00 \times 10^2$ km/h, and $w = 1.00 \times 10^2$ km/h, calculate the round-trip flight times for parts (a), (b), and (c). Which case gives the slowest round-trip?

3–67 An elevator is moving upward at a constant speed of 2.50 m/s. A bolt in the elevator ceiling 3.00 m above the elevator floor works loose and falls. a) How long does it take for the bolt to fall to the elevator floor? What is the speed of the bolt just as it hits the elevator floor b) according to an observer in the elevator? c) according to an observer standing on one of the floor landings of the building? d) According to the observer in part (c), what distance did the bolt travel between the ceiling and floor of the elevator?

CHALLENGE PROBLEMS

3–68 A man is riding on a flatcar traveling with a constant speed of 9.10 m/s (Fig. 3–41). He wishes to throw a ball through a stationary hoop 4.90 m above the height of his hands in such a manner that the ball will be moving horizontally as it passes through the hoop. He throws the ball with a speed of 10.8 m/s with respect to himself. a) What must be the vertical component of the initial velocity of the ball? b) How many seconds after he releases the ball will it pass through the hoop? c) At what horizontal distance in front of the hoop must he release the ball? d) When the ball leaves the man's hands, what is the direction of its velocity relative to the frame of reference of the flatcar? Relative to the frame of reference of an observer standing on the ground?

4.90 m

$v = 9.10$ m/s

FIGURE 3–41 Challenge Problem 3–68.

3–69 A shotgun fires a large number of pellets upward, with some pellets traveling very nearly vertically and others as much as $1.0°$ from the vertical. Assume that the initial speed of the pellets is uniformly 150 m/s, and ignore air resistance. a) Within what radius from the point of firing will the pellets land? b) If there are 1000 pellets, and they fall in a uniform distribution over a circle with the radius calculated in part (a), what is the probability that at least one pellet will fall on the head of the person who fires the shotgun? Assume that his head has a radius of 10 cm. c) Air resistance in fact has several effects. It slows down the rising pellets, decreases their horizontal component of velocity, and limits the speed with which they fall. Which of these effects will tend to make the radius greater than calculated in part (a), and which will tend to make it less? What do you think the overall effect of air resistance will be? (The effect of air resistance on a velocity component increases as the magnitude of the component increases.)

3–70 A projectile is thrown from a point P. It moves in such a way that its distance from P is always increasing. Find the maximum angle above the horizontal with which the projectile could have been thrown. Ignore air resistance.

3–71 A baseball is given an initial velocity with magnitude v_0 at an angle ϕ above the surface of an incline, which is in turn inclined at an angle θ above the horizontal (Fig. 3–42). a) Calculate the distance, measured along the incline, from the launch point to where the baseball strikes the incline. Your answer will be in terms of v_0, g, θ, and ϕ. b) What angle ϕ gives the maximum range, measured along the incline? (*Note:* You might be interested in the three different methods of solution presented by I. R. Lapidus in *Amer. Jour. of Phys.*, Vol. 51 (1983), pp. 806 and 847. See also H. A. Buckmaster in *Amer. Jour. of Phys.*, Vol. 53 (1985), pp. 638–641, for a thorough study of this and some similar problems.)

3–72 Refer to Challenge Problem 3–71. a) An archer on ground of constant upward slope of $30.0°$ aims at a target 60.0 m farther up the incline. The arrow in the bow and the bull's-eye at the center of the target are each 1.50 m above the ground. The initial velocity of the arrow just after it leaves the bow has magnitude 32.0 m/s. At what angle above the *horizontal* should the archer aim to hit the bull's-eye? If there are two such angles, calculate the smaller of the two. You might have to solve the equation for the angle by iteration, that is, by trial and error. How does the angle compare to that required when the ground is level, with zero slope? b) Repeat the above for ground that has a constant *downward* slope of $30.0°$.

3–73 For no apparent reason, a poodle is running at a constant speed of $v = 5.00$ m/s in a circle with radius $R = 2.50$ m. Let $\vec{v}_1$ be the velocity vector at time t_1, and let $\vec{v}_2$ be the velocity vector at time t_2. Consider $\Delta\vec{v} = \vec{v}_2 - \vec{v}_1$ and $\Delta t = t_2 - t_1$. Recall that $\vec{a}_{av} = \Delta\vec{v}/\Delta t$. For $\Delta t = 0.5$ s, 0.1 s, and 0.05 s, calculate the magnitude (to four significant figures) and direction (relative to $\vec{v}_1$) of the average acceleration $\vec{a}_{av}$. Compare your results to the general expression for the instantaneous acceleration $\vec{a}$ for uniform circular motion that is derived in the text.

3–74 A rocket designed to place small payloads into orbit is carried to an altitude of 12.0 km above sea level by a converted airliner. When the airliner is flying in a straight line at a constant speed of 850 km/h, the rocket is dropped. After the drop, the airliner maintains the same altitude and speed and continues to fly in a straight line. The rocket falls for a brief time, after which its rocket motor turns on. Once its rocket motor is on, the combined effects of thrust and gravity give the rocket a constant acceleration of magnitude $3.00g$ directed at an angle of $30.0°$ above the horizontal. For reasons of safety, the rocket should be at least 1.00 km in front of the airliner when it climbs through the airliner's altitude. Your job is to determine the minimum time that the rocket must fall before its engine starts. Ignore air resistance. Your answer should include i) a diagram showing the flight paths of both the rocket and the airliner, labeled at several points with vectors for their velocities and accelerations; ii) an *x-t* graph showing the motions of both the rocket and the airliner; and iii) a *y-t* graph showing the motions of both the rocket and the airliner. On the diagram and the graphs, indicate when the rocket is dropped, when the rocket motor turns on, and when the rocket climbs through the altitude of the airliner.

3–75 Two students are canoeing on a river. While heading upstream they accidentally drop an empty bottle overboard. They then continue paddling for 60 minutes, reaching a point 2.0 km further upstream. At this point they realize that the bottle is missing and, driven by ecological awareness, they turn around and head downstream. They catch up with and retrieve the bottle (which has been moving along with the current) 5.0 km downstream from the turn-around point. a) Assuming a constant paddling effort throughout, how fast is the river flowing? b) What would be the canoe speed in a still lake for the same paddling effort?

v_0

ϕ

θ

FIGURE 3–42 Challenge Problem 3–71.

4

Newton's Laws of Motion

To hit a home run, a baseball player requires an intuitive understanding of Newton's laws. He must swing the bat so that it hits the ball with a force of sufficient magnitude and duration and in just the right direction. If the batter does everything just right, the ball will be accelerated into an upward trajectory that sends it flying into the stands.

4–1 INTRODUCTION

How can a tugboat push a cruise ship that's much heavier than the tug? Why does it take a long distance to stop the ship once it is in motion? Why does your foot hurt more when you kick a big rock than when you kick a pebble? Why is it harder to control a car on wet ice than on dry concrete? The answers to these and similar questions take us into the subject of **dynamics,** the relationship of motion to the forces that cause it. In the two preceding chapters we studied *kinematics,* the language for *describing* motion. Now we are ready to think about what makes bodies move the way they do.

In this chapter we will use the kinematic quantities displacement, velocity, and acceleration along with two new concepts, *force* and *mass,* to analyze the principles of dynamics. These principles can be wrapped up in a neat package of three statements called **Newton's laws of motion.** The first law states that when the net force on a body is zero, its motion doesn't change. The second law relates force to acceleration when the net force is *not* zero. The third law is a relation between the forces that two interacting bodies exert on each other. These laws, based on experimental studies of moving bodies, are *fundamental* in two ways. First, they cannot be deduced or proved from other principles. Second, they make it possible to understand most familiar kinds of motion; they are the foundation of **classical mechanics** (also called **Newtonian mechanics**).

Newton's laws are not universal, however; they require modification at very high speeds (near the speed of light) and for very small sizes (such as within the atom).

The laws of motion were clearly stated for the first time by Sir Isaac Newton (1642–1727), who published them in 1687 in his *Philosophiae Naturalis Principia Mathematica* ("Mathematical Principles of Natural Philosophy"). Many other scientists before Newton contributed to the foundations of mechanics, including Copernicus, Brahe, Kepler, and especially Galileo Galilei (1564–1642), who died the same year Newton was born. Indeed, Newton himself said, "If I have been able to see a little farther than other men, it is because I have stood on the shoulders of giants." Now it's your turn to stand on the shoulders of Newton and use his laws to understand how the physical world works.

Newton's laws are very simple to state, yet many students find these laws difficult to grasp and to work with. The reason is that before studying physics, you've spent years walking, throwing balls, pushing boxes, and doing dozens of things that involve motion. Along the way, you've developed a set of "common sense" ideas about motion and its causes. But many of these "common sense" ideas, while they seem to work well enough in everyday life, don't stand up to logical analysis or comparison with experiment. A big part of the job of this chapter—and of the rest of our study of physics—is helping you to recognize when "common sense" ideas need to be replaced by other kinds of analysis.

4–2 FORCE AND INTERACTIONS

In everyday language, a **force** is a push or pull. The concept of force gives us a quantitative description of the interaction between two bodies or between a body and its environment. When you push on a car that is stuck in the snow, you exert a force on it. A locomotive exerts a force on the train it is pulling or pushing, a steel cable exerts a force on the beam it is hoisting at a construction site, and so on.

When a force involves direct contact between two bodies, we call it a **contact force.** Contact forces include the pushes or pulls you exert with your hand, the force of a rope pulling on a block to which it is tied, and the friction force that the ground exerts on a ball player sliding into home. There are also forces, called **long-range forces,** that act even when the bodies are separated by empty space. You've experienced long-range forces if you've ever played with a pair of magnets. Gravity, too, is a long-range force; the sun exerts a gravitational pull on the earth, even over a distance of 150 million kilometers, that keeps the earth in orbit. The force of gravitational attraction that the earth exerts on your body is called your **weight.**

Force is a vector quantity; you can push or pull a body in different directions. Thus to describe a force, we need to describe the *direction* in which it acts as well as its *magnitude,* the quantity that describes "how much" or "how hard" the force pushes or pulls. The SI unit of the magnitude of force is the *newton,* abbreviated N. (We'll give a precise definition of the newton in Section 4–4.) Table 4–1 lists some typical force magnitudes.

A common instrument for measuring forces is the *spring balance.* It consists of a coil spring, enclosed in a case for protection, with a pointer attached to one end. When forces are applied to the ends of the spring, it stretches; the amount of stretch depends on the force. We can make a scale for the pointer and calibrate it by using a number of identical bodies with weights of exactly 1 N each. When two, three, or more of these are suspended simultaneously from the balance, the total force stretching the spring is 2 N, 3 N, and so on, and we can label the corresponding positions of the pointer 2 N, 3 N, and so on. Then we can use this instrument to measure the magnitude of an unknown force. We can also make a similar instrument that measures pushes instead of pulls.

Suppose we slide a box along the floor, applying a force to it by pulling it with a string or pushing it with a stick (Fig. 4–1). In each case we draw a vector to represent the force applied. The labels indicate the magnitude and direction of the force, and the length of the arrow (drawn to some scale, such as 1 cm = 10 N) also shows the magnitude.

When *two* forces $\vec{F}_1$ and $\vec{F}_2$ act at the same time at a point A of a body (Fig. 4–2), experiment shows that the effect on the body's motion is the same as the effect of a single force $\vec{R}$ equal to the *vector sum* of the original forces: $\vec{R} = \vec{F}_1 + \vec{F}_2$. More generally,

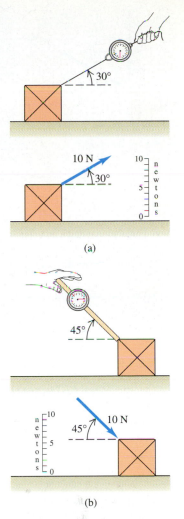

(a)

(b)

4–1 A force may be exerted on the box by either (a) pulling it or (b) pushing it. A force diagram illustrates each case.

4–2 Two forces $\vec{F}_1$ and $\vec{F}_2$ acting simultaneously on a body have the same effect as a single force $\vec{R}$ equal to the vector sum (resultant) of $\vec{F}_1$ and $\vec{F}_2$.

TABLE 4–1

TYPICAL FORCE MAGNITUDES

Sun's gravitational force on the earth	3.5×10^{22} N
Thrust of the *Energia* rocket	3.9×10^{7} N
Weight of a large blue whale	1.9×10^{6} N
Maximum pulling force of a locomotive	8.9×10^{5} N
Weight of a 250-lb linebacker	1.1×10^{3} N
Weight of a medium apple	1 N
Weight of smallest insect eggs	2×10^{-6} N
Electric attraction between the proton and the electron in a hydrogen atom	8.2×10^{-8} N
Weight of a very small bacterium	1×10^{-18} N
Weight of a hydrogen atom	1.6×10^{-26} N
Weight of an electron	8.9×10^{-30} N
Gravitational attraction between the proton and the electron in a hydrogen atom	3.6×10^{-47} N

(a)

(b)

4–3 The force $\vec{F}$, which acts at an angle θ from the x-axis, may be replaced by its rectangular component vectors $\vec{F}_x$ and $\vec{F}_y$. The x- and y-components of $\vec{F}$ are $F_x = F \cos \theta$ and $F_y = F \sin \theta$.

4–4 F_x and F_y are the components of $\vec{F}$ parallel and perpendicular to the sloping surface of the inclined plane.

the effect of any number of forces applied at a point on a body is the same as the effect of a single force equal to the vector sum of the forces. This important principle goes by the name **superposition of forces.**

The experimental discovery that forces combine according to vector addition is of the utmost importance. We will use this fact many times throughout our study of physics. It allows us to replace a force by its component vectors, as we did with displacements in Section 1–9. For example, in Fig. 4–3a, force $\vec{F}$ acts on a body at point O. The component vectors of $\vec{F}$ in the directions Ox and Oy are $\vec{F}_x$ and $\vec{F}_y$. When $\vec{F}_x$ and $\vec{F}_y$ are applied simultaneously, as in Fig. 4–3b, the effect is exactly the same as the effect of the original force $\vec{F}$. *Any force can be replaced by its component vectors, acting at the same point.*

It's frequently more convenient to describe a force $\vec{F}$ in terms of its x- and y-components F_x and F_y rather than by its component vectors (recall from Section 1–9 that *component vectors* are vectors, but *components* are just numbers). For the case shown in Fig. 4–3, both F_x and F_y are positive; for other orientations of the force $\vec{F}$, either F_x or F_y can be negative or zero.

There is no law that says our coordinate axes have to be vertical and horizontal. Figure 4–4 shows a stone block being pulled up a ramp by a force $\vec{F}$, represented by its components F_x and F_y parallel and perpendicular to the sloping surface of the ramp.

CAUTION ▶ In Fig. 4–4 we draw a wiggly line through the force vector $\vec{F}$ to show that we have replaced it by its x- and y-components. Otherwise, the diagram would include the same force twice. We will draw such a wiggly line in any force diagram where a force is replaced by its components. ◀

We will often need to find the vector sum (resultant) of *all* the forces acting on a body. We will call this the **net force** acting on the body. We will use the Greek letter Σ (capital "sigma," equivalent to the Roman S) as a shorthand notation for a sum. If the forces are labeled $\vec{F}_1$, $\vec{F}_2$, $\vec{F}_3$, and so on, we abbreviate the sum as

$$\vec{R} = \vec{F}_1 + \vec{F}_2 + \vec{F}_3 + \cdots = \Sigma \vec{F}, \qquad (4\text{–}1)$$

where $\Sigma \vec{F}$ is read as "the vector sum of the forces" or "the net force." The component version of Eq. (4–1) is the pair of component equations

$$R_x = \Sigma F_x, \qquad R_y = \Sigma F_y, \qquad (4\text{–}2)$$

where ΣF_x is the sum of the x-components, and so on. Each component may be positive or negative, so be careful with signs when you evaluate the sums in Eqs. (4–2).

Once we have R_x and R_y, we can find the magnitude and direction of the net force $\vec{R} = \Sigma \vec{F}$ acting on the body. The magnitude is

$$R = \sqrt{R_x{}^2 + R_y{}^2},$$

and the angle θ between $\vec{R}$ and the +x-axis can be found from the relation $\tan \theta = R_y/R_x$. The components R_x and R_y may be positive, negative, or zero, and the angle θ may be in any of the four quadrants.

In three-dimensional problems, forces may also have z-components; then we add the equation $R_z = \Sigma F_z$ to Eqs. (4–2). The magnitude of the net force is then

$$R = \sqrt{R_x{}^2 + R_y{}^2 + R_z{}^2}.$$

EXAMPLE 4–1

Superposition of forces Three customers are fighting over the same bargain basement coat. They apply the three horizontal forces to the coat that are shown in Fig. 4–5, where the coat is located at the origin. Find the x- and y-components of the net force on the coat, and find the magnitude and direction of the net force.

SOLUTION This is just a problem in vector addition, which we attack using the component method. The angles between the three forces $\vec{F}_1$, $\vec{F}_2$, and $\vec{F}_3$ and the +x-axis are $\theta_1 = 30°$, $\theta_2 = 180° - 45° = 135°$, and $\theta_3 = 180° + 53° = 233°$. The x- and y-components of the three forces are

$$F_{1x} = (200 \text{ N}) \cos 30° = 173 \text{ N},$$
$$F_{1y} = (200 \text{ N}) \sin 30° = 100 \text{ N},$$
$$F_{2x} = (300 \text{ N}) \cos 135° = -212 \text{ N},$$
$$F_{2y} = (300 \text{ N}) \sin 135° = 212 \text{ N},$$
$$F_{3x} = (155 \text{ N}) \cos 233° = -93 \text{ N},$$
$$F_{3y} = (155 \text{ N}) \sin 233° = -124 \text{ N}.$$

From Eqs. (4–2) the net force $\vec{R} = \Sigma \vec{F}$ has components

$$R_x = \Sigma F_x = F_{1x} + F_{2x} + F_{3x}$$
$$= 173 \text{ N} + (-212 \text{ N}) + (-93 \text{ N}) = -132 \text{ N},$$
$$R_y = \Sigma F_y = F_{1y} + F_{2y} + F_{3y}$$
$$= 100 \text{ N} + 212 \text{ N} + (-124 \text{ N}) = 188 \text{ N}.$$

4–5 Three forces acting on a coat. What is the net force?

The net force has a negative x-component and a positive y-component, so it points to the left and toward the top of the page in Fig. 4–5 (that is, in the second quadrant).

The magnitude of the net force $\vec{R} = \Sigma \vec{F}$ is

$$R = \sqrt{R_x{}^2 + R_y{}^2} = \sqrt{(-132 \text{ N})^2 + (188 \text{ N})^2} = 230 \text{ N}.$$

To find the angle between the net force and the +x-axis, we use the relation $\tan \theta = R_y/R_x$, or

$$\theta = \arctan \frac{R_y}{R_x} = \arctan \left(\frac{188 \text{ N}}{-132 \text{ N}} \right) = \arctan (-1.42).$$

The two possible solutions are $\theta = -55°$ or $\theta = -55° + 180° = 125°$. Since the net force lies in the second quadrant, as mentioned above, the correct answer is $\theta = 125°$.

4–3 NEWTON'S FIRST LAW

We have discussed some of the properties of forces, but so far have said nothing about how forces affect motion. To begin with, let's consider what happens when the net force on a body is *zero*. You would almost certainly agree that if a body is at rest, and if no net force acts on it (that is, no net push or pull), that body will remain at rest. But what if there is zero net force acting on a body in *motion?*

To see what happens in this case, suppose you slide a hockey puck along a horizontal table top, applying a horizontal force to it with your hand (Fig. 4–6a). After you stop pushing, the puck *does not* continue to move indefinitely; it slows down and stops. To keep it moving, you have to keep pushing (that is, applying a force). You might come to the "common sense" conclusions that bodies in motion naturally come to rest and that a force is required to sustain motion.

But now imagine pushing the puck across the smooth surface of a freshly waxed floor (Fig. 4–6b; see next page). After you quit pushing, the puck will slide a lot farther

(a)

4–6 A hockey puck is given an initial velocity. (a) It stops in a short distance on a table top.

(b)

(c)

4–6 (cont.) (b) A freshly waxed surface decreases the friction force, and the puck slides farther. (c) On an air-hockey table the friction force is practically zero, so the puck continues with almost constant velocity.

(a)

(b)

4–7 (a) A hockey puck accelerates in the direction of a net applied force $\vec{F}_1$. (b) When the net force is zero, the acceleration is zero, and the puck is in equilibrium.

before it stops. Put it on an air-hockey table, where it floats on a thin cushion of air, and it moves still farther (Fig. 4–6c). In each case, what slows the puck down is *friction*, an interaction between the lower surface of the puck and the surface on which it slides. Each surface exerts a frictional force on the puck that resists the puck's motion; the difference between the three cases is the magnitude of the frictional force. The slippery waxed floor exerts less friction than the table top, so the puck travels farther. The gas molecules of the air-hockey table exert the least friction of all. If we could eliminate friction completely, the puck would never slow down, and we would need no force at all to keep the puck moving once it had been started. Thus the "common sense" idea that a force is required to sustain motion is *incorrect*.

Experiments like the ones we've just described show that when no net force acts on a body, the body either remains at rest or moves with constant velocity in a straight line. Once a body has been set in motion, no net force is needed to keep it moving. In other words:

> **A body acted on by no net force moves with constant velocity (which may be zero) and zero acceleration.**

This is **Newton's first law of motion.**

The tendency of a body to keep moving once it is set in motion results from a property called **inertia.** You use this when you try to get ketchup out of a bottle by shaking it. First you start the bottle (and the ketchup inside) moving forward; when you jerk the bottle back, the ketchup tends to keep moving forward and, you hope, ends up on your burger. The tendency of a body at rest to remain at rest is also due to inertia. You may have seen a tablecloth yanked out from under the china without breaking anything. The force on the china isn't great enough to make it move appreciably during the short time it takes to pull the tablecloth away.

It's important to note that the *net* force is what matters in Newton's first law. For example, a physics book at rest on a horizontal table top has two forces acting on it: the downward force of the earth's gravitational attraction (a long-range force that acts even if the table top is elevated above the ground) and an upward supporting force exerted by the table top (a contact force). The upward push of the surface is just as great as the downward pull of gravity, so the *net* force acting on the book (that is, the vector sum of the two forces) is zero. In agreement with Newton's first law, if the book is at rest on the table top, it remains at rest. The same principle applies to a hockey puck sliding on a horizontal, frictionless surface: The vector sum of the upward push of the surface and the downward pull of gravity is zero. Once the puck is in motion, it continues to move with constant velocity because the *net* force acting on it is zero. (The upward supporting force of the surface is called a **normal force** because it is *normal*, or perpendicular, to the surface of contact. We'll discuss the normal force in more detail in Chapter 5.)

Here's another example. Suppose a hockey puck rests on a horizontal surface with negligible friction, such as an air-hockey table or a slab of wet ice. If the puck is initially at rest and a single horizontal force $\vec{F}_1$ acts on it (Fig. 4–7a), the puck starts to move. If the puck is in motion to begin with, the force changes its speed, its direction, or both, depending on the direction of the force. In this case the net force is equal to $\vec{F}_1$, which is *not* zero. (There are also two vertical forces: the earth's gravitational attraction and the upward contact force exerted by the surface. But as we mentioned earlier, these two forces cancel.)

Now suppose we apply a second force $\vec{F}_2$ (Fig. 4–7b), equal in magnitude to $\vec{F}_1$ but opposite in direction. The two forces are negatives of each other, $\vec{F}_2 = -\vec{F}_1$, and their vector sum is zero:

$$\Sigma \vec{F} = \vec{F}_1 + \vec{F}_2 = \vec{F}_1 + (-\vec{F}_1) = 0.$$

Again, we find that if the body is at rest at the start, it remains at rest; if it is initially moving, it continues to move in the same direction with constant speed. These results show that in Newton's first law, *zero net force is equivalent to no force at all.* This is just the principle of superposition of forces that we saw in Section 4–2.

When a body is acted on by no forces, or by several forces such that their vector sum (resultant) is zero, we say that the body is in **equilibrium.** In equilibrium, a body is either at rest or moving in a straight line with constant velocity. For a body in equilibrium, the net force is zero:

$$\Sigma \vec{F} = 0 \quad \text{(body in equilibrium)}. \tag{4–3}$$

For this to be true, each component of the net force must be zero, so

$$\Sigma F_x = 0, \qquad \Sigma F_y = 0 \quad \text{(body in equilibrium)}. \tag{4–4}$$

When Eqs. (4–4) are satisfied, the body is in equilibrium.

We are assuming that the body can be represented adequately as a point particle. When the body has finite size, we also have to consider *where* on the body the forces are applied. We will return to this point in Chapter 11.

EXAMPLE 4–2

Zero net force means constant velocity In the classic 1950 science fiction film *Rocketship X-M,* a spaceship is moving in the vacuum of outer space, far from any planet, when its engine dies. As a result, the spaceship slows down and stops. What does Newton's first law say about this event?

SOLUTION In this situation there are no forces acting on the spaceship, so according to Newton's first law, it will *not* stop. It continues to move in a straight line with constant speed. This example of science fiction contains more fiction than science.

EXAMPLE 4–3

Constant velocity means zero net force You are driving your Porsche 911 Carrera along a straight testing track at a constant speed of 150 km/h. You pass a 1971 Volkswagen Beetle doing a constant 75 km/h. For which car is the net force greater?

SOLUTION The key word in this question is "net." Both cars are in equilibrium because their velocities are both constant; therefore the *net* force on each car is *zero.*

This conclusion seems to contradict the "common sense" idea that the faster car must have a greater force pushing it. It's true that there is a forward force on both cars, and it's true that

the forward force on your Porsche is much greater than that on the Volkswagen (thanks to your Porsche's high-power engine). But there is also a *backward* force acting on each car due to friction with the road and air resistance. The only reason these cars need their engines running at all is to counteract this backward force so that the vector sum of the forward and backward forces will be zero and the car will travel with constant velocity. The backward force on your Porsche is greater because of its greater speed, which is why your engine has to be more powerful than the Volkswagen's.

INERTIAL FRAMES OF REFERENCE

In discussing relative velocity in Section 3–6, we introduced the concept of *frame of reference.* This concept is central to Newton's laws of motion. Suppose you are in an airplane that is accelerating down the runway during takeoff. If you could stand in the aisle on roller skates, you would start moving *backward* relative to the plane as the pilot opens the throttle. If instead the airplane was landing, you would start moving forward down the aisle as the airplane slowed to a stop. In either case, it looks as though Newton's first law is not obeyed; there is no net force acting on you, yet your velocity changes. What's wrong?

The point is that the airplane is accelerating with respect to the earth and is *not* a suitable frame of reference for Newton's first law. This law is valid in some frames of reference, and is not valid in others. A frame of reference in which Newton's first law *is* valid is called an **inertial frame of reference.** The earth is at least approximately an

4–8 Riding in an accelerating vehicle. (a) If you and the vehicle are initially at rest, you tend to remain at rest as the vehicle accelerates around you. (b) If you and the vehicle are initially moving, you tend to continue moving with constant velocity as the vehicle slows down. (c) You tend to continue moving in a straight line as the vehicle turns.

inertial frame of reference, but the airplane is not. (The earth is not a completely inertial frame, owing to the acceleration associated with its rotation and its motion around the sun. These effects are quite small, however; see Exercises 3–21 and 3–24.) Because Newton's first law is used to define what we mean by an inertial frame of reference, it is sometimes called the *law of inertia*.

Figure 4–8 shows how we can use Newton's first law to understand what you experience when riding in a vehicle that's accelerating. In Fig. 4–8a a vehicle is initially at rest, then begins to accelerate to the right. A passenger on roller skates has virtually no net force acting on her, since the skates nearly eliminate the effects of friction; hence she tends to remain at rest relative to the inertial frame of the earth, in accordance with Newton's first law. As the vehicle accelerates around her, she moves backwards relative to the vehicle. In the same way, a passenger in a vehicle that is slowing down tends to continue moving with constant velocity relative to the earth (Fig. 4–8b). This passenger moves forward relative to the vehicle. A vehicle is also accelerating if it moves at a constant speed but is turning (Fig. 4–8c). In this case a passenger tends to continue moving relative to the earth at constant speed in a straight line; relative to the vehicle, the passenger moves to the side of the vehicle on the outside of the turn.

In each case shown in Fig. 4–8, an observer in the vehicle's frame of reference might be tempted to conclude that there *is* a net force acting on the passenger in each case, since the passenger's velocity *relative to the vehicle* changes in each case. This conclusion is simply wrong; the net force on the passenger is indeed zero. The vehicle observer's mistake is in trying to apply Newton's first law in the vehicle's frame of reference, which is *not* an inertial frame and in which Newton's first law isn't valid. In this book we will use *only* inertial frames of reference.

We've mentioned only one (approximately) inertial frame of reference: the earth's surface. But there are many inertial frames. If we have an inertial frame of reference *A*, in which Newton's first law is obeyed, then any second frame of reference *B* will also be inertial if it moves relative to A with constant velocity $\vec{v}_{B/A}$. To prove this, we use the relative velocity relation Eq. (3–36) from Section 3–6:

$$\vec{v}_{P/A} = \vec{v}_{P/B} + \vec{v}_{B/A}$$

Suppose that P is a body that moves with constant velocity $\vec{v}_{P/A}$ with respect to an inertial frame A. By Newton's first law the net force on this body is zero. The velocity of P relative to another frame B has a different value, $\vec{v}_{P/B} = \vec{v}_{P/A} - \vec{v}_{B/A}$. But if the relative velocity $\vec{v}_{B/A}$ of the two frames is constant, then $\vec{v}_{P/B}$ will be constant as well. Thus B is also an inertial frame; the velocity of P in this frame is constant, and the net force on P is zero, so Newton's first law is obeyed in B. Observers in frames A and B will disagree about the velocity of P, but they will agree that P has a constant velocity (zero acceleration) and has zero net force acting on it.

There is no single inertial frame of reference that is preferred over all others for formulating Newton's laws. If one frame is inertial, then every other frame moving relative to it with constant velocity is also inertial. Viewed in this light, the state of rest and the state of motion with constant velocity are not very different; both occur when the vector sum of forces acting on the body is zero.

From the frame of reference of this car, it seems as though a force is pushing the crash test dummies forward as the car comes to a sudden stop. But there is really no such force: as the car stops, the dummies keep moving forward as a consequence of Newton's first law.

4-4 NEWTON'S SECOND LAW

In discussing Newton's first law, we have seen that when a body is acted on by no force or zero net force, it moves with constant velocity and zero acceleration. In Fig. 4–9a a hockey puck is sliding to the right on wet ice, so there is negligible friction. There are no horizontal forces acting on the puck; the downward force of gravity and the upward contact force exerted by the ice surface sum to zero. So the net force $\Sigma \vec{F}$ acting on the puck is zero, the puck has zero acceleration, and its velocity is constant.

But what happens when the net force is *not* zero? In Fig. 4–9b we apply a constant horizontal force to a sliding puck in the same direction that the puck is moving. Then $\Sigma \vec{F}$ is constant and in the same horizontal direction as $\vec{v}$. We find that during the time the force is acting, the velocity of the puck changes at a constant rate; that is, the puck moves with constant acceleration. The speed of the puck increases, so the acceleration $\vec{a}$ is in the same direction as $\vec{v}$ and as $\Sigma \vec{F}$.

(a)

(b)

(c)

4–9 The acceleration of a body is in the same direction as the net force acting on the body (in this case, a hockey puck on a frictionless surface). (a) If $\Sigma \vec{F} = 0$, the puck is in equilibrium; the velocity is constant and the acceleration is zero. (b) If $\Sigma \vec{F}$ is to the right, the acceleration is to the right. (c) If $\Sigma \vec{F}$ is to the left, the acceleration is to the left.

4–10 A top view of a hockey puck in uniform circular motion on a frictionless horizontal surface. At any point in the motion the acceleration $\vec{a}$ and the net force $\Sigma\vec{F}$ are in the same direction, toward the center of the circle.

(a)

(b)

(c)

4–11 (a) Acceleration $\vec{a}$ is proportional to the net force $\Sigma\vec{F}$. (b) Doubling the net force doubles the acceleration. (c) Halving the net force halves the acceleration.

Figure 4–9c shows another experiment, in which we reverse the direction of the force on the puck so that $\Sigma\vec{F}$ acts in the direction opposite to $\vec{v}$. In this case as well, the puck has an acceleration; the puck moves more and more slowly to the right. If the leftward force continues to act, the puck eventually stops and begins to move more and more rapidly to the left. The acceleration $\vec{a}$ in this experiment is to the left, in the same direction as $\Sigma\vec{F}$. As in the previous case, experiment shows that the acceleration is constant if $\Sigma\vec{F}$ is constant.

We conclude that the presence of a net force acting on a body causes the body to accelerate. The direction of the acceleration is the same as that of the net force. If the magnitude of the net force is constant, as in Figs. 4–9b and 4–9c, then so is the magnitude of the acceleration.

These conclusions about net force and acceleration also apply to a body moving along a curved path. For example, Fig. 4–10 shows a hockey puck moving in a horizontal circle on an ice surface of negligible friction. A string attaching the puck to the ice exerts a force of constant magnitude toward the center of the circle. The result is acceleration that is constant in magnitude and directed toward the center of the circle. The speed of the puck is constant, so this is uniform circular motion, as discussed in Section 3–5.

Figure 4–11a shows another experiment to explore the relationship between the acceleration of a body and the net force acting on that body. We apply a constant horizontal force to a puck on a frictionless horizontal surface, using the spring balance described in Section 4–2 with the spring stretched a constant amount. As in Figs. 4–9b and 4–9c, this horizontal force equals the net force on the puck. If we change the magnitude of the net force, the acceleration changes in the same proportion. Doubling the net force doubles the acceleration (Fig. 4–11b), halving the force halves the acceleration (Fig. 4–11c), and so on. Many such experiments show that for any given body the magnitude of the acceleration is directly proportional to the magnitude of the net force acting on the body.

For a given body the ratio of the magnitude $|\Sigma\vec{F}|$ of the net force to the magnitude $a = |\vec{a}|$ of the acceleration is constant, regardless of the magnitude of the net force. We call this ratio the inertial mass, or simply the **mass**, of the body and denote it by m. That is, $m = |\Sigma\vec{F}|/a$, or

$$|\Sigma\vec{F}| = ma. \tag{4–5}$$

MASS AND FORCE

Mass is a quantitative measure of inertia, which we discussed in Section 4–3. The greater its mass, the more a body "resists" being accelerated. The concept is easy to relate to everyday experience. If you hit a table-tennis ball and then a basketball with the same force, the basketball has much smaller acceleration because it has much greater mass. If a force causes a large acceleration, the mass of the body is small; if the same force causes only a small acceleration, the mass of the body is large.

The SI unit of mass is the **kilogram.** We mentioned in Section 1–4 that the kilogram is officially defined to be the mass of a chunk of platinum-iridium alloy kept in a vault near Paris. We can use this standard kilogram, along with Eq. (4–5), to define the **newton:**

> **One newton is the amount of net force that gives an acceleration of one meter per second squared to a body with a mass of one kilogram.**

We can use this definition to calibrate the spring balances and other instruments used to measure forces. Because of the way we have defined the newton, it is related to the units of mass, length, and time. For Eq. (4–5) to be dimensionally consistent, it must be true that

$$1 \text{ newton} = (1 \text{ kilogram})(1 \text{ meter per second squared}),$$

or

$$1 \text{ N} = 1 \text{ kg} \cdot \text{m/s}^2.$$

We will use this relation many times in the next few chapters, so keep it in mind.

We can also use Eq. (4–5) to compare a mass with the standard mass and thus to *measure* masses. Suppose we apply a constant net force F to a body having a known mass m_1 and we find an acceleration of magnitude a_1. We then apply the same force to another body having an unknown mass m_2, and we find an acceleration of magnitude a_2. Then, according to Eq. (4–5),

$$m_1 a_1 = m_2 a_2,$$

$$\frac{m_2}{m_1} = \frac{a_1}{a_2} \qquad \text{(same net force).} \qquad (4\text{–}6)$$

The ratio of the masses is the inverse of the ratio of the accelerations. Figure 4–12 shows the inverse proportionality between mass and acceleration. In principle we could use Eq. (4–6) to measure an unknown mass m_2, but it is usually easier to determine mass indirectly by measuring the body's *weight*. We'll return to this point in Section 4–5.

When two bodies with masses m_1 and m_2 are fastened together, we find that the mass of the composite body is always $m_1 + m_2$ (Fig. 4–12c). This additive property of mass may seem obvious, but it has to be verified experimentally. Ultimately, the mass of a body is related to the number of protons, electrons, and neutrons it contains. This wouldn't be a good way to *define* mass because there is no practical way to count these particles. But the concept of mass is the most fundamental way to characterize the quantity of matter in a body.

NEWTON'S SECOND LAW

We've been careful to state that the *net* force on a body is what causes that body to accelerate. Experiment shows that if a combination of forces $\vec{F}_1, \vec{F}_2, \vec{F}_3, \ldots$ is applied to a body, the body will have the same acceleration (magnitude and direction) as when only a single force is applied, if that single force is equal to the vector sum $\vec{F}_1 + \vec{F}_2 + \vec{F}_3 + \ldots$. In other words, the principle of superposition of forces also holds true when the net force is not zero and the body is accelerating.

Equation (4–5) relates the magnitude of the net force on a body to the magnitude of the acceleration that it produces. We have also seen that the direction of the net force is the same as the direction of the acceleration, whether the body's path is straight or curved. Newton wrapped up all these relationships and experimental results in a single concise statement that we now call **Newton's second law of motion:**

> **If a net external force acts on a body, the body accelerates. The direction of acceleration is the same as the direction of the net force. The net force vector is equal to the mass of the body times the acceleration of the body.**

In symbols,

$$\sum \vec{F} = m\vec{a} \qquad \text{(Newton's second law).} \qquad (4\text{–}7)$$

An alternative statement is that the acceleration of a body (the rate of change of its velocity) is equal to the vector sum (resultant) of all forces acting on the body, divided by its mass. The acceleration has the same direction as the net force. Newton's second law is a fundamental law of nature, the basic relation between force and motion. Most

(a)

(b)

(c)

4–12 For a constant net force $\sum \vec{F}$, acceleration is inversely proportional to the mass of a body. Masses add like ordinary scalars. (a) $\vec{a}_1 = \sum \vec{F}/m_1$. (b) With a greater mass m_2, the acceleration has smaller magnitude: $\vec{a}_2 = \sum \vec{F}/m_2$. (c) If the two masses are combined, the acceleration is $\vec{a}_3 = \sum \vec{F}/(m_1 + m_2)$.

ActivPhysics

2.1.3
Tension Change

2.1.4
Sliding on an Incline

of the remainder of this chapter and all of the next are devoted to learning how to apply this principle in various situations.

Though Eq. (4–7) may be new to you, you've actually been using it all your life to measure your body's acceleration. In your inner ear, microscopic hair cells sense the magnitude and direction of the force that they must exert to cause small membranes to accelerate along with the rest of your body. By Newton's second law, the acceleration of the membranes—and hence that of your body as a whole—is proportional to this force and has the same direction. In this way, you can sense the magnitude and direction of your acceleration even with your eyes closed!

There are at least four aspects of Newton's second law that deserve special attention. First, Eq. (4–7) is a *vector* equation. Usually, we will use it in component form, with a separate equation for each component of force and the corresponding acceleration:

$$\Sigma F_x = ma_x, \qquad \Sigma F_y = ma_y, \qquad \Sigma F_z = ma_z \qquad \text{(Newton's second law).} \qquad (4\text{--}8)$$

This set of component equations is equivalent to the single vector equation (4–7). Each component of total force equals the mass times the corresponding component of acceleration.

Second, the statement of Newton's second law refers to *external* forces. By this we mean forces exerted on the body by other bodies in its environment. It's impossible for a body to affect its own motion by exerting a force on itself; if it was possible, you could lift yourself to the ceiling by pulling up on your belt! That's why only external forces are included in the sum $\Sigma \vec{F}$ in Eqs. (4–7) and (4–8).

Third, Eqs. (4–7) and (4–8) are valid only when the mass m is *constant*. It's easy to think of systems whose masses change, such as a leaking tank truck, a rocket ship, or a moving railroad car being loaded with coal. But such systems are better handled by using the concept of momentum; we'll get to that in Chapter 8.

Finally, Newton's second law is valid only in inertial frames of reference, just like the first law. Thus it is not valid in the reference frame of any of the accelerating vehicles in Fig. 4–8; relative to any of these frames, the passenger accelerates even though the net force on the passenger is zero. We will usually assume that the earth is an adequate approximation to an inertial frame, although because of its rotation and orbital motion it is not precisely inertial.

CAUTION ▶ Please note that the quantity $m\vec{a}$ is *not* a force. All that Eqs. (4–7) and (4–8) say is that the vector $m\vec{a}$ is equal in magnitude and direction to the vector sum $\Sigma \vec{F}$ of all the forces acting on the body. It's incorrect to think of acceleration as a force; rather, acceleration is a result of a nonzero net force. It's "common sense" to think that there is a "force of acceleration" that pushes you back into your seat when your car accelerates forward from rest. But *there is no such force;* instead, your inertia causes you to tend to stay at rest relative to the earth, and the car accelerates around you. The "common sense" confusion arises from trying to apply Newton's second law in a frame of reference where it isn't valid, like the non-inertial reference frame of an accelerating car. We will always examine motion relative to *inertial* frames of reference only. ◀

In learning how to use Newton's second law, we will begin in this chapter with examples of straight-line motion. Then in Chapter 5 we will consider more general cases and develop more detailed problem-solving strategies for applying Newton's laws of motion.

The design of high-performance motorcycles depends fundamentally on Newton's second law. To maximize the forward acceleration, the designer makes the motorcycle as light as possible (that is, minimizes the mass) and uses the most powerful engine possible (thus maximizing the forward force).

EXAMPLE 4-4

Determining acceleration from force A worker applies a constant horizontal force with magnitude 20 N to a box with mass 40 kg resting on a level surface with negligible friction. What is the acceleration of the box?

SOLUTION The first steps in *any* problem involving forces are to choose a coordinate system and to identify all the forces acting on the body in question. We take the +x-axis in the direction of the horizontal force and the +y-axis to be upward (Fig. 4–13).

4-13 (a) A 20-N force accelerates the box to the right. (b) Vector diagram of forces acting on the box, considered as a particle. The tiles under the box are freshly waxed, so friction is negligible.

The forces acting on the box are (i) the horizontal force $\vec{F}$ exerted by the worker; (ii) the weight $\vec{w}$ of the box, that is, the downward force of the earth's gravitational attraction; and (iii) the upward supporting force $\vec{n}$ exerted by the surface. As in Section 4-3, we call the force $\vec{n}$ a *normal* force because it is normal (perpendicular) to the surface of contact. (We use a script letter n to avoid confusion with the abbreviation N for newton.) We are told that friction is negligible, so no friction force is present. The acceleration is given by Newton's second law, Eqs. (4-8). There is no vertical acceleration, so we know that the two vertical forces must sum to zero. There is only one horizontal force, and we have

$$\Sigma F_x = ma_x,$$

$$a_x = \frac{\Sigma F_x}{m} = \frac{20 \text{ N}}{40 \text{ kg}} = \frac{20 \text{ kg} \cdot \text{m/s}^2}{40 \text{ kg}} = 0.50 \text{ m/s}^2.$$

The acceleration is in the $+x$-direction, the same direction as the net force. The net force is constant, so the acceleration is also constant. If we are given the initial position and velocity of the box, we can find the position and velocity at any later time from the equations of motion with constant acceleration we derived in Chapter 2.

EXAMPLE 4-5

Determining force from acceleration A waitress shoves a ketchup bottle with mass 0.45 kg toward the right along a smooth, level lunch counter. As the bottle leaves her hand, it has an initial velocity of 2.8 m/s. As it slides, it slows down because of the constant horizontal friction force exerted on it by the counter top. It slides a distance of 1.0 m before coming to rest. What are the magnitude and direction of the friction force acting on it?

SOLUTION As in the previous example, the first step is to choose a coordinate system. Suppose the bottle slides in the $+x$-direction, leaving her hand at the point $x_0 = 0$ with the given initial velocity (Fig. 4-14a). The forces acting on the bottle are shown in Fig. 4-14b. The friction force $\vec{f}$ acts to slow the bottle down, so its direction must be opposite the direction of velocity (see Fig. 4-9c).

To find the magnitude of $\vec{f}$ using Eqs. (4-8), we must first

4-14 (a) As the bottle slides to the right, the friction force slows it down. (b) Vector diagram of forces acting on the bottle, considered as a particle.

determine the acceleration of the bottle. We are given that the friction force is constant. The acceleration is then constant also, so we may use the constant-acceleration formula Eq. (2–13):

$$v^2 = v_0{}^2 + 2a(x - x_0),$$

$$a = \frac{v^2 - v_0{}^2}{2(x - x_0)} = \frac{(0 \text{ m/s})^2 - (2.8 \text{ m/s})^2}{2(1.0 \text{ m} - 0 \text{ m})} = -3.9 \text{ m/s}^2.$$

The negative sign means that the acceleration is toward the *left;* the velocity is in the opposite direction to acceleration, as it must be, since the bottle is slowing down. The net force in the x-direction is the x-component –f of the friction force, so

$$\Sigma F_x = -f = ma = (0.45 \text{ kg})(-3.9 \text{ m/s}^2)$$

$$= -1.8 \text{ kg} \cdot \text{m/s}^2 = -1.8 \text{ N}.$$

Again the negative sign shows that the force on the bottle is directed toward the left. The magnitude of the friction force is f = 1.8 N. Remember that magnitudes are *always* positive!

SOME NOTES ON UNITS

A few words about units are in order. In the cgs metric system (not used in this book), the unit of mass is the gram, equal to 10^{-3} kg, and the unit of distance is the centimeter, equal to 10^{-2} m. The corresponding unit of force is called the *dyne:*

$$1 \text{ dyne} = 1 \text{ g} \cdot \text{cm/s}^2.$$

One dyne is equal to 10^{-5} N. In the British system, the unit of force is the *pound* (or pound-force) and the unit of mass is the *slug.* The unit of acceleration is one foot per second squared, so

$$1 \text{ pound} = 1 \text{ slug} \cdot \text{ft/s}^2.$$

The official definition of the pound is

$$1 \text{ pound} = 4.448221615260 \text{ newtons.}$$

It is handy to remember that a pound is about 4.4 N and a newton is about 0.22 pound. Next time you want to order a "quarter-pounder," try asking for a "one-newtoner" and see what happens. Another useful fact: A body with a mass of 1 kg has a weight of about 2.2 lb at the earth's surface.

The units of force, mass, and acceleration in the three systems are summarized in Table 4–2.

Despite its name, the English unit of mass has nothing to do with the type of slug shown here. A common garden slug has a mass of about 15 grams, or about 10^{-3} slug.

TABLE 4–2

UNITS OF FORCE, MASS, AND ACCELERATION

SYSTEM OF UNITS	FORCE	MASS	ACCELERATION
SI	newton (N)	kilogram (kg)	m/s^2
cgs	dyne (dyn)	gram (g)	cm/s^2
British	pound (lb)	slug	ft/s^2

4–5 MASS AND WEIGHT

The *weight* of a body is a familiar force. It is the force of the earth's gravitational attraction for the body. We will study gravitational interactions in detail in Chapter 12, but we need some preliminary discussion now. The terms *mass* and *weight* are often misused and interchanged in everyday conversation. It is absolutely essential for you to understand clearly the distinctions between these two physical quantities.

Mass characterizes the *inertial* properties of a body. Mass is what keeps the china on the table when you yank the tablecloth out from under it. The greater the mass, the greater the force needed to cause a given acceleration; this is reflected in Newton's second law, $\Sigma \vec{F} = m\vec{a}$. Weight, on the other hand, is a *force* exerted on a body by the pull of the earth or some other large body. Everyday experience shows us that bodies having large mass also have large weight. A large stone is hard to throw because of its large *mass,* and hard to lift off the ground because of its large *weight.* On the moon the stone would be just as hard to throw horizontally, but it would be easier to lift. So what exactly *is* the relationship between mass and weight?

The answer to this question, according to legend, came to Newton as he sat under an apple tree watching the apples fall. A freely falling body has an acceleration equal to g, and because of Newton's second law, a force must act to produce this acceleration. If a 1-kg body falls with an acceleration of 9.8 m/s², the required force has magnitude

$$F = ma = (1 \text{ kg})(9.8 \text{ m/s}^2) = 9.8 \text{ kg} \cdot \text{m/s}^2 = 9.8 \text{ N}.$$

But the force that makes the body accelerate downward is the gravitational pull of the earth, that is, the *weight* of the body. Any body near the surface of the earth that has a mass of 1 kg *must* have a weight of 9.8 N to give it the acceleration we observe when it is in free fall. More generally, a body with mass m must have weight with magnitude w given by

$$w = mg \quad \text{(magnitude of the weight of a body of mass } m\text{)}. \tag{4–9}$$

The weight of a body is a force, a vector quantity, and we can write Eq. (4–9) as a vector equation:

$$\vec{w} = m\vec{g}. \tag{4–10}$$

Remember that g is the *magnitude* of $\vec{g}$, the acceleration due to gravity, so g is always a positive number, by definition. Thus w, given by Eq. (4–9), is the *magnitude* of the weight and is also always positive.

It is important to understand that the weight of a body acts on the body *all the time*, whether it is in free fall or not. When a 10-kg flowerpot hangs suspended from a chain, it is in equilibrium, and its acceleration is zero. But its weight, given by Eq. (4–10), is still pulling down on it. In this case the chain pulls up on the pot, applying an upward force. The *vector sum* of the forces is zero, and the pot is in equilibrium.

2.11
Pole Vaulter Vaults

EXAMPLE 4–6

Net force and acceleration in free fall In Example 2–6 (Section 2–6) a 500-lira coin was dropped from rest from the Leaning Tower of Pisa. If the coin falls freely, so that the effects of the air are negligible, how does the net force on the coin vary as it falls?

SOLUTION In free fall, the acceleration $\vec{a}$ of the coin is constant and equal to $\vec{g}$. Hence by Newton's second law the net force $\Sigma \vec{F} = m\vec{a}$ is also constant and equal to $m\vec{g}$, which is the weight $\vec{w}$ of the coin (Fig. 4–15). The velocity of the coin changes as it falls, but the net force acting on it remains constant. If this result surprises you, it's because you still believe in the erroneous "common sense" idea that greater speed implies greater force. If so, you should reread Example 4–3.

The net force on a freely-falling coin is constant even if you initially toss it upward. The force that your hand exerts on the coin is a contact force, and it disappears the instant that the coin loses contact with your hand. From then on, the only force acting on the coin is its weight $\vec{w}$.

$$\vec{a} = \vec{g}$$

$$\Sigma \vec{F} = \vec{w}$$

4–15 The acceleration of a freely-falling object is constant, and so is the net force acting on the object.

VARIATION OF g WITH LOCATION

We will use $g = 9.80$ m/s² for problems on the earth. In fact, the value of g varies somewhat from point to point on the earth's surface, from about 9.78 to 9.82 m/s², because the earth is not perfectly spherical and because of effects due to its rotation and orbital

(a)

(b)

4–16 (a) A standard kilogram weighs about 9.8 N on earth. (b) The same kilogram weighs only about 1.6 N on the moon.

motion. At a point where $g = 9.80 \text{ m/s}^2$, the weight of a standard kilogram is $w = 9.80$ N. At a different point, where $g = 9.78 \text{ m/s}^2$, the weight is $w = 9.78$ N but the mass is still 1 kg. The weight of a body varies from one location to another; the mass does not. If we take a standard kilogram to the surface of the moon, where the acceleration of free fall (equal to the value of g at the moon's surface) is 1.62 m/s^2, its weight is 1.62 N, but its mass is still 1 kg (Fig. 4–16). An 80.0-kg astronaut has a weight on earth of $(80.0 \text{ kg})(9.80 \text{ m/s}^2) = 784$ N, but on the moon the astronaut's weight would be only $(80.0 \text{ kg})(1.62 \text{ m/s}^2) = 130$ N.

The following excerpt from Astronaut Buzz Aldrin's book *Men from Earth,* describing his experiences during the Apollo 11 mission to the moon in 1969, offers some interesting insights into the distinction between mass and weight:

> Our portable life-support system backpacks looked simple, but they were hard to put on and tricky to operate. On Earth the portable life-support system and space suit combination weighed 190 pounds, but on the moon it was only 30. Combined with my own body weight, that brought me to a total lunar-gravity weight of around 60 pounds.
>
> One of my tests was to jog away from the lunar module to see how maneuverable an astronaut was on the surface. I remembered what Isaac Newton had taught us two centuries before: Mass and weight are not the same. I weighed only 60 pounds, but my *mass* was the same as it was on Earth. Inertia was a problem. I had to plan ahead several steps to bring myself to a stop or to turn without falling.

MEASURING MASS AND WEIGHT

In Section 4–4 we described a way to compare masses by comparing their accelerations when subjected to the same net force. Usually, however, the easiest way to measure the mass of a body is to measure its weight, often by comparing with a standard. Because of Eq. (4–9), two bodies that have the same weight at a particular location also have the same mass. We can compare weights very precisely; the familiar equal-arm balance (Fig. 4–17) can determine with great precision (up to 1 part in 10^6) when the weights of two bodies are equal and hence when their masses are equal. This method doesn't work in the apparent "zero-gravity" environment of outer space. Instead, we have to use Newton's second law directly. We apply a known force to the body, measure its acceleration, and compute the mass as the ratio of force to acceleration. This method, or a variation of it, is used to measure the masses of astronauts in orbiting space stations as well as the masses of atomic and subatomic particles.

The concept of mass plays two rather different roles in mechanics. The weight of a body (the gravitational force acting on it) is proportional to its mass; we may call the property related to gravitational interactions *gravitational mass*. On the other hand, we can call the inertial property that appears in Newton's second law the *inertial mass*. If these two quantities were different, the acceleration due to gravity might well be different for different bodies. However, extraordinarily precise experiments have established with a precision of better than one part in 10^{12} that in fact the two *are* the same.

CAUTION ▶ The SI units for mass and weight are often misused in everyday life. Incorrect expressions such as "This box weighs 6 kg" are nearly universal. What is meant is that the *mass* of the box, probably determined indirectly by *weighing,* is 6 kg. This usage is so common that there is probably no hope of straightening things out, but be sure you recognize that the term *weight* is often used when *mass* is meant. Be careful to avoid this kind of mistake in your own work! In SI units, weight (a force) is measured in newtons, while mass is measured in kilograms. ◀

w_{unknown} w_{known}

4–17 An equal-arm balance determines the mass of a body by comparing its weight to a known weight.

EXAMPLE 4–7

A 1.96×10^4 N Lincoln Town Car traveling in the +x-direction makes a fast stop; the x-component of the net force acting on it is -1.50×10^4 N. What is its acceleration?

SOLUTION Because the newton is a unit of force, we know that 1.96×10^4 N is the weight, not the mass, of the car. Its mass m is

$$m = \frac{w}{g} = \frac{1.96 \times 10^4 \text{ N}}{9.80 \text{ m/s}^2} = \frac{1.96 \times 10^4 \text{ kg} \cdot \text{m/s}^2}{9.80 \text{ m/s}^2},$$

$$= 2000 \text{ kg}.$$

Then $\Sigma F_x = ma_x$ gives

$$a_x = \frac{\Sigma F_x}{m} = \frac{-1.50 \times 10^4 \text{ N}}{2000 \text{ kg}} = \frac{-1.50 \times 10^4 \text{ kg} \cdot \text{m/s}^2}{2000 \text{ kg}}$$

$$= -7.5 \text{ m/s}^2.$$

This acceleration can be written as $-0.77g$. Note that -0.77 is also the ratio of -1.50×10^4 N (the x-component of the net force) to 1.96×10^4 N (the weight).

4–6 NEWTON'S THIRD LAW

A force acting on a body is always the result of its interaction with another body, so forces always come in pairs. You can't pull on a doorknob without the doorknob pulling back on you. When you kick a football, the forward force that your foot exerts on the ball launches it into its trajectory, but you also feel the force the ball exerts back on your foot. If you kick a boulder, the pain you feel is due to the force that the boulder exerts on your foot.

In each of these cases the force that you exert on the other body is in the opposite direction to the force that body exerts on you. Experiments show that whenever two bodies interact, the two forces that they exert on each other are always *equal in magnitude* and *opposite in direction*. This fact is called **Newton's third law of motion.** In Fig. 4–18, $\vec{F}_{A \text{ on } B}$ is the force applied *by* body A (first subscript) *on* body B (second subscript), and $\vec{F}_{B \text{ on } A}$ is the force applied *by* body B (first subscript) *on* body A (second subscript). The mathematical statement of Newton's third law is

$$\vec{F}_{A \text{ on } B} = -\vec{F}_{B \text{ on } A}. \qquad (4\text{–}11)$$

Expressed in words,

> If body A exerts a force on body B (an "action"), then body B exerts a force on body A (a "reaction"). These two forces have the same magnitude but are opposite in direction. These two forces act on *different* bodies.

In this statement, "action" and "reaction" are the two opposite forces; we sometimes refer to them as an **action-reaction pair.** This is not meant to imply any cause-and-effect relationship; we can consider either force as the "action" and the other as the "reaction." We often say simply that the forces are "equal and opposite," meaning that they have equal magnitudes and opposite directions.

We stress that the two forces described in Newton's third law act on *different* bodies. This is important in problems involving Newton's first or second law, which involve the forces that act *on* a body. For instance, the net force acting on the football in Fig. 4–18 is the vector sum of the weight of the football and the force $\vec{F}_{A \text{ on } B}$ exerted by the kicker. You would not include the force $\vec{F}_{B \text{ on } A}$ because this force acts on the *kicker*, not on the football.

In Fig. 4–18 the action and reaction forces are *contact* forces that are present only when the two bodies are touching. But Newton's third law also applies to *long-range* forces that do not require physical contact, such as the force of gravitational attraction. A table-tennis ball exerts an upward gravitational force on the earth that's equal in magnitude to the downward gravitational force the earth exerts on the ball. When you drop the ball, both the ball and the earth accelerate toward each other. The net force on each body has the same magnitude, but the earth's acceleration is microscopically small because its mass is so great. Nevertheless, it does move!

4–18 If body A exerts a force $\vec{F}_{A \text{ on } B}$ on body B, then body B exerts a force $\vec{F}_{B \text{ on } A}$ on body A that is equal in magnitude and opposite in direction: $\vec{F}_{A \text{ on } B} = -\vec{F}_{B \text{ on } A}$.

EXAMPLE 4–8

After your sports car breaks down, you start to push it to the nearest repair shop. While the car is starting to move, how does the force you exert on the car compare to the force the car exerts on you? How do these forces compare when you are pushing the car along at a constant speed?

SOLUTION In *both* cases, the force you exert on the car is equal in magnitude and opposite in direction to the force the car exerts on you. It's true that you have to push harder to get the car going than to keep it going. But no matter how hard you push on the car, the car pushes just as hard back on you. Newton's third law gives the same result whether the two bodies are at rest, moving with constant velocity, or accelerating.

EXAMPLE 4–9

An apple sits on a table in equilibrium. What forces act on it? What is the reaction force to each of the forces acting on the apple? What are the action-reaction pairs?

SOLUTION Figure 4–19a shows the apple on the table, and Fig. 4–19b shows the forces acting on the apple. The apple is A, the table T, and the earth E. In the diagram, $\vec{F}_{\text{E on A}}$ is the weight of the apple; that is, the downward gravitational force exerted *by* the earth E (first subscript) *on* the apple A (second subscript). Similarly, $\vec{F}_{\text{T on A}}$ is the upward force exerted *by* the table T (first subscript) *on* the apple A (second subscript).

As the earth pulls down on the apple, the apple exerts an equally strong upward pull $\vec{F}_{\text{A on E}}$ on the earth, as shown in Fig.

4–19d. $\vec{F}_{\text{A on E}}$ and $\vec{F}_{\text{E on A}}$ are an action-reaction pair, representing the mutual interaction of the apple and the earth, so

$$\vec{F}_{\text{A on E}} = -\vec{F}_{\text{E on A}}.$$

Also, as the table pushes up on the apple with force $\vec{F}_{\text{T on A}}$, the corresponding reaction is the downward force $\vec{F}_{\text{A on T}}$ exerted by the apple on the table (Fig. 4–19c). So we have

$$\vec{F}_{\text{A on T}} = -\vec{F}_{\text{T on A}}.$$

The two forces acting on the apple are $\vec{F}_{\text{T on A}}$ and $\vec{F}_{\text{E on A}}$. Are they an action-reaction pair? No, they aren't, despite being equal and opposite. They do not represent the mutual interaction of two bodies; they are two different forces acting on the *same*

(b) $\vec{F}_{\text{T on A}}$ $\vec{F}_{\text{E on A}} = \vec{w}$

(c) $\vec{F}_{\text{A on T}} = -\vec{F}_{\text{T on A}}$

(a)

(d) $\vec{F}_{\text{A on E}} = -\vec{F}_{\text{E on A}}$

(e) $\vec{F}_{\text{E on A}}$ $\vec{F}_{\text{A on E}}$ $\vec{F}_{\text{T on A}} = 0 = \vec{F}_{\text{A on T}}$

4–19 The two forces in an action-reaction pair always act on different bodies.

body. *The two forces in an action-reaction pair **never** act on the same body.* Here's another way to look at it. Suppose we suddenly yank the table out from under the apple (Fig. 4–19e). The two forces $\vec{F}_{\text{A on T}}$ and $\vec{F}_{\text{T on A}}$ then become zero, but $\vec{F}_{\text{A on E}}$ and

$\vec{F}_{\text{E on A}}$ are still there (the gravitational interaction is still present). Since $\vec{F}_{\text{T on A}}$ is now zero, it can't be the negative of $\vec{F}_{\text{E on A}}$, and these two forces can't be an action-reaction pair.

EXAMPLE 4–10

A stonemason drags a marble block across a floor by pulling on a rope attached to the block (Fig. 4–20a). The block may or may not be in equilibrium. How are the various forces related? What are the action-reaction pairs?

SOLUTION Figure 4–20b shows the horizontal forces acting on each body, the block (B), the rope (R), and the mason (M). We'll use subscripts on all the forces to help explain things. Vector $\vec{F}_{\text{M on R}}$ represents the force exerted *by* the mason *on* the rope. Its reaction is the equal and opposite force $\vec{F}_{\text{R on M}}$ exerted *by* the rope *on* the mason. Vector $\vec{F}_{\text{R on B}}$ represents the force exerted by the rope on the block. The reaction to it is the equal and opposite force $\vec{F}_{\text{B on R}}$ exerted by the block on the rope:

$$\vec{F}_{\text{R on M}} = -\vec{F}_{\text{M on R}} \quad \text{and} \quad \vec{F}_{\text{B on R}} = -\vec{F}_{\text{R on B}}.$$

Be sure you understand that the forces $\vec{F}_{\text{M on R}}$ and $\vec{F}_{\text{B on R}}$ are *not* an action-reaction pair. Both of these forces act on the *same* body (the rope); an action and its reaction *must* always act on *different* bodies. Furthermore, the forces $\vec{F}_{\text{M on R}}$ and $\vec{F}_{\text{B on R}}$ are not necessarily equal in magnitude. Applying Newton's second law to the rope, we get

$$\Sigma\vec{F} = \vec{F}_{\text{M on R}} + \vec{F}_{\text{B on R}} = m_{\text{rope}}\vec{a}_{\text{rope}}.$$

If the block and rope are accelerating (that is, speeding up or slowing down), the rope is not in equilibrium, and $\vec{F}_{\text{M on R}}$ must have a different magnitude than $\vec{F}_{\text{B on R}}$. By contrast, the action-

reaction forces $\vec{F}_{\text{M on R}}$ and $\vec{F}_{\text{R on M}}$ are always equal in magnitude, as are $\vec{F}_{\text{R on B}}$ and $\vec{F}_{\text{B on R}}$. Newton's third law holds whether or not the bodies are accelerating.

In the special case in which the rope is in equilibrium, the forces $\vec{F}_{\text{M on R}}$ and $\vec{F}_{\text{B on R}}$ are equal in magnitude. But this is an example of Newton's *first* law, not his *third*. Another way to look at this is that in equilibrium, $\vec{a}_{\text{rope}} = 0$ in the above equation. Then $\vec{F}_{\text{B on R}} = -\vec{F}_{\text{M on R}}$ because of Newton's first or second law.

This is also true if the rope is accelerating but has negligibly small mass compared to the block or the mason. In this case, $m_{\text{rope}} = 0$ in the above equation, so again $\vec{F}_{\text{B on R}} = -\vec{F}_{\text{M on R}}$. Since $\vec{F}_{\text{B on R}}$ *always* equals $-\vec{F}_{\text{R on B}}$ by Newton's third law (they are an action-reaction pair), in these same special cases, $\vec{F}_{\text{R on B}}$ also equals $\vec{F}_{\text{M on R}}$. In other words, in these cases the force of the rope on the block equals the force of the mason on the rope, and we can then think of the rope as "transmitting" to the block, without change, the force the person exerts on the rope (Fig. 4–20c). This is a useful point of view, but you have to remember that it is valid *only* when the rope has negligibly small mass or is in equilibrium.

If you feel as though you're drowning in subscripts at this point, take heart. Go over this discussion again, comparing the symbols with the vector diagrams, until you're sure you see what's going on.

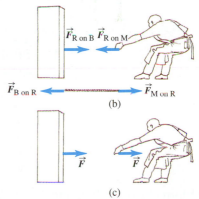

4–20 (a) A mason pulls on a rope attached to a block. (b) Separate diagrams showing the force of the rope on the block, the force of the rope on the mason, and the forces of the block and the mason on the rope. (c) If the rope is not accelerating or if its mass can be neglected, it can be considered to transmit an undiminished force from the mason to the block, and vice versa.

A body, such as the rope in Fig. 4–20, that has pulling forces applied at its ends is said to be in **tension**. The tension at any point is the magnitude of force acting at that point. In Fig. 4–20b the tension at the right-hand end of the rope is the magnitude of $\vec{F}_{\text{M on R}}$ (or of $\vec{F}_{\text{R on M}}$), and the tension at the left-hand end equals the magnitude of $\vec{F}_{\text{B on R}}$ (or of $\vec{F}_{\text{R on B}}$). If the rope is in equilibrium and if no forces act except at its ends, the tension is the same at both ends and throughout the rope. Thus if the magnitudes of $\vec{F}_{\text{B on R}}$ and $\vec{F}_{\text{M on R}}$ are 50 N each, the tension in the rope is 50 N (*not* 100 N). The *total* force vector $\vec{F}_{\text{B on R}} + \vec{F}_{\text{M on R}}$ acting on the rope in this case is zero!

Finally, we emphasize once more a fundamental truth: The two forces in an action-reaction pair *never* act on the same body. Remembering this simple fact can often help you avoid confusion about action-reaction pairs and Newton's third law.

4–7 USING NEWTON'S LAWS

Newton's three laws of motion contain all the basic principles we need to solve a wide variety of problems in mechanics. These laws are very simple in form, but the process of applying them to specific situations can pose real challenges. This section introduces some useful techniques. You'll learn others in Chapter 5, which also extends the use of Newton's laws to cover more complex situations.

When you use Newton's first law, $\Sigma \vec{F} = 0$, for an equilibrium situation, or Newton's second law, $\Sigma \vec{F} = m\vec{a}$, for a non-equilibrium situation, you must apply it to some specific body. It is absolutely essential to decide at the beginning what body you are talking about. This may sound trivial, but it isn't. Once you have chosen a body, then you have to identify all the forces acting on it. Don't get confused between the forces acting on a body and the forces exerted by that body on some other body. Only the forces acting on the body you've chosen go into $\Sigma \vec{F}$.

To help identify the relevant forces, draw a **free-body diagram.** What's that? It is a diagram showing the chosen body by itself, "free" of its surroundings, with vectors drawn to show the magnitudes and directions of all the forces applied to the body by the various other bodies that interact with it. We have already shown some free-body diagrams in Figs. 4–13, 4–14, and 4–19. Be careful to include all the forces acting *on* the body, but be equally careful *not* to include any forces that the body exerts on any other body. In particular, the two forces in an action-reaction pair must *never* appear in the same free-body diagram because they never act on the same body. Furthermore, forces that a body exerts on itself are never included, since these can't affect the body's motion.

CAUTION ▶ When you have a complete free-body diagram, you *must* be able to answer for each force the question "What other body is applying this force?" If you can't answer that question, you may be dealing with a nonexistent force. Be especially on your guard to avoid nonexistent forces such as "the force of acceleration" or "the $m\vec{a}$ force," discussed in Section 4–4. ◀

When a problem involves more than one body, you have to take the problem apart and draw a separate free-body diagram for each body. For example, Fig. 4–20b shows a separate free-body diagram for the rope in the case in which the rope is considered massless (so that no gravitational force acts on it). This figure also shows diagrams for the block and the mason, but these are *not* complete free-body diagrams because they don't show all the forces acting on each body. Which forces are missing in these diagrams?

The simple act of walking depends crucially on Newton's third law. To start moving forward, you push backward on the ground with your foot. As a reaction, the ground pushes forward on your foot (and hence on your body as a whole) with a force of the same magnitude. This *external* force provided by the ground is what accelerates your body forward.

2.1.1
Force Magnitudes

Problem–Solving Strategy

NEWTON'S LAWS

1. Define your coordinate system. A diagram showing the location of the origin and the positive axis direction is always helpful. If you know the direction of the acceleration, it is often convenient to take that as your positive direction.

2. Be consistent with signs. Once you define the *x*-axis and its positive direction, then components in that direction of velocity, acceleration, and force are also positive.

3. In applying Newton's first or second law, always concentrate on a specific body. Draw a free-body diagram showing all the forces acting *on* this body, but do not include forces that the body exerts on any other body. The acceleration of the body is determined by the forces acting on it, not by the forces it exerts on anything else. In your free-body diagram, you can usually represent the body as a particle. Using a colored pencil or pen for the

force vectors may help (Fig. 4–21). Make sure that you can answer the question "What other body is applying this force?" for each force in your diagram. To avoid confusion, the only vector quantities in your diagram should be forces; don't include other vector quantities such as velocity and acceleration.

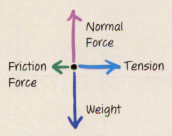

4–21 A free-body diagram. We will use different colors to distinguish certain force vectors in free-body diagrams: dark blue for weight, purple for normal forces, and green for friction forces. All other forces will appear in bright blue. We'll use this color scheme for force vectors throughout the remainder of this book.

4. Identify the known and unknown quantities, and give each unknown quantity an algebraic symbol. If you know the direction of a force at the start, use a symbol to represent the *magnitude* of the force, always a positive quantity. Keep in mind that the *component* of this force along a particular axis direction may still be either positive or negative.

5. Write Newton's first law (for a problem with zero acceleration) or Newton's second law (for problems where there is an acceleration) in component form, using the coordinate system you defined in Step 1. Solve these equations for the unknown quantities.

6. Always check for unit consistency; when appropriate, use the conversion $1 \, \text{N} = 1 \, \text{kg} \cdot \text{m/s}^2$.

   In Chapter 5 we will expand this strategy to deal with more complex problems, but it is important for you to use it consistently from the very start, to develop good habits in the systematic analysis of problems.

EXAMPLE 4–11

Tension in a massless chain To improve the acoustics in an auditorium, a sound reflector with a mass of 200 kg is suspended by a chain from the ceiling (Fig. 4–22a). What is the weight of the reflector? What force (magnitude and direction) does the chain exert on it? What is the tension in the chain? Assume that the mass of the chain itself is negligible.

SOLUTION The reflector is in equilibrium, so we use Newton's first law, $\Sigma \vec{F} = 0$. We draw separate free-body diagrams for the reflector (Fig. 4–22b) and the chain (Fig. 4–22c). We take the positive y-axis to be upward, as shown. Each force has only a y-component. The magnitude of the weight of the reflector is given by Eq. (4–9):

$$w = mg = (200 \, \text{kg})(9.80 \, \text{m/s}^2) = 1960 \, \text{N}.$$

This force points in the negative y-direction, so its y-component is (–1960 N). The upward force exerted by the chain has unknown magnitude T. Because the reflector is in equilibrium, the sum of the y-components of force acting on it must be zero:

$$\Sigma F_y = T + (-1960 \, \text{N}) = 0, \quad \text{so} \quad T = 1960 \, \text{N}.$$

The chain pulls *up* on the reflector with a force T of magnitude 1960 N. By Newton's third law, the reflector pulls *down* on the chain with a force of magnitude 1960 N.

   The chain is also in equilibrium. We have assumed that it is weightless, so an upward force of 1960 N must act on it at its

top end to make the vector sum of forces on it equal to zero. The tension in the chain is 1960 N.

   Note that we have defined T to be the *magnitude* of a force, so it is always positive. But the y-component of force acting on the chain at its lower end is $-T = -1960 \, \text{N}$.

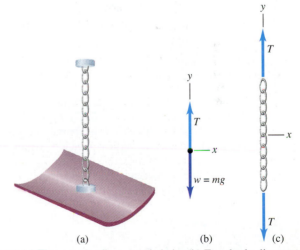

4–22 (a) The sound reflector and chain. (b) Free-body diagram of the reflector. (c) Free-body diagram of the chain, assuming its weight to be negligible.

EXAMPLE 4–12

Tension in a chain with mass In Example 4–11, suppose the mass of the chain is not negligible but is 10.0 kg. Find the forces at the ends of the chain.

SOLUTION The chain's weight is $w_C = m_C g =$

$(10.0 \, \text{kg})(9.8 \, \text{m/s}^2) = 98 \, \text{N}$. Again we draw separate free-body diagrams for the reflector and the chain; each is in equilibrium. The free-body diagrams, shown in Figs. 4–23a and 4–23b, differ from those in Fig. 4–22 because the magnitudes of the forces at the two ends of the chain are no longer equal. We label them T_1

and T_2, as shown. Note that the two forces labeled T_2 form an action-reaction pair; that's how we know they have the same magnitude. The equilibrium condition $\Sigma F_y = 0$ for the reflector is

$$T_2 + (-1960 \text{ N}) = 0, \quad \text{so} \quad T_2 = 1960 \text{ N}.$$

There are now three forces acting on the chain: its weight and the forces at the two ends. The equilibrium condition $\Sigma F_y = 0$ for the chain is

$$T_1 + (-T_2) + (-w_C) = 0.$$

Note that the y-component of T_1 is positive, because it points in the $+y$-direction, but the y-components of both T_2 and w_C are negative. When we substitute the values $T_2 = 1960$ N and $w_C = 98$ N and solve for T_1, we find

$$T_1 = T_2 + w_C = 2058 \text{ N}.$$

Alternatively, we could draw a free-body diagram for the composite body consisting of the reflector and the chain together (Fig. 4–23c). The two forces on this composite object are the upward force T_1 at the top of the chain and the total weight, with magnitude $w_R + w_C = 1960$ N + 98 N = 2058 N. Again we find $T_1 = 2058$ N. Note that we cannot find T_2 by this method.

4–23 (a) Free-body diagram for the reflector (weight w_R). (b) Free-body diagram for the chain (weight w_C). (c) Free-body diagram for the reflector and chain, considered as a single particle.

EXAMPLE 4–13

The cutting-blade assembly on a radial-arm saw has a mass of 5.0 kg. It is pulled along a pair of frictionless horizontal rails aligned with the x-axis by a force F_x (Fig. 4–24a). The position of the blade assembly as a function of time is

$$x = (0.18 \text{ m/s}^2)t^2 - (0.030 \text{ m/s}^3)t^3.$$

Find the net force acting on the blade assembly as a function of time. What is the force at time $t = 3.0$ s? For what times is the force positive? Negative? Zero?

SOLUTION Figure 4–24b shows the free-body diagram and coordinate axes. The forces are the horizontal force F_x, the weight w, and the upward force n that the rails exert to support the cutting head. There is no acceleration in the vertical direction, so the sum of the vertical components of force must be zero. The net force has only a horizontal component, equal to F_x; from Newton's second law, this is equal to ma_x. To determine F_x, we

first find the acceleration a_x by taking the second derivative of x. The second derivative of t^2 is 2, and the second derivative of t^3 is $6t$, so

$$a_x = \frac{d^2x}{dt^2} = 0.36 \text{ m/s}^2 - (0.18 \text{ m/s}^3)t.$$

Then, from Newton's second law,

$$F_x = ma_x = (5.0 \text{ kg})[0.36 \text{ m/s}^2 - (0.18 \text{ m/s}^3)t]$$

$$= 1.80 \text{ N} - (0.90 \text{ N/s})t.$$

At time $t = 3.0$ s the force is 1.80 N − (0.90 N/s)(3.0 s) = −0.90 N. The force is zero when $a_x = 0$, that is, when 0.36 m/s² − (0.18 m/s³)t = 0. This happens when $t = 2.0$ s. When $t < 2.0$ s, F_x is positive, and the blade assembly is being pulled to the right. When $t > 2.0$ s, F_x is negative, and the horizontal force on the assembly is to the left. Figure 4–25 shows graphs of F_x and a_x as functions of time. You shouldn't be surprised that the net force and acceleration are directly proportional; according to Newton's second law, this is *always* true.

4–24 (a) A radial-arm saw cutting blade. (b) Free-body diagram of blade assembly. Both the magnitude and direction of the net force are functions of time.

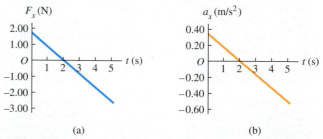

4–25 (a) The net force on the cutting-blade assembly is directly proportional to (b) its acceleration.

4–8 FREE-BODY DIAGRAMS VISUALIZED

Drawing a correct free-body diagram is the first step in analyzing almost any physics or engineering problem involving the motion of a body. In this section we present some real-life situations and the corresponding free-body diagrams. Note that while the situations depicted in the following figures occur in very different surroundings, the free-body diagrams are remarkably similar.

Immersed in water, a person's body experiences an upward force $\vec{B}$ due to buoyancy. This is balanced by the downward force of the diver's weight $\vec{w}$. In this situation the diver's motion depends on the force with which the water pushes on her, due either to the water currents or to reaction forces to the swimmer's arm and leg movements.

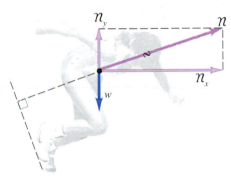

A sprinter gains a large forward acceleration at the start of a race by kicking back hard against the angled starting blocks. The blocks exert an equally large reaction force $\vec{n}$ on the sprinter. This force has a large forward component n_x that springs her into motion, as well as a smaller upward component n_y. If this upward component has the same magnitude as the sprinter's weight w, the net vertical force acting on her is zero and she does not accelerate vertically.

A basketball player jumps by pushing with his feet against the floor. The forces acting on him are the reaction force with which the floor pushes back up and his weight $\vec{w}$. Once the player is in the air, the only force acting on him is his weight; his acceleration is downward, even as he rises up. His stationary opponent is acted on by his own weight and by the upward normal force $\vec{n}$ exerted on him by the floor.

SUMMARY

KEY TERMS

dynamics, 92

Newton's laws of motion, 92

classical (Newtonian) mechanics, 92

force, 93

contact force, 93

long-range forces, 93

weight, 93

superposition of forces, 94

net force, 94

Newton's first law of motion, 96

inertia, 96

normal force, 96

equilibrium, 97

inertial frame of reference, 97

mass, 100

kilogram, 100

newton, 100

Newton's second law of motion, 101

Newton's third law of motion, 107

action-reaction pair, 107

tension, 109

free-body diagram, 110

■ Force, a vector quantity, is a quantitative measure of the interaction between two bodies. When several forces act on a body, the effect on its motion is the same as when a single force, equal to the vector sum (resultant) of the forces, acts on the body.

■ Newton's first law states that when no force acts on a body, or when the vector sum of all forces acting on it (the *net force*) is zero, the body is in equilibrium. If the body is initially at rest, it remains at rest; if it is initially in motion, it continues to move with constant velocity. This law is valid only in inertial frames of reference.

■ The inertial properties of a body are characterized by its *mass*. The acceleration of a body under the action of a given set of forces is directly proportional to the vector sum of the forces (the *net force*) and inversely proportional to the mass of the body. This relationship is Newton's second law:

$$\Sigma \vec{F} = m\vec{a}. \tag{4-7}$$

Like the first law, this is valid only in inertial frames of reference.

■ The unit of force is defined in terms of the units of mass and acceleration. In SI units the unit of force is the newton (N), equal to $1 \text{ kg} \cdot \text{m/s}^2$.

■ The weight of a body is the gravitational force exerted on it by the earth (or whatever other body exerts the gravitational force). Weight is a force and is therefore a vector quantity. The magnitude of the weight of a body at any specific location is equal to the product of its mass m and the magnitude of the acceleration due to gravity g at that location:

$$w = mg. \tag{4-9}$$

The weight of a body depends on its location, but the mass is independent of location.

■ Newton's third law states that "action equals reaction"; when two bodies interact, they exert forces on each other that are equal in magnitude and opposite in direction. Each force in an action-reaction pair acts on only one of the two bodies; the action and reaction forces never act on the same body.

DISCUSSION QUESTIONS

Q4–1 Can a body be in equilibrium when only one force acts on it? Explain.

Q4–2 A ball thrown straight up has zero velocity at its highest point. Is the ball in equilibrium at this point? Why or why not?

Q4–3 A helium balloon hovers in midair, neither ascending nor descending. Is it in equilibrium? What forces act on it?

Q4–4 When you fly in an airplane at night in smooth air, there is no sensation of motion, even though the plane may be moving at 800 km/h (500 mi/h). Why is this?

Q4–5 If the two ends of a rope in equilibrium are pulled with forces of equal magnitude and opposite direction, why is the total tension in the rope not zero?

Q4–6 You tie a brick to the end of a rope and whirl the brick around you in a horizontal circle. Describe the path of the brick after you suddenly let go of the rope.

Q4–7 When a car stops suddenly, the passengers tend to move forward relative to their seats. Why? When a car makes a sharp turn, the passengers tend to slide to one side of the car. Why?

Q4–8 Some people say that the "force of inertia" (or "force of momentum") throws the passengers forward when a car brakes sharply. What is wrong with this explanation?

Q4–9 A passenger in a moving bus with no windows notices that a ball that has been at rest in the aisle suddenly starts to move toward the rear of the bus. Think of two different possible explanations, and devise a way to decide which is correct.

Q4–10 Suppose the fundamental SI units had been chosen to be force, length, and time instead of mass, length, and time. What would be the units of mass in terms of those fundamental units?

Q4–11 Inertia is *not* a force that keeps things in place or that keeps them moving. How do we know this?

Q4–12 Why is the earth only approximately an inertial reference frame?

Q4–13 Does Newton's second law hold true for an observer in a van as it speeds up, slows down, or rounds a corner? Explain.

Q4–14 Some students refer to the quantity $m\vec{a}$ as "the force of acceleration." Is it correct to refer to this quantity as a force? If so, what exerts this force? If not, what is a better description of this quantity?

Q4–15 The acceleration of a falling body is measured in an elevator traveling upward at a constant speed of 9.8 m/s. What result is obtained?

Q4–16 You can play catch with a softball in a bus moving with constant speed on a straight road, just as though the bus were at rest. Is this still possible when the bus is making a turn at constant speed on a level road? Why or why not?

Q4–17 Is the coat of Example 4–1 in equilibrium? Explain.

Q4–18 The head of a hammer begins to come loose from its wooden handle. How should you strike the handle on a concrete sidewalk to reset the head? Why does this work?

Q4–19 Why can it hurt your foot more to kick a big rock than a small pebble? *Must* the big rock hurt more? Explain.

Q4–20 "It's not the fall that hurts you; it's the sudden stop at the bottom." Translate this saying into the language of Newton's laws of motion.

Q4–21 A person can dive into water from a height of 10 m without injury, but a person who jumps off the roof of a 10-m-tall building and lands on a concrete street is likely to be seriously injured. Why is there a difference?

Q4–22 Why are cars designed to crumple up in front and back for safety? Why not for side collisions and rollovers?

Q4–23 When a bullet is fired from a gun, what is the origin of the force that accelerates the bullet?

Q4–24 When a heavy weight is lifted by a string that is barely strong enough, the weight can be lifted by a steady pull; but if the string is jerked, it will break. Explain in terms of Newton's laws of motion.

Q4–25 A large crate is suspended from the end of a vertical rope. Is the tension in the rope larger when the crate is at rest or when it is moving upward at constant speed? If the crate is traveling upward, is the tension in the rope greatest when the crate is speeding up or when it is slowing down? In each case explain in terms of Newton's laws of motion.

Q4–26 Automotive engineers, in discussing the motion of an automobile, call the rate of change of the acceleration the "jerk." Why is this a useful quantity in characterizing the riding qualities of an automobile?

Q4–27 Why can't we correctly say that 1.0 kg *equals* 2.2 lb?

Q4–28 A horse is hitched to a wagon. Since the wagon pulls back on the horse just as hard as the horse pulls on the wagon, why doesn't the wagon remain in equilibrium, no matter how hard the horse pulls?

Q4–29 A 450-N girl from the south slaps a 800-N boy from the north. Her fingers exert a force of 30 N to the west on his cheek. There may be other reactions, but by Newton's third law, what is the reaction *force* to the slap?

Q4–30 A large truck and a small compact car have a head-on collision. During the collision the truck exerts a force $\vec{F}_{\text{T on C}}$ on the car, and the car exerts a force $\vec{F}_{\text{C on T}}$ on the truck. Which force has the larger magnitude, or are they the same? Does your answer depend on how fast each vehicle was moving before the collision? Why or why not?

Q4–31 If you ask most people what force makes a car accelerate forward, they will say "the force of the engine." But what force is *directly* responsible for making the car accelerate?

Q4–32 A small compact car is giving a push to a large van that has broken down, and they travel along the road with equal velocities and accelerations. While the car is speeding up, is the force it exerts on the van larger than, smaller than, or the same magnitude as the force the van exerts on it? Which object, the car or the van, has the largest net force on it, or are the net forces the same? Explain.

Q4–33 Consider a tug-of-war between two people who pull in opposite directions on the ends of a rope. By Newton's third law, the force that *A* exerts on *B* is just as great as the force that *B* exerts on *A*. So what determines who wins? (*Hint:* Draw a free-body diagram showing all the forces that act on each person.)

Q4–34 On the moon, $g = 1.62$ m/s². If a 2-kg brick drops onto your foot from a height of 2 m, will this hurt more, or less, or the same if it happens on the moon instead of on the earth? Explain. If a 2-kg brick is thrown and hits you when it is moving horizontally at 6 m/s, will this hurt more, less, or the same if it happens on the moon instead of on the earth? Explain. (On the moon, assume that you are inside a pressurized structure, so you are not wearing a spacesuit.)

Q4–35 A manual for student pilots contains the following passage: "When an airplane flies at a steady altitude, neither climbing nor descending, the upward lift force from the wings equals the airplane's weight. When the airplane is climbing at a steady rate, the upward lift is greater than the weight; when the airplane is descending at a steady rate, the upward lift is less than the weight." Are these statements correct? Explain.

EXERCISES

SECTION **4–2** **FORCE AND INTERACTIONS**

4–1 Two forces have the same magnitude *F*. What is the angle between the two vectors if their sum has a magnitude of a) 2*F*? b) $\sqrt{2}\,F$? c) zero? Sketch the three vectors in each case.

4–2 Instead of using the *x*- and *y*-axes of Fig. 4–5 to analyze the situation of Example 4–1, use axes rotated 30.0° counter-clockwise, so the *x*-axis is parallel to the 200-N force. a) For these axes find the *x*- and *y*-components of the net force on the coat. b) From the components computed in part (a) find the magnitude and direction of the net force. Compare your results to Example 4–1.

4–3 A warehouse worker pushes a crate along the floor as in Fig. 4–1b by a force of 10 N that points downward at an angle of

45° below the horizontal. Find the horizontal and vertical components of the force.

4–4 A man is dragging a trunk up the loading ramp of a mover's truck. The ramp has a slope angle of 20.0°, and the man pulls upward with a force $\vec{F}$ whose direction makes an angle of 30.0° with the ramp (Fig. 4–26). a) How large a force $\vec{F}$ is necessary for the component F_x parallel to the ramp to be 60.0 N? b) How large will the component F_y then be?

FIGURE 4–26 Exercise 4–4.

4–5 Two dogs pull horizontally on ropes attached to a post; the angle between the ropes is 60.0°. If dog A exerts a force of 270 N and dog B exerts a force of 300 N, find the magnitude of the resultant force and the angle it makes with dog A's rope.

4–6 Two forces, $\vec{F_1}$ and $\vec{F_2}$, act at a point. The magnitude of $\vec{F_1}$ is 9.00 N, and its direction is 60.0° above the x-axis in the second quadrant. The magnitude of $\vec{F_2}$ is 6.00 N, and its direction is 53.1° below the x-axis in the third quadrant. a) What are the x- and y-components of the resultant force? b) What is the magnitude of the resultant force?

SECTION 4–4 NEWTON'S SECOND LAW

4–7 If a net horizontal force of 132 N is applied to a person with mass 60 kg who is resting on the edge of a swimming pool, what horizontal acceleration is produced?

4–8 What magnitude of net force is required to give a 135-kg refrigerator an acceleration of magnitude 1.40 m/s²?

4–9 A box rests on a frozen pond, which serves as a frictionless horizontal surface. If a fisherman applies a horizontal force with magnitude 48.0 N to the box and produces an acceleration of magnitude 3.00 m/s², what is the mass of the box?

4–10 A dockworker applies a constant horizontal force of 80.0 N to a block of ice on a smooth horizontal floor. The frictional force is negligible. The block starts from rest and moves 11.0 m in 5.00 s. a) What is the mass of the block of ice? b) If the worker stops pushing at the end of 5.00 s, how far does the block move in the next 5.00 s?

4–11 A hockey puck with mass 0.160 kg is at rest at the origin ($x = 0$) on the horizontal frictionless surface of the rink. At time $t = 0$ a player applies a force of 0.250 N to the puck, parallel to

the x-axis; he continues to apply this force until $t = 2.00$ s. a) What are the position and speed of the puck at $t = 2.00$ s? b) If the same force is again applied at $t = 5.00$ s, what are the position and speed of the puck at $t = 7.00$ s?

4–12 A crate with mass 32.5 kg initially at rest on a warehouse floor is acted on by a net horizontal force of 140 N. a) What acceleration is produced? b) How far does the crate travel in 10.0 s? c) What is its speed at the end of 10.0 s?

4–13 A hockey puck moves from point A to point B at constant velocity while under the influence of several forces. a) What can you say about the forces? b) Graph the puck's path from A to B. c) On the graph, continue the path to point C if a new constant force is added to the puck at point B, where the new force is perpendicular to the puck's velocity at point B. d) Continue the path to point D if at point C the constant force added at B is replaced by one of constant magnitude but a direction always perpendicular to the path of the puck.

4–14 An electron (mass = 9.11×10^{-31} kg) leaves one end of a TV picture tube with zero initial speed and travels in a straight line to the accelerating grid, which is 1.80 cm away. It reaches the grid with a speed of 3.00×10^6 m/s. If the accelerating force is constant, compute a) the acceleration; b) the time to reach the grid; c) the net force, in newtons. (The gravitational force on the electron may be neglected.)

SECTION 4–5 MASS AND WEIGHT

4–15 Superman throws a 2400-N boulder at an adversary. What horizontal force must Superman apply to the boulder to give it a horizontal acceleration of 12.0 m/s²?

4–16 A bowling ball weighs 71.2 N (16.0 lb). The bowler applies a horizontal force of 160 N (36.0 lb) to the ball. What is the magnitude of the horizontal acceleration of the ball?

4–17 At the surface of Jupiter's moon Io the acceleration due to gravity is $g = 1.81$ m/s². A watermelon weighs 44.0 N at the surface of the earth. a) What is its mass on the earth's surface? b) What are its mass and weight on the surface of Io?

4–18 a) What is the mass of a book that weighs 3.20 N at a point where $g = 9.80$ m/s²? b) At the same location, what is the weight of a dog whose mass is 14.0 kg?

SECTION 4–6 NEWTON'S THIRD LAW

4–19 World-class sprinters can accelerate out of the starting blocks with an acceleration that is nearly horizontal and has magnitude 15 m/s². How much horizontal force must a 55-kg sprinter exert on the starting blocks during a start to produce this acceleration? Which body exerts the force that propels the sprinter: the blocks or the sprinter herself?

4–20 Imagine that you are holding a book weighing 4 N at rest on the palm of your hand. Complete the following sentences: a) A downward force of magnitude 4 N is exerted on the book by _____. b) An upward force of magnitude _____ is exerted on _____ by the hand. c) Is the upward force in part (b) the reaction to the downward force in part (a)? d) The reaction to the force in part (a) is a force of magnitude _____, exerted on _____ by _____. Its direction is _____. e) The reaction to the force in part (b) is a force of magnitude _____, exerted on _____ by _____. Its direction is _____. f) The forces in parts (a) and (b) are equal and opposite because of Newton's _____ law. g) The forces in parts (b) and (e) are equal and opposite because of Newton's _____ law. Now suppose that you exert an upward force of magnitude 5 N on the book. h) Does the book remain in equilibrium? i) Is the force exerted on the book by your hand equal and opposite to the force exerted on the book by the earth? j) Is the force exerted on the book by the earth equal and opposite to the force exerted on the earth by the book? k) Is the force exerted on the book by your hand equal and opposite to the force exerted on your hand by the book? Finally, suppose that you snatch your hand away while the book is moving upward. l) How many forces then act on the book? m) Is the book in equilibrium?

4–21 A bottle is given a push along a table top and slides off the edge of the table. Do *not* neglect air resistance. a) What forces are exerted on the bottle while it is falling from the table to the floor? b) What is the reaction to each force; that is, on what body and by what body is the reaction exerted?

4–22 The upward normal force exerted by the floor is 620 N on an elevator passenger who weighs 650 N. What are the reaction forces to these two forces? Is the passenger accelerating? If so, what are the magnitude and direction of the acceleration?

4–23 A student with mass 45 kg jumps off a high diving board. Using 6.0×10^{24} kg for the mass of the earth, what is the acceleration of the earth toward her as she accelerates toward the earth with an acceleration of 9.8 m/s²? Assume that the net force on the earth is the force of gravity she exerts on it.

SECTION 4–7 USING NEWTON'S LAWS

4–24 An astronaut is tethered by a strong cable to a spacecraft. The astronaut and her spacesuit have a total mass of 105 kg, while the mass of the cable is negligible. The mass of the spacecraft is 9.05×10^4 kg. The spacecraft is far from any large astronomical bodies, so we can neglect the gravitational forces on it and the astronaut. We also assume that both the spacecraft and the astronaut are initially at rest in an inertial reference frame. The astronaut then pulls on the cable with a force of 80.0 N. a) What force does the cable exert on the astronaut? b) Since $\Sigma \vec{F} = m\vec{a}$, how can a "massless" ($m = 0$) cable exert a force? c) What is the astronaut's acceleration? d) What force does the cable exert on the spacecraft? e) What is the acceleration of the spacecraft?

4–25 A 4.80-kg bucket of water is accelerated upward by a cord of negligible mass whose breaking strength is 75.0 N. Find the maximum upward acceleration that can be given to the bucket without breaking the cord.

4–26 An elevator with mass m is moving upward and has an acceleration of magnitude $|\vec{a}|$. The mass of the supporting cable can be neglected. What is the tension in the supporting cable a) if the elevator is speeding up as it rises? b) if the elevator is slowing down as it rises?

4–27 Two crates, one with mass 4.00 kg and the other with mass 6.00 kg, sit on the frictionless surface of a frozen pond, connected by a light rope (Fig. 4–27). A woman wearing golf shoes (so she can get traction on the ice) pulls horizontally on the 6.00-kg crate with a force F that gives the crate an acceleration of 2.50 m/s². a) What is the magnitude of the force F? b) What is the tension T in the rope connecting the two crates?

4–28 Refer to Fig. 4–27. The crates are on a horizontal frictionless surface. The woman (still wearing golf shoes for traction) applies a horizontal force $F = 50.0$ N to the 6.00-kg crate. The ropes are of negligible mass. a) Draw a free-body diagram for the 4.00-kg crate, a free-body diagram for the 6.00-kg crate, and a free-body diagram for the woman. For each force, indicate what body exerts that force. b) What is the magnitude of the acceleration of the 6.00-kg crate? c) What is the tension T in the rope connecting the two crates?

FIGURE 4–27 Exercises 4–27 and 4–28.

4–29 A parachutist relies on air resistance (mainly on her parachute) to decrease her downward velocity. If she and her parachute have a mass of 55.0 kg and air resistance exerts a total upward force of 620 N on her and her parachute, what is her acceleration?

4–30 The position of a 2.75×10^5 N training helicopter under test is given by $\vec{r} = (0.020 \text{ m/s}^3)t^3\,\hat{\imath} + (2.2 \text{ m/s})t\,\hat{\jmath} - (0.060 \text{ m/s}^2)t^2\hat{k}$. Find the net force on the helicopter at $t = 5.0$ s.

4–31 An object with mass m moves along the x-axis. Its position as a function of time is given by $x(t) = At - Bt^3$, where A and B are constants. Calculate the net force on the object as a function of time.

PROBLEMS

4–32 A .22 rifle bullet, traveling at 350 m/s, strikes a block of soft wood, which it penetrates to a depth of 0.130 m. The mass of the bullet is 1.80 g. Assume a constant retarding force. a) How much time is required for the bullet to stop? b) What force, in newtons, does the wood exert on the bullet?

4–33 Two horses pull horizontally on ropes attached to a stump. The two forces $\vec{F}_1$ and $\vec{F}_2$ that they apply to the stump are such that the net (resultant) force $\vec{R}$ has a magnitude equal to that of $\vec{F}_1$ and makes an angle of 90° with $\vec{F}_1$ (Fig. 4–28; see next page). Let $F_1 = 1300$ N and $R = 1300$ N also. Find the magnitude of $\vec{F}_2$ and its direction (relative to $\vec{F}_1$).

4–34 A proud angler hangs her catch from a spring balance, which is supported from the roof of an elevator. a) If the elevator has an upward acceleration of 2.45 m/s² and the balance reads

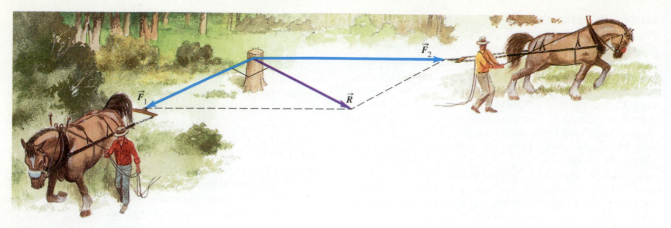

FIGURE 4–28 Problem 4–33.

50.0 N, what is the true weight of the fish? b) Under what circumstances will the balance read 30.0 N? c) What will the balance read if the elevator cable breaks?

4–35 Two adults and a child want to push a wheeled cart in the direction marked x in Fig. 4–29. The two adults push with horizontal forces $\vec{F}_1$ and $\vec{F}_2$ as shown in the figure. a) Find the magnitude and direction of the *smallest* force that the child should exert. Ignore the effects of friction. b) If the child exerts the minimum force found in part (a), the cart accelerates at 2.0 m/s^2 in the +x-direction. What is the weight of the cart?

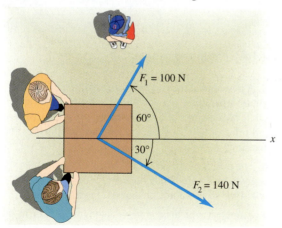

FIGURE 4–29 Problem 4–35.

4–36 An oil tanker's engines have broken down, and the wind is blowing the tanker straight toward a reef at a constant speed of 1.5 m/s (Fig. 4–30). When the tanker is 500 m from the reef, the wind dies down just as the engineer gets the engines going again. The rudder is stuck, so the only choice is to try to

FIGURE 4–30 Problem 4–36.

accelerate straight backward away from the reef. The mass of the tanker and cargo is 3.6×10^7 kg, and the engines produce a net horizontal force of 8.0×10^4 N on the tanker. Will the ship hit the reef? If it does, will the oil be safe? The hull can withstand an impact at a speed of 0.2 m/s or less. Neglect the retarding force of the water on the tanker's hull.

4–37 A Standing Vertical Jump. Basketball player Darrell Griffith is on record as attaining a standing vertical jump of 1.2 m (4 ft). (This means that he moved upward by 1.2 m after his feet left the floor.) If Griffith weighed 890 N (200 lb) and the time of the jump before his feet left the ground was 0.300 s, what was the average force he applied to the ground?

4–38 An advertisement claims that a particular automobile can "stop on a dime." What net force would actually be necessary to stop a 850-kg automobile traveling initially at 45.0 km/h in a distance equal to the diameter of a dime, which is 1.8 cm?

4–39 To study damage to aircraft that collide with large birds, you design a test gun that will accelerate chicken-sized objects so that their displacement along the gun barrel is given by $x = (9.0 \times 10^3 \text{ m/s}^2)t^2 - (8.0 \times 10^4 \text{ m/s}^3)t^3$. The object leaves the end of the barrel at $t = 0.025$ s. a) How long must the gun barrel be? b) What will be the speed of the objects as they leave the end of the barrel? c) What net force must be exerted on a 1.50-kg object at i) $t = 0$? ii) $t = 0.025$ s?

4–40 A spacecraft descends vertically near the surface of Planet X. An upward thrust of 25.0 kN from its engines slows it down at a rate of 1.20 m/s^2, but it speeds up at a rate of 0.80 m/s^2 with an upward thrust of 10.0 kN. What is the spacecraft's weight near the surface of Planet X?

4–41 A train (an engine plus four cars) running on level tracks is speeding up with an acceleration of magnitude $|\vec{a}|$. If each car has a mass m and if each car has negligible frictional forces acting on it, what is a) the force of the engine on the first car? b) the force of the first car on the second car? c) the force of the second car on the third car? d) the force of the fourth car on the third car? e) What would these same four forces be if the train were slowing down with an acceleration of magnitude $|\vec{a}|$? Your answer to each question must include a clearly labeled free-body diagram.

4–42 A gymnast of mass m climbs a vertical rope attached to the ceiling. The weight of the rope can be neglected. Calculate the tension in the rope if the gymnast a) climbs at a constant rate; b) hangs motionless on the rope; c) accelerates up the rope with an acceleration of magnitude $|\vec{a}|$; d) slides down the rope with a downward acceleration of magnitude $|\vec{a}|$.

4–43 A loaded elevator with very worn cables has a total mass of 2200 kg, and the cables can withstand a maximum tension of 28,000 N. a) What is the maximum upward acceleration for the elevator if the cables are not to break? b) What would be the answer to part (a) if the elevator were on the moon, where $g = 1.62$ m/s²?

4–44 Jumping to the Ground. A 75.0-kg man steps off a platform 3.10 m above the ground. He keeps his legs straight as he falls, but at the moment his feet touch the ground his knees begin to bend, and, treated as a particle, he moves an additional 0.60 m before coming to rest. a) What is his speed at the instant his feet touch the ground? b) Treating him as a particle, what is his acceleration as he slows down, if the acceleration is assumed to be constant? c) What force do his feet exert on the ground while he slows down? Express this force in newtons and also as a multiple of his weight.

4–45 A 4.9-N hammer head is stopped from an initial downward velocity of 3.2 m/s in a distance of 0.45 cm by a nail in a pine board. In addition to its weight, there is a 15-N downward force on the hammer head applied by the person using the hammer. Assume that the acceleration of the hammer head is constant while it is in contact with the nail and moving downward. a) Draw a free-body diagram for the hammer head. Identify the reaction force to each force in the diagram. b) Calculate the downward force $\vec{F}$ exerted by the hammer head on the nail while the hammer head is in contact with the nail and moving downward. c) Suppose the nail is in hardwood and the distance the hammer head travels in coming to rest is only 0.12 cm. The downward forces on the hammer head are the same as on part (b). What then is the force $\vec{F}$ exerted by the hammer head on the nail while the hammer head is in contact with the nail and moving downward?

4–46 A uniform cable of weight w hangs vertically downward, supported by an upward force of magnitude w at its top end. What is the tension in the cable a) at its top end? b) at its bottom end? c) at its middle? Your answer to each part must include a free-body diagram. (*Hint:* For each question choose the body to be analyzed to be a section of the cable or a point along the cable.) d) Graph the tension in the rope versus the distance from its top end.

4–47 The two blocks in Fig. 4–31 are connected by a heavy uniform rope with a mass of 4.00 kg. An upward force of 200 N is applied as shown. a) Draw three free-body diagrams, one for the 6.00-kg block, one for the 4.00-kg rope, and another one for the 5.00-kg block. For each force, indicate what body exerts that force. b) What is the acceleration of the system? c) What is the tension at the top of the heavy rope? d) What is the tension at the midpoint of the rope?

FIGURE 4–31 Problem 4–47.

4–48 If a beach ball with mass 0.0900 kg is thrown vertically upward in a vacuum, so there is no drag force on it, it reaches a height of 5.0 m. If the ball is thrown upward with the same initial velocity but in air instead of vacuum, its maximum height is 3.8 m. What is the average force exerted on the ball by the air during its upward motion?

*4–49 An object with mass m initially at rest is acted on by a force $\vec{F} = k_1\hat{\imath} + k_2 t^3\hat{\jmath}$, where k_1 and k_2 are constants. Calculate the velocity $\vec{v}(t)$ of the object as a function of time.

CHALLENGE PROBLEMS

*4–50 If we know $F(t)$, the force as a function of time, for straight-line motion, Newton's second law gives us $a(t)$, the acceleration as a function of time. We can then integrate $a(t)$ to find $v(t)$ and $x(t)$. However, suppose we know $F(v)$ instead. a) The net force on a body moving along the x-axis equals $-Cv^2$. Use Newton's second law written as $\Sigma F = m \, dv/dt$ and two integrations to show that $x - x_0 = (m/C) \ln (v_0/v)$. b) Show that Newton's second law can be written as $\Sigma F = mv \, dv/dx$. Derive the same expression as in part (a) using this form of the second law and one integration.

*4–51 An object of mass m is at rest in equilibrium at the origin. At $t = 0$ a new force $\vec{F}(t)$ is applied that has components

$$F_x(t) = k_1 + k_2 y, \qquad F_y(t) = k_3 t,$$

where k_1, k_2, and k_3 are constants. Calculate the position $\vec{r}(t)$ and velocity $\vec{v}(t)$ vectors as functions of time.

5

Applications of Newton's Laws

A bird glides by making use of Newton's first and third laws. The wings push downward on the air as it flows past the bird, and the air reacts by pushing upward on the wings with a force— called *lift*—of equal magnitude but opposite direction. In a steady glide the aerodynamic forces on the bird exactly balance the downward force of gravity, so that the net external force on the bird is zero.

5–1 INTRODUCTION

Newton's three laws of motion, the foundation of classical mechanics, can be stated very simply, as we have seen. But *applying* these laws to situations such as an iceboat skating across a frozen lake, a toboggan sliding down a hill, a jet fighter making a steep turn, or even a weight hanging from the ceiling requires analytical skills and problem-solving technique. In this chapter we'll help you to extend the problem-solving skills you began to develop in the previous chapter.

We begin with equilibrium problems, in which a body is at rest or moving with constant velocity. Then we'll generalize our problem-solving techniques to include bodies that are not in equilibrium, for which we need to deal precisely with the relationships between forces and motion. We'll learn how to describe and analyze the contact force acting on a body when it rests or slides on a surface. Next we'll study the important case of uniform circular motion, in which a body moves in a circle with constant speed. We'll use the concept of force throughout our study of physics, and we close the chapter with a brief look at the fundamental nature of force and the classes of forces found in nature.

5–2 USING NEWTON'S FIRST LAW: PARTICLES IN EQUILIBRIUM

We learned in Chapter 4 that a body is in *equilibrium* when it is at rest or moving with constant velocity in an inertial frame of reference. A hanging lamp, a rope and pulley setup for hoisting heavy loads, a suspension bridge—all are examples of equilibrium situations. In this section we consider only equilibrium of a body that can be modeled as a particle. (In Chapter 11 we'll see what to do when the body can't be represented adequately as a particle.) The essential physical principle is Newton's first law: When a particle is at rest or is moving with constant velocity in an inertial frame of reference, the net force acting on it—that is, the vector sum of all the forces acting on it—must be zero:

$$\Sigma \vec{F} = 0 \qquad \text{(particle in equilibrium).} \qquad (5\text{–}1)$$

We will usually use this in component form:

$$\Sigma F_x = 0, \qquad \Sigma F_y = 0 \qquad \text{(particle in equilibrium).} \qquad (5\text{–}2)$$

This section is about using Newton's first law to solve problems dealing with bodies in equilibrium. The important thing to remember is that *all* such problems are done in the same way. The following strategy details the steps you need to follow for any and all such problems. Study this strategy carefully, look at how it's applied in the worked-out examples, and try to apply it yourself when you solve problems.

Problem–Solving Strategy

EQUILIBRIUM OF A PARTICLE

1. Draw a simple sketch of the physical situation, showing dimensions and angles.

2. Choose some body that is in equilibrium, and draw a free-body diagram of this body. For the present we will consider it as a particle, so a large dot will do to represent the body. In your free-body diagram, *do not* include the other bodies that interact with it, such as a surface it may be resting on, or a rope pulling on it.

3. Now ask yourself what is interacting with the body by touching it or in any other way. On your free-body diagram, draw a force vector for each of the interactions. If you know the angle at which a force is directed, draw the angle accurately and label it. A surface in contact with the body exerts a normal force perpendicular to the surface and possibly a friction force parallel to the surface. Remember that a rope or chain can't push on a body, but can only pull in a direction along its length. Be sure to include the body's weight, except in the case in which the body has negligible mass (and hence negligible weight). If the mass is given, use $w = mg$ to find the weight. Label each force with a symbol representing the *magnitude* of the force and the numerical value if it's given. Make sure you can answer the question "What other body causes that force?" for each force. If you can't answer that question, you may be imagining a force that isn't there.

4. *Do not* show in the free-body diagram any forces exerted *by* the body on any other body. The sums in Eqs. (5–1) and (5–2) include only forces that act *on* the body.

5. Choose a set of coordinate axes and represent each force acting on the body by its components along these axes. Draw a wiggly line through each force vector that you replace by its components so that you don't count it twice. Often you can simplify the problem by your choice of coordinate axes. For example, when a body rests or slides on a plane surface, it's usually simplest to take the axes in the directions parallel and perpendicular to this surface, even when the plane is tilted.

6. Set the algebraic sum of all x-components of force equal to zero. In a separate equation, set the algebraic sum of all y-components equal to zero. (*Never* add x- and y-components in a single equation.) You can then solve these equations for up to two unknown quantities, which may be force magnitudes, components, or angles.

7. If there are two or more bodies, repeat Steps 2 through 6 for each body. If the bodies interact with each other, use Newton's third law to relate the forces they exert on each other. You need to find as many independent equations as the number of unknown quantities. Then solve these equations to obtain the unknowns. This part is algebra, not physics, but it's an essential step.

8. Look at your results and ask whether they make sense. When the result is a symbolic expression or formula, try to think of special cases (particular values or extreme cases for the various quantities) for which you can guess what the results ought to be. Check to see that your formula works in these particular cases.

EXAMPLE 5–1

One-dimensional equilibrium A gymnast has just begun climbing up a rope hanging from a gymnasium ceiling (Fig. 5–1a). She stops, suspended from the lower end of the rope by her hands. Her weight is 500 N, and the weight of the rope is 100 N. Analyze the forces on the gymnast and on the rope. (This example is similar to Example 4–12 in Section 4–7; our purpose is to emphasize the steps in the problem-solving strategy presented above.)

SOLUTION Figure 5–1b is a free-body diagram for the gymnast. The forces acting on her are her weight (magnitude $w_G = 500$ N) and the upward tension force exerted on her by the rope, of magnitude $T_{R \, on \, G}$ (force exerted *by* the rope *on* the gymnast). We don't include the downward force she exerts on the rope because it isn't a force that acts *on* her. We take the y-axis vertically upward and the x-axis horizontal; there are no x-components of force. The rope pulls upward (in the positive y-direction), and the corresponding y-component of force is just the magnitude $T_{R \, on \, G}$, a positive (scalar) quantity. But the weight acts in the negative y-direction, and its y-component is the *negative* of the magnitude, that is, $-w_G = -500$ N. The algebraic sum of

y-components is $T_{R \, on \, G} + (-w_G)$, and from the equilibrium condition, Eq. (5–2), we have

$$\Sigma F_y = T_{R \, on \, G} + (-w_G) = 0,$$
$$T_{R \, on \, G} = w_G = 500 \text{ N}.$$

The tension at the bottom of the rope equals the gymnast's weight, as you probably expected.

The two forces acting on the gymnast have equal magnitude 500 N and are opposite in direction, but they are *not* an action-reaction pair. The reason is that the weight and the tension force in Fig. 5–1b both act on the gymnast, whereas action and reaction forces always act on different bodies. These two forces are equal and opposite because of Newton's first law, $\Sigma F_y = 0$, not Newton's third law. The gymnast's weight is the attractive (downward) force exerted on the gymnast by the earth. Its reaction force is the equal and opposite (upward) attractive force exerted on the *earth* by the *gymnast*. This force acts on the earth, not on the gymnast, so it does not appear in the free-body diagram for the gymnast. Compare the discussion of the apple in Example 4–9 (Section 4–6).

Figure 5–1c shows a free-body diagram for the rope. The

reaction to the upward force of magnitude 500 N exerted by the rope on the gymnast is a downward force exerted by the gymnast on the rope, as shown. According to Newton's third law, the magnitude $T_{\text{G on R}}$ of this downward force is also 500 N. The other forces on the rope are its own weight (magnitude 100 N) and the upward force (magnitude $T_{\text{C on R}}$) exerted by the ceiling on the upper end of the rope.

The y-component of force acting on the top end of the rope is $+T_{\text{C on R}}$, the y-component acting on the bottom end is $-T_{\text{G on R}} = -500$ N, and the y-component of the weight is $-w_R = -100$ N.

The equilibrium condition $\Sigma F_y = 0$ for the rope gives

$$\Sigma F_y = T_{\text{C on R}} + (-T_{\text{G on R}}) + (-w_R) = 0,$$

$$T_{\text{C on R}} = T_{\text{G on R}} + w_R = 500 \text{ N} + 100 \text{ N} = 600 \text{ N}.$$

This result makes sense; the tension is 100 N greater at the top of the rope (where it must support the combined weight of both the rope and the gymnast) than at the bottom (where it supports only the gymnast). Figure 5–1d is a partial free-body diagram for the ceiling, showing that the rope exerts a downward force $F_{\text{R on C}}$ on the ceiling. From Newton's third law the magnitude of this force is 600 N, the same as $T_{\text{C on R}}$.

(a)   (b)   (c)   (d)

5–1 (a) A gymnast hanging at rest from the end of a vertical rope. (b) Free-body diagram for the gymnast. (c) Free-body diagram for the rope. (d) The force exerted on the ceiling by the rope. (This is not a complete free-body diagram because it does not show the other forces acting on the ceiling.)

EXAMPLE 5–2

Two-dimensional equilibrium In Fig. 5–2a a car engine with weight w hangs from a chain that is linked at point O to two other chains, one fastened to the ceiling and the other to the wall. Find the tensions in these three chains, assuming that w is given and the weights of the chains themselves are negligible.

SOLUTION It may seem strange that we neglect the weight of the chains in this example, while in Example 5–1 we did *not* neglect the weight of a mere rope. The reason is that the weight of the chains is very small compared to the weight of the massive engine. By contrast, in Example 5–1 the weight of the rope was a reasonable fraction of the gymnast's weight (100 N compared to 500 N).

Figure 5–2b is a free-body diagram for the engine. The two forces acting on the engine are its weight and the upward force exerted by the vertical chain. Without further ceremony we can conclude that $T_1 = w$. The horizontal and slanted chains do not exert forces on the engine itself, because they are not attached to it, but they do exert forces on the ring where three chains join. So let's consider the *ring* as a particle in equilibrium; the weight of the ring itself is negligible.

In the free-body diagram for the ring (Fig. 5–2c), remember that T_1, T_2, and T_3 are the *magnitudes* of the forces; their directions are shown by the vectors on the diagram. An x-y coordinate axis system is also shown, and the force with magnitude T_3 has been resolved into its x- and y-components. Note that the vertical chain exerts forces of the same magnitude T_1 at both of its ends, upward on the engine in Fig. 5–2b and downward on the ring in

Fig. 5–2c. This is because the weight of the chain is negligible (see Example 4–11 in Section 4–7). If the weight were not negligible, these two forces would have different magnitudes, as was the case for the rope in Example 5–1.

We now apply the equilibrium conditions for the ring, writing separate equations for the x- and y-components. (Note that x- and y-components are *never* added together in a single equation.) We find

$$\Sigma F_x = 0: \quad T_3 \cos 60° + (-T_2) = 0;$$

$$\Sigma F_y = 0: \quad T_3 \sin 60° + (-T_1) = 0.$$

Because $T_1 = w$, we can rewrite the second equation as

$$T_3 = \frac{T_1}{\sin 60°} = \frac{w}{\sin 60°} = 1.155w.$$

We can now use this result in the first equation:

$$T_2 = T_3 \cos 60° = w\frac{\cos 60°}{\sin 60°} = 0.577w.$$

So we can express all three tensions as multiples of the weight w of the engine, which we assume is known. To summarize,

$$T_1 = w,$$

$$T_2 = 0.577w,$$

$$T_3 = 1.155w.$$

If the engine's weight is $w = 2200$ N (about 500 lb), then

$$T_1 = 2200 \text{ N},$$

$$T_2 = (0.577)(2200 \text{ N}) = 1270 \text{ N},$$

$$T_3 = (1.155)(2200 \text{ N}) = 2540 \text{ N}.$$

The chain attached to the ceiling exerts a force on the ring of

magnitude T_3, which is greater than the weight of the engine. If this seems strange, note that the *vertical* component of this force is equal to T_1, which in turn is equal to w. But since this force also has a horizontal component, its magnitude T_3 must be some-what *larger* than w. Hence the chain attached to the ceiling is under the greatest tension and is the one most susceptible to breaking.

(a)

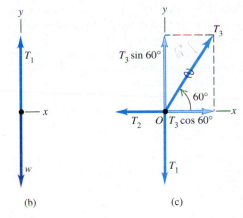

(b)   (c)

5–2 (a) A car engine with weight w is suspended from a chain linked at O to two other chains. The chains and ring are considered to be massless. (b) Free-body diagram for the engine. (c) Free-body diagram for the ring with $\vec{T}_3$ replaced by its components.

EXAMPLE 5–3

An inclined plane   A car rests on the slanted tracks of a ramp leading to a car-transporter trailer (Fig. 5–3a). The car's brakes and transmission lock are released; only a cable attached to the car and to the frame of the trailer prevents the car from rolling down the ramp. If the car's weight is w, find the tension in the cable and the force with which the tracks push on the car tires.

SOLUTION   Figure 5–3b shows a free-body diagram for the car. The three forces acting on the car are its weight (magnitude w), the tension in the cable (magnitude T), and the forces exerted on the wheels by the tracks; we have lumped the forces on the wheels together as a single force with magnitude n. This force must be normal (perpendicular) to the tracks, and hence we call it a *normal force* (we first used this term in Section 4–3). We use a special script letter n to avoid confusion with the abbreviation N for newton. (If this force had a component *along* the tracks, it would tend to prevent the car from moving along the tracks. We have assumed that there is no such effect.)

We take our coordinate axes parallel and perpendicular to the tracks, as shown. To find the components of the weight, we first note that the angle α between the plane and the horizontal is equal to the angle α between the weight vector $\vec{w}$ and the normal

(a)

(b)   (c)

5–3 (a) A cable holds the car on the ramp. (b) Free-body diagram for the car. (c) The weight component $w \sin \alpha$ acts down the plane, and $w \cos \alpha$ is perpendicular to the plane.

to the plane of the ramp. The angle α is *not* measured from the $+x$-axis toward the $+y$-axis, so we *cannot* use Eqs. (1–7) directly to find the components. (You may want to review Section 1–9 to make sure that you understand this important point.) Instead, consider the right triangles in Fig. 5–3c. The sine of α is the magnitude of the x-component of $\vec{w}$ (that is, the side of the triangle opposite the angle α) divided by the magnitude w (the hypotenuse of the triangle). Similarly, the cosine of α is the magnitude of the y-component (the side of the triangle adjacent to α) divided by w. Both components are negative, so $w_x = -w \sin \alpha$ and $w_y = -w \cos \alpha$.

We draw a wiggly line through the original vector representing the weight to remind us not to count it twice. The equilibrium conditions then give us

$$\Sigma F_x = 0: \qquad T + (-w \sin \alpha) = 0;$$
$$\Sigma F_y = 0: \qquad n + (-w \cos \alpha) = 0.$$

Be sure you understand how these signs are related to our choice of coordinates. Remember that by definition, T, w, and n are all *magnitudes* of vectors and are therefore all positive.

Solving these equations for T and n, we find

$$T = w \sin \alpha,$$
$$n = w \cos \alpha.$$

In general, the magnitude n of the normal force is *not* equal to the weight w.

To check some special cases, first note that if the angle α is zero, then $\sin \alpha = 0$ and $\cos \alpha = 1$. In this case the tracks are horizontal; no cable tension T is needed to hold the car, and the total normal force n is equal in magnitude to the weight. If the angle is 90°, then $\sin \alpha = 1$ and $\cos \alpha = 0$. Then the cable tension T equals the weight w, and the normal force n is zero. Are these the results you would expect for these particular cases?

As a final note, ask yourself how the answers for T and n would be affected if the car were not stationary but were being pulled up the ramp at a constant speed. You should immediately realize that this, too, is an equilibrium situation, since the car's velocity is constant. So the calculation is exactly the same, and T and n have the same values as when the car is at rest. (It's true that T must be greater than $w \sin \alpha$ to *start* the car moving up the ramp, but that's not what we asked.)

EXAMPLE 5-4

Tension over a frictionless pulley Blocks of granite are being hauled up a 15° slope out of a quarry. For environmental reasons, dirt is also being dumped into the quarry to fill up old holes. You have been asked to find a way to use this dirt to move the granite out more easily. You design a system in which a granite block on a cart with steel wheels (weight w_1, including the cart) is pulled uphill on steel rails by a bucket of dirt (weight w_2, including the bucket) dropping vertically into the quarry (Fig. 5–4a). Ignoring friction in the pulley and wheels and the weight of the cable, determine how the weights w_1 and w_2 must be related in order for the system to move with constant speed.

SOLUTION Our idealized model for the system is shown in Fig. 5–4b. Figure 5–4c is the free-body diagram for the dirt and bucket, and Fig. 5–4d is the free-body diagram for the granite block and cart. We have drawn an axis system for each body. Note that we are at liberty to orient the axes differently for each body; the choices shown are the most convenient ones. We have represented the weight of the granite block in terms of its components in the chosen axis system. These components are found in the same way as in Example 5–3. We assume the weight of the cable to be negligible so that the tension T is the same throughout the cable.

5–4 (a) A bucket of dirt pulls up a cart carrying a granite block. (b) Idealized model of the system. (c) Free-body diagram for the dirt and bucket. (d) Free-body diagram for the block and cart.

Applying $\Sigma F_y = 0$ to the dirt and bucket in Fig. 5–4c, we find

$$T + (-w_2) = 0, \qquad T = w_2.$$

Applying $\Sigma F_x = 0$ to the block and cart in Fig. 5–4d, we get

$$T + (-w_1 \sin 15°) = 0, \qquad T = w_1 \sin 15°.$$

Equating the two expressions for T, we find

$$w_2 = w_1 \sin 15° = 0.26w_1.$$

If the weight of dirt and bucket totals 26% of the weight of the granite block and cart, the system can move with constant speed in either direction (our analysis doesn't depend on the direction of motion, only on the velocity being constant). Notice that we didn't need to apply the equation $\Sigma F_y = 0$ to the cart and block; this would be useful only if we wanted to find the value of n. Can you show that $n = w_1 \cos 15°$?

5–3 USING NEWTON'S SECOND LAW: DYNAMICS OF PARTICLES

ActivPhysics

2.1.5
Car Race

2.2
Lifting a Crate

2.3
Lowering a Crate

We are now ready to discuss *dynamics* problems, in which we apply Newton's second law to bodies that are accelerating and hence are *not* in equilibrium. In this case the net force on the body is not zero, but is equal to the mass of the body times its acceleration:

$$\Sigma \vec{F} = m\vec{a} \qquad \text{(Newton's second law).} \qquad (5–3)$$

We will usually use this relation in component form:

$$\Sigma F_x = ma_x, \qquad \Sigma F_y = ma_y \qquad \text{(Newton's second law).} \qquad (5–4)$$

The following problem-solving strategy is very similar to our strategy for equilibrium problems in Section 5–2. We urge you to study it carefully, watch how it's applied in our examples, and use it when you tackle the end-of-chapter problems. Remember that *all* dynamics problems can be solved by using this strategy.

CAUTION ▶ We emphasize again that the quantity $m\vec{a}$ is *not* a force; it's not a push or a pull exerted by anything in the body's environment. All that Eqs. (5–3) and (5–4) say is that the acceleration $\vec{a}$ is proportional to the net force $\Sigma \vec{F}$. When you draw the free-body diagram for a body not in equilibrium, make sure you *never* include "the $m\vec{a}$ force," because *there is no such force* (Fig. 5–5). Sometimes we will draw the acceleration vector $\vec{a}$ *alongside* a free-body diagram; the acceleration will *never* be drawn with its tail touching the body (a position reserved for the forces acting on the body). ◀

5–5 The quantity $m\vec{a}$ is *not* a force. It must never be included in free-body diagrams.

Problem–Solving Strategy

NEWTON'S SECOND LAW

1. Sketch the physical situation. Identify one or more moving bodies to which you will apply Newton's second law.

2. Draw a free-body diagram for each chosen body. Be sure to include all the forces acting *on* the body, but be equally careful *not* to include any force exerted *by* the body on some other body. Never include the quantity $m\vec{a}$ in your free-body diagram; it's not a force! Label the magnitude of each force with an algebraic symbol and the numerical value if it's given. Usually, one of the forces will be the body's weight; it is usually best to label this as $w = mg$. If a numerical value of mass is given, you can compute the corresponding weight.

3. Show your coordinate axes explicitly in the free-body diagram, and then determine components of forces with reference to these axes. If you know the direction of the acceleration, it is usually best to take that direction as one of the axes. When you represent a force in terms of its components, draw a wiggly line through the original

force vector to remind you not to include it twice. When there are two or more bodies, you can use a separate axis system for each body; you don't have to use the same axis system for all the bodies. But in the equations for each body, the signs of the components *must* be consistent with the axes you have chosen for that body.

4. Write the equations for Newton's second law, Eqs. (5–4), using a separate equation for each component.

5. If more than one body is involved, repeat Steps 2 through 4 for each body. There may be relationships among the motions of the bodies; for example, they may be connected by a rope. Express any such relationships in algebraic form as relations between the accelerations of the various bodies. Then solve the equations to find the required unknowns.

6. Check particular values or extreme cases of quantities, when possible, and compare the results with your intuitive expectations. Ask, "Does this result make sense?"

EXAMPLE 5–5

Acceleration in one dimension An iceboat is at rest on a frictionless horizontal surface (Fig. 5–6a). What constant horizontal force F do we need to apply (along the direction of the runners) to give it a velocity of 6.0 m/s (about 22 km/h, or 13 mi/h) at the end of 4.0 s? The mass of iceboat and rider is 200 kg.

SOLUTION The forces acting on the iceboat and rider are the weight, the normal force exerted by the surface, and the horizontal force F. Figure 5–6b shows a free-body diagram and a coordinate system. We can find the unknown force by using Eqs. (5–4) if we can find the acceleration, so let's start there. The y-component of acceleration is zero, and we can get the x-component from the velocity data. The forces are all constant, so a_x is also constant, and we can use one of the constant-acceleration formulas, Eq. (2–8). Since the iceboat starts from rest, we find

$$a_x = \frac{v - v_0}{t} = \frac{6.0 \text{ m/s} - 0 \text{ m/s}}{4.0 \text{ s}} = 1.5 \text{ m/s}^2.$$

The sum of the x-components of force is simply

$$\Sigma F_x = F,$$

and Newton's second law gives

$$\Sigma F_x = F = ma_x,$$
$$F = (200 \text{ kg})(1.5 \text{ m/s}^2) = 300 \text{ kg} \cdot \text{m/s}^2$$
$$= 300 \text{ N}   \text{(about 75 lb).}$$

Note that we did not need the y-components at all to find F. We can use these to find the normal force n:

$$a_y = 0,$$
$$\Sigma F_y = n + (-mg) = ma_y = 0,$$
$$n = mg = (200 \text{ kg})(9.8 \text{ m/s}^2) = 1960 \text{ N}   \text{(about 440 lb).}$$

In this situation the magnitude n of the normal force is equal to the weight of the iceboat and rider because the vertical acceleration is zero, the surface is horizontal, and these are the only vertical forces acting. As we will see in Examples 5–8 and 5–9, there are many common situations in which the normal force can be quite different from the weight.

(a)   (b)

5–6 (a) An iceboat starting from rest. (b) Free-body diagram for the iceboat and rider with no friction.

EXAMPLE 5–6

In the situation of Example 5–5, suppose the motion of the iceboat is opposed by a constant horizontal friction force with magnitude 100 N. Now what force F must we apply to give the iceboat a velocity of 6.0 m/s at the end of 4.0 s?

SOLUTION The acceleration is the same as before, $a_x = 1.5$ m/s^2. The new free-body diagram is shown in Fig. 5–7; the difference between it and Fig. 5–6b is the addition of the friction force $\vec{f}$. (Note that its *magnitude*, $f = 100$ N, is a positive quantity but that its *component* in the x-direction is negative, equal to $-f$ or -100 N.) Now Newton's second law gives

$$\Sigma F_x = F + (-f) = ma_x,$$
$$F = ma_x + f = (200 \text{ kg})(1.5 \text{ m/s}^2) + (100 \text{ N}) = 400 \text{ N}.$$

We need 100 N to overcome friction and 300 N more to give the iceboat the necessary acceleration.

5–7 Free-body diagram for the iceboat and rider with a frictional force $\vec{f}$ opposing the motion.

EXAMPLE 5–7

Tension in an elevator cable An elevator and its load have a total mass of 800 kg (Fig. 5–8a). The elevator is originally moving downward at 10.0 m/s; it is brought to rest with constant acceleration in a distance of 25.0 m. Find the tension T in the supporting cable while the elevator is being brought to rest.

SOLUTION The only forces acting on the elevator are its weight and the tension force of the cable. The acceleration vector is also shown in Fig. 5–8b; we draw it off to one side because it's not a force. We can use Eqs. (5–4) to find T if we can first find the acceleration. The simplest way to do this is to use the constant-acceleration formula $v^2 = v_0^2 + 2a_y(y - y_0)$. Taking the positive y-axis upward, we have $v_0 = -10.0$ m/s, $v = 0$, and $y - y_0 = -25.0$ m. Then

$$a_y = \frac{v^2 - v_0^2}{2(y - y_0)} = \frac{(0)^2 - (-10.0 \text{ m/s})^2}{2(-25.0 \text{ m})} = +2.00 \text{ m/s}^2.$$

Note that the velocity is downward and the acceleration is upward, corresponding to downward motion with decreasing speed.

Now we are ready to use Newton's second law:

$$\Sigma F_y = T + (-w) = ma_y,$$
$$T = w + ma_y = mg + ma_y = m(g + a_y)$$
$$= (800 \text{ kg})(9.80 \text{ m/s}^2 + 2.00 \text{ m/s}^2) = 9440 \text{ N}.$$

The tension must be 1600 N *greater* than the weight ($w = mg$ = 7840 N) to stop the elevator in the required distance.

Can you see that we would get the same answer for T if the elevator were moving upward and gaining speed at a rate of 2.00 m/s² ?

(a)   (b)

5–8 (a) A loaded elevator moving downward being brought to rest. (b) Free-body diagram for the elevator. Note that we draw the acceleration vector (in yellow) to one side of the free-body diagram; unlike tension and weight, acceleration is *not* a force.

EXAMPLE 5–8

Apparent weight in an accelerating elevator A 50.0-kg woman stands on a bathroom scale while riding in the elevator in Example 5–7 (Fig. 5–9a). What is the reading on the scale?

SOLUTION The scale reads the magnitude of the downward force exerted *by* the passenger *on* the scale; by Newton's third law, this equals the magnitude of the upward normal force exerted *by* the scale *on* the passenger. Hence we can solve the problem by finding the magnitude n of the normal force.

Figure 5–9b shows a free-body diagram for the passenger. The forces acting on her are her weight $w = mg$ = (50.0 kg)(9.80 m/s²) = 490 N and the normal force n exerted by the scale. The passenger's acceleration is the same as the elevator's, and Newton's second law gives

$$\Sigma F_y = n + (-mg) = ma_y,$$
$$n = m(g + a_y) = (50.0 \text{ kg})(9.80 \text{ m/s}^2 + 2.00 \text{ m/s}^2)$$
$$= 590 \text{ N}.$$

While the elevator is stopping, the scale pushes up on the passenger with a force of 590 N. By Newton's third law, she pushes down on the scale with the same force; so the scale reads 590 N, which is 100 N more than her actual weight. The scale reading is called the passenger's **apparent weight**. The passen-

ger *feels* a greater strain in her legs and feet than when the elevator is stationary or moving with constant velocity because the floor is pushing up harder on her feet. You can feel this yourself; try taking a few steps in an elevator that is coming to a stop after descending.

(a)   (b)

5–9 (a) A passenger in the elevator as it slows down. (b) Free-body diagram for the passenger.

Astronauts in orbit feel as though they are weightless because they have the same acceleration as their spacecraft—*not* because they are "outside the pull of the earth's gravity." (If they were, the astronauts and their spacecraft could not stay in orbit, but would fly off into deep space.)

Let's generalize the result of Example 5–8. A passenger with mass m rides in an elevator with y-acceleration a_y. We saw above that a scale shows the passenger's apparent weight to be

$$n = m(g + a_y).$$

When a_y is positive, the elevator is accelerating upward (moving upward with increasing speed, or moving downward with decreasing speed) and n is greater than the passenger's weight $w = mg$. When the elevator is accelerating downward (moving upward with decreasing speed, or moving downward with increasing speed), a_y is negative and n is less than the weight. If the passenger doesn't know the elevator is accelerating, she may feel as though her weight is changing; indeed, this is just what the scale shows. With upward acceleration the apparent weight is greater than the true weight, and the passenger feels heavier; with downward acceleration she feels lighter.

The extreme case occurs when the elevator has a downward acceleration $a_y = -g$, that is, when it is in free fall. In that case, $n = 0$ and the passenger *seems* to be weightless. Similarly, an astronaut orbiting the earth in a spacecraft experiences *apparent* weightlessness. In each case the person is not really weightless because there is still a gravitational force acting. But the effect of this free-fall condition is just the same as though the body were in outer space with no gravitational force at all. In both cases the person and their vehicle (elevator or spacecraft) are falling together with the same acceleration g, so nothing pushes the person against the floor or walls of the vehicle.

The physiological effect of prolonged apparent weightlessness is an interesting medical problem that is being actively explored. Gravity plays an important role in blood distribution in the body; one reaction to apparent weightlessness is a decrease in blood volume through increased excretion of water. In some cases, astronauts returning to earth have experienced temporary impairment of their sense of balance and a greater tendency toward motion sickness.

EXAMPLE 5–9

Acceleration down a hill A toboggan loaded with vacationing students (total weight w) slides down a long, snow-covered slope (Fig. 5–10a). The hill slopes at a constant angle α, and the toboggan is so well waxed that there is virtually no friction. What is the toboggan's acceleration?

SOLUTION The only forces acting on the toboggan are its weight w and the normal force n exerted by the hill (Fig. 5–10b). We take axes parallel and perpendicular to the surface of the hill and resolve the weight into x- and y-components: $w_x = w \sin \alpha$ and $w_y = -w \cos \alpha$. (Compare to Example 5–3, in which the

5–10 (a) A loaded toboggan slides down a frictionless hill. (b) The free-body diagram shows that the x-component of the weight, $w \sin \alpha$, accelerates the toboggan down the hill.

x-component of weight was $-w \sin \alpha$. The difference is that the positive *x*-axis is uphill in Example 5–3, while in Fig. 5–10b it is downhill.) Newton's second law in the *x*-direction then gives

$$\Sigma F_x = w \sin \alpha = ma_x,$$

and since $w = mg$,

$$a_x = g \sin \alpha.$$

The mass does not appear in the final result. This means that *any* toboggan, regardless of its mass or number of passengers, slides down a frictionless hill with an acceleration of $g \sin \alpha$. If the plane is horizontal, $\alpha = 0$ and $a_x = 0$ (the toboggan does not accelerate); if the plane is vertical, $\alpha = 90°$ and $a_x = g$ (the toboggan is in free fall).

Once again, we did not need the *y*-components to find the acceleration; let's use them to find the normal force n exerted on the toboggan by the surface of the hill. We know that $a_y = 0$, so $\Sigma F_y = 0$ and

$$n = mg \cos \alpha.$$

This force is *not* equal to the toboggan's weight (compare Example 5–3 in Section 5–2). The moral is that you should *never* automatically assume that the normal force is equal to the weight. Very often, as in this example, their values are different.

CAUTION ▶ Figure 5–11 shows a common *incorrect* way to draw the free-body diagram for the toboggan. This diagram is wrong for two reasons: the normal force should be drawn perpendicular to the surface, and it's completely bogus to include an "$m\vec{a}$ force." If you remember that "normal" means "perpendicular" and that the mass times the acceleration is *not* itself a force, you won't make these mistakes. ◀

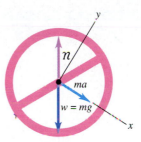

5–11 The *wrong* way to draw the free-body diagram for the toboggan. The normal force points in the wrong direction, and $m\vec{a}$ shouldn't be included because it isn't a force.

EXAMPLE 5–10

Two bodies with the same acceleration A robot arm pulls a 4.0-kg cart along a horizontal frictionless track with a 0.50-kg rope, applying a horizontal force with magnitude $F = 9.0$ N to the rope (Fig. 5–12a; see page 130). Find the acceleration of the system and the tension at the point where the rope is fastened to the cart. (On earth the rope would sag a little; to avoid this complication, suppose the robot arm is operating in a zero-gravity space station.)

SOLUTION We choose the *x*-axis in the direction of acceleration, so only *x*-components are relevant. The cart and the rope move together in the same direction and so have the same acceleration a in the *x*-direction. We can proceed in either of two ways.

Method 1: We write Newton's second law for the cart and a separate equation for Newton's second law for the rope, using a and T for the unknown acceleration and tension (Fig. 5–12b). For the cart,

$$\Sigma F_x = T = m_{cart} a,$$

and for the rope,

$$\Sigma F_x = F + (-T) = m_{rope} a.$$

Note the role of Newton's third law in equating the magnitudes of the two forces labeled T; one is the force on the cart exerted by the rope, the other is the force on the rope exerted by the cart. We now have two simultaneous equations for T and a. An easy way to solve them is to add the two equations; this eliminates T, giving

$$F = m_{cart} a + m_{rope} a = (m_{cart} + m_{rope})a,$$

and

$$a = \frac{F}{m_{cart} + m_{rope}} = \frac{9.0 \text{ N}}{4.0 \text{ kg} + 0.50 \text{ kg}} = 2.0 \text{ m/s}^2.$$

Substituting this back into the first equation,

$$T = m_{cart} a = (4.0 \text{ kg})(2.0 \text{ m/s}^2) = 8.0 \text{ N}.$$

The tension is different at the two ends of the rope; $T = 8.0$ N on the right, $F = 9.0$ N on the left. This is because the rope has mass, and so there must be a net force acting on the rope in order to accelerate it. This net force is equal to $F - T = 9.0$ N $- 8.0$ N $= 1.0$ N, exactly enough to accelerate a 0.50-kg rope at 2.0 m/s².

Method 2: Take the cart and rope as a composite body with a total mass $m = m_{cart} + m_{rope} = 4.5$ kg (Fig. 5–12c). The only force acting on this composite body in the *x*-direction is F, so

$$a = \frac{F}{m} = \frac{9.0 \text{ N}}{4.5 \text{ kg}} = 2.0 \text{ m/s}^2.$$

Then, looking at the 4.0-kg cart by itself, we see that to give it an acceleration of 2.0 m/s² requires a force

$$T = m_{cart} a = (4.0 \text{ kg})(2.0 \text{ m/s}^2) = 8.0 \text{ N}.$$

CAUTION ▶ The method of treating the two bodies as a single composite body works here because the two bodies have the same magnitude *and* direction of acceleration. If the accelerations are different, you should treat the two bodies separately, as in the following example. ◀

5–12 (a) A robot arm pulling a cart along a frictionless track in a zero-gravity space station. (b) Individual free-body diagrams for the rope and cart. (c) Free-body diagram for the rope and cart as a composite body.

EXAMPLE 5–11

Two bodies with the same magnitude of acceleration In Fig. 5–13a an air-track glider with mass m_1 moves on a level, frictionless air track in the physics lab. It is connected to a lab weight with mass m_2 by a light, flexible, nonstretching string that passes over a small frictionless pulley. Find the acceleration of each body and the tension in the string.

SOLUTION The two bodies have different motions—one horizontal, one vertical—so we won't consider them together as we did the bodies in Example 5–10. We use a separate free-body diagram and $\Sigma \vec{F} = m\vec{a}$ equations for each. Figures 5–13b and 5–13c show a free-body diagram and a coordinate system for each body. There is no friction in the pulley, and we consider the string to be massless, so the tension T in the string is the same throughout; it applies a force of magnitude T to each body. The weights are m_1g and m_2g. For the glider on the track, Newton's second law gives

$$\Sigma F_x = T = m_1 a_{1x},$$
$$\Sigma F_y = n + (-m_1 g) = m_1 a_{1y} = 0.$$

For the lab weight it's convenient to take the $+y$-direction as downward so that both bodies accelerate in positive axis directions. Then for the lab weight,

$$\Sigma F_y = m_2 g + (-T) = m_2 a_{2y}.$$

We note again that it is perfectly all right to use different coordinate axes for the two bodies.

If the string doesn't stretch, the two bodies must move equal distances in equal times, and so their speeds at any instant must be equal. When the speeds change, they change by equal amounts in a given time, so the accelerations of the two bodies must have the same magnitude a. We can express this relation as

$$a_{1x} = a_{2y} = a.$$

The *directions* of the two accelerations are different, but their *magnitudes* are the same. The two equations for Newton's second law are then

$$T = m_1 a \qquad \text{(glider)},$$
$$m_2 g + (-T) = m_2 a \qquad \text{(lab weight)}.$$

These are two simultaneous equations for the unknowns T and a. We add the two equations to eliminate T, getting

$$m_2 g = m_1 a + m_2 a = (m_1 + m_2)a$$

and

$$a = \frac{m_2}{m_1 + m_2}\, g.$$

The acceleration is less than g, as you might expect. Substituting this back into the first equation, we get

$$T = \frac{m_1 m_2}{m_1 + m_2} \, g.$$

We see that the tension T is *not* equal to the weight $m_2 g$ of mass m_2, but is *less* by a factor of $m_1/(m_1 + m_2)$. If T *were* equal to $m_2 g$, then m_2 would be in equilibrium, and it isn't.

CAUTION▶ It's a common mistake to assume that if an object is attached to a vertical string, the string tension must be equal to the object's weight. That would certainly be wrong in

this example! The only safe approach is to treat the tension as a variable, as we have done here. ◀

Now let's check some special cases. If $m_1 = 0$, then m_2 would fall freely and there would be no tension in the string. The equations do give $T = 0$ and $a = g$ when $m_1 = 0$. Also, if $m_2 = 0$, we expect no tension and no acceleration; for this case the equations give $T = 0$ and $a = 0$. Thus in these two special cases the results agree with our intuitive expectations.

(a)

(b)

(c)

5–13 (a) A lab weight accelerates a glider along an air track. (b) Free-body diagram for the glider (mass m_1). (c) Free-body diagram for the lab weight (mass m_2).

EXAMPLE 5–12

A simple accelerometer Figure 5–14a shows a lead fish-line sinker hanging from a string attached to point P on the ceiling of a car. When the system has an acceleration a toward the right, the string makes an angle β with the vertical. In a practical instrument, some form of damping would be needed to keep the string from swinging when the acceleration changes. For example, the system might hang in a tank of oil. (You can make a primitive version of this device by taping a thread to the ceiling light in a car, tying a key or nearly any other small object to the other end, and using a protractor. Let someone else drive the car while you are doing the experiment.) The problem is this: Given m and β, what is the acceleration a?

SOLUTION As shown in the free-body diagram, Fig. 5–14b, two forces act on the body: its weight $w = mg$ and the tension T in the string. In finding the components of the tension, note that the angle β is not measured from the $+x$-axis. Referring to the right triangles in Fig. 5–14b and recalling the definitions of the sine and cosine functions, we find the components shown (compare to Example 1–6 in Section 1–9). The sum of the horizontal components of force is

$$\Sigma F_x = T \sin \beta,$$

and the sum of the vertical components is

$$\Sigma F_y = T \cos \beta + (-mg).$$

The x-acceleration is the acceleration a of the system, and the y-acceleration is zero, so

$$T \sin \beta = ma, \qquad T \cos \beta = mg.$$

When we divide the first equation by the second and use $(\sin \beta)/(\cos \beta) = \tan \beta$, we get

$$a = g \tan \beta.$$

The acceleration a is proportional to the tangent of the angle β.

When $\beta = 0$, the sinker hangs vertically and the acceleration is zero; when $\beta = 45°$, $a = g$, and so on. We note that β can never be 90° because that would require an infinite acceleration. Also note that this result doesn't depend on the mass of the sinker.

(a)   (b)

5–14 (a) A simple accelerometer. (b) Free-body diagram for the lead sinker.

5-4 FRICTIONAL FORCES

We have seen several problems where a body rests or slides on a surface that exerts forces on the body, and we have used the terms *normal force* and *friction force* to describe these forces. Whenever two bodies interact by direct contact (touching) of their surfaces, we call the interaction forces *contact forces*. Normal and friction forces are both contact forces.

Our concern in this section is with friction, an important force in many aspects of everyday life. The oil in a car engine minimizes friction between moving parts, but without friction between the tires and the road we couldn't drive or turn the car. Air drag—the frictional force exerted by the air on a body moving through it—decreases automotive fuel economy but makes parachutes work. Without friction, nails would pull out, light bulbs would unscrew effortlessly, and riding a bicycle would be hopeless.

KINETIC AND STATIC FRICTION

Let's consider a body sliding across a surface. When you try to slide a heavy box of books across the floor, the box doesn't move at all unless you push with a certain minimum force. Then the box starts moving, and you can usually keep it moving with less force than you needed to get it started. If you take some of the books out, you need less force than before to get it started or keep it moving. What general statements can we make about this behavior?

First, when a body rests or slides on a surface, we can always represent the contact force exerted by the surface on the body in terms of components of force perpendicular and parallel to the surface. We call the perpendicular component vector the *normal force,* denoted by $\vec{n}$. (Recall that *normal* is a synonym for *perpendicular.*) The component vector parallel to the surface is the **friction force,** denoted by $\vec{f}$. By definition, $\vec{n}$ and $\vec{f}$ are always perpendicular to each other. We use script symbols for these quantities to emphasize their special role in representing the contact force. If the surface is frictionless, then the contact force has *only* a normal component, and $\vec{f}$ is zero. (Frictionless surfaces are an unattainable idealization, but we can approximate a surface as frictionless if the effects of friction are negligibly small.) The direction of the friction force is always such as to oppose relative motion of the two surfaces.

The kind of friction that acts when a body slides over a surface is called a **kinetic friction force** $\vec{f}_k$. The adjective "kinetic" and the subscript "k" remind us that the two surfaces are moving relative to each other. The *magnitude* of the kinetic friction force usually increases when the normal force increases. It takes more force to slide a box full of books across the floor than to slide the same box when it is empty. This principle is also used in automotive braking systems; the harder the brake pads are squeezed against the rotating brake disks, the greater the braking effect. In many cases the magnitude of the kinetic friction force f_k is found experimentally to be approximately *proportional* to the magnitude n of the normal force. In such cases we can write

$$f_k = \mu_k n \quad \text{(magnitude of kinetic friction force),} \tag{5-5}$$

where μ_k (pronounced "mu-sub-k") is a constant called the **coefficient of kinetic friction.** The more slippery the surface, the smaller the coefficient of friction. Because it is a quotient of two force magnitudes, μ_k is a pure number without units.

CAUTION ▶ Remember, the friction force and the normal force are always perpendicular. Equation (5-5) is *not* a vector equation, but a *scalar* relation between the *magnitudes* of the two perpendicular forces. ◀

Equation (5-5) is only an approximate representation of a complex phenomenon. On a microscopic level, friction and normal forces result from the intermolecular forces (fundamentally electrical in nature) between two rough surfaces at points where they come into contact (Fig. 5-15). The actual area of contact is usually much smaller than

The sport of ice hockey depends crucially on having just the right amount of friction between a player's skates and the ice. If there were too much friction, the players would move much more slowly; if there were too little friction, they could hardly keep from falling over.

5-15 The normal and friction forces arise from interactions between molecules at high points on the surfaces of the block and the floor.

the total surface area. As a box slides over the floor, bonds between the two surfaces form and break, and the total number of such bonds varies; hence the kinetic friction force is not perfectly constant. Smoothing the surfaces can actually increase friction, since more molecules are able to interact and bond; bringing two smooth surfaces of the same metal together can cause a "cold weld." Lubricating oils work because an oil film between two surfaces (such as the pistons and cylinder walls in a car engine) prevents them from coming into actual contact.

Table 5–1 shows a few representative values of μ_k. Although these values are given with two significant figures, they are only approximate, since friction forces can also depend on the *speed* of the body relative to the surface. We'll ignore this effect and assume that μ_k and f_k are independent of speed so that we can concentrate on the simplest cases. Table 5–1 also lists coefficients of *static* friction; we'll define these shortly.

TABLE 5–1
APPROXIMATE COEFFICIENTS OF FRICTION

MATERIALS	STATIC, μ_s	KINETIC, μ_k
Steel on steel	0.74	0.57
Aluminum on steel	0.61	0.47
Copper on steel	0.53	0.36
Brass on steel	0.51	0.44
Zinc on cast iron	0.85	0.21
Copper on cast iron	1.05	0.29
Glass on glass	0.94	0.40
Copper on glass	0.68	0.53
Teflon on Teflon	0.04	0.04
Teflon on steel	0.04	0.04
Rubber on concrete (dry)	1.0	0.8
Rubber on concrete (wet)	0.30	0.25

Friction forces may also act when there is *no* relative motion. If you try to slide that box of books across the floor, the box may not move at all because the floor exerts an equal and opposite friction force on the box. This is called a **static friction force** $\vec{f}_s$. In Fig. 5–16a the box is at rest in equilibrium under the action of its weight $\vec{w}$ and the upward normal force $\vec{n}$, which is equal in magnitude to the weight and exerted on the box by the floor. Now we tie a rope to the box (Fig. 5–16b) and gradually increase the tension T in the rope. At first the box remains at rest because, as T increases, the force of static friction f_s also increases (staying equal in magnitude to T).

At some point, T becomes greater than the maximum static friction force f_s the surface can exert. Then the box "breaks loose" (the tension T is able to break the bonds between molecules in the surfaces of the box and floor) and starts to slide. Figure 5–16c is the force diagram when T is at this critical value. If T exceeds this value, the box is no longer in equilibrium. For a given pair of surfaces the maximum value of f_s depends on the normal force. Experiment shows that in many cases this maximum value, called $(f_s)_{max}$, is approximately *proportional* to n; we call the proportionality factor μ_s (pronounced "mu-sub-s") the **coefficient of static friction.** Some representative values of μ_s are shown in Table 5–1. In a particular situation, the actual force of static friction can have any magnitude between zero (when there is no other force parallel to the surface) and a maximum value given by $\mu_s n$. In symbols,

$$f_s \le \mu_s n \qquad \text{(magnitude of static friction force).} \qquad (5\text{–}6)$$

Like Eq. (5–5), this is a relation between magnitudes, *not* a vector relation. The

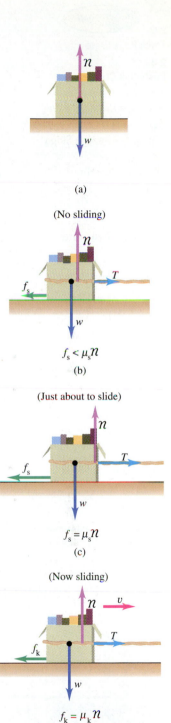

(a)

(No sliding)

$f_s < \mu_s n$
(b)

(Just about to slide)

$f_s = \mu_s n$
(c)

(Now sliding)

$f_k = \mu_k n$
(d)

5–16 (a), (b), (c) When there is no relative motion of the surfaces, the magnitude of the static friction force f_s is less than or equal to $\mu_s n$. (d) When there is relative motion, the magnitude of the kinetic friction force f_k equals $\mu_k n$.

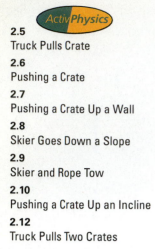
equality sign holds only when the applied force T, parallel to the surface, has reached the critical value at which motion is about to start (Fig. 5–16c). When T is less than this value (Fig. 5–16b), the inequality sign holds. In that case we have to use the equilibrium conditions ($\Sigma \vec{F} = 0$) to find f_s. If there is no applied force ($T = 0$) as in Fig. 5–16a, then there is no static friction force either ($f_s = 0$).

As soon as sliding starts (Fig. 5–16d), the friction force usually *decreases*; it's easier to keep the box moving than to start it moving. Hence the coefficient of kinetic friction is usually *less* than the coefficient of static friction for any given pair of surfaces, as shown in Table 5–1. If we start with no applied force ($T = 0$) at time $t = 0$ and gradually increase the force, the friction force varies somewhat, as shown in Fig. 5–17.

In some situations the surfaces will alternately stick (static friction) and slip (kinetic friction). This is what causes the horrible squeak made by chalk held at the wrong angle while writing on the blackboard. Another stick-slip phenomenon is the squeaky noise your windshield-wiper blades make when the glass is nearly dry; still another is the outraged shriek of tires sliding on asphalt pavement. A more positive example is the motion of a violin bow against the string.

When a body slides on a layer of gas, friction can be made very small. In the linear air track used in physics laboratories, the gliders are supported on a layer of air. The frictional force is velocity-dependent, but at typical speeds the effective coefficient of friction is of the order of 0.001. A similar device is the air table, where the pucks are supported by an array of small air jets about 2 cm apart.

5–17 In response to an externally applied force, the friction force increases to $(f_s)_{\text{max}}$. Then the surfaces begin to slide across one another, and the frictional force drops back to a nearly constant value f_k. The kinetic friction force varies somewhat as intermolecular bonds form and break.

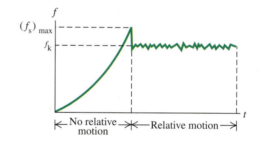

EXAMPLE 5-13

Friction in horizontal motion   A delivery company has just unloaded a 500-N crate full of home exercise equipment on the sidewalk in front of your house (Fig. 5–18a). You find that to get it started moving toward your front door, you have to pull with a horizontal force of magnitude 230 N. Once it "breaks loose" and starts to move, you can keep it moving at constant velocity with only 200 N. What are the coefficients of static and kinetic friction?

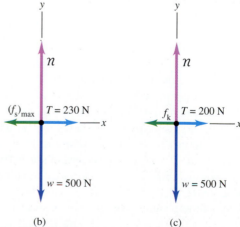

(a)   (b)   (c)

5–18 (a) Pulling a crate with a horizontal force. (b) Free-body diagram for the crate as it starts to move. (c) Free-body diagram for the crate moving at constant velocity.

SOLUTION The state of rest and the state of motion with constant velocity are both equilibrium conditions, so we use Eqs. (5–2). An instant before the crate starts to move, the static friction force has its maximum possible value, $(f_s)_{max} = \mu_s n$. The appropriate force diagram is Fig. 5–18b. We find

$$\Sigma F_x = T + (-(f_s)_{max}) = 230 \text{ N} - (f_s)_{max} = 0, \qquad (f_s)_{max} = 230 \text{ N},$$
$$\Sigma F_y = n + (-w) = n - 500 \text{ N} = 0, \qquad\qquad n = 500 \text{ N},$$
$$(f_s)_{max} = \mu_s n \quad \text{(motion about to start)},$$
$$\mu_s = \frac{(f_s)_{max}}{n} = \frac{230 \text{ N}}{500 \text{ N}} = 0.46.$$

After the crate starts to move, the forces are as shown in Fig. 5–18c, and we have

$$\Sigma F_x = T + (-f_k) = 200 \text{ N} - f_k = 0, \qquad f_k = 200 \text{ N},$$
$$\Sigma F_y = n + (-w) = n - 500 \text{ N} = 0, \qquad n = 500 \text{ N},$$
$$f_k = \mu_k n \quad \text{(motion occurs)},$$
$$\mu_k = \frac{f_k}{n} = \frac{200 \text{ N}}{500 \text{ N}} = 0.40.$$

It's easier to keep the crate moving than to start it moving, and so the coefficient of kinetic friction is less than the coefficient of static friction.

EXAMPLE 5–14

In Example 5–13, what is the friction force if the crate is at rest on the surface and a horizontal force of 50 N is applied to it?

SOLUTION From the equilibrium conditions we have

$$\Sigma F_x = T + (-f_s) = 50 \text{ N} - f_s = 0,$$
$$f_s = 50 \text{ N}.$$

In this case, f_s is less than the maximum value $(f_s)_{max} = \mu_s n$. The frictional force can prevent motion for any horizontal applied force up to 230 N.

EXAMPLE 5–15

In Example 5–13, suppose you try to move your crate full of exercise equipment by tying a rope around it and pulling upward on the rope at an angle of 30° above the horizontal (Fig. 5–19a). How hard do you have to pull to keep the crate moving with constant velocity? Is this easier or harder than pulling horizontally? Assume that $w = 500$ N and $\mu_k = 0.40$.

SOLUTION Figure 5–19b is a free-body diagram showing the forces on the crate. The kinetic friction force f_k is still equal to $\mu_k n$, but now the normal force n is *not* equal in magnitude to the weight of the crate. The force exerted by the rope has an additional vertical component that tends to lift the crate off the floor.

The crate is in equilibrium, since its velocity is constant, so

$$\Sigma F_x = T \cos 30° + (-f_k) = T \cos 30° - 0.40n = 0,$$
$$\Sigma F_y = T \sin 30° + n + (-500 \text{ N}) = 0.$$

These are two simultaneous equations for the two unknown quantities T and n. To solve them, we can eliminate one unknown and solve for the other. There are many ways to do this; here is one way. Rearrange the second equation to the form

$$n = 500 \text{ N} - T \sin 30°.$$

Substitute this expression for n back into the first equation:

$$T \cos 30° - 0.40(500 \text{ N} - T \sin 30°) = 0.$$

(a)   (b)

5–19 (a) Pulling a crate with a force applied at an upward angle. (b) Free-body diagram for the crate moving at constant velocity.

Finally, solve this equation for T, then substitute the result back into either of the original equations to obtain n. The results are

$$T = 188 \text{ N}, \qquad n = 406 \text{ N}.$$

Note that the normal force is *less* than the weight of the box ($w = 500$ N) because the vertical component of tension pulls upward on the crate. Despite this, the tension required is a little less than the 200-N force needed when you pulled horizontally in Example 5–13. Try pulling at 22°; you'll find that you need even less force (see Challenge Problem 5–111).

EXAMPLE 5-16

Toboggan ride with friction I Let's go back to the toboggan that we studied in Example 5–9 (Section 5–3). The wax has worn off, and there is now a coefficient of kinetic friction μ_k. The slope has just the right angle to make the toboggan slide with constant speed. Derive an expression for the slope angle in terms of w and μ_k.

SOLUTION Figure 5–20 shows the free-body diagram. The slope angle is α. The forces on the toboggan are its weight w and the normal and frictional components of the contact force exerted on it by the sloping surface. We take axes perpendicular and parallel to the surface and represent the weight in terms of its components in these two directions, as shown. The toboggan is in equilibrium because its velocity is constant, and the equilibrium conditions are

$$\Sigma F_x = w \sin \alpha + (-f_k) = w \sin \alpha - \mu_k n = 0,$$
$$\Sigma F_y = n + (-w \cos \alpha) = 0.$$

Rearranging, we get

$$\mu_k n = w \sin \alpha, \qquad n = w \cos \alpha.$$

Just as in Example 5–9, the normal force n is *not* equal to the weight w. When we divide the first of these equations by the second, we find

$$\mu_k = \frac{\sin \alpha}{\cos \alpha} = \tan \alpha.$$

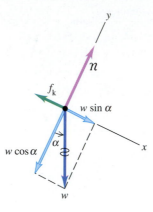

5–20 Free-body diagram for the toboggan with friction.

The weight w doesn't appear in this expression. *Any* toboggan, regardless of its weight, slides down an incline with constant speed if the coefficient of kinetic friction equals the tangent of the slope angle of the incline. The steeper the slope, the greater the coefficient of friction has to be for the toboggan to slide with constant velocity. This is just what we should expect.

EXAMPLE 5-17

Toboggan ride with friction II What if we have the same toboggan and coefficient of friction as in Example 5–16, but a steeper hill? This time the toboggan accelerates, although not as much as in Example 5–9, when there was no friction. Derive an expression for the acceleration in terms of g, α, μ_k, and w.

SOLUTION The free-body diagram (Fig. 5–21) is almost the same as for Example 5–16, but the body is no longer in equilibrium; a_y is still zero, but a_x is not. Using $w = mg$, Newton's second law gives us the two equations

$$\Sigma F_x = mg \sin \alpha + (-f_k) = ma_x,$$
$$\Sigma F_y = n + (-mg \cos \alpha) = 0.$$

From the second equation and Eq. (5–5), $f_k = \mu_k n$, we get an expression for f_k:

$$n = mg \cos \alpha,$$
$$f_k = \mu_k n = \mu_k mg \cos \alpha.$$

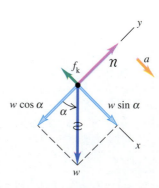

5–21 Free-body diagram for the toboggan with friction, going down a steeper hill.

We substitute this back into the *x*-component equation. The result is

$$mg \sin \alpha + (-\mu_k mg \cos \alpha) = ma_x,$$

$$a_x = g(\sin \alpha - \mu_k \cos \alpha).$$

Does this result make sense? Here are some special cases we can check. First, if the hill is *vertical*, $\alpha = 90°$; then $\sin \alpha = 1$, $\cos \alpha = 0$, and $a_x = g$. This is free fall, just what we would expect. Second, on a hill at angle α with *no* friction, $\mu_k = 0$. Then $a_x = g \sin \alpha$. The situation is the same as in Example 5–9, and we get the same result; that's encouraging! Next, suppose that there is just enough friction to make the toboggan move with constant velocity. In that case, $a_x = 0$, and our result gives

$$\sin \alpha = \mu_k \cos \alpha \qquad \text{and} \qquad \mu_k = \tan \alpha.$$

This agrees with our result from Example 5–16; good! Finally, note that there may be so much friction that $\mu_k \cos \alpha$ is actually greater than $\sin \alpha$. In that case, a_x is negative; if we give the

toboggan an initial downhill push to start it moving, it will slow down and eventually stop.

We have pretty much beaten the toboggan problem to death, but there is an important lesson to be learned. We started out with a simple problem and then extended it to more and more general situations. Our most general result, found in this example, includes *all* the previous ones as special cases, and that's a nice, neat package! Don't memorize this package; it is useful only for this one set of problems. But do try to understand how we obtained it and what it means.

One final variation that you may want to try out is the case in which we give the toboggan an initial push *up* the hill. The direction of the friction force is now reversed, so the acceleration is different from the downhill value. It turns out that the expression for a_x is the same as for downhill motion except that the minus sign becomes plus. Can you prove this?

ROLLING FRICTION

It's a lot easier to move a loaded filing cabinet across a horizontal floor by using a cart with wheels than to slide it. How much easier? We can define a **coefficient of rolling friction** μ_r, which is the horizontal force needed for constant speed on a flat surface divided by the upward normal force exerted by the surface. Transportation engineers call μ_r the *tractive resistance*. Typical values of μ_r are 0.002 to 0.003 for steel wheels on steel rails and 0.01 to 0.02 for rubber tires on concrete. These values show one reason why railroad trains are in general much more fuel-efficient than highway trucks.

EXAMPLE 5–18

Motion with rolling friction A typical car weighs about 12,000 N (about 2700 lb). If the coefficient of rolling friction is $\mu_r = 0.010$, what horizontal force must you apply to push the car at constant speed on a level road? Neglect air resistance.

SOLUTION The normal force n is equal to the weight w, because the road surface is horizontal and there are no other vertical forces. From the definition of μ_r, the rolling friction force f_r is

$$f_r = \mu_r n = (0.010)(12{,}000 \text{ N}) = 120 \text{ N} \qquad \text{(about 27 lb).}$$

From Newton's first law, a forward force with this magnitude is needed to keep the car moving with constant speed.

We invite you to apply this analysis to your crate of exercise equipment (Example 5–13). If the delivery company brings it on a rubber-wheeled dolly with $\mu_r = 0.02$, only a 10-N force is needed to keep it moving at constant velocity. Can you verify this?

FLUID RESISTANCE AND TERMINAL SPEED

Sticking your hand out the window of a fast-moving car will convince you of the existence of **fluid resistance,** the force that a fluid (a gas or liquid) exerts on a body moving through it. The moving body exerts a force on the fluid to push it out of the way. By Newton's third law, the fluid pushes back on the body with an equal and opposite force.

The *direction* of the fluid resistance force acting on a body is always opposite the direction of the body's velocity relative to the fluid. The *magnitude* of the fluid resistance force usually increases with the speed of the body through the fluid. Contrast this behavior with that of the kinetic friction force between two surfaces in contact, which we can usually regard as independent of speed. For low speeds, the magnitude f of the resisting force of the fluid is approximately proportional to the body's speed v:

$$f = kv \qquad \text{(fluid resistance at low speed),} \qquad (5\text{–}7)$$

5-22 Free-body diagram for a body falling through a fluid.

5-23 Graphs of acceleration, velocity, and position versus time for a body falling with fluid resistance proportional to v, shown as dark color curves. The light color curves show the corresponding relations if there is *no* fluid resistance.

where k is a proportionality constant that depends on the shape and size of the body and the properties of the fluid. In motion through air at the speed of a tossed tennis ball or faster, the resisting force is approximately proportional to v^2 rather than to v. It is then called **air drag** or simply *drag*. Airplanes, falling raindrops, and cars moving at high speed all experience air drag. In this case we replace Eq. (5–7) by

$$f = Dv^2 \qquad \text{(fluid resistance at high speed).} \qquad (5\text{--}8)$$

Because of the v^2 dependence, air drag increases rapidly with increasing speed. The air drag on a typical car is negligible at low speeds but comparable to or greater than the force of rolling resistance at highway speeds. The value of D depends on the shape and size of the body and on the density of the air.

We invite you to show that the units of the constant k in Eq. (5–7) are $N \cdot s/m$ or kg/s and that the units of the constant D in Eq. (5–8) are $N \cdot s^2/m^2$ or kg/m.

Because of the effects of fluid resistance, an object falling in a fluid will *not* have a constant acceleration. To describe its motion, we can't use the constant-acceleration relationships from Chapter 2; instead, we have to start over, using Newton's second law. Let's consider the following situation. You release a rock at the surface of a deep pond, and it falls to the bottom. The fluid resistance force in this situation is given by Eq. (5–7). What are the acceleration, velocity, and position of the rock as functions of time?

The free-body diagram is shown in Fig. 5–22. We take the positive direction to be downward and neglect any force associated with buoyancy in the water. There are no x-components, and Newton's second law gives

$$\Sigma F_y = mg + (-kv) = ma.$$

When the rock first starts to move, $v = 0$, the resisting force is zero, and the initial acceleration is $a = g$. As its speed increases, the resisting force also increases until finally it is equal in magnitude to the weight. At this time, $mg - kv = 0$, the acceleration becomes zero, and there is no further increase in speed. The final speed v_t, called the **terminal speed,** is given by $mg - kv_t = 0$, or

$$v_t = \frac{mg}{k} \qquad \text{(terminal speed, fluid resistance } f = kv). \qquad (5\text{--}9)$$

Figure 5–23 shows how the acceleration, velocity, and position vary with time. As time goes by, the acceleration approaches zero, and the velocity approaches v_t (remember that we chose the positive y-direction to be down). The slope of the graph of y versus t becomes constant as the velocity becomes constant.

To see how the graphs in Fig. 5–23 are derived, we must find the relation between speed and time during the interval before the terminal speed is reached. We go back to Newton's second law, which we rewrite as

$$m\frac{dv}{dt} = mg - kv.$$

After rearranging terms and replacing mg/k by v_t, we integrate both sides, noting that $v = 0$ when $t = 0$:

$$\int_0^v \frac{dv}{v - v_t} = -\frac{k}{m} \int_0^t dt,$$

which integrates to

$$\ln \frac{v_t - v}{v_t} = -\frac{k}{m} t, \qquad \text{or} \qquad 1 - \frac{v}{v_t} = e^{-(k/m)t},$$

and finally

$$v = v_t\left[1 - e^{-(k/m)t}\right].$$ (5–10)

2.1.2
Skyvider

Note that v becomes equal to the terminal speed v_t only in the limit that $t \to \infty$; the rock cannot attain terminal speed in any finite length of time.

The derivative of v gives a as a function of time, and the integral of v gives y as a function of time. We leave the derivations for you to complete (see Exercise 5–40); the results are

$$a = ge^{-(k/m)t},$$ (5–11)

$$y = v_t\left[t - \frac{m}{k}(1 - e^{-(k/m)t})\right].$$ (5–12)

Now look again at Fig. 5–23, which shows graphs of these three relations.

In deriving the terminal speed in Eq. (5–9), we assumed that the fluid resistance force was proportional to the speed. For an object falling through the air at high speeds, so that the fluid resistance is proportional to v^2 as in Eq. (5–8), we invite you to show that the terminal speed v_t is given by

$$v_t = \sqrt{\frac{mg}{D}} \qquad \text{(terminal speed, fluid resistance } f = Dv^2\text{).}$$ (5–13)

This expression for terminal speed explains the observation that heavy objects in air tend to fall faster than light objects. Two objects with the same physical size but different mass (say, a table-tennis ball and a lead ball with the same radius) have the same value of D but different values of m. The more massive object has a larger terminal speed and falls faster. The same idea explains why a sheet of paper falls faster if you first crumple it into a ball; the mass m is the same, but the smaller size makes D smaller (less air drag for a given speed) and v_t larger.

By changing the positions of their arms and legs while falling, sky-divers can change the value of the constant D in Eq. (5–8) and hence adjust the terminal speed of their fall (Eq. (5–13)).

EXAMPLE 5-19

Terminal speed of a sky diver For a human body falling through air in a spread-eagle position, the numerical value of the constant D in Eq. (5–8) is about 0.25 kg/m. For an 80-kg sky diver the terminal velocity is, from Eq. (5–13),

$$v_t = \sqrt{\frac{mg}{D}} = \sqrt{\frac{(80 \text{ kg})(9.8 \text{ m/s}^2)}{0.25 \text{ kg/m}}}$$

$$= 56 \text{ m/s} \qquad \text{(about 200 km/h, or 125 mi/h).}$$

When the sky diver deploys the parachute, the value of D increases greatly, and the terminal speed of the sky diver and parachute is (thankfully) much less than 56 m/s.

5-5 DYNAMICS OF CIRCULAR MOTION

We talked about uniform circular motion in Section 3–5. We showed that when a particle moves in a circular path with constant speed, the particle's acceleration is always directed toward the center of the circle (perpendicular to the instantaneous velocity). The magnitude a_{rad} of the acceleration is constant and is given in terms of the speed v and the radius R of the circle by

$$a_{rad} = \frac{v^2}{R}.$$ (5–14)

5–24 In uniform circular motion, both the acceleration and the net force are directed toward the center of the circle.

The subscript "rad" is a reminder that at each point the acceleration is radially inward toward the center of the circle, perpendicular to the instantaneous velocity. We explained in Section 3–5 why this acceleration is sometimes called *centripetal acceleration*.

We can also express the centripetal acceleration a_{rad} in terms of the *period T*, the time for one revolution:

$$T = \frac{2\pi R}{v}. \tag{5–15}$$

In terms of the period, a_{rad} is

$$a_{rad} = \frac{4\pi^2 R}{T^2}. \tag{5–16}$$

Uniform circular motion, like all other motion of a particle, is governed by Newton's second law. The particle's acceleration toward the center of the circle must be caused by a force, or several forces, such that the vector sum $\Sigma \vec{F}$ is a vector directed always toward the center (Fig. 5–24). The magnitude of the acceleration is constant, so the magnitude F_{net} of the net inward radial force must also be constant. If an ice skater whirls his partner around in a circle on the ice, he must constantly pull on her inward toward the center of the circle. If he lets go, then the inward force no longer acts, and his partner flies off in a straight line tangent to the circle (Fig. 5–25).

5–25 If the inward radial force suddenly ceases to act on a body in circular motion, the body flies off in a straight line with constant velocity (as it must, since the net force acting is zero).

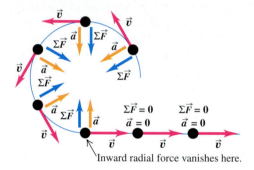

Inward radial force vanishes here.

The magnitude of the radial acceleration is given by $a_{rad} = v^2/R$, so the magnitude of the net inward radial force F_{net} on a particle with mass m must be

$$F_{net} = ma_{rad} = m\frac{v^2}{R}. \tag{5–17}$$

Uniform circular motion can result from *any* combination of forces, just so the net force $\Sigma \vec{F}$ is always directed toward the same point at the center of the circle and has a constant magnitude.

CAUTION ▶ In doing problems involving uniform circular motion, you may be tempted to include an extra outward force of magnitude $m(v^2/R)$ to "keep the body out there" or to "keep it in equilibrium"; this outward force is usually called *"centrifugal force"* ("centrifugal" means "fleeing from the center"). Resist this temptation, because this approach is simply *wrong!* First, the body does *not* "stay out there"; it is in constant motion around its circular path. Its velocity is constantly changing in direction, so it accelerates and is *not* in equilibrium. Second, if there *were* an additional outward ("centrifugal") force to balance the inward force, there would then be *no* net inward force to cause the circular motion, and the body would move in a straight line, not a circle (Fig. 5–25). Third, the quantity $m(v^2/R)$ is *not* a force; it corresponds to the $m\vec{a}$ side of $\Sigma \vec{F} = m\vec{a}$ and *does not* appear in $\Sigma \vec{F}$. It's certainly true that a passenger riding in a car going around a circular path on a level road tends to slide to the outside of the turn, as though responding to a "centrifugal force." But such a passenger is in an accelerating, non-inertial frame of reference in which Newton's first and second laws don't apply. As we discussed in Section 4–3, what really happens is that the passenger tends to keep moving in a straight line, and the outer side of the car "runs into" the passenger as the car turns (Fig. 4–8c). *In an inertial frame of reference there is no such thing as a "centrifugal force" acting on a body.* We promise not to mention this term again, and we strongly advise you to avoid using it as well. ◄

EXAMPLE 5–20

Force in uniform circular motion A small plastic box with a mass of 0.300 kg revolves uniformly in a circle on a horizontal frictionless surface such as an air-hockey table (Fig. 5–26a). The box is attached by a cord 0.140 m long to a pin set in the surface. If the box makes two complete revolutions per second, find the force F exerted on it by the cord.

SOLUTION To calculate the force F using Newton's second law, we first need to find the magnitude of the centripetal acceleration. The period is $T = (1.00 \text{ s})/(2 \text{ rev}) = 0.500$ s, so from Eq. (5–16) the acceleration is

$$a_{rad} = \frac{4\pi^2 R}{T^2} = \frac{4\pi^2(0.140 \text{ m})}{(0.500 \text{ s})^2} = 22.1 \text{ m/s}^2.$$

Alternatively, we can first use Eq. (5–15) to find the speed v:

$$v = \frac{2\pi R}{T} = \frac{2\pi(0.140 \text{ m})}{0.500 \text{ s}} = 1.76 \text{ m/s}.$$

Then, using Eq. (5–14),

$$a_{rad} = \frac{v^2}{R} = \frac{(1.76 \text{ m/s})^2}{0.140 \text{ m}} = 22.1 \text{ m/s}^2.$$

The body has no vertical acceleration, so the vertical forces $\vec{n}$ and $\vec{w}$ in Fig. 5–26b add to zero. The only force toward the center of the circle is the tension $\vec{F}$ in the cord, and its magnitude is

$$F = ma_{rad} = (0.300 \text{ kg})(22.1 \text{ m/s}^2)$$
$$= 6.63 \text{ kg} \cdot \text{m/s}^2 = 6.63 \text{ N}.$$

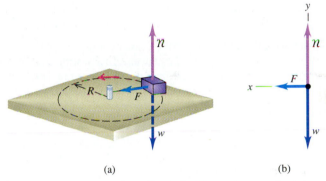

(a)   (b)

5–26 (a) A box moving in uniform circular motion on a horizontal frictionless surface. (b) Free-body diagram for the box. The +x-direction is toward the center of the circle.

EXAMPLE 5–21

The conical pendulum An inventor who dares to be different proposes to make a pendulum clock using a pendulum bob with mass m at the end of a thin wire of length L. Instead of swinging back and forth, the bob moves in a horizontal circle with constant speed v, with the wire making a constant angle β with the vertical direction (Fig. 5–27a). Assuming that the time T for one revolution (that is, the period) is known, find the tension F in the wire and the angle β.

SOLUTION A free-body diagram for the bob and a coordinate system are shown in Fig. 5–27b. The center of the circular path is at the center of the shaded area, *not* at the top end of the wire. The forces on the bob in the position shown are the weight mg and the tension F in the wire. The tension has a horizontal component $F \sin \beta$ and a vertical component $F \cos \beta$. (This free-body diagram is exactly like the one in Fig. 5–14b, but in this case the acceleration a is the *radial* acceleration, v^2/R.)

The system has no vertical acceleration, so the sum of the vertical components of force is zero. The horizontal force is directed toward the center of the circle; it must equal the mass m times the radial acceleration $a_{rad} = v^2/R$. Thus the $\Sigma \vec{F} = m\vec{a}$ equations are

$$\Sigma F_x = F \sin \beta = m \frac{v^2}{R}, \qquad \Sigma F_y = F \cos \beta + (-mg) = 0.$$

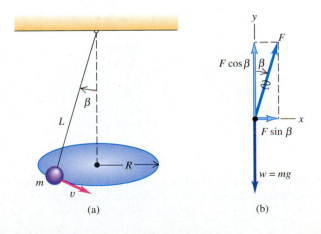

(a)   (b)

5–27 (a) The bob on the end of the wire moves in uniform circular motion. (b) Free-body diagram for the bob. The +x-direction is toward the center of the circle.

These are two simultaneous equations for the unknowns F and β. The equation for ΣF_y gives $F = mg/\cos\beta$; substituting this result into the equation for ΣF_x, we find

$$\tan\beta = \frac{v^2}{gR}.$$

We can rewrite this expression to relate β to the period T. The radius of the circle is $R = L\sin\beta$, and the speed is the circumference $2\pi L\sin\beta$ divided by the period T:

$$v = \frac{2\pi L\sin\beta}{T}.$$

Substituting this into $\tan\beta = v^2/gR$, we obtain

$$\cos\beta = \frac{gT^2}{4\pi^2 L},$$

or

$$T = 2\pi\sqrt{\frac{L\cos\beta}{g}}.$$

Once we know β, we can find the tension F from $F = mg/\cos\beta$. For a given length L, as the period T becomes smaller, $\cos\beta$ decreases, the angle β increases, and the tension $F = mg/\cos\beta$ increases. The angle can never be 90°, however; this would require that $T = 0$, $F = \infty$, and $v = \infty$. This system is called a *conical pendulum* because the suspending wire traces out a cone. It would not make a very good clock because the period depends on β in such a direct way.

EXAMPLE 5–22

Rounding a flat curve The BMW Z3 roadster in Example 3–11 (Section 3–5) is rounding a flat, unbanked curve with radius R (Fig. 5–28a). If the coefficient of friction between tires and road is μ_s, what is the maximum speed v_{max} at which the driver can take the curve without sliding?

SOLUTION Figure 5–28b shows a free-body diagram for the car. The acceleration v^2/R toward the center of the curve must be caused by the friction force f, and there is no vertical acceleration. Thus we have

$$\Sigma F_x = f = m\frac{v^2}{R}, \qquad \Sigma F_y = n - mg = 0.$$

The first equation shows that the friction force *needed* to keep the car moving in its circular path increases with the car's speed. But the maximum friction force *available* is $f_{max} = \mu_s n = \mu_s mg$, which is constant, and this determines the car's maximum speed. Substituting f_{max} for f and v_{max} for v in the ΣF_x equation, we find

$$\mu_s mg = m\frac{v_{max}^2}{R},$$

so the maximum speed is

$$v_{max} = \sqrt{\mu_s gR}.$$

If $\mu_s = 0.87$ and $R = 230$ m, then

$$v_{max} = \sqrt{(0.87)(9.8 \text{ m/s}^2)(230 \text{ m})} = 44 \text{ m/s},$$

or about 160 km/h (100 mi/h). This is the maximum speed for this radius. Note that the maximum centripetal acceleration (called the "lateral acceleration" in Example 3–11) is equal to $\mu_s g$. Why do we use the coefficient of *static* friction for a moving car?

If we take the curve at a slower speed than $\sqrt{\mu_s gR}$, the friction force is *less* than the maximum possible value $f_{max} = \mu_s mg$. If we try to take the curve going *faster* than the maximum speed, the car can still go in a circle without skidding, but the radius will be larger and the car will run off the road.

(a)   (b)

5–28 (a) A sports car rounding a curve on a level road. (b) Free-body diagram for the car. The $+x$-direction is toward the center of the circle.

EXAMPLE 5–23

Rounding a banked curve For a car traveling at a certain speed, it is possible to bank a curve at just the right angle so that *no* friction at all is needed. Then a car can round the curve even on wet ice with Teflon tires. Bobsled racing depends on this same idea. An engineer proposes to rebuild the curve in Example 5–22 so that a car moving at speed v can safely make the turn even if there is no friction (Fig. 5–29a). At what angle β should the curve be banked?

SOLUTION The free-body diagram is shown in Fig. 5–29b. The normal force $\vec{n}$ is no longer vertical but is perpendicular to the roadway at an angle β with the vertical. Thus it has a vertical component $n \cos \beta$ and a horizontal component $n \sin \beta$, as shown. We want to get around the curve without relying on friction, so no friction force is included. The horizontal component of $\vec{n}$ must now cause the acceleration v^2/R; there is no vertical acceleration. Thus

$$\Sigma F_x = n \sin \beta = \frac{mv^2}{R}, \qquad \Sigma F_y = n \cos \beta + (-mg) = 0.$$

From the ΣF_y equation, $n = mg/\cos \beta$. Substituting this into the ΣF_x equation gives an expression for the banking angle:

$$\tan \beta = \frac{v^2}{gR}.$$

The banking angle depends on the speed and the radius. For a given radius, no one angle is correct for all speeds. In the design of highways and railroads, curves are often banked for the *average* speed of the traffic over them. If $R = 230$ m and $v = 25$ m/s (a highway speed of 88 km/h, or 55 mi/h), then

$$\beta = \arctan \frac{(25 \text{ m/s})^2}{(9.8 \text{ m/s}^2)(230 \text{ m})} = 15°.$$

This is within the range of banking angles actually used in highways. With the same radius and $v = 44$ m/s, as in Example 5–22, $\beta = 41°$; such steeply banked curves are found at automobile raceways.

Note that the expression for the banking angle β is the same as that found for the angle of a conical pendulum in Example 5–21. In fact, the free-body diagrams of Figs. 5–27b and 5–29b are very similar; the normal force acting on the car plays the role of the tension acting on the pendulum bob.

The same considerations apply to the correct banking angle of an airplane when it makes a turn in level flight. The free-body diagram for the pilot of the airplane is exactly as shown in Fig. 5–29b; the normal force $n = mg/\cos \beta$ is exerted on the pilot by the seat. As in Example 5–8, n is equal to the *apparent weight* of the pilot, which is greater than his true weight mg. For a jet fighter making a tight turn (small R) at high speed (large v), $\tan \beta = v^2/gR$ will be large, and the required angle β will approach 90°. The pilot's apparent weight will then be tremendous, $5.8mg$ at $\beta = 80°$ and $9.6mg$ at $\beta = 84°$. Pilots black out in such tight turns because the apparent weight of their blood increases by the same factor, and the human heart isn't strong enough to pump such apparently "heavy" blood to the brain.

(a)

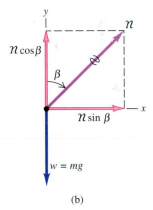

(b)

5–29 (a) A sports car rounding a banked curve. If the curve is banked at the correct angle β, no friction is needed to round the curve at speed v. (b) Free-body diagram for the car. The +x-direction is toward the center of the circle.

MOTION IN A VERTICAL CIRCLE

In all the preceding examples the body moved in a horizontal circle. Motion in a vertical circle is no different in principle, but the weight of the body has to be treated carefully. The following example shows what we mean.

4.7

Car Circles a Track

EXAMPLE 5-24

Uniform circular motion in a vertical circle A passenger on a carnival Ferris wheel moves in a vertical circle of radius R with constant speed v. Assuming that the seat remains upright during the motion, derive expressions for the force the seat exerts on the passenger at the top of the circle and at the bottom.

SOLUTION Figure 5–30 is a diagram of the situation, with free-body diagrams for the two positions. We take the positive y-direction as upward in both cases. Let n_T be the upward normal force the seat applies to the passenger at the top of the circle, and let n_B be the normal force at the bottom. At the top the acceleration has magnitude v^2/R, but its vertical component is negative because its direction is downward, toward the center of the circle. So the $\Sigma F_y = ma_y$ equation at the top is

$$\Sigma F_y = n_T + (-mg) = m\left(-\frac{v^2}{R}\right), \quad \text{or}$$

$$n_T = m\left(g - \frac{v^2}{R}\right).$$

This means that at the top of the Ferris wheel, the upward force the seat applies to the passenger is *smaller* in magnitude than the passenger's weight. If the ride goes fast enough that $g - v^2/R$ becomes zero, the seat applies *no* force, and the passenger is about to become airborne. If v becomes still larger, n_T becomes negative; this means that a *downward* force (such as from a seat belt) is needed to keep the passenger in the seat.

At the bottom the acceleration is upward, and Newton's second law is

$$\Sigma F_y = n_B + (-mg) = m\left(+\frac{v^2}{R}\right), \quad \text{or}$$

$$n_B = m\left(g + \frac{v^2}{R}\right).$$

At this point the upward force provided by the seat is always *greater than* the passenger's weight. You feel the seat pushing up on you more firmly than when you are at rest.

We recognize that n_T and n_B are the values of the passenger's *apparent weight* at the top and bottom of the circle, respectively. As a specific example, suppose the passenger's mass is 60.0 kg, the radius of the circle is $R = 8.00$ m, and the wheel makes one revolution in 10.0 s. From Eq. (5–16),

$$a_{rad} = \frac{4\pi^2 R}{T^2} = \frac{4\pi^2(8.00 \text{ m})}{(10.0 \text{ s})^2} = 3.16 \text{ m/s}^2.$$

The two values of the apparent weight are

$$n_T = (60.0 \text{ kg})(9.80 \text{ m/s}^2 - 3.16 \text{ m/s}^2) = 398 \text{ N},$$

$$n_B = (60.0 \text{ kg})(9.80 \text{ m/s}^2 + 3.16 \text{ m/s}^2) = 778 \text{ N}.$$

The passenger's weight is 588 N, and n_T and n_B are about 30% less than and greater than the weight, respectively.

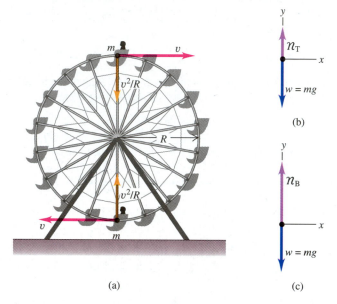

(a)

(b)

(c)

5–30 (a) The passenger's acceleration vector has the same magnitude at the top and bottom of the Ferris wheel and in both cases points toward the center of the circle. (b) Free-body diagram for a passenger at the top of the circle. (c) Free-body diagram for a passenger at the bottom of the circle.

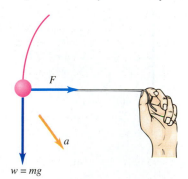

5–31 A ball moving in a vertical circle. At a typical point in the circle, there is a component of the net force (and hence a component of acceleration) tangent to the circle. The ball is in *non-uniform* circular motion.

When we tie a string to an object and whirl it in a vertical circle, the preceding analysis isn't directly applicable. The reason is that v is *not* constant in this case; at every point in the circle except the top and bottom, the net force (and hence the acceleration) does *not* point toward the center of the circle (Fig. 5–31). So both $\Sigma \vec{F}$ and $\vec{a}$ have a component tangent to the circle, and this is a case of *non-uniform* circular motion (see Section 3–5). Even worse, we can't use the constant-acceleration formulas to relate the speeds at various points because *neither* the magnitude nor the direction of the acceleration is constant. The speed relations that we need are best obtained by using the concept of energy. We'll consider such problems in Chapter 7.

*5-6 THE FUNDAMENTAL FORCES OF NATURE

We have discussed several different kinds of forces—including weight, tension, friction, fluid resistance, and the normal force—and we will encounter others as we continue our study of physics. But just how many different kinds of forces are there? Our current understanding is that all forces are expressions of just four distinct classes of *fundamental* forces, or interactions between particles (Fig. 5–32). Two are familiar in everyday experience. The other two involve interactions between subatomic particles that we cannot observe with the unaided senses.

Of the two familiar classes, **gravitational interactions** were the first to be studied in detail. The *weight* of a body results from the earth's gravitational attraction acting on it. The sun's gravitational attraction for the earth keeps the earth in its nearly circular orbit around the sun. Newton recognized that the motions of the planets around the sun and the free fall of objects on earth are both the result of gravitational forces. In Chapter 12 we will study gravitational interactions in greater detail, and we will analyze their vital role in the motions of planets and satellites.

The second familiar class of forces, **electromagnetic interactions,** includes electric and magnetic forces. If you run a comb through your hair, you can then use the comb to pick up bits of paper or fluff; this interaction is the result of electric charge on the comb. All atoms contain positive and negative electric charge, so atoms and molecules can exert electric forces on each other. Contact forces, including the normal force, friction, and fluid resistance, are the combination of all such forces exerted on the atoms of a body by atoms in its surroundings. *Magnetic* forces occur in interactions between magnets or between a magnet and a piece of iron. These may seem to form a different category, but magnetic interactions are actually the result of electric charges in motion. In an electromagnet an electric current in a coil of wire causes magnetic interactions. We will study electric and magnetic interactions in detail in the second half of this book.

These two interactions differ enormously in their strength. The electrical repulsion between two protons at a given distance is stronger than their gravitational attraction by a factor of the order of 10^{35}. Gravitational forces play no significant role in atomic or molecular structure. But in bodies of astronomical size, positive charge and negative charge are usually present in nearly equal amounts, and the resulting electrical interactions nearly cancel each other out. Gravitational interactions are thus the dominant influence in the motion of planets and in the internal structure of stars.

The other two classes of interactions are less familiar. One, the **strong interaction,** is responsible for holding the nucleus of an atom together. Nuclei contain electrically

(a)

(b)

5–32 Examples of the fundamental interactions in nature.
(a) Gravitational forces cause the orbital motions of the earth around the sun and of the moon around the earth. (b) Electromagnetic forces act between atoms to form molecules, as in this atomic force micrograph of bacterial plasmid DNA. The smallest details seen here are 3 to 4 nm in size.
(c) Strong forces between nuclear particles are responsible for thermonuclear reactions in the center of the sun; the released energy reaches us as sunlight. (d) Weak forces, characteristic of interactions involving subatomic particles called neutrinos, play a crucial role when a star (shown on the left by an arrow) explodes and becomes a supernova (shown on the right).

(c)

(d)

neutral neutrons and positively charged protons. The charged protons repel each other, and a nucleus could not be stable if it were not for the presence of an attractive force of a different kind that counteracts the repulsive electrical interactions. In this context the strong interaction is also called the *nuclear force.* It has much shorter range than electrical interactions, but within its range it is much stronger. The strong interaction is also responsible for the creation of unstable particles in high-energy particle collisions.

Finally, there is the **weak interaction.** It plays no direct role in the behavior of ordinary matter, but it is of vital importance in interactions among fundamental particles. The weak interaction is responsible for a common form of radioactivity called beta decay, in which a neutron in a radioactive nucleus is transformed into a proton while ejecting an electron and an essentially massless particle called an antineutrino. The weak interaction between the antineutrino and ordinary matter is so feeble that an antineutrino could easily penetrate a wall of lead a million kilometers thick!

During the past several decades a unified theory of the electromagnetic and weak interactions has been developed. We now speak of the *electroweak* interaction, and in a sense this reduces the number of classes of interactions from four to three. Similar attempts have been made to understand strong, electromagnetic, and weak interactions on the basis of a single unified theory called a *grand unified theory* (GUT), and the first tentative steps have been taken toward a possible unification of all interactions into a *theory of everything* (TOE). Such theories are still speculative, and there are many unanswered questions in this very active field of current research.

5-7 PROJECTILE MOTION WITH AIR RESISTANCE

A Case Study in Computer Analysis

In our study of projectile motion in Chapter 3, we assumed that air-resistance effects are negligibly small. But in fact air resistance (often called *air drag,* or simply *drag*) has a major effect on the motion of many objects, including tennis balls, bicycle riders, and airplanes. In Section 5–4 we considered how a fluid resistance force affected a body falling straight down. We'd now like to extend this analysis to a projectile moving in a plane.

It's not difficult to include the force of air resistance in the equations for a projectile, but solving them for the position and velocity as functions of time, or the shape of the path, can get quite complex. Fortunately, it is fairly easy to make quite precise numerical approximations to these solutions, using a computer. That's what this section is about.

When we omitted air drag, the only force acting on a projectile with mass m was its weight $\vec{w} = m\vec{g}$. The components of the projectile's acceleration were simply

$$a_x = 0, \qquad a_y = -g.$$

The $+x$-axis is horizontal, and the $+y$-axis is vertically upward. We must now include the air drag force acting on the projectile. At the speed of a tossed tennis ball or faster, the magnitude f of the air drag force is approximately proportional to the square of the projectile's speed relative to the air, as discussed in Section 5–4:

$$f = Dv^2, \qquad\qquad (5\text{–}8)$$

where $v^2 = v_x^2 + v_y^2$. We'll assume that the air is still, so $\vec{v}$ is the velocity of the projectile relative to the ground as well as to the air (Fig. 5–33a). The direction of $\vec{f}$ is opposite the direction of $\vec{v}$, so we can write $\vec{f} = -Dv\vec{v}$ and the components of $\vec{f}$ are

$$f_x = -Dvv_x, \qquad f_y = -Dvv_y.$$

v

(a)

5-33 (a) A projectile in flight with velocity $\vec{v}$ relative to the ground.

Note that each component is opposite in direction to the corresponding component of velocity and that $f = \sqrt{f_x^2 + f_y^2}$. The free-body diagram is shown in Fig. 5–33b. Newton's second law gives

$$\Sigma F_x = -Dvv_x = ma_x, \qquad \Sigma F_y = -mg - Dvv_y = ma_y,$$

and the components of acceleration including the effects of both gravity and air drag are

$$a_x = -(D/m)vv_x, \qquad a_y = -g - (D/m)vv_y. \qquad (5\text{–}18)$$

The constant D depends on the density ρ of air, the silhouette area A of the body (its area as seen from the front), and a dimensionless constant C called the **drag coefficient** that depends on the shape of the body. Typical values of C for baseballs, tennis balls, and the like are in the range from 0.2 to 1.0. In terms of these quantities, D is given by

$$D = \frac{\rho C A}{2}. \qquad (5\text{–}19)$$

Now comes the basic idea of our numerical calculation. The acceleration components a_x and a_y are constantly changing as the velocity components change. But over a sufficiently short time interval Δt, we can regard the acceleration as essentially constant. If we know the coordinates and velocity components at some time t, we can find these quantities at a slightly later time $t + \Delta t$ using the formulas for constant acceleration. Here's how we do it. During a time interval Δt, the average x-component of acceleration is $a_x = \Delta v_x / \Delta t$ and the x-velocity v_x changes by an amount $\Delta v_x = a_x \Delta t$. Similarly, v_y changes by an amount $\Delta v_y = a_y \Delta t$. So the values of the x-velocity and y-velocity at the end of the interval are

$$v_x + \Delta v_x = v_x + a_x \Delta t, \qquad v_y + \Delta v_y = v_y + a_y \Delta t. \qquad (5\text{–}20)$$

While this is happening, the projectile is moving, so the coordinates are also changing. The average x-velocity during the time interval Δt is the average of the value v_x (at the beginning of the interval) and $v_x + \Delta v_x$ (at the end of the interval), or $v_x + \Delta v_x / 2$. During Δt the coordinate x changes by an amount

$$\Delta x = (v_x + \Delta v_x / 2)\Delta t = v_x \Delta t + \frac{1}{2} a_x (\Delta t)^2,$$

and similarly for y. (Compare this to Eq. (2–12) for motion with constant acceleration; it's exactly the same expression.) So the coordinates of the projectile at the end of the interval are

$$x + \Delta x = x + v_x \Delta t + \frac{1}{2} a_x (\Delta t)^2, \qquad y + \Delta y = y + v_y \Delta t + \frac{1}{2} a_y (\Delta t)^2. \qquad (5\text{–}21)$$

We have to specify the starting conditions, that is, the initial values of x, y, v_x, and v_y. Then we can step through the calculation to find the position and velocity at the end of each interval in terms of their values at the beginning, and thus to find the values at the end of any number of intervals. It would be a lot of work to do all the calculations for 100 or more steps with a hand calculator, but the computer does it for us quickly and accurately.

Exactly how you implement this plan depends on whether you use a computer language such as C++, BASIC, FORTRAN, or Pascal or use a spreadsheet or a numerical-analysis software package such as MathCad. Here's a sketch of a general algorithm using a programming language.

Step 1: Identify the parameters of the problem, m, A, C, and ρ, and evaluate D.

Step 2: Choose the time interval Δt and the initial values of x, y, v_x, v_y and t. You may want to express the initial velocity components in terms of the magnitude v_0 and direction α_0 of the initial velocity.

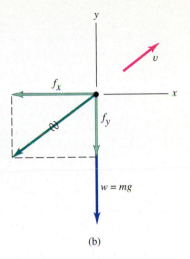

5–33 (b) Free-body diagram for the projectile. In still air, the air drag force $\vec{f}$ is always directed opposite $\vec{v}$.

Step 3: Choose the maximum number of intervals N or the maximum time $t_{max} = N\Delta t$ for which you want to get the numerical solution.

Step 4: Loop (or iterate) Steps 5 through 9 while $n < N$ or $t < t_{max}$:

Step 5: Calculate the acceleration components.

$$a_x = -(D/m)vv_x, \qquad a_y = -g - (D/m)vv_y.$$

Step 6: Print or plot x, y, v_x, v_y, a_x, and a_y.

Step 7: Calculate the new velocity components from Eqs. (5–20):

$$v_x + \Delta v_x = v_x + a_x\Delta t, \qquad v_y + \Delta v_y = v_y + a_y\Delta t.$$

Step 8: Calculate the new coordinates from Eqs. (5–21):

$$x + \Delta x = x + v_x\Delta t + \frac{1}{2}a_x(\Delta t)^2, \qquad y + \Delta y = y + v_y\Delta t + \frac{1}{2}a_y(\Delta t)^2.$$

Step 9: Increment the time by Δt:

$$t = t + \Delta t.$$

Step 10: Stop.

If you are using a spreadsheet, the following notation may be useful. Let $x(n), y(n), v_x(n)$, and $v_y(n)$ be the values at the end of the nth interval. These values appear across the nth line of the spreadsheet. The initial values are $x(1), y(1), v_x(1)$, and $v_y(1)$. Then Eqs. (5–20) and (5–21) become

$$v_x(n + 1) = v_x(n) + a_x(n)\Delta t, \qquad x(n + 1) = x(n) + v_x(n)\Delta t + \frac{1}{2}a_x(n)(\Delta t)^2 \quad (5\text{–}22)$$

$$v_y(n + 1) = v_y(n) + a_y(n)\Delta t, \qquad y(n + 1) = y(n) + v_y(n)\Delta t + \frac{1}{2}a_y(n)(\Delta t)^2. \quad (5\text{–}23)$$

In Eqs. (5–22) and (5–23) we compute $a_x(n)$ and $a_y(n)$ by substituting the values $v_x(n)$ and $v_y(n)$ into Eqs. (5–18).

Figure 5–34 shows the trajectory of a baseball with and without air drag. The radius of the baseball is $r = 0.0366$ m, and $A = \pi r^2$. The mass is $m = 0.145$ kg, and we have chosen the drag coefficient to be $C = 0.5$ (appropriate for a batted ball or pitched fastball) and the density of air to be $\rho = 1.2$ kg/m^3 (appropriate for a ballpark at sea level). In this example the baseball was given an initial velocity of 50 m/s at an angle of 35° above the $+x$-axis. You can see that both the range of the baseball and the maximum height reached are substantially less than the zero-drag calculation would lead you to believe. Hence the baseball trajectory that we calculated in Example 3–7 (Section 3–4) is quite unrealistic, since there we ignored air drag completely. Air drag is an important part of the game of baseball!

5–34 Computer-generated trajectories of a baseball with and without drag. Note that different scales are used on the horizontal and vertical axes.

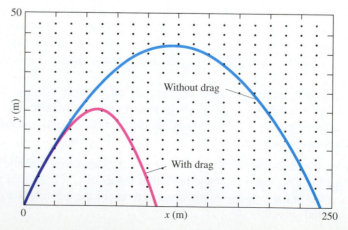

SUMMARY

- When a body is in equilibrium in an inertial frame of reference, the vector sum of forces acting on it must be zero. In component form,

$$\Sigma F_x = 0, \qquad \Sigma F_y = 0 \qquad \text{(particle in equilibrium)}. \qquad (5\text{–}2)$$

Free-body diagrams are useful in identifying the forces acting on the body being considered. Newton's third law is also frequently needed in equilibrium problems. The two forces in an action-reaction pair *never* act on the same body.

- When the vector sum of forces on a body is not zero, the body has an acceleration determined by Newton's second law:

$$\Sigma \vec{F} = m\vec{a}. \qquad (5\text{–}3)$$

In component form,

$$\Sigma F_x = ma_x, \qquad \Sigma F_y = ma_y. \qquad (5\text{–}4)$$

- The contact force between two bodies can always be represented in terms of a normal force $\vec{n}$ perpendicular to the interaction surface and a frictional force $\vec{f}$ parallel to the surface. When sliding occurs, f_k is approximately proportional to n, and the proportionality constant is μ_k, the coefficient of kinetic friction:

$$f_k = \mu_k n \qquad \text{(magnitude of kinetic friction force)}. \qquad (5\text{–}5)$$

When there is no relative motion, the maximum possible friction force is approximately proportional to the normal force. The proportionality constant is μ_s, the coefficient of static friction:

$$f_s \leq \mu_s n \qquad \text{(magnitude of static friction force)}. \qquad (5\text{–}6)$$

The actual static friction force may be anything from zero to this maximum value, depending on the situation. Usually, μ_k is less than μ_s for a given pair of surfaces.

- In uniform circular motion, the acceleration vector is directed toward the center of the circle and has magnitude v^2/R. The motion is governed by $\Sigma \vec{F} = m\vec{a}$, just as for any other dynamics problem.

- The fundamental forces of nature are the gravitational, electromagnetic, strong, and weak interactions. The electromagnetic and weak interactions have been unified into a single interaction, the electroweak interaction.

KEY TERMS

apparent weight, 127
friction force, 132
kinetic friction force, 132
coefficient of kinetic friction, 132
static friction force, 133
coefficient of static friction, 133
coefficient of rolling friction, 137
fluid resistance, 137
air drag, 138
terminal speed, 138
gravitational interaction, 145
electromagnetic interaction, 145
strong interaction, 145
weak interaction, 146
drag coefficient, 147

DISCUSSION QUESTIONS

Q5–1 A man sits in a seat that is suspended from a rope. The rope passes over a pulley suspended from the ceiling, and the man holds the other end of the rope in his hands. What is the tension in the rope, and what force does the seat exert on the man? Draw a free-body force diagram for the man.

Q5–2 "In general, the normal force is not equal to the weight." Give an example where the two forces are equal in magnitude, and at least two examples where they are not.

Q5–3 A clothesline is hung between two poles. No matter how tightly the line is stretched, it always sags a little at the center. Explain why.

Q5–4 A car is being driven up a steep hill at constant speed. Discuss all the forces acting on the car. What pushes it up the hill?

Q5–5 When you tighten a nut on a bolt, how are you increasing the frictional force? How does a lock washer work?

Q5–6 In pushing a box up a ramp, is the force required smaller if you push horizontally or if you push parallel to the ramp? Why?

Q5–7 A block rests on an inclined plane with enough friction to prevent its sliding down. To start the block moving, is it easier to push it up the plane, down the plane, or sideways? Why?

Q5–8 A crate of books rests on a level floor. To move it along the floor at a constant velocity, why do you exert a smaller force if you pull it at an angle θ above the horizontal than if you push it at the same angle below the horizontal?

Q5–9 In stopping a car on an icy road, is it better to push the brake pedal hard enough to "lock" the wheels and make them slide or to push gently so that the wheels continue to roll? Why?

Q5–10 Can the force of kinetic friction acting on an object ever cause that object to *speed up?* If it cannot, explain why not. If it can, give at least one example. Repeat for the force of *static* friction.

Q5–11 When you stand with bare feet in a wet bathtub, the grip feels fairly secure, and yet a catastrophic slip is quite possible. Explain this in terms of the two coefficients of friction.

Q5–12 For medical reasons it is important for astronauts in outer space to determine their body mass at regular intervals. Devise a scheme for measuring body mass in an apparently weightless environment.

Q5–13 A woman in an elevator lets go of her briefcase but it does not fall to the floor. How is the elevator moving?

Q5–14 Scales for weighing objects can be classified as those that use springs and those that use standard masses to balance unknown masses. Which group would be the more accurate when used in an accelerating elevator? When used on the moon? Does it matter whether you are trying to determine mass or weight?

Q5–15 Fast pitchers can make a baseball move faster than its 43 m/s (96 mi/h) terminal speed. How is this possible?

Q5–16 Because of air resistance, two objects of unequal mass do *not* fall at precisely the same rate. If two bodies of identical shape but unequal mass are dropped simultaneously from the same height, which one reaches the ground first? Explain.

Q5–17 A tennis ball is dropped from rest at the top of a tall glass cylinder, first with the air pumped out of the cylinder so there is no air resistance and then a second time after the air has been readmitted to the cylinder. You examine multiflash photographs (like Fig. 2–18) of the two drops. From these photos how can you tell which one is which, or can you?

Q5–18 Roller Coaster Road near Tucson, Arizona, was so named because it resembled an amusement park ride. If you drove on it at constant speed, where was the normal force the greatest and the least? Why would it have been especially poor road design to put an unbanked right or left curve at the top of one of the hills?

Q5–19 An automotive magazine calls decreasing-radius curves "the bane of the Sunday driver." Explain.

Q5–20 If you hang fuzzy dice from your rear-view mirror and drive through a banked curve, how can you tell whether you are traveling less than, equal to, or greater than the speed used to calculate the banking angle?

Q5–21 If there is a net force on a particle in uniform circular motion, why doesn't the particle's speed change?

Q5–22 A curve in a road has the banking angle calculated for 80 km/h. However, the road is covered with ice, and you plan to creep around the highest lane at 20 km/h. What may happen to your car? Why?

Q5–23 If you swing a ball on the end of a lightweight string in a horizontal circle at constant speed, the string at any instant does not lie exactly along the radius vector from the center of the path to the ball's location. Does the string lie above or below the horizontal? In relation to the direction the ball is moving, is the string in front of or behind the radius vector? Draw a free-body force diagram for the ball and use it to explain your answers. (Note that air resistance may be a factor.)

Q5–24 The centrifugal force is not included in the free-body diagrams of Figs. 5–28b and 5–29b. Explain why not.

Q5–25 A professor swings a rubber stopper in a horizontal circle on the end of a string in front of his class. He tells Caroline in the first row that he is going to let the string go when the stopper is directly in front of her face. Should Caroline worry?

Q5–26 To keep the forces on the riders within allowable limits, loop-the-loop roller coaster rides are often designed so the loop, rather than being a perfect circle, has a larger radius of curvature at the bottom than at the top. Explain.

Q5–27 You throw a baseball straight upward. If air resistance is *not* neglected, how does the time required for the ball to go from the height at which it was thrown up to its maximum height compare to the time required for it to fall from its maximum height back down to the height from which it was thrown? Explain your answer.

Q5–28 Will the frictional force always be opposite a baseball's velocity even if the wind is blowing? Explain.

Q5–29 When can a baseball in flight have an acceleration with a positive upward component? Explain in terms of the forces on the ball and also in terms of the velocity components compared to the terminal speed. Do *not* neglect air resistance.

Q5–30 When a batted baseball moves with air drag, does it spend more time climbing to its maximum height, or descending from its maximum height back to the ground? Or is the time the same for both? Explain in terms of the forces acting on the ball.

Q5–31 When a batted baseball moves with air drag, does it travel a greater horizontal distance while climbing to its maximum height, or while descending from its maximum height back to the ground? Or is the horizontal distance traveled the same for both? Explain in terms of the forces acting on the ball.

Q5–32 "A ball is thrown from the edge of a high cliff. No matter what the angle at which it is thrown, due to air resistance the ball will eventually end up moving vertically downward." Justify this statement.

EXERCISES

SECTION 5–2 USING NEWTON'S FIRST LAW: PARTICLES IN EQUILIBRIUM

5–1 Two 25.0-N weights are suspended at opposite ends of a rope that passes over a light, frictionless pulley. The pulley is attached to a chain that goes to the ceiling. a) What is the tension in the rope? b) What is the tension in the chain?

5–2 In Fig. 5–35 each of the suspended blocks has weight w. The pulleys are frictionless and the ropes have negligible weight. Calculate in each case the tension T in the rope in terms of the weight w. In each case, include the free-body diagram or diagrams you used to determine the answer.

FIGURE 5–35 Exercise 5–2.

5–3 An adventurous archaeologist crosses between two rock cliffs by slowly going hand-over-hand along a rope stretched between the cliffs. He stops to rest at the middle of the rope (Fig. 5–36). The rope will break if the tension in it exceeds 2.50×10^4 N, and our hero's mass is 90.0 kg. a) If the angle θ is 10.0°, find the tension in the rope. b) What is the smallest value the angle θ can have if the rope is not to break?

FIGURE 5–36 Exercise 5–3.

5–4 A picture frame hung against a wall is suspended by two wires attached to its upper corners. If the two wires make the same angle with the vertical, what must this angle be if the tension in each wire is equal to 0.75 of the weight of the frame? (Neglect any friction between the wall and the picture frame.)

5–5 Solve the problem in Example 5–3 using coordinate axes where the y-axis is vertical and the x-axis is horizontal. Do you get the same answers as obtained in the example using this different set of axes?

5–6 Certain streets in San Francisco make an angle of 17.5° with the horizontal. What force parallel to the street surface is required to keep a loaded 1967 Corvette of mass 1390 kg from rolling down such a street?

5–7 A large wrecking ball is held in place by two light steel cables (Fig. 5–37). If the mass m of the wrecking ball is 4090 kg, what is a) the tension T_B in the cable that makes an angle of 40° with the vertical? b) the tension T_A in the horizontal cable?

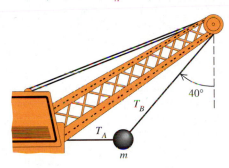

FIGURE 5–37 Exercise 5–7.

5–8 Find the tension in each cord in Fig. 5–38 if the weight of the suspended object is w.

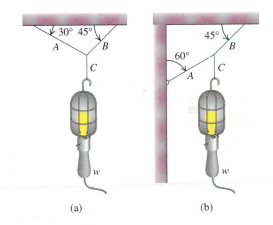

FIGURE 5–38 Exercise 5–8.

5–9 At a certain part of your drive from home to school, your car (mass 1600 kg) will coast in neutral at a constant speed of 72 km/h if there is no wind. Examination of a topographical map shows that for this stretch of straight highway the elevation decreases 200 m for each 6000 m of road. What is the total resistive force (friction plus air resistance) that acts on your car when it is traveling at 72 km/h?

5–10 A man pushes on a piano with mass 180 kg so that it slides at constant velocity down a ramp that is inclined at 11.0° above the horizontal. Neglect friction acting on the piano. If the force applied by the man is parallel to the incline, calculate the magnitude of this force.

5–11 In Fig. 5–39 the weight w is 60.0 N. a) What is the tension in the diagonal string? b) Find the magnitudes of the horizontal forces $\vec{F}_1$ and $\vec{F}_2$ that must be applied to hold the system in the position shown.

FIGURE 5–39 Exercise 5–11.

5-12 A tether ball leans against the post to which it is attached (Fig. 5–40). If the string to which the ball is attached is 1.40 m long and the ball has a radius of 0.110 m and a mass of 0.270 kg, what are the tension in the rope and the force the pole exerts on the ball? Neglect any friction between the ball and the pole. (The string is attached to the ball such that a line along the string passes through the center of the ball.)

FIGURE 5–40 Exercise 5–12.

5–13 Two blocks, each with weight w, are held in place on a frictionless incline (Fig. 5–41). In terms of w and the angle α of the incline, calculate the tension in a) the rope connecting the blocks; b) the rope that connects block A to the wall. c) Calculate the magnitude of the force that the incline exerts on each block. d) Interpret your answers for the cases $\alpha = 0$ and $\alpha = 90°$.

FIGURE 5–41 Exercise 5–13.

5–14 An airplane is in level flight at a constant speed. There are four forces acting on it: its weight $w = mg$, a forward force F provided by the engine (the *thrust*), the air resistance or *drag*

force f which acts opposite to the direction of flight, and a lift force L provided by the wings, which acts perpendicular to the direction of flight. The drag force f is proportional to the square of the speed. a) Show that $F = f$ and $L = w$. b) Suppose the pilot pushes the throttle forward, doubling the value of F, while maintaining a constant altitude. The airplane will eventually attain a new, faster, constant speed. At this faster speed, how does the new value of f compare to the previous value? c) How much faster than the original speed is the airplane going now?

SECTION 5–3 USING NEWTON'S SECOND LAW: DYNAMICS OF PARTICLES

5–15 Atwood's Machine. A 15.0-kg load of bricks hangs from one end of a rope that passes over a small, frictionless pulley. A 28.0-kg counterweight is suspended from the other end of the rope, as shown in Fig. 5–42. The system is released from rest. a) Draw two free-body diagrams, one for the load of bricks and one for the counterweight. b) What is the magnitude of the upward acceleration of the load of bricks? c) What is the tension in the rope while the load is moving? How does the tension compare to the weight of the load of bricks? To the weight of the counterweight?

FIGURE 5–42 Exercise 5–15.

5–16 A 8.00-kg block of ice released from rest at the top of a 1.50-m-long frictionless ramp slides downhill, reaching a speed of 2.50 m/s at the bottom. What is the angle between the ramp and the horizontal?

5–17 A light rope is attached to a block with a mass of 4.00 kg that rests on a horizontal, frictionless surface. The horizontal rope passes over a frictionless, massless pulley and a block of mass m is suspended from the other end. When the blocks are released, the tension in the rope is 10.0 N. a) Draw two free-body diagrams, one for the 4.00-kg block and one for the block of mass m. What is b) the acceleration of either block? c) the mass m of the hanging block? d) How does the tension compare to the weight of the hanging block?

5–18 A transport plane takes off from a level landing field with two gliders in tow, one behind the other. The mass of each glider is 700 kg, and the total resistance (air drag plus friction with the runway) on each may be assumed constant and equal to 2500 N. The tension in the towrope between the transport plane and the first glider is not to exceed 12,000 N. a) If a speed of 40 m/s is required for takeoff, what minimum length of runway is needed? b) What is the tension in the towrope between the two gliders while they are accelerating for the takeoff?

5–19 A 550-N physics student stands on a bathroom scale in an elevator. As the elevator starts moving, the scale reads 450 N. a) Find the acceleration of the elevator (magnitude and direction). b) What is the acceleration if the scale reads 670 N? c) If the scale reads zero, should the student worry? Explain.

5–20 A physics student playing with an air hockey table (a frictionless surface) finds that if she gives the puck a velocity of 3.80 m/s along the length (1.75 m) of the table at one end, by the time it has reached the other end the puck has drifted 2.50 cm to the right but still has a velocity component along the length of 3.80 m/s. She concludes correctly that the table is not level and correctly calculates its inclination from the above information. What is the angle of inclination?

5–21 a) Which way will the accelerometer in Fig. 5–14 deflect if the car is at rest on a sloping surface? b) Suppose the car is given a push up an icy (and hence frictionless) hill. After being pushed, the car moves uphill, slows to a halt, then moves back downhill. What is the direction of the accelerometer deflection in each stage of the motion? Explain your answers.

5–22 Which way will the accelerometer in Fig. 5–14 deflect under the following conditions? a) The car is moving toward the left and speeding up. b) The car is moving toward the left and slowing down. c) The car is moving toward the right and speeding up. Explain your answers.

SECTION 5–4 FRICTIONAL FORCES

5–23 **Free-Body Diagrams.** The first two steps in the solution of Newton's second law problems are to select an object for analysis and then to draw free-body diagrams for that object. Draw free-body diagrams for the following situations: a) a mass M sliding down a frictionless inclined plane of angle α; b) a mass M sliding up a frictionless inclined plane of angle α; c) a mass M sliding up an inclined plane of angle α with kinetic friction present; d) masses M and m sliding down an inclined plane of angle α with friction present, as shown in Fig. 5–43a. Here draw free-body diagrams for both m and M. Identify the forces that are action-reaction pairs. e) Draw free-body diagrams for masses m and M shown in Fig. 5–43b. Identify all action-reaction pairs. There is a frictional force between all surfaces in contact. The pulley is frictionless and massless. In all cases be sure you have the correct direction of the forces and are completely clear on what object is causing each force in your free-body diagram.

5–24 a) A large rock rests upon a rough horizontal surface. A bulldozer pushes on the rock with a horizontal force T that is slowly increased, starting from zero. In a graph with T along the x-axis and the friction force f along the y-axis, start at $T = 0$ and show the region of no motion, the point at which the rock is ready to move, and the region where the rock is moving. b) A block with weight w rests on a rough horizontal plank. The slope

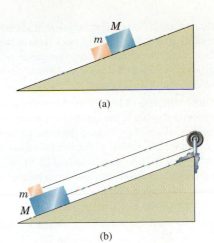

(a)

(b)

FIGURE 5–43 Exercise 5–23.

angle of the plank θ is gradually increased until the block starts to slip. Draw two graphs, both with θ along the x-axis. In one graph, show the ratio of the normal force to the weight, n/w, as a function of θ. In the second graph, show the ratio of the friction force to the weight, f/w. Indicate the region of no motion, the point at which the block is ready to move, and the region where the block is moving.

5–25 A stockroom worker pushes a box with mass 11.2 kg on a horizontal surface with a constant speed of 3.50 m/s. The coefficient of kinetic friction between the box and the surface is 0.20. a) What horizontal force must be applied by the worker to maintain the motion? b) If the force calculated in part (a) is removed, how far does the box slide before coming to rest?

5–26 A box of bananas weighing 40.0 N rests on a horizontal surface. The coefficient of static friction between the box and the surface is 0.40, and the coefficient of kinetic friction is 0.20. a) If no horizontal force is applied to the box and the box is at rest, how large is the friction force exerted on the box? b) What is the magnitude of the friction force if a monkey applies a horizontal force of 6.0 N to the box and the box is initially at rest? c) What minimum horizontal force must the monkey apply to start the box in motion? d) What minimum horizontal force must the monkey apply to keep the box moving at constant velocity once it has been started? e) If the monkey applies a horizontal force of 18.0 N, what is the magnitude of the friction force?

5–27 In a physics lab experiment, a 6.00-kg box is pushed across a flat table by a horizontal force $\vec{F}$. a) If the box is moving at a constant speed of 0.350 m/s and the coefficient of kinetic friction is 0.12, what is the magnitude of $\vec{F}$? b) What is the magnitude of $\vec{F}$ if the box is speeding up with a constant acceleration of 0.180 m/s²? c) How would your answers to parts (a) and (b) change if the experiments were performed on the moon, where $g = 1.62$ m/s²?

5–28 An 85-N box of oranges is being pushed across a horizontal floor. As it moves, it is slowing at a constant rate of 0.90 m/s each second. The push force has a horizontal component of 20 N and a vertical component of 25 N downward. Calculate the coefficient of kinetic friction between the box and floor.

5–29 A safe with mass 260 kg is to be lowered at constant speed down skids 20.0 m long, from a truck 2.00 m high. a) If the coefficient of kinetic friction between safe and skids is 0.25, does the safe need to be pulled down or held back? b) How great a force parallel to the skids is needed?

5–30 a) If the coefficient of kinetic friction between tires and dry pavement is 0.80, what is the shortest distance in which an automobile can be stopped by locking the brakes when travel-ing at 28.7 m/s (about 65 mi/h)? b) On wet pavement the coefficient of kinetic friction may be only 0.25. How fast should you drive on wet pavement in order to be able to stop in the same distance as in part (a)? (*Note:* Locking the brakes is *not* the safest way to stop.)

5–31 A clean brass washer slides along a horizontal clean steel surface until it stops. Using the values from Table 5–1, how many times farther would it have slid with the same initial speed if the washer had been Teflon-coated?

5–32 Consider the system shown in Fig. 5–44. Block A has weight w_A and block B has weight w_B. Once block B is set into downward motion, it descends at a constant speed. a) Calculate the coefficient of kinetic friction between block A and the table top. b) A cat, also of weight w_A, falls asleep on top of block A. If block B is now set into downward motion, what is its accelera-tion (magnitude and direction)?

FIGURE 5–44 Exercise 5–32, Exercise 5–35, and Problem 5–68.

5–33 Two crates connected by a rope lie on a horizontal surface (Fig. 5–45). Crate A has mass m_A, and crate B has mass m_B. The coefficient of kinetic friction between each crate and the surface is μ_k. The crates are pulled to the right at constant velocity by a horizontal force $\vec{F}$. In terms of m_A, m_B, and μ_k, calculate a) the magnitude of the force $\vec{F}$; b) the tension in the rope connecting the blocks. Include the free-body diagram or diagrams you used to determine each answer.

FIGURE 5–45 Exercise 5–33.

5–34 Two bicycle tires are set rolling with the same initial speed of 3.50 m/s along a long, straight road, and the distance each travels before its speed is reduced by half is measured. One tire is inflated to a pressure of 40 psi and goes 18.1 m; the other is at 105 psi and goes 92.9 m. What is the coefficient of rolling friction μ_r for each? Assume that the net horizontal force is due to rolling friction only.

5–35 As shown in Fig. 5–44, block A (mass 2.25 kg) rests on a table top. It is connected by a horizontal cord passing over a light, frictionless pulley to a hanging block B (mass 1.30 kg). The coefficient of kinetic friction between block A and the table top is 0.450. After the blocks are released from rest, find a) the speed of each block after moving 3.00 cm; b) the tension in the cord. Include the free-body diagram or diagrams you used to determine the answers.

5–36 A 25.0-kg box of textbooks rests on a loading ramp that makes an angle α with the horizontal. The coefficient of kinetic friction is 0.25, and the coefficient of static friction is 0.35. a) As the angle α is increased, find the minimum angle at which the box starts to slip. b) At this angle, find the acceleration once the box has begun to move. c) At this angle, how fast will the box be moving after it has slid 5.0 m along the inclined plane?

5–37 A large crate with mass m rests on a horizontal floor. The coefficients of friction between the crate and the floor are μ_s and μ_k. A woman pushes downward at an angle θ below the horizon-tal on the crate with a force $\vec{F}$. a) What magnitude of force $\vec{F}$ is required to keep the crate moving at constant velocity? b) If μ_s is larger than some critical value, the woman cannot start the crate moving no matter how hard she pushes. Calculate this critical value of μ_s.

5–38 A box with mass m is dragged across a level floor having a coefficient of kinetic friction μ_k by a rope that is pulled upward at an angle θ above the horizontal with a force of magnitude F. a) In terms of m, μ_k, θ, and g, obtain an expression for the mag-nitude of force required to move the box with constant speed. b) Knowing that you are studying physics, a CPR instructor asks you how much force it would take to slide a 90-kg patient across a floor at constant speed by pulling on him at an angle of 25° above the horizontal. By dragging some weights wrapped in an old pair of pants down the hall with a spring balance, you find that $\mu_k = 0.35$. Use the result of part (a) to answer the instructor's question.

5–39 Blocks A, B, and C are placed as in Fig. 5–46 and connected by ropes of negligible mass. Both A and B weigh 25.0 N each, and the coefficient of kinetic friction between each block and the sur-face is 0.35. Block C descends with constant velocity. a) Draw two separate free-body diagrams showing the forces acting on A and on B. b) Find the tension in the rope connecting blocks A and B. c) What is the weight of block C? d) If the rope connecting A and B were cut, what would be the acceleration of C?

FIGURE 5–46 Exercise 5–39.

5–40 Starting from Eq. (5–10), derive Eqs. (5–11) and (5–12).

5–41 a) In Example 5–19, what value of D is required to make $v_t = 42$ m/s for this sky diver? b) If the sky diver's daughter, whose mass is 45 kg, is falling through the air and has the same D (0.25 kg/m) as her father, what is the daughter's terminal speed?

5–42 A baseball is thrown straight up. The drag force is proportional to v^2. In terms of g, what is the y-component of the ball's acceleration when its speed is half its terminal speed and a) it is moving up? b) it is moving back down?

SECTION 5–5 DYNAMICS OF CIRCULAR MOTION

5–43 A stone with a mass of 0.80 kg is attached to one end of a string 0.90 m long. The string will break if its tension exceeds 600 N. (This is called the *breaking strength* of the string.) The stone is whirled in a horizontal circle on a frictionless table top; the other end of the string is kept fixed. Find the maximum speed the stone can attain without breaking the string.

5–44 A flat (unbanked) curve on a highway has a radius of 220 m. A car rounds the curve at a speed of 25.0 m/s. What is the minimum coefficient of friction that will prevent sliding?

5–45 Aircraft experience a lift force (due to the air) that is perpendicular to the plane of the wings. Light airplanes are designed so that their wings can safely provide a lift force of 3.8 times the weight of the airplane. A greater force can damage the structure of the wing. (Aerobatic airplanes and jet fighters are designed with much higher limits.) a) What is the maximum banking angle that a pilot can safely maintain in level flight without threatening the safety of the airplane (and her own safety)? b) Does your answer to part (a) depend on the speed of the airplane? Explain why or why not.

5–46 The "Giant Swing" at a county fair consists of a vertical central shaft with a number of horizontal arms attached at its upper end (Fig. 5–47). Each arm supports a seat suspended from a cable 5.00 m long, the upper end of the cable being fastened to the arm at a point 3.00 m from the central shaft. a) Find the time of one revolution of the swing if the cable supporting a seat makes an angle of 30.0° with the vertical. b) Does the angle depend on the weight of the passenger for a given rate of revolution?

FIGURE 5–47 Exercise 5–46.

5–47 Aircraft experience a lift force (due to the air) that is perpendicular to the plane of the wings. A small airplane is flying at a constant speed of 240 km/h. At what angle from the horizontal must the wings of the airplane be tilted for the plane to execute a horizontal turn from east to north with a turning radius of 1200 m?

5–48 A small button placed on a horizontal rotating platform with diameter 0.320 m will revolve with the platform when it is brought up to a speed of 40.0 rev/min, provided the button is no more than 0.150 m from the axis. a) What is the coefficient of static friction between the button and the platform? b) How far from the axis can the button be placed, without slipping, if the platform rotates at 60.0 rev/min?

5–49 Rotating Space Stations. One problem for humans living in outer space is that they are apparently weightless. One way around this problem is to design a space station that spins about its center at a constant rate. This creates "artificial gravity" at the outside rim of the station. a) If the diameter of the space station is 800 m, how many revolutions per minute are needed in order for the "artificial gravity" acceleration to be 9.8 m/s^2? b) If the space station is a waiting area for travelers going to Mars, it might be desirable to simulate the acceleration due to gravity on the Martian surface (3.70 m/s^2). How many revolutions per minute are needed in this case?

5–50 The Cosmoclock 21 Ferris wheel in Yokohama City, Japan, has a diameter of 100 m. Its name comes from its 60 arms, each of which can function as a second hand (so that it makes one revolution every 60.0 s). a) Find the speed of the passengers when the Ferris wheel is rotating at this rate. b) A passenger weighs 882 N at the weight-guessing booth on the ground. What is his apparent weight at the highest and at the lowest point on the Ferris wheel? c) What would be the time for one revolution if the passenger's apparent weight at the highest point were zero? d) What then would be the passenger's apparent weight at the lowest point?

5–51 An airplane flies in a loop (a circular path in a vertical plane) of radius 150 m. The pilot's head always points toward the center of the loop. The speed of the airplane is not constant; the airplane goes slowest at the top of the loop and fastest at the bottom. a) At the top of the loop, the pilot feels weightless. What is the speed of the airplane at this point? b) At the bottom of the loop, the speed of the airplane is 280 km/h. What is the apparent weight of the pilot at this point? His true weight is 700 N.

5–52 A 50.0-kg stunt pilot who has been diving her airplane vertically pulls out of the dive by changing her course to a circle in a vertical plane. a) If the plane's speed at the lowest point of the circle is 95.0 m/s, what is the minimum radius of the circle for the acceleration at this point not to exceed $4.00g$? b) What is the apparent weight of the pilot at the lowest point of the pullout?

5–53 A cord is tied to a pail of water, and the pail is swung in a vertical circle of radius 0.600 m. What must be the minimum speed of the pail at the highest point of the circle if no water is to spill from it?

5–54 A bowling ball weighing 71.2 N (16.0 lb) is attached to the ceiling by a 3.80-m rope. The ball is pulled to one side and released; it then swings back and forth as a pendulum. As the rope swings through the vertical, the speed of the bowling ball is 4.20 m/s. a) What is the acceleration of the bowling ball, in magnitude and direction, at this instant? b) What is the tension in the rope at this instant?

SECTION 5–7 PROJECTILE MOTION WITH AIR RESISTANCE: A CASE STUDY IN COMPUTER ANALYSIS

5–55 a) Implement the algorithm in Section 5–7 on a computer, and use it to reproduce the graph labeled "with drag" in Fig. 5–34. b) The home field for the Colorado Rockies is in Denver (altitude 1 mile), where the air density ρ is only 1.0 kg/m^3. For $m = 0.145$ kg, $r = 0.0366$ m, and $C = 0.5$, how much farther does a ball go than at sea level (where $\rho = 1.2$ kg/m^3) if $v_0 = 50$ m/s and $\alpha_0 = 35°$?

5–56 A baseball has $m = 0.145$ kg, $r = 0.0366$ m, and $C = 0.5$. At what angle should the ball be hit or thrown to maximize the range? Assume $v_0 = 50$ m/s and $\rho = 1.2$ kg/m^3.

5–57 Bo Jackson made a flat-footed 300-ft throw (no bounces) from left field to home plate to put out a tenth-inning tying run. Assuming that $\alpha_0 = 40°$, with what speed did he throw the ball? Use $m = 0.145$ kg, $r = 0.0366$ m, $C = 0.5$, and $\rho = 1.2$ kg/m^3. How does this compare to the 100-mi/h throwing speed of a good fastball pitcher?

5–58 In tennis, 100 mi/h is a very good serving speed. How fast is the ball moving when it crosses the baseline in the opposite court (24 m distant)? Use $m = 0.055$ kg, $r = 0.031$ m, $C = 0.75$, and $\rho = 1.2$ kg/m^3. The ball leaves the racquet moving horizontally and does not hit the ground until after it crosses the opposite court baseline.

5–59 Estimate the maximum distance a human being can throw a ping-pong ball. A ping-pong ball has $r = 0.019$ m and $m = 0.0024$ kg; use $C = 0.5$ and $\rho = 1.2$ kg/m^3. The fastest pitchers can throw at about 100 mi/h. Why can a baseball be thrown much farther than a ping-pong ball?

PROBLEMS

5–60 In Fig. 5–48 a worker lifts a weight w by pulling down on a rope with a force $\vec{F}$. The upper pulley is attached to the ceiling by a chain, and the lower pulley is attached to the weight by another chain. In terms of w, find the tension in each chain and the magnitude of the force $\vec{F}$ if the weight is lifted at constant speed. Include the free-body diagram or diagrams you used to determine your answers. Assume that the rope, pulleys, and chains all have negligible weights.

FIGURE 5–48 Problem 5–60.

5–61 A man is pushing a refrigerator up a ramp at constant speed. The ramp is inclined at an angle α above the horizontal, but the man applies a *horizontal* force $\vec{F}$. Calculate the magnitude of $\vec{F}$ in terms of α and the refrigerator mass m. Neglect friction on the refrigerator.

5–62 A Rope with Mass. In most problems in this book, the ropes, cords, or cables have so little mass compared to other objects in the problem that their mass can safely be neglected. But if the rope is the *only* object in the problem, then clearly its mass cannot be neglected. For example, suppose we have a clothesline attached to two poles (Fig. 5–49). The clothesline has a mass M, and each end makes an angle θ with the horizontal.

What is a) the tension at the ends of the clothesline? b) the tension at the lowest point? c) Why can't we have $\theta = 0$? (See Discussion Question Q5–3.) The curve of the clothesline, or of any flexible cable hanging under its own weight, is called a *catenary*. For a more advanced treatment of this curve, see K. R. Symon, *Mechanics*, 3rd ed. (Addison–Wesley, Reading, MA, 1971), pp. 237–241.

FIGURE 5–49 Problem 5–62.

5–63 A block of mass M is attached to the lower end of a vertical, uniform rope of mass m and length L. A constant upward force $\vec{F}$ is applied to the top of the rope, causing the rope and block to accelerate upward. Find the tension in the rope at a distance x from the top end of the rope, where x can have any value from 0 to L.

5–64 A block with mass m_1 is placed on an inclined plane with slope angle α and is connected to a second hanging block that has mass m_2 by a cord passing over a small, frictionless pulley (Fig. 5–50). The coefficient of static friction is μ_s, and the coefficient of kinetic friction is μ_k. a) Find the mass m_2 for which block m_1 moves up the plane at constant speed once it has been set in motion. b) Find the mass m_2 for which block m_1 moves down the plane at constant speed once it has been set in motion. c) For what range of values of m_2 will the blocks remain at rest if they are released from rest?

5–65 a) Block A in Fig. 5–51 weighs 60.0 N. The coefficient of static friction between the block and the surface on which it rests is 0.25. The weight w is 12.0 N, and the system is in equilibrium. Find the friction force exerted on block A. b) Find the maximum weight w for which the system will remain in equilibrium.

FIGURE 5–50 Problem 5–64.

FIGURE 5–51 Problem 5–65.

5–66 Block A in Fig. 5–52 weighs 1.20 N, and block B weighs 3.60 N. The coefficient of kinetic friction between all surfaces is 0.300. Find the magnitude of the horizontal force $\vec{F}$ necessary to drag block B to the left at constant speed a) if A rests on B and moves with it (Fig. 5–52a); b) if A is held at rest (Fig. 5–52b).

(a)   (b)

FIGURE 5–52 Problem 5–66.

5–67 A window washer pushes his scrub brush up a vertical window at constant speed by applying a force $\vec{F}$ as shown in Fig. 5–53. The brush weighs 12.0 N, and the coefficient of kinetic friction is $\mu_k = 0.150$. Calculate a) the magnitude of the force $\vec{F}$; b) the normal force exerted by the window on the brush.

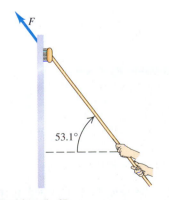

FIGURE 5–53 Problem 5–67.

5–68 In the system shown in Fig. 5–44, block A has mass m_A, block B has mass m_B, and the rope connecting them has a *nonzero* mass m_{rope}. The rope has a total length L and the pulley has a very small radius. Ignore any sag in the horizontal part of the string. a) If there is no friction between block A and the table top, find the acceleration of the blocks at an instant when a length d of rope hangs vertically between the pulley and block B. As block B falls, will the magnitude of the acceleration of the system increase, decrease, or remain constant? Explain. b) Let $m_A = 2.00$ kg, $m_B = 0.400$ kg, $m_{rope} = 0.160$ kg, and $L = 1.00$ m. If there is friction between block A and the table top, with $\mu_k = 0.200$ and $\mu_s = 0.250$, find the minimum value of the distance d such that the blocks will start to move if they are initially at rest. c) Repeat part (b) for the case $m_{rope} = 0.040$ kg. Will the blocks move in this case?

5–69 If the coefficient of static friction between a table and a uniform massive rope is μ_s, what fraction of the rope can hang over the edge of the table without the rope sliding?

5–70 A woman attempts to push a box of books that has mass m up a ramp inclined at an angle α above the horizontal. The coefficients of friction between the ramp and the box are μ_s and μ_k. The force $\vec{F}$ applied by the woman is *horizontal*. a) If μ_s is greater than some critical value, the woman cannot start the box moving up the ramp no matter how hard she pushes. Calculate this critical value of μ_s. b) Assume that μ_s is less than this critical value. What magnitude of force must the woman apply to keep the box moving up the ramp at constant speed?

5–71 A 30.0-kg packing case is initially at rest on the floor of a 1500-kg pickup truck. The coefficient of static friction between the case and the truck floor is 0.30, and the coefficient of kinetic friction is 0.20. Before each acceleration given below, the truck is traveling due north at constant speed. Find the magnitude and direction of the friction force acting on the case a) when the truck accelerates at 2.20 m/s^2 northward; b) when it accelerates at 3.40 m/s^2 southward.

5–72 The truck in Problem 5–71 is traveling at a constant speed on a country road where the speed limit is 72 km/h (45 mi/h). Spotting a stop sign ahead, the driver applies the brakes and comes to a halt in a distance of 47.0 m. Without warning a policeman emerges from behind a bush and hands the driver a speeding ticket. When the driver protests that he was going less than the speed limit, the policeman replies that "I saw that packing case slide from the back to the front of your truck's cargo area. To make it slide like that, you must have had to brake very hard, which means you had to have been speeding." Will the policeman's argument hold up in traffic court? (Assume that the judge, like you, knows physics.)

5–73 Troy uses a frayed rope to pull a crate across a level floor. The maximum tension in the rope without breaking is T_{max} and the coefficient of kinetic friction is μ_k. a) Show that the maximum weight that he can pull at constant speed is $T_{max}/(\sin\theta)$, where $\theta = \arctan\mu_k$ is the angle of the rope above the horizontal. b) The answer to part (a) suggests that Troy can pull a weight that approaches infinity with a bit of spiderweb as the coefficient of kinetic friction approaches zero. Explain.

5–74 The airplane in Exercise 5–14 has an engine failure (so $F = 0$), and the airplane glides at constant speed toward a safe landing. The direction of flight is a constant angle α (called the *glide angle*) below the horizontal (Fig. 5–54). a) Find the magnitude of the lift force L (which acts perpendicular to the direction of flight) and of the drag force f in terms of w and α. b) Show that $\alpha = \arctan (f/L)$. c) A fully loaded Cessna 182 (a single-engine airplane) weighs 12,900 N and has 1300 N of drag at a speed of 130 km/h. If this airplane has an engine failure at an altitude of 2500 m, what maximum distance over the ground can it glide while searching for a safe landing spot? d) Justify the statement that "it's drag, not gravity, that makes the airplane go down."

FIGURE 5–54 Problem 5–74.

5–75 The pilot of the Cessna 182 airplane in part (c) of Problem 5–74 manages to get the engine restarted. He applies full throttle, and the airplane climbs along a straight line angled above the horizontal. The airplane is flying at a constant speed of 130 km/h, weighs 12,900 N, and has 1300 N of drag. The rate-of-climb indicator on the instrument panel shows that the airplane is gaining altitude at a steady rate of 5.00 m/s (300 m/min, or about 980 feet/min). Determine the magnitude of the thrust (the forward force produced by the engine). (*Hint:* The thrust acts in the direction that the airplane is moving.)

5–76 A 12.0-kg box rests on the flat floor of a truck. The coefficients of friction between the box and floor are $\mu_s = 0.19$ and $\mu_k = 0.15$. The truck stops at a stop sign and then starts to move with an acceleration of 2.20 m/s². If the box is 1.80 m from the rear of the truck when the truck starts, how much time elapses before the box falls off the truck? How far does the truck travel in this time?

5–77 Block A in Fig. 5–55 weighs 1.40 N, and block B weighs 4.20 N. The coefficient of kinetic friction between all surfaces is 0.30. Find the magnitude of the horizontal force $\vec{F}$ necessary to drag block B to the left at constant speed if A and B are connected by a light, flexible cord passing around a fixed, frictionless pulley.

FIGURE 5–55 Problem 5–77.

5–78 A 30.0-kg crate is released from an airplane traveling due east at an altitude of 1200 m with a ground speed of 70.0 m/s. The wind applies a constant force on the crate of 180 N directed horizontally in the opposite direction to the plane's flight path. Where and when (with respect to the release location and time) does the crate hit the ground?

5–79 Block A in Fig. 5–56 has a mass of 4.00 kg, and block B has a mass of 12.0 kg. The coefficient of kinetic friction between block B and the horizontal surface is 0.25. a) What is the mass of block C if block B is moving to the right and speeding up with an acceleration 2.00 m/s²? b) What is the tension in each cord when block B has this acceleration?

FIGURE 5–56 Problem 5–79.

5–80 Two blocks connected by a cord passing over a small, frictionless pulley rest on frictionless planes (Fig. 5–57). a) Which way will the system move when the blocks are released from rest? b) What is the acceleration of the blocks? c) What is the tension in the cord?

FIGURE 5–57 Problem 5–80.

5–81 In terms of m_1, m_2, and g, find the accelerations of each block in Fig. 5–58. There is no friction anywhere in the system.

FIGURE 5–58 Problem 5–81.

5–82 Block B, of mass m_B, rests on block A, of mass m_A, which in turn is on a horizontal table top (Fig. 5–59). The coefficient of kinetic friction between block A and the table top is μ_k and the coefficient of static friction between block A and block B is μ_s. A light string attached to block A passes over a frictionless, massless pulley and block C is suspended from the other end of the string. What is the largest mass m_C that block C can have so that blocks A and B still slide together when the system is released from rest?

FIGURE 5–59 Problem 5–82.

5–83 Two objects with masses of 5.00 kg and 2.00 kg hang 0.600 m above the floor from the ends of a cord 6.00 m long passing over a frictionless pulley. Both objects start from rest. Find the maximum height reached by the 2.00-kg object.

5–84 Friction in an Elevator. You are riding in an elevator on the way to the eighteenth floor of your dormitory. The elevator is accelerating upward with $a = 1.90$ m/s^2. Beside you is the box containing your new computer; box and contents have a total mass of 28.0 kg. While the elevator is accelerating upward, you push horizontally on the box to slide it at constant speed toward the elevator door. If the coefficient of kinetic friction between the box and elevator floor is $\mu_k = 0.32$, what magnitude of force must you apply?

5–85 What acceleration must the cart in Fig. 5–60 have in order that block A does not fall? The coefficient of static friction between the block and the cart is μ_s. How would the behavior of the block be described by an observer on the cart?

FIGURE 5–60 Problem 5–85.

5–86 Two blocks with masses of 4.00 kg and 8.00 kg are connected by a string and slide down a 30.0° inclined plane (Fig. 5–61). The coefficient of kinetic friction between the 4.00-kg block and the plane is 0.25; that between the 8.00-kg block and the plane is 0.35. a) Calculate the acceleration of each block. b) Calculate the tension in the string. c) What happens if the positions of the blocks are reversed, so the 4.00-kg block is above the 8.00-kg block?

5–87 Block A, with weight $3w$, slides down an inclined plane S of slope angle 36.9° at a constant speed while plank B, with weight w, rests on top of A. The plank is attached by a cord to the

FIGURE 5–61 Problem 5–86.

top of the plane (Fig. 5–62). a) Draw a diagram of all the forces acting on block A. b) If the coefficient of kinetic friction is the same between A and B and between S and A, determine its value.

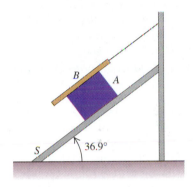

FIGURE 5–62 Problem 5–87.

5–88 A man with mass 70.0 kg stands on a platform with mass 25.0 kg. He pulls on the free end of a rope that runs over a pulley on the ceiling and has its other end fastened to the platform. The mass of the rope and the mass of the pulley can be neglected, and the pulley is frictionless. The rope is vertical on either side of the pulley. a) With what force does he have to pull so that he and the platform have an upward acceleration of 1.80 m/s^2? b) What is the acceleration of the rope relative to him?

5–89 Two blocks, of mass m_1 and m_2, are stacked as shown in Fig. 5–63 and placed on a frictionless horizontal surface. There is friction between the two blocks. An external force of magnitude F is applied to the top block at an angle α below the horizontal. a) If the two blocks move together, find their acceleration. b) Show that the two blocks will move together only if

$$F \leq \frac{\mu_s m_1 (m_1 + m_2) g}{m_2 \cos \alpha - \mu_s (m_1 + m_2) \sin \alpha},$$

where μ_s is the coefficient of static friction between the two blocks.

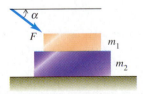

FIGURE 5–63 Problem 5–89.

5–90 A curve with a 120-m radius on a level road is banked at the correct angle for a speed of 20 m/s. If an automobile rounds this curve at 30 m/s, what is the minimum coefficient of static friction between tires and road needed to prevent skidding?

5–91 Consider a wet roadway banked as in Example 5–23 (Section 5–5), where there is a coefficient of static friction of 0.30 and a coefficient of kinetic friction of 0.25 between the tires and the roadway. The radius of the curve is $R = 50$ m. a) If the banking angle is $\beta = 25°$, what is the *maximum* speed the automobile can have before sliding *up* the banking? b) What is the *minimum* speed the automobile can have before sliding *down* the banking?

5–92 You are riding in a school bus. As the bus rounds a flat curve at constant speed, a lunch box with a mass of 0.500 kg suspended from the ceiling of the bus by a string 1.80 m long is found to hang at rest relative to the bus when the string makes an angle of 30.0° with the vertical. In this position the lunch box is 50.0 m from the center of curvature of the curve. What is the speed v of the bus?

5–93 The Monkey and Bananas Problem. A 20-kg monkey has a firm hold on a light rope that passes over a frictionless pulley and is attached to a 20-kg bunch of bananas (Fig. 5–64). The monkey looks upward, sees the bananas, and starts to climb the rope to get them. a) As the monkey climbs, do the bananas move up, down, or remain at rest? b) As the monkey climbs, does the distance between the monkey and the bananas decrease, increase, or remain constant? c) The monkey releases her hold on the rope. What happens to the distance between the monkey and the bananas while she is falling? d) Before reaching the ground, the monkey grabs the rope to stop her fall. What do the bananas do?

20 kg

20 kg

FIGURE 5–64 Problem 5–93.

5–94 A rock is thrown downward into water with a speed of $3mg/k$, where k is the coefficient in Eq. (5–7). Assume that the rela-

tion between fluid resistance and speed is as given in Eq. (5–7), and calculate the speed of the rock as a function of time.

5–95 A rock with mass $m = 3.00$ kg falls from rest in a viscous medium. The rock is acted on by a net constant downward force of 18.0 N (a combination of gravity and the buoyant force exerted by the medium) and by a fluid resistance force $f = kv$, where v is the speed in m/s and $k = 2.20$ N $\cdot$ s/m. (See Section 5–4.) a) Find the initial acceleration, a_0. b) Find the acceleration when the speed is 3.00 m/s. c) Find the speed when the acceleration equals $0.1a_0$. d) Find the terminal speed, v_t. e) Find the coordinate, speed, and acceleration 2.00 s after the start of the motion. f) Find the time required to reach a speed $0.9v_t$.

5–96 The 4.00-kg block in Fig. 5–65 is attached to a vertical rod by means of two strings. When the system rotates about the axis of the rod, the strings are extended as shown in the diagram and the tension in the upper string is 80.0 N. a) What is the tension in the lower cord? b) How many revolutions per minute does the system make? c) Find the number of revolutions per minute at which the lower cord just goes slack. d) Explain what happens if the number of revolutions per minute is less than in part (c).

1.25 m

2.00 m   4.00 kg

1.25 m

FIGURE 5–65 Problem 5–96.

5–97 Two identical twins, Jena and Jackie, are playing one December on a large merry-go-round (a disk mounted parallel to the ground on a vertical axle through its center) in their school playground in northern Minnesota. Each twin has a mass of 30.0 kg. The icy coating on the merry-go-round surface makes it frictionless. The merry-go-round revolves at a constant rate as the twins ride on it. Jena, sitting 1.80 m from the center of the merry-go-round, must hold onto one of the metal posts attached to the merry-go-round with a horizontal force of 60.0 N to keep from sliding off. Jackie is sitting at the edge, 3.60 m from the center. a) With what horizontal force must Jackie hold on to keep from falling off? b) If Jackie falls off, what will be her horizontal velocity when she becomes airborne?

5–98 Consider a passenger on a Ferris wheel like that of Example 5–24. a) What is the passenger's apparent weight when the passenger's velocity is vertical, in the $\pm y$-direction? b) Where in the rotation are the passenger's apparent weight and true weight equal in magnitude?

5–99 In the ride "Spindletop" at the amusement park Six Flags Over Texas, people stand against the inner wall of a hollow vertical cylinder with radius 2.5 m. The cylinder starts to rotate, and when it reaches a constant rotation rate of 0.60 rev/s, the floor on which people are standing drops about 0.5 m. The people remain pinned against the wall. a) Draw a force diagram for a person in this ride, after the floor has dropped. b) What minimum coefficient of static friction is required if the person in the ride is not to slide downward to the new position of the floor? c) Does your answer in part (b) depend on the mass of the passenger? (*Note:* When the ride is over, the cylinder is slowly brought to rest. As it slows down, people slide down the walls to the floor.)

5–100 A physics major is working to pay his college tuition by performing in a traveling carnival. He rides a motorcycle inside a hollow transparent plastic sphere. After gaining sufficient speed, he travels in a vertical circle with a radius of 13.0 m. The physics major has a mass of 70.0 kg, and his motorcycle has a mass of 40.0 kg. a) What minimum speed must he have at the top of the circle if the tires of the motorcycle are not to lose contact with the sphere? b) At the bottom of the circle his speed is twice the value calculated in part (a). What is the magnitude of the normal force exerted on the motorcycle by the sphere at this point?

5–101 You are driving a classic 1954 Nash Ambassador with a friend who is sitting to your right on the passenger side of the front seat. The Ambassador has flat bench seats. You would like to be closer to your friend and decide to use physics to achieve your romantic goal by making a quick turn. a) Which way (to the left or to the right) should you turn the car to get your friend to slide closer to you? b) If the coefficient of static friction between your friend and the car seat is 0.35 and you keep driving at a constant speed of 20 m/s, what is the maximum radius you could make your turn and still have your friend slide your way?

5–102 A small block with mass m rests on a frictionless horizontal tabletop a distance r from a hole in the center of the table (Fig. 5–66). A string tied to the small block passes down through the hole, and a larger block with mass M is suspended from the free end of the string. The small block is set into uniform circular motion with radius r and speed v. What must v be if the large block is to remain motionless when released?

FIGURE 5–66 Problem 5–102.

5–103 A small bead can slide without friction on a circular hoop that is in a vertical plane and has a radius of 0.100 m. The hoop rotates at a constant rate of 4.00 rev/s about a vertical diameter (Fig. 5–67). a) Find the angle β at which the bead is in vertical equilibrium. (Of course, it has a radial acceleration toward the axis.) b) Is it possible for the bead to "ride" at the

same elevation as the center of the hoop? c) What will happen if the hoop rotates at 1.00 rev/s?

FIGURE 5–67 Problem 5–103.

5–104 A model airplane with mass 2.20 kg moves in the xy-plane such that its x- and y-coordinates vary in time according to $x(t) = \alpha - \beta t^3$ and $y(t) = \gamma t - \delta t^2$, where $\alpha = 1.50$ m, $\beta = 0.120$ m/s^3, $\gamma = 3.00$ m/s, and $\delta = 1.00$ m/s^2. a) Calculate the x- and y-components of the net force on the plane as functions of time. b) Sketch the trajectory of the airplane between $t = 0$ and $t = 3.00$ s, and draw on your sketch vectors showing the net force on the airplane at $t = 0$, $t = 1.00$ s, $t = 2.00$ s, and $t = 3.00$ s. For each of these times, relate the direction of the net force to the direction that the airplane is turning, and to whether the airplane is speeding up or slowing down (or neither). c) What are the magnitude and direction of the net force at $t = 3.00$ s?

5–105 A particle moves on a frictionless surface along a path as shown in Fig. 5–68. (The figure gives a view looking down on the surface.) The particle is initially at rest at point A, then begins to move toward B as it gains speed at a constant rate. From B to C the particle moves along a circular path at a constant speed. The speed remains constant along the straight-line path from C to D. From D to E the particle moves along a circular path, but now its speed is decreasing at a constant rate. The speed continues to decrease at a constant rate as the particle moves from E to F; the particle comes to a halt at F. (The time intervals between the marked points are not equal.) At each point marked with a dot, draw arrows to represent the velocity, the acceleration, and the net force acting on the particle. Draw a longer or shorter arrow to represent vectors of larger or smaller magnitude.

FIGURE 5–68 Problem 5–105.

5–106 A small remote-control car with a mass of 1.60 kg moves at a constant speed of $v = 12.0$ m/s in a vertical circle inside a hollow metal cylinder that has a radius of 5.00 m (Fig. 5–69). What is the magnitude of the normal force exerted on the car by the walls of the cylinder at a) point A (at the bottom of the vertical circle)? b) point B (at the top of the vertical circle)?

FIGURE 5–69 Problem 5–106.

5–107 A small block with mass m is placed inside an inverted cone that is rotating about a vertical axis such that the time for one revolution of the cone is T (Fig. 5–70). The walls of the cone make an angle β with the vertical. The coefficient of static friction between the block and the cone is μ_s. If the block is to remain at a constant height h above the apex of the cone, what are the maximum and minimum values of T?

FIGURE 5–70 Problem 5–107.

CHALLENGE PROBLEMS

5–108 A wedge with mass M rests on a frictionless horizontal table top. A block with mass m is placed on the wedge (Fig. 5–71a). There is no friction between the block and the wedge. The system is released from rest. a) Calculate the acceleration of the wedge and the horizontal and vertical components of the acceleration of the block. b) Do your answers to part (a) reduce to the correct results when M is very large? c) As seen by a stationary observer, what is the shape of the trajectory of the block?

5–109 A wedge with mass M rests on a frictionless horizontal table top. A block with mass m is placed on the wedge, and a horizontal force $\vec{F}$ is applied to the wedge (Fig. 5–71b). What must be the magnitude of $\vec{F}$ if the block is to remain at a constant height above the table top?

(a)   (b)

FIGURE 5–71 Challenge Problems 5–108 and 5–109.

5–110 A box of weight w is accelerated up a ramp by a rope that exerts a tension T. The ramp makes an angle α with the horizontal and the rope makes an angle θ above the ramp. The coefficient of kinetic friction between the box and the ramp is μ_k. Show that no matter what the value of α, the acceleration is maximum if $\theta = \arctan \mu_k$ (as long as the box remains in contact with the ramp).

5–111 A box with weight w is pulled at constant speed along a level floor by a force $\vec{F}$ that is at an angle θ above the horizontal. The coefficient of kinetic friction between the floor and box is μ_k. a) In terms of θ, μ_k, and w, calculate F. b) For $w = 400$ N and $\mu_k = 0.25$, calculate F for θ ranging from 0° to 90° in increments of 10°. Graph F versus θ. c) From the general expression in part (a), calculate the value of θ for which the value of F required to maintain constant speed is a minimum. (*Hint:* At a point where a function is minimum, what are the first and second derivatives of the function? Here F is a function of θ.) For the special case of $w = 400$ N and $\mu_k = 0.25$, evaluate this optimal θ and compare your result to the graph you constructed in part (b).

5–112 A baseball is dropped from the roof of a tall building. As the ball falls, the air exerts a drag force proportional to the square of the ball's speed ($f = Dv^2$). a) In a diagram, show the direction of motion and indicate with the aid of vectors all of the forces acting on the ball. b) Apply Newton's second law and infer from the resulting equation the general properties of the motion. c) Show that the ball acquires a terminal speed that is as given in Eq. (5–13). d) Derive the equation for the speed at any time. (*Note:*

$$\int \frac{dx}{a^2 - x^2} = \frac{1}{a} \operatorname{arctanh}\left(\frac{x}{a}\right),$$

where

$$\tanh(x) = \frac{e^x - e^{-x}}{e^x + e^{-x}} = \frac{e^{2x} - 1}{e^{2x} + 1}$$

defines the hyperbolic tangent.)

5–113 Double Atwood's Machine. In Fig. 5–72 masses m_1 and m_2 are connected by a light string A over a light frictionless pulley B. The axle of pulley B is connected by a second light string C over a second light frictionless pulley D to a mass m_3. Pulley D is suspended from the ceiling by an attachment to its axle. The system is released from rest. In terms of m_1, m_2, m_3, and g, what is a) the acceleration of block m_3? b) the acceleration of pulley B? c) the acceleration of block m_1? d) the acceleration of block m_2? e) the tension in string A? f) the tension in string C? g) What do your expressions give for the special case of $m_1 = m_2$ and $m_3 = m_1 + m_2$? Is this sensible?

FIGURE 5–72 Challenge Problem 5–113.

5–114 The masses of blocks A and B in Fig. 5–73 are 20.0 kg and 10.0 kg, respectively. The blocks are initially at rest on the floor and are connected by a massless string passing over a massless and frictionless pulley. An upward force $\vec{F}$ is applied to the pulley. Find the accelerations $\vec{a}_1$ of block A and $\vec{a}_2$ of block B when F is a) 124 N; b) 294 N; c) 424 N.

FIGURE 5–73 Challenge Problem 5–114.

5–115 A ball is held at rest at position A in Fig. 5–74 by two light strings. The horizontal string is cut, and the ball starts swinging as a pendulum. Point B is the farthest to the right the ball goes as it swings back and forth. What is the ratio of the tension in the supporting string in position B to its value at A before the horizontal string was cut?

FIGURE 5–74 Challenge Problem 5–115.

6 Work and Kinetic Energy

To prepare an arrow for flight, an archer does work—that is, exerts a force over a distance—to draw back the arrow and stretch the bow and bowstring. The bow in turn does work on the arrow when it is released, giving the arrow kinetic energy, or energy of motion. This kinetic energy is used to do work on the target as the arrow penetrates into the bull's-eye.

6-1 INTRODUCTION

Some problems are harder than they look. Suppose you try to find the speed of an arrow that has been shot from a bow. You apply Newton's laws and all the problem-solving techniques that we've learned, but you run across a major stumbling block: After the archer releases the arrow, the bow string exerts a *varying* force that depends on the arrow's position. As a result, the simple methods that we've learned aren't enough to calculate the speed. Never fear; we aren't by any means finished with mechanics, and there are other methods for dealing with such problems.

The new method that we're about to introduce uses the ideas of *work* and *energy*. The applications of these ideas go far beyond mechanics, however. The importance of the energy idea stems from the *principle of conservation of energy:* Energy is a quantity that can be converted from one form to another but cannot be created or destroyed. In an automobile engine, chemical energy stored in the fuel is converted partially to the energy of the automobile's motion and partially to thermal energy. In a microwave oven, electromagnetic energy obtained from your power company is converted to thermal energy of the food being cooked. In these and all other processes, the *total* energy the sum of all energy present in all different forms—remains the same. No exception has ever been found.

We'll use the energy idea throughout the rest of this book to study a tremendous range of physical phenomena.

This idea will help you understand why a sweater keeps you warm, how a camera's flash unit can produce a short burst of light, and the meaning of Einstein's famous equation $E = mc^2$.

In this chapter, though, our concentration will be on mechanics. We'll learn about one important form of energy called *kinetic energy,* or energy of motion, and how it relates to the concept of *work.* We'll also consider *power,* which is the time rate of doing work. In Chapter 7 we'll expand the ideas of work and kinetic energy into an understanding of the general concept of energy, and we'll see how the conservation of energy arises.

6-2 WORK

You'd probably agree that it's hard work to pull a heavy sofa across the room, to lift a stack of encyclopedias from the floor to a high shelf, or to push a stalled car off the road. Indeed, all of these examples agree with the everyday meaning of *work*—any activity that requires muscular or mental effort.

In physics, work has a much more precise definition. By making use of this definition we'll find that in any motion, no matter how complicated, the total work done on a particle by all forces that act on it equals the change in its *kinetic energy*—a quantity that's related to the particle's speed. This relationship holds even when the forces acting on the particle aren't constant, a situation that can be difficult or impossible to handle with the techniques you learned in Chapters 4 and 5. So the ideas of work and kinetic energy will enable us to solve problems in mechanics that we could not have attempted before.

We'll develop the relationship between work and kinetic energy in Section 6–3, and see what to do with non-constant forces in Section 6–4. Meanwhile, let's see how work is defined and how to calculate work in a variety of situations involving *constant* forces. Even though we already know how to solve problems in which the forces are constant, the idea of work is still a useful one in such problems.

The three examples of work described above—pulling a sofa, lifting encyclopedias, and pushing a car—have something in common. In each case you do work by exerting a *force* on a body while that body *moves* from one place to another, that is, undergoes a *displacement*. You do more work if the force is greater (you pull harder on the sofa) or if the displacement is greater (you pull the sofa farther across the room).

The physicist's definition of work is based on these observations. Consider a body that undergoes a displacement of magnitude s along a straight line. (For now, we'll assume that any body we discuss can be treated as a particle so that we can ignore any rotation or changes in shape of the body.) While the body moves, a constant force with magnitude F acts on it in the same direction as the displacement (Fig. 6–1). We define the **work** W done by a constant force F acting on the body under these conditions as

$$W = Fs \quad \text{(constant force in direction of straight-line displacement).} \quad (6\text{--}1)$$

The work done on the body is greater if either the force F or the displacement s is greater, in agreement with our observations above.

CAUTION ▶ Don't confuse W (work) with w (weight). Though the symbols may be almost the same, work and weight are different quantities. ◀

The SI unit of work is the **joule** (abbreviated J, pronounced "jewel," and named in honor of the nineteenth century English physicist James Prescott Joule). From Eq. (6–1) we see that in any system of units, the unit of work is the unit of force multiplied by the unit of distance. In SI units the unit of force is the newton and the unit of distance is the meter, so one joule is equivalent to one *newton-meter* (N · m):

$$1 \text{ joule} = (1 \text{ newton})(1 \text{ meter}) \quad \text{or} \quad 1 \text{ J} = 1 \text{ N} \cdot \text{m}.$$

In the British system the unit of force is the pound (lb), the unit of distance is the foot, and the unit of work is the *foot-pound* (ft · lb). The following conversions are useful:

$$1 \text{ J} = 0.7376 \text{ ft} \cdot \text{lb}, \quad 1 \text{ ft} \cdot \text{lb} = 1.356 \text{ J}.$$

6–1 When a constant force $\vec{F}$ acts in the same direction as the displacement $\vec{s}$, the work done by the force is $W = Fs$.

EXAMPLE 6–1

Work done by a force in the direction of motion Steve is trying to impress Elayne with his new car, but the engine dies in the middle of an intersection. While Elayne steers, Steve pushes the car 19 m to clear the intersection. If he pushes in the direction of motion with a constant force of 210 N (about 47 lb), how much work does he do on the car?

SOLUTION From Eq. (6–1),

$$W = Fs = (210 \text{ N})(19 \text{ m}) = 4.0 \times 10^3 \text{ J}.$$

In Example 6–1, Steve pushed the car in the direction he wanted it to go. What if he had pushed at an angle ϕ with the car's displacement (Fig. 6–2)? Only the component of force in the direction of the car's motion, (210 N) cos ϕ, would be effective in moving the car. (Other forces must act on the car so that it moves along $\vec{s}$, not in the direction of $\vec{F}$. But we're interested only in the work done by Steve, so we'll consider only the force

6–2 When a constant force $\vec{F}$ acts at an angle ϕ to the displacement $\vec{s}$, the work done by the force is $(F \cos \phi) s = Fs \cos \phi$.

he exerts.) When the force $\vec{F}$ and the displacement $\vec{s}$ have different directions, we take the component of $\vec{F}$ in the direction of the displacement $\vec{s}$, and we define the work as the product of this component and the magnitude of the displacement. The component of $\vec{F}$ in the direction of $\vec{s}$ is $F \cos \phi$, so

$$W = Fs \cos \phi \qquad \text{(constant force, straight-line displacement).} \qquad (6–2)$$

We are assuming that F and ϕ are constant during the displacement. If $\phi = 0$, so that $\vec{F}$ and $\vec{s}$ are in the same direction, then $\cos \phi = 1$ and we are back to Eq. (6–1).

Equation (6–2) has the form of the *scalar product* of two vectors, introduced in Section 1–11: $\vec{A} \cdot \vec{B} = AB \cos \phi$. You may want to review that definition. Using this, we can write Eq. (6–2) more compactly as

$$W = \vec{F} \cdot \vec{s} \qquad \text{(constant force, straight-line displacement).} \qquad (6–3)$$

It's important to understand that work is a *scalar* quantity, even though it's calculated by using two vector quantities (force and displacement). A 5-N force toward the east acting on a body that moves 6 m to the east does exactly the same work as a 5-N force toward the north acting on a body that moves 6 m to the north.

It's also important to realize that work can be positive, negative, or zero. This is the essential way in which work as defined in physics differs from the "everyday" definition of work. When the force has a component in the *same direction* as the displacement (ϕ between zero and 90°), $\cos \phi$ in Eq. (6–2) is positive and the work W is *positive* (Fig. 6–3a). When the force has a component *opposite* to the displacement (ϕ between 90° and 180°), $\cos \phi$ is negative and the work is *negative* (Fig. 6–3b). When the force is *perpendicular* to the displacement, $\phi = 90°$ and the work done by the force is *zero* (Fig. 6–3c). The cases of zero work and negative work bear closer examination, so let's look at some examples.

There are many situations in which forces act but do zero work. You might think it's "hard work" to hold this book motionless at arm's length for five minutes, but you actually aren't doing any work at all on the book because there is no displacement. You get tired because the muscle fibers in your arm do work as they continually contract and relax. But this is work done by one part of your arm exerting a force on another part, *not* on the book. (We'll say more in Section 6–3 about work done by one part of a body on another part.) Even when you walk with constant velocity on a level floor while carrying the book, you still do no work on it. The book now has a displacement, but the

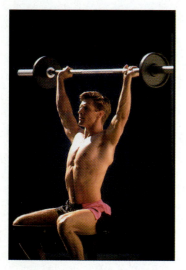

A weight lifter exerts an upward force as he raises the barbell upward, and so does *positive* work on it. But when the weight lifter lowers the barbell to the ground, he does *negative* work because the upward force he exerts on the barbell is opposite to its displacement.

6–3 (a) W is positive because $\vec{F}$ has a component in the direction of $\vec{s}$. (b) W is negative because $\vec{F}$ has a component opposite the direction of $\vec{s}$. (c) W is zero because $\vec{F}$ has no component in the direction of $\vec{s}$.

(a)   (b)   (c)

6–4 (a) When an outfielder catches a baseball, his hand and the ball have the same displacement $\vec{s}$. (b) The work done on his hand by the baseball is positive because the force $\vec{F}_{B \text{ on } H}$ is in the same direction as the displacement $\vec{s}$. (c) The outfielder's hand exerts an equal and opposite force $\vec{F}_{H \text{ on } B} = -\vec{F}_{B \text{ on } H}$ on the baseball. The work done on the baseball by the outfielder's hand is negative because $\vec{F}_{H \text{ on } B}$ and $\vec{s}$ are in opposite directions.

(vertical) supporting force that you exert on the book has no component in the direction of the (horizontal) motion. Then $\phi = 90°$ in Eq. (6–2), and $\cos \phi = 0$. When a body slides along a surface, the work done on the body by the normal force is zero; and when a ball on a string moves in uniform circular motion, the work done on the ball by the tension in the string is also zero. In both cases the work is zero because the force has no component in the direction of motion.

What does it really mean to do *negative* work? The answer comes from Newton's third law of motion. When a baseball outfielder catches a fly ball as in Fig. 6–4a, the outfielder's hand and the ball move together with the same displacement $\vec{s}$ (Fig. 6–4b). The ball exerts a force $\vec{F}_{B \text{ on } H}$ on the outfielder's hand in the same direction as the hand's displacement, so the work done on the outfielder's *hand* by the *ball* is *positive*. But by Newton's third law the outfielder's hand exerts an equal and opposite force $\vec{F}_{H \text{ on } B} = -\vec{F}_{B \text{ on } H}$ on the ball (Fig. 6–4c). This force, which slows the ball to a stop, acts opposite to the displacement of the ball. Thus the work done on the *ball* by the outfielder's *hand* is *negative*. Since the outfielder's hand and the ball have the same displacement, the work done on the ball by the hand is just the negative of the work done on the hand by the ball. In general, when one body does negative work on a second body, the second body does an equal amount of *positive* work on the first body.

CAUTION ► We always speak of work done *on* a particular body *by* a specific force. Always be sure to specify exactly what force is doing the work you are talking about. When you lift a book, you exert an upward force on the book and the book's displacement is upward, so the work done by the lifting force on the book is positive. But the work done by the *gravitational* force (weight) on a book being lifted is *negative* because the downward gravitational force is opposite to the upward displacement. ◄

How do we calculate work when several forces act on a body? One way is to use Eq. (6–2) or (6–3) to compute the work done by each separate force. Then, because work is a scalar quantity, the *total* work W_{tot} done on the body by all the forces is the algebraic sum of the quantities of work done by the individual forces. An alternative way to find the total work W_{tot} is to compute the vector sum of the forces (that is, the net force) and then use this vector sum as $\vec{F}$ in Eq. (6–2) or (6–3).

5.1
Work Calculations

EXAMPLE 6–2

Work done by several forces Farmer Johnson hitches his tractor to a sled loaded with firewood and pulls it a distance of 20 m along level frozen ground (Fig. 6–5). The total weight of sled and load is 14,700 N. The tractor exerts a constant 5000-N force at an angle of 36.9° above the horizontal, as shown. There is a 3500-N friction force opposing the motion. Find the work done

by each force acting on the sled and the total work done by all the forces.

SOLUTION Let's do the easy parts first. The work W_w done by the weight is zero because its direction is perpendicular to the displacement. (The angle between the force of gravity and the

(a)   (b)

6–5 (a) A tractor pulling a sled of firewood. (b) Free-body diagram for the sled and its load, treated as a particle.

displacement is 90°, and the cosine of the angle is zero.) For the same reason the work W_n done by the normal force is also zero. So $W_w = W_n = 0$. (Incidentally, the normal force is *not* equal in magnitude to the weight; see Example 5–15 of Section 5–4, in which the free-body diagram is very similar to Fig. 6–5b.)

That leaves the force F_T exerted by the tractor and the friction force f. From Eq. (6–2) the work W_T done by the tractor is

$$W_T = F_T s \cos \phi = (5000\ \text{N})(20\ \text{m})(0.800) = 80,000\ \text{N} \cdot \text{m}$$
$$= 80\ \text{kJ}.$$

The friction force $\vec{f}$ is opposite to the displacement, so for this force $\phi = 180°$ and $\cos \phi = -1$. The work W_f done by the friction force is

$$W_f = f s \cos 180° = (3500\ \text{N})(20\ \text{m})(-1) = -70,000\ \text{N} \cdot \text{m}$$
$$= -70\ \text{kJ}.$$

The total work W_{tot} done on the sled by all forces is the *algebraic* sum of the work done by the individual forces:

$$W_{\text{tot}} = W_w + W_n + W_T + W_f = 0 + 0 + 80\ \text{kJ} + (-70\ \text{kJ})$$
$$= 10\ \text{kJ}.$$

In the alternative approach, we first find the *vector* sum of all the forces (the net force) and then use it to compute the total work. The vector sum is best found by using components. From Fig. 6–5b,

$$\Sigma F_x = F_T \cos \phi + (-f) = (5000\ \text{N}) \cos 36.9° - 3500\ \text{N}$$
$$= 500\ \text{N},$$
$$\Sigma F_y = F_T \sin \phi + n + (-w)$$
$$= (5000\ \text{N}) \sin 36.9° + n - 14,700\ \text{N}.$$

We don't really need the second equation; we know that the y-component of force is perpendicular to the displacement, so it does no work. Besides, there is no y-component of acceleration, so ΣF_y has to be zero anyway. The total work is therefore the work done by the total x-component:

$$W_{\text{tot}} = \left(\Sigma \vec{F} \right) \cdot \vec{s} = (\Sigma F_x)s = (500\ \text{N})(20\ \text{m}) = 10,000\ \text{J}$$
$$= 10\ \text{kJ}.$$

This is the same result that we found by computing the work done by each force separately.

EXAMPLE 6–3

Total work when velocity is constant An electron moves in a straight line toward the east with a constant speed of 8×10^7 m/s. It has electric, magnetic, and gravitational forces acting on it. Calculate the total work done on the electron during a 1-m displacement.

SOLUTION The electron moves with constant velocity, so its acceleration is zero, and by Newton's second law the vector sum of forces is also zero. Therefore the total work done by all the forces (equal to the work done by the vector sum of all the forces) must be *zero*. Individual forces may do nonzero work, but that's not what the problem asks for.

6–3 WORK AND KINETIC ENERGY

The total work done on a body by external forces is related to the body's displacement, that is, to changes in its position. But the total work is also related to changes in the *speed*

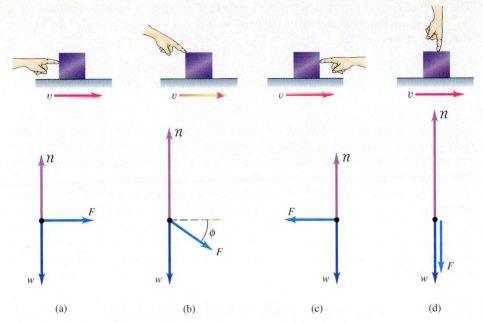

6-6 A block sliding on a frictionless table. (a) The net force causes the speed to increase and does positive work. (b) Again the net force causes the speed to increase and does positive work. (c) The net force opposes the displacement, causes the speed to decrease, and does negative work. (d) The net force is zero and does no work, and the speed is constant.

of the body. To see this, consider Fig. 6–6, which shows several examples of a block sliding on a frictionless table. The forces acting on the block are its weight $\vec{w}$, the normal force $\vec{n}$, and the force $\vec{F}$ exerted on it by the hand.

In Fig. 6–6a the net force on the block is in the direction of its motion. From Newton's second law, this means that the block speeds up; from Eq. (6–1), this also means that the total work W_{tot} done on the block is positive. The total work is also positive in Fig. 6–6b, but only the component $F \cos \phi$ contributes to W_{tot}. The block again speeds up, and this same component $F \cos \phi$ is what causes the acceleration. The total work is *negative* in Fig. 6–6c because the net force opposes the displacement; in this case the block slows down. The net force is zero in Fig. 6–6d, so the speed of the block stays the same and the total work done on the block is zero. We can conclude that when a particle undergoes a displacement, it speeds up if $W_{tot} > 0$, slows down if $W_{tot} < 0$, and maintains the same speed if $W_{tot} = 0$.

Let's make these observations more quantitative. Consider a particle with mass m moving along the x-axis under the action of a constant net force with magnitude F directed along the positive x-axis (Fig. 6–1). The particle's acceleration is constant and given by Newton's second law, $F = ma$. Suppose the speed changes from v_1 to v_2 while the particle undergoes a displacement $s = x_2 - x_1$ from point x_1 to x_2. Using a constant-acceleration equation, Eq. (2–13), and replacing v_0 by v_1, v by v_2, and $(x - x_0)$ by s, we have

$$v_2{}^2 = v_1{}^2 + 2as,$$

$$a = \frac{v_2{}^2 - v_1{}^2}{2s}.$$

When we multiply this equation by m and equate ma to the net force F, we find

$$F = ma = m\frac{v_2{}^2 - v_1{}^2}{2s}$$

and

$$Fs = \frac{1}{2}mv_2{}^2 - \frac{1}{2}mv_1{}^2. \qquad (6\text{–}4)$$

The product Fs is the work done by the net force F and thus is equal to the total work W_{tot} done by all the forces acting on the particle. The quantity $\frac{1}{2}mv^2$ is called the **kinetic energy** K of the particle:

$$K = \frac{1}{2}mv^2 \qquad \text{(definition of kinetic energy).} \qquad (6-5)$$

Like work, the kinetic energy of a particle is a scalar quantity; it depends only on the particle's mass and speed, not its direction of motion. A car (viewed as a particle) has the same kinetic energy when going north at 10 m/s as when going east at 10 m/s. Kinetic energy can never be negative, and it is zero only when the particle is at rest.

We can now interpret Eq. (6–4) in terms of work and kinetic energy. The first term on the right side of Eq. (6–4) is $K_2 = \frac{1}{2}mv_2{}^2$, the final kinetic energy of the particle (that is, after the displacement). The second term is the initial kinetic energy, $K_1 = \frac{1}{2}mv_1{}^2$, and the difference between these terms is the *change* in kinetic energy. So Eq. (6–4) says that **the work done by the net force on a particle equals the change in the particle's kinetic energy:**

$$W_{tot} = K_2 - K_1 = \Delta K \qquad \text{(work-energy theorem).} \qquad (6-6)$$

This result is the **work-energy theorem.**

The work-energy theorem agrees with our observations about the block in Fig. 6–6. When W_{tot} is *positive*, K_2 is greater than K_1, the kinetic energy *increases,* and the particle is going faster at the end of the displacement than at the beginning. When W_{tot} is *negative,* the kinetic energy *decreases* and the speed is less after the displacement. When $W_{tot} = 0$, the initial and final kinetic energies K_1 and K_2 are the same and the speed is unchanged. We stress that the work-energy theorem by itself tells us only about changes in *speed,* not velocity, since the kinetic energy carries no information about the direction of motion.

From Eq. (6–4) or (6–6), kinetic energy and work must have the same units. Hence the joule is the SI unit of both work and kinetic energy (and, as we will see later, of all kinds of energy). To verify this, note that in SI units the quantity $K = \frac{1}{2}mv^2$ has units $\text{kg} \cdot (\text{m/s})^2$ or $\text{kg} \cdot \text{m}^2/\text{s}^2$; we recall that $1\,\text{N} = 1\,\text{kg} \cdot \text{m/s}^2$, so

$$1\,\text{J} = 1\,\text{N} \cdot \text{m} = 1\,(\text{kg} \cdot \text{m/s}^2) \cdot \text{m} = 1\,\text{kg} \cdot \text{m}^2/\text{s}^2.$$

In the British system the unit of kinetic energy and of work is

$$1\,\text{ft} \cdot \text{lb} = 1\,\text{ft} \cdot \text{slug} \cdot \text{ft/s}^2 = 1\,\text{slug} \cdot \text{ft}^2/\text{s}^2.$$

Because we used Newton's laws in deriving the work-energy theorem, we can use it only in an inertial frame of reference. The speeds that we use to compute the kinetic energies and the distances that we use to compute work *must* be measured in an inertial frame. Note also that the work-energy theorem is valid in *any* inertial frame, but the values of W_{tot} and $K_2 - K_1$ may differ from one inertial frame to another (because the displacement and speed of a body may be different in different frames).

We have derived the work-energy theorem for the special case of straight-line motion with constant forces, and in the following examples we'll apply it to this special case only. We'll find in the next section that the theorem is valid in general, even when the forces are not constant and the particle's trajectory is curved.

When a billiards player hits a cue ball at rest, the ball's kinetic energy after being hit is equal to the work that was done on it by the cue. The greater the force exerted by the cue and the greater the distance the ball moves while in contact with it, the greater the ball's kinetic energy.

Problem–Solving Strategy

CALCULATIONS WITH WORK AND KINETIC ENERGY

1. Choose the initial and final positions of the body, and draw a free-body diagram showing all the forces that act on the body. List the forces, and calculate the work done by each force. In some cases, one or more forces may be unknown; represent the unknowns by algebraic symbols. Be sure to check signs. When a force has a component in the same direction as the displacement, its work is positive; when it has a component opposite to the displacement, the work is negative. When force and displacement are perpendicular, the work is zero.

2. Add the amounts of work done by the separate forces to find the total work. Again be careful with signs. Sometimes it may be easier to calculate the vector sum of the forces (the net force) first and then find the work done by the net force.

3. Write expressions for the initial and final kinetic energies K_1 and K_2. If a quantity such as v_1 or v_2 is unknown, express it in terms of the corresponding algebraic symbol. When you calculate kinetic energies, make sure you use the *mass* of the body, not its *weight*.

4. Use the relationship $W_{tot} = K_2 - K_1 = \Delta K$; insert the results from the above steps, and solve for whatever unknown is required. Remember that kinetic energy can never be negative. If you come up with a negative K, you've made a mistake. Maybe you interchanged subscripts 1 and 2 or made a sign error in one of the work calculations.

EXAMPLE 6–4

Using work and energy to calculate speed Let's look again at the sled in Fig. 6–5 and the numbers at the end of Example 6–2. The free-body diagram is shown again in Fig. 6–7. We found that the total work done by all the forces is 10,000 J = 10 kJ, so the kinetic energy of the sled must increase by 10 kJ. The mass of the sled and its load is $m = w/g$, equal to $(14,700 \text{ N})/(9.8 \text{ m/s}^2) = 1500 \text{ kg}$. Suppose the initial speed v_1 is 2.0 m/s. What is the final speed?

SOLUTION Steps 1 and 2 of the problem-solving strategy were done in Example 6–2, where we found $W_{tot} = 10$ kJ. The initial kinetic energy K_1 is

$$K_1 = \frac{1}{2} mv_1{}^2 = \frac{1}{2} (1500 \text{ kg})(2.0 \text{ m/s})^2 = 3000 \text{ kg} \cdot \text{m}^2/\text{s}^2$$

$$= 3000 \text{ J}.$$

The final kinetic energy K_2 is

$$K_2 = \frac{1}{2} mv_2{}^2 = \frac{1}{2} (1500 \text{ kg})v_2{}^2,$$

where v_2 is the unknown speed we want to find. Equation (6–6) gives

$$K_2 = K_1 + W_{tot} = 3000 \text{ J} + 10,000 \text{ J} = 13,000 \text{ J}.$$

Setting these two expressions for K_2 equal, substituting $1 \text{ J} = 1 \text{ kg} \cdot \text{m}^2/\text{s}^2$, and solving for v_2, we find

$$v_2 = 4.2 \text{ m/s}.$$

The total work is positive, so the kinetic energy increases $(K_2 > K_1)$ and the speed increases $(v_2 > v_1)$.

This problem can also be done without the work-energy theorem. We can find the acceleration from $\Sigma \vec{F} = m\vec{a}$ and then use the equations of motion for constant acceleration to find v_2. Since the acceleration is along the x-axis,

$$a = a_x = \frac{\Sigma F_x}{m} = \frac{(5000 \text{ N}) \cos 36.9° - 3500 \text{ N}}{1500 \text{ kg}}$$

$$= 0.333 \text{ m/s}^2;$$

then

$$v_2{}^2 = v_1{}^2 + 2as = (2.0 \text{ m/s})^2 + 2(0.333 \text{ m/s}^2)(20 \text{ m})$$

$$= 17.3 \text{ m}^2/\text{s}^2,$$

$$v_2 = 4.2 \text{ m/s}.$$

This is the same result that we obtained with the work-energy approach, but there we avoided the intermediate step of finding the acceleration. You will find several other examples in this chapter and the next that *can* be done without using energy considerations but that are easier when energy methods are used. When a problem can be done by two different methods, doing it by both methods (as we did in this example) is a very good way to check your work.

6–7 Free-body diagram for the sled and its load in Example 6–2.

EXAMPLE 6–5

Forces on a hammerhead In a pile driver, a steel hammerhead with mass 200 kg is lifted 3.00 m above the top of a vertical I-beam being driven into the ground (Fig. 6–8a). The hammer is then dropped, driving the I-beam 7.4 cm farther into the ground. The vertical rails that guide the hammerhead exert a constant 60-N friction force on the hammerhead. Use the work-energy theorem to find a) the speed of the hammerhead just as it hits the I-beam and b) the average force the hammerhead exerts on the I-beam. Ignore the effects of the air.

SOLUTION Figure 6–8b is a free-body diagram showing the vertical forces on the falling hammerhead. Because the displacement is vertical, any horizontal forces that may be present do no work.

a) Let point 1 be the initial position of the hammerhead, and let point 2 be where it just hits the I-beam. The vertical forces are the downward weight $w = mg = (200 \text{ kg})(9.8 \text{ m/s}^2) = 1960$ N and the upward friction force $f = 60$ N. Thus the net downward force is $w - f = 1900$ N. The displacement of the hammerhead from point 1 to point 2 is downward and equal to $s_{12} = 3.00$ m. The total work done on the hammerhead as it moves from point 1 to point 2 is then

$$W_{\text{tot}} = (w - f)s_{12} = (1900 \text{ N})(3.00 \text{ m}) = 5700 \text{ J}.$$

At point 1 the hammerhead is at rest, so its initial kinetic energy K_1 is zero. Equation (6–6) gives

$$W_{\text{tot}} = K_2 - K_1 = \frac{1}{2}mv_2^2 - 0,$$

$$v_2 = \sqrt{\frac{2W_{\text{tot}}}{m}} = \sqrt{\frac{2(5700 \text{ J})}{200 \text{ kg}}} = 7.55 \text{ m/s}.$$

This is the hammerhead's speed at point 2, just as it hits the beam.

b) Let point 3 be where the hammerhead finally comes to rest. Then $K_3 = 0$. As Fig. 6–8c shows, there is now an additional force, the upward normal force n that the beam exerts on the hammerhead during the additional downward displacement $s_{23} = 7.4$ cm. This force actually varies as the hammerhead comes to a halt, but for simplicity we'll treat n as a constant; our result for n will then be an *average* value of the upward force during the motion. The total work done on the hammerhead during the 7.4-cm displacement is

$$W_{\text{tot}} = (w - f - n)s_{23}.$$

This is equal to the change in kinetic energy $K_3 - K_2$, which is negative because the kinetic energy of the hammerhead decreases. So we have

$$(w - f - n)s_{23} = K_3 - K_2,$$

$$n = w - f - \frac{(K_3 - K_2)}{s_{23}}$$

$$= 1960 \text{ N} - 60 \text{ N} - \frac{(0 \text{ J} - 5700 \text{ J})}{0.074 \text{ m}}$$

$$= 79{,}000 \text{ N}.$$

The force that the *hammerhead* exerts on the *I-beam* during this part of the motion is the equal and opposite reaction force of 79,000 N (about 9 tons) downward—more than 40 times the weight of the hammerhead.

The total change in the hammerhead's kinetic energy during the whole process is zero; a relatively small net force does

(a)

(b)   (c)

6–8 (a) A pile driver pounds an I-beam into the ground. (b) Free-body diagram for the hammerhead while falling. (c) Free-body diagram for the hammerhead while driving the I-beam. Vector lengths are not to scale.

positive work over a large distance, and then a much larger net force does negative work over a much smaller distance. The same thing happens if you speed your car up gradually and then drive it into a brick wall. The very large force needed to reduce the kinetic energy to zero over a short distance is what does the damage to your car—and possibly to you.

THE MEANING OF KINETIC ENERGY

Example 6–5 gives insight into the physical meaning of kinetic energy. The hammerhead is dropped from rest, and its kinetic energy when it hits the I-beam equals the total work done on it up to that point by the net force. This result is true in general: to accelerate a particle of mass m from rest (zero kinetic energy) up to a speed v, the total work done on it must equal the change in kinetic energy from zero to $K = \frac{1}{2}mv^2$;

$$W_{tot} = K - 0 = K.$$

So *the kinetic energy of a particle is equal to the total work that was done to accelerate it from rest to its present speed.* The definition $K = \frac{1}{2}mv^2$, Eq. (6–5), wasn't chosen at random; it's the *only* definition that agrees with this interpretation of kinetic energy.

In the second part of Example 6–5 the kinetic energy of the hammerhead was used to do work on the I-beam and drive it into the ground. This gives us another interpretation of kinetic energy: *the kinetic energy of a particle is equal to the total work that particle can do in the process of being brought to rest.* This is why a baseball outfielder pulls his hand and arm back when catching a fly ball (Fig. 6–4). As the ball comes to rest, it does an amount of work (force times distance) on the outfielder's hand equal to the ball's initial kinetic energy. By pulling his hand back, the outfielder maximizes the distance over which the force acts and so minimizes the force exerted on his hand.

EXAMPLE 6–6

Comparing kinetic energies Two iceboats like the one in Example 5–5 (Section 5–3) hold a race on a frictionless horizontal lake (Fig. 6–9). The two iceboats have masses m and $2m$, respectively. Each iceboat has an identical sail, so the wind exerts the same constant force $\vec{F}$ on each iceboat. The two iceboats start from rest and cross the finish line a distance s away. Which iceboat crosses the finish line with greater kinetic energy?

SOLUTION If you simply use the mathematical definition of kinetic energy, $K = \frac{1}{2}mv^2$ (Eq. (6–5)), the answer to this problem isn't immediately obvious. The iceboat of mass $2m$ has greater mass, so you might guess that the larger iceboat attains a greater

kinetic energy at the finish line. But the smaller iceboat, of mass m, will cross the finish line with a greater speed, and you might guess that *this* iceboat will have the greater kinetic energy. How can we decide?

The correct way to approach this problem is to remember that *the kinetic energy of a particle is just equal to the total work done to accelerate it from rest.* Both iceboats travel the same distance s, and only the horizontal force F in the direction of motion does work on either iceboat. Hence the total work done between the starting line and the finish line is the *same* for each iceboat, $W_{tot} = Fs$. At the finish line, each iceboat has a kinetic energy equal to the work W_{tot} done on it, because each iceboat started from rest. So both iceboats have the *same* kinetic energy at the finish line!

6–9 A race between iceboats.

You might think this is a "trick" question, but it isn't. If you really understand the physical meanings of quantities such as kinetic energy, you can solve problems more easily and with better insight into the physics.

Notice that we didn't need to say anything about how much time each iceboat took to reach the finish line. This is because the work-energy theorem makes no direct reference to time, only to displacement. In fact the iceboat of mass m takes less time to reach the finish line than does the larger iceboat of mass $2m$; we leave the calculation to you (Exercise 6–12).

6–10 The external forces acting on a skater pushing off a wall. The work done by these forces is zero, but his kinetic energy changes nonetheless.

WORK AND KINETIC ENERGY IN COMPOSITE SYSTEMS

You may have noticed that in this section we've been careful to apply the work-energy theorem only to bodies that we can represent as *particles,* that is, as moving point masses. The reason is that new subtleties appear for more complex systems that have to be represented in terms of many particles with different motions. We can't go into these subtleties in detail in this chapter, but here's an example.

Consider a man standing on frictionless roller skates on a level surface, facing a rigid wall (Fig. 6–10). He pushes against the wall, setting himself in motion to the right. The forces acting on him are his weight $\vec{w}$, the upward normal forces $\vec{n}_1$ and $\vec{n}_2$ exerted by the ground on his skates, and the horizontal force $\vec{F}$ exerted on him by the wall. There is no vertical displacement, so $\vec{w}$, $\vec{n}_1$, and $\vec{n}_2$ do no work. The force $\vec{F}$ is the horizontal force that accelerates him to the right, but the parts of his body where that force is applied (the man's hands) do not move. Thus the force $\vec{F}$ also does no work. So where does the man's kinetic energy come from?

The difficulty is that it's simply not adequate to represent the man as a single point mass. For the motion to occur as we've described it, different parts of the man's body must have different motions; his hands are stationary against the wall while his torso is moving away from the wall. The various parts of his body interact with each other, and one part can exert forces and do work on another part. Therefore the *total* kinetic energy of this composite system of body parts can change, even though no work is done by forces applied by bodies (such as the wall) that are outside the system. This would not be possible with a system that can be represented as a single point particle. In Chapter 8 we'll consider further the motion of a collection of particles that interact with each other. We'll discover that just as for the man in this example, the total kinetic energy of such a system can change even when no work is done on any part of the system by anything outside it.

6–4 WORK AND ENERGY WITH VARYING FORCES

So far in this chapter we've considered work done by *constant forces* only. But what happens when you stretch a spring? The more you stretch it, the harder you have to pull, so the force you exert is *not* constant as the spring is stretched. We've also restricted our discussion to *straight-line* motion. You can think of many situations in which a force that varies in magnitude, direction, or both acts on a body moving along a curved path. We need to be able to compute the work done by the force in these more general cases. Fortunately, we'll find that the work-energy theorem holds true even when varying forces are considered and when the body's path is not straight.

To add only one complication at a time, let's consider straight-line motion with a force that is directed along the line but with an x-component F that may change as the body moves. For example, imagine a train moving on a straight track with the engineer constantly changing the locomotive's throttle setting or applying the brakes. Suppose a particle moves along the x-axis from point x_1 to x_2. Figure 6–11a is a graph of the

6–11 (a) Curve showing how a certain force F varies with x.

x-component of force as a function of the particle's coordinate x. To find the work done by this force, we divide the total displacement into small segments Δx_a, Δx_b, and so on (Fig. 6–11b). We approximate the work done by the force during segment Δx_a as the average force F_a in that segment multiplied by the displacement Δx_a. We do this for each segment and then add the results for all the segments. The work done by the force in the total displacement from x_1 to x_2 is approximately

$$W = F_a \Delta x_a + F_b \Delta x_b + \cdots.$$

As the number of segments becomes very large and the width of each becomes very small, this sum becomes (in the limit) the *integral* of F from x_1 to x_2:

$$W = \int_{x_1}^{x_2} F\,dx \quad \text{(varying x-component of force, straight-line displacement).}$$

$$(6\text{–}7)$$

Note that $F_a \Delta x_a$ represents the *area* of the first vertical strip in Fig. 6–11b and that the integral in Eq. (6–7) represents the area under the curve of Fig. 6–11a between x_1 and x_2. *On a graph of force as a function of position, the total work done by the force is represented by the area under the curve between the initial and final positions.* An alternative interpretation of Eq. (6–7) is that the work W equals the average force that acts over the entire displacement, multiplied by the displacement.

Equation (6–7) also applies if F, the x-component of the force, is constant. In that case, F may be taken outside the integral:

$$W = \int_{x_1}^{x_2} F\,dx = F \int_{x_1}^{x_2} dx = F(x_2 - x_1).$$

But $x_2 - x_1 = s$, the total displacement of the particle. So in the case of a constant force F, Eq. (6–7) says that $W = Fs$, in agreement with Eq. (6–1). The interpretation of work as the area under the curve of F as a function of x also holds for a constant force; $W = Fs$ is the area of a rectangle of height F and width s (Fig. 6–12).

Now let's apply what we've learned to the stretched spring. To keep a spring stretched beyond its unstretched length by an amount x, we have to apply a force with magnitude F at each end (Fig. 6–13). If the elongation x is not too great, we find that F is directly proportional to x:

$$F = kx \quad \text{(force required to stretch a spring),} \quad (6\text{–}8)$$

where k is a constant called the **force constant** (or spring constant) of the spring. Equation (6–8) shows that the units of k are force divided by distance, N/m in SI units and lb/ft in British units. A floppy toy spring such as a Slinky™ has a force constant of about 1 N/m; for the much stiffer springs in an automobile's suspension, k is about 10^5 N/m. The observation that elongation is directly proportional to force for elongations that are not too great was made by Robert Hooke in 1678 and is known as **Hooke's law.** It really shouldn't be called a "law," since it's a statement about a specific device and not a fundamental law of nature. Real springs don't always obey Eq. (6–8) precisely, but it's still a useful idealized model. We'll discuss Hooke's law more fully in Chapter 11.

To stretch a spring, we must do work. We apply equal and opposite forces to the ends of the spring and gradually increase the forces. We hold the left end stationary, so the force we apply at this end does no work. The force at the moving end *does* do work. Figure 6–14 is a graph of F as a function of x, the elongation of the spring. The work done by F when the elongation goes from zero to a maximum value X is

$$W = \int_{0}^{X} F\,dx = \int_{0}^{X} kx\,dx = \frac{1}{2}kX^2. \quad (6\text{–}9)$$

(b)

6–11 (b) If the area is partitioned into small rectangles, the sum of their areas approximates the total work done during the displacement; the greater the number of rectangles used, the closer the approximation. The total area under the curve equals the work done by the force as the particle moves from x_1 to x_2.

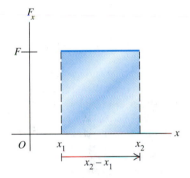

6–12 The work done by a constant force F in the x-direction as a particle moves from x_1 to x_2 is equal to the rectangular area under the graph of force versus displacement.

6–13 The force needed to stretch an ideal spring is proportional to its elongation: $F = kx$.

6–14 The work done to stretch a spring by a length X is equal to the triangular area under the graph of force versus displacement.

We can also obtain this result graphically. The area of the shaded triangle in Fig. 6–14, representing the total work done by the force, is equal to half the product of the base and altitude, or

$$W = \frac{1}{2}(X)(kX) = \frac{1}{2}kX^2.$$

This equation also says that the work is the *average* force $kX/2$ multiplied by the total displacement X. We see that the total work is proportional to the *square* of the final elongation X. To stretch an ideal spring by 2 cm, you must do four times as much work as is needed to stretch it by 1 cm.

Equation (6–9) assumes that the spring was originally unstretched. If initially the spring is already stretched a distance x_1, the work we must do to stretch it to a greater elongation x_2 is

$$W = \int_{x_1}^{x_2} F\,dx = \int_{x_1}^{x_2} kx\,dx = \frac{1}{2}kx_2^2 - \frac{1}{2}kx_1^2. \qquad (6\text{--}10)$$

If the spring has spaces between the coils when it is unstretched, then it can also be compressed, and Hooke's law holds for compression as well as stretching. In this case the force F and displacement x are in the opposite directions from those shown in Fig. 6–13, and so F and x in Eq. (6–8) will both be negative. Since both F and x are reversed, the force again is in the same direction as the displacement, and the work done by F is again positive. So the total work is still given by Eq. (6–9) or (6–10), even when X is negative or either or both of x_1 and x_2 are negative.

EXAMPLE 6–7

Work done on a spring scale   A woman weighing 600 N steps on a bathroom scale containing a stiff spring (Fig. 6–15). In equilibrium the spring is compressed 1.0 cm under her weight. Find the force constant of the spring and the total work done on it during the compression.

SOLUTION   In equilibrium the net force on the woman is zero, so the woman's weight and the spring force acting on her have the same 600-N magnitude but opposite directions. We take positive values of x to correspond to elongation, so $x = -1.0$ cm $= -0.010$ m, and the force that the woman applies to the spring is $F = -600$ N. From Eq. (6–8) the force constant k is

$$k = \frac{F}{x} = \frac{-600 \text{ N}}{-0.010 \text{ m}} = 60{,}000 \text{ N/m}.$$

Then, from Eq. (6–9),

$$W = \frac{1}{2}kX^2 = \frac{1}{2}(60{,}000 \text{ N/m})(-0.010 \text{ m})^2 = 3.0 \text{ J}.$$

6–15 Compressing a spring in a bathroom scale. A negative force causes a negative displacement but does a positive amount of work.

WORK-ENERGY THEOREM FOR STRAIGHT-LINE MOTION, VARYING FORCES

In Section 6–3 we derived the work-energy theorem, $W_{\text{tot}} = K_2 - K_1$, for the special case of straight-line motion with a constant net force. We can now prove that this theorem is true even when the force varies with position. As in Section 6–3, let's consider a particle that undergoes a displacement x while being acted on by a net force with x-component F, which we now allow to vary. Just as in Fig. 6–11, we divide the total displacement x into a large number of small segments Δx. We can apply the work-energy theorem, Eq. (6–6), to each segment because the value of F in each small segment is approximately constant. The change in kinetic energy in segment Δx_a is equal to the work $F_a \Delta x_a$, and so on. The total change of kinetic energy is the sum of the changes in the individual segments, and thus is equal to the total work done on the particle during the entire displacement. So $W_{tot} = \Delta K$ holds for varying forces as well as for constant ones.

Here's an alternative derivation of the work-energy theorem for a force that may vary with position. It involves making a change of variable from x to v in the work integral. As a preliminary, we note that the acceleration a of the particle can be expressed in various ways, using $a = dv/dt$, $v = dx/dt$, and the chain rule for derivatives:

$$a = \frac{dv}{dt} = \frac{dv}{dx}\frac{dx}{dt} = v\frac{dv}{dx}. \qquad (6\text{–}11)$$

Using this result, Eq. (6–7) tells us that the total work done by the *net* force F is

$$W_{tot} = \int_{x_1}^{x_2} F\,dx = \int_{x_1}^{x_2} ma\,dx = \int_{x_1}^{x_2} mv\,\frac{dv}{dx}\,dx. \qquad (6\text{–}12)$$

Now $(dv/dx)\,dx$ is the change in velocity dv during the displacement dx, so in Eq. (6–12) we can substitute dv for $(dv/dx)\,dx$. This changes the integration variable from x to v, so we change the limits from x_1 and x_2 to the corresponding velocities v_1 and v_2 at these points. This gives us

$$W_{tot} = \int_{v_1}^{v_2} mv\,dv.$$

The integral of $v\,dv$ is just $v^2/2$. Substituting the upper and lower limits, we finally find

$$W_{tot} = \frac{1}{2}mv_2{}^2 - \frac{1}{2}mv_1{}^2. \qquad (6\text{–}13)$$

This is the same result as Eq. (6–6) but without the assumption that the net force F is constant. So the work-energy theorem is valid even when F varies during the displacement.

EXAMPLE 6-8

Motion with a varying force An air-track glider of mass 0.100 kg is attached to the end of a horizontal air track by a spring with force constant 20.0 N/m (Fig. 6–16a). Initially the spring is unstretched and the glider is moving at 1.50 m/s to the right. Find the maximum distance d that the glider moves to the right a) if the air track is turned on so that there is no friction, and b) if the air is turned off so that there is kinetic friction with coefficient $\mu_k = 0.47$.

(a)

SOLUTION Figures 6–16b and 6–16c show the free-body diagrams for the glider without and with friction, respectively. The force exerted by the spring is not constant, so we *cannot* use the constant-acceleration formulas of Chapter 2 to solve this

(b)   (c)

6–16 (a) A glider attached to an air track by a spring. (b) Free-body diagram for the glider without friction. (c) Free-body diagram for the glider with kinetic friction.

problem. Instead, we'll use the work-energy theorem.

a) The glider moves only horizontally, so only the horizontal spring force does work. As it moves a distance d to the right, the glider stretches the spring by d and does an amount of work on the spring equal to $\frac{1}{2}kd^2$. The *spring* does an amount of work on the *glider* equal to the negative of this, or $-\frac{1}{2}kd^2$. The spring stretches until the glider comes instantaneously to rest, so the glider's final kinetic energy is zero. The initial kinetic energy of the glider is $\frac{1}{2}mv_1^2$. Using the work-energy theorem, we find

$$-\frac{1}{2}kd^2 = 0 - \frac{1}{2}mv_1^2,$$

so the distance the glider moves is

$$d = v_1\sqrt{\frac{m}{k}} = (1.50 \text{ m/s})\sqrt{\frac{0.100 \text{ kg}}{20.0 \text{ N/m}}}$$

$$= 0.106 \text{ m} = 10.6 \text{ cm}.$$

The stretched spring pulls the glider back to the left, so the glider is at rest only instantaneously.

b) If the air is turned off, we must also include the work done by the constant force of kinetic friction. The normal force n is equal in magnitude to the weight of the glider, since the track is horizontal and there are no other vertical forces. The magnitude of the kinetic friction force is then $f_k = \mu_k n = \mu_k mg$. The friction force is directed opposite to the displacement, so the work done by friction is

$$W_{\text{fric}} = f_k d \cos 180° = -f_k d = -\mu_k mgd.$$

The total work is the sum of W_{fric} and the work done by the spring, $-\frac{1}{2}kd^2$. Hence

$$-\mu_k mgd - \frac{1}{2}kd^2 = 0 - \frac{1}{2}mv_1^2,$$

$$-(0.47)(0.100 \text{ kg})(9.8 \text{ m/s}^2)d - \frac{1}{2}(20.0 \text{ N/m})d^2$$

$$= -\frac{1}{2}(0.100 \text{ kg})(1.50 \text{ m/s})^2,$$

$$(10.0 \text{ N/m})d^2 + (0.461 \text{ N})d - (0.113 \text{ N} \cdot \text{m}) = 0.$$

This is a quadratic equation for d. The solutions are

$$d = \frac{-(0.461 \text{ N}) \pm \sqrt{(0.461 \text{ N})^2 - 4(10.0 \text{ N/m})(-0.113 \text{ N} \cdot \text{m})}}{2(10.0 \text{ N/m})}$$

$$= 0.086 \text{ m} \quad \text{or} \quad -0.132 \text{ m}.$$

We have used d as the symbol for a positive displacement, so only the positive value of d makes sense. Thus with friction the glider moves a distance

$$d = 0.086 \text{ m} = 8.6 \text{ cm}.$$

With friction present, the glider goes a shorter distance and the spring stretches less, as you might expect. Again the glider stops instantaneously, and again the spring force pulls the glider to the left; whether it moves or not depends on how great the *static* friction force is. How large would the coefficient of static friction μ_s have to be to keep the glider from springing back to the left?

WORK-ENERGY THEOREM FOR MOTION ALONG A CURVE

We can generalize our definition of work further to include a force that varies in direction as well as magnitude, and a displacement that lies along a curved path. Suppose a particle moves from point P_1 to P_2 along a curve, as shown in Fig. 6–17a. We divide the portion of the curve between these points into many infinitesimal vector displacements, and we call a typical one of these $d\vec{l}$. Each $d\vec{l}$ is tangent to the path at its position. Let $\vec{F}$ be the force at a typical point along the path, and let ϕ be the angle between $\vec{F}$ and $d\vec{l}$ at this point. Then the small element of work dW done on the particle during the displacement $d\vec{l}$ may be written as

$$dW = F \cos \phi \, dl = F_\parallel \, dl = \vec{F} \cdot d\vec{l},$$

where $F_\parallel = F \cos \phi$ is the component of $\vec{F}$ in the direction parallel to $d\vec{l}$ (Fig. 6–17b). The total work done by $\vec{F}$ on the particle as it moves from P_1 to P_2 is then

$$W = \int_{P_1}^{P_2} F \cos \phi \, dl = \int_{P_1}^{P_2} F_\parallel \, dl = \int_{P_1}^{P_2} \vec{F} \cdot d\vec{l} \quad \text{(work done on a curved path).}$$

(6–14)

6–17 (a) A particle moves along a curved path from point P_1 to P_2, acted on by a force $\vec{F}$ that varies in magnitude and direction. During an infinitesimal displacement $d\vec{l}$ (a small segment of the path), the work dW done by the force is given by $dW = \vec{F} \cdot d\vec{l} = F \cos \phi \, dl$. (b) The force contributing to the work is the component of force parallel to the displacement, $F_\parallel = F \cos \phi$.

We can now show that the work-energy theorem, Eq. (6–6), holds true even with varying forces and a displacement along a curved path. The force $\vec{F}$ is essentially constant over any given infinitesimal segment $d\vec{l}$ of the path, so we can apply the work-energy theorem for straight-line motion to that segment. Thus the change in the particle's kinetic energy K over that segment equals the work $dW = F_\parallel \, dl = \vec{F} \cdot d\vec{l}$ done on the particle. Adding up these infinitesimal quantities of work from all the segments along the whole path gives the total work done, Eq. (6–14), and this equals the total change in kinetic energy over the whole path. So $W_{\text{tot}} = \Delta K = K_2 - K_1$ is true *in general*, no matter what the path and no matter what the character of the forces. This can be proved more rigorously by using steps like those in Eqs. (6–11) through (6–13) (see Challenge Problem 6–90).

Note that only the component of the net force parallel to the path, $F_\parallel$, does work on the particle, so only this component can change the speed and kinetic energy of the particle. The component perpendicular to the path, $F_\perp = F \sin \phi$, has no effect on the particle's speed; it only acts to change the particle's direction.

The integral in Eq. (6–14) is called a *line integral*. To evaluate this integral in a specific problem, we need some sort of detailed description of the path and of how $\vec{F}$ varies along the path. We usually express the line integral in terms of some scalar variable, as in the following example.

EXAMPLE 6–9

Motion on a curved path At a family picnic you are appointed to push your obnoxious cousin Throckmorton in a swing (Fig. 6–18a). His weight is w, the length of the chains is R, and you push Throcky until the chains make an angle θ_0 with the vertical. To do this, you exert a varying horizontal force $\vec{F}$ that starts at zero and gradually increases just enough so that Throcky and the swing move very slowly and remain very nearly in equilibrium. What is the total work done on Throcky by all forces? What is the work done by the tension T in the chains? What is the work you do by exerting the force $\vec{F}$? (Neglect the weight of chains and seat.)

SOLUTION The free-body diagram is shown in Fig. 6–18b. We have replaced the tensions in the two chains with a single tension T. Because Throcky is in equilibrium at every point, the net force on him is zero, and the total work done on him by all

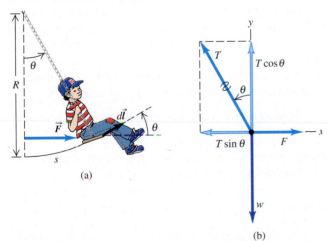

(a)

(b)

6–18 (a) Pushing cousin Throckmorton in a swing. (b) Free-body diagram for Throcky, considered as a particle and neglecting the small weight of chains and seat.

forces is zero. At any point during the motion the chain force on Throcky is perpendicular to each $d\vec{l}$, so the angle between the chain force and the displacement is always 90°. Therefore the work done by the chain tension is zero.

To compute the work done by $\vec{F}$, we have to find out how it varies with the angle θ. Throcky is in equilibrium at every point, so from $\Sigma F_x = 0$ we get

$$F + (-T \sin \theta) = 0,$$

and from $\Sigma F_y = 0$ we find

$$T \cos \theta + (-w) = 0.$$

By eliminating T from these two equations, we obtain

$$F = w \tan \theta.$$

The point where $\vec{F}$ is applied swings through the arc s. The arc length s equals the radius R of the circular path multiplied by the length θ (in radians), so $s = R\theta$. Therefore the displacement $d\vec{l}$ corresponding to a small change of angle $d\theta$ has a magnitude $dl = ds = R \, d\theta$. The work done by $\vec{F}$ is

$$W = \int \vec{F} \cdot d\vec{l} = \int F \cos \theta \, ds.$$

Now we express everything in terms of the varying angle θ:

$$W = \int_0^{\theta_0} (w \tan \theta) \cos \theta \, (R \, d\theta) = wR \int_0^{\theta_0} \sin \theta \, d\theta$$

$$= wR (1 - \cos \theta_0).$$

If $\theta_0 = 0$, there is no displacement; in that case, $\cos \theta_0 = 1$ and $W = 0$, as we should expect. If $\theta_0 = 90°$, then $\cos \theta_0 = 0$ and $W = wR$. In that case the work you do is the same as if you had lifted Throcky straight up a distance R with a force equal to his weight w. In fact, the quantity $R (1 - \cos \theta_0)$ is the increase in his height above the ground during the displacement, so for any value of θ_0 the work done by force $\vec{F}$ is the change in height multiplied by the weight. This is an example of a more general result that we'll prove in Section 7–2.

6–5 POWER

The definition of work makes no reference to the passage of time. If you lift a barbell weighing 400 N through a vertical distance of 0.5 m at constant velocity, you do

Propeller-driven airliners of the 1950s had maximum speeds of about 600 km/h (370 mi/h). By contrast, modern jet airliners are much larger and fly much faster—about 900 km/h (560 mi/h). This is due in large part to a nearly 20-fold increase in engine power, from 3400 hp (2.5×10^6 W) for a large 1950's piston engine to over 66,000 hp (4.9×10^7 W) for a modern jet engine (see Example 6–10).

(400 N)(0.5 m) = 200 J of work whether it takes you 1 second, 1 hour, or 1 year to do it. But often we need to know how quickly work is done. We describe this in terms of *power.* In ordinary conversation the word "power" is often synonymous with "energy" or "force." In physics we use a much more precise definition: **power** is the time *rate* at which work is done. Like work and energy, power is a scalar quantity.

When a quantity of work ΔW is done during a time interval Δt, the average work done per unit time or **average power** P_{av} is defined to be

$$P_{av} = \frac{\Delta W}{\Delta t} \qquad \text{(average power)}. \tag{6–15}$$

The rate at which work is done might not be constant. Even when it varies, we can define **instantaneous power** P as the limit of the quotient in Eq. (6–15) as Δt approaches zero:

$$P = \lim_{\Delta t \to 0} \frac{\Delta W}{\Delta t} = \frac{dW}{dt} \qquad \text{(instantaneous power)}. \tag{6–16}$$

The SI unit of power is the **watt** (W), named for the Scottish inventor James Watt. One watt equals one joule per second (1 W = 1 J/s). The kilowatt (1 kW = 10^3 W) and the megawatt (1 MW = 10^6 W) are also commonly used. In the British system, work is expressed in foot-pounds, and the unit of power is the foot-pound per second. A larger unit called the *horsepower* (hp) is also used:

$$1 \text{ hp} = 550 \text{ ft} \cdot \text{lb/s} = 33{,}000 \text{ ft} \cdot \text{lb/min}.$$

That is, a 1-hp motor running at full load does 33,000 ft · lb of work every minute. A useful conversion factor is

$$1 \text{ hp} = 746 \text{ W} = 0.746 \text{ kW},$$

or 1 horsepower equals about $\frac{3}{4}$ of a kilowatt.

The watt is a familiar unit of *electrical* power; a 100-W light bulb converts 100 J of electrical energy into light and heat each second. But there's nothing inherently electrical about a watt or a kilowatt. A light bulb could be rated in horsepower, and some automobile manufacturers rate their engines in kilowatts rather than horsepower.

The units of power can be used to define new units of work or energy. The *kilowatt-hour* (kWh) is the usual commercial unit of electrical energy. One kilowatt-hour is the total work done in 1 hour (3600 s) when the power is 1 kilowatt (10^3 J/s), so

$$1 \text{ kWh} = (10^3 \text{ J/s})(3600 \text{ s}) = 3.6 \times 10^6 \text{ J} = 3.6 \text{ MJ}.$$

The kilowatt-hour is a unit of *work* or *energy,* not power.

In mechanics we can also express power in terms of force and velocity. Suppose that a force $\vec{F}$ acts on a body while it undergoes a vector displacement $\Delta \vec{s}$. If $F_{\parallel}$ is the component of $\vec{F}$ tangent to the path (parallel to $\Delta \vec{s}$), then the work done by the force is $\Delta W = F_{\parallel} \Delta s$, and the average power is

$$P_{av} = \frac{F_{\parallel} \Delta s}{\Delta t} = F_{\parallel} \frac{\Delta s}{\Delta t} = F_{\parallel} v_{av}. \tag{6–17}$$

Instantaneous power P is the limit of this as $\Delta t \to 0$:

$$P = F_{\parallel} v, \tag{6–18}$$

where v is the magnitude of the instantaneous velocity. We can also express Eq. (6–18) in terms of the scalar product:

$$P = \vec{F} \cdot \vec{v} \qquad \text{(instantaneous rate at which force } \vec{F} \text{ does work on a particle)}. \tag{6–19}$$

EXAMPLE 6-10

Each of the two jet engines on a Boeing 767 airliner develops a thrust (a forward force on the airplane) of 197,000 N (44,300 lb). When the airplane is flying at 250 m/s (900 km/h, or roughly 560 mi/h), what horsepower does each engine develop?

SOLUTION The force is in the same direction as the velocity, so $F = F_{\parallel}$. From Eq. (6-18),

$$P = Fv = (1.97 \times 10^5 \text{ N})(250 \text{ m/s}) = 4.93 \times 10^7 \text{ W}$$

$$= (4.93 \times 10^7 \text{ W}) \frac{1 \text{ hp}}{746 \text{ W}} = 66,000 \text{ hp}.$$

EXAMPLE 6-11

As part of a charity fund-raising drive, a Chicago marathon runner with mass 50.0 kg runs up the stairs to the top of the 443-m-tall Sears Tower, the tallest building in the United States (Fig. 6-19). In order to lift herself to the top in 15.0 minutes, what must be her average power output in watts? In kilowatts? In horsepower?

SOLUTION We'll treat the runner as a particle of mass m. As in Example 6-9, lifting a mass m against gravity requires an

6-19 How much power is required to run up the stairs of Chicago's Sears Tower in 15 minutes?

amount of work equal to the weight mg multiplied by the height h it is lifted. Hence the work she must do is

$$W = mgh = (50.0 \text{ kg})(9.80 \text{ m/s}^2)(443 \text{ m})$$

$$= 2.17 \times 10^5 \text{ J}.$$

The time is 15.0 min = 900 s, so from Eq. (6-15) the average power is

$$P_{av} = \frac{2.17 \times 10^5 \text{ J}}{900 \text{ s}} = 241 \text{ W} = 0.241 \text{ kW} = 0.323 \text{ hp}.$$

An alternative approach is to use Eq. (6-17). The force exerted is vertical, and the average vertical component of velocity is (443 m)/(900 s) = 0.492 m/s, so the average power is

$$P_{av} = F_{\parallel} v_{av} = (mg) v_{av}$$

$$= (50.0 \text{ kg})(9.80 \text{ m/s}^2)(0.492 \text{ m/s}) = 241 \text{ W}.$$

In fact the runner's *total* power output will be several times greater than we have calculated. The reason is that the runner isn't really a particle but a collection of parts that exert forces on each other and do work, such as the work done to inhale and exhale and to make her arms and legs swing. What we've calculated is only the part of her power output that goes into lifting her to the top of the building.

6-6 AUTOMOTIVE POWER

A Case Study in Energy Relations

The power requirements of a gasoline-powered automobile are an important and practical example of the concepts in this chapter. If roads were flat and frictionless and air resistance didn't exist, there would be no need for an automobile to have an engine. All you'd need to go for a drive would be a few strong friends to give you a push to get started and a few other friends at your destination to stop you. (Steering on frictionless roads would be a problem, though.) In the real world, however, a moving car without an engine slows down because of forces that resist its motion. The engine's function is to continuously provide power to overcome this resistance. So to understand how much power is required from a car's engine, we must analyze the forces that act on the car.

Two forces oppose the motion of an automobile: rolling friction and air resistance. We described rolling friction in Section 5-4 in terms of a coefficient of rolling friction μ_r. A typical value of μ_r for properly inflated tires on hard pavement is 0.015. A Porsche 911 Carrera has a mass of 1251 kg and a weight of (1251 kg)(9.80 m/s^2) = 12,260 N, and so the resisting force of rolling friction on a level road (where the normal force $n = mg$) is

$$F_{roll} = \mu_r n = (0.015)(12{,}260 \text{ N}) = 180 \text{ N}.$$

This force is nearly independent of car speed.

The air resistance force F_{air} is approximately proportional to the square of the speed and can be expressed by the equation

$$F_{air} = \frac{1}{2} CA\rho v^2, \tag{6–20}$$

where A is the silhouette area of the car (seen from the front), ρ is the density of air (about 1.2 kg/m^3 at sea level at ordinary temperatures), v is the car's speed, and C is a dimensionless constant called the *drag coefficient* that depends on the shape of the moving body. Typical values of C for cars range from 0.35 to 0.50; for the 911 Carrera, $C = 0.38$ and $A = 1.77$ m^2. For this car the air-resistance force is

$$F_{air} = \frac{1}{2}(0.38)(1.77 \text{ m}^2)(1.2 \text{ kg/m}^3)v^2 = (0.40 \text{ N} \cdot \text{s}^2/\text{m}^2)v^2.$$

In a residential speed zone, where $v = 10$ m/s (36 km/h, or about 22 mi/h), the air-resistance force is about

$$F_{air} = (0.40 \text{ N} \cdot \text{s}^2/\text{m}^2)(10 \text{ m/s})^2 = 40 \text{ N}.$$

At a moderate speed of 15 m/s (54 km/h or 34 mi/h), F_{air} is 90 N, and at a highway speed of 30 m/s (110 km/h or 67 mi/h) it is 360 N. Thus at slow speeds, air resistance is less important than rolling friction. At moderate speeds they are comparable, and at highway speeds air resistance dominates.

What does this mean in terms of the *power* needed from the engine? In constant-speed driving on a level road, the sum of F_{roll} and F_{air} must be just balanced by the forward force $F_{forward}$ supplied by the drive wheels. (The drive wheels push backward on the pavement, and the pavement pushes forward on the drive wheels.) The power is just this forward force multiplied by the speed v. For the 911 Carrera the power needed for constant speed v is

$$P = F_{forward}v = (F_{roll} + F_{air})v = [180 \text{ N} + (0.40 \text{ N} \cdot \text{s}^2/\text{m}^2)v^2]v.$$

For the three speeds mentioned above, you can do the arithmetic yourself to find the following results:

v (m/s)	F_{roll} (N)	F_{air} (N)	$F_{forward}$ (N)	P (kW)	P (hp)
10	180	40	220	2.2	2.9
15	180	90	270	4.1	5.5
30	180	360	540	16	22

How much fuel must be consumed in the engine to provide this power? Burning one liter (1 L) of gasoline releases about 3.5×10^7 J of energy. But not all of this is converted into useful work. The laws of thermodynamics, which we'll encounter in Chapters 17 and 18, impose fundamental limits on the efficiency of converting heat to work. In a typical car engine, about 65% of the heat released from gasoline combustion is wasted in the cooling system and the exhaust. Another 20% or so is converted to work that does nothing to propel the car; this includes work done to oppose friction in the drive train and to run accessories such as the air conditioner and power steering. This leaves only

External antenna Driveshaft Stand-out mirrors Flush-mounted windshield with built-in antenna

Non-aerodynamic steel body Aerodynamic composite body Faired rear-view mirrors

Iron V-8 engine

Chrome grille & bumper Steel box frame Trailing arm suspension

Belt-driven cooling fan Differential & rear axle Thermostatically controlled electric cooling fan

Transverse 4-cylinder fuel-injected computer-controlled aluminum front wheel drive engine Air dam

(a) (b)

6–20 (a) Cars designed in the early 1970s did not feature aerodynamic styling and used heavy materials such as iron for engines and steel for body panels. Rear-wheel drive required a heavy drive train as well. (b) By the 1990s, economics forced new ideas in car designs. Body shapes had lower drag coefficients, engines were built of aluminum, and body panels were often made of plastic. These changes doubled the typical fuel efficiency of cars.

about 15% of the energy to do work against the rolling friction and air resistance we described above. The available energy per liter is then

$$(0.15)(3.5 \times 10^7 \text{ J/L}) = 5.3 \times 10^6 \text{ J/L}. \qquad (6\text{–}21)$$

Let's look at the fuel consumption in the 15-m/s case. The power required is 4.1 kW = 4100 J/s. In 1 hour (3600 s) the total energy required is

$$(4100 \text{ J/s})(3600 \text{ s}) = 1.5 \times 10^7 \text{ J},$$

and during that hour the car travels a distance of

$$(15 \text{ m/s})(3600 \text{ s}) = 5.4 \times 10^4 \text{ m} = 54 \text{ km}.$$

From Eq. (6–21) the amount of fuel consumed in 1 hour, traveling 54 km at 15 m/s, is

$$\frac{1.5 \times 10^7 \text{ J}}{5.3 \times 10^6 \text{ J/L}} = 2.8 \text{ L}.$$

That amount of gasoline moves the car 54 km, so the distance traveled per liter of fuel is (54 km)/(2.8 L) = 19 km/L, or about 45 miles per gallon. (Figure 6–20 shows some of the features of contemporary car design that have improved fuel efficiency.)

The power required for a steady 15 m/s on level ground is 4.1 kW, but the power required for acceleration and hill climbing may be much greater. The 911 Carrera is advertised as going from zero to 60 mi/h (27 m/s) in 6.1 s. The final kinetic energy is then

$$K = \frac{1}{2}mv^2 = \frac{1}{2}(1251 \text{ kg})(27 \text{ m/s})^2 = 4.6 \times 10^5 \text{ J}.$$

The average additional power required for the acceleration is

$$P_{av} = \frac{4.6 \times 10^5 \text{ J}}{6.1 \text{ s}} = 7.5 \times 10^4 \text{ W} = 75 \text{ kW} = 100 \text{ hp}.$$

This rapid acceleration requires about 18 times as much power as cruising at a steady 15 m/s (not including the power to overcome friction in the drive train). For the record, the 911 Carrera advertises a maximum horsepower of 214 hp at an engine speed of 5900 rpm.

What about hill climbing? A 5% grade, about the maximum found on most interstate highways, rises 5 meters for every 100 meters of horizontal distance. A car moving at

30 m/s up a 5% grade is gaining elevation at the rate of $(0.05)(30 \text{ m/s}) = 1.5$ m/s. The 911 Carrera weighs 12,260 N, so lifting it at this rate requires a power of

$$P = Fv = (12{,}260 \text{ N})(1.5 \text{ m/s}) = 1.8 \times 10^4 \text{ J/s} = 18 \text{ kW} = 24 \text{ hp}.$$

The *total* power required is this amount plus the 16 kW needed to maintain 30 m/s on a level road, that is,

$$P_{\text{tot}} = 18 \text{ kW} + 16 \text{ kW} = 34 \text{ kW} = 46 \text{ hp}.$$

Finally, let's compare the energy requirements of an automobile to your energy requirements while walking. A 70-kg man needs about 2.0×10^5 J of energy (released from food) to walk 1 km at 5 km/h $= 1.4$ m/s. If the 3.5×10^7 J obtained from 1 L of gasoline could somehow be made available to the man's body, he could travel a distance of $(3.5 \times 10^7 \text{ J})/(2.0 \times 10^5 \text{ J/km}) = 170$ km. Yet this same amount of energy can propel our car only 19 km. Of course, the car travels at over ten times the speed of a walking man (15 m/s versus 1.4 m/s). But this example shows that the speed and convenience of traveling by automobile come only at the cost of greatly increased energy consumption.

SUMMARY

KEY TERMS

work, 165
joule, 165
kinetic energy, 170
work-energy theorem, 170
force constant, 175
Hooke's law, 175
power, 180
average power, 180
instantaneous power, 180
watt, 180

- When a constant force $\vec{F}$ acts on a particle that undergoes a straight-line displacement $\vec{s}$, the work W done by the force is defined as

$$W = Fs \cos\phi = \vec{F} \cdot \vec{s}, \qquad (6\text{-}2), (6\text{-}3)$$

where ϕ is the angle between the directions of $\vec{F}$ and $\vec{s}$. The unit of work in SI units is 1 joule $= 1$ newton $\cdot$ meter ($1 \text{ J} = 1 \text{ N} \cdot \text{m}$). Work is a scalar quantity; it has an algebraic sign (positive or negative) but no direction in space.

- The kinetic energy K of a particle equals the amount of work required to accelerate the particle from rest to speed v; it is also equal to the amount of work the particle can do in the process of being brought to rest. The kinetic energy of a particle with mass m and speed v is

$$K = \frac{1}{2}mv^2. \qquad (6\text{-}5)$$

Kinetic energy is a scalar quantity that has no direction in space; it is always positive or zero. Its units are the same as the units of work: $1 \text{ J} = 1 \text{ N} \cdot \text{m} = 1 \text{ kg} \cdot \text{m}^2/\text{s}^2$.

- When forces act on a particle while it undergoes a displacement, the particle's kinetic energy changes by an amount equal to the total work W_{tot} done on the particle by all the forces:

$$W_{\text{tot}} = K_2 - K_1 = \Delta K. \qquad (6\text{-}6)$$

This relation is called the work-energy theorem. It is valid whether the forces are constant or varying and whether the path followed by the particle is straight or curved. It is applicable only to bodies that can be treated as a particle.

- When the force varies during a straight-line displacement, and the force is in the same direction as the displacement, the work done by the force is given by

$$W = \int_{x_1}^{x_2} F\, dx. \qquad (6\text{-}7)$$

If the force makes an angle ϕ with the displacement, the work done by the force is

$$W = \int_{P_1}^{P_2} F \cos \phi \, dl = \int_{P_1}^{P_2} F_\parallel \, dl = \int_{P_1}^{P_2} \vec{F} \cdot d\vec{l}. \qquad (6\text{–}14)$$

This expression is valid even if the path is curved and the angle ϕ varies during the displacement.

- Power is the time rate of doing work. If an amount of work ΔW is done in a time Δt, the average power P_{av} is

$$P_{av} = \frac{\Delta W}{\Delta t}. \qquad (6\text{–}15)$$

The instantaneous power is defined as

$$P = \lim_{\Delta t \to 0} \frac{\Delta W}{\Delta t} = \frac{dW}{dt}. \qquad (6\text{–}16)$$

When a force $\vec{F}$ acts on a particle moving with velocity $\vec{v}$, the instantaneous power or rate at which the force does work is

$$P = \vec{F} \cdot \vec{v}. \qquad (6\text{–}19)$$

Like work and kinetic energy, power is a scalar quantity. The unit of power in SI units is 1 watt = 1 joule/second (1 W = 1 J/s).

DISCUSSION QUESTIONS

Q6–1 For a constant force in the direction of the displacement, how can twice the work be done by a force of half the magnitude?

Q6–2 An elevator is hoisted by its cables at constant speed. Is the total work done on the elevator positive, negative, or zero? Explain.

Q6–3 A rope tied to a body is pulled, causing the body to accelerate. But according to Newton's third law, the body pulls back on the rope with an equal and opposite force. Is the total work done then zero? If so, how can the body's kinetic energy change? Explain.

Q6–4 When you use an automobile jack, you lift a car by exerting a force much smaller in magnitude than the weight of the car. Does this mean that you do less work on the car than if you had lifted the car directly? Explain.

Q6–5 If there is a net force on a moving object that is nonzero and constant in magnitude, is it possible for the total work done on the object to be zero? Explain, with an example that illustrates your answer.

Q6–6 In Example 5–4 (Section 5–2), how does the work done on the bucket by the tension in the cable compare to the work done on the cart by the tension in the cable?

Q6–7 In Example 5–20 (Section 5–5), does the force $\vec{F}$ do work on the box? Do any of the forces on the box do work on it? Is the speed of the box constant? Is the velocity of the box constant? Explain your answers.

Q6–8 In Example 5–9 (Section 5–3), gravity does work on the toboggan as it travels a distance d down the hill, where d is measured parallel to the slope. The magnitude of the gravity force depends on the mass of the toboggan and its contents but is

independent of the hill's slope angle α. For a constant distance d, does the *work* done by gravity depend on α? Explain.

Q6–9 A force $\vec{F}$ is in the x-direction and has a magnitude that depends on x. Sketch a possible graph of F versus x such that the force does zero work on an object that moves from x_1 to x_2, even though the force magnitude is not zero at all x in this range.

Q6–10 Does the kinetic energy of a car change more when it speeds up from 10 to 15 m/s or from 15 to 20 m/s? Explain.

Q6–11 A falling brick has a mass of 1.5 kg and is moving straight downward with a speed of 5.0 m/s. A 1.5-kg physics book is sliding across the floor with a speed of 5.0 m/s. A 1.5-kg melon is traveling with a horizontal velocity component 3.0 m/s to the right and a vertical component 4.0 m/s upward. Do these objects all have the same velocity? Do these objects all have the same kinetic energy? For each question, give the reasoning behind your answer.

Q6–12 Can the *total* work done on an object during a displacement be negative? Explain. If the total work is negative, can its magnitude be larger than the initial kinetic energy of the object? Explain.

Q6–13 A net force acts on an object and accelerates it from rest to a speed v_1. In doing so the force does an amount of work W_1. By what factor must the work done on the object be increased to produce three times the final speed, with the object again starting from rest?

Q6–14 A truck speeding down the highway has a lot of kinetic energy relative to a stopped state trooper, but no kinetic energy relative to the truck driver. In these two frames of reference, is the same amount of work required to stop the truck? Explain.

Q6–15 A vertical spring has one end attached to the floor. A force $\vec{F}$ is applied to the other end of the spring, slowly stretching it. Is the net work done on the spring equal to the change in its kinetic energy? Explain.

Q6–16 When a book slides along a table top, the force of friction does negative work on it. Can friction ever do *positive* work? Explain. (*Hint:* Think of a box in the back of an accelerating truck, with friction between the box and the truck floor.)

Q6–17 Time yourself while running up a flight of steps and compute the average rate at which you do work against the force of gravity. Express your answer in watts and in horsepower.

Q6–18 When a constant force is applied to a body moving with constant acceleration, is the power of the force constant? If not, how would the force have to vary with speed for the power to be constant?

Q6–19 An advertisement for a portable electrical generating unit claims that the unit's diesel engine produces 28,000 hp to drive an electrical generator that produces 30 MW of electrical power. Is this possible? Explain.

Q6–20 A car speeds up while the engine delivers constant power. Is the acceleration greater at the beginning of this process or at the end? Explain.

Q6–21 Consider a graph of instantaneous power versus time, with the vertical P axis starting at $P = 0$. What is the physical significance of the area under the P versus t curve between vertical lines at t_1 and t_2? How could you find the average power from the graph? Draw a P versus t curve that consists of two straight-line sections and for which the peak power is equal to twice the average power.

Q6–22 A traffic engineer claims that timing the signal lights so that motorists can travel long distances at constant speeds is one way to improve the air quality of a city. Explain the physics that supports this statement.

EXERCISES

SECTION 6–2 WORK

6–1 You push your physics book 1.50 m along a horizontal tabletop with a horizontal force of 2.40 N. The opposing force of friction is 0.600 N. a) How much work does your 2.40-N force do on the book? b) What is the work done on the book by the friction force? c) What is the total work done on the book?

6–2 An old oaken bucket of mass 6.75 kg hangs in a well at the end of a rope. The rope passes over a frictionless pulley at the top of the well, and you pull horizontally on the end of the rope to raise the bucket slowly a distance of 4.00 m. a) How much work do you do on the bucket in pulling it up? b) How much work is done by the gravitational force acting on the bucket? c) What is the total work done on the bucket?

6–3 A fisherman reels in 12.0 m of line while pulling in a fish that exerts a constant resisting force of 25.0 N. If the fish is pulled in at constant velocity, how much work is done on it by the tension in the line?

6–4 A factory worker pushes a 30.0-kg crate a distance of 4.5 m along a level floor at constant velocity by pushing horizontally on it. The coefficient of kinetic friction between the crate and floor is 0.25. a) What magnitude of force must the worker apply? b) How much work is done on the crate by this force? c) How much work is done on the crate by friction? d) How much work is done by the normal force? By gravity? e) What is the total work done on the crate?

6–5 Suppose the worker in Exercise 6–4 pushes downward at an angle of 30° below the horizontal. a) What magnitude of force must the worker apply to move the crate at constant velocity? b) How much work is done on the crate by this force when the crate is pushed a distance of 4.5 m? c) How much work is done on the crate by friction during this displacement? d) How much work is done by the normal force? By gravity? e) What is the total work done on the crate?

6–6 A water skier is pulled behind a boat by a horizontal tow rope. She skis off to the side, so the rope makes an angle of 15.0° with her direction of motion, and then continues in a straight line. The tension in the rope is 180 N. How much work is done on the skier by the rope during a displacement of 300 m?

6–7 Two tugboats pull a disabled supertanker. Each tug exerts a constant force of 1.80×10^6 N, one 14° west of north and the other 14° east of north, as they pull the tanker 0.75 km toward the north. What is the total work they do on the supertanker?

SECTION 6–3 WORK AND KINETIC ENERGY

6–8 a) Compute the kinetic energy, in joules, of a 1600-kg automobile traveling at 50.0 km/h. b) By what factor does the kinetic energy change if the speed is doubled?

6–9 The dinosaur *Tyrannosaurus rex* is thought to have had a mass of about 7000 kg. a) Treat the dinosaur as a particle and estimate its kinetic energy at a walking speed of 4.0 km/h. b) With what speed would a 70-kg person have to move to have the same kinetic energy as a walking *T. rex?*

6–10 A car is stopped by a constant friction force that is independent of the car's speed. By what factor is the stopping distance changed if the car's initial speed is doubled? (Solve using work-energy methods.)

6–11 A baseball leaves a pitcher's hand at a speed of 32.0 m/s. The mass of the baseball is 0.145 kg. Ignore air resistance. How much work has the pitcher done on the ball in throwing it?

6–12 In Example 6–6 (Section 6–3), call the iceboat with mass m boat A and the other iceboat, of mass $2m$, boat B. a) At the finish line what is the ratio v_A/v_B of the speeds of the two iceboats? b) Let t_A be the elapsed time it takes boat A to reach the finish line and t_B the elapsed time for boat B. What is the ratio t_B/t_A of these two times?

6–13 A moving electron has kinetic energy K_1. After a net amount of work W has been done on it, the electron is moving one-quarter as fast in the opposite direction. a) Find W in terms of K_1. b) Does your answer depend on the final direction of the electron's motion?

6–14 A sled with a mass of 8.00 kg moves in a straight line on a frictionless horizontal surface. At one point in its path, its speed is 4.00 m/s; after it has traveled 2.50 m beyond this point, its speed is 6.00 m/s. Use the work-energy theorem to find the force acting on the sled, assuming that this force is constant and that it acts in the direction of the sled's motion.

6–15 A soccer ball of mass 0.420 kg is initially moving with speed 2.00 m/s. A soccer player kicks the ball, exerting a constant force of magnitude 40.0 N in the same direction as the ball's motion. Over what distance must her foot be in contact with the ball to increase the ball's speed to 6.00 m/s?

6–16 A 12–pack of Omni-Cola (mass 4.30 kg) is initially at rest on a horizontal floor. It is then pushed in a straight line for 1.20 m by a trained dog who exerts a horizontal force with magnitude 36.0 N. Use the work-energy theorem to find the final speed of the 12-pack if a) there is no friction between the 12-pack and the floor; b) the coefficient of kinetic friction between the 12-pack and the floor is 0.30.

6–17 A baseball with mass 0.145 kg is thrown straight upward with an initial speed of 25.0 m/s. a) How much work has gravity done on the baseball when it reaches a height of 20.0 m above the pitcher's hand? b) Use the work-energy theorem to calculate the speed of the baseball at a height of 20.0 m above the pitcher's hand. Ignore air resistance. c) Does the answer to part (b) depend on whether the baseball is moving upward or downward at a height of 20.0 m? Explain.

6–18 A 4.80-kg watermelon is dropped (zero initial speed) from the roof of a 25.0-m-tall building. a) Calculate the work done by gravity on the watermelon during its displacement from the roof to the ground. b) What is the kinetic energy of the watermelon just before it strikes the ground? Ignore air resistance.

6–19 A little red wagon with a mass of 7.00 kg moves in a straight line on a frictionless horizontal surface. It has an initial speed of 4.00 m/s and then is pushed 3.0 m in the direction of the initial velocity by a force with a magnitude of 10.0 N. a) Use the work-energy theorem to calculate the wagon's final speed. b) Calculate the acceleration produced by the force. Use this acceleration in the kinematic relations of Chapter 2 to calculate the wagon's final speed. Compare this result to that calculated in part (a).

6–20 A block of ice with mass 2.00 kg slides 0.750 m down an inclined plane that slopes downward at an angle of 36.9° below the horizontal. If the block of ice starts from rest, what is its final speed? Friction can be neglected.

6–21 A car is traveling on a level road with speed v_0 at the instant when the brakes lock, so that the tires slide rather than roll. a) Use the work-energy theorem to calculate the minimum stopping distance of the car in terms of v_0, g, and the coefficient of kinetic friction μ_k between the tires and the road. b) The car stops in a distance of 91.2 m if $v_0 = 80.0$ km/h. What is the stopping distance if $v_0 = 60.0$ km/h? Assume that the value of μ_k remains the same.

SECTION 6–4 WORK AND ENERGY WITH VARYING FORCES

6–22 To stretch a spring 3.00 cm from its unstretched length, 12.0 J of work must be done. How much work must be done to compress this spring 4.00 cm from its unstretched length?

6–23 A force of 160 N stretches a spring 0.050 m beyond its unstretched length. a) What magnitude of force is required to stretch the spring 0.015 m beyond its unstretched length? To compress the spring 0.020 m? b) How much work must be done to stretch the spring 0.015 m beyond its unstretched length? How much work to compress the spring 0.020 m from its unstretched length?

6–24 A child applies a force $\vec{F}$ parallel to the x-axis to a 10.0-kg sled moving on the frozen surface of a small pond. As the child controls the speed of the sled, the x-component of the force she applies varies with the x-coordinate of the sled as shown in Fig. 6–21. Calculate the work done by the force $\vec{F}$ when the sled moves a) from $x = 0$ to $x = 8.0$ m; b) from $x = 8.0$ m to $x = 12.0$ m; c) from $x = 0$ to 12.0 m.

6–25 Suppose the sled in Exercise 6–24 is initially at rest at $x = 0$. Use the work-energy theorem to find the speed of the sled at a) $x = 8.0$ m; b) $x = 12.0$ m. Neglect friction between the sled and the surface of the pond.

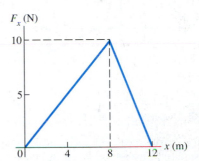

FIGURE 6–21 Exercises 6–24 and 6–25.

6–26 Leg Presses. As part of your daily workout, you lie on your back and push with your feet against a platform attached to two stiff springs arranged side by side so that they are parallel to each other. When you push the platform, you compress the springs. You do 80.0 J of work when you compress the springs 0.200 m from their uncompressed length. a) What magnitude of force must you apply to hold the platform in this position? b) How much *additional* work must you do to move the platform 0.200 m *farther*, and what maximum force must you apply?

6–27 a) In Example 6–8 (Section 6–4) it is calculated that with the air track turned off the glider travels 8.6 cm before it stops instantaneously. How large would the coefficient of static friction μ_s have to be to keep the glider from springing back to the left? b) If the coefficient of static friction between the glider and the track is $\mu_s = 0.60$, what is the maximum initial speed v_1 that the glider can be given and remain at rest after it stops instantaneously? With the air track turned off, the coefficient of kinetic friction is $\mu_k = 0.47$.

6–28 A 4.00-kg block of ice is placed against a horizontal spring that has force constant $k = 200$ N/m and is compressed 0.025 m. The spring is released and accelerates the block along a horizontal surface. Friction and the mass of the spring can be neglected. a) Calculate the work done on the block by the spring during the motion of the block from its initial position to where the spring has returned to its uncompressed length. b) What is the speed of the block after it leaves the spring?

6–29 A force $\vec{F}$ is applied to a 2.0-kg radio-controlled model car parallel to the x-axis as it moves along a straight track. The x-component of the force varies with the x-coordinate of the car as shown in Fig. 6–22. Calculate the work done by the force $\vec{F}$ when the car moves from a) $x = 0$ to $x = 3.0$ m; b) $x = 3.0$ m to $x = 4.0$ m; c) $x = 4.0$ m to $x = 7.0$ m; d) $x = 0$ to $x = 7.0$ m; e) $x = 7.0$ m to $x = 2.0$ m.

6–30 Suppose the 2.0-kg model car in Exercise 6–29 is initially at rest at $x = 0$ and $\vec{F}$ is the net force acting on the car. Use the work-energy theorem to find the speed of the car at a) $x = 3.0$ m; b) $x = 4.0$ m; c) $x = 7.0$ m.

FIGURE 6–22 Exercises 6–29 and 6–30.

6–31 At a waterpark, sleds with riders are sent along a slippery horizontal surface by the release of a large compressed spring. The spring with a force constant $k = 4000$ N/m and negligible mass rests on the frictionless horizontal surface. One end is in contact with a stationary wall. A sled and rider with total mass 70.0 kg are pushed against the other end, compressing the spring 0.375 m. The sled is then released with zero initial velocity. What is the sled's speed when the spring a) returns to its uncompressed length? b) is still compressed 0.200 m?

6–32 In Example 6–9 (Section 6–4), instead of applying a varying horizontal force $\vec{F}$ that maintains Throcky very nearly in equilibrium, you apply a constant horizontal force with magnitude $F = 2w$, where w is Throcky's weight. As in Example 6–9, consider Throcky to be a particle and neglect the small weight of the chains and seat. You push Throcky until the chains make an angle θ_0 with the vertical. a) Use Eq. (6–14) to calculate the work done on Throcky by the force $\vec{F}$ that you apply. b) At the angle θ_0, how does the magnitude of $\vec{F}$ in this exercise compare to that in Example 6–9? c) How does the work done by $\vec{F}$ in this exercise compare to that in Example 6–9?

6–33 A small glider is placed against a compressed spring at the bottom of an air track that slopes upward at an angle of 40.0° above the horizontal. The glider has mass 0.0900 kg. The spring has $k = 640$ N/m and negligible mass. When the spring is

released, the glider travels a maximum distance of 1.80 m along the air track before sliding back down. Before reaching this maximum distance, the glider loses contact with the spring. a) What distance was the spring originally compressed? b) When the glider has traveled along the air track 0.80 m from its initial position against the compressed spring, is it still in contact with the spring? What is the kinetic energy of the glider at this point?

6–34 An ingenious bricklayer builds a device for shooting bricks up to the top of the wall where he is working. He places a brick on a vertical compressed spring with force constant $k = 450$ N/m and negligible mass. When the spring is released, the brick is propelled upward. If the brick has mass 1.80 kg and is to reach a maximum height of 3.6 m above its initial position on the compressed spring, what distance must the spring be compressed initially? (The brick loses contact with the spring when the spring has returned to its uncompressed length. Why?)

SECTION 6–5 POWER

6–35 A tandem (two-person) bicycle team must overcome a force of 165 N to maintain a speed of 9.00 m/s. Find the power required per rider, assuming that each contributes equally. Express your answer in watts and in horsepower.

6–36 The total consumption of electrical energy in the United States is about 1.0×10^{19} joules per year. a) What is the average rate of electrical energy consumption in watts? b) If the population of the United States is 260 million, what is the average rate of electrical energy consumption per person? c) The sun transfers energy to the earth by radiation at a rate of approximately 1.0 kW per square meter of surface. If this energy could be collected and converted to electrical energy with 40% efficiency, how great an area (in square kilometers) would be required to collect the electrical energy used by the United States?

6–37 When its 75-kW (100-hp) engine is generating full power, a small single-engine airplane with a mass of 700 kg gains altitude at a rate of 2.5 m/s (150 m/min, or 500 ft/min). What fraction of the engine power is being used to make the airplane climb? (The remainder is used to overcome the effects of air resistance and of inefficiencies in the propeller and engine.)

6–38 Your job is to lift 30-kg crates a vertical distance of 0.90 m from the ground onto the bed of a truck. a) How many crates would you have to load onto the truck in one minute for your average power output that goes into lifting the crates to be 0.50 hp? b) How many crates for an average power output of 100 W?

6–39 An elevator has a mass of 600 kg, not including passengers. The elevator is designed to ascend at constant speed a vertical distance of 20.0 m (five floors) in 16.0 s, and is driven by a motor that can provide up to 40 hp to the elevator. What is the maximum number of passengers that can ride in the elevator? Assume that an average passenger has a mass of 65.0 kg.

6–40 The hammer of a pile driver weighs 3800 N and must be lifted 2.80 m vertically at constant speed in 4.00 s. What horsepower must the engine provide to the hammer?

6–41 The aircraft carrier *John F. Kennedy* has a mass of 7.4×10^7 kg. When its engines are developing their full power of 280,000 hp, *John F. Kennedy* travels at its top speed of 35 knots (65 km/h). If 70% of the power output of the engines is applied to pushing the ship through the water, what is the magnitude of the force of water resistance that opposes the carrier's motion at this speed?

6–42 A ski tow is operated on a 15.0° slope of length 300 m. The rope moves at 12.0 km/h and power is provided for 50 riders at one time, with an average mass per rider of 70.0 kg. Estimate the power required to operate the tow.

6–43 A particle is accelerated from rest by a constant net force. a) Show that the instantaneous power supplied by the net force is ma^2t. b) To triple the acceleration at any given time, by what factor must the power be increased? c) At $t = 5.0$ s the power supplied by the net force is 36 W. What must the power be at $t = 15.0$ s to maintain constant acceleration?

6–44 Show that the instantaneous power P supplied by the net force acting on a particle is related to the particle's kinetic energy K by $P = dK/dt$.

SECTION 6–6 AUTOMOTIVE POWER: A CASE STUDY IN ENERGY RELATIONS

6–45 Consider the Porsche 911 Carrera of Section 6–6. a) Verify that the power needed for a constant speed of 30 m/s on a level road is 16 kW. b) If 15% of the 3.5×10^7 J of energy obtained by burning each liter of gasoline is available to propel the car, what volume of gasoline is consumed in 1.0 h at this speed? c) Calculate the fuel consumption per unit distance in L/km and in gal/mi.

6–46 A truck engine transmits 28.0 kW (37.5 hp) to the driving wheels when the truck is traveling at a constant velocity of magnitude 60.0 km/h (37.3 mi/h) on a level road. a) What is the resisting force acting on the truck? b) Assume that 65% of the resisting force is due to rolling friction, and that the remainder is due to air resistance. If the force of rolling friction is independent of speed and the force of air resistance is proportional to the square of the speed, what power will drive the truck at 30.0 km/h? At 120.0 km/h? Give your answers in kilowatts and in horsepower.

6–47 a) If 8.00 hp are required to drive a 1800-kg automobile at 60.0 km/h on a level road, what is the total retarding force due to friction, air resistance, and so on? b) What power is necessary to drive the car at 60.0 km/h up a 10.0% grade (a hill rising 10.0 m vertically in 100 m horizontally)? c) What power is necessary to drive the car at 60.0 km/h *down* a 1.00% grade? d) Down what percent grade would the car coast at 60.0 km/h?

6–48 Adding a 75-kg passenger to the Porsche of Section 6–6 increases the mass by 6%. By what percent does this increase the required power at a) 10 m/s? b) 30 m/s?

PROBLEMS

6–49 A luggage handler pulls a 20.0-kg suitcase up a ramp inclined at 25.0° above the horizontal by a force $\vec{F}$ of magnitude 140 N that acts parallel to the ramp. The coefficient of kinetic friction between the ramp and the incline is $\mu_k = 0.300$. If the suitcase travels 3.80 m along the ramp, calculate a) the work done on the suitcase by the force $\vec{F}$; b) the work done on the suitcase by the gravitational force; c) the work done on the suitcase by the normal force; d) the work done on the suitcase by the friction force; e) the total work done on the suitcase. f) If the speed of the suitcase is zero at the bottom of the ramp, what is its speed after it has traveled 3.80 m along the ramp?

6–50 While doing a chin-up, a man's body is lifted 0.40 m. a) How much work must the man do per kilogram of body mass? b) The muscles involved in doing a chin-up can generate about 70 J of work per kilogram of muscle mass. If the man can just barely do a 0.40-m chin-up, what percentage of his body's mass do these muscles constitute? (For comparison, the *total* percentage of muscle in a typical 70-kg man with 14% body fat is about 43%.) c) Repeat part (b) for the man's young son, who has arms half as long as his father's but whose muscles can also generate 70 J of work per kilogram of muscle mass. d) Adults and children have about the same percentage of muscle in their bodies. Explain why children can commonly do chin-ups more easily than their fathers.

6–51 Ramps for the disabled are used because a large weight w can be raised by relatively small force equal to $w \sin \alpha$ plus the small friction force. Such inclined planes are an example of a class of devices called *simple machines*. An input force F_{in} is applied to the system and results in an output force F_{out} applied to the object being moved. For a simple machine the ratio of these forces, F_{out}/F_{in}, is called the actual mechanical advantage (AMA). The inverse ratio of the distances that the points of application of these forces move through during the motion of the object, s_{in}/s_{out}, is called the ideal mechanical advantage (IMA). a) Find the IMA for an inclined plane. b) What can we say about the relation between the work supplied to the machine, W_{in}, and the work output of the machine, W_{out}, if AMA = IMA? c) Sketch a single pulley arranged to give IMA = 2. d) We define the efficiency e of a simple machine to equal the ratio of the output work to the input work, $e = W_{out}/W_{in}$. Show that $e = $ AMA/IMA.

6–52 A woman stands in an elevator that has a constant upward acceleration while the elevator travels upward a distance of 18.0 m. During the 18.0-m displacement, the normal force exerted by the elevator floor does 8.25 kJ of work on her and gravity does −7.35 kJ of work on her. a) What is the mass of the woman? b) What is the normal force that the elevator floor exerts on her? c) What is the acceleration of the elevator?

6–53 The space shuttle *Endeavour*, with mass 86,400 kg, is in a circular orbit of radius 6.66×10^6 m around the earth. It takes 90.1 min for the shuttle to complete each orbit. On a repair mission, it is cautiously moving 1.00 m closer to a disabled satellite every 3.00 s. Calculate the shuttle's kinetic energy a) relative to the earth; b) relative to the satellite.

6–54 A 5.00-kg package slides 1.50 m down a long ramp that is inclined at 12.0° below the horizontal. The coefficient of kinetic friction between the package and the ramp is $\mu_k = 0.310$. Calculate a) the work done on the package by friction; b) the work done on the package by gravity; c) the work done on the package by the normal force; d) the total work done on the package. e) If the package has a speed of 2.20 m/s at the top of the ramp, what is its speed after sliding 1.50 m down the ramp?

6-55 The package in Problem 6–54 has a speed of 2.20 m/s at the top of the ramp. Use the work-energy theorem to find how far down the ramp the package slides before it stops.

6-56 An object is attracted toward the origin with a force given by $F_x = -k/x^2$. (Gravitational and electrical forces have this distance dependence.) a) Calculate the work done by the force F_x when the object moves in the x-direction from x_1 to x_2. If $x_2 > x_1$, is the work done by F_x positive or negative? b) The only other force acting on the object is a force that you exert with your hand to move the object slowly from x_1 to x_2. How much work do you do? If $x_2 > x_1$, is the work you do positive or negative? c) Explain the similarities and differences between your answers to parts (a) and (b).

6-57 An object that can move along the x-axis is attracted toward the origin with a force of magnitude $F = \alpha x^3$, where $\alpha = 4.00 \text{ N/m}^3$. What is the force F when the object is a) at $x = 1.00$ m? b) at $x = 2.00$ m? c) How much work is done by the force F when the object moves from $x = 1.00$ m to $x = 2.00$ m? Is this work positive or negative?

6-58 Consider a spring that does not obey Hooke's law very faithfully. One end of the spring is held fixed. To keep the spring stretched or compressed an amount x, a force along the x-axis with x-component $F_x = kx - bx^2 + cx^3$ must be applied to the free end. Here $k = 100$ N/m, $b = 700 \text{ N/m}^2$, and $c = 12,000 \text{ N/m}^3$. Note that $x > 0$ when the spring is stretched and $x < 0$ when it is compressed. a) How much work must be done to stretch this spring by 0.050 m from its unstretched length? b) How much work must be done to *compress* this spring by 0.050 m from its unstretched length? c) Is it easier to stretch or compress this spring? Explain why in terms of the dependence of F_x on x. (Many real springs behave qualitatively in the same way.)

6-59 A small block with a mass of 0.120 kg is attached to a cord passing through a hole in a horizontal frictionless surface (Fig. 6–23). The block is originally revolving at a distance of 0.40 m from the hole with a speed of 0.70 m/s. The cord is then pulled from below, shortening the radius of the circle in which the block revolves to 0.10 m. At this new distance the speed of the block is observed to be 2.80 m/s. a) What is the tension in the cord in the original situation when the block has speed $v = 0.70$ m/s? b) What is the tension in the cord in the final situation when the block has speed $v = 2.80$ m/s? c) How much work was done by the person who pulled on the cord?

FIGURE 6–23 Problem 6–59.

6-60 Proton Bombardment. A proton with a mass of 1.67×10^{-27} kg is propelled at an initial speed of 3.00×10^5 m/s directly toward a uranium nucleus 5.00 m away. The proton is repelled by the uranium nucleus with a force of magnitude $F = \alpha/x^2$, where x is the separation between the two objects and $\alpha = 2.12 \times 10^{-26} \text{ N} \cdot \text{m}^2$. Assume that the uranium nucleus remains at rest. a) What is the speed of the proton when it is 8.00×10^{-10} m from the uranium nucleus? b) As the proton approaches the uranium nucleus, the repulsive force slows down the proton until it comes momentarily to rest, after which the proton moves away from the uranium nucleus. How close to the uranium nucleus does the proton get? c) What is the speed of the proton when it is again 5.00 m away from the uranium nucleus?

6-61 A block of ice with mass 6.00 kg is initially at rest on a frictionless horizontal surface. A worker then applies a horizontal force $\vec{F}$ to it. As a result, the block moves along the x-axis such that its position as a function of time is given by $x(t) = \alpha t^2 + \beta t^3$, where $\alpha = 0.200 \text{ m/s}^2$ and $\beta = 0.0200 \text{ m/s}^3$. a) Calculate the velocity of the object when $t = 4.00$ s. b) Calculate the magnitude of $\vec{F}$ when $t = 4.00$ s. c) Calculate the work done by the force $\vec{F}$ during the first 4.00 s of the motion.

6-62 A force with magnitude $(5.00 \text{ N/m}^2)x^2$ and directed at a constant 31.0° angle with the $+x$-axis acts on a 0.250-kg object. Other forces with only y-components also act so that the object moves parallel to the x-axis. What is the object's speed at $x = 1.50$ m if its speed at $x = 1.00$ m is 4.00 m/s?

6-63 You and your bicycle have a combined mass of 80.0 kg. When you reach the base of an bridge, you are traveling along the road at 5.00 m/s (Fig. 6–24). At the top of the bridge you have climbed a vertical distance of 5.20 m and have slowed to 1.50 m/s. Ignore work done by friction and any inefficiency in the bike or your legs. a) What is the total work done on you and your bicycle when you go from the base to the top of the bridge? b) How much work have you done with the force you apply to the pedals?

FIGURE 6–24 Problem 6–63.

6-64 A force in the $+x$-direction has magnitude $F = b/x^n$, where b and n are constants. a) For $n > 1$, calculate the work done on a particle by this force when the particle moves along the x-axis from $x = x_0$ to infinity. b) Show that for $0 < n < 1$, even though F becomes zero as x becomes very large, an infinite amount of work is done by F when the particle moves from $x = x_0$ to infinity.

6–65 You are asked to design spring bumpers for the walls of a parking garage. A freely rolling 1200-kg car moving at 0.65 m/s is to compress the spring no more than 0.070 m before stopping. What should be the force constant of the spring? Assume that the spring has negligible mass.

6–66 The spring of a spring gun has force constant $k = 400$ N/m and negligible mass. The spring is compressed 6.00 cm and a ball with mass 0.0300 kg is placed in the horizontal barrel against the compressed spring. The spring is then released, and the ball is propelled out the barrel of the gun. The barrel is 6.00 cm long, so the ball leaves the barrel at the same point that it loses contact with the spring. The gun is held so the barrel is horizontal. a) Calculate the speed with which the ball leaves the barrel if friction can be neglected. b) Calculate the speed of the ball as it leaves the barrel if a constant resisting force of 6.00 N acts on the ball as it moves along the barrel. c) For the situation in part (b), at what position along the barrel does the ball have the greatest speed, and what is that speed? (In this case the maximum speed does not occur at the end of the barrel.)

6–67 A 2.50-kg textbook is forced against a horizontal spring of negligible mass and force constant 250 N/m, compressing the spring a distance of 0.250 m. When released, the textbook slides on a horizontal table top with coefficient of kinetic friction $\mu_k = 0.30$. Use the work-energy theorem to find how far the textbook moves from its initial position before coming to rest.

6–68 Your cat "Ms." (mass 7.00 kg) is trying to make it to the top of a frictionless ramp 2.00 m long and inclined upward at 30.0° above the horizontal. Since the poor cat can't get any traction on the ramp, you push her up the entire length of the ramp by exerting a constant 100-N force parallel to the ramp. If Ms. takes a running start so that she is moving at 2.40 m/s at the bottom of the ramp, what is her speed when she reaches the top of the incline? Use the work-energy theorem.

6–69 A student proposes a design for an automobile crash barrier in which a 1700-kg sport utility vehicle moving at 20.0 m/s crashes into a spring of negligible mass that slows it to a stop. To avoid injuring the passengers, the acceleration of the vehicle as it slows can be no more than $5.00g$. a) Find the required spring constant k, and find the distance the spring will compress in slowing the vehicle to a stop. In your calculation disregard any deformation or crumpling of the vehicle and the friction between the vehicle and the ground. b) What disadvantages are there to this design?

6–70 A physics professor is pushed up a ramp inclined upward at 30.0° above the horizontal as he sits in his desk chair that slides on frictionless rollers. The combined mass of the professor and chair is 85.0 kg. He is pushed 2.50 m along the incline by a group of students who together exert a constant horizontal force of 600 N. The professor's speed at the bottom of the ramp is 2.00 m/s. Use the work-energy theorem to find his speed at the top of the ramp.

6–71 A 5.00-kg block is moving at $v_0 = 6.00$ m/s along a frictionless horizontal surface toward a spring with force constant $k = 500$ N/m that is attached to a wall (Fig. 6–25). The spring has negligible mass. a) Find the maximum distance the spring

will be compressed. b) If the spring is to compress by no more than 0.150 m, what should be the maximum value of v_0?

FIGURE 6–25 Problem 6–71.

6–72 Consider the system shown in Fig. 6–26. The rope and pulley have negligible mass, and the pulley is frictionless. The coefficient of kinetic friction between the 8.00-kg block and the table top is $\mu_k = 0.250$. The blocks are released from rest. Use energy methods to calculate the speed of the 6.00-kg block after it has descended 1.50 m.

6–73 Consider the system shown in Fig. 6–26. The rope and pulley have negligible mass, and the pulley is frictionless. Initially the 6.00-kg block is moving downward and the 8.00-kg block is moving to the right, both with a speed of 0.900 m/s. The blocks come to rest after moving 2.00 m. Use the work-energy theorem to calculate the coefficient of kinetic friction between the 8.00-kg block and the table top.

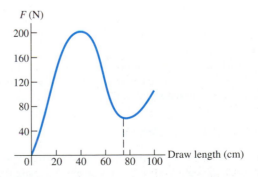

FIGURE 6–26 Problems 6–72 and 6–73.

6–74 Figure 6–27 shows how the force exerted by the string of a compound bow on an arrow varies as a function of how far back the arrow is pulled (the *draw length*). Assume that the same force is exerted on the arrow as it moves forward after being released. Full draw for this bow is at a draw length of 75.0 cm. If the bow shoots a 0.0250-kg arrow from full draw, what is the speed of the arrow as it leaves the bow?

FIGURE 6–27 Problem 6–74.

6–75 A pump is required to lift 800 kg of water (about 210 gallons) per minute from a well 14.0 m deep and eject it with a speed of 18.0 m/s. a) How much work is done per minute in lifting the water? b) How much in giving the water the kinetic energy it has when ejected? c) What must be the power output of the pump?

6–76 Find the power output of the worker in Problem 6–61 as a function of time. What is the numerical value of the power (in watts) at $t = 4.00$ s?

6–77 A physics student spends part of her day walking between classes or for recreation, during which time she expends energy at an average rate of 280 W. The remainder of the day she is sitting in class, studying, or resting; during these activities, she expends energy at an average rate of 100 W. If she expends a total of 1.1×10^7 J of energy in a 24-hour day, how much of the day did she spend walking?

6–78 All birds, independent of their size, must maintain a power output of 10 to 25 watts per kilogram of body mass in order to fly by flapping their wings. (a) The Andean giant hummingbird (*Patagona gigas*) has a mass of 70 g and flaps its wings ten times per second while hovering. Estimate the amount of work done by such a hummingbird in each wingbeat. (b) A 70-kg athlete can maintain a power output of 1.4 kW for no more than a few seconds; the *steady* power output of a typical athlete is only 500 W or so. Is it possible for a human-powered aircraft to fly for extended periods by flapping its wings? Explain.

6–79 The Grand Coulee Dam is 1270 m long and 170 m high. The electrical power output from generators at its base is approximately 2000 MW. How many cubic meters of water must flow from the top of the dam per second to produce this amount of power if 92% of the work done on the water by gravity is converted to electrical energy? (Each cubic meter of water has a mass of 1000 kg.)

*6–80 The engine of a car of mass m supplies a constant power P to the wheels to accelerate the car. Rolling friction and air resistance can be neglected. The car is initially at rest. a) Show that the speed of the car is given as a function of time by $v = (2Pt/m)^{1/2}$. b) Show that the acceleration of the car is not constant, but is given as a function of time by $a = (P/2mt)^{1/2}$. c) Show that the displacement as a function of time is given by $x - x_0 = (8P/9m)^{1/2} t^{3/2}$.

6–81 Power of the Human Heart. The human heart is a powerful and extremely reliable pump. Each day it takes in and discharges about 7500 L of blood. Assume that the work done by the heart is equal to the work required to lift this amount of blood a height equal to that of the average American female (1.63 m). The density (mass per unit volume) of blood is 1.05×10^3 kg/m^3. a) How much work does the heart do in a day? b) What is its power output in watts?

6–82 Six diesel units in series can provide 13.4 MW of power to the lead car of a freight train. The diesel units have a total mass of 1.10×10^6 kg. The average car in the train has a mass of 8.2×10^4 kg and requires a horizontal pull of 2.8 kN to move at a constant 27 m/s on level tracks. a) How many cars can be in the train under these conditions? b) However, this would leave no power for accelerating or climbing hills. Show that the extra force needed to accelerate the train is about the same for a 0.10 m/s^2 acceleration or a 1.0% slope (slope angle $\alpha =$ arctan 0.010). c) With the 1.0% slope, show that an extra 2.9 MW of power is needed to maintain the 27 m/s speed of the diesel units. d) With 2.9 MW less power available, how many cars can the six diesel units pull up a 1.0% slope at a constant 27 m/s?

6–83 It takes a force of 53 kN on the lead car of a 16-car passenger train with a mass of 9.1×10^5 kg to pull it at a constant 45 m/s (101 mi/h) on level tracks. a) What power must the locomotive provide to the lead car? b) How much more power to the lead car than calculated in part (a) would be needed to give the train an acceleration of 1.5 m/s^2, at the instant that the train has a speed of 45 m/s on level tracks? c) How much more power to the lead car than that calculated in part (a) would be needed to move the train up a 1.5% grade (slope angle $\alpha =$ arctan 0.015) at a constant 45 m/s?

6–84 An object has several forces acting on it. One of these forces is $\vec{F} = \alpha xy\hat{\imath}$, a force in the x-direction whose magnitude depends on the position of the object, with $\alpha = 2.50$ N/m^2. Calculate the work done on the object by this force for the following displacements of the object: a) The object starts at the point $x = 0$, $y = 3.00$ m and moves parallel to the x-axis to the point $x = 2.00$ m, $y = 3.00$ m. b) The object starts at the point $x = 2.00$ m, $y = 0$ and moves in the y-direction to the point $x = 2.00$ m, $y = 3.00$ m. c) The object starts at the origin and moves on the line $y = 1.5x$ to the point $x = 2.00$ m, $y = 3.00$ m.

6–85 For a touring bicyclist the drag coefficient is 1.00, the frontal area is 0.463 m^2, and the coefficient of rolling friction is 0.0045. The rider has a mass of 50.0 kg, and her bike has a mass of 12.0 kg. a) To maintain a speed of 12.0 m/s (about 27 mi/h) on a level road, what must the rider's power output to the rear wheel be? b) For racing, the same rider uses a different bike with a coefficient of rolling friction 0.0030 and a mass of 9.00 kg. She also crouches down, reducing her drag coefficient to 0.88 and reducing her frontal area to 0.366 m^2. What must her power output to the rear wheel be then to maintain a speed of 12.0 m/s? c) For the situation in part (b), what power output is required to maintain a speed of 6.0 m/s? Note the great drop in power requirement when the speed is only halved. (For more on aerodynamic speed limitations for a wide variety of human-powered vehicles see "The Aerodynamics of Human-Powered Land Vehicles," *Scientific American,* December, 1983.)

CHALLENGE PROBLEMS

6–86 On a winter's day in Maine, a warehouse worker is shoving boxes up a rough plank inclined at an angle α above the horizontal. The plank is partially covered with ice, with more ice near the bottom of the plank than near the top, so that the coefficient of friction increases with the distance x along the plank: $\mu = Ax$, where A is a positive constant and the bottom of the plank is at $x = 0$. (For this plank the coefficients of kinetic and static friction are equal: $\mu_k = \mu_s = \mu$.) A box is shoved up the plank so that it leaves the bottom of the plank moving at speed v_0. Show that when the box first comes to rest, it will remain at rest if

$$v_0^2 \geq \frac{3g \sin^2 \alpha}{A \cos \alpha}$$

6–87 A Spring with Mass. We usually ignore the kinetic energy of the moving coils of a spring, but let's try to get a reasonable approximation to this. Consider a spring of mass M, equilibrium length L_0, and spring constant k. The work done to stretch or compress the spring by a distance L is $\frac{1}{2}kX^2$, where $X = L - L_0$. a) Consider a spring as described above that has one end fixed and the other end moving with speed v. Assume that the speed of points along the length of the spring varies linearly with distance l from the fixed end. Assume also that the mass M of the spring is distributed uniformly along the length of the spring. Calculate the kinetic energy of the spring in terms of M and v. (*Hint:* Divide the spring into pieces of length dl; find the speed of each piece in terms of l, v, and L; find the mass of each piece in terms of dl, M, and L; and integrate from 0 to L. The result is *not* $\frac{1}{2}Mv^2$, since not all of the spring moves with the same speed.) In a spring gun, a spring of mass 0.243 kg and force constant 3200 N/m is compressed 2.50 cm from its unstretched length. When the trigger is pulled, the spring pushes horizontally on a 0.053-kg ball. The work done by friction is negligible. Calculate the ball's speed when the spring reaches its uncompressed length b) neglecting the mass of the spring; c) including, using the results of part (a), the mass of the spring. d) In part (c), what is the final kinetic energy of the ball and of the spring?

6–88 An airplane in flight is subject to an air resistance force proportional to the square of its speed v, just as in Eq. (6–20). But there is an additional resistive force because the airplane has wings. Air flowing over the wings is pushed down and slightly forward, so from Newton's third law the air exerts a force on the wings and airplane that is up and slightly backward (Fig. 6–28). The upward force is the lift force that keeps the airplane aloft, and the backward force is called *induced drag*. At flying speeds induced drag is inversely proportional to v^2, so that the total air resistance force can be expressed by $F_{\text{air}} = \alpha v^2 + \beta/v^2$, where α and β are positive constants that depend on the shape and size of the airplane and the density of the air. For a Cessna 150, a small single-engine airplane, $\alpha = 0.30$ N $\cdot$ s²/m² and $\beta = 3.5 \times 10^5$ N $\cdot$ m²/s². In steady flight, the engine must provide a forward force that exactly balances the air resistance force. a) Calculate the speed (in km/h) at which this airplane will have the maximum *range* (that is, travel the greatest distance) for a given quantity of fuel. b) Calculate the speed (in km/h) for which the airplane will have the maximum *endurance* (that is, will remain in the air the longest time).

6–89 Figure 6–29 shows the oxygen consumption rate of men walking and running at different speeds. The vertical axis shows the volume of oxygen (in cm³) that a man consumes per kilogram of body mass per minute. Note the transition from walking to running that occurs naturally at about 9 km/h. The metabolism of 1 cm³ of oxygen releases about 20 J of energy. Using the data in the graph, calculate the energy required for a 70-kg man to travel 1 km on foot at each of the following speeds a) 5 km/h (walking); b) 10 km/h (running); c) 15 km/h (running). d) Which speed is the most efficient, that is, requires the least energy to travel 1 km?

FIGURE 6–29 Challenge Problem 6–89.

6–90 General Proof of the Work-Energy Theorem.
Consider a particle that moves along a curved path in space from (x_1, y_1, z_1) to (x_2, y_2, z_2). At the initial point the particle has velocity $\vec{v} = v_{1x}\hat{i} + v_{1y}\hat{j} + v_{1z}\hat{k}$. The path that the particle follows may be divided into infinitesimal segments $d\vec{l} = dx\hat{i} + dy\hat{j} + dz\hat{k}$. As the particle moves, it is acted on by a net force $\vec{F} = F_x\hat{i} + F_y\hat{j} + F_z\hat{k}$. The force components F_x, F_y, and F_z are in general functions of position. By the same sequence of steps used in Eqs. (6–11) through (6–13), prove the work-energy theorem for this general case. That is, prove that

$$W_{\text{tot}} = K_2 - K_1,$$

where

$$W_{\text{tot}} = \int_{(x_1,y_1,z_1)}^{(x_2,y_2,z_2)} \vec{F} \cdot d\vec{l} = \int_{(x_1,y_1,z_1)}^{(x_2,y_2,z_2)} (F_x\,dx + F_y\,dy + F_z\,dz).$$

FIGURE 6–28 Challenge Problem 6–88.

7

Potential Energy and Energy Conservation

As water cascades down a waterfall, gravitational potential energy is converted into kinetic energy (so that the water gains speed) and thermal energy (manifested as an increase in the water's temperature). Careful measurements show that no energy is lost in this process—instead, the energy merely changes form.

7–1 INTRODUCTION

When a diver jumps off a high board into a swimming pool, she hits the water moving pretty fast, with a lot of kinetic energy. Where does that energy come from? The answer we learned in Chapter 6 was that the gravitational force (her weight) does work on the diver as she falls. The diver's kinetic energy—energy associated with her *motion*—increases by an amount equal to the work done.

But there is a very useful alternative way to think about work and kinetic energy. This new approach is based on the concept of *potential energy,* which is energy associated with the *position* of a system rather than its motion. In this approach, there is *gravitational potential energy* even while the diver is standing on the high board. Energy is not added to the earth-diver system as the diver falls, but rather a storehouse of energy is *transformed* from one form (potential energy) to another (kinetic energy) as she falls. In this chapter we'll see how this transformation can be understood from the work-energy theorem.

If the diver bounces on the end of the board before she jumps, the bent board stores a second kind of potential energy called *elastic potential energy.* We'll discuss elastic potential energy of simple systems such as a stretched or compressed spring. (An important third kind of potential energy is associated with the positions of electrically charged particles relative to each other. We'll encounter this in Chapter 24.)

We will prove that in some cases the sum of a system's kinetic and potential energy, called the *total mechanical energy* of the system, is constant during the motion of the system. This will lead us to the general statement of the *law of conservation of energy,* one of the most fundamental and far-reaching principles in all of science.

7–2 GRAVITATIONAL POTENTIAL ENERGY

A particle gains or loses kinetic energy because it interacts with other objects that exert forces on it. We learned in Chapter 6 that during any interaction the change in a particle's kinetic energy is equal to the total work done on the particle by the forces that act on it.

In many situations it seems as though energy has been stored in a system, to be recovered later. It's like a savings account in a bank; you deposit money, and then you can draw it out later. For example, you must do work on the hammerhead of a pile driver to lift it. It seems reasonable that in hoisting the hammerhead into the air you are storing energy in the system, energy that is later converted into kinetic energy as the hammerhead falls. Or consider your cousin Throckmorton on a swing. Suppose you give him a shove, which gives him some kinetic energy, and then you let him swing freely back and forth. He stops momentarily when he reaches the front and back end points of his arc, so at these points he has no kinetic energy. But he regains his kinetic energy as he passes through the low point in the arc. It seems as though at the high points the energy is stored in some other form, related to his height above the ground, and is converted back to kinetic energy as he swings toward the low point.

Both these examples point to the idea of an energy associated with the *position* of bodies in a system. This kind of energy is a measure of the *potential* or *possibility* for work to be done; when a hammerhead is raised into the air, there is a potential for work to be done on it by the gravitational force, but only if the hammerhead is allowed to fall to the ground. For this reason, energy associated with position is called **potential energy.** Our discussion suggests that there is potential energy associated with a body's weight and its height above the ground. We call this *gravitational potential energy.*

When a body falls without air resistance, gravitational potential energy decreases and the body's kinetic energy increases. But in Chapter 6 we used the work-energy theorem to conclude that the kinetic energy of a falling body increases because the force of the earth's gravity (the body's weight) does work on the body. Let's use the work-energy theorem to show that these two descriptions of a falling body are equivalent and to derive the expression for gravitational potential energy.

Let's consider a body with mass m that moves along the (vertical) y-axis, as in Fig. 7–1. The forces acting on it are its weight, with magnitude $w = mg$, and possibly some other forces; we call the vector sum (resultant) of all the other forces $\vec{F}_{other}$. We'll assume that the body stays close enough to the earth's surface that the weight is constant. (We'll find in Chapter 12 that weight decreases with altitude.) We want to find the work done by the weight when the body drops from a height y_1 above the origin to a lower height y_2 (Fig. 7–1a). The weight and displacement are in the same direction, so the work W_{grav} done on the body by its weight is positive;

$$W_{grav} = Fs = w(y_1 - y_2) = mgy_1 - mgy_2. \qquad (7\text{–}1)$$

This expression also gives the correct work when the body moves upward and y_2 is greater than y_1 (Fig. 7–1b). In that case the quantity $(y_1 - y_2)$ is negative, and W_{grav} is negative because the weight and displacement are opposite in direction.

Equation (7–1) shows that we can express W_{grav} in terms of the values of the quantity mgy at the beginning and end of the displacement. This quantity, the product of the weight mg and the height y above the origin of coordinates, is called the **gravitational potential energy,** U:

$$U = mgy \qquad \text{(gravitational potential energy).} \qquad (7\text{–}2)$$

Its initial value is $U_1 = mgy_1$, and its final value is $U_2 = mgy_2$. The change in U is the final value minus the initial value, or $\Delta U = U_2 - U_1$. We can express the work W_{grav} done by the gravitational force during the displacement from y_1 to y_2 as

$$W_{grav} = U_1 - U_2 = -(U_2 - U_1) = -\Delta U. \qquad (7\text{–}3)$$

The negative sign in front of ΔU is *essential.* When the body moves up, y increases, the work done by the gravitational force is negative, and the gravitational potential energy increases ($\Delta U > 0$). When the body moves down, y decreases, the gravitational force does positive work, and the gravitational potential energy decreases ($\Delta U < 0$). It's like drawing money out of the bank (decreasing U) and spending it (doing positive work). As Eq. (7–3) shows, the unit of potential energy is the joule (J), the same unit as is used for work.

CAUTION ▶ Although you might be tempted to do so, it is *not* correct to call $U = mgy$ the "gravitational potential energy of the body." The reason is that gravitational potential energy is a *shared* property of the body and the earth. Gravitational potential energy

7–1 Work is done by the gravitational force $\vec{w}$ during the vertical motion of a body from an initial height y_1 to a final height y_2. The vertical displacement can be either (a) downward or (b) upward.

increases if the earth stays fixed and the height of the body increases; it also increases if the body is fixed in space and the earth is moved away from the body. Notice that the formula $U = mgy$ involves characteristics of both the body (its mass m) and the earth (the value of g). ◄

CONSERVATION OF MECHANICAL ENERGY (GRAVITATIONAL FORCES ONLY)

To see what gravitational potential energy is good for, suppose the body's weight is the *only* force acting on it, so $\vec{F}_{\text{other}} = 0$. The body is then falling freely with no air resistance, and can be moving either up or down. Let its speed at point y_1 be v_1, and let its speed at y_2 be v_2. The work-energy theorem, Eq. (6–6), says that the total work done on the body equals the change in the body's kinetic energy; $W_{\text{tot}} = \Delta K = K_2 - K_1$. If gravity is the only force that acts, then from Eq. (7–3), $W_{\text{tot}} = W_{\text{grav}} = -\Delta U = U_1 - U_2$. Putting these together, we get

$$\Delta K = -\Delta U \qquad \text{or} \qquad K_2 - K_1 = U_1 - U_2,$$

which we can rewrite as

$$K_1 + U_1 = K_2 + U_2 \qquad \text{(if only gravity does work)}, \qquad (7\text{–}4)$$

or

$$\frac{1}{2}mv_1^2 + mgy_1 = \frac{1}{2}mv_2^2 + mgy_2 \qquad \text{(if only gravity does work)}. \qquad (7\text{–}5)$$

We now define the sum $K + U$ of kinetic and potential energy to be E, the **total mechanical energy of the system.** By "system" we mean the body of mass m *and* the earth considered together, because gravitational potential energy U is a shared property of both bodies. Then $E_1 = K_1 + U_1$ is the total mechanical energy at y_1, and $E_2 = K_2 + U_2$ is the total mechanical energy at y_2. Equation (7–4) says that when the body's weight is the only force doing work on it, $E_1 = E_2$. That is, E is constant; it has the same value at y_1 and y_2. But since the positions y_1 and y_2 are arbitrary points in the motion of the body, the total mechanical energy E has the same value at *all* points during the motion;

$$E = K + U = \text{constant} \qquad \text{(if only gravity does work).}$$

A quantity that always has the same value is called a *conserved* quantity. *When only the force of gravity does work, the total mechanical energy is constant, that is, is conserved.* This is our first example of the **conservation of mechanical energy.**

When we throw a ball into the air, its speed decreases on the way up as kinetic energy is converted to potential energy; $\Delta K < 0$ and $\Delta U > 0$. On the way back down, potential energy is converted back to kinetic energy and the ball's speed increases; $\Delta K > 0$ and $\Delta U < 0$. But the *total* mechanical energy (kinetic plus potential) is the same at every point in the motion, provided that no force other than gravity does work on the ball (that is, air resistance must be negligible). It's still true that the gravitational force does work on the body as it moves up or down, but we no longer have to calculate work directly; keeping track of changes in the value of U takes care of this completely.

An important point about gravitational potential energy is that it doesn't matter what height we choose to be $y = 0$, the origin of coordinates. If we shift the origin for y, the values of y_1 and y_2 both change, but the *difference* $y_2 - y_1$ stays the same. It follows that although U_1 and U_2 depend on where we place the origin, the difference $U_2 - U_1 = mg(y_2 - y_1)$ does not. The physically significant quantity is not the value of U at a particular point, but only the *difference* in U between two points. So we can define U to be zero at whatever point we choose without affecting the physics of the situation. This is shown in the following example.

During the time this athlete is in midair, only gravity does work on him (if we neglect the minor effects of air resistance). Mechanical energy—the sum of kinetic and gravitational potential energy—is conserved.

EXAMPLE 7-1

Height of a baseball from energy conservation You throw a 0.145-kg baseball straight up in the air, giving it an initial upward velocity of magnitude 20.0 m/s. Use conservation of energy to find how high it goes, ignoring air resistance.

SOLUTION After the ball leaves your hand, the only force doing work on the ball is its weight, so we can use Eq. (7–4). Let's take the origin at the point where the ball leaves your hand (point 1). Then $y_1 = 0$ (Fig. 7–2), and the potential energy at point 1 is $U_1 = mgy_1 = 0$. At this point, $v_1 = 20.0$ m/s. We want to find the height y_2 at point 2, where it stops and begins to fall back to earth. At this point, $v_2 = 0$, and the kinetic energy is $K_2 = \frac{1}{2}mv_2^2 = 0$. Equation (7–4) says that $K_1 + U_1 = K_2 + U_2$; since $U_1 = 0$ and $K_2 = 0$,

$$K_1 = U_2.$$

As the energy bar graphs in Fig. 7–2 show, the kinetic energy of the ball at point 1 is completely converted into gravitational potential energy at point 2. At point 1 the kinetic energy is

$$K_1 = \frac{1}{2}mv_1^2 = \frac{1}{2}(0.145 \text{ kg})(20.0 \text{ m/s})^2 = 29.0 \text{ J}.$$

This equals the gravitational potential energy $U_2 = mgy_2$ at point 2, so

$$y_2 = \frac{U_2}{mg} = \frac{29.0 \text{ J}}{(0.145 \text{ kg})(9.80 \text{ m/s}^2)} = 20.4 \text{ m}.$$

We can also solve the equation $K_1 = U_2$ algebraically for y_2:

$$\frac{1}{2}mv_1^2 = mgy_2,$$

$$y_2 = \frac{v_1^2}{2g} = \frac{(20.0 \text{ m/s})^2}{2(9.80 \text{ m/s}^2)} = 20.4 \text{ m}.$$

The mass divides out, as we should expect; we learned in Chapter 2 that the motion of a body in free fall doesn't depend

7–2 After a baseball leaves your hand, the only force acting on it is its weight (neglecting air resistance), so mechanical energy $E = K + U$ is conserved. The energy bar graphs show the values of E, K, and U at $y_1 = 0$ and y_2.

on its mass. Indeed, we could have derived the result $y_2 = v_1^2/2g$ using Eq. (2–13).

In carrying out the calculation above, we chose the origin to be at point 1, so $y_1 = 0$ and $U_1 = 0$. What happens if we make a different choice? As an example, suppose we choose the origin to be 5.0 m below point 1, so $y_1 = 5.0$ m. With this choice the total mechanical energy at point 1 is part kinetic and part potential, while at point 2 it's purely potential energy. If you work through the calculation again with this choice of origin, you'll find $y_2 = 25.4$ m; this is 20.4 m above point 1, just as with the first choice of origin. In every problem, it's up to you to choose the height at which $U = 0$; don't agonize over the choice, however, because the physics of the answer doesn't depend on your choice.

EFFECT OF OTHER FORCES

When other forces act on the body in addition to its weight, then $\vec{F}_{\text{other}}$ in Fig. 7–1 is *not* zero. For the pile driver described in Example 6–5 (Section 6–3) the force applied by the hoisting cable and the friction with the vertical guide rails are examples of forces that might be included in $\vec{F}_{\text{other}}$. The gravitational work W_{grav} is still given by Eq. (7–3), but the total work W_{tot} is then the sum of W_{grav} and the work done by $\vec{F}_{\text{other}}$. We will call this additional work W_{other}, so the total work done by all forces is $W_{\text{tot}} = W_{\text{grav}} + W_{\text{other}}$. Equating this to the change in kinetic energy, we have

$$W_{\text{other}} + W_{\text{grav}} = K_2 - K_1. \tag{7–6}$$

Also, from Eq. (7–3), $W_{\text{grav}} = U_1 - U_2$, so

$$W_{\text{other}} + U_1 - U_2 = K_2 - K_1,$$

which we can rearrange in the form

$$K_1 + U_1 + W_{\text{other}} = K_2 + U_2 \quad \text{(if forces other than gravity do work).} \tag{7–7}$$

5.3
Sliding on Snow and Grass

5.5
Upward-Moving Elevator Stops

5.6
Stopping a Downward-Moving Elevator

5.9
Skier Speed

5.10
Skier Rope Tow

5.13
Bricklayer Perils

Finally, using the appropriate expressions for the various energy terms, we obtain

$$\frac{1}{2}mv_1{}^2 + mgy_1 + W_{\text{other}} = \frac{1}{2}mv_2{}^2 + mgy_2 \qquad \text{(if forces other than gravity do work).}$$

(7–8)

The meaning of Eqs. (7–7) and (7–8) is this: *The work done by all forces* other than the gravitational force *equals the change in the total mechanical energy* $E = K + U$ *of the system, where U is the gravitational potential energy.* When W_{other} is positive, E increases, and $K_2 + U_2$ is greater than $K_1 + U_1$. When W_{other} is negative, E decreases. In the special case in which no forces other than the body's weight do work, $W_{\text{other}} = 0$. The total mechanical energy is then constant, and we are back to Eq. (7–4) or (7–5).

Problem–Solving Strategy

PROBLEMS USING MECHANICAL ENERGY

1. First decide whether the problem should be solved by energy methods, by using $\Sigma \vec{F} = m\vec{a}$ directly, or by a combination. The energy approach is particularly useful when the problem involves motion with varying forces, motion along a curved path (discussed later in this section), or both. But if the problem involves elapsed time, the energy approach is usually *not* the best choice because this approach doesn't involve time directly.

2. When using the energy approach, first decide what the initial and final states (the positions and velocities) of the system are. Use the subscript 1 for the initial state and the subscript 2 for the final state. It helps to draw sketches showing the initial and final states.

3. Define your coordinate system, particularly the level at which $y = 0$. You will use this to compute gravitational potential energies. Equation (7–2) assumes that the positive direction for y is upward; we suggest that you use this choice consistently.

4. List the initial and final kinetic and potential energies, that is, K_1, K_2, U_1, and U_2. In general, some of these will be known and some will be unknown. Use algebraic symbols for any unknown coordinates or velocities.

5. Identify all nongravitational forces that do work. A free-body diagram is always helpful. Calculate the work W_{other} done by all these forces. If some of the quantities you need are unknown, represent them by algebraic symbols.

6. Relate the kinetic and potential energies and the nongravitational work W_{other} using Eq. (7–7). If there is no nongravitational work, this becomes Eq. (7–4). It's helpful to draw bar graphs showing the initial and final values of K, U, and $E = K + U$. Then solve to find whatever unknown quantity is required.

7. Keep in mind, here and in later sections, that the work done by each force must be represented either in $U_1 - U_2 = -\Delta U$ or as W_{other}, but *never* in both places. If the gravitational work is included in ΔU, do not include it again in W_{other}.

EXAMPLE 7–2

Work and energy throwing a baseball In Example 7–1, suppose your hand moves up 0.50 m while you are throwing the ball, which leaves your hand with an upward velocity of 20.0 m/s. Again ignore air resistance. a) Assuming that your hand exerts a constant upward force on the ball, find the magnitude of that force. b) Find the speed of the ball at a point 15.0 m above the point where it leaves your hand.

SOLUTION Figure 7–3 shows a diagram of the situation, including a free-body diagram for the ball while it is being thrown. a) Let's use Eq. (7–8) to solve for W_{other}, the work done by the upward force $\vec{F}$ that your hand applies to the ball as you throw it. We'll be able to determine the magnitude F from this. Let point 1 be the point where your hand first starts to move, and let point 2 be the point where the ball leaves your hand. With the same

coordinate system as in Example 7–1, we have $y_1 = -0.50$ m and $y_2 = 0$. Then

$K_1 = 0$,

$U_1 = mgy_1 = (0.145 \text{ kg})(9.80 \text{ m/s}^2)(-0.50 \text{ m}) = -0.71 \text{ J}$,

$K_2 = \frac{1}{2}mv_2{}^2 = \frac{1}{2}(0.145 \text{ kg})(20.0 \text{ m/s})^2 = 29.0 \text{ J}$,

$U_2 = mgy_2 = (0.145 \text{ kg})(9.80 \text{ m/s}^2)(0) = 0$.

The initial potential energy U_1 is negative because the ball was initially below the origin. According to Eq. (7–7), $K_1 + U_1 + W_{\text{other}} = K_2 + U_2$, so

$W_{\text{other}} = (K_2 - K_1) + (U_2 - U_1)$

$= (29.0 \text{ J} - 0) + (0 - (-0.71 \text{ J})) = 29.7 \text{ J}$.

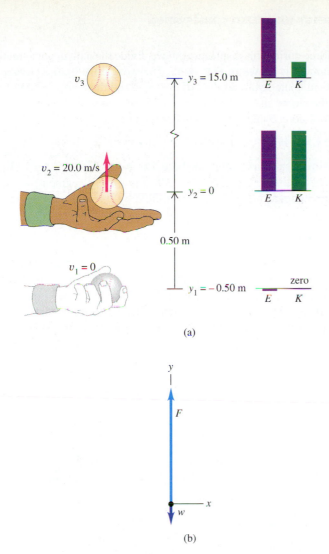

(a)

(b)

7–3 (a) Throwing a ball vertically upward. (b) The free-body diagram for the ball as the force $\vec{F}$ applied by the hand does the work W_{other} on the ball. The force $\vec{F}$ and gravity both act between y_1 and y_2. From y_2 to y_3, only gravity acts on the ball.

The kinetic energy of the ball increases by $K_2 - K_1 = 29.0$ J, and the potential energy increases by $U_2 - U_1 = 0.71$ J; the sum is $E_2 - E_1$, the change in total mechanical energy, which is equal to W_{other}.

Assuming that the upward force $\vec{F}$ that your hand applies is constant, the work W_{other} done by this force is equal to the magnitude F of the force multiplied by the upward displacement $y_2 - y_1$ over which it acts:

$$W_{other} = F(y_2 - y_1),$$

$$F = \frac{W_{other}}{y_2 - y_1} = \frac{29.7 \text{ J}}{0.50 \text{ m}} = 59 \text{ N}.$$

This is about 40 times greater than the weight of the ball.
b) Let point 3 be at the 15-m height. Then $y_3 = 15$ m, and we want to find the speed v_3 at this point. Between points 2 and 3, total mechanical energy is conserved; the force of your hand no longer acts, so $W_{other} = 0$. We can then find the kinetic energy at point 3 using Eq. (7–4);

$$K_2 + U_2 = K_3 + U_3,$$

$$U_3 = mgy_3 = (0.145 \text{ kg})(9.80 \text{ m/s}^2)(15.0 \text{ m}) = 21.3 \text{ J},$$

$$K_3 = (K_2 + U_2) - U_3 = (29.0 \text{ J} + 0 \text{ J}) - 21.3 \text{ J} = 7.7 \text{ J}.$$

Since $K_3 = \frac{1}{2}mv_3^2$, where v_3 is the y-component of the ball's velocity at point 3, we have

$$v_3 = \pm\sqrt{\frac{2K_3}{m}} = \pm\sqrt{\frac{2(7.7 \text{ J})}{0.145 \text{ kg}}} = \pm10 \text{ m/s}.$$

The significance of the plus-or-minus sign is that the ball passes point 3 *twice*, once on the way up and again on the way down. The total mechanical energy E is constant and equal to 29.0 J while the ball is in free fall, and the potential energy at point 3 is $U_3 = 21.3$ J whether the ball is moving up or down. So at point 3, the ball's kinetic energy K_3 and *speed* don't depend on the direction the ball is moving. The velocity is positive (+10 m/s) when the ball is moving up and negative (−10 m/s) when it is moving down; the speed is 10 m/s in either case.

GRAVITATIONAL POTENTIAL ENERGY FOR MOTION ALONG A CURVED PATH

In our first two examples the body moved along a straight vertical line. What happens when the path is slanted or curved (Fig. 7–4a)? The body is acted on by the gravitational force $\vec{w} = m\vec{g}$ and possibly by other forces whose resultant we call $\vec{F}_{other}$. To find the work

(a)

(b)

7–4 (a) A displacement along a curved path. (b) The work done by the gravitational force $\vec{w} = m\vec{g}$ depends only on the vertical component of displacement Δy (in this figure, Δy is negative).

5.14
Cart Leaves Circular Hill

done by the gravitational force during this displacement, we divide the path up into small segments $\Delta \vec{s}$; a typical segment is shown in Fig. 7–4b. The work done by the gravitational force over this segment is the scalar product of the force and the displacement. In terms of unit vectors, the force is $\vec{w} = m\vec{g} = -mg\hat{\jmath}$ and the displacement is $\Delta \vec{s} = \Delta x \hat{\imath} + \Delta y \hat{\jmath}$, so the work done by the gravitational force is

$$\vec{w} \cdot \Delta \vec{s} = -mg\hat{\jmath} \cdot (\Delta x \hat{\imath} + \Delta y \hat{\jmath}) = -mg\Delta y.$$

The work done by gravity is the same as though the body had been displaced vertically a distance Δy, with no horizontal displacement. This is true for every segment, so the *total* work done by the gravitational force is $-mg$ multiplied by the *total* vertical displacement $(y_2 - y_1)$:

$$W_{grav} = -mg(y_2 - y_1) = mgy_1 - mgy_2 = U_1 - U_2.$$

This is the same as Eq. (7–1) or (7–3), in which we assumed a purely vertical path. So even if the path a body follows between two points is curved, the total work done by the gravitational force depends only on the difference in height between the two points of the path. This work is unaffected by any horizontal motion that may occur. So *we can use the same expression for gravitational potential energy whether the body's path is curved or straight.*

EXAMPLE 7–3

Energy in projectile motion A batter hits two identical fly balls with the same initial speed but different initial angles. Prove that at a given height h, both balls have the same speed if air resistance can be neglected.

SOLUTION If there is no air resistance, the only force acting on each ball after it is hit is its weight. Hence the total mechanical energy for each ball is constant. Figure 7–5 shows the trajectories of two balls batted at the same height with the same initial speed, and thus the same total mechanical energy, but with different initial angles. At all points at the same height the potential energy is the same. Thus the kinetic energy at this height must be the same for both balls, and the speeds are the same.

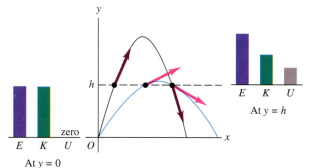

7–5 For the same initial speed and initial height the speed of a projectile at a given elevation h is always the same, neglecting air resistance.

EXAMPLE 7–4

Maximum height of a projectile, using energy methods In Example 3–9 (Section 3–4) we derived an expression for the maximum height h of a projectile launched with initial speed v_0 at initial angle α_0:

$$h = \frac{v_0^2 \sin^2 \alpha_0}{2g}.$$

Derive this expression using energy considerations.

SOLUTION We neglect air resistance, so just as in Example 7–3, total mechanical energy is conserved. We take point 1 to be the

launch point and point 2 to be the highest point on the trajectory (Fig. 7–6). We choose $y = 0$ at launch level; then $U_1 = 0$ and $U_2 = mgh$. We can express the kinetic energy at each point in terms of the components of velocity, using $v^2 = v_x^2 + v_y^2$:

$$K_1 = \frac{1}{2} m(v_{1x}^2 + v_{1y}^2),$$

$$K_2 = \frac{1}{2} m(v_{2x}^2 + v_{2y}^2).$$

Conservation of energy then gives $K_1 + U_1 = K_2 + U_2$, so

$$\frac{1}{2} m(v_{1x}^2 + v_{1y}^2) + 0 = \frac{1}{2} m(v_{2x}^2 + v_{2y}^2) + mgh.$$

To simplify this, we multiply through by $2/m$ to obtain

$$v_{1x}^2 + v_{1y}^2 = v_{2x}^2 + v_{2y}^2 + 2gh.$$

Now for the *coup de grace:* We recall that in projectile motion the x-component of acceleration is zero, so the x-component of velocity is constant and $v_{1x} = v_{2x}$. Also, because point 2 is the highest point, the vertical component of velocity is zero at that point: $v_{2y} = 0$. Subtracting the v_x^2 terms from both sides, we get

$$v_{1y}^2 = 2gh.$$

But v_{1y} is just the y-component of initial velocity, which is equal to $v_0 \sin \alpha_0$. Making this substitution and solving for h, we find

$$h = \frac{v_0^2 \sin^2 \alpha_0}{2g},$$

which agrees with the result of Example 3–9.

7–6 Trajectory of a projectile.

EXAMPLE 7–5

Calculating speed along a vertical circle Your cousin Throckmorton skateboards down a curved playground ramp. Treating Throcky and his skateboard as a particle, his center moves through a quarter-circle with radius R (Fig. 7–7). The total mass of Throcky and his skateboard is 25.0 kg. He starts from rest, and there is no friction. (a) Find his speed at the bottom of the ramp. (b) Find the normal force that acts on him at the bottom of the curve.

SOLUTION a) We can't use the equations of motion with constant acceleration; the acceleration isn't constant because the slope decreases as Throcky descends. Instead, we'll use the energy approach. Since there is no friction, the only force other than Throcky's weight is the normal force $\vec{n}$ exerted by the ramp (Fig. 7–7b). Although this force acts all along the path, it does *zero* work because $\vec{n}$ is perpendicular to Throcky's velocity at every point. Thus $W_{\text{other}} = 0$, and total mechanical energy is conserved.

7–7 (a) Throcky skateboarding down a frictionless circular ramp. The total mechanical energy is constant. (b) Free-body diagrams for Throcky and his skateboard at various points on the ramp.

Take point 1 at the starting point and point 2 at the bottom of the curved ramp, and let $y = 0$ be at the bottom of the ramp (Fig. 7–7a). Then $y_1 = R$ and $y_2 = 0$. Throcky starts from rest at the top, so $v_1 = 0$. The various energy quantities are

$$K_1 = 0, \qquad U_1 = mgR,$$

$$K_2 = \frac{1}{2}mv_2{}^2, \qquad U_2 = 0.$$

From conservation of energy,

$$K_1 + U_1 = K_2 + U_2,$$

$$0 + mgR = \frac{1}{2}mv_2{}^2 + 0,$$

$$v_2 = \sqrt{2gR}.$$

The speed is the same as if Throcky had fallen vertically through a height R, and it is independent of his mass.

As a numerical example, let $R = 3.00$ m. Then

$$v_2 = \sqrt{2(9.80 \text{ m/s}^2)(3.00 \text{ m})} = 7.67 \text{ m/s}.$$

This example shows a general rule about the role of forces in problems in which we use energy techniques: What matters is not simply whether a force acts, but whether that force does work. If the force does no work, like the normal force $\vec{n}$ in this example, then it does not appear at all in Eq. (7–7), $K_1 + U_1 + W_{\text{other}} = K_2 + U_2$.

b) We want the magnitude n of the normal force at point 2. Since n doesn't appear at all in the energy equation, we'll use Newton's second law. The free-body diagram at point 2 is shown in Fig. 7–7b. At point 2, Throcky is moving at speed $v_2 = \sqrt{2gR}$ in a circle of radius R; his acceleration is radial and has magnitude

$$a_{\text{rad}} = \frac{v_2{}^2}{R} = \frac{2gR}{R} = 2g.$$

If we take the positive y-direction upward, the y-component of Newton's second law is

$$\Sigma F_y = n + (-w) = ma_{\text{rad}} = 2mg,$$

$$n = w + 2mg = 3mg.$$

At point 2 the normal force is three times Throcky's weight. This result is independent of the radius of the circular ramp. We learned in Example 5–8 (Section 5–3) and Example 5–24 (Section 5–5) that the magnitude of n is the *apparent weight,* so Throcky will feel as though he weighs three times his true weight mg. But as soon as he reaches the horizontal part of the ramp to the right of point 2, the normal force will decrease to $w = mg$, and Throcky will feel normal again. Can you see why?

Notice we had to use *both* the energy approach and Newton's second law to solve this problem; energy conservation gave us the speed, and $\Sigma \vec{F} = m\vec{a}$ gave us the normal force. For each part of the problem we used the technique that most easily led to the answer.

EXAMPLE 7–6

A vertical circle with friction In Example 7–5, suppose that the ramp is not frictionless and that Throcky's speed at the bottom is only 6.00 m/s. What work was done by the friction force acting on him? Use $R = 3.00$ m.

SOLUTION We choose the same initial and final points and reference level as in Example 7–5 (Fig. 7–8). Again the normal force does no work, but now there is a friction force $\vec{f}$ that *does* do work. In this case, $W_{\text{other}} = W_f$. The energy quantities are

$$K_1 = 0,$$

$$U_1 = mgR = (25.0 \text{ kg})(9.80 \text{ m/s}^2)(3.00 \text{ m}) = 735 \text{ J},$$

$$K_2 = \frac{1}{2}mv_2{}^2 = \frac{1}{2}(25.0 \text{ kg})(6.00 \text{ m/s})^2 = 450 \text{ J},$$

$$U_2 = 0.$$

From Eq. (7–7),

$$W_f = K_2 + U_2 - K_1 - U_1$$

$$= 450 \text{ J} + 0 - 0 - 735 \text{ J} = -285 \text{ J}.$$

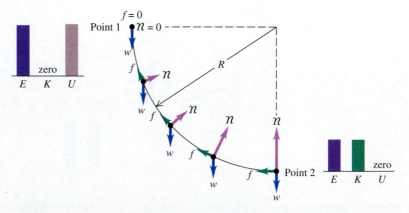

7–8 Free-body diagram for Throcky skateboarding down a ramp with friction. The total mechanical energy decreases as Throcky moves down the ramp.

The work done by the friction force is −285 J, and the total mechanical energy *decreases* by 285 J. Do you see why W_f has to be negative?

Throcky's motion is determined by Newton's second law, $\Sigma \vec{F} = m\vec{a}$. But it would be very difficult to apply the second law directly to this problem because the normal and friction forces and the acceleration are continuously changing in both magnitude and direction as Throcky skates down. The energy approach, by contrast, relates the motions at the top and bottom of the ramp without involving the details of what happens in between. Many problems are easy if energy considerations are used but very complex if we try to use Newton's laws directly.

EXAMPLE 7–7

An inclined plane with friction A 12-kg crate sits on the floor. We want to load it into a truck by sliding it up a ramp 2.5 m long, inclined at 30°. A worker, giving no thought to friction, calculates that he can get the crate up the ramp by giving it an initial speed of 5.0 m/s at the bottom and letting it go. But friction is *not* negligible; the crate slides 1.6 m up the ramp, stops, and slides back down (Fig. 7–9). a) Assuming that the friction force acting on the crate is constant, find its magnitude. b) How fast is the crate moving when it reaches the bottom of the ramp?

SOLUTION a) As in Example 7–2, we'll find the magnitude of the nongravitational force that does work (in this case, friction) by using the energy approach. Let point 1 be the bottom of the ramp, and let point 2 be where the crate stops (Fig. 7–9a). If we take $U = 0$ at floor level, we have $y_1 = 0$, $y_2 = (1.6 \text{ m}) \sin 30°$ = 0.80 m. The energy quantities are

$$K_1 = \frac{1}{2}(12 \text{ kg})(5.0 \text{ m/s})^2 = 150 \text{ J}, \qquad U_1 = 0,$$

$$K_2 = 0, \qquad U_2 = (12 \text{ kg})(9.8 \text{ m/s}^2)(0.80 \text{ m}) = 94 \text{ J},$$

$$W_{\text{other}} = -fs,$$

where f is the unknown magnitude of the friction force and $s = 1.6$ m. Using Eq. (7–7), we find

$$K_1 + U_1 + W_{\text{other}} = K_2 + U_2,$$
$$W_{\text{other}} = -fs = (K_2 + U_2) - (K_1 + U_1),$$
$$f = -\frac{(K_2 + U_2) - (K_1 + U_1)}{s}$$
$$= -\frac{[(0 + 94 \text{ J}) - (150 \text{ J} + 0)]}{1.6 \text{ m}} = 35 \text{ N}.$$

The friction force of 35 N, acting over 1.6 m, causes the mechanical energy of the crate to decrease from 150 J to 94 J. b) The crate returns to point 3 at the bottom of the ramp; $y_3 = 0$ and $U_3 = 0$ (Fig. 7–9b). On the way down, the friction force and the displacement both reverse direction but have the same magnitudes, so the frictional work has the same negative value for each half of the trip. The total work done by friction between points 1 and 3 is

$$W_{\text{other}} = W_{\text{fric}} = -2fs = -2(35 \text{ N})(1.6 \text{ m}) = -112 \text{ J}.$$

From part (a), $K_1 = 150$ J and $U_1 = 0$. Eq. (7–7) then gives

$$K_1 + U_1 + W_{\text{other}} = K_3 + U_3,$$
$$K_3 = K_1 + U_1 - U_3 + W_{\text{other}}$$
$$= 150 \text{ J} + 0 - 0 + (-112 \text{ J}) = 38 \text{ J}.$$

7–9 (a) A crate slides part way up the ramp, stops, and (b) slides back down. (c) Energy bar graphs for points 1, 2, and 3.

The crate returns to the bottom of the ramp with only 38 J of the original 150 J of mechanical energy. Using $K_3 = \frac{1}{2}mv_3^2$, we get

$$v_3 = \sqrt{\frac{2K_3}{m}} = \sqrt{\frac{2(38 \text{ J})}{12 \text{ kg}}} = 2.5 \text{ m/s}.$$

Alternatively, we could have applied Eq. (7–7) to points 2 and 3, considering the second half of the trip by itself. Try it and see whether you get the same result.

7–3 ELASTIC POTENTIAL ENERGY

When a railroad car runs into a spring bumper at the end of the track, the spring is compressed as the car is brought to a stop. If there is no friction, the bumper springs back and the car moves away with its original speed in the opposite direction. During the interaction with the spring, the car's kinetic energy has been "stored" in the elastic deformation of the spring. Something very similar happens in a rubber-band slingshot. Work is done on the rubber band by the force that stretches it, and that work is stored in the rubber band until you let it go. Then the rubber band gives kinetic energy to the projectile.

This is the same pattern we saw with the pile-driver in Section 7–2: do work on the system to store energy, which can later be converted to kinetic energy. We'll describe the process of storing energy in a deformable body such as a spring or rubber band in terms of *elastic potential energy*. A body is called *elastic* if it returns to its original shape and size after being deformed. To be specific, we'll consider storing energy in an ideal spring like the ones we discussed in Section 6–4. To keep such an ideal spring stretched by a distance x, we must exert a force $F = kx$, where k is the force constant of the spring. The ideal spring is a useful idealization because many other elastic bodies show this same direct proportionality between force F and displacement x, provided that x is sufficiently small.

We proceed just as we did for gravitational potential energy. We begin with the work done by the elastic (spring) force and then combine this with the work-energy theorem. The difference is that gravitational potential energy is a shared property of a body and the earth, but elastic potential energy is stored just in the spring (or other deformable body).

Figure 7–10 shows the ideal spring from Fig. 6–13, with its left end held stationary and its right end attached to a block with mass m that can move along the x-axis. In Fig. 7–10a the body is at $x = 0$ when the spring is neither stretched nor compressed. We move the block to one side, thereby stretching or compressing the spring, and then let it go. As the block moves from one position x_1 to another position x_2, how much work does the elastic (spring) force do on the block?

The Achilles tendon, which runs along the back of the ankle to the heel bone, acts like a natural spring. When it stretches and then relaxes, this tendon stores and then releases elastic potential energy. This spring action reduces the amount of work your leg muscles must do as you run.

7–10 (a) A block attached to a spring in equilibrium ($x = 0$) on a horizontal surface. (b) When the block moves from positive x_1 to positive $x_2 > x_1$, the spring does negative work as it stretches. (c) When the block moves from positive x_1 to positive $x_2 < x_1$, the spring relaxes and does positive work. (d) When the block moves from negative x_1 to a less negative x_2, the compressed spring relaxes and again does positive work.

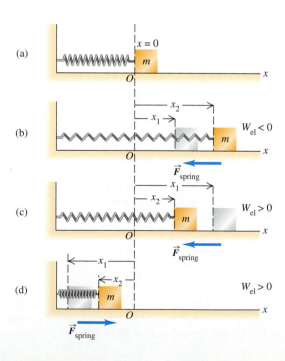

We found in Section 6–4 that the work we must do *on* the spring to move one end from an elongation x_1 to a different elongation x_2 is

$$W = \frac{1}{2} kx_2^2 - \frac{1}{2} kx_1^2 \qquad \text{(work done *on* a spring),}$$

where k is the force constant of the spring. If we stretch the spring further, we do positive work on the spring; if we let the spring relax while holding one end, we do negative work on it. We also saw that this expression for work is still correct if the spring is compressed, not stretched, so that x_1 or x_2 or both are negative. Now we need to find the work done *by* the spring. From Newton's third law the two quantities of work are just negatives of each other. Changing the signs in this equation, we find that in a displacement from x_1 to x_2 the spring does an amount of work W_{el} given by

$$W_{el} = \frac{1}{2} kx_1^2 - \frac{1}{2} kx_2^2 \qquad \text{(work done *by* a spring).}$$

The subscript "el" stands for *elastic*. When x_1 and x_2 are both positive and $x_2 > x_1$ (Fig. 7–10b), the spring does negative work on the block, which moves in the $+x$-direction while the spring pulls on it in the $-x$-direction. The spring stretches further, and the block slows down. When x_1 and x_2 are both positive and $x_2 < x_1$ (Fig. 7–10c), the spring does positive work as it relaxes and the block speeds up. If the spring can be compressed as well as stretched, x_1 or x_2 or both may be negative, but the expression for W_{el} is still valid. In Fig. 7–10d, both x_1 and x_2 are negative, but x_2 is less negative than x_1; the compressed spring does positive work as it relaxes, speeding the block up.

Just as for gravitational work, we can express the work done by the spring in terms of a given quantity at the beginning and end of the displacement. This quantity is $\frac{1}{2}kx^2$, and we define it to be the **elastic potential energy**:

$$U = \frac{1}{2} kx^2 \qquad \text{(elastic potential energy).} \qquad (7-9)$$

Figure 7–11 is a graph of Eq. (7–9). The unit of U is the joule (J), the unit used for *all* energy and work quantities; to see this from Eq. (7–9), recall that the units of k are N/m and that $1 \text{ N} \cdot \text{m} = 1 \text{ J}$.

We can use Eq. (7–9) to express the work W_{el} done on the block by the elastic force in terms of the change in potential energy:

$$W_{el} = \frac{1}{2} kx_1^2 - \frac{1}{2} kx_2^2 = U_1 - U_2 = -\Delta U. \qquad (7-10)$$

When a stretched spring is stretched further, as in Fig. 7–10b, W_{el} is negative and U *increases;* a greater amount of elastic potential energy is stored in the spring. When a stretched spring relaxes, as in Fig. 7–10c, x decreases, W_{el} is positive, and U *decreases;* the spring loses elastic potential energy. Negative values of x refer to a compressed spring. But, as Fig. 7–11 shows, U is positive for both positive and negative x, and Eqs. (7–9) and (7–10) are valid for both cases. Thus when a compressed spring is compressed further, $W_{el} < 0$ and U increases; when a compressed spring relaxes (Fig. 7–10d), $W_{el} > 0$ and U decreases. The more a spring is compressed *or* stretched, the greater its elastic potential energy.

CAUTION ▶ An important difference between gravitational potential energy $U = mgy$ and elastic potential energy $U = \frac{1}{2}kx^2$ is that we do *not* have the freedom to choose $x = 0$ to be wherever we wish. To be consistent with Eq. (7–9), $x = 0$ *must* be the position at which the spring is neither stretched nor compressed. At that position, its elastic potential energy and the force that it exerts are both zero. ◀

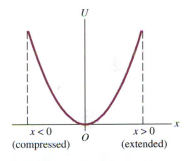

7–11 The graph of elastic potential energy for an ideal spring is a parabola: $U = \frac{1}{2}kx^2$, where x is the extension or compression of the spring. For extension (stretching), x is positive. For compression (when that is possible), x is negative. Elastic potential energy U is never negative.

The work-energy theorem says that $W_{tot} = K_2 - K_1$, no matter what kind of forces are acting on a body. If the elastic force is the *only* force that does work on the body, then

$$W_{tot} = W_{el} = U_1 - U_2.$$

The work-energy theorem $W_{tot} = K_2 - K_1$ then gives us

$$K_1 + U_1 = K_2 + U_2 \qquad \text{(if only the elastic force does work).} \qquad (7\text{--}11)$$

Here U is given by Eq. (7–9), so

$$\frac{1}{2}mv_1^2 + \frac{1}{2}kx_1^2 = \frac{1}{2}mv_2^2 + \frac{1}{2}kx_2^2 \qquad \text{(if only the elastic force does work).}$$
$$(7\text{--}12)$$

In this case the total mechanical energy $E = K + U$ (the sum of kinetic and elastic potential energy) is *conserved*. An example of this is the motion of the block in Fig. 7–10, provided that the horizontal surface is frictionless so that no force does work other than that exerted by the spring.

For Eq. (7–12) to be strictly correct, the ideal spring that we've been discussing must also be *massless*. If the spring has a mass, it will also have kinetic energy as the coils of the spring move back and forth. We can neglect the kinetic energy of the spring if its mass is much less than the mass m of the body attached to the spring. For instance, a typical automobile has a mass of 1200 kg or more. The springs in its suspension have masses of only a few kilograms, so their mass can be neglected if we want to study how a car bounces on its suspension.

If forces other than the elastic force also do work on the body, we call their work W_{other}, as before. Then the total work is $W_{tot} = W_{el} + W_{other}$, and the work-energy theorem gives

$$W_{el} + W_{other} = K_2 - K_1.$$

The work done by the spring is still $W_{el} = U_1 - U_2$, so again

$$K_1 + U_1 + W_{other} = K_2 + U_2 \qquad \text{(if forces other than the elastic force do work)}$$
$$(7\text{--}13)$$

and

$$\frac{1}{2}mv_1^2 + \frac{1}{2}kx_1^2 + W_{other} = \frac{1}{2}mv_2^2 + \frac{1}{2}kx_2^2$$
$$(7\text{--}14)$$

(if forces other than the elastic force do work).

This equation shows that *the work done by all forces* other than the elastic force *equals the change in the total mechanical energy* $E = K + U$ *of the system, where* U *is the elastic potential energy.* The "system" is made up of the body of mass m and the spring of force constant k. When W_{other} is positive, E increases; when W_{other} is negative, E decreases. You should compare Eq. (7–14) to Eq. (7–8), which describes situations in which there is gravitational potential energy but no elastic potential energy.

SITUATIONS WITH BOTH GRAVITATIONAL AND ELASTIC POTENTIAL ENERGY

Equations (7–11), (7–12), (7–13), and (7–14) are valid when the only potential energy in the system is elastic potential energy. What happens when we have *both* gravitational and elastic forces, such as a block attached to the lower end of a vertically hanging spring? We can still use Eq. (7–13), but now U_1 and U_2 are the initial and final values of the *total* potential energy, including both gravitational and elastic potential energies. That is, $U = U_{grav} + U_{el}$. Thus *the most general statement* of the relationship between kinetic energy, potential energy, and work done by other forces is

$$K_1 + U_{grav,\,1} + U_{el,\,1} + W_{other} = K_2 + U_{grav,\,2} + U_{el,\,2} \qquad \text{(valid in general).} \quad (7\text{--}15)$$

Work is done on this bungee jumper by both gravity and the elastic force of the bungee cord. Mechanical energy is not conserved, however, because work is also done by frictional forces within the bungee cord and by air resistance. This is a good thing—if mechanical energy were conserved, the bungee jumper would keep bouncing up and down forever!

That is, **the work done by all forces other than the gravitational force or elastic force equals the change in the total mechanical energy $E = K + U$ of the system, where U is the sum of the gravitational potential energy and elastic potential energy.** If the gravitational and elastic forces are the *only* forces that do work on the body, then $W_{other} = 0$ and the total mechanical energy (including both gravitational and elastic potential energy) is conserved. Pole vaulting is a familiar example of transformations between kinetic energy, elastic potential energy, and gravitational potential energy. The athlete's initial kinetic energy is partly stored as elastic potential energy in the flexed pole. Most of this elastic potential energy is then used to help give the increase in gravitational potential energy needed to clear the bar.

The strategy outlined in Section 7–2 is equally useful in solving problems that involve elastic forces as well as gravitational forces. The only new wrinkle is that the potential energy U now includes the elastic potential energy $U_{el} = \frac{1}{2}kx^2$, where x is the displacement of the spring *from its unstretched length*. The work done by the gravitational and elastic forces is accounted for by their potential energies; the work of the other forces, W_{other}, has to be included separately.

ActivPhysics

5.2
Ejector Pad

5.7
Inverse Bungee Jumper

5.8
Spring-Launched Bowler

5.11
Spring-Launched Ski Lift

EXAMPLE 7–8

Motion with elastic potential energy In Fig. 7–12a a glider with mass $m = 0.200$ kg sits on a frictionless horizontal air track, connected to a spring with force constant $k = 5.00$ N/m. You pull on the glider, stretching the spring 0.100 m, and then release it with no initial velocity (Fig. 7–12b). The glider begins to move back toward its equilibrium position ($x = 0$). What is its x-velocity when $x = 0.080$ m?

SOLUTION Because the spring force varies with position, this problem *cannot* be done by using the equations for motion with

constant acceleration. The energy method, by contrast, offers a simple and elegant solution.

As the glider starts to move, elastic potential energy is converted into kinetic energy. (The glider remains at the same height throughout the motion, so *gravitational* potential energy is not a factor.) The spring force is the only force doing work on the glider, so $W_{other} = 0$, and we may use Eq. (7–11). Let point 1 be where the glider is released (Fig. 7–12b), and let point 2 be at $x = 0.080$ m (Fig. 7–12c). The energy quantities are

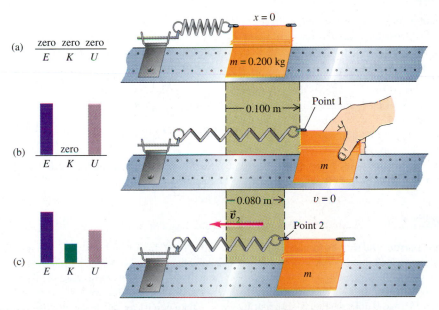

7–12 (a) An air-track glider attached to a spring. (b) Elastic potential energy is added to the system by stretching the spring. (c) Elastic potential energy is transformed to kinetic energy as the glider moves back toward its equilibrium position.

$$K_1 = \frac{1}{2}(0.200 \text{ kg})(0)^2 = 0,$$

$$U_1 = \frac{1}{2}(5.00 \text{ N/m})(0.100 \text{ m})^2 = 0.0250 \text{ J},$$

$$K_2 = \frac{1}{2}mv_2{}^2,$$

$$U_2 = \frac{1}{2}(5.00 \text{ N/m})(0.080 \text{ m})^2 = 0.0160 \text{ J},$$

Then from Eq. (7–11),

$$K_2 = K_1 + U_1 - U_2 = 0 + 0.0250 \text{ J} - 0.0160 \text{ J} = 0.0090 \text{ J},$$

$$v_2 = \pm\sqrt{\frac{2K_2}{m}} = \pm\sqrt{\frac{2(0.0090 \text{ J})}{0.200 \text{ kg}}} = \pm 0.30 \text{ m/s}.$$

We choose the negative root because the glider is moving in the $-x$-direction; the answer we want is $v_2 = -0.30$ m/s.

EXAMPLE 7–9

Motion with elastic potential energy and work done by other forces For the system of Example 7–8, suppose the glider is initially at rest at $x = 0$, with the spring unstretched. You then apply a constant force $\vec{F}$ in the $+x$-direction with magnitude 0.610 N to the glider. What is the glider's speed when it has moved to $x = 0.100$ m?

SOLUTION Again, the acceleration of the glider isn't constant. (Do you see why not?) Total mechanical energy is not conserved because of the work done by the force $\vec{F}$, but we can still use the energy relation of Eq. (7–13). Now let point 1 be at $x = 0$ and point 2 at $x = 0.100$ m (these are different from the points labeled in Fig. 7–12). The energy quantities are

$$K_1 = 0, \qquad\qquad U_1 = 0,$$

$$K_2 = \frac{1}{2}mv_2{}^2, \qquad U_2 = \frac{1}{2}(5.00 \text{ N/m})(0.100 \text{ m})^2$$

$$= 0.0250 \text{ J},$$

$$W_{other} = (0.610 \text{ N})(0.100 \text{ m}) = 0.0610 \text{ J}.$$

(To calculate W_{other} we multiplied the magnitude of the force by the displacement, since both are in the $+x$-direction.) Initially, the total mechanical energy is zero; the work done by the force $\vec{F}$ increases the total mechanical energy to 0.0610 J, of which 0.0250 J is elastic potential energy. The remainder is kinetic energy. From Eq. (7–13),

$$K_1 + U_1 + W_{other} = K_2 + U_2,$$

$$K_2 = K_1 + U_1 + W_{other} - U_2$$

$$= 0 + 0 + 0.0610 \text{ J} - 0.0250 \text{ J} = 0.0360 \text{ J},$$

$$v_2 = \sqrt{\frac{2K_2}{m}} = \sqrt{\frac{2(0.0360 \text{ J})}{0.200 \text{ kg}}} = 0.60 \text{ m/s}.$$

EXAMPLE 7–10

Motion with elastic potential energy after other forces have ceased In Example 7–9, suppose the force $\vec{F}$ is removed when the glider reaches the point $x = 0.100$ m. How much farther does the glider move before coming to rest?

SOLUTION After $\vec{F}$ is removed, the spring force is the only force doing work. Hence for this part of the motion, total mechanical energy $E = K + U$ is conserved. We saw in Example 7–9 that the kinetic energy at the 0.100-m point (point 2) is $K_2 = 0.0360$ J and the potential energy at this point is $U_2 = 0.0250$ J. The total mechanical energy at and beyond this point is therefore 0.0610 J (equal to the work done by the force $\vec{F}$ between points 1 and 2).

When the body comes to rest at x_{max} (point 3), its kinetic energy K_3 is zero, and its potential energy U_3 is equal to the total mechanical energy 0.0610 J. We can also see this from $K_2 + U_2 = K_3 + U_3$:

$$U_3 = K_2 + U_2 - K_3 = 0.0360 \text{ J} + 0.0250 \text{ J} - 0 = 0.0610 \text{ J}.$$

But $U_3 = \frac{1}{2}kx_{max}{}^2$, so

$$x_{max} = \sqrt{\frac{2U_3}{k}} = \sqrt{\frac{2(0.0610 \text{ J})}{5.00 \text{ N/m}}} = 0.156 \text{ m}.$$

So the body moves an additional 0.056 m after the force is removed at $x = 0.100$ m.

EXAMPLE 7–11

Motion with gravitational, elastic, and friction forces In a "worst-case" design scenario, a 2000-kg elevator with broken cables is falling at 25 m/s when it first contacts a cushioning spring at the bottom of the shaft. The spring is supposed to stop the elevator, compressing 3.00 m as it does so (Fig. 7–13). During the motion a safety clamp applies a constant 17,000-N frictional force to the elevator. As a design consultant, you are asked to determine what the force constant of the spring should be.

SOLUTION In this problem there is both gravitational and elastic potential energy, so we must use Eq. (7–15). Total mechanical energy is not conserved because the friction force does negative work W_{other} on the elevator.

Take point 1 as the elevator's position when it initially contacts the spring, and take point 2 as its position when it is at rest. We choose the origin to be at point 1, so $y_1 = 0$ and $y_2 = -3.00$ m. With this choice the coordinate of the upper end of the spring is

7–13 The fall of an elevator is stopped by a spring and by a constant friction force.

the same as the coordinate of the elevator, so the elastic potential energy at any point between point 1 and point 2 is $U_{el} = \frac{1}{2}ky^2$. The elevator's initial speed is $v_1 = 25$ m/s, so the initial kinetic energy is

$$K_1 = \frac{1}{2}mv_1^2 = \frac{1}{2}(2000 \text{ kg})(25 \text{ m/s})^2 = 625,000 \text{ J}.$$

The elevator stops at point 2, so $K_2 = 0$. The potential energy at point 1, U_1, is zero; U_{grav} is zero because $y_1 = 0$, and $U_{el} = 0$ because the spring is not yet compressed. At point 2 there is both gravitational and elastic potential energy, so

$$U_2 = mgy_2 + \frac{1}{2}ky_2^2;$$

the gravitational potential energy at point 2 is

$$mgy_2 = (2000 \text{ kg})(9.80 \text{ m/s}^2)(-3.00 \text{ m}) = -58,800 \text{ J}.$$

The other force is the 17,000-N friction force, acting opposite to the 3.00-m displacement, so

$$W_{other} = -(17,000 \text{ N})(3.00 \text{ m}) = -51,000 \text{ J}.$$

Putting these into $K_1 + U_1 + W_{other} = K_2 + U_2$, we have

$$K_1 + 0 + W_{other} = 0 + (mgy_2 + \frac{1}{2}ky_2^2),$$

so the force constant of the spring is

$$k = \frac{2(K_1 + W_{other} - mgy_2)}{y_2^2}$$

$$= \frac{2(625,000 \text{ J} + (-51,000 \text{ J}) - (-58,800 \text{ J}))}{(-3.00 \text{ m})^2}$$

$$= 1.41 \times 10^5 \text{ N/m}.$$

This is comparable to the springs in an automobile suspension.

Let's notice what might seem to be a paradox in this problem. The elastic potential energy in the spring at point 2 is

$$\frac{1}{2}ky_2^2 = \frac{1}{2}(1.41 \times 10^5 \text{ N/m})(-3.00 \text{ m})^2 = 632,800 \text{ J}.$$

This is *more* than the total mechanical energy at point 1,

$$E_1 = K_1 + U_1 = 625,000 \text{ J} + 0 = 625,000 \text{ J}.$$

But the friction force caused the mechanical energy of the system to *decrease* by 51,000 J between point 1 and point 2. Does this mean that energy appeared from nowhere? Don't panic; there is no paradox. At point 2 there is also *negative* gravitational potential energy $mgy_2 = -58,800$ J because point 2 is below the origin. The total mechanical energy at point 2 is

$$E_2 = K_2 + U_2 = 0 + \frac{1}{2}ky_2^2 + mgy_2$$

$$= 632,800 \text{ J} + (-58,800 \text{ J}) = 574,000 \text{ J}.$$

This is just the initial mechanical energy of 625,000 J, minus 51,000 J lost to friction.

Your next job as design consultant would be to tell your client that the elevator won't stay at the bottom of the shaft. Instead, it will bounce back up. The reason is that at point 2 the compressed spring exerts an upward force of magnitude $F_{spring} = (1.41 \times 10^5 \text{ N/m})(3.00 \text{ m}) = 422,000$ N. The weight of the elevator is only $w = mg = (2000 \text{ kg})(9.80 \text{ m/s}^2) = 19,600$ N, so the net force will be upward. The elevator bounces back upward even if the safety clamp now exerts a *downward* friction force of magnitude $f = 17,000$ N; the spring force is greater than the sum of f and mg. The elevator will bounce again and again until enough mechanical energy has been removed by friction for it to stop.

Can you also show that the acceleration of the falling elevator when it hits the spring is unacceptably high?

7–4 CONSERVATIVE AND NONCONSERVATIVE FORCES

In our discussions of potential energy we have talked about "storing" kinetic energy by converting it to potential energy. We always have in mind that later we may retrieve it again as kinetic energy. When you throw a ball up in the air, it slows down as kinetic energy is converted into potential energy. But on the way down, the conversion is reversed, and the ball speeds up as potential energy is converted back to kinetic energy. If there is no air resistance, the ball is moving just as fast when you catch it as when you threw it.

If a glider moving on a frictionless horizontal air track runs into a spring bumper at the end of the track, the spring compresses and the glider stops. But then it bounces back, and if there is no friction, the glider has the same speed and kinetic energy it had before the collision. Again, there is a two-way conversion from kinetic to potential energy and back. In both cases we find that we can define a potential-energy function so that the total mechanical energy, kinetic plus potential, is constant or *conserved* during the motion.

A force that offers this opportunity of two-way conversion between kinetic and potential energies is called a **conservative force.** We have seen two examples of conservative forces: the gravitational force and the spring force. An essential feature of conservative forces is that their work is always *reversible*. Anything that we deposit in the energy "bank" can later be withdrawn without loss. Another important aspect of conservative forces is that a body may move from point 1 to point 2 by various paths, but the work done by a conservative force is the same for all of these paths (Fig. 7–14). Thus if a body stays close to the surface of the earth, the gravitational force $m\vec{g}$ is independent of height, and the work done by this force depends only on the change in height. If the body moves around a closed path, ending at the same point where it started, the *total* work done by the gravitational force is always zero.

The work done by a conservative force *always* has these properties:

1. It can always be expressed as the difference between the initial and final values of a *potential energy* function.
2. It is reversible.
3. It is independent of the path of the body and depends only on the starting and ending points.
4. When the starting and ending points are the same, the total work is zero.

When the only forces that do work are conservative forces, the total mechanical energy $E = K + U$ is constant.

Not all forces are conservative. Consider the friction force acting on the crate sliding on a ramp in Example 7–7 (Section 7–2). When the body slides up and then back down to the starting point, the total work done on it by the friction force is *not* zero. When the direction of motion reverses, so does the friction force, and friction does *negative* work in *both* directions. When a car with its brakes locked skids across the pavement with decreasing speed (and decreasing kinetic energy), the lost kinetic energy cannot be recovered by reversing the motion or in any other way, and mechanical energy is *not* conserved. There is no potential-energy function for the friction force.

In the same way, the force of fluid resistance (Section 5–4) is not conservative. If you throw a ball up in the air, air resistance does negative work on the ball while it's rising *and* while it's descending. The ball returns to your hand with less speed and less

7–14 For any conservative force, the work done by that force depends only on the end points, not on the path taken. Thus the gravitational force, which is conservative, does the same work on the hummingbird no matter what path it flies from point 1 to point 2.

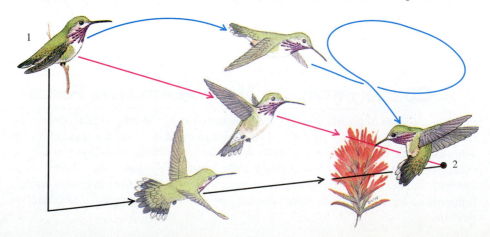

kinetic energy than when it left, and there is no way to get back the lost mechanical energy.

A force that is not conservative is called a **nonconservative force.** The work done by a nonconservative force *cannot* be represented by a potential-energy function. Some nonconservative forces, like kinetic friction or fluid resistance, cause mechanical energy to be lost or dissipated; a force of this kind is called a **dissipative force.** There are also nonconservative forces that *increase* mechanical energy. The fragments of an exploding firecracker fly off with very large kinetic energy, thanks to a chemical reaction of gunpowder with oxygen. The forces unleashed by this reaction are nonconservative because the process is not reversible. Imagine the fragments spontaneously reassembling themselves into a complete firecracker!

EXAMPLE 7–12

Frictional work depends on the path You are rearranging your furniture and wish to move a 40.0-kg futon 2.50 m across the room (Fig. 7–15). However, the straight-line path is blocked by a heavy coffee table that you don't want to move. Instead, you slide the futon in a dogleg path over the floor; the doglegs are 2.00 m and 1.50 m long. Compared to the straight-line path, how much more work must you do to push the futon in the dogleg path? The coefficient of kinetic friction is 0.200.

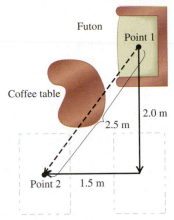

7–15 A view from above of furniture being moved. How much more work must you do to move the futon on a dogleg path?

SOLUTION The initial and final points are shown in Fig. 7–15. The futon is at rest at both point 1 and point 2, so $K_1 = K_2 = 0$. The gravitational potential energy does not change because the futon moves only horizontally; to be specific, we'll say $U_1 = U_2 = 0$. From Eq. (7–7) it follows that $W_{other} = 0$. The other work done on the futon is the sum of the positive work you do, W_{you}, and the negative work W_{fric} done by the kinetic friction force. Since the sum of these is zero, we have

$$W_{you} = -W_{fric}.$$

Because the floor is horizontal, the normal force on the futon equals its weight mg, and the magnitude of the friction force is $f_k = \mu_k n = \mu_k mg$. The work you must do over each path is then

$$W_{you} = -W_{fric} = -(-f_k s) = +\mu_k mgs$$

$$= (0.200)(40.0 \text{ kg})(9.80 \text{ m/s}^2)(2.50 \text{ m})$$

$$= 196 \text{ J} \quad \text{(straight-line path)},$$

$$W_{you} = -W_{fric} = (0.200)(40.0 \text{ kg})(9.80 \text{ m/s}^2)(2.00 \text{ m} + 1.50 \text{ m})$$

$$= 274 \text{ J} \quad \text{(dogleg path)}.$$

The extra work you must do is 274 J − 196 J = 78 J.

The work done by friction is $W_{fric} = -W_{you} = -196$ J on the straight-line path, and −274 J on the dogleg. Friction is a nonconservative force, so the work done by friction depends on the path taken.

EXAMPLE 7–13

Conservative or nonconservative? In a certain region of space the force on an electron is $\vec{F} = Cx\hat{\jmath}$, where C is a positive constant. The electron moves in a counterclockwise direction around a square loop in the xy-plane (Fig. 7–16). The corners of the square are at $(x, y) = (0, 0)$, $(L, 0)$, (L, L) and $(0, L)$. Calculate the work done on the electron by the force $\vec{F}$ during one complete trip around the square. Is this force conservative or nonconservative?

SOLUTION The force $\vec{F}$ is not constant and in general is not in the same direction as the displacement. So we'll use the more general expression for work, Eq. (6–14):

$$W = \int_{P_1}^{P_2} \vec{F} \cdot d\vec{l},$$

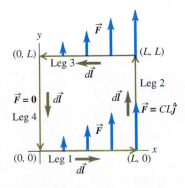

7–16 An electron moving around a square loop while being acted on by the force $\vec{F} = Cx\hat{\jmath}$.

where $d\vec{l}$ is an infinitesimal displacement. Let's calculate the work done by the force $\vec{F}$ on each leg of the square and then add the results to find the work done on the round trip.

On the first leg, from (0, 0) to (L, 0), the force varies but is everywhere perpendicular to the displacement. So $\vec{F} \cdot d\vec{l} = 0$, and the work done on the first leg is $W_1 = 0$. The force has the same value $\vec{F} = CL\hat{j}$ everywhere on the second leg from (L, 0) to (L, L). The displacement on this leg is in the +y-direction, so $d\vec{l} = dy\hat{j}$ and

$$\vec{F} \cdot d\vec{l} = CL\hat{j} \cdot dy\hat{j} = CL\,dy.$$

The work done on the second leg is then

$$W_2 = \int_{(L,0)}^{(L,L)} \vec{F} \cdot d\vec{l} = \int_{y=0}^{y=L} CL\,dy = CL \int_0^L dy = CL^2.$$

On the third leg, from (L, L) to (0, L), $\vec{F}$ is again perpendicular to the displacement so $W_3 = 0$. The force is zero on the final leg, from (0, L) to (0, 0), so no work is done and $W_4 = 0$. The work done by the force $\vec{F}$ on the round trip is

$$W = W_1 + W_2 + W_3 + W_4 = 0 + CL^2 + 0 + 0 = CL^2.$$

The starting and ending points are the same, but the total work done by $\vec{F}$ is not zero. This is a nonconservative force; it cannot be represented by a potential energy function.

Because W is positive, the mechanical energy of the electron *increases* as it goes around the loop. This is not a mathematical curiosity; it's a description of what happens in an electrical generating plant. A loop of wire is moved through a magnetic field, which gives rise to a nonconservative force similar to the one in this example. Electrons in the wire gain energy as they move around the loop, and this energy is carried via transmission lines to the consumer. (We'll discuss how this works in detail in Chapter 30.) All the electrical energy used in the home and in industry comes from work done by nonconservative forces!

How would the value of W change if the electron went around the loop clockwise instead of counterclockwise? The force $\vec{F}$ would be unaffected, but the direction of each infinitesimal displacement $d\vec{l}$ would reverse. Thus the sign of work would also reverse, and the work for a clockwise round trip would be $W = -CL^2$. This is a different behavior than the nonconservative friction force. When a body slides over a stationary surface with friction, the work done by friction is always negative, no matter what the direction of motion (see Example 7–7).

THE LAW OF CONSERVATION OF ENERGY

Nonconservative forces cannot be represented in terms of potential energy. But we can describe the effects of these forces in terms of kinds of energy other than kinetic and potential energy. When a car with locked brakes skids to a stop, the tires and the road surface both become hotter. The energy associated with this change in the state of the materials is called **internal energy.** Raising the temperature of a body increases its internal energy; lowering the body's temperature decreases its internal energy.

To see the significance of internal energy, let's consider a block sliding on a rough surface. Friction does *negative* work on the block as it slides, and the change in internal energy of the block and surface (both of which get hotter) is *positive*. Careful experiments show that the increase in the internal energy is *exactly* equal to the absolute value of the work done by friction. In other words,

$$\Delta U_{\text{int}} = -W_{\text{other}},$$

where ΔU_{int} is the change in internal energy. If we substitute this into Eq. (7–7), (7–13), or (7–15), we find

$$K_1 + U_1 - \Delta U_{\text{int}} = K_2 + U_2.$$

Writing $\Delta K = K_2 - K_1$ and $\Delta U = U_2 - U_1$, we can finally express this as

$$\Delta K + \Delta U + \Delta U_{\text{int}} = 0 \qquad \text{(law of conservation energy)}. \qquad (7\text{–}16)$$

This remarkable statement is the general form of the **law of conservation of energy.** In a given process, the kinetic energy, potential energy, and internal energy of a system may all change. But the *sum* of those changes is always zero. If there is a decrease in one form of energy, it is made up for by an increase in the other forms. When we expand our definition of energy to include internal energy, Eq. (7–16) says that *energy is never created or destroyed; it only changes form.* No exception to this rule has ever been found.

Notice that the concept of work has been banished from Eq. (7–16). Instead, this equation invites us to think purely in terms of the conversion of energy from one form to another. When you throw a baseball straight up, you convert a portion of the internal energy of your molecules into kinetic energy of the baseball. This is converted into grav-

When a liter of gasoline is burned in an automotive engine, it releases 3.3×10^7 J of internal energy. Hence $\Delta U_{\text{int}} = -3.3 \times 10^7$ J, where the minus sign means that the amount of energy stored in the gasoline has decreased. This energy can be converted into kinetic energy (making the car go faster) or into potential energy (enabling the car to climb uphill).

itational potential energy as the ball climbs and back to kinetic energy as the ball falls. If there is air resistance, part of the energy is used to heat up the air and the ball and increase their internal energy. Energy is converted back into the kinetic form as the ball falls. If you catch the ball in your hand, whatever energy was not lost to the air once again becomes internal energy; the ball and your hand are now warmer than they were at the beginning.

In a hydroelectric generating station, falling water is used to drive turbines ("water wheels"), which in turn run electric generators. Fundamentally, gravitational potential energy is released as the water falls, and the generating station converts this into electrical energy. Even if we know nothing about the details of how this is done, we can still use the law of conservation of energy, Eq. (7–16), to draw an important conclusion: the amount of electrical energy produced can be no greater than the amount of gravitational potential energy lost. (Because of friction, some of the potential energy goes into raising the temperatures of the water and the mechanism.)

In later chapters we will study the relation of internal energy to temperature changes, heat, and work. This is the heart of the area of physics called *thermodynamics*.

5.4
Inventing Processes

5.12
Modified Atwood Machine

EXAMPLE 7-14

Work done by friction Let's look again at Example 7–6 in Section 7–2, in which your cousin Throcky skateboards down a curved ramp. He starts with zero kinetic energy and 735 J of potential energy, and at the bottom he has 450 J of kinetic energy and zero potential energy. So $\Delta K = +450$ J, and $\Delta U = -735$ J. The work $W_{other} = W_{fric}$ done by the nonconservative friction forces is -285 J, so the change in internal energy is

$\Delta U_{int} = -W_{other} = +285$ J. The wheels, the bearings, and the ramp all get a little warmer as Throcky rolls down. In accordance with Eq. (7–16) the sum of the energy changes equals zero:

$$\Delta K + \Delta U + \Delta U_{int} = +450 \text{ J} + (-735 \text{ J}) + 285 \text{ J} = 0.$$

The total energy of the system (including nonmechanical forms of energy) is conserved.

7-5 FORCE AND POTENTIAL ENERGY

For the two kinds of conservative forces (gravitational and elastic) we have studied, we started with a description of the behavior of the force and derived from that an expression for the potential energy. For a body with mass m in a uniform gravitational field, the gravitational force is $F_y = -mg$. We found that the corresponding potential energy is $U(y) = mgy$ (Fig. 7–17a). To stretch an ideal spring by a distance x, we exert a force equal to $+kx$. By Newton's third law the force that an ideal spring exerts on a body is opposite this, or $F_x = -kx$. The corresponding potential energy function is $U(x) = \frac{1}{2}kx^2$ (Fig. 7–17b).

We can reverse this procedure: If we are given a potential-energy expression, we can find the corresponding force. Here's how we do it. First let's consider motion along a straight line, with coordinate x. We denote the x-component of force, a function of x, by $F_x(x)$, and the potential energy as $U(x)$. This notation reminds us that both F_x and U are *functions* of x. Now we recall that in any displacement the work W done by a conservative force equals the negative of the change ΔU in potential energy:

$$W = -\Delta U.$$

Let's apply this to a small displacement Δx. The work done by the force $F_x(x)$ during this displacement is approximately equal to $F_x(x) \Delta x$. We have to say "approximately" because $F_x(x)$ may vary a little over the interval Δx. But it is at least approximately true that

$$F_x(x)\Delta x = -\Delta U \qquad \text{and} \qquad F_x(x) = -\frac{\Delta U}{\Delta x}.$$

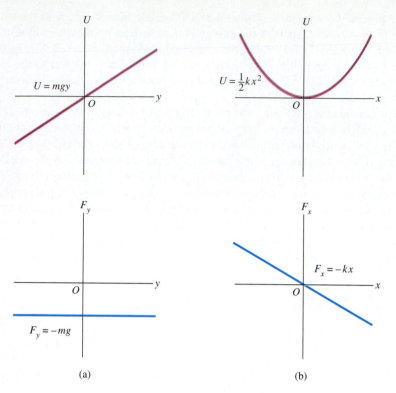

7–17 Graphs of potential energy and force versus position for (a) the gravitational force and (b) the spring force. In each case the force is the negative derivative of the potential energy.

You can probably see what's coming. We take the limit as $\Delta x \to 0$; in this limit, the variation of F_x becomes negligible, and we have the exact relation

$$F_x(x) = -\frac{dU(x)}{dx} \quad \text{(force from potential energy, one dimension).} \quad (7-17)$$

This result makes sense; in regions where $U(x)$ changes most rapidly with x (that is, where $dU(x)/dx$ is large), the greatest amount of work is done during a given displacement, and this corresponds to a large force magnitude. Also, when $F_x(x)$ is in the positive x-direction, $U(x)$ *decreases* with increasing x. So $F_x(x)$ and $dU(x)/dx$ should indeed have opposite signs. The physical meaning of Eq. (7–17) is that *a conservative force always acts to push the system toward lower potential energy.*

As a check, let's consider the function for elastic potential energy, $U(x) = \frac{1}{2}kx^2$. Substituting this into Eq. (7–17) yields

$$F_x(x) = -\frac{d}{dx}\left(\frac{1}{2}kx^2\right) = -kx.$$

which is the correct expression for the force exerted by an ideal spring.

EXAMPLE 7–15

An electrical force and its potential energy An electrically charged particle is held at rest at the point $x = 0$, while a second particle with equal charge is free to move along the positive x-axis. The potential energy of the system is

$$U(x) = \frac{C}{x},$$

where C is a positive constant that depends on the magnitude of the charges. Derive an expression for the x-component of force acting on the movable charge, as a function of its position.

SOLUTION We are given the potential-energy function $U(x)$, so we can use Eq. (7–17). The derivative with respect to x of the function $1/x$ is $-1/x^2$. So the force on the movable charge for $x > 0$ is

$$F_x(x) = -\frac{dU(x)}{dx} = -C\left(-\frac{1}{x^2}\right) = \frac{C}{x^2}.$$

The x-component of force is positive, corresponding to a repulsive interaction between like electric charges. The potential energy is very large for small x and approaches zero as x

becomes large; the force pushes the movable charge toward large positive values of x, for which the potential energy is less. The force varies as $1/x^2$; it is small when the particles are far apart (large x) but becomes large when the particles are close together (small x). This is an example of Coulomb's law for electrical interactions, which we will study in greater detail in Chapter 22.

FORCE AND POTENTIAL ENERGY IN THREE DIMENSIONS

We can extend this analysis to three dimensions, where the particle may move in the x-, y-, or z-direction, or all at once, under the action of a conservative force that has components F_x, F_y, and F_z. Each component of force may be a function of the coordinates x, y, and z. The potential-energy function U is also a function of all three space coordinates. We can now use Eq. (7–17) to find each component of force. The potential-energy change ΔU when the particle moves a small distance Δx in the x-direction is again given by $-F_x \Delta x$; it doesn't depend on F_y and F_z, which represent force components that are perpendicular to the displacement and do no work. So we again have the approximate relation

$$F_x = -\frac{\Delta U}{\Delta x}.$$

The y- and z-components of force are determined in exactly the same way:

$$F_y = -\frac{\Delta U}{\Delta y}, \qquad F_z = -\frac{\Delta U}{\Delta z}.$$

To make these relations exact, we need to take the limits $\Delta x \to 0$, $\Delta y \to 0$, and $\Delta z \to 0$ so that these ratios become derivatives. Because U may be a function of all three coordinates, we need to remember that when we calculate each of these derivatives, only one coordinate changes at a time. We compute the derivative of U with respect to x by assuming that y and z are constant and only x varies, and so on. Such a derivative is called a *partial derivative*. The usual notation is $\partial U/\partial x$ and so on; the symbol ∂ is a modified d to remind us of the nature of this operation. So we write

$$F_x = -\frac{\partial U}{\partial x}, \qquad F_y = -\frac{\partial U}{\partial y}, \qquad F_z = -\frac{\partial U}{\partial z} \qquad \text{(force from potential energy)}.$$

$$(7–18)$$

We can use unit vectors to write a single compact vector expression for the force $\vec{F}$:

$$\vec{F} = -\left(\frac{\partial U}{\partial x}\hat{\imath} + \frac{\partial U}{\partial y}\hat{\jmath} + \frac{\partial U}{\partial z}\hat{k}\right) \qquad \text{(force from potential energy).} \qquad (7–19)$$

The expression inside the parentheses represents a particular operation on the function U, in which we take the partial derivative of U with respect to each coordinate, multiply by the corresponding unit vector, and take the vector sum. This operation is called the **gradient** of U and is often abbreviated as $\vec{\nabla}U$. Thus the force is the negative of the gradient of the potential-energy function:

$$\vec{F} = -\vec{\nabla}U. \qquad (7–20)$$

As a check, let's substitute into Eq. (7–20) the function $U = mgy$ for gravitational potential energy:

$$\vec{F} = -\vec{\nabla}(mgy) = -\left(\frac{\partial(mgy)}{\partial x}\hat{\imath} + \frac{\partial(mgy)}{\partial y}\hat{\jmath} + \frac{\partial(mgy)}{\partial z}\hat{k}\right) = (-mg)\hat{\jmath}.$$

This is just the familiar expression for the gravitational force.

EXAMPLE 7–16

Force and potential energy in two dimensions A puck slides on a level, frictionless air-hockey table. The coordinates of the puck are x and y. It is acted on by a conservative force described by the potential-energy function

$$U(x,\ y) = \frac{1}{2} k(x^2 + y^2).$$

Derive an expression for the force acting on the puck.

SOLUTION We can use Eqs. (7–18) to find the components of force:

$$F_x = -\frac{\partial U}{\partial x} = -kx, \qquad F_y = -\frac{\partial U}{\partial y} = -ky.$$

From Eq. (7–19) this corresponds to the vector expression

$$\vec{F} = -k(x\hat{\imath} + y\hat{\jmath}).$$

Now $x\hat{\imath} + y\hat{\jmath}$ is just the position vector $\vec{r}$ of the particle, so we can rewrite this as $\vec{F} = -k\vec{r}$. This represents a force that at each point is opposite in direction to the position vector of the point, that is, a force that at each point is directed toward the origin.

The potential energy is minimum at the origin, so again the force pushes in the direction of decreasing potential energy.

Furthermore, the *magnitude* of the force at any point is

$$F = k\sqrt{x^2 + y^2} = kr,$$

where r is the particle's distance from the origin. This is the force exerted by a spring that obeys Hooke's law and has a negligibly small length (compared to the other distances in the problem) when it is not stretched. So the motion of the puck is the same as if it were attached to one end of an ideal spring of negligible unstretched length; the other end is attached to the air-hockey table at the origin.

The potential-energy function can also be expressed as $U = \frac{1}{2}kr^2$, so the radial component of force F_r can be expressed as

$$F_r = -\frac{\partial U}{\partial r} = -\frac{\partial\left(\frac{1}{2}kr^2\right)}{\partial r} = -kr.$$

The minus sign indicates that the force is radially inward (toward the origin).

7–6 ENERGY DIAGRAMS

When a particle moves along a straight line under the action of a conservative force, we can get a lot of insight into the possible motions by looking at the graph of the potential-energy function $U(x)$. Figure 7–18a shows a glider with mass m that moves along the x-axis on an air track. The spring exerts on the glider a force with x-component $F_x = -kx$. Figure 7–18b is a graph of the corresponding potential-energy function $U(x) = \frac{1}{2}kx^2$. If the elastic force of the spring is the *only* horizontal force acting on the glider, the total mechanical energy $E = K + U$ is constant, independent of x. A graph of E as a function of x is thus a straight horizontal line.

The vertical distance between the U and E graphs at each point represents the difference $E - U$, equal to the kinetic energy K at that point. We see that K is greatest at $x = 0$. It is zero at the values of x where the two graphs cross, labeled A and $-A$ in the diagram. Thus the speed v is greatest at $x = 0$, and it is zero at $x = \pm A$, the points of *maximum* possible displacement from $x = 0$ for a given value of the total energy E. The potential energy U can never be greater than the total energy E; if it were, K would be negative, and that's impossible. The motion is a back-and-forth oscillation between the points $x = A$ and $x = -A$.

At each point, the force F_x on the glider is equal to the negative of the slope of the $U(x)$ curve: $F_x = -dU/dx$. When the particle is at $x = 0$, the slope and the force are zero, so this is an *equilibrium* position. When x is positive, the slope of the $U(x)$ curve is positive and the force F_x is negative, directed toward the origin. When x is negative, the slope is negative and F_x is positive, again toward the origin. Such a force is sometimes called a *restoring force;* when the glider is displaced to either side of $x = 0$, the resulting force tends to "restore" it back to $x = 0$. An analogous situation is a marble rolling around in a round-bottomed salad bowl. We say that $x = 0$ is a point of **stable equilibrium.** More generally, *any minimum in a potential-energy curve is a stable equilibrium position.*

Figure 7–19a shows a hypothetical but more general potential-energy function $U(x)$. Figure 7–19b shows the corresponding force $F_x = -dU/dx$. Points x_1 and x_3 are stable

(a)

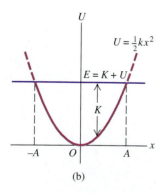

(b)

7–18 (a) A glider on an air track. The spring exerts a force $F_x = -kx$. (b) The potential-energy function. The limits of the motion are the points where the U curve intersects the horizontal line representing the total mechanical energy E.

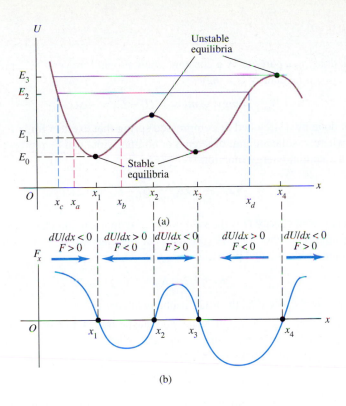

7-19 (a) A hypothetical potential-energy function $U(x)$. (b) The corresponding force $F_x = -dU/dx$. Maxima and minima of $U(x)$ correspond to points where $F_x = 0$.

equilibrium points. At each, F_x is zero because the slope of the $U(x)$ curve is zero. When the particle is displaced to either side, the force pushes back toward the equilibrium point. The slope of the $U(x)$ curve is also zero at points x_2 and x_4, and these are also equilibrium points. But when the particle is displaced a little to the right of either point, the slope of the $U(x)$ curve becomes negative, corresponding to a positive F_x that tends to push the particle still farther from the point. When the particle is displaced a little to the left, F_x is negative, again pushing away from equilibrium. This is analogous to a marble rolling on the top of a bowling ball. Points x_2 and x_4 are called **unstable equilibrium** points; *any maximum in a potential-energy curve is an unstable equilibrium position.*

CAUTION ▶ The direction of the force on a body is *not* determined by the sign of the potential energy U. Rather, it's the sign of $F_x = -dU/dx$ that matters. As we discussed in Section 7-2, the physically significant quantity is the *difference* in the value of U between two points, which is just what the derivative $F_x = -dU/dx$ measures. This means that you can always add a constant to the potential-energy function without changing the physics of the situation. ◀

If the total energy is E_1 and the particle is initially near x_1, it can move only in the region between x_a and x_b determined by the intersection of the E_1 and U graphs (Fig. 7-19a). Again, U cannot be greater than E_1 because K can't be negative. We speak of the particle as moving in a *potential well*, and x_a and x_b are the *turning points* of the particle's motion (since at these points, the particle stops and reverses direction). If we increase the total energy to the level E_2, the particle can move over a wider range, from x_c to x_d. If the total energy is greater than E_3, the particle can "escape" and move to indefinitely large values of x. At the other extreme, E_0 represents the least possible total energy the system can have.

SUMMARY

- The work done on a particle by a constant gravitational force can be represented in terms of a potential energy $U = mgy$:

$$W_{grav} = mgy_1 - mgy_2 = U_1 - U_2 = -\Delta U. \qquad (7-1), (7-3)$$

- The work done by a stretched or compressed spring that exerts a force $F_x = -kx$ on a particle, where x is the amount of stretch or compression, can be represented in terms of a potential-energy function $U = \frac{1}{2}kx^2$:

$$W_{el} = \frac{1}{2}kx_1^2 - \frac{1}{2}kx_2^2 = U_1 - U_2 = -\Delta U. \qquad (7-10)$$

- The total potential energy U is the sum of the gravitational and elastic potential energy. If no forces other than the gravitational and elastic forces do work on a particle, the sum of kinetic and potential energy is conserved:

$$K_1 + U_1 = K_2 + U_2. \qquad (7-11)$$

The sum $K + U = E$ is called the total mechanical energy.

- When forces other than the gravitational and elastic forces do work on a particle, the work W_{other} done by these other forces equals the change in total mechanical energy (kinetic energy plus total potential energy):

$$K_1 + U_1 + W_{other} = K_2 + U_2. \qquad (7-13)$$

- All forces are either conservative or nonconservative. A conservative force is one for which the work-kinetic energy relation is completely reversible. The work of a conservative force can always be represented by a potential-energy function, but the work of a nonconservative force cannot.

- The work done by nonconservative forces manifests itself as changes in the internal energy of bodies. The sum of kinetic, potential, and internal energy is always conserved:

$$\Delta K + \Delta U + \Delta U_{int} = 0. \qquad (7-16)$$

- For motion along a straight line, a conservative force $F_x(x)$ and its associated potential energy $U(x)$ are related by

$$F_x(x) = -\frac{dU}{dx}. \qquad (7-17)$$

In three dimensions, where U is a function of x, y, and z, the components of force are

$$F_x = -\frac{\partial U}{\partial x}, \qquad F_y = -\frac{\partial U}{\partial y}, \qquad F_z = -\frac{\partial U}{\partial z}. \qquad (7-18)$$

or, in vector form,

$$\vec{F} = -\left(\frac{\partial U}{\partial x}\hat{\imath} + \frac{\partial U}{\partial y}\hat{\jmath} + \frac{\partial U}{\partial z}\hat{k}\right). \qquad (7-19)$$

DISCUSSION QUESTIONS

Q7-1 A baseball is thrown straight up with initial speed v_1, as in Example 7-1 (Section 7-2). If air resistance is *not* neglected, when the ball returns to its initial height its speed is less than v_1. Explain, using energy concepts.

Q7-2 In Fig. 7-5 the projectile has the same initial kinetic energy in each case. Why does it not rise to the same maximum height in each case?

Q7–3 In Example 7–5 (Section 7–2), does Throckmorton's speed at the bottom depend on the shape of the ramp or just the difference in height of points 1 and 2? Explain. Answer this same question if the ramp is not frictionless, as in Example 7–6 (Section 7–2).

Q7–4 An egg is released from rest at the roof of a building and falls to the ground. Its fall is observed by a student on the roof of the building who uses coordinates with origin at the roof and by a student on the ground who uses coordinates with origin at the ground. Do the two students assign the same or different values to the initial gravitational potential energy, the final gravitational potential energy, the change in gravitational potential energy, and the kinetic energy of the egg just before it strikes the ground? Explain.

Q7–5 A physics teacher had a bowling ball suspended from a very long rope attached to the high ceiling of a large lecture hall. To illustrate his faith in conservation of energy, he would back up to one side of the stage, pull the ball far to one side until the taut rope brought it just to the end of his nose, and then release it. The massive ball would swing in a mighty arc across the stage and then return to stop momentarily just in front of the nose of the stationary, unflinching teacher. However, one day after the demonstration he looked up just in time to see a student at the other side of the stage *push* the ball away from his nose as he tried to duplicate the demonstration. Tell the rest of the story and explain the reason for the potentially tragic outcome.

Q7–6 When a returning space shuttle touches down on the runway, it has lost almost all the kinetic energy it had in orbit. The gravitational potential energy has also decreased considerably. Where did all that energy go?

Q7–7 Is it possible for a frictional force to *increase* the mechanical energy of a system? If so, give examples.

Q7–8 A woman bounces on a trampoline, going a little higher with each bounce. Explain how she increases the total mechanical energy.

Q7–9 In a siphon, water is lifted above its original level during its flow from one container to another. Where does it get the needed potential energy?

Q7–10 A compressed spring is clamped in its compressed position and then is dissolved in acid. What becomes of its potential energy?

Q7–11 When an object moves away from the earth the gravitational potential energy increases; when it moves closer the potential energy decreases. But the potential energy for a spring increases both when the spring is stretched and when it is compressed. Explain the reason for the difference in behavior of these two potential energies.

Q7–12 Since only changes in potential energy are important in any problem, a student decides to let the elastic potential energy of a spring be zero when the spring is stretched a distance x_1. The student decides therefore to let $U = \frac{1}{2}k(x - x_1)^2$. Is this correct? Explain.

Q7–13 Figure 7–17b shows the potential energy function for the force $F_x = -kx$. Sketch the potential energy function for the force $F_x = +kx$. For this force, is $x = 0$ a point of equilibrium? Is this equilibrium stable or unstable? Explain.

Q7–14 Figure 7–17a shows the potential energy function associated with the gravitational force between an object and the earth. Use this graph to explain why objects always fall toward the earth when they are released.

Q7–15 For a system of two particles we often let the potential energy for the force between the particles approach zero as the separation of the particles approaches infinity. If this choice is made, explain why the potential energy at noninfinite separation is positive if the particles repel one another and negative if they attract.

Q7–16 Explain why the points $x = A$ and $x = -A$ in Fig. 7–18b are called *turning points*. How are the values of E and U related at a turning point?

Q7–17 A particle is in *neutral equilibrium* if the net force on it is zero and remains zero if the particle is displaced slightly in any direction. Sketch the potential energy function near a point of neutral equilibrium, for the case of one-dimensional motion. Give an example of an object in neutral equilibrium.

Q7–18 The net force on a particle of mass m has the potential-energy function graphed in Figure 7–19a. If the total energy is E_1, graph the speed v of the particle versus its position x. At what value of x is the speed greatest? Sketch v versus x if the total energy is E_2.

EXERCISES

SECTION 7–2 GRAVITATIONAL POTENTIAL ENERGY

7–1 What is the potential energy for an 800-kg elevator at the top of Chicago's Sears Tower, 440 m above street level? Let the potential energy be zero at street level.

7–2 A 5.00-kg sack of flour is lifted vertically at a constant speed of 3.50 m/s through a height of 15.0 m. a) How great a force is required? b) How much work is done on the sack by the lifting force? What becomes of this work?

7–3 Repeat part (a) of Example 6–5 (Section 6–3) using Eq. (7–7).

7–4 A mail bag with a mass of 120 kg is suspended by a vertical rope 6.0 m long. a) What horizontal force is necessary to hold the bag in a position displaced sideways 3.0 m from its initial position? b) How much work is done by the worker in moving the bag to this position?

7–5 A baseball is thrown from the roof of a 22.0-m-tall building with an initial velocity of magnitude 12.0 m/s and directed at an angle of 53.1° above the horizontal. a) What is the speed of the ball just before it strikes the ground? Use energy methods and ignore air resistance. b) What is the answer for part (a) if the initial velocity is at an angle of 53.1° *below* the horizontal? c) If the effects of air resistance are included, will part (a) or (b) give the higher speed?

7–6 a) In Example 7–7 (Section 7–2), calculate the minimum initial speed required for the crate to reach the top of the ramp. b) If the initial speed of the crate in Example 7–7 is 11.0 m/s, what is its speed at the top of the ramp?

7–7 Answer part (b) of Example 7–7 (Section 7–2) by applying Eq. (7–7) to points 2 and 3, rather than to points 1 and 3 as was done in the example.

7–8 An empty crate slides down a ramp, starting with a speed v_0 and reaching the bottom with a speed v and kinetic energy K. Some books are placed in the box, so that the total mass is quadrupled. The coefficient of kinetic friction is constant and air resistance is negligible. Again starting with v_0 at the top of the ramp, what are the speed and kinetic energy at the bottom? Explain the reasoning behind your answers.

7–9 A small rock with a mass of 0.20 kg is released from rest at point A, which is at the top edge of a large hemispherical bowl with radius $R = 0.50$ m (Fig. 7–20). Assume that the size of the rock is small compared to R, so that the rock can be treated as a particle, and assume that the rock slides rather than rolling. The work done by friction on the rock when it moves from point A to point B at the bottom of the bowl is –0.22 J. What is the speed of the rock when it reaches point B?

FIGURE 7–20 Exercise 7–9.

7–10 Tarzan, in one tree, sights Jane in another tree. He grabs the end of a vine of length 20 m that makes an angle of 45° with the vertical, steps off his tree limb, and swings down and then up to Jane's open arms. When he arrives, his vine makes an angle of 30° with the vertical. Determine whether he gives her a tender embrace or knocks her off her limb by calculating Tarzan's speed just before he reaches Jane. Ignore air resistance and the mass of the vine.

7–11 A 10.0-kg microwave oven is pushed 8.00 m up the sloping surface of a loading ramp inclined at an angle of 36.9° above the horizontal, by a constant force $\vec{F}$ with a magnitude of 110 N and acting parallel to the ramp. The coefficient of kinetic friction between the oven and the ramp is 0.250. a) What is the work done on the oven by the force $\vec{F}$? b) What is the work done on the oven by the friction force? c) Compute the increase in potential energy for the oven. d) Use your answers to parts (a), (b), and (c) to calculate the increase in the oven's kinetic energy. e) Use $\Sigma \vec{F} = m\vec{a}$ to calculate the acceleration of the oven. Assuming that the oven is initially at rest, use the acceleration to calculate the oven's speed after traveling 8.00 m. Compute from this the increase in the oven's kinetic energy, and compare to the answer you got in part (d).

7–12 A small rock with mass 0.12 kg is fastened to a massless string with length 0.80 m to form a pendulum. The pendulum is swinging so as to make a maximum angle of 45° with the vertical. Air resistance is negligible. a) What is the speed of the rock when the string passes through the vertical position? b) What is the tension in the string when it makes an angle of 45° with the vertical? c) What is the tension in the string as it passes through the vertical?

SECTION 7–3 ELASTIC POTENTIAL ENERGY

7–13 A force of 800 N stretches a certain spring a distance of 0.200 m. a) What is the potential energy of the spring when it is stretched 0.200 m? b) What is its potential energy when it is compressed 5.00 cm?

7–14 A force of 720 N stretches a certain spring a distance of 0.150 m. What is the potential energy of the spring when a 60.0-kg mass hangs vertically from it?

7–15 A spring of negligible mass has force constant $k = 1600$ N/m. a) How far must the spring be compressed for 3.20 J of potential energy to be stored in it? b) You place the spring vertically with one end on the floor. You then drop a 1.20-kg book onto it from a height of 0.80 m above the top of the spring. Find the maximum distance the spring will be compressed.

7–16 A slingshot will shoot a 10-g pebble 22.0 m straight up. a) How much potential energy is stored in the slingshot's rubber band? b) With the same potential energy stored in the rubber band, how high can the slingshot shoot a 25-g pebble? c) What physical effects did you ignore in solving this problem?

7–17 A 1.20-kg piece of cheese is placed on a vertical spring of negligible mass and force constant $k = 1800$ N/m that is compressed 15.0 cm. When the spring is released, how high does the cheese rise from this initial position? (The cheese and the spring are *not* attached.)

7–18 Consider the glider of Example 7–8 (Section 7–3) and Fig. 7–12. As in the example, the glider is released from rest with the spring stretched 0.100 m. What is the displacement x of the glider from its equilibrium position when its speed is 0.20 m/s? (You should get more than one answer. Explain why.)

7–19 Consider the glider of Example 7–8 (Section 7–3) and Fig. 7–12. a) As in the example, the glider is released from rest with the spring stretched 0.100 m. What is the speed of the glider when it returns to $x = 0$? b) What must be the initial displacement of the glider if its maximum speed in the subsequent motion is to be 2.50 m/s?

7–20 Consider the glider of Example 7–8 (Section 7–3) and Fig. 7–12. As in the example, the glider is released from rest with the spring stretched 0.100 m. But now the air track is turned off so that there is a kinetic friction force acting on the glider. a) If the coefficient of kinetic friction between the air track and the glider is $\mu_k = 0.050$, what is the speed of the glider when it has traveled 0.020 m, so that $x = 0.080$ m? b) If $\mu_k = 0.050$, what is the speed of the glider when it has traveled 0.100 m, so that $x = 0$? c) What must be the value of μ_k so that the glider reaches the $x = 0$ position with zero speed?

7–21 a) For the elevator of Example 7–11 (Section 7–3), what is the speed of the elevator after it has moved downward 1.00 m from point 1 in Fig. 7–13? b) When the elevator is 1.00 m below point 1 in Fig. 7–13, what is its acceleration?

7–22 You are asked to design a spring that will give a 1160-kg satellite a speed of 2.50 m/s relative to an orbiting space shuttle. Your spring is to give the satellite a maximum acceleration of 5.00g. The spring's mass, the recoil kinetic energy of the shuttle, and changes in gravitational potential energy will all be negligible. a) What must be the force constant of the spring? b) What distance must the spring be compressed?

SECTION 7–4 CONSERVATIVE AND NONCONSERVATIVE FORCES

7–23 A 0.75-kg book moves vertically upward a distance of 16 m, and then vertically downward 16 m, returning to its initial position. a) How much work is done by gravity during the upward motion of the book? b) How much work is done by gravity during the downward motion of the book? c) What is the total work done on the book by gravity during the complete round trip? d) On the basis of your answer to part (c), would you say that the gravitational force is conservative or nonconservative? Explain.

7–24 A 0.050-kg rock moves from the origin to the point (3.0 m, 5.0 m) in a coordinate system in which the positive y-direction is upward. a) The rock first moves horizontally from the origin to the point (3.0 m, 0). Then it moves vertically from (3.0 m, 0) to (3.0 m, 5.0 m). Sketch the path of the rock in the xy-plane. How much work is done on the rock by gravity during the displacement? b) Instead of the path in part (a), let the rock first move vertically from the origin to (0, 5.0 m) and then horizontally from (0, 5.0 m) to (3.0 m, 5.0 m). Sketch the path of the rock in the xy-plane. How much work is done on the rock by gravity during the displacement? c) Compare the answers to parts (a) and (b). From your results would you say that the gravitational force is conservative or nonconservative? Explain.

7–25 In an experiment, one of the forces exerted on a proton is $\vec{F} = -\alpha x^2 \hat{\imath}$, where $\alpha = 12$ N/m². a) How much work does $\vec{F}$ do when the proton moves along the straight-line path from the point (0.10 m, 0) to the point (0.10 m, 0.40 m)? b) Along the straight-line path from the point (0.10 m, 0) to the point (0.30 m, 0)? c) Along the straight-line path from the point (0.30 m, 0) to the point (0.10 m, 0)? d) Is the force $\vec{F}$ conservative? Explain. If $\vec{F}$ is conservative, what is the potential energy function for it? Let $U = 0$ when $x = 0$.

7–26 Consider the electron and force $\vec{F}$ of Example 7–13 (Section 7–4). a) The electron travels from the point (0, 0) to the point (L, L) along the straight line from (0, 0) to (0, L) and then along the straight line from (0, L) to (L, L). What is the work done by the force $\vec{F}$ during this displacement? b) The electron travels from the point (0, 0) to the point (L, L) along the straight line from (0, 0) to (L, 0) and then along the straight line from (L, 0) to (L, L). What is the work done by the force $\vec{F}$ during this displacement? c) The electron travels from the point (0, 0) to the point (L, L) along the straight line from (0, 0) to (L, L). What is the work done by the force $\vec{F}$ during this displacement? d) Compare the answers in parts (a), (b), and (c) and explain the results of this comparison.

7–27 A 0.60-kg book slides on a horizontal table. The kinetic friction force on the book has magnitude 1.2 N. a) How much work is done on the book by friction during a displacement of 3.0 m to the left? b) The book now slides 3.0 m to the right, returning to its starting point. During this second 3.0-m displacement, how much work is done on the book by friction? c) What is the total work done on the book by friction during the complete round trip? d) On the basis of your answer to part (c), would you say that the friction force is conservative or nonconservative? Explain.

7–28 A 30.0-kg packing crate in a warehouse is pushed to the loading dock by a worker who applies a horizontal force. The coefficient of kinetic friction between the crate and the floor is 0.20. The loading dock is 15.0 m southwest of the initial position of the crate. a) If the crate is pushed 10.6 m south and then 10.6 m west, what is the total work done on the crate by friction? b) If the crate is pushed along a straight-line path to the dock, so that it travels 15.0 m southwest, what is the work done on the crate by friction? c) Sketch the paths along which the crate moves in parts (a) and (b). Based on your answers to parts (a) and (b), would you say that the friction force is conservative or nonconservative? Explain.

7–29 You and three friends stand at the corners of a square whose sides are 8.0 m long in the middle of the gym floor, as shown in Fig. 7–21. You take your physics book and push it from one person to the other. The book has a mass of 1.5 kg, and the coefficient of kinetic friction between the book and the floor is $\mu_k = 0.25$. a) The book slides from you to Beth and then from Beth to Carlos, along the lines connecting these people. What is the work done by friction during this displacement? b) You slide the book from you to Carlos along the diagonal of the square. What is the work done by friction during this displacement? c) You slide the book to Kim who then slides it back to you. What is the total work done by friction during this motion of the book? d) Is the friction force on the book conservative or nonconservative? Explain.

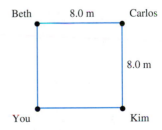

FIGURE 7–21 Exercise 7–29.

7–30 A block of mass m is attached to an ideal spring that has force constant k. a) The block moves from x_1 to x_2, where $x_2 > x_1$. How much work does the spring force do during this displacement? b) The block moves from x_1 to x_2 and then from x_2 to x_1. How much work does the spring force do during the displacement from x_2 to x_1? What is the total work done by the spring during the entire $x_1 \to x_2 \to x_1$ displacement? Explain why you obtained the answer you did. c) The block moves from x_1 to x_3, where $x_3 > x_2$. How much work does the spring force do during this displacement? The block then moves from x_3 to x_2. How much work does the spring force do during this displacement?

What is the total work done by the spring force during the $x_1 \rightarrow x_3 \rightarrow x_2$ displacement? Compare your answer to the answer to part (a), where the starting and ending points are the same but the path is different.

SECTION 7–5 FORCE AND POTENTIAL ENERGY

7–31 The potential energy of a pair of hydrogen atoms separated by a large distance x is given by $U(x) = -C_6/x^6$, where C_6 is a positive constant. What is the force that one atom exerts on the other? Is this force attractive or repulsive?

7–32 A force parallel to the x-axis acts on a particle moving along the x-axis. This force produces a potential energy $U(x)$ given by $U(x) = \alpha x^4$, where $\alpha = 1.20 \ \text{J/m}^4$. What is the force (magnitude and direction) when the particle is at $x = -0.800 \ \text{m}$?

7–33 An experimental device moving in the xy-plane is acted on by a conservative force described by the potential-energy function $U(x, y) = k(x^2 + y^2) + k'xy$, where k and k' are positive constants. Derive an expression for the force expressed in terms of the unit vectors $\hat{\imath}$ and $\hat{\jmath}$.

7–34 An object moving in the xy-plane is acted on by a conservative force described by the potential-energy function $U(x, y) = \alpha (1/x^2 + 1/y^2)$, where α is a positive constant. Derive an expression for the force expressed in terms of the unit vectors $\hat{\imath}$ and $\hat{\jmath}$.

SECTION 7–6 ENERGY DIAGRAMS

7–35 The potential energy of two atoms in a diatomic molecule is approximated by $U(r) = a/r^{12} - b/r^6$, where r is the spacing between atoms and a and b are positive constants. a) Find the

force $F(r)$ on one atom as a function of r. Make two graphs, one of $U(r)$ versus r and one of $F(r)$ versus r. b) Find the equilibrium distance between the two atoms. Is this equilibrium stable? c) Suppose the distance between the two atoms is equal to the equilibrium distance found in part (b). What minimum energy must be added to the molecule to *dissociate* it, that is, to separate the two atoms to an infinite distance apart? This is called the *dissociation energy* of the molecule. d) For the molecule CO, the equilibrium distance between the carbon and oxygen atoms is 1.13×10^{-10} m and the dissociation energy is 1.54×10^{-18} J per molecule. Find the values of the constants a and b.

7–36 A marble moves along the x-axis. The potential-energy function is shown in Fig. 7–22. a) At which of the labeled x-coordinates is the force on the marble zero? b) Which of the labeled x-coordinates is a position of stable equilibrium? c) Which of the labeled x-coordinates is a position of unstable equilibrium?

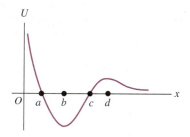

FIGURE 7–22 Exercise 7–36.

PROBLEMS

7–37 A man with mass 70.0 kg sits on a platform suspended from a movable pulley as shown in Fig. 7–23 and raises himself at constant speed by a rope passing over a fixed pulley. The platform and the pulleys have negligible mass. Assume that there are no friction losses. a) Find the force he must exert. b) Find the

FIGURE 7–23 Problem 7–37.

increase in the energy of the system when he raises himself 1.20 m. (Answer by calculating the increase in potential energy and also by computing the product of the force on the rope and the length of the rope passing through his hands.)

7–38 A 2.00-kg block is pushed against a spring with negligible mass and force constant $k = 400 \ \text{N/m}$, compressing it 0.220 m. When the block is released, it moves along a frictionless, horizontal surface and then up a frictionless incline with slope $37.0°$ (Fig. 7–24). a) What is the speed of the block as it slides along the horizontal surface after having left the spring? b) How far does the block travel up the incline before starting to slide back down?

FIGURE 7–24 Problem 7–38.

$k = 400 \ \text{N/m}$ $m = 2.00 \ \text{kg}$ $37.0°$ 0.220 m

7–39 A block with mass 0.50 kg is forced against a horizontal spring of negligible mass, compressing the spring a distance of 0.20 m (Fig. 7–25). When released, the block moves on a horizontal table top for 1.00 m before coming to rest. The spring constant k is 100 N/m. What is the coefficient of kinetic friction, μ_k, between the block and the table?

FIGURE 7–25 Problem 7–39.

7–40 Riding a Loop-the-loop. A car in an amusement park ride rolls without friction around the track shown in Fig. 7–26. It starts from rest at point A at a height h above the bottom of the loop. Treat the car as a particle. a) What is the minimum value of h (in terms of R) such that the car moves around the loop without falling off at the top (point B)? b) If $h = 3.50R$ and $R = 20.0$ m, compute the speed, radial acceleration, and tangential acceleration of the passengers when the car is at point C, which is at the end of a horizontal diameter. Show these acceleration components in a diagram, approximately to scale.

FIGURE 7–26 Problem 7–40.

7–41 The Great Sandini is a 60-kg circus performer who is shot from a cannon (actually a spring gun). You don't find many men of his caliber, so you help him design a new gun. This new gun has a very large spring with a very small mass and a force constant of 1100 N/m that he will compress with a force of 4400 N. The inside of the gun barrel is coated with Teflon, so the average friction force will be only 40 N during the 4.0 m he moves in the barrel. At what speed will he emerge from the end of the barrel, 2.5 m above his initial rest position?

7–42 You are designing a delivery ramp for crates containing exercise equipment. The 1470-N crates will move at 1.8 m/s at the top of a ramp that slopes downward at 22.0°. The ramp exerts a 550-N kinetic friction force on each crate, and the maximum static friction force also has this value. Each crate will compress a spring at the bottom of the ramp and will come to rest after traveling a total distance of 8.0 m along the ramp. Once stopped, a crate must not rebound back up the ramp. Calculate the force constant of the spring that will need to be compressed the least in order to meet the design criteria.

7–43 A system of two paint buckets connected by a lightweight rope is released from rest with the 12.0-kg bucket 2.00 m above

the floor (Fig. 7–27). Use the principle of conservation of energy to find the speed with which this bucket strikes the floor. Neglect friction and the inertia of the pulley.

FIGURE 7–27 Problem 7–43.

7–44 a) For the elevator of Example 7–11 (Section 7–3), how much potential energy is stored in the spring when the elevator is at point 2 in Fig. 7–13? b) What maximum distance does the elevator travel upward from point 2 before it starts to slide down the shaft again? c) As the elevator slides back down, what is its speed just before it reaches the spring? d) When the elevator compresses the spring the second time, what is the maximum energy that is stored in the spring and what is the force that the spring exerts on the elevator?

7–45 Redesign the elevator safety system of Example 7–11 (Section 7–3) so that the elevator does not bounce but stays at rest the first time its speed becomes zero. The mass of the elevator is 2000 kg and its speed when it first touches the spring is 25 m/s. There is a kinetic friction force of 17,000 N, and the maximum static friction force on the elevator is also 17,000 N. The mass of the spring can be neglected. a) What spring constant is required, and what distance is the spring compressed when the elevator is stopped? Do you think this design is practical? Explain. b) What is the maximum magnitude of the acceleration of the elevator?

7–46 A wooden rod of negligible mass and length 80.0 cm is pivoted about a horizontal axis through its center. A white rat of mass 0.500 kg clings to one end of the stick, and a mouse of mass 0.200 kg clings to the other end. The system is released from rest with the rod horizontal. If the animals can manage to hold on, what are their speeds as the rod swings through a vertical position?

7–47 A 0.100-kg potato is tied to a string with length 2.50 m, and the other end of the string is tied to a rigid support. The potato is held straight out horizontally from the point of support, with the string pulled taut, and is then released. a) What is the speed of the potato at the lowest point of its motion? b) What is the tension in the string at this point?

7–48 The following is data from a computer simulation for a batted baseball of mass 0.145 kg, including air resistance:

t	x	y	v_x	v_y
0	0	0	30.0 m/s	40.0 m/s
3.05 s	70.2 m	53.6 m	18.6 m/s	0
6.59 s	124.4 m	0	11.9 m/s	−28.7 m/s

a) How much work was done by the air on the baseball as it moved from its initial position to its maximum height? b) How much work was done by the air on the baseball as it moved from its maximum height back to the starting elevation? c) Explain why the magnitude of the answer to part (b) is smaller than the magnitude of the answer in part (a).

7–49 A fireman of mass m slides a distance d down a pole. He starts from rest. He moves as fast at the bottom as if he had stepped off a platform a distance $h \leq d$ above the ground and descended with negligible air resistance. a) What average friction force did the fireman exert on the pole? Does your answer make sense in the special cases of $h = d$ and $h = 0$? b) Find a numerical value for the average friction force a 75-kg fireman exerts, for $d = 2.5$ m and $h = 1.0$ m. c) In terms of g, h, and d, what is the speed of the fireman when he is a distance y above the bottom of the pole?

7–50 A 60.0-kg skier starts from rest at the top of a ski slope 65.0 m high. a) If frictional forces do -10.5 kJ of work on her as she descends, how fast is she going at the bottom of the slope? b) Now moving horizontally, the skier crosses a patch of soft snow, where $\mu_k = 0.20$. If the patch is 82.0 m wide and the average force of air resistance on the skier is 160 N, how fast is she going after crossing the patch? c) The skier hits a snowdrift and penetrates 2.5 m into it before coming to a stop. What is the average force exerted on her by the snowdrift as it stops her?

7–51 A skier starts at the top of a very large frictionless snowball, with a very small initial speed, and skis straight down the side (Fig. 7–28). At what point does she lose contact with the snowball and fly off at a tangent? That is, at the instant she loses contact with the snowball, what angle α does a radial line from the center of the snowball to the skier make with the vertical?

FIGURE 7–28 Problem 7–51.

7–52 A rock is tied to a cord, and the other end of the cord is held fixed. The rock is given an initial tangential velocity that causes it to rotate in a vertical circle. Prove that the tension in the cord at the lowest point exceeds that at the highest point by six times the weight of the rock.

7–53 In a truck-loading station at a post office a small 0.200-kg package is released from rest at point A on a track that is one-quarter of a circle with radius 1.60 m (Fig. 7–29). The size of the package is much less than 1.60 m, so the package can be treated as a particle. It slides down the track and reaches point B with a speed of 4.80 m/s. From point B it slides on a level surface a distance of 3.00 m to point C, where it comes to rest. a) What is the coefficient of kinetic friction on the horizontal surface? b) How much work is done on the package by friction as it slides down the circular arc from A to B?

FIGURE 7–29 Problem 7–53.

7–54 A truck of mass m has a brake failure while going down an icy mountain road of constant downward slope angle α (Fig. 7–30). Initially the truck is moving downhill at speed v_0. After careening downhill a distance L with negligible friction, the truck driver steers the runaway vehicle onto a runaway truck ramp of constant upward slope angle β. The truck ramp has a soft sand surface for which the coefficient of rolling friction is μ_r. What is the distance that the truck moves up the ramp before coming to a halt? Use energy methods.

FIGURE 7–30 Problem 7–54.

7–55 A certain spring is found *not* to obey Hooke's law; it exerts a restoring force $F_x(x) = -\alpha x - \beta x^2$ if it is stretched or compressed, where $\alpha = 60.0$ N/m and $\beta = 18.0$ N/m^2. The mass of the spring is negligible. a) Calculate the potential energy function $U(x)$ for this spring. Let $U = 0$ when $x = 0$. b) An object with a mass of 0.900 kg on a frictionless horizontal surface is attached to this spring, pulled a distance 1.00 m to the right (the $+x$-direction) to stretch the spring, and released. What is the speed of the object when it is 0.50 m to the right of the $x = 0$ equilibrium position?

7–56 A variable force $\vec{F}$ is maintained tangent to a frictionless semicircular surface (Fig. 7–31). By slowly varying the force, a block with weight w is moved, and the spring to which it is attached is stretched from position 1 to position 2. The spring has negligible mass and force constant k. The end of the spring moves in an arc of radius a. Calculate the work done by the force $\vec{F}$.

FIGURE 7–31 Problem 7–56.

7–57 A 0.150-kg block of ice is placed against a horizontal compressed spring mounted on a horizontal tabletop that is 1.20 m above the floor. The spring has force constant 1900 N/m and is initially compressed 0.045 m. The mass of the spring is negligible. The spring is released, and the block slides along the table, goes off the edge, and travels to the floor. If there is negligible friction between the ice and the table, what is the speed of the block of ice when it reaches the floor?

7–58 An 80.0-kg man jumps from a height of 2.50 m onto a platform mounted on springs. As the springs compress, the platform is pushed down a maximum distance of 0.240 m below its initial position, and then it rebounds. The platform and springs have negligible mass. a) What is the man's speed at the instant the platform is depressed 0.120 m? b) If the man had just stepped gently onto the platform, what maximum distance would it have been pushed down?

7–59 An experimental apparatus of mass m is placed on a vertical spring of negligible mass and pushed down until the spring has been compressed a distance x. The apparatus is then released and reaches its maximum height at a distance h above the point where it was released. The apparatus is not attached to the spring, and at its maximum height it is no longer in contact with the spring. The maximum magnitude of acceleration the apparatus can have without being damaged is a, where $a > g$. a) What should be the force constant of the spring? b) What distance x must the spring be compressed initially?

7–60 If a fish is attached to a vertical spring and slowly lowered to its equilibrium position, it is found to stretch the spring by an amount d. If the same fish is attached to the end of the unstretched spring and then allowed to fall from rest, through what maximum distance does it stretch the spring? (*Hint:* Calculate the force constant of the spring in terms of the distance d and the mass m of the fish.)

7–61 A wooden block with mass 1.50 kg is placed against a compressed spring at the bottom of an incline of slope 30.0° (point A). When the spring is released, it projects the block up the incline. At point B, a distance of 6.00 m up the incline from A, the block is moving up the incline at 7.00 m/s and is no longer in contact with the spring. The coefficient of kinetic friction between the block and incline is $\mu_k = 0.50$. The mass of the spring is negligible. Calculate the amount of potential energy that was initially stored in the spring.

7–62 A 2.00-kg package is released on a 53.1° incline, 4.00 m from a long spring with force constant 120 N/m that is attached at the bottom of the incline (Fig. 7–32). The coefficients of friction between the package and the incline are $\mu_s = 0.40$ and $\mu_k = 0.20$. The mass of the spring is negligible. a) What is the speed of the package just before it reaches the spring? b) What is the maximum compression of the spring? c) The package rebounds back up the incline. How close does it get to its initial position?

FIGURE 7–32 Problem 7–62.

7–63 A 0.500-kg block attached to a spring with length 0.60 m and force constant 40.0 N/m is at rest with the back of the block at point A on a frictionless, horizontal air table (Fig. 7–33). The mass of the spring is negligible. You move the block to the right along the surface by pulling with a constant 20.0-N horizontal force. a) What is the block's speed when the back of the block reaches point B, which is 0.25 m to the right of point A? b) When the back of the block reaches point B, you let go of the block. In the subsequent motion, how close does the block get to the wall where the left end of the spring is attached?

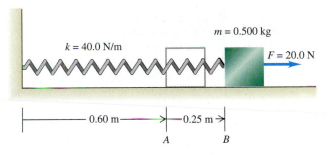

FIGURE 7–33 Problem 7–63.

7–64 A box of mass m is pushed against a spring of negligible mass and force constant k, compressing it a distance x. The box is then released and travels up a ramp that is at an angle α above the horizontal. The coefficient of kinetic friction between the box and the ramp is μ_k, where $\mu_k < 1$. The box is still moving up the ramp after traveling a distance $s > |x|$ along the ramp. Calculate the angle α for which the speed of the box after traveling distance s is a minimum. Explain why the minimum speed doesn't occur when $\alpha = 90°$, even though for this α there is the maximum increase in gravitational potential energy.

7–65 The brothers of Iota Eta Pi fraternity build a platform supported at all four corners by vertical springs in the basement of their frat house. A brave fraternity brother wearing a football helmet stands in the middle of the platform; his weight compresses the springs by 0.18 m. Then four of his fraternity brothers, pushing down at the corners of the platform, compress the springs another 0.53 m until the top of the brave brother's helmet is 0.90 m below the basement ceiling. They then simultaneously release the platform. Neglect the masses of the springs and platform. a) When the dust clears, the fraternity asks you to calculate their fraternity brother's speed just before his helmet hit the flimsy ceiling. b) Without the ceiling, how high would he have gone? c) In discussing their probation, the dean of students suggests that the next time they try this, they do it outdoors on another planet. Would the answer to part (b) be the same if this stunt were performed on a planet with a different value of g? Assume that the fraternity brothers push the platform down 0.53 m as before. Explain your reasoning.

7–66 a) Is the force $\vec{F} = Cy^2\hat{j}$, where C is a negative constant with units of N/m^2, conservative or nonconservative? Justify your answer. b) Is the force $\vec{F} = Cy^2\hat{i}$, where C is a negative constant with units of N/m^2, conservative or nonconservative? Justify your answer.

7–67 A cutting tool under microprocessor control has several forces acting on it. One force is $\vec{F} = -\alpha xy^2\hat{j}$, a force in the negative y-direction whose magnitude depends on the position of the tool. The constant is $\alpha = 2.50$ N/m^3. Consider the displacement of the tool from the origin to the point $x = 3.00$ m, $y = 3.00$ m. a) Calculate the work done on the tool by $\vec{F}$ if this displacement is along the straight line $y = x$ that connects these two points. b) Calculate the work done on the tool by $\vec{F}$ if the tool is first moved out along the x-axis to the point $x = 3.00$ m, $y = 0$ and then moved parallel to the y-axis to $x = 3.00$ m, $y = 3.00$ m. c) Compare the work done by $\vec{F}$ along these two paths. Is $\vec{F}$ conservative or nonconservative? Explain.

7–68 An object has several forces acting on it. One force is $\vec{F} = \alpha xy\hat{i}$, a force in the x-direction whose magnitude depends on the position of the object. (See Problem 6–84.) The constant is $\alpha = 2.00$ N/m^2. The object moves along the following path: (1) it starts at the origin and moves along the y-axis to the point $x = 0$, $y = 1.50$ m; (2) it moves parallel to the x-axis to the point $x = 1.50$ m, $y = 1.50$ m; (3) it moves parallel to the y-axis to the point $x = 1.50$ m, $y = 0$; (4) it moves parallel to the x-axis back to

the origin. a) Sketch this path in the xy-plane. b) Calculate the work done on the object by $\vec{F}$ for each leg of the path and for the complete round trip. c) Is $\vec{F}$ conservative or nonconservative? Explain.

7–69 A Hooke's law force $-kx$ and a constant conservative force F in the $+x$-direction act on an atomic ion. a) Show that a possible potential-energy function for this combination of forces is $U(x) = \frac{1}{2}kx^2 - Fx - F^2/2k$. Is this the *only* possible function? Explain. b) Find the stable equilibrium position. c) Graph $U(x)$ (in units of F^2/k) versus x (in units of F/k) for values of x between $-5F/k$ to $5F/k$. d) Are there any unstable equilibrium positions? e) If the total energy is $E = F^2/k$, what are the maximum and minimum values of x that the ion reaches in its motion? f) If the ion has mass m, find its maximum speed if the total energy is $E = F^2/k$. For what value of x is the speed maximum?

7–70 A particle moves along the x-axis while acted on by a single conservative force parallel to the x-axis. The force corresponds to the potential-energy function graphed in Fig. 7–34. The particle is released from rest at point A. a) What is the direction of the force on the particle when it is at point A? b) At point B? c) At what value of x is the kinetic energy of the particle a maximum? d) What is the force on the particle when it is at point C? e) What is the largest value of x reached by the particle during its motion? f) What value or values of x correspond to points of stable equilibrium? g) Of unstable equilibrium?

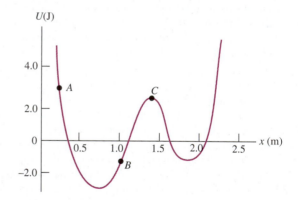

FIGURE 7–34 Problem 7–70.

CHALLENGE PROBLEM

7–71 A proton with mass m moves in one dimension. The potential-energy function is $U(x) = \alpha/x^2 - \beta/x$, where α and β are positive constants. The proton is released from rest at $x_0 = \alpha/\beta$. a) Show that $U(x)$ can be written as

$$U(x) = \frac{\alpha}{x_0^2}\left[\left(\frac{x_0}{x}\right)^2 - \frac{x_0}{x}\right].$$

Graph $U(x)$. Calculate $U(x_0)$ and thereby locate the point x_0 on the graph. b) Calculate $v(x)$, the speed of the proton as a func-

tion of position. Graph $v(x)$, and give a qualitative description of the motion. c) For what value of x is the speed of the proton a maximum? What is the value of that maximum speed? d) What is the force on the proton at the point in part (c)? e) Let the proton be released instead at $x_1 = 3\alpha/\beta$. Locate the point x_1 on the graph of $U(x)$. Calculate $v(x)$, and give a qualitative description of the motion. f) For each release point ($x = x_0$ and $x = x_1$), what are the maximum and minimum values of x reached during the motion?

Momentum, Impulse, and Collisions

Between July 16 and July 22, 1994, 23 kilometer-sized fragments of a comet collided with the planet Jupiter. This artist's conception shows the chain of fragments streaming toward the giant planet, as viewed from Jupiter's moon Io. Though the energy released in each impact was equivalent to more than 10^{14} tons of TNT, the fragments had so little momentum compared to Jupiter that there was no noticeable effect on the planet's orbit.

8–1 INTRODUCTION

When an eighteen-wheeler collides head-on with a compact car, what determines which way the wreckage goes after the collision? Why are the occupants of the car much more likely to be injured than those of the truck? How do you decide how to aim the cue ball in pool to knock the eight ball into the pocket?

A common theme of all these questions is that they can't be answered by directly applying Newton's second law, $\Sigma \vec{F} = m\vec{a}$, because there are forces acting about which we know very little: the forces acting between the car and the eighteen-wheeler, or between two pool balls. Remarkably, we will find in this chapter that we don't have to know *anything* about these forces in order to answer questions of this kind!

Our approach will use two new concepts, *momentum* and *impulse,* and a new conservation law, *conservation of momentum.* This conservation law is every bit as important as that of conservation of energy. The law of conservation of momentum is valid even in situations in which Newton's laws are inadequate, such as bodies moving at very high speeds (near the speed of light) or objects on a very small scale (such as the constituents of atoms). Within the domain of Newtonian mechanics, conservation of momentum enables us to analyze many situations that would be very difficult if we tried to use Newton's laws directly. Among these are *collision* problems, in which

two bodies collide and can exert very large forces on each other for a short time.

8–2 MOMENTUM AND IMPULSE

In Chapter 6 we re-expressed Newton's second law for a particle, $\Sigma \vec{F} = m\vec{a}$, in terms of the work-energy theorem. This theorem helped us to tackle a great number of physics problems and led us to the law of conservation of energy. Let's now return to $\Sigma \vec{F} = m\vec{a}$ and see yet another useful way in which this fundamental law can be restated.

Let's consider a particle of constant mass m. (Later in this chapter we'll see how to deal with situations in which the mass of a body changes.) Because $\vec{a} = d\vec{v}/dt$, we can write Newton's second law for this particle as

$$\Sigma \vec{F} = m \frac{d\vec{v}}{dt} = \frac{d}{dt}(m\vec{v}). \qquad (8\text{–}1)$$

We can take the mass m inside the derivative because it is constant. Thus Newton's second law says that the net force $\Sigma \vec{F}$ acting on a particle equals the time rate of change of the combination $m\vec{v}$, the product of the particle's mass and velocity. We'll call this combination the **momentum,** or **linear momentum,** of the particle. Using the symbol $\vec{p}$ for momentum, we have

$$\vec{p} = m\vec{v} \qquad \text{(definition of momentum)}. \qquad (8\text{–}2)$$

Momentum is a vector quantity that has a magnitude (*mv*) and a direction (the same as the velocity vector $\vec{v}$). The momentum of a car driving north at 20 m/s is different from the momentum of the same car driving east at the same speed. A fast ball thrown by a major-league pitcher has greater magnitude of momentum than the same ball thrown by a child because the speed is greater. An eighteen-wheeler going 65 mi/h has greater magnitude of momentum than a Saturn automobile with the same speed because the truck's mass is greater. The units of the magnitude of momentum are units of mass times speed; the SI units of momentum are kg · m/s. The plural of momentum is "momenta."

Substituting Eq. (8–2) into Eq. (8–1), we have

$$\Sigma \vec{F} = \frac{d\vec{p}}{dt} \qquad \text{(Newton's second law in terms of momentum).} \qquad (8\text{–}3)$$

The net force (vector sum of all forces) acting on a particle equals the time rate of change of momentum of the particle. This, not $\Sigma \vec{F} = m\vec{a}$, is the form in which Newton originally stated his second law (though he called momentum the "quantity of motion"). It is valid only in inertial frames of reference.

According to Eq. (8–3), a rapid change in momentum requires a large net force, while a gradual change in momentum requires less net force. This principle is used in the design of automobile safety devices such as air bags. The driver of a fast-moving automobile has a large momentum (the product of the driver's mass and velocity). If the car stops suddenly in a collision, the driver's momentum becomes zero. An air bag causes the driver to lose momentum more gradually than would an abrupt collision with the steering wheel, reducing the force exerted on the driver (and the possibility for injury). The same principle applies to the padding used to package fragile objects for shipping.

We will often express the momentum of a particle in terms of its components. If the particle has velocity components v_x, v_y, and v_z, its momentum components p_x, p_y, and p_z are given by

$$p_x = mv_x, \qquad p_y = mv_y, \qquad p_z = mv_z. \qquad (8\text{–}4)$$

These three component equations are equivalent to Eq. (8–2).

A particle's momentum $\vec{p} = m\vec{v}$ and its kinetic energy $K = \frac{1}{2}mv^2$ both depend on the mass and velocity of the particle. What is the fundamental difference between these two quantities? A purely mathematical answer is that momentum is a vector whose magnitude is proportional to speed, while kinetic energy is a scalar proportional to the speed squared. But to see the *physical* difference between momentum and kinetic energy, we must first define a quantity closely related to momentum called *impulse*.

Let's first consider a particle acted on by a *constant* net force $\Sigma \vec{F}$ during a time interval Δt from t_1 to t_2. (We'll look at the case of varying forces shortly.) The **impulse** of the net force, denoted by $\vec{J}$, is defined to be the product of the net force and the time interval:

$$\vec{J} = \Sigma \vec{F}(t_2 - t_1) = \Sigma \vec{F} \Delta t \qquad \text{(assuming constant net force).} \qquad (8\text{–}5)$$

Impulse is a vector quantity; its direction is the same as the net force $\Sigma \vec{F}$. Its magnitude is the product of the magnitude of the net force and the length of time that the net force acts. The SI unit of impulse is the newton-second (N · s). Because 1 N = 1 kg · m/s^2, an alternative set of units for impulse is kg · m/s, the same as the units of momentum.

To see what impulse is good for, let's go back to Newton's second law as restated in terms of momentum, Eq. (8–3). If the net force $\Sigma \vec{F}$ is constant, then $d\vec{p}/dt$ is also constant. In that case, $d\vec{p}/dt$ is equal to the *total* change in momentum $\vec{p}_2 - \vec{p}_1$ during the time interval $t_2 - t_1$, divided by the interval:

$$\Sigma \vec{F} = \frac{\vec{p}_2 - \vec{p}_1}{t_2 - t_1}.$$

Multiplying this equation by $(t_2 - t_1)$, we have

$$\Sigma \vec{F}(t_2 - t_1) = \vec{p}_2 - \vec{p}_1.$$

Comparing to Eq. (8–5), we end up with a result called the **impulse-momentum theorem:**

$$\vec{J} = \vec{p}_2 - \vec{p}_1 \qquad \text{(impulse-momentum theorem).} \qquad (8\text{–}6)$$

The change in momentum of a particle during a time interval equals the impulse of the net force that acts on the particle during that interval.

The impulse-momentum theorem also holds when forces are not constant. To see this, we integrate both sides of Newton's second law $\Sigma \vec{F} = d\vec{p}/dt$ over time between the limits t_1 and t_2:

$$\int_{t_1}^{t_2} \Sigma \vec{F} \, dt = \int_{t_1}^{t_2} \frac{d\vec{p}}{dt} \, dt = \int_{\vec{p}_1}^{\vec{p}_2} d\vec{p} = \vec{p}_2 - \vec{p}_1.$$

The integral on the left is defined to be the impulse $\vec{J}$ of the net force $\Sigma \vec{F}$ during this interval:

$$\vec{J} = \int_{t_1}^{t_2} \Sigma \vec{F} \, dt \qquad \text{(general definition of impulse).} \qquad (8\text{–}7)$$

With this definition the impulse-momentum theorem $\vec{J} = \vec{p}_2 - \vec{p}_1$, Eq. (8–6), is valid even when the net force $\Sigma \vec{F}$ varies with time.

We can define an *average* net force $\vec{F}_{av}$ such that even when $\Sigma \vec{F}$ is not constant, the impulse $\vec{J}$ is given by

$$\vec{J} = \vec{F}_{av}(t_2 - t_1). \qquad (8\text{–}8)$$

When $\Sigma \vec{F}$ is constant, $\Sigma \vec{F} = \vec{F}_{av}$ and Eq. (8–8) reduces to Eq. (8–5).

Figure 8–1 shows a graph of the x-component of net force ΣF_x as a function of time during a collision. This might represent the force on a soccer ball that is in contact with a player's foot from time t_1 to t_2. The x-component of impulse during this interval is represented by the blue area under the curve between t_1 and t_2. This area is equal to the rectangular area bounded by t_1, t_2, and $(F_{av})_x$, so $(F_{av})_x(t_2 - t_1)$ is equal to the impulse of the actual time-varying force during the same interval.

Impulse and momentum are both vector quantities, and Eqs. (8–5) through (8–8) are all vector equations. In specific problems it is often easiest to use them in component form:

$$J_x = \int_{t_1}^{t_2} \Sigma F_x \, dt = (F_{av})_x(t_2 - t_1) = p_{2x} - p_{1x} = mv_{2x} - mv_{1x},$$

$$(8\text{–}9)$$

$$J_y = \int_{t_1}^{t_2} \Sigma F_y \, dt = (F_{av})_y(t_2 - t_1) = p_{2y} - p_{1y} = mv_{2y} - mv_{1y},$$

and similarly for the z-component.

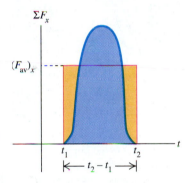

8–1 The x-component of the impulse of ΣF_x between t_1 and t_2 equals the blue area under the ΣF_x–t curve, which also equals the area under the rectangle with height $(F_{av})_x$.

MOMENTUM AND KINETIC ENERGY COMPARED

We can now see the fundamental difference between momentum and kinetic energy. The impulse-momentum theorem $\vec{J} = \vec{p}_2 - \vec{p}_1$ says that changes in a particle's momentum are due to impulse, which depends on the *time* over which the net force acts. By contrast, the work-energy theorem $W_{tot} = K_2 - K_1$ tells us that kinetic energy changes when work is done on a particle; the total work depends on the *distance* over which the net force

6.1
Momentum and Energy Change

acts. Consider a particle that starts from rest at t_1 so that $\vec{v}_1 = 0$. Its initial momentum is $\vec{p}_1 = m\vec{v}_1 = 0$, and its initial kinetic energy is $K_1 = \frac{1}{2}mv_1{}^2 = 0$. Now let a constant net force equal to $\vec{F}$ act on that particle from time t_1 until time t_2. During this interval the particle moves a distance s in the direction of the force. From Eq. (8–6), the particle's momentum at time t_2 is

$$\vec{p}_2 = \vec{p}_1 + \vec{J} = \vec{J},$$

where $\vec{J} = \vec{F}(t_2 - t_1)$ is the impulse that acts on the particle. So *the momentum of a particle equals the impulse that accelerated it from rest to its present speed;* impulse is the product of the net force that accelerated the particle and the *time* required for the acceleration. By comparison the kinetic energy of the particle at t_2 is $K_2 = W_{\text{tot}} = Fs$, the total *work* done on the particle to accelerate it from rest. The total work is the product of the net force and the *distance* required to accelerate the particle.

Here's an application of the distinction between momentum and kinetic energy. Suppose you have a choice between catching a 0.50-kg ball moving at 4.0 m/s or a 0.10-kg ball moving at 20 m/s. Which will be easier to catch? Both have the same magnitude of momentum, $p = mv = (0.50\text{ kg})(4.0\text{ m/s}) = (0.10\text{ kg})(20\text{ m/s}) = 2.0\text{ kg}\cdot\text{m/s}$, but the two balls have very different values of kinetic energy $K = \frac{1}{2}mv^2$; the large, slow-moving ball has $K = 4.0$ J, while the small, fast-moving ball has $K = 20$ J. Since the momentum is the same for both balls, both require the same *impulse* in order to be brought to rest. But stopping the 0.10-kg ball with your hand requires five times more *work* than stopping the 0.50-kg ball, because the smaller ball has five times as much kinetic energy. Hence for a given force that you exert with your hand, it takes the same amount of time (the duration of the catch) to stop either ball, but your hand and arm will be pushed back five times farther if you choose to catch the small, fast-moving ball. To minimize arm strain, you should choose to catch the 0.50-kg ball with its lower kinetic energy. (Kinetic energy may not be the only factor to consider, however. A pitched baseball and a .22 bullet fired from a rifle have roughly the same kinetic energy, and the bullet has less momentum than does the baseball. Nonetheless, you'd probably prefer to catch the baseball rather than the bullet. Why?)

Both the impulse-momentum and work-energy theorems are relationships between force and motion, and both rest on the foundation of Newton's laws. They are *integral* principles, relating the motion at two different times separated by a finite interval. By contrast, Newton's second law itself (in either of the forms $\Sigma \vec{F} = m\vec{a}$ or $\Sigma \vec{F} = d\vec{p}/dt$) is a *differential* principle, relating the forces to the rate of change of velocity or momentum at each instant.

The momentum of a pitched baseball is equal to the impulse imparted to it by the pitcher, while the kinetic energy of the ball is equal to the work the pitcher does on it. Hence the momentum is related to the *time* it took to bring the ball up to speed, while the kinetic energy depends on the *distance* the ball moved while being accelerated.

EXAMPLE 8–1

Momentum vs. kinetic energy Consider again the race described in Example 6–6 (Section 6–3) between two iceboats on a frictionless frozen lake. The iceboats have masses m and $2m$ respectively, and the wind exerts the same constant horizontal force $\vec{F}$ on each iceboat (Fig. 6–9). The two iceboats start from rest and cross the finish line a distance s away. Which iceboat crosses the finish line with greater momentum?

SOLUTION In Example 6–6 we asked how the kinetic energies of the iceboats compare when they cross the finish line. The way to determine this was not by using the formula $K = \frac{1}{2}mv^2$, but rather by remembering that a body's kinetic energy is equal to the total work done to accelerate it from rest. Both iceboats started from rest, and the total work done between the starting and finish lines was the same for both iceboats (because the net

force and displacement were the same for both). Hence both iceboats cross the finish line with the same kinetic energy.

Similarly, the best way to compare the momenta of the iceboats is *not* to use the formula for momentum, $\vec{p} = m\vec{v}$. By itself this formula isn't enough to decide which iceboat has greater momentum at the finish line. The iceboat of mass $2m$ has greater mass, which suggests greater momentum; but this iceboat crosses the finish line going slower than the other one, which suggests less momentum.

Instead, we use the idea that the momentum of each iceboat equals the impulse that accelerated it from rest. For each iceboat the downward force of gravity and the upward normal force add to zero, so the net force equals the constant horizontal wind force $\vec{F}$. Let Δt be the time an iceboat takes to reach the finish line, so that the impulse on the iceboat during that time is

$\vec{J} = \vec{F}\,\Delta t$. Since the iceboat starts from rest, this just equals the iceboat's momentum $\vec{p}$ at the finish line:

$$\vec{p} = \vec{F}\,\Delta t.$$

Both iceboats are subjected to the same force $\vec{F}$, but they do *not* take the same amount of time Δt to reach the finish line. The more massive iceboat of mass $2m$ accelerates more slowly and takes a longer time to travel the distance s; thus there is a greater impulse on this iceboat between the starting and finish lines. So the iceboat of mass $2m$ crosses the finish line with a greater magnitude of momentum than the iceboat of mass m (but with the same kinetic energy). Can you show that the iceboat of mass $2m$ has $\sqrt{2}$ times as much momentum at the finish line as the iceboat of mass m?

EXAMPLE 8–2

A ball hits a wall Suppose you throw a ball with a mass of 0.40 kg against a brick wall. It hits the wall moving horizontally to the left at 30 m/s and rebounds horizontally to the right at 20 m/s. a) Find the impulse of the net force on the ball during its collision with the wall. b) If the ball is in contact with the wall for 0.010 s, find the average horizontal force that the wall exerts on the ball during the impact.

8–2 A ball moving to the left, hitting a wall, and rebounding to the right.

SOLUTION a) Take the x-axis as horizontal and the positive direction to the right (Fig. 8–2). Then the initial x-component of momentum of the ball is

$$p_1 = mv_1 = (0.40\ \text{kg})(-30\ \text{m/s}) = -12\ \text{kg} \cdot \text{m/s}.$$

The final x-component of momentum is

$$p_2 = mv_2 = (0.40\ \text{kg})(+20\ \text{m/s}) = +8.0\ \text{kg} \cdot \text{m/s}.$$

The *change* in the x-component of momentum is

$$p_2 - p_1 = mv_2 - mv_1 = 8.0\ \text{kg} \cdot \text{m/s} - (-12\ \text{kg} \cdot \text{m/s})$$

$$= 20\ \text{kg} \cdot \text{m/s}.$$

According to Eqs. (8–9), this equals the x-component of impulse of the net force on the ball, so $J_x = 20\ \text{kg} \cdot \text{m/s} = 20\ \text{N} \cdot \text{s}$.

The time variation of the net horizontal force may be similar to one of the curves in Fig. 8–3. The net horizontal force is zero before impact, rises to a maximum, and then decreases to zero when the ball loses contact with the wall. If the ball is relatively rigid, like a baseball or a golf ball, the collision lasts a short time and the maximum force is large, as in curve (a). If the ball is softer, like a tennis ball, the collision time is longer and the maximum force is less, as in curve (b). In any case the *area* under the curve represents the impulse.

b) If the collision time is $\Delta t = 0.010$ s, then from Eq. (8–9),

$$J_x = (F_{av})_x\,\Delta t, \qquad (F_{av})_x = \frac{J_x}{\Delta t} = \frac{20\ \text{N} \cdot \text{s}}{0.010\ \text{s}} = 2000\ \text{N}.$$

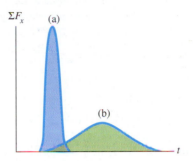

8–3 (a) A "hard" collision, such as that of a golf ball. (b) A "soft" collision, such as that of a tennis ball. The impulse is the same for both collisions, but the maximum force is less for the longer-lasting collision.

This average force is represented by the horizontal line $(F_{av})_x$ in Fig. 8–1. The horizontal force is exerted on the ball by the wall itself. The magnitude of this force has to be very large to cause the ball to change momentum in such a short time. Other forces that act on the ball during the collision are very weak by comparison; for instance, the gravitational force is only 3.9 N. Thus during the brief time that the collision lasts, we can ignore all other forces on the ball to a very good approximation. Figure 8–4 is a photograph showing the impact of a tennis ball and racket.

8–4 Typically, a tennis ball is in contact with the racket for approximately 0.01 s. The ball flattens noticeably due to the tremendous force exerted by the racket.

EXAMPLE 8–3

Kicking a soccer ball   A soccer ball has a mass of 0.40 kg. Initially, it is moving to the left at 20 m/s, but then it is kicked and given a velocity at 45° upward and to the right, with a magnitude of 30 m/s (Fig. 8–5a). Find the impulse of the net force and the average net force, assuming a collision time $\Delta t = 0.010$ s.

SOLUTION   The initial and final velocities are not along the same line, and we have to be careful to treat momentum and impulse as vector quantities, using their x- and y-components. Taking the x-axis horizontally to the right and the y-axis vertically upward, we find the following velocity components:

$$v_{1x} = -20 \text{ m/s}, \qquad v_{1y} = 0,$$

$$v_{2x} = v_{2y} = (0.707)(30 \text{ m/s}) = 21.2 \text{ m/s}$$

$$\text{(since } \cos 45° = \sin 45° = 0.707\text{).}$$

The x-component of impulse is equal to the x-component of momentum change, and the same is true for the y-components:

$$J_x = p_{2x} - p_{1x} = m(v_{2x} - v_{1x})$$

$$= (0.40 \text{ kg})[21.2 \text{ m/s} - (-20 \text{ m/s})] = 16.5 \text{ kg} \cdot \text{m/s},$$

$$J_y = p_{2y} - p_{1y} = m(v_{2y} - v_{1y})$$

$$= (0.40 \text{ kg})(21.2 \text{ m/s} - 0) = 8.5 \text{ kg} \cdot \text{m/s}.$$

The components of the average net force on the ball are

$$(F_{av})_x = \frac{J_x}{\Delta t} = 1650 \text{ N}, \qquad (F_{av})_y = \frac{J_y}{\Delta t} = 850 \text{ N}.$$

The magnitude and direction of the average force are

$$F_{av} = \sqrt{(1650 \text{ N})^2 + (850 \text{ N})^2} = 1.9 \times 10^3 \text{ N},$$

$$\theta = \arctan \frac{850 \text{ N}}{1650 \text{ N}} = 27°,$$

where θ is measured upward from the $+x$-axis (Fig. 8–5b). Note that because the ball was not initially at rest, the ball's final velocity does *not* have the same direction as the average force acting on it.

The average net force $\vec{F}_{av}$ includes the effects of the force of gravity, though these are small; the weight of the ball is only 3.9 N. As in Example 8–2, the average force acting during the collision is exerted almost entirely by the object that the ball hit (in this case the soccer player's foot).

8–5 (a) A soccer ball (1) before and (2) after being kicked. (b) Finding the average force from its components.

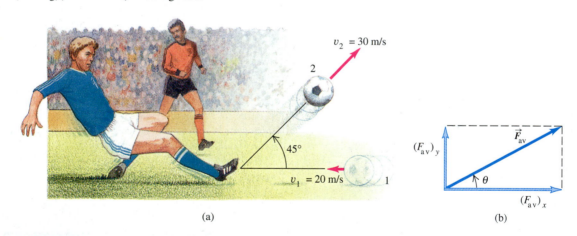

(a)   (b)

8-3 CONSERVATION OF MOMENTUM

The concept of momentum is particularly important in situations in which we have two or more interacting bodies. Let's consider first an idealized system consisting of two bodies that interact with each other but not with anything else—for example, two astronauts who touch each other as they float freely in the zero-gravity environment of outer space (Fig. 8–6a). Think of the astronauts as particles. Each particle exerts a force on the other; according to Newton's third law, the two forces are always equal in magnitude and opposite in direction. Hence the *impulses* that act on the two particles will be equal and opposite, and the changes in momentum of the two particles will be equal and opposite.

Let's go over that again with some new terminology. For any system, the forces that the particles of the system exert on each other are called **internal forces.** Forces exerted

(a)

(b)   (c)

8–6 (a) Two astronauts push each other as they float freely in the zero-gravity environment of space. (b) Free-body diagram for astronaut A. (c) Free-body diagram for astronaut B. The only forces that act are internal forces; they don't change the total momentum of the two astronauts.

on any part of the system by some object outside it are called **external forces.** For the system we have described, the internal forces are $\vec{F}_{B \text{ on } A}$, exerted by particle B on particle A, and $\vec{F}_{A \text{ on } B}$, exerted by particle A on particle B (Figs. 8–6b, c). There are *no* external forces; when this is the case, we have an **isolated system.**

The net force on particle A is $\vec{F}_{B \text{ on } A}$, and the net force on particle B is $\vec{F}_{A \text{ on } B}$, so from Eq. (8–3) the rates of change of the momenta of the two particles are

$$\vec{F}_{B \text{ on } A} = \frac{d\vec{p}_A}{dt}, \qquad \vec{F}_{A \text{ on } B} = \frac{d\vec{p}_B}{dt}. \qquad (8\text{–}10)$$

The momentum of each particle changes, but these changes are not independent; according to Newton's third law, the two forces $\vec{F}_{B \text{ on } A}$ and $\vec{F}_{A \text{ on } B}$ are always equal in magnitude and opposite in direction. That is, $\vec{F}_{B \text{ on } A} = -\vec{F}_{A \text{ on } B}$, so $\vec{F}_{B \text{ on } A} + \vec{F}_{A \text{ on } B} = 0$. Adding together the two equations in Eq. (8–10), we have

$$\vec{F}_{B \text{ on } A} + \vec{F}_{A \text{ on } B} = \frac{d\vec{p}_A}{dt} + \frac{d\vec{p}_B}{dt} = \frac{d(\vec{p}_A + \vec{p}_B)}{dt} = 0. \qquad (8\text{–}11)$$

The rates of change of the two momenta are equal and opposite, so that the rate of change of the vector sum $\vec{p}_A + \vec{p}_B$ is zero. We now define the **total momentum** $\vec{P}$ of the system of two particles as the vector sum of the momenta of the individual particles. That is,

$$\vec{P} = \vec{p}_A + \vec{p}_B. \qquad (8\text{–}12)$$

Then Eq. (8–11) becomes, finally,

$$\vec{F}_{B \text{ on } A} + \vec{F}_{A \text{ on } B} = \frac{d\vec{P}}{dt} = 0. \qquad (8\text{–}13)$$

(a)

(b)   (c)

8–7 (a) Two ice skaters push each other as they skate on a friction-less, horizontal surface. (b) Free-body diagram for skater A. (c) Free-body diagram for skater B. The normal forces and gravitational forces are external forces, but the vector sum of the external forces is zero and total momentum is conserved. (Compare to Fig. 8–6.)

6.3
Momentum Conservation and Collisions

6.7
Explosion Problems

6.8
Block Sliding Down an Incline

6.11
Pendulum Person-Projectile Bowling

The time rate of change of the *total* momentum $\vec{P}$ is zero. Hence the total momentum of the system is constant, even though the individual momenta of the particles that make up the system can change.

If external forces are also present, they must be included on the left side of Eq. (8–13) along with the internal forces. Then the total momentum is, in general, not constant. But if the vector sum of the external forces is zero, as in Fig. 8–7, these forces don't contribute to the sum, and $d\vec{P}/dt$ is again zero. Thus we have the following general result:

If the vector sum of the external forces on a system is zero, the total momentum of the system is constant.

This is the simplest form of the **principle of conservation of momentum.** This principle is a direct consequence of Newton's third law. What makes this principle useful is that it doesn't depend on the detailed nature of the internal forces that act between members of the system. This means that we can apply conservation of momentum even if (as is often the case) we know very little about the internal forces. We have used Newton's second law to derive this principle, so we have to be careful to use it only in inertial frames of reference.

We can generalize this principle for a system containing any number of particles A, B, C, … interacting only with each other. The total momentum of such a system is

$$\vec{P} = \vec{p}_A + \vec{p}_B + \cdots = m_A\vec{v}_A + m_B\vec{v}_B + \cdots \qquad \text{(total momentum of a system of particles).} \qquad (8\text{–}14)$$

We make the same argument as before; the total rate of change of momentum of the system due to each action-reaction pair of internal forces is zero. Thus the total rate of change of momentum of the entire system is zero whenever the vector sum of the external forces acting on it is zero. The internal forces can change the momenta of individual particles in the system but not the *total* momentum of the system.

CAUTION ▶ When you apply the conservation of momentum to a system, it's absolutely essential to remember that momentum is a *vector* quantity. Hence you *must* use vector addition to compute the total momentum of a system. Using components is usually the simplest method. If p_{Ax}, p_{Ay}, and p_{Az} are the components of momentum of particle A, and similarly for the other particles, then Eq. (8–14) is equivalent to the component equations

$$P_x = p_{Ax} + p_{Bx} + \cdots,$$
$$P_y = p_{Ay} + p_{By} + \cdots, \qquad (8\text{–}15)$$
$$P_z = p_{Az} + p_{Bz} + \cdots.$$

If the vector sum of the external forces on the system is zero, then P_x, P_y, and P_z are all constant. ◀

In some ways the principle of conservation of momentum is more general than the principle of conservation of mechanical energy. For example, mechanical energy is conserved only when the internal forces are *conservative*—that is, when the forces allow two-way conversion between kinetic and potential energy—but conservation of momentum is valid even when the internal forces are *not* conservative. In this chapter we will analyze situations in which both momentum and mechanical energy are conserved, and others in which only momentum is conserved. These two principles play a fundamental role in all areas of physics, and we will encounter them throughout our study of physics.

Problem–Solving Strategy

CONSERVATION OF MOMENTUM

1. Before applying conservation of momentum to a problem, you must first decide whether or not momentum *is* conserved! This will be true *only* if the vector sum of the external forces acting on the system of particles is zero. If this is not the case, you can't use conservation of momentum.

2. Define a coordinate system. Make a sketch showing the coordinate axes, including the positive direction for each. Often it is easiest to choose the x-axis to have the direction of one of the initial velocities. Make sure you are using an inertial frame of reference. Most of the problems in this chapter deal with two-dimensional situations, in which the vectors have only x- and y-components; all the following statements can be generalized to include z-components when necessary.

3. Treat each body as a particle. Draw "before" and "after" sketches, and include vectors on each to represent all known velocities. Label the vectors with magnitudes, angles, components, or whatever information is given, and give each unknown magnitude, angle, or component an algebraic symbol. You may find it helpful to use the subscripts 1 and 2 for velocities before and after the inter-

action, respectively; if you use these subscripts, use letters (not numbers) to label each particle.

4. Write an equation in terms of symbols equating the total *initial* x-component of momentum (that is, before the interaction) to the total *final* x-component of momentum (that is, after the interaction), using $p_x = mv_x$ for each particle. Write another equation for the y-components, using $p_y = mv_y$ for each particle. Remember that the x- and y-components of velocity or momentum are *never* added together in the same equation! Even when all the velocities lie along a line (such as the x-axis), the components of velocity along this line can be positive or negative; be careful with signs!

5. Solve these equations to determine whatever results are required. In some problems you will have to convert from the x- and y-components of a velocity to its magnitude and direction, or the reverse.

6. In some problems, energy considerations give additional relationships among the various velocities, as we will see later in this chapter.

EXAMPLE 8-4

Recoil of a rifle A marksman holds a rifle of mass $m_R = 3.00$ kg loosely in his hands, so as to let it recoil freely when fired. He fires a bullet of mass $m_B = 5.00$ g horizontally with a velocity relative to the ground of $v_B = 300$ m/s (Fig. 8–8). What is the recoil velocity v_R of the rifle? What are the final momentum and kinetic energy of the bullet? Of the rifle?

SOLUTION We consider an idealized model in which the horizontal forces the marksman exerts on the rifle are negligible. Then there is no net horizontal force on the system (the bullet and rifle) during the firing of the rifle, and so the total horizontal momentum of the system is the same before and after the rifle is fired (i.e., is conserved).

Take the positive x-axis to be the direction the rifle is aimed. Initially, both the rifle and bullet are at rest, so the initial x-component of total momentum is zero. After the bullet is fired, its x-component of momentum is $m_B v_B$, and that of the rifle is $m_R v_R$. Conservation of the x-component of total momentum gives

$$P_x = 0 = m_B v_B + m_R v_R,$$

$$v_R = -\frac{m_B}{m_R} v_B = -\left(\frac{0.00500 \text{ kg}}{3.00 \text{ kg}}\right)(300 \text{ m/s}) = -0.500 \text{ m/s}.$$

The negative sign means that the recoil is in the direction oppo-

v_R $v_B = 300$ m/s $+x$ $m_B = 5.00$ g $m_R = 3.00$ kg

8–8 The ratio of bullet speed to recoil speed is the inverse of the ratio of bullet mass to rifle mass.

site to that of the bullet. If the butt of a rifle were to hit your shoulder traveling at this speed, you'd feel it. That's what the "kick" of a rifle is all about. It's more comfortable to hold the rifle tightly against your shoulder when you fire it; then m_R is replaced by the sum of your mass and the rifle's mass, and the recoil speed is much less.

The momentum and kinetic energy of the bullet are

$$p_B = m_B v_B = (0.00500 \text{ kg})(300 \text{ m/s}) = 1.50 \text{ kg} \cdot \text{m/s},$$

$$K_B = \frac{1}{2} m_B v_B{}^2 = \frac{1}{2}(0.00500 \text{ kg})(300 \text{ m/s})^2 = 225 \text{ J};$$

for the rifle, the momentum and the kinetic energy are

$$p_R = m_R v_R = (3.00 \text{ kg})(-0.500 \text{ m/s}) = -1.50 \text{ kg} \cdot \text{m/s},$$

$$K_R = \frac{1}{2} m_R v_R{}^2 = \frac{1}{2}(3.00 \text{ kg})(-0.500 \text{ m/s})^2 = 0.375 \text{ J}.$$

The bullet and the rifle have equal and opposite momenta after the interaction because they were subjected to equal and opposite interaction forces for the same amount of *time* (that is, equal and opposite impulses). But the bullet acquires much greater kinetic energy than the rifle because the bullet travels a much greater *distance* than the rifle during the interaction. Thus the force on the bullet does more work than the force on the rifle. The ratio of the two kinetic energies, 600:1, is equal to the inverse ratio of the masses; in fact, it can be shown that this always happens in recoil situations. We leave the proof as a problem (see Exercise 8–21).

Notice that our calculation doesn't depend at all on the details of how the rifle works. In a real rifle, the bullet is propelled forward by an explosive charge; if instead a very stiff spring were used to give the same bullet velocity, the answers would have been exactly the same.

EXAMPLE 8-5

Collision along a straight line Two gliders move toward each other on a frictionless linear air track (Fig. 8–9a). After they collide, glider B moves away with a final velocity of $+2.0$ m/s (Fig. 8–9b). What is the final velocity of glider A? How do the changes in momentum and in velocity compare for the two gliders?

SOLUTION The total vertical force on each glider is zero; the net force on each glider is the horizontal force exerted on it by the other glider. The net *external* force on the two gliders together is zero, so the total momentum is constant. Take the x-axis to lie along the air track, with the positive direction to the right. The masses of the gliders and their initial x-components of velocity are shown in the figure. The x-component of total momentum before the collision is

$$P_x = m_A v_{A1} + m_B v_{B1}$$

$$= (0.50 \text{ kg})(2.0 \text{ m/s}) + (0.30 \text{ kg})(-2.0 \text{ m/s}) = 0.40 \text{ kg} \cdot \text{m/s}.$$

8–9 Two gliders (a) before and (b) after their collision.

O

$v_{A1} = 2.0$ m/s $v_{B1} = -2.0$ m/s

x

A B

$m_A = 0.50$ kg $m_B = 0.30$ kg

(a)

O

v_{A2} $v_{B2} = 2.0$ m/s

x

A B

(b)

This is positive (to the right in Fig. 8–9) because glider A has a greater magnitude of momentum before the collision than does glider B. The x-component of total momentum has the same value after the collision, so

$$P_x = m_A v_{A2} + m_B v_{B2}.$$

Solving this equation for v_{A2}, the final velocity of A, we find

$$v_{A2} = \frac{P_x - m_B v_{B2}}{m_A} = \frac{0.40 \text{ kg} \cdot \text{m/s} - (0.30 \text{ kg})(2.0 \text{ m/s})}{0.50 \text{ kg}}$$

$$= -0.40 \text{ m/s}.$$

The change in momentum of glider A is

$$m_A v_{A2} - m_A v_{A1} = (0.50 \text{ kg})(-0.40 \text{ m/s}) - (0.50 \text{ kg})(2.0 \text{ m/s})$$

$$= -1.2 \text{ kg} \cdot \text{m/s},$$

and the change in momentum of glider B is

$$m_B v_{B2} - m_B v_{B1} = (0.30 \text{ kg})(2.0 \text{ m/s}) - (0.30 \text{ kg})(-2.0 \text{ m/s})$$

$$= +1.2 \text{ kg} \cdot \text{m/s}.$$

The two interacting gliders undergo changes in momentum that are equal in magnitude and opposite in direction. The same is not true of their changes in *velocity*, however. For A, $v_{A2} - v_{A1}$ $= (-0.40 \text{ m/s}) - 2.0 \text{ m/s} = -2.4 \text{ m/s}$; for B, $v_{B2} - v_{B1} = 2.0 \text{ m/s} -$ $(-2.0 \text{ m/s}) = +4.0 \text{ m/s}$. The magnitude of the interaction force is the same for both bodies, but from Newton's second law the magnitude of acceleration (and hence of velocity change) is greater for the less massive body (glider B). In a collision between an eighteen-wheeler truck and a compact car, the car's occupants are subjected to greater acceleration (and greater chance of injury) than the truck driver is. If the truck collides with an insect, the truck driver won't notice the resulting acceleration at all, but the insect surely will!

EXAMPLE 8–6

Collision in a horizontal plane Figure 8–10a shows two chunks of ice sliding on a frictionless frozen pond. Chunk A, with mass $m_A = 5.0$ kg, moves with initial velocity $v_{A1} = 2.0$ m/s parallel to the x-axis. It collides with chunk B, which has mass $m_B = 3.0$ kg and is initially at rest. After the collision, the velocity of chunk A is found to be $v_{A2} = 1.0$ m/s in a direction making an angle $\alpha = 30°$ with the initial direction. What is the final velocity of chunk B?

SOLUTION There are no horizontal (x or y) external forces, so the total horizontal momentum of the system is the same before and after the collision. The velocities are not all along a single line, so we have to treat momentum as a *vector* quantity, using components of each momentum in the x- and y-directions. Then momentum conservation requires that the sum of the x-components *before* the collision (subscript 1) must equal the sum *after* the collision (subscript 2), and similarly for the y-components. Just as with force equilibrium problems, we write a separate equation for each component. For the x-components (before and after) we have

$$m_A v_{A1x} + m_B v_{B1x} = m_A v_{A2x} + m_B v_{B2x},$$

$$v_{B2x} = \frac{m_A v_{A1x} + m_B v_{B1x} - m_A v_{A2x}}{m_B}$$

$$= \frac{\begin{bmatrix} (5.0 \text{ kg})(2.0 \text{ m/s}) + (3.0 \text{ kg})(0) \\ -(5.0 \text{ kg})(1.0 \text{ m/s})(\cos 30°) \end{bmatrix}}{3.0 \text{ kg}}$$

$$= 1.89 \text{ m/s}.$$

Conservation of the y-component of total momentum gives

$$m_A v_{A1y} + m_B v_{B1y} = m_A v_{A2y} + m_B v_{B2y},$$

$$v_{B2y} = \frac{m_A v_{A1y} + m_B v_{B1y} - m_A v_{A2y}}{m_B}$$

$$= \frac{\begin{bmatrix} (5.0 \text{ kg})(0) + (3.0 \text{ kg})(0) \\ -(5.0 \text{ kg})(1.0 \text{ m/s})(\sin 30°) \end{bmatrix}}{3.0 \text{ kg}}$$

$$= -0.83 \text{ m/s}.$$

After the collision, chunk B moves in the positive x-direction and the negative y-direction (Fig. 8–10b). The magnitude of v_{B2} is

$$v_{B2} = \sqrt{(1.89 \text{ m/s})^2 + (-0.83 \text{ m/s})^2} = 2.1 \text{ m/s},$$

and the angle of its direction from the positive x-axis is

$$\beta = \arctan \frac{-0.83 \text{ m/s}}{1.89 \text{ m/s}} = -24°.$$

(a)

(b)

8–10 Views from above of the velocities (a) before and (b) after the collision.

8–11 Model of an elastic collision: Glider A and glider B approach each other on a frictionless surface. Each glider has a steel spring bumper on the end to ensure an elastic collision.

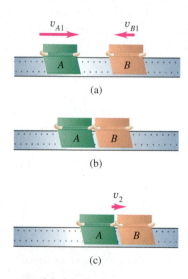

8–12 Model of a completely inelastic collision. The spring bumpers on the gliders are replaced by putty, so the gliders stick together after collision.

6.4.1
Collision Problems
6.9
Skier and Cart

8-4 INELASTIC COLLISIONS

To the average person on the street the term *collision* is likely to mean some sort of automotive disaster. We'll use it in that sense, but we'll also broaden the meaning to include any strong interaction between bodies that lasts a relatively short time. So we include not only car accidents but also balls colliding on a billiard table, neutrons hitting atomic nuclei in a nuclear reactor, a bowling ball striking the pins, the impact of a meteor on the Arizona desert, and a close encounter of a spacecraft with the planet Saturn.

If the forces between the bodies are much larger than any external forces, as is the case in most collisions, we can neglect the external forces entirely and treat the bodies as an *isolated* system. Then momentum is conserved in the collision, and the total momentum of the system has the same value before and after the collision. Two cars colliding at an icy intersection provide a good example. Even two cars colliding on dry pavement can be treated as an isolated system during the collision if the forces between the cars are much larger than the friction forces of pavement against tires.

If the forces between the bodies are also *conservative,* so that no mechanical energy is lost or gained in the collision, the total *kinetic energy* of the system is the same after the collision as before. Such a collision is called an **elastic collision.** A collision between two marbles or two billiard balls is almost completely elastic. Figure 8–11 shows a model for an elastic collision. When the bodies collide, the springs are momentarily compressed and some of the original kinetic energy is momentarily converted to elastic potential energy. Then the bodies bounce apart, the springs expand, and this potential energy is converted back to kinetic energy.

A collision in which the total kinetic energy after the collision is *less* than that before the collision is called an **inelastic collision.** A meatball landing on a plate of spaghetti and a bullet embedding itself in a block of wood are examples of inelastic collisions. An inelastic collision in which the colliding bodies stick together and move as one body after the collision is often called a **completely inelastic collision.** An example is shown in Fig. 8–12; we have replaced the spring bumper in Fig. 8–11 with a ball of putty that squashes and sticks the two bodies together.

CAUTION ▶ It's a common misconception that the *only* inelastic collisions are those in which the colliding bodies stick together. In fact, inelastic collisions include many situations in which the bodies do *not* stick. If two cars bounce off each other in a "fender-bender," the work done to deform the fenders cannot be recovered as kinetic energy of the cars, and so the collision is inelastic. ◀

Remember the following rule: **In any collision in which external forces can be neglected, momentum is conserved and the total momentum before equals the total momentum after; in elastic collisions *only,* the total kinetic energy before equals the total kinetic energy after.**

COMPLETELY INELASTIC COLLISIONS

Let's look at what happens to momentum and kinetic energy in a *completely* inelastic collision of two bodies (A and B), as in Fig. 8–12. Because the two bodies stick together after the collision, their final velocities must be equal:

$$\vec{v}_{A2} = \vec{v}_{B2} = \vec{v}_2.$$

Conservation of momentum gives the relation

$$m_A \vec{v}_{A1} + m_B \vec{v}_{B1} = (m_A + m_B)\vec{v}_2 \qquad \text{(completely inelastic collision).} \quad (8\text{-}16)$$

If we know the masses and initial velocities, we can compute the common final velocity $\vec{v}_2$.

Suppose, for example, that a body with mass m_A and initial x-component of velocity v_1 collides inelastically with a body with mass m_B that is initially at rest ($v_{B1} = 0$). From Eq. (8–16) the common x-component of velocity v_2 of both bodies after the collision is

$$v_2 = \frac{m_A}{m_A + m_B} v_1 \quad \text{(completely inelastic collision, } B \text{ initially at rest).} \quad (8\text{–}17)$$

Let's verify that the total kinetic energy after this completely inelastic collision is less than before the collision. The kinetic energies K_1 and K_2 before and after the collision, respectively, are

$$K_1 = \frac{1}{2} m_A v_1^{\,2},$$

$$K_2 = \frac{1}{2}(m_A + m_B)v_2^{\,2} = \frac{1}{2}(m_A + m_B)\left(\frac{m_A}{m_A + m_B}\right)^2 v_1^{\,2}.$$

The ratio of final to initial kinetic energy is

$$\frac{K_2}{K_1} = \frac{m_A}{m_A + m_B} \quad \text{(completely inelastic collision, } B \text{ initially at rest).} \quad (8\text{–}18)$$

The right side is always less than unity because the denominator is always greater than the numerator. Even when the initial velocity of m_B is not zero, it is not hard to verify that the kinetic energy after a completely inelastic collision is always less than before.

Please note: We don't recommend memorizing Eqs. (8–17) or (8–18). We derived them only to prove that kinetic energy is always lost in a completely inelastic collision.

Automobile collisions are intended to be inelastic, so that the structure of the car absorbs as much of the energy of the collision as possible. This absorbed energy cannot be recovered, since it goes into a permanent deformation of the car.

EXAMPLE 8–7

A completely inelastic collision Suppose that in the collision described in Example 8–5 (Section 8–3), the two gliders do not bounce but stick together after the collision. The masses and initial velocities are as shown in Fig. 8–9a. Find the common final velocity v_2, and compare the initial and final kinetic energies.

SOLUTION Take the x-axis to lie along the direction of motion, as before. From conservation of the x-component of momentum,

$$m_A v_{A1} + m_B v_{B1} = (m_A + m_B)v_2,$$

$$v_2 = \frac{m_A v_{A1} + m_B v_{B1}}{m_A + m_B}$$

$$= \frac{(0.50 \text{ kg})(2.0 \text{ m/s}) + (0.30 \text{ kg})(-2.0 \text{ m/s})}{(0.50 \text{ kg} + 0.30 \text{ kg})}$$

$$= 0.50 \text{ m/s}.$$

Because v_2 is positive, the gliders move together to the right (the +x-direction) after the collision.

Before the collision, the kinetic energies of gliders A and B are

$$K_A = \frac{1}{2} m_A v_{A1}^{\,2} = \frac{1}{2}(0.50 \text{ kg})(2.0 \text{ m/s})^2 = 1.0 \text{ J},$$

$$K_B = \frac{1}{2} m_B v_{B1}^{\,2} = \frac{1}{2}(0.30 \text{ kg})(-2.0 \text{ m/s})^2 = 0.60 \text{ J}.$$

(Note that the kinetic energy of glider B is positive, even though the x-components of its velocity v_{B1} and momentum mv_{B1} are both negative.) The *total* kinetic energy before the collision is 1.6 J. The kinetic energy after the collision is

$$\frac{1}{2}(m_A + m_B)v_2^{\,2} = \frac{1}{2}(0.50 \text{ kg} + 0.30 \text{ kg})(0.50 \text{ m/s})^2 = 0.10 \text{ J}.$$

The final kinetic energy is only $\frac{1}{16}$ of the original; $\frac{15}{16}$ is converted from mechanical energy to various other forms. If there is a ball of chewing gum between the gliders, it squashes and becomes warmer. If the gliders couple together like two freight cars, the energy goes into elastic vibrations that eventually dissipate. If there is a spring between the gliders that is compressed as they lock together, then the energy is stored as potential energy of the spring. In all of these cases the *total* energy of the system is conserved, although *kinetic* energy is not. However, in an isolated system, momentum is *always* conserved, whether the collision is elastic or not.

EXAMPLE 8-8

The ballistic pendulum Figure 8–13 shows a ballistic pendulum, a system for measuring the speed of a bullet. The bullet, with mass m, is fired into a block of wood with mass M, suspended like a pendulum, and makes a completely inelastic collision with it. After the impact of the bullet, the block swings up to a maximum height y. Given the values of y, m, and M, what is the initial speed v of the bullet?

8–13 A ballistic pendulum.

SOLUTION We analyze this event in two stages. The first is the embedding of the bullet in the block, and the second is the subsequent swinging of the block on its strings. As we will see, momentum is conserved in the first stage, mechanical energy in the second.

During the first stage, the bullet embeds itself in the block so quickly that the block has no time to swing appreciably away from its straight-down position. So during the impact the supporting strings remain very nearly vertical, there is negligible external horizontal force acting on the system, and the horizontal component of momentum is conserved. Let V be the x-component of velocity of bullet and block just after the collision. Momentum conservation gives

$$mv = (m + M)V, \qquad v = \frac{m + M}{m} V.$$

The kinetic energy of the system just after the collision is $K = \frac{1}{2}(m + M)V^2$. As in Eq. (8–18), this is less than the kinetic energy before the collision; the collision is inelastic. Momentum is conserved during the first stage, but mechanical energy is not.

In the second stage, after the collision, the block and bullet move as a unit. The only forces are the weight (a conservative force) and the string tensions (which do no work). So as the pendulum swings upward and to the right, mechanical energy is conserved. Momentum is *not* conserved during this stage because there is a net external force. The pendulum comes to rest for an instant at a height y, where its kinetic energy $\frac{1}{2}(m + M)V^2$ has all become potential energy $(m + M)gy$; then it swings back down. Energy conservation gives

$$\frac{1}{2}(m + M)V^2 = (m + M)gy,$$

$$V = \sqrt{2gy}.$$

Now we substitute this into the momentum equation to find v:

$$v = \frac{m + M}{m}\sqrt{2gy}.$$

By measuring m, M, and y, we can compute the original speed v of the bullet. For example, if $m = 5.00$ g $= 0.00500$ kg, $M = 2.00$ kg, and $y = 3.00$ cm $= 0.0300$ m,

$$v = \frac{2.00 \text{ kg} + 0.00500 \text{ kg}}{0.00500 \text{ kg}} \sqrt{2(9.80 \text{ m/s}^2)(0.0300 \text{ m})}$$

$$= 307 \text{ m/s}.$$

The velocity V of the block just after impact is

$$V = \sqrt{2gy} = \sqrt{2(9.80 \text{ m/s}^2)(0.0300 \text{ m})}$$

$$= 0.767 \text{ m/s}.$$

The kinetic energy of the bullet just before impact is $\frac{1}{2}(0.00500 \text{ kg})(307 \text{ m/s})^2 = 236$ J. Just after impact the kinetic energy of the bullet and block is $\frac{1}{2}(2.005 \text{ kg})(0.767 \text{ m/s})^2 = 0.589$ J. Nearly all the kinetic energy disappears as the wood splinters and the bullet and block become hotter.

EXAMPLE 8-9

Analysis of an automobile collision A small compact car with a mass of 1000 kg is traveling north on Morewood Avenue at 15 m/s. At the intersection of Morewood and Fifth Avenue it collides with an enormous luxury car with a mass of 2000 kg, traveling east on Fifth Avenue at 10 m/s (Fig. 8–14a). Fortunately, all occupants are wearing seat belts and there are no injuries, but the two cars are thoroughly tangled and move away from the impact point as one mass. a) Treating each car as a particle, find the total momentum just before the collision. b) The insurance adjustor has asked you to find the velocity of the wreckage just after impact. What do you tell her?

SOLUTION a) In the figure, we have drawn coordinate axes and have labeled the small car A and the large car B. From Eqs.

(8–15) the components of the total momentum $\vec{P}$ are

$$P_x = p_{Ax} + p_{Bx} = m_A v_{Ax} + m_B v_{Bx}$$

$$= (1000 \text{ kg})(0) + (2000 \text{ kg})(10 \text{ m/s})$$

$$= 2.0 \times 10^4 \text{ kg} \cdot \text{m/s},$$

$$P_y = p_{Ay} + p_{By} = m_A v_{Ay} + m_B v_{By}$$

$$= (1000 \text{ kg})(15 \text{ m/s}) + (2000 \text{ kg})(0)$$

$$= 1.5 \times 10^4 \text{ kg} \cdot \text{m/s}.$$

The magnitude of $\vec{P}$ is

$$P = \sqrt{(2.0 \times 10^4 \text{ kg} \cdot \text{m/s})^2 + (1.5 \times 10^4 \text{ kg} \cdot \text{m/s})^2}$$

$$= 2.5 \times 10^4 \text{ kg} \cdot \text{m/s},$$

and its direction is given by the angle θ shown in Fig. 8–14b,

(a)

(b)

8–14 (a) View from a traffic heli-copter overhead of the start of the collision at Morewood and Fifth. (b) Finding the total momentum just before the collision.

where

$$\tan \theta = \frac{P_y}{P_x} = \frac{1.5 \times 10^4 \text{ kg} \cdot \text{m/s}}{2.0 \times 10^4 \text{ kg} \cdot \text{m/s}} = 0.75, \qquad \theta = 37°.$$

b) To find out what happens *after* the collision, we'll assume that we can treat the cars as an isolated system during the collision. This may seem implausible, so let's look at some numbers. The mass of the large car is 2000 kg, its weight is about 20,000 N, and if the coefficient of friction is 0.5, the friction force when it slides across the pavement is about 10,000 N. Its kinetic energy just before the impact is $\frac{1}{2}(2000 \text{ kg})(10 \text{ m/s})^2 = 1.0 \times 10^5$ J. The car may crumple 0.2 m or so; to do -1.0×10^5 J of work on it to stop the car in 0.2 m would require a force of 5.0×10^5 N, which is 50 times greater than the friction force.

Compared to the large forces that act during such a collision, it's not unreasonable to neglect the friction forces on both cars and assume that the cars form an isolated system during the col-lision. In that case the total momentum just after the collision is

the same as just before, which we found to be 2.5×10^4 kg · m/s in a direction 37° north of due east. Assuming that no parts fall off, the total mass of wreckage is $M = 3000$ kg. Calling the final velocity $\vec{V}$, we have $\vec{P} = M\vec{V}$. The direction of the velocity is the same as that of the momentum, and its magnitude is

$$V = \frac{P}{M} = \frac{2.5 \times 10^4 \text{ kg} \cdot \text{m/s}}{3000 \text{ kg}} = 8.3 \text{ m/s}.$$

This is an inelastic collision, so we expect the total kinetic energy to be less after the collision than before. We invite you to carry out the calculations; you will find that the initial kinetic energy is 2.1×10^5 J and the final value is 1.0×10^5 J. Over half of the initial kinetic energy is converted to other forms.

CAUTION ▶ You may have been tempted to try to find the final velocity by taking the vector sum of the initial velocities. But there is *no* "law of conservation of velocities." The conserved quantity is the total momentum of the system, and the above analysis is the only correct way to do the problem. ◀

8–5 ELASTIC COLLISIONS

As discussed in Section 8–4, an *elastic collision* in an isolated system is a collision in which kinetic energy (as well as momentum) is conserved. Elastic collisions occur when the forces between the colliding bodies are *conservative*. When two steel balls collide, they squash a little near the surface of contact, but then they spring back. Some of the kinetic energy is stored temporarily as elastic potential energy, but at the end it is recon-verted to kinetic energy.

Let's look at an elastic collision between two bodies A and B. We start with a one-dimensional collision, in which all the velocities lie along the same line; we choose this line to be the x-axis. Each momentum and velocity then has only an x-component. We call the x-components of velocity before the collision v_{A1} and v_{B1}, and those after the col-lision v_{A2} and v_{B2}. From conservation of kinetic energy we have

$$\frac{1}{2} m_A v_{A1}^2 + \frac{1}{2} m_B v_{B1}^2 = \frac{1}{2} m_A v_{A2}^2 + \frac{1}{2} m_B v_{B2}^2,$$

and conservation of momentum gives

$$m_A v_{A1} + m_B v_{B1} = m_A v_{A2} + m_B v_{B2}.$$

6.2
Collisions and Elasticity

6.4.2 and 6.4.3
Collision Problems

6.5
Car Collision: Two Dimensions

6.10
Pendulum Bashes Box

If the masses m_A and m_B and the initial velocities v_{A1} and v_{B1} are known, these two equations can be solved simultaneously to find the two final velocities v_{A2} and v_{B2}.

The general solution is a little complicated, so we will concentrate on the particular case in which body B is at rest before the collision. Think of it as a target for body A to hit. We can then simplify the velocity notation; we let v be the initial x-component of velocity of A, and v_A and v_B be the final x-components for A and B. Then the kinetic energy and momentum conservation equations are, respectively.

$$\frac{1}{2}m_A v^2 = \frac{1}{2}m_A v_A{}^2 + \frac{1}{2}m_B v_B{}^2, \tag{8-19}$$

$$m_A v = m_A v_A + m_B v_B. \tag{8-20}$$

We may solve for v_A and v_B in terms of the masses and the initial velocity v. This involves some fairly strenuous algebra, but it's worth it. No pain, no gain! The simplest approach is somewhat indirect, but along the way it uncovers an additional interesting feature of elastic collisions.

First rearrange Eqs. (8–19) and (8–20) as follows:

$$m_B v_B{}^2 = m_A(v^2 - v_A{}^2) = m_A(v - v_A)(v + v_A), \tag{8-21}$$

$$m_B v_B = m_A(v - v_A). \tag{8-22}$$

Now divide Eq. (8–21) by Eq. (8–22) to obtain

$$v_B = v + v_A. \tag{8-23}$$

Substitute this back into Eq. (8–22) to eliminate v_B, and then solve for v_A:

$$m_B(v + v_A) = m_A(v - v_A),$$

$$v_A = \frac{m_A - m_B}{m_A + m_B}v. \tag{8-24}$$

Finally, substitute this result back into Eq. (8–23) to obtain

$$v_B = \frac{2m_A}{m_A + m_B}v. \tag{8-25}$$

Now we can interpret the results. Suppose body A is a ping-pong ball and body B is a bowling ball. Then we expect A to bounce off after the collision with a velocity nearly equal to its original value but in the opposite direction (Fig. 8–15a), and we expect B's velocity to be much smaller. That's just what the equations predict. When m_A is much smaller than m_B, the fraction in Eq. (8–24) is approximately equal to (-1), so v_A is approximately equal to $-v$. The fraction in Eq. (8–25) is much smaller than unity, so v_B is much less than v. Figure 8–15b shows the opposite case, in which A is the bowling ball and B the ping-pong ball and m_A is much larger than m_B. What do you expect to happen then? Check your predictions against Eqs. (8–24) and (8–25).

Billiard balls deform very little when they collide, and quickly spring back from any deformation they do undergo. Hence the force of interaction between the balls is almost perfectly conservative, and the collision is elastic.

8–15 Collisions between a ping-pong ball and a bowling ball.

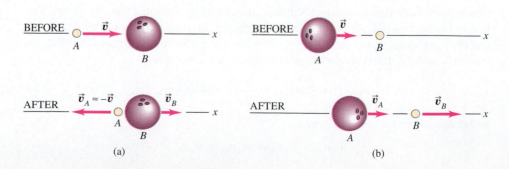

(a)   (b)

Another interesting case occurs when the masses are equal (Fig. 8–16). If $m_A = m_B$, then Eqs. (8–24) and (8–25) give $v_A = 0$ and $v_B = v$. That is, the body that was moving stops dead; it gives all its momentum and kinetic energy to the body that was at rest. This behavior is familiar to all pool players and marbles shooters.

Let's return to the general case in which A and B have different masses. Equation (8–23) can be rewritten as

$$v = v_B - v_A. \qquad (8\text{–}26)$$

Here $v_B - v_A$ is the velocity of B relative to A *after* the collision; from Eq. (8–26), this equals v, which is the *negative* of the velocity of B relative to A before the collision. (We discussed relative velocity in Section 3–6.) The relative velocity has the same magnitude, but opposite sign, before and after the collision. The sign changes because A and B are approaching before the collision but receding from each other after the collision. If we view this collision from a second coordinate system moving with constant velocity relative to the first, the velocities of the bodies are different but the *relative* velocities are the same. Hence our statement about relative velocities holds for *any* straight-line elastic collision, even when neither body is at rest initially. *In a straight-line elastic collision of two bodies, the relative velocities before and after the collision have the same magnitude but opposite sign.* This can be expressed using our original notation as

$$v_{B2} - v_{A2} = -(v_{B1} - v_{A1}). \qquad (8\text{–}27)$$

It turns out that a *vector* relationship similar to Eq. (8–27) is a general property of *all* elastic collisions, even when both bodies are moving initially and the velocities do not all lie along the same line. This result provides an alternative and equivalent definition of an elastic collision: *In an elastic collision the relative velocity of the two bodies has the same magnitude before and after the collision.* Whenever this condition is satisfied, the total kinetic energy is also conserved.

When an elastic two-body collision isn't head-on, the velocities don't all lie along a single line. If they all lie in a plane, each final velocity has two unknown components, and there are four unknowns in all. Conservation of energy and conservation of the x- and y-components of momentum give only three equations. To determine the final velocities uniquely, we need additional information, such as the direction or magnitude of one of the final velocities.

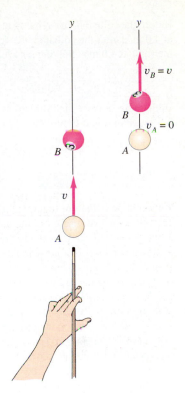

8–16 In a one-dimensional elastic collision between bodies of equal mass, if one body is at rest, it receives all the momentum and kinetic energy of the moving body.

EXAMPLE 8–10

An elastic straight-line collision The situation in Fig. 8–17 is the same as for Example 8–5 in Section 8–3, but now we've added ideal spring bumpers to the gliders so that the collision is elastic. What are the velocities of A and B after the collision?

SOLUTION From conservation of momentum,

$$m_A v_{A1} + m_B v_{B1} = m_A v_{A2} + m_B v_{B2},$$

$$(0.50 \text{ kg})(2.0 \text{ m/s}) + (0.30 \text{ kg})(-2.0 \text{ m/s})$$

$$= (0.50 \text{ kg})v_{A2} + (0.30 \text{ kg})v_{B2},$$

$$0.50v_{A2} + 0.30v_{B2} = 0.40 \text{ m/s}.$$

(In the last equation we have divided through by the unit "kg.") From Eq. (8–27), the relative velocity relation for an elastic collision,

$$v_{B2} - v_{A2} = -(v_{B1} - v_{A1})$$

$$= -(-2.0 \text{ m/s} - 2.0 \text{ m/s}) = 4.0 \text{ m/s}.$$

8–17 Air-track gliders (a) before and (b) after an elastic collision.

Before the collision, the velocity of B relative to A is to the left at 4.0 m/s; after the collision, the velocity of B relative to A is to the right at 4.0 m/s. Solving these equations simultaneously, we find

$$v_{A2} = -1.0 \text{ m/s}, \qquad v_{B2} = 3.0 \text{ m/s}.$$

Both bodies reverse their directions of motion; A moves to the left at 1.0 m/s, and B moves to the right at 3.0 m/s. This is differ-

ent from the result of Example 8–5 because that collision was *not* elastic. The total kinetic energy after the elastic collision is

$$\frac{1}{2}(0.50 \text{ kg})(-1.0 \text{ m/s})^2 + \frac{1}{2}(0.30 \text{ kg})(3.0 \text{ m/s})^2 = 1.6 \text{ J}.$$

As expected, this equals the total kinetic energy before the collision (which we calculated in Example 8–7, Section 8–4).

EXAMPLE 8–11

Moderator in a nuclear reactor High-speed neutrons are produced in a nuclear reactor during nuclear fission processes. Before a neutron can trigger additional fissions, it has to be slowed down by collisions with nuclei in the *moderator* of the reactor. The first nuclear reactor (built in 1942 at the University of Chicago) and the reactor involved in the 1986 Chernobyl accident both used carbon (graphite) as the moderator material. A neutron (mass 1.0 u) traveling at 2.6×10^7 m/s undergoes a head-on elastic collision with a carbon nucleus (mass 12 u) initially at rest. The external forces during the collision are negligible. What are the velocities after the collision? (1 u is the *atomic mass unit,* equal to 1.66×10^{-27} kg.)

SOLUTION Because one body is initially at rest, we can use Eqs. (8–24) and (8–25) with

$$m_A = 1.0 \text{ u}, \qquad m_B = 12 \text{ u}, \qquad \text{and} \qquad v = 2.6 \times 10^7 \text{ m/s}.$$

We'll let you do the arithmetic; the results are

$$v_A = -2.2 \times 10^7 \text{ m/s}, \qquad v_B = 0.4 \times 10^7 \text{ m/s}.$$

The neutron ends up with $\frac{11}{13}$ of its initial speed, and the speed of the recoiling carbon nucleus is $\frac{2}{13}$ of the neutron's initial speed. Kinetic energy is proportional to speed squared, so the neutron's final kinetic energy is $\left(\frac{11}{13}\right)^2$, or about 0.72 of its original value. If it makes a second such collision, its kinetic energy is $(0.72)^2$, or about half its original value, and so on. After several collisions the neutron will be moving quite slowly.

EXAMPLE 8–12

The gravitational slingshot effect Figure 8–18 shows the planet Saturn moving in the negative x-direction at its orbital speed (with respect to the sun) of 9.6 km/s. The mass of Saturn is 5.69×10^{26} kg. A spacecraft with mass 825 kg approaches Saturn, moving initially in the $+x$-direction at 10.4 km/s. The gravitational attraction of Saturn (a conservative force) causes the spacecraft to swing around it and head off in the opposite direction. Find the final speed of the spacecraft after it is far enough away to be nearly free of Saturn's gravitational pull.

SOLUTION Here the "collision" is not an impact but a gravitational interaction. Let the spacecraft be body A, and let Saturn be body B. We can assume that Saturn's speed is essentially constant during the interaction because its mass is so much greater than that of the spacecraft. So we have a one-dimensional elastic collision in which $v_{B1} = v_{B2} = -9.6$ km/s and $v_{A1} = 10.4$ km/s. The relative velocity of the two bodies has the same magnitude before and after the interaction, as shown by Eq. (8–27). Being very careful with signs, we find

$$v_{A2} = v_{B2} + v_{B1} - v_{A1}$$

$$= (-9.6 \text{ km/s}) + (-9.6 \text{ km/s}) - 10.4 \text{ km/s} = -29.6 \text{ km/s}.$$

The spacecraft's final speed (relative to the sun) is nearly three times as great as before the encounter with Saturn. Not

only does the spacecraft get a close look at Saturn, but its kinetic energy increases by a factor of $(29.6/10.4)^2 = 8.1$ in the process. The reason for this seemingly mysterious increase is that Saturn is *moving* in its orbit around the sun. If Saturn were initially at rest, the spacecraft would reverse direction but would undergo hardly any change in speed; a baseball behaves similarly when it bounces off a bat that's held stationary (a "bunt"). But when a baseball collides with a bat that's swinging toward it, the baseball leaves the bat with a much greater speed, perhaps enough to make a home run. In our example, Saturn takes the place of the swinging bat, and the spacecraft gains speed much like a batted baseball.

This is a simplified version of the *gravitational slingshot effect,* used to give an extra boost to spacecraft. (The actual motion of a spacecraft is never purely one-dimensional as we have assumed here. We'll discuss the trajectories of spacecraft and planets in detail in Chapter 12.) The *Voyager 2* spacecraft, launched in 1977, used the slingshot effect in flying past Jupiter, Saturn, and Uranus. Thanks to the kinetic energy gained in this way, *Voyager 2* reached the planet Neptune in 1989; without the slingshot effect, *Voyager 2* would not have arrived at Neptune until 2008.

In this bit of celestial gymnastics, does the spacecraft get something for nothing? Where does its extra energy come from?

$v_{A2} = -29.6$ km/s

$v_{A1} = 10.4$ km/s

$v_B = -9.6$ km/s

8–18 The gravitational slingshot effect: a simplified model of a spacecraft making a non-impact "collision" with the planet Saturn.

EXAMPLE 8–13

A two-dimensional elastic collision Figure 8–19 shows an elastic collision of two pucks on a frictionless air table. Puck A has mass $m_A = 0.500$ kg, and puck B has mass $m_B = 0.300$ kg. Puck A has an initial velocity of 4.00 m/s in the positive x-direction and a final velocity of 2.00 m/s in an unknown direction. Puck B is initially at rest. Find the final speed v_{B2} of puck B and the angles α and β in the figure.

SOLUTION Because the collision is elastic, the initial and final kinetic energies are equal:

$$\frac{1}{2} m_A v_{A1}^2 = \frac{1}{2} m_A v_{A2}^2 + \frac{1}{2} m_B v_{B2}^2,$$

$$v_{B2}^2 = \frac{m_A v_{A1}^2 - m_A v_{A2}^2}{m_B}$$

$$= \frac{(0.500 \text{ kg})(4.00 \text{ m/s})^2 - (0.500 \text{ kg})(2.00 \text{ m/s})^2}{0.300 \text{ kg}},$$

$$v_{B2} = 4.47 \text{ m/s}.$$

Conservation of the x-component of total momentum gives

$$m_A v_{A1x} = m_A v_{A2x} + m_B v_{B2x},$$

$$(0.500 \text{ kg})(4.00 \text{ m/s}) = (0.500 \text{ kg})(2.00 \text{ m/s})(\cos \alpha)$$
$$+ (0.300 \text{ kg})(4.47 \text{ m/s})(\cos \beta),$$

and conservation of the y-component gives

$$0 = m_A v_{A2y} + m_B v_{B2y},$$

$$0 = (0.500 \text{ kg})(2.00 \text{ m/s})(\sin \alpha) - (0.300 \text{ kg})(4.47 \text{ m/s})(\sin \beta).$$

These are two simultaneous equations for α and β. The simplest solution is to eliminate β as follows: We solve the first equation for $\cos \beta$ and the second for $\sin \beta$; we then square each equation and add. Since $\sin^2 \beta + \cos^2 \beta = 1$, this eliminates β and leaves an equation that we can solve for $\cos \alpha$ and hence for α. We can then substitute this value back into either of the two equations and solve the result for β. We leave the details for you to work out as a problem; the results are

$$\alpha = 36.9°, \qquad \beta = 26.6°.$$

y

A $v_{A1} = 4.00$ m/s $m_B = 0.300$ kg

O $m_A = 0.500$ kg B

$v_{B1} = 0$

x

(a)

y

A

$v_{A2} = 2.00$ m/s

α

O x

β

B v_{B2}

(b)

8–19 (a) Before and (b) after an elastic collision that isn't head-on.

The examples in this and the preceding sections show that we can classify collisions according to energy considerations. A collision in which kinetic energy is conserved is called *elastic*. A collision in which the total kinetic energy decreases is called *inelastic*. When the two bodies have a common final velocity, we say that the collision is *completely inelastic*. There are also cases in which the final kinetic energy is *greater* than the initial value. Rifle recoil, discussed in Example 8–4 (Section 8–3), is an example.

Finally, we emphasize again that we can sometimes use momentum conservation even when there are external forces acting on the system, if the net external force acting on the colliding bodies is small in comparison to the internal forces during the collision.

8-6 CENTER OF MASS

We can restate the principle of conservation of momentum in a useful way by using the concept of **center of mass.** Suppose we have several particles, with masses m_1, m_2, and so on. Let the coordinates of m_1 be (x_1, y_1), those of m_2 be (x_2, y_2), and so on. We define the center of mass of the system as the point having coordinates (x_{cm}, y_{cm}) given by

$$x_{cm} = \frac{m_1 x_1 + m_2 x_2 + m_3 x_3 + \cdots}{m_1 + m_2 + m_3 + \cdots} = \frac{\sum_i m_i x_i}{\sum_i m_i},$$

$$\text{(center of mass)} \quad (8\text{–}28)$$

$$y_{cm} = \frac{m_1 y_1 + m_2 y_2 + m_3 y_3 + \cdots}{m_1 + m_2 + m_3 + \cdots} = \frac{\sum_i m_i y_i}{\sum_i m_i}.$$

The position vector $\vec{r}_{cm}$ of the center of mass can be expressed in terms of the position vectors $\vec{r}_1, \vec{r}_2, \ldots$ of the particles as

$$\vec{r}_{cm} = \frac{m_1 \vec{r}_1 + m_2 \vec{r}_2 + m_3 \vec{r}_3 + \cdots}{m_1 + m_2 + m_3 + \cdots} = \frac{\sum_i m_i \vec{r}_i}{\sum_i m_i} \quad \text{(center of mass).} \quad (8\text{–}29)$$

In statistical language the center of mass is a *mass-weighted average* position of the particles.

EXAMPLE 8-14

Center of mass of the water molecule Figure 8–20 shows a simple model of the structure of a water molecule. The separation between atoms is $d = 9.57 \times 10^{-11}$ m. Each hydrogen atom has mass 1.0 u, and the oxygen atom has mass 16.0 u. We represent each atom as a point because nearly all the mass of each atom is concentrated in its nucleus, which is only about 10^{-5} times the overall radius of the atom. Using the coordinate system shown, find the position of the center of mass.

SOLUTION The x-coordinate of each hydrogen atom is $d \cos (105°/2)$; the y-coordinates of the upper and lower hydrogen atoms are $+d \sin (105°/2)$ and $-d \sin (105°/2)$, respectively. The coordinates of the oxygen atom are $x = 0$, $y = 0$. From Eqs. (8–28) the x-coordinate of the center of mass is

$$x_{cm} = \frac{(1.0 \text{ u})(d \cos 52.5°) + (1.0 \text{ u})(d \cos 52.5°) + (16.0 \text{ u})(0)}{1.0 \text{ u} + 1.0 \text{ u} + 16.0 \text{ u}}$$

$$= 0.068d,$$

and the y-coordinate is

$$y_{cm} = \frac{(1.0 \text{ u})(d \sin 52.5°) + (1.0 \text{ u})(-d \sin 52.5°) + (16.0 \text{ u})(0)}{1.0 \text{ u} + 1.0 \text{ u} + 16.0 \text{ u}}$$

$$= 0.$$

8–20 Where is the center of mass of a water molecule?

The actual value of d for the water molecule is 9.57×10^{-11} m, so we find

$$x_{cm} = (0.068)(9.57 \times 10^{-11} \text{ m}) = 6.5 \times 10^{-12} \text{ m.}$$

The center of mass is much closer to the oxygen atom than to either hydrogen atom because the oxygen atom is much more massive. Notice that the center of mass lies along the x-axis, which is the *axis of symmetry* of this molecule. If the molecule is rotated by 180° around this axis, it looks exactly the same as before. The position of the center of mass can't be affected by this rotation, so it must lie on the axis of symmetry.

For solid bodies, in which we have (at least on a macroscopic level) a continuous distribution of matter, the sums in Eqs. (8–28) have to be replaced by integrals. The calculations can get quite involved, but we can say some general things about such problems. First, whenever a homogeneous body has a geometric center, such as a billiard ball, a sugar cube, or a can of frozen orange juice, the center of mass is at the geometric center. Second, whenever a body has an axis of symmetry, such as a wheel or a pulley, the center of mass always lies on that axis. Third, there is no law that says the center of mass has to be within the body. For example, the center of mass of a doughnut is right in the middle of the hole. You can probably think of other examples.

We'll talk a little more about locating the center of mass in Chapter 11 in connection with the related concept *center of gravity*.

MOTION OF THE CENTER OF MASS

To see the significance of the center of mass of a collection of particles, we must ask what happens to the center of mass when the particles move. The x- and y-components of velocity of the center of mass, $v_{cm\text{-}x}$ and $v_{cm\text{-}y}$, are the time derivatives of x_{cm} and y_{cm}. Also, dx_1/dt is the x-component of velocity of particle 1, and so on, so $dx_1/dt = v_{1x}$, and so on. Taking time derivatives of Eqs. (8–28), we get

$$v_{cm\text{-}x} = \frac{m_1 v_{1x} + m_2 v_{2x} + m_3 v_{3x} + \cdots}{m_1 + m_2 + m_3 + \cdots},$$

$$v_{cm\text{-}y} = \frac{m_1 v_{1y} + m_2 v_{2y} + m_3 v_{3y} + \cdots}{m_1 + m_2 + m_3 + \cdots}. \qquad (8\text{–}30)$$

These equations are equivalent to the single vector equation obtained by taking the time derivative of Eq. (8–29):

$$\vec{v}_{cm} = \frac{m_1 \vec{v}_1 + m_2 \vec{v}_2 + m_3 \vec{v}_3 + \cdots}{m_1 + m_2 + m_3 + \cdots} \qquad (8\text{–}31)$$

We denote the *total* mass $m_1 + m_2 + \ldots$ by M. We can then rewrite Eq. (8–31) as

$$M\vec{v}_{cm} = m_1 \vec{v}_1 + m_2 \vec{v}_2 + m_3 \vec{v}_3 + \cdots = \vec{P}. \qquad (8\text{–}32)$$

The right side is simply the total momentum $\vec{P}$ of the system. Thus we have proved that the *total momentum is equal to the total mass times the velocity of the center of mass.* When you catch a baseball, you are really catching a collection of a very large number of molecules of masses $m_1, m_2, m_3, \ldots$. The impulse you feel is due to the total momentum of this entire collection. But this impulse is the same as if you were catching a single particle of mass $M = m_1 + m_2 + m_3 + \cdots$ moving with velocity $\vec{v}_{cm}$, the velocity of the collection's center of mass. So Eq. (8–32) helps to justify representing an extended body as a particle.

For a system of particles on which the net external force is zero, so that the total momentum $\vec{P}$ is constant, the velocity of the center of mass $\vec{v}_{cm} = \vec{P}/M$ is also constant. Suppose we mark the center of mass of an adjustable wrench, which lies at a point partway down the handle. We then slide the wrench with a spinning motion across a smooth, horizontal table top (Fig. 8–21). The overall motion appears complicated, but the center of mass follows a straight line, as though all the mass were concentrated at that point.

8–21 The center of mass of this wrench is marked with a white dot. The total external force acting on the wrench is almost zero. As it spins on a smooth horizontal surface, the center of mass moves in a straight line with nearly constant velocity.

EXAMPLE 8-15

A tug-of-war on the ice James and Ramon are standing 20.0 m apart on the slippery surface of a frozen pond. Ramon has mass 60.0 kg, and James has mass 90.0 kg. Midway between the two men a mug of their favorite beverage sits on the ice (Fig. 8–22). They pull on the ends of a light rope that is stretched between them. When James has moved 6.0 m toward the mug, how far and in what direction has Ramon moved?

SOLUTION The frozen surface is horizontal and essentially frictionless, so the net external force on the system of James, Ramon, and the rope is zero. Hence their total momentum is conserved. Initially, there is no motion, so the total momentum is zero; thus the velocity of the center of mass is zero, and the center of mass remains at rest. Let's take the origin at the position of the mug, and let the +x-axis extend from the mug toward Ramon. Since the rope is light, we can neglect its mass in calculating the position of the center of mass. The initial x-coordinates of James and Ramon are −10.0 m and +10.0 m, respectively, so the x-coordinate of the center of mass is

$$x_{cm} = \frac{(90.0 \text{ kg})(-10.0 \text{ m}) + (60.0 \text{ kg})(10.0 \text{ m})}{90.0 \text{ kg} + 60.0 \text{ kg}} = -2.0 \text{ m}.$$

When James moves 6.0 m toward the mug, his new x-coordinate is −4.0 m; we'll call Ramon's new x-coordinate x_2. The center of mass doesn't move, so

$$x_{cm} = \frac{(90.0 \text{ kg})(-4.0 \text{ m}) + (60.0 \text{ kg})x_2}{90.0 \text{ kg} + 60.0 \text{ kg}} = -2.0 \text{ m},$$

$$x_2 = 1.0 \text{ m}.$$

8–22 No matter how hard James or Ramon pulls, their center of mass will not move.

James has moved 6.0 m and is still 4.0 m from the mug, but Ramon has moved 9.0 m and is only 1.0 m from it. The ratio of how far each man moved, (6.0 m)/(9.0 m) = $\frac{2}{3}$, equals the inverse ratio of their masses. Can you see why? If the two men keep moving (and if the surface is frictionless, they will!), Ramon will reach the mug first. This result is completely independent of how hard either person pulls; pulling harder just helps Ramon quench his thirst sooner.

EXTERNAL FORCES AND CENTER-OF-MASS MOTION

If the net external force on a system of particles is not zero, then total momentum is not conserved and the velocity of the center of mass will change. Let's look at the relation between the motion of the center of mass and the forces acting on the system.

Equations (8–31) and (8–32) give the *velocity* of the center of mass in terms of the velocities of the individual particles. Proceeding one additional step, we take the time derivatives of these equations to show that the accelerations are related in the same way. Let $\vec{a}_{cm} = d\vec{v}_{cm}/dt$ be the acceleration of the center of mass; then we find

$$M\vec{a}_{cm} = m_1\vec{a}_1 + m_2\vec{a}_2 + m_3\vec{a}_3 + \cdots \qquad (8\text{–}33)$$

Now $m_1\vec{a}_1$ is equal to the vector sum of forces on the first particle, and so on, so the right side of Eq. (8–33) is equal to the vector sum $\Sigma\vec{F}$ of *all* the forces on *all* the particles. Just as we did in Section 8–3, we may classify each force as *internal* or *external*. The sum of all forces on all the particles is then

$$\Sigma\vec{F} = \Sigma\vec{F}_{ext} + \Sigma\vec{F}_{int} = M\vec{a}_{cm}.$$

Because of Newton's third law, the internal forces all cancel in pairs, and $\Sigma\vec{F}_{int} = 0$. What survives on the left side is only the sum of the *external* forces, and we have

$$\Sigma\vec{F}_{ext} = M\vec{a}_{cm} \qquad \text{(body or collection of particles)}. \qquad (8\text{–}34)$$

When a body or a collection of particles is acted on by external forces, the center of

(a)

(b)

8–23 (a) A shell explodes into two parts in flight. If air resistance is ignored, the fragments follow individual parabolic paths, but the center of mass continues on the same trajectory as the shell's path before exploding—until a fragment hits the ground. (b) The same effect occurs with exploding fireworks.

mass moves just as though all the mass were concentrated at that point and it were acted on by a net force equal to the sum of the external forces on the system.

This result may not sound very impressive, but in fact it is central to the whole subject of mechanics. In fact, we've been using this result all along; without it, we would not be able to represent an extended body as a point particle when we apply Newton's laws. It explains why only *external* forces can affect the motion of an extended body. If you pull upward on your belt, your belt exerts an equal downward force on your hands; these are *internal* forces that cancel and have no effect on the overall motion of your body.

Suppose a cannon shell traveling in a parabolic trajectory (neglecting air resistance) explodes in flight, splitting into two fragments with equal mass (Fig. 8–23a). The fragments follow new parabolic paths, but the center of mass continues on the original parabolic trajectory, just as though all the mass were still concentrated at that point. A skyrocket exploding in air (Fig. 8–23b) is a spectacular example of this effect.

This property of the center of mass is important when we analyze the motion of rigid bodies. We describe the motion of an extended body as a combination of translational motion of the center of mass and rotational motion about an axis through the center of mass. We will return to this topic in Chapter 10. This property also plays an important role in the motion of astronomical objects. It's not correct to say that the moon orbits the earth; rather, the earth and moon both move in orbits around their center of mass.

There's one more useful way to describe the motion of a system of particles. Using $\vec{a}_{cm} = d\vec{v}_{cm}/dt$, we can rewrite Eq. (8–33) as

$$M\vec{a}_{cm} = M\frac{d\vec{v}_{cm}}{dt} = \frac{d(M\vec{v}_{cm})}{dt} = \frac{d\vec{P}}{dt}. \qquad (8\text{–}35)$$

The total system mass M is constant, so we're allowed to take it inside the derivative. Substituting Eq. (8–35) into Eq. (8–34), we find

$$\Sigma\vec{F}_{ext} = \frac{d\vec{P}}{dt} \qquad \text{(extended body or system of particles)}. \qquad (8\text{–}36)$$

This equation looks like Eq. (8–3). The difference is that Eq. (8–36) describes a *system* of particles, such as an extended body, while Eq. (8–3) describes a single particle. The interactions between the particles that make up the system can change the individual momenta of the particles, but the *total* momentum $\vec{P}$ of the system can be changed only by external forces acting from outside the system.

Finally, we note that if the net external force is zero, Eq. (8–34) shows that the acceleration $\vec{a}_{cm}$ of the center of mass is zero. So the center-of-mass velocity $\vec{v}_{cm}$ is constant, as for the wrench in Fig. 8–21. From Eq. (8–36) the total momentum $\vec{P}$ is also constant. This reaffirms our statement in Section 8–3 of the principle of conservation of momentum.

6.6
Saving an Astronaut

*8-7 ROCKET PROPULSION

Momentum considerations are particularly useful for analyzing a system in which the masses of parts of the system change with time. In such cases we can't use Newton's second law $\Sigma \vec{F} = m\vec{a}$ directly because m changes. Rocket propulsion offers a typical and interesting example of this kind of analysis. A rocket is propelled forward by rearward ejection of burned fuel that initially was in the rocket. The forward force on the rocket is the reaction to the backward force on the ejected material. The total mass of the system is constant, but the mass of the rocket itself decreases as material is ejected.

As a simple example, we consider a rocket fired in outer space, where there is no gravitational force and no air resistance. We choose our x-axis to be along the rocket's direction of motion. Figure 8–24a shows the rocket at a time t, when its mass is m and the x-component of its velocity relative to our coordinate system is v. The x-component of total momentum at this instant is $P_1 = mv$. In a short time interval dt the mass of the rocket changes by an amount dm. This is an inherently negative quantity because the rocket's mass m *decreases* with time. During dt a *positive* mass $-dm$ of burned fuel is ejected from the rocket. Let v_{ex} be the exhaust *speed* of this material *relative to the rocket;* the burned fuel is ejected opposite the direction of motion, so its x-component of *velocity* relative to the rocket is $-v_{ex}$. The x-component of velocity v_{fuel} of the burned fuel relative to our coordinate system is then

$$v_{fuel} = v + (-v_{ex}) = v - v_{ex},$$

and the x-component of momentum of the ejected mass $(-dm)$ is

$$(-dm)v_{fuel} = (-dm)(v - v_{ex}).$$

As shown in Fig. 8–24b, at the end of the time interval dt, the x-component of velocity of the rocket and unburned fuel has increased to $v + dv$, and its mass has decreased to $m + dm$ (remember that dm is negative). The rocket's momentum at this time is

$$(m + dm)(v + dv).$$

Thus the *total* x-component of momentum P_2 of rocket plus ejected fuel at time $t + dt$ is

$$P_2 = (m + dm)(v + dv) + (-dm)(v - v_{ex}).$$

According to our initial assumption, the rocket and fuel are an isolated system. Thus momentum is conserved, and the total x-component of momentum of the system must be the same at time t and at time $t + dt$: $P_1 = P_2$. Hence

$$mv = (m + dm)(v + dv) + (-dm)(v - v_{ex}).$$

8–24 (a) A rocket moving in gravity-free outer space with mass m and x-component of velocity v at time t. (b) At time $t + dt$ the rocket's mass has decreased, so dm is negative. The mass of the rocket (including the unburned fuel) is $m + dm$, and its velocity component is $v + dv$. The mass of burned fuel in the exhaust is $-dm$, and its velocity relative to the coordinate system shown is $v_{fuel} = v - v_{ex}$.

+x direction

v

m

$v_{fuel} = v - v_{ex}$ $v + dv$

$-dm$ $m + dm$

(a)   (b)

This can be simplified to

$$m\,dv = -dm\,v_{ex} - dm\,dv.$$

We can neglect the term $(-dm\,dv)$ because it is a product of two small quantities and thus is much smaller than the other terms. Dropping this term, dividing by dt, and rearranging, we find

$$m\frac{dv}{dt} = -v_{ex}\frac{dm}{dt}. \tag{8–37}$$

Now dv/dt is the acceleration of the rocket, so the left side of this equation (mass times acceleration) equals the net force F, or *thrust*, on the rocket,

$$F = -v_{ex}\frac{dm}{dt}. \tag{8–38}$$

The thrust is proportional both to the relative speed v_{ex} of the ejected fuel and to the mass of fuel ejected per unit time, $-dm/dt$. (Remember that dm/dt is negative because it is the rate of change of the rocket's mass.)

The x-component of acceleration of the rocket is

$$a = \frac{dv}{dt} = -\frac{v_{ex}}{m}\frac{dm}{dt}. \tag{8–39}$$

This is positive because v_{ex} is positive (remember, it's the exhaust *speed*) and dm/dt is negative. The rocket's mass m decreases continuously while the fuel is being consumed. If v_{ex} and dm/dt are constant, the acceleration increases until all the fuel is gone.

An effective rocket burns fuel at a rapid rate (large $-dm/dt$) and ejects the burned fuel at a high relative speed (large v_{ex}), as in Fig. 8–25. In the early days of rocket propulsion, people who didn't understand conservation of momentum thought that a rocket couldn't function in outer space because "it doesn't have anything to push against." On the contrary, rockets work *best* in outer space, where there is no air resistance! The rocket in Fig. 8–25 is *not* "pushing against the ground" to get into the air.

8–25 Launch of a space shuttle, a dramatic example of rocket propulsion. In order to provide enough upward thrust to overcome gravity, space launch vehicles can consume over 10,000 kg/s of fuel and eject the burned fuel at speeds of over 4000 m/s.

If the exhaust speed v_{ex} is constant, we can integrate Eq. (8–39) to find a relation between the velocity v at any time and the remaining mass m. At time $t = 0$, let the mass be m_0 and the velocity v_0. Then we rewrite Eq. (8–39) as

$$dv = -v_{ex} \frac{dm}{m}.$$

We change the integration variables to v' and m', so we can use v and m as the upper limits (the final speed and mass). Then we integrate both sides, using limits v_0 to v and m_0 to m, and take the constant v_{ex} outside the integral:

$$\int_{v_0}^{v} dv' = -\int_{m_0}^{m} v_{ex} \frac{dm'}{m'} = -v_{ex} \int_{m_0}^{m} \frac{dm'}{m'},$$

$$v - v_0 = -v_{ex} \ln \frac{m}{m_0} = v_{ex} \ln \frac{m_0}{m}. \tag{8–40}$$

The ratio m_0/m is the original mass divided by the mass after the fuel has been exhausted. In practical spacecraft this ratio is made as large as possible to maximize the speed gain, which means that the initial mass of the rocket is almost all fuel. The final velocity of the rocket will be greater in magnitude (and is often *much* greater) than the relative speed v_{ex} if $\ln (m_0/m) > 1$, that is, if $m_0/m > e = 2.71828 \ldots$.

We've assumed throughout this analysis that the rocket is in gravity-free outer space. However, gravity must be taken into account when a rocket is launched from the surface of a planet, as in Fig. 8–25 (see Problem 8–90).

EXAMPLE 8–16

Acceleration of a rocket A rocket is in outer space, far from any planet, when the rocket engine is turned on. In the first second of firing, the rocket ejects $\frac{1}{120}$ of its mass with a relative speed of 2400 m/s. What is the rocket's initial acceleration?

SOLUTION We are given that the initial rate of change of mass is

$$\frac{dm}{dt} = -\frac{m_0/120}{1 \text{ s}} = -\frac{m_0}{120 \text{ s}},$$

where m_0 is the initial ($t = 0$) mass of the rocket. From Eq. (8–39) the initial acceleration is

$$a = -\frac{v_{ex}}{m_0} \frac{dm}{dt} = -\frac{2400 \text{ m/s}}{m_0} \left(-\frac{m_0}{120 \text{ s}}\right) = 20 \text{ m/s}^2.$$

Note that the answer didn't depend on the value of m_0. If v_{ex} is the same, the initial acceleration is the same for a 120,000-kg spacecraft that ejects 1000 kg/s as for a 60-kg astronaut equipped with a small rocket that ejects 0.5 kg/s.

EXAMPLE 8–17

Speed of a rocket Suppose that $\frac{3}{4}$ of the initial mass m_0 of the rocket in Example 8–16 is fuel, so the final mass is $m = m_0/4$, and that the fuel is completely consumed at a constant rate in a total time $t = 90$ s. If the rocket starts from rest in our coordinate system, find its speed at the end of this time.

SOLUTION We have $m_0/m = 4$, so from Eq. (8–40),

$$v = v_0 + v_{ex} \ln \frac{m_0}{m} = 0 + (2400 \text{ m/s})(\ln 4) = 3327 \text{ m/s}.$$

At the start of the flight, when the velocity of the rocket is zero, the ejected fuel is moving to the left, relative to our coordi-

nate system, at 2400 m/s. At the end of the first second ($t = 1$ s) the rocket is moving at 20 m/s, and the fuel's speed relative to our system is 2380 m/s. During the next second the acceleration, given by Eq. (8–39), is a little greater. At $t = 2$ s the rocket is moving at a little over 40 m/s, and the fuel's speed is a little less than 2360 m/s. Detailed calculation shows that at about $t = 75.6$ s, the rocket's velocity v in our coordinate system equals 2400 m/s. The burned fuel ejected after this time moves *forward*, not backward, in our system. Since the final velocity of the rocket is 3327 m/s and the relative velocity is 2400 m/s, the last portion of the ejected fuel has a forward velocity (relative to our frame) of $(3327 - 2400)$ m/s $= 927$ m/s. (We are using more figures than are significant to illustrate our point.)

8–8 THE NEUTRINO

A Case Study in Modern Physics

The laws of conservation of energy and momentum, vitally important in all areas of physical science, rest on a very solid foundation of experimental evidence. Nevertheless, the decay of radioactive nuclei by a process called *beta decay* led the great physicist Niels Bohr to suggest seriously in 1930 that these laws might not be obeyed in nuclear processes. Most physicists disagreed; they believed so strongly in these conservation laws that they accepted instead an alternative hypothesis proposed by Wolfgang Pauli that an unseen "ghost particle" is emitted in beta decay. Not until 25 years later was this particle, the *neutrino,* actually observed experimentally.

To see how this challenge to the conservation laws arose, let's consider an isolated system consisting of two particles, with masses m_1 and m_2, initially at rest. The system might be the bullet and rifle of Example 8–4 (Section 8–3), or two balls tied together with a compressed spring between them, or an unstable nucleus that comes apart into two fragments. When the two bodies fly apart, with speeds v_1 and v_2, some total quantity of energy Q is divided between them:

$$Q = K_1 + K_2 = \frac{1}{2}m_1 v_1^2 + \frac{1}{2}m_2 v_2^2. \qquad (8\text{–}41)$$

In Example 8–4, the value of Q was 225 J. We noted in that example that the ratio of the kinetic energies of the two particles equals the inverse ratio of their masses. With a 600:1 mass ratio the recoiling rifle and the bullet get kinetic energies of $\frac{1}{601}Q$ and $\frac{600}{601}Q$, respectively. The energy always divides this way, with the lighter particle getting more kinetic energy than the heavier one.

To prove this, we note that the system is isolated, so the total momentum is zero both before and after it flies apart. Thus the final momenta of the two particles must be equal and opposite:

$$m_1 v_1 = -m_2 v_2.$$

When we square this equation and divide by 2, we get

$$\frac{1}{2}m_1^2 v_1^2 = \frac{1}{2}m_2^2 v_2^2,$$

$$m_1 K_1 = m_2 K_2, \qquad \text{or} \qquad \frac{K_2}{K_1} = \frac{m_1}{m_2}.$$

When we combine this with Eq. (8–41) to get expressions for K_1 and K_2, we find

$$K_1 = \frac{m_2}{m_1 + m_2}Q \qquad \text{and} \qquad K_2 = \frac{m_1}{m_1 + m_2}Q. \qquad (8\text{–}42)$$

The essential point is that when there are only two final particles, the total available energy Q *must* divide this way every time. If there are three or more final particles, the energy can be divided up in many different ways, depending on the directions in which the particles move.

EXAMPLE 8–18

Energy in radioactive decay The radioactive isotope ^{222}Rn, the inert gas radon, is naturally present in the air. It undergoes a process called *alpha decay,* the end products of which are a ^{218}Po nucleus (mass $= 3.62 \times 10^{-25}$ kg) and an alpha particle (mass $= 6.65 \times 10^{-27}$ kg). (This radiation, along with that pro-duced by further decay events, is responsible for the health hazards of radon.) The quantity of energy released in the decay of ^{222}Rn is $Q = 9.0 \times 10^{-13}$ J. What are the kinetic energies of the emitted alpha particle and the recoiling ^{218}Po nucleus?

SOLUTION Let m_1 be the alpha particle, and let m_2 be the ^{218}Po nucleus. We can use Eqs. (8–42) with $m_1 = 6.65 \times 10^{-27}$ kg and $m_2 = 3.62 \times 10^{-25}$ kg. The energy of the alpha particle is

$$K_1 = \frac{m_2}{m_1 + m_2} Q = \frac{3.62 \times 10^{-25} \text{ kg}}{3.69 \times 10^{-25} \text{ kg}} (9.0 \times 10^{-13} \text{ J})$$

$$= 8.8 \times 10^{-13} \text{ J}.$$

The amount left for the ^{218}Po nucleus is

$$K_2 = Q - K_1 = 2 \times 10^{-14} \text{ J}.$$

About 98% of the energy released is given to the alpha particle, leaving only 2% for kinetic energy of the ^{218}Po nucleus.

8–26 A graph of the numbers of alpha particles (N) emitted with various energies K in the decay of ^{222}Rn to ^{218}Po. The graph consists of a single sharp peak at $K = 8.8 \times 10^{-13}$ J.

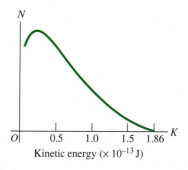

8–27 A graph of the number N of electrons emitted with various energies K in the β^- decay of ^{210}Bi to ^{210}Po. The graph shows a continuous distribution of electron energies, with a maximum value of 1.86×10^{-13} J.

Experimental measurements of ^{222}Rn decay give data on the number of alpha particles emitted with various kinetic energies (Fig. 8–26). The graph shows that *every* alpha particle emitted from *every* ^{222}Rn nucleus at rest has the same kinetic energy, within experimental error. (The alpha particles are said to be *monoenergetic*.) If there are only two final particles, then this is to be expected. If all ^{222}Rn nuclei are identical, then Q is the same for all, and K_1 must be the same. Hence these measurements confirm the two-particle picture of the decay process.

Another type of radioactive decay process, called beta decay, involves an electron or its antiparticle, the positron. Electrons are sometimes called beta-minus (β^-) particles, and positrons are called beta-plus (β^+) particles. An example of β^- decay occurs in the unstable nucleus ^{210}Bi, which decays into an electron and a ^{210}Po nucleus. Suppose this is a two-particle decay, like alpha emission. The rest mass of the ^{210}Po nucleus is 383,000 times the rest mass of the electron; a calculation similar to the one in Example 8–18 predicts that virtually all the released energy $Q = 1.86 \times 10^{-13}$ J should go to the electron and that the electrons from all decays should be monoenergetic. (The calculation requires relativistic expressions for kinetic energy and momentum, developed in Chapter 39, because the electrons emitted in ^{210}Bi decay can move as fast as 95% of the speed of light.)

Henri Becquerel, who discovered radioactivity in 1896, found indications by 1900 that the electrons emitted in β^- decay were *not* monoenergetic. Measurements made in 1914 by James Chadwick, the discoverer of the neutron, suggested this even more strongly. In the case of ^{210}Bi decay, later experimenters found a whole range of electron energies, with a maximum value of 1.86×10^{-13} J (Fig. 8–27). Since conservation of energy and momentum predicted that all the electrons should have the *same* energy, these results cast doubt on the validity of these conservation laws.

Looking for a way to save the conservation laws, Wolfgang Pauli suggested in 1931 that an unseen, electrically neutral third particle might be emitted during β^- decay. If so, it could carry off some of the energy and momentum. With three particles the energy and momentum can be shared in many different ways that are consistent with the conservation laws. With two particles (Fig. 8–28a) the momenta must be equal and opposite, and this requirement determines the energy of each particle uniquely. But with three particles, zero total momentum is represented by a triangle of momentum vectors (Figs. 8–28b, c, d) that allows various divisions of the momentum and energy, provided only that the kinetic energies add up to Q. So if there is a third particle, the electron can have a whole range of kinetic energies from zero to almost Q, as is observed.

Since Pauli's hypothetical particle was neutral, the Italian physicist Enrico Fermi first wanted to call it a neutron, but Chadwick had already claimed that name for a massive neutral particle. So Fermi christened the particle *neutrino* ("little neutral one"). (In present-day nomenclature the particle emitted in β^- decay is called an *antineutrino*, the antiparticle of the neutrino.) Subsequent measurements of electron energies in β^- decay were consistent with the existence of this unseen neutral particle. Because it interacts with other particles only extremely weakly, it is very difficult to detect. Not until 1956 were

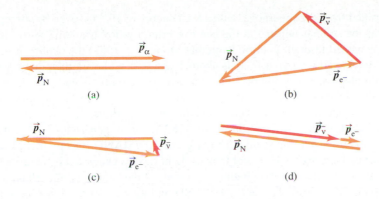

8-28 (a) Two-particle decay. The momenta are equal in magnitude and opposite in direction. (b) In the case of three particles the electron's kinetic energy will be somewhere between zero and Q. (c) When the momentum and energy of the antineutrino are small, the electron's kinetic energy is nearly Q. (d) When the antineutrino's energy is nearly Q, the electron's kinetic energy is nearly zero.

neutrinos observed directly. But in the intervening years, physicists had such confidence in the conservation laws that hardly anyone doubted that the neutrino would be found.

The neutrino's importance in modern physics goes far beyond helping to confirm conservation laws. Neutrinos have helped us understand what happens in a *supernova,* one of the most dramatic phenomena in nature, in which a star many times more massive than the sun explodes cataclysmically. The most recent nearby supernova, SN 1987A, was observed on February 24, 1987 (Fig. 8-29). At the same time, two laboratories on earth (one in Japan, the other under Lake Erie) detected a total of 19 neutrinos coming from the direction of SN 1987A. Because of the extremely weak interactions of neutrinos, even detecting this many neutrinos required enormous detectors made up of millions of kilograms of purified water. (Collisions of neutrinos with atomic nuclei in the water produce fast-moving secondary particles, which in turn produce brief flashes of light. It is these light flashes that are detected.) Analysis of the data showed that during the first second of the supernova explosion, approximately 10^{58} neutrinos

8-29 Before (left) and after (right) photographs of Supernova 1987A. Less than 0.1% of the energy released in the supernova appeared as visible light; most of the remainder was in the form of neutrinos.

were emitted from the supernova, with a total energy of 10^{46} J—over 100 times as much energy as the sun has emitted in the last five billion years! (About 5×10^{14} supernova neutrinos passed through your body on that Monday in 1987; the chance that even one of them interacted with your body is about 1 in 5,000, and even then there would have been no biological effect on you.) These neutrinos were produced by nuclear reactions deep within the core of the exploding supernova; detecting even 19 supernova neutrinos gives astrophysicists direct information about these reactions.

Spurred on by the observation of SN 1987A, a new generation of "neutrino telescopes" is coming into operation. The largest is DUMAND (Deep Underwater Muon and Neutrino Detector), which will look for light flashes generated by neutrinos passing through the clear water 4800 m below the surface of the Pacific off the island of Hawaii. By scanning 10^8 tons of seawater, DUMAND will look for ultra-high-energy neutrinos that are predicted to be emitted by material falling into supermassive black holes. Neutrino physics continues to be a vital and exciting field of research.

SUMMARY

KEY TERMS

momentum (linear momentum), 227

impulse, 228

impulse-momentum theorem, 229

internal force, 232

external force, 233

isolated system, 233

total momentum, 233

principle of conservation of momentum, 234

elastic collision, 238

inelastic collision, 238

completely inelastic collision, 238

center of mass, 246

■ The momentum $\vec{p}$ of a particle with mass m moving with velocity $\vec{v}$ is defined as the vector quantity

$$\vec{p} = m\vec{v}. \tag{8–2}$$

In terms of momentum, Newton's second law for a particle may be expressed as

$$\Sigma \vec{F} = \frac{d\vec{p}}{dt}. \tag{8–3}$$

■ The impulse $\vec{J}$ of a constant net force $\Sigma \vec{F}$ acting for a time interval from t_1 to t_2 is the vector quantity

$$\vec{J} = \Sigma \vec{F}(t_2 - t_1) = \Sigma \vec{F} \Delta t. \tag{8–5}$$

If the net force varies with time, the impulse is

$$\vec{J} = \int_{t_1}^{t_2} \Sigma \vec{F} \, dt. \tag{8–7}$$

The change in momentum of a particle in any time interval equals the impulse of the net force that acts on the particle during that interval:

$$\vec{J} = \vec{p}_2 - \vec{p}_1. \tag{8–6}$$

The momentum of a particle equals the impulse that accelerated it from rest to its present speed.

■ An internal force is a force exerted by one part of a system on another. An external force is a force exerted on a part of a system by something outside the system. An isolated system is one with no external forces.

■ The total momentum of a system of particles A, B, C, … is the vector sum of the momenta of the individual particles:

$$\vec{P} = \vec{p}_A + \vec{p}_B + \cdots = m_A \vec{v}_A + m_B \vec{v}_B + \cdots. \tag{8–14}$$

■ If the net external force on a system is zero, the total momentum of the system is constant (conserved); each component of total momentum is separately conserved.

■ Collisions can be classified according to energy relations and final velocities. In an elastic collision between two bodies, the initial and final total kinetic energies are equal and the initial and final relative velocities have the same magnitude. In an inelastic two-body collision the final total kinetic energy is less than the initial total kinetic energy. If the two bodies have the same final velocity, the collision is completely inelastic.

■ The coordinates x_{cm} and y_{cm} of the center of mass of a system of particles are defined as

$$x_{cm} = \frac{m_1 x_1 + m_2 x_2 + m_3 x_3 + \cdots}{m_1 + m_2 + m_3 + \cdots} = \frac{\sum_i m_i x_i}{\sum_i m_i},$$

$$y_{cm} = \frac{m_1 y_1 + m_2 y_2 + m_3 y_3 + \cdots}{m_1 + m_2 + m_3 + \cdots} = \frac{\sum_i m_i y_i}{\sum_i m_i}. \qquad (8\text{–}28)$$

In terms of the position vectors $\vec{r}_1, \vec{r}_2, \ldots$ of the particles, the position vector $\vec{r}_{cm}$ of the center of mass is

$$\vec{r}_{cm} = \frac{m_1 \vec{r}_1 + m_2 \vec{r}_2 + m_3 \vec{r}_3 + \cdots}{m_1 + m_2 + m_3 + \cdots} = \frac{\sum_i m_i \vec{r}_i}{\sum_i m_i}. \qquad (8\text{–}29)$$

■ The total momentum $\vec{P}$ of a system equals its total mass M multiplied by the velocity $\vec{v}_{cm}$ of its center of mass:

$$M\vec{v}_{cm} = m_1 \vec{v}_1 + m_2 \vec{v}_2 + m_3 \vec{v}_3 + \cdots = \vec{P}. \qquad (8\text{–}32)$$

■ The center of mass of a system moves as though all the mass M was concentrated at the center of mass. If the net external force on the system is zero, the center of mass velocity $\vec{v}_{cm}$ is constant. If the net external force is not zero, the center of mass accelerates:

$$\Sigma\vec{F}_{ext} = M\vec{a}_{cm}. \qquad (8\text{–}34)$$

■ In rocket propulsion the mass of a rocket changes as the fuel is burned and exhausted. Analysis of the motion of the rocket must include the momentum carried away by the fuel as well as the momentum of the rocket itself.

DISCUSSION QUESTIONS

Q8–1 In splitting logs with a hammer and wedge, is a heavy hammer more effective than a lighter hammer? Why?

Q8–2 Suppose you catch a baseball and then someone invites you to catch a bowling ball with either the same momentum or with the same kinetic energy as the baseball. Which would you choose? Explain.

Q8–3 When rain falls from the sky, what becomes of its momentum as it hits the ground? Is your answer also valid for Newton's famous apple?

Q8–4 A car has the same kinetic energy when it is traveling south at 30 m/s as when it is traveling northwest at 30 m/s. Is the momentum of the car the same in both cases? Explain.

Q8–5 A truck is accelerating as it speeds down the highway. One inertial frame of reference is attached to the ground with its origin at a fence post. A second frame of reference is attached to a police car that is traveling down the highway at constant velocity. Is the momentum of the truck the same in these two reference frames? Explain. Is the rate of change of the momentum of the truck the same in these two frames? Explain.

Q8–6 When a large, heavy truck collides with a passenger car, the occupants of the car are more likely to be hurt than the truck driver. Why?

Q8–7 A woman holding a large rock stands on a frictionless horizontal sheet of ice. She throws the rock with speed v_0 at an angle α above the horizontal. Consider the system consisting of the woman plus the rock. Is the momentum of the system conserved? Why or why not? Is any component of the momentum of the system conserved? Again, why or why not?

Q8–8 In Example 8–7 (Section 8–4), where the two gliders in Fig. 8–9a stick together after the collision, the collision is inelastic since $K_2 < K_1$. In Example 8–5 (Section 8–3), is the collision inelastic? Explain.

Q8–9 In a completely inelastic collision between two objects, where the objects stick together after the collision, is it possible for the final kinetic energy of the system to be zero? If so, give an example in which this would occur. If the final kinetic energy is zero, what must be the initial momentum of the system? Is the initial kinetic energy of the system zero? Explain.

Q8–10 Since for a particle the kinetic energy is given by $K = \frac{1}{2} mv^2$ and the momentum by $\vec{p} = m\vec{v}$, it is easy to show that $K = p^2/2m$. How, then, is it possible to have an event during which the total momentum of the system is constant but the total kinetic energy changes?

Q8–11 In each of Examples 8–10, 8–11, 8–12, and 8–13 (Section 8–5), verify that the relative velocity vector of the two bodies has the same magnitude before and after the collision. In each case what happens to the *direction* of the relative velocity vector?

Q8–12 A glass dropped on the floor is more likely to break if the floor is concrete than if it is wood. Why? (Refer to Fig. 8–3.)

Q8–13 In Fig. 8–18 the kinetic energy of the spacecraft is larger after its interaction with Saturn than before. Where does the extra energy come from? Describe the event in terms of conservation of energy.

Q8–14 A machine gun is fired at a steel plate. Is the average force on the plate from the bullet impact greater if the bullets bounce off or if they are squashed and stick to the plate? Explain.

Q8–15 A net force of 4 N acts on an object initially at rest for 0.25 s and gives it a final speed of 5 m/s. How could a net force of 2 N produce the same final speed?

Q8–16 A net force with x-component ΣF_x acts on an object from time t_1 to time t_2. The x-component of the momentum of the object is the same at t_1 as it is at t_2, but ΣF_x is not zero at all times between t_1 and t_2. What can you say about the graph of ΣF_x versus t?

Q8–17 A tennis player hits a tennis ball with a racket. Consider the system made up of the ball and the racket. Is the total momentum of the system the same just before and just after the hit? Is the total momentum just after the hit the same as two seconds later, when the ball is in midair at the high point of its trajectory? Explain any differences between the two cases.

Q8–18 In Example 8–4 (Section 8–3) consider the system consisting of the rifle plus the bullet. What is the speed of the center of mass of the system after the rifle is fired? Explain.

Q8–19 An egg is released from rest at the roof of a building and falls to the ground. As the egg falls, what happens to the momentum of the system of the egg plus the earth?

Q8–20 A woman stands in the middle of a perfectly smooth, frictionless frozen lake. She can set herself in motion by throwing things, but suppose she has nothing to throw. Can she propel herself to shore *without* throwing anything?

Q8–21 In a zero-gravity environment, can a rocket-propelled spaceship ever attain a speed greater than the relative speed with which the burnt fuel is exhausted?

Q8–22 It is estimated that Supernova 1987A, a distance of 170,000 light years from the earth, emitted 10^{58} neutrinos. But two huge detectors on earth detected only 19 of them. Give at least two reasons why the number detected was so much smaller than the number emitted.

EXERCISES

SECTION 8–2 MOMENTUM AND IMPULSE

8–1 a) What is the magnitude of the momentum of a 10,000-kg truck whose speed is 12.0 m/s? b) What speed would a 2,000-kg sport utility vehicle have to attain in order to have i) the same momentum? ii) the same kinetic energy?

8–2 In Example 8–1 (Section 8–2), show that the iceboat of mass $2m$ has $\sqrt{2}$ times as much momentum at the finish line as does the iceboat of mass m.

8–3 a) Show that the kinetic energy K and the momentum magnitude p of a particle of mass m are related by $K = p^2/2m$. b) A 0.040-kg cardinal (*Richmondena cardinalis*) and a 0.145-kg baseball have the same kinetic energy. Which has the greater magnitude of momentum? What is the ratio of the cardinal's magnitude of momentum to the baseball's? c) A 700-N man and a 450-N woman have the same momentum. Which has the greater kinetic energy? What is the ratio of the man's kinetic energy to that of the woman?

8–4 A soccer ball of mass 0.420 kg is traveling with a speed of 4.50 m/s at an angle of 20.0° counterclockwise from the +x-axis (Fig. 8–30). What are the ball's x- and y-components of momentum?

8–5 A baseball of mass 0.145 kg is moving in the +y-direction with a speed of 1.30 m/s, and a tennis ball of mass 0.0570 kg is moving in the −y-direction with a speed of 7.80 m/s. What are

4.50 m/s

20.0°

$m = 0.420$ kg

FIGURE 8–30 Exercise 8–4.

the magnitude and direction of the total momentum of the system consisting of the two balls?

8–6 A golf ball of mass 0.045 kg is moving in the +x-direction with a speed of 9.00 m/s, and a baseball of mass 0.145 kg is moving in the −y-direction with a speed of 7.00 m/s. What are the magnitude and direction of the total momentum of the system consisting of the two balls?

8–7 Force of a Golf Swing. A 0.0450-kg golf ball initially at rest is given a speed of 25.0 m/s when it is struck by a club. If the club and ball are in contact for 2.00 ms, what average force acts on the ball? Is the effect of the ball's weight during the time of contact significant? Why or why not?

8–8 Force of a Baseball Swing. A baseball has a mass of 0.145 kg. a) If the velocity of a pitched ball has a magnitude of 45.0 m/s and the batted ball's velocity is 55.0 m/s in the opposite direction, find the magnitude of the change in momentum of the ball and of the impulse applied to it by the bat. b) If the ball remains in contact with the bat for 2.00 ms, find the magnitude of the average force applied by the bat.

8–9 A 0.160-kg hockey puck is moving on an icy, frictionless, horizontal surface. At $t = 0$ the puck is moving to the right at 3.00 m/s. a) Calculate the velocity of the puck (magnitude and direction) after a force of 25.0 N directed to the right has been applied for 0.050 s. b) If instead a force of 12.0 N directed to the left is applied from $t = 0$ to $t = 0.050$ s, what is the final velocity of the puck?

8–10 An engine of the orbital maneuvering system (OMS) on a space shuttle exerts a force of $(26{,}700 \text{ N})\hat{j}$ for 3.90 s, exhausting a negligible mass of fuel relative to the 95,000-kg mass of the shuttle. a) What is the impulse of the force for this 3.90 s? b) What is the shuttle's change in momentum from this impulse? c) What is the shuttle's change in velocity from this impulse? d) Why can't we find the resulting change in kinetic energy of the shuttle?

8–11 Coach Johnson's bat exerts a horizontal force on a 0.145-kg baseball of $\vec{F} = [(1.60 \times 10^{7} \text{ N/s})t - (6.00 \times 10^{9} \text{ N/s}^{2})t^{2}]\hat{i}$ between $t = 0$ and $t = 2.50$ ms. At $t = 0$ the baseball's velocity is $-(40.0\hat{i} + 5.0\hat{j})$ m/s. a) Calculate the impulse exerted by the bat on the ball during the 2.50 ms that they are in contact. b) Calculate the impulse exerted by gravity on the ball during this time interval. c) Calculate the average force exerted by the bat on the ball during this time interval. d) Calculate the momentum and the velocity of the baseball at $t = 2.50$ ms.

8–12 A 0.145-kg baseball is struck by a bat. Just before impact, the ball is traveling horizontally to the right at 50.0 m/s, and it leaves the bat traveling to the left at an angle of 30° above horizontal with a speed of 65.0 m/s. If the ball and bat are in contact for 1.75 ms, find the horizontal and vertical components of the average force on the ball.

8–13 A net force $\Sigma F_x(t) = A + Bt^2$ in the +x-direction is applied to a girl on roller skates. The girl has mass m. The force starts at $t_1 = 0$ and continues until $t = t_2$. a) What is the impulse J_x of the force? b) If the girl is initially at rest, what is her speed at time t_2?

SECTION 8–3 CONSERVATION OF MOMENTUM

8–14 Frustrated by having the goalie block his shots, a 75.0-kg hockey player standing on ice throws a 0.160-kg puck horizontally at the net with a speed of 20.0 m/s. With what speed and in what direction will the hockey player begin to move if there is no friction between his feet and the ice?

8–15 You are standing on a sheet of ice that covers the football stadium parking lot in Buffalo; there is negligible friction between your feet and the ice. A friend throws you a 0.400-kg ball that is traveling horizontally at 10.0 m/s. Your mass is 70.0 kg. a) If you catch the ball, with what speed do you and the ball move afterwards? b) If the ball hits you and bounces off your chest, so afterwards it is moving horizontally at

8.0 m/s in the opposite direction, what is your speed after the collision?

8–16 On a frictionless horizontal air table, puck A (with mass 0.250 kg) is moving toward puck B (with mass 0.350 kg), which is initially at rest. After the collision, puck A has a velocity of 0.120 m/s to the left, and puck B has velocity 0.650 m/s to the right. a) What was the speed of puck A before the collision? b) Calculate the change in the total kinetic energy of the system that occurs during the collision.

8–17 Energy Change During a Hip Check. Ice hockey star Wayne Gretzky is skating at 13.0 m/s toward a defender, who in turn is skating at 5.00 m/s toward Gretzky (Fig. 8–31). Gretzky's weight is 756 N; that of the defender is 900 N. Immediately after the collision Gretzky is moving at 1.50 m/s in his original direction. Neglect external horizontal forces applied by the ice to the skaters during the collision. a) What is the velocity of the defender immediately after the collision? b) Calculate the change in the combined kinetic energy of the two players.

13.0 m/s

5.0 m/s

$w = 756$ N

$w = 900$ N

FIGURE 8–31 Exercise 8–17.

8–18 The expanding gases that leave the muzzle of a rifle also contribute to the recoil. A .30 caliber bullet has a mass of 0.00720 kg and a speed of 601 m/s relative to the muzzle when fired from a rifle that has a mass of 2.80 kg. The loosely held rifle recoils at a speed of 1.85 m/s relative to the earth. Find the momentum of the propellant gases in a coordinate system attached to the earth as they leave the muzzle of the rifle.

8–19 Block A in Fig. 8–32 (see next page) has a mass of 1.00 kg, and block B has a mass of 3.00 kg. The blocks are forced together, compressing a spring S between them; then the system is released from rest on a level, frictionless surface. The spring, which has negligible mass, is not fastened to either block and drops to the surface after it has expanded. Block B acquires a speed of 1.20 m/s. a) What is the final speed of block A? b) How much potential energy was stored in the compressed spring?

8–20 One of James Bond's adversaries is standing on a frozen lake; there is no friction between his feet and the ice. He throws his steel-lined hat with a velocity of 22.0 m/s at 36.9° above the horizontal, hoping to hit James. If his mass is 120 kg and that of his hat is 4.50 kg, what is the magnitude of his horizontal recoil velocity?

FIGURE 8-32 Exercise 8-19.

8-21 A ceramic penguin on top of your television set suddenly bursts apart into two pieces. One piece, mass m_A, travels off to the left with speed v_A. The other piece, mass m_B, travels off to the right with speed v_B. a) Use conservation of momentum to solve for v_B in terms of m_A, m_B, and v_A. b) Use the results of part (a) to show that $K_A/K_B = m_B/m_A$, where K_A and K_B are the kinetic energies of the two pieces.

8-22 Two ice skaters, Daniel (mass 65.0 kg) and Rebecca (mass 45.0 kg), are practicing. Daniel stops to tie his shoelace and, while at rest, is struck by Rebecca, who is moving at 13.0 m/s before she collides with him. After the collision Rebecca has a velocity of magnitude 8.00 m/s at an angle of 53.1° from her initial direction. Both skaters move on the frictionless horizontal surface of the rink. a) What are the magnitude and direction of Daniel's velocity after the collision? b) What is the change in total kinetic energy of the two skaters as a result of the collision?

8-23 Ken and Kim are skating together on a rink at 3.00 m/s. Ken keeps asking Kim how much she weighs. Annoyed, Kim pushes away from Ken so that she speeds up to 4.00 m/s and he slows down to 2.25 m/s in the same direction. Friction, in the physics sense, is negligible in this drama. If Ken weighs 700 N, what does Kim weigh?

8-24 An open-topped freight car of mass 24,000 kg is coasting without friction along a level track. It is raining very hard, and the rain is falling vertically downward. The car is originally empty and moving with a speed of 4.00 m/s. What is the speed of the car after it has collected 3000 kg of rainwater?

8-25 A hockey puck B rests on a smooth ice surface and is struck by a second puck A, which was originally traveling at 40.0 m/s and which is deflected 30.0° from its original direction (Fig. 8-33). Puck B acquires a velocity at a 45.0° angle to the original direction of A. The pucks have the same mass. a) Compute the speed of each puck after the collision. b) What fraction of the original kinetic energy of puck A is dissipated during the collision?

FIGURE 8-33 Exercise 8-25.

SECTION **8-4 INELASTIC COLLISIONS**

8-26 On a greasy, essentially frictionless lunch counter, a 0.500-kg submarine sandwich moving 3.00 m/s to the left col-

lides with an 0.250-kg grilled cheese sandwich moving 1.20 m/s to the right. a) If the two sandwiches stick together, what is the final velocity? b) How much mechanical energy is dissipated in the collision?

8-27 Your 1050-kg sports car, parked on a hill without the parking brake being set, has rolled to the bottom of the hill and is moving westbound on a level road at a speed of 15.0 m/s. A truck driver driving east on the same road sees the sports car coming, and decides to stop the car by running his 6320-kg truck into it. The two vehicles remain locked together after the collision. a) If the truck is moving at 10.0 m/s when it hits your sports car head-on, what is the velocity (magnitude and direction) of the two vehicles just after the collision? b) At what speed should the truck be moving so that the truck and car are both stopped in the collision? c) Find the change in kinetic energy of the system of two vehicles for the situation of part (a) and the situation of part (b). For which situation is the change in kinetic energy greater in magnitude?

8-28 On a very muddy football field, a 110-kg linebacker tackles an 85-kg halfback. Immediately before the collision, the linebacker is slipping with a velocity of 8.8 m/s north and the halfback is sliding with a velocity of 7.2 m/s east. What is the velocity (magnitude and direction) at which the two players move together immediately after the collision?

8-29 In Dallas the morning after a winter ice storm, a 1400-kg automobile going westward on Chestnut Street at 35.0 km/h collides with a 2800-kg truck going southward *across* Chestnut Street at 50.0 km/h. If they become coupled on collision, what are the magnitude and direction of their velocity after colliding? Friction forces between the vehicles and the icy road can be neglected.

8-30 At the intersection of Texas Avenue and University Drive, a blue subcompact car of mass 950 kg traveling east on University collides with a maroon pickup truck of mass 1900 kg that is traveling north on Texas and ran a red light (Fig. 8-34). The two vehicles stick together as a result of the collision, and after the collision the wreckage is sliding at 16.0 m/s in the direction 24.0° east of north. Calculate the speed of each vehicle before the collision. The collision occurs during a heavy rainstorm; friction forces between the vehicles and the wet road can be neglected.

FIGURE 8-34 Exercise 8-30.

8–31 A 5.00-g bullet is fired horizontally into a 1.20-kg wooden block resting on a horizontal surface. The coefficient of kinetic friction between block and surface is 0.20. The bullet remains embedded in the block, which is observed to slide 0.230 m along the surface before stopping. What was the initial speed of the bullet?

8–32 **A Ballistic Pendulum.** A 12.0-g rifle bullet is fired with a speed of 380 m/s into a ballistic pendulum with a mass of 6.00 kg, suspended from a cord 70.0 cm long. (See Example 8–8 in Section 8–4.) Compute a) the vertical height through which the pendulum rises; b) the initial kinetic energy of the bullet; c) the kinetic energy of the bullet and pendulum immediately after the bullet becomes embedded in the pendulum.

SECTION 8–5 ELASTIC COLLISIONS

8–33 A 0.150-kg glider is moving to the right on a horizontal, frictionless air track with a speed of 0.80 m/s. It makes a head-on collision with a 0.300-kg glider that is moving to the left with a speed of 2.20 m/s. Find the final velocity (magnitude and direction) of each glider if the collision is elastic.

8–34 A 10.0-g marble slides to the left with a velocity of magnitude 0.400 m/s on the frictionless horizontal surface of an icy New York sidewalk and makes a head-on elastic collision with a larger 30.0-g marble sliding to the right with a velocity of magnitude 0.200 m/s (Fig. 8–35). a) Find the velocity of each marble (magnitude and direction) after the collision. (Since the collision is head-on, all the motion is along a line.) b) Calculate the *change in momentum* (that is, the momentum after the collision minus the momentum before the collision) for each marble. Compare the values you get for each marble. c) Calculate the *change in kinetic energy* (that is, the kinetic energy after the collision minus the kinetic energy before the collision) for each marble. Compare the values you get for each marble.

FIGURE 8–35 Exercise 8–34.

8–35 Supply the details of the calculation of α and β in Example 8–13 (Section 8–5).

8–36 Canadian nuclear reactors use *heavy water* moderators in which elastic collisions occur between the neutrons and deuterons of mass 2.0 u. (See Example 8–11 in Section 8–5.) a) What is the speed of a neutron, expressed as a fraction of its original speed, after a head-on elastic collision with a deuteron which is initially at rest? b) What is its kinetic energy, expressed as a fraction of its original kinetic energy? c) How many such successive collisions will reduce the speed of a neutron to 1/59,000 of its original value?

8–37 You are at the controls of a particle accelerator, sending a beam of 1.50×10^7 m/s protons (mass m) at a gas target of an unknown element. Your detector tells you that some protons bounce straight back after a collision with one of the nuclei of the unknown element. All such protons rebound with a speed of 1.20×10^7 m/s. Assume that the initial speed of the target nucleus is negligible and that the collision is elastic. a) Find the mass of one nucleus of the unknown element. Express your answer in terms of the proton mass m. b) What is the speed of the unknown nucleus immediately after such a collision?

SECTION 8–6 CENTER OF MASS

8–38 Three odd-shaped blocks of chocolate have the following masses and center-of-mass coordinates: (1) 0.300 kg, (0.200 m, 0.300 m); (2) 0.400 kg, (0.100 m, –0.400 m); (3) 0.200 kg, (–0.300 m, 0.600 m). Find the coordinates of the center of mass of the system of three chocolate blocks.

8–39 Find the position of the center of mass of the system of the sun and Jupiter. (Since Jupiter is more massive than the rest of the planets put together, this is essentially the position of the center of mass of the solar system.) Does the center of mass lie inside or outside the sun? Use the data in Appendix F.

8–40 A 1200-kg station wagon is moving along a straight highway at 12.0 m/s. Another car, with mass 1800 kg and speed 20.0 m/s, has its center of mass 40.0 m ahead of the center of mass of the station wagon (Fig. 8–36). a) Find the position of the center of mass of the system consisting of the two automobiles. b) Find the magnitude of the total momentum of the system from the above data. c) Find the speed of the center of mass of the system. d) Find the total momentum of the system, using the speed of the center of mass. Compare your result with that of part (b).

FIGURE 8–36 Exercise 8–40.

8–41 At one instant, the center of mass of a system of two particles is located on the x-axis at $x = 2.0$ m and has a velocity of $(5.0$ m/s$)\hat{\imath}$. One of the particles is at the origin. The other particle has a mass of 0.10 kg and is at rest on the x-axis at $x = 8.0$ m. a) What is the mass of the particle at the origin? b) Calculate the total momentum of this system. c) What is the velocity of the particle at the origin?

8–42 In Example 8–15 (Section 8–6) Ramon pulls on the rope to give himself a speed of 0.70 m/s. What is the speed of James?

8–43 A system consists of two particles. At $t = 0$ one particle is at the origin; the other, which has a mass of 0.50 kg, is on the y-axis at $y = 6.0$ m. At $t = 0$ the center of mass of the system is on the y-axis at $y = 2.4$ m. The velocity of the center of mass is given by $(0.75$ m/s$^3)t^2\hat{\imath}$. a) Find the total mass of the system. b) Find the acceleration of the center of mass at any time t. c) Find the net external force acting on the system at $t = 3.0$ s.

8–44 A radio-controlled model airplane has a momentum given by $[(-0.75$ kg $\cdot$ m/s$^3)t^2 + (3.0$ kg $\cdot$ m/s$)]\hat{\imath} + (0.25$ kg $\cdot$ m/s$^2)t\hat{\jmath}$. a) What are the x-, y-, and z-components of the net force on the airplane? b) At what time t is the x-component of the net force on the airplane equal to zero?

*SECTION 8–7 ROCKET PROPULSION

*8–45 A small rocket burns 0.0500 kg of fuel per second, ejecting it as a gas with a velocity relative to the rocket of magnitude 1600 m/s. a) What is the thrust of the rocket? b) Would the rocket operate in outer space where there is no atmosphere? If so, how would you steer it? Could you brake it?

*8–46 A 70-kg astronaut floating in space in a 110-kg MMU (manned maneuvering unit) experiences an acceleration of 0.029 m/s^2 when he fires one of the MMU's thrusters. a) If the speed of the escaping N$_2$ gas relative to the astronaut is 490 m/s, how much gas is used by the thruster in 5.0 s? b) What is the thrust of the thruster?

*8–47 A rocket is fired in deep space, where gravity is negligible. If the rocket has an initial mass of 6000 kg and ejects gas at a relative velocity of magnitude 2000 m/s, how much gas must it eject in the first second to have an initial acceleration of 25.0 m/s^2?

*8–48 A rocket is fired in deep space, where gravity is negligible. In the first second it ejects $\frac{1}{160}$ of its mass as exhaust gas and has an acceleration of 15.0 m/s^2. What is the speed of the exhaust gas relative to the rocket?

*8–49 A C 6-5 model rocket engine has an impulse of 10.0 N $\cdot$ s for 1.70 s, while burning 0.0125 kg of propellant. It has a maximum thrust of 13.3 N. The initial mass of the engine plus propellant is 0.0258 kg. a) What fraction of the maximum thrust is the average thrust? b) Calculate the relative speed of the exhaust gases, assuming it is constant. c) Assuming that the relative speed of the exhaust gases is constant, find the final speed of the engine if it was attached to a very light frame and fired from rest in gravity-free outer space.

*8–50 A single-stage rocket is fired from rest from a deep-space platform, where gravity is negligible. If the rocket burns its fuel in a time of 50.0 s and the relative speed of the exhaust gas is $v_{ex} = $ 2100 m/s, what must the mass ratio m_0/m for a final speed v of 8.00 km/s (about equal to the orbital speed of an earth satellite)?

*8–51 Obviously rockets can be made to go very fast, but what is a reasonable top speed? Assume that a rocket is fired from rest at a space station in deep space, where gravity is negligible. a) If the rocket ejects gas at a relative speed of 2000 m/s and you want the rocket's speed eventually to be $1.00 \times 10^{-3}\ c$, where c is the speed of light, what fraction of the initial mass of the rocket and fuel is *not* fuel? b) What is this fraction if the final speed is to be 3000 m/s?

SECTION 8–8 THE NEUTRINO: A CASE STUDY IN MODERN PHYSICS

8–52 A ^{232}Th (thorium) nucleus at rest decays to a ^{228}Ra (radon) nucleus with the emission of an alpha particle. The total kinetic energy of the decay fragments is 6.54×10^{-13} J. An alpha particle has 1.76% of the mass of a ^{228}Ra nucleus. Calculate the kinetic energy of a) the recoiling ^{228}Ra nucleus; b) the alpha particle.

8–53 In a certain alpha decay the kinetic energy of the alpha particle is 9.650×10^{-13} J and the Q value for the decay is 9.850×10^{-13} J. What is the mass of the recoiling nucleus?

8–54 A ^{210}Bi (bismuth) nucleus at rest undergoes beta decay to ^{210}Po (polonium). Suppose that the emitted electron moves to the right with a momentum (calculated using the theory of relativity) of 5.60×10^{-22} kg $\cdot$ m/s. The ^{210}Po nucleus, with a mass of 3.50×10^{-25} kg, recoils to the left at a speed of 1.14×10^3 m/s. Calculate the magnitude and direction of the momentum of the antineutrino that is emitted in the decay. (The ^{210}Po nucleus moves at much less than the speed of light, so the theory of relativity is not needed to calculate its momentum. Instead, the nonrelativistic expression in Eq. (8–2) can be used.)

8–55 A ^{210}Bi (bismuth) nucleus at rest undergoes β^- decay to ^{210}Po (polonium). In a particular decay event the electron and antineutrino are emitted at right angles to one another. The magnitudes of their momenta are 3.60×10^{-22} kg $\cdot$ m/s for the electron and 5.20×10^{-22} kg $\cdot$ m/s for the antineutrino. The ^{210}Po nucleus has a mass of 3.50×10^{-25} kg. Calculate a) the magnitude of the momentum of the recoiling ^{210}Po nucleus; b) the kinetic energy of the recoiling ^{210}Po nucleus.

PROBLEMS

8–56 A steel ball of mass 40.0 g is dropped from a height of 2.00 m onto a horizontal steel slab. The ball rebounds to a height of 1.60 m. a) Calculate the impulse delivered to the ball during impact. b) If the ball is in contact with the slab for 2.00 ms, find the average force on the ball during impact.

8–57 The net force acting on a 2.00-kg discus while it is being thrown is $(\alpha t^2)\hat{i} + (\beta + \gamma t)\hat{j}$, where $\alpha = 25.0$ N/s^2, $\beta = 30.0$ N, and $\gamma = 5.0$ N/s. If the discus was originally at rest, what is its velocity after the net force has acted for 0.500 s? Express your answer in terms of the unit vectors $\hat{i}$ and $\hat{j}$.

8–58 Just before being struck by a racket, a tennis ball weighing 0.560 N has a velocity of $(20.0\ \text{m/s})\hat{i} - (4.0\ \text{m/s})\hat{j}$. During the 3.00 ms that the racket and ball are in contact, the net force on the ball is constant and equal to $-(380\ \text{N})\hat{i} + (110\ \text{N})\hat{j}$. a) What are the x- and y-components of the impulse of the net force applied to the ball? b) What are the x- and y-components of the final velocity of the ball?

8–59 Three coupled railroad cars roll along and couple with a fourth car, which is initially at rest. These four roll along and couple to a fifth car initially at rest. This process continues until the speed of the final collection of railroad cars is one-fifth the speed of the initial three railroad cars. All the cars are identical. Ignoring friction, how many cars are in the final collection?

8–60 A 1500-kg blue convertible is traveling south, and a 2000-kg red sports utility vehicle is traveling west. If the total momentum of the system consisting of the two cars is 8000 kg $\cdot$ m/s directed at 60.0° west of south, what is the speed of each car?

8–61 Three identical pucks on a horizontal air table have repelling magnets. They are held together, then released simultaneously. Each has the same speed at any instant. One puck moves due west. What is the direction of the velocity of each of the other two pucks?

8–62 Spheres A (mass 0.020 kg), B (mass 0.030 kg), and C (mass 0.050 kg), are each approaching the origin as they slide on a frictionless air table (Fig. 8–37). The initial velocities of A and B are given in the figure. All three spheres arrive at the origin at the same time and stick together. a) What must be the x- and y-components of the initial velocity of C if all three objects are to end up moving at 0.50 m/s in the $+x$-direction after the collision? b) If C has the velocity found in part (a), what is the change in the kinetic energy of the system of three spheres as a result of the collision?

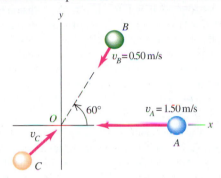

FIGURE 8–37 Problem 8–62.

8–63 A railroad handcar is moving along straight frictionless tracks with negligible air resistance. In the following cases, the car initially has a total mass (car and contents) of 200 kg and is traveling east with a velocity of magnitude 5.00 m/s. Find the *final velocity* of the car in each case, assuming that the handcar does not leave the tracks. a) A 25.0-kg mass is thrown sideways out of the car with a velocity of magnitude 2.00 m/s relative to the car's initial velocity. b) A 25.0-kg mass is thrown backward out of the car with a velocity of 5.00 m/s relative to the initial motion of the car. c) A 25.0-kg mass is thrown into the car with a velocity of 6.00 m/s relative to the ground and opposite in direction to the initial velocity of the car.

8–64 A railroad hopper car filled with sand is rolling with an initial speed of 15.0 m/s on straight horizontal tracks. Ignore frictional forces on the railroad car. The total mass of the car plus sand is 85,000 kg. The hopper door is not fully closed so sand leaks out the bottom. After 20 minutes, 13,000 kg of sand has leaked out. What then is the speed of the railroad car? (Compare your analysis to that used to solve Exercise 8–24.)

8–65 At a classic auto show, a 840-kg 1955 Nash Metropolitan motors by at 9.0 m/s, followed by a 1620-kg 1957 Packard Clipper purring past at 5.0 m/s. a) Which car has the greater kinetic energy? What is the ratio of the kinetic energy of the Nash to that of the Packard? b) Which car has the greater magnitude of momentum? What is the ratio of the magnitude of momentum of the Nash to that of the Packard? c) Let F_N be the net force required to stop the Nash in time t, and let F_P be the net force required to stop the Packard in the same time. Which is larger, F_N or F_P? What is the ratio F_N/F_P of these two forces? d) Now let F_N be the net force required to stop the Nash in a distance d, and let F_P be the net force required to stop the Packard in the same distance. Which is larger, F_N or F_P? What is the ratio F_N/F_P?

8–66 A soldier on a firing range fires an 8-shot burst from an assault weapon at a full automatic rate of 1000 rounds per minute. Each bullet has a mass of 7.45 g and a speed of 293 m/s relative to the ground as it leaves the barrel of the weapon. Calculate the average recoil force exerted on the weapon during that burst.

8–67 A 0.150-kg frame, when suspended from a coil spring, stretches the spring 0.050 m. A 0.200-kg lump of putty is dropped from rest onto the frame from a height of 30.0 cm (Fig. 8–38). Find the maximum distance the frame moves downward from its initial position.

FIGURE 8–38 Problem 8–67.

8–68 A rifle bullet of mass 8.00 g strikes and embeds itself in a block of mass 0.992 kg that rests on a frictionless horizontal surface and is attached to a coil spring (Fig. 8–39). The impact compresses the spring 15.0 cm. Calibration of the spring shows that a force of 0.750 N is required to compress the spring 0.250 cm. a) Find the magnitude of the velocity of the block just after impact. b) What was the initial speed of the bullet?

FIGURE 8–39 Problem 8–68.

8–69 A Ricocheting Bullet. A 0.100-kg stone rests on a frictionless horizontal surface. A bullet of mass 6.00 g, traveling horizontally at 350 m/s, strikes the stone and rebounds horizontally at right angles to its original direction with a speed of 250 m/s. a) Compute the magnitude and direction of the velocity of the stone after it is struck. b) Is the collision perfectly elastic?

8–70 A movie stuntman (mass 80.0 kg) stands on a window ledge 5.0 m above the floor (Fig. 8–40). Grabbing a rope attached to a chandelier, he swings down to grapple with the movie's villain (mass 70.0 kg), who is standing directly under the chandelier. (Assume that the stuntman's center of mass moves downward 5.0 m. He releases the rope just as he reaches the villain.) a) With what speed do the entwined foes start to slide across the floor? b) If the coefficient of kinetic friction of their bodies with the floor is $\mu_k = 0.250$, how far do they slide?

5.0 m

$m = 80.0$ kg

$m = 70.0$ kg

FIGURE 8–40 Problem 8–70.

8–71 A 4.00-g bullet traveling horizontally with a velocity of magnitude 400 m/s is fired into a wooden block with a mass of 0.800 kg, initially at rest on a level surface. The bullet passes through the block and emerges with its speed reduced to 120 m/s. The block slides a distance of 45.0 cm along the surface from its initial position. a) What is the coefficient of kinetic friction between block and surface? b) What is the decrease in kinetic energy of the bullet? c) What is the kinetic energy of the block at the instant after the bullet passes through it?

8–72 A 5.00-g bullet is shot *through* a 1.00-kg wood block suspended on a string 2.000 m long. The center of mass of the block rises a distance of 0.45 cm. Find the speed of the bullet as it emerges from the block if its initial speed is 450 m/s.

8–73 A neutron of mass m makes a head-on elastic collision with a nucleus of mass M, which is initially at rest. a) Show that if the neutron's initial kinetic energy is K_0, the kinetic energy that it loses during the collision is $4mMK_0/(M + m)^2$. b) For what value of M does the incident neutron lose the most energy? c) When M has the value calculated in part (b), what is the speed of the neutron after the collision?

8–74 A blue puck of mass 0.0400 kg, sliding with a velocity of magnitude 0.200 m/s on a frictionless, horizontal air table, makes a perfectly elastic head-on collision with a red puck of mass m, initially at rest. After the collision the velocity of the blue puck is 0.050 m/s in the same direction as its initial velocity. Find a) the velocity (magnitude and direction) of the red puck after the collision; b) the mass m of the red puck.

8–75 Two asteroids with masses m_A and m_B are moving with velocities $\vec{v}_A$ and $\vec{v}_B$ with respect to an astronomer in a space vehicle. a) Show that the total kinetic energy as measured by the astronomer is

$$K = \frac{1}{2} M v_{cm}^2 + \frac{1}{2}(m_A v'_A{}^2 + m_B v'_B{}^2),$$

with $\vec{v}_{cm}$ and M defined as in Section 8–6, $\vec{v}'_A = \vec{v}_A - \vec{v}_{cm}$ and $\vec{v}'_B = \vec{v}_B - \vec{v}_{cm}$. In this expression the total kinetic energy of the two

asteroids is the energy associated with their center of mass plus that associated with the internal motion relative to the center of mass. b) If the asteroids collide with each other, what is the *minimum* possible kinetic energy they can have after the collision, as measured by the astronomer? Explain.

8–76 Suppose you hold a small ball in contact with and directly over the center of a large ball. If you then drop the small ball a short time after dropping the large ball, the small ball rebounds with surprising speed. To show the extreme case, neglect air resistance and suppose the large ball makes an elastic collision with the floor, then rebounds to make an elastic collision with the still-descending small ball. Just before the collision between the two balls, the large ball is moving upward with velocity $\vec{v}$, and the small ball has velocity $-\vec{v}$. (Do you see why?) Assume the large ball has a much greater mass than the small ball. a) What is the velocity of the small ball immediately after its collision with the large ball? b) From the answer to part (a), what is the ratio of the small ball's rebound distance to the distance it fell before the collision?

8–77 Jack and Jill are standing on a crate at rest on the frictionless horizontal surface of a frozen pond. Jack has mass 75.0 kg, Jill has mass 45.0 kg, and the crate has mass 15.0 kg. They remember that they must fetch a pail of water, so each jumps horizontally from the top of the crate. Just after each jumps, that person is moving away from the crate with a speed of 4.00 m/s relative to the crate. a) What is the final speed of the crate if both Jack and Jill jump simultaneously and in the same direction? (*Hint:* Use an inertial coordinate system attached to the ground.) b) What is the final speed of the crate if Jack jumps first and then a few seconds later Jill jumps in the same direction? c) What is the final speed of the crate if Jill jumps first and then Jack, again in the same direction?

8–78 A proton moving with speed v_{A1} in the +x-direction makes an elastic, off-center collision with an identical proton originally at rest. After impact the first proton moves with speed v_{A2} in the first quadrant at an angle α with the x-axis, and the second moves with speed v_{B2} in the fourth quadrant at an angle β with the x-axis (Fig. 8–10). a) Write the equations expressing conservation of linear momentum in the x- and y-directions. b) Square the equations from part (a) and add them. c) At this point, introduce the fact that the collision is elastic. d) Prove that $\alpha + \beta = \pi/2$. (You have shown that this equation is obeyed in any elastic off-center collision between objects of equal mass when one object is initially at rest.)

8–79 Hockey puck B rests on a smooth ice surface and is struck by a second puck A, which has the same mass. Puck A is initially traveling at 15.0 m/s and is deflected 25.0° from its initial direction. Assume that the collision is perfectly elastic. Find the final speed of each puck and the direction of B's velocity after the collision. (*Hint:* Use the relation derived in part (d) of Problem 8–78.)

8–80 Jonathan and Jane are sitting in a sleigh that is at rest on frictionless ice. Jonathan's weight is 800 N, Jane's weight is 600 N, and that of the sleigh is 1000 N. They see a poisonous spider on the floor of the sleigh and immediately jump off. Jonathan jumps to the left with a velocity (relative to the ice) of 5.00 m/s at 30.0° above the horizontal, and Jane jumps to the right at 7.00 m/s at 36.9° above the horizontal (relative to the ice). Calculate the sleigh's horizontal velocity (magnitude and direction) after they jump out.

8–81 The objects in Fig. 8–41 are constructed of uniform wire bent into the shapes shown. Find the position of the center of mass of each.

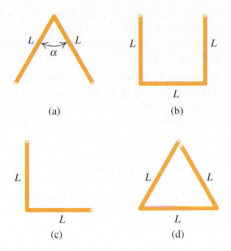

FIGURE 8–41 Problem 8–81.

8–82 A 45.0-kg woman stands up in a 60.0-kg canoe of length 5.00 m. She walks from a point 1.00 m from one end to a point 1.00 m from the other end (Fig. 8–42). If resistance to motion of the canoe in the water can be neglected, how far does the canoe move during this process?

FIGURE 8–42 Problem 8–82.

8–83 You are standing on a concrete slab, which in turn is resting on a frozen lake. Assume that there is no friction between the slab and the ice. The slab has a weight five times your weight. If you begin walking forward at 2.00 m/s, relative to the ice, with what speed relative to the ice does the slab move?

8–84 A 20.0-kg projectile is fired at an angle of 60.0° above the horizontal and with a speed of 80.0 m/s. At the highest point of its trajectory the projectile explodes into two fragments with equal mass, one of which falls vertically with zero initial speed. Ignore air resistance. a) How far from the point of firing does the other fragment strike if the terrain is level? b) How much energy is released during the explosion?

8–85 A Nuclear Reaction. Fission, the process that supplies energy in nuclear power plants, occurs when a heavy nucleus is split into two medium-sized nuclei. One such reaction occurs when a neutron colliding with a ^{235}U (uranium) nucleus splits that nucleus into a ^{141}Ba (barium) nucleus and a ^{92}Kr (krypton) nucleus. In this reaction, two neutrons also are split off from the original ^{235}U. Before the collision we have the arrangement in Fig. 8–43a. After the collision we have the ^{141}Ba nucleus moving in the +z-direction and the ^{92}Kr nucleus in the −z-direction. The three neutrons are moving in the xy-plane as shown in Fig. 8–43b. If the incoming neutron has an initial velocity of magnitude 3.0×10^3 m/s and a final velocity of magnitude 2.0×10^3 m/s in the directions shown, what are the speeds of the other two neutrons, and what can you say about the speeds of the ^{141}Ba and ^{92}Kr nuclei? (The mass of the ^{141}Ba nucleus is approximately 2.3×10^{-25} kg, and that of ^{92}Kr is about 1.5×10^{-25} kg.)

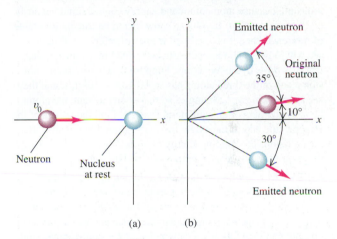

FIGURE 8–43 Problem 8–85.

8–86 Center-of-Mass Coordinate System. Puck A (mass m_A) is moving on a frictionless horizontal air table in the +x-direction with velocity $\vec{v}_{A1}$ and makes an elastic head-on collision with puck B (mass m_B) that is initially at rest. After the collision both pucks are moving along the x-axis. a) Calculate the velocity of the center of mass of the two-puck system before the collision. b) Consider a coordinate system whose origin is at the center of mass and moves with it. Is this an inertial reference frame? c) What are the initial velocities $\vec{u}_{A1}$ and $\vec{u}_{B1}$ of the two pucks in this center-of-mass reference frame? What is the total momentum in this frame? d) Use conservation of momentum and energy, applied in the center-of-mass reference frame, to relate the final momentum of each puck to its initial momentum and thus the final velocity of each puck to its initial velocity. Your results should show that a one-dimensional elastic collision has a very simple description in center-of-mass coordinates. e) Let $m_A = 0.400$ kg, $m_B = 0.200$ kg, and $v_{A1} = 6.00$ m/s. Find the center-of-mass velocities $\vec{u}_{A1}$ and $\vec{u}_{B1}$, apply the simple result found in part (d), and transform back to velocities in a stationary frame to find the final velocities of the pucks. Does your result agree with Eqs. (8–24) and (8–25)?

8–87 The *coefficient of restitution* ϵ for a head-on collision is defined as the ratio of the relative speed after the collision to the relative speed before. a) What is ϵ for a completely inelastic collision? b) What is ϵ for an elastic collision? c) A ball is dropped from a height h onto a stationary surface and rebounds back to a height H_1. Show that $\epsilon = \sqrt{H_1/h}$. d) A properly inflated basketball should have a coefficient of restitution 0.85. When dropped from a height of 1.2 m above a solid wood floor, to what height should a properly inflated basketball bounce? e) The height of the first bounce is H_1. If ϵ is constant, show that the height of the nth bounce is $H_n = \epsilon^{2n}h$. f) If ϵ is constant, what is the height of the eighth bounce of a properly inflated basketball dropped from 1.2 m?

8–88 Binding Energy of the Hydrogen Molecule. When two hydrogen atoms of mass m combine to form a diatomic hydrogen molecule (H_2), the potential energy of the system after they combine is $-\Delta$, where Δ is a positive quantity called the *binding energy* of the molecule. a) Show that in a collision that only involves two hydrogen atoms, it is *impossible* to form an H_2 molecule because momentum and energy cannot simultaneously be conserved. (*Hint:* If you can show this to be true in one frame of reference, then it is true in all frames of reference. Can you see why?) b) An H_2 molecule can be formed in a collision that involves *three* hydrogen atoms. Suppose that before such a collision, each of the three atoms has speed 1.00×10^3 m/s, and they are approaching at 120° angles so that at any instant, the atoms lie at the corners of an equilateral triangle. Find the speeds of the H_2 molecule and of the single hydrogen atom that remains after the collision. The binding energy of H_2 is $\Delta = 7.23 \times 10^{-19}$ J, and the mass of the hydrogen atom is 1.67×10^{-27} kg.

8–89 A wagon with two boxes of gold and having total mass 300 kg is cut loose from the horses by an outlaw when the wagon is at rest 50 m up a 6.0° slope (Fig. 8–44). The outlaw plans to have the wagon roll down the slope and across the level ground, and then fall into a canyon where his confederates wait. But in a tree 40 m from the canyon edge wait the Lone Ranger (mass 75.0 kg) and Tonto (mass 60.0 kg). They drop vertically into the wagon as it passes beneath them. a) If they require 5.0 s to grab the gold and jump out, will they make it before the wagon goes over the edge? The wagon rolls with negligible friction. b) When the two heroes drop into the wagon, is the kinetic energy of the system of the heroes plus the wagon conserved? If not, does it increase or decrease, and by how much?

FIGURE 8–44 Problem 8–89.

*8–90 In Section 8–7 we considered a rocket fired in outer space where there is no air resistance and where gravity is negligible. Suppose instead that the rocket is accelerating vertically upward from rest on the earth's surface. Continue to neglect air resistance and consider only that part of the motion

where the altitude of the rocket is small so that g may be assumed to be constant. a) How is Eq. (8–37) modified by the presence of the gravity force? b) Derive an expression for the acceleration a of the rocket, analogous to Eq. (8–39). c) What is the acceleration of the rocket in Example 8–16 (Section 8–7) if it is near the earth's surface rather than in outer space? Neglect air resistance. d) Find the speed of the rocket in Example 8–17 (Section 8–7) after 90 s if the rocket is fired from the earth's surface rather than in outer space. Neglect air resistance. How does your answer compare to the rocket speed calculated in Example 8–17?

*8–91 A Multistage Rocket. Suppose the first stage of a two-stage rocket has a total mass of 12,000 kg, of which 9000 kg is fuel. The total mass of the second stage is 1000 kg, of which 700 kg is fuel. Assume that the relative speed v_{ex} of ejected material is constant, and neglect any effect of gravity. (The latter effect is small during the firing period if the rate of fuel consumption is large.) a) Suppose the entire fuel supply carried by the two-stage rocket is utilized in a single-stage rocket of the same total mass of 13,000 kg. In terms of v_{ex}, what is the speed of the rocket, starting from rest, when its fuel is exhausted? b) For the two-stage rocket, what is the speed when the fuel of the first stage is exhausted if the first stage carries the second stage with it to this point? This speed then becomes the initial speed of the second stage. At this point the second stage separates from the first stage. c) What is the final speed of the second stage? d) What value of v_{ex} is required to give the second stage of the above rocket a speed of 7.00 km/s?

*8–92 The equation $F = -v_{ex}(dm/dt)$ for the thrust on a rocket, Eq. (8–38), can also be applied to an airplane propeller. In fact, there are *two* contributions to the thrust: one positive and one negative. The positive contribution comes from air pushed backward, away from the propeller (so $dm/dt < 0$), at a speed v_{ex} relative to the propeller. The negative contribution comes from this same quantity of air flowing into the front of the propeller (so $dm/dt > 0$) at speed v, equal to the speed of the airplane through the air. a) Write an equation for the *net* thrust developed by an aircraft propeller in terms of v, v_{ex}, and the absolute value $|dm/dt|$. b) For a Cessna 182 (a single-engine airplane) flying at 130 km/h, 150 kg of air flows through the propeller each second and the propeller develops a net thrust of 1300 N. Determine the speed increase (in km/h) that the propeller imparts to the air.

*8–93 For the rocket described in Examples 8–16 and 8–17 (Section 8–7), the mass of the rocket as a function of time is

$$m(t) = \begin{cases} m_0, & \text{for } t < 0 \\ m_0\left(1 - \dfrac{t}{120 \text{ s}}\right), & \text{for } 0 \le t \le 90 \text{ s} \\ m_0/4, & \text{for } t \ge 90 \text{ s}. \end{cases}$$

a) Calculate and graph the velocity of the rocket as a function of time from $t = 0$ s to $t = 100$ s. b) Calculate and graph the acceleration of the rocket as a function of time from $t = 0$ s to $t = 100$ s. c) A 75-kg astronaut lies on a reclined chair during the firing of the rocket. What is the maximum net force on the astronaut exerted by the chair during the firing? How does your answer compare to her weight on earth?

CHALLENGE PROBLEMS

8–94 You are going to entertain your Aunt Emma by pulling a tablecloth out from under the dishes at her birthday party. The birthday cake is resting on a tablecloth at the center of a round table of radius $r = 0.90$ m. The tablecloth is the same size as the table top. You grab the edge of the tablecloth and pull sharply. The tablecloth and cake are in contact for time t after you start pulling. Then the sliding cake is stopped (you hope) by the friction between the cake and table top. The coefficient of kinetic friction between the cake and tablecloth is $\mu_{k1} = 0.30$, and that between the cake and table top is $\mu_{k2} = 0.40$. Apply the impulse-momentum relation (Eq. 8–9) and the work-energy theorem (Eq. 6–6) to calculate the maximum value of t if the cake is not to end up on the floor. (*Hint:* Assume that the cake moves a distance d while still on the tablecloth and therefore a distance $r - d$ while sliding on the table top. Assume that the friction forces are independent of the relative speed of the sliding surfaces. You can easily try this trick yourself by pulling a sheet of paper out from under a glass of water, but have a mop handy just in case!)

8–95 In Section 8–6 we calculated the center of mass by considering objects composed of a *finite* number of point masses or objects that by symmetry could be represented by a finite number of point masses. For a solid object whose mass distribution does not allow for a simple determination of the center of mass by symmetry, the sums of Eqs. (8–28) must be generalized to integrals:

$$x_{cm} = \frac{1}{M} \int x \, dm, \qquad y_{cm} = \frac{1}{M} \int y \, dm,$$

where x and y are the coordinates of the small piece of the object that has mass dm. The integration is over the whole of the object. Consider a thin rod of length L, mass M, and cross-sectional area A. Let the origin of coordinates be at the left end of the rod and the positive x-axis lie along the rod. a) If the density $\rho = M/V$ of the object is uniform, perform the integration described above to show that the x-coordinate of the center of mass of the rod is at its geometrical center. b) If the density of the object varies linearly with x—that is, $\rho = \alpha x$, where α is a positive constant—calculate the x-coordinate of the rod's center of mass.

8–96 Use the methods of Challenge Problem 8–95 to calculate the x- and y-coordinates of the center of mass of a semicircular metal plate with uniform density ρ and thickness t. Let the radius of the plate be a. The mass of the plate is thus $M = \frac{1}{2} \rho \pi a^2 t$. Use the coordinate system indicated in Fig. 8–45.

8–97 One-fourth of a rope of length l is hanging down over the edge of a frictionless table. The rope has a uniform linear density (mass per unit length) λ ("lambda"), and the end already on the table is held by a person. How much work does the person do when she pulls on the rope to slowly raise the rest of the rope onto the table? Do the problem in two ways as follows. a) Find the force that the person must exert to raise the rope and from this the work done. Note that this is a variable force because at different times different amounts of rope are hanging over the edge. b) Suppose the segment of the rope initially hanging over the edge of the table has all of its mass concentrated at its center of mass. Find the work necessary to raise this to table height. You will probably find this approach simpler than that of part (a). How do the answers compare, and why is this so?

*8–98 A Variable-Mass Raindrop. In a rocket-propulsion problem the mass is variable. Another such problem is a raindrop falling through a cloud of small water droplets. Some of these small droplets adhere to the raindrop, thereby *increasing* its mass as it falls. The force on the raindrop is

$$F_{ext} = \frac{dp}{dt} = m\frac{dv}{dt} + v\frac{dm}{dt}.$$

Suppose the mass of the raindrop depends on the distance x that it has fallen. Then $m = kx$, where k is a constant, and $dm/dt = kv$. This gives, since $F_{ext} = mg$,

$$mg = m\frac{dv}{dt} + v(kv).$$

Or, dividing by k,

$$xg = x\frac{dv}{dt} + v^2.$$

This is a differential equation that has a solution of the form $v = at$, where a is the acceleration and is constant. Take the initial velocity of the raindrop to be zero. a) Using the proposed solution for v, find the acceleration a. b) Find the distance the raindrop has fallen in $t = 3.00$ s. c) Given that $k = 2.00$ g/m, find the mass of the raindrop at $t = 3.00$ s. For many more intriguing aspects of this problem see K. S. Krane, *Amer. Jour. Phys.,* Vol. 49 (1981), pp. 113–117.

FIGURE 8–45 Challenge Problem 8–96.

9

Rotation of Rigid Bodies

No matter where you sit on a rotating amusement park ride, you have the same angular velocity and angular acceleration. But your *linear* velocity and *linear* acceleration depend on how far you sit from the axis of rotation. In general, the farther you are from the axis, the greater the linear velocity and acceleration and the wilder the ride.

9-1 INTRODUCTION

What do the motions of a compact disc, a Ferris wheel, a circular saw blade, and a ceiling fan have in common? None of these can be represented adequately as a moving *point;* each involves a body that *rotates* about an axis that is stationary in some inertial frame of reference. Rotation occurs at all scales, from the motion of electrons in atoms to the motions of entire galaxies. We need to develop some general methods for analyzing the motion of a rotating body. In this chapter and the next we consider bodies that have definite size and definite shape, and that in general can have rotational as well as translational motion.

Real-world bodies can be even more complicated; the forces that act on them can deform them—stretching, twisting, and squeezing them. We'll neglect these deformations for now and assume that the body has a perfectly definite and unchanging shape and size. We call this idealized model a **rigid body.** This chapter and the next are mostly about rotational motion of a rigid body. We begin with kinematic language for *describing* rotational motion. Next we look at the kinetic energy of rotation, the key to using energy methods for rotational motion. Then in Chapter 10 we'll develop dynamic principles that relate the forces on a body to its rotational motion.

9-2 ANGULAR VELOCITY AND ACCELERATION

In analyzing rotational motion, let's think first about a rigid body that rotates about a fixed axis. By *fixed axis* we mean an axis that is at rest in some inertial frame of reference and does not change direction relative to that frame. The rotating rigid body might be a motor shaft, a chunk of roast beef on a barbecue skewer, or a merry-go-round.

Figure 9–1 shows a rigid body rotating about a fixed axis that passes through point O and is perpendicular to the plane of the diagram, which we choose to call the xy-plane. One way to describe the rotation of this body would be to choose a particular point P on the body and to keep track of the x- and y-coordinates of this point. This isn't a terribly convenient method, since it takes two numbers (the two coordinates x and y) to specify the rotational position of the body. Instead, we notice that the line OP is fixed in the body and rotates with it. The angle θ that this line makes with the $+x$-axis describes the rotational position of the body; we will use this single quantity θ as a *coordinate* for rotation.

The angular coordinate θ of a rigid body rotating around a fixed axis can be positive or negative. If we choose positive angles to be measured counterclockwise from the positive x-axis, then the angle θ in Fig. 9–1 is positive. If we instead choose the positive rotation direction to be clockwise, then θ in Fig. 9–1 is negative. When we considered the motion of a particle along a straight line, it was essential to specify the direction of positive displacement along that line; in discussing rotation around a fixed axis, it's just as essential to specify the direction of positive rotation.

In describing rotational motion, the most natural way to measure the angle θ is not in degrees, but in **radians.** As shown in Fig. 9–2a, one radian (1 rad) is the angle subtended at the center of a circle by an arc with a length equal to the radius of the circle. In Fig. 9–2b an angle θ is subtended by an arc of length s on a circle of radius r. The value of θ (in radians) is equal to s divided by r:

$$\theta = \frac{s}{r}, \quad \text{or} \quad s = r\theta. \tag{9–1}$$

An angle in radians is the ratio of two lengths, so it is a pure number, without dimensions. If $s = 3.0$ m and $r = 2.0$ m, then $\theta = 1.5$, but we will often write this as 1.5 rad to distinguish it from an angle measured in degrees or revolutions.

The circumference of a circle (that is, the arc length all the way around the circle) is 2π times the radius, so there are 2π (about 6.283) radians in one complete revolution (360°). Therefore

$$1 \text{ rad} = \frac{360°}{2\pi} = 57.3°.$$

Similarly, $180° = \pi$ rad, $90° = \pi/2$ rad, and so on. If we had insisted on measuring the angle θ in degrees, we would have needed to include an extra factor of $(2\pi/360)$ on the right-hand side of $s = r\theta$ in Eq. (9–1). By measuring angles in radians, we keep the relationship between angle and distance along an arc as simple as possible.

ANGULAR VELOCITY

The coordinate θ shown in Fig. 9–1 specifies the rotational position of a rigid body at a given instant. We can describe the rotational *motion* of such a rigid body in terms of the rate of change of θ. We'll do this in an analogous way to our description of straight-line motion in Chapter 2. In Fig. 9–3a a reference line OP in a rotating body makes an angle θ_1 with the +x-axis at time t_1. At a later time t_2 the angle has changed to θ_2. We define the **average angular velocity** ω_{av} (the Greek letter "omega") of the body in the time interval $\Delta t = t_2 - t_1$ as the ratio of the *angular displacement* $\Delta\theta = \theta_2 - \theta_1$ to Δt:

$$\omega_{av} = \frac{\theta_2 - \theta_1}{t_2 - t_1} = \frac{\Delta\theta}{\Delta t}. \tag{9–2}$$

The **instantaneous angular velocity** ω is the limit of ω_{av} as Δt approaches zero, that is, the derivative of θ with respect to t:

$$\omega = \lim_{\Delta t \to 0} \frac{\Delta\theta}{\Delta t} = \frac{d\theta}{dt} \quad \text{(definition of angular velocity).} \tag{9–3}$$

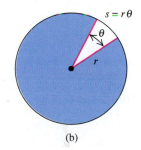

9–1 A rigid body rotating counterclockwise about a fixed axis that passes through O and is perpendicular to the page.

9–2 An angle θ in radians is defined as the ratio of the arc length s to the radius r.

(a)

(b)

9–3 (a) Angular displacement $\Delta\theta$ of a rotating body. (b) Every part of a rigid body has the same angular velocity.

When we refer simply to "angular velocity," we mean the instantaneous angular velocity, not the average angular velocity.

Different points on a rotating rigid body move different distances in a given time interval, depending on how far the point lies from the rotation axis. But because the body is rigid, *all* points rotate through the same angle in the same time (Fig. 9–3b). Hence *at any instant, every part of a rotating rigid body has the same angular velocity.* The angular velocity is positive if the body is rotating in the direction of increasing θ and negative if rotating in the direction of decreasing θ.

If the angle θ is in radians, the unit of angular velocity is the radian per second (rad/s). Other units, such as the revolution per minute (rev/min or rpm), are often used. Since 1 rev = 2π rad, two useful conversions are

$$1 \text{ rev/s} = 2\pi \text{ rad/s} \quad \text{and} \quad 1 \text{ rev/min} = 1 \text{ rpm} = \frac{2\pi}{60} \text{ rad/s}.$$

That is, 1 rad/s is about 10 rpm.

EXAMPLE 9-1

Calculating angular velocity Figure 9–4 shows the flywheel in a car engine under test. The angular position θ of the flywheel is given by

$$\theta = (2.0 \text{ rad/s}^3)t^3.$$

The diameter of the flywheel is 0.36 m. a) Find the angle θ, in radians and in degrees, at times $t_1 = 2.0$ s and $t_2 = 5.0$ s. b) Find the distance that a particle on the rim moves during that time interval. c) Find the average angular velocity, in rad/s and in rev/min (rpm), between $t_1 = 2.0$ s and $t_2 = 5.0$ s. d) Find the instantaneous angular velocity at time $t = 3.0$ s.

SOLUTION (a) We substitute the values of t into the given equation:

$$\theta_1 = (2.0 \text{ rad/s}^3)(2.0 \text{ s})^3 = 16 \text{ rad}$$

$$= (16 \text{ rad}) \frac{360°}{2\pi \text{ rad}} = 920°,$$

$$\theta_2 = (2.0 \text{ rad/s}^3)(5.0 \text{ s})^3 = 250 \text{ rad}$$

$$= (250 \text{ rad}) \frac{360°}{2\pi \text{ rad}} = 14,000°.$$

b) The flywheel turns through an angular displacement of $\theta_2 - \theta_1 = 250 \text{ rad} - 16 \text{ rad} = 234 \text{ rad}$. The radius r is half the diameter, or 0.18 m. Equation (9–1) gives

$$s = r\theta = (0.18 \text{ m})(234 \text{ rad}) = 42 \text{ m}.$$

Notice that in order to use Eq. (9–1), the angle *must* be expressed in radians. We drop "radians" from the unit for s because θ is really a dimensionless pure number; s is a distance and is measured in meters, the same unit as r.

Flywheel
$d = 0.36$ m

9–4 Rotating flywheel in a car engine.

c) In Eq. (9–2) we have

$$\omega_{av} = \frac{\theta_2 - \theta_1}{t_2 - t_1} = \frac{250 \text{ rad} - 16 \text{ rad}}{5.0 \text{ s} - 2.0 \text{ s}} = 78 \text{ rad/s}$$

$$= \left(78 \frac{\text{rad}}{\text{s}}\right)\left(\frac{1 \text{ rev}}{2\pi \text{ rad}}\right)\left(\frac{60 \text{ s}}{1 \text{ min}}\right) = 740 \text{ rev/min}.$$

d) We use Eq. (9–3):

$$\omega = \frac{d\theta}{dt} = \frac{d}{dt}[(2.0 \text{ rad/s}^3)t^3] = (2.0 \text{ rad/s}^3)(3t^2)$$

$$= (6.0 \text{ rad/s}^3)t^2.$$

At time $t = 3.0$ s,

$$\omega = (6.0 \text{ rad/s}^3)(3.0 \text{ s})^2 = 54 \text{ rad/s}.$$

ANGULAR ACCELERATION

When the angular velocity of a rigid body changes, it has an *angular acceleration.* If ω_1 and ω_2 are the instantaneous angular velocities at times t_1 and t_2, we define the **average angular acceleration** α_{av} over the interval $\Delta t = t_2 - t_1$ as the change in angular velocity divided by Δt:

$$\alpha_{av} = \frac{\omega_2 - \omega_1}{t_2 - t_1} = \frac{\Delta \omega}{\Delta t}. \qquad (9\text{–}4)$$

The **instantaneous angular acceleration** α is the limit of α_{av} as $\Delta t \rightarrow 0$:

$$\alpha = \lim_{\Delta t \rightarrow 0} \frac{\Delta \omega}{\Delta t} = \frac{d\omega}{dt} \qquad \text{(definition of angular acceleration).} \qquad (9\text{–}5)$$

The usual unit of angular acceleration is the radian per second per second, or rad/s^2. Henceforth we will use the term "angular acceleration" to mean the instantaneous angular acceleration rather than the average angular acceleration.

Because $\omega = d\theta/dt$, we can also express angular acceleration as the second derivative of the angular coordinate:

$$\alpha = \frac{d}{dt}\frac{d\theta}{dt} = \frac{d^2\theta}{dt^2}. \qquad (9\text{–}6)$$

You have probably noticed that we are using Greek letters for angular kinematic quantities: θ for angular position, ω for angular velocity, and α for angular acceleration. These are analogous to x for position, v for velocity, and a for acceleration, respectively, in straight-line motion. In each case, velocity is the rate of change of position with respect to time and acceleration is the rate of change of velocity with respect to time. We will sometimes use the terms *linear* velocity and *linear* acceleration for the familiar quantities we defined in Chapters 2 and 3 to make sure we distinguish clearly between these and the *angular* quantities introduced in this chapter.

In rotational motion, if the angular acceleration α is positive, the angular velocity ω is increasing; if α is negative, ω is decreasing. The rotation is speeding up if α and ω have the same sign and slowing down if α and ω have opposite signs. (These relationships are exactly the same as those between *linear* acceleration a and *linear* velocity v for straight-line motion; see Section 2–4.)

The pilot of a helicopter can change the angular velocity of the rotor—that is, give it an angular acceleration—by adjusting a control in the cockpit.

EXAMPLE 9–2

Calculating angular acceleration In Example 9–1 we found that the instantaneous angular velocity ω of the flywheel at any time t is given by

$$\omega = (6.0 \text{ rad/s}^3)t^2.$$

a) Find the average angular acceleration between $t_1 = 2.0$ s and $t_2 = 5.0$ s. b) Find the instantaneous angular acceleration at time $t = 3.0$ s.

SOLUTION (a) The values of ω at the two times are

$$\omega_1 = (6.0 \text{ rad/s}^3)(2.0 \text{ s})^2 = 24 \text{ rad/s},$$

$$\omega_2 = (6.0 \text{ rad/s}^3)(5.0 \text{ s})^2 = 150 \text{ rad/s}.$$

From Eq. (9–4) the average angular acceleration is

$$\alpha_{av} = \frac{150 \text{ rad/s} - 24 \text{ rad/s}}{5.0 \text{ s} - 2.0 \text{ s}} = 42 \text{ rad/s}^2.$$

b) From Eq. (9–5) the instantaneous angular acceleration at any time t is

$$\alpha = \frac{d\omega}{dt} = \frac{d}{dt}[(6.0 \text{ rad/s}^3)(t^2)] = (6.0 \text{ rad/s}^3)(2t)$$

$$= (12 \text{ rad/s}^3)t.$$

At time $t = 3.0$ s,

$$\alpha = (12 \text{ rad/s}^3)(3.0 \text{ s}) = 36 \text{ rad/s}^2.$$

Note that the angular acceleration is *not* constant in this situation. The angular velocity ω is always increasing because α is always positive. Furthermore, the rate at which the angular velocity increases is itself increasing, since α increases with time.

In straight-line motion, velocity v and acceleration a are the *components* of the vectors $\vec{v}$ and $\vec{a}$, respectively. In fixed-axis rotation problems it is likewise useful to think of ω and α as components of vector quantities. The angular velocity *vector* $\vec{\omega}$ is directed

(a)

(b)

Speeding up   Slowing down

(c)

9–5 (a) When the fingers of the right hand curl in the direction of rotation, the thumb position is the direction of the angular velocity vector. The direction is also the direction of advance of an ordinary screw with a right-hand thread. (b) Reversing the direction of rotation makes the angular velocity vector point in the opposite direction. (c) When the rotation axis is fixed, the angular acceleration and angular velocity vectors both lie along that axis.

7.8
Rotational Kinematics

along the axis of rotation. As Fig. 9–5 shows, the direction is given by the right-hand rule we used to define the vector product in Section 1–11. We define the angular acceleration vector $\vec{\alpha}$ as the time derivative of the angular velocity vector. If the direction of the axis is fixed, then $\vec{\alpha}$ is also along the axis; it points in the same direction as $\vec{\omega}$ if the rotation is speeding up and in the opposite direction if the rotation is slowing down (Fig. 9–5c). The vector formulation is especially useful in situations in which the direction of the axis *changes;* we'll examine such situations briefly at the end of Chapter 10.

9–3 ROTATION WITH CONSTANT ANGULAR ACCELERATION

In Chapter 2 we found that straight-line motion is particularly simple when the acceleration is constant. This is also true of rotational motion about a fixed axis. When the angular acceleration is constant, we can derive equations for angular velocity and angular position using exactly the same procedure that we used for straight-line motion in Section 2–5. In fact, the equations we are about to derive are identical to Eqs. (2–8), (2–12), (2–13), and (2–14) if we replace x with θ, v with ω, and a with α. We suggest that you review Section 2–5 before continuing.

Let ω_0 be the angular velocity of a rigid body at time $t = 0$, and let ω be its angular velocity at any later time t. The angular acceleration α is constant and equal to the average value for any interval. Using Eq. (9–4) with the interval from 0 to t, we find

$$\alpha = \frac{\omega - \omega_0}{t - 0}, \quad \text{or}$$

$$\omega = \omega_0 + \alpha t \quad \text{(constant angular acceleration only)}. \qquad (9\text{–}7)$$

The product αt is the total change in ω between $t = 0$ and the later time t; the angular velocity ω at time t is the sum of the initial value ω_0 and this total change.

With constant angular acceleration the angular velocity changes at a uniform rate, so its average value between 0 and t is the average of the initial and final values:

$$\omega_{\text{av}} = \frac{\omega_0 + \omega}{2}. \qquad (9\text{–}8)$$

We also know that ω_{av} is the total angular displacement $(\theta - \theta_0)$ divided by the time interval $(t - 0)$;

$$\omega_{\text{av}} = \frac{\theta - \theta_0}{t - 0}. \qquad (9\text{–}9)$$

When we equate Eqs. (9–8) and (9–9) and multiply the result by t, we get

$$\theta - \theta_0 = \frac{1}{2}(\omega_0 + \omega)t \quad \text{(constant angular acceleration only)}. \qquad (9\text{–}10)$$

To obtain a relation between θ and t that doesn't contain ω, we substitute Eq. (9–7) into Eq. (9–10):

$$\theta - \theta_0 = \frac{1}{2}[\omega_0 + (\omega_0 + \alpha t)]t,$$

$$\theta = \theta_0 + \omega_0 t + \frac{1}{2}\alpha t^2 \quad \text{(constant angular acceleration only)}. \qquad (9\text{–}11)$$

That is, if at the initial time $t = 0$ the body is at angular position θ_0 and has angular velocity ω_0, then its angular position θ at any later time t is the sum of three terms: its initial

TABLE 9–1

COMPARISON OF LINEAR AND ANGULAR MOTION WITH CONSTANT ACCELERATION

STRAIGHT-LINE MOTION WITH CONSTANT LINEAR ACCELERATION	FIXED-AXIS ROTATION WITH CONSTANT ANGULAR ACCELERATION
$a = \text{constant}$	$\alpha = \text{constant}$
$v = v_0 + at$	$\omega = \omega_0 + \alpha t$
$x = x_0 + v_0 t + \frac{1}{2}at^2$	$\theta = \theta_0 + \omega_0 t + \frac{1}{2}\alpha t^2$
$v^2 = v_0^2 + 2a(x - x_0)$	$\omega^2 = \omega_0^2 + 2\alpha(\theta - \theta_0)$
$x - x_0 = \frac{1}{2}(v + v_0)t$	$\theta - \theta_0 = \frac{1}{2}(\omega + \omega_0)t$

angular position θ_0, plus the rotation $\omega_0 t$ it would have if the angular velocity were constant, plus an additional rotation $\frac{1}{2}\alpha t^2$ caused by the changing angular velocity.

Following the same procedure as for straight-line motion in Section 2–5, we can combine Eqs. (9–7) and (9–11) to obtain a relation between θ and ω that does not contain t. We invite you to work out the details, following the same procedure we used to get Eq. (2–13). (See Exercise 9–10.) In fact, because of the perfect analogy between straight-line and rotational quantities, we can simply take Eq. (2–13) and replace each straight-line quantity by its rotational analog. We get

$$\omega^2 = \omega_0^2 + 2\alpha(\theta - \theta_0) \qquad \text{(constant angular acceleration only).} \qquad (9–12)$$

CAUTION ▶ Keep in mind that all of these results are valid *only* when the angular acceleration α is *constant;* be careful not to try to apply them to problems in which α is *not* constant. Table 9–1 shows the analogy between Eqs. (9–7), (9–10), (9–11), and (9–12) for fixed-axis rotation with constant angular acceleration and the corresponding equations for straight-line motion with constant linear acceleration. ◀

EXAMPLE 9–3

Rotation with constant angular acceleration A bicycle wheel is being tested at a repair shop. The angular velocity of the wheel is 4.00 rad/s at time $t = 0$, and its angular acceleration is constant and equal to -1.20 rad/s^2. A spoke OP on the wheel coincides with the $+x$-axis at time $t = 0$ (Fig. 9–6). a) What is the wheel's angular velocity at $t = 3.00$ s? b) What angle does the spoke OP make with the $+x$-axis at this time?

SOLUTION We can use Eqs. (9–7) and (9–11) to find ω and θ at any time in terms of the given initial conditions.
a) From Eq. (9–7), $\omega = \omega_0 + \alpha t$. At time $t = 3.00$ s,

$$\omega = 4.00 \text{ rad/s} + (-1.20 \text{ rad/s}^2)(3.00 \text{ s}) = 0.40 \text{ rad/s}.$$

The angular velocity has decreased because α is negative.
b) The angle θ is given as a function of time by Eq. (9–11):

$$\theta = \theta_0 + \omega_0 t + \frac{1}{2}\alpha t^2$$

$$= 0 + (4.00 \text{ rad/s})(3.00 \text{ s}) + \frac{1}{2}(-1.20 \text{ rad/s}^2)(3.00 \text{ s})^2$$

$$= 6.60 \text{ rad} = 6.60 \text{ rad}\left(\frac{1 \text{ rev}}{2\pi \text{ rad}}\right) = 1.05 \text{ rev.}$$

The wheel has turned through one complete revolution plus an additional 0.05 rev, that is, through an additional angle of $(0.05 \text{ rev})(2\pi \text{ rad/rev}) = 0.32 \text{ rad} = 18°$. The spoke OP is at an angle of 18° with the $+x$-axis.

9–6 The spoke OP of a rotating wheel at $t = 0$. What angle does this spoke make with the $+x$-axis at $t = 3.00$ s?

We can also determine θ by rearranging Eq. (9–12), $\omega^2 = \omega_0^2 + 2\alpha(\theta - \theta_0)$, so that it reads

$$\theta = \theta_0 + \frac{\omega^2 - \omega_0^2}{2\alpha}$$

$$= 0 + \frac{(0.40 \text{ rad/s})^2 - (4.00 \text{ rad/s})^2}{2(-1.20 \text{ rad/s}^2)} = 6.60 \text{ rad.}$$

Can you show that the wheel will stop instantaneously at $t = 3.33$ s after having rotated through 6.67 rad? If the angular acceleration continues to remain constant after this time, the wheel will reverse direction and start rotating clockwise.

9–7 A rigid body rotating about a fixed axis through point O. The distance s that point P on the body moves through equals $r\theta$ if θ is measured in radians. The linear speed v of point P equals $r\omega$ if the angular speed ω is measured in rad/s.

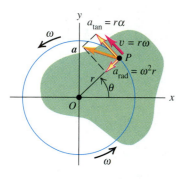

9–8 The component of acceleration of point P parallel to $\vec{v}$ is $a_{\text{tan}} = r\alpha$; the component perpendicular to $\vec{v}$, radially inward toward the rotation axis, is $a_{\text{rad}} = \omega^2 r$. The rigid body shown here is speeding up.

9–4 RELATING LINEAR AND ANGULAR KINEMATICS

How do we find the velocity and acceleration of a particular point in a rotating rigid body? We need to answer this question in order to proceed with our study of rotation. For example, to find the kinetic energy of a rotating body, we have to start from $K = \frac{1}{2}mv^2$ for a particle, and this requires knowing v for each particle in the body. So it's worthwhile to develop general relations between the *angular* velocity and acceleration of a rigid body rotating about a fixed axis and the *linear* velocity and acceleration of a specific point or particle in the body.

When a rigid body rotates about a fixed axis, every particle in the body moves in a circular path. The circle lies in a plane perpendicular to the axis and is centered on the axis. The speed of a particle is directly proportional to the body's angular velocity; the faster the body rotates, the greater the speed of each particle. In Fig. 9–7, point P is a constant distance r from the axis of rotation, so it moves in a circle of radius r. At any time the angle θ (in radians) and the arc length s are related by

$$s = r\theta.$$

We take the time derivative of this, noting that r is constant for any specific particle, and take the absolute value of both sides:

$$\left|\frac{ds}{dt}\right| = r\left|\frac{d\theta}{dt}\right|.$$

Now $|ds/dt|$ is the absolute value of the rate of change of arc length, which is equal to the instantaneous *linear* speed v of the particle. Analogously, $|d\theta/dt|$, the absolute value of the rate of change of the angle, is the instantaneous **angular speed** ω—that is, the magnitude of the instantaneous angular velocity in rad/s. Thus

$$v = r\omega \qquad \text{(relation between linear and angular speed).} \tag{9–13}$$

The farther a point is from the axis, the greater its linear speed. The *direction* of the linear velocity *vector* is tangent to its circular path at each point (Fig. 9–7).

We can represent the acceleration of a particle moving in a circle in terms of its centripetal and tangential components, a_{rad} and a_{tan} (Fig. 9–8), as we did in Section 3–5. It would be a good idea to review that section now. We found that the **tangential component of acceleration** a_{tan}, the component parallel to the instantaneous velocity, acts to change the *magnitude* of the particle's velocity (i.e., the speed) and is equal to the rate of change of speed. Taking the derivative of Eq. (9–13), we find

$$a_{\text{tan}} = \frac{dv}{dt} = r\frac{d\omega}{dt} = r\alpha \qquad \text{(tangential acceleration of a point on a rotating body).} \tag{9–14}$$

This component of a particle's acceleration is always tangent to the circular path of the particle.

The component of the particle's acceleration directed toward the rotation axis, the **centripetal component of acceleration** a_{rad}, is associated with the change of *direction* of the particle's velocity. In Section 3–5 we worked out the relation $a_{\text{rad}} = v^2/r$. We can express this in terms of ω by using Eq. (9–13):

$$a_{\text{rad}} = \frac{v^2}{r} = \omega^2 r \qquad \text{(centripetal acceleration of a point on a rotating body).} \tag{9–15}$$

This is true at each instant *even when ω and v are not constant.* The centripetal component always points toward the axis of rotation.

The vector sum of the centripetal and tangential components of acceleration of a particle in a rotating body is the linear acceleration $\vec{a}$ (Fig. 9–8).

CAUTION ▶ It's important to remember that Eq. (9–1), $s = r\theta$, is valid *only* when θ is measured in radians. The same is true of any equation derived from this, including

Eqs. (9–13), (9–14), and (9–15). When you use these equations, you *must* express the angular quantities in radians, not revolutions or degrees. ◄

Equations (9–1), (9–13), and (9–14) also apply to any particle that has the same tangential velocity as a point in a rotating rigid body. For example, when a rope wound around a circular cylinder unwraps without stretching or slipping, its speed and acceleration at any instant are equal to the speed and tangential acceleration of the point at which it is tangent to the cylinder. The same principle holds for situations such as bicycle chains and sprockets, belts and pulleys that turn without slipping, and so on. We will have several opportunities to use these relations later in this chapter and in Chapter 10. Note that Eq. (9–15) for the centripetal component a_{rad} is applicable to the rope or chain *only* at points that are in contact with the cylinder or sprocket. Other points do not have the same acceleration toward the center of the circle that points on the cylinder or sprocket have.

EXAMPLE 9–4

Throwing a discus A discus thrower turns with angular acceleration $\alpha = 50$ rad/s^2, moving the discus in a circle of radius 0.80 m. We model the thrower's arm as a rigid body, so r is constant. Find the tangential and centripetal components of acceleration of the discus and the magnitude of its acceleration at the instant when the angular velocity is 10 rad/s.

SOLUTION We model the discus as a particle moving in a circular path (Fig. 9–9). The acceleration components are given by Eqs. (9–14) and (9–15):

$$a_{tan} = r\alpha = (0.80 \text{ m})(50 \text{ rad/s}^2) = 40 \text{ m/s}^2,$$

$$a_{rad} = \omega^2 r = (10 \text{ rad/s})^2 (0.80 \text{ m}) = 80 \text{ m/s}^2.$$

The magnitude of the acceleration vector is

$$a = \sqrt{a_{rad}^2 + a_{tan}^2} = 89 \text{ m/s}^2,$$

or about nine times the acceleration due to gravity. Note that we have dropped the unit "radian" from our results for a_{tan}, a_{rad},

9–9 Treating a discus thrower's arm as a rigid body.

and a; we can do this because "radian" is a dimensionless quantity.

EXAMPLE 9–5

Designing a propeller You are asked to design an airplane propeller to turn at 2400 rpm. The forward air speed of the plane is to be 75.0 m/s (270 km/h, or about 168 mi/h), and the speed of the tips of the propeller blades through the air must not exceed 270 m/s. (This is about 0.80 times the speed of sound in air. If the propeller tips were to move at the speed of sound, they would produce a tremendous amount of noise. Keeping the speed comfortably below this keeps the noise level acceptably low.) a) What is the maximum radius the propeller can have? b) With this radius, what is the acceleration of the propeller tip?

SOLUTION First we convert ω to rad/s:

$$\omega = 2400 \text{ rpm} = \left(2400 \frac{\text{rev}}{\text{min}}\right)\left(\frac{2\pi \text{ rad}}{1 \text{ rev}}\right)\left(\frac{1 \text{ min}}{60 \text{ s}}\right)$$

$$= 251 \text{ rad/s}.$$

a) The tangential velocity $v_P = r\omega$ of the propeller tip is perpendicular to the forward velocity v_A through the air (Fig. 9–10; see page 276). The magnitude of the vector sum of these velocities

is v_{total}, where

$$v_{total}^2 = v_A^2 + v_P^2 = v_A^2 + r^2\omega^2,$$

so

$$r^2 = \frac{v_{total}^2 - v_A^2}{\omega^2} \quad \text{and} \quad r = \frac{\sqrt{v_{total}^2 - v_A^2}}{\omega}.$$

If $v_{total} = 270$ m/s, the propeller radius is

$$r = \frac{\sqrt{(270 \text{ m/s})^2 - (75.0 \text{ m/s})^2}}{251 \text{ rad/s}} = 1.03 \text{ m}.$$

b) The angular velocity of the propeller is constant, so the acceleration of the propeller tip is purely centripetal:

$$a_{rad} = \omega^2 r$$

$$= (251 \text{ rad/s})^2 (1.03 \text{ m}) = 6.5 \times 10^4 \text{ m/s}^2.$$

From $\Sigma \vec{F} = m\vec{a}$, the propeller must exert a force of 6.5×10^4 N on each kilogram of material at its tip! This is why propellers are made out of tough material (usually aluminum alloy).

9–10 Designing an airplane propeller.

EXAMPLE 9–6

Bicycle gears How are the angular speeds of the two bicycle sprockets in Fig. 9–11 related to the number of teeth on each sprocket?

SOLUTION The chain does not slip or stretch, so it moves at the same tangential speed on both sprockets. Letting v and ω denote the linear and angular *speeds,* we have

$$v = r_1 \omega_1 = r_2 \omega_2, \qquad \text{so} \qquad \frac{\omega_2}{\omega_1} = \frac{r_1}{r_2}.$$

The angular speed is inversely proportional to the radius. This relationship also holds for pulleys connected by a belt, provided that the belt doesn't slip. For chain sprockets the teeth must be equally spaced on the circumferences of both sprockets in order for the chain to mesh properly with both. Let N_1 and N_2 be the numbers of teeth; the condition that the tooth spacing is the same on both sprockets is

$$\frac{2\pi r_1}{N_1} = \frac{2\pi r_2}{N_2}, \qquad \text{or} \qquad \frac{r_1}{r_2} = \frac{N_1}{N_2}.$$

Combining this with the other equation, we get

$$\frac{\omega_2}{\omega_1} = \frac{N_1}{N_2}.$$

The angular speed of each sprocket is inversely proportional to the number of teeth. On a multi-speed bike, you get the highest

9–11 How are the angular velocities of the two sprockets related to their number of teeth?

angular speed ω_2 of the rear wheel for a given pedaling rate ω_1 when the ratio N_1/N_2 is maximum; this means using the largest-radius front sprocket (largest N_1) and the smallest-radius rear sprocket (smallest N_2).

9–5 ENERGY IN ROTATIONAL MOTION

A rotating rigid body consists of mass in motion, so it has kinetic energy. We can express this kinetic energy in terms of the body's angular velocity and a new quantity called *moment of inertia* that we will define below. To develop this relationship, we think of the body as being made up of a large number of particles, with masses $m_1, m_2, \ldots$, at distances $r_1, r_2, \ldots$ from the axis of rotation. We label the particles with the index i: The mass of the ith particle is m_i, and its distance from the axis of rotation is r_i. The particles don't necessarily all lie in the same plane, so we specify that r_i is the *perpendicular* distance from the axis to the ith particle.

When a rigid body rotates about a fixed axis, the speed v_i of the ith particle is given by Eq. (9–13), $v_i = r_i \omega$, where ω is the body's angular speed. Different particles have different values of r, but ω is the same for all (otherwise, the body wouldn't be rigid).

The kinetic energy of the ith particle can be expressed as

$$\frac{1}{2}m_i v_i^2 = \frac{1}{2}m_i r_i^2 \omega^2.$$

The *total* kinetic energy of the body is the sum of the kinetic energies of all its particles:

$$K = \frac{1}{2}m_1 r_1^2 \omega^2 + \frac{1}{2}m_2 r_2^2 \omega^2 + \cdots = \sum_i \frac{1}{2}m_i r_i^2 \omega^2.$$

Taking the common factor $\omega^2/2$ out of this expression, we get

$$K = \frac{1}{2}(m_1 r_1^2 + m_2 r_2^2 + \cdots)\omega^2 = \frac{1}{2}(\sum_i m_i r_i^2)\omega^2.$$

The quantity in parentheses, obtained by multiplying the mass of each particle by the square of its distance from the axis of rotation and adding these products, is denoted by I and is called the **moment of inertia** of the body for this rotation axis:

$$I = m_1 r_1^2 + m_2 r_2^2 + \cdots = \sum_i m_i r_i^2 \qquad \text{(definition of moment of inertia)}. \quad (9\text{--}16)$$

(a)

The word "moment" means that I depends on how the body's mass is distributed in space; it has nothing to do with a "moment" of time. For a body with a given rotation axis and a given total mass, the greater the distance from the axis to the particles that make up the body, the greater the moment of inertia. In a rigid body, the distances r_i are all constant and I is independent of how the body is rotating around the given axis. The SI unit of moment of inertia is the kilogram-meter^2 (kg · m^2).

In terms of moment of inertia I, the **rotational kinetic energy** K of a rigid body is

$$K = \frac{1}{2}I\omega^2 \qquad \text{(rotational kinetic energy of a rigid body)}. \quad (9\text{--}17)$$

(b)

The kinetic energy given by Eq. (9–17) is not a new form of energy; it's the sum of the kinetic energies of the individual particles that make up the rigid body, written in a compact and convenient form in terms of the moment of inertia. When using Eq. (9–17), ω *must* be measured in radians per second, not revolutions or degrees per second, to give K in joules; this is because we used $v_i = r_i\omega$ in our derivation.

Equation (9–17) gives a simple physical interpretation of moment of inertia: the greater the moment of inertia, the greater the kinetic energy of a rigid body rotating with a given angular speed ω. We learned in Chapter 6 that the kinetic energy of a body equals the amount of work done to accelerate that body from rest. So the greater a body's moment of inertia, the harder it is to start the body rotating if it's at rest and the harder it is to stop its rotation if it's already rotating (Fig. 9–12). For this reason, I is also called the *rotational inertia*.

Example 9–7 shows how *changing* the rotation axis can affect the value of I.

9–12 An apparatus free to rotate around a vertical axis. The two cylinders of mass m can be locked into any position on the horizontal shaft. (a) If the two cylinders are located close to the rotation axis, the moment of inertia is small and it's easy to start the apparatus rotating. (b) If the cylinders are further from the rotation axis, the moment of inertia is greater and it's more difficult to start or stop the rotation.

EXAMPLE 9–7

Moments of inertia for different rotation axes An engineer is designing a one-piece machine part consisting of three heavy connectors linked by light molded struts (Fig. 9–13). The connectors can be considered as massive particles connected by massless rods. a) What is the moment of inertia of this body about an axis through point A, perpendicular to the plane of the diagram? b) What is the moment of inertia about an axis coinciding with rod BC? c) If the body rotates about an axis through A perpendicular to the plane of the diagram, with angular speed $\omega = 4.0$ rad/s, what is its kinetic energy?

9–13 A strangely shaped machine part.

SOLUTION a) The particle at point A lies *on* the axis. Its distance r from the axis is zero, so it contributes nothing to the moment of inertia. Equation (9–16) gives

$$I = \Sigma m_i r_i^2 = (0.10 \text{ kg})(0.50 \text{ m})^2 + (0.20 \text{ kg})(0.40 \text{ m})^2$$

$$= 0.057 \text{ kg} \cdot \text{m}^2.$$

b) The particles at B and C both lie *on* the axis, so for them $r = 0$, and neither contributes to the moment of inertia. Only A contributes, and we have

$$I = \Sigma m_i r_i^2 = (0.30 \text{ kg})(0.40 \text{ m})^2 = 0.048 \text{ kg} \cdot \text{m}^2.$$

Since this moment of inertia is less than in part (a), it's easier to make the body rotate about this axis than about the axis through point A.

c) From Eq. (9–17),

$$K = \frac{1}{2} I \omega^2 = \frac{1}{2} (0.057 \text{ kg} \cdot \text{m}^2)(4.0 \text{ rad/s})^2 = 0.46 \text{ J}.$$

CAUTION ▶ The results of parts (a) and (b) of Example 9–7 show that the moment of inertia of a body depends on the location and orientation of the axis. It's not enough to just say, "The moment of inertia of this body is $0.048 \text{ kg} \cdot \text{m}^2$." We have to be specific and say, "The moment of inertia of this body *about axis BC* is $0.048 \text{ kg} \cdot \text{m}^2$." ◀

In Example 9–7 we represented the body as several point masses, and we evaluated the sum in Eq. (9–16) directly. When the body is a *continuous* distribution of matter, such as a solid cylinder or plate, the sum becomes an integral, and we need to use calculus to calculate the moment of inertia. We will give several examples of such calculations in Section 9–7; meanwhile, Table 9–2 gives moments of inertia for several familiar shapes in terms of the masses and dimensions. Each body shown in Table 9–2 is *uniform;* that is, the density has the same value at all points within the solid parts of the body.

7.7

Rotational Inertia

TABLE 9–2

MOMENTS OF INERTIA OF VARIOUS BODIES

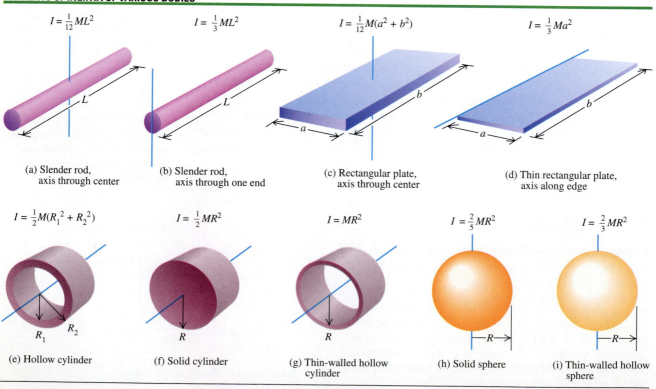

$I = \frac{1}{12} ML^2$

(a) Slender rod, axis through center

$I = \frac{1}{3} ML^2$

(b) Slender rod, axis through one end

$I = \frac{1}{12} M(a^2 + b^2)$

(c) Rectangular plate, axis through center

$I = \frac{1}{3} Ma^2$

(d) Thin rectangular plate, axis along edge

$I = \frac{1}{2} M(R_1^2 + R_2^2)$

(e) Hollow cylinder

$I = \frac{1}{2} MR^2$

(f) Solid cylinder

$I = MR^2$

(g) Thin-walled hollow cylinder

$I = \frac{2}{5} MR^2$

(h) Solid sphere

$I = \frac{2}{3} MR^2$

(i) Thin-walled hollow sphere

You may be tempted to try to compute the moment of inertia of a body by assuming that all the mass is concentrated at the center of mass and then multiplying the total mass by the square of the distance from the center of mass to the axis. Resist that temptation; it doesn't work! For example, when a uniform thin rod of length L and mass M is pivoted about an axis through one end, perpendicular to the rod, the moment of inertia is $I = ML^2/3$ (case (b) in Table 9–2). If we took the mass as concentrated at the center, a distance $L/2$ from the axis, we would obtain the *incorrect* result $I = M(L/2)^2 = ML^2/4$.

Now that we know how to calculate the kinetic energy of a rotating rigid body, we can apply the energy principles of Chapter 7 to rotational motion. Here are some points of strategy and some examples.

Problem–Solving Strategy

ROTATIONAL ENERGY

1. You can use work-energy relations and conservation of energy to find relations involving position and motion of a rotating body. The problem-solving strategy outlined for energy problems in Section 7–2 is equally useful for rotating rigid bodies; the only difference is that rotational kinetic energy is expressed in terms of the moment of inertia I and angular speed ω of the body ($K = \frac{1}{2}I\omega^2$) instead of its mass m and speed v.

2. Many problems involve a rope or cable wrapped around a rotating rigid body that functions as a pulley. In these situations, remember that the point on the pulley that contacts the rope has the same linear speed as the rope, provided that the rope doesn't slip on the pulley. You can then take advantage of Eqs. (9–13) and (9–14), which relate the linear speed and tangential acceleration of a point on a rigid body to the angular speed and angular acceleration of the body. Examples 9–8 and 9–9 illustrate this point.

EXAMPLE 9–8

An unwinding cable A light, flexible, nonstretching cable is wrapped several times around a winch drum, a solid cylinder of mass 50 kg and diameter 0.120 m, which rotates about a stationary horizontal axis held by frictionless bearings (Fig. 9–14). The free end of the cable is pulled with a constant force of magnitude 9.0 N for a distance of 2.0 m. It unwinds without slipping, turning the cylinder as it does so. If the cylinder is initially at rest, find its final angular speed and the final speed of the cable.

SOLUTION There is friction between the cable and the cylinder; this is what makes the cylinder rotate when the cable is pulled. But because the cable doesn't slip, there is no sliding of the cable relative to the cylinder, and no mechanical energy is lost due to friction. The change in kinetic energy of the cylinder is

equal to the work $W = Fs$ done by the force $F = 9.0$ N acting over a distance $s = 2.0$ m; hence $W = (9.0\ \text{N})(2.0\ \text{m}) = 18$ J. From Table 9–2 the moment of inertia is

$$I = \frac{1}{2}MR^2 = \frac{1}{2}(50\ \text{kg})(0.060\ \text{m})^2 = 0.090\ \text{kg} \cdot \text{m}^2.$$

(The radius R is just half the diameter of the cylinder.) Since the cylinder is initially at rest ($\omega_0 = 0$), the work-energy theorem gives

$$W = K_2 - K_1$$

$$= \frac{1}{2}I\omega^2 - \frac{1}{2}I\omega_0^2 = \frac{1}{2}I\omega^2,$$

$$\omega = \sqrt{\frac{2W}{I}} = \sqrt{\frac{2(18\ \text{J})}{0.090\ \text{kg} \cdot \text{m}^2}}$$

$$= 20\ \text{rad/s}.$$

The final speed of the cable is equal to the final tangential speed v of the cylinder, which is given by Eq. (9–13):

$$v = r\omega = (0.060\ \text{m})(20\ \text{rad/s}) = 1.2\ \text{m/s}.$$

9–14 A cable unwinds from a cylinder. The force F does work and increases the kinetic energy of the cylinder.

EXAMPLE 9–9

An unwinding cable II In a lab experiment to test conservation of energy in rotational motion, we wrap a light, flexible cable around a solid cylinder with mass M and radius R. The cylinder rotates with negligible friction about a stationary horizontal axis (Fig. 9–15). We tie the free end of the cable to an object of mass m and release the object with no initial velocity at a distance h above the floor. As the object falls, the cable unwinds without stretching or slipping, turning the cylinder. Find the speed of the falling object and the angular speed of the cylinder just as the object strikes the floor.

SOLUTION Initially, the system has no kinetic energy ($K_1 = 0$). We take the potential energy to be zero when the object is at

9–15 As the cylinder rotates counterclockwise, the cable unwinds, the hanging object falls, and gravitational potential energy is transformed into kinetic energy.

floor level; then $U_1 = mgh$ and $U_2 = 0$. (We can ignore the gravitational potential energy for the rotating cylinder, since its height doesn't change.) As in the previous example, no work is done by friction, so $W_{other} = 0$ and $K_1 + U_1 = K_2 + U_2$. (The cable does no net work; at one end the force and displacement are in the same direction, and at the other end they are in opposite directions. So the total work done by both ends of the cable is zero.) Just before the falling object hits the floor, both this object and the cylinder have kinetic energy. The total kinetic energy K_2 at that instant is

$$K_2 = \frac{1}{2} mv^2 + \frac{1}{2} I\omega^2.$$

From Table 9–2 the moment of inertia of the cylinder is $I = \frac{1}{2}MR^2$. Also, v and ω are related by $v = R\omega$, since the speed of the falling mass must be equal to the tangential speed at the outer surface of the cylinder. Using these relations and equating the initial and final total energies, we find

$$0 + mgh = \frac{1}{2} mv^2 + \frac{1}{2}\left(\frac{1}{2} MR^2\right)\left(\frac{v}{R}\right)^2 = \frac{1}{2}\left(m + \frac{1}{2} M\right)v^2,$$

$$v = \sqrt{\frac{2gh}{1 + M/2m}}.$$

The final angular speed ω is obtained from $\omega = v/R$.

Let's check some particular cases. When M is much larger than m, v is very small, as we would expect. When M is much smaller than m, v is nearly equal to $\sqrt{2gh}$, which is the speed of a body that falls freely from an initial height h. Does it surprise you that v doesn't depend on the radius of the cylinder?

7.14
Woman and Flywheel Elevator—
Energy Approach

7.15
Rotoride—Energy Approach

GRAVITATIONAL POTENTIAL ENERGY FOR AN EXTENDED BODY

In Example 9–9 the cable was of negligible mass, so we could ignore its kinetic energy as well as the gravitational potential energy associated with it. If the mass is *not* negligible, we need to know how to calculate the *gravitational potential* energy associated with such an extended body. If the acceleration of gravity g is the same at all points on the body, the gravitational potential energy is the same as though all the mass were concentrated at the center of mass of the body. Suppose we take the y-axis vertically upward. Then for a body with total mass M, the gravitational potential energy U is simply

$$U = Mgy_{cm} \qquad \text{(gravitational potential energy for an extended body),} \qquad (9\text{–}18)$$

where y_{cm} is the y-coordinate of the center of mass. This expression applies to any extended body, whether it is rigid or not.

To prove Eq. (9–18), we again represent the body as a collection of mass elements m_i. The potential energy for element m_i is $m_i g y_i$, so the total potential energy is

$$U = m_1 g y_1 + m_2 g y_2 + \cdots = (m_1 y_1 + m_2 y_2 + \cdots)g.$$

But from Eq. (8–28), which defines the coordinates of the center of mass,

$$m_1 y_1 + m_2 y_2 + \cdots = (m_1 + m_2 + \cdots)y_{cm} = My_{cm},$$

where $M = m_1 + m_2 + \cdots$ is the total mass. Combining this with the above expression for U, we find $U = Mgy_{cm}$, in agreement with Eq. (9–18).

We leave the application of Eq. (9–18) to the problems. We'll make use of this relationship in Chapter 10 in the analysis of rigid-body problems where the axis of rotation moves.

9–6 PARALLEL-AXIS THEOREM

We pointed out in Section 9–5 that a body doesn't have just one moment of inertia. In fact, it has infinitely many, because there are infinitely many axes about which it might rotate. But there is a simple relationship between the moment of inertia I_{cm} of a body of mass M about an axis through its center of mass and the moment of inertia I_P about any other axis parallel to the original one but displaced from it by a distance d. This relationship, called the **parallel-axis theorem,** states that

$$I_P = I_{cm} + Md^2 \qquad \text{(parallel-axis theorem)}. \tag{9–19}$$

In a technique called the "Fosbury flop" after its innovator, this athlete arches his body as he passes over the bar in the high jump. As a result, his center of mass actually passes *under* the bar. This technique requires a smaller increase in gravitational potential energy (Eq. (9–18)) than the older method of straddling the bar.

To prove this theorem, we consider two axes, both parallel to the z-axis, one through the center of mass and the other through a point P (Fig. 9–16). First we take a very thin slice of the body, parallel to the xy-plane and perpendicular to the z-axis. We take the origin of our coordinate system to be at the center of mass of the body; the coordinates of the center of mass are then $x_{cm} = y_{cm} = z_{cm} = 0$. The axis through the center of mass passes through this thin slice at point O, and the parallel axis passes through point P, whose x- and y-coordinates are (a, b). The distance of this axis from the axis through the center of mass is d, where $d^2 = a^2 + b^2$.

We can write an expression for the moment of inertia I_P about the axis through point P. Let m_i be a mass element in our slice, with coordinates (x_i, y_i, z_i). Then the moment of inertia I_{cm} of the slice about the axis through the center of mass (at O) is

$$I_{cm} = \sum_i m_i(x_i^2 + y_i^2).$$

The moment of inertia of the slice about the axis through P is

$$I_P = \sum_i m_i[(x_i - a)^2 + (y_i - b)^2].$$

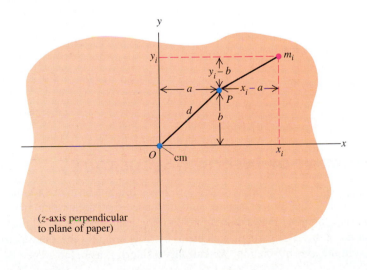

(z-axis perpendicular to plane of paper)

9–16 The mass element m_i has coordinates (x_i, y_i) with respect to an axis of rotation through the center of mass and perpendicular to the plane of the figure. The mass element has coordinates $(x_i - a, y_i - b)$ with respect to the parallel axis through point P.

These expressions don't involve the coordinates z_i measured perpendicular to the slices, so we can extend the sums to include *all* particles in *all* slices. Then I_p becomes the moment of inertia of the *entire* body for an axis through P. We then expand the squared terms and regroup, and obtain

$$I_P = \sum_i m_i(x_i^2 + y_i^2) - 2a \sum_i m_i x_i - 2b \sum_i m_i y_i + (a^2 + b^2) \sum_i m_i.$$

The first sum is I_{cm}. From Eq. (8–28), the definition of the center of mass, the second and third sums are proportional to x_{cm} and y_{cm}; these are zero because we have taken our origin to be the center of mass. The final term is d^2 multiplied by the total mass, or Md^2. This completes our proof that $I_P = I_{cm} + Md^2$.

As Eq. (9–19) shows, a rigid body has a lower moment of inertia about an axis through its center of mass than about any other parallel axis. Thus it's easier to start a body rotating if the rotation axis passes through the center of mass. This suggests that it's somehow most natural for a rotating body to rotate about an axis through its center of mass; we'll make this idea more quantitative in Chapter 10.

EXAMPLE 9–10

Using the parallel-axis theorem I A part of a mechanical linkage (Fig. 9–17) has a mass of 3.6 kg. We measure its moment of inertia about an axis 0.15 m from its center of mass to be $I_P = 0.132$ kg · m². What is the moment of inertia I_{cm} about a parallel axis through the center of mass?

SOLUTION We rearrange Eq. (9–19) and substitute in the values:

$$I_{cm} = I_P - Md^2 = 0.132 \text{ kg} \cdot \text{m}^2 - (3.6 \text{ kg})(0.15 \text{ m})^2$$
$$= 0.051 \text{ kg} \cdot \text{m}^2.$$

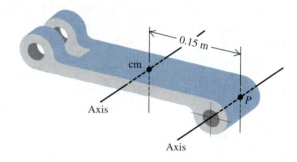

9–17 Calculating I_{cm} from a measurement of I_P.

EXAMPLE 9–11

Using the parallel-axis theorem II Find the moment of inertia of a thin uniform disk of mass M and radius R about an axis perpendicular to its plane at the edge.

SOLUTION From Table 9–2, $I_{cm} = MR^2/2$, and in this case $d = R$. The parallel-axis theorem gives

$$I_P = I_{cm} + Md^2 = \frac{MR^2}{2} + MR^2 = \frac{3MR^2}{2}.$$

This is a very simple solution of a problem that would be fairly complicated if we tried to solve it using the integration techniques to be discussed in the next section.

*9–7 MOMENT OF INERTIA CALCULATIONS

When a rigid body cannot be represented by a few point masses but is a continuous distribution of mass, the sum of masses and distances that defines moment of inertia (Eq. (9–16)) becomes an integral. Imagine dividing the body into small mass elements dm so that all points in a particular element are at essentially the same perpendicular distance from the axis of rotation. We call this distance r, as before. Then the moment

of inertia is

$$I = \int r^2 \, dm. \tag{9–20}$$

To evaluate the integral, we have to represent r and dm in terms of the same integration variable. When we have an effectively one-dimensional object, such as the slender rods (a) and (b) in Table 9–2, we can use a coordinate x along the length and relate dm to an increment dx. For a three-dimensional object it is usually easiest to express dm in terms of an element of volume dV and the *density* ρ of the body. Density is mass per unit volume, $\rho = dm/dV$, so we may also write Eq. (9–20) as

$$I = \int r^2 \, \rho \, dV.$$

If the body is uniform in density, then we may take ρ outside the integral:

$$I = \rho \int r^2 \, dV. \tag{9–21}$$

To use this equation, we have to express the volume element dV in terms of the differentials of the integration variables, such as $dV = dx \, dy \, dz$. The element dV must always be chosen so that all points within it are at very nearly the same distance from the axis of rotation. The limits on the integral are determined by the shape and dimensions of the body. For regularly shaped bodies this integration can often be carried out quite easily.

By measuring small variations in the orbits of satellites, geophysicists can measure the earth's moment of inertia. This tells us how our planet's mass is distributed within its interior. The same technique has been used with interplanetary spacecraft to explore the internal structure of other worlds.

EXAMPLE 9–12

Uniform thin rod, axis perpendicular to length Figure 9–18 shows a slender uniform rod with mass M and length L. It might be a baton held by a twirler in a marching band (less the rubber end caps). Compute its moment of inertia about an axis through O, at an arbitrary distance h from one end.

SOLUTION We choose as an element of mass a short section of rod with length dx at a distance x from point O. The ratio of the mass dm of this element to the total mass M is equal to the ratio of its length dx to the total length L:

$$\frac{dm}{M} = \frac{dx}{L}.$$

We solve this for dm, substitute into Eq. (9–20), and add the appropriate integration limits on x to obtain

$$I = \int x^2 \, dm = \frac{M}{L} \int_{-h}^{L-h} x^2 \, dx$$

$$= \left[\frac{M}{L} \left(\frac{x^3}{3} \right) \right]_{-h}^{L-h} = \frac{1}{3} M(L^2 - 3Lh + 3h^2).$$

From this general expression we can find the moment of inertia about an axis through any point on the rod. For example, if the axis is at the left end, $h = 0$ and

$$I = \frac{1}{3} ML^2.$$

9–18 Finding the moment of inertia of a thin rod about an axis through O. The mass element is a segment with length dx.

If the axis is at the right end, we should get the same result. Putting $h = L$, we again get

$$I = \frac{1}{3} ML^2.$$

If the axis passes through the center, the usual place for a twirled baton, then $h = L/2$ and

$$I = \frac{1}{12} ML^2.$$

These results agree with the expressions in Table 9–2.

EXAMPLE 9-13

Hollow or solid cylinder, rotating about axis of symmetry Figure 9–19 shows a hollow, uniform cylinder with length L, inner radius R_1, and outer radius R_2. This might be a steel cylinder in a printing press or a sheet-steel rolling mill. Find the moment of inertia about the axis of symmetry of the cylinder.

SOLUTION We choose as a volume element a thin cylindrical shell of radius r, thickness dr, and length L. All parts of this element are at very nearly the same distance from the axis. The volume of the element is very nearly equal to that of a flat sheet with thickness dr, length L, and width $2\pi r$ (the circumference of the shell). Then

$$dm = \rho \, dV = \rho \, (2\pi r L \, dr)$$

The moment of inertia is given by

$$I = \int r^2 \, dm = \int_{R_1}^{R_2} r^2 \, \rho \, (2\pi r L \, dr)$$

$$= 2\pi \rho L \int_{R_1}^{R_2} r^3 \, dr$$

$$= \frac{2\pi \rho L}{4} (R_2{}^4 - R_1{}^4)$$

$$= \frac{\pi \rho L}{2} (R_2{}^2 - R_1{}^2)(R_2{}^2 + R_1{}^2).$$

It is usually more convenient to express the moment of inertia in terms of the total mass M of the body, which is its density ρ multiplied by the total volume V. The volume is

$$V = \pi L (R_2{}^2 - R_1{}^2),$$

so the total mass M is

$$M = \rho V = \pi L \rho (R_2{}^2 - R_1{}^2),$$

and the moment of inertia is

$$I = \frac{1}{2} M (R_1{}^2 + R_2{}^2).$$

This is as shown in Table 9–2, case (e).

If the cylinder is solid, such as a lawn roller, $R_1 = 0$. Calling the outer radius R_2 simply R, we find that the moment of inertia

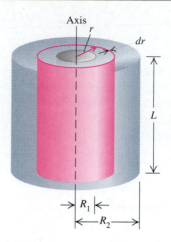

9-19 Finding the moment of inertia of a hollow cylinder about its symmetry axis. The mass element is a cylindrical shell with radius r and thickness dr.

of a solid cylinder of radius R is

$$I = \frac{1}{2} MR^2.$$

If the cylinder has a very thin wall (like a stovepipe), R_1 and R_2 are very nearly equal; if R represents this common radius,

$$I = MR^2.$$

We could have predicted this last result; in a thin-walled cylinder, all the mass is the same distance $r = R$ from the axis, so $I = \int r^2 \, dm = R^2 \int dm = MR^2$.

Note that the moment of inertia of a cylinder about an axis coinciding with its axis of symmetry depends on its mass and radii but not on its length L. Two hollow cylinders with the same inner and outer radii, one of wood and one of brass, but having the same mass M have equal moments of inertia even though the length of the wood cylinder is much greater. The moment of inertia depends only on the *radial* distribution of mass, not on its distribution along the axis. Thus the above results hold also for a thin disk, such as a washer or a compact disc (CD).

EXAMPLE 9-14

Uniform sphere with radius *R*, axis through center The object in Fig. 9–20 might be a billiard ball, a steel ball bearing, or the world's largest ball of string. Find the moment of inertia about an axis through the center of the sphere.

SOLUTION We begin by dividing the sphere into thin disks of thickness dx, whose moment of inertia we know from Example 9–13. The radius r of the disk shown is

$$r = \sqrt{R^2 - x^2}.$$

Its volume is

$$dV = \pi r^2 \, dx = \pi (R^2 - x^2) \, dx,$$

and its mass is

$$dm = \rho \, dV = \pi \rho (R^2 - x^2) \, dx.$$

From Example 9–13, the moment of inertia of a disk of radius r and mass dm is

$$dI = \frac{1}{2} r^2 \, dm = \frac{1}{2} \left(\sqrt{R^2 - x^2} \right)^2 [\pi \rho (R^2 - x^2) \, dx]$$

$$= \frac{\pi \rho}{2} (R^2 - x^2)^2 \, dx.$$

Integrating this expression from $x = 0$ to $x = R$ gives the moment of inertia of the right hemisphere. From symmetry, the total I for the entire sphere is just twice this:

$$I = (2)\frac{\pi\rho}{2}\int_0^R (R^2 - x^2)^2\,dx.$$

Carrying out the integration, we obtain

$$I = \frac{8\pi\rho}{15}R^5.$$

The mass M of the sphere of volume $V = 4\pi R^3/3$ is

$$M = \rho V = \frac{4\pi\rho R^3}{3}.$$

By comparing the expressions for I and M, we find

$$I = \frac{2}{5}MR^2.$$

This agrees with the expression in Table 9–2, case (h). Note that the moment of inertia of a solid sphere of mass M and R is less than the moment of inertia of a solid *cylinder* of the same mass and radius, $I = \frac{1}{2}MR^2$. The reason is that more of the sphere's mass is located close to the axis.

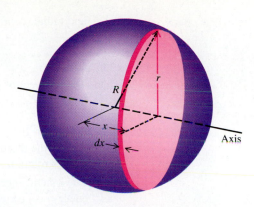

9–20 Finding the moment of inertia of a sphere about an axis through its center. The mass element is a disk with thickness dx.

SUMMARY

- When a rigid body rotates about a fixed axis, its position is described by an angular coordinate θ. The angular velocity ω is defined as the derivative of the angular coordinate θ:

$$\omega = \lim_{\Delta t \to 0}\frac{\Delta\theta}{\Delta t} = \frac{d\theta}{dt}. \tag{9–3}$$

The angular acceleration α is defined as the derivative of the angular velocity ω, or the second derivative of the angular coordinate θ:

$$\alpha = \lim_{\Delta t \to 0}\frac{\Delta\omega}{\Delta t} = \frac{d\omega}{dt} = \frac{d^2\theta}{dt^2}. \tag{9–6}$$

- When a rigid body rotates about a fixed axis with constant angular acceleration, the angular position, velocity, and acceleration are related by

$$\theta = \theta_0 + \omega_0 t + \frac{1}{2}\alpha t^2, \tag{9–11}$$

$$\theta - \theta_0 = \frac{1}{2}(\omega_0 + \omega)t, \tag{9–10}$$

$$\omega = \omega_0 + \alpha t, \tag{9–7}$$

$$\omega^2 = \omega_0^2 + 2\alpha(\theta - \theta_0), \tag{9–12}$$

where θ_0 and ω_0 are the initial values of angular position and angular velocity, respectively.

- A particle in a rigid body rotating with angular speed ω has a speed v given by

$$v = r\omega, \tag{9–13}$$

and an acceleration $\vec{a}$ with tangential component

$$a_{\text{tan}} = \frac{dv}{dt} = r\frac{d\omega}{dt} = r\alpha \tag{9–14}$$

KEY TERMS

rigid body, 268
radian, 269
average angular velocity, 269
instantaneous angular velocity, 269
average angular acceleration, 270
instantaneous angular acceleration, 271
angular speed, 274
tangential component of acceleration, 274
centripetal component of acceleration, 274
moment of inertia, 277
rotational kinetic energy, 277
parallel-axis theorem, 281

and radial (centripetal) component

$$a_{\text{rad}} = \frac{v^2}{r} = \omega^2 r. \tag{9–15}$$

■ The moment of inertia I of a body about a given axis is defined as

$$I = m_1 r_1{}^2 + m_2 r_2{}^2 + \cdots = \sum_i m_i r_i{}^2, \tag{9–16}$$

where r_i is the perpendicular distance of mass m_i from the axis of rotation. The greater the moment of inertia, the more difficult it is to change the state of the body's rotation.

■ The kinetic energy of a rigid body rotating about a fixed axis (rotational kinetic energy) is given by

$$K = \frac{1}{2} I \omega^2, \tag{9–17}$$

where I is the moment of inertia for that rotation axis.

■ The moment of inertia I_{cm} of a body of mass M about an axis through the center of mass is related to the moment of inertia I_P about a parallel axis at a distance d from the first axis by

$$I_P = I_{\text{cm}} + Md^2. \tag{9–19}$$

DISCUSSION QUESTIONS

Q9–1 When rewinding an audio or video tape, why does the tape wind up fastest at the end?

Q9–2 Does a body rotating about a fixed axis have to be *perfectly* rigid for all points on the body to have the same angular velocity and the same angular acceleration? Explain.

Q9–3 What is the difference between tangential and radial acceleration for a point on a rotating body?

Q9–4 In Fig. 9–11, all points on the chain have the same linear speed v. Is the magnitude a of the linear acceleration also the same for all points on the chain? How are the angular accelerations of the two sprockets related? Explain.

Q9–5 In Fig. 9–11, how are the radial accelerations of points at the teeth of the two sprockets related? Explain the reasoning behind your answer.

Q9–6 A flywheel rotates with constant angular velocity. Does a point on its rim have a tangential acceleration? A radial acceleration? Are these accelerations constant in magnitude? In direction? In each case give the reasoning behind your answer.

Q9–7 What is the purpose of the spin cycle of a washing machine? Explain in terms of acceleration components.

Q9–8 Although angular velocity and angular acceleration can be treated as vectors, the angular displacement θ, despite having a magnitude and a direction, cannot. This is because θ does not follow the commutative law of vector addition (Eq. 1–4). Prove this to yourself in the following way. Lay your physics textbook flat on the desk in front of you with the cover side up and where you can read the writing on it. Rotate the farthest edge toward you by 90° about a horizontal axis. Call this angular displacement θ_1. Then rotate the left edge toward you by 90° about a vertical axis. Call this angular displacement θ_2. The spine of the

book should now face you, with the writing on it oriented so that it can be read. Now start over again but carry out the two rotations in the reverse order. Do you get a different result? That is, is $\theta_1 + \theta_2$ equal to $\theta_2 + \theta_1$? Now repeat this experiment but this time with an angle of 1° rather than 90°. Do you think that the infinitesimal displacement $d\vec{\theta}$ obeys the commutative law of addition and hence qualifies as a vector? If so, how is the direction of $d\vec{\theta}$ related to the direction of $\vec{\omega}$?

Q9–9 Can you think of a body that has the same moment of inertia for all possible axes? If so, give an example, and if not, explain why this is not possible. Can you think of a body that has the same moment of inertia for all axes passing through a certain point? If so, give an example and indicate where the point is located.

Q9–10 To maximize the moment of inertia of a flywheel while minimizing its weight, what shape and distribution of mass should it have? Explain.

Q9–11 How might you determine experimentally the moment of inertia of an irregularly shaped body about a given axis?

Q9–12 A cylindrical body has mass M and radius R. Can the mass be distributed within the body in such a way that its moment of inertia about its axis of symmetry is greater than MR^2? Explain.

Q9–13 Describe how part (b) of Table 9–2 could be used to derive the result in part (d).

Q9–14 The moment of inertia I of a rigid body for an axis through its center of mass is I_{cm}. Is there any other axis parallel to this axis for which I is less than I_{cm}? Explain.

Q9–15 For the equations for I given in parts (a) and (b) of Table 9–2 to be valid, must the rod have a circular cross section?

Is there any restriction on the size of the cross section in order for these equations to apply? Explain.

Q9–16 In part (d) of Table 9–2, the thickness of the plate must be much less than a for the expression given for I to apply. But in part (c), the expression given for I applies no matter how thick the plate is. Explain.

Q9–17 In Fig. 5–26a, use the expressions $K = \frac{1}{2}I\omega^2$ and $K = \frac{1}{2}mv^2$ to calculate the kinetic energy of the box (treated as a single particle). How do these two results compare? Explain.

Q9–18 Equation (9–18) says to use y_{cm} to calculate U for an extended rigid body. But in Example 9–9 (Section 9–5), y is measured to the bottom of the hanging mass rather than to the center of mass. Is this a mistake? Explain.

Q9–19 Any angular measure—radians, degrees, or revolutions—can be used in some of the equations in Chapter 9, but only radian measure can be used in others. Identify those for which using radians is necessary and those for which it is not, and in each case give the reasoning behind your answer.

EXERCISES

SECTION 9–2 ANGULAR VELOCITY AND ACCELERATION

9–1 a) What angle in radians is subtended by an arc 1.50 m in length on the circumference of a circle of radius 2.50 m? What is this angle in degrees? b) An arc 14.0 cm in length on the circumference of a circle subtends an angle of 128°. What is the radius of the circle? c) The angle between two radii of a circle with radius 1.50 m is 0.700 rad. What length of arc is intercepted on the circumference of the circle by the two radii?

9–2 An airplane propeller is rotating at 1900 rev/min. a) Compute the propeller's angular velocity in rad/s. b) How long in seconds does it take for the propeller to turn through 35°?

9–3 Consider the flywheel of Examples 9–1 and 9–2 (Section 9–2). a) Calculate the instantaneous angular acceleration at $t = 3.5$ s. Explain why your result is equal to the average angular acceleration for the time interval from 2.0 s to 5.0 s. b) Calculate the instantaneous angular velocity at $t = 3.5$ s. Explain why your result is *not* equal to the average angular velocity for the 2.0 s to 5.0 s time interval, even though 3.5 s is at the middle of this time interval.

9–4 A fan blade rotates with angular velocity given by $\omega(t) = \gamma - \beta t^2$, where $\gamma = 5.00$ rad/s and $\beta = 0.800$ rad/s^3. (a) Calculate the angular acceleration as a function of time. (b) Calculate the instantaneous angular acceleration α at $t = 3.00$ s and the average angular acceleration α_{av} for the time interval $t = 0$ to $t = 3.00$ s. How do these two quantities compare? If they are different, why are they different?

9–5 A child is pushing a merry-go-round. The angle through which the merry-go-round has turned varies with time according to $\theta(t) = \gamma t + \beta t^3$, where $\gamma = 0.400$ rad/s and $\beta = 0.0120$ rad/s^3. a) Calculate the angular velocity of the merry-go-round as a function of time. b) What is the initial value of the angular velocity? c) Calculate the instantaneous value of the angular velocity ω at $t = 5.00$ s and the average angular velocity ω_{av} for the time interval $t = 0$ to $t = 5.00$ s. Show that ω_{av} is *not* equal to the average of the instantaneous angular velocities at $t = 0$ and $t = 5.00$ s, and explain why it is not.

9–6 At $t = 0$ the current to a dc electric motor is reversed, resulting in an angular displacement of the motor shaft given by $\theta(t) = (250 \text{ rad/s})t - (20.0 \text{ rad/s}^2)t^2 - (1.50 \text{ rad/s}^3)t^3$. a) At what time is the angular velocity of the motor shaft zero? b) Calculate the angular acceleration at the instant that the motor shaft has zero angular velocity. c) How many revolutions does the motor shaft turn through between the time when the current is reversed and the instant when the angular velocity is zero? d) How fast was the motor shaft rotating at $t = 0$, when the current was

reversed? e) Calculate the average angular velocity for the time period from $t = 0$ to the time calculated in part (a).

9–7 The angle θ through which a bicycle wheel turns is given by $\theta(t) = a + bt^2 - ct^3$, where a, b, and c are positive constants such that for t in seconds, θ will be in radians. a) Calculate the angular acceleration of the wheel as a function of time. b) At what time is the angular velocity of the wheel instantaneously not changing?

SECTION 9–3 ROTATION WITH CONSTANT ANGULAR ACCELERATION

9–8 A bicycle wheel has an initial angular velocity of 1.50 rad/s. a) If its angular acceleration is constant and equal to 0.300 rad/s^2, what is its angular velocity at $t = 2.50$ s? b) Through what angle has the wheel turned between $t = 0$ and $t = 2.50$ s?

9–9 An electric fan is turned off, and its angular velocity decreases uniformly from 500 rev/min to 200 rev/min in 4.00 s. a) Find the angular acceleration in rev/s^2 and the number of revolutions made by the motor in the 4.00-s interval. b) How many more seconds are required for the fan to come to rest if the angular acceleration remains constant at the value calculated in part (a)?

9–10 a) Derive Eq. (9–12) by combining Eqs. (9–7) and (9–11) to eliminate t. b) The angular velocity of an airplane propeller increases from 12.0 rad/s to 16.0 rad/s while turning through 7.00 rad. What is the angular acceleration in rad/s^2?

9–11 The rotating blade of a blender turns with constant angular acceleration 1.50 rad/s^2. a) How much time does it take to reach an angular velocity of 36.0 rad/s, starting from rest? b) Through how many revolutions does the blade turn in this time interval?

9–12 A flywheel requires 4.00 s to rotate through 162 rad. Its angular velocity at the end of this time is 108 rad/s. Find a) the angular velocity at the beginning of the 4.00-s interval; b) the constant angular acceleration.

9–13 Emilie's potter's wheel rotates with a constant 2.25-rad/s^2 angular acceleration. After 4.00 s the wheel has rotated through an angle of 60.0 rad. What was the angular velocity of the wheel at the beginning of the 4.00-s interval?

9–14 A circular saw blade 0.200 m in diameter starts from rest. In 6.00 s it accelerates with constant angular acceleration to an angular velocity of 140 rad/s. Find the angular acceleration and the angle through which the blade has turned.

9–15 A safety device brings the blade of a power mower from an initial angular velocity of ω_1 to rest in 1.00 revolution. At the same constant acceleration, how many revolutions would it take the blade to come to rest from an initial angular velocity ω_3 that was three times as great, $\omega_3 = 3\omega_1$?

9–16 A straight piece of reflecting tape extends from the center of a wheel to its rim. You darken the room and use a camera and strobe unit that flashes once every 0.050 s to take pictures of the wheel as it rotates counterclockwise. You trigger the strobe so that the first flash ($t = 0$) occurs when the tape is horizontal to the right at an angular displacement of zero. For the following situations draw a sketch of the photo you will get for the time exposure over five flashes (at $t = 0$, 0.050 s, 0.100 s, 0.150 s, and 0.200 s) and graph θ versus t and ω versus t for $t = 0$ to $t = 0.200$ s. a) The angular velocity is constant at 10.0 rev/s. b) The wheel starts from rest with a constant angular acceleration of 25.0 rev/s². c) The wheel is rotating at 10.0 rev/s at $t = 0$ and changes angular velocity at a constant rate of −50.0 rev/s².

9–17 At $t = 0$ a grinding wheel has an angular velocity of 24.0 rad/s. It has a constant angular acceleration of 30.0 rad/s² until a circuit breaker trips at $t = 2.00$ s. From then on, it turns through 432 rad as it coasts to a stop at constant angular acceleration. a) Through what total angle did the wheel turn between $t = 0$ and the time it stopped? b) At what time did it stop? c) What was its acceleration as it slowed down?

9–18 a) Derive a constant-angular-acceleration equation that gives $\theta - \theta_0$ in terms of ω, α, and t (no ω_0 in the equation). b) At $t = 8.0$ s a gear is rotating about a fixed axis at 4.50 rad/s. During the preceding 8.0 s, it has turned through 40.0 rad. Use the result of part (a) to calculate the gear's constant angular acceleration. c) What was the angular velocity of the gear at $t = 0$?

SECTION 9–4 RELATING LINEAR AND ANGULAR KINEMATICS

9–19 The main rotor of a helicopter is turning in a horizontal plane at 90.0 rev/min. The distance from the center of the rotor shaft to each blade tip is 5.00 m. Calculate the speed through the air of the blade tip if a) the helicopter is sitting on the ground; b) the helicopter is rising vertically at 4.00 m/s.

9–20 A compact disc (CD) stores music in a coded pattern of tiny pits 10^{-7} m deep. The pits are arranged in a track that spirals outward toward the rim of the disc; the inner and outer radii of this spiral are 25.0 mm and 58.0 mm, respectively. As the disc spins inside a CD player, the track is scanned at a constant *linear* speed of 1.25 m/s. a) What is the angular speed of the CD when scanning the innermost part of the track? The outermost part of the track? b) The maximum playing time of a CD is 74.0 min. What would be the length of the track on such a maximum-duration CD if it was stretched out in a straight line? c) What is the average angular acceleration of a maximum-duration CD during its 74.0-min playing time? Take the direction of rotation of the disc to be positive.

9–21 A wheel rotates with a constant angular velocity of 6.00 rad/s. a) Compute the radial acceleration of a point 0.500 m from the axis, using the relation $a_{rad} = \omega^2 r$. b) Find the tangential speed of the point and compute its radial acceleration from the relation $a_{rad} = v^2/r$.

9–22 Find the required angular velocity (in rev/min) of an ultracentrifuge for the radial acceleration of a point 2.50 cm from the axis to equal 400,000g (that is, 400,000 times the acceleration due to gravity).

9–23 A flywheel with a radius of 0.300 m starts from rest and accelerates with a constant angular acceleration of 0.600 rad/s². Compute the magnitude of the tangential acceleration, the radial acceleration and the resultant acceleration of a point on its rim a) at the start; b) after it has turned through 60.0°; c) after it has turned through 120.0°.

9–24 An electric ceiling fan with blades 0.750 m in diameter is rotating about a fixed axis with an initial angular velocity of 0.250 rev/s. The angular acceleration is 0.900 rev/s². a) Compute the angular velocity after 0.200 s. b) Through how many revolutions has the blade turned in this time interval? c) What is the tangential speed of a point on the tip of the blade at $t = 0.200$ s? d) What is the magnitude of the resultant acceleration of a point on the tip of the blade at $t = 0.200$ s?

9–25 An ad claims that a centrifuge takes up only 0.127 m of bench space but can produce a radial acceleration of 3000g at 5000 rev/min. Calculate the required radius of the centrifuge. Is the claim a realistic one?

9–26 a) Derive an equation for the radial acceleration that includes v and ω but not r. b) You are designing a merry-go-round for which a point of the rim will have a radial acceleration of 0.500 m/s² when the tangential velocity of that point has magnitude 2.00 m/s. What angular velocity is required to achieve these values?

9–27 **A Drill Problem.** When drilling a 12.7-mm-diameter hole in wood, plastic, or aluminum, a shop manual recommends a drill speed of 1250 rev/min. For a 12.7-mm-diameter drill bit turning at a constant 1250 rev/min, find a) the maximum linear speed of any part of the bit; b) the maximum radial acceleration of any part of the bit.

9–28 At $t = 3.00$ s a point on the rim of a 0.200-m-radius wheel has a tangential speed of 50.0 m/s as the wheel slows down with a tangential acceleration of constant magnitude 10.0 m/s². a) Calculate the wheel's constant angular acceleration. b) Calculate the angular velocities at $t = 3.00$ s and $t = 0$. c) Through what angle did the wheel turn between $t = 0$ and $t = 3.00$ s? d) At what time will the radial acceleration equal g?

9–29 The spin cycles of a washing machine have two angular speeds, 423 rev/min and 640 rev/min. The internal diameter of the drum is 0.470 m. a) What is the ratio of the maximum radial force on the laundry for the higher angular speed to that for the lower speed? b) What is the ratio of the maximum tangential speed of the laundry for the higher angular speed to that for the lower speed? c) Find the laundry's maximum tangential speed and the maximum radial acceleration, in terms of g.

SECTION 9–5 ENERGY IN ROTATIONAL MOTION

9–30 Small blocks, each with mass m, are clamped at the ends and at the center of a light rod of length L. Compute the moment of inertia of the system about an axis perpendicular to the rod and passing through a point $\frac{1}{4}$ of the length from one end. Neglect the moment of inertia of the light rod.

9–31 A twirler's baton is made of a slender metal cylinder of mass M and length L. Each end has a rubber cap of mass m and each cap can be accurately treated as a particle in this problem. Find the total moment of inertia of the baton about the usual twirling axis (perpendicular to the baton through its center).

9–32 Find the moment of inertia about each of the following axes for a rod that is 0.300 cm in diameter and 1.50 m long and has a mass of 0.0420 kg. Use the formulas of Table 9–2. a) About an axis perpendicular to the rod and passing through its center. b) About an axis perpendicular to the rod and passing through one end. c) About a longitudinal axis passing through the center of the rod.

9–33 Four small spheres, each of which can be regarded as a point of mass 0.200 kg, are arranged in a square 0.400 m on a side and connected by light rods (Fig. 9–21). Find the moment of inertia of the system about an axis a) through the center of the square, perpendicular to its plane (an axis through point O in the figure); b) bisecting two opposite sides of the square (an axis along the line AB in the figure); c) that passes through the centers of the upper left and lower right spheres and through point O.

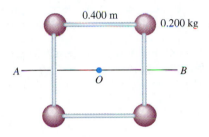

FIGURE 9–21 Exercise 9–33.

9–34 How I Scales. If we multiply all design dimensions of an object by a scaling factor f, its volume and mass will be multiplied by f^3. a) By what factor will its moment of inertia be multiplied? b) If a $\frac{1}{48}$-scale model has a rotational kinetic energy of 2.5 J, what will be the kinetic energy for the full-scale object of the same material rotating at the same angular velocity?

9–35 A wagon wheel is constructed as in Fig. 9–22. The radius of the wheel is 0.300 m, and the rim has a mass of 1.40 kg. Each of the eight spokes, which lie along a diameter and are 0.300 m long, has a mass of 0.280 kg. What is the moment of inertia of the wheel about an axis through its center and perpendicular to the plane of the wheel? (Use the formulas given in Table 9–2.)

FIGURE 9–22 Exercise 9–35.

9–36 An airplane propeller is 2.08 m in length (from tip to tip) and has a mass of 117 kg. The propeller is rotating at 2400 rev/min about an axis through its center. a) What is its rotational kinetic energy? Treat the propeller as a slender rod. b) If it were not rotating, how far would it have to drop in free fall to acquire the same kinetic energy?

9–37 a) Show that $\frac{1}{2}I\omega^2$ has units equivalent to joules. Explain why "rads" need not be included in the units. b) Often ω is expressed in rev/min rather than rad/s. Write an expression for rotational kinetic energy such that if I is in kg · m² and ω is in rev/min, the energy will be in joules.

9–38 An antique phonograph turntable has a kinetic energy of 0.0250 J when turning at 45.0 rev/min. What is the moment of inertia of the turntable about the rotation axis?

9–39 The flywheel of a gasoline engine is required to give up 500 J of kinetic energy while its angular velocity decreases from 650 rev/min to 520 rev/min. What moment of inertia is required?

9–40 A light flexible rope is wrapped several times around a *hollow* cylinder with a weight of 40.0 N and a radius of 0.25 m, which rotates without friction about a fixed horizontal axis. The cylinder is attached to the axle by spokes of negligible moment of inertia. The cylinder is initially at rest. The free end of the rope is pulled with a constant force P for a distance of 5.00 m, at which point the end of the rope is moving at 6.00 m/s. If the rope does not slip on the cylinder, what is the value of P?

9–41 Energy is to be stored in a 70.0-kg flywheel in the shape of a uniform solid disk with radius $R = 1.20$ m. To prevent structural failure of the flywheel, the maximum allowed radial acceleration of a point on its rim is 3500 m/s². What is the maximum kinetic energy that can be stored in the flywheel?

9–42 Suppose the solid cylinder in the apparatus described in Example 9–9 (Section 9–5) is replaced by a thin-walled hollow cylinder of the same mass M and radius R. The cylinder is attached to the axle by spokes of negligible moment of inertia. a) Find the speed of the hanging mass m just as it strikes the floor. b) Is the answer to part (a) greater than, equal to, or less than the speed found in Example 9–9? Explain why, using energy concepts.

9–43 Rate of Kinetic Energy Loss. A rigid body with moment of inertia I spins once every T seconds. The rotation is slowing down, so that $dT/dt > 0$. a) Express the kinetic energy of rotation of the body in terms of I and T. b) Express the rate of change of the kinetic energy of rotation in terms of I, T, and dT/dt. c) A large flywheel has $I = 8.0$ kg · m². What is the kinetic energy of the flywheel when the period of rotation is 1.5 s? d) What is the rate of change of the kinetic energy of the flywheel in part (c) at the instant when the period is 1.5 s and is changing at the rate of $dT/dt = 0.0060$?

9–44 A uniform rope 10.0 m long and with mass 3.00 kg is hanging with one end attached to a gymnasium ceiling and the other end just touching the floor. The upper end of the rope is released, and the rope falls to the floor. What is the change in the gravitational potential energy if the rope ends up flat on the floor (not coiled up)?

9–45 Center of Mass of an Extended Object. How much work must a wrestler do to raise the center of mass of his 120-kg opponent a vertical distance of 0.700 m?

SECTION 9–6 PARALLEL-AXIS THEOREM

9–46 Find the moment of inertia of a hoop (a thin-walled hollow ring) of mass M and radius R about an axis perpendicular to the hoop's plane at an edge.

9–47 About what axis will a uniform balsa-wood sphere have the same moment of inertia as for a thin-walled hollow lead sphere of the same mass and radius with the axis along a diameter?

9–48 Use the parallel-axis theorem to show that the moments of inertia given in parts (a) and (b) of Table 9–2 are consistent.

9–49 A thin, rectangular sheet of metal has mass M and sides of length a and b. Use the parallel-axis theorem to calculate the moment of inertia of the sheet for an axis that is perpendicular to the plane of the sheet and that passes through one corner of the sheet.

9–50 a) For the thin rectangular plate shown in part (d) of Table 9–2, find the moment of inertia about an axis that lies in the plane of the plate, passes through the center of the plate, and is parallel to the axis shown in the figure. b) Find the moment of inertia of the plate for an axis that lies in the plane of the plate, passes through the center of the plate, and is perpendicular to the axis in part (a).

SECTION 9–7 MOMENT OF INERTIA CALCULATIONS

*9–51 Using the information in Table 9–2 and the parallel-axis theorem, find the moment of inertia of the slender rod with mass M and length L shown in Fig. 9–18 about an axis through O, at an arbitrary distance h from one end. Compare your result to that found by integration in Example 9–12 (Section 9–7).

*9–52 Use Eq. (9–20) to calculate the moment of inertia of a uniform solid disk with mass M and radius R for an axis perpendicular to the plane of the disk and passing through its center.

*9–53 Use Eq. (9–20) to calculate the moment of inertia of a slender, uniform rod with mass M and length L about an axis at one end, perpendicular to the rod.

*9–54 A slender rod with length L has a mass per unit length that varies with distance from the left-hand end, where $x = 0$, according to $dm/dx = \gamma x$, where γ has units of kg/m^2. a) Calculate the total mass of the rod in terms of γ and L. b) Use Eq. (9–20) to calculate the moment of inertia of the rod for an axis at the left-hand end, perpendicular to the rod. Use the expression you derived in part (a) to express I in terms of M and L. How does your result compare to that for a uniform rod? Explain this comparison. c) Repeat part (b) for an axis at the right-hand end of the rod. How do the results for parts (b) and (c) compare? Explain this result.

PROBLEMS

9–55 Sketch a wheel lying in the plane of your paper and rotating counterclockwise. Choose a point on the rim and draw a vector $\vec{r}$ from the center of the wheel to that point. a) What is the direction of $\vec{\omega}$? b) Show that the velocity of the point is $\vec{v} = \vec{\omega} \times \vec{r}$. c) Show that the radial acceleration of the point is $\vec{a}_{\text{rad}} = \vec{\omega} \times \vec{v} = \vec{\omega} \times (\vec{\omega} \times \vec{r})$. (See Exercise 9–26.)

9–56 a) Prove that when an object starts from rest and rotates about a fixed axis with constant angular acceleration, the radial acceleration of a point in the object is directly proportional to its angular displacement. b) Through what angle has the object turned at the instant when the resultant acceleration of a point makes an angle of 36.9° with the radial direction?

9–57 A roller in a printing press turns through an angle $\theta(t)$ given by $\theta(t) = \gamma t^2 - \beta t^3$, where $\gamma = 3.20$ rad/s^2 and $\beta = 0.500$ rad/s^3. a) Calculate the angular velocity of the roller as a function of time. b) Calculate the angular acceleration of the roller as a function of time. c) What is the maximum positive angular velocity, and at what value of t does it occur?

*9–58 A bicycle wheel with a radius of 0.33 m turns with angular acceleration $\alpha(t) = \gamma - \beta t$, where $\gamma = 1.80$ rad/s^2 and $\beta = 0.25$ rad/s^3. It is at rest at $t = 0$. a) Calculate the angular velocity and angular displacement as functions of time. b) Calculate the maximum positive angular velocity and maximum positive angular displacement of the wheel. (*Hint:* See Section 2–7.)

9–59 When a toy car is rapidly scooted across the floor, it stores energy in a flywheel. The car has mass 0.180 kg, and its flywheel has a moment of inertia of 4.00×10^{-5} kg $\cdot$ m^2. The car is 15.0 cm long. An advertisement claims that the car can be made to travel at a scale speed of up to 700 km/h (440 mi/h). The scale speed is the speed of the toy car multiplied by the ratio of the length of an actual car to the length of the toy. Assume a length of 3.0 m for a real car. a) For a scale speed of 700 km/h, what is the actual translational speed of the car? b) If all the kinetic energy that is initially in the flywheel is converted to the translational kinetic energy of the toy, how much energy is originally stored in the flywheel? c) What initial angular velocity of the flywheel was needed to store the amount of energy calculated in part (b)?

9–60 A classic 1957 Chevrolet Corvette of mass 1240 kg starts from rest and speeds up with a constant tangential acceleration of 3.00 m/s^2 on a circular test track of radius 60.0 m. Treat the car as a particle. a) What is its angular acceleration? b) What is its angular velocity 6.00 s after it starts? c) What is its radial acceleration at this time? d) Sketch a view from above showing the circular track, the car, the velocity vector, and the acceleration component vectors 6.00 s after the car starts. e) What are the magnitudes of the total acceleration and net force for the car at this time? f) What angle do the total acceleration and net force make with the car's velocity at this time?

9–61 The flywheel of a punch press has a moment of inertia of 16.0 kg $\cdot$ m^2 and runs at 300 rev/min. The flywheel supplies all the energy needed in a quick punching operation. a) Find the speed in rev/min to which the flywheel will be reduced by a sudden punching operation requiring 4000 J of work. b) What must be the constant power supply to the flywheel (in watts) to bring it back to its initial speed in a time of 5.00 s?

9–62 A totally inedible cafeteria meatball of mass 40.0 g is attached to the free end of a 2.50-m piece of string that is attached to the ceiling. The meatball is pulled to the left so that the string makes a 36.9° angle with the vertical and is then released. a) What is the magnitude and direction of the angular velocity of the meatball the first time the angular acceleration is zero? b) What is it the second time that $\alpha = 0$? c) At the times described in parts (a) and (b), what is the magnitude and direction of the meatball's radial acceleration? d) Show that the answer to part (c) is independent of the length of the string.

9–63 A vacuum cleaner belt is looped over a shaft of radius 0.45 cm and a wheel of radius 2.00 cm. The arrangement of the belt, shaft, and wheel is similar to that of the chain and sprockets in Fig. 9–11. The motor turns the shaft at 60.0 rev/s and the moving belt turns the wheel, which in turn is connected by another shaft to the roller that beats the dirt out of the rug being vacuumed. Assume that the belt doesn't slip on either the shaft or the wheel. a) What is the speed of a point on the belt? b) What is the angular velocity of the wheel in rad/s?

9–64 The motor of a table saw is rotating at 3450 rev/min. A pulley attached to the motor shaft drives a second pulley of half the diameter by means of a V-belt. A circular saw blade of diameter 0.208 m is mounted on the same rotating shaft as the second pulley. a) The operator is careless and the blade catches and throws back a small piece of wood. This piece of wood moves with linear speed equal to the tangential speed of the rim of the blade. What is this speed? b) Calculate the radial acceleration of points on the outer edge of the blade to see why sawdust doesn't stick to its teeth.

9–65 A wheel changes its angular velocity with a constant angular acceleration while rotating about a fixed axis through its center. a) Show that the change in the magnitude of the radial acceleration during any time interval of a point on the wheel is twice the product of the angular acceleration, the angular displacement, and the perpendicular distance of the point from the axis. b) The radial acceleration of a point on the wheel that is 0.250 m from the axis changes from 25.0 m/s² to 85.0 m/s² as the wheel rotates through 15.0 rad. Calculate the tangential acceleration of this point. c) Show that the change in the wheel's kinetic energy during any time interval is the product of the moment of inertia about the axis, the angular acceleration, and the angular displacement. d) During the 15.0-rad angular displacement of part (b), the kinetic energy of the wheel increases from 20.0 J to 45.0 J. What is the moment of inertia of the wheel about the rotation axis?

9–66 The three uniform objects shown in Fig. 9–23 have the same mass m. Object A is a solid cylinder with radius R. Object B is a hollow, thin cylinder with radius R. Object C is a solid cube whose sides are of length $2R$. The axis of rotation of each object is perpendicular to the page and through the object's center of mass. a) Which object has the smallest moment of inertia? Explain. b) Which object has the largest moment of inertia? Explain. c) Where would the moment of inertia of a uniform solid sphere rank if its radius is R, its mass is m, and the axis of rotation is along a diameter of the sphere? Explain.

9–67 The earth, which is not a uniform sphere, has moment of inertia of $0.3308MR^2$ about an axis through its north and south poles. It takes the earth 86,164 s to spin once about this axis. Use Appendix F to calculate a) the earth's kinetic energy due to its rotation about this axis and b) the earth's kinetic energy due to its orbital motion around the sun. c) Explain how the value of the earth's moment of inertia tells us that the mass of the earth is concentrated toward the planet's center.

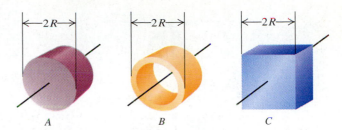

FIGURE 9–23 Problem 9–66.

9–68 A uniform solid disk with mass m and radius R is pivoted about a horizontal axis through its center. A small object of the same mass m is glued to the rim of the disk. If the disk is released from rest with the small object at the end of a horizontal radius, find the angular speed when the small object is directly below the axis.

9–69 A meter stick with a mass of 0.160 kg is pivoted about one end so it can rotate without friction about a horizontal axis. The meter stick is held in a horizontal position and released. As it swings through the vertical, calculate a) the change in gravitational potential energy that has occurred; b) the angular speed of the stick; c) the linear speed of the end of the stick opposite the axis. d) Compare the answer in part (c) to the speed of a particle that has fallen 1.00 m, starting from rest.

9–70 Exactly one turn of a flexible rope of mass m is wrapped around a uniform cylinder of mass M and radius R. The cylinder rotates without friction about a horizontal axle along the cylinder axis. One end of the rope is attached to the cylinder. The cylinder starts with an angular speed ω_0. After one revolution of the cylinder the rope has unwrapped and at this instant hangs vertically down, tangent to the cylinder. Find the angular speed of the cylinder and the linear speed of the lower end of the rope at this time. Neglect the thickness of the rope. (*Hint:* Use Eq. (9–18).)

9–71 The pulley in Fig. 9–24 has radius R and moment of inertia I. The rope does not slip over the pulley, and the pulley spins on a frictionless axle. The coefficient of kinetic friction between block A and the tabletop is μ_k. The system is released from rest, and block B descends. Block A has mass m_A and block B has mass m_B. Use energy methods to calculate the speed of block B as a function of the distance d that it has descended.

FIGURE 9–24 Problem 9–71.

9–72 The pulley in Fig. 9–25 has radius 0.160 m and moment of inertia 0.480 kg · m². The rope does not slip on the pulley rim. Use energy methods to calculate the speed of the 4.00-kg block just before it strikes the floor.

4.00 kg

5.00 m

2.00 kg

FIGURE 9–25 Problem 9–72.

9–73 You hang a thin hoop of radius R over a nail at the rim of the hoop. You displace it to the side through an angle β from its equilibrium position and let it go. What is its angular speed when it returns to its equilibrium position? (*Hint:* Use Eq. (9–18).)

9–74 A passenger bus in Zurich, Switzerland, derived its motive power from the energy stored in a large flywheel. The wheel was brought up to speed periodically, when the bus stopped at a station, by an electric motor, which could then be attached to the electric power lines. The flywheel was a solid cylinder of mass 1000 kg and diameter 1.80 m; its top angular speed was 3000 rev/min. a) At this angular speed, what is the kinetic energy of the flywheel? b) If the average power required to operate the bus is 1.86×10^4 W, how long could it operate between stops?

9–75 Two metal disks, one with radius $R_1 = 2.50$ cm and mass $M_1 = 0.80$ kg and the other with radius $R_2 = 5.00$ cm and mass $M_2 = 1.60$ kg, are welded together and mounted on a frictionless axis through their common center (Fig. 9–26). a) What is the total moment of inertia of the two disks? b) A light string is wrapped around the edge of the smaller disk, and a 1.50-kg block suspended from the free end of the string. If the block is released from rest a distance of 2.00 m above the floor, what is its speed just before it strikes the floor? c) Repeat the calculation of part (b), this time with the string wrapped around the edge of the larger disk. In which case is the final speed of the block the greatest? Explain why this should be so.

9–76 In the cylinder and mass combination described in Example 9–9 (Section 9–5), suppose the falling mass m is made of ideal rubber, so that no mechanical energy is lost when the mass hits the ground. a) If the cylinder is originally not rotating and the mass m is released from rest at a height h above the ground, to what height will this mass rebound if it bounces straight back up from the floor? b) Explain, in terms of energy, why the answer to (a) is *less* than h.

9–77 A thin flat uniform disk has mass M and radius R. A circular hole of radius $R/4$, centered at a point $R/2$ from the disk's center, is then punched in the disk. a) Find the moment

R_1 R_2

1.50 kg

FIGURE 9–26 Problem 9–75.

of inertia of the disk with the hole about an axis through the original center of the disk, perpendicular to the plane of the disk. (*Hint:* Find the moment of inertia of the piece punched from the disk.) b) Find the moment of inertia of the disk with the hole about an axis through the center of the hole, perpendicular to the plane of the disk.

9–78 A pendulum is made of a uniform solid sphere of mass M and radius R suspended from the end of a light rod. The distance from the pivot at the upper end of the rod to the center of the sphere is L. The pendulum's moment of inertia I_P for rotation about the pivot is usually approximated as ML^2. a) Use the parallel-axis theorem to show that if R is 5% of L and the mass of the rod is neglected, I_P is only 0.1% larger than ML^2. b) If the mass of the rod is 1% of M and R is 5% of L, what is the ratio of I_{rod} for an axis at the pivot to ML^2?

9–79 Perpendicular-Axis Theorem. Consider a rigid body that is a thin plane sheet of arbitrary shape. Take the body to lie in the xy-plane and let the origin O of coordinates be located at any point within or outside the body. Let I_x and I_y be the moments of inertia about the x- and y-axes and let I_O be the moment of inertia about an axis through O perpendicular to the plane. a) By considering mass elements m_i with coordinates (x_i, y_i), show that $I_x + I_y = I_O$. This is called the perpendicular-axis theorem. Note that point O does not have to be the center of mass. b) For a thin washer of mass M and with inner and outer radii R_1 and R_2, use the perpendicular-axis theorem to find the moment of inertia about an axis that is in the plane of the washer and that passes through its center. You may use the information in Table 9–2. c) Use the perpendicular-axis theorem to show that for a thin, square sheet of mass M and side L, the moment of inertia about *any* axis in the plane of the sheet that passes through the center of the sheet is $\frac{1}{12}ML^2$. You may use the information in Table 9–2.

9–80 A thin uniform rod is bent into a square of side length a. If the total mass is M, find the moment of inertia about an axis through the center and perpendicular to the plane of the square. (*Hint:* Use the parallel-axis theorem.)

*9–81 A cylinder of radius R and mass M has density that increases linearly with distance r from the cylinder axis, $\rho = \alpha r$, where α is a positive constant. a) Calculate the moment of inertia of the cylinder about a longitudinal axis through its center in terms of M and R. b) Is your answer greater or smaller than the moment of inertia of a cylinder of the same mass and radius but of uniform density? Explain why this result makes qualitative sense.

9–82 Neutron Stars and Supernova Remnants. The Crab Nebula is a cloud of glowing gas about 10 light years across, located about 6500 light years from the earth (Fig. 9–27). It is the remnant of a star that underwent a *supernova explosion*, seen on earth in 1054 A.D. Energy is released by the Crab Nebula at a rate of about 5×10^{31} W, about 10^5 times the rate at which the sun radiates energy. The Crab Nebula obtains its energy from the rotational kinetic energy of a rapidly spinning *neutron star* at its center. This object rotates once every 0.0331 s, and this period is increasing by 4.22×10^{-13} s for each second of time that elapses. a) If the rate at which energy is lost by the neutron star is equal to the rate at which energy is released by the nebula, find the moment of inertia of the neutron star. (Use the expression derived in Exercise 9–43.) b) Theories of supernovae predict that the neutron star in the Crab Nebula has a mass of about 1.4 times that of the sun. Modeling the neutron star as a solid uniform sphere, calculate its radius in kilometers. c) What is the linear speed of a point on the equator of the neutron star? Compare to the speed of light. d) Assume that the neutron star is uniform and calculate its density. Compare to the density of ordinary rock (3000 kg/m^3) and to the density of an atomic nucleus (about 10^{17} kg/m^3). Justify the statement that a neutron star is essentially a large atomic nucleus.

FIGURE 9–27 Problem 9–82.

CHALLENGE PROBLEMS

9–83 The moment of inertia of a sphere with uniform density about an axis through its center is $\frac{2}{5}MR^2 = 0.400MR^2$. Satellite observations show that the earth's moment of inertia is $0.3308MR^2$. Geophysical data suggests the earth consists of five main regions: the inner core ($r = 0$ to $r = 1220$ km) of average density 12,900 kg/m^3, the outer core ($r = 1220$ km to $r = 3480$ km) of average density 10,900 kg/m^3, the lower mantle ($r = 3480$ km to $r = 5700$ km) of average density 4900 kg/m^3, the upper mantle ($r = 5700$ km to $r = 6350$ km) of average density 3600 kg/m^3, and the outer crust and oceans ($r = 6350$ km to $r = 6370$ km) of average density 2400 kg/m^3. a) Show that the moment of inertia about a diameter of a uniform spherical shell of inner radius R_1, outer radius R_2, and density ρ is $I = \rho(8\pi/15)(R_2^5 - R_1^5)$. (*Hint:* Form the shell by superposition of a sphere of density ρ and a smaller sphere of density $-\rho$.) b) Check the given data by using them to calculate the mass of the earth. c) Use the given data to calculate the earth's moment of inertia in terms of MR^2.

*9–84 Calculate the moment of inertia of a uniform solid cone about an axis through its center (Fig. 9–28). The cone has mass M and altitude h. The radius of its circular base is R.

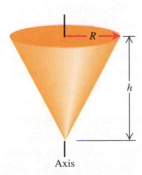

FIGURE 9–28 Challenge Problem 9–84.

9–85 On a compact disc (CD), music is coded in a pattern of tiny pits arranged in a track that spirals outward toward the rim of the disc. As the disc spins inside a CD player, the track is scanned at a constant *linear* speed of $v = 1.25$ m/s. Because the radius of the track varies as it spirals outward, the *angular* speed of the disc must change as the CD is played. (See Exercise 9–20.) Let's see what angular acceleration is required to keep v constant. The equation of a spiral is $r(\theta) = r_0 + \beta\theta$, where r_0 is the radius of the spiral at $\theta = 0$ and β is a constant. On a CD, r_0 is the inner radius of the spiral track. If we take the rotation direction of the CD to be positive, β must be positive so that r increases as the disc turns and θ increases. a) When the disc rotates through a small angle $d\theta$, the distance scanned along the track is $ds = r\, d\theta$. Using the above expression for $r(\theta)$, integrate ds to find the total distance s scanned along the track as a function of the total angle θ through which the disc has rotated. b) Since the track is scanned at a constant linear speed v, the distance s found in part (a) is equal to vt. Use this to find θ as a function of time. There will be two solutions for θ; choose the positive one, and explain why this is the solution to choose. c) Use your expression for $\theta(t)$ to find the angular velocity ω and angular acceleration α as functions of time. Is α constant? d) On a CD, the inner radius of the track is 25.0 mm, the track radius increases by 1.55 μm per revolution, and the playing time is 74.0 min. Find the values of r_0 and β, and find the total number of revolutions made during the playing time. e) Using your results from parts (c) and (d), make graphs of ω (in rad/s) versus t and α (in rad/s^2) versus t between $t = 0$ and $t = 74.0$ min.

10

Dynamics of Rotational Motion

The swirling air in a hurricane both circles around and moves inward toward the low-pressure region at the center. As the air moves inward, it gains speed in accordance with the law of conservation of angular momentum. As a result, the fastest and most destructive winds are found just outside the "eye," a region of calm air at the very center of the hurricane.

10–1 INTRODUCTION

When you use a wrench to loosen the lug nuts while changing a flat tire, the whole wheel starts to turn unless you figure out a way to hold it. What is it about the force you apply to the wrench handle that makes the wheel turn? More generally, what is it that gives a rotating body an *angular* acceleration? A force is a push or a pull, but to produce rotational motion, we need a twisting or turning action.

In this chapter we will define a new physical quantity, *torque,* that describes the twisting or turning effort of a force. We'll find that the net torque acting on a rigid body determines its angular acceleration, in the same way that the net force on a body determines its linear acceleration. We'll also look at work and power in rotational motion in order to understand such problems as how energy is transmitted by the rotating drive shaft in a car. Finally, we will develop a new conservation principle, *conservation of angular momentum,* that is tremendously useful for understanding the rotational motion of both rigid and nonrigid bodies. We'll finish this chapter by studying *gyroscopes,* rotating devices that seemingly defy common sense and don't fall over when you might think they should—but that actually behave in perfect accordance with the dynamics of rotational motion.

10–2 TORQUE

What is it about a force that determines how effective it is in causing or changing rotational motion? The magnitude and direction of the force are important, but so is the position of the point where the force is applied. When you try to swing a heavy door open, it's a lot more effective to push farther away from the axis of rotation (near the knob) than close to the axis (near the hinges). In Fig. 10–1 a wrench is being used to loosen a tight bolt. Force $\vec{F}_b$, applied near the end of the handle, is more effective than an equal force $\vec{F}_a$ applied near the bolt. Force $\vec{F}_c$ doesn't do any good at all; it's applied at the same point and has the same magnitude as $\vec{F}_b$, but it's directed along the length of the handle.

The quantitative measure of the tendency of a force to cause or change the rotational motion of a body is called *torque.* Figure 10–2 shows a body that can rotate about an axis that passes through point O and is perpendicular to the plane of the figure. The body is acted on by three forces, $\vec{F}_1$, $\vec{F}_2$, and $\vec{F}_3$, in the plane of the figure. The tendency of $\vec{F}_1$ to cause a rotation about point O depends on its magnitude F_1. It also depends on the *perpendicular* distance l_1 between the **line of action** of the force (that is, the line along which the force vector lies) and point O. We call the distance l_1 the **lever arm** (or **moment arm**) of force $\vec{F}_1$ about O. The twisting effort is directly proportional to both F_1 and l_1. We define the **torque** (or *moment*) of the force $\vec{F}_1$ with respect to point O as the product $F_1 l_1$. We will use the Greek letter τ ("tau") for torque. For a force of magnitude F whose line of action is a perpendicular distance l from

the point O, the torque is

$$\tau = Fl. \qquad (10\text{–}1)$$

Physicists usually use the term "torque," while engineers usually use "moment" (unless they are talking about a rotating shaft). Both groups use the term "lever arm" or "moment arm" for the distance l.

The lever arm of $\vec{F}_1$ in Fig. 10–2 is the perpendicular distance OA or l_1, and the lever arm of $\vec{F}_2$ is the perpendicular distance OB or l_2. The line of action of $\vec{F}_3$ passes through the reference point O, so the lever arm for $\vec{F}_3$ is zero and its torque with respect to point O is zero. In the same way, force $\vec{F}_c$ in Fig. 10–1 has zero torque with respect to point O. On the other hand, $\vec{F}_b$ has a greater torque than $\vec{F}_a$ because its lever arm is greater.

CAUTION ▶ Note that torque is always defined with reference to a specific point, often (but not always) the origin of a coordinate system. If we shift the position of this point, the torque of each force may also change. For example, the torque of force $\vec{F}_3$ in Fig. 10–2 is zero with respect to point O, but is *not* zero about point A or B. When describing the torque of a certain force, it's not enough to call it "the torque of $\vec{F}$"; you must say "the torque of $\vec{F}$ with respect to point X" or "the torque of $\vec{F}$ about point X." ◀

10–1 Which of these three forces is most likely to loosen the tight bolt?

7.1
Calculating Torques

Force $\vec{F}_1$ in Fig. 10–2 tends to cause *counterclockwise* rotation about O, while $\vec{F}_2$ tends to cause *clockwise* rotation. To distinguish between these two possibilities, we will choose a positive sense of rotation. With the choice that *counterclockwise torques are positive and clockwise torques are negative*, the torques of $\vec{F}_1$ and $\vec{F}_2$ about O are

$$\tau_1 = +F_1 l_1, \qquad \tau_2 = -F_2 l_2.$$

We will often use the symbol

to indicate our choice of the positive sense of rotation.

The SI unit of torque is the newton-meter. In our discussion of work and energy we called this combination the joule. But torque is *not* work or energy, and torque should be expressed in newton-meters, *not* joules.

Figure 10–3 shows a force $\vec{F}$ applied at a point P described by a position vector $\vec{r}$ with respect to the chosen point O. There are several ways to calculate the torque of this force. One is to find the lever arm l and use $\tau = Fl$. Or we can determine the angle ϕ between the vectors $\vec{r}$ and $\vec{F}$; the lever arm is $r \sin \phi$, so $\tau = rF \sin \phi$. A third method is to represent $\vec{F}$ in terms of a radial component F_{rad} along the direction of $\vec{r}$ and a

10–2 The torque or moment of a force about a point is the product of the force magnitude and the lever arm.

10–3 The torque of the force $\vec{F}$ about the point O is defined by $\vec{\tau} = \vec{r} \times \vec{F}$. The magnitude of $\vec{\tau}$ is $rF \sin \phi$. In this figure, $\vec{r}$ and $\vec{F}$ are in the plane of the paper; by the right-hand rule for the vector product, $\vec{\tau}$ points out of the page toward you.

tangential component F_{tan} at right angles, perpendicular to $\vec{r}$. (We call this a tangential component because if the body rotates, the point where the force acts moves in a circle, and this component is tangent to that circle.) Then $F_{tan} = F \sin \phi$, and $\tau = r(F \sin \phi) = F_{tan} r$. The component F_{rad} has no torque with respect to O because its lever arm with respect to that point is zero (compare to forces $\vec{F}_c$ in Fig. 10–1 and $\vec{F}_3$ in Fig. 10–2). Summarizing these three expressions for torque, we have

$$\tau = Fl = rF \sin \phi = F_{tan} r \qquad \text{(magnitude of torque).} \qquad (10\text{-}2)$$

We saw in Section 9–2 that angular velocity and angular acceleration can be represented as vectors; the same is true for torque. To see how to do this, note that the quantity $rF \sin \phi$ in Eq. (10–2) is the magnitude of the *vector product* $\vec{r} \times \vec{F}$ that we defined in Section 1–11. You should review that definition. We now generalize the definition of torque as follows: When a force $\vec{F}$ acts at a point having a position vector $\vec{r}$ with respect to an origin O, as in Fig. 10–3, the torque $\vec{\tau}$ of the force with respect to O is the *vector* quantity

$$\vec{\tau} = \vec{r} \times \vec{F} \qquad \text{(definition of torque vector).} \qquad (10\text{-}3)$$

The torque as defined in Eq. (10–2) is just the magnitude of the torque vector $\vec{r} \times \vec{F}$. The direction of $\vec{\tau}$ is perpendicular to both $\vec{r}$ and $\vec{F}$. In particular, if both $\vec{r}$ and $\vec{F}$ lie in a plane perpendicular to the axis of rotation, as in Fig. 10–3, then the torque vector $\vec{\tau} = \vec{r} \times \vec{F}$ is directed along the axis of rotation, with a sense given by the right-hand rule (Fig. 1–20). The direction relationships are shown in Fig. 10–4.

In the following sections we will usually be concerned with rotation of a body about an axis oriented in a specified constant direction. In that case, only the component of torque along that axis is of interest, and we often call that component the torque with respect to the specified axis.

10–4 The torque, $\vec{\tau} = \vec{r} \times \vec{F}$, is directed along the axis of the bolt. When the fingers of the right hand curl from the direction of $\vec{r}$ into the direction of $\vec{F}$—that is, in the direction of rotation that the torque tends to cause—the outstretched right thumb points in the direction of $\vec{\tau}$ (perpendicular to both $\vec{r}$ and $\vec{F}$).

EXAMPLE 10–1

A weekend plumber, unable to loosen a pipe fitting, slips a piece of scrap pipe (a "cheater") over his wrench handle. He then applies his full weight of 900 N to the end of the cheater by standing on it. The distance from the center of the fitting to the point where the weight acts is 0.80 m, and the wrench handle and cheater make an angle of 19° with the horizontal (Fig. 10–5a). Find the magnitude and direction of the torque he applies about the center of the pipe fitting.

SOLUTION From Fig. 10–5b, the angle ϕ between $\vec{r}$ and $\vec{F}$ is 109°, and the lever arm l is

$$l = (0.80 \text{ m})(\sin 109°) = (0.80 \text{ m})(\sin 71°) = 0.76 \text{ m}.$$

10–5 (a) A weekend plumber tries to loosen a pipe fitting by standing on a "cheater." (b) Vector diagram to find the torque about O.

(a)

(b)

We can find the magnitude of the torque from Eq. (10–1):

$$\tau = Fl = (900\ \text{N})(0.76\ \text{m}) = 680\ \text{N}\cdot\text{m}.$$

Or, from Eq. (10–2),

$$\tau = rF\sin\phi = (0.80\ \text{m})(900\ \text{N})(\sin 109°) = 680\ \text{N}\cdot\text{m}.$$

Alternatively, we can find F_{tan}, the tangential component of $\vec{F}$. This is the component that acts perpendicular to $\vec{r}$ (that is, per-

pendicular to the "cheater"). The vector $\vec{r}$ is oriented 19° from the horizontal, so the perpendicular to $\vec{r}$ is oriented 19° from the vertical. Since $\vec{F}$ is vertical, this means $F_{\text{tan}} = F(\cos 19°)$ = (900 N)(cos 19°) = 851 N. Then the torque is

$$\tau = F_{\text{tan}}r = (851\ \text{N})(0.80\ \text{m}) = 680\ \text{N}\cdot\text{m}.$$

The force tends to produce a clockwise rotation about O, and the direction of the torque vector $\vec{\tau}$ is *into* the plane of the figure.

10–3 TORQUE AND ANGULAR ACCELERATION FOR A RIGID BODY

We are now ready to develop the fundamental relation for the rotational dynamics of a rigid body. We will show that the angular acceleration of a rotating rigid body is directly proportional to the sum of the torque components along the axis of rotation. The proportionality factor is the moment of inertia.

To develop this relation, we again imagine the body as being made up of a large number of particles. We choose the axis of rotation to be the y-axis; the first particle has mass m_1 and distance r_1 from this axis (Fig. 10–6). The *net force* acting on this particle has a component $F_{1,\text{rad}}$ along the radial direction, a component $F_{1,\text{tan}}$ that is tangent to the circle of radius r_1 in which the particle moves as the body rotates, and a component F_{1y} along the axis of rotation. Newton's second law for the tangential component is

$$F_{1,\text{tan}} = m_1 a_{1,\text{tan}}. \tag{10–4}$$

We can express the tangential acceleration of the first particle in terms of the angular acceleration α, using Eq. (9–14): $a_{1,\text{tan}} = r_1\alpha$. Using this relation and multiplying both sides of Eq. (10–4) by r_1, we obtain

$$F_{1,\text{tan}}r_1 = m_1 r_1^2\alpha. \tag{10–5}$$

From Eq. (10–2), $F_{1,\text{tan}}r_1$ is just the magnitude of the *torque* τ_1 of the net force with respect to the rotation axis (equal to the component of the torque vector along the rotation axis). Neither of the components $F_{1,\text{rad}}$ or F_{1y} contributes to the torque about the y-axis, since neither tends to change the particle's rotation about that axis. So $\tau_1 = F_{1,\text{tan}}r_1$ is the total torque acting on the particle with respect to the rotation axis. Also, $m_1 r_1^2$ is I_1, the moment of inertia of the particle about the rotation axis. With this in mind, we rewrite Eq. (10–5) as

$$\tau_1 = I_1\alpha = m_1 r_1^2\alpha.$$

We write an equation like this for every particle in the body and then add all these equations:

$$\tau_1 + \tau_2 + \cdots = I_1\alpha + I_2\alpha + \cdots = m_1 r_1^2\alpha + m_2 r_2^2\alpha + \cdots,$$

or

$$\Sigma\tau_i = (\Sigma m_i r_i^2)\alpha.$$

The left side of this equation is the sum of all the torques about the rotation axis that act on all the particles. The right side is $I = \Sigma m_i r_i^2$, the total moment of inertia about the rotation axis, multiplied by the angular acceleration α. Note that α is the same for every particle because this is a *rigid* body. Thus for the entire rigid body we have the *rotational analog of Newton's second law*:

$$\Sigma\tau = I\alpha \quad \text{(rotational analog of Newton's second law for a rigid body).} \tag{10–6}$$

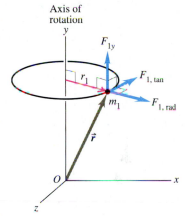

10–6 Three components of the net force acting on one of the particles of a rigid body. Only $F_{1,\text{tan}}$ has a y-component of torque about O.

Loosening or tightening a screw requires giving it an angular acceleration and hence applying a torque. This is made easier by using a screwdriver with a large-radius handle, which provides a large lever arm for the force you apply with your hand.

10–7 Particle 1 and particle 2 in a rigid body exert equal and opposite forces on each other. If these forces act along the line from one particle to another, the lever arms are the same, and the torques due to the two forces are equal and opposite. Only external torques affect the rotation of a rigid body.

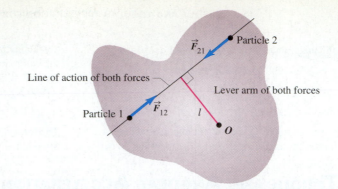

Line of action of both forces

Lever arm of both forces

Particle 1

Particle 2

$\vec{F}_{21}$

$\vec{F}_{12}$

l

O

7.10
Rotoride—Dynamics Approach

7.11
Falling Ladder

7.12
Woman and Flywheel Elevator—Dynamics Approach

Just as Newton's second law says that the net force on a particle equals the particle's mass times its acceleration, Eq. (10–6) says that the net torque on a rigid body equals the body's moment of inertia about the rotation axis times its angular acceleration.

We emphasize that Eq. (10–6) is valid *only* for *rigid* bodies. If the body is not rigid, like a rotating tank of water or a swirling tornado of air, the angular acceleration α is different for different particles in the body, and the derivation of Eq. (10–6) isn't valid. Also note that since our derivation used Eq. (9–14), $a_{\text{tan}} = r\alpha$, α must be measured in rad/s^2.

The torque on each particle is due to the net force on that particle, which is the vector sum of external and internal forces (defined in Section 8–3). According to Newton's third law, the *internal* forces that any pair of particles in the rigid body exert on each other are equal and opposite (Fig. 10–7). If these forces act along the line joining the two particles, their lever arms with respect to any axis are also equal. So the torques for each such pair are equal and opposite, and add to zero. Indeed, *all* the internal torques add to zero, so the sum $\Sigma\tau$ in Eq. (10–6) includes only the torques of the *external* forces.

Often, one of the important external forces acting on a body is its *weight*. This force is not concentrated at a single point; it acts on every particle in the entire body. Nevertheless, it turns out that if the value of $\vec{g}$ is the same at all points, we always get the correct torque (about any specified axis) if we assume that all the weight is concentrated at the *center of mass* of the body. We will prove this statement in Chapter 11, but meanwhile we will use it for some of the problems in this chapter.

Problem–Solving Strategy

ROTATIONAL DYNAMICS

Our strategy for solving problems in rotational dynamics is very similar to that used in Section 5–3 for applications of Newton's laws:

1. Draw a sketch of the situation, and select the body or bodies to be analyzed.

2. For each body, draw a free-body diagram isolating the body and including all the forces (and *only* those) that act on the body, including its weight. Label unknown quantities with algebraic symbols. A new consideration is that you must show the *shape* of the body accurately, including all dimensions and angles you will need for torque calculations.

3. Choose coordinate axes for each body, and indicate a positive sense of rotation for each rotating body. If there is a linear acceleration, it's usually simplest to pick a positive axis in its direction. If you know the sense of α in advance, picking that as the positive sense of rotation simplifies the calculations. When you represent a force in terms of its components, cross out the original force so that you don't include it twice.

4. Some problems will include bodies that have translational motion, others that have rotational motion, and some that have both. Depending on the behavior of the body in question, apply $\Sigma \vec{F} = m\vec{a}$ or $\Sigma\tau = I\alpha$ or both.

5. If more than one body is involved, carry out Steps 2 through 4 for each body. Write a separate equation of motion for each body. There may also be *geometrical* relations between the motions of two or more bodies, as with a string that unwinds from a pulley while turning it or a wheel that rolls without slipping (to be discussed in Section 10–4). Express these in algebraic form, usually as relations between two linear accelerations or a linear acceleration and an angular acceleration. Then solve the equations to find the unknown quantities. Often this involves solving a set of simultaneous equations.

6. Check the results for special cases or extreme values of quantities when possible, and compare with your intuitive expectations. Ask yourself: "Does this result make sense?"

EXAMPLE 10–2

An unwinding cable Figure 10–8a shows the same situation that we analyzed in Example 9–8 (Section 9–5) using energy methods. A cable is wrapped several times around a uniform solid cylinder that can rotate about its axis. The cylinder has diameter 0.120 m and mass 50 kg. The cable is pulled with a force of 9.0 N. Assuming that the cable unwinds without stretching or slipping, what is its acceleration?

SOLUTION Figure 10–8b shows the free-body diagram. We take the positive sense of rotation for the cylinder to be clockwise. The net force on the cylinder must be zero because its center of mass remains at rest. The weight (magnitude Mg) and the normal force (magnitude n) exerted by the bearing act along lines through the rotation axis and thus have no torque with respect to that axis. The only torque about the rotation axis is due to the force F exerted by the cable on the cylinder: $\tau = Fl$ = (9.0 N)(0.060 m) = 0.54 N · m. From Example 9–8, the moment of inertia is $I = \frac{1}{2}MR^2 = \frac{1}{2}(50 \text{ kg})(0.060 \text{ m})^2 = 0.090 \text{ kg} \cdot \text{m}^2$. The angular acceleration α is given by Eq. (10–6):

$$\alpha = \frac{\tau}{I} = \frac{0.54 \text{ N} \cdot \text{m}}{0.090 \text{ kg} \cdot \text{m}^2} = 6.0 \text{ rad/s}^2.$$

We invite you to check the units in this equation and make sure they come out right.

Now we need a kinematic relation. We remarked in Section 9–4 that the acceleration of a cable unwinding from a cylinder is

10–8 (a) Cylinder and cable. (b) The free-body diagram for the cylinder.

the same as the tangential component of acceleration of a point on the surface of the cylinder where the cable is tangent to it. This tangential acceleration is given by Eq. (9–14):

$$a = r\alpha = (0.060 \text{ m})(6.0 \text{ rad/s}^2) = 0.36 \text{ m/s}^2.$$

Can you use this result, together with an equation from Chapter 2, to determine the speed of the cable after it has been pulled 2.0 m? Try it, and compare your result with Example 9–8, in which we found this speed using work and energy considerations.

EXAMPLE 10–3

An unwinding cable II Figure 10–9a shows the same situation that we analyzed in Example 9–9 (Section 9–5) using energy methods. Find the acceleration of the object of mass m and the angular acceleration of the cylinder.

SOLUTION We have to treat the two bodies separately. Figure 10–9b shows a free-body diagram for each body. We take the positive sense of rotation for the cylinder to be counterclockwise and the positive direction of the y-coordinate for the object to be downward. For the object, Newton's second law gives

$$\Sigma F_y = mg + (-T) = ma.$$

The weight Mg and the normal force n (exerted by the bearing) have no torques with respect to the rotation axis because they act along lines through that axis, just as in Example 10–2. The only torque is that due to the cable tension T. Applying Eq. (10–6) to the cylinder gives

$$\Sigma \tau = RT = I\alpha = \frac{1}{2} MR^2 \alpha.$$

As in Example 10–2, the acceleration of the cable is the same as the tangential acceleration of a point on the cylinder rim. According to Eq. (9–14), this acceleration is given by $a = a_{\text{tan}} = R\alpha$. We use this to replace ($R\alpha$) with a in the cylinder equation above, and then divide by R; the result is

$$T = \frac{1}{2} Ma.$$

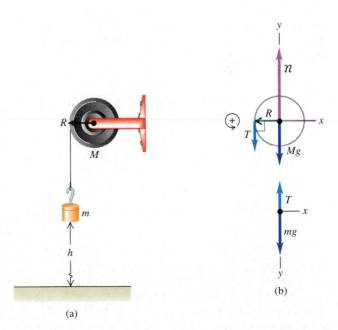

10–9 (a) Cylinder, object, and cable. (b) Free-body diagrams for the cylinder and for the object hanging from the cable. The cable is assumed to have negligible mass.

Now we substitute this expression for T into Newton's second law for the object and solve for the acceleration a:

$$mg - \frac{1}{2}Ma = ma,$$

$$a = \frac{g}{1 + M/2m}.$$

Finally, we can substitute this back into Newton's second law for the object to find T:

$$T = mg - ma = mg - m\left(\frac{g}{1 + M/2m}\right) = \frac{mg}{1 + 2m/M}.$$

We note that the tension in the cable is *not* equal to the weight mg of the object; if it were, the object could not accelerate.

Let's check some particular cases. When M is much larger than m, the tension is nearly equal to mg, and the acceleration is correspondingly much less than g. When M is zero, $T = 0$ and $a = g$; the object then falls freely. If the object starts from a height h above the floor with an initial speed v_0, its speed v when it strikes the ground is given by $v^2 = v_0^2 + 2ah$. If it starts from rest, $v_0 = 0$ and

$$v = \sqrt{2ah} = \sqrt{\frac{2gh}{1 + M/2m}}.$$

This is the same result we obtained from energy considerations in Example 9–9.

EXAMPLE 10-4

In Fig. 10–10a a glider of mass m_1 slides without friction on a horizontal air track. It is attached to an object of mass m_2 by a massless string. The pulley is a thin cylindrical shell (with massless spokes) with mass M and radius R, and the string turns the pulley without slipping or stretching. Find the acceleration of each body, the angular acceleration of the pulley, and the tension in each part of the string.

SOLUTION Figure 10–10b shows the free-body diagrams and coordinate systems for the three bodies.

CAUTION ▶ We considered a similar situation in Example 5–11 (Section 5–3). In that example the string slid without friction over a fixed pulley, and the tension was the same throughout the massless string. But with a rotating pulley, and with friction between the pulley and string to prevent slipping, the two tensions T_1 and T_2 *cannot* be equal. If they were, the pulley could not have an angular acceleration. Labeling the tension in both parts of the string as simply T would be a serious error. Be on your guard against this error in any problem that involves a rotating pulley. ◀

The equations of motion for the glider and the object are

$$\Sigma F_x = T_1 = m_1 a_1, \tag{10-7}$$

$$\Sigma F_y = m_2 g + (-T_2) = m_2 a_2. \tag{10-8}$$

The unknown normal force n_2 acts on a line through the pulley's axis of rotation, so it has no lever arm or torque with respect to

that axis. From Table 9–2 the moment of inertia of the pulley about this axis is $I = MR^2$. We take the positive sense of rotation as clockwise (the sense of the pulley's actual angular acceleration). Then the pulley's equation of motion is

$$\Sigma \tau = T_2 R + (-T_1 R) = I\alpha = (MR^2)\alpha. \tag{10-9}$$

Because the string does not stretch or slip, we have the additional *kinematic* relations

$$a_1 = a_2 = R\alpha. \tag{10-10}$$

(The accelerations of the glider and the object have different directions but the same magnitude.)

Equations (10–7) through (10–10) are five simultaneous equations for the five unknowns a_1, a_2, α, T_1, and T_2. (Equation (10–10) is actually two equations.) This may sound like an awesome problem, but we have as many equations as unknowns, and solving them is a lot easier than you might imagine. First we use Eqs. (10–10) to eliminate a_2 and α from Eqs. (10–7) through (10–9). We then have three equations for the three unknowns T_1, T_2, and a_1:

$$T_1 = m_1 a_1,$$

$$m_2 g - T_2 = m_2 a_1,$$

$$T_2 - T_1 = Ma_1.$$

The easiest way to solve these is simply to *add* the three equations, which eliminates T_1 and T_2, and then to solve for a_1:

$$a_1 = \frac{m_2 g}{m_1 + m_2 + M}.$$

10–10 (a) An air-track glider pulled by an object hanging over a pulley. (b) Free-body diagrams for all three bodies.

(a)

(b)

We can then substitute this back into Eqs. (10–7) and (10–8) to find the tensions. The results are

$$T_1 = \frac{m_1 m_2 g}{m_1 + m_2 + M}, \qquad T_2 = \frac{(m_1 + M)m_2 g}{m_1 + m_2 + M}.$$

Let's check some special cases to see whether these results make sense. First, if either m_1 or M is much larger than m_2, the

accelerations are very small, and T_2 is approximately $m_2 g$. But if m_2 is much *larger* than either m_1 or M, the acceleration is approximately g. Both of these results are what we should expect. When $M = 0$, do we get the same result as Example 5–11 (Section 5–3)? Why or why not? Can you think of other special cases to check?

10–4 RIGID-BODY ROTATION ABOUT A MOVING AXIS

We can extend our analysis of the dynamics of rotational motion to some cases in which the axis of rotation moves. When that happens, the motion of the body is **combined translation and rotation.** The key to understanding such situations is this: every possible motion of a rigid body can be represented as a combination of *translational motion of the center of mass* and *rotation about an axis through the center of mass.* This is true even when the center of mass accelerates, so that it is not at rest in any inertial frame. A graphic example is the motion of a hammer thrown upward (Fig. 10–11). The center of mass of the hammer follows a parabolic curve, as though the hammer were a particle located at the center of mass. At the same time, the hammer rotates with constant angular velocity about the center of mass (compare the motion of the wrench in Fig. 8–21). The translation of the center of mass and the rotation about the center of mass can be treated as separate but related problems. Other examples of such motion are a ball rolling down a hill and a yo-yo unwinding at the end of a string.

10–11 The motion of a rigid body like this thrown hammer is a combination of translational motion of the center of mass and rotation around the center of mass.

COMBINED TRANSLATION AND ROTATION: ENERGY RELATIONS

It's beyond the scope of this book to prove that the motion of a rigid body can always be divided into translation of the center of mass and rotation about the center of mass. But we can show that this is true for the *kinetic energy* of a rigid body that has both translational and rotational motion. In this case, the body's kinetic energy is the sum of a part $\frac{1}{2}Mv_{cm}^2$ associated with motion of the center of mass and a part $\frac{1}{2}I_{cm}\omega^2$ associated with rotation about an axis through the center of mass:

$$K = \frac{1}{2} M v_{cm}^2 + \frac{1}{2} I_{cm} \omega^2 \qquad \text{(rigid body with both translation and rotation).} \qquad (10-11)$$

To prove this relation, we again imagine the rigid body to be made up of particles. Consider a typical particle with mass m_i as shown in Fig. 10–12. The velocity $\vec{v}_i$ of this particle relative to an inertial frame is the vector sum of the velocity $\vec{v}_{cm}$ of the center of mass and the velocity $\vec{v}_i'$ of the particle *relative to* the center of mass:

$$\vec{v}_i = \vec{v}_{cm} + \vec{v}_i'. \qquad (10-12)$$

The kinetic energy K_i of this particle in the inertial frame is $\frac{1}{2}m_i v_i^2$, which we can also express as $\frac{1}{2}m_i(\vec{v}_i \cdot \vec{v}_i)$. Substituting Eq. (10–12) into this, we get

$$K_i = \frac{1}{2} m_i (\vec{v}_{cm} + \vec{v}_i') \cdot (\vec{v}_{cm} + \vec{v}_i')$$

$$= \frac{1}{2} m_i (\vec{v}_{cm} \cdot \vec{v}_{cm} + 2\vec{v}_{cm} \cdot \vec{v}_i' + \vec{v}_i' \cdot \vec{v}_i')$$

$$= \frac{1}{2} m_i (v_{cm}^2 + 2\vec{v}_{cm} \cdot \vec{v}_i' + v_i'^2).$$

The total kinetic energy is the sum ΣK_i for all the particles making up the body.

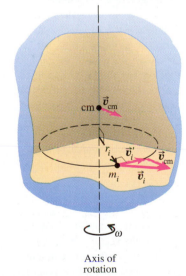

10–12 A rigid body with both translational and rotational motion. The velocity $\vec{v}_i$ of a typical particle in the rigid body is the sum of the center-of-mass velocity $\vec{v}_{cm}$ and the particle's velocity $\vec{v}_i'$ relative to the center of mass. The angular speed of the rigid body about the center of mass is ω, so $v_i' = r_i \omega$.

7.13

Race Between a Block and a Disk

Expressing the three terms in this equation as separate sums, we get

$$K = \Sigma K_i = \Sigma \left(\frac{1}{2} m_i v_{cm}{}^2 \right) + \Sigma (m_i \vec{v}_{cm} \cdot \vec{v}_i{}') + \Sigma \left(\frac{1}{2} m_i v_i{}'^2 \right).$$

The first and second terms have common factors that can be taken outside the sum:

$$K = \frac{1}{2} (\Sigma m_i) v_{cm}{}^2 + \vec{v}_{cm} \cdot (\Sigma m_i \vec{v}_i{}') + \Sigma \left(\frac{1}{2} m_i v_i{}'^2 \right). \tag{10–13}$$

Now comes the reward for our effort. In the first term, Σm_i is the total mass M. The second term is zero because $\Sigma m_i \vec{v}_i{}'$ is M times the velocity of the center of mass *relative to the center of mass,* and this is zero by definition. The last term is the sum of kinetic energies of the particles computed by using their speeds with respect to the center of mass; this is just the kinetic energy of rotation around the center of mass. Using the same steps that led to Eq. (9–17) for the rotational kinetic energy of a rigid body, we can write this last term as $\frac{1}{2} I_{cm} \omega^2$, where I_{cm} is the moment of inertia with respect to the axis through the center of mass and ω is the angular speed. So Eq. (10–13) becomes Eq. (10–11):

$$K = \frac{1}{2} M v_{cm}{}^2 + \frac{1}{2} I_{cm} \omega^2.$$

Rolling without slipping is the norm for automobile tires, but is not always the case. If the car slides on ice, the tires will spin too fast for the car's speed and $R\omega$ will be greater than v_{cm}. If the driver applies the brakes too heavily so that the car skids, the tires will spin too slowly (if at all) and $R\omega$ will be less than v_{cm}.

An important case of combined translation and rotation is **rolling without slipping,** such as the motion of the wheel shown in Fig. 10–13. The wheel is symmetrical, so its center of mass is at its geometric center. We view the motion in an inertial frame of reference in which the surface on which the wheel rolls is at rest. In this frame, the point on the wheel that contacts the surface must be instantaneously *at rest* so that it does not slip. Hence the velocity $\vec{v}_1{}'$ of the point of contact relative to the center of mass must have the same magnitude but opposite direction as the center-of-mass velocity $\vec{v}_{cm}$. If the radius of the wheel is R and its angular speed about the center of mass is ω, then the magnitude of $\vec{v}_1{}'$ is $R\omega$; hence we must have

$$v_{cm} = R\omega \quad \text{(condition for rolling without slipping).} \tag{10–14}$$

As Fig. 10–13 shows, the velocity of a point on the wheel is the vector sum of the velocity of the center of mass and the velocity of the point relative to the center of mass. Thus while point 1, the point of contact, is instantaneously at rest, point 3 at the top of the wheel is moving forward *twice as fast* as the center of mass, and points 2 and 4 at the sides have velocities at 45° to the horizontal.

At any instant we can think of the wheel as rotating about an "instantaneous axis" of rotation that passes through the point of contact with the ground. The angular velocity ω is the same for this axis as for an axis through the center of mass; an observer at the center of mass sees the rim make the same number of revolutions per second as does an observer at the rim watching the center of mass spin around him. If we think of the

10–13 As seen from the inertial frame of the surface, the velocity $\vec{v}$ of a point on the rim of a rolling wheel is the vector sum of the velocity $\vec{v}_{cm}$ of the center of mass and the velocity $\vec{v}'$ of the point relative to the center of mass due to the rotation of the wheel. The magnitude of $\vec{v}'$ is equal to $R\omega$. When the wheel rolls without slipping, $v' = v_{cm}$. The rolling wheel is instantaneously at rest at the point where it contacts the ground. The instantaneous velocity is greatest at the top of the wheel.

motion of the rolling wheel in Fig. 10–13 in this way, the kinetic energy of the wheel is $K = \frac{1}{2}I_1\omega^2$, where I_1 is the moment of inertia of the wheel about an axis through point 1. But by the parallel-axis theorem, Eq. (9–19), $I_1 = I_{cm} + MR^2$, where M is the total mass of the wheel and I_{cm} is the moment of inertia with respect to an axis through the center of mass. Using Eq. (10–14), the kinetic energy of the wheel is

$$K = \frac{1}{2}I_1\omega^2 = \frac{1}{2}I_{cm}\omega^2 + \frac{1}{2}MR^2\omega^2 = \frac{1}{2}I_{cm}\omega^2 + \frac{1}{2}Mv_{cm}^2,$$

which is the same as Eq. (10–11).

If a rigid body changes height as it moves, we must also consider gravitational potential energy. As we discussed in Section 9–5, the gravitational potential energy associated with any extended body of mass M, rigid or not, is the same as if we replace the body by a particle of mass M located at the body's center of mass. That is,

$$U = Mgy_{cm}.$$

EXAMPLE 10–5

A rolling cylindrical shell A hollow cylindrical shell with mass M and radius R rolls without slipping with speed v_{cm} on a flat surface. What is its kinetic energy?

SOLUTION The moment of inertia is $I_{cm} = MR^2$, and from Eq. (10–14) the angular speed ω is equal to v_{cm}/R. Substituting these expressions into Eq. (10–11), we get

$$K = \frac{1}{2}Mv_{cm}^2 + \frac{1}{2}(MR^2)\left(\frac{v_{cm}}{R}\right)^2$$
$$= Mv_{cm}^2.$$

The kinetic energy is twice as great as it would be if the cylindrical shell were sliding at speed v_{cm} without rolling. Half of the total kinetic energy is translational, and half is rotational.

EXAMPLE 10–6

Speed of a primitive yo-yo A primitive yo-yo is made by wrapping a string several times around a solid cylinder with mass M and radius R (Fig. 10–14). You hold the end of the string stationary while releasing the cylinder with no initial motion. The string unwinds but does not slip or stretch as the cylinder drops and rotates. Use energy considerations to find the speed v_{cm} of the center of mass of the solid cylinder after it has dropped a distance h.

SOLUTION The upper end of the string is held fixed, not pulled upward, so the hand in Fig. 10–14 does no work on the system of string and cylinder. As in Example 9–8 (Section 9–5), there is friction between the string and the cylinder, but because the string never slips on the surface of the cylinder, no mechanical energy is lost. Thus we can use conservation of mechanical energy. The potential energies are $U_1 = Mgh$ and $U_2 = 0$. The string has no kinetic energy because it's massless. The initial kinetic energy of the cylinder is $K_1 = 0$, and its final kinetic energy K_2 is given by Eq. (10–11):

$$K_2 = \frac{1}{2}Mv_{cm}^2 + \frac{1}{2}I_{cm}\omega^2.$$

But $\omega = v_{cm}/R$ from Eq. (10–14), and $I_{cm} = \frac{1}{2}MR^2$, so

$$K_2 = \frac{1}{2}Mv_{cm}^2 + \frac{1}{2}\left(\frac{1}{2}MR^2\right)\left(\frac{v_{cm}}{R}\right)^2$$
$$= \frac{3}{4}Mv_{cm}^2.$$

10–14 Calculating the speed of a primitive yo-yo.

Finally, conservation of energy gives

$$K_1 + U_1 = K_2 + U_2,$$
$$0 + Mgh = \frac{3}{4}Mv_{cm}^2 + 0,$$

and

$$v_{cm} = \sqrt{\frac{4}{3}gh}.$$

This is less than the speed $\sqrt{2gh}$ that a dropped object would have, because one third of the potential energy released as the cylinder falls appears as rotational kinetic energy.

EXAMPLE 10-7

Race of the rolling bodies In a physics lecture demonstration, an instructor "races" various round rigid bodies by releasing them from rest at the top of an inclined plane (Fig. 10–15). What shape should a body have to reach the bottom of the incline first?

SOLUTION We can again use conservation of energy because there is no sliding of the rigid bodies over the inclined plane. No work is done by kinetic friction if the bodies roll without slipping. We can also ignore the effects of *rolling friction,* introduced in Section 5–4, provided that the bodies and the surface on which they roll are perfectly rigid. (Later in this section we'll explain why this is so.) Each body starts from rest at the top of an incline with height h, so $K_1 = 0$, $U_1 = Mgh$, and $U_2 = 0$. From Eq. (10–11),

$$K_2 = \frac{1}{2} M v_{cm}^2 + \frac{1}{2} I_{cm} \omega^2.$$

If the bodies roll without slipping, $\omega = v_{cm}/R$. The moments of inertia of all the round bodies in Table 9–2 (about axes through their centers of mass) can be expressed as $I_{cm} = cMR^2$, where c is a pure number less than or equal to 1 that depends on the shape of the body. Then from conservation of energy,

$$K_1 + U_1 = K_2 + U_2,$$

$$0 + Mgh = \frac{1}{2} M v_{cm}^2 + \frac{1}{2} cMR^2 \left(\frac{v_{cm}}{R} \right)^2$$

$$= \frac{1}{2} (1 + c) M v_{cm}^2,$$

10–15 Which body rolls down the plane fastest, and why?

so the speed at the bottom of the incline is

$$v_{cm} = \sqrt{\frac{2gh}{1 + c}}.$$

This is a fairly amazing result; the speed doesn't depend on either the mass M of the body or its radius R. All uniform solid cylinders have the same speed at the bottom, even if their masses and radii are different, because they have the same c. All solid spheres have the same speed, and so on. The smaller the value of c, the faster the body is moving at the bottom (and at any point on the way down). Small-c bodies always beat large-c bodies, because they have less of their kinetic energy tied up in rotation and have more available for translation. Reading the values of c from Table 9–2, we see that the order of finish is as follows: any solid sphere, any solid cylinder, any thin-walled hollow sphere, and any thin-walled hollow cylinder.

COMBINED TRANSLATION AND ROTATION: DYNAMICS

We can also analyze the combined translational and rotational motion of a rigid body from the standpoint of dynamics. We showed in Section 8–6 that for a body with total mass M, the acceleration $\vec{a}_{cm}$ of the center of mass is the same as that of a point mass M acted on by all the external forces on the actual body:

$$\Sigma \vec{F}_{ext} = M \vec{a}_{cm}. \tag{10–15}$$

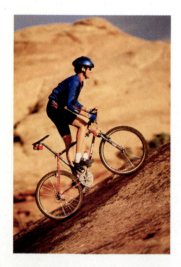

The axle of a bicycle wheel passes through the wheel's center of mass and is an axis of symmetry. Hence the rotation of the wheel is described by Eq. (10–16), provided the bicycle doesn't turn or tilt to one side (which would change the orientation of the axle).

The rotational motion about the center of mass is described by the rotational analog of Newton's second law, Eq. (10–6):

$$\Sigma \tau = I_{cm} \alpha, \tag{10–16}$$

where I_{cm} is the moment of inertia with respect to an axis through the center of mass and the sum $\Sigma \tau$ includes all external torques with respect to this axis. It's not immediately obvious that Eq. (10–16) should apply to the motion of a translating rigid body; after all, our derivation of $\Sigma \tau = I\alpha$ in Section 10–3 assumed that the axis of rotation was stationary. But in fact, Eq. (10–16) is valid *even when the axis of rotation moves,* provided the following two conditions are met:

1. The axis through the center of mass must be an axis of symmetry.
2. The axis must not change direction.

Note that in general this moving axis of rotation is *not* at rest in an inertial frame of reference.

We can now solve dynamics problems involving a rigid body that undergoes translational and rotational motion at the same time, provided that the rotation axis satisfies the two conditions just mentioned. The problem-solving strategy outlined in Section 10–3 is equally useful here, and you should review it now. Keep in mind that when a body undergoes translational and rotational motion at the same time, we need two separate equations of motion *for the same body*. One of these, Eq. (10–15), describes the translational motion of the center of mass. The other equation of motion, Eq. (10–16), describes the rotational motion about the axis through the center of mass.

EXAMPLE 10-8

Acceleration of a primitive yo-yo For the primitive yo-yo in Example 10–6, find the downward acceleration of the cylinder and the tension in the string.

SOLUTION Figure 10–16 shows a free-body diagram for the yo-yo, including the choice of positive coordinate directions. The equation for the translational motion of the center of mass is

$$\Sigma F_y = Mg + (-T) = Ma_{cm}. \qquad (10\text{--}17)$$

The moment of inertia for an axis through the center of mass is $I_{cm} = \frac{1}{2}MR^2$. Only the tension force has a torque with respect to the axis through the center of mass, so the equation for rotational motion about this axis is

$$\Sigma \tau = TR = I_{cm}\alpha = \frac{1}{2}MR^2\alpha. \qquad (10\text{--}18)$$

The string unwinds without slipping, so $v_{cm} = R\omega$ from Eq. (10–14); the derivative of this relation with respect to time is

$$a_{cm} = R\alpha. \qquad (10\text{--}19)$$

We now use Eq. (10–19) to eliminate α from Eq. (10–18) and then solve Eqs. (10–17) and (10–18) simultaneously for T and a_{cm}. The results are amazingly simple:

$$a_{cm} = \frac{2}{3}g, \qquad T = \frac{1}{3}Mg.$$

10–16 Free-body diagram for a primitive yo-yo.

Using the constant-acceleration formula $v_{cm}^2 = v_0^2 + 2a_{cm}h$, you can show that the speed of the yo-yo after it has fallen a distance h is $v_{cm} = \sqrt{\frac{4}{3}gh}$, just as we found in Example 10–6.

From the standpoint of dynamics, the tension force is essential; it causes the yo-yo's acceleration to be less than g, and its torque is what causes the yo-yo to turn. Yet when we analyzed this situation using energy methods in Example 10–6, we didn't have to consider the tension force at all! Because no mechanical energy was lost or gained, from the energy standpoint the string is significant only as a way of helping to convert some of the gravitational potential energy into rotational kinetic energy.

EXAMPLE 10-9

Acceleration of a rolling sphere A solid bowling ball rolls without slipping down the return ramp at the side of the alley (Fig. 10–17a). The ramp is inclined at an angle β to the horizontal. What is the ball's acceleration? Treat the ball as a uniform solid sphere, ignoring the finger holes.

SOLUTION Figure 10–17b shows a free-body diagram, with positive coordinate directions indicated. From Table 9–2 the moment of inertia of a solid sphere is $I_{cm} = \frac{2}{5}MR^2$. The equations of motion for translation and for rotation about the axis through

10–17 (a) A bowling ball rolling down a ramp. (b) The free-body diagram for the bowling ball.

the center of mass, respectively, are

$$\Sigma F_x = Mg \sin \beta + (-f) = Ma_{cm}, \qquad (10\text{–}20)$$

$$\Sigma \tau = fR = I_{cm}\alpha = \left(\frac{2}{5}MR^2\right)\alpha. \qquad (10\text{–}21)$$

Note that only the friction force f has a torque with respect to the axis through the center of mass. If the ball rolls without slipping, we have the same kinematic relation $a_{cm} = R\alpha$ as in Example 10–8. We use this to eliminate α from Eq. (10–21):

$$fR = \frac{2}{5}MRa_{cm}.$$

This equation and Eq. (10–20) are two equations for two unknowns, a_{cm} and f. We solve Eq. (10–20) for f, substitute the expression into the above equation to eliminate f, and then solve for a_{cm} to obtain

$$a_{cm} = \frac{5}{7}g \sin \beta.$$

The acceleration is just $\frac{5}{7}$ as large as it would be if the ball could *slide* without friction down the slope, like the toboggan in Example 5–9 (Section 5–3). Finally, we substitute this back into Eq. (10–20) and solve for f:

$$f = \frac{2}{7}Mg \sin \beta.$$

Because the ball does not slip at the instantaneous point of contact with the ramp, the friction force f is a *static* friction force; it prevents slipping and gives the ball its angular acceleration. We can derive an expression for the minimum coefficient of static friction μ_s needed to prevent slipping. The normal force is $n = Mg \cos \beta$. The maximum force of static friction equals $\mu_s n$, so the coefficient of friction must be at least as great as

$$\mu_s = \frac{f}{n} = \frac{\frac{2}{7}Mg \sin \beta}{Mg \cos \beta} = \frac{2}{7}\tan \beta.$$

If the plane is tilted only a little, β is small, and only a small value of μ_s is needed to prevent slipping. But as the angle increases, the required value of μ_s increases, as we might expect intuitively. If the ball begins to slip, Eqs. (10–20) and (10–21) are both still valid, but it's no longer true that $v_{cm} = R\omega$ or $a_{cm} = R\alpha$; we have only two equations for three unknowns (a_{cm}, α, and f). To solve the problem of rolling *with* slipping requires taking *kinetic* friction into account (see Challenge Problem 10–85).

If the bowling ball descends a vertical distance h as it moves down the ramp, the displacement along the ramp is $h/\sin \beta$. You should be able to show that the speed of the ball at the bottom of the ramp would be $v_{cm} = \sqrt{\frac{10}{7}gh}$, which is just the result you found in Example 10–7 with $c = \frac{2}{5}$.

If the ball were rolling *uphill,* the force of friction would still be directed uphill as in Fig. 10–17b. Can you see why?

ActivPhysics

7.9
Car "Peels Out"

9.14
The Sweet Spot

ROLLING FRICTION

In Example 10–7 we said that we can ignore rolling friction if both the rolling body and the surface over which it rolls are perfectly rigid. In Fig. 10–18a a perfectly rigid sphere is rolling down a perfectly rigid incline. The line of action of the normal force passes through the center of the sphere, so its torque is zero; there is no sliding at the point of contact, so the friction force does no work. Figure 10–18b shows a more realistic situation, in which the surface "piles up" in front of the sphere and the sphere rides in a shallow trench. Because of these deformations, the contact forces on the sphere no longer act along a single point, but over an area; the forces are concentrated on the front

10–18 (a) Forces on a perfectly rigid sphere rolling down a perfectly rigid incline. (b) If the sphere or incline is deformable, the contact forces act at different positions. The normal force produces a counterclockwise torque about the rotation axis, opposing the clockwise rotation. The deformation is shown greatly exaggerated.

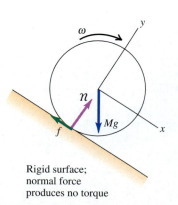

Rigid surface;
normal force
produces no torque

(a)

Deformable surface;
normal force produces
torque opposing rotation

(b)

of the sphere as shown. As a result, the normal force now exerts a torque that opposes the rotation. In addition, there is some sliding of the sphere over the surface due to the deformation, causing mechanical energy to be lost. The combination of these two effects is the phenomenon of *rolling friction*. Rolling friction also occurs if the rolling body is deformable, such as an automobile tire. Often the rolling body and surface are rigid enough that rolling friction can be ignored, as we have assumed in all the examples in this section.

10–5 WORK AND POWER IN ROTATIONAL MOTION

When you pedal a bicycle, you apply forces to a rotating body and do work on it. Similar things happen in many other real-life situations, such as a rotating motor shaft driving a power tool or a car engine propelling the vehicle. We can express this work in terms of torque and angular displacement.

Suppose a tangential force $\vec{F}_{tan}$ acts at the rim of a pivoted disk—for example, a child running while pushing on a playground merry-go-round (Fig. 10–19a). The disk rotates through an infinitesimal angle $d\theta$ about a fixed axis during an infinitesimal time interval dt (Fig. 10–19b). The work dW done by the force $\vec{F}_{tan}$ while a point on the rim moves a distance ds is $dW = F_{tan}\,ds$. If $d\theta$ is measured in radians, then $ds = R\,d\theta$ and

$$dW = F_{tan}R\,d\theta.$$

Now $F_{tan}R$ is the *torque* τ due to the force $\vec{F}_{tan}$, so

$$dW = \tau\,d\theta. \tag{10–22}$$

The total work W done by the torque during an angular displacement from θ_1 to θ_2 is

$$W = \int_{\theta_1}^{\theta_2} \tau\,d\theta \qquad \text{(work done by a torque).} \tag{10–23}$$

If the torque is *constant* while the angle changes by a finite amount $\Delta\theta = \theta_2 - \theta_1$, then

$$W = \tau(\theta_2 - \theta_1) = \tau\,\Delta\theta \qquad \text{(work done by a constant torque).} \tag{10–24}$$

The work done by a *constant* torque is the product of torque and the angular displacement. If torque is expressed in newton-meters (N · m) and angular displacement in radians, the work is in joules. Equation (10–24) is the rotational analog of Eq. (6–1), $W = Fs$, and Eq. (10–23) is the analog of Eq. (6–7), $W = \int F\,dx$, for the work done by a force in a straight-line displacement.

If the force in Fig. 10–19 had an axial or radial component, that component would do no work because the displacement of the point of application has only a tangential component. An axial or radial component of force would also make no contribution to the torque about the axis of rotation, so Eqs. (10–23) and (10–24) are correct for any force, no matter what its components.

When a torque does work on a rotating rigid body, the kinetic energy changes by an amount equal to the work done. We can prove this by using exactly the same procedure that we used in Eqs. (6–11) through (6–13) for a particle. We first let τ represent the *net* torque on the body so that from Eq. (10–6), $\tau = I\alpha$. By using this equation, we are assuming that the body is rigid so that the moment of inertia I is constant. We then transform the integrand in Eq. (10–23) into an integral on ω as follows:

$$\tau\,d\theta = (I\alpha)\,d\theta = I\frac{d\omega}{dt}\,d\theta = I\frac{d\theta}{dt}\,d\omega = I\omega\,d\omega.$$

Since τ is the net torque, the integral in Eq. (10–23) is the *total* work done on the

(a)

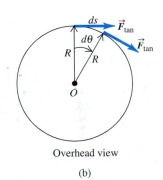

Overhead view

(b)

10–19 A tangential force acting on a rotating body does work.

rotating rigid body. This equation then becomes

$$W_{\text{tot}} = \int_{\omega_1}^{\omega_2} I\omega \, d\omega = \frac{1}{2} I\omega_2^2 - \frac{1}{2} I\omega_1^2. \qquad (10\text{--}25)$$

The change in the rotational kinetic energy of a *rigid* body equals the work done by forces exerted from outside the body. This equation is analogous to Eq. (6–13), the work-energy theorem for a particle.

What about the *power* associated with work done by a torque acting on a rotating body? When we divide both sides of Eq. (10–22) by the time interval dt during which the angular displacement occurs, we find

$$\frac{dW}{dt} = \tau \frac{d\theta}{dt}.$$

But dW/dt is the rate of doing work, or *power P*, and $d\theta/dt$ is angular velocity ω, so

$$P = \tau\omega. \qquad (10\text{--}26)$$

When a torque τ (with respect to the axis of rotation) acts on a body that rotates with angular velocity ω, its power (rate of doing work) is the product of τ and ω. This is the analog of the relation $P = \vec{F} \cdot \vec{v}$ that we developed in Section 6–5 for particle motion.

EXAMPLE 10–10

The power output of an automobile engine is advertised to be 200 hp at 6000 rpm. What is the corresponding torque?

SOLUTION First we have to convert the power to watts and the angular velocity to rad/s:

$$P = 200 \text{ hp} = 200 \text{ hp} \left(\frac{746 \text{ W}}{1 \text{ hp}} \right) = 1.49 \times 10^5 \text{ W},$$

$$\omega = 6000 \text{ rev/min} = \left(\frac{6000 \text{ rev}}{1 \text{ min}} \right)\left(\frac{2\pi \text{ rad}}{1 \text{ rev}} \right)\left(\frac{1 \text{ min}}{60 \text{ s}} \right)$$

$$= 628 \text{ rad/s}.$$

From Eq. (10–26),

$$\tau = \frac{P}{\omega} = \frac{1.49 \times 10^5 \text{ N} \cdot \text{m/s}}{628 \text{ rad/s}} = 237 \text{ N} \cdot \text{m}.$$

You could apply this much torque by using a wrench 0.25 m long and applying a force of 948 N (213 lb) to the end of its handle. Could you do it?

EXAMPLE 10–11

An electric motor exerts a constant torque of $\tau = 10 \text{ N} \cdot \text{m}$ on a grindstone mounted on its shaft. The moment of inertia of the grindstone is $I = 2.0 \text{ kg} \cdot \text{m}^2$. If the system starts from rest, find the work done by the motor in 8.0 seconds and the kinetic energy at the end of this time. What was the average power delivered by the motor?

SOLUTION From $\tau = I\alpha$, the angular acceleration is 5.0 rad/s². The angular velocity and kinetic energy after 8.0 s are

$$\omega = \alpha t = (5.0 \text{ rad/s}^2)(8.0 \text{ s}) = 40 \text{ rad/s},$$

$$K = \frac{1}{2} I\omega^2 = \frac{1}{2}(2.0 \text{ kg} \cdot \text{m}^2)(40 \text{ rad/s})^2 = 1600 \text{ J}.$$

The total angle through which the system turns in 8.0 s is

$$\Delta\theta = \frac{1}{2} \alpha t^2 = \frac{1}{2}(5.0 \text{ rad/s}^2)(8.0 \text{ s})^2 = 160 \text{ rad},$$

and the total work done by the torque is

$$W = \tau\Delta\theta = (10 \text{ N} \cdot \text{m})(160 \text{ rad}) = 1600 \text{ J}$$

This equals the total kinetic energy, as it must.

The average power is the total work divided by the time interval:

$$P_{\text{av}} = \frac{1600 \text{ J}}{8.0 \text{ s}} = 200 \text{ J/s} = 200 \text{ W}.$$

The instantaneous power, given by $P = \tau\omega$, is not constant because ω increases continuously. But we can compute the total work by taking the time integral of P, as follows:

$$W = \int_{t_1}^{t_2} P \, dt = \int_{t_1}^{t_2} \tau\omega \, dt = \int_{t_1}^{t_2} \tau(\alpha t) \, dt$$

$$= \int_0^{8.0 \text{ s}} (10 \text{ N} \cdot \text{m})(5 \text{ rad/s}^2) t \, dt = \left[(50 \text{ J/s}^2) \frac{t^2}{2} \right]_0^{8.0 \text{ s}}$$

$$= 1600 \text{ J},$$

as before. The instantaneous power P increases from zero at $t = 0$ to $(10 \text{ N} \cdot \text{m})(40 \text{ rad/s}) = 400 \text{ W}$ at $t = 8.0 \text{ s}$. The angular velocity and the power increase uniformly with time, so the *average* power is just half this maximum value, or 200 W.

10-6 ANGULAR MOMENTUM

Every rotational quantity that we have encountered in Chapters 9 and 10 is the analog of some quantity in the translational motion of a particle. The analog of *momentum* of a particle is **angular momentum,** a vector quantity denoted as $\vec{L}$. Its relation to momentum $\vec{p}$ (which we will sometimes call *linear momentum* for clarity) is exactly the same as the relation of torque to force, $\vec{\tau} = \vec{r} \times \vec{F}$. For a particle with constant mass m, velocity $\vec{v}$, momentum $\vec{p} = m\vec{v}$, and position vector $\vec{r}$ relative to the origin O of an inertial frame, we define angular momentum $\vec{L}$ as

$$\vec{L} = \vec{r} \times \vec{p} = \vec{r} \times m\vec{v} \qquad \text{(angular momentum of a particle).} \qquad (10\text{--}27)$$

The value of $\vec{L}$ depends on the choice of origin O, since it involves the position vector of the particle relative to the origin. The units of angular momentum are $\text{kg} \cdot \text{m}^2/\text{s}$.

In Fig. 10-20 a particle moves in the xy-plane; its position vector $\vec{r}$ and momentum $\vec{p} = m\vec{v}$ are shown. The angular momentum vector $\vec{L}$ is perpendicular to the xy-plane. The right-hand rule for vector products shows that its direction is along the $+z$-axis, and its magnitude is

$$L = mvr \sin\phi = mvl, \qquad (10\text{--}28)$$

where l is the perpendicular distance from the line of $\vec{v}$ to O. This distance plays the role of "lever arm" for the momentum vector.

When a net force $\vec{F}$ acts on a particle, its velocity and momentum change, so its angular momentum may also change. We can show that the *rate of change* of angular momentum is equal to the torque of the net force. We take the time derivative of Eq. (10-27), using the rule for the derivative of a product:

$$\frac{d\vec{L}}{dt} = \left(\frac{d\vec{r}}{dt} \times m\vec{v} \right) + \left(\vec{r} \times m\frac{d\vec{v}}{dt} \right) = (\vec{v} \times m\vec{v}) + (\vec{r} \times m\vec{a}).$$

The first term is zero because it contains the vector product of the vector $\vec{v} = d\vec{r}/dt$ with itself. In the second term we replace $m\vec{a}$ with the net force $\vec{F}$, obtaining

$$\frac{d\vec{L}}{dt} = \vec{r} \times \vec{F} = \vec{\tau} \qquad \text{(for a particle acted on by net force } \vec{F}\text{).} \qquad (10\text{--}29)$$

The rate of change of angular momentum of a particle equals the torque of the net force acting on it. Compare this result to Eq. (8-3), which states that the rate of change $d\vec{p}/dt$ of the *linear* momentum of a particle equals the net force that acts on it.

We can use Eq. (10-28) to find the total angular momentum of a rigid body rotating about the z-axis with angular speed ω. First consider a thin slice of the body lying in the xy-plane (Fig. 10-21). Each particle in the slice moves in a circle centered at the origin, and at each instant its velocity $\vec{v}_i$ is perpendicular to its position vector $\vec{r}_i$, as shown. Hence in Eq. (10-28), $\phi = 90°$ for every particle. A particle with mass m_i at a distance r_i from O has a speed v_i equal to $r_i\omega$. From Eq. (10-28) the magnitude L_i of its angular momentum is

$$L_i = m_i(r_i\omega)r_i = m_i r_i{}^2\omega. \qquad (10\text{--}30)$$

The direction of each particle's angular momentum, as given by the right-hand rule for the vector product, is along the $+z$-axis.

The *total* angular momentum of the slice of the body lying in the xy-plane is the sum ΣL_i of the angular momenta L_i of the particles. Summing Eq. (10-30), we have

$$L = \Sigma L_i = (\Sigma m_i r_i{}^2)\omega = I\omega,$$

where I is the moment of inertia of the slice about the z-axis.

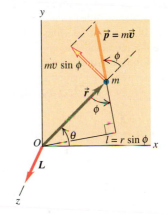

10-20 A particle with mass m moving in the xy-plane. The angular momentum of the particle is $\vec{L} = \vec{r} \times m\vec{v} = \vec{r} \times \vec{p}$, and its magnitude is $L = mvl$.

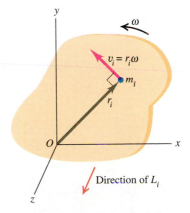

10-21 Calculating the angular momentum of a rotating rigid body. Every particle in the body moves in a circle about the rotation axis (in this case, the z-axis) with the same angular speed.

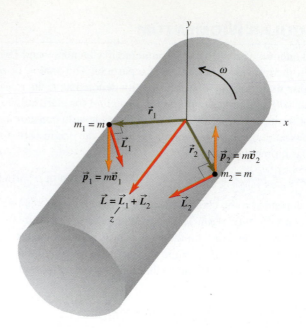

10–22 Two particles of the same mass ($m_1 = m_2 = m$) located symmetrically on either side of the axis of rotation (the z-axis). The angular momentum vectors $\vec{L}_1$ and $\vec{L}_2$ of the individual particles do not lie along the rotation axis, but their vector sum $\vec{L} = \vec{L}_1 + \vec{L}_2$ does.

We can do this same calculation for the other slices of the body, all parallel to the xy-plane. For points that do not lie in the xy-plane, a complication arises because the $\vec{r}$ vectors have components in the z-direction as well as the x- and y-directions; this gives the angular momentum of each particle a component perpendicular to the z-axis. But *if the z-axis is an axis of symmetry,* the perpendicular components for particles on opposite sides of this axis add up to zero (Fig. 10–22). So when a body rotates about an axis of symmetry, its angular momentum vector $\vec{L}$ lies along the symmetry axis, and its magnitude is $L = I\omega$.

The angular velocity vector $\vec{\omega}$ also lies along the rotation axis, as we discussed at the end of Section 9–2. Hence for a rigid body rotating around an axis of symmetry, $\vec{L}$ and $\vec{\omega}$ are in the same direction (Fig. 10–23). So we have the *vector* relationship

$$\vec{L} = I\vec{\omega} \qquad \text{(for a rigid body rotating around a symmetry axis).} \qquad (10\text{–}31)$$

From Eq. (10–29) the rate of change of angular momentum of a particle equals the torque of the net force acting on the particle. For any system of particles (including both rigid and nonrigid bodies) the rate of change of the *total* angular momentum equals the sum of the torques of all forces acting on all the particles. The torques of the *internal* forces add to zero if these forces act along the line from one particle to another, as in Fig. 10–7, and so the sum of torques includes only the torques of the *external* forces. (A similar cancellation occurred in our discussion of center-of-mass motion in Section 8–6.) If the total angular momentum of the system of particles is $\vec{L}$ and the sum of external torques is $\Sigma\vec{\tau}$, then

$$\Sigma\vec{\tau} = \frac{d\vec{L}}{dt} \qquad \text{(for any system of particles).} \qquad (10\text{–}32)$$

10–23 For rotation about an axis of symmetry, $\vec{\omega}$ and $\vec{L}$ are parallel and along the axis. The directions are given by the same right-hand rule as in Fig. 10–4.

Finally, if the system of particles is a rigid body rotating about a symmetry axis, then $L = I\omega$ and I is constant. If this axis has a fixed direction in space, then the vectors $\vec{L}$ and $\vec{\omega}$ change only in magnitude, not in direction. In that case, $dL/dt = I\,d\omega/dt = I\alpha$, or

$$\Sigma\tau = I\alpha,$$

which is again our basic relation for the dynamics of rigid-body rotation. If the body is not rigid, I may change, and in that case, L changes even when ω is constant. For a non-rigid body, Eq. (10–32) is still valid, even though Eq. (10–6) is not.

When the axis of rotation is *not* a symmetry axis, the angular momentum is in general *not* parallel to the axis (Fig. 10–24). As the body turns, the angular momentum vector $\vec{L}$ traces out a cone around the rotation axis. Because $\vec{L}$ changes, there must be a net external torque acting on the body even though the angular velocity magnitude ω may be constant. If the body is an unbalanced wheel on a car, this torque is provided by friction in the bearings, which causes the bearings to wear out. "Balancing" a wheel means distributing the mass so that the rotation axis is an axis of symmetry; then $\vec{L}$ points along the rotation axis, and no net torque is required to keep the wheel turning.

In fixed-axis rotation we often use the term "angular momentum of the body" to refer just to the *component* of $\vec{L}$ along the rotation axis of the body (the z-axis in Fig. 10–24), with a sign to indicate the sense of rotation just as with angular velocity.

10–24 If the rotation axis of a rigid body is not a symmetry axis, the angular momentum vector $\vec{L}$ does not in general lie along the rotation axis. Even if ω is constant, the direction of $\vec{L}$ changes, and a net torque is required to maintain rotation.

EXAMPLE 10–12

A turbine fan in a jet engine (Fig. 10–25) has a moment of inertia of 2.5 kg · m² about its axis of rotation. As the turbine is starting up, its angular velocity as a function of time is

$$\omega = (400 \text{ rad/s}^3)t^2.$$

a) Find the fan's angular momentum as a function of time, and find its value at time $t = 3.0$ s. b) Find the net torque acting on the fan as a function of time, and find the torque at time $t = 3.0$ s.

SOLUTION a) Just like an electric fan, the turbine fan rotates around an axis of symmetry, so we may use Eq. (10–31) to relate the magnitudes L and ω:

$$L = I\omega = (2.5 \text{ kg} \cdot \text{m}^2)(400 \text{ rad/s}^3)t^2 = (1000 \text{ kg} \cdot \text{m}^2/\text{s}^3)t^2.$$

At time $t = 3.0$ s,

$$L = (1000 \text{ kg} \cdot \text{m}^2/\text{s}^3)(3.0 \text{ s})^2 = 9000 \text{ kg} \cdot \text{m}^2/\text{s}.$$

b) From Eq. (10–32),

$$\tau = \frac{dL}{dt} = (1000 \text{ kg} \cdot \text{m}^2/\text{s}^2)(2t) = (2000 \text{ kg} \cdot \text{m}^2/\text{s}^3)t.$$

10–25 A turbine fan is used in turbojet engines to force air into the engine at high speeds.

At time $t = 3.0$ s,

$$\tau = (2000 \text{ kg} \cdot \text{m}^2/\text{s}^3)(3.0 \text{ s}) = 6000 \text{ kg} \cdot \text{m}^2/\text{s}^2 = 6000 \text{ N} \cdot \text{m}.$$

10–7 CONSERVATION OF ANGULAR MOMENTUM

We have just seen that angular momentum can be used for an alternative statement of the basic dynamic principle for rotational motion. It also forms the basis for the **principle of conservation of angular momentum.** Like conservation of energy and of linear momentum, this principle is a universal conservation law, valid at all scales from atomic and nuclear systems to the motions of galaxies. This principle follows directly from Eq. (10–32): $\Sigma\vec{\tau} = d\vec{L}/dt$. If $\Sigma\vec{\tau} = 0$, then $d\vec{L}/dt = 0$, and $\vec{L}$ is constant. **When the net external torque acting on a system is zero, the total angular momentum of the system is constant (conserved).**

A circus acrobat, a diver, and an ice skater pirouetting on the toe of one skate all take advantage of this principle. Suppose an acrobat has just left a swing with arms and legs extended and rotating counterclockwise about her center of mass. When she pulls her arms and legs in, her moment of inertia I_{cm} with respect to her center of mass changes from a large value I_1 to a much smaller value I_2. The only external force acting on her is her weight, which has no torque with respect to an axis through her center of mass. So her angular momentum $L = I_{cm}\omega$ remains constant, and her angular velocity ω increases

ActivPhysics

7.16
Ball Hits Bat

7.17
Scattering

as I_{cm} decreases. That is,

$$I_1\omega_1 = I_2\omega_2 \qquad \text{(zero net external torque).} \qquad (10\text{--}33)$$

When a skater or ballerina spins with arms outstretched and then pulls her arms in, her angular velocity increases as her moment of inertia decreases. In each case there is conservation of angular momentum in a system in which the net external torque is zero.

When a system has several parts, the internal forces that the parts exert on each other cause changes in the angular momenta of the parts, but the *total* angular momentum doesn't change. Here's an example. Consider two bodies A and B that interact with each other but not with anything else, such as the astronauts we discussed in Section 8–3 (Fig. 8–6). Suppose body A exerts a force $\vec{F}_{A\text{ on }B}$ on body B; the corresponding torque (with respect to whatever point we choose) is $\vec{\tau}_{A\text{ on }B}$. According to Eq. (10–32), this torque is equal to the rate of change of angular momentum of B:

$$\vec{\tau}_{A\text{ on }B} = \frac{d\vec{L}_B}{dt}.$$

At the same time, body B exerts a force $\vec{F}_{B\text{ on }A}$ on body A, with a corresponding torque $\vec{\tau}_{B\text{ on }A}$, and

$$\vec{\tau}_{B\text{ on }A} = \frac{d\vec{L}_A}{dt}.$$

From Newton's third law, $\vec{F}_{B\text{ on }A} = -\vec{F}_{A\text{ on }B}$. Furthermore, if the forces act along the same line, as in Fig. 10–7, their lever arms with respect to the chosen axis are equal. Thus the *torques* of these two forces are equal and opposite, and $\vec{\tau}_{B\text{ on }A} = -\vec{\tau}_{A\text{ on }B}$. So if we add the two previous equations, we find

$$\frac{d\vec{L}_A}{dt} + \frac{d\vec{L}_B}{dt} = 0,$$

or, because $\vec{L}_A + \vec{L}_B$ is the *total* angular momentum $\vec{L}$ of the system,

$$\frac{d\vec{L}}{dt} = 0 \qquad \text{(zero net external torque).} \qquad (10\text{--}34)$$

That is, the total angular momentum of the system is constant. The torques of the internal forces can transfer angular momentum from one body to the other, but they can't change the *total* angular momentum of the system.

A falling cat twists different parts of its body in different directions so that it lands feet first. At all times during this process the angular momentum of the cat as a whole remains zero.

EXAMPLE 10-13

Anyone can be a ballerina An acrobatic physics professor stands at the center of a turntable, holding his arms extended horizontally with a 5.0-kg dumbbell in each hand (Fig. 10–26). He is set rotating about a vertical axis, making one revolution in 2.0 s. Find the prof's new angular velocity if he pulls the dumbbells in to his stomach, and discuss how this affects the kinetic energy. His moment of inertia (without the dumbbells) is 3.0 kg · m² when his arms are outstretched, dropping to 2.2 kg · m² when his hands are at his stomach. The dumbbells are 1.0 m from the axis initially and 0.20 m from it at the end. Treat the dumbbells as particles.

SOLUTION If we neglect friction in the turntable, no external torques act about the vertical axis, so the angular momentum

about this axis is constant, and we can use Eq. (10–33). The moment of inertia of the system is $I = I_{prof} + I_{dumbbells}$, so

$$I_1 = 3.0 \text{ kg} \cdot \text{m}^2 + 2(5.0 \text{ kg})(1.0 \text{ m})^2 = 13 \text{ kg} \cdot \text{m}^2,$$

$$I_2 = 2.2 \text{ kg} \cdot \text{m}^2 + 2(5.0 \text{ kg})(0.20 \text{ m})^2 = 2.6 \text{ kg} \cdot \text{m}^2,$$

$$\omega_1 = \frac{1 \text{ rev}}{2.0 \text{ s}} = 0.5 \text{ rev/s}.$$

From conservation of angular momentum the new angular velocity is

$$\omega_2 = \frac{I_1}{I_2}\omega_1 = \frac{13 \text{ kg} \cdot \text{m}^2}{2.6 \text{ kg} \cdot \text{m}^2}(0.5 \text{ rev/s}) = 2.5 \text{ rev/s}.$$

That is, the angular velocity increases by a factor of five while the angular momentum remains constant. Note that we didn't have to change "revolutions" to "radians" in this calculation. Why not?

To calculate the kinetic energy, we must express ω_1 and ω_2 in rad/s. (Why?) We have $\omega_1 = (0.5 \text{ rev/s})(2\pi \text{ rad/rev}) = 3.14 \text{ rad/s}$ and $\omega_2 = (2.5 \text{ rev/s})(2\pi \text{ rad/rev}) = 15.7 \text{ rad/s}$. The initial kinetic energy is

$$K_1 = \frac{1}{2} I_1 \omega_1{}^2 = \frac{1}{2}(13 \text{ kg} \cdot \text{m}^2)(3.14 \text{ rad/s})^2 = 64 \text{ J},$$

and the final kinetic energy is

$$K_2 = \frac{1}{2} I_2 \omega_2{}^2 = \frac{1}{2}(2.6 \text{ kg} \cdot \text{m}^2)(15.7 \text{ rad/s})^2 = 320 \text{ J}.$$

Where did the extra energy come from?

10–26 Fun with conservation of angular momentum—if you don't get dizzy.

EXAMPLE 10–14

Figure 10–27 shows two disks, one an engine flywheel, the other a clutch plate attached to a transmission shaft. Their moments of inertia are I_A and I_B; initially, they are rotating with constant angular velocities ω_A and ω_B, respectively. We then push the disks together with forces acting along the axis, so as not to apply any torque on either disk. The disks rub against each other and eventually reach a common final angular speed ω. Derive an expression for ω.

SOLUTION The only torque acting on either disk is the torque applied by the other disk; there are no external torques. Thus the total angular momentum of the system of two disks is the same before and after they are pushed together. At the end they rotate together as one body with total moment of inertia $I = I_A + I_B$ and angular speed ω. Conservation of angular momentum gives

$$I_A \omega_A + I_B \omega_B = (I_A + I_B)\omega,$$

$$\omega = \frac{I_A \omega_A + I_B \omega_B}{I_A + I_B}.$$

10–27 When the net external torque is zero, angular momentum is conserved. The forces shown are along the axis of rotation and thus exert no torque about any point on it.

EXAMPLE 10–15

In Example 10–14, suppose flywheel A has a mass of 2.0 kg, a radius of 0.20 m, and an initial angular speed of 50 rad/s (about 500 rpm) and that clutch plate B has a mass of 4.0 kg, a radius of 0.10 m, and an initial angular speed of 200 rad/s. Find the common final angular speed ω after the disks are pushed into contact. Is kinetic energy conserved during this process?

SOLUTION The moments of inertia of the two disks are

$$I_A = \frac{1}{2} m_A r_A{}^2 = \frac{1}{2}(2.0 \text{ kg})(0.20 \text{ m})^2 = 0.040 \text{ kg} \cdot \text{m}^2,$$

$$I_B = \frac{1}{2} m_B r_B{}^2 = \frac{1}{2}(4.0 \text{ kg})(0.10 \text{ m})^2 = 0.020 \text{ kg} \cdot \text{m}^2.$$

From Example 10–14 we have

$$\omega = \frac{I_A \omega_A + I_B \omega_B}{I_A + I_B}$$

$$= \frac{(0.040 \text{ kg} \cdot \text{m}^2)(50 \text{ rad/s}) + (0.020 \text{ kg} \cdot \text{m}^2)(200 \text{ rad/s})}{0.040 \text{ kg} \cdot \text{m}^2 + 0.020 \text{ kg} \cdot \text{m}^2}$$

$$= 100 \text{ rad/s}.$$

The initial kinetic energy is

$$K_1 = \frac{1}{2} I_A \omega_A{}^2 + \frac{1}{2} I_B \omega_B{}^2$$

$$= \frac{1}{2}(0.040 \text{ kg} \cdot \text{m}^2)(50 \text{ rad/s})^2 + \frac{1}{2}(0.020 \text{ kg} \cdot \text{m}^2)(200 \text{ rad/s})^2$$

$$= 450 \text{ J}.$$

The final kinetic energy is

$$K_2 = \frac{1}{2}(I_A + I_B)\omega^2$$

$$= \frac{1}{2}(0.040 \text{ kg} \cdot \text{m}^2 + 0.020 \text{ kg} \cdot \text{m}^2)(100 \text{ rad/s})^2 = 300 \text{ J}.$$

One third of the initial kinetic energy was lost during this "angular collision," the rotational analog of a completely inelastic collision. We shouldn't expect kinetic energy to be conserved, even though the resultant external force and torque are zero, because nonconservative (frictional) internal forces act while the two disks rub together and gradually approach a common angular velocity.

EXAMPLE 10–16

Angular momentum in a crime bust A door 1.0 m wide, of mass 15 kg, is hinged at one side so that it can rotate without friction about a vertical axis. It is unlatched. A police officer fires a bullet with a mass of 10 g and a speed of 400 m/s into the exact center of the door, in a direction perpendicular to the plane of the door (Fig. 10–28). Find the angular speed of the door just after the bullet imbeds itself in the door. Is kinetic energy conserved?

SOLUTION We consider the door and bullet together as a system. There is no external torque about the axis defined by the hinges, so angular momentum about this axis is conserved. The initial angular momentum of the bullet is given by Eq. (10–28):

$$L = mvl = (0.010 \text{ kg})(400 \text{ m/s})(0.50 \text{ m}) = 2.0 \text{ kg} \cdot \text{m}^2/\text{s}.$$

The final angular momentum is $I\omega$, where $I = I_{\text{door}} + I_{\text{bullet}}$. From Table 9–2, for a door of width d,

$$I_{\text{door}} = \frac{Md^2}{3} = \frac{(15 \text{ kg})(1.0 \text{ m})^2}{3} = 5.0 \text{ kg} \cdot \text{m}^2.$$

The moment of inertia of the bullet (with respect to the axis along the hinges) is

$$I_{\text{bullet}} = ml^2 = (0.010 \text{ kg})(0.50 \text{ m})^2 = 0.0025 \text{ kg} \cdot \text{m}^2.$$

Conservation of angular momentum requires that $mvl = I\omega$, or

$$\omega = \frac{mvl}{I} = \frac{2.0 \text{ kg} \cdot \text{m}^2/\text{s}}{5.0 \text{ kg} \cdot \text{m}^2 + 0.0025 \text{ kg} \cdot \text{m}^2} = 0.40 \text{ rad/s}.$$

10–28 Shooting open a door (top view). The bullet imbeds itself in the center of the door.

The collision of bullet and door is inelastic because nonconservative forces act during the impact. Thus we do not expect kinetic energy to be conserved. To check, we calculate initial and final kinetic energies:

$$K_1 = \frac{1}{2}mv^2 = \frac{1}{2}(0.010 \text{ kg})(400 \text{ m/s})^2 = 800 \text{ J},$$

$$K_2 = \frac{1}{2}I\omega^2 = \frac{1}{2}(5.0025 \text{ kg} \cdot \text{m}^2)(0.40 \text{ rad/s})^2$$

$$= 0.40 \text{ J}.$$

The final kinetic energy is only 1/2000 of the initial value!

10–8 GYROSCOPES AND PRECESSION

In all of the situations we've looked at so far in this chapter, the axis of rotation either has stayed fixed or has moved and kept the same direction (such as rolling without slipping). But a variety of new physical phenomena, some quite unexpected, can occur when the axis of rotation can change direction. For example, consider a toy gyroscope that's supported at one end (Fig. 10–29). If we hold it with the flywheel axis horizontal and let go, the free end of the axis simply drops owing to gravity—*if* the flywheel isn't spinning. But if the flywheel *is* spinning, what happens is quite different. One possible motion is a steady circular motion of the axis in a horizontal plane, combined with the spin motion of the flywheel about the axis. This surprising, non-intuitive motion of the axis is called **precession.** Precession is found in nature, as well as in rotating machines such as gyroscopes. As you read these words, the earth itself is precessing; its spin axis (through the north and south poles) slowly changes direction, going through a complete cycle of precession every 26,000 years.

   To study this strange phenomenon of precession, we must remember that angular velocity, angular momentum, and torque are all *vector* quantities. In particular, we need the general relation between the net torque $\Sigma\vec{\tau}$ that acts on a body and the rate of change of the body's angular momentum $\vec{L}$, given by Eq. (10–32), $\Sigma\vec{\tau} = d\vec{L}/dt$. Let's first apply this equation to the case in which the flywheel is *not* spinning (Fig. 10–30a). We take the origin O at the pivot and assume that the flywheel is symmetrical, with mass M and moment of inertia I about the flywheel axis. The flywheel axis is initially along the *x*-axis. The only external forces on the gyroscope are the normal force $\vec{n}$ acting at the pivot (assumed to be frictionless) and the weight $\vec{w}$ of the flywheel that acts at its center of mass, a distance r from the pivot. The normal force has zero torque with respect to the pivot, and the weight has a torque $\vec{\tau}$ in the *y*-direction as shown in Fig. 10–30a. Initially, there is no rotation, and the initial angular momentum $\vec{L}_i$ is zero. From Eq. (10–32) the *change* $d\vec{L}$ in angular momentum in a short time interval dt following this is

$$d\vec{L} = \vec{\tau}\,dt. \qquad (10\text{–}35)$$

This is in the *y*-direction because $\vec{\tau}$ is. As each additional time interval dt elapses, the angular momentum changes by additional increments $d\vec{L}$ in the *y*-direction because the direction of the torque is constant (Fig. 10–30b). The steadily increasing horizontal angular momentum means that the gyroscope rotates downward faster and faster around the *y*-axis until it hits either the stand or the table on which it sits.

Now let's see what happens if the flywheel *is* spinning initially, so the initial angular momentum $\vec{L}_i$ is not zero (Fig. 10–31a). Since the flywheel rotates around its symmetry axis, $\vec{L}_i$ lies along the axis. But each change in angular momentum $d\vec{L}$ is perpendicular to the axis, because the torque $\vec{\tau} = \vec{r} \times \vec{w}$ is perpendicular to the axis (Fig. 10–31b). This causes the *direction* of $\vec{L}$ to change, but not its magnitude. The changes $d\vec{L}$ are always in the horizontal *xy*-plane, so the angular momentum vector and the flywheel axis with which it moves are always horizontal. In other words, the axis doesn't fall—it just precesses.

If this still seems mystifying to you, think about a ball attached to a string. If the ball is initially at rest and you pull the string toward you, the ball will move toward you also.

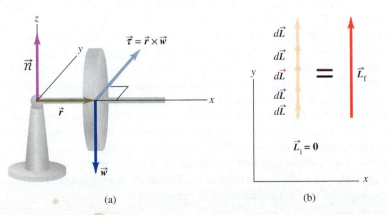

(a)   (b)

10-30 (a) The flywheel is initially not spinning. The torque $\vec{\tau}$ is due to the weight $\vec{w}$. (b) Looking down from directly over the gyroscope. In each successive time interval dt the torque produces a change $d\vec{L}$ in the angular momentum; this change is in the same direction as $\vec{\tau}$. The flywheel has zero initial angular momentum, so all the $d\vec{L}$'s are parallel, and the final angular momentum $\vec{L}_f$ also has the same direction as $\vec{\tau}$.

10–31 (a) The flywheel is spinning initially with angular momentum $\vec{L}_i$. The free-body diagram is the same as in Fig. 10–30, so the forces are not shown.
(b) With an initial angular momentum, each change $d\vec{L}$ is perpendicular to $\vec{L}$; this continuously changes the direction of $\vec{L}$, causing precession. The magnitude of the angular momentum is constant: $|\vec{L}_i| = |\vec{L}_f|$.

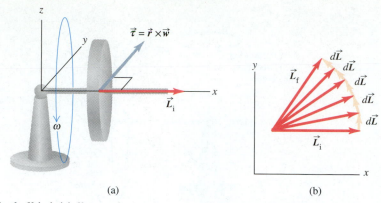

(a)

(b)

But if the ball is initially moving and you continuously pull the string in a direction perpendicular to the ball's motion, the ball moves in a circle around your hand; it does not approach your hand at all. In the first case the ball has zero linear momentum $\vec{p}$ to start with; when you apply a force $\vec{F}$ toward you for a time dt, the ball acquires a momentum $d\vec{p} = \vec{F}\,dt$, which is also toward you. But if the ball already has linear momentum $\vec{p}$, a change in momentum $d\vec{p}$ that's perpendicular to $\vec{p}$ will change the direction of motion, not the speed. Replace $\vec{p}$ with $\vec{L}$ and $\vec{F}$ with $\vec{\tau}$ in this argument, and you'll see that precession is simply the rotational analog of uniform circular motion.

At the instant shown in Fig. 10–31a, the gyroscope has angular momentum $\vec{L}$. A short time interval dt later, the angular momentum is $\vec{L} + d\vec{L}$; the infinitesimal change in angular momentum is $d\vec{L} = \vec{\tau}\,dt$, which is perpendicular to $\vec{L}$. As the vector diagram in Fig. 10–32 shows, this means that the flywheel axis of the gyroscope has turned through a small angle $d\phi$ given by $d\phi = |d\vec{L}|/|\vec{L}|$. The rate at which the axis moves, $d\phi/dt$, is called the **precession angular speed;** denoting this quantity by Ω, we find

$$\Omega = \frac{d\phi}{dt} = \frac{|d\vec{L}|/|\vec{L}|}{dt} = \frac{\tau}{L} = \frac{wr}{I\omega}. \qquad (10\text{–}36)$$

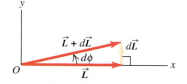

10–32 In a time dt the angular momentum vector and the flywheel axis precess together through an angle $d\phi$.

Thus the precession angular speed is *inversely* proportional to the angular speed of spin about the axis. A rapidly spinning gyroscope precesses slowly; if friction in its bearings causes the flywheel to slow down, the precession angular speed *increases!* The precession angular speed of the earth is very slow (1 rev/26,000 yr) because its spin angular momentum L is large and the torque τ, due to the gravitational influences of the moon and sun, is relatively small.

As a gyroscope precesses, its center of mass moves in a circle with radius r in a horizontal plane. Its vertical component of acceleration is zero, so the upward normal force $\vec{n}$ exerted by the pivot must be just equal in magnitude to the weight. The circular motion of the center of mass with angular speed Ω requires a force $\vec{F}$ directed toward the center of the circle, with magnitude $F = M\Omega^2 r$. This force must also be supplied by the pivot.

One key assumption that we made in our analysis of the gyroscope was that the angular momentum vector $\vec{L}$ is associated only with the spin of the flywheel and is purely horizontal. But there will also be a vertical component of angular momentum associated with the precessional motion of the gyroscope. By ignoring this, we've tacitly assumed that the precession is *slow,* that is, that the precession angular speed Ω is very much less than the spin angular speed ω. As Eq. (10–36) shows, a large value of ω automatically gives a small value of Ω, so this approximation is reasonable. When the precession is not slow, additional effects show up, including an up-and-down wobble or *nutation* of the flywheel axis that's superimposed on the precessional motion. You can see nutation occurring in a gyroscope as its spin slows down, so that Ω increases and the vertical component of $\vec{L}$ can no longer be ignored.

EXAMPLE 10–17

Figure 10–33a shows a top view of a cylindrical gyroscope wheel that has been set spinning by an electric motor. The pivot is at O, and the mass of the axle is negligible. a) As seen from above, is the precession clockwise or counterclockwise? b) If the gyro takes 4.0 s for one revolution of precession, at what angular speed does the wheel spin?

SOLUTION a) The right-hand rule shows that $\vec{\omega}$ and $\vec{L}$ are to the left (Fig. 10–33b). The weight $\vec{w}$ points into the page in this top view and acts at the center of mass (denoted by an $\times$); the torque $\vec{\tau} = \vec{r} \times \vec{w}$ is toward the top of the page; and $d\vec{L}/dt$ is also toward the top of the page. Adding a small $d\vec{L}$ to the $\vec{L}$ that we have initially changes the direction of $\vec{L}$ as shown, so the precession is clockwise as seen from above.

b) Be careful not to confuse ω and Ω! We are given $\Omega = (1\ \text{rev})/(4.0\ \text{s}) = (2\pi\ \text{rad})/(4.0\ \text{s}) = 1.57\ \text{rad/s}$. The weight is equal to mg, and the moment of inertia about its symmetry axis of a solid cylinder with radius R is $I = \frac{1}{2}mR^2$. Solving Eq. (10–36) for ω, we find

$$\omega = \frac{wr}{I\Omega} = \frac{mgr}{(mR^2/2)\Omega} = \frac{2gr}{R^2\Omega}$$

$$= \frac{2(9.8\ \text{m/s}^2)(2.0 \times 10^{-2}\ \text{m})}{(3.0 \times 10^{-2}\ \text{m})^2(1.57\ \text{rad/s})} = 280\ \text{rad/s} = 2600\ \text{rev/min}.$$

The precession angular speed Ω is very much less than the spin angular speed ω, so this is an example of slow precession.

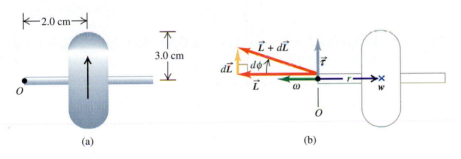

(a)　　　　　(b)

10–33 In which direction does this gyroscope precess?

SUMMARY

KEY TERMS

- When a force $\vec{F}$ acts on a body, the torque τ of that force with respect to a point O is given by

$$\tau = Fl, \tag{10–2}$$

where l is the lever arm. A common sign convention is "counterclockwise positive, clockwise negative." A generalized definition of torque as a vector quantity $\vec{\tau}$ is

$$\vec{\tau} = \vec{r} \times \vec{F}, \tag{10–3}$$

where $\vec{r}$ is the position vector of the point at which the force acts.

- The angular acceleration α of a rigid body rotating about a stationary axis is related to the net torque $\Sigma\tau$ and the moment of inertia I of the body by

$$\Sigma\tau = I\alpha. \tag{10–6}$$

- If a rigid body has both translational and rotational motion, its kinetic energy may be expressed as the sum of translational kinetic energy due to motion of the center of mass and rotational kinetic energy about an axis through the center of mass:

$$K = \frac{1}{2}Mv_{cm}^2 + \frac{1}{2}I_{cm}\omega^2. \tag{10–11}$$

In the special case of a rigid body with radius R that rolls without slipping, $v_{cm} = R\omega$. Combined translation and rotation may also be described by $\Sigma\vec{F}_{ext} = M\vec{a}_{cm}$ (for translation of the center of mass) and $\Sigma\tau = I_{cm}\alpha$ (for rotation about the center of mass), provided that the rotation axis is a symmetry axis that does not change direction as the body moves.

line of action, 294

lever arm (moment arm), 294

torque, 294

combined translation and rotation, 301

rolling without slipping, 302

angular momentum, 309

principle of conservation of angular momentum, 311

precession, 314

precession angular speed, 316

■ When a torque τ acts on a rigid body that undergoes an angular displacement from θ_1 to θ_2, the work W done by the torque is

$$W = \int_{\theta_1}^{\theta_2} \tau \, d\theta. \qquad (10\text{–}23)$$

If the torque is *constant,* then

$$W = \tau(\theta_2 - \theta_1) = \tau \, \Delta\theta. \qquad (10\text{–}24)$$

The work-energy theorem for rotational motion of a rigid body is

$$W_{\text{tot}} = \frac{1}{2} I\omega_2{}^2 - \frac{1}{2} I\omega_1{}^2. \qquad (10\text{–}25)$$

When the body rotates with angular velocity ω, the power P (rate at which the torque does work) is

$$P = \tau\omega. \qquad (10\text{–}26)$$

■ The angular momentum $\vec{L}$, with respect to point O, of a particle with mass m, velocity $\vec{v}$, and linear momentum $\vec{p} = m\vec{v}$, is

$$\vec{L} = \vec{r} \times \vec{p} = \vec{r} \times m\vec{v}, \qquad (10\text{–}27)$$

where $\vec{r}$ is the particle's position vector with respect to point O. When a symmetric rigid body with moment of inertia I rotates with angular velocity $\vec{\omega}$ about a stationary axis of symmetry, its angular momentum is given by

$$\vec{L} = I\vec{\omega}. \qquad (10\text{–}31)$$

If the body is not symmetric or the rotation axis is not an axis of symmetry, the component of angular momentum along the axis of rotation is equal to $I\omega$.

■ In terms of vector torque $\vec{\tau}$ and angular momentum $\vec{L}$, the basic dynamic relation for the rotational motion of any system can be restated as

$$\Sigma\vec{\tau} = \frac{d\vec{L}}{dt}. \qquad (10\text{–}32)$$

■ If the net torque associated with the external forces acting on a system is zero, the total angular momentum of the system is constant (conserved).

DISCUSSION QUESTIONS

Q10–1 In tightening cylinder-head bolts in an automobile engine, the critical quantity is the *torque* applied to the bolts. Why is this more important than the actual *force* applied to the wrench handle?

Q10–2 Can a single force applied to a body change both its translational and rotational motion? Explain.

Q10–3 Suppose you could use wheels of any type in the design of a soapbox-derby racer (an unpowered four-wheel vehicle that coasts from rest down a hill). Within the rules on the total weight of the vehicle and rider, should you design with large massive wheels or small light wheels? Should you use solid wheels or wheels with most of the mass at the rim? Explain.

Q10–4 Unless $\vec{r}$ and $\vec{F}$ are perpendicular, there are always two angles between their directions that give the same torque for given magnitudes of $\vec{r}$ and $\vec{F}$. Explain why. Illustrate your answer with a sketch.

Q10–5 The crankshaft of an automobile engine has a flywheel to increase the moment of inertia about the rotation axis. Why is this desirable?

Q10–6 The harder you hit the brakes while driving forward, the more the front end of your car will move down (and the rear end move up). Why? What happens when accelerating? Why don't drag racers use front-wheel drive only?

Q10–7 When an acrobat walks on a tightrope, she extends her arms straight out from her sides. She does this to make it easier for her to catch herself if she should tip to one side or the other. Explain how this works. (*Hint:* Think about Eq. (10–6).)

Q10–8 When an electric motor is turned on, it takes longer to come up to final speed if there is a grinding wheel attached to the shaft. Why?

Q10–9 Experienced cooks can tell whether an egg is raw or hard-boiled by rolling it down a slope (taking care to catch it at the bottom). How is this possible? What are they looking for?

Q10–10 An electric grinding wheel coasts for a minute or more after the power is turned off, but an electric drill coasts for only a few seconds. Why is there a difference?

Q10–11 The hammer in Fig. 10–11 is acted on by the force of gravity, and forces produce torques that cause a body's angular velocity to change. Why, then, is the angular velocity of the hammer in the figure constant?

Q10–12 In Example 10–6 (Section 10–4) suppose you pulled your hand and the end of the string upward. Would mechanical energy be conserved in this situation? Why or why not?

Q10–13 A wheel is rolling without slipping on a horizontal surface. In an inertial frame of reference in which the surface is at rest, is there any point on the wheel that has a velocity that is purely vertical? Is there any point that has a horizontal velocity component opposite to the velocity of the center of mass? Explain. Do your answers change if the wheel is slipping as it rolls? Why or why not?

Q10–14 Part of the kinetic energy of a moving automobile is in the rotational motion of its wheels. When the brakes are applied hard on an icy street, the wheels "lock," and the car starts to slide. What becomes of the rotational kinetic energy?

Q10–15 You are standing at the center of a large horizontal turntable in a carnival funhouse. The turntable is set rotating on frictionless bearings, and rotates freely (that is, there is no motor driving the turntable). As you walk toward the edge of the turntable, what happens to the combined angular momentum of you and the turntable? What happens to the rotation speed of the turntable? Explain your answer.

Q10–16 A point particle travels in a circle at constant speed. With respect to the origin at the center of the circle, is there a net torque acting on the particle? A net force? What if the speed of the particle is changing? In each case, explain your answers.

Q10–17 A point particle travels in a straight line at constant speed and the closest distance it comes to the origin of coordinates is a distance l. With respect to this origin, does the particle have nonzero angular momentum? As the particle moves along its straight-line path, does its angular momentum with respect to origin change?

Q10–18 In Example 10–13 (Section 10–7) the angular speed ω changes and this must mean that there is nonzero angular acceler-

ation. But there is no torque about the rotation axis if the forces the professor applies to the weights are directly radially inward. Then by Eq. (10–6) α must be zero. Explain what is wrong with this reasoning that leads to this apparent contradiction.

Q10–19 In Example 10–13 (Section 10–7) the rotational kinetic energy of the professor and dumbbells increases. But since there are no external torques, no work is being done to change the rotational kinetic energy. Then, by Eq. (10–25) the kinetic energy must remain the same! Explain what is wrong with this reasoning that leads to this apparent contradiction. Where *does* the extra kinetic energy come from?

Q10–20 As discussed in Section 10–7, the angular momentum of a circus acrobat is conserved as she tumbles through the air. Is her *linear* momentum conserved? Why or why not?

Q10–21 If you stop a spinning raw egg for the shortest instant you can and then release it, the egg will start spinning again. If you do the same to a hard-boiled egg, it will remain stopped. Try it. Explain it.

Q10–22 A helicopter has a large main rotor that rotates in a horizontal plane and provides lift. There is also a small rotor on the tail that rotates in a vertical plane. What is the purpose of the tail rotor? (*Hint:* if there were no tail rotor, what would happen when the pilot changed the angular speed of the main rotor?) Some helicopters have no tail rotor, but instead have two large main rotors that rotate in a horizontal plane. Why is it important that the two main rotors rotate in opposite directions?

Q10–23 In a common design for a gyroscope, the flywheel and flywheel axis are enclosed in a light, spherical frame with the flywheel at the center of the frame. The gyroscope is then balanced on top of a pivot so that the flywheel is directly above the pivot. Does the gyroscope precess if it is released while the flywheel is spinning? Explain.

Q10–24 A gyroscope takes 3.8 s to precess 1.0 revolution about a vertical axis. Two minutes later it takes only 1.9 s to precess 1.0 revolution. No one has touched the gyroscope. Explain.

Q10–25 A gyroscope is precessing as in Fig. 10–29. What happens if you gently add some weight to the end of the flywheel axis farthest from the pivot?

EXERCISES

SECTION 10–2 TORQUE

10–1 Calculate the torque (magnitude and direction) about point O due to the force $\vec{F}$ in each of the situations sketched in Fig. 10–34. In each case, the force $\vec{F}$ and the yellow rod both lie in the plane of the page, the rod has length 4.00 m, and the force has magnitude $F = 10.0$ N.

10–2 Calculate the net torque about point O for the two forces applied as in Fig. 10–35. The rod and both forces are in the plane of the page.

10–3 A square metal plate 0.180 m on each side is pivoted about an axis through point O at its center and perpendicular to the plate (Fig. 10–36). Calculate the net torque about this axis due to the three forces shown in the figure if the magnitudes of

the forces are $F_1 = 18.0$ N, $F_2 = 26.0$ N, and $F_3 = 14.0$ N. The plate and all forces are in the plane of the page.

10–4 Forces $F_1 = 7.50$ N and $F_2 = 5.30$ N are applied tangentially to a wheel with radius 0.330 m, as shown in Fig. 10–37. What is the net torque on the wheel due these two forces for an axis perpendicular to the wheel and passing through its center?

10–5 One force acting on a machine part is $\vec{F} = (-5.00 \text{ N})\hat{\imath} + (4.00 \text{ N})\hat{\jmath}$. The vector from the origin to the point where the force is applied is $\vec{r} = (-0.450 \text{ m})\hat{\imath} + (0.150 \text{ m})\hat{\jmath}$. a) In a sketch show $\vec{r}$, $\vec{F}$, and the origin. b) Use the right-hand rule to determine the direction of the torque. c) Calculate the vector torque produced by this force. Verify that the direction of the torque is the same as you obtained in (b).

(a) 90.0° F

(b) 120.0° F

(c) 30.0° F 60.0°

(d) O 2.00 m

(e) F 60.0° O

(f) F

FIGURE 10–34 Exercise 10–1.

$F_2 = 12.0$ N $F_1 = 8.00$ N

30.0°

O

|←2.00 m→|←3.00 m→|

FIGURE 10–35 Exercise 10–2.

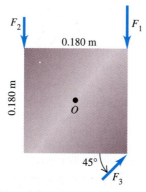

F_2 F_1

0.180 m

0.180 m

O

45° F_3

FIGURE 10–36 Exercise 10–3.

F_1

0.330 m

F_2

FIGURE 10–37 Exercise 10–4.

SECTION 10–3 TORQUE AND ANGULAR ACCELERATION FOR A RIGID BODY

10–6 The flywheel of an engine has moment of inertia 2.50 kg · m² about its rotation axis. a) What constant torque is required to bring it up to an angular speed of 400 rev/min in 8.00 s, starting from rest? b) What is its final kinetic energy?

10–7 Using the value of a calculated in Example 10–2 (Section 10–3), what is the speed of the cable after it has been pulled 2.0 m? Compare your result with Example 9–8 (Section 9–5).

10–8 A cord is wrapped around the rim of a wheel 0.250 m in radius, and a steady pull of 40.0 N is exerted on the cord. The wheel is mounted on frictionless bearings on a horizontal shaft through its center. The moment of inertia of the wheel about this shaft is 5.00 kg · m². Compute the angular acceleration of the wheel.

10–9 a) Find the magnitude n of the normal force for the situation described in Example 10–3 (Section 10–3). b) Is your answer to part (a) less than, equal to, or greater than the total weight $(M + m)g$ of the cylinder and mass? Explain why this should be so. c) Suppose the cylinder is initially rotating clockwise so that the hanging mass m is initially moving upward at speed v_0 (the cable remains taut). What effect does this have on the tension T and normal force n? Explain.

10–10 a) In the situation described in Example 10–2 of Section 10–3 (Fig. 10–8), the normal force $\vec{n}$ exerted on the cylinder by the bearing is directed up and to the left. Why must the normal force be in this direction? b) Calculate the magnitude and direction of $\vec{n}$.

10–11 A grindstone in the shape of a solid disk with a diameter of 0.520 m and a mass of 50.0 kg is rotating at 850 rev/min. You press an ax against the rim with a normal force of 160 N (Fig. 10–38), and the grindstone comes to rest in 7.50 s. Find the coefficient of friction between the ax and the grindstone. Neglect friction in the bearings.

ω

$F = 160$ N

$m = 50.0$ kg

FIGURE 10–38 Exercise 10–11 and Problem 10–44.

10–12 A 15.0-kg bucket of water is suspended by a rope wrapped around a windlass, which is a solid cylinder 0.300 m in diameter with mass 12.0 kg. The cylinder is pivoted on a frictionless axle through its center. The bucket is released from rest at the top of a well and falls 10.0 m to the water. Neglect the weight of the rope. a) What is the tension in the rope while the bucket is falling? b) With what speed does the bucket strike the water? c) What is the time of fall? d) While the bucket is falling, what is the force exerted on the cylinder by the axle?

10–13 A 2.00-kg textbook rests on a frictionless horizontal surface. A cord attached to the book passes over a pulley whose diameter is 0.150 m to a hanging book of mass 3.00 kg. The system is released from rest, and the books are observed to move 1.20 m in 0.800 s. a) What is the tension in each part of the cord? b) What is the moment of inertia of the pulley about its rotation axis?

10–14 A thin, horizontal rod with length l and mass M is pivoted about a vertical axis at one end. A force with constant magnitude F is applied to the other end, causing the rod to rotate in a horizontal plane. The force is maintained perpendicular to the rod and to the axis of rotation. Calculate the magnitude of the angular acceleration of the rod.

SECTION 10–4 RIGID-BODY ROTATION ABOUT A MOVING AXIS

10–15 A string is wrapped several times around the rim of a small hoop of radius 0.0800 m and mass 0.180 kg. If the free end of the string is held in place and the hoop is released from rest (Fig. 10–39), calculate a) the tension in the string while the hoop descends as the string unwinds; b) the time it takes the hoop to descend 0.750 m; c) the angular speed of the rotating hoop after it has descended 0.750 m.

0.0800 m

FIGURE 10–39 Exercise 10–15 and Problem 10–63.

10–16 Repeat part (c) of Exercise 10–15, this time using energy considerations.

10–17 In Example 10–5 (Section 10–4) we found that for a hollow cylindrical shell rolling without slipping on a horizontal surface, half of the total kinetic energy is translational and half is rotational. What fraction of the total kinetic energy is rotational for the following objects rolling without slipping on a horizontal surface? a) A uniform solid cylinder. b) A uniform sphere. c) A thin-walled hollow sphere. d) A hollow cylinder of outer radius R and inner radius $R/2$.

10–18 A hollow spherical shell of mass 2.00 kg rolls without slipping down a 38.0° slope. a) Find the acceleration, the friction force, and the minimum coefficient of friction needed to prevent slipping. b) How would your answers to part (a) change if the mass were doubled to 4.00 kg?

10–19 A 392-N wheel comes off a moving truck and rolls without slipping along a highway. At the bottom of a hill it is rotating at 25.0 rad/s. The radius of the wheel is 0.600 m and its moment of inertia about its rotation axis is $0.800MR^2$. Friction does work on the wheel as it rolls up the hill to a stop, a height h above the bottom of the hill; this work has absolute value 3500 J. Calculate h.

10–20 A Ball Rolling Uphill. A bowling ball rolls without slipping up a ramp that slopes upward at an angle β to the horizontal. (See Example 10–9 in Section 10–4.) Treat the ball as a uniform solid sphere, ignoring the finger holes. a) Draw the free-body diagram for the ball. Explain why the friction force must be directed *uphill*. b) What is the acceleration of the center of mass of the ball? c) What minimum coefficient of static friction is needed to prevent slipping?

SECTION 10–5 WORK AND POWER IN ROTATIONAL MOTION

10–21 A playground merry-go-round has radius 2.40 m and moment of inertia 2100 kg · m^2 about a vertical axle through its center, and turns with negligible friction. a) A child applies an 18.0-N force tangentially to the edge of the merry-go-round for 15.0 s. If the merry-go-round is initially at rest, what is its angular speed after this 15.0-s interval? b) How much work did the child do on the merry-go-round? c) What is the average power supplied by the child?

10–22 Example 9–5 (Section 9–4) described the design of an aircraft propeller. The engine delivers 175 hp to the propeller at 2400 rev/min. a) How much torque does the aircraft engine provide? b) How much work does the engine do in one revolution of the propeller?

10–23 A 1.50-kg grinding wheel is in the form of a solid cylinder of radius 0.100 m. a) What constant torque will bring it from rest to an angular speed of 1200 rev/min in 2.5 s? b) Through what angle has it turned during that time? c) Use Eq. (10–24) to calculate the work done by the torque. d) What is the grinding wheel's kinetic energy when it is rotating at 1200 rev/min? Compare your answer to the result in part (c).

10–24 What is the power output in horsepower of an electric motor turning at 4800 rev/min and developing a torque of 4.30 N · m?

10–25 The carbide tips of the cutting teeth of a circular saw are 8.6 cm from the axis of rotation. a) The no-load speed of the saw, when it is not cutting anything, is 4800 rev/min. Why is its no-load power output negligible? b) While cutting lumber, the angular speed of the saw slows to 2400 rev/min and the power output is 1.9 hp. What is the tangential force that the wood exerts on the carbide tips?

10–26 An airplane propeller is 2.08 m in length (from tip to tip) and has a mass of 117 kg. When the airplane's engine is first started, it applies a constant torque of 1950 N · m to the propeller, which starts from rest. a) What is the angular acceleration of the propeller? Treat the propeller as a slender rod. (*Hint:* See Table 9–2.) b) What is the propeller's angular speed after making 5.00 rev? c) How much work is done by the engine during the first 5.00 rev? d) What is the average power output of the engine during the first 5.00 rev? e) What is the instantaneous power output of the motor at the instant that the propeller has turned through 5.00 rev?

10–27 a) Compute the torque developed by an industrial motor whose output is 150 kW at an angular speed of 4000 rev/min. b) A drum of negligible mass, 0.400 m in diameter, is attached to the motor shaft and the power output of the motor is used to raise a weight hanging from a rope wrapped around the drum. How heavy a weight can be lifted at constant speed? c) At what constant speed will the weight rise?

SECTION 10–6 ANGULAR MOMENTUM

10–28 A woman of mass 50 kg is standing on the rim of a large disk that is rotating at 0.50 rev/s about an axis through its center. The disk has a mass of 110 kg and a radius of 4.0 m. Calculate the magnitude of the total angular momentum of the woman-plus-disk system. (Assume that the woman can be treated as a point.)

10–29 A 2.00-kg rock has a horizontal velocity of magnitude 12.0 m/s when it is at point P in Fig. 10–40. a) At this instant, what is the magnitude and direction of its angular momentum relative to point O? b) If the only force acting on the rock is its weight, what is the rate of change (magnitude and direction) of its angular momentum at this instant?

$v = 12.0$ m/s
P
8.00 m
36.9°
O

FIGURE 10–40 Exercise 10–29.

10–30 a) Calculate the magnitude of the angular momentum of the earth considered as a particle orbiting the sun. The mass of the earth is 5.97×10^{24} kg. Treat it as moving in a circular orbit of radius 1.50×10^{11} m at a speed of 2.98×10^4 m/s. b) Calculate the magnitude of the angular momentum of the earth due its rotation around an axis through the north and south poles. Treat the earth as a uniform sphere of radius 6.38×10^6 m that makes one revolution in 24.0 hours.

10–31 Find the magnitude of the angular momentum of the second hand on a clock about an axis through the center of the clock face. The clock hand has a length of 15.0 cm and a mass of 6.00 g. Take the second hand to be a slender rod rotating with constant angular velocity about one end.

SECTION 10–7 CONSERVATION OF ANGULAR MOMENTUM

10–32 Under some circumstances a star can collapse into an extremely dense object made mostly of neutrons and called a *neutron star*. The density of a neutron star is roughly 10^{14} times as great as that of ordinary solid matter. Suppose we represent the star as a uniform solid rigid sphere, both before and after the collapse. The star's initial radius was 7.0×10^5 km (comparable to our sun); its final radius is 16 km. If the original star rotated once in 30 days, find the angular speed of the neutron star.

10–33 A small block on a frictionless horizontal surface has a mass of 0.0250 kg. It is attached to a massless cord passing through a hole in the surface (Fig. 10–41). The block is originally revolving at a distance of 0.300 m from the hole with an angular speed of 1.75 rad/s. The cord is then pulled from below, shortening the radius of the circle in which the block revolves to 0.150 m. The block may be treated as a particle. a) Is angular momentum conserved? Why or why not? b) What is the new angular speed? c) Find the change in kinetic energy of the block. d) How much work was done in pulling the cord?

FIGURE 10–41 Exercise 10–33, Problem 10–76, and Challenge Problem 10–87.

10–34 **The Spinning Figure Skater.** The outstretched hands and arms of a figure skater preparing for a spin can be considered a slender rod pivoting about an axis through its center (Fig. 10–42). When his hands and arms are brought in and wrapped around his body to execute the spin, the hands and arms can be considered a thin-walled hollow cylinder. His hands and arms have a combined mass of 8.0 kg. When outstretched, they span 1.8 m; when wrapped, they form a cylinder of radius 25 cm. The moment of inertia about the rotation axis of the remainder of his body is constant and equal to 0.40 kg · m². If his original angular speed is 0.40 rev/s, what is his final angular speed?

FIGURE 10–42 Exercise 10–34.

10–35 A diver comes off a board with arms straight up and legs straight down, giving her a moment of inertia about her rotation axis of 18 kg · m². She then tucks into a small ball, decreasing this moment of inertia to 3.6 kg · m². While tucked, she makes two complete revolutions in 1.0 s. If she hadn't tucked at all, how many revolutions would she have made in the 1.5 s from board to water?

10–36 A large turntable rotates about a fixed vertical axis, making one revolution in 6.00 s. The moment of inertia of the turntable about this axis is 1200 kg · m². A child of mass 40.0 kg, initially standing at the center of the turntable, runs out along a radius. What is the angular speed of the turntable when the child is 2.00 m from the center? (Assume that the child can be treated as a particle.)

10–37 A large wooden turntable in the shape of a flat disk has a radius of 2.00 m and a total mass of 120 kg. The turntable is initially rotating at 3.00 rad/s about a vertical axis through its center. Suddenly, a 70-kg parachutist makes a soft landing on the turntable at a point near the outer edge. a) Find the angular speed of the turntable after the parachutist lands. (Assume that the parachutist can be treated as a particle.) b) Compute the kinetic energy of the system before and after the parachutist lands. Why are these kinetic energies not equal?

10–38 A solid wood door 1.00 m wide and 2.00 m high is hinged along one side and has a total mass of 40.0 kg. Initially open and at rest, the door is struck at its center by a handful of sticky mud of mass 0.500 kg, traveling perpendicular to the door at 12.0 m/s just before impact. Find the final angular speed of the door. Does the mud make a significant contribution to the moment of inertia?

SECTION 10–8 GYROSCOPES AND PRECESSION

10–39 Draw a top view of the gyroscope in Fig. 10–29. a) Draw labeled arrows on your sketch for $\vec{\omega}$, $\vec{L}$, and $\vec{\tau}$. Draw $d\vec{L}$ produced by $\vec{\tau}$. Draw $\vec{L} + d\vec{L}$. Determine the sense of precession by examining the directions of $\vec{L}$ and $\vec{L} + d\vec{L}$. b) Reverse the direction of the spin angular velocity of the rotor and repeat all steps in part (a). c) Move the pivot to the other end of the shaft, with the same direction of spin angular velocity as in part (b), and repeat all the steps. d) Keeping the pivot as in part (c), reverse the spin angular velocity of the rotor and repeat all steps.

10–40 The rotor (flywheel) of a toy gyroscope has a mass of 0.140 kg. Its moment of inertia about its axis is 1.20×10^{-4} kg · m². The mass of the frame is 0.0250 kg. The gyroscope is supported on a single pivot (Fig. 10–43) with its center of mass a horizontal distance of 4.00 cm from the pivot. The gyroscope is precessing in a horizontal plane at the rate of one revolution in 2.20 s. a) Find the upward force exerted by the pivot. b) Find the angular speed with which the rotor is spinning about its axis, expressed in rev/min. c) Copy the diagram, and draw vectors to show the angular momentum of the rotor and the torque acting on it.

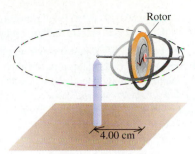

Rotor

4.00 cm

FIGURE 10–43 Exercise 10–40.

10–41 **A Gyro Stabilizer.** The stabilizing gyroscope of a ship is a solid disk of mass 60,000 kg; its radius is 2.00 m, and it rotates about a vertical axis with an angular speed of 500 rev/min. a) How much time is required to bring it up to speed, starting from rest, with a constant power input of 7.46×10^4 W? b) Find the torque needed to cause the axis to precess in a vertical fore-and-aft plane at the rate of 1.00°/s.

10–42 A gyroscope is precessing about a vertical axis. Describe what happens to the precession angular speed if the following changes in the variables are made, with all other variables remaining the same: a) the angular speed of the spinning flywheel is doubled; b) the total weight is doubled; c) the moment of inertia about the axis of the spinning flywheel is doubled; d) the distance from the pivot to the center of gravity is doubled. e) What happens if all four of the variables in parts (a) through (d) are doubled?

10–43 The earth precesses once every 26,000 years, and spins on its axis once a day. Estimate the magnitude of the torque that causes the precession of the earth. You may need some data from Appendix F. Make the estimate by assuming i) the earth is a uniform sphere and ii) the precession of the earth is like that of the gyroscope shown in Fig. 10–30. In this model, the precession axis and rotation axis are perpendicular. Actually, the angle between these two axes for the earth is only $23\frac{1}{2}°$; this affects the calculated torque by about a factor of 2.

PROBLEMS

10–44 A 55.0-kg grindstone is a solid disk 0.520 m in diameter. You press an ax down on the rim with a normal force of 160 N (Fig. 10–38). The coefficient of kinetic friction between the blade and the stone is 0.60, and there is a constant friction torque of 6.50 N · m between the axle of the stone and its bearings. a) How much force must be applied tangentially at the end of a crank handle 0.500 m long to bring the stone from rest to 120 rev/min in 9.00 s? b) After the grindstone attains an angular speed of 120 rev/min, what tangential force at the end of the handle is needed to maintain a constant angular speed of 120 rev/min? c) How much time does it take the grindstone to come from 120 rev/min to rest if it is acted on by the axle friction alone?

10–45 An experimental bicycle wheel is placed on a test stand so that it is free to turn on its axle. If a constant net torque of 5.00 N · m is applied to the tire for 2.00 s, the angular speed of the tire increases from zero to 100 rev/min. The external torque is then removed, and the wheel is brought to rest by friction in its bearings in 125 s. Compute a) the moment of inertia of the wheel

about the rotation axis; b) the friction torque; c) the total number of revolutions made by the wheel in the 125-s time interval.

10–46 A flywheel 0.600 m in diameter is pivoted on a horizontal axis. A rope is wrapped around the outside of the flywheel, and a steady pull of 40.0 N is exerted on the rope. The flywheel starts from rest, and 5.00 m of rope are unwound in 2.00 s. a) What is the angular acceleration of the flywheel? b) What is its final angular speed? c) What is its final kinetic energy? d) What is its moment of inertia about the rotation axis?

10–47 A wheel starts from rest and rotates with constant angular acceleration about a fixed axis. a) Prove that the power at any given time is proportional to the square of the net torque about the axis. b) If the power at $t = 3.00$ s is 500 W with a constant net torque of 20.0 N · m, what would the power have been at $t = 3.00$ s with a constant net torque of 60.0 N · m? c) Prove that the power at any given angular displacement is proportional to the $\frac{3}{2}$ power of the total torque about the axis at that angular displacement. d) If the power after having rotated 37.5 rad with a torque of 20.0 N · m is 500 W, what would the power have been after having rotated

37.5 rad with a torque of 60.0 N · m? e) Do the answers to parts (a) and (b) contradict the answers to parts (c) and (d)? Why or why not?

10–48 A beam of length l lies on the $+x$-axis with its left end at the origin. A cable pulls on the beam in the $+y$-direction with a force $\vec{F}$ with a magnitude that depends on where along the beam it is exerted, $F = F_0(1 - x/l)$, where F_0 is a constant equal to the force magnitude when it is applied at the left end of the beam. a) What is the direction of the torque due to $\vec{F}$? b) Graph F versus x from $x = 0$ to $x = l$. Express F in terms of F_0 and x in terms of l. c) Graph the torque versus x from $x = 0$ to $x = l$. Express the torque in terms of F_0l, and express x in terms of l. d) Where along the beam should the force be applied to produce maximum torque and what is that maximum torque?

10–49 While exploring a castle, Exena the Exterminator is spotted by a dragon who chases her down a hallway. Exena runs into a room and attempts to swing the heavy door shut before the dragon gets her. The door is initially perpendicular to the wall, so it must be turned through 90° to close. The door is 3.00 m tall and 1.25 m wide, and it weighs 750 N. The friction at the hinges can be neglected. If Exena applies a force of 220 N at the edge of the door and perpendicular to it, how much time does it take her to close the door?

10–50 A thin rod of length l lies on the $+x$-axis with its left end at the origin. A string pulls on the rod with a force $\vec{F}$ directed toward a point P a distance h above the rod. Where along the rod should you attach the string to get the greatest torque about the origin if point P is a) above the right end of the rod? b) above the left end of the rod? c) above the center of the rod?

10–51 Balancing Act. Attached to one end of a long, thin, uniform rod of length L and mass M is a small blob of clay of the same mass M. a) Locate the position of the center of mass of the system of rod and clay. Note this position on a drawing of the rod. b) You carefully balance the rod on a frictionless table top so that it is standing vertically, with the end without the clay touching the table. If the rod is now tipped so that it is a small angle θ away from the vertical, determine its angular acceleration at this instant. Assume that the end without the clay remains in contact with the table top. (*Hint:* See Table 9–2.) c) You again balance the rod on the frictionless table top so that it is standing vertically, but now the end of the rod *with* the clay is touching the table. If the rod is again tipped so that it is a small angle θ away from the vertical, determine its angular acceleration at this instant. Assume that the end with the clay remains in contact with the table top. How does this compare to the angular acceleration in part (b)? d) A pool cue is a tapered wooden rod that is thick at one end and thin at the other. You can easily balance a pool cue vertically on one finger if the thin end is in contact with your finger; this is quite a bit harder to do if the thick end is in contact with your finger. Explain why there is a difference.

10–52 You connect a light string to a point on the edge of a uniform vertical disk of radius R and mass M. The disk is free to rotate without friction about a stationary horizontal axis through its center. Initially, the disk is at rest with the string connection at the highest point on the disk. You pull the string with a constant horizontal force $\vec{F}$ until the wheel has made exactly one-quarter revolution about a horizontal axis through its center, then let go. a) Use Eq. (10–23) to find the work done by the

string. b) Use Eq. (6–14) to find the work done by the string. Do you obtain the same result as in part (a)? c) Find the final angular speed of the disk. d) Find the maximum tangential acceleration of a point on the disk. e) Find the maximum radial (centripetal) acceleration of a point on the disk.

10–53 The mechanism shown in Fig. 10–44 is used to raise a crate of supplies from a ship's hold. The crate has a total mass of 50 kg. A rope is wrapped around a wooden cylinder that turns on a metal axle. The cylinder has radius 0.25 m and moment of inertia $I = 2.9$ kg · m² about the axle. The crate is suspended from the free end of the rope. One end of the axle is pivoted on frictionless bearings; a crank handle is attached to the other end. When the crank is turned, the end of the handle rotates about the axle in a vertical circle of radius 0.12 m, the cylinder turns, and the crate is raised. What magnitude of the force $\vec{F}$ applied tangentially to the rotating crank is required to raise the crate with an acceleration of 0.80 m/s²? (The mass of the rope can be neglected, as can the moments of inertia of the axle and the crank.)

FIGURE 10–44 Problem 10–53.

10–54 A large 16.0-kg roll of paper with radius $R = 18.0$ cm rests against the wall and is held in place by a bracket attached to a rod through the center of the roll (Fig. 10–45). The rod turns without friction in the bracket, and the moment of inertia of the paper and rod about the axis is 0.260 kg · m². The other end of the bracket is attached by a frictionless hinge to the wall such that the bracket makes an angle of 30.0° with the wall. The weight of the bracket is negligible. The coefficient of kinetic friction between the paper and the wall is $\mu_k = 0.25$. A constant vertical force $F = 40.0$ N is applied to the paper, and the paper unrolls. a) What is the magnitude of the force that the rod exerts on the paper as it unrolls? b) What is the angular acceleration of the roll?

10–55 A block with mass $m = 5.00$ kg slides down a surface inclined 36.9° to the horizontal (Fig. 10–46). The coefficient of kinetic friction is 0.25. A string attached to the block is wrapped around a flywheel on a fixed axis at O. The flywheel has mass 25.0 kg and moment of inertia 0.500 kg · m² with respect to the axis of rotation. The string pulls without slipping at a perpendicular distance of 0.200 m from that axis. a) What is the acceleration of the block down the plane? b) What is the tension in the string?

10–56 Two metal disks, one with radius $R_1 = 2.50$ cm and mass $M_1 = 0.80$ kg and the other with radius $R_2 = 5.00$ cm and mass

FIGURE 10–47 Problem 10–58.

FIGURE 10–45 Problem 10–54.

FIGURE 10–46 Problem 10–55.

$M_2 = 1.60$ kg, are welded together and mounted on a frictionless axis through their common center, as in Problem 9–75. a) A light string is wrapped around the edge of the smaller disk and a 1.50-kg block is suspended from the free end of the string. What is the magnitude of the downward acceleration of the block after it is released? b) Repeat the calculation of part (a), this time with the string wrapped around the edge of the larger disk. In which case is the acceleration of the block the greatest? Does your answer make sense?

10–57 A lawn roller in the form of a thin-walled hollow cylinder of mass M is pulled horizontally with a constant horizontal force F applied by a handle attached to the axle. If it rolls without slipping, find the acceleration and the friction force.

10–58 Atwood's Machine. Figure 10–47 shows an Atwood's machine. Find the linear accelerations of blocks A and B, the angular acceleration of the wheel C, and the tension in each side of the cord if there is no slipping between the cord and the surface of the wheel. The masses of the blocks A and B are m_A and m_B, respectively, the moment of inertia of the wheel about its axis is I, and the radius of the semicircle in which the string moves is R.

10–59 A solid disk is rolling without slipping on a level surface at a constant speed of 2.50 m/s. a) If the disk rolls up a 30.0° ramp, how far along the ramp will it move before it stops? b) Explain why your answer to (a) does not depend on either the mass or the radius of the disk.

10–60 The Yo-yo. A yo-yo is made from two uniform disks, each with mass m and radius R, connected by a light axle of radius b. A light, thin string is wound several times around the axle and

then held stationary while the yo-yo is released from rest, dropping as the string unwinds. Find the linear acceleration and angular acceleration of the yo-yo and the tension in the string.

10–61 A uniform marble of radius r starts from rest with its center of mass a height h above the lowest point of a loop-the-loop track of radius R, a track shaped like the one in Fig. 7–26. It rolls without slipping. Rolling friction and air resistance are negligible. a) What is the minimum value of h such that the marble will not leave the track at the top of the loop? (*Note:* The radius r is *not* negligible compared to the radius R.) b) What is the answer to part (a) if the track is well lubricated, making friction negligible?

10–62 Figure 10–48 shows three identical yo-yos initially at rest on a horizontal surface. For each yo-yo the string is pulled in the direction shown. In each case there is sufficient friction for the yo-yo to roll without slipping. Draw the free-body diagram for each yo-yo. In what direction will each yo-yo rotate? (Try it!) Explain your answers.

FIGURE 10–48 Problem 10–62.

10–63 As shown in Fig. 10–39, a string is wrapped several times around the rim of a small hoop of radius 0.0800 m and mass 0.180 kg. The free end of the string is pulled upward in just the right way so that the hoop does not move vertically as the string unwinds. a) Find the tension in the string as the string unwinds. b) Find the angular acceleration of the hoop as the string unwinds. c) Find the upward acceleration of the hand that pulls on the free end of the string. d) How would your answers be different if the hoop were replaced by a solid disk of the same mass and radius?

10–64 In a lab experiment you let a uniform ball roll down a curved track. The ball starts from rest and rolls without slipping. While on the track, the ball descends a vertical distance h. The lower end of the track is horizontal and extends over the edge of the lab table; the ball leaves the track traveling horizontally. While free falling after leaving the track, the ball moves a horizontal distance x and a vertical distance y. a) Calculate x in terms of h and y, ignoring the work done by friction. b) Would the answer to part (a) be any different on the moon? c) In doing the experiment very carefully, your measured value of x is consistently a bit smaller than that calculated in part (a). Why? d) What would x be for the same h and y as in part (a) if you let a silver dollar roll down the track? Neglect the work done by friction.

10–65 In a spring gun, a spring of force constant 400 N/m is compressed 0.15 m. When fired, 80.0% of the elastic potential energy stored in the spring is eventually converted into kinetic energy of a 0.0590-kg uniform ball that is rolling without slipping at the base of a ramp. The ball continues to roll without slipping up the ramp with 90.0% of the kinetic energy at the bottom converted into an increase in gravitational potential energy at the instant it stops. a) What is the speed of the ball's center of mass at the base of the ramp? b) At this position, what is the speed of a point at the top of the ball? c) At this position, what is the speed of a point at the bottom of the ball? d) What maximum vertical height up the ramp does the ball move?

10–66 If a wheel rolls along a horizontal surface at constant speed, the coordinates of a certain point on the rim of the wheel are $x(t) = R[(2\pi t/T) - \sin(2\pi t/T)]$ and $y(t) = R[1 - \cos(2\pi t/T)]$, where R and T are constants. a) Sketch the trajectory of the point from $t = 0$ to $t = 2T$. A curve with this shape is called a *cycloid*. b) What are the meanings of the constants R and T? c) Find the x- and y-components of the velocity and of the acceleration of the point at any time t. d) Find the times at which the point is instantaneously at rest. What are the x- and y-components of the acceleration at these times? e) Find the magnitude of the acceleration of the point. Does it depend on time? Compare to the magnitude of the acceleration of a particle in uniform circular motion, $a_{\text{rad}} = 4\pi^2 R/T^2$. Explain your result for the magnitude of the acceleration of the point on the rolling wheel using the idea that rolling is a combination of rotational and translational motion.

10–67 A child rolls a 0.600-kg basketball up a long ramp. The basketball can be considered a thin-walled hollow sphere. When the child releases the basketball at the bottom of the ramp, it has a speed of 8.0 m/s. When the ball returns to her after rolling up the ramp and then rolling back down, it has a speed of 4.0 m/s. Assume the work done by friction on the basketball is the same when the ball moves up or down the ramp and that the basketball rolls without slipping. Find the maximum vertical height increase of the ball as it rolls up the ramp.

10–68 A wheel rotates from rest about a fixed axis through its center of mass such that $\theta = bt^3$, where b is a positive constant with units rad/s^3. a) Use Eq. (10–23) to show that the work done by the net torque on the wheel when it has turned through an angle θ is $\frac{9}{2}I_{\text{cm}}b^{2/3}\theta^{4/3}$. b) Use Eq. (9–3) to calculate the angular speed of the wheel when it has turned through an angle θ. c) Use the result of part (b) to calculate the kinetic energy of the wheel when it has turned through an angle θ. Is the work-energy theorem, Eq. (10–25), obeyed? Explain.

10–69 A uniform solid cylinder of mass M and radius $2R$ rests on a horizontal table top. A string is attached by a yoke to a frictionless axle through the center of the cylinder so that the cylinder can rotate about the axle. The string runs over a disk-shaped pulley of mass M and radius R that is mounted on a

frictionless axle through its center. A block of mass M is suspended from the free end of the string (Fig. 10–49). The string doesn't slip over the pulley surface, and the cylinder rolls without slipping on the table top. Find the magnitude of the acceleration of the block after the system is released from rest.

FIGURE 10–49 Problem 10–69.

10–70 A uniform 0.0300-kg rod of length 0.400 m rotates in a horizontal plane about a fixed axis through its center and perpendicular to the rod. Two small rings, each of mass 0.0200 kg, are mounted so that they can slide along the rod. They are initially held by catches at positions 0.0500 m on each side of the center of the rod, and the system is rotating at 30.0 rev/min. Without otherwise changing the system, the catches are released, and the rings slide outward along the rod and fly off at the ends. a) What is the angular speed of the system at the instant when the rings reach the ends of the rod? b) What is the angular speed of the rod after the rings leave it?

10–71 A uniform rod of length L rests on a frictionless horizontal surface. The rod is pivoted about a fixed frictionless axis at one end. The rod is initially at rest. A bullet traveling parallel to the horizontal surface and perpendicular to the rod with speed v strikes the rod at its center and becomes embedded in it. The mass of the bullet is one-fourth the mass of the rod. a) What is the final angular speed of the rod? b) What is the ratio of the kinetic energy of the system after the collision to the kinetic energy of the bullet before the collision?

10–72 The solid wood door of a gymnasium is 1.00 m wide and 2.00 m high, has a total mass of 35.0 kg, and is hinged along one side. The door is open and at rest when a stray basketball hits the center of the door head-on, applying an average force of 1500 N to the door for 8.00 ms. Find the angular speed of the door after the impact. (*Hint:* Integrating Eq. (10–32) yields $\Delta L = \int_{t_1}^{t_2} (\Sigma\,\tau)\,dt = (\Sigma\,\tau)_{\text{av}}\Delta t$. The quantity $\int_{t_1}^{t_2} (\Sigma\,\tau)\,dt$ is called the *angular impulse*.)

10–73 A target in a shooting gallery consists of a vertical square wooden board, 0.250 m on a side and of mass 0.750 kg, that is pivoted on a horizontal axis along its top edge. The board is struck face-on at its center by a bullet of mass 1.90 g that is traveling at 360 m/s and that remains embedded in the board. a) What is the angular speed of the board just after the bullet's impact? b) What maximum height above the equilibrium position does the center of the board reach before starting to swing down again? c) What minimum bullet speed would be required for the board to swing all the way over after impact?

10–74 Neutron Star Glitches. Occasionally a rotating neutron star (Exercise 10–32) undergoes a sudden and unexpected speedup called a *glitch*. One explanation is that a glitch occurs when the crust of the neutron star settles slightly, decreasing the moment of inertia about the rotation axis. A neutron star with angular speed $\omega_0 = 70.4$ rad/s underwent such a glitch in October 1975 that increased its angular speed to $\omega = \omega_0 + \Delta\omega$, where $\Delta\omega/\omega_0 = 2.01 \times 10^{-6}$. If the radius of the neutron star before the glitch was 11 km, by how much did its radius decrease in the starquake? Assume that the neutron star is a uniform sphere.

10–75 Disks A and B are mounted on a shaft SS and may be connected or disconnected by a clutch C (Fig. 10–50). Disk A is made of a lighter material than disk B so that the moment of inertia of disk A about the shaft is one-third that of disk B. The moments of inertia of the shaft and clutch are negligible. With the clutch disconnected, A is brought up to an angular speed ω_0. The accelerating torque is then removed from A, and A is coupled to disk B by the clutch. (Bearing friction may be neglected.) It is found that 2400 J of thermal energy is developed in the clutch when the connection is made. What was the original kinetic energy of disk A?

FIGURE 10–50 Problem 10–75.

10–76 A small block of mass 0.250 kg is attached to a string passing through a hole in a frictionless horizontal surface (see Fig. 10–41). The block is originally revolving in a circle with a radius of 0.800 m about the hole with a tangential speed of 4.00 m/s. The string is then pulled slowly from below, shortening the radius of the circle in which the block revolves. The breaking strength of the string is 30.0 N. What is the radius of the circle when the string breaks?

10–77 A horizontal plywood disk of mass 7.00 kg and diameter 1.00 m is pivoted on frictionless bearings about a vertical axis through its center. You attach a circular model railroad track of negligible mass and average diameter 0.95 m to the disk. A 1.20-kg, battery-driven model train rests on the tracks. To demonstrate conservation of angular momentum, you switch on the train's engine. The train moves counterclockwise, soon attaining a constant speed of 0.600 m/s relative to the tracks. Find the magnitude and direction of the angular velocity of the disk relative to the earth.

10–78 One particle of mass m moves at a constant speed v in a circle of radius R that is a distance R above the xz-plane. Another particle of mass m moves the same way at the same speed in another circle of radius R that is a distance R below the xz-plane. The two particles are one-half a revolution apart, so that when one is at (x, R, z), the other is at $(-x, -R, -z)$. Thus their center of mass is at the origin but their rotation axis (the y-axis) is not an axis of symmetry. a) Sketch these two particles at the instant they are at $(R, R, 0)$ and $(-R, -R, 0)$, and show their position, velocity, and angular momentum vectors with respect to the origin. b) Show that at any time the two particles have equal angular momentum. c) What angle do $\vec{\omega}$ (the angular velocity vector of the system of two particles) and the total angular momentum of the system make

with each other? d) Show that the y-component of the system's total angular momentum is constant and equal to $L_y = 2mvR$. e) What is the y-component of the total torque acting on the system? f) Find the magnitude of the net force acting on each particle and the magnitude of the total torque acting on the system. g) Show, using your sketch from part (a), the direction of the net torque on the system and that this torque rotates parallel to the xz-plane.

10–79 In a physics laboratory you do the following ballistic pendulum experiment. You shoot a ball of mass m horizontally from a spring gun with a speed v. The ball is immediately caught a distance r below a frictionless pivot by a pivoted catcher assembly of mass M. The moment of inertia of this assembly about its rotation axis through the pivot is I. The distance r is much greater than the radius of the ball. a) Use conservation of angular momentum to show that the angular speed of the ball and catcher just after the ball is caught is $\omega = mvr/(mr^2 + I)$. b) After the ball is caught, the center of mass of the ball-catcher assembly system swings up with a maximum height increase h. Use conservation of energy to show that $\omega = \sqrt{2(M + m)gh/(mr^2 + I)}$. c) Your lab partner says that linear momentum is conserved in the collision and derives the expression $mv = (m + M)V$, where V is the speed of the ball immediately after the collision. She then uses conservation of energy to derive that $V = \sqrt{2gh}$, so that $mv = (m + M)\sqrt{2gh}$. Use the results of parts (a) and (b) to show that this equation is satisfied only for the special case when r is given by $I = Mr^2$.

10–80 A 55-kg runner runs around the edge of a horizontal turntable mounted on a vertical frictionless axis through its center. The runner's velocity relative to the earth has magnitude 2.8 m/s. The turntable is rotating in the opposite direction with an angular velocity of magnitude 0.20 rad/s relative to the earth. The radius of the turntable is 3.0 m, and its moment of inertia about the axis of rotation is 80 kg $\cdot$ m^2. Find the final angular velocity of the system if the runner comes to rest relative to the turntable. (The runner may be treated as a particle.)

10–81 A Falling Bicycle. The moment of inertia of the front wheel of a bicycle about its axle is 0.085 kg $\cdot$ m^2, its radius is 0.33 m, and the forward speed of the bicycle is 6.00 m/s. With what angular velocity must the front wheel be turned about a vertical axis to counteract the capsizing torque due to a mass of 50.0 kg located 0.040 m horizontally to the right or left of the line of contact of wheels and ground? (Bicycle riders: Compare your own experience and decide whether your answer seems reasonable.)

10–82 Center of Percussion. A baseball bat rests on a horizontal frictionless surface. The bat has a length of 0.900 m, a mass of 0.800 kg, and its center of mass is 0.600 m from the handle end of the bat (Fig. 10–51). The moment of inertia of the bat about its center of mass is 0.0530 kg $\cdot$ m^2. The bat is struck by a baseball traveling perpendicular to the bat. The impact applies an impulse $J = \int_{t_1}^{t_2} F\, dt$ at a point a distance x from the handle end of the bat. What must x be so that the handle end of the bat remains at rest as the bat begins to move? (*Hint:* Consider the motion of the center of mass and the rotation about the center of mass. Find x so that these two motions combine to give $v = 0$ for the end of the bat just after the collision. Also, note that integration of Eq. (10–32) gives $\Delta L = \int_{t_1}^{t_2} (\Sigma \tau)\, dt$ (Problem 10–72).) The point on the bat you have located is called the *center of percussion*. Hitting a pitched ball at the center of percussion of the bat minimizes the "sting" the batter experiences on the hands.

FIGURE 10–51 Problem 10–82.

10–83 Consider a gyroscope with an axis that is not horizontal, but is inclined from the horizontal by an angle β. Show that the precession angular frequency does not depend on the value of β, but is given by Eq. (10–36).

CHALLENGE PROBLEMS

10–84 A uniform ball of radius R rolls without slipping between two rails such that the horizontal distance is d between the two contact points of the rails to the ball. a) In a sketch, show that at any instant $v_{cm} = \omega\sqrt{R^2 - d^2/4}$. Discuss this expression in the limits $d = 0$ and $d = 2R$. b) For a uniform ball starting from rest and descending a vertical distance h while rolling without slipping down a ramp, $v_{cm} = \sqrt{10gh/7}$. Replacing the ramp with the two rails, show that

$$v_{cm} = \sqrt{\frac{10gh}{5 + 2/(1 - d^2/4R^2)}}.$$

In each case, the work done by friction has been neglected. c) Which speed in part (b) is smaller? Why? Answer in terms of how the loss of potential energy is shared between the gain in translational and rotational kinetic energies. d) For what value of the ratio d/R do the two expressions for the speed in part (b) differ by 5.0%? By 0.50%?

10–85 When an object is rolling without slipping, the rolling friction force is much less than the friction force when the object is sliding; a silver dollar will roll on its edge much farther than it will slide on its flat side. (See Section 5–4.) When an object is rolling without slipping on a horizontal surface, we can approximate the friction force to be zero, so that a and α are approximately zero and v and ω are approximately constant. Rolling without slipping means $v = r\omega$ and $a = r\alpha$. If an object is set into motion on a surface *without* these equalities, sliding (kinetic) friction will act on the object as it slips until rolling without slipping is established. A solid cylinder with mass M and radius R, rotating with angular velocity ω_0 about an axis through its center, is set on a horizontal surface for which the kinetic friction coefficient is μ_k. a) Draw a free-body diagram for the cylinder on the surface. Think carefully about the direction of the kinetic friction force on the cylinder. Calculate the accelerations a of the center of mass and α of rotation about the center of mass. b) The cylinder is initially slipping completely, as initially $\omega = \omega_0$ but $v = 0$. Rolling without slipping sets in when $v = R\omega$. Calculate the *distance* the cylinder rolls before slipping stops. c) Calculate the work done by the friction force on the cylinder as it moves from where it was set down to where it begins to roll without slipping.

10–86 A demonstration gyroscope wheel is constructed by removing the tire from a bicycle wheel 0.650 m in diameter, wrapping lead wire around the rim, and taping it in place. The shaft projects 0.200 m at each side of the wheel, and a woman holds the ends of the shaft in her hands. The mass of the system is 8.00 kg; its entire mass may be assumed to be located at its rim. The shaft is horizontal and the wheel is spinning about the shaft at 5.00 rev/s. Find the magnitude and direction of the force each hand exerts on the shaft a) when the shaft is at rest; b) when the shaft is rotating in a horizontal plane about its center at 0.050 rev/s; c) when the shaft is rotating in a horizontal plane about its center at 0.300 rev/s. d) At what rate must the shaft rotate in order that it may be supported at one end only?

10–87 A block with mass m is revolving with linear speed v_1 in a circle of radius r_1 on a frictionless horizontal surface (see Fig. 10–41). The string is slowly pulled from below until the radius of the circle in which the block is revolving is reduced to r_2. a) Calculate the tension T in the string as a function of r, the distance of the block from the hole. Your answer will be in terms of the initial velocity v_1 and the radius r_1. b) Use $W = \int_{r_1}^{r_2} \vec{T}(r) \cdot d\vec{r}$ to calculate the work done by $\vec{T}$ when r changes from r_1 to r_2. c) Compare the results of part (b) to the change in the kinetic energy of the block.

11

11

Equilibrium and Elasticity

To help submarines maintain structural integrity while diving deep in the ocean, their hulls are built with a cylindrical cross section. This shape stabilizes the hull against compressive stresses due to the inward water pressure. The rounded shape also minimizes the possibility of stress fractures in the hull, which would tend to form at sharp corners.

11–1 INTRODUCTION

We've devoted a good deal of effort to understanding why and how bodies accelerate in response to the forces that act on them. But very often we're interested in making sure that bodies *don't* accelerate. Any building, from the World Trade Center to the humblest shed, must be designed so that it won't topple over. Similar concerns occur with a suspension bridge, a ladder leaning against a wall, or a crane hoisting a bucket full of concrete.

A body that can be modeled as a *particle* is in equilibrium whenever the vector sum of the forces acting on it is zero. But for the situations we've just described, that condition isn't enough. If forces act at different points on an extended body, an additional requirement must be satisfied to ensure that the body has no tendency to *rotate:* the sum of the *torques* about any point must be zero. This requirement is based on the principles of rotational dynamics developed in Chapter 10. We can compute the torque due to the weight of a body using the concept of center of mass from Section 8–6 and the related concept of center of gravity, which we introduce in this chapter.

Rigid bodies don't bend, stretch, or squash when forces act on them. But the rigid body is an idealization; all real materials are *elastic* and do deform to some extent. Elastic properties of materials are tremendously important. You want the wings of an airplane to be able to bend a little, but you'd rather not have them break off. The steel frame of an earthquake-resistant building has to be able to flex, but not

too much. Many of the necessities of everyday life, from rubber bands to suspension bridges, depend on the elastic properties of materials. In this chapter we'll introduce the concepts of *stress, strain,* and *elastic modulus* and a simple principle called *Hooke's law* that helps us predict what deformations will occur when forces are applied to a real (not perfectly rigid) body.

11–2 CONDITIONS FOR EQUILIBRIUM

We learned in Sections 4–3 and 5–2 that a particle is in equilibrium—that is, the particle does not accelerate—in an inertial frame of reference if the vector sum of all the forces acting on the particle is zero, $\Sigma \vec{F} = 0$. The equivalent statement for an *extended* body is that the center of mass of the body has zero acceleration if the vector sum of all external forces acting on the body is zero, as discussed in Section 8–6. This is often called the **first condition for equilibrium.** In terms of components,

$$\Sigma F_x = 0, \quad \Sigma F_y = 0, \quad \Sigma F_z = 0 \quad \begin{array}{l} \text{(first condition} \\ \text{for equilibrium),} \end{array} \quad (11\text{–}1)$$

where the sum includes *external* forces only.

A second condition for an extended body to be in equilibrium is that the body must have no tendency to *rotate.* This condition is based on the dynamics of rotational motion in exactly the same way that the first condition is based on Newton's first law. A rigid body that, in an inertial frame, is not rotating about a certain point has zero angular momentum about that point, $\vec{L} = 0$. If it is not to start rotating about that point, the rate of change of angular momentum $d\vec{L}/dt$ must *also* be zero. From the discussion in Section 10–6, particularly Eq. (10–32), this means that the sum of torques $\Sigma \vec{\tau}$ due to all the external forces acting on the body must be zero. A rigid body in equilibrium can't have any tendency to start rotating about *any* point, so the sum of external torques must be zero about any point. This is the **second condition for equilibrium:**

$$\Sigma \vec{\tau} = 0 \quad \text{about any point} \qquad \text{(second condition for equilibrium).} \qquad (11\text{–}2)$$

The sum of the torques due to all external forces acting on the body, with respect to any specified point, must be zero.

   CAUTION ▶ Although the choice of reference point is arbitrary, once you choose a point you must use the *same* point to calculate *all* of the torques on a body. An important element of problem-solving strategy is to pick the point so as to simplify the calculations as much as possible. ◀

   In this chapter we will apply the first and second conditions for equilibrium to situations in which a rigid body is at rest (no translation or rotation). Such a body is said to be in **static equilibrium.** But the same conditions apply to a rigid body in uniform *translational* motion (without rotation), such as an airplane in flight with constant speed, direction, and altitude. Such a body is in equilibrium but is not static.

11–3 CENTER OF GRAVITY

In most equilibrium problems, one of the forces acting on the body is its weight. We need to be able to calculate the *torque* of this force. The weight doesn't act at a single point; it is distributed over the entire body. But we can always calculate the torque due to the body's weight by assuming that the entire force of gravity (weight) is concentrated at a point called the **center of gravity** (abbreviated "cg"). The acceleration due to gravity $\vec{g}$ decreases with altitude; but if we can ignore this variation over the vertical dimension of the body, then the body's center of gravity is identical with its *center of mass,* which we defined in Section 8–6. We stated this result without proof in Section 10–3, and now we'll prove it.

   First let's review the definition of the center of mass. For a collection of particles with masses $m_1, m_2, \ldots$ and coordinates $(x_1, y_1, z_1), (x_2, y_2, z_2), \ldots$, the coordinates x_{cm}, y_{cm}, and z_{cm} of the center of mass are given by

$$x_{cm} = \frac{m_1 x_1 + m_2 x_2 + m_3 x_3 + \cdots}{m_1 + m_2 + m_3 + \ldots} = \frac{\sum_i m_i x_i}{\sum_i m_i},$$

$$y_{cm} = \frac{m_1 y_1 + m_2 y_2 + m_3 y_3 + \cdots}{m_1 + m_2 + m_3 + \cdots} = \frac{\sum_i m_i y_i}{\sum_i m_i}, \qquad \text{(center of mass)} \qquad (11\text{–}3)$$

$$z_{cm} = \frac{m_1 z_1 + m_2 z_2 + m_3 z_3 + \cdots}{m_1 + m_2 + m_3 + \cdots} = \frac{\sum_i m_i z_i}{\sum_i m_i}.$$

Also, x_{cm}, y_{cm}, and z_{cm} are the components of the position vector $\vec{r}_{cm}$ of the center of mass, so Eqs. (11–3) are equivalent to the vector equation

$$\vec{r}_{cm} = \frac{m_1 \vec{r}_1 + m_2 \vec{r}_2 + \cdots}{m_1 + m_2 + \cdots} = \frac{\sum_i m_i \vec{r}_i}{\sum_i m_i}. \qquad (11\text{–}4)$$

   Now let's consider the gravitational torque on a body of arbitrary shape (Fig. 11–1). We assume that the acceleration due to gravity $\vec{g}$ has the same magnitude and direction

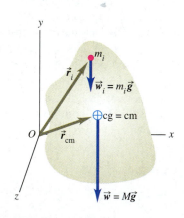

11–1 The gravitational torque about any point can be found by assuming that all the weight of the body acts as its center of gravity (cg), which is identical to its center of mass (cm) if $\vec{g}$ is the same at all points on the body.

at every point in the body. Every particle in the body experiences a gravitational force, and the total weight of the body is the vector sum of a large number of parallel forces. A typical particle has mass m_i and weight $\vec{w}_i = m_i\vec{g}$. If $\vec{r}_i$ is the position vector of this particle with respect to an arbitrary origin O, the torque vector $\vec{\tau}_i$ of the weight $\vec{w}_i$ with respect to O is, from Eq. (10–3),

$$\vec{\tau}_i = \vec{r}_i \times \vec{w}_i = \vec{r}_i \times m_i\vec{g}.$$

The *total* torque due to the gravitational forces on all the particles is

$$\vec{\tau} = \Sigma\vec{\tau}_i = \vec{r}_1 \times m_1\vec{g} + \vec{r}_2 \times m_2\vec{g} + \cdots$$
$$= (m_1\vec{r}_1 + m_2\vec{r}_2 + \cdots) \times \vec{g}$$
$$= \left(\sum_i m_i\vec{r}_i\right) \times \vec{g}.$$

When we multiply and divide this by the total mass of the body,

$$M = m_1 + m_2 + \cdots = \sum_i m_i,$$

we get

$$\vec{\tau} = \frac{m_1\vec{r}_1 + m_2\vec{r}_2 + \cdots}{m_1 + m_2 + \cdots} \times M\vec{g} = \frac{\sum_i m_i\vec{r}_i}{\sum_i m_i} \times M\vec{g}.$$

The fraction in this equation is just the position vector $\vec{r}_{\text{cm}}$ of the center of mass, with components x_{cm}, y_{cm}, and z_{cm}, as given by Eq. (11–4), and $M\vec{g}$ is equal to the total weight $\vec{w}$ of the body. Thus

$$\vec{\tau} = \vec{r}_{\text{cm}} \times M\vec{g} = \vec{r}_{\text{cm}} \times \vec{w}. \qquad (11–5)$$

The total gravitational torque, given by Eq. (11–5), is the same as though the total weight $\vec{w}$ were acting on the position $\vec{r}_{\text{cm}}$ of the center of mass, which we also call the *center of gravity*. **If $\vec{g}$ has the same value at all points on a body, its center of gravity is identical to its center of mass.** Note, however, that the center of mass is defined independently of any gravitational effect. In this chapter we'll assume that the center of mass and the center of gravity are identical unless explicitly stated otherwise.

FINDING AND USING THE CENTER OF GRAVITY

We can often use symmetry considerations to locate the center of gravity of a body, just as we did for center of mass. The center of gravity of a homogeneous sphere, cube, circular sheet, or rectangular plate is at its geometric center. The center of gravity of a right circular cylinder or cone is on its axis of symmetry.

For a body of more complex shape, we can sometimes locate the center of gravity by thinking of the body as being made of symmetrical pieces. For example, we could approximate the human body as a collection of solid cylinders, with a sphere for the head. Then we can compute the coordinates of the center of gravity of the combination from Eqs. (11–3), letting m_1, m_2, ... be the masses of the individual pieces and (x_1, y_1, z_1), (x_2, y_2, z_2), ... be the coordinates of their centers of gravity.

When a body acted on by gravity is supported or suspended at a single point, the center of gravity is always at or directly above or below the point of suspension. If it were anywhere else, the weight would have a torque with respect to the point of suspension, and the body could not be in rotational equilibrium. This fact can be used to determine experimentally the location of the center of gravity of an irregular body. To see how this works, hold one corner of a piece of paper gently between your thumb and forefinger so that the paper hangs below your hand. Draw a line on the paper that extends vertically downward from the point of suspension; the center of gravity must lie somewhere along this line. Then hold the paper by a different corner and draw a line vertically downward

The center of gravity and center of mass of an orbiting space station are not at exactly the same place. This is because $\vec{g}$, the acceleration due to gravity, can vary measurably over the dimensions of the station. On the earth's surface this variation is so small that the center of mass and center of gravity do coincide for all but the largest structures.

11–2 In (a) the center of gravity is within the area bounded by the supports, and the car is in equilibrium. The car in (b) and the truck in (c) will tip over because their centers of gravity lie outside the area of support.

Area of support   Area of support   Area of support
(a)   (b)   (c)

from the new point of suspension. The point where the lines intersect is the paper's center of gravity. (Try this with a coffee cup suspended first from the upper part of its handle, then from the lower part of its handle.)

Using the same reasoning, we can see that a body supported at several points must have its center of gravity somewhere within the area bounded by the supports. This explains why a car can drive on a straight but slanted road if the slant angle is relatively small (Fig. 11–2a) but will tip over if the angle is too great (Fig. 11–2b). The truck in Fig. 11–2c has a higher center of gravity than the car and will tip over on a shallower incline; when a truck overturns on a highway and blocks traffic for hours, it's the high center of gravity that's to blame.

The lower the center of gravity and the larger the area of support, the more difficult it is to overturn a body. Four-legged animals such as deer and horses have a large area of support bounded by their legs; hence they are naturally stable and need only small feet or hooves. Animals that walk erect on two legs, such as humans and birds, need relatively large feet to give them a reasonable area of support. If a two-legged animal holds its body approximately horizontal, like a chicken or the dinosaur *Tyrannosaurus rex,* it must perform a delicate balancing act as it walks to keep its center of gravity over the foot that is on the ground. A chicken does this by moving its head; *T. rex* probably did it by moving its massive tail.

7.2
A Tilted Beam: Torques and Equilibrium

7.3
Arm Levers

EXAMPLE 11–1

Walking the plank A massive uniform plank of length $L = 6.0$ m and mass $M = 90$ kg rests on top of two sawhorses separated by $D = 1.5$ m, located equal distances from the center of the plank (Fig. 11–3). Your cousin Throckmorton tries to stand on the right-hand end of the plank. If the plank is to remain at rest, how massive can Throckmorton be?

SOLUTION If the system of plank and Throckmorton is just in balance, the center of gravity of this system will be directly over the right-hand sawhorse. We take the origin at C, the geometrical center and center of gravity of the uniform plank. Then the x-coordinates of the centers of gravity of the plank and Throckmorton are $x_P = 0$ and $x_T = L/2 = 3.0$ m, respectively. If we let m be Throcky's mass, then from Eqs. (11–3) the center of gravity of the system of plank and Throcky is at

$$x_{cg} = \frac{M(0) + m(L/2)}{M + m} = \frac{m}{M + m}\frac{L}{2}.$$

$L = 6.0$ m

$D = 1.5$ m

$\dfrac{L}{2} - \dfrac{D}{2}$

Plank, mass
$M = 90$ kg

$\oplus$ cg

Throckmorton,
mass m

C S

11–3 Throckmorton and the plank, just barely in equilibrium.

Setting this equal to $D/2$, the x-coordinate of the right-hand sawhorse, we have

$$\frac{m}{M+m}\frac{L}{2} = \frac{D}{2},$$

$$mL = (M+m)D,$$

$$m = M\frac{D}{L-D} = (90 \text{ kg})\frac{1.5 \text{ m}}{6.0 \text{ m} - 1.5 \text{ m}}$$

$$= 30 \text{ kg}.$$

We can get the same result even more easily if we take the origin at S, the position of the right-hand sawhorse, so that $x_{cg} = 0$. The centers of gravity of the plank and Throcky are now at $x_P = -D/2$ and $x_T = (L/2) - (D/2)$ respectively, so

$$x_{cg} = \frac{M(-D/2) + m((L/2) - (D/2))}{M+m} = 0,$$

$$m = \frac{MD/2}{(L/2) - (D/2)} = M\frac{D}{L-D} = 30 \text{ kg}.$$

A 60-kg child could only stand halfway between the right-hand sawhorse and the end of the plank. Can you see why?

11–4 SOLVING RIGID-BODY EQUILIBRIUM PROBLEMS

There are just two key principles of rigid-body equilibrium: the vector sum of the forces on the body must be zero, and the sum of torques about any point must be zero. To keep things simple, we'll restrict our attention to situations in which we can treat all forces as acting in a single plane, which we'll call the xy-plane. Then we can ignore the condition $\Sigma F_z = 0$ in Eq. (11–1), and in Eq. (11–2) we need only consider the z-components of torque (perpendicular to the plane). The first and second conditions for equilibrium are then

$$\Sigma F_x = 0 \quad \text{and} \quad \Sigma F_y = 0 \qquad \text{(first condition for equilibrium, forces in } xy\text{-plane),}$$

$$\Sigma \tau_z = 0 \qquad \text{(second condition for equilibrium, forces in } xy\text{-plane).}$$

(11–6)

ActivPhysics

7.4
Two Painters on a Beam

7.5
The Back

7.6
Lecturing from a Beam

The challenge is to apply these simple principles to specific problems. The following strategy is very similar to the suggestions in Section 5–3 for equilibrium of a particle. You should compare it with the strategy for rotational dynamics problems given in Section 10–3.

Problem–Solving Strategy

EQUILIBRIUM OF A RIGID BODY

1. Draw a sketch of the physical situation, including dimensions, and select the body in equilibrium to be analyzed.

2. Draw a free-body diagram showing the forces acting *on* the selected body and no others. *Do not* include forces exerted *by* this body on other bodies. Be careful to show correctly the point at which each force acts; this is crucial for correct torque calculations. You can't represent a rigid body as a point.

3. Choose coordinate axes and specify a positive sense of rotation for torques. Represent forces in terms of their components with respect to the axes you have chosen; when you do this, cross out the original force so that you don't include it twice.

4. In choosing a point to compute torques, note that if a force has a line of action that goes *through* a particular point, the torque of the force with respect to that point is zero. You can often eliminate unknown forces or components from the torque equation by a clever choice of point for your calculation. The body doesn't actually have to be pivoted about an axis through the chosen point.

5. Write equations expressing the equilibrium conditions. Remember that $\Sigma F_x = 0$, $\Sigma F_y = 0$, and $\Sigma \tau_z = 0$ are always separate equations; *never* add x- and y-components in a single equation. Also remember that when a force is represented in terms of its components, you can compute the torque of that force by finding the torque of each component separately, each with its appropriate lever arm and sign, and adding the results. This is often easier than determining the lever arm of the original force.

6. You always need as many equations as you have unknowns. Depending on the number of unknowns, you may need to compute torques with respect to two or more axes to obtain enough equations. Often, there are several equally good sets of force and torque equations for a particular problem; there is usually no single "right" combination of equations. When you have as many independent equations as unknowns, you can solve the equations simultaneously. We will illustrate this point in some of the following examples by using various sets of equations.

EXAMPLE 11-2

Weight distribution for a car An auto magazine reports, "The Nissan 240SX (Fig. 11–4a) has 53% of its weight on its front wheels and 47% on its rear wheels, with a 2.46-m wheelbase." This means that the total normal force on the front wheels is $0.53w$ and that on the rear wheels is $0.47w$, where w is the total weight. The wheelbase is the distance between front and rear axles. How far in front of the rear axle is the 240SX's center of gravity?

SOLUTION Figure 11–4b shows a free-body diagram for the car, along with the x- and y-axes and our convention that counterclockwise torques are positive. We can see that the first condition for equilibrium in Eqs. (11–6) is satisfied; $\Sigma F_x = 0$ because there aren't any x-components of force, and $\Sigma F_y = 0$ because $0.47w + 0.53w + (-w) = 0$. So far, so good. We have drawn the weight w as acting at the center of gravity, and the distance we want is L_{cg}. This is the lever arm of the weight with respect to the rear axle R, so it is reasonable to take torques with respect to R. Note that the torque due to the weight is negative because it tends to cause a clockwise rotation about R. The torque due to the upward normal force at the front wheels is positive because it tends to cause a counterclockwise rotation about R. The torque equation is

$$\Sigma \tau_R = 0.47w(0) - wL_{cg} + 0.53w(2.46 \text{ m}) = 0,$$

$$L_{cg} = 1.30 \text{ m}.$$

Can you show that if f is the fraction of the weight on the front

11–4 (a) The Nissan 240SX at rest on a horizontal surface. (b) Free-body diagram for the car.

wheels and d is the wheelbase, then the center of gravity is a distance fd in front of the rear wheels? In terms of the center of gravity, why do owners of rear-wheel-drive vehicles put bags of sand in their trunks to improve traction on snow and ice? Would this help with a front-wheel-drive car? With a four-wheel-drive car?

EXAMPLE 11-3

A heroic rescue Sir Lancelot is trying to rescue the Lady Elayne from Castle Von Doom by climbing a uniform ladder that is 5.0 m long and weighs 180 N. Lancelot, who weighs 800 N, stops a third of the way up the ladder (Fig. 11–5a). The bottom of the ladder rests on a horizontal stone ledge and leans across the moat in equilibrium against a vertical wall that is frictionless

because of a thick layer of moss. The ladder makes an angle of 53.1° with the horizontal, conveniently forming a 3–4–5 right triangle. a) Find the normal and friction forces on the ladder at its base. b) Find the minimum coefficient of static friction needed to prevent slipping at the base. c) Find the magnitude and direction of the contact force on the ladder at the base.

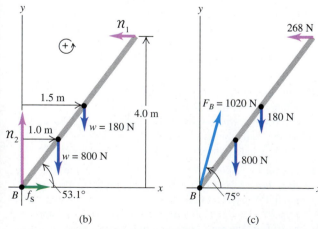

11–5 (a) Sir Lancelot pauses a third of the way up the ladder, fearing it will slip. (b) Free-body diagram for the ladder.

SOLUTION a) Figure 11–5b shows the free-body diagram for the ladder. The ladder is described as "uniform," so its center of gravity is at its geometrical center, halfway between the base and the wall. Lancelot's weight pushes down on the ladder at a point one-third of the way from the base toward the wall. The forces at the base are the upward normal force n_2 and the static friction force f_s, which must point to the right to prevent slipping. The frictionless wall exerts only a normal force n_1 at the top of the ladder.

From Eqs. (11–6), the first condition for equilibrium gives

$$\Sigma F_x = f_s + (-n_1) = 0,$$

$$\Sigma F_y = n_2 + (-800 \text{ N}) + (-180 \text{ N}) = 0.$$

These are two equations for the three unknowns n_1, n_2, and f_s. The first equation tells us that the two horizontal forces must be equal and opposite, and the second equation gives

$$n_2 = 980 \text{ N}.$$

The ground pushes up with a force of 980 N to balance the total (downward) weight (800 N + 180 N).

We don't yet have enough equations, but now we can use the second condition for equilibrium. We can take torques about any point we choose. The smart choice is the point that will give us the fewest terms and fewest unknowns in the torque equation. With this in mind, we choose point B, at the base of the ladder. The two forces n_2 and f_s have no torque about that point. From Fig. 11–5b we see that the lever arm for the ladder's weight is 1.5 m, the lever arm for Lancelot's weight is 1.0 m, and the lever arm for n_1 is 4.0 m. The torque equation for point B is

$$\Sigma \tau_B = n_1(4.0 \text{ m}) - (180 \text{ N})(1.5 \text{ m}) - (800 \text{ N})(1.0 \text{ m})$$
$$+ n_2(0) + f_s(0) = 0.$$

Solving for n_1, we get $n_1 = 268$ N. We now substitute this back into the $\Sigma F_x = 0$ equation to get

$$f_s = 268 \text{ N}.$$

b) The static friction force f_s cannot exceed $\mu_s n_2$, so the *minimum* coefficient of static friction to prevent slipping is

$$(\mu_s)_{\min} = \frac{f_s}{n_2} = \frac{268 \text{ N}}{980 \text{ N}} = 0.27.$$

c) The components of the contact force $\vec{F}_B$ at the base are the static friction force f_s and the normal force n_2, so

$$\vec{F}_B = f_s \hat{\imath} + n_2 \hat{\jmath} = (268 \text{ N})\hat{\imath} + (980 \text{ N})\hat{\jmath}.$$

The magnitude and direction of $\vec{F}_B$ (Fig. 11–5c) are then

$$F_B = \sqrt{(268 \text{ N})^2 + (980 \text{ N})^2} = 1020 \text{ N},$$

$$\theta = \arctan \frac{980 \text{ N}}{268 \text{ N}} = 75°.$$

As Fig. 11–5c shows, the contact force $\vec{F}_B$ is *not* directed along the length of the ladder. You may be surprised by this, but there's really no good reason why the two directions should be the same. Can you show that if $\vec{F}_B$ were directed along the ladder, there would be a net counterclockwise torque with respect to the top of the ladder, and equilibrium would be impossible?

Here are a few final comments. First, as Lancelot climbs higher on the ladder, the lever arm and torque of his weight about B increase; this increases the values of n_1, f_s, and $(\mu_s)_{\min}$. At the top, his lever arm would be nearly 3 m, giving a minimum coefficient of static friction of nearly 0.7. The value of μ_s would not be this large for a present-day aluminum ladder on a wood floor, which is why such ladders are usually equipped with nonslip rubber pads. Lancelot's medieval ladder is unlikely to have this feature, and the ladder is likely to slip as he climbs.

Second, a larger ladder angle would decrease the lever arms with respect to B of the weights of the ladder and Lancelot and increase the lever arm of n_1, all of which would decrease the required friction force. The R. D. Werner Ladder Co. recommends that its ladders be used at an angle of 75°. (Why not 90°?)

Finally, if we had assumed friction on the wall as well as on the floor, the problem would be impossible to solve by using the equilibrium conditions alone. (Try it!) Such a problem is said to be *statically indeterminate*. The difficulty is that it's no longer adequate to treat the body as being perfectly rigid. Another simple example of such a problem is a four-legged table; there is no way to use the equilibrium conditions alone to find the force on each separate leg. In a sense, four legs are one too many; three, properly located, are sufficient for stability.

EXAMPLE 11–4

Another rescue attempt After Lancelot falls into the moat, Sir Gawain tosses a grappling hook through an open upstairs window. In Fig. 11–6a (page 336) he is resting partway up the rope. Gawain weighs 700 N, his body makes an angle of 60° with the wall, and his center of gravity is 0.85 m from his feet. The rope force acts 1.30 m from his feet, and the rope makes an angle of 20° with the vertical. Find the tension in the rope and the forces exerted on his feet by the moss-free region of the wall.

SOLUTION Figure 11–6b shows the free-body diagram for Gawain. The forces on his feet include a (horizontal) normal component n and a (vertical) frictional component f. The rope tension is T. The angles needed to find the torques are shown.

Because two unknown forces (n and f) act at point B at Gawain's feet, we take torques about B, obtaining a torque equa-

tion in which T is the only unknown. The easiest way to find the torque of each force is to use $\tau = rF \sin \phi$, where ϕ is the angle between the position vector $\vec{r}$ (from B to the point of application of the force) and the force vector $\vec{F}$. Taking counterclockwise torques as positive, we obtain the torque equation

$$\Sigma \tau_B = (1.30 \text{ m})T (\sin 140°)$$
$$- (0.85 \text{ m})(700 \text{ N}) (\sin 60°) + (0)n + (0)f = 0.$$

We solve this equation for T; the result is

$$T = 617 \text{ N}.$$

We note that the rope tension is *less than* Gawain's weight.

To find the components of force n and f at Gawain's feet, we use the first equilibrium condition. From Fig. 11–6b, T_x is negative and T_y is positive, so $T_x = -T \sin 20°$ and $T_y = T \cos 20°$.

(a)   (b)

11–6 (a) Gawain pausing to catch his breath. (b) Free-body diagram for Gawain, considered as a rod.

Then

$$\Sigma F_x = n + (-T \sin 20°) = 0,$$

$$\Sigma F_y = f + T \cos 20° + (-700 \text{ N}) = 0.$$

When we substitute the value of T into these equations, we get

$$n = 211 \text{ N} \quad \text{and} \quad f = 120 \text{ N}.$$

The sum of the vertical friction force ($f = 120$ N) and the vertical component of the tension ($T \cos 20° = 580$ N) equals Gawain's 700-N weight.

EXAMPLE 11–5

Equilibrium and pumping iron Figure 11–7a shows a human arm lifting a dumbbell, and Fig. 11–7b is a free-body diagram for the forearm. The forearm is in equilibrium under the action of the weight w of the dumbbell, the tension T in the tendon connected to the biceps muscle, and the force E exerted on the forearm by the upper arm at the elbow joint. For clarity the point A where the tendon is attached is drawn farther away from the elbow than its actual position. The weight w and the angle θ are given; we want to find the tendon tension and the two components of force at the elbow (three unknown scalar quantities in all). We neglect the weight of the forearm itself.

SOLUTION First we represent the tendon force in terms of its components T_x and T_y, using the given angle θ and the unknown magnitude T:

$$T_x = T \cos \theta, \qquad T_y = T \sin \theta.$$

We also represent the force at the elbow in terms of its components E_x and E_y. We'll guess that the directions of these components are as shown in Fig. 11–7b; there's no need to agonize over this guess, since the results for E_x and E_y will tell us the actual directions. Next we note that if we take torques about the elbow joint, the resulting torque equation does not contain

(a)   (b)

11–7 (a) Building up the biceps. (b) Free-body diagram for the forearm. The weight of the forearm is neglected, and the distance D is greatly exaggerated for clarity.

E_x, E_y, or T_x because the lines of action of all these forces pass through this point. The torque equation is then simply

$$\Sigma \tau_E = Lw - DT_y = 0.$$

From this we find

$$T_y = \frac{Lw}{D} \quad \text{and} \quad T = \frac{Lw}{D \sin \theta}.$$

To find E_x and E_y, we use the first conditions for equilibrium, $\Sigma F_x = 0$ and $\Sigma F_y = 0$:

$$\Sigma F_x = T_x + (-E_x) = 0,$$

$$E_x = T_x = T \cos \theta = \frac{Lw}{D \sin \theta} \cos \theta = \frac{Lw}{D} \cot \theta$$

$$= \frac{Lw}{D} \frac{D}{h} = \frac{Lw}{h};$$

$$\Sigma F_y = T_y + E_y + (-w) = 0,$$

$$E_y = w - \frac{Lw}{D} = -\frac{(L-D)w}{D}.$$

The negative sign shows that our guess for the direction of E_y, shown in Fig. 11–7b, was wrong; it is actually vertically *downward*.

An alternative way to find E_x and E_y is to use two more torque equations. We take torques about the tendon attach point, A:

$$\Sigma \tau_A = (L-D)w + DE_y = 0 \quad \text{and} \quad E_y = -\frac{(L-D)w}{D}.$$

Finally, we take torques about point B in the figure:

$$\Sigma \tau_B = Lw - hE_x = 0 \quad \text{and} \quad E_x = \frac{Lw}{h}.$$

Notice how much we have simplified these calculations by choosing the point for calculating torques so as to eliminate one or more of the unknown quantities. In the last step, the force T has no torque about point B; thus when the torques of T_x and T_y are computed separately, they must add to zero. We invite you to check out this statement.

As a specific example, suppose $w = 200$ N, $D = 0.050$ m, $L = 0.30$ m, and $\theta = 80°$. Then from $\tan \theta = h/D$, we find

$$h = D \tan \theta = (0.050 \text{ m})(5.67) = 0.28 \text{ m.}$$

From the previous general results we find

$$T = \frac{Lw}{D \sin \theta} = \frac{(0.30 \text{ m})(200 \text{ N})}{(0.050 \text{ m})(0.98)} = 1220 \text{ N,}$$

$$E_y = -\frac{(L-D)w}{D} = -\frac{(0.30 \text{ m} - 0.050 \text{ m})(200 \text{ N})}{0.050 \text{ m}}$$

$$= -1000 \text{ N,}$$

$$E_x = \frac{Lw}{h} = \frac{(0.30 \text{ m})(200 \text{ N})}{0.28 \text{ m}} = 210 \text{ N.}$$

The magnitude of the force at the elbow is

$$E = \sqrt{E_x{}^2 + E_y{}^2} = 1020 \text{ N.}$$

In our alternative determination of E_x and E_y, we didn't explicitly use the first condition for equilibrium (that the vector sum of the forces is zero). As a check, compute ΣF_x and ΣF_y to verify that they really *are* zero. Consistency checks are always a good idea!

In view of the magnitudes of our results, neglecting the weight of the forearm itself, which may be 20 N or so, will cause only relatively small errors in our results.

11–5 STRESS, STRAIN, AND ELASTIC MODULI

The rigid body is a useful idealized model, but the stretching, squeezing, and twisting of real bodies when forces are applied are often too important to ignore. Figure 11–8 shows three examples. We want to study the relation between the forces and deformations for each case.

(a)   (b)   (c)

11–8 Three types of stress. (a) A cable under *tension,* being stretched by forces acting at its ends. (b) A submersible under *compression,* being squeezed from all sides by the force of the water pressure. (c) A drive shaft under *shear,* being twisted by forces at its ends that cause torques about its axis.

For each kind of deformation we will introduce a quantity called **stress** that characterizes the strength of the forces causing the stretch, squeeze, or twist, usually on a "force per unit area" basis. Another quantity, **strain,** describes the resulting deformation. When the stress and strain are small enough, we often find that the two are directly proportional, and we call the proportionality constant an **elastic modulus.** The harder you pull on something, the more it stretches; and the more you squeeze it, the more it compresses. The general pattern that emerges can be formulated as

$$\frac{\text{Stress}}{\text{Strain}} = \text{Elastic modulus} \qquad \text{(Hooke's law)}. \qquad (11\text{--}7)$$

The proportionality of stress and strain (under certain conditions) is called **Hooke's law,** after Robert Hooke (1635–1703), a contemporary of Newton. We used one form of Hooke's law in Sections 6–4 and 7–3: the elongation of an ideal spring is proportional to the stretching force. Remember that Hooke's law is not really a general law but an experimental finding that is valid only over a limited range. The last section of this chapter discusses what this limited range is.

TENSILE AND COMPRESSIVE STRESS AND STRAIN

The simplest elastic behavior to understand is the stretching of a bar, rod, or wire when its ends are pulled. Figure 11–9a shows a bar with uniform cross-section area A, with equal and opposite forces F pulling at its ends. We say that the bar is in **tension.** We've already talked a lot about tension in ropes and strings; it's the same concept here. Figure 11–9b shows a cross section through the bar. The part of the bar to the right of the section pulls on the part to the left with a force $F_\perp$, and vice versa. We use the notation $F_\perp$ as a reminder that the force acts in a direction perpendicular to the cross section. We will assume that the forces at every cross section are distributed uniformly over the section, as shown by the short arrows in Fig. 11–9b. (This is the case if the bar is uniform and the forces at the *ends* are uniformly distributed.)

We define the **tensile stress** at the cross section as the ratio of the force $F_\perp$ to the cross-section area A:

$$\text{Tensile stress} = \frac{F_\perp}{A}. \qquad (11\text{--}8)$$

This is a *scalar* quantity because $F_\perp$ is the *magnitude* of the force. The SI unit of stress is the **pascal** (abbreviated Pa, and named for the 17th-century French scientist and philosopher Blaise Pascal). Equation (11–8) shows that one pascal equals one newton per square meter (N/m^2):

$$1 \text{ pascal} = 1 \text{ Pa} = 1 \text{ N/m}^2.$$

In the British system the logical unit of stress would be the pound per square foot, but the pound per square inch (lb/in.^{2} or psi) is more commonly used. The conversion factors are

$$1 \text{ psi} = 6895 \text{ Pa} \qquad \text{and} \qquad 1 \text{ Pa} = 1.450 \times 10^{-4} \text{ psi}.$$

This rock climber exerts a stress on the climbing rope due to his weight. The strain of the rope is the fractional amount that it stretches as a result of this stress.

11–9 (a) A bar in tension. (b) The stress at a perpendicular section equals $F_\perp/A$, the ratio of the force to the cross-section area.

(a)   (b)

The units of stress are the same as those of *pressure,* which we will encounter often in later chapters. Air pressure in automobile tires is typically around 3×10^5 Pa = 300 kPa, and steel cables are commonly required to withstand tensile stresses of the order of 10^8 Pa.

The fractional change of length (stretch) of a body under a tensile stress is called the **tensile strain.** Figure 11–10 shows a bar with unstretched length l_0 that stretches to a length $l = l_0 + \Delta l$ when equal and opposite forces F are applied to its ends. The elongation Δl does not occur only at the ends; every part of the bar stretches in the same proportion. The tensile strain is defined as the ratio of the elongation Δl to the original length l_0:

$$\text{Tensile strain} = \frac{l - l_0}{l_0} = \frac{\Delta l}{l_0}. \qquad (11-9)$$

Tensile strain is stretch per unit length. It is a ratio of two lengths, always measured in the same units, and so is a pure (dimensionless) number with no units.

Experiment shows that for a sufficiently small tensile stress, stress and strain are proportional, as in Eq. (11–7). The corresponding elastic modulus is called **Young's modulus,** denoted by Y:

$$Y = \frac{\text{Tensile stress}}{\text{Tensile strain}},$$

$$Y = \frac{F_\perp/A}{\Delta l/l_0} = \frac{F_\perp}{A}\frac{l_0}{\Delta l} \qquad \text{(Young's modulus).} \qquad (11-10)$$

Since strain is a pure number, the units of Young's modulus are the same as those of stress: force per unit area. Some typical values are listed in Table 11–1. (This table also gives values of two other elastic moduli that we will discuss later in this chapter.) A material with a large value of Y is relatively unstretchable; a large stress is required for a given strain. For example, the value of Y for cast steel (2×10^{11} Pa) is much larger than that for rubber (5×10^8 Pa).

TABLE 11–1
APPROXIMATE ELASTIC MODULI

MATERIAL	YOUNG'S MODULUS, Y (Pa)	BULK MODULUS, B (Pa)	SHEAR MODULUS, S (Pa)
Aluminum	7.0×10^{10}	7.5×10^{10}	2.5×10^{10}
Brass	9.0×10^{10}	6.0×10^{10}	3.5×10^{10}
Copper	11×10^{10}	14×10^{10}	4.4×10^{10}
Crown glass	6.0×10^{10}	5.0×10^{10}	2.5×10^{10}
Iron	21×10^{10}	16×10^{10}	7.7×10^{10}
Lead	1.6×10^{10}	4.1×10^{10}	0.6×10^{10}
Nickel	21×10^{10}	17×10^{10}	7.8×10^{10}
Steel	20×10^{10}	16×10^{10}	7.5×10^{10}

When the forces on the ends of a bar are pushes rather than pulls (Fig. 11–11), the bar is in **compression,** and the stress is a **compressive stress.** At the cross section shown,

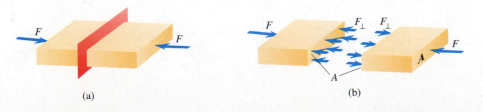

(a)   (b)

11–10 The tensile strain is defined as $\Delta l/l_0$. The elongation Δl is exaggerated for clarity.

11–11 (a) A bar in compression. (b) The stress at a perpendicular section also equals $F_\perp/A$.

(a)

(b)   (c)

11-12 (a) A beam supported at both ends is under both compression and tension. (b) The cross-section shape of an I-beam minimizes both stress and weight. (c) A hollow pole or tube is resistant to bending in all directions. (d) At a height of 553 m, the CN Tower in Toronto is the world's tallest self-supporting structure. In cross section the tower forms a Y, which gives the 1.3×10^8-kg structure tremendous rigidity against twisting or bending. In a 200-km/h wind, the Skypod (the seven-story building two-thirds of the way up the tower) would sway back and forth only 25 cm.

each side pushes rather than pulls on the other. The **compressive strain** of a bar in compression is defined in the same way as tensile strain, but Δl has the opposite direction. Hooke's law and Eq. (11–10) are valid for compression as well as tension if the compressive stress is not too great. For many materials, Young's modulus has the same value for both tensile and compressive stresses; composite materials such as concrete are an exception.

In many situations, bodies can experience both tensile and compressive stresses at the same time. As an example, a horizontal beam supported at each end sags under its own weight. As a result, the top of the beam is under compression, while the bottom of the beam is under tension (Fig. 11–12a). To minimize the stress and hence the bending strain, the top and bottom of the beam are given a large cross-section area. There is neither compression nor tension along the centerline of the beam, so this part can have a small cross section; this helps to keep the weight of the bar to a minimum and further helps to reduce the stress. The result is an I-beam of the familiar shape used in building construction (Fig. 11–12b).

The vertical pole that supports a traffic light or a highway sign has a circular cross section because it must be resistant to bending in all directions due to wind or earthquakes. A hollow circular pole is more resistant to bending than a solid circular pole of the same mass but smaller radius, since it has a greater effective cross-section area (Fig. 11–12c). The CN Tower in Toronto (Fig. 11–12d) is basically a hollow circular pole with the sides pinched in to give some of the natural stability of a tripod. Sections of the tower closer to the ground must support a larger fraction of the tower's weight, so the cross-section area is greater near the ground to keep the stress relatively constant.

Bridges are under tremendous stress due to the weight they must support. The design of bridges is an exercise in imparting this stress to the supporting ground. A suspension bridge supports its load primarily through tension in the cables and compression in the

(d)

11-13 (a) In a suspension bridge, the cables are under tension and the towers are under compression. Tension structures are most efficient for supporting light or moderate loads over long distances; the world's longest suspension bridge, the Humber Bridge in England, carries cars and trucks (but no trains) across a center span of 1410 m. (b) An arched bridge is under compression. Compression structures are very efficient for supporting large loads over short distances. An example is the Sydney Harbour Bridge in Australia, which has a 500-m span and was built to carry four railway lines and a 17-m-wide road.

(a)   (b)

towers (Fig. 11–13a). The downward force on the towers due to tension is balanced by the upward force provided by the ground under the towers. The ground must be strong enough to bear this stress without fracturing. An arched bridge supports its load primarily through compression (Fig. 11–13b). The compressive stress is borne by the ground at the ends of the arch.

EXAMPLE 11–6

A steel rod 2.0 m long has a cross-section area of 0.30 cm². A 550-kg milling machine is now hung from the rod. Determine the stress, the strain, and the elongation of the rod.

SOLUTION We use the definitions of stress, strain, and Young's modulus given by Eqs. (11–8), (11–9), and (11–10), respectively, and the value of Y for steel from Table 11–1:

$$\text{Stress} = \frac{F_\perp}{A} = \frac{(550 \text{ kg})(9.8 \text{ m/s}^2)}{3.0 \times 10^{-5} \text{ m}^2} = 1.8 \times 10^8 \text{ Pa};$$

$$\text{Strain} = \frac{\Delta l}{l_0} = \frac{\text{Stress}}{Y} = \frac{1.8 \times 10^8 \text{ Pa}}{20 \times 10^{10} \text{ Pa}} = 9.0 \times 10^{-4};$$

$$\text{Elongation} = \Delta l = (\text{Strain}) \times l_0 = (9.0 \times 10^{-4})(2.0 \text{ m})$$
$$= 0.0018 \text{ m} = 1.8 \text{ mm}.$$

The small size of this elongation, which results from a load of over half a ton, is a testament to the stiffness of steel.

11–6 BULK STRESS AND STRAIN

When a submersible plunges deep into the ocean, the water exerts nearly uniform pressure everywhere on its surface and squeezes the submersible to a slightly smaller volume (Fig. 11–8b). This is a different situation from the tensile and compressive stresses and strains we have discussed. The stress is now a uniform pressure on all sides, and the resulting deformation is a volume change. We use the terms **bulk stress** (or **volume stress**) and **bulk strain** (or **volume strain**) to describe these quantities. Another familiar example is the compression of a gas under pressure, such as the air in a car's tire.

If we choose an arbitrary cross section within a fluid (liquid or gas) at rest, the force acting on each side of the section is always *perpendicular* to it. If we tried to exert a force parallel to a section, the fluid would slip sideways to counteract the effort. When a solid is immersed in a fluid and both are at rest, the forces that the fluid exerts on the surface of the solid are always perpendicular to the surface at each point. The force $F_\perp$ per unit area A on such a surface is called the **pressure** p in the fluid:

$$p = \frac{F_\perp}{A} \qquad \text{(pressure in a fluid).} \tag{11–11}$$

When we apply pressure to the surface of a fluid in a container, such as the cylinder and piston shown in Fig. 11–14, the pressure is transmitted through the fluid and also acts on the surface of any body immersed in the fluid. This principle is called *Pascal's law*. If pressure differences due to differences in depth within the fluid can be neglected, the pressure is the same at every point in the fluid and at every point on the surface of any submerged body.

Pressure has the same units as stress; commonly used units include 1 Pa (= 1 N/m²) and 1 lb/in.² (1 psi). Also in common use is the **atmosphere,** abbreviated atm. One atmosphere is the approximate average pressure of the earth's atmosphere at sea level:

$$1 \text{ atmosphere} = 1 \text{ atm} = 1.013 \times 10^5 \text{ Pa} = 14.7 \text{ lb/in.}^2.$$

Pressure is a scalar quantity, not a vector quantity; it has no direction.

11–14 Every element dA of the surface of a submerged body experiences an inwardly normal force $dF_\perp = p \, dA$ from the surrounding fluid.

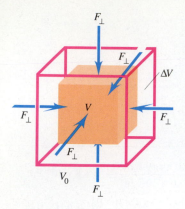

11–15 Under bulk stress, the cube occupies the shaded volume. Without the stress, the cube would occupy the outlined volume. The bulk strain is the fractional change in volume, $\Delta V/V_0$. The volume change ΔV is exaggerated for clarity.

TABLE 11–2

COMPRESSIBILITIES OF LIQUIDS

	COMPRESSIBILITY, k	
LIQUID	Pa^{-1}	atm^{-1}
Carbon disulfide	93×10^{-11}	94×10^{-6}
Ethyl alcohol	110×10^{-11}	111×10^{-6}
Glycerine	21×10^{-11}	21×10^{-6}
Mercury	3.7×10^{-11}	3.8×10^{-6}
Water	45.8×10^{-11}	46.4×10^{-6}

Pressure plays the role of stress in a volume deformation. The corresponding strain is the fractional change in volume (Fig. 11–15), that is, the ratio of the volume change ΔV to the original volume V_0:

$$\text{Bulk (volume) strain} = \frac{\Delta V}{V_0}. \tag{11-12}$$

Volume strain is the change in volume per unit volume. Like tensile or compressive strain, it is a pure number, without units.

When Hooke's law is obeyed, an increase in pressure (bulk stress) produces a *proportional* bulk strain (fractional change in volume). The corresponding elastic modulus (ratio of stress to strain) is called the **bulk modulus,** denoted by B. When the pressure on a body changes by a small amount Δp, from p_0 to $p_0 + \Delta p$, and the resulting bulk strain is $\Delta V/V_0$, Hooke's law takes the form

$$B = -\frac{\Delta p}{\Delta V/V_0} \qquad \text{(bulk modulus)}. \tag{11-13}$$

We include a minus sign in this equation because an *increase* of pressure always causes a *decrease* in volume. In other words, if Δp is positive, ΔV is negative. The bulk modulus B itself is a positive quantity.

For small pressure changes in a solid or a liquid, we consider B to be constant. The bulk modulus of a *gas,* however, depends on the initial pressure p_0. Table 11–1 includes values of the bulk modulus for several solid materials. Its units, force per unit area, are the same as those of pressure (and of tensile or compressive stress).

The reciprocal of the bulk modulus is called the **compressibility** and is denoted by k. From Eq. (11–13),

$$k = \frac{1}{B} = -\frac{\Delta V/V_0}{\Delta p} = -\frac{1}{V_0}\frac{\Delta V}{\Delta p} \qquad \text{(compressibility)}. \tag{11-14}$$

Compressibility is the fractional decrease in volume, $-\Delta V/V_0$, per unit increase Δp in pressure. The units of compressibility are those of *reciprocal pressure,* Pa^{-1} or atm^{-1}.

Values of compressibility k for several liquids are listed in Table 11–2. From this table the compressibility of water is 46.4×10^{-6} atm^{-1}. This means that for each atmosphere increase in pressure, the volume of water decreases by 46.4 parts per million. Materials with small bulk modulus and large compressibility are easy to compress; those with larger bulk modulus and smaller compressibility compress less with the same pressure increase.

EXAMPLE 11–7

A hydraulic press contains 0.25 m^3 (250 L) of oil. Find the decrease in volume of the oil when it is subjected to a pressure increase $\Delta p = 1.6 \times 10^7$ Pa (about 160 atm or 2300 psi). The bulk modulus of the oil is $B = 5.0 \times 10^9$ Pa (about 5.0×10^4 atm), and its compressibility is $k = 1/B = 20 \times 10^{-6}$ atm^{-1}.

SOLUTION To find the volume change ΔV, we solve Eq. (11–13) for ΔV and then substitute in the given values:

$$\Delta V = -\frac{V_0 \Delta p}{B} = -\frac{(0.25 \text{ m}^3)(1.6 \times 10^7 \text{ Pa})}{5.0 \times 10^9 \text{ Pa}}$$

$$= -8.0 \times 10^{-4} \text{ m}^3 = -0.80 \text{ L}.$$

Alternatively, we can use Eq. (11–14). Solving for ΔV and using the approximate unit conversions given above, we get

$$\Delta V = -kV_0 \ \Delta p = -(20 \times 10^{-6} \text{ atm}^{-1})(0.25 \text{ m}^3)(160 \text{ atm})$$

$$= -8.0 \times 10^{-4} \text{ m}^3.$$

Even though the pressure increase is very large, the fractional change in volume is very small:

$$\Delta V/V_0 = (-8.0 \times 10^{-4} \text{ m}^3)/(0.25 \text{ m}^3) = -0.0032, \text{ or } -0.32\%.$$

11-7 SHEAR STRESS AND STRAIN

The third kind of stress-strain situation is called *shear*. The drive shaft in Fig. 11–8c and the body in Fig. 11–16 are under **shear stress**; the result will be a twisting or deformation of the body. We define the shear stress as the force $F_\parallel$ *tangent* to a material surface, divided by the area A on which it acts:

$$\text{Shear stress} = \frac{F_\parallel}{A}. \qquad (11-15)$$

This is the sort of stress provided by a pair of scissors or shears. Shear stress, like the other two types of stress, is a force per unit area.

A shear deformation is shown in Fig. 11–17. The black outline *abcd* represents an unstressed block of material, such as a book, a wall, or a chunk of the earth's crust. The area *a' b' c' d'* shaded in yellow shows the same block under shear stress. In Fig. 11–17a the centers of the stressed and unstressed block coincide. The deformation in Fig. 11–17b is the same as that in Fig. 11–17a, but the figure has been shifted and rotated to make the edges *ad* and *a' d'* coincide. In shear stress the lengths of the faces remain nearly constant; all dimensions parallel to the diagonal *ac* increase in length, and those parallel to the diagonal *bd* decrease in length. We define **shear strain** as the ratio of the displacement x of corner b to the transverse dimension h:

$$\text{Shear strain} = \frac{x}{h} = \tan\phi. \qquad (11-16)$$

In real-life situations, x is nearly always much smaller than h, $\tan\phi$ is very nearly equal to ϕ (in radians), and the strain is simply the angle ϕ measured in radians. Like all strains, shear strain is a dimensionless number; it is a ratio of two lengths.

If the forces are small enough that Hooke's law is obeyed, the shear strain is *proportional* to the shear stress. The corresponding elastic modulus (ratio of shear stress to shear strain) is called the **shear modulus**, denoted by S:

$$S = \frac{\text{Shear stress}}{\text{Shear strain}} = \frac{F_\parallel/A}{x/h} = \frac{F_\parallel}{A}\frac{h}{x} = \frac{F_\parallel/A}{\phi} \quad \text{(shear modulus)}, \qquad (11-17)$$

with x and h defined as in Fig. 11–17.

Several values of shear modulus are given in Table 11–1. For a given material, S is usually $\frac{1}{3}$ to $\frac{1}{2}$ as large as Young's modulus Y for tensile stress. Keep in mind that the concepts of shear stress, shear strain, and shear modulus apply to *solid* materials only. Here's why: the shear forces in Fig. 11–17a are required to deform the solid block, and the block tends to return to its original shape if the shear forces are removed. By contrast, gases and liquids do not have a definite shape.

11-16 A body under shear stress. The area A is the edge area on which each force $F_\parallel$ acts.

Scissors and knives work by applying shear stress on the object to be cut. Cutting takes place when the strain caused by the stress is so large that chemical bonds between molecules in the object break apart.

(a)

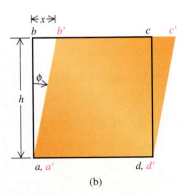

(b)

11-17 (a) A block changes shape under shear stress. (b) The deformed block has been shifted and rotated so that its bottom edge coincides with that of the undeformed block to show the angle ϕ clearly. The shear strain, defined as x/h, is very nearly equal to ϕ in radians.

EXAMPLE 11-8

Suppose the body in Fig. 11–17 is the brass base plate of an out-door sculpture; it experiences shear forces as a result of an earthquake. The frame is 0.80 m square and 0.50 cm thick. How large a force $F_\parallel$ must be exerted on each of its edges if the displacement x (see Fig. 11–17b) is 0.16 mm?

SOLUTION From Table 11–1 the shear modulus of brass is 3.5×10^{10} Pa. The shear strain is

$$\text{Shear strain} = \frac{x}{h} = \frac{1.6 \times 10^{-4}\,\text{m}}{0.80\,\text{m}} = 2.0 \times 10^{-4}.$$

From Eq. (11–17) the shear stress equals the shear strain multiplied by the shear modulus S,

$$\text{Stress} = (\text{Shear strain}) \times S$$
$$= (2.0 \times 10^{-4})(3.5 \times 10^{10}\,\text{Pa}) = 7.0 \times 10^{6}\,\text{Pa},$$

and the force at each edge is the shear stress multiplied by the area of the edge:

$$F_\parallel = (\text{Shear stress}) \times A$$
$$= (7.0 \times 10^{6}\,\text{Pa})(0.80\,\text{m})(0.0050\,\text{m}) = 2.8 \times 10^{4}\,\text{N}.$$

This is a force of over three tons.

11–8 ELASTICITY AND PLASTICITY

Hooke's law, the proportionality of stress and strain in elastic deformations, has a limited range of validity. In the preceding three sections we used phrases such as "provided that the forces are small enough that Hooke's law is obeyed." Just what *are* the limitations of Hooke's law? We know that if you pull, squeeze, or twist *anything* hard enough, it will bend or break. Can we be more precise than that?

Let's look at tensile stress and strain again. Suppose we plot a graph of stress as a function of strain. If Hooke's law is obeyed, the graph is a straight line with a slope equal to Young's modulus. Figure 11–18 shows a typical stress-strain graph for a metal such as copper or soft iron. The strain is shown as the *percent* elongation; the horizontal scale is not uniform beyond the first portion of the curve, up to a strain of less than 1%. The first portion is a straight line, indicating Hooke's law behavior with stress directly proportional to strain. This straight-line portion ends at point *a*; the stress at this point is called the *proportional limit.*

From *a* to *b*, stress and strain are no longer proportional, and Hooke's law is *not* obeyed. If the load is gradually removed, starting at any point between *O* and *b*, the curve is retraced until the material returns to its original length. The deformation is *reversible,* and the forces are conservative; the energy put into the material to cause the deformation is recovered when the stress is removed. In region *Ob* we say that the material shows *elastic behavior.* Point *b*, the end of this region, is called the *yield point;* the stress at the yield point is called the *elastic limit.*

11–18 Typical stress-strain diagram for a ductile metal under tension.

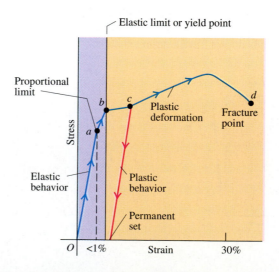

When we increase the stress beyond point b, the strain continues to increase. But now when we remove the load at some point beyond b, say c, the material does not come back to its original length. Instead, it follows the red line in Fig. 11–18. The length at zero stress is now greater than the original length; the material has undergone an irreversible deformation and has acquired what we call a *permanent set*. Further increase of load beyond c produces a large increase in strain for a relatively small increase in stress, until a point d is reached at which *fracture* takes place. The behavior of the material from b to d is called *plastic flow* or *plastic deformation*. A plastic deformation is irreversible; when the stress is removed, the material does not return to its original state.

For some materials a large amount of plastic deformation takes place between the elastic limit and the fracture point. Such a material is said to be *ductile*. But if fracture occurs soon after the elastic limit is passed, the material is said to be *brittle*. A soft iron wire that can have considerable permanent stretch without breaking is ductile, while a steel piano string that breaks soon after its elastic limit is reached is brittle.

Figure 11–19 shows a stress-strain curve for vulcanized rubber that has been stretched by over seven times its original length. The stress is not proportional to the strain, but the behavior is elastic because when the load is removed, the material returns to its original length. However, the material follows different curves for increasing and decreasing stress. This is called *elastic hysteresis*. The work done by the material when it returns to its original shape is less than the work required to deform it; there are nonconservative forces associated with internal friction. Rubber with large elastic hysteresis is very useful for absorbing vibrations, such as in engine mounts and shock-absorber bushings for cars.

The stress required to cause actual fracture of a material is called the *breaking stress,* the *ultimate strength,* or (for tensile stress) the *tensile strength.* Two materials, such as two types of steel, may have very similar elastic constants but vastly different breaking stresses. Table 11–3 gives a few typical values of breaking stress for several materials in tension.

The conversion factor 6.9×10^8 Pa = 100,000 psi will help put these numbers in perspective. For example, if the breaking stress of a particular steel is 6.9×10^8 Pa, then a bar with 1-in.^{2} cross section has a breaking strength of 100,000 lb.

11–19 Typical stress-strain diagram for vulcanized rubber, showing elastic hysteresis.

TABLE 11–3

APPROXIMATE BREAKING STRESSES OF MATERIALS

MATERIAL	BREAKING STRESS (Pa or N/m^2)
Aluminum	2.2×10^8
Brass	4.7×10^8
Glass	10×10^8
Iron	3.0×10^8
Phosphor bronze	5.6×10^8
Steel	$5–20 \times 10^8$

SUMMARY

- For a rigid body to be in equilibrium, two conditions must be satisfied. First, the vector sum of forces must be zero. Second, the sum of torques about any point must be zero. The first condition, in terms of components, is

$$\Sigma F_x = 0, \qquad \Sigma F_y = 0, \qquad \Sigma F_z = 0. \qquad (11\text{–}1)$$

The second condition is

$$\Sigma \vec{\tau} = 0 \qquad \text{about } any \text{ point.} \qquad (11\text{–}2)$$

- The torque due to the weight of a body can be obtained by assuming the entire weight to be concentrated at the center of gravity. If $\vec{g}$ has the same value at all points, the coordinates of the center of gravity are the same as those of the center of mass:

$$x_{cm} = \frac{m_1 x_1 + m_2 x_2 + \cdots}{m_1 + m_2 + \cdots} = \frac{\sum\limits_i m_i x_i}{\sum\limits_i m_i}, \qquad y_{cm} = \frac{m_1 y_1 + m_2 y_2 + \cdots}{m_1 + m_2 + \cdots} = \frac{\sum\limits_i m_i y_i}{\sum\limits_i m_i},$$

$$z_{cm} = \frac{m_1 z_1 + m_2 z_2 + \cdots}{m_1 + m_2 + \cdots} = \frac{\sum\limits_i m_i z_i}{\sum\limits_i m_i}. \qquad (11\text{–}3)$$

KEY TERMS

first condition for equilibrium, 329

second condition for equilibrium, 330

static equilibrium, 330

center of gravity, 330

stress, 338

strain, 338

elastic modulus, 338

Hooke's law, 338

tension, 338

tensile stress, 338

pascal, 338

tensile strain, 339

Young's modulus, 339

compression, 339

compressive stress, 339

- The torque due to a force can be computed by finding the torque of each force component, using its appropriate lever arm and sign, and adding these.

- Hooke's law states that in elastic deformations, stress is proportional to strain:

$$\frac{\text{Stress}}{\text{Strain}} = \text{Elastic modulus.} \tag{11–7}$$

- Tensile stress is tensile force per unit area, $F_\perp/A$. Tensile strain is fractional change in length, $\Delta l/l_0$. Young's modulus Y is the ratio of tensile stress to tensile strain:

$$Y = \frac{F_\perp/A}{\Delta l/l_0} = \frac{F_\perp}{A} \frac{l_0}{\Delta l}. \tag{11–10}$$

Compressive stress and strain are defined the same way as tensile stress and strain. For many materials, Young's modulus has the same value for both tension and compression.

- Pressure in a fluid is force per unit area:

$$p = \frac{F_\perp}{A}. \tag{11–11}$$

- The bulk modulus B is the negative of the ratio of pressure change Δp (bulk stress) to fractional volume change $\Delta V/V_0$:

$$B = -\frac{\Delta p}{\Delta V/V_0}. \tag{11–13}$$

Compressibility k is the reciprocal of bulk modulus: $k = 1/B$.

- Shear stress is force per unit area $F_\parallel/A$ for a force applied parallel to a surface. Shear strain is the angle ϕ shown in Fig. 11–17. The shear modulus S is the ratio of shear stress to shear strain:

$$S = \frac{\text{Shear stress}}{\text{Shear strain}} = \frac{F_\parallel/A}{x/h} = \frac{F_\parallel}{A} \frac{h}{x} = \frac{F_\parallel/A}{\phi}. \tag{11–17}$$

- The proportional limit is the maximum stress for which stress and strain are proportional. Beyond the proportional limit, Hooke's law is not valid. The elastic limit is the stress beyond which irreversible deformation occurs. The breaking stress, or ultimate strength, is the stress at which the material breaks.

DISCUSSION QUESTIONS

Q11–1 Does a rigid body in uniform rotation about a fixed axis satisfy the first and second conditions for equilibrium? Why or why not? An alternative definition of equilibrium states: "For a rigid body to be in equilibrium, every part of the body must be in equilibrium." On the basis of this definition, is a rigid body in uniform rotation about a fixed axis in equilibrium? Explain.

Q11–2 Car tires are sometimes "balanced" on a machine that pivots the tire and wheel about the center. Weights are placed around the wheel rim until it does not tip from the horizontal plane. Discuss this procedure in terms of the center of gravity.

Q11–3 Does the center of gravity of a solid body always lie within the material of the body? If not, give a counterexample.

Q11–4 In Section 11–3 we always assumed that the value of g was the same at all points on the body. This is *not* a good approximation if the dimensions of the body are great enough, because the value of g decreases with altitude. If this is taken into account, will the center of gravity of a long vertical rod be above, below, or at its center of mass? Explain how this can be used to keep the long axis of an orbiting spacecraft pointed toward the earth. (This would be useful for a weather satellite that must always keep its camera lens trained on the earth.) The moon is not exactly spherical, but is somewhat elongated. Explain why this same effect is responsible for keeping the same face of the moon pointed toward the earth at all times.

Q11–5 You are balancing a wrench by suspending it at a single point. Is the equilibrium stable, unstable, or neutral if the point is above, at, or below the wrench's center of gravity? In each case give the reasoning behind your answer. (For rotation a rigid body is in *stable* equilibrium if a small rotation of the body produces a torque that tends to return the body to equilibrium, it is in *unstable* equilibrium if a small rotation produces a torque that tends to take the body farther from equilibrium, and it is in *neutral* equilibrium if a small rotation produces no torque.)

Q11–6 You can probably stand flatfooted on the floor, then rise up and balance on your tiptoes. Why can't you do it if your toes are touching the wall of your room? (Try it!)

Q11–7 You freely pivot a horseshoe from a horizontal nail through one of its nail holes. You then hang a long string with a weight at its bottom from the same nail, so that the string hangs vertically in front of the horseshoe without touching it. How do you know that the horseshoe's center of gravity is along the line behind the string? How can you locate the center of gravity by repeating the process at another nail hole? Will the center of gravity be within the solid material of the horseshoe?

Q11–8 A stunt driver drives a car in a straight line at constant speed while balanced on the two right-hand wheels. Where is the car's center of gravity?

Q11–9 When four-legged animals walk, they always have three of their legs on the ground at any instant. Explain why, using the concept of center of gravity.

Q11–10 Why must a water skier moving with constant velocity lean backwards? What determines how far back she must lean? Draw a free-body diagram for the water skier to justify your answers.

Q11–11 In pioneer days, when a Conestoga wagon was stuck in the mud, people would grasp the wheel spokes and try to turn the wheels, rather than simply pushing the wagon. Why?

Q11–12 A physical therapist says a physicist-patient has weak stomach muscles because he can't do a partial sit-up without his feet being held down. The physicist, who has short slender legs and a long heavy upper torso, correctly disagrees. Explain, using the concept of center of gravity.

Q11–13 Why is it easier to hold a 10-kg dumbbell in your hand at your side than to hold it with your arm extended horizontally?

Q11–14 Figure 11–2 shows that an object at rest on an incline will tip over if its center of gravity is not within the area of support. Explain why this result does *not* depend on the value of the coefficient of static friction between the object and the surface of the incline. Draw a free-body diagram as part of your explanation.

Q11–15 In Example 11–3 (Section 11–4), explain why n_1 is negative in the $\Sigma F_x = 0$ equation but gives a positive torque in the $\Sigma \tau_B = 0$ equation.

Q11–16 Why is a tapered water glass with a narrow base easier to tip over than one with straight sides? Does it matter whether the glass is full or empty?

Q11–17 When a tall, heavy refrigerator is pushed across a rough floor, what determines whether it slides or tips?

Q11–18 Climbing ropes used by mountaineers are usually made of nylon. Would a steel cable of equal strength be just as good? What advantages and disadvantages would it have in comparison to nylon?

Q11–19 Why is concrete with steel reinforcing rods embedded in it stronger than plain concrete?

Q11–20 Electric power lines are sometimes made by using wires with steel core and copper jacket or with strands of copper and steel twisted together. Why?

Q11–21 Compare the mechanical properties of a steel cable, made by twisting many thin wires together, with those of a solid steel rod of the same diameter. What advantages does each have?

Q11–22 Two vases have flat bottoms, the same weight, and centers of gravity at the same distance above their bottoms. The one that exerts the greater pressure on the table is likely to be the one that tips over easier. Why?

Q11–23 Is the bulk modulus of steel greater or less than that of air? By roughly what factor?

Q11–24 Is the shear modulus of steel greater or less than that of jelly? By roughly what factor?

Q11–25 Why does Fig. 11–16 show two pairs of antiparallel forces, all of the same magnitude? Wouldn't just one pair result in shear strain and equilibrium?

Q11–26 The material in human bones and elephant bones is essentially the same, but an elephant has much thicker legs. Explain why, in terms of breaking stress.

Q11–27 There is a small but appreciable amount of elastic hysteresis in the large tendon at the back of a horse's leg. Explain how this can cause damage to the tendon if a horse runs too hard for too long a time.

Q11–28 When rubber mounting blocks are used to absorb machine vibrations through elastic hysteresis, as mentioned in Section 11–8, what becomes of the energy associated with the vibrations?

EXERCISES

SECTION **11–3 CENTER OF GRAVITY**

11–1 A ball with radius $r_1 = 0.080$ m and mass 1.00 kg is attached by a light rod 0.400 m in length to a second ball with radius $r_2 = 0.100$ m and mass 2.00 kg (Fig. 11–20). Where is the center of gravity of this system?

11–2 Suppose the rod in Exercise 11–1 is uniform and has a mass of 1.50 kg. Where is the system's center of gravity?

11–3 For the plank of Example 11–1 (Section 11–3), show

FIGURE 11–20 Exercise 11–1.

that a 60-kg child could stand no farther to the right than halfway between the right-hand sawhorse and the end of the plank without the plank tipping over.

SECTION 11–4 SOLVING RIGID-BODY EQUILIBRIUM PROBLEMS

11–4 A uniform 300-N trapdoor in a ceiling is hinged at one side. Find the net upward force needed to begin to open it and the total force exerted on the door by the hinges, a) if the upward force is applied at the center; b) if the upward force is applied at the center of the edge opposite the hinges.

11–5 Raising a Ladder. A ladder carried by a fire truck is 20.0 m long. The ladder weighs 2800 N and its center of gravity is at its center. The ladder is pivoted at one end (*A*) about a pin (Fig. 11–21); the friction torque at the pin can be neglected. The ladder is raised into position by a force applied by a hydraulic piston at *C*. Point *C* is 8.0 m from *A*, and the force $\vec{F}$ exerted by the piston makes an angle of 40° with the ladder. What magnitude must $\vec{F}$ have to just lift the ladder off the support bracket at *B*?

FIGURE 11–21 Exercise 11–5.

11–6 Two people are carrying a uniform wooden board that is 3.00 m long and weighs 160 N. If one person applies an upward force equal to 60 N at one end, at what point does the other person lift?

11–7 Two people carry a heavy electric motor by placing it on a light board 2.00 m long. One person lifts at one end with a force of 400 N, and the other lifts the opposite end with a force of 600 N. What is the weight of the motor, and where along the board is its center of gravity located?

11–8 Suppose the board in Exercise 11–7 is not light but weighs 200 N, with its center of gravity at its center. The two people each exert the same forces as before. What is the weight of the motor in this case, and where is its center of gravity located?

11–9 Show that the torques about point *B* due to T_x and T_y in Example 11–5 (Section 11–4) sum to zero. Do this for the general case, rather than for the specific numerical values given at the end of the example.

11–10 A uniform ladder 5.0 m long rests against a frictionless vertical wall with its lower end 3.0 m from the wall. The ladder weighs 160 N. The coefficient of static friction between the foot of the ladder and the ground is 0.40. A man weighing 740 N climbs slowly up the ladder. a) What is the maximum frictional force that the ground can exert on the ladder at its lower end? b) What is the actual frictional force when the man has climbed 1.0 m along the ladder? c) How far along the ladder can the man climb before the ladder starts to slip?

11–11 A diving board 3.00 m long is supported at a point 1.00 m from the end, and a diver weighing 500 N stands at the free end (Fig. 11–22). The diving board is of uniform cross section and weighs 280 N. Find a) the force at the support point; b) the force at the end that is held down.

FIGURE 11–22 Exercise 11–11.

11–12 A uniform aluminum beam 9.00 m long, weighing 300 N, rests symmetrically on two supports 5.00 m apart (Fig. 11–23). A boy weighing 600 N starts at point *A* and walks toward the right. a) In the same diagram construct two graphs showing the upward forces F_A and F_B exerted on the beam at points *A* and *B*, as functions of the coordinate *x* of the boy. Let 1 cm = 100 N vertically, and 1 cm = 1.00 m horizontally. b) From your diagram, how far beyond point *B* can the boy walk before the beam tips? c) How far from the right end of the beam should support *B* be placed so that the boy can walk just to the end of the beam without causing it to tip?

FIGURE 11–23 Exercise 11–12.

11–13 Find the tension *T* in each cable and the magnitude and direction of the force exerted on the strut by the pivot in each of the arrangements in Fig. 11–24. In each case let *w* be the weight of the suspended crate full of priceless art objects. The strut is uniform and also has weight *w*.

FIGURE 11–24 Exercise 11–13.

11–14 The horizontal beam in Fig. 11–25 weighs 150 N, and its center of gravity is at its center. Find a) the tension in the cable; b) the horizontal and vertical components of the force exerted on the beam at the wall.

FIGURE 11–25 Exercise 11–14.

11–15 A door 1.00 m wide and 2.00 m high weighs 280 N and is supported by two hinges, one 0.50 m from the top and the other 0.50 m from the bottom. Each hinge supports half the total weight of the door. Assuming that the door's center of gravity is at its center, find the horizontal components of force exerted on the door by each hinge.

11–16 The boom in Fig. 11–26 weighs 2600 N. The boom is attached with a frictionless pivot at its lower end. It is not uniform; the distance of its center of gravity from the pivot is 35% of its length. a) Find the tension in the guy wire and the horizontal and vertical components of the force exerted on the boom at its lower end. b) Does the line of action of the force exerted on the boom at its lower end lie along the boom?

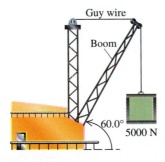

FIGURE 11–26 Exercise 11–16.

11–17 A 3.00-m-long, 240-N uniform rod at the zoo is held in a horizontal position by two ropes at its ends (Fig. 11–27). The left rope makes an angle of 150° with the rod and the right rope

FIGURE 11–27 Exercise 11–17.

makes an angle θ with the horizontal. A 90-N howler monkey (*Alouatta seniculus*) hangs motionless 0.50 m from the right end of the rod as he carefully studies you. Calculate the tensions in the two ropes and the angle θ.

11–18 A non-uniform beam 4.50 m long and weighing 1.00 kN makes an angle of 25.0° with the horizontal. It is held in position by a frictionless pivot at its upper right end and by a cable 3.00 m farther down the beam and perpendicular to it. The center of gravity of the beam is 2.00 m down the beam from the pivot. Lighting equipment exerts a 5.00-kN downward force on the lower left end of the beam. Find the tension T in the cable and the horizontal and vertical components of the force exerted on the beam by the pivot.

11–19 A Couple. Two forces equal in magnitude and opposite in direction acting on an object at two different points form what is called a *couple*. Two antiparallel forces with equal magnitudes $F_1 = F_2 = 8.00$ N are applied to a rod as shown in Fig. 11–28. a) What should be the distance l between the forces if they are to provide a net torque of 6.40 N · m about the left-hand end of the rod? b) Is the sense of this torque clockwise or counterclockwise? c) Repeat parts (a) and (b) for a pivot at the point on the rod where $\vec{F_2}$ is applied.

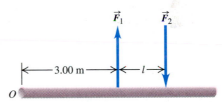

FIGURE 11–28 Exercise 11–19.

SECTION 11–5 STRESS, STRAIN, AND ELASTIC MODULI

11–20 A relaxed biceps muscle requires a force of 25.0 N for an elongation of 3.0 cm; the same muscle under maximum tension requires a force of 500 N for the same elongation. Find Young's modulus for the muscle tissue under each of these conditions if the muscle is assumed to be a uniform cylinder with length 0.200 m and cross-section area 50.0 cm^2.

11–21 A circular steel wire 2.00 m long is to stretch no more than 0.25 cm when a tensile force of 400 N is applied to each end of the wire. What minimum diameter is required for the wire?

11–22 Two round rods, one steel and the other copper, are joined end to end. Each rod is 0.750 m long and 1.50 cm in diameter. The combination is subjected to a tensile force with magnitude 4000 N. For each rod, what is a) the strain? b) the elongation?

11–23 A metal rod that is 4.00 m long and 0.50 cm^2 in cross-section area is found to stretch 0.20 cm under a tension of 5000 N. What is Young's modulus for this metal?

11–24 Stress on a Mountaineer's Rope. A nylon rope used by mountaineers elongates 1.10 m under the weight of an 65.0-kg climber. If the rope is 45.0 m in length and 7.0 mm in diameter, what is Young's modulus for this material?

11–25 In constructing a large mobile, an artist hangs an aluminum sphere of mass 6.0 kg from a vertical steel wire 0.50 m long and 2.5×10^{-3} cm^2 in cross-section area. On the bottom of the sphere he attaches a similar steel wire, from which he hangs a brass cube of mass 10.0 kg. For each wire, compute a) the tensile strain; b) the elongation.

11–26 A vertical solid steel post 25 cm in diameter and 2.50 m long is required to support a load of 8000 kg. The weight of the post can be neglected. What is a) the stress in the post? b) the strain in the post? c) the change in post's length when the load is applied?

SECTION 11–6 BULK STRESS AND STRAIN

11–27 Outside a house 1.0 km from ground zero of a 100-kiloton nuclear bomb explosion, the pressure will rapidly rise to as high as 2.8 atm while the pressure inside the house remains 1.0 atm. If the area of the front of the house is 50 m^2, what is the resulting net force exerted by the air on the front of the house?

11–28 A solid gold bar is pulled up from the hold of the sunken RMS *Titanic*. a) What happens to its volume as it goes from the pressure at the ship to the lower pressure at the ocean's surface? b) The pressure difference is proportional to the depth. How many times greater would the volume change have been had the ship been twice as deep? c) The bulk modulus of lead is one-fourth that of gold. Find the ratio of the volume change of a solid lead bar to that of an equal volume gold bar for the same pressure change.

11–29 A petite young woman distributes her 500 N weight equally over the heels of her high-heel shoes. Each heel has an area of 0.750 cm^2. a) What is the pressure exerted on the floor by each heel? b) With the same pressure, how much weight could be supported by two flat-bottomed sandals, each of area 200 cm^2?

11–30 In the Challenger Deep of the Marianas Trench the depth of seawater is 10.9 km and the pressure is 1.16×10^8 Pa (about 1.15×10^3 atm). a) If a cubic meter of water is taken from the surface to this depth, what is the change in its volume? (Normal atmospheric pressure is about 1.0×10^5 Pa. Assume that k for seawater is the same as the freshwater value given in Table 11–2.) b) What is the density of seawater at this depth? (At the surface, seawater has a density of 1.03×10^3 kg/m^3.)

11–31 A specimen of oil having an initial volume of 600 cm^3 is subjected to a pressure increase of 3.6×10^6 Pa, and the

volume is found to decrease by 0.45 cm^3. What is the bulk modulus of the material? The compressibility?

SECTION 11–7 SHEAR STRESS AND STRAIN

11–32 Suppose the object in Fig. 11–17 is a square steel plate, 10.0 cm on a side and 0.500 cm thick. a) Find the shear strain that results if a force of magnitude 9.0×10^5 N is applied to each of the four sides as shown in Fig. 11–17a. b) Find the deformation angle ϕ in degrees and the displacement x in centimeters (see Fig. 11–17b).

11–33 Shear forces are applied to a rectangular solid. The same forces are applied to another rectangular solid of the same material, but with three times each edge length. In each case the forces are small enough that Hooke's law is obeyed. What is the ratio of the shear strain for the larger object to that of the smaller object?

11–34 Two strips of brass, each 3.00 cm wide and 45.0 cm long, are placed with their ends overlapping by 1.00 cm. The overlapping ends are then riveted together by four rivets, each with diameter 0.250 cm. Tests show that when a tension of at least 1.20×10^4 N is applied to the ends of the riveted strip, the rivets fail and shear apart. What is the shearing stress on each rivet when it fails? Assume that each rivet carries one quarter of the load.

SECTION 11–8 ELASTICITY AND PLASTICITY

11–35 In a materials testing laboratory, a metal wire made from a new alloy is found to break when a tensile force of 90.8 N is applied perpendicular to each end. If the diameter of the wire is 1.84 mm, what is the breaking stress of the alloy?

11–36 A 4.0-m-long steel wire has a cross-section area of 0.050 cm^2. Its proportional limit has a value of 0.0016 times its Young's modulus (Table 11–1). Its breaking stress has a value of 0.0065 times its Young's modulus. The wire is fastened at its upper end and hangs vertically. a) How great a weight can be hung from the wire without exceeding the proportional limit? b) How much will the wire stretch under this load? c) What is the maximum weight that can be supported?

11–37 A steel cable with cross-section area 3.00 cm^2 has an elastic limit of 2.40×10^8 Pa. Find the maximum upward acceleration that can be given a 1200-kg elevator supported by the cable if the stress is not to exceed one-third of the elastic limit.

11–38 A brass wire is to withstand a tensile force of 350 N without breaking. What minimum diameter must the wire have?

PROBLEMS

11–39 Center of Gravity of a Car. For an automobile with weight w the front wheels support a fraction f of the weight, so the normal force exerted at the front wheels is fw, and at the rear wheels the normal force is $(1 - f)w$. The distance between the front and real axles is d. a) Show that the center of gravity of the car is a distance fd in front of the rear wheels. b) Show that the general result derived in part (a) reproduces the numerical answer of Example 11–2 (Section 11–4).

11–40 Sir Lancelot rides slowly out of the castle at Camelot and onto the 12.0-m-long drawbridge that passes over the moat (Fig. 11–29). Unbeknownst to him, his enemies have partially severed the vertical cable holding up the front end of the bridge so that it will break under a tension of 5.80×10^3 N. The bridge has mass 200 kg and its center of gravity is at its center. Lancelot, his lance, his armor, and his horse together have a combined mass of 600 kg. Will the cable break before

Lancelot reaches the end of the drawbridge? If so, how far from the castle end of the bridge will the center of gravity of the horse plus rider be when the cable breaks?

FIGURE 11–29 Problem 11–40.

11–41 Three vertical forces act on an airplane when it is flying at a constant altitude and with a constant velocity. These are the weight of the airplane, an aerodynamic force on the wing of the airplane, and an aerodynamic force on the airplane's horizontal tail. (The aerodynamic forces are exerted by the surrounding air, and are reactions to the forces that the wing and tail exert on the air as the airplane flies through it.) For a particular light airplane with a weight of 6700 N, the center of gravity is 0.30 m in front of the point where the wing's vertical aerodynamic force acts and 3.66 m in front of the point where the tail's vertical aerodynamic force acts. Determine the magnitude and direction (upward or downward) of each of the two vertical aerodynamic forces.

11–42 A pickup truck has a wheelbase of 3.00 m. Ordinarily, 10,780 N rests on the front wheels and 8820 N on the rear wheels when the truck is parked on a level road. a) A box weighing 3600 N is now placed on the tailgate, 1.00 m behind the rear axle. How much total weight now rests on the front wheels? On the rear wheels? b) How much weight would need to be placed on the tailgate to make the front wheels come off the ground?

11–43 A uniform seesaw is 4.00 m long and weighs 240 N. The seesaw is pivoted about a point on its bottom surface and the location of this pivot along the length of the seesaw can be adjusted as far as 0.20 m from the center of the seesaw. Little Karen, who weighs 420 N, has her center of gravity over the right end, with the seesaw set so she exerts the most torque about the pivot. The seesaw balances horizontally when Elwood has his center of gravity 0.15 m from the left end. Both have their feet off the ground. a) What does Elwood weigh? b) Is this equilibrium stable, unstable, or neutral? (*Hint:* See Question Q11–5.)

11–44 A piece of steel will not balance with the tip of one end just barely resting on the edge of a table (point O in Fig. 11–30). But suppose we fasten on two other pieces as shown in Fig. 11–30. All pieces have the same square cross section and density. Neglect the weight of the fasteners. The system can now balance if we find the right length l of the third piece. a) What is the length l for which the system will balance with the first and third pieces horizontal, as shown in the figure?

b) How would your answer change if the pieces were made of copper rather than steel? Explain.

FIGURE 11–30 Problem 11–44.

11–45 Several external forces $\vec{F}_i$ are applied to a rigid body. With respect to an origin O, force $\vec{F}_1$ is applied at the point located at $\vec{r}_1$, force $\vec{F}_2$ at $\vec{r}_2$, and so on. Point P is located at $\vec{r}_P$. If the body is in translational equilibrium, so that $\Sigma \vec{F}_i = 0$, prove that the sum of the torques about point P equals the sum of torques about point O. (A special case of this result is when $\Sigma \vec{\tau}_O = 0$; for a body in translational equilibrium, if the resultant torque about one point is zero, the resultant torque about any point is zero.)

11–46 A claw hammer is used to pull a nail out of a board (Fig. 11–31). The nail is at an angle of 60° to the board, and a force $\vec{F}_1$ of magnitude 500 N applied to the nail is required to pull it from the board. The hammer head contacts the board at point A, which is 0.080 m from where the nail enters the board. A horizontal force $\vec{F}_2$ is applied to the hammer handle at a distance of 0.300 m above the board. What magnitude of force $\vec{F}_2$ is required to apply the required 500-N force (F_1) to the nail? (Neglect the weight of the hammer.)

FIGURE 11–31 Problem 11–46.

11–47 End A of the bar AB in Fig. 11–32 rests on a frictionless horizontal surface, and end B is hinged. A horizontal force $\vec{F}$ of magnitude 120 N is exerted on end A. Neglect the weight of the

FIGURE 11–32 Problem 11–47.

bar. What are the horizontal and vertical components of the force exerted by the bar on the hinge at B?

11–48 A single additional force is to be applied to the bar in Fig. 11–33 to maintain it in equilibrium in the position shown. The weight of the bar can be neglected. a) What are the horizontal and vertical components of the required force? b) What is the angle the force must make with the bar? c) What is the magnitude of the required force? d) Where should the force be applied?

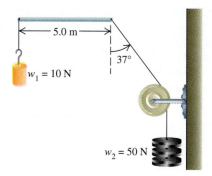

FIGURE 11–33 Problem 11–48.

11–49 Your dog Nikita is 0.90 m long (nose to hindlegs). Her forelegs are 0.18 m horizontally from her nose, her center of gravity is 0.28 m horizontally from her hindlegs, and she weighs 190 N. a) How much force does a level floor exert on each of Nikita's forefeet, and on each of her hindfeet? b) If Nikita picks up a 25-N bone and holds it in her mouth (directly under her nose), what is the force exerted by the floor on each of her forefeet and each of her hindfeet?

11–50 A Truck on a Drawbridge. A loaded cement mixer drives onto an old drawbridge, where it stalls with its center of gravity three-quarters of the way across the span. The truck driver radios for help, sets the hand-brake, and waits. Meanwhile, a boat approaches, so the drawbridge is raised by means of a cable attached to its end opposite the hinge (Fig. 11–34). The drawbridge span is 40.0 m long and has a mass of 12,000 kg; its center of gravity is at its midpoint. The cement mixer, with driver, has a mass of 30,000 kg. When the drawbridge has been raised to an angle of 30° above the horizontal, the cable makes an angle of 70° with the surface of the bridge. a) What is the tension T in the cable when the drawbridge is held in this position? b) What are the horizontal and vertical components of the force the hinge exerts on the span?

FIGURE 11–34 Problem 11–50.

11–51 Couples. a) A force $\vec{F}_1$ is applied to a rod at a distance x from some point P on the rod. A second force $\vec{F}_2$ is applied at a distance $x + l$ from point P. The forces are equal in magnitude and opposite in direction and hence form a couple (Exercise 11–19). The angle between the rod and the direction of each force is ϕ. Derive an expression for the net torque about P produced by these two forces and show that it is independent of x. This shows that a couple produces the same torque about any pivot point. b) Two forces with magnitude $F_1 = F_2 = 14.0$ N are applied to a rod as shown in Fig. 11–35, with $\phi = 37°$. Calculate the net torque about point O due to these two forces by calculating the torque due to each separate force. Calculate the net torque about point P due to these two forces by calculating the torque due to each separate force. Compare your results to the general result derived in part (a).

FIGURE 11–35 Problem 11–51.

11–52 A non-uniform fire escape ladder is 6.0 m long when extended to the icy alley below. It is held at the top by a frictionless pivot, and there is negligible frictional force from the icy surface at the bottom. The ladder weighs 250 N, and its center of gravity is 2.0 m along the ladder from its bottom. A mother and child of total weight 750 N are on the ladder 1.5 m from the pivot. The ladder makes an angle θ with the horizontal. Find the magnitude and direction of a) the force exerted by the icy alley on the ladder and b) the force exerted by the ladder on the pivot.

11–53 A uniform strut of mass m makes an angle θ with the horizontal. It is supported by a frictionless pivot located at one-third its length from its lower left end and a horizontal rope at its upper right end. A cable and package of total weight w hang from its upper right end. a) Find the vertical and horizontal components V and H of the pivot's force on the strut as well as the tension T in the rope. b) If the maximum safe tension in the rope is 700 N and the mass of the strut is 20.0 kg, find the maximum safe weight of the cable and package when the strut makes an angle of 55.0° with the horizontal. c) For what angle θ can no weight be safely suspended from the right end of the strut?

11–54 a) In Fig. 11–36 a 6.00-m-long uniform beam is hanging from a point 1.00 m to the right of its center. The beam weighs 140 N and makes an angle of 30.0° with the vertical. At the right-hand end of the beam a 100-N weight is hung; an unknown weight w hangs at the other end. If the system is in equilibrium, what is w? Ignore the thickness of the beam. b) If the beam makes instead an angle of 45.0° with the vertical, what is w?

FIGURE 11–36 Problem 11–54.

11–55 A uniform horizontal flagpole 5.00 m in length and with a weight of 200 N is hinged to a vertical wall at one end. A 600-N stuntwoman hangs from its other end. The flagpole is supported by a guy wire running from its outer end to a point on the wall directly above the pole. a) If the tension in this wire is not to exceed 1000 N, what is the minimum height above the pole at which it may be fastened to the wall? b) If the flagpole remains horizontal, by how many newtons would the tension be increased if the wire were fastened 0.50 m below this point?

11–56 A holiday decoration consists of two shiny glass spheres with masses 0.0240 kg and 0.0360 kg suspended as shown in Fig. 11–37 from a uniform rod with mass 0.120 kg and length 1.00 m. The rod is suspended from the ceiling by a vertical cord at each end, so that it is horizontal. Calculate the tension in each of the cords A through F.

FIGURE 11–37 Problem 11–56.

11–57 The sign for Lew's Bait Shop and Sushi Restaurant is a uniform rectangle, 1.50 m wide and 0.800 m tall, weighing 300 N. It is supported by a wire at its upper-right corner and a pivot at its lower-left corner. a) Late one night Lew's competitor from across the street cuts the wire. The sign stays in place because the pivot has rusted tight. After the wire has been cut, what are the magnitude and direction of the force and of the torque exerted by the pivot on the sign? b) When the sign was brand new, the pivot was frictionless and the wire was at the angle that minimized the tension in it. What was the tension in the wire and the magnitude and direction of the force exerted on the sign by the pivot?

11–58 When you stretch a wire, rope, or rubber band, it gets thinner as well as longer. When Hooke's law holds, the fractional decrease in width is proportional to the tensile strain. If w_0 is the original width and Δw is the change in width, then $\Delta w/w_0 = -\sigma \Delta l/l_0$, where the minus sign reminds us that width decreases when length increases. The dimensionless constant, σ, different for different materials, is called *Poisson's ratio.* a) If the steel cable of Example 11–6 (Section 11–5) has a circular cross section and a Poisson's ratio of 0.23, what is its change in diameter when the milling machine is hung from it? b) A cylinder made of nickel (Poisson's ratio = 0.42) has radius 2.0 cm. What tensile force $F_\perp$ must be applied perpendicular to each end of the cylinder to cause its radius to decrease by 0.10 mm? Assume that the breaking stress and proportional limit for the metal are extremely large and are not exceeded.

11–59 A circular disk 0.500 m in diameter, pivoted about a horizontal axis through its center, has a cord wrapped around its rim. The cord passes over a frictionless pulley P and is attached to an object that weighs 240 N. A uniform rod 2.00 m long is fastened to the disk, with one end at the center of the disk. The apparatus is in equilibrium, with the rod horizontal (Fig. 11–38). a) What is the weight of the rod? b) What is the new equilibrium direction of the rod when a second object weighing 20.0 N is suspended from the other end of the rod, as shown by the broken line? That is, what angle does the rod then make with the horizontal?

FIGURE 11–38 Problem 11–59.

11–60 One end of a uniform meter stick is placed against a vertical wall (Fig. 11–39). The other end is held by a lightweight cord making an angle θ with the stick. The coefficient of static friction between the end of the meter stick and the wall is 0.40. a) What is the maximum value the angle θ can have if the stick is to remain in equilibrium? b) Let the angle θ be 15°. A block of the same weight as the meter stick is suspended from the stick as

FIGURE 11–39 Problem 11–60.

shown, at a distance x from the wall. What is the minimum value of x for which the stick will remain in equilibrium? c) When $\theta = 15°$, how large must the coefficient of static friction be so that the block can be attached 10 cm from the left end of the stick without causing it to slip?

11–61 Two friends are carrying a 200-kg crate up a flight of stairs. The crate is 1.25 m long and 0.50 m high, and its center of gravity is at its center. The stairs make a 45.0° angle with respect to the floor. The crate also is carried at a 45.0° angle, so that its bottom side is parallel to the slope of the stairs (Fig. 11–40). If the force each person applies is vertical, what is the magnitude of each of these forces? Is it best to be the person above or below on the stairs?

FIGURE 11–40 Problem 11–61.

11–62 You are trying to raise a bicycle wheel of mass m and radius R up over a curb of height h. To do this, you apply a horizontal force $\vec{F}$. What is the least magnitude of the force $\vec{F}$ that

FIGURE 11–41 Problem 11–62.

will succeed in raising the wheel onto the curb when the force is applied a) at the center of the wheel (Fig. 11–41)? b) at the top of the wheel? c) In which case is less force required?

11–63 The Farmyard Gate. A gate 4.00 m wide and 2.00 m high weighs 500 N. Its center of gravity is at its center, and it is hinged at A and B. To relieve the strain on the top hinge, a wire CD is connected as shown in Fig. 11–42. The tension in CD is increased until the horizontal force at hinge A is zero. a) What is the tension in the wire CD? b) What is the magnitude of the horizontal component of the force at hinge B? c) What is the combined vertical force exerted by hinges A and B?

FIGURE 11–42 Problem 11–63.

11–64 If you put a uniform block at the edge of a table, the center of the block must be over the table for the block not to fall off. a) If you stack two identical blocks at the table edge, the center of the top block must be over the bottom block, and the center of gravity of the two blocks together must be over the table. In terms of the length L of each block, what is the maximum overhang possible (Fig. 11–43)? b) Repeat part (a) for three identical blocks and for four identical blocks. c) Is it possible to make a stack of blocks such that the uppermost block is not directly over the table at all? How many blocks would it take to do this? (Try this with your friends using copies of this book.)

FIGURE 11–43 Problem 11–64.

11–65 An engineer is designing a conveyor system for loading hay bales into a wagon (Fig. 11–44). Each bale is 0.25 m wide, 0.50 m high, and 0.80 m long (the dimension perpendicular to the plane of the figure), and has a mass of 30.0 kg. The center

FIGURE 11-44 Problem 11-65.

of gravity of each bale is at its geometrical center. The coefficient of static friction between a bale and the conveyor belt is 0.60, and the belt moves with constant speed. a) The angle β of the conveyor is slowly increased. At some critical angle a bale will tip (if it doesn't slip first), and at some different critical angle it will slip (if it doesn't tip first). Find the two critical angles, and determine which happens at the smaller angle. b) Would the outcome of part (a) be different if the coefficient of friction were 0.40?

11-66 The hay bale of Problem 11-65 is dragged along a horizontal surface with constant speed by a force $\vec{F}$ (Fig. 11-45). The coefficient of kinetic friction is 0.35. a) Find the magnitude of the force $\vec{F}$. b) Find the value of h for which the bale just begins to tip.

FIGURE 11-45 Problem 11-66.

11-67 A garage door is mounted on an overhead rail (Fig. 11-46). The wheels at A and B have rusted so that they do not roll, but rather slide along the track. The coefficient of kinetic friction is 0.52. The distance between the wheels is 2.00 m, and each is 0.50 m from the vertical sides of the door. The door is uniform and weighs 950 N. It is pushed to the left at constant speed by a horizontal force $\vec{F}$. a) If the distance h is 1.60 m, what is the vertical component of the force exerted on each wheel by the track? b) Find the maximum value h can have without causing one wheel to leave the track.

FIGURE 11-46 Problem 11-67.

11-68 A horizontal boom is supported at its left end by a frictionless pivot. It is held in place by a cable attached to the right-hand end of the boom. A chain and crate of total weight w hang from somewhere along the boom. The boom's weight w_b cannot be neglected and the boom may or may not be uniform. a) Show that the tension in the cable is the same whether the cable makes an angle θ or an angle $180° - \theta$ with the horizontal, and that the horizontal force component exerted on the boom by the pivot has equal magnitude but opposite direction for the two angles. b) Show that the cable cannot be horizontal. c) Show that the tension in the cable is a minimum when the cable is vertical, pulling upward on the right end of the boom. d) Show that when the cable is vertical, the force exerted by the pivot on the boom is vertical.

11-69 Prior to being placed in its hole, a 5700-N, 9.0-m-long uniform utility pole makes some nonzero angle with the vertical. A vertical cable attached 2.0 m below its upper end holds it in place while its lower end rests on the ground. a) Find the tension in the cable and the magnitude and direction of the force exerted by the ground on the pole. b) Why don't we need the angle the pole makes with the vertical, as long as it is not zero?

11-70 Hooke's law for a tensile stress can be written as $F = kx$, where x is the object's change in length and k is the force constant. a) What is the force constant of a rod of length l_0, cross-section area A, and Young's modulus Y? b) In terms of l_0, A, and Y, how much work is required to stretch the object a distance x?

11-71 A 12.0-kg mass, fastened to the end of an aluminum wire with an unstretched length of 0.50 m, is whirled in a vertical circle with a constant angular speed of 120 rev/min. The cross-section area of the wire is 0.014 cm². Calculate the elongation of the wire when the mass is a) at the lowest point of the path; b) at the highest point of its path.

11-72 A metal wire 3.50 m long and 0.70 mm in diameter was given the following test. A load weighing 20 N was originally hung from the wire to keep it taut. The position of the lower end of the wire was read on a scale as load was added.

Added load (N)	Scale reading (cm)
0	3.02
10	3.07
20	3.12
30	3.17
40	3.22
50	3.27
60	3.32
70	4.27

a) Graph these values, plotting the increase in length horizontally and the added load vertically. b) Calculate the value of Young's modulus. c) The proportional limit occurred at a scale reading of 3.34 cm. What was the stress at this point?

11-73 A 1.05-m-long rod of negligible weight is supported at its ends by wires A and B of equal length (Fig. 11-47). The cross-section area of A is 2.00 mm² and that of B is 4.00 mm². Young's modulus for wire A is 1.80×10^{11} Pa; that for B is 1.20×10^{11} Pa. At what point along the rod should a weight w

be suspended to produce a) equal stresses in A and B? b) equal strains in A and B?

FIGURE 11–47 Problem 11–73.

11–74 An amusement park ride consists of airplane-shaped cars attached to steel rods (Fig. 11–48). Each rod has a length of 15.0 m and a cross-section area of 8.00 cm². a) How much is the rod stretched when the ride is at rest? (Assume that each car plus two people seated in it has a total weight of 1900 N.) b) When operating, the ride has a maximum angular speed of 8.0 rev/min. How much is the rod stretched then?

FIGURE 11–48 Problem 11–74.

11–75 A brass rod with a length of 1.40 m and a cross-section area of 2.00 cm² is fastened end to end to a nickel rod with length L and cross-section area 1.00 cm². The compound rod is subjected to equal and opposite pulls of magnitude 4.00×10^4 N at its ends. a) Find the length L of the nickel rod if the elongations of the two rods are equal. b) What is the stress in each rod? c) What is the strain in each rod?

11–76 Stress on the Shin Bone. Compressive strength of our bones is important in everyday life. Young's modulus for bone is about 1.4×10^{10} Pa. Bone can take only about a 1.0% change in its length before fracturing. a) What is the maximum force that can be applied to a bone whose minimum cross-section area is 3.0 cm²? (This is approximately the cross-section area of a tibia, or shin bone, at its narrowest point.) b) Estimate the maximum

height from which a 70-kg man could jump and not fracture the tibia. Take the time between when he first touches the floor and when he has stopped to be 0.030 s, and assume that the stress is distributed equally between his two legs.

11–77 You hang a floodlamp from the end of a vertical steel wire. The floodlamp stretches the wire 0.18 mm and the stress is proportional to the strain. How much would it have stretched a) if the wire were twice as long? b) if the wire had the same length but twice the diameter? c) for a copper wire of the original length and diameter?

11–78 A moonshiner produces pure ethanol (ethyl alcohol) late at night and stores it in a stainless steel tank in the form of a cylinder 0.300 m in diameter with a tight-fitting piston at the top. The total volume of the tank is 250 L (0.250 m³). In an attempt to squeeze a little more into the tank, the moonshiner piles 1420 kg of lead bricks on top of the piston. What additional volume of ethanol can the moonshiner squeeze into the tank? (Assume that the wall of the tank is perfectly rigid.)

11–79 A bar with cross-section area A is subjected to equal and opposite tensile forces $\vec{F}$ at its ends. Consider a plane through the bar making an angle θ with a plane at right angles to the bar (Fig. 11–49). a) What is the tensile (normal) stress at this plane in terms of F, A, and θ? b) What is the shear (tangential) stress at the plane in terms of F, A, and θ? c) For what value of θ is the tensile stress a maximum? d) For what value of θ is the shear stress a maximum?

FIGURE 11–49 Problem 11–79.

11–80 A horizontal uniform copper rod has an original length l_0, a cross-section area A, a Young's modulus Y, and a mass m. It is supported by a frictionless pivot at its right end and by a cable at its left end. Both pivot and cable are attached so that they exert their forces uniformly over the rod's cross section. The cable makes an angle θ with the rod and compresses it. a) Find the stress exerted by the cable and pivot on the rod. b) Find the change in length of the rod due to this stress. c) The mass of the rod equals $\rho A l_0$, where ρ is the density. Show that the answers to parts (a) and (b) are independent of the cross-section area of the rod. d) The density of copper is 8900 kg/m³. Take Y for compression as given for copper in Table 11–1. Find the stress and change in length for an original length of 1.8 m and an angle of 30°. e) By how much would you multiply the answers of part (d) if the rod were twice as long?

CHALLENGE PROBLEMS

11–81 A bookcase weighing 1500 N rests on a horizontal surface for which the coefficient of static friction is $\mu_s = 0.40$. The bookcase is 1.80 m tall and 2.00 m wide; its center of gravity is

at its geometrical center. The bookcase rests on four short legs that are each 0.10 m from the edge of the bookcase. A person pulls on a rope attached to an upper corner of the bookcase with

FIGURE 11–50 Challenge Problem 11–81.

a force $\vec{F}$ that makes an angle θ with the bookcase (Fig. 11–50).
a) If $\theta = 90°$, so $\vec{F}$ is horizontal, show that as F is increased from
zero, the bookcase will start to slide before it tips, and calculate
the magnitude of $\vec{F}$ that will start the bookcase sliding. b) If
$\theta = 0°$, so $\vec{F}$ is vertical, show that the bookcase will tip over
rather than slide, and calculate the magnitude of $\vec{F}$ that will
cause the bookcase to start to tip. c) Calculate as a function of θ
the magnitude of $\vec{F}$ that will cause the bookcase to start to slide
and the magnitude that will cause it to start to tip. What is the
smallest value that θ can have so that the bookcase will still start
to slide before it starts to tip?

11–82 Knocking Over a Post. One end of a post weighing
400 N and with height h rests on a rough horizontal surface with
$\mu_s = 0.30$. The upper end is held by a rope fastened to the surface
and making an angle of 36.9° with the post (Fig. 11–51). A horizontal force $\vec{F}$ is exerted on the post as shown. a) If the force $\vec{F}$ is
applied at the midpoint of the post, what is the largest value it
can have without causing the post to slip? b) How large can the
force be without causing the post to slip if its point of application is six-tenths of the way from the ground to the top of the
post? c) Show that if the point of application of the force is too
high, the post cannot be made to slip, no matter how great the
force. Find the critical height for the point of application.

FIGURE 11–51 Challenge Problem 11–82.

11–83 Minimizing the Tension. A heavy horizontal girder of
length L has several objects suspended from it. It is supported by
a frictionless pivot at its left end and a cable of negligible weight
that is attached to an I-beam at a point a distance h directly
above the girder's center. Where should the other end of the
cable be attached to the girder so that the cable's tension is a

minimum? (*Hint:* In evaluating and presenting your answer,
don't forget that the maximum distance of the point of attachment from the pivot is the length L of the beam.)

11–84 Two ladders, 4.00 m and 3.00 m long, are hinged at
point A and tied together by a horizontal rope 0.90 m above the
floor (Fig. 11–52). The ladders weigh 480 N and 360 N, respectively, and the center of gravity of each is at its center. Assume
that the floor is freshly waxed and frictionless. a) Find the
upward force at the bottom of each ladder. b) Find the tension
in the rope. c) Find the magnitude of the force one ladder exerts
on the other at point A. d) If an 800-N painter stands at point A,
find the tension in the horizontal rope.

FIGURE 11–52 Challenge Problem 11–84.

11–85 The compressibility of sodium is to be measured by
observing the displacement of the piston in Fig. 11–14 when a
force is applied. The sodium is immersed in an oil that fills the
cylinder below the piston. Assume that the piston and walls of
the cylinder are perfectly rigid and that there is no friction and
no oil leak. Compute the compressibility of the sodium in terms
of the applied force F, the piston displacement x, the piston
area A, the initial volume of the oil V_O, the initial volume of
the sodium V_S, and the compressibility of the oil k_O.

11–86 Bulk Modulus of an Ideal Gas. The equation of state
(the equation relating pressure, volume, and temperature) for an
ideal gas is $pV = nRT$, where n and R are constants. a) Show that
if the gas is compressed while the temperature T is held constant, the bulk modulus is equal to the pressure. b) When an
ideal gas is compressed without the transfer of any heat into or
out of it, the pressure and volume are related by pV^γ = constant,
where γ is a constant having different values for different gases.
Show that in this case the bulk modulus is given by $B = \gamma p$.

11–87 An angler hangs a 4.50-kg fish from a vertical steel
wire 1.50 m long and 5.00×10^{-3} cm^2 in cross-section area.
The upper end of the wire is securely fastened to a support.
a) Calculate the amount the wire is stretched by the hanging
fish. The angler now applies a force $\vec{F}$ to the fish, pulling it very
slowly downward by 0.500 mm from its equilibrium position.
For this downward motion, calculate b) the work done by gravity; c) the work done by the force $\vec{F}$; d) the work done by the
force the wire exerts on the fish; and e) the change in the elastic
potential energy (the potential energy associated with the tensile
stress in the wire). Compare the answers in parts (d) and (e).
(*Hint:* See Problem 11–70.)

The great galaxy M31 in the constellation Andromeda is more than 10^{18} km (100,000 light years) across. It lies some 2.9×10^6 light years from earth. All the matter in this galaxy—stars, glowing gas, interstellar dust, and other material not visible in this image—is kept in orbit around the galaxy's center by its mutual gravitational attraction.

12

Gravitation

12–1 INTRODUCTION

Why are planets, moons, and the sun all nearly spherical? Why do some earth satellites circle the earth in 90 minutes, while the moon takes 27 days for the trip? And why don't satellites fall back to earth? The study of gravitation provides the answers for these and many related questions.

As we remarked in Chapter 5, gravitation is one of the four classes of interactions found in nature, and it was the earliest of the four to be studied extensively. Newton discovered in the seventeenth century that the same interaction that makes an apple fall out of a tree also keeps the planets in their orbits around the sun. This was the beginning of *celestial mechanics,* the study of the dynamics of objects in space. Today, our knowledge of celestial mechanics allows us to determine how to put a satellite into any desired orbit around the earth or to choose just the right trajectory to send a spacecraft to another planet.

In this chapter you will learn the basic law that governs gravitational interactions. This law is *universal:* gravity acts in the same fundamental way between the earth and your body, between the sun and a planet, and between a planet and one of its moons. We'll apply the law of gravitation to phenomena such as the variation of weight with altitude, the orbits of satellites around the earth, and the orbits of planets around the sun.

12–2 NEWTON'S LAW OF GRAVITATION

The example of gravitational attraction that's probably most familiar to you is your *weight,* the force that attracts you toward the earth. During his study of the motions of the planets and of the moon, Newton discovered the fundamental character of the gravitational attraction between *any* two bodies. Along with his three laws of motion, Newton published the **law of gravitation** in 1687. It may be stated as follows:

Every particle of matter in the universe attracts every other particle with a force that is directly proportional to the product of the masses of the particles and inversely proportional to the square of the distance between them.

Translating this into an equation, we have

$$F_g = \frac{Gm_1m_2}{r^2} \qquad \text{(law of gravitation)}, \qquad (12\text{--}1)$$

where F_g is the magnitude of the gravitational force on either particle, m_1 and m_2 are their masses, r is the distance between them (Fig. 12–1), and G is a fundamental physical constant called the **gravitational constant.** The numerical value of G depends on the system of units used.

CAUTION ▶ Because the symbols g and G are almost the same, it's common to confuse the two gravitational quantities that these symbols represent. Lowercase g is the acceleration due to gravity, which relates the weight w of a body to its mass m: $w = mg$. The value of g is different at different locations on the earth's surface and on the surfaces of different planets. By contrast, capital G relates the gravitational force between any two bodies to their masses and the distance between them. We call G a *universal* constant because it has the same value for any two bodies, no matter where in space they are located. In the next section we'll see how the values of g and G are related, but remember that they're two very different quantities! ◄

Gravitational forces always act along the line joining the two particles, and they form an action-reaction pair. Even when the masses of the particles are different, the two interaction forces have equal magnitude. The attractive force that your body exerts on the earth has the same magnitude as the force that it exerts on you. When you fall from a diving board into a swimming pool, the entire earth rises up to meet you! (Why don't you notice this? The earth's mass is greater than yours by a factor of about 10^{23}, so its acceleration is only 10^{-23} as great as yours.)

We have stated the law of gravitation in terms of the interaction between two *particles*. It turns out that the gravitational interaction of any two bodies having *spherically symmetric* mass distributions (such as solid spheres or spherical shells) is the same as though we concentrated all the mass of each at its center, as in Fig. 12–2. Thus if we model the earth as a spherically symmetric body with mass m_E, the force it exerts on a particle or a spherically symmetric body with mass m, at a distance r between centers, is

$$F_g = \frac{Gm_E m}{r^2}, \qquad (12\text{–}2)$$

provided that the body lies outside the earth. A force of the same magnitude is exerted *on* the earth by the body. (We will prove these statements in Section 12–7.)

At points *inside* the earth the situation is different. If we could drill a hole to the center of the earth and measure the gravitational force on a body at various depths, we would find that toward the center of the earth the force *decreases*, rather than increasing as $1/r^2$. As the body enters the interior of the earth (or other spherical body), some of the earth's mass is on the side of the body opposite from the center and pulls in the opposite direction. Exactly at the center, the earth's gravitational force on the body is zero.

Spherically symmetric bodies are an important case because moons, planets, and stars all tend to be spherical (Fig. 12–3a). Since all particles in a body gravitationally attract each other, the particles tend to move to minimize the distance between them. As a result, the body naturally tends to assume a spherical shape, just as a lump of clay forms into a sphere if you squeeze it with equal forces on all sides. This effect is greatly reduced in celestial bodies of low mass since the gravitational attraction is less, and these bodies tend *not* to be spherical (Fig. 12–3b).

12–1 Two particles, separated by a distance r, exert attractive gravitational forces on each other. The forces on each particle are of equal magnitude even if their masses are quite different.

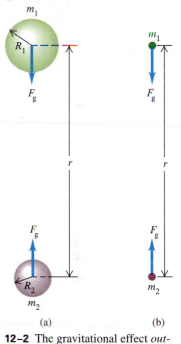

(a)   (b)

12–2 The gravitational effect *outside* any spherically symmetric mass distribution is the same as though all the mass of the sphere were concentrated at its center.

12–3 (left) The sun's mass of 1.99×10^{30} kg is great enough that gravitation pulls it into a nearly spherical shape of radius 696,000 km. (right) The asteroid Gaspra is an irregularly shaped body about 20 km long. The mass of Gaspra is only about 10^{16} kg, and the gravitational attraction between its parts isn't great enough to overcome the interatomic forces that maintain Gaspra's solid shape.

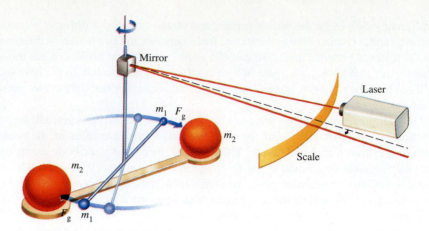

12–4 The principle of the Cavendish balance, used for determining the value of G. The angle of deflection has been exaggerated for clarity.

DETERMINING THE VALUE OF G

To determine the value of the gravitational constant G, we have to *measure* the gravitational force between two bodies of known masses m_1 and m_2 at a known distance r. The force is extememly small for bodies that are small enough to be brought into the laboratory, but it can be measured with an instrument called a *torsion balance*, which Sir Henry Cavendish used in 1798 to determine G.

A modern version of the Cavendish torsion balance is shown in Fig. 12–4. A light, rigid rod shaped like an inverted T is supported by a very thin vertical quartz fiber. Two small spheres, each of mass m_1, are mounted at the ends of the horizontal arms of the T. When we bring two large spheres, each of mass m_2, to the positions shown, the attractive gravitational forces twist the T through a small angle. To measure this angle, we shine a beam of light on a mirror fastened to the T. The reflected beam strikes a scale, and as the T twists, the reflected beam moves along the scale.

After calibrating the Cavendish balance, we can measure gravitational forces and thus determine G. The presently accepted value (in SI units) is

$$G = 6.67259(85) \times 10^{-11} \text{ N} \cdot \text{m}^2/\text{kg}^2.$$

To three significant figures, $G = 6.67 \times 10^{-11}$ N $\cdot$ m^2/kg^2. Because 1 N = 1 kg $\cdot$ m/s^2, the units of G can also be expressed (in fundamental SI units) as m^3/(kg $\cdot$ s^2).

Gravitational forces combine vectorially. If each of two masses exerts a force on a third, the *total* force on the third mass is the vector sum of the individual forces of the first two. Example 12–3 makes use of this property, which is often called *superposition of forces*.

EXAMPLE 12–1

Calculating gravitational force The mass m_1 of one of the small spheres of a Cavendish balance is 0.0100 kg, the mass m_2 of one of the large spheres is 0.500 kg, and the center-to-center distance between each large sphere and the nearer small one is 0.0500 m. Find the gravitational force F_g on each sphere due to the nearest other sphere.

SOLUTION The magnitude of each force is

$$F_g = \frac{(6.67 \times 10^{-11} \text{ N} \cdot \text{m}^2/\text{kg}^2)(0.0100 \text{ kg})(0.500 \text{ kg})}{(0.0500 \text{ m})^2}$$

$$= 1.33 \times 10^{-10} \text{ N}.$$

This is a very small force. *Reminder:* The two bodies experience the *same* magnitude of force, even though their masses are very different.

EXAMPLE 12-2

Acceleration due to gravitational attraction Suppose one large sphere and one small sphere are detached from the apparatus in Example 12–1 and placed 0.0500 m (between centers) from each other at a point in space far removed from all other bodies. What is the magnitude of the acceleration of each, relative to an inertial system?

SOLUTION The force on each sphere has the same magnitude that we found in Example 12–1. The acceleration a_1 of the smaller sphere has magnitude

$$a_1 = \frac{F_g}{m_1} = \frac{1.33 \times 10^{-10} \text{ N}}{0.0100 \text{ kg}} = 1.33 \times 10^{-8} \text{ m/s}^2.$$

The acceleration a_2 of the larger sphere has magnitude

$$a_2 = \frac{F_g}{m_2} = \frac{1.33 \times 10^{-10} \text{ N}}{0.500 \text{ kg}} = 2.66 \times 10^{-10} \text{ m/s}^2.$$

Although the forces on the bodies are equal in magnitude, the magnitudes of the two accelerations are *not* equal. Also, the accelerations are not constant; the gravitational forces increase as the spheres start to move toward each other.

EXAMPLE 12-3

Superposition of gravitational forces Three spheres are arranged at the vertices of a 45° right triangle (Fig. 12–5). Find the magnitude and direction of the total gravitational force exerted on the small sphere by both large ones.

SOLUTION We use the principle of superposition: the total force on the small sphere is the vector sum of the two forces due to each large sphere. We can compute the vector sum by using components, but first we find the magnitudes of the forces. The magnitude F_1 of the force on the small mass due to the upper large one is

$$F_1 = \frac{(6.67 \times 10^{-11} \text{ N} \cdot \text{m}^2/\text{kg}^2)(0.500 \text{ kg})(0.0100 \text{ kg})}{(0.200 \text{ m})^2 + (0.200 \text{ m})^2}$$

$$= 4.17 \times 10^{-12} \text{ N}.$$

The magnitude F_2 of the force due to the lower large mass is

$$F_2 = \frac{(6.67 \times 10^{-11} \text{ N} \cdot \text{m}^2/\text{kg}^2)(0.500 \text{ kg})(0.0100 \text{ kg})}{(0.200 \text{ m})^2}$$

$$= 8.34 \times 10^{-12} \text{ N}.$$

The x- and y-components of these forces are

$$F_{1x} = (4.17 \times 10^{-12} \text{ N})(\cos 45°) = 2.95 \times 10^{-12} \text{ N},$$

$$F_{1y} = (4.17 \times 10^{-12} \text{ N})(\sin 45°) = 2.95 \times 10^{-12} \text{ N},$$

$$F_{2x} = 8.34 \times 10^{-12} \text{ N},$$

$$F_{2y} = 0.$$

The components of the total force on the small mass are

$$F_x = F_{1x} + F_{2x} = 11.3 \times 10^{-12} \text{ N},$$

$$F_y = F_{1y} + F_{2y} = 2.95 \times 10^{-12} \text{ N}.$$

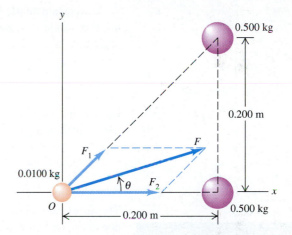

12–5 The total gravitational force on the 0.0100-kg sphere is the vector sum of the forces exerted on it by the two 0.500-kg spheres.

The magnitude of this force is

$$F = \sqrt{F_x^2 + F_y^2} = \sqrt{(11.3 \times 10^{-12} \text{ N})^2 + (2.95 \times 10^{-12} \text{ N})^2}$$

$$= 1.17 \times 10^{-11} \text{ N},$$

and its direction relative to the x-axis is

$$\theta = \arctan \frac{F_y}{F_x} = \arctan \frac{2.95 \times 10^{-12} \text{ N}}{11.3 \times 10^{-12} \text{ N}} = 14.6°.$$

Can you show that this force is *not* directed toward the center of mass of the two large masses? (See Problem 12–42.)

Perhaps the most remarkable aspect of the gravitational force of one body on another is that it acts *at a distance,* without direct contact between the two bodies. Electric and magnetic forces have this same remarkable property, as do the strong and weak interactions that we discussed in Section 5–6. A useful way to describe forces that act at a distance is in terms of a *field.* One body sets up a disturbance or field at all points in space, and the force that acts on a second body at a particular point is its response to the first body's field at that point. There is a field associated with each force that acts at a distance, and so we refer to gravitational fields, electric fields, magnetic fields, and so on. We won't need the field concept for our study of gravitation in this chapter, so we won't discuss it further here. But in later chapters we'll find that the field description is an extraordinarily powerful tool for describing electric and magnetic interactions.

12–3 WEIGHT

We defined the *weight* of a body in Section 4–5 as the attractive gravitational force exerted on it by the earth. We can now broaden our definition. **The weight of a body is the total gravitational force exerted on the body by all other bodies in the universe.** When the body is near the surface of the earth, we can neglect all other gravitational forces and consider the weight as just the earth's gravitational attraction. At the surface of the *moon* we consider a body's weight to be the gravitational attraction of the moon, and so on.

If we again model the earth as a spherically symmetric body with radius R_E and mass m_E, the weight w of a small body of mass m at the earth's surface (distance R_E from its center) is

$$w = F_g = \frac{Gm_E m}{R_E{}^2} \qquad \text{(weight of a body of mass } m \text{ at the earth's surface).} \qquad (12\text{–}3)$$

But we also know from Section 4–5 that the weight w of a body is the force that causes the acceleration g of free fall, so by Newton's second law, $w = mg$. Equating this with Eq. (12–3) and dividing by m, we find

$$g = \frac{Gm_E}{R_E{}^2} \qquad \text{(acceleration due to gravity at the earth's surface).} \qquad (12\text{–}4)$$

The acceleration due to gravity g is independent of the mass m of the body because m doesn't appear in this equation. We already knew that, but we can now see how it follows from the law of gravitation.

We can *measure* all the quantities in Eq. (12–4) except for m_E, so this relation allows us to compute the mass of the earth. Solving Eq. (12–4) for m_E and using $R = 6380$ km $= 6.38 \times 10^6$ m and $g = 9.80$ m/s², we find

$$m_E = \frac{gR_E{}^2}{G} = 5.98 \times 10^{24} \text{ kg,}$$

very close to the currently accepted value of 5.974×10^{24} kg. Once Cavendish had measured G, he computed the mass of the earth in just this way.

At a point above the earth's surface a distance r from the center of the earth (a distance $r - R_E$ above the surface), the weight of a body is given by Eq. (12–3) with R_E replaced by r:

$$w = F_g = \frac{Gm_E m}{r^2}. \qquad (12\text{–}5)$$

The weight of a body decreases inversely with the square of its distance from the earth's center. Figure 12–6 shows how the weight varies with height above the earth for an astronaut who weighs 700 N at the earth's surface.

The *apparent* weight of a body on earth differs slightly from the earth's gravitational force because the earth rotates and is therefore not precisely an inertial frame of refer-

In an airliner at high altitude, you are farther from the center of the earth than when on the ground and you actually weigh less. The effect is quite small, though measurable. Can you show that at an altitude of 10 km above the surface, you weigh just 0.3% less than you do on the ground?

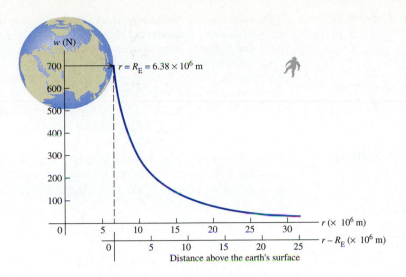

$r = R_E = 6.38 \times 10^6$ m

r ($\times\ 10^6$ m)

$r - R_E$ ($\times\ 10^6$ m)

Distance above the earth's surface

12–6 An astronaut weighing 700 N at the earth's surface experiences less gravitational attraction when above that surface (10^6 m = 1000 km = 620 mi). The astronaut's distance from the *center* of the earth is r; her distance from the *surface* of the earth is $r - R_E = r - 6.38 \times 10^6$ m.

ence. We have ignored this effect in the above discussion and have assumed that the earth *is* an inertial system. We will return to the effect of the earth's rotation in Section 12–8.

In our discussion of weight, we've used the fact that the earth is an approximately spherically symmetric distribution of mass. But this does *not* mean that the earth is uniform. To demonstrate that it cannot be uniform, let's first calculate the average *density*, or mass per unit volume, of the earth. If we assume a spherical earth, the volume is

$$V_E = \frac{4}{3}\pi R_E^{\ 3} = \frac{4}{3}\pi(6.38 \times 10^6 \text{ m})^3 = 1.09 \times 10^{21} \text{ m}^3.$$

The average density ρ (the Greek letter "rho") of the earth is the total mass divided by the total volume:

$$\rho = \frac{m_E}{V_E} = \frac{5.97 \times 10^{24} \text{ kg}}{1.09 \times 10^{21} \text{ m}^3}$$

$$= 5500 \text{ kg/m}^3 = 5.5 \text{ g/cm}^3.$$

(For comparison, the density of water is 1000 kg/m³ = 1.00 g/cm³.) If the earth were uniform, we would expect the density of individual rocks near the earth's surface to have this same value. In fact, the density of igneous surface rocks such as granite or gneiss is about 3000 kg/m³ = 3 g/cm³, although some basaltic rock has a density of about 5000 kg/m³ = 5 g/cm³. So the earth *cannot* be uniform, and the interior of the earth must be much more dense than the surface in order that the *average* density be 5500 kg/m³ = 5.5 g/cm³. According to geophysical models of the earth's interior, the maximum density at the center is about 13,000 kg/m³ = 13 g/cm³. Figure 12–7 is a graph of density as a function of distance from the center.

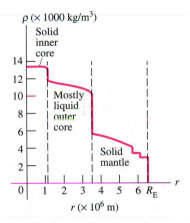

ρ ($\times$ 1000 kg/m³)

Solid inner core

Mostly liquid outer core

Solid mantle

r ($\times\ 10^6$ m)

12–7 The density of the earth decreases with increasing distance from its center.

EXAMPLE 12–4

Gravity on Mars You're involved in the design of a mission carrying humans to the surface of the planet Mars, which has a radius $r_M = 3.40 \times 10^6$ m and a mass $m_M = 6.42 \times 10^{23}$ kg. The earth weight of the Mars lander is 39,200 N. Calculate its weight F_g and the acceleration g_M due to the gravity of Mars: a) 6.0×10^6 m above the surface of Mars (the distance at which the moon Phobos orbits Mars); b) at the surface of Mars.

Neglect the gravitational effects of the (very small) moons of Mars.

SOLUTION a) In Eq. (12–5) we replace m_E with m_M. The value of G is the same everywhere in the universe; it is a fundamental physical constant. The distance r from the *center* of Mars is

$$r = (6.0 \times 10^6 \text{ m}) + (3.40 \times 10^6 \text{ m}) = 9.4 \times 10^6 \text{ m}.$$

The mass m of the lander is its earth weight w divided by the acceleration of gravity g on earth:

$$m = \frac{w}{g} = \frac{39{,}200 \text{ N}}{9.8 \text{ m/s}^2} = 4000 \text{ kg}.$$

The mass is the same whether the lander is on the earth, on Mars, or in between. From Eq. (12–5),

$$F_g = \frac{Gm_M m}{r^2}$$

$$= \frac{(6.67 \times 10^{-11} \text{ N} \cdot \text{m}^2/\text{kg}^2)(6.42 \times 10^{23} \text{ kg})(4000 \text{ kg})}{(9.4 \times 10^6 \text{ m})^2}$$

$$= 1940 \text{ N}.$$

The acceleration due to the gravity of Mars at this point is

$$g_M = \frac{F_g}{m} = \frac{1940 \text{ N}}{4000 \text{ kg}} = 0.48 \text{ m/s}^2.$$

This is also the acceleration experienced by Phobos in its orbit, 6.0×10^6 m above the surface of Mars.

b) To find F_g and g_M at the surface, we repeat the above calculations, replacing $r = 9.4 \times 10^6$ m with $R_M = 3.40 \times 10^6$ m. Alternatively, because F_g and g_M are inversely proportional to $1/r^2$ (at any point outside the planet), we can multiply the results of (a) by the factor

$$\left(\frac{9.4 \times 10^6 \text{ m}}{3.40 \times 10^6 \text{ m}} \right)^2.$$

We invite you to complete the calculation by both methods and show that at the surface $F_g = 15{,}000$ N and $g_M = 3.7$ m/s^2. That is, on the surface of Mars, F_g and g are roughly 40% as large as at the surface of the earth.

12–4 GRAVITATIONAL POTENTIAL ENERGY

When we first developed the concept of gravitational potential energy in Section 7–2, we assumed that the gravitational force on a body is constant in magnitude and direction. This led to the expression $U = mgy$. But we now know that the earth's gravitational force on a body of mass m at any point outside the earth is given more generally by Eq. (12–2), $F_g = Gm_E m/r^2$, where m_E is the mass of the earth and r is the distance of the body from the earth's center. For problems in which r changes enough that the gravitational force can't be considered constant, we need a more general expression for gravitational potential energy.

To find this expression, we follow the same basic sequence of steps as in Section 7–2. We consider a body of mass m outside the earth, and first compute the work W_{grav} done by the gravitational force when the body moves directly away from or toward the center of the earth from $r = r_1$ to $r = r_2$ as in Fig. 12–8. This is given by

$$W_{grav} = \int_{r_1}^{r_2} F_r \, dr, \tag{12–6}$$

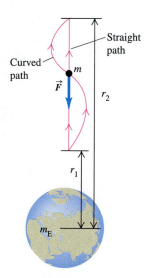

12–8 Work done by the gravitational force $\vec{F}$ as a body moves from radial coordinate r_1 to r_2. The work done by $\vec{F}$ is the same whether the body moves on a straight path or a curved one.

where F_r is the radial component of the gravitational force $\vec{F}$, that is, the component in the direction *outward* from the center of the earth. Because $\vec{F}$ points directly *inward* toward the center of the earth, F_r is negative. It differs from Eq. (12–2), the magnitude of the gravitational force, by a minus sign:

$$F_r = -\frac{Gm_E m}{r^2}. \tag{12–7}$$

Substituting Eq. (12–7) into Eq. (12–6), we see that W_{grav} is given by

$$W_{grav} = -Gm_E m \int_{r_1}^{r_2} \frac{dr}{r^2} = \frac{Gm_E m}{r_2} - \frac{Gm_E m}{r_1}. \tag{12–8}$$

The path doesn't have to be a straight line; it could also be a curve like the one in Fig. 12–8. By an argument similar to that in Section 7–2, this work depends only on the initial and final values of r, not on the path taken. This also proves that the gravitational force is always *conservative*.

We now define the corresponding potential energy U so that $W_{grav} = U_1 - U_2$, as in Eq. (7–3). Comparing this with Eq. (12–8), we see that the appropriate definition for

gravitational potential energy is

$$U = -\frac{Gm_E m}{r} \qquad \text{(gravitational potential energy).} \qquad (12\text{--}9)$$

Figure 12–9 shows how the gravitational potential energy depends on the distance r between the body of mass m and the center of the earth. When the body moves away from the earth, r increases, the gravitational force does negative work, and U increases (i.e., becomes less negative). When the body "falls" toward earth, r decreases, the gravitational work is positive, and the potential energy decreases (i.e., becomes more negative).

You may be troubled by Eq. (12–9) because it states that gravitational potential energy is always negative. But in fact you've seen negative values of U before. In using the formula $U = mgy$ in Section 7–2, we found that U was negative whenever the body of mass m was at a value of y below the arbitrary height we chose to be $y = 0$—that is, whenever the body and the earth were closer together than some certain arbitrary distance. (See, for instance, Example 7–2 in Section 7–2.) In defining U by Eq. (12–9), we have chosen U to be zero when the body of mass m is infinitely far away from the earth ($r = \infty$). As the body moves toward the earth, gravitational potential energy decreases and so becomes negative. If we wanted, we could make $U = 0$ at the surface of the earth, where $r = R_E$, by simply adding the quantity $Gm_E m/R_E$ to Eq. (12–9). This would make U positive when $r > R_E$, but at the price of making the expression for U more complicated. This added term would not affect the *difference* in U between any two points, which is the only physically significant quantity; that is why we omit this term and use Eq. (12–9) for potential energy.

CAUTION ▶ Be careful not to confuse the expressions for gravitational force, Eq. (12–7), and gravitational potential energy, Eq. (12–9). The force F_r is proportional to $1/r^2$, while potential energy U is proportional to $1/r$. ◀

Armed with Eq. (12–9), we can now use general energy relations for problems in which the $1/r^2$ behavior of the earth's gravitational force has to be included. If the gravitational force on the body is the only force that does work, the total mechanical energy of the system is constant, or *conserved*. In the following example we'll use this principle to calculate **escape speed**, the speed required for a body to escape completely from a planet.

12–9 Graph of gravitational potential energy U versus distance from the center of the earth r. Note that U is always negative, but U becomes less negative with increasing r.

EXAMPLE 12–5

"From the earth to the moon" In Jules Verne's 1865 story with this title, three men were shot to the moon in a shell fired from a giant cannon sunk in the earth in Florida. a) Find the muzzle speed needed to shoot the shell straight up to a height above the earth equal to the earth's radius. (b) Find the *escape speed*—that is, the muzzle speed that would allow the shell to escape from the earth completely. Neglect air resistance, the earth's rotation, and the gravitational pull of the moon. The earth's radius is $R_E = 6380$ km $= 6.38 \times 10^6$ m, and its mass is $m_E = 5.97 \times 10^{24}$ kg (see Appendix F).

SOLUTION a) The only force doing work is the conservative gravitational force, so mechanical energy is conserved: $K_1 + U_1 = K_2 + U_2$. Let point 1 be the starting point, and let point 2 be the point of maximum height, so the speed at point 2 is $v_2 = 0$. If the radius of the earth is R_E, then $r_1 = R_E$ and $r_2 = 2R_E$ (Fig. 12–10). The mass of the shell (with passengers) is m. We can determine v_1 from the energy-conservation equation

$K_1 + U_1 = K_2 + U_2$:

$$\frac{1}{2}mv_1^2 + \left(-\frac{Gm_E m}{R_E}\right) = 0 + \left(-\frac{Gm_E m}{2R_E}\right).$$

Rearranging this, we find that

$$v_1 = \sqrt{\frac{Gm_E}{R_E}}$$

$$= \sqrt{\frac{(6.67 \times 10^{-11} \text{ N} \cdot \text{m}^2/\text{kg}^2)(5.97 \times 10^{24} \text{ kg})}{6.38 \times 10^6 \text{ m}}}$$

$$= 7900 \text{ m/s} \ (= 28{,}400 \text{ km/h} = 17{,}700 \text{ mi/h}).$$

b) Again mechanical energy is conserved, so $K_1 + U_1 = K_2 + U_2$. We want the shell barely to be able to "reach" $r_2 = \infty$, with no kinetic energy left over, so $K_2 = 0$. When the shell is infinitely far from the earth, the potential energy $U_2 = 0$ as well, so the total mechanical energy $K_2 + U_2$ must be zero if the shell is just barely to escape to infinity. When the shell is fired, its positive kinetic energy K_1 and negative potential energy U_1 must also add to zero:

$$\frac{1}{2}mv_1^2 + \left(-\frac{Gm_E m}{R_E}\right) = 0,$$

$$v_1 = \sqrt{\frac{2Gm_E}{R_E}}$$

$$= \sqrt{\frac{2(6.67 \times 10^{-11}\ \text{N}\cdot\text{m}^2/\text{kg}^2)(5.97 \times 10^{24}\ \text{kg})}{6.38 \times 10^6\ \text{m}}}$$

$$= 1.12 \times 10^4\ \text{m/s} \quad (= 40{,}200\ \text{km/h} = 25{,}000\ \text{mi/h}).$$

This result does not depend on the mass of the shell, nor does it depend on the direction in which the shell is launched. Modern spacecraft launched from Florida must attain essentially this same speed to escape the earth. A spacecraft on the ground at Cape Canaveral is already moving at 410 m/s to the east because of the earth's rotation; by launching to the east, the spacecraft takes advantage of this "free" contribution toward escape speed.

Generalizing our result, the initial speed v_1 needed for a body to escape from the surface of a spherical mass M with radius R (ignoring air resistance) is

$$v_1 = \sqrt{\frac{2GM}{R}} \qquad \text{(escape speed)}.$$

12–10 A projectile fired from the surface of the earth to a height equal to the earth's radius.

You can use this result to compute the escape speed for other bodies. You will find 5.02×10^3 m/s for Mars, 5.95×10^4 m/s for Jupiter, and 6.18×10^5 m/s for the sun.

MORE ON GRAVITATIONAL FORCE AND POTENTIAL ENERGY

The relationship between the gravitational force on a body, given by Eq. (12–7), and its gravitational potential energy, given by Eq. (12–9), can be expressed in a different way by using the methods of Section 7–5. Equations (7–17) and (7–18) show that the component of force in a given direction equals the negative of the derivative of U with respect to the corresponding coordinate. For motion along the x-axis,

$$F_x = -\frac{dU}{dx}. \tag{12–10}$$

The gravitational force has a component only in the radial direction, so we replace x by r in Eq. (12–10). We find

$$F_r = -\frac{dU}{dr} = -\frac{d}{dr}\left(-\frac{Gm_E m}{r}\right) = -\frac{Gm_E m}{r^2}. \tag{12–11}$$

This agrees with the F_r we started with in Eq. (12–7). As we remarked before, F_r is negative, showing that the force points in the opposite direction from the direction of increasing r.

As a final note, let's show that when we are close to the earth's surface, Eq. (12–9) reduces to the familiar $U = mgy$ from Chapter 7. We first rewrite Eq. (12–8) as

$$W_{\text{grav}} = Gm_E m\,\frac{r_1 - r_2}{r_1 r_2}.$$

If the body stays close to the earth, then in the denominator we may replace r_1 and r_2 by R_E, the earth's radius, so

$$W_{\text{grav}} = Gm_E m\,\frac{r_1 - r_2}{R_E^{\,2}}.$$

According to Eq. (12–4), $g = Gm_E/R_E^2$, so

$$W_{\text{grav}} = mg(r_1 - r_2).$$

If we replace the r's by y's, this is just Eq. (7–1) for the work done by a constant gravitational force. In Section 7–2 we used this equation to derive Eq. (7–2), $U = mgy$, so we may consider this expression for gravitational potential energy to be a special case of the more general Eq. (12–9).

12–5 THE MOTION OF SATELLITES

Artificial satellites orbiting the earth are a familiar fact of contemporary life. But how do they stay in orbit, and what determines the properties of their orbits? We can use Newton's laws and the law of gravitation to provide the answers. We'll see in the next section that the motion of planets can be analyzed in the same way.

To begin, think back to the discussion of projectile motion in Section 3–4. In Example 3–6 a motorcycle rider rides horizontally off the edge of a cliff, launching himself into a parabolic path that ends on the flat ground at the base of the cliff. If he survives and repeats the experiment with increased launch speed, he will land farther from the starting point. We can imagine him launching himself with great enough speed that the earth's curvature becomes significant. As he falls, the earth curves away beneath him. If he is going fast enough, and if his launch point is high enough that he clears the mountain tops, he may be able to go right on around the earth without ever landing.

Figure 12–11 shows a variation on this theme. We launch a projectile from point A in the direction AB, tangent to the earth's surface. Trajectories (1) to (7) show the effect of increasing the initial speed. In trajectories (3) through (5) the projectile misses the earth and becomes an earth satellite. If there is no retarding force, the projectile's speed when it returns to point A is the same as its initial speed and it repeats its motion indefinitely.

Trajectories (1) through (5) close on themselves and are called **closed orbits.** All closed orbits are ellipses or segments of ellipses; trajectory (4) is a circle, a special case

Many satellites, including this gamma-ray telescope, have been placed in orbit by NASA's fleet of space shuttles.

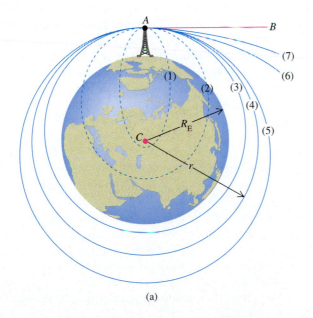

(a)

12–11 Trajectories of a body projected from point A in the direction AB with different initial velocities. Orbits (1) and (2) would be completed as shown if the earth were a point mass at C.

12-12 The force F_g due to the earth's gravitational attraction provides the centripetal acceleration that keeps a satellite in orbit. Compare this figure to Fig. 5–24.

4.8

Satellites Orbit

of an ellipse. Trajectories (6) and (7) are **open orbits.** For these the projectile never returns to its starting point but travels ever farther away from the earth.

A circular orbit like trajectory (4) in Fig. 12–11 is the simplest case, so we'll analyze it in detail. It is also an important case, since many artificial satellites have nearly circular orbits and the orbits of the planets around the sun are also fairly circular. The only force acting on a satellite in circular orbit around the earth is the earth's gravitational attraction, which is directed toward the center of the earth and hence toward the center of the orbit (Fig. 12–12). As we discussed in Section 5–5, this means that the satellite is in *uniform* circular motion and its speed is constant. The satellite isn't falling *toward* the earth; rather, it's constantly falling *around* the earth, and in a circular orbit the speed is just right to keep the distance from the satellite to the center of the earth constant.

How can we find the constant speed v of a satellite in a circular orbit? The radius of the orbit is r, measured from the *center* of the earth; the acceleration of the satellite has magnitude $a_{\text{rad}} = v^2/r$ and is always directed toward the center of the circle. By the law of gravitation, the net force (gravitational force) on the satellite of mass m has magnitude $F_g = Gm_{\text{E}}m/r^2$ and is in the same direction as the acceleration. Newton's second law ($\Sigma \vec{F} = m\vec{a}$) then tells us that

$$\frac{Gm_{\text{E}}m}{r^2} = \frac{mv^2}{r}.$$

Solving this for v, we find

$$v = \sqrt{\frac{Gm_{\text{E}}}{r}} \qquad \text{(circular orbit).} \qquad (12\text{–}12)$$

This relation shows that we can't choose the orbit radius r and the speed v independently; for a given radius r, the speed v for a circular orbit is determined.

Equation (12–12) also shows that the satellite's motion does not depend on its mass m because m doesn't appear in the equation. If we could cut a satellite in half without changing its speed, each half would continue on with the original motion. An astronaut on board a space shuttle is herself a satellite of the earth, held by the earth's gravitational attraction in the same orbit as the shuttle. The astronaut has the same velocity and acceleration as the shuttle, so nothing is pushing her against the floor or walls of the shuttle. She is in a state of *apparent weightlessness,* as in a freely falling elevator; see the discussion following Example 5–8 in Section 5–3. (*True* weightlessness would occur only if the astronaut were infinitely far from any other masses, so that the gravitational force on her would be zero.) Indeed, every part of her body is apparently weightless; she feels nothing pushing her stomach against her intestines or her head against her shoulders.

Apparent weightlessness is not just a feature of circular orbits; it occurs whenever gravity is the only force acting on a spacecraft. Hence it occurs for orbits of any shape, including open orbits such as (6) and (7) in Fig. 12–11.

We can derive a relation between the radius r of a circular orbit and the period T, the time for one revolution. The speed v is the distance $2\pi r$ traveled in one revolution, divided by the period:

$$v = \frac{2\pi r}{T}. \tag{12–13}$$

To get an expression for T, we solve Eq. (12–13) for T and substitute v from Eq. (12–12):

$$T = \frac{2\pi r}{v} = 2\pi r\sqrt{\frac{r}{Gm_E}} = \frac{2\pi r^{3/2}}{\sqrt{Gm_E}} \quad \text{(circular orbit).} \tag{12–14}$$

Equations (12–12) and (12–14) show that larger orbits correspond to slower speeds and longer periods.

It's interesting to compare Eq. (12–12) to the calculation of escape speed in Example 12–5. We see that the escape speed from a spherical body with radius R is just $\sqrt{2}$ times as large as the speed of a satellite in a circular orbit at that radius. If our spacecraft is in circular orbit around *any* planet, we have to multiply our speed by a factor of $\sqrt{2}$ to escape to infinity, regardless of the planet's mass.

Since the speed v in a circular orbit is determined by Eq. (12–12) for a given orbit radius r, the total mechanical energy $E = K + U$ is determined as well. Using Eqs. (12–9) and (12–12), we have

$$E = K + U = \frac{1}{2}mv^2 + \left(-\frac{Gm_E m}{r}\right) = \frac{1}{2}m\left(\frac{Gm_E}{r}\right) - \frac{Gm_E m}{r},$$
$$\tag{12–15}$$

$$E = -\frac{Gm_E m}{2r} \quad \text{(circular orbit).}$$

The total mechanical energy in a circular orbit is negative and equal to one-half the potential energy. Increasing the orbit radius r means increasing the mechanical energy (that is, making E less negative). If the satellite is in a relatively low orbit that encounters the outer fringes of earth's atmosphere, mechanical energy will decrease due to negative work done by the force of air resistance; as a result, the orbit radius will decrease until the satellite hits the ground or burns up in the atmosphere.

We have talked mostly about earth satellites, but we can apply the same analysis to the circular motion of *any* body under its gravitational attraction to a stationary body. Other examples include the earth's moon, the moons of other planets, and the rings of Saturn. If Saturn's rings were a rotating rigid body, all parts of the rings would have the same orbital period T, and by Eq. (12–13) the outer regions of the rings (large r) would have a greater speed v than the inner regions (small r). But since the rings are actually composed of many individual orbiting particles, Eqs. (12–12) and (12–14) tell us that particles in the outer regions of the rings move at a slower speed and have a longer period than do particles in the inner regions.

The rings of Saturn are made up of countless numbers of icy particles, each of which is an independent satellite of Saturn.

EXAMPLE 12–6

Suppose you want to place a 1000-kg weather satellite into a circular orbit 300 km above the earth's surface. a) What speed, period, and radial acceleration must it have? b) How much work had to be done to place this satellite in orbit? c) How much additional work would have to be done to make this satellite escape the earth? The earth's radius is $R_E = 6380$ km, and its mass is $m_E = 5.97 \times 10^{24}$ kg.

SOLUTION a) The radius of the satellite's orbit is

$$r = 6380 \text{ km} + 300 \text{ km} = 6680 \text{ km} = 6.68 \times 10^6 \text{ m}.$$

From Eq. (12–12),

$$v = \sqrt{\frac{Gm_E}{r}} = \sqrt{\frac{(6.67 \times 10^{-11} \text{ N} \cdot \text{m}^2/\text{kg}^2)(5.97 \times 10^{24} \text{ kg})}{6.68 \times 10^6 \text{ m}}}$$

$$= 7720 \text{ m/s}.$$

From Eq. (12–14),

$$T = \frac{2\pi r}{v} = \frac{2\pi(6.68 \times 10^6 \text{ m})}{7720 \text{ m/s}}$$

$$= 5440 \text{ s} = 90.6 \text{ min}.$$

The radial acceleration is

$$a_{rad} = \frac{v^2}{r} = \frac{(7720 \text{ m/s})^2}{6.68 \times 10^6 \text{ m}}$$

$$= 8.92 \text{ m/s}^2.$$

This is the value of g at a height of 300 km above the earth's surface; it is somewhat less than the value of g at the surface. b) The work required is the difference between E_2, the total mechanical energy when the satellite is in orbit, and E_1, the orig-

inal mechanical energy when the satellite was at rest on the launch pad back on earth. In orbit, if we use Eq. (12–15), the energy is

$$E_2 = -\frac{Gm_E m}{2r} = -\frac{(6.67 \times 10^{-11} \text{ N} \cdot \text{m}^2/\text{kg}^2)(5.97 \times 10^{24} \text{ kg})(1000 \text{ kg})}{2(6.68 \times 10^6 \text{ m})}$$

$$= -2.99 \times 10^{10} \text{ J}.$$

At rest on the earth's surface the energy is purely potential:

$$E_1 = K_1 + U_1 = 0 + \left(-\frac{Gm_E m}{R_E} \right)$$

$$= -\frac{(6.67 \times 10^{-11} \text{ N} \cdot \text{m}^2/\text{kg}^2)(5.97 \times 10^{24} \text{ kg})(1000 \text{ kg})}{6.38 \times 10^6 \text{ m}}$$

$$= -6.25 \times 10^{10} \text{ J},$$

and so

$$W_{required} = E_2 - E_1 = -2.99 \times 10^{10} \text{ J} - (-6.25 \times 10^{10} \text{ J})$$

$$= 3.26 \times 10^{10} \text{ J}.$$

Notice that we ignored the initial kinetic energy due to the rotation of the earth. How much difference does this make? (See Example 12–5.)
c) We saw in part (b) of Example 12–5 that for a satellite to escape to infinity, the total mechanical energy must be zero. The total mechanical energy in the circular orbit is $E_2 = -2.99 \times 10^{10}$ J; to increase this to zero, an amount of work equal to 2.99×10^{10} J would have to be done. This extra energy could be supplied by rocket engines attached to the satellite.

12–6 THE MOTION OF PLANETS

The name *planet* comes from a Greek word meaning "wanderer," and indeed the planets continually change their positions in the sky relative to the background of stars. One of the great intellectual accomplishments of the sixteenth and seventeenth centuries was the threefold realization that the earth is also a planet, that all planets orbit the sun, and that the apparent motions of the planets as seen from the earth can be used to precisely determine the orbits of the planets.

The first and second of these ideas were published by Nicolaus Copernicus in Poland in 1543. The determination of planetary orbits was carried out between 1601 and 1619 by the German astronomer and mathematician Johannes Kepler, using a voluminous set of precise data on apparent planetary motions compiled by his mentor, the Danish astronomer Tycho Brahe. By trial and error, Kepler discovered three empirical laws that accurately described the motions of the planets:

1. **Each planet moves in an elliptical orbit, with the sun at one focus of the ellipse.**
2. **A line from the sun to a given planet sweeps out equal areas in equal times.**
3. **The periods of the planets are proportional to the $\frac{3}{2}$ powers of the major axis lengths of their orbits.**

Kepler did not know *why* the planets moved in this way. Three generations later, when Newton turned his attention to the motion of the planets, he discovered that each of Kepler's laws can be *derived;* they are consequences of Newton's laws of motion and the law of gravitation. Let's examine each of Kepler's laws in turn and see how each law arises.

Let's first consider the elliptical orbits described in Kepler's first law. Figure 12–13 shows the geometry of the ellipse. The longest dimension is the *major axis,* with half-length a; this half-length is called the **semi-major axis.** The sum of the distances from S to P and from S' to P is the same for all points on the curve. S and S' are the *foci* (plural of focus). The sun is at S, and the planet is at P; we think of them both as points because the size of each is very small in comparison to the distance between them. There is nothing at the other focus S'.

The distance of each focus from the center of the ellipse is ea, where e is a dimensionless number between 0 and 1 called the **eccentricity.** If $e = 0$, the ellipse is a circle. The actual orbits of the planets are fairly circular; their eccentricities range from 0.007 for Venus to 0.248 for Pluto (the earth's orbit has $e = 0.017$). The point in the planet's orbit closest to the sun is the *perihelion,* and the point most distant from the sun is the *aphelion.*

Newton was able to show that for a body acted on by an attractive force proportional to $1/r^2$, the only possible closed orbits are a circle or an ellipse; he also showed that open orbits (trajectories (6) and (7) in Fig. 12–11) must be parabolas or hyperbolas. These results can be derived by a straightforward application of Newton's laws and the law of gravitation, together with a lot more differential equations than we're ready for.

Kepler's second law is shown in Fig. 12–14. In a small time interval dt, the line from the sun S to the planet P turns through an angle $d\theta$. The area swept out is the colored triangle with height r, base length $r\,d\theta$, and area $dA = \frac{1}{2}r^2\,d\theta$. The rate at which area is swept out, dA/dt, is called the *sector velocity:*

$$\frac{dA}{dt} = \frac{1}{2}r^2\frac{d\theta}{dt}. \qquad (12\text{--}16)$$

The essence of Kepler's second law is that the sector velocity has the same value at all points in the orbit. When the planet is close to the sun, r is small and $d\theta/dt$ is large; when the planet is far from the sun, r is large and $d\theta/dt$ is small.

To see how Kepler's second law follows from Newton's laws, we express dA/dt in terms of the velocity vector $\vec{v}$ of the planet P. The component of $\vec{v}$ perpendicular to the radial line is $v_\perp = v\sin\phi$. From Fig. 12–14b the displacement along the direction of $v_\perp$ during time dt is $r\,d\theta$, so we also have $v_\perp = r\,d\theta/dt$. Using this relation in Eq. (12–16), we find

$$\frac{dA}{dt} = \frac{1}{2}rv\sin\phi \qquad \text{(sector velocity)}. \qquad (12\text{--}17)$$

Now $rv\sin\phi$ is the magnitude of the vector product $\vec{r}\times\vec{v}$, which in turn is $1/m$ times the angular momentum $\vec{L} = \vec{r}\times m\vec{v}$ of the planet with respect to the sun. So we have

$$\frac{dA}{dt} = \frac{1}{2m}|\vec{r}\times m\vec{v}| = \frac{L}{2m}. \qquad (12\text{--}18)$$

Thus Kepler's second law, that sector velocity is constant, means that angular momentum is constant!

It is easy to see why the angular momentum of the planet *must* be constant. According to Eq. (10–29), the rate of change of $\vec{L}$ equals the torque of the gravitational force $\vec{F}$ acting on the planet:

$$\frac{d\vec{L}}{dt} = \vec{\tau} = \vec{r}\times\vec{F}.$$

In our situation, $\vec{r}$ is the vector from the sun to the planet, and the force $\vec{F}$ is directed from the planet to the sun. So these vectors always lie along the same line, and their vector

12–13 Geometry of an ellipse. The sum of the distances SP and $S'P$ is the same for every point on the curve. The sizes of the sun and planet are exaggerated for clarity.

(a)

(b)

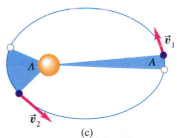

(c)

12–14 (a) The planet P moves about the sun (at S) in an elliptical orbit. (b) In a time dt the line from the sun to the planet sweeps out a triangular area $dA = \frac{1}{2}(r\,d\theta)\,r = \frac{1}{2}r^2\,d\theta$. (c) The planet's velocity varies so that the ratio of area dA to time interval dt is constant, regardless of the planet's position in its orbit.

product $\vec{r} \times \vec{F}$ is zero. Hence $d\vec{L}/dt = 0$. This conclusion does not depend on the $1/r^2$ behavior of the force; angular momentum is conserved for *any* force that acts always along the line joining the particle to a fixed point. Such a force is called a *central force*. (Kepler's first and third laws are valid *only* for a $1/r^2$ force.)

Conservation of angular momentum also explains why the orbit lies in a plane. The vector $\vec{L} = \vec{r} \times m\vec{v}$ is always perpendicular to the plane of the vectors $\vec{r}$ and $\vec{v}$; since $\vec{L}$ is constant in magnitude *and* direction, $\vec{r}$ and $\vec{v}$ always lie in the same plane, which is just the plane of the planet's orbit.

We have already derived Kepler's third law for the particular case of circular orbits. Equation (12–14) shows that the period of a satellite or planet in a circular orbit is proportional to the $\frac{3}{2}$ power of the orbit radius. Newton was able to show that this same relationship holds for an elliptical orbit, with the orbit radius r replaced by the semimajor axis a:

$$T = \frac{2\pi a^{3/2}}{\sqrt{Gm_S}} \qquad \text{(elliptical orbit around the sun).} \qquad (12\text{–}19)$$

Since the planet orbits the sun, not the earth, we have replaced the earth's mass m_E in Eq. (12–14) by the sun's mass m_S. Note that the period does not depend on the eccentricity e. An asteroid in an elongated elliptical orbit with semi-major axis a will have the same orbital period as a planet in a circular orbit of radius a. The key difference is that the asteroid moves at different speeds at different points in its elliptical orbit (Fig. 12–14c), while the planet's speed is constant around its circular orbit.

EXAMPLE 12–7

Kepler's third law Using the data in Appendix F, check whether the orbital semi-major axes and periods of Uranus and Saturn obey Kepler's third law.

SOLUTION The ratio of the orbital semi-major axis of Uranus to that of Saturn is

$$\frac{a_U}{a_S} = \frac{2.88 \times 10^{12} \text{ m}}{1.43 \times 10^{12} \text{ m}} = 2.01.$$

According to Kepler's third law, the ratio of the periods should be the $\frac{3}{2}$ power of this, or 2.85. The actual ratio is

$$\frac{T_U}{T_S} = \frac{83.75 \text{ y}}{29.42 \text{ y}} = 2.85.$$

As a historical note, Uranus wasn't discovered until 1781, a century and a half after Kepler's publication of his third law. While Kepler deduced his three laws from the motions of the five planets (other than the earth) known in his time, these laws have proven to apply equally well to all of the planets, asteroids, and comets subsequently discovered to be orbiting the sun.

EXAMPLE 12–8

Halley's comet Halley's comet moves in an elongated elliptical orbit around the sun (Fig. 12–15). At perihelion, Halley's comet is 8.75×10^7 km from the sun; at aphelion it is 5.26×10^9 km from the sun. a) At what point in the orbit does the comet have the greatest speed? b) Find the semi-major axis, eccentricity, and period of the orbit.

SOLUTION a) Mechanical energy is conserved as the comet moves around its orbit. The comet's kinetic energy $K = \frac{1}{2}mv^2$ will be maximum when the potential energy $U = -Gm_Sm/r$ is minimum (that is, the most negative), which occurs when r is a minimum. Hence the speed v is maximum at perihelion. Your intuition about falling bodies is helpful here; as the comet falls inward toward the sun, it picks up speed, and its speed is maximum when closest to the sun. By the same reasoning, the comet

slows down as it moves away from the sun, and its speed is minimum at aphelion.

b) From Fig. 12–13 the length of the major axis equals the sum of the comet-sun distance at perihelion plus the comet-sun distance at aphelion. The length of the major axis is $2a$, so

$$a = \frac{8.75 \times 10^7 \text{ km} + 5.26 \times 10^9 \text{ km}}{2} = 2.67 \times 10^9 \text{ km}.$$

Further inspection of Fig. 12–13 shows that the comet-sun distance at perihelion is given by

$$a - ea = a(1 - e).$$

Since we are given that this distance is 8.75×10^7 km, the eccentricity is

$$e = 1 - \frac{8.75 \times 10^7 \text{ km}}{a} = 1 - \frac{8.75 \times 10^7 \text{ km}}{2.67 \times 10^9 \text{ km}} = 0.967.$$

12–15 (a) The orbit of Halley's comet. (b) Halley's comet as it appeared in 1986. At the heart of the comet is an icy body, called the nucleus, that is about 10 km across. When the comet's orbit carries it close to the sun, the heat of sunlight causes the nucleus to partially evaporate. The evaporated material forms the tail, which can be tens of millions of kilometers long.

This is very close to 1, so this is a very elongated orbit.

The period is given by Eq. (12–19):

$$T = \frac{2\pi a^{3/2}}{\sqrt{Gm_s}} = \frac{2\pi(2.67 \times 10^9 \text{ km})^{3/2}}{\sqrt{(6.67 \times 10^{-11} \text{ N} \cdot \text{m}^2/\text{kg}^2)(1.99 \times 10^{30} \text{ kg})}}$$

$$= 2.38 \times 10^9 \text{ s} = 75.5 \text{ years}.$$

Halley's comet was at perihelion in early 1986; it will next reach perihelion one period later, in 2061.

We have assumed that as a planet or comet orbits the sun, the sun remains absolutely stationary. Of course, this can't be correct; because the sun exerts a gravitational force on the planet, the planet exerts a gravitational force on the sun of the same magnitude but opposite direction. In fact, *both* the sun and the planet orbit around their common center of mass. We've made only a small transgression by ignoring this effect, however; the sun's mass is about 750 times the total mass of all the planets combined, so the center of mass of the solar system is not far from the center of the sun. But in binary star systems, in which two stars of comparable mass orbit each other, these effects are not small, and we must account for the motions of both stars about the center of mass.

Newton's analysis of planetary motions was a complete success, and his results are used on a daily basis by modern-day astronomers. But the most remarkable result of Newton's work is that the motions of bodies in the heavens obey the *same* laws of motion as do bodies on the earth. This *Newtonian synthesis,* as it has come to be called, is one of the great unifying principles of science. It has had profound effects on the way that humanity looks at the universe—not as a realm of impenetrable mystery, but as a direct extension of our everyday world, subject to scientific study and calculation. The Newtonian synthesis was an astonishing leap in understanding, made by a giant intellect.

*12–7 SPHERICAL MASS DISTRIBUTIONS

We have used without proof the statement that the gravitational interaction between two spherically symmetric mass distributions is the same as though all the mass of each were concentrated at its center. Now we're ready to prove this statement. Newton searched for

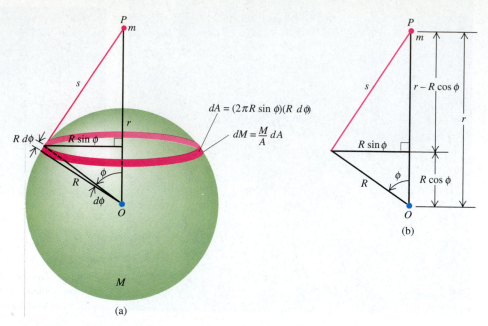

12–16 (a) To calculate the gravitational effects *outside* a spherical shell, all the mass M can be considered concentrated at the center. (b) The distance s is the hypotenuse of a right triangle with sides $(r - R \cos \phi)$ and $R \sin \phi$.

a proof for several years, and he delayed publication of the law of gravitation until he found one.

Here's our program. Rather than starting with two spherically symmetric masses, we'll tackle the simpler problem of a point mass m interacting with a thin spherical shell with total mass M. We will show that when m is outside the sphere, the *potential energy* associated with this gravitational interaction is the same as though M were all concentrated at the center of the sphere. According to Eq. (12–10), the force is the negative derivative of the potential energy, so the *force* on m is also the same as for a point mass M. Any spherically symmetric mass distribution can be thought of as being made up of many concentric spherical shells, so our result will also hold for *any* spherically symmetric M.

We start by considering a ring on the surface of the shell (Fig. 12–16a), centered on the line from the center of the shell to m. The reason we do this is that all of the particles that make up the ring are the same distance s from the point mass m. From Eq. (12–9) the potential energy of interaction between the earth (mass m_E) and a point mass m, separated by a distance r, is $U = -Gm_E m/r$. By changing notation in this expression, we see that in the situation shown in Fig. 12–16a, the potential energy of interaction between the point mass m and a particle of mass m_i within the ring is given by

$$U_i = -\frac{Gmm_i}{s}.$$

To find the potential energy of interaction between m and the entire ring of mass $dM = \sum_i m_i$, we sum this expression for U_i over all particles in the ring. Calling this potential energy dU, we find

$$dU = \sum_i U_i = \sum_i \left(-\frac{Gmm_i}{s} \right) = -\frac{Gm}{s} \sum_i m_i = -\frac{Gm\, dM}{s}. \qquad (12\text{–}20)$$

To proceed, we need to know the mass dM of the ring. We can find this with the aid of a little geometry. The radius of the shell is R, so in terms of the angle ϕ shown in the figure, the radius of the ring is $R \sin \phi$, and its circumference is $2\pi R \sin \phi$. The width of the ring is $R\, d\phi$, and its area dA is approximately equal to its width times its circumference:

$$dA = 2\pi R^2 \sin \phi\, d\phi.$$

The ratio of the ring mass dM to the total mass M of the shell is equal to the ratio of the area dA of the ring to the total area $A = 4\pi R^2$ of the shell:

$$\frac{dM}{M} = \frac{2\pi R^2 \sin\phi \; d\phi}{4\pi R^2} = \frac{1}{2}\sin\phi \; d\phi. \tag{12–21}$$

Now we solve Eq. (12–21) for dM and substitute the result into Eq. (12–20) to find the potential energy of interaction between the point mass m and the ring:

$$dU = -\frac{GMm \sin\phi \; d\phi}{2s}. \tag{12–22}$$

The total potential energy of interaction between the point mass and the *shell* is the integral of Eq. (12–22) over the whole sphere as ϕ varies from 0 to π (*not* 2π!) and s varies from $r - R$ to $r + R$. To carry out the integration, we have to express the integrand in terms of a single variable; we choose s. To express ϕ and $d\phi$ in terms of s, we have to do a little more geometry. Figure 12–16b shows that s is the hypotenuse of a right triangle with sides $(r - R\cos\phi)$ and $R\sin\phi$, so the Pythagorean theorem gives

$$s^2 = (r - R\cos\phi)^2 + (R\sin\phi)^2$$

$$= r^2 - 2rR\cos\phi + R^2. \tag{12–23}$$

We take differentials of both sides:

$$2s \; ds = 2rR\sin\phi \; d\phi.$$

Next we divide this by $2rR$ and substitute the result into Eq. (12–22):

$$dU = -\frac{GMm}{2s}\frac{s \; ds}{rR} = -\frac{GMm}{2rR}\;ds. \tag{12–24}$$

We can now integrate Eq. (12–24), recalling that s varies from $r - R$ to $r + R$:

$$U = -\frac{GMm}{2rR}\int_{r-R}^{r+R} ds = -\frac{GMm}{2rR}[(r+R)-(r-R)]. \tag{12–25}$$

Finally, we have

$$U = -\frac{GmM}{r} \qquad \text{(point mass } m \text{ outside spherical shell } M\text{).} \tag{12–26}$$

This is equal to the potential energy of two point masses m and M at a distance r. So we have proved that the gravitational potential energy of the spherical shell M and the point mass m at any distance r is the same as though they were point masses. Because the force is given by $F_r = -dU/dr$, the force is also the same.

Any spherically symmetric mass distribution can be thought of as a combination of concentric spherical shells. Because of the principle of superposition of forces, what is true of one shell is also true of the combination. So we have proved half of what we set out to prove, that the gravitational interaction between any spherically symmetric mass distribution and a point mass is the same as though all the mass of the spherically symmetric distribution were concentrated at its center.

The other half is to prove that *two* spherically symmetric mass distributions interact as though they were both points. That's easier. In Fig. 12–16 the forces the two bodies exert on each other are an action-reaction pair, and they obey Newton's third law. So we have also proved that the force that m exerts *on* the sphere M is the same as though M were a point. But now if we replace m with a spherically symmetric mass distribution

12–17 When a point mass m is *inside* a uniform spherical shell of mass M, the potential energy is the same no matter where inside the shell the point mass is located. The force from the masses' mutual gravitational interaction is zero.

centered at m's location, the resulting gravitational force on any part of M is the same as before, and so is the total force. This completes our proof.

A POINT MASS INSIDE A SPHERICAL SHELL

We assumed at the beginning that the point mass m was outside the spherical shell, so our proof is valid only when m is outside a spherically symmetric mass distribution. When m is *inside* a spherical shell, the geometry is as shown in Fig. 12–17. The entire analysis goes just as before; Eqs. (12–20) through (12–24) are still valid. But when we get to Eq. (12–25), the limits of integration have to be changed to $R - r$ and $R + r$. We then have

$$ U = -\frac{GMm}{2rR} \int_{R-r}^{R+r} ds = -\frac{GMm}{2rR}[(R + r) - (R - r)], \qquad (12\text{–}27) $$

and the final result is

$$ U = -\frac{GmM}{R} \qquad \text{(point mass } m \text{ inside spherical shell } M\text{).} \qquad (12\text{–}28) $$

Compare this result to Eq. (12–26): instead of having r, the distance between m and the center of M, in the denominator, we have R, the radius of the shell. This means that U in Eq. (12–28) doesn't depend on r and thus has the same value everywhere inside the shell. When m moves around inside the shell, no work is done on it, so the force on m at any point inside the shell must be zero.

More generally, at any point in the interior of any spherically symmetric mass distribution (not necessarily a shell), at a distance r from its center, the gravitational force on a point mass m is the same as though we removed all the mass at points farther than r from the center and concentrated all the remaining mass at the center.

EXAMPLE 12–9

"Journey to the center of the earth" Suppose we drill a hole through the earth (radius R_E, mass m_E) along a diameter and drop a mail pouch (mass m) down the hole. Derive an expression for the gravitational force on the pouch as a function of its distance r from the center. Assume that the density of the earth is uniform (not a very realistic model; see Fig. 12–7).

SOLUTION According to the statements above, the gravitational force at distance r from the center is determined only by the mass M within a sphere of radius r (Fig. 12–18). With uniform

density the mass is proportional to the volume of the sphere, which is $\frac{4}{3}\pi r^3$ for this sphere and $\frac{4}{3}\pi R_E^{\ 3}$ for the entire earth. So we have

$$ \frac{M}{m_E} = \frac{\frac{4}{3}\pi r^3}{\frac{4}{3}\pi R_E^{\ 3}} = \frac{r^3}{R_E^{\ 3}}. $$

The magnitude of the gravitational force on m is given by

$$ F_g = \frac{GMm}{r^2} = \frac{Gm_E m}{r^2}\frac{r^3}{R_E^{\ 3}} = \frac{Gm_E m}{R_E^{\ 3}}r. $$

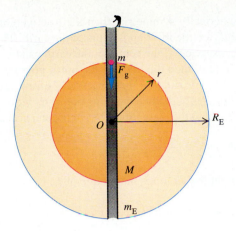

At points inside the earth, F_g is *directly proportional* to the distance r from the center, rather than proportional to $1/r^2$ as it is outside the sphere. Right at the surface, where $r = R_E$, the above expression gives $F_g = Gm_Em/R_E^2$, as we should expect. In the next chapter we'll learn how to compute the time it would take for the mail pouch to emerge on the other side of the earth. Meanwhile, keep in mind that the assumption of uniform density of the earth is, like the hole, pure fantasy. We discussed the variation of density in the earth in Section 12–3.

12–18 A hole through the center of the earth (assumed to be uniform). When an object is at a distance r from the center, only the mass inside a sphere of radius r exerts a net gravitational force on it.

*12–8 APPARENT WEIGHT AND THE EARTH'S ROTATION

Because the earth rotates on its axis, it is not precisely an inertial frame of reference. For this reason the apparent weight of a body on earth is not precisely equal to the earth's gravitational attraction, which we will call the **true weight** $\vec{w}_0$ of the body. Figure 12–19 is a cutaway view of the earth, showing three observers. Each one holds a spring balance with a body of mass m hanging from it. Each balance applies a tension force $\vec{F}$ to the body hanging from it, and the reading on each balance is the magnitude F of this force. If the observers are unaware of the earth's rotation, each one *thinks* that the scale reading equals the weight of the body because he thinks the body on his spring balance is in equilibrium. So each observer thinks that the tension $\vec{F}$ must be opposed by an equal and opposite force $\vec{w}$, which we call the **apparent weight.** But if the bodies are rotating with the earth, they are *not* precisely in equilibrium. Our problem is to find the relation between the apparent weight $\vec{w}$ and the true weight $\vec{w}_0$.

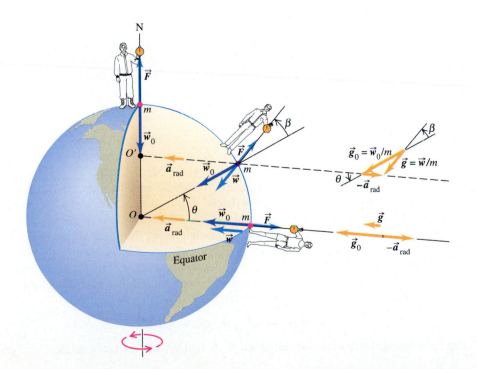

12–19 Except at the poles, the scale reads less than the gravitational force of attraction because a net force is needed to provide the centripetal acceleration. The sketch exaggerates the difference for clarity. On the earth the maximum angle β between the true and apparent weight vectors is 0.1°, and the magnitude of the apparent weight is smaller than that of the true weight by less than 0.35%.

If we assume that the earth is spherically symmetric, then the true weight $\vec{w}_0$ has magnitude $Gm_{\mathrm{E}}m/R_{\mathrm{E}}^2$, where m_{E} and R_{E} are the mass and radius of the earth. This value is the same for all points on the earth's surface. If the center of the earth can be taken as the origin of an inertial coordinate system, then the body at the north pole really *is* in equilibrium in an inertial system, and the reading on that observer's spring balance is equal to w_0. But the body at the equator is moving in a circle of radius R_{E} with speed v, and there must be a net inward force equal to the mass times the centripetal acceleration:

$$w_0 - F = \frac{mv^2}{R_{\mathrm{E}}}.$$

So the magnitude of the apparent weight (equal to the magnitude of F) is

$$w = w_0 - \frac{mv^2}{R_{\mathrm{E}}} \qquad \text{(at the equator).} \tag{12–29}$$

If the earth were not rotating, the body when released would have a free-fall acceleration $g_0 = w_0/m$. Since the earth *is* rotating, the falling body's actual acceleration relative to the observer at the equator is $g = w/m$. Dividing Eq. (12–29) by m and using these relations, we find

$$g = g_0 - \frac{v^2}{R_{\mathrm{E}}} \qquad \text{(at the equator).}$$

To evaluate v^2/R_{E}, we note that in 86,164 s a point on the equator moves a distance equal to the earth's circumference, $2\pi R_{\mathrm{E}} = 2\pi(6.38 \times 10^6 \text{ m})$. (The solar day, 86,400 s, is $\frac{1}{365}$ longer than this because in one day the earth also completes $\frac{1}{365}$ of its orbit around the sun.) Thus we find

$$v = \frac{2\pi(6.38 \times 10^6 \text{ m})}{86,164 \text{ s}} = 465 \text{ m/s},$$

$$\frac{v^2}{R_{\mathrm{E}}} = \frac{(465 \text{ m/s})^2}{6.38 \times 10^6 \text{ m}} = 0.0339 \text{ m/s}^2.$$

So for a spherically symmetric earth the acceleration due to gravity should be about 0.03 m/s^2 less at the equator than at the poles.

At locations intermediate between the equator and the poles, the true weight $\vec{w}_0$ and the centripetal acceleration are not along the same line, and we need to write a vector equation corresponding to Eq. (12–29). From Fig. 12–19 we see that the appropriate equation is

$$\vec{w} = \vec{w}_0 - m\vec{a}_{\mathrm{rad}} = m\vec{g}_0 - m\vec{a}_{\mathrm{rad}}. \tag{12–30}$$

The difference in the magnitudes of g and g_0 lies between zero and 0.0339 m/s^2. As shown in Fig. 12–19, the *direction* of the apparent weight differs from the direction toward the center of the earth by a small angle β, which is 0.1° or less.

Table 12–1 gives the values of g at several locations, showing variations with latitude. There are also small additional variations due to the lack of perfect spherical symmetry of the earth, local variations in density, and differences in elevation.

Our discussion of apparent weight can also be applied to the phenomenon of apparent weightlessness in orbiting spacecraft, which we described in Section 12–5. Bodies in an orbiting spacecraft are *not* weightless; the earth's gravitational attraction continues to act on them just as though they were at rest relative to the earth. The apparent weight of a body in a spacecraft is again given by Eq. (12–30):

$$\vec{w} = \vec{w}_0 - m\vec{a}_{\mathrm{rad}} = m\vec{g}_0 - m\vec{a}_{\mathrm{rad}}.$$

But for a spacecraft in orbit, as well as any body inside the spacecraft, the acceleration

TABLE 12–1

VARIATIONS OF g WITH LATITUDE AND ELEVATION

STATION	NORTH LATITUDE	ELEVATION (m)	$g\,(\text{m/s}^2)$	$g\,(\text{ft/s}^2)$
Canal Zone	9°	0	9.78243	32.0944
Jamaica	18°	0	9.78591	32.1059
Bermuda	32°	0	9.79806	32.1548
Denver	40°	1638	9.79609	32.1393
Pittsburgh, Pa.	40.5°	235	9.80118	32.1561
Cambridge, Mass.	42°	0	9.80398	32.1652
Greenland	70°	0	9.82534	32.2353

$\vec{a}_{\text{rad}}$ toward the earth's center is equal to the value of the acceleration of gravity $\vec{g}_0$ at the position of the spacecraft. Hence

$$\vec{g}_0 = \vec{a}_{\text{rad}},$$

and the apparent weight is

$$\vec{w} = 0.$$

This is what we mean when we say that an astronaut or other body in the spacecraft is apparently weightless. Note that we didn't make any assumptions about the shape of the orbit; as we mentioned in Section 12–5, an astronaut will be apparently weightless no matter what the orbit.

This orbiting astronaut is acted on by the earth's gravity, but *feels* weightless because his acceleration is equal to $\vec{g}$.

12–9 BLACK HOLES

A Case Study in Modern Physics

The concept of a black hole is one of the most interesting and startling products of modern gravitational theory, yet the basic idea can be understood on the basis of Newtonian principles. Think first about the properties of our own sun. Its mass $M = 1.99 \times 10^{30}$ kg and radius $R = 6.96 \times 10^8$ m are much larger than those of any planet, but compared to other stars, our sun is not exceptionally massive.

What's the average *density* ρ of the sun? You can find it the same way we found the average density of the earth in Section 12–3:

$$\rho = \frac{M}{V} = \frac{M}{\frac{4}{3}\pi R^3} = \frac{1.99 \times 10^{30}\ \text{kg}}{\frac{4}{3}\pi (6.96 \times 10^8\ \text{m})^3}$$

$$= 1410\ \text{kg/m}^3.$$

The sun's temperatures range from 5800 K (about 5500°C, or 10,000°F) at the surface up to 1.5×10^7 K (about 2.7×10^7 °F) in the interior, so it surely contains no solids or liquids. Yet gravitational attraction pulls the sun's gas atoms together until the sun is, on average, 41% denser than water and about 1200 times as dense as the air we breathe.

Now think about the escape speed for a body at the surface of the sun. In Example 12–5 (Section 12–4) we found that the escape speed from the surface of a spherical mass M with radius R is $v = \sqrt{2GM/R}$. We can relate this to the average density. Substituting $M = \rho V = \rho(\frac{4}{3}\pi R^3)$ into the expression for escape speed gives

$$v = \sqrt{\frac{2GM}{R}} = \sqrt{\frac{8\pi G \rho}{3}} R. \qquad (12\text{–}31)$$

Using either form of this equation, you can show that the escape speed for a body at the surface of our sun is $v = 6.18 \times 10^5$ m/s (about 2.2 million km/h, or 1.4 million mi/h).

This value, roughly 1/500 the speed of light, is independent of the mass of the escaping body; it depends only on the mass and radius (or average density and radius) of the sun.

Now consider various stars with the same average density ρ and different radii R. Equation (12–31) shows that for a given value of density ρ, the escape speed v is directly proportional to R. In 1783 the Rev. John Mitchell, an amateur astronomer, noted that if a body with the same average density as the sun had about 500 times the radius of the sun, the magnitude of its escape velocity would be greater than the speed of light c. With his statement that "all light emitted from such a body would be made to return toward it," Mitchell became the first person to suggest the existence of what we now call a **black hole.**

The first expression for escape speed in Eq. (12–31) also suggests that a body of mass M will act as a black hole if its radius R is less than or equal to a certain critical radius. How can we determine this critical radius? You might think that you can find the answer by simply setting $v = c$ in Eq. (12–31). As a matter of fact, this does give the correct result, but only because of two compensating errors. The kinetic energy of light is *not* $mc^2/2$, and the gravitational potential energy near a black hole is *not* given by Eq. (12–9). In 1916, Karl Schwarzschild used Einstein's general theory of relativity (in part a generalization and extension of Newtonian gravitation theory) to derive an expression for the critical radius R_S, now called the **Schwarzschild radius.** The result turns out to be the same as though we had set $v = c$ in Eq. (12–31), so

$$c = \sqrt{\frac{2GM}{R_S}}.$$

Solving for the Schwarzschild radius R_S, we find

$$R_S = \frac{2GM}{c^2}. \tag{12–32}$$

If a spherical, nonrotating body with mass M has a radius less than R_S, then *nothing* (not even light) can escape from the surface of the body, and the body functions as a black hole (Fig. 12–20). In this case, any other body within a distance R_S of the center of the black hole is trapped by the gravitational attraction of the black hole and cannot escape from it.

The surface of the sphere with radius R_S surrounding a black hole is called the **event horizon** because, since light can't escape from within that sphere, we can't see events occurring inside. All that an observer outside the event horizon can know about a black hole is its mass (from its gravitational effects on other bodies), its electric charge (from

12–20 (a) When the radius R of a body is greater than the Schwarzschild radius R_S, light can escape from the surface of the body. As it travels away, it is "red-shifted" to longer wavelengths. (b) When the body is within the event horizon (radius R_S), it is a black hole with an escape speed greater than the speed of light. In that case we can obtain very little information about it.

(a)

(b)

the electric forces it exerts on other charged bodies), and its angular momentum (because a rotating black hole tends to drag space—and everything in that space—around with it). All other information about the body is irretrievably lost when it collapses inside its event horizon.

EXAMPLE 12–10

Current astrophysical theory suggests that a burned-out star can collapse under its own gravity to form a black hole when its mass is as small as two solar masses. If it does, what is the radius of its event horizon?

SOLUTION The radius is R_S, and "two solar masses" means $M = 2(1.99 \times 10^{30} \text{ kg}) = 4.0 \times 10^{30}$ kg. From Eq. (12–32),

$$R_S = \frac{2GM}{c^2} = \frac{2(6.67 \times 10^{-11} \text{ N} \cdot \text{m}^2/\text{kg}^2)(4.0 \times 10^{30} \text{ kg})}{(3.00 \times 10^8 \text{ m/s})^2}$$

$$= 5.9 \times 10^3 \text{ m} = 5.9 \text{ km},$$

or less than 4 miles.

If the radius of such an object is just equal to the Schwarzschild radius, the average density has the incredibly large value

$$\rho = \frac{M}{\frac{4}{3}\pi R^3} = \frac{4.0 \times 10^{30} \text{ kg}}{\frac{4}{3}\pi(5.9 \times 10^3 \text{ m})^3}$$

$$= 4.6 \times 10^{18} \text{ kg/m}^3.$$

This is of the order of 10^{15} times as great as the density of familiar matter on earth and is comparable to the densities of atomic nuclei. In fact, once the body collapses to a radius of R_S, nothing can prevent it from collapsing further. All of the mass ends up being crushed down to a single point called a *singularity* at the center of the event horizon. This point has zero volume and so has *infinite* density.

At points far from a black hole, its gravitational effects are the same as those of any normal body with the same mass. If the sun collapsed to form a black hole, the orbits of the planets would be unaffected. But things get dramatically different close to the black hole. If you decided to become a martyr for science and jump into a black hole, those you left behind would notice several odd effects as you moved toward the event horizon, most of them associated with effects of general relativity. If you carried a radio transmitter to send back your comments on what was happening, they would have to retune their receiver continuously to lower and lower frequencies, an effect called the *gravitational red shift*. Consistent with this shift, they would observe that your clocks (electronic or biological) would appear to run more and more slowly, an effect called *time dilation*. In fact, during their lifetimes they would never see you make it to the event horizon. In your frame of reference, you would make it to the event horizon in a rather short time but in a rather disquieting way. As you fell feet first into the black hole, the gravitational pull on your feet would be greater than that on your head, which would be slightly farther away from the black hole. The *differences* in gravitational force on different parts of your body would be great enough to stretch you along the direction toward the black hole and compress you perpendicular to it. These effects (called *tidal forces*) would rip you to atoms, and then rip your atoms apart, before you reached the event horizon.

If light cannot escape from a black hole and if black holes are as small as Example 12–10 suggests, how can we know that such things exist in space? Here's the idea. Any gas or dust near the black hole tends to be pulled into an *accretion disk* that swirls around and into the black hole, rather like a whirlpool (Fig. 12–21). Friction within the accretion disk's material causes it to lose mechanical energy and spiral into the black hole; as it moves inward, it is compressed together. This causes heating of the material, just as air compressed in a bicycle pump gets hotter. Temperatures in excess of 10^6 K can occur in the accretion disk, so hot that the disk emits not just visible light (as do bodies that are "red-hot" or "white-hot") but x rays. Astronomers look for these x rays (emitted *before* the material crosses the event horizon) to signal the presence of a black hole. Several promising candidates have been found, and astronomers now express considerable confidence in the existence of black holes.

Black holes in binary star systems like the one depicted in Fig. 12–21 have masses a few times greater than the sun's mass. There is also mounting evidence for the

12–21 A binary star system in which an ordinary star and a black hole orbit each other. Matter is pulled from the ordinary star to form an accretion disk around the black hole. The gas in the accretion disk is compressed and heated to such high temperatures that it becomes an intense source of x rays.

existence of much larger *supermassive black holes.* One example is believed to occur in the galaxy M87, a collection of about 10^{12} stars located 50 million light years away. Figure 12–22 shows a 6000-light-year-long jet of material being ejected from the center of M87 at speeds of about one tenth the speed of light. At the galaxy's center itself, shown in the inset, a spiral-shaped disk of hot gas 300 light years across is swirling about a massive unseen object at the center of the disk. The most plausible, albeit bizarre, explanation for both of these phenomena is that there is a black hole at the center of the disk, with about 3×10^9 times the mass of the sun. In this model, material in the disk spirals in to form a hot accretion disk within about 10^{11} km of the black hole. But the Schwarzschild radius of the black hole is only about 10^{10} km (roughly the same size as the orbit of the planet Pluto around the sun), and not all of the infalling material hits this relatively small target. The material that misses is swirled outward at high speed

12–22 Two Hubble Space Telescope images of the galaxy M87. At the center of M87, shown in the inset, gas clouds orbit at speeds in excess of 550 km/s around an invisible object that is presumed to be a supermassive black hole.

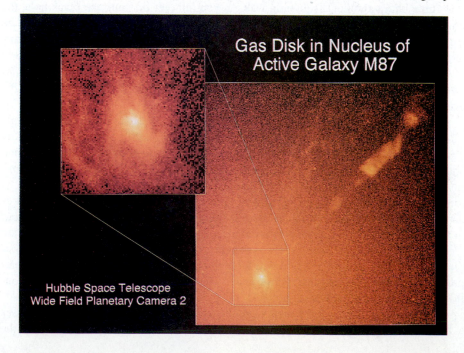

along the rotation axis of the accretion disk, and magnetic forces cause this material to form jets like the one shown in Fig. 12–22.

Astronomical observations like those for M87 indicate that there are supermassive black holes at the center of a number of galaxies. In fact, there is now compelling evidence that a black hole with a mass more than 2×10^6 times that of the sun lies at the center of our own Milky Way galaxy. Observational and theoretical studies of such black holes continue to be an exciting area of research in contemporary physics and astronomy.

SUMMARY

- Newton's law of gravitation states that two particles with masses m_1 and m_2, a distance r apart, attract each other with forces of magnitude

$$F_g = G \frac{m_1 m_2}{r^2}. \qquad (12\text{–}1)$$

These forces form an action-reaction pair and obey Newton's third law. When two or more bodies exert gravitational forces on a particular body, the total gravitational force on that particular body is the vector sum of the forces exerted by the other bodies individually.

- The weight w of a body is the total gravitational force exerted on it by all other bodies in the universe. Near the surface of the earth (mass m_E and radius R_E), this is essentially equal to the gravitational force of the earth alone. The weight w of a body with mass m is then

$$w = F_g = \frac{G m_E m}{R_E^{\;2}}, \qquad (12\text{–}3)$$

and the acceleration due to gravity g is

$$g = \frac{G m_E}{R_E^{\;2}}. \qquad (12\text{–}4)$$

- The gravitational potential energy U of two spherically symmetric bodies of masses m and m_E with their centers separated by a distance r is

$$U = - \frac{G m_E m}{r}. \qquad (12\text{–}9)$$

This potential energy is never positive; it is zero only when the two bodies are infinitely far apart.

- When a satellite moves in a circular orbit, the centripetal acceleration is provided by the gravitational attraction of the earth. The speed v and period T of a satellite in a circular orbit with radius r are

$$v = \sqrt{\frac{G m_E}{r}}, \qquad (12\text{–}12)$$

$$T = \frac{2\pi r}{v} = 2\pi r \sqrt{\frac{r}{G m_E}} = \frac{2\pi r^{3/2}}{\sqrt{G m_E}}. \qquad (12\text{–}14)$$

- Kepler's three laws describe characteristics of the elliptical orbits of planets around the sun or satellites around a planet.
- The gravitational interaction of any spherically symmetric mass distribution, at points outside the distribution, is the same as though all the mass were concentrated at the center.

KEY TERMS

law of gravitation, 358
gravitational constant, 358
gravitational potential energy, 365
escape speed, 365
closed orbit, 367
open orbit, 368
semi-major axis, 371
eccentricity, 371
true weight, 377
apparent weight, 377
black hole, 380
Schwarzschild radius, 380
event horizon, 380

- Because of the earth's rotation, a body's apparent weight at the equator is about 0.3% less than its true weight, and the free-fall acceleration g is about the same percentage less than the value it would have without rotation.

- If a nonrotating spherical mass distribution with total mass M has a radius less than $R_S = 2\,GM/c^2$, called the Schwarzschild radius, then the gravitational interaction prevents anything, including light, from escaping from within a sphere with radius R_S. Such a body is called a black hole.

DISCUSSION QUESTIONS

Q12–1 A student wrote: "The only reason an apple falls downward to meet the earth instead of the earth rising upward to meet the apple is that the earth is much more massive and so exerts a much greater pull." Please comment.

Q12–2 To use the Cavendish balance (Fig. 12–4) to determine G, you must first know the masses m_1 and m_2. How could you determine these masses *without* weighing them?

Q12–3 If all planets had the same average density, how would the acceleration due to gravity at the surface of a planet depend on its radius?

Q12–4 Is a pound of butter on earth the same amount as a pound of butter on Mars? What about a kilogram of butter? Explain.

Q12–5 Example 12–2 (Section 12–2) shows that the acceleration of each sphere caused by the gravitational force is inversely proportional to the mass of that sphere. So why does the force of gravity give all masses the same acceleration when dropped near the surface of the earth?

Q12–6 When will you attract the sun more: today at noon, or tonight at midnight? Explain.

Q12–7 Since the moon is constantly attracted toward the earth by the gravitational interaction, why doesn't it crash into the earth?

Q12–8 Imagine that the sun somehow became twice as massive as it is now. What effect would this have on your weight on earth (as measured by standing on a scale)? Explain your answer.

Q12–9 The sun pulls on the moon with a force that is more than twice the magnitude of the force with which the earth attracts the moon. Why, then, doesn't the sun take the moon away from the earth?

Q12–10 As defined in Chapter 7, gravitational potential energy is $U = mgy$ and is positive for a body of mass m above the earth's surface (which is at $y = 0$). But in this chapter, gravitational potential energy is $U = -Gm_Em/r^2$, which is *negative* for a body of mass m above the earth's surface (which is at $r = R_E$). How can you reconcile these seemingly incompatible descriptions of gravitational potential energy?

Q12–11 A planet is moving at constant speed in a circular orbit around a star. In one complete orbit, what is the net amount of work done on the planet by the star's gravitational force: positive, negative, or zero? What if the planet's orbit is an ellipse, so that the speed is not constant? Explain your answers.

Q12–12 Does the escape speed for an object at the earth's surface depend on the direction in which it is launched? Explain.

Does your answer depend on whether or not you include the effects of air resistance?

Q12–13 If a projectile is fired straight up from the earth's surface, what would happen if the total mechanical energy (kinetic plus potential) is a) less than zero? c) greater than zero? In each case, ignore air resistance and the gravitational effects of the sun, the moon and the other planets.

Q12–14 Discuss whether the following statement is correct: "In the absence of air resistance, the trajectory of a projectile thrown near the earth's surface is an *ellipse*, not a parabola."

Q12–15 The earth is closer to the sun in November than in May. In which month does it move faster in its orbit? Explain.

Q12–16 A communications firm wants to place a satellite in orbit so that it is always directly above the earth's 45th parallel (latitude 45° north). This means that the plane of the orbit will not pass through the center of the earth. Is such an orbit possible? Why or why not?

Q12–17 At what point in an elliptical orbit is the acceleration maximum? At what point is it minimum? Justify your answers.

Q12–18 Which takes more fuel, a voyage from the earth to the moon or from the moon to the earth? Explain.

Q12–19 What would Kepler's third law be for circular orbits if an amendment to Newton's law of gravitation made the gravitational force inversely proportional to r^3? Would this change affect Kepler's other two laws? Explain.

Q12–20 In the elliptical orbit of Halley's comet, shown in Fig. 12–15a, the sun's gravity is responsible for making the comet fall inward from aphelion to perihelion. But what is responsible for making the comet move from perihelion back outward to aphelion?

Q12–21 Many people believe that orbiting astronauts feel weightless because they are "beyond the pull of the earth's gravity." How far from the earth would a spacecraft have to travel to be truly beyond the earth's gravitational influence? If a spacecraft were really unaffected by the earth's gravity, would it remain in orbit? Explain. What is the real reason why astronauts in orbit feel weightless?

Q12–22 As part of their training before going into orbit, astronauts ride in an airliner that is flown along the same parabolic trajectory as a freely falling projectile. Explain why this gives the same experience of apparent weightlessness as being in orbit.

EXERCISES

SECTION 12–2 NEWTON'S LAW OF GRAVITATION

12–1 What is the ratio of the gravitational pull of the sun on the moon to that of the earth on the moon? Use the data in Appendix F. Is it more accurate to say that the moon orbits the earth, or that the moon orbits the sun?

12–2 A 2150-kg satellite used in a cellular telephone network is in a circular orbit at a height of 780 km above the surface of the earth. What is the gravitational force on the satellite? What fraction is this of its weight at the surface of the earth?

12–3 A set of two uniform spheres of masses m_1 and m_2 have their centers separated by a distance r_{12}. They exert a gravitational force of magnitude F_{12} on each other. What magnitude of gravitational force does a second set of uniform spheres exert on one another if they have masses nm_1 and nm_2 and a center-to-center separation of nr_{12}, where n is any positive number?

12–4 Two uniform spheres, each of mass M and radius R, touch one another. What is the magnitude of their gravitational force of attraction?

12–5 An interplanetary spaceship passes through the point in space where the gravitational forces from the sun and the earth on the ship exactly cancel. a) How far from the center of the earth is it? Use the data in Appendix F. b) What, if anything, happens to the spaceship when it passes through the point described in part (a)? Explain.

12–6 a) In Fig. 12–23 what are the magnitude and direction of the net gravitational force exerted on the 0.100-kg uniform sphere by the other two uniform spheres? The centers of all three spheres are on the same line. b) According to Newton's third law, does the 0.100-kg sphere exert forces of the same magnitude as your answer to part (a), but opposite direction, on *each* of the other two spheres?

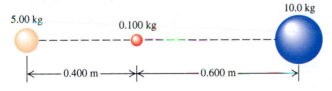

FIGURE 12–23 Exercise 12–6.

12–7 The typical adult human has a mass of about 70 kg. What force does a full moon exert on such a human when it is directly above with its center 378,000 km away?

12–8 The sun has a mass 333,000 times that of the earth. For a person on earth, the average distance to the center of the sun is 23,500 times the distance to the center of the earth. In magnitude, what is the ratio of the sun's gravitational force on you to the earth's gravitational force on you?

12–9 Calculate the magnitude and direction of the net gravitational force on the moon due to the earth and the sun when the moon is in each of the positions shown in Fig. 12–24. (Note that the figure is *not* drawn to scale. Assume that the sun is in the

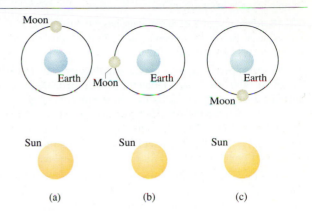

FIGURE 12–24 Exercise 12–9

plane of the earth-moon orbit, even though this is usually not the case.) Use the data in Appendix F.

12–10 Determine the magnitude and direction of the gravitational force on a particle of mass m that is midway between two spherically symmetric objects, one of mass m_1 and the other of mass m_2, where $m_2 > m_1$. The centers of the two spherically symmetric objects are a distance d apart.

12–11 Two uniform spheres, each of mass 0.260 kg, are fixed at points A and B (Fig. 12–25). Find the magnitude and direction of the initial acceleration of a uniform sphere with mass 0.010 kg if released from rest at point P and acted on only by forces of gravitational attraction of the spheres at A and B.

FIGURE 12–25 Exercise 12–11.

SECTION 12–3 WEIGHT

12–12 Use the mass and radius of the planet Pluto given in Appendix F to calculate the acceleration due to gravity at the surface of Pluto.

12–13 At what distance above the surface of the earth is the acceleration due to the earth's gravity 0.980 m/s^2 if the acceleration due to gravity at the surface has magnitude 9.80 m/s^2?

12–14 The mass of Venus is 81.5% that of the earth, and its radius is 94.9% that of the earth. a) Compute the acceleration due to gravity on the surface of Venus from these data. b) What is the weight of a 5.00-kg rock on the surface of Venus?

12–15 Titania, the largest moon of the planet Uranus, has $\frac{1}{8}$ the radius of the earth and $\frac{1}{1700}$ the mass of the earth. a) What is the acceleration due to gravity at the surface of Titania? b) What is the average density of Titania? (This is less than the density of rock, which is one piece of evidence that Titania is made primarily of ice.)

12–16 Rhea, one of Saturn's moons, has a radius of 765 km and an acceleration due to gravity of 0.278 m/s² at its surface. Calculate its mass and average density.

12–17 An experiment using the Cavendish balance to measure the gravitational constant G found that a uniform 0.400-kg sphere attracts another 0.00300-kg sphere with a force of 8.00×10^{-10} N when the distance between the centers of the spheres is 0.0100 m. The acceleration due to gravity at the earth's surface is 9.80 m/s², and the radius of the earth is 6380 km. Compute the mass of the earth from these data.

12–18 In Example 12–4 (Section 12–3) we neglected the gravitational effects of the moons of Mars. The larger of the two moons, Phobos, has a radius of approximately 12 km and an average density of 2000 kg/m³. a) Calculate the gravitational force of Phobos on the Mars lander of Example 12–4 if the Mars lander were on the surface of Phobos. b) Discuss whether the approximation made in Example 12–4 was appropriate.

SECTION **12–4 GRAVITATIONAL POTENTIAL ENERGY**

12–19 The asteroid Dactyl, discovered in 1993, has a radius of only about 700 m and a mass of about 3.6×10^{12} kg. Use the results of Example 12–5 (Section 12–4) to calculate the escape speed for an object at the surface of Dactyl. Could a person reach this speed just by walking?

12–20 When in orbit, a communication satellite attracts the earth with a force of 19.0 kN and the earth-satellite gravitational potential energy (relative to zero at infinite separation) is -1.39×10^{11} J. a) Find the satellite's altitude above the earth's surface. b) Find the mass of the satellite.

12–21 Use the results of Example 12–5 (Section 12–4) to calculate the escape speed for a spacecraft a) from the surface of Mars; b) from the surface of Jupiter. Use the data in Appendix F. c) Why is the escape speed for a spacecraft independent of the spacecraft's mass?

12–22 Ten days after it was launched toward Mars in December 1998, the *Mars Climate Orbiter* spacecraft (mass 629 kg) was 2.87×10^6 km from the earth and traveling at 1.20×10^4 km/h relative to the earth. At this time, what were a) the spacecraft's kinetic energy relative to the earth and b) the potential energy of the earth-spacecraft system?

SECTION **12–5 THE MOTION OF SATELLITES**

12–23 An earth satellite moves in a circular orbit with an orbital speed of 6200 m/s. a) Find the time of one revolution. b) Find the radial acceleration of the satellite in its orbit.

12–24 What is the period of revolution of a satellite of mass m that orbits the earth in a circular path of radius 7880 km (about 1500 km above the surface of the earth)?

12–25 For a satellite to be in a circular orbit 780 km above the surface of the earth, what must be its orbital speed?

SECTION **12–6 THE MOTION OF PLANETS**

12–26 For a time in the nineteenth century it was thought (incorrectly) that there was a planet called Vulcan orbiting the sun in a circular orbit with a period of 19.7 days. If Vulcan had turned out to be real, what would have been the radius of its orbit? (The name "Vulcan" may have been the inspiration for the planet of the same name in *Star Trek*.)

12–27 The star Rho¹ Cancri is 57 light years from the earth and has a mass 0.85 times that of our sun. A planet has been detected in a circular orbit around Rho¹ Cancri with an orbital radius equal to 0.11 times the radius of the earth's orbit around the sun. What is a) the orbital speed and b) the orbital period of the planet of Rho¹ Cancri?

12–28 Venus orbits the sun in a nearly circular orbit. Use the data in Appendix F for the orbital radius and period of Venus to calculate the mass of the sun.

12–29 a) Use Fig. 12–13 to show that the sun-planet distance at perihelion is $(1 - e)a$, the sun-planet distance at aphelion is $(1 + e)a$, and therefore that the sum of these two distances is $2a$. b) Pluto is usually called the outermost planet, but at perihelion in 1989 it was almost 100 million km closer to the sun than Neptune. The semi-major axes of the orbits of Pluto and Neptune are 5.92×10^{12} m and 4.50×10^{12} m, respectively, and the eccentricities are 0.248 and 0.010. Find Pluto's closest distance and Neptune's farthest distance from the sun. c) How many years after being at perihelion in 1989 will Pluto again be at perihelion?

12–30 The star 70 Virginis is 59 light years from earth and has a mass of 1.9×10^{30} kg. a) A large planet of mass 1.3×10^{28} kg is known to orbit this star. The planet is attracted to the star by a force of 3.3×10^{26} N when their centers are separated by a distance equal to the semi-major axis of the planet's orbit. Calculate this distance (in km). b) What is the period (in days) of this planet's orbit? c) Other as yet undiscovered planets may orbit 70 Virginis. If a second planet is in a circular orbit around 70 Virginis with an orbital period equal to eight times the value found in part (b), what is the radius (in km) of the second planet's orbit?

12–31 The *Helios B* spacecraft had a speed of 71 km/s when it was 4.3×10^7 km from the sun. a) Prove that it was not in a circular orbit about the sun. b) Prove that its orbit about the sun was closed and therefore elliptical.

*SECTION **12–7 SPHERICAL MASS DISTRIBUTIONS**

***12–32** A thin, uniform rod has length L and mass M. A small uniform sphere of mass m is placed a distance x from one end of the rod, along the axis of the rod (Fig. 12–26). a) Calculate the gravitational potential energy of the rod-sphere system. Take the potential energy to be zero when the rod and sphere are infinitely far apart. Show that your answer reduces to the expected result when x is much larger than L. (*Hint:* Use the power series expansion for $\ln(1 + x)$ given in Appendix B.) b) Use $F_x = -dU/dx$ to find the magnitude and direction of the gravitational force exerted on the sphere by the rod. Show that your answer reduces to the expected result when x is much larger than L.

FIGURE 12–26 Exercise 12–32, Problem 12–71.

*12–33 Consider the ring-shaped body of Fig. 12–27. A particle of mass m is placed a distance x from the center of the ring, along the line through the center of the ring and perpendicular to its plane. a) Calculate the gravitational potential energy U of this system. Take the potential energy to be zero when the two objects are far apart. b) Show that your answer to part (a) reduces to the expected result when x is much larger than the radius a of the ring. c) Use $F_x = -dU/dx$ to find the magnitude and direction of the force on the particle. d) Show that your answer to part (c) reduces to the expected result when x is much larger than a. e) What are the values of U and F_x when $x = 0$? Explain why these results make sense.

FIGURE 12–27 Exercise 12–33, Problem 12–70.

*SECTION 12–8 APPARENT WEIGHT AND THE EARTH'S ROTATION

*12–34 The weight of a person at the equator, as determined by a spring balance, is 690 N. By how much does this differ from the true force of gravitational attraction at the same point? Assume that the earth is spherically symmetric.

*12–35 The acceleration due to gravity at the north pole of Neptune is approximately 10.7 m/s^2. Neptune has mass 1.0×10^{26} kg and radius 2.5×10^4 km and rotates once around its axis in about 16 h. a) What is the gravitational force on a 5.0-kg object at the north pole of Neptune? b) What is the apparent weight of this same object at Neptune's equator? (Note that Neptune's "surface" is gaseous, not solid, so it is impossible to stand on it.)

SECTION 12–9 BLACK HOLES: A CASE STUDY IN MODERN PHYSICS

12–36 a) Show that a black hole attracts an object of mass m with a force of $mc^2 R_S/(2r^2)$, where r is the distance between the object and the center of the black hole. b) Calculate the magnitude of the gravitational force exerted by a black hole of Schwarzschild radius 14.0 mm on a 5.00-kg mass 3000 km from it. c) What is the mass of this black hole?

12–37 At the Galaxy's Core. Astronomers have observed a small, massive object at the center of our Milky Way galaxy. A ring of material orbits this massive object; the ring has a diameter of about 15 light years and an orbital speed of about 200 km/s. a) Determine the mass of the massive object at the center of the Milky Way galaxy. Give your answer both in kilograms and in solar masses (one solar mass is the mass of the sun). b) Observations of stars, as well as theories of the structure of stars, suggest that it is impossible for a single star to have a mass of more than about 50 solar masses. Can this massive object be a single, ordinary star? c) Many astronomers believe that the massive object at the center of the Milky Way galaxy is a black hole. If so, what must the Schwarzschild radius of this black hole be? Would a black hole of this size fit inside the earth's orbit around the sun?

12–38 Derive the relation $R_S = (3.0 \text{ km/solar mass})M$, where M is in solar masses. (One solar mass is the mass of the sun.) How accurate is this relation?

12–39 To what fraction of its current radius would the earth have to be compressed to become a black hole?

PROBLEMS

12–40 Three uniform spheres are fixed at positions shown in Fig. 12–28. a) What is the magnitude and direction of the force on a 0.0150-kg particle placed at P? b) If the spheres are in deep outer space and a 0.0150-kg particle is released from rest 300 m from the origin along a line 45° below the $-x$-axis, what will the particle's speed be when it reaches the origin?

FIGURE 12–28 Problem 12–40.

12–41 A uniform sphere with mass 60.0 kg is held with its center at the origin and a second uniform sphere with mass 80.0 kg is held with its center at the point $x = 0$, $y = 3.00$ m. a) What are the magnitude and direction of the net gravitational force due to these objects on a third uniform sphere with mass 0.500 kg placed at the point $x = 4.00$ m, $y = 0$? b) Where, other than infinitely far away, could the third sphere be placed such that the net gravitational force acting on it from the other two spheres is equal to zero?

12–42 a) Show that the gravitational force on the small mass due to the two large masses in Example 12–3 (Section 12–2) is *not* directed toward the point midway between the two large masses. b) Consider the two large masses as making up a single rigid body. (That is, the two masses are joined by a rod of negligible mass). Calculate the torque exerted by the small mass on the rigid body for a pivot at its center of mass. c) Explain how the result in part (b) shows that the center of mass does not coincide with the center of gravity. Why is this the case in this situation?

12–43 At a certain instant the earth, the moon, and a stationary 1250-kg spacecraft lie at the vertices of an equilateral triangle whose sides are 3.84×10^5 km in length. a) Find the magnitude and direction of the net gravitational force exerted on the spacecraft by the earth and moon. State the direction as an angle measured from a line connecting the earth and the spacecraft. In a sketch, show the earth, the moon, the spacecraft, and the force vector. b) What is the minimum amount of work that you would have to do to move the spacecraft to a point far from the earth and moon? Ignore any gravitational effects due to the other planets or the sun.

12–44 An experiment is performed in deep space with two uniform spheres of equal radii, one with mass 25.0 kg and the other with mass 100.0 kg. The spheres are released from rest with their centers 40.0 m apart. They accelerate toward each other because of their mutual gravitational attraction. Neglect all gravitational forces other than that between the two spheres. a) Explain why linear momentum is conserved. b) When their centers are 20.0 m apart, find i) the speed of each sphere and ii) the magnitude of the relative velocity with which one sphere is approaching the other. c) How far from the initial position of the center of the 25.0-kg sphere do the surfaces of the two spheres collide?

12–45 Assume that the moon orbits the earth in a circular orbit. From the observed orbital period of 27.3 d, calculate the distance of the moon from the center of the earth. Assume that the motion of the moon is determined solely by the gravitational force exerted on it by the earth, and use the mass of the earth given in Appendix F.

12–46 Gravitational Field. The *gravitational field* $\vec{g}$ produced by an object is defined as the gravitational force $\vec{F}_g$ that the object exerts on a small test particle of mass m divided by m. That is, $\vec{g} = \vec{F}_g/m$. What are the magnitude and direction of the gravitational field produced by a 20.0-kg spherically symmetric object at a point 1.50 m directly above the center of the object?

12–47 Geosynchronous Satellites. Many satellites are moving in a circle in the earth's equatorial plane. They are at such a height above the earth's surface that they remain always above the same point. a) Find the altitude of these satellites above the earth's surface. (Such an orbit is said to be *geosynchronous*.) b) Explain, with a sketch, why the radio signals from these satellites cannot directly reach receivers on earth that are north of 81.3° N latitude.

12–48 Suppose an object is to be placed in a circular orbit around the asteroid Dactyl (see Exercise 12–19), with the radius of its orbit just slightly larger than the asteroid's radius. What is the speed of the object? Could you launch yourself into orbit around Dactyl just by walking?

12–49 What is the escape speed from a 300-km-diameter asteroid with a density of 2500 kg/m³?

12–50 a) Asteroids have average densities of about 2500 kg/m³ and radii from 470 km down to less than a kilometer. Assuming that the asteroid has a spherically symmetric mass distribution, estimate the radius of the largest asteroid from which you could escape simply by jumping off. (*Hint*: You can estimate your jump speed by relating it to the maximum height that you can jump on earth.) b) Europa, one of Jupiter's four large moons, has a radius of 1570 km. The acceleration due to gravity at its surface is 1.33 m/s². Calculate its average density.

12–51 There are two equations from which a change in the gravitational potential energy U of the system of a mass m and the earth can be calculated. One is $U = mgy$ (Eq. 7–2). The other

is $U = -Gm_Em/r$ (Eq. 12–9). As shown in Section 12–4, the first equation is correct only if the gravitational force is a constant over the change in height Δy. The second is always correct. Actually, the gravitational force is never exactly constant over any change in height, but if the variation is small we can ignore it. Consider the difference in U between a mass at the earth's surface and a distance h above it using both equations, and find the value of h for which Eq. (7–2) is in error by 1%. Express this value of h as a fraction of the earth's radius, and also obtain a numerical value for it.

12–52 In Example 12–5 (Section 12–4) we ignored the gravitational effects of the moon on a spacecraft en route from the earth to the moon. In fact, we must include the gravitational potential energy due to the moon as well. For this problem, ignore the motion of the earth and moon. a) If the moon has radius R_M and the distance between the centers of the earth and the moon is R_{EM}, find the total gravitational potential energy of the particle-earth and particle-moon systems when a particle of mass m is between the earth and the moon and a distance r from the center of the earth. Take the gravitational potential energy to be zero when the objects are far from each other. b) There is a point along a line between the earth and the moon where the net gravitational force is zero. Use the expression derived in part (a) and numerical values from Appendix F to find the distance of this point from the center of the earth. With what speed must a spacecraft be launched from the surface of the earth to just barely reach this point? c) If a spacecraft were launched from the earth's surface toward the moon with an initial speed of 11.2 km/s, with what speed would it impact the moon?

12–53 An unmanned spacecraft is in a circular orbit around the moon, observing the lunar surface from an altitude of 50.0 km. (See Appendix F.) To the dismay of scientists on earth, an electrical fault causes an on-board thruster to fire, decreasing the speed of the spacecraft by 20.0 m/s. If nothing is done to correct its orbit, with what speed (in km/h) will the spacecraft crash into the lunar surface?

*12–54 What would be the length of a day (that is, the time required for one rotation of the earth on its axis) if the rate of rotation of the earth were such that $g = 0$ at the equator?

12–55 A hammer with mass m is dropped from rest from a height h above the earth's surface. This height is not necessarily small compared with the radius R_E of the earth. If air resistance is neglected, derive an expression for the speed v of the hammer when it reaches the surface of the earth. Your expression should involve h, R_E, and m_E, the mass of the earth.

12–56 a) Calculate how much work is required to launch a spacecraft of mass m from the surface of the earth (mass m_E, radius R_E) and place it in a circular *low earth orbit*, that is, an orbit whose altitude above the earth's surface is much less than R_E. (As an example, the International Space Station is in low earth orbit at an altitude of about 400 km, much less than $R_E = 6380$ km.) Ignore the kinetic energy that the spacecraft has on the ground due to the earth's rotation. b) Calculate the minimum amount of additional work required to move the spacecraft from low earth orbit to a very great distance from the earth. Ignore the gravitational effects of the sun, the moon, and the other planets. c) Justify the statement that "in terms of energy, low earth orbit is halfway to the edge of the universe."

12–57 A spacecraft is to be launched from the surface of the earth so that it will escape from the solar system altogether. a) Find the speed relative to the center of the earth with which it must be launched. Take into consideration the gravitational effects of both the earth and the sun and include the effects of the earth's orbital speed, but ignore air resistance. b) The rotation of the earth can help this spacecraft achieve escape speed. Find the speed that the spacecraft must have relative to the earth's *surface* if the spacecraft is launched from Florida at the point shown in Fig. 12–29. The rotation and orbital motions of the earth are in the same direction. The launch facilities at Cape Canaveral are 28.5° north of the equator. c) The European Space Agency (ESA) uses launch facilities in French Guiana (immediately north of Brazil), 5.15° north of the equator. What speed relative to the earth's surface would a spacecraft need to escape the solar system if launched from French Guiana?

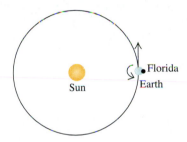

FIGURE 12–29 Problem 12–57.

12–58 We can define a useful quantity called the *gravitational potential* of the earth (denoted by ϕ) by requiring that the product of its value at any point and the mass of a particle placed at that point gives us the gravitational potential energy of that earth-particle system. (We will use an analogous quantity, the electric potential V, in the discussion of electrical interactions in Chapter 24.) a) Find the SI units for the gravitational potential and show they are equivalent to m^2/s^2. b) What is the equation for the gravitational potential at any point outside the earth, if it is taken to be zero infinitely far from the earth? c) What is the increase in gravitational potential between the earth's surface and the International Space Station's altitude of 400 km? d) Use the result of part (c) to calculate the amount of work that must be done against the earth's gravitational force to move a 15,000-kg payload from Cape Canaveral to an altitude of 400 km.

12–59 Kirkwood Gaps. Hundreds of thousands of asteroids orbit the sun within the *asteroid belt*, which extends from about 3×10^8 km to about 5×10^8 km from the sun. a) Find the orbital period (in years) of i) an asteroid at the inside of the belt and ii) an asteroid at the outside of the belt. Assume circular orbits. b) In 1867 the American astronomer Daniel Kirkwood pointed out that several gaps exist in the asteroid belt where relatively few asteroids are found. It is now understood that these *Kirkwood gaps* are caused by the gravitational attraction of Jupiter, largest of the planets, which orbits the sun once every

11.86 years. As an example, if an asteroid has an orbital period half that of Jupiter, or 5.93 years, on every other orbit this asteroid would be at its closest to Jupiter and feel a strong attraction toward the planet. This attraction, acting over and over on successive orbits, could sweep asteroids out of the Kirkwood gap. Use this hypothesis to determine the orbital radius for this Kirkwood gap. c) One of several other Kirkwood gaps appears at a distance from the sun where the orbital period is 0.400 that of Jupiter. Explain why this happens, and find the orbital radius for this Kirkwood gap.

12–60 If a satellite is in a sufficiently low orbit, it will encounter air drag from the earth's atmosphere. Since air drag does negative work (the force of air drag is directed opposite the motion), the mechanical energy will decrease. According to Eq. (12–15), if E decreases (becomes more negative), the radius r of the orbit will decrease. If air drag is relatively small, the satellite can be considered to be in a circular orbit of continually decreasing radius. a) According to Eq. (12–12), if the radius of a satellite's circular orbit decreases, the satellite's orbital speed v increases. How can you reconcile this with the statement that the mechanical energy *decreases*? (*Hint:* Is air drag the only force that does work on the satellite as the orbital radius decreases?) b) Due to air drag, the radius of a satellite's circular orbit decreases from r to $r - \Delta r$, where the positive quantity Δr is much less than r. The mass of the satellite is m. Show that the increase in orbital speed is $\Delta v = +(\Delta r/2) \sqrt{Gm_E/r^3}$; that the change in kinetic energy is $\Delta K = -W = +(Gm_E m/2r^2)\Delta r$; that the change in gravitational potential energy is $\Delta U = 2W = -2\Delta K = -(Gm_E m/r^2)\Delta r$; and that the amount of work done by the force of air drag is $W = -(Gm_E m/2r^2)\Delta r$. Interpret these results in light of your comments in part (a). c) A satellite of mass 3000 kg is initially in a circular orbit 300 km above the earth's surface. Due to air drag, the satellite's altitude decreases to 250 km. Calculate the initial orbital speed; the increase in orbital speed; the initial mechanical energy; the change in kinetic energy; the change in gravitational potential energy; the change in mechanical energy; and the work done by the force of air drag. d) Eventually a satellite will descend to a low enough altitude in the atmosphere that the satellite burns up and the debris falls to earth. What becomes of the initial mechanical energy?

12–61 Binary Star—Equal Masses. Two identical stars of mass M orbit around their center of mass. Each orbit is circular and has radius R, so that the two stars are always on opposite sides of the circle. a) Find the gravitational force of one star on the other. b) Find the orbital speed of each star and the period of the orbit. c) How much energy would be required to separate the two stars to infinity?

12–62 Binary Star—Different Masses. Two stars, of mass M_1 and M_2, are in circular orbits around their center of mass. The star of mass M_1 has an orbit of radius R_1; the star of mass M_2 has an orbit of radius R_2. a) Show that the ratio of the orbital radii of the two stars equals the reciprocal of the ratio of their masses, that is, $R_1/R_2 = M_2/M_1$. b) Explain why the two stars have the same orbital period, and show that the period T is given by $T = 2\pi(R_1 + R_2)^{3/2}/\sqrt{G(M_1 + M_2)}$. c) The two stars in a certain binary star system move in circular orbits. The first star, Alpha, has an orbital speed of 36.0 km/s. The other star, Beta, has an orbital speed of 12.0 km/s. The orbital period is 137 d. What are the masses of each of the two stars? d) One of the best candidates for a black hole is found in the binary system called A0620-0090. The two objects in the binary system are an orange star, V616 Monocerotis, and a compact object believed to be a black hole (Fig. 12–21). The orbital period of A0620-0090 is 7.75 hours, the mass of V616 Monocerotis is estimated to be 0.67 times the mass of the sun, and the mass of the black hole is estimated to be 3.8 times the mass of the sun. Assuming that the orbits are circular, find the radius of each object's orbit and the orbital speed of each object. Compare these answers to the orbital radius and orbital speed of the earth in its orbit around the sun.

12–63 Comets travel around the sun in elliptical orbits with large eccentricities. If a comet has speed 2.0×10^4 m/s when at a distance of 2.5×10^{11} m from the center of the sun, what is its speed when at a distance of 5.0×10^{10} m?

12–64 As Mars orbits the sun in its elliptical orbit, its distance of closest approach to the center of the sun (at perihelion) is 2.067×10^{11} m, and its maximum distance from the center of the sun (at aphelion) is 2.492×10^{11} m. If the orbital speed of Mars at aphelion is 2.198×10^4 m/s, what is its orbital speed at perihelion? (Neglect the influence of the other planets.)

12–65 Consider a spacecraft in an elliptical orbit around the earth. At the low point, or perigee of its orbit, it is 400 km above the earth's surface; at the high point, or apogee, it is 4000 km above the earth's surface. a) What is the period of the spacecraft's orbit? b) Using conservation of angular momentum, find the ratio of the spacecraft's speed at perigee to its speed at apogee. c) Using conservation of energy, find the speed at perigee and the speed at apogee. d) It is desired to have the spacecraft escape from the earth completely. If the spacecraft's rockets are fired at perigee, by how much would the speed have to be increased to achieve this? What if the rockets were fired at apogee? Which point in the orbit is the more efficient to use?

12–66 The planet Uranus has a radius of 25,560 km and a surface acceleration due to gravity of 11.1 m/s^2 at its poles. Its moon Miranda (discovered by Kuiper in 1948) is in a circular orbit about Uranus at an altitude of 104,000 km above the planet's surface. Miranda has a mass of 6.6×10^{19} kg and a radius of 235 km. a) Calculate the mass of Uranus from the above data. b) Calculate the magnitude of Miranda's acceleration due to its orbital motion about Uranus. c) Calculate the acceleration due to Miranda's gravity at the surface of Miranda. d) Do the answers to parts (b) and (c) mean that an object released 1 m above Miranda's surface on the side toward Uranus will fall *up* relative to Miranda? Explain.

12–67 A 3000-kg spacecraft is in a circular orbit 2000 km above the surface of Mars. How much work must the spacecraft engines perform to move the spacecraft to a circular orbit 4000 above the surface?

12–68 In Example 12–8 (Section 12–6) the orbital period of Halley's comet was calculated to be 75.5 years. Many other comets have orbits that extend much farther from the sun, and hence have much longer periods. a) At aphelion, a typical long-period comet is about 8×10^{12} km from the sun; at perihelion, it passes inside the earth's orbit. Estimate the orbital period of such a comet. Give your answer in years. b) Estimate the speed of a typical long-period comet at perihelion. (*Hint:* At aphelion, the comet is moving very slowly.) c) The massive nucleus of a comet contains about 10^{15} kg of material. If the earth were to be struck by a long-period comet, estimate the kinetic energy the comet would have just before impact. (*Hint:* The motion of the comet is due primarily to the gravitational attraction of the sun, not the earth. Why?) Compare this to the energy released in a very large volcanic eruption, about 6×10^{18} J, and to the energy that would be released by burning all the fossil fuel on earth, about 2×10^{23} J. An impact of this kind apparently occurred 65 million years ago in the Yucatan, and is implicated in the demise of the dinosaurs as well as many other ancient species.

12–69 One of the brightest comets of the twentieth century was Comet Hyakutake, which passed close to the sun in early 1996. The orbital period of this comet is estimated to be about 30,000 years. Find the semi-major axis of this comet's orbit. Compare to the average sun-Pluto distance and to the distance to Alpha Centauri, the nearest star to the sun, which is 4.3 light years distant.

*12–70 An object in the shape of a thin ring has radius a and mass M. A uniform sphere with a mass m and radius R is placed with its center at a distance x to the right of the center of the ring along a line through the center of the ring and perpendicular to its plane (Fig. 12–27). What is the gravitational force that the sphere exerts on the ring-shaped object? Show that your result reduces to the expected result when x is much larger than a.

*12–71 A thin uniform rod has length L and mass M. Calculate the magnitude of the gravitational force the rod exerts on a particle of mass m that is at a point along the axis of the rod a distance x from one end (Fig. 12–26). Show that your result reduces to the expected result when x is much larger than L.

*12–72 A shaft is drilled from the surface to the center of the earth (Fig. 12–18). As in Example 12–9 (Section 12–7), make the rather unrealistic assumption that the density of the earth is uniform. With this approximation, the gravitational force on an object of mass m that is inside the earth at a distance r from the center has magnitude $F_g = Gm_E mr/R_E^3$ (as shown in Example 12–9) and points toward the center of the earth. a) Derive an expression for the gravitational potential energy $U(r)$ of the object-and-earth system as a function of the object's distance from the center of the earth. Take the potential energy to be zero when the object is at the center of the earth. b) If an object is released in the shaft at the surface of the earth, what speed will it have when it reaches the center of the earth?

CHALLENGE PROBLEMS

12-73 a) When an object is in a circular orbit of radius r around the earth (mass m_E), the period of the orbit is T (given by Eq. (12–14)) and the orbital speed is v (given by Eq. (12–12)). Show that when the object is moved into a circular orbit of slightly larger radius $r + \Delta r$, where $\Delta r \ll r$, its new period is $T + \Delta T$ and its new orbital speed is $v - \Delta v$, where Δr, ΔT, and Δv are all positive quantities and

$$\Delta T = \frac{3\pi \Delta r}{v} \quad \text{and} \quad \Delta v = \frac{\pi \Delta r}{T}$$

(*Hint:* Use the expression $(1 + x)^n \approx 1 + nx$, valid for $|x| \ll 1$.) b) The International Space Station (ISS) is in a nearly circular orbit at an altitude of 398.00 km above the surface of the earth. A maintenance crew is about to arrive on a space shuttle that is also in a circular orbit in the same orbital plane as the ISS, but with an altitude of 398.10 km. The crew has come to remove a faulty 125-m electrical cable, one end of which is attached to the ISS and the other end of which is floating free in space. The plan is for the shuttle to snag the free end just at the moment that the shuttle, the ISS, and the center of the earth all lie along the same line. The cable will then break free from the ISS when it becomes taut. How long after the free end is caught by the space shuttle will it detach from the ISS? Give your answer in minutes. c) If the shuttle misses catching the cable, show that the crew must wait a time $t \approx T^2/\Delta T$ before they have a second chance. Find the numerical value of t and explain whether it would be worth the wait.

12-74 Interplanetary Navigation. The most efficient way to send a spacecraft from the earth to another planet is by using a *Hohmann transfer orbit* (Fig. 12–30). If the orbits of the departure and destination planets are circular, the Hohmann transfer orbit is an elliptical orbit whose perihelion and aphelion are tangent to the orbits of the two planets. The rockets are fired briefly at the departure planet to put the spacecraft into the transfer orbit; the spacecraft then coasts until it reaches the destination planet. The rockets are then fired again to put the spacecraft into the same orbit about the sun as the destination planet. a) For a flight from earth to Mars, in what direction must the rockets be fired at the earth and at Mars: in the direction of motion, or opposite the direction of motion? What about from a flight from Mars to the earth? b) How long does a one-way trip from the

earth to Mars take, between the firings of the rockets? c) To reach Mars from earth, the launch must be timed so that Mars will be at the right spot when the spacecraft reaches Mars's orbit around the sun. At launch, what must be the angle between a sun-Mars line and a sun-earth line? Use data from Appendix F.

12-75 Tidal Forces Near a Black Hole. An astronaut inside a spacecraft, which protects her from harmful radiation, is orbiting a black hole at a distance of 120 km from its center. The black hole is 5.00 times the mass of the sun and has a Schwarzschild radius of 15.0 km. The astronaut is positioned inside the spaceship such that one of her 0.030-kg ears is 6.0 cm farther from the black hole than the center of mass of the spacecraft and the other ear is 6.0 cm closer. a) What is the tension between her ears? Would the astronaut find it difficult to keep from being torn apart by the gravitational forces? (Since her whole body orbits with the same angular velocity, one ear is moving too slowly for the radius of its orbit and the other is moving too fast. Hence her head must exert forces on her ears to keep them in their orbits.) b) Is the center of gravity of her head at the same point as the center of mass? Explain.

***12-76** Mass M is distributed uniformly over a disk of radius a. Find the gravitational force (magnitude and direction) between this disk-shaped mass and a particle of mass m located a distance x above the center of the disk (Fig. 12–31). Does your result reduce to the correct expression as x becomes very large? (*Hint:* Divide the disk into infinitesimally thin concentric rings, use the expression derived in Problem 12–70 for the gravitational force due to each ring, and integrate to find the total force.)

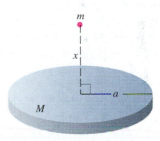

FIGURE 12–31 Challenge Problem 12–76.

***12-77** Mass M is distributed uniformly along a line of length $2L$. A particle of mass m is at a point that is a distance a above the center of the line on its perpendicular bisector (point P in Fig. 12–32). For the gravitational force that the line exerts on the particle, calculate the components perpendicular and parallel to the line. Does your result reduce to the correct expression as a becomes very large?

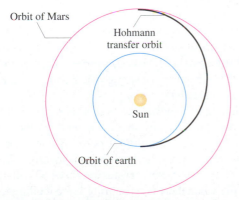

FIGURE 12–30 Challenge Problem 12–74.

FIGURE 12–32 Challenge Problem 12–77.

13 Periodic Motion

An electric toothbrush is an everyday example of periodic motion, or motion that repeats itself. The brush oscillates through 31,000 up-and-down motions per second, dislodging harmful plaque that an ordinary manual toothbrush might not remove. Using such a toothbrush can make it less likely that your teeth will have to encounter another rapidly moving object—a dentist's drill.

13–1 INTRODUCTION

The vibration of a quartz crystal in a watch, the swinging pendulum of a grandfather clock, the sound vibrations produced by a clarinet or an organ pipe, the back-and-forth motion of the pistons in a car engine—all these are examples of motion that repeats itself over and over. We call this **periodic motion** or **oscillation,** and it is the subject of this chapter. Understanding periodic motion will be essential for our later study of waves, sound, alternating electric currents, and light.

A body that undergoes periodic motion always has a stable equilibrium position. When it is moved away from this position and released, a force or torque comes into play to pull it back toward equilibrium. But by the time it gets there, it has picked up some kinetic energy, so it overshoots, stopping somewhere on the other side, and is again pulled back toward equilibrium. Picture a ball rolling back and forth in a round bowl or a pendulum that swings back and forth past its straight-down position.

In this chapter we will concentrate on two simple examples of systems that can undergo periodic motions: spring-mass systems and pendulums. We will also study why oscillations often tend to die out with time and why some oscillations can build up to greater and greater displacements from equilibrium when periodically varying forces act.

13–2 THE CAUSES OF OSCILLATION

One of the simplest systems that can have periodic motion is shown in Fig. 13–1a. A body with mass m moves on a frictionless horizontal guide system, such as a linear air track, so it moves only along the x-axis. The body is attached to a spring of negligible mass that can be either stretched or compressed. The left end of the spring is held fixed, and the right end is attached to the body. The spring force is the only horizontal force acting on the body; the vertical normal and gravitational forces always add to zero. The quantities x, v, a, and F refer to the x-components of the position, velocity, acceleration, and force vectors, respectively. Like all components, they can be positive, negative, or zero.

It's simplest to define our coordinate system so that the origin O is at the equilibrium position, where the spring is neither stretched nor compressed. Then x is the x-component of the **displacement** of the body from equilibrium and is also the change in length of the spring. The x-component of acceleration a is given by $a = F/m$.

Figure 13–1b shows free-body diagrams for three different positions of the body. Whenever the body is displaced from its equilibrium position, the spring force tends to restore it to the equilibrium position. We call a force with this character a **restoring force.** Oscillation can occur only when there is a restoring force tending to return the system to equilibrium.

13-1 Model for periodic motion. (a) When the body is to the left of the equilibrium position, the compressed spring exerts a force to the right. When the body is at the equilibrium position, the spring is neither compressed nor stretched and exerts no horizontal force. When the body is to the right of the equilibrium position, the stretched spring exerts a force to the left. (b) Free-body diagrams for the three positions. If the spring force F is proportional to the displacement from equilibrium, the motion is simple harmonic.

Let's analyze how oscillation occurs in this system. If we displace the body to the right to $x = A$ and then let go, the net force and the acceleration are to the left. The speed increases as the body approaches the equilibrium position O. When the body is at O, the net force acting on it is zero, but because of its motion it *overshoots* the equilibrium position. On the other side of the equilibrium position the velocity is to the left but the acceleration is to the right; the speed decreases until the body comes to a stop. We will show later that with an ideal spring, the stopping point is at $x = -A$. The body then accelerates to the right, overshoots equilibrium again, and stops at the starting point $x = A$, ready to repeat the whole process. The body is oscillating! If there is no friction or other force to remove mechanical energy from the system, this motion repeats itself forever; the restoring force perpetually draws the body back toward the equilibrium position, only to have the body overshoot time after time.

In different situations the force may depend on the displacement x from equilibrium in different ways. But oscillation will *always* occur if the force is a *restoring* force that tends to return the system to equilibrium.

Here are some terms that we'll use in discussing periodic motions of all kinds:

The **amplitude** of the motion, denoted by A, is the maximum magnitude of displacement from equilibrium; that is, the maximum value of $|x|$. It is always positive. If the spring in Fig. 13–1 is an ideal one, the total overall range of the motion is $2A$. The SI unit of A is the meter. A complete vibration, or **cycle,** is one complete round trip, say from A to $-A$ and back to A, or from O to A, back through O to $-A$, and back to O. Note that motion from one side to the other (say, $-A$ to A) is a half-cycle, not a whole cycle.

The **period,** T, is the time for one cycle. It is always positive. The SI unit is the second, but it is sometimes expressed as "seconds per cycle."

The **frequency,** f, is the number of cycles in a unit of time. It is always positive. The SI unit of frequency is the hertz:

$$1 \text{ hertz} = 1 \text{ Hz} = 1 \text{ cycle/s} = 1 \text{ s}^{-1}.$$

This unit is named in honor of the German physicist Heinrich Hertz (1857–1894), a pioneer in investigating electromagnetic waves.

The **angular frequency**, ω, is 2π times the frequency:

$$\omega = 2\pi f.$$

We'll learn shortly why ω is a useful quantity. It represents the rate of change of an angular quantity (not necessarily related to a rotational motion) that is always measured in radians, so its units are rad/s. Since f is in cycle/s, we may regard the number 2π as having units rad/cycle.

From the definitions of period T and frequency f we see that each is the reciprocal of the other:

$$f = \frac{1}{T}, \qquad T = \frac{1}{f} \qquad \text{(relationships between frequency and period).} \qquad (13\text{-}1)$$

Also, from the definition of ω,

$$\omega = 2\pi f = \frac{2\pi}{T} \qquad \text{(angular frequency).} \qquad (13\text{-}2)$$

EXAMPLE 13-1

An ultrasonic transducer (a kind of loudspeaker) used for medical diagnosis oscillates at a frequency of 6.7 MHz = 6.7×10^6 Hz. How much time does each oscillation take, and what is the angular frequency?

SOLUTION The period T is given by Eq. (13-1):

$$T = \frac{1}{f} = \frac{1}{6.7 \times 10^6 \text{ Hz}} = 1.5 \times 10^{-7} \text{ s} = 0.15 \ \mu s.$$

We get ω from Eq. (13-2):

$$\omega = 2\pi f = 2\pi(6.7 \times 10^6 \text{ Hz})$$
$$= (2\pi \text{ rad/cycle})(6.7 \times 10^6 \text{ cycle/s})$$
$$= 4.2 \times 10^7 \text{ rad/s}.$$

A very rapid vibration corresponds to large f and ω and small T; a slow vibration corresponds to small f and ω and large T.

13-3   SIMPLE HARMONIC MOTION

The very simplest kind of oscillation occurs when the restoring force F is *directly proportional* to the displacement from equilibrium x. This happens if the spring in Fig. 13-1 is an ideal one that obeys Hooke's law. The constant of proportionality between F and x is the force constant k. (You may want to review Hooke's law and the definition of the force constant in Section 6-4.) On either side of the equilibrium position, F and x always have opposite signs. In Section 6-4 we represented the force acting *on* a stretched ideal spring as $F = kx$. The x-component of force the spring exerts *on the body* is the negative of this, so the x-component of force F on the body is

$$F = -kx \qquad \text{(restoring force exerted by an ideal spring).} \qquad (13\text{-}3)$$

This equation gives the correct magnitude and sign of the force, whether x is positive, negative, or zero. The force constant k is always positive and has units of N/m (a useful alternative set of units is kg/s^2). We are assuming that there is no friction, so Eq. (13-3) gives the *net* force on the body.

When the restoring force is directly proportional to the displacement from equilibrium, as given by Eq. (13-3), the oscillation is called **simple harmonic motion,** abbreviated **SHM.** The acceleration $a = d^2x/dt^2 = F/m$ of a body in SHM is given by

$$a = \frac{d^2x}{dt^2} = -\frac{k}{m}x \qquad \text{(simple harmonic motion).} \qquad (13\text{-}4)$$

The minus sign means the acceleration and displacement always have opposite signs. This acceleration is *not* constant, so don't even think of using the constant-acceleration equations from Chapter 2. We'll see shortly how to solve this equation to find the displacement x as a function of time. A body that undergoes simple harmonic motion is called a **harmonic oscillator.**

Why is simple harmonic motion important? Keep in mind that not all periodic motions are simple harmonic; in periodic motion in general, the restoring force depends on displacement in a more complicated way than in Eq. (13–3). But in many systems the restoring force is *approximately* proportional to displacement if the displacement is sufficiently small (Fig. 13–2). That is, if the amplitude is small enough, the oscillations of such systems will be approximately simple harmonic and therefore approximately described by Eq. (13–4). Thus we can use SHM as an approximate model for many different periodic motions, such as the vibration of the quartz crystal in a watch, the motion of a tuning fork, the electric current in an alternating-current circuit, and the vibrations of atoms in molecules and solids.

13–2 In most real oscillations the restoring force is not directly proportional to the displacement, and is different in magnitude for displacements on either side of the equilibrium position. But $F = -kx$ is often a good approximation to the force if the displacement x is sufficiently small.

EQUATIONS OF SIMPLE HARMONIC MOTION

To explore the properties of simple harmonic motion, we must express the displacement x of the oscillating body as a function of time, $x(t)$. The second derivative of this function, d^2x/dt^2, must be equal to $(-k/m)$ times the function itself, as required by Eq. (13–4). As we mentioned above, the formulas for constant acceleration from Section 2–5 are no help because the acceleration changes constantly as the displacement x changes. Instead, we'll find $x(t)$ by noticing a striking similarity between SHM and another form of motion that we've already studied in detail.

Figure 13–3 shows a top view of a horizontal disk of radius A with a ball attached to its rim at point Q. The disk rotates with constant angular speed ω (measured in rad/s), so the ball moves in uniform circular motion. A horizontal light beam shines on the rotating disk and casts a shadow of the ball on a screen. The shadow at point P oscillates back and forth as the ball moves in a circle. We then arrange a body attached to an ideal spring, like the combination shown in Fig. 13–1, so that the body oscillates parallel to the shadow. We will prove that the motion of the body and the motion of the ball's shadow are *identical* if the amplitude of the body's oscillation is equal to the disk radius A, and if the angular frequency $2\pi f$ of the oscillating body is equal to the angular speed ω of the rotating disk. That is, *simple harmonic motion is the projection of uniform circular motion onto a diameter.*

We can verify this remarkable statement by finding the acceleration of the shadow at P and comparing it to the acceleration of a body undergoing SHM, given by Eq. (13–4). The circle in which the ball moves so that its projection matches the motion of the oscillating body is called the **circle of reference**; we will call the point Q the *reference point*. We take the circle of reference to lie in the xy-plane, with the origin O at the center of the circle (Fig. 13–4a). At time t the vector OQ from the origin to the reference point Q makes an angle θ with the positive x-axis. As the point Q moves around the

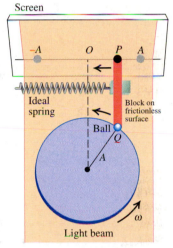

13–3 The ball at point Q rotates counterclockwise in uniform circular motion. Its shadow at point P, projected onto the screen by a beam of light, moves in exactly the same way as a body oscillating on an ideal spring, that is, in simple harmonic motion.

13–4 (a) The coordinate of the ball's shadow P along the x-axis changes with time as the ball Q rotates counterclockwise in uniform circular motion. (b) and (c) The velocity and acceleration of point P are the x-components of the velocity and acceleration vectors, respectively, of point Q.

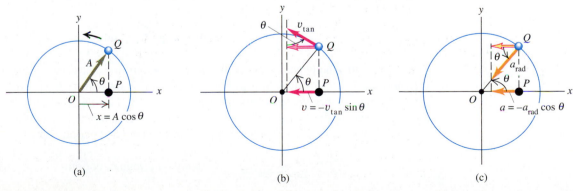

reference circle with constant angular speed ω, the vector OQ rotates with the same angular speed. Such a rotating vector is called a **phasor.** (This term was in use long before the invention of the similarly named Star Trek stun gun. The phasor method for analyzing oscillations is useful in many areas of physics. We'll use phasors when we study alternating-current circuits in Chapter 32 and the interference of light in Chapters 37 and 38.)

The x-component of the phasor at time t is just the x-coordinate of the point Q:

$$x = A \cos \theta. \qquad (13\text{–}5)$$

This is also the x-coordinate of the shadow P, which is the *projection* of Q onto the x-axis. Hence the acceleration of the shadow P along the x-axis is equal to the x-component of the acceleration vector of the reference point Q (Fig. 13–4c). Since point Q is in uniform circular motion, its acceleration vector $\vec{a}_Q$ is always directed toward O. Furthermore, the magnitude of $\vec{a}_Q$ is constant and given by the angular velocity squared times the radius of the circle (see Section 3–5):

$$a_Q = \omega^2 A. \qquad (13\text{–}6)$$

Figure 13–4c show that the x-component of $\vec{a}_Q$ is $a = -a_Q \cos \theta$. Combining this with Eqs. (13–5) and (13–6), we get that the acceleration of point P is

$$a = -a_Q \cos \theta = -\omega^2 A \cos \theta, \qquad (13\text{–}7)$$

or

$$a = -\omega^2 x. \qquad (13\text{–}8)$$

The acceleration of the point P is directly proportional to the displacement x and always has the opposite sign. These are precisely the hallmarks of simple harmonic motion.

Equation (13–8) will be *exactly* the same as Eq. (13–4) for the acceleration of a harmonic oscillator, provided that the angular speed ω of the reference point Q is related to the force constant k and mass m of the oscillating body by

$$\omega^2 = \frac{k}{m}, \qquad \text{or} \qquad \omega = \sqrt{\frac{k}{m}}. \qquad (13\text{–}9)$$

We have been using the same symbol ω for the angular *speed* of the reference point Q and the angular *frequency* of the oscillating point P. The reason is that these quantities are equal! If point Q makes one complete revolution in time T, then point P goes through one complete cycle of oscillation in the same time; hence T is the period of the oscillation. During time T the point Q moves through 2π radians, so its angular speed is $\omega = 2\pi/T$. But this is just the same as Eq. (13–2) for the angular frequency of the point P, which verifies our statement about the two interpretations of ω. This is why we introduced angular frequency in Section 13–2; it's this quantity that makes the connection between oscillation and circular motion. So we re-interpret Eq. (13–9) as an expression for the angular frequency of simple harmonic motion for a body of mass m, acted on by a restoring force with force constant k:

$$\omega = \sqrt{\frac{k}{m}} \qquad \text{(simple harmonic motion).} \qquad (13\text{–}10)$$

When you start a body oscillating in SHM, the value of ω is not yours to choose; it is predetermined by the values of k and m. The units of k are N/m or kg/s^2, so k/m is in (kg/s^2)/kg = s^{-2}. When we take the square root in Eq. (13–10), we get s^{-1}, or more properly rad/s because this is an *angular* frequency (recall that a radian is not a true unit).

According to Eqs. (13–1) and (13–2), the frequency f and period T are

$$f = \frac{\omega}{2\pi} = \frac{1}{2\pi} \sqrt{\frac{k}{m}} \qquad \text{(simple harmonic motion),} \qquad (13\text{–}11)$$

$$T = \frac{1}{f} = \frac{2\pi}{\omega} = 2\pi\sqrt{\frac{m}{k}} \qquad \text{(simple harmonic motion).} \qquad (13-12)$$

We see from Eq. (13–12) that a larger mass m, with its greater inertia, will have less acceleration, move more slowly, and take a longer time for a complete cycle. In contrast, a stiffer spring (one with a larger force constant k) exerts a greater force at a given deformation x, causing greater acceleration, higher speeds, and a shorter time T per cycle.

CAUTION► You can run into trouble if you don't make the distinction between frequency f and angular frequency $\omega = 2\pi f$. Frequency tells you how many cycles of oscillation occur per second, while angular frequency tells you how many radians per second this corresponds to on the reference circle. In solving problems, pay careful attention to whether the goal is to find f or ω. ◄

Equations (13–11) and (13–12) show that the period and frequency of simple harmonic motion are completely determined by the mass m and the force constant k. *In simple harmonic motion the period and frequency do not depend on the amplitude A.* For given values of m and k the time of one complete oscillation is the same whether the amplitude is large or small. Equation (13–3) shows why we should expect this. Larger A means that the body reaches larger values of $|x|$ and is subjected to larger restoring forces. This increases the average speed of the body over a complete cycle; this exactly compensates for having to travel a larger distance, so the same total time is involved.

The oscillations of a tuning fork are essentially simple harmonic motion, which means that it always vibrates with the same frequency, independent of amplitude. This is why a tuning fork can be used as a standard for musical pitch. If it were not for this characteristic of simple harmonic motion, it would be impossible to make familiar types of mechanical and electronic clocks run accurately or to play most musical instruments in tune. If you encounter an oscillating body with a period that *does* depend on the amplitude, the oscillation is *not* simple harmonic motion.

Equation (13–11) explains why tuning forks of different sizes produce sounds of different pitch. The smaller the mass m in the tuning fork's tines, the higher the frequency of oscillation and the higher the pitch of the sound the tuning fork produces.

EXAMPLE 13-2

Angular frequency, frequency, and period in SHM   A spring is mounted horizontally, with its left end held stationary. By attaching a spring balance to the free end and pulling toward the right (Fig. 13–5), we determine that the stretching force is proportional to the displacement and that a force of 6.0 N causes a displacement of 0.030 m. We remove the spring balance and attach a 0.50-kg body to the end, pull it a distance of 0.020 m, release it, and watch it oscillate in SHM. a) Find the force constant of the spring. b) Find the angular frequency, frequency, and period of the oscillation.

SOLUTION   a) When $x = 0.030$ m, the force the spring exerts on the spring balance is $F = -6.0$ N. From Eq. (13–3),

$$k = -\frac{F}{x} = -\frac{-6.0 \text{ N}}{0.030 \text{ m}} = 200 \text{ N/m}.$$

b) We are given $m = 0.50$ kg and $k = 200$ N/m $= 200$ kg/s^2. Using Eq. (13–10), we find

$$\omega = \sqrt{\frac{k}{m}} = \sqrt{\frac{200 \text{ kg/s}^2}{0.50 \text{ kg}}} = 20 \text{ rad/s}.$$

The frequency f is

$$f = \frac{\omega}{2\pi} = \frac{20 \text{ rad/s}}{2\pi \text{ rad/cycle}} = 3.2 \text{ cycle/s} = 3.2 \text{ Hz}.$$

13–5 The force exerted *on* the spring is 6.0 N. The force exerted *by* the spring is −6.0 N.

The period T is the reciprocal of the frequency f:

$$T = \frac{1}{f} = \frac{1}{3.2 \text{ cycle/s}} = 0.31 \text{ s}.$$

A period is usually stated in "seconds" rather than "seconds per cycle."

The amplitude of the oscillation is 0.020 m, the distance to the right that we pulled the body attached to the spring before releasing it. We didn't need to use this information to find the angular frequency, frequency, or period because in SHM, none of these quantities depend on the amplitude.

13-6 Graph of *x* versus *t* [Eq. (13–13)] for simple harmonic motion. The case shown has $\phi = 0$.

(a)

(b)

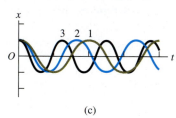

(c)

13-7 Variations of simple harmonic motion. All cases shown have $\phi = 0$. (a) Amplitude *A* increases from curve 1 to 2 to 3, while mass *m* and force constant *k* are unchanged. Changing the amplitude has no effect on the period. (b) Mass *m* increases from 1 to 2 to 3, while *A* and *k* are unchanged. (c) Force constant *k* increases from 1 to 2 to 3, while *A* and *m* are unchanged.

13-8 Position as a function of time for simple harmonic motion with the same frequency and amplitude but with three different phase angles: $\phi = 0$ (green), $\phi = \pi/4$ (blue), and $\phi = \pi/2$ (black).

DISPLACEMENT, VELOCITY, AND ACCELERATION IN SHM

We still need to find the displacement *x* as a function of time for a harmonic oscillator. Equation (13–4) for a body in simple harmonic motion along the *x*-axis is identical to Eq. (13–8) for the *x*-coordinate of the reference point in uniform circular motion with constant angular speed $\omega = \sqrt{k/m}$. It follows that Eq. (13–5), $x = A \cos \theta$, describes the coordinate *x* for both of these situations. If at $t = 0$ the phasor *OQ* makes an angle ϕ ("phi") with the positive *x*-axis, then at any later time *t* this angle is $\theta = \omega t + \phi$. We substitute this into Eq. (13–5) to obtain

$$x = A \cos (\omega t + \phi) \qquad \text{(displacement in SHM)}, \qquad (13\text{–}13)$$

where $\omega = \sqrt{k/m}$. Figure 13–6 shows a graph of Eq. (13–13) for the particular case $\phi = 0$. The displacement *x* is a periodic function of time, as expected for SHM. We could also have written Eq. (13–13) in terms of a sine function rather than a cosine by using the identity $\cos \alpha = \sin (\alpha + \pi/2)$. *In simple harmonic motion the position is a periodic, sinusoidal function of time.* There are many other periodic functions, but none so smooth and simple as a sine or cosine function.

The value of the cosine function is always between −1 and 1, so in Eq. (13–13), *x* is always between −*A* and *A*. This confirms that *A* is the amplitude of the motion. Figure 13–7a shows the graph of *x* versus *t* for various values of *A*.

The period *T* is the time for one complete cycle of oscillation. The cosine function repeats itself whenever the quantity in parentheses in Eq. (13–13) increases by 2π radians. Thus if we start at time $t = 0$, the time *T* to complete one cycle is given by

$$\omega T = \sqrt{\frac{k}{m}}\, T = 2\pi, \qquad \text{or} \qquad T = 2\pi \sqrt{\frac{m}{k}},$$

which is just Eq. (13–12). Changing either *m* or *k* changes the period of oscillation, as shown in Figs. 13–7b and 13–7c.

The constant ϕ in Eq. (13–13) is called the **phase angle.** It tells us at what point in the cycle the motion was at $t = 0$ (equivalent to where around the circle the point *Q* was at $t = 0$). We denote the position at $t = 0$ by x_0. Putting $t = 0$ and $x = x_0$ in Eq. (13–13), we get

$$x_0 = A \cos \phi. \qquad (13\text{–}14)$$

If $\phi = 0$, then $x_0 = A \cos 0 = A$, and the body starts at its maximum positive displacement. If $\phi = \pi$, then $x_0 = A \cos \pi = -A$, and the particle starts at its maximum *negative* displacement. If $\phi = \pi/2$, then $x_0 = A \cos(\pi/2) = 0$, and the particle is initially at the origin. Figure 13–8 shows the displacement *x* versus time for different phase angles.

We find the velocity *v* and acceleration *a* as functions of time for a harmonic oscillator by taking derivatives of Eq. (13–13) with respect to time:

$$v = \frac{dx}{dt} = -\omega A \sin (\omega t + \phi) \qquad \text{(velocity in SHM)}, \qquad (13\text{–}15)$$

$$a = \frac{dv}{dt} = \frac{d^2 x}{dt^2} = -\omega^2 A \cos (\omega t + \phi) \qquad \text{(acceleration in SHM)}. \qquad (13\text{–}16)$$

The velocity *v* oscillates between $v_{max} = +\omega A$ and $-v_{max} = -\omega A$, and the acceleration *a* oscillates between $a_{max} = +\omega^2 A$ and $-a_{max} = -\omega^2 A$ (Fig. 13–9). Comparing Eq. (13–16) with Eq. (13–13) and recalling that $\omega^2 = k/m$ from Eq. (13–9), we see that

$$a = -\omega^2 x = -\frac{k}{m} x,$$

which is just Eq. (13–4) for simple harmonic motion. This confirms that Eq. (13–13) for *x* as a function of time is correct.

We actually derived Eq. (13–16) earlier in a geometrical way by taking the x-component of the acceleration vector of the reference point Q. This was done in Fig. 13–4c and Eq. (13–7) (recall that $\theta = \omega t + \phi$). In the same way, we could have derived Eq. (13–15) by taking the x-component of the velocity vector of Q, as shown in Fig. 13–4b. We'll leave the details for you to work out (see Problem 13–67).

Note that the sinusoidal graph of displacement versus time (Fig. 13–9a) is shifted by one-quarter period from the graph of velocity versus time (Fig. 13–9b) and by one-half period from the graph of acceleration versus time (Fig. 13–9c). When the body is passing through the equilibrium position so that the displacement is zero, the velocity equals either v_{max} or $-v_{max}$ (depending on which way the body is moving) and the acceleration is zero. When the body is at either its maximum positive displacement, $x = +A$, or its maximum negative displacement, $x = -A$, the velocity is zero and the body is instantaneously at rest. At these points, the restoring force $F = -kx$ and the acceleration of the body have their maximum magnitudes. At $x = +A$ the acceleration is negative and equal to $-a_{max}$. At $x = -A$ the acceleration is positive: $a = +a_{max}$.

If we are given the initial position x_0 and initial velocity v_0 for the oscillating body, we can determine the amplitude A and the phase angle ϕ. Here's how to do it. The initial velocity v_0 is the velocity at time $t = 0$; putting $v = v_0$ and $t = 0$ in Eq. (13–15), we find

$$v_0 = -\omega A \sin \phi. \tag{13–17}$$

To find ϕ, divide Eq. (13–17) by Eq. (13–14). This eliminates A and gives an equation that we can solve for ϕ:

$$\frac{v_0}{x_0} = \frac{-\omega A \sin \phi}{A \cos \phi} = -\omega \tan \phi,$$

$$\phi = \arctan\left(-\frac{v_0}{\omega x_0}\right) \quad \text{(phase angle in SHM).} \tag{13–18}$$

It is also easy to find the amplitude A if we are given x_0 and v_0. We'll sketch the derivation, and you can fill in the details. Square Eq. (13–14); divide Eq. (13–17) by ω, square it, and add to the square of Eq. (13–14). The right side will be $A^2(\sin^2 \phi + \cos^2 \phi)$, which is equal to A^2. The final result is

$$A = \sqrt{x_0^2 + \frac{v_0^2}{\omega^2}} \quad \text{(amplitude in SHM).} \tag{13–19}$$

Note that when the body has both an initial displacement x_0 and a nonzero initial velocity v_0, the amplitude A is *not* equal to the initial displacement. That's reasonable; if you start the body at a positive x_0 but give it a positive velocity v_0, it will go *farther* than x_0 before it turns and comes back.

13–9 (a) Graph of x versus t for simple harmonic motion. In this graph, $\phi = \pi/3$. (b) Graph of v versus t for the same motion. Note that this graph is shifted by $\frac{1}{4}$ cycle from the graph of x versus t. (c) Graph of a versus t for the same motion. This graph is shifted by $\frac{1}{4}$ cycle from the graph of v versus t and by $\frac{1}{2}$ cycle from the graph of x versus t.

ActivPhysics

9.1
Position Graphs and Equations

9.2
Describing Vibrational Motion

9.5
Ape Drops Tarzan

Problem–Solving Strategy

SIMPLE HARMONIC MOTION I

1. Be careful to distinguish between two different kinds of quantities: those that represent basic physical properties of the system and those that describe a particular motion that occurs when the system is set in motion in a specific way. The physical properties include the mass m, the force constant k, and quantities derived from these, including the period T, the frequency f, and the angular frequency $\omega = 2\pi f$. In some problems, m or k or both can be determined from other information given about the

system. Quantities that describe a particular motion include the amplitude A, the maximum velocity v_{max}, the phase angle ϕ, and the displacement, velocity, or acceleration at a particular time.

2. When detailed information is required about displacements, velocities, and accelerations at various times, you will need to use Eqs. (13–13), (13–15), and (13–16). If the initial position x_0 and initial velocity v_0 are both given,

you can determine the phase angle and amplitude from Eqs. (13–18) and (13–19). If the body is given an initial positive displacement x_0 but no initial velocity ($v_0 = 0$), then the amplitude is $A = x_0$ and the phase angle is $\phi = 0$. If it has an initial positive velocity v_0 but no initial displacement ($x_0 = 0$), the amplitude is $A = v_0/\omega$ and the phase angle is $\phi = -\pi/2$.

EXAMPLE 13–3

Let's return to the horizontal spring we considered in Example 13–2. The force constant is $k = 200$ N/m, and the spring is attached to a body with mass $m = 0.50$ kg. This time we give the body an initial displacement of $+0.015$ m and an initial velocity of $+0.40$ m/s. a) Find the period, amplitude, and phase angle of the motion. b) Write equations for the displacement, velocity, and acceleration as functions of time.

SOLUTION a) The period is exactly the same as in Example 13–2, $T = 0.31$ s. In simple harmonic motion the period does not depend on the amplitude, only on the values of k and m.

In Example 13–2 we found that $\omega = 20$ rad/s. So from Eq. (13–19),

$$A = \sqrt{x_0^2 + \frac{v_0^2}{\omega^2}}$$

$$= \sqrt{(0.015 \text{ m})^2 + \frac{(0.40 \text{ m/s})^2}{(20 \text{ rad/s})^2}}$$

$$= 0.025 \text{ m}.$$

To find the phase angle ϕ, we use Eq. (13–18):

$$\phi = \arctan\left(\frac{-v_0}{\omega x_0}\right)$$

$$= \arctan\left(\frac{-0.40 \text{ m/s}}{(20 \text{ rad/s})(0.015 \text{ m})}\right) = -53° = -0.93 \text{ rad}.$$

b) The displacement at any time is given by Eq. (13–13); substituting the appropriate values, we find

$$x = (0.025 \text{ m}) \cos [(20 \text{ rad/s})t - 0.93 \text{ rad}].$$

The velocity at any time is given by Eq. (13–15); substituting the values, we find

$$v = -(0.50 \text{ m/s}) \sin [(20 \text{ rad/s})t - 0.93 \text{ rad}].$$

The velocity varies sinusoidally between -0.50 m/s and $+0.50$ m/s. The acceleration is given by Eq. (13–16); substituting the values, we get

$$a = -(10 \text{ m/s}^2) \cos [(20 \text{ rad/s})t - 0.93 \text{ rad}].$$

The acceleration varies between -10 m/s^2 and $+10$ m/s^2.

13–4 ENERGY IN SIMPLE HARMONIC MOTION

ActivPhysics

9.3
Vibrational Energy

9.4
Two Ways to Measure Young Tarzan's Mass

9.6
Releasing a Vibrating Skier I

9.7
Releasing a Vibrating Skier II

We can learn even more about simple harmonic motion by using energy considerations. Take another look at the body oscillating on the end of a spring in Fig. 13–1. We've already noted that the spring force is the only horizontal force on the body. The force exerted by an ideal spring is a conservative force, and the vertical forces do no work, so the total mechanical energy of the system is *conserved*. We will also assume that the mass of the spring itself is negligible.

The kinetic energy of the body is $K = \frac{1}{2}mv^2$, and the potential energy of the spring is $U = \frac{1}{2}kx^2$, just as in Section 7–3. (It would be helpful to review that section.) There are no nonconservative forces that do work, so the total mechanical energy $E = K + U$ is conserved:

$$E = \frac{1}{2}mv^2 + \frac{1}{2}kx^2 = \text{constant}. \tag{13–20}$$

The total mechanical energy E is also directly related to the amplitude A of the motion. When the body reaches the point $x = A$, its maximum displacement from equilibrium, it momentarily stops as it turns back toward the equilibrium position. That is, when $x = A$ (or $-A$), $v = 0$. At this point the energy is entirely potential, and $E = \frac{1}{2}kA^2$. Because E is constant, this quantity equals E at any other point. Combining this expression with Eq. (13–20), we get

$$E = \frac{1}{2}mv^2 + \frac{1}{2}kx^2 = \frac{1}{2}kA^2 = \text{constant} \qquad \text{(total mechanical energy in SHM).} \tag{13–21}$$

$v = 0$ $v = \pm\sqrt{3/4}\, v_{max}$ $v = \pm v_{max}$ $v = \pm\sqrt{3/4}\, v_{max}$ $v = 0$

a_{max} $a_{max}/2$ $a = 0$ $-a_{max}/2$ $-a_{max}$

$-A$ $-A/2$ O $A/2$ A

O

E K U   E K U   E K U   E K U   E K U

9.8
One- and Two-Spring Vibrating Systems

9.9
Vibro-Ride

13–10 In simple harmonic motion the total mechanical energy E is constant, continuously transforming from potential energy U to kinetic energy K and back again as the body oscillates.

We can verify this equation by substituting x and v from Eqs. (13–13) and (13–15) and using $\omega^2 = k/m$ from Eq. (13–9):

$$E = \frac{1}{2}mv^2 + \frac{1}{2}kx^2 = \frac{1}{2}m(-\omega A \sin(\omega t + \phi))^2 + \frac{1}{2}k(A\cos(\omega t + \phi))^2$$

$$= \frac{1}{2}kA^2 \sin^2(\omega t + \phi) + \frac{1}{2}kA^2 \cos^2(\omega t + \phi)$$

$$= \frac{1}{2}kA^2.$$

(Recall that $\sin^2\alpha + \cos^2\alpha = 1$.) Hence our expressions for displacement and velocity in SHM are consistent with energy conservation, as they must be.

We can use Eq. (13–21) to solve for the velocity v of the body at a given displacement x:

$$v = \pm\sqrt{\frac{k}{m}}\sqrt{A^2 - x^2}. \qquad (13\text{–}22)$$

The $\pm$ sign means that at a given value of x the body can be moving in either direction. For example, when $x = \pm A/2$,

$$v = \pm\sqrt{\frac{k}{m}}\sqrt{A^2 - \left(\pm\frac{A}{2}\right)^2} = \pm\sqrt{\frac{3}{4}}\sqrt{\frac{k}{m}}A.$$

Equation (13–22) also shows that the *maximum* speed v_{max} occurs at $x = 0$. Using Eq. (13–10), $\omega = \sqrt{k/m}$, we find that

$$v_{max} = \sqrt{\frac{k}{m}}A = \omega A. \qquad (13\text{–}23)$$

This agrees with Eq. (13–15), which showed that v oscillates between $-\omega A$ and $+\omega A$.

Figure 13–10 shows the energy quantities E, K, and U at $x = 0$, $x = \pm A/2$, and $x = \pm A$. Figure 13–11 is a graphical display of Eq. (13–21); energy (kinetic, potential, and total) is plotted vertically and the coordinate x is plotted horizontally. The parabolic curve in Fig. 13–11a represents the potential energy $U = \frac{1}{2}kx^2$. The horizontal line represents the total mechanical energy E, which is constant and does not vary with x. This line intersects the potential-energy curve at $x = -A$ and $x = A$, where the energy is entirely potential. At any value of x between $-A$ and A, the vertical distance from the x-axis to the parabola is U; since $E = K + U$, the remaining vertical distance up to the horizontal line is K. Figure 13–11b shows both K and U as functions of x. As the body oscillates between $-A$ and A, the energy is continuously transformed from potential to kinetic and back again.

Figure 13–11a shows the connection between the amplitude A and the corresponding total mechanical energy $E = \frac{1}{2}kA^2$. If we tried to make x greater than A (or less than $-A$), U would be greater than E, and K would have to be negative. But K can never be negative, so x can't be greater than A or less than $-A$.

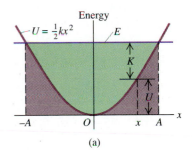

Energy

$U = \frac{1}{2}kx^2$ E

K

U

$-A$ O x A x

(a)

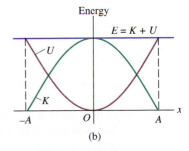

Energy

$E = K + U$

U

K

$-A$ O A x

(b)

13–11 Kinetic energy K, potential energy U, and total mechanical energy E as functions of position for SHM. At each value of x the sum of the values of K and U equals the constant value of E.

Problem–Solving Strategy

SIMPLE HARMONIC MOTION II

The energy equation, Eq. (13–21), is a useful alternative relation between velocity and position, especially when energy quantities are also required. If the problem involves a relation among position, velocity, and acceleration without reference to time, it is usually easier to use Eq. (13–4) (from Newton's second law) or (13–21) (from energy conservation) than to use the general expression for x, v, and a as functions of time

[Eqs. (13–13), (13–15), and (13–16), respectively]. Because the energy equation involves x^2 and v^2, it cannot tell you the sign of x or of v; you have to infer the sign from the situation. For instance, if the body is moving from the equilibrium position toward the point of greatest positive displacement, then x is positive and v is positive.

EXAMPLE 13–4

In the oscillation described in Example 13–2, $k = 200$ N/m, $m = 0.50$ kg, and the oscillating mass is released from rest at $x = 0.020$ m. a) Find the maximum and minimum velocities attained by the oscillating body. b) Compute the maximum acceleration. c) Determine the velocity and acceleration when the body has moved halfway to the center from its original position. d) Find the total energy, potential energy, and kinetic energy at this position.

SOLUTION a) The velocity v at any displacement x is given by Eq. (13–22):

$$v = \pm \sqrt{\frac{k}{m}} \sqrt{A^2 - x^2}.$$

The maximum velocity occurs when the body is moving to the right through the equilibrium position, where $x = 0$:

$$v = v_{max} = \sqrt{\frac{k}{m}} A = \sqrt{\frac{200 \text{ N/m}}{0.50 \text{ kg}}} (0.020 \text{ m}) = 0.40 \text{ m/s}.$$

The minimum (i.e., most negative) velocity occurs when the body is moving to the left through $x = 0$; its value is $-v_{max} = -0.40$ m/s.

b) From Eq. (13–4),

$$a = -\frac{k}{m} x.$$

The maximum (most positive) acceleration occurs at the most negative value of x, that is, $x = -A$; therefore

$$a_{max} = -\frac{k}{m}(-A) = -\frac{200 \text{ N/m}}{0.50 \text{ kg}}(-0.020 \text{ m}) = 8.0 \text{ m/s}^2.$$

The minimum (most negative) acceleration is -8.0 m/s^2, occurring at $x = +A = +0.020$ m.

c) At a point halfway to the center from the initial position, $x = A/2 = 0.010$ m. From Eq. (13–22),

$$v = -\sqrt{\frac{200 \text{ N/m}}{0.50 \text{ kg}}} \sqrt{(0.020 \text{ m})^2 - (0.010 \text{ m})^2} = -0.35 \text{ m/s}.$$

We choose the negative square root because the body is moving from $x = A$ toward $x = 0$. From Eq. (13–4),

$$a = -\frac{200 \text{ N/m}}{0.50 \text{ kg}}(0.010 \text{ m}) = -4.0 \text{ m/s}^2.$$

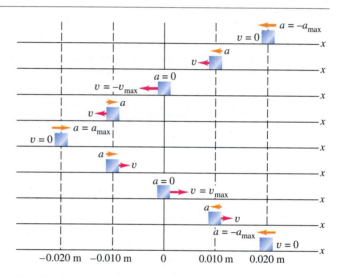

13–12 A body is attached to a spring, pulled 0.020 m from the equilibrium position, and released. The velocity and acceleration are shown at $x = 0$, $x = \pm A/2$, and $x = \pm A$. The velocity of the body is not constant, so these images of the body at equally spaced points in its motion are *not* equally spaced in time.

At this point the velocity and the acceleration have the same sign, so the speed is increasing. The conditions at $x = 0$, $\pm A/2$, and $\pm A$ are shown in Fig. 13–12.

d) The total energy has the same value at all points during the motion:

$$E = \frac{1}{2} kA^2 = \frac{1}{2}(200 \text{ N/m})(0.020 \text{ m})^2 = 0.040 \text{ J}.$$

The potential energy is

$$U = \frac{1}{2} kx^2 = \frac{1}{2}(200 \text{ N/m})(0.010 \text{ m})^2 = 0.010 \text{ J},$$

and the kinetic energy is

$$K = \frac{1}{2} mv^2 = \frac{1}{2}(0.50 \text{ kg})(-0.35 \text{ m/s})^2 = 0.030 \text{ J}.$$

At this point, E is one fourth potential energy and three fourths kinetic energy.

EXAMPLE 13–5

Energy and momentum in SHM A block with mass M attached to a horizontal spring with force constant k is moving with simple harmonic motion having amplitude A_1. At the instant when the block passes through its equilibrium position, a lump of putty with mass m is dropped vertically onto the block from a very small height and sticks to it (Fig. 13–13a). a) Find the new amplitude and period. b) Repeat part (a) for the case in which the putty is dropped on the block when it is at one end of its path (Fig. 13–13b).

SOLUTION a) The problem involves the motion at a given position, not a given time, so let's use energy methods. Before the putty lands on the block, the mechanical energy of the oscillating block and spring is constant. When the putty lands on the block, it's a completely inelastic collision (Section 8–4); the x-component of momentum is conserved, but mechanical energy decreases. Once the collision is over, the mechanical energy remains constant at its new, lower value. Let's look at these three stages in turn—before, during, and after the collision.

Before the collision the total mechanical energy of the block and spring is $E_1 = \frac{1}{2}kA_1^2$. Since the block is at the equilibrium position, $U = 0$, and the energy is purely kinetic. If we let v_1 be the speed of the block at the equilibrium position, we have

$$E_1 = \frac{1}{2}Mv_1^2 = \frac{1}{2}kA_1^2, \quad \text{so} \quad v_1 = \sqrt{\frac{k}{M}}A_1.$$

During the collision the x-component of momentum of the system of block and putty is conserved. (Why?) Just before the collision this component is the sum of Mv_1 (for the block) and zero (for the putty). Just after the collision the block and putty move together with speed v_2, and their combined x-component of momentum is $(M + m)v_2$. From conservation of momentum,

$$Mv_1 + 0 = (M + m)v_2, \quad \text{so} \quad v_2 = \frac{M}{M + m}v_1.$$

The collision lasts a very short time, so just after the collision the block and putty are still at the equilibrium position. The energy is still purely kinetic, but is *less* than before the collision:

$$E_2 = \frac{1}{2}(M + m)v_2^2 = \frac{1}{2}\frac{M^2}{M + m}v_1^2 = \frac{M}{M + m}\left(\frac{1}{2}Mv_1^2\right)$$

$$= \left(\frac{M}{M + m}\right)E_1.$$

(The lost mechanical energy goes into raising the temperature of the block and the putty.) Since E_2 equals $\frac{1}{2}kA_2^2$, where A_2 is the amplitude after the collision, we have

$$\frac{1}{2}kA_2^2 = \left(\frac{M}{M + m}\right)\frac{1}{2}kA_1^2,$$

$$A_2 = A_1\sqrt{\frac{M}{M + m}}.$$

The larger the putty mass m, the smaller the final amplitude.

Finding the period of oscillation after the collision is the easy

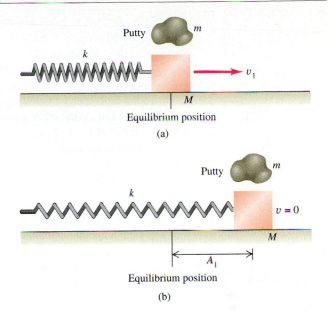

13–13 (a) Dropping a lump of putty on an oscillating block as it passes through equilibrium. (b) Dropping a lump of putty on the block at $x = A_1$.

part. Using Eq. (13–12), we have

$$T_2 = 2\pi\sqrt{\frac{M + m}{k}}.$$

Dropping the putty on the block at the equilibrium position makes each cycle take a longer time and cover a smaller total distance.

b) When the putty drops on the block, the block is instantaneously at rest; all the mechanical energy is stored in the spring as potential energy. Again the x-component of momentum is conserved during the collision, but now this component is zero both before and after the collision. The block had zero kinetic energy just before the collision, and the block and putty have zero kinetic energy just after the collision. So in this case, adding the extra mass of the putty has *no effect* on the mechanical energy. That is,

$$E_2 = E_1 = \frac{1}{2}kA_1^2,$$

and the amplitude is still A_1. The period still changes when the putty is added, though; its value doesn't depend on how the mass is added, only on what the total mass is. So T_2 is the same as we found in part (a), $T_2 = 2\pi\sqrt{(M + m)/k}$. Adding the mass of the putty in this way slows down the oscillation but doesn't change the amplitude.

13-5 APPLICATIONS OF SIMPLE HARMONIC MOTION

So far, we've looked at a grand total of *one* situation in which simple harmonic motion (SHM) occurs: a body attached to an ideal horizontal spring. But SHM can occur in any system in which there is a restoring force that is directly proportional to the displacement from equilibrium, as given by Eq. (13–3), $F = -kx$. The restoring force will originate in different ways in different situations, so the force constant k has to be found for each case by examining the net force on the system. Once this is done, it's straightforward to find the angular frequency ω, frequency f, and period T; we just substitute the value of k into Eqs. (13–10), (13–11), and (13–12). Let's use these ideas to examine several examples of simple harmonic motion.

VERTICAL SHM

Suppose we hang a spring with force constant k (Fig. 13–14a) and suspend from it a body with mass m. Oscillations will now be vertical; will they still be SHM? In Fig.

13-14 (a) A hanging spring. (b) A body suspended from the spring is in equilibrium when the upward spring force has the same magnitude as the body's weight. (c) If the body is displaced from equilibrium, the restoring force is proportional to the coordinate measured from the equilibrium position. Hence the oscillations of the body are simple harmonic.

(a)   (b)   (c)

13–14b the body hangs at rest, in equilibrium. In this position the spring is stretched an amount Δl just great enough that the spring's upward vertical force $k\Delta l$ on the body balances its weight mg:

$$k\Delta l = mg.$$

Take $x = 0$ to be this equilibrium position, and take the positive x-direction to be upward. When the body is a distance x *above* its equilibrium position (Fig. 13–14c), the extension of the spring is $\Delta l - x$. The upward force it exerts on the body is then $k(\Delta l - x)$, and the net x-component of force on the body is

$$F_{\text{net}} = k(\Delta l - x) + (-mg) = -kx,$$

that is, a net downward force of magnitude kx. Similarly, when the body is *below* the equilibrium position, there is a net upward force with magnitude kx. In either case there is a restoring force with magnitude kx. If the body is set in vertical motion, it oscillates in SHM with the same angular frequency as though it were horizontal, $\omega = \sqrt{k/m}$. So vertical SHM doesn't differ in any essential way from horizontal SHM. The only real change is that the equilibrium position $x = 0$ no longer corresponds to the point at which the spring is unstretched. The same ideas hold if a body with weight mg is placed atop a compressible spring (Fig. 13–15) and compresses it a distance Δl.

13-15 If the weight mg compresses the spring a distance Δl, the force constant is $k = mg/\Delta l$ and the angular frequency for vertical SHM is $\omega = \sqrt{k/m}$.

EXAMPLE 13–6

Vertical SHM in an old car The shock absorbers in an old car with mass 1000 kg are completely worn out. When a 980-N person climbs slowly into the car to its center of gravity, the car sinks 2.8 cm. When the car, with the person aboard, hits a bump, the car starts oscillating up and down in SHM. Model the car and person as a single body on a single spring, and find the period and frequency of the oscillation.

SOLUTION When the force increases by 980 N, the spring compresses an additional 0.028 m, and the coordinate x of the car changes by -0.028 m. Hence the effective force constant (including the effect of the entire suspension) is

$$k = -\frac{F}{x} = -\frac{980 \text{ N}}{-0.028 \text{ m}} = 3.5 \times 10^4 \text{ kg/s}^2.$$

The person's mass is $w/g = (980 \text{ N})/(9.8 \text{ m/s}^2) = 100$ kg. The *total* oscillating mass is $m = 1000$ kg $+ 100$ kg $= 1100$ kg. The period T is

$$T = 2\pi\sqrt{\frac{m}{k}} = 2\pi\sqrt{\frac{1100 \text{ kg}}{3.5 \times 10^4 \text{ kg/s}^2}} = 1.11 \text{ s},$$

and the frequency is

$$f = \frac{1}{T} = \frac{1}{1.11 \text{ s}} = 0.90 \text{ Hz}.$$

If you have ridden in a car suffering from this affliction, the persistent oscillation may have left your stomach in an altered state.

ANGULAR SHM

Figure 13–16 shows the balance wheel of a mechanical watch. The wheel has a moment of inertia I about its axis. A coil spring (called the *hairspring*) exerts a restoring torque τ that is proportional to the angular displacement θ from the equilibrium position. We write $\tau = -\kappa\theta$, where κ (the Greek letter "kappa") is a constant called the *torsion constant*. Using the rotational analog of Newton's second law for a rigid body, $\Sigma\tau = I\alpha = I\, d^2\theta/dt^2$, the equation of motion is

$$-\kappa\theta = I\alpha, \qquad \text{or} \qquad \frac{d^2\theta}{dt^2} = -\frac{\kappa}{I}\theta.$$

13–16 The balance wheel of a mechanical watch. The spring exerts a restoring torque that is proportional to the angular displacement from the equilibrium position. Therefore the motion is angular SHM.

The form of this equation is exactly the same as Eq. (13–4) for the acceleration in simple harmonic motion, with x replaced by θ and k/m replaced by κ/I. So we are dealing with a form of *angular* simple harmonic motion. The angular frequency ω and frequency f are given by Eqs. (13–10) and (13–11), respectively, with the same replacement:

$$\omega = \sqrt{\frac{\kappa}{I}} \qquad \text{and} \qquad f = \frac{1}{2\pi}\sqrt{\frac{\kappa}{I}} \qquad \text{(angular SHM).} \qquad (13\text{–}24)$$

The motion is described by the function

$$\theta = \Theta \cos(\omega t + \phi),$$

where Θ (capital "theta") plays the role of an angular amplitude.

It's a good thing that the motion of a balance wheel *is* simple harmonic. If it weren't, the frequency might depend on the amplitude, and the watch would run too fast or too slow as the spring ran down.

*VIBRATIONS OF MOLECULES

The following discussion of the vibrations of molecules uses the binomial theorem. If you aren't familiar with this theorem, you should read about it in the appropriate section of your math textbook.

When two atoms are separated from each other by a few atomic diameters, they can exert attractive forces on each other. But if the atoms are so close to each other that their electron shells overlap, the forces between the atoms are repulsive. Between these limits, there can be an equilibrium separation distance at which two atoms form a *molecule*. If these atoms are displaced slightly from equilibrium, they will oscillate. Let's see whether these oscillations can be simple harmonic.

13–17 (a) Two atoms with centers separated by r. (b) Potential energy U in the van der Waals interaction as a function of r. At the equilibrium separation $r = R_0$, U is minimum. Near $r = R_0$, U can be approximated by a parabola. (c) Force F on the right-hand atom as function of r. At the equilibrium separation $r = R_0$, F is zero. Near $r = R_0$, F can be approximated by a straight line.

As an example, we'll consider one type of interaction between atoms called the *van der Waals interaction*. Our immediate task here is specifically to study oscillations, so we won't go into the details of how this interaction arises. Let the center of one atom be at the origin, and let the center of the other atom be a distance r away (Fig. 13–17a); the equilibrium distance between centers is $r = R_0$. Experiment shows that the van der Waals interaction can be described by the potential energy function

$$U = U_0\left[\left(\frac{R_0}{r}\right)^{12} - 2\left(\frac{R_0}{r}\right)^6\right], \qquad (13–25)$$

where U_0 is a positive constant with units of joules. When the two atoms are very far apart, $U = 0$; when they are separated by the equilibrium distance $r = R_0$, $U = -U_0$. The force on the second atom is the negative derivative of Eq. (13–25),

$$F = -\frac{dU}{dr} = U_0\left[\frac{12R_0^{12}}{r^{13}} - 2\frac{6R_0^6}{r^7}\right] = 12\frac{U_0}{R_0}\left[\left(\frac{R_0}{r}\right)^{13} - \left(\frac{R_0}{r}\right)^7\right]. \qquad (13–26)$$

The potential energy and force are plotted in Figs. 13–17b and 13–17c, respectively. The force is positive for $r < R_0$ and negative for $r > R_0$, so this is a *restoring* force.

To study small-amplitude oscillations around the equilibrium separation $r = R_0$, we introduce the quantity x to represent the displacement from equilibrium:

$$x = r - R_0, \qquad \text{so} \qquad r = R_0 + x.$$

In terms of x, the force F in Eq. (13–26) becomes

$$F = 12\frac{U_0}{R_0}\left[\left(\frac{R_0}{R_0 + x}\right)^{13} - \left(\frac{R_0}{R_0 + x}\right)^7\right] = 12\frac{U_0}{R_0}\left[\frac{1}{(1 + x/R_0)^{13}} - \frac{1}{(1 + x/R_0)^7}\right]. \qquad (13–27)$$

This looks nothing like Hooke's law, $F = -kx$, so we might be tempted to conclude that molecular oscillations cannot be SHM. But let us restrict ourselves to *small-amplitude* oscillations so that the absolute value of the displacement x will be small in comparison to R_0 and the absolute value of the ratio x/R_0 will be much less than 1. We can then simplify Eq. (13–27) by using the *binomial theorem*:

$$(1 + u)^n = 1 + nu + \frac{n(n - 1)}{2!}u^2 + \frac{n(n - 1)(n - 2)}{3!}u^3 + \cdots. \qquad (13–28)$$

If $|u|$ is much less than 1, each successive term in Eq. (13–28) is much smaller than the one it follows, and we can safely approximate $(1 + u)^n$ by just the first two terms. In Eq.

(13–27), u is replaced by x/R_0 and n equals -13 or -7, so

$$\frac{1}{(1 + x/R_0)^{13}} = (1 + x/R_0)^{-13} \approx 1 + (-13)\frac{x}{R_0},$$

$$\frac{1}{(1 + x/R_0)^{7}} = (1 + x/R_0)^{-7} \approx 1 + (-7)\frac{x}{R_0},$$

$$F \approx 12\frac{U_0}{R_0}\left[\left(1 + (-13)\frac{x}{R_0}\right) - \left(1 + (-7)\frac{x}{R_0}\right)\right] = -\left(\frac{72U_0}{R_0{}^2}\right)x. \qquad (13\text{–}29)$$

This is just Hooke's law, with force constant $k = 72U_0/R_0{}^2$. (Note that k has the correct units, J/m^2 or N/m.) So oscillations of molecules bound by the van der Waals interaction can be simple harmonic motion, provided that the amplitude is small in comparison to R_0 so that the approximation $|x/R_0| \ll 1$ used in the derivation of Eq. (13–29) is valid.

You can also show that the potential energy U, Eq. (13–25), can be written as $U \approx \frac{1}{2}kx^2 + C$, where $C = -U_0$ and k is again equal to $72U_0/R_0{}^2$. Adding a constant to the potential energy has no effect on the physics, so the system of two atoms is fundamentally no different from a mass attached to a horizontal spring for which $U = \frac{1}{2}kx^2$. The proof is left as an exercise.

EXAMPLE 13–7

Two argon atoms can form a weakly bound molecule, Ar_2, held together by a van der Waals interaction with $U_0 = 1.68 \times 10^{-21}$ J and $R_0 = 3.82 \times 10^{-10}$ m. Find the frequency for small oscillations of one of the atoms about its equilibrium position.

SOLUTION Since the oscillations are small, we can use Eq. (13–11) for the frequency of simple harmonic motion. From Eq. (13–29) the force constant is

$$k = \frac{72U_0}{R_0{}^2} = \frac{72(1.68 \times 10^{-21}\text{ J})}{(3.82 \times 10^{-10}\text{ m})^2} = 0.829\text{ J/m}^2 = 0.829\text{ N/m}.$$

This is comparable to the force constant of a loose, floppy toy spring like a Slinky™.

From the periodic table of the elements (Appendix D), the average atomic mass of argon is $(39.948\text{ u})(1.66 \times 10^{-27}\text{ kg}/1\text{ u}) = 6.63 \times 10^{-26}$ kg. If one of the argon atoms is fixed and the other atom oscillates, the frequency of oscillation is

$$f = \frac{1}{2\pi}\sqrt{\frac{k}{m}} = \frac{1}{2\pi}\sqrt{\frac{0.829\text{ N/m}}{6.63 \times 10^{-26}\text{ kg}}} = 5.63 \times 10^{11}\text{ Hz}.$$

The oscillating mass is very small, so even a loose spring causes very rapid oscillations.

Our answer for f isn't quite right, however. If there is no net external force acting on the molecule, the center of mass of the molecule (located halfway between the two atoms) doesn't accelerate. To ensure this, *both* atoms must oscillate with the same amplitude in opposite directions. It turns out that we can account for this by replacing m with $m/2$ in the expression for f. (See Problem 13–69.) This makes f larger by a factor of $\sqrt{2}$, so $f = \sqrt{2}(5.63 \times 10^{11}\text{ Hz}) = 7.96 \times 10^{11}$ Hz. An additional complication is that on the atomic scale we must use *quantum mechanics*, not Newtonian mechanics, to describe oscillation and other motions; happily, the frequency has the same value in quantum mechanics.

13–6 THE SIMPLE PENDULUM

A **simple pendulum** is an idealized model consisting of a point mass suspended by a massless, unstretchable string. When the point mass is pulled to one side of its straight-down equilibrium position and released, it oscillates about the equilibrium position. Familiar situations such as a wrecking ball on a crane's cable, the plumb bob on a surveyor's transit, and a child on a swing can be modeled as simple pendulums.

The path of the point mass (sometimes called a pendulum bob) is not a straight line but the arc of a circle with radius L equal to the length of the string (Fig. 13–18). We use as our coordinate the distance x measured along the arc. If the motion is simple harmonic, the restoring force must be directly proportional to x or (because $x = L\theta$) to θ. Is it?

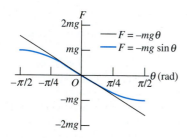

13-18 The forces on the bob of a simple pendulum.

13-19 The restoring force for a simple pendulum, $F = -mg \sin \theta$ (shown in blue), can be approximated by $F = -mg\theta$ (shown in black) for small values of θ. Hence for small angles the restoring force is approximately proportional to θ, and the oscillations are simple harmonic.

ActivPhysics

9.10
Pendulum Frequency

9.11
Pendulum Timer

9.12
Risky Pendulum Walk

In Fig. 13–18 we represent the forces on the mass in terms of tangential and radial components. The restoring force F is the tangential component of the net force:

$$F = -mg \sin \theta. \tag{13–30}$$

The restoring force is provided by gravity; the tension T merely acts to make the point mass move in an arc. The restoring force is proportional *not* to θ but to $\sin \theta$, so the motion is *not* simple harmonic. However, *if the angle θ is small*, $\sin \theta$ is very nearly equal to θ in radians (Fig. 13–19). For example, when $\theta = 0.1$ rad (about 6°), $\sin \theta = 0.0998$, a difference of only 0.2%. With this approximation, Eq. (13–30) becomes

$$F = -mg\theta = -mg \frac{x}{L}, \quad \text{or}$$

$$F = -\frac{mg}{L} x. \tag{13–31}$$

The restoring force is then proportional to the coordinate *for small displacements*, and the force constant is $k = mg/L$. From Eq. (13–10) the angular frequency ω of a simple pendulum with small amplitude is

$$\omega = \sqrt{\frac{k}{m}} = \sqrt{\frac{mg/L}{m}} = \sqrt{\frac{g}{L}} \quad \text{(simple pendulum, small amplitude).} \tag{13–32}$$

The corresponding frequency and period relations are

$$f = \frac{\omega}{2\pi} = \frac{1}{2\pi} \sqrt{\frac{g}{L}} \quad \text{(simple pendulum, small amplitude),} \tag{13–33}$$

$$T = \frac{2\pi}{\omega} = \frac{1}{f} = 2\pi \sqrt{\frac{L}{g}} \quad \text{(simple pendulum, small amplitude).} \tag{13–34}$$

Note that these expressions do not involve the *mass* of the particle. This is because the restoring force, a component of the particle's weight, is proportional to m. Thus the mass appears on *both* sides of $\Sigma \vec{F} = m\vec{a}$ and cancels out. (This is the same physics that explains why bodies of different masses fall with the same acceleration in a vacuum.) For small oscillations the period of a pendulum for a given value of g is determined entirely by its length.

The dependence on L and g in Eqs. (13–32) through (13–34) is just what we should expect. A long pendulum has a longer period than a shorter one. Increasing g increases the restoring force, causing the frequency to increase and the period to decrease.

We emphasize again that the motion of a pendulum is only *approximately* simple harmonic. When the amplitude is not small, the departures from simple harmonic motion can be substantial. But how small is "small"? The period can be expressed by an infinite series; when the maximum angular displacement is Θ, the period T is given by

$$T = 2\pi \sqrt{\frac{L}{g}} \left(1 + \frac{1^2}{2^2} \sin^2 \frac{\Theta}{2} + \frac{1^2 \cdot 3^2}{2^2 \cdot 4^2} \sin^4 \frac{\Theta}{2} + \cdots \right). \tag{13–35}$$

We can compute the period to any desired degree of precision by taking enough terms in the series. We invite you to check that when $\Theta = 15°$ (on either side of the central position), the true period is longer than that given by the approximate Eq. (13–34) by less than 0.5%.

The usefulness of the pendulum as a timekeeper depends on the period being *very nearly* independent of amplitude, provided that the amplitude is small. Thus as a pendulum clock runs down and the amplitude of the swings decreases a little, the clock still keeps very nearly correct time.

A simple pendulum, or a variation thereof, is also a precise and convenient method for measuring the acceleration of gravity g, since L and T can be measured easily and precisely. Such measurements are often used in geophysics. Local deposits of ore or oil affect the local value of g because their density differs from that of their surroundings. Precise measurements of this quantity over an area being surveyed often furnish valuable information about the nature of underlying deposits.

EXAMPLE 13–8

Find the period and frequency of a simple pendulum 1.000 m long at a location where $g = 9.800$ m/s^2.

SOLUTION From Eq. (13–34),

$$T = 2\pi\sqrt{\frac{L}{g}} = 2\pi\sqrt{\frac{1.000 \text{ m}}{9.800 \text{ m/s}^2}} = 2.007 \text{ s}.$$

Then

$$f = \frac{1}{T} = 0.4983 \text{ Hz}.$$

The period is almost exactly 2 s. In fact, when the metric system was first established, the second was defined as half the period of a one-meter pendulum. This wasn't a very good standard for time, however, because the value of g varies from place to place. We discussed more modern time standards in Section 1–4.

13–7 THE PHYSICAL PENDULUM

A **physical pendulum** is any *real* pendulum, using a body of finite size, as contrasted to the idealized model of the *simple* pendulum with all the mass concentrated at a single point. For small oscillations, analyzing the motion of a real, physical pendulum is almost as easy as for a simple pendulum. Figure 13–20 shows a body of irregular shape pivoted so that it can turn without friction about an axis through point O. In the equilibrium position the center of gravity is directly below the pivot; in the position shown in the figure, the body is displaced from equilibrium by an angle θ, which we use as a coordinate for the system. The distance from O to the center of gravity is d, the moment of inertia of the body about the axis of rotation through O is I, and the total mass is m. When the body is displaced as shown, the weight mg causes a restoring torque

$$\tau = -(mg)(d\sin\theta). \tag{13–36}$$

The negative sign shows that the restoring torque is clockwise when the displacement is counterclockwise, and vice versa.

When the body is released, it oscillates about its equilibrium position. The motion is not simple harmonic because the torque τ is proportional to $\sin\theta$ rather than to θ itself. However, if θ is small, we can again approximate $\sin\theta$ by θ in radians, and the motion is *approximately* simple harmonic. With this approximation,

$$\tau = -(mgd)\theta.$$

The equation of motion is $\Sigma\tau = I\alpha$, so

$$-(mgd)\theta = I\alpha = I\frac{d^2\theta}{dt^2},$$

$$\frac{d^2\theta}{dt^2} = -\frac{mgd}{I}\theta. \tag{13–37}$$

Comparing this with Eq. (13–4), we see that the role of (k/m) for the mass-spring system is played here by the quantity (mgd/I). Thus the angular frequency is given by

$$\omega = \sqrt{\frac{mgd}{I}} \qquad \text{(physical pendulum, small amplitude).} \tag{13–38}$$

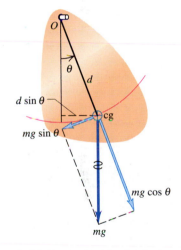

13–20 Dynamics of a physical pendulum.

9.13
Physical Pendulum

The frequency f is $1/2\pi$ times this, and the period T is

$$T = 2\pi\sqrt{\frac{I}{mgd}} \qquad \text{(physical pendulum, small amplitude).} \qquad (13\text{–}39)$$

Equation (13–39) is the basis of a common method for experimentally determining the moment of inertia of a body with a complicated shape. The center of gravity of the body is first located by balancing. The body is then suspended so that it is free to oscillate about an axis, and the period T of small-amplitude oscillations is measured. By using Eq. (13–39), the moment of inertia I of the body about this axis can then be calculated from T, the body's mass m, and the distance d from the axis to the center of gravity (see Exercise 13–42). Biomechanics researchers use this method to find the moments of inertia of an animal's limbs. This information is important for analyzing how an animal walks, as we'll see in the second of the two following examples.

EXAMPLE 13–9

Physical pendulum versus simple pendulum Suppose the body in Fig. 13–20 is a uniform rod with length L, pivoted at one end. Find the period of its motion.

SOLUTION From Chapter 9 the moment of inertia of a uniform rod about an axis through one end is $I = \frac{1}{3}ML^2$. The distance from the pivot to the center of gravity is $d = L/2$. From Eq. (13–39),

$$T = 2\pi\sqrt{\frac{\frac{1}{3}ML^2}{MgL/2}} = 2\pi\sqrt{\frac{2L}{3g}}.$$

If the rod is a meter stick ($L = 1.00$ m) and $g = 9.80$ m/s^2, then

$$T = 2\pi\sqrt{\left(\frac{2}{3}\right)(1.00 \text{ m})/(9.80 \text{ m/s}^2)} = 1.64 \text{ s}.$$

The period is smaller by a factor of $\sqrt{2/3} = 0.816$ than the period of a simple pendulum with the same length, calculated in Example 13–8 (Section 13–6).

EXAMPLE 13–10

***Tyrannosaurus rex* and the physical pendulum** All walking animals, including humans, have a natural walking pace, a number of steps per minute that is more comfortable than a faster or slower pace. Suppose this natural pace is equal to the period of the leg, viewed as a uniform rod pivoted at the hip joint. a) How does the natural walking pace depend on the length L of the leg, measured from hip to foot? b) Fossil evidence shows that *Tyrannosaurus rex*, a two-legged dinosaur that lived about 65 million years ago at the end of the Cretaceous period, had a leg length $L = 3.1$ m and a stride length (the distance from one footprint to the next print of the same foot) $S = 4.0$ m (Fig. 13–21). Estimate the walking speed of *Tyrannosaurus rex*.

SOLUTION a) The period of oscillation of the leg is given by the expression found in Example 13–9, $T = 2\pi\sqrt{2L/3g}$, which is proportional to $\sqrt{L}$. Each period (a complete back-and-forth swing of the leg) corresponds to *two* steps, so the walking pace in steps per unit time is just twice the oscillation frequency $f = 1/T$. Hence the walking pace is proportional to $1/\sqrt{L}$. Animals with short legs (small values of L), such as mice or Chihuahuas, have rapid walking paces; humans, giraffes, and other animals with long legs (large values of L) walk at slower paces.

b) According to our model for the natural walking pace, the elapsed time for one stride of a walking *Tyrannosaurus rex* is

$$T = 2\pi\sqrt{\frac{2L}{3g}} = 2\pi\sqrt{\frac{2(3.1 \text{ m})}{3(9.8 \text{ m/s}^2)}} = 2.9 \text{ s}.$$

The distance moved during this time is the stride length S, so the walking speed is

$$v = \frac{S}{T} = \frac{4.0 \text{ m}}{2.9 \text{ s}} = 1.4 \text{ m/s} = 5.0 \text{ km/h} = 3.1 \text{ mi/h}.$$

This is about the same as a typical human walking speed! Our estimate must be somewhat in error, however, because a uniform rod isn't a very good model for a leg. The legs of many animals, including *T. rex* as well as humans, are tapered; there is a lot more mass between the knee and the hip than between the knee and the foot. Thus the center of mass is less than $L/2$ from the hip; a reasonable guess would be about $L/4$. The moment of inertia is *considerably* less than $ML^2/3$, probably somewhere around $ML^2/15$. Try these numbers out with the analysis of Example 13–9; you'll get a shorter oscillation period and an even faster walking speed for *T. rex*.

13-21 The walking speed of *Tyrannosaurus rex* can be estimated from its leg length L and its stride length S.

13-8 DAMPED OSCILLATIONS

The idealized oscillating systems we have discussed thus far are frictionless. There are no nonconservative forces, the total mechanical energy is constant, and a system set into motion continues oscillating forever with no decrease in amplitude.

Real-world systems always have some dissipative forces, however, and oscillations do die out with time unless we provide some means for replacing the dissipated mechanical energy. A mechanical pendulum clock continues to run because potential energy stored in the spring or a hanging weight system replaces the mechanical energy lost due to friction in the pivot and the gears. But eventually the spring runs down or the weights reach the bottom of their travel. Then no more energy is available, and the pendulum swings decrease in amplitude and stop.

The decrease in amplitude caused by dissipative forces is called **damping,** and the corresponding motion is called **damped oscillation.** The simplest case to analyze in detail is a simple harmonic oscillator with a frictional damping force that is directly proportional to the *velocity* of the oscillating body. This behavior occurs in friction involving viscous fluid flow, such as in shock absorbers or sliding between oil-lubricated surfaces. We then have an additional force on the body due to friction, $F = -bv$, where $v = dx/dt$ is the velocity and b is a constant that describes the strength of the damping force. The negative sign shows that the force is always opposite in direction to the velocity. The *net* force on the body is then

$$\Sigma F = -kx - bv, \qquad (13-40)$$

and Newton's second law for the system is

$$-kx - bv = ma, \qquad \text{or} \qquad -kx - b\frac{dx}{dt} = m\frac{d^2x}{dt^2}. \qquad (13-41)$$

A disturbance such as an earthquake would make this chandelier oscillate back and forth. The oscillations would be damped, however, by air resistance and friction at the point of suspension, and would eventually die out.

13–22 Graph of Eq. (13–42), showing damped harmonic motion, in the case in which the phase angle $\phi = 0$. The period when there is no damping ($b = 0$) is T_0. The blue curve shows the motion when $b = 0.1 \sqrt{km}$, and the red curve shows the motion when $b = 0.4 \sqrt{km}$. The amplitude decreases more rapidly for the larger value of b. Close inspection of the points where the curves cross the t-axis also reveals that the period increases slightly with increasing b. The critical-damping condition is $b = 2 \sqrt{km}$.

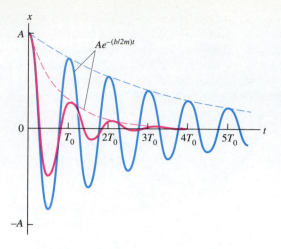

Equation (13–41) is a differential equation for x; it would be the same as Eq. (13–4), the equation for the acceleration in SHM, except for the added term $-b \, dx/dt$. Solving this equation is a straightforward problem in differential equations, but we won't go into the details here. If the damping force is relatively small, the motion is described by

$$x = Ae^{-(b/2m)t} \cos{(\omega' t + \phi)} \qquad \text{(oscillator with little damping).} \qquad (13\text{–}42)$$

The angular frequency of oscillation ω' is given by

$$\omega' = \sqrt{\frac{k}{m} - \frac{b^2}{4m^2}} \qquad \text{(oscillator with little damping).} \qquad (13\text{–}43)$$

You can verify that Eq. (13–42) is a solution of Eq. (13–41) by calculating the first and second derivatives of x, substituting them into Eq. (13–41), and checking whether the left and right sides are equal. This is a straightforward but slightly tedious procedure.

The motion described by Eq. (13–42) differs from the undamped case in two ways. First, the amplitude $Ae^{-(b/2m)t}$ is not constant but decreases with time because of the decreasing exponential factor $e^{-(b/2m)t}$. Figure 13–22 is a graph of Eq. (13–42) for the case $\phi = 0$; it shows that the larger the value of b, the more quickly the amplitude decreases.

Second, the angular frequency ω', given by Eq. (13–43), is no longer equal to $\omega = \sqrt{k/m}$ but is somewhat smaller. It becomes zero when b becomes so large that

$$\frac{k}{m} - \frac{b^2}{4m^2} = 0, \qquad \text{or} \qquad b = 2\sqrt{km}. \qquad (13\text{–}44)$$

When Eq. (13–44) is satisfied, the condition is called **critical damping.** The system no longer oscillates but returns to its equilibrium position without oscillation when it is displaced and released.

If b is greater than $2\sqrt{km}$, the condition is called **overdamping.** Again there is no oscillation, but the system returns to equilibrium more slowly than with critical damping. For the overdamped case the solutions of Eq. (13–41) have the form

$$x = C_1 e^{-a_1 t} + C_2 e^{-a_2 t},$$

where C_1 and C_2 are constants that depend on the initial conditions and a_1 and a_2 are constants determined by m, k, and b.

When b is less than the critical value, as in Eq. (13–42), the condition is called **underdamping.** The system oscillates with steadily decreasing amplitude.

In a vibrating tuning fork or guitar string, it is usually desirable to have as little damping as possible. By contrast, damping plays a beneficial role in the oscillations of an automobile's suspension system. The shock absorbers provide a velocity-dependent damping force so that when the car goes over a bump, it doesn't continue bouncing forever (Fig. 13–23). For optimal passenger comfort, the system should be critically damped or slightly underdamped. As the shocks get old and worn, the value of b decreases and the bouncing persists longer. Not only is this nauseating, it is bad for steering because the front wheels have less positive contact with the ground. Thus damping is an advantage in this system. Too much damping would be counterproductive, however; if b is excessively large, the system is overdamped and the suspension returns to equilibrium more slowly. If the suspension is overdamped and the car hits a second bump just after the first one, the springs in the suspension will still be compressed somewhat from the first bump and will be unable to fully absorb the impact.

In damped oscillations the damping force is nonconservative; the mechanical energy of the system is not constant but decreases continuously, approaching zero after a long time. To derive an expression for the rate of change of energy, we first write an expression for the total mechanical energy E at any instant:

$$E = \frac{1}{2}mv^2 + \frac{1}{2}kx^2.$$

To find the rate of change of this quantity, we take its time derivative:

$$\frac{dE}{dt} = mv\frac{dv}{dt} + kx\frac{dx}{dt}.$$

But $dv/dt = a$ and $dx/dt = v$, so

$$\frac{dE}{dt} = v(ma + kx).$$

From Eq. (13–41), $ma + kx = -b\,dx/dt = -bv$, so

$$\frac{dE}{dt} = v(-bv) = -bv^2 \qquad \text{(damped oscillations)}. \qquad (13\text{–}45)$$

The right side of Eq. (13–45) is always negative, whether v is positive or negative. This shows that E continuously decreases, though not at a uniform rate. The term $-bv^2 = (-bv)v$ (force times velocity) is the rate at which the damping force does (negative) work on the system (that is, the damping *power*). This equals the rate of change of the total mechanical energy of the system.

Similar behavior occurs in electric circuits containing inductance, capacitance, and resistance. There is a natural frequency of oscillation, and the resistance plays the role of the damping constant b. In such circuits it is usually desirable to minimize damping, but damping can never be prevented completely. We will study these circuits in detail in Chapters 31 and 32.

Upper cylinder remains relatively stationary

— Piston

— Viscous fluid

Lower cylinder moves with wheel and axle

13–23 A car shock absorber. The top part, connected to the piston, is attached to the car's frame; the bottom part, connected to the lower cylinder, is attached to the axle. The viscous fluid causes a damping force that depends on the relative velocity of the two ends of the unit. This helps to control wheel bounce and jounce.

13-9 FORCED OSCILLATIONS, RESONANCE, AND CHAOS

A damped oscillator left to itself will eventually stop moving altogether. But we can maintain a constant-amplitude oscillation by applying a force that varies with time in a periodic or cyclic way, with a definite period and frequency. As an example, consider your cousin Throckmorton on a playground swing. You can keep him swinging with constant amplitude by giving him a little push once each cycle. We call this additional force a **driving force.**

If we apply a periodically varying driving force with angular frequency ω_d to a damped harmonic oscillator, the motion that results is called a **forced oscillation** or a *driven oscillation*. It is different from the motion that occurs when the system is simply displaced from equilibrium and then left alone, in which case the system oscillates with a **natural angular frequency** ω' determined by m, k, and b, as in Eq. (13–43). In a forced oscillation, however, the angular frequency with which the mass oscillates will be equal to the driving angular frequency ω_d. This does *not* have to be equal to the angular frequency ω' with which the system would oscillate without a driving force. If you grab the ropes of Throckmorton's swing, you can force the swing to oscillate with any frequency you like.

Suppose we force the oscillator to vibrate with an angular frequency ω_d that is nearly *equal* to the angular frequency ω' it would have with no driving force. What happens? The oscillator is naturally disposed to oscillate at $\omega = \omega'$, so we expect the amplitude of the resulting oscillation to be larger than when the two frequencies are very different. Detailed analysis and experiment shows that this is just what happens. The easiest case to analyze is a *sinusoidally* varying force, say, $F(t) = F_{max} \cos \omega_d t$. If we vary the frequency ω_d of the driving force, the amplitude of the resulting forced oscillation varies in an interesting way (Fig. 13–24). When there is very little damping (small b), the amplitude goes through a sharp peak as the driving angular frequency ω_d nears the natural oscillation angular frequency ω'. When the damping is increased (larger b), the peak becomes broader and smaller in height and shifts toward lower frequencies.

We could work out an expression that shows how the amplitude A of the forced oscillation depends on the frequency of a sinusoidal driving force, with maximum value F_{max}. That would involve more differential equations than we're ready for, but here is the result:

$$A = \frac{F_{max}}{\sqrt{(k - m\omega_d^{\,2})^2 + b^2\omega_d^{\,2}}} \qquad \text{(amplitude of a driven oscillator).} \quad (13\text{–}46)$$

When $k - m\omega_d^{\,2} = 0$, the first term under the radical is zero, so A has a maximum near $\omega_d = \sqrt{k/m}$. The height of the curve at this point is proportional to $1/b$; the less damping, the higher the peak. At the low-frequency extreme, when $\omega_d = 0$, we get $A = F_{max}/k$. This corresponds to a *constant* force F_{max} and a constant displacement $A = F_{max}/k$ from equilibrium, as we might expect.

13–24 Graph of the amplitude A of forced oscillation of a damped harmonic oscillator as a function of the frequency ω_d of the driving force. The horizontal axis shows the ratio of ω_d to the angular frequency $\omega = \sqrt{k/m}$ of an undamped oscillator. Each curve is labeled with the value of the dimensionless quantity $b/\sqrt{km}$, which characterizes the amount of damping. The highest curve has $b = 0.2\sqrt{km}$, the next has $b = 0.4\sqrt{km}$, and so on. As b increases, the peak becomes broader and less sharp and shifts toward lower frequencies. When b is larger than $\sqrt{2km}$, the peak disappears completely.

The fact that there is an amplitude peak at driving frequencies close to the natural frequency of the system is called **resonance.** Physics is full of examples of resonance; building up the oscillations of a child on a swing by pushing with a frequency equal to the swing's natural frequency is one. A vibrating rattle in a car that occurs only at a certain engine speed or wheel-rotation speed is an all too familiar example. Inexpensive loudspeakers often have an annoying boom or buzz when a musical note happens to coincide with the resonant frequency of the speaker cone or the speaker housing. A tuned circuit in a radio or television receiver responds strongly to waves having frequencies near its resonant frequency, and this fact is used to select a particular station and reject the others. We will study resonance in electric circuits in detail in Chapter 32.

Resonance in mechanical systems can be destructive. A company of soldiers once destroyed a bridge by marching across it in step; the frequency of their steps was close to a natural vibration frequency of the bridge, and the resulting oscillation had large enough amplitude to tear the bridge apart. Ever since, marching soldiers have been ordered to break step before crossing a bridge. Some years ago, vibrations of the engines of a particular airplane had just the right frequency to resonate with the natural frequencies of its wings. Large oscillations built up, and occasionally the wings fell off.

Nearly everyone has seen the film of the collapse of the Tacoma Narrows suspension bridge in 1940 (Fig. 13–25). This is usually cited as an example of resonance driven by the wind, but there's some doubt as to whether it should be called that. The wind didn't have to vary *periodically* with a frequency close to a natural frequency of the bridge. The air flow past the bridge was turbulent, and vortices were formed in the air with a regular frequency that depended on the flow speed. It is conceivable that this frequency may have coincided with a natural frequency of the bridge. But the cause may well have been something more subtle called a *self-excited oscillation*, in which the aerodynamic forces caused by a *steady* wind blowing on the bridge tended to displace it farther from equilibrium at times when it was already moving away from equilibrium. It is as though we had a damping force such as the $-bv$ term in Eq. (13–40) but with the sign reversed. Instead of draining mechanical energy away from the system, this anti-damping force pumps energy into the system, building up the oscillations to destructive amplitudes. The approximate differential equation is Eq. (13–41) with the sign of the b term reversed, and the oscillating solution is Eq. (13–42) with a *positive* sign in the exponent. You can see that we're headed for trouble. Engineers have learned how to stabilize suspension bridges, both structurally and aerodynamically, to prevent such disasters.

13–25 The Tacoma Narrows Bridge collapsed four months and six days after it was opened for traffic. The main span was 2800 ft long and 39 ft wide, with 8-ft-high steel stiffening girders on both sides. The maximum amplitude of the torsional vibrations was 35°; the frequency was about 0.2 Hz.

CHAOTIC OSCILLATIONS

If a sinusoidally varying driving force $F(t) = F_{max} \cos \omega_d t$ is applied to a harmonic oscillator, the resulting forced oscillation has the same angular frequency ω_d as the driving force. But what happens if the oscillator is *not* a harmonic one?

Here's an example. Suppose you have a pendulum that is made up of a mass m attached to a lightweight but rigid rod of length L and is free to rotate around a pivot at the top of the rod, as in Fig. 13–18. We saw in Section 13–6 that large-amplitude oscillations of a simple pendulum are *not* simple harmonic motion, because the period depends on the amplitude. To apply a sinusoidally varying driving force to the pendulum, you hold the pivot in your hand and oscillate it up and down with angular frequency ω_d. To be specific, let's suppose that ω_d is twice the angular frequency $\omega = \sqrt{g/L}$ at which the pendulum would oscillate by itself and that there is friction in the pivot.

As you increase the amplitude F_{max} of the driving force that you apply to the pivot, the amplitude of oscillation of the pendulum increases as well. If F_{max} isn't too large, the pendulum oscillates at the same angular frequency as the driving force, just like a driven harmonic oscillator. But if you increase F_{max} to a value close to the weight mg of the pendulum, after the transient behavior has died down, you will notice something odd: the motion of the pendulum during one up-and-down cycle of the pivot is slightly different from the oscillation during the next up-and-down cycle. The pendulum's motion repeats itself only after *two* up-and-down cycles. So if the period of the driving force is $T_d = 2\pi/\omega_d$, the period of the actual motion of the pendulum is $T = 2T_d$, a phenomenon called *period doubling*. If you increase the value of F_{max} to about $1.5mg$, the pendulum will make complete circles around the pivot, moving somewhat differently during one circle than during the following circle, so again $T = 2T_d$. Further increases in F_{max} will cause the period of the pendulum to double again, to $T = 4T_d$; the pendulum's motion repeats on only every fourth cycle of the pivot's motion. As you continue to increase F_{max}, T becomes a larger and larger multiple of T_d. At a finite value of F_{max} (around $2mg$ for the pendulum we've described) the period of the motion becomes infinite. This means that the motion *never* repeats!

Although the motion of the pendulum is completely determined by its initial position, its initial velocity, and the forces that act on it (through Newton's second law, $\Sigma \vec{F} = m\vec{a}$), the motion appears to be completely random and unpredictable. Such *seemingly* random motion is called **chaotic motion,** and its behavior is called **chaos.** Many physical systems in nature and technology can display chaotic behavior. Examples include water dripping from a faucet, a mass bouncing on top of a vertically oscillating piston, and turbulent fluid flow. Chaos also occurs in such situations outside of physics such as the growth and decay of animal populations and fluctuations in stock prices. The study of chaotic motion in mechanical systems can give tremendous insight into chaotic behavior in biology, economics, and other fields. Chaos is an active field of research in many fields of natural and social science.

SUMMARY

KEY TERMS

periodic motion, 392
oscillation, 392
displacement, 392
restoring force, 392
amplitude, 393
cycle, 393

■ Periodic motion is motion that repeats itself in a definite cycle. It occurs whenever a body has a stable equilibrium position and a restoring force or torque that acts when it is displaced from equilibrium. Period T is the time for one cycle. Frequency f is the number of cycles per unit time:

$$f = \frac{1}{T}, \qquad T = \frac{1}{f}. \tag{13–1}$$

Angular frequency ω is

$$\omega = 2\pi f = \frac{2\pi}{T}. \tag{13-2}$$

- If the net force is a restoring force F that is directly proportional to the displacement x, the motion is called simple harmonic motion (SHM). In that case,

$$F = -kx, \tag{13-3}$$

$$a = \frac{F}{m} = -\frac{k}{m}x. \tag{13-4}$$

In many cases the restoring force F is given approximately by Eq. (13–3) if the displacement from equilibrium is small.

- The circle of reference construction uses a rotating vector called a phasor, having a length equal to the amplitude of the motion. Its projection on the horizontal axis represents the actual motion of a body in simple harmonic motion.

- The angular frequency, frequency, and period in SHM do not depend on the amplitude, but depend only on the mass m and force constant k:

$$\omega = \sqrt{\frac{k}{m}}, \tag{13-10}$$

$$f = \frac{\omega}{2\pi} = \frac{1}{2\pi}\sqrt{\frac{k}{m}}, \tag{13-11}$$

$$T = \frac{1}{f} = 2\pi\sqrt{\frac{m}{k}}. \tag{13-12}$$

- In SHM the displacement, velocity, and acceleration are sinusoidal functions of time. The displacement x is given as a function of time t by

$$x = A\cos(\omega t + \phi), \tag{13-13}$$

where $\omega = \sqrt{k/m}$. The amplitude A and phase angle ϕ are determined by the initial position and velocity of the body.

- Conservation of energy in SHM leads to the relation

$$E = \frac{1}{2}mv^2 + \frac{1}{2}kx^2 = \frac{1}{2}kA^2 = \text{constant}. \tag{13-21}$$

- In angular simple harmonic motion the frequency and angular frequency are related to the moment of inertia I and the torsion constant κ by

$$\omega = \sqrt{\frac{\kappa}{I}} \quad \text{and} \quad f = \frac{1}{2\pi}\sqrt{\frac{\kappa}{I}}. \tag{13-24}$$

- A simple pendulum consists of a point mass m at the end of a massless string of length L. Its motion is approximately simple harmonic for sufficiently small amplitude; the angular frequency, frequency, and period are then given by

$$\omega = \sqrt{\frac{g}{L}}, \tag{13-32}$$

$$f = \frac{\omega}{2\pi} = \frac{1}{2\pi}\sqrt{\frac{g}{L}}, \tag{13-33}$$

$$T = \frac{2\pi}{\omega} = \frac{1}{f} = 2\pi\sqrt{\frac{L}{g}}. \tag{13-34}$$

■ A physical pendulum is a body suspended from an axis of rotation a distance d from its center of gravity. If the moment of inertia about the axis of rotation is I, the angular frequency and period for small-amplitude oscillations are

$$\omega = \sqrt{\frac{mgd}{I}}, \tag{13-38}$$

$$T = 2\pi\sqrt{\frac{I}{mgd}}. \tag{13-39}$$

■ When a damping force $F = -bv$ proportional to velocity is added to a simple harmonic oscillator, the motion is called a damped oscillation. For a relatively small damping force,

$$x = Ae^{-(b/2m)t} \cos \omega' t. \tag{13-42}$$

The angular frequency of oscillation ω' is given by

$$\omega' = \sqrt{\frac{k}{m} - \frac{b^2}{4m^2}}. \tag{13-43}$$

This motion occurs when $b < 2\sqrt{km}$, a condition called underdamping. When $b = 2\sqrt{km}$, the system is critically damped and no longer oscillates. When $b > 2\sqrt{km}$, the system is overdamped.

■ When a sinusoidally varying driving force is added to a damped harmonic oscillator, the resulting motion is called a forced oscillation. The amplitude is given as a function of driving angular frequency ω_d by

$$A = \frac{F_{max}}{\sqrt{(k - m\omega_d^2)^2 + b^2\omega_d^2}}. \tag{13-46}$$

The amplitude reaches a peak at a driving frequency close to the natural oscillation frequency of the system. This behavior is called resonance.

DISCUSSION QUESTIONS

Q13-1 Your schedule of classes, study time, meals, and recreation is essentially the same from one week to the next. Is this a form of periodic motion? Of simple harmonic motion? Explain.

Q13-2 Think of several examples in everyday life of motion that is at least approximately simple harmonic. In what respects does each differ from SHM?

Q13-3 Does a tuning fork or similar tuning instrument undergo simple harmonic motion? Why is this a crucial question for musicians?

Q13-4 A highly elastic "superball" bouncing on a hard floor has a motion that is approximately periodic. In what ways is the motion similar to SHM? In what ways is it different?

Q13-5 If a uniform spring is cut in half, what is the force constant of each half? Justify your answer. How would the frequency of SHM using a half-spring differ from that using the same mass and the entire spring?

Q13-6 The analysis of SHM in this chapter neglected the mass of the spring. How does the spring mass change the characteristics of the motion?

Q13-7 Suppose the stationary end of the spring in Fig. 13–1 is connected instead to another identical body that is free to slide along the same line. Could such a system undergo SHM? Explain. How would the period compare with that of the original system? Explain.

Q13-8 For the system shown in Fig. 13–1, is there any point during the motion at which the body is in equilibrium? Explain.

Q13-9 The system shown in Fig. 13–14 is mounted in an elevator, which accelerates upward with constant acceleration. Does the period increase, decrease, or remain the same? Justify your answer.

Q13-10 Why is the "springiness" of a diving board adjusted for different dives and different weights of divers? How is the adjustment made?

Q13-11 A pendulum is mounted in an elevator that accelerates upward with constant acceleration. Does the period increase, decrease, or remain the same? Explain your answer.

Q13-12 What should you do to the length of the string of a simple pendulum to a) double its frequency? b) double its period? c) double its angular frequency?

Q13–13 If a pendulum clock is taken to a mountain top, does it gain or lose time, assuming it is correct at a lower elevation? Explain your answer.

Q13–14 When the amplitude of a simple pendulum increases, should its period increase or decrease? Give a qualitative argument; do not rely on Eq. (13–35). Is your argument also valid for a physical pendulum?

Q13–15 Why do short dogs (like Chihuahuas) walk with quicker strides than do tall dogs (like Great Danes)?

Q13–16 At what point in the motion of a simple pendulum is the string tension greatest? Least? In each case give the reasoning behind your answer.

Q13–17 Could a standard of time be based on the period of a certain standard pendulum? What advantages and disadvantages would such a standard have compared to the actual present-day standard discussed in Section 1–4?

Q13–18 The frequency with which dogs pant is the natural frequency of their respiratory system. Why do they choose this frequency?

Q13–19 In designing structures in an earthquake-prone region, how should the natural frequencies of oscillation of a structure relate to typical earthquake frequencies? Why? Should the structure have a large or small amount of damping?

EXERCISES

SECTION 13–2 THE CAUSES OF OSCILLATION

13–1 A piano string sounds a middle A by vibrating primarily at 220 Hz. a) Calculate its period and angular frequency. b) Calculate the period and angular frequency for a soprano singing a "high A," two octaves up, which is four times the frequency of the piano string.

13–2 The body in Fig. 13–1 is displaced 0.120 m from its equilibrium position and released with zero initial speed. After 0.800 s its displacement is found to be 0.120 m on the opposite side, and it has passed the equilibrium position once during this interval. Find a) the amplitude; b) the period; c) the frequency.

13–3 The tip of a tuning fork goes through 440 complete vibrations in 0.500 s. Find the angular frequency and the period of the motion.

SECTION 13–3 SIMPLE HARMONIC MOTION

13–4 In physics lab, you attach a 0.200-kg air-track glider to the end of an ideal spring of negligible mass as in Fig. 13–1 and start it oscillating. The elapsed time from when the glider first moves through the equilibrium point to the second time it moves through that point is 2.60 s. Find the spring's force constant.

13–5 A body of unknown mass is attached to an ideal spring with force constant 120 N/m in the arrangement shown in Fig. 13–1. It is found to vibrate with a frequency of 6.00 Hz. Find a) the period; b) the angular frequency; c) the mass of the body.

13–6 A harmonic oscillator is made by using a 0.600-kg frictionless block and an ideal spring of unknown force constant. The oscillator is found to have a period of 0.150 s. Find the force constant of the spring.

13–7 A harmonic oscillator has a mass of 0.500 kg and an ideal spring with force constant 140 N/m. Find a) the period; b) the frequency; c) the angular frequency.

13–8 Substitute the following equations, in which A, ω, and β are constant, into Eq. (13–4) to see if they describe SHM. If so, what must ω equal? a) $x = A \sin (\omega t + \beta)$. b) $x = A\omega t^2 + \beta$. c) $x = Ae^{i(\omega t + \beta)}$, where $i = \sqrt{-1}$.

13–9 A guitar string vibrates at a frequency of 440 Hz. A point at its center moves in SHM with an amplitude of 3.0 mm and a phase angle of zero. a) Write an equation for the position of the center of the string as a function of time. b) What are the maximum values of the magnitudes of the velocity and acceleration of the center of the string? c) The derivative of the acceleration with respect to time is a quantity called the *jerk*. Write an equation for the jerk of the center of the string as a function of time, and find the maximum value of the magnitude of the jerk.

13–10 A 2.00-kg frictionless block is attached to an ideal spring with force constant 300 N/m. At $t = 0$ the spring is neither stretched nor compressed and the block is moving in the negative direction at 12.0 m/s. Find a) the amplitude and b) the phase angle. c) Write an equation for the position as a function of time.

13–11 Repeat Exercise 13–10, but assume that at $t = 0$ the block has velocity −4.00 m/s and displacement +0.200 m.

13–12 The point of the needle of a sewing machine moves in SHM along the x-axis with a frequency of 2.5 Hz. At $t = 0$ its position and velocity components are +1.1 cm and −15 cm/s. a) Find the acceleration component of the needle at $t = 0$. b) Write equations giving the position, velocity, and acceleration components of the point as a function of time.

13–13 An object is undergoing simple harmonic motion with period 1.200 s and amplitude 0.600 m. At $t = 0$ the object is at $x = 0$. How far is the object from the equilibrium position when $t = 0.480$ s?

13–14 An object is undergoing simple harmonic motion with period 0.300 s and amplitude 6.00 cm. At $t = 0$ the object is instantaneously at rest at $x = 6.00$ cm. Calculate the time it takes the object to go from $x = 6.00$ cm to $x = -1.50$ cm.

SECTION 13–4 ENERGY IN SIMPLE HARMONIC MOTION

13–15 Derive Eq. (13–19) a) from Eqs. (13–14) and (13–17); b) from conservation of energy, Eq. (13–21).

13–16 A tuning fork labeled 392 Hz has the tip of each of its two prongs vibrating with an amplitude of 0.600 mm. a) What is the maximum speed of the tip of a prong? b) A housefly (*Musca domestica*) of mass 0.0270 g is holding on to the tip of one of the prongs. As the prong vibrates, what is the fly's maximum kinetic energy? Assume that the fly's mass has a negligible effect on the frequency of oscillation.

13–17 A harmonic oscillator has angular frequency ω and amplitude A. a) What are the magnitudes of the displacement and velocity when the elastic potential energy is equal to the kinetic energy? (Assume that $U = 0$ at equilibrium.) b) How often does this occur each cycle? What is the time between occurrences? c) At an instant when the displacement is equal to $A/2$, what fraction of the total energy of the system is kinetic and what fraction is potential?

13–18 An ideal spring with force constant $k = 450$ N/m is mounted as in Fig. 13–1. A 0.500-kg glider attached to the end undergoes simple harmonic motion with an amplitude 0.040 m. Compute a) the maximum speed of the glider; b) the speed of the glider when it is at $x = -0.015$ m; c) the magnitude of the maximum acceleration of the glider; d) the acceleration of the glider at $x = -0.015$ m; e) the total mechanical energy of the glider at any point in its motion.

13–19 A cheerleader waves her pom-pom in SHM with an amplitude of 18.0 cm and a frequency of 0.850 Hz. Find a) the maximum magnitude of the acceleration and of the velocity; b) the acceleration and speed when the pom-pom's coordinate is $x = +9.0$ cm; c) the time required to move from the equilibrium position directly to a point 12.0 cm away. d) Which of the quantities asked for in parts (a), (b), and (c) can be found using the energy approach used in Section 13–4, and which cannot? Explain.

13–20 For the situation described in part (a) of Example 13–5, what should be the value of the putty mass m so that the amplitude after the collision is one-half the original amplitude? For this value of m, what fraction of the original mechanical energy is converted into heat?

13–21 A 0.150-kg toy is undergoing simple harmonic motion on the end of a horizontal spring with force constant $k = 300$ N/m. When the object is 0.012 m from its equilibrium position, it is observed to have a speed of 0.300 m/s. What is a) the total energy of the object at any point of its motion? b) the amplitude of the motion? c) the maximum speed attained by the object during its motion?

SECTION 13–5 APPLICATIONS OF SIMPLE HARMONIC MOTION

13–22 The scale of a spring balance reading from zero to 200 N is 12.5 cm long. A fish hanging from the bottom of the spring oscillates vertically at 2.60 Hz. What is the mass of the fish? Neglect the mass of the spring.

13–23 A proud deep-sea fisherman hangs a 65.0-kg fish from an ideal spring having negligible mass. The fish stretches the spring 0.120 m. a) What is the force constant of the spring? b) What is the period of oscillation of the fish if it is pulled down and released?

13–24 A thrill-seeking cat with a mass of 4.00 kg is attached by a harness to an ideal spring of negligible mass and oscillates vertically in simple harmonic motion. The amplitude is 0.050 m, and at the highest point of the motion the spring has its natural unstretched length. Calculate the elastic potential energy of the spring (take it to be zero for the unstretched spring), the kinetic energy of the cat, the gravitational potential energy of the system relative to the lowest point of the motion, and the sum of these three energies when the cat is a) at its highest point; b) at its lowest point; c) at its equilibrium position.

13–25 A 2.00-kg block of Cheddar cheese hangs from an ideal spring of negligible mass. When displaced from equilibrium and released, the cheese oscillates with a period of 0.400 s. How much is the spring stretched when the cheese hangs in equilibrium (at rest)?

13–26 Your boss at the Cut-Rate Cuckoo Clock Company asks you what would happen to the frequency of the angular SHM of the balance wheel if it had the same density and the same coil spring (thus the same torsion constant), but all the balance wheel dimensions were made one-third as great to save material. a) What do you tell her? b) By what factor would the torsion constant need to be changed to make the smaller balance wheel oscillate at the original frequency?

13–27 A certain alarm clock ticks four times each second, each tick representing half a period. The balance wheel consists of a thin rim with radius 0.55 cm, connected to the balance staff by thin spokes of negligible mass. The total mass of the balance wheel is 0.90 g. a) What is the moment of inertia of the balance wheel about its shaft? b) What is the torsion constant of the hairspring?

13–28 A thin metal disk of mass 2.00×10^{-3} kg and radius 2.20 cm is attached at its center to a long fiber (Fig. 13–26). The disk, when twisted and released, oscillates with a period of 1.00 s. Find the torsion constant of the fiber.

FIGURE 13–26 Exercise 13–28.

13–29 You want to find the moment of inertia of a complicated machine part about an axis through its center of mass. You suspend it from a wire along this axis. The wire has a torsion constant of 0.450 N · m/rad. You twist the part a small amount about this axis and let it go, timing 125 oscillations in 265 s. What is the desired moment of inertia?

13–30 The balance wheel of a watch vibrates with an angular amplitude Θ, angular frequency ω, and phase angle $\phi = 0$. a) Find expressions for the angular velocity $d\theta/dt$ and angular acceleration $d^2\theta/dt^2$ as functions of time. b) Find its angular velocity and angular acceleration when its angular displacement is Θ; when its angular displacement is $\Theta/2$ and θ is decreasing. (*Hint:* Sketch a graph of θ versus t.)

*13–31 For the van der Waals interaction with potential energy function given by Eq. (13–25), show that when the magnitude of the displacement x from equilibrium ($r = R_0$) is small the potential energy can be written approximately as $U \approx \frac{1}{2}kx^2 - U_0$. (*Hint:* In Eq. (13–25), let $r = R_0 + x$ and $u = x/R_0$. Then approximate $(1 + u)^n$ by the first *three* terms in Eq. (13–28).) How does k in this equation compare with the force constant in Eq. (13–29) for the force?

*13-32 When displaced from equilibrium, the two hydrogen atoms in an H_2 molecule are acted on by a restoring force $F = -kx$ with $k = 580$ N/m. Calculate the oscillation frequency of the H_2 molecule. (*Hint:* The mass of a hydrogen atom is 1.008 atomic mass units, or 1 u; see Appendix E. As in Example 13–7 of Section 13–5, use $m/2$ instead of m in the expression for f.)

SECTION 13-6 THE SIMPLE PENDULUM

13-33 A Pendulum on Mars. A certain simple pendulum has a period on earth of 1.60 s. What is its period on the surface of Mars, where $g = 3.71$ m/s^2?

13-34 You pull a simple pendulum of length 0.240 m to the side through an angle of 3.50° and release it. a) How much time does it take the pendulum bob to reach its highest speed? b) How much time does it take if the pendulum is released at an angle of 1.75° instead of 3.50°?

13-35 An apple weighs 1.00 N. When you hang it from the end of a long spring of force constant 1.50 N/m and negligible mass, it bounces up and down in SHM. If you stop the bouncing and let the apple swing from side to side through a small angle, the frequency of this simple pendulum is half the bounce frequency. (Because the angle is small, the back and forth swings do not cause any appreciable change in the length of the spring.) What is the unstretched length of the spring?

13-36 A small sphere of mass m is attached to a massless rod of length L that is pivoted at the top, forming a simple pendulum. The pendulum is pulled to one side so that the rod is at an angle Θ from the vertical, and released from rest. a) In a diagram, show the pendulum just after it is released. On the diagram, draw vectors representing the *forces* acting on the small sphere and the *acceleration* of the sphere. Accuracy counts! At this point, what is the linear acceleration of the sphere? b) Repeat part (a) for the instant when the pendulum rod is at an angle $\Theta/2$ from the vertical. c) Repeat part (a) for the instant when the pendulum rod is vertical. At this point, what is the linear speed of the sphere?

13-37 After landing on an unfamiliar planet, a space explorer constructs a simple pendulum of length 50.0 cm. She finds that the pendulum makes 100 complete swings in 136 s. What is the value of g on this planet?

SECTION 13-7 THE PHYSICAL PENDULUM

13-38 We want to support a thin hoop by a horizontal nail and have the hoop make one complete small-angle oscillation each 2.0 s. What must the hoop's radius be?

13-39 Show that the expression for the period of a physical pendulum reduces to that of a simple pendulum if the physical pendulum consists of a particle of mass m on the end of a massless string of length L.

13-40 A 1.80-kg monkey wrench is pivoted 0.250 m from its center of mass and allowed to swing as a physical pendulum. The period for small-angle oscillations is 0.940 s. a) What is the moment of inertia of the wrench about an axis through the pivot? b) If the wrench is initially displaced 0.400 rad from its equilibrium position, what is the angular speed of the wrench as it passes through the equilibrium position?

13-41 A holiday ornament in the shape of a hollow sphere with mass $M = 0.015$ kg and radius $R = 0.050$ m is hung from a tree limb by a small loop of wire attached to the surface of the sphere. If the ornament is displaced a small distance and released, it swings back and forth as a physical pendulum. Calculate its period. (Neglect friction at the pivot. The moment of inertia of the sphere about the pivot at the tree limb is $5MR^2/3$.)

13-42 A 1.80-kg connecting rod from a car engine is pivoted about a horizontal knife edge as shown in Fig. 13–27. The center of gravity of the rod was located by balancing and is 0.200 m from the pivot. When it is set into small amplitude oscillation, the rod makes 100 complete swings in 120 s. Calculate the moment of inertia of the rod about the rotation axis through the pivot.

$d = 0.200$ m

cg

FIGURE 13-27 Exercise 13–42.

SECTION 13-8 DAMPED OSCILLATIONS

13-43 An unhappy 0.300-kg mouse moving on the end of a spring with force constant $k = 2.50$ N/m is acted on by a damping force $F_x = -bv$. a) If the constant b has the value 0.900 kg/s, what is the frequency of oscillation of the mouse? b) For what value of the constant b will the motion be critically damped?

13-44 A 50.0-g hard-boiled egg moves on the end of a spring with force constant $k = 25.0$ N/m. Its initial displacement is 0.300 m. A damping force $F_x = -bv$ acts on the egg, and the amplitude of the motion decreases to 0.100 m in 5.00 s. Calculate the magnitude of the damping constant b.

13-45 The motion of an underdamped oscillator is described by Eq. (13–42). Let the phase angle ϕ be zero. a) According to this equation, what is the value of x at $t = 0$? b) What are the magnitude and direction of the velocity at $t = 0$? What does the result tell you about the slope of the graph of x versus t near $t = 0$? c) Obtain an expression for the acceleration a at $t = 0$. For what value or range of values of the damping constant b (in terms of k and m) is the acceleration at $t = 0$ negative, zero, and positive? Discuss each case in terms of the shape of the graph of x versus t near $t = 0$.

SECTION 13-9 FORCED OSCILLATIONS, RESONANCE, AND CHAOS

13-46 A sinusoidally varying driving force is applied to a damped harmonic oscillator of force constant k and mass m. If the damping constant has a value b_1, the amplitude is A_1 when the driving angular frequency equals $\sqrt{k/m}$. In terms of A_1, what is the amplitude for the same driving frequency and the same driving force amplitude F_{max} if the damping constant is a) $3b_1$? b) $b_1/2$?

13–47 A sinusoidally varying driving force is applied to a damped harmonic oscillator. a) What are the units of the damping constant b? b) Show that the quantity $\sqrt{km}$ has the same units as b. c) In terms of F_{max} and k, what is the amplitude for $\omega_d = \sqrt{k/m}$ when i) $b = 0.2\sqrt{km}$? ii) $b = 0.4\sqrt{km}$? Compare your results to Fig. 13–24.

13–48 An experimental package and its support structure to be placed on board the International Space Station act as an underdamped spring-mass system with a force constant of 2.1×10^6 N/m and a mass of 108 kg. A NASA requirement is that resonance for forced oscillations not occur for any frequency below 35 Hz. Does this package meet the requirement?

PROBLEMS

13–49 SHM in a Car Engine. The motion of the piston of an automobile engine (Fig. 13–28) is approximately simple harmonic. a) If the stroke of an engine (twice the amplitude) is 0.100 m and the engine runs at 3500 rev/min, compute the acceleration of the piston at the endpoint of its stroke. b) If the piston has a mass of 0.450 kg, what net force must be exerted on it at this point? c) What are the speed and kinetic energy of the piston at the midpoint of its stroke? d) What average power is required to accelerate the piston from rest to the speed found in part (c)? e) If the engine runs at 7000 rev/min, what are the answers to parts (b), (c), and (d)?

FIGURE 13–28 Problem 13–49.

13–50 Four passengers whose combined mass is 250 kg compress the springs of a car with worn-out shock absorbers by 4.00 cm when they enter it. Model the car and passengers as a single body on a single ideal spring. If the loaded car has a period of vibration of 1.08 s, what is the period of vibration of the empty car?

13–51 A glider is oscillating in SHM on an air track with an amplitude A_1. You slow it so that its amplitude is halved. What happens to its a) period, frequency, and angular frequency? b) total mechanical energy? c) maximum speed? d) speed at $x = \pm A_1/4$? e) potential and kinetic energies at $x = \pm A_1/4$?

13–52 You hang an unknown weight from the end of a spring, moving your hand down slowly so that when the weight comes to equilibrium it has stretched the spring a distance L. If the spring has negligible mass, prove that the weight can oscillate in SHM with the same period as a simple pendulum of length L.

13–53 A child with poor table manners is sliding his 250-g dinner plate back and forth in SHM with an amplitude of 0.100 m on a horizontal surface. At a point 0.060 m away from equilibrium, the speed of the plate is 0.300 m/s. a) What is the period?

b) What is the displacement when the speed is 0.160 m/s? c) In the center of the dinner plate is a 10.0-g carrot slice. If the carrot slice is just on the verge of slipping at the end point of the path, what is the coefficient of static friction between the carrot slice and the plate?

13–54 A mechanical bucking horse moves vertically in SHM with an amplitude of 0.250 m and a frequency of 1.50 Hz, that are the same whether there is a rider or not. A drugstore cowboy on the horse decides the macho way to ride is to not hold on in any way. a) He leaves the saddle when it is moving upward. What is the magnitude of the downward acceleration of the saddle when he loses contact with it? b) How high is the top surface of the saddle above its equilibrium position when he first gets airborne? c) How fast is he moving upward when he leaves the saddle? d) He is in free fall until he returns to the saddle. Show that this happens 0.538 s later. e) What is his speed relative to the saddle at the instant he returns?

13–55 A block of mass M rests on a frictionless surface and is connected to a horizontal spring of force constant k. The other end of the spring is attached to a wall (Fig. 13–29). A second block of mass m rests on top of the first block. The coefficient of static friction between the blocks is μ_s. Find the *maximum* amplitude of oscillation such that the top block will not slip on the bottom block.

FIGURE 13–29 Problem 13–55.

13–56 An object with a mass of 0.200 kg is acted on by an elastic restoring force with force constant 10.0 N/m. a) Graph elastic potential energy U as a function of displacement x over a range of x from −0.300 m to +0.300 m. On your graph, let 1 cm = 0.05 J vertically and 1 cm = 0.05 m horizontally. The object is set into oscillation with an initial potential energy of 0.140 J and an initial kinetic energy of 0.060 J. Answer the following questions by referring to the graph. b) What is the amplitude of oscillation? c) What is the potential energy when the displacement is one-half the amplitude? d) At what displacement are the kinetic and potential energies equal? e) What is the value of the phase angle ϕ if the initial velocity is positive and the initial displacement is negative?

13–57 Figure 13–14 shows a body of mass m hanging from a spring of force constant k. The positive x-axis is upward and $x = 0$

is the equilibrium position of the body. a) Show that when the body is at coordinate x, the elastic potential energy in the spring is $U_{el} = \frac{1}{2}k(\Delta l - x)^2$. b) Let $x = x_0$ be the coordinate at which the gravitational potential energy is zero. Show that the total potential energy is $U = \frac{1}{2}kx^2 + \frac{1}{2}k(\Delta l)^2 - mgx_0$. c) The expression for the total potential energy in part (b) is of the form $U = \frac{1}{2}kx^2 + C$, where the constant C equals $\frac{1}{2}k(\Delta l)^2 - mgx_0$. Explain why the behavior of the system does not depend on the value of this constant, so that vertical SHM is fundamentally no different than horizontal SHM, for which $U = \frac{1}{2}kx^2$.

13–58 A hanging wire is 1.80 m in length. When a 60.0-kg steel ball is suspended from the wire, the wire stretches by 2.00 mm. If the ball is pulled down a small additional distance and released, at what frequency will it vibrate? Assume that the stress on the wire is less than the proportional limit (see Section 11–8).

13–59 A 5.00-kg partridge is suspended from a pear tree by an ideal spring of negligible mass. When the partridge is pulled down 0.100 m below its equilibrium position and released, it vibrates with a period of 4.20 s. a) What is its speed as it passes through the equilibrium position? b) What is its acceleration when it is 0.050 m above the equilibrium position? c) When it is moving upward, how much time is required for it to move from a point 0.050 m below its equilibrium position to a point 0.050 m above it? d) The motion of the partridge is stopped, and then it is removed from the spring. How much does the spring shorten?

13–60 A 0.0200-kg bolt moves with simple harmonic motion that has an amplitude of 0.240 m and a period of 1.500 s. The displacement of the bolt is +0.240 m when $t = 0$. Compute a) the displacement of the bolt when $t = 0.500$ s; b) the magnitude and direction of the force acting on the bolt when $t = 0.500$ s; c) the minimum time required for the bolt to move from its initial position to the point where $x = -0.180$ m; d) the speed of the bolt when $x = -0.180$ m.

13–61 SHM of a Butcher's Scale. A spring of negligible mass and force constant $k = 400$ N/m is hung vertically, and a 0.200-kg pan is suspended from its lower end. A butcher drops a 2.2-kg steak onto the pan from a height of 0.40 m. The steak makes a totally inelastic collision with the pan and sets the system into vertical simple harmonic motion. What is a) the speed of the pan and steak immediately after the collision? b) the amplitude of the subsequent motion? c) the period of that motion?

13–62 A 40.0-N force stretches a vertical spring 0.250 m. a) What mass must be suspended from the spring so that the system will oscillate with a period of 1.00 s? b) If the amplitude of the motion is 0.050 m and the period is that specified in part (a), where is the object and in what direction is it moving 0.35 s after it has passed the equilibrium position, moving downward? c) What force (magnitude and direction) does the spring exert on the object when it is 0.030 m below the equilibrium position, moving upward?

13–63 A small excursion boat with a flat deck bobs up and down executing simple harmonic motion due to the waves on a lake. The amplitude of the motion is 0.200 m and the period is 2.80 s. A stable dock is next to the boat at a level equal to the highest level of the deck. People wish to step off the boat onto the dock but can do so comfortably only if the level of the deck is within 0.100 m of the dock level. How much time do the people have to get off comfortably during each period of the simple harmonic motion?

13–64 An interesting though highly impractical example of oscillation is the motion of an object dropped down a hole that extends from one side of the earth, through its center, to the other side. With the assumption (not very realistic) that the earth is a sphere of uniform density, prove that the motion is simple harmonic and find the period. (*Note:* The gravitational force on the object as a function of the object's distance r from the center of the earth was derived in Example 12–9 (Section 12–7). The motion is simple harmonic if the acceleration a and the displacement from equilibrium x are related by Eq. (13–8), and the period is then $T = 2\pi/\omega$.)

13–65 Let t_1 be the time it takes a body undergoing SHM to move from $x = 0$ (when $t = 0$) to $x = A$. Obtain an equation for t_1 in the following way. In Eq. (13–22) replace v by dx/dt. Separate variables by writing all factors containing x on one side and all those containing t on the other side so that each side can be integrated. Integrate the equation with limits on t of 0 to t_1 and limits on x of 0 to A and thereby obtain an equation for t_1. How does t_1 compare to the period T?

13–66 For a certain oscillator the net force on the body of mass m is given by $F = -cx^3$. a) What is the potential energy function for this oscillator, if we take $U = 0$ at $x = 0$? b) One-quarter of a period is the time for the body to move from $x = 0$ to $x = A$. Calculate this time and hence the period. (*Hint:* Begin with Eq. (13–20), modified to include the potential energy function you found in part (a), and solve for the velocity v as a function of x. Then replace v by dx/dt and follow the steps outlined in Problem 13–65. In the x-integral make the change of variable $u = x/A$. The resulting integral can be evaluated by numerical methods on a computer and has the value $\int_0^1 du/\sqrt{1 - u^4} = 1.31$.) c) According to the result you obtained in part (b), does the period depend on the amplitude A of the motion? Are the oscillations simple harmonic?

13–67 Consider the circle of reference shown in Fig. 13–4. The x-component of the velocity of Q is the velocity of P. Compute this component and show that the velocity of P is as given by Eq. (13–15).

13–68 To measure g in an unorthodox manner, a student places a ball bearing on the concave side of a lens (Fig. 13–30). She attaches the lens to a simple harmonic oscillator (actually a small stereo speaker) whose amplitude is A and whose frequency f can be varied. She can measure both A and f with a strobe light. a) If the ball bearing has mass m, find the normal force exerted by the lens on the ball bearing as a function of time. Your result should be in terms of A, f, m, g, and a phase angle ϕ. b) The frequency is slowly increased. When it reaches a value f_b, the ball is heard to bounce. What is g in terms of A and f_b?

*13–69 Two identical atoms in a diatomic molecule vibrate as harmonic oscillators. However, their center of mass, midway between them, remains at rest. a) Show that at any instant the momenta of the atoms relative to the center of mass are $\vec{p}$ and $-\vec{p}$. b) Show that the total kinetic energy K of the two atoms

FIGURE 13–30 Problem 13–68.

at any instant is the same as that of a single object of mass $m/2$ with a momentum of magnitude p. (Use $K = p^2/2m$.) This result shows why $m/2$ should be used in the expression for f in Example 13–7 (Section 13–5). c) If the atoms are not identical, but have masses m_1 and m_2, show that the result of part (a) still holds and the single object's mass in part (b) is $m_1m_2/(m_1 + m_2)$. The quantity $m_1m_2/(m_1 + m_2)$ is called the *reduced mass* of the system.

*13–70 An approximation for the potential energy of a KCl molecule is $U = A[(R_0^7/8r^8) - 1/r]$, where $R_0 = 2.67 \times 10^{-10}$ m and $A = 2.31 \times 10^{-28}$ J $\cdot$ m. Using this approximation: a) Show that the radial component of the force on each atom is $F = A[(R_0^7/r^9) - 1/r^2]$. b) Show that R_0 is the equilibrium separation. c) Find the minimum potential energy. d) Use $r = R_0 + x$ and the first two terms of the binomial theorem (Eq. 13–28) to show that $F \approx -(7A/R_0^3)x$ so that the molecule's force constant is $k = 7A/R_0^3$. e) With both the K and Cl atoms vibrating in opposite directions on opposite sides of the molecule's center of mass, $m_1m_2/(m_1 + m_2) = 3.06 \times 10^{-26}$ kg is the mass to use in calculating the frequency. (See Problem 13–69.) Calculate the frequency of small-amplitude vibrations.

*13–71 Three positively charged particles are maintained in a straight line. The end particles are identical and are held fixed at a distance $2R_0$ apart. The potential energy for the force acting on the center particle can be written as $U = A[1/r - 1/(r - 2R_0)]$, where A is a constant and r is the distance from the left-hand particle to the center particle. a) Show that the force on the center particle is $F = A[1/r^2 - 1/(r - 2R_0)^2]$. b) Show that $r = R_0$ at equilibrium. c) Use $r = R_0 + x$ and the first two terms of the binomial theorem (Eq. 13–28) to show that $F \approx -(4A/R_0^3)x$ so that the force constant is $k = 4A/R_0^3$ d) What is the frequency of small-amplitude oscillation of the center particle if it has mass m?

13–72 Many real springs are more easily stretched than compressed. We can represent this by using different spring constants for $x > 0$ and for $x < 0$. As an example, consider a spring that exerts the following restoring force:

$$F = \begin{cases} -kx & \text{for } x > 0, \\ -2kx & \text{for } x < 0. \end{cases}$$

A mass m on a frictionless horizontal surface is attached to this spring, displaced to $x = A$ by stretching the spring, and released. a) Find the period of the motion. Does the period depend on A? Are the oscillations simple harmonic? b) What is the most negative value of x that the mass m reaches? Is the oscillation symmetric about the point $x = 0$?

13–73 Two solid cylinders connected along their common axis by a short, light rod have a radius R and a total mass M and rest

on a horizontal table top. A spring with force constant k has one end attached to a clamp and the other end attached to a frictionless ring at the center of mass of the cylinders (Fig. 13–31). The cylinders are pulled to the left a distance x, which stretches the spring, and released. There is sufficient friction between the table top and the cylinders for the cylinders to roll without slipping as they move back and forth on the end of the spring. Show that the motion of the center of mass of the cylinders is simple harmonic, and calculate its period in terms of M and k. (*Hint:* The motion is simple harmonic if a and x are related by Eq. (13–8), and the period then is $T = 2\pi/\omega$. Apply $\Sigma \, \tau = I_{cm}\alpha$ and $\Sigma \, F = Ma_{cm}$ to the cylinders in order to relate a_{cm} and the displacement x of the cylinders from their equilibrium position.)

FIGURE 13–31 Problem 13–73.

13–74 A slender, uniform metal rod with mass M is pivoted without friction about an axis through its midpoint and perpendicular to the rod. A horizontal spring with force constant k is attached to the lower end of the rod, with the other end of the spring attached to a rigid support. If the rod is displaced by a small angle Θ from the vertical (Fig. 13–32) and released, show that it moves in angular simple harmonic motion and calculate the period. (*Hint:* Assume that the angle Θ is small enough for the approximations $\sin \Theta \approx \Theta$ and $\cos \Theta \approx 1$ to be valid. The motion is simple harmonic if $d^2\theta/dt^2 = -\omega^2\theta$, and the period is then $T = 2\pi/\omega$.)

FIGURE 13–32 Problem 13–74.

13–75 The Silently Ringing Bell Problem. A large bell is hung from a wooden beam so it can swing back and forth with negligible friction. The center of mass of the bell is 0.60 m below the pivot, the bell's mass is 34.0 kg, and the moment of inertia of the bell about an axis at the pivot is 18.0 kg $\cdot$ m^2. The clapper is a small 1.8-kg mass attached to one end of a slender rod that has length L and negligible mass. The other end of the rod is attached to the inside of the bell so it can swing freely about the same axis as the bell. What should be the length L of the clapper rod for the bell to ring silently, that is, for the period of oscillation for the bell to equal that for the clapper?

13–76 Two identical thin rods, each of mass m and length L, are joined at right angles to form an L-shaped object. This object is balanced on top of a sharp edge (Fig. 13–33). If the L-shaped object is deflected slightly, it oscillates. Find the frequency of oscillation.

FIGURE 13–33 Problem 13–76.

13–77 You want to construct a pendulum with a period of 4.00 s at a location where $g = 9.80$ m/s^2. a) What is the length of a *simple* pendulum having this period? b) Suppose the pendulum must be mounted in a case that is not over 0.50 m high. Can you devise a pendulum having a period of 4.00 s that will satisfy this requirement?

13–78 Show that $x(t)$ as given in Eq. (13–42) is a solution of Newton's second law for damped oscillations, Eq. (13–41), if ω' is as defined in Eq. (13–43).

13–79 A uniform rod of length L oscillates through small angles about a point a distance x from its center. a) Prove that its angular frequency is $\sqrt{gx/[(L^2/12) + x^2]}$. b) Show that its maximum angular frequency occurs when $x = L/\sqrt{12}$. c) What is the length of the rod if the maximum angular frequency is 2π rad/s?

13–80 When Is It Simple? To make a pendulum, you hang a uniform solid sphere of radius R on the end of a light string. The distance from the pivot to the center of the sphere is L. a) Prove that the period of your pendulum is $T = T_{sp}\sqrt{1 + 2R^2/5L^2}$, where T_{sp} is the period of a simple pendulum of length L. b) At what value of L is T only 0.10% larger than T_{sp}? c) For a sphere of diameter 2.540 cm, what is the value of L in part (b)?

CHALLENGE PROBLEMS

13–81 Two springs, each with unstretched length 0.200 m but with different force constants k_1 and k_2, are attached to opposite ends of a block with mass m on a level, frictionless surface. The outer ends of the springs are now attached to two pins P_1 and P_2, 0.100 m from the original positions of the ends of the springs (Fig. 13–34). Let $k_1 = 2.00$ N/m, $k_2 = 6.00$ N/m, and $m = 0.100$ kg. a) Find the length of each spring when the block is in its new equilibrium position after the springs have been attached to the pins. b) Find the period of vibration of the block if it is slightly displaced from its new equilibrium position and released.

FIGURE 13–34 Challenge Problem 13–81.

13–82 The Effective Force Constant of Two Springs. Two springs with the same unstretched length but different force constants k_1 and k_2 are attached to a block of mass m on a level, frictionless surface. Calculate the effective force constant k_{eff} in each of the three cases (a), (b), and (c) depicted in Fig. 13–35. (The effective force constant is defined by $\Sigma F = -k_{eff}x$.) d) An object with mass m, suspended from a uniform spring with a

force constant k, vibrates with a frequency f_1. When the spring is cut in half and the same object is suspended from one of the halves, the frequency is f_2. What is the ratio f_2/f_1?

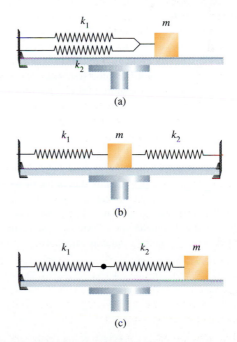

FIGURE 13–35 Challenge Problem 13–82.

13–83 a) What is the change ΔT in the period of a simple pendulum when the acceleration of gravity g changes by Δg? (*Hint:* The new period $T + \Delta T$ is obtained by substituting $g + \Delta g$ for g:

$$T + \Delta T = 2\pi\sqrt{\frac{L}{g + \Delta g}}.$$

To obtain an approximate expression, expand the factor $(g + \Delta g)^{-1/2}$ using the binomial theorem (Appendix B) and keeping only the first two terms:

$$(g + \Delta g)^{-1/2} = g^{-1/2} - \frac{1}{2}g^{-3/2}\Delta g + \cdots.$$

The other terms contain higher powers of Δg and are very small if Δg is small.) Express your result as the *fractional* change in period $\Delta T/T$ in terms of the fractional change $\Delta g/g$. b) A pendulum clock keeps correct time at a point where $g = 9.8000$ m/s^2, but is found to lose 4.0 s each day at a higher elevation. Use the result of part (a) to find the approximate value of g at this new location.

13–84 A Spring with Mass. The preceding problems in this chapter have assumed that the springs had negligible mass. But of course no spring is completely massless. To find the effect of the spring's mass, consider a spring of mass M, equilibrium length L_0, and spring constant k. When stretched or compressed to a length L, the potential energy is $\frac{1}{2}kx^2$, where $x = L - L_0$. a) Consider a spring as described above that has one end fixed and the other end moving with speed v. Assume that the speed of points along the length of the spring varies linearly with distance l from the fixed end. Assume also that the mass M of the spring is distributed uniformly along the length of the spring. Calculate the kinetic energy of the spring in terms of M and v. (*Hint:* Divide the spring into pieces of length dl; find the speed of each piece in terms of l, v, and L; find the mass of each piece in terms of dl, M, and L; and integrate from 0 to L. The result is *not* $\frac{1}{2}Mv^2$, since not all of the spring moves with the same speed.) b) Take the time derivative of the conservation of energy equation, Eq. (13–21), for a mass m moving on the end of a *massless* spring. By comparing your results to Eq. (13–8), which defines ω, show that the angular frequency of oscillation is $\omega = \sqrt{k/m}$. c) Apply the procedure of part (b) to obtain the angular frequency of oscillation ω of the spring considered in part (a). If the *effective mass* M' of the spring is defined by $\omega = \sqrt{k/M'}$, what is M' in terms of M?

13–85 A uniform 1.00-m stick hangs from a horizontal axis at one end and oscillates as a physical pendulum. An object of small dimensions and of mass equal to that of the meter stick can be clamped to the stick at a distance y below the axis. Let T represent the period of the system with the body attached and T_0 the period of the meter stick alone. a) Find the ratio T/T_0. Evaluate your expression for y ranging from 0 to 1.0 m in steps

of 0.1 m, and graph T/T_0 versus y. b) Is there any value of y, other than $y = 0$, for which $T = T_0$? If so, find it and explain why the period is unchanged when y has this value.

13–86 You measure the period of a physical pendulum about one pivot point to be T. Then you find another pivot point on the opposite side of the center of mass that gives the same period. The two points are separated by a distance L. Use the parallel-axis theorem to show that $g = L(2\pi/T)^2$. (This result shows a way that you can measure g without knowing the mass or any moments of inertia of the physical pendulum.)

13–87 Resonance in a Mechanical System. A mass m is attached to one end of a massless spring with a force constant k and an unstretched length l_0. The other end of the spring is free to turn about a nail driven into a frictionless horizontal surface (Fig. 13–36). The mass is made to revolve in a circle with an angular frequency of revolution ω'. a) Calculate the length l of the spring as a function of ω'. b) What happens to the result in part (a) when ω' approaches the natural frequency $\omega = \sqrt{k/m}$ of the mass-spring system? (If your result bothers you, remember that massless springs and frictionless surfaces don't exist as such, but are only approximate descriptions of real springs and surfaces. Also, Hooke's law is itself only an approximation to the way real springs behave; the greater the elongation of the spring, the greater the deviation from Hooke's law.)

FIGURE 13–36 Challenge Problem 13–87.

*13–88 Vibration of a Covalently Bonded Molecule. Many diatomic (two-atom) molecules are bound together by *covalent bonds* that are much stronger than the van der Waals interaction. Examples include H_2, O_2, and N_2. Experiment shows that for many such molecules, the interaction can be described by a force of the form

$$F = A\left[e^{-2b(r - R_0)} - e^{-b(r - R_0)}\right],$$

where A and b are positive constants, r is the center-to-center separation of the atoms, and R_0 is the equilibrium separation. For the hydrogen molecule (H_2), $A = 2.97 \times 10^{-8}$ N, $b = 1.95 \times 10^{10}$ m^{-1}, and $R_0 = 7.4 \times 10^{-11}$ m. Find the force constant for small oscillations around equilibrium. (*Hint:* Use the expansion for e^x given in Appendix B.) Compare your result to the value given in Exercise 13–32.

14

Fluid Mechanics

The beach is a wonderful place to watch fluid mechanics in action. Water moves in response to gravity and the pressure of the surrounding fluid; the flow of water changes from laminar to turbulent as the waves break; and surface tension makes the water in a breaking wave form into drops.

14–1 INTRODUCTION

Fluids play a vital role in many aspects of everyday life. We drink them, breathe them, swim in them. They circulate through our bodies and control our weather. Airplanes fly through them; ships float in them. A fluid is any substance that can flow; we use the term for both liquids and gases. We usually think of a gas as easily compressed and a liquid as nearly incompressible, although there are exceptional cases.

We begin our study with **fluid statics,** the study of fluids at rest in equilibrium situations. Like other equilibrium situations, it is based on Newton's first and third laws. We will explore the key concepts of density, pressure, buoyancy, and surface tension. **Fluid dynamics,** the study of fluids in motion, is much more complex; indeed, it is one of the most complex branches of mechanics. Fortunately, we can analyze many important situations using simple idealized models and familiar principles such as Newton's laws and conservation of energy. Even so, we will barely scratch the surface of this broad and interesting topic.

14–2 DENSITY

An important property of any material is its **density,** defined as its mass per unit volume. A homogeneous mate-

rial such as ice or iron has the same density throughout. We use the Greek letter ρ ("rho") for density. If a mass m of homogeneous material has volume V, the density ρ is

$$\rho = \frac{m}{V} \qquad \text{(definition of density).} \qquad (14\text{–}1)$$

The density of some materials varies from point to point within the material; some examples are the earth's atmosphere (which is less dense at high altitude) and oceans (which are denser at greater depths). For these materials, Eq. (14–1) describes the *average* density. In general, the density of a material depends on environmental factors such as temperature and pressure.

The SI unit of density is the kilogram per cubic meter (1 kg/m^3). The cgs unit, the gram per cubic centimeter (1 g/cm^3), is also widely used. The conversion factor

$$1 \text{ g/cm}^3 = 1000 \text{ kg/m}^3$$

is useful. Densities of several common substances at ordinary temperatures are given in Table 14–1. Note the wide range of magnitudes. The densest material found on earth is the metal osmium ($\rho = 22{,}500$ kg/m^3), but this pales by comparison to the densities of exotic astronomical objects such as white dwarf stars and neutron stars.

The **specific gravity** of a material is the ratio of its density to the density of water at 4.0°C, 1000 kg/m^3; it is a pure number without units. For example, the specific

TABLE 14–1
DENSITIES OF SOME COMMON SUBSTANCES

MATERIAL	DENSITY (kg/m³)*	MATERIAL	DENSITY (kg/m³)*
Air (1 atm, 20° C)	1.20	Iron, steel	7.8×10^3
Ethanol	0.81×10^3	Brass	8.6×10^3
Benzene	0.90×10^3	Copper	8.9×10^3
Ice	0.92×10^3	Silver	10.5×10^3
Water	1.00×10^3	Lead	11.3×10^3
Seawater	1.03×10^3	Mercury	13.6×10^3
Blood	1.06×10^3	Gold	19.3×10^3
Glycerin	1.26×10^3	Platinum	21.4×10^3
Concrete	2×10^3	White dwarf star	10^{10}
Aluminum	2.7×10^3	Neutron star	10^{18}

*To obtain the densities in grams per cubic centimeter, simply divide by 10^3.

The price of gold is quoted by weight (say, in dollars per ounce). Because gold is one of the densest of the metals, a small fortune in gold can be stored in a small volume.

gravity of aluminum is 2.7. "Specific gravity" is a poor term, since it has nothing to do with gravity; "relative density" would have been better.

Measuring density is an important analytical technique. For example, we can determine the charge condition of a storage battery by measuring the density of its electrolyte, a sulfuric acid solution. As the battery discharges, the sulfuric acid (H_2SO_4) combines with lead in the battery plates to form insoluble lead sulfate ($PbSO_4$), decreasing the concentration of the solution. The density decreases from about 1.30×10^3 kg/m³ for a fully charged battery to 1.15×10^3 kg/m³ for a discharged battery.

Another automotive example is permanent-type antifreeze, which is usually a solution of ethylene glycol ($\rho = 1.12 \times 10^3$ kg/m³) and water. The freezing point of the solution depends on the glycol concentration, which can be determined by measuring the density. Such measurements of fluid density are performed routinely in service stations with a device called a hydrometer, which we'll discuss in Section 14–4.

EXAMPLE 14–1

The weight of a roomful of air Find the mass and weight of the air in a living room with a 4.0 m × 5.0 m floor and a ceiling 3.0 m high. What is the mass and weight of an equal volume of water?

SOLUTION The densities of air and water are given in Table 14–1. The volume of the room is $V = (3.0 \text{ m})(4.0 \text{ m})(5.0 \text{ m}) = 60$ m³. The mass m_{air} of air is given by Eq. (14–1):

$$m_{air} = \rho_{air}V = (1.2 \text{ kg/m}^3)(60 \text{ m}^3) = 72 \text{ kg}.$$

The weight of the air is

$$w_{air} = m_{air}g = (72 \text{ kg})(9.8 \text{ m/s}^2) = 700 \text{ N} = 160 \text{ lb}.$$

Does it surprise you that a roomful of air weighs this much? The mass of an equal volume of water is

$$m_{water} = \rho_{water}V = (1000 \text{ kg/m}^3)(60 \text{ m}^3) = 6.0 \times 10^4 \text{ kg}.$$

The weight is

$$w_{water} = m_{water}g = (6.0 \times 10^4 \text{ kg})(9.8 \text{ m/s}^2) = 5.9 \times 10^5 \text{ N}$$
$$= 1.3 \times 10^5 \text{ lb} = 66 \text{ tons}.$$

This much weight would certainly collapse the floor of an ordinary house.

14–3 PRESSURE IN A FLUID

When a fluid (either liquid or gas) is at rest, it exerts a force perpendicular to any surface in contact with it, such as a container wall or a body immersed in the fluid. This is the force that you feel pressing on your legs when you dangle them in a swimming pool. While the fluid as a whole is at rest, the molecules that make up the fluid are in motion; the force exerted by the fluid is due to molecules colliding with their surroundings.

If we think of an imaginary surface *within* the fluid, the fluid on the two sides of the surface exerts equal and opposite forces on the surface. (Otherwise, the surface would accelerate and the fluid would not remain at rest.) Consider a small surface of area dA centered on a point in the fluid; the normal force exerted by the fluid on each side is $dF_\perp$ (Fig. 14–1). We define the **pressure** p at that point as the normal force per unit area, that is, the ratio of $dF_\perp$ to dA:

$$p = \frac{dF_\perp}{dA} \quad \text{(definition of pressure).} \tag{14–2}$$

If the pressure is the same at all points of a finite plane surface with area A, then

$$p = \frac{F_\perp}{A}, \tag{14–3}$$

where $F_\perp$ is the net normal force on one side of the surface. The SI unit of pressure is the **pascal,** where

$$1 \text{ pascal} = 1 \text{ Pa} = 1 \text{ N/m}^2.$$

We introduced the pascal in Chapter 11. Two related units, used principally in meteorology, are the *bar,* equal to 10^5 Pa, and the *millibar,* equal to 100 Pa.

Atmospheric pressure p_a is the pressure of the earth's atmosphere, the pressure at the bottom of this sea of air in which we live. This pressure varies with weather changes and with elevation. Normal atmospheric pressure at sea level (an average value) is 1 *atmosphere* (atm), defined to be exactly 101,325 Pa. To four significant figures,

$$(p_a)_{av} = 1 \text{ atm} = 1.013 \times 10^5 \text{ Pa}$$

$$= 1.013 \text{ bar} = 1013 \text{ millibar} = 14.70 \text{ lb/in.}^2$$

CAUTION ▶ In everyday language the words "pressure" and "force" mean pretty much the same thing. In fluid mechanics, however, these words describe distinct quantities with different characteristics. Fluid pressure acts perpendicular to any surface in the fluid, no matter how that surface is oriented. Hence pressure has no intrinsic direction of its own; it's a scalar. By contrast, force is a vector with a definite direction. Remember, too, that pressure is force per unit area. ◀

14–1 An imaginary small area dA within a fluid has equal normal forces $dF_\perp$ exerted on each side. The pressure acting over the area is $p = dF_\perp/dA$.

EXAMPLE 14–2

In the room described in Example 14–1, what is the total downward force on the floor surface due to air pressure of 1.00 atm?

SOLUTION The floor area is $A = (4.0 \text{ m})(5.0 \text{ m}) = 20 \text{ m}^2$. The pressure is uniform, so from Eq. (14–3) the total downward force is

$$F = pA = (1.013 \times 10^5 \text{ N/m}^2)(20 \text{ m}^2)$$

$$= 2.0 \times 10^6 \text{ N} = 4.5 \times 10^5 \text{ lb} = 225 \text{ tons}.$$

As in Example 14–1, this is more than enough force to collapse the floor. So why doesn't it collapse? Because there is an upward force on the underside of the floor. If the house has a basement, this upward force is exerted by the air underneath the floor. In this case, if we neglect the thickness of the floor, the *net* force due to air pressure is zero.

PRESSURE, DEPTH, AND PASCAL'S LAW

If the weight of the fluid can be neglected, the pressure in a fluid is the same throughout its volume. We used that approximation in our discussion of bulk stress and strain in Section 11–6. But often the fluid's weight is *not* negligible. Atmospheric pressure is less at high altitude than at sea level, which is why an airplane cabin has to be pressurized when flying at 35,000 feet. When you dive into deep water, your ears tell you that the pressure increases rapidly with increasing depth below the surface.

We can derive a general relation between the pressure p at any point in a fluid at rest and the elevation y of the point. We'll assume that the density ρ and the acceleration due to gravity g are the same throughout the fluid. If the fluid is in equilibrium, every

(a)

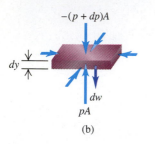

(b)

14–2 The forces on an element of fluid in equilibrium.

14–3 The pressure at a depth h in a fluid is greater than the surface pressure p_0 by ρgh.

14–4 The pressure in the fluid is the same at all points having the same elevation. The shape of the container does not affect the pressure.

volume element is in equilibrium. Consider a thin element of fluid, with height dy (Fig. 14–2). The bottom and top surfaces each have area A, and they are at elevations y and $y + dy$ above some reference level where $y = 0$. The volume of the fluid element is $dV = A\,dy$, its mass is $dm = \rho\,dV = \rho A\,dy$, and its weight is $dw = dm\,g = \rho gA\,dy$.

What are the other forces on this fluid element? Call the pressure at the bottom surface p; the total y-component of upward force on this surface is pA. The pressure at the top surface is $p + dp$, and the total y-component of (downward) force on the top surface is $-(p + dp)A$. The fluid element is in equilibrium, so the total y-component of force, including the weight and the forces at the bottom and top surfaces, must be zero:

$$\Sigma F_y = 0, \qquad \text{so} \qquad pA - (p + dp)A - \rho gA\,dy = 0.$$

When we divide out the area A and rearrange, we get

$$\frac{dp}{dy} = -\rho g. \tag{14–4}$$

This equation shows that when y increases, p decreases; that is, as we move upward in the fluid, pressure decreases, as we expect. If p_1 and p_2 are the pressures at elevations y_1 and y_2, respectively, and if ρ and g are constant, then

$$p_2 - p_1 = -\rho g(y_2 - y_1) \qquad \text{(pressure in a fluid of uniform density).} \tag{14–5}$$

It's often convenient to express Eq. (14–5) in terms of the *depth* below the surface of a fluid (Fig. 14–3). Take point 1 at any level in the fluid and let p represent the pressure at this point. Take point 2 at the *surface* of the fluid, where the pressure is p_0 (subscript zero for zero depth). The depth of point 1 below the surface is $h = y_2 - y_1$, and Eq. (14–5) becomes

$$p_0 - p = -\rho g(y_2 - y_1) = -\rho gh,$$

$$p = p_0 + \rho gh \qquad \text{(pressure in a fluid of uniform density).} \tag{14–6}$$

The pressure p at a depth h is greater than the pressure p_0 at the surface by an amount ρgh. Note that the pressure is the same at any two points at the same level in the fluid. The *shape* of the container does not matter (Fig. 14–4).

Equation (14–6) shows that if we increase the pressure p_0 at the top surface, possibly by using a piston that fits tightly inside the container to push down on the fluid surface, the pressure p at any depth increases by exactly the same amount. This fact was recognized in 1653 by the French scientist Blaise Pascal (1623–1662) and is called **Pascal's law: Pressure applied to an enclosed fluid is transmitted undiminished to every portion of the fluid and the walls of the containing vessel.**

The hydraulic lift shown schematically in Fig. 14–5 illustrates Pascal's law. A piston with small cross-section area A_1 exerts a force F_1 on the surface of a liquid such as oil. The applied pressure $p = F_1/A_1$ is transmitted through the connecting pipe to a larger piston of area A_2. The applied pressure is the same in both cylinders, so

$$p = \frac{F_1}{A_1} = \frac{F_2}{A_2} \qquad \text{and} \qquad F_2 = \frac{A_2}{A_1}\,F_1. \tag{14–7}$$

The hydraulic lift is a force-multiplying device with a multiplication factor equal to the ratio of the areas of the two pistons. Dentist's chairs, car lifts and jacks, many elevators, and hydraulic brakes all use this principle.

For gases the assumption that the density ρ is uniform is realistic only over short vertical distances. In a room with a ceiling height of 3.0 m filled with air of uniform density 1.2 kg/m^3, the difference in pressure between floor and ceiling, given by Eq. (14–6), is

$$\rho gh = (1.2\ \text{kg/m}^3)(9.8\ \text{m/s}^2)(3.0\ \text{m}) = 35\ \text{Pa},$$

or about 0.00035 atm, a very small difference. But between sea level and the summit of Mount Everest (8882 m) the density of air changes by nearly a factor of three, and in this case we cannot use Eq. (14–6). Liquids, by contrast, are nearly incompressible, and it is usually a very good approximation to regard their density as independent of pressure. A pressure of several hundred atmospheres will cause only a few percent increase in the density of most liquids.

ABSOLUTE PRESSURE, GAUGE PRESSURE, AND PRESSURE GAUGES

If the pressure inside a car tire is equal to atmospheric pressure, the tire is flat. The pressure has to be *greater* than atmospheric to support the car, so the significant quantity is the *difference* between the inside and outside pressures. When we say that the pressure in a car tire is "32 pounds" (actually 32 lb/in.^{2}, equal to 220 kPa or 2.2×10^5 Pa), we mean that it is *greater* than atmospheric pressure (14.7 lb/in.^{2} or 1.01×10^5 Pa) by this amount. The *total* pressure in the tire is then 47 lb/in.^{2} or 320 kPa. The excess pressure above atmospheric pressure is usually called **gauge pressure,** and the total pressure is called **absolute pressure.** Engineers use the abbreviations psig and psia for "pounds per square inch gauge" and "pounds per square inch absolute," respectively. If the pressure is *less* than atmospheric, as in a partial vacuum, the gauge pressure is negative.

14–5 Principle of the hydraulic lift, an application of Pascal's law. The size of the fluid-filled container is exaggerated for clarity.

EXAMPLE 14–3

Finding absolute and gauge pressures A solar water-heating system uses solar panels on the roof, 12.0 m above the storage tank. The water pressure at the level of the panels is one atmosphere. What is the absolute pressure in the tank? The gauge pressure?

SOLUTION From Eq. (14–6), the absolute pressure is

$$p = p_0 + \rho g h$$
$$= (1.01 \times 10^5 \text{ Pa}) + (1000 \text{ kg/m}^3)(9.80 \text{ m/s}^2)(12.0 \text{ m})$$
$$= 2.19 \times 10^5 \text{ Pa} = 2.16 \text{ atm} = 31.8 \text{ lb/in.}^2.$$

The gauge pressure is

$$p - p_0 = (2.19 - 1.01) \times 10^5 \text{ Pa} = 1.18 \times 10^5 \text{ Pa}$$
$$= 1.16 \text{ atm} = 17.1 \text{ lb/in.}^2.$$

If such a tank has a pressure gauge, it is usually calibrated to read gauge pressure rather than absolute pressure. As we have mentioned, the variation in *atmospheric* pressure over this height is negligibly small.

The simplest pressure gauge is the open-tube *manometer* (Fig. 14–6a). The U-shaped tube contains a liquid of density ρ, often mercury or water. One end of the tube

(a)

(b)

14–6 Pressure gauges. (a) The open-tube manometer. (b) The mercury barometer.

is connected to the container where the pressure p is to be measured, and the other end is open to the atmosphere at pressure $p_0 = p_a$. The pressure at the bottom of the tube due to the fluid in the left column is $p + \rho g y_1$, and the pressure at the bottom due to the fluid in the right column is $p_a + \rho g y_2$. These pressures are measured at the same point, so they must be equal:

$$p + \rho g y_1 = p_a + \rho g y_2,$$
$$p - p_a = \rho g(y_2 - y_1) = \rho g h. \qquad (14-8)$$

In Eq. (14-8), p is the *absolute pressure,* and the difference $p - p_a$ between absolute and atmospheric pressure is the gauge pressure. Thus the gauge pressure is proportional to the difference in height $(y_2 - y_1)$ of the liquid columns.

Another common pressure gauge is the **mercury barometer.** It consists of a long glass tube, closed at one end, that has been filled with mercury and then inverted in a dish of mercury (Fig. 14-6b). The space above the mercury column contains only mercury vapor; its pressure is negligibly small, so the pressure p_0 at the top of the mercury column is practically zero. From Eq. (14-6),

$$p_a = p = 0 + \rho g(y_2 - y_1) = \rho g h. \qquad (14-9)$$

Thus the mercury barometer reads the atmospheric pressure p_a directly from the height of the mercury column.

In many applications, pressures are still commonly described in terms of the height of the corresponding mercury column, as so many "inches of mercury" or "millimeters of mercury" (abbreviated mm Hg). A pressure of 1 mm Hg is called *one torr,* after Evangelista Torricelli, inventor of the mercury barometer. But these units depend on the density of mercury, which varies with temperature, and on the value of g, which varies with location, so the pascal is the preferred unit of pressure.

EXAMPLE 14-4

Compute the atmospheric pressure p_a on a day when the height of mercury in a barometer is 760 mm.

SOLUTION We use Eq. (14-9). Assuming that $g = 9.80$ m/s^2 and

$\rho = 13.6 \times 10^3$ kg/m^3 for mercury, we find

$$p_a = \rho g h = (13.6 \times 10^3 \text{ kg/m}^3)(9.80 \text{ m/s}^2)(0.760 \text{ m})$$
$$= 1.01 \times 10^5 \text{ N/m}^2 = 1.01 \times 10^5 \text{ Pa} = 1.01 \text{ bar.}$$

One common type of blood-pressure gauge, called a *sphygmomanometer,* uses a mercury-filled manometer. Blood-pressure readings, such as 130/80, refer to the maximum and minimum gauge pressures in the arteries, measured in mm Hg or torrs. Blood pressure varies with height; the standard reference point is the upper arm, level with the heart.

Many types of pressure gauges use a flexible sealed vessel (Fig. 14-7). A change in the pressure either inside or outside the vessel causes a change in its dimensions. This change is detected optically, electrically, or mechanically.

14-4 BUOYANCY

Buoyancy is a familiar phenomenon: a body immersed in water seems to weigh less than when it is in air. When the body is less dense than the fluid, it floats. The human body usually floats in water, and a helium-filled balloon floats in air.

Archimedes' principle states: **When a body is completely or partially immersed in a fluid, the fluid exerts an upward force on the body equal to the weight of the fluid displaced by the body.** To prove this principle, we consider an arbitrary portion of

Pressure p being measured

14-7 A Bourdon pressure gauge. The spiral metal tube is attached to a pointer. When the pressure inside the tube increases, the tube straightens out a little, deflecting the pointer on the scale.

(a)   (b)

14–8 (a) An element of fluid in equilibrium under its weight and the buoyant force of the surrounding fluid. (b) The element is replaced by a body with identical shape. The buoyant force is the same and is equal to the weight of the fluid that the body replaced, no matter what the weight of the body that replaces the fluid.

fluid at rest. In Fig. 14–8a the irregular outline is the surface bounding this portion of fluid. The arrows represent the forces exerted on the boundary surface by the surrounding fluid.

The entire fluid is in equilibrium, so the sum of all the y-components of force on this portion of fluid is zero. Hence the sum of the y-components of the *surface* forces must be an upward force equal in magnitude to the weight mg of the fluid inside the surface. Also, the sum of the torques on the portion of fluid must be zero, so the line of action of the resultant y-component of surface force must pass through the center of gravity of this portion of fluid.

Now we remove the fluid inside the surface and replace it with a solid body having exactly the same shape (Fig. 14–8b). The pressure at every point is exactly the same as before. So the total upward force exerted on the body by the fluid is also the same, again equal in magnitude to the weight mg of the fluid displaced to make way for the body. We call this upward force the **buoyant force** on the solid body. The line of action of the buoyant force again passes through the center of gravity of the displaced fluid (which doesn't necessarily coincide with the center of gravity of the body).

When a balloon floats in equilibrium in air, its weight (including the gas inside it) must be the same as the weight of the air displaced by the balloon. When a submerged submarine is in equilibrium, its weight must equal the weight of the water it displaces. A body whose average density is less than that of a liquid can float partially submerged at the free upper surface of the liquid. The greater the density of the liquid, the less of the body is submerged. When you swim in seawater (density 1030 kg/m^3), your body floats higher than in fresh water (1000 kg/m^3). Unlikely as it may seem, lead floats in mercury. Very flat-surfaced "float glass" is made by floating molten glass on molten tin and then letting it cool.

Another familiar example is the hydrometer, used to measure the density of liquids (Fig. 14–9a). The calibrated float sinks into the fluid until the weight of the fluid it displaces is exactly equal to its own weight. The hydrometer floats *higher* in denser liquids than in less dense liquids. It is weighted at its bottom end so that the upright position is stable, and a scale in the top stem permits direct density readings. Figure 14–9b shows a type of hydrometer that is commonly used to measure the density of battery acid or antifreeze. The bottom of the large tube is immersed in the liquid; the bulb is squeezed to expel air and is then released, like a giant medicine dropper. The liquid rises into the outer tube, and the hydrometer floats in this sample of the liquid.

(a)   (b)

14–9 (a) A simple hydrometer. (b) Use of a hydrometer as a tester for battery acid or antifreeze.

EXAMPLE 14–5

A 15.0-kg solid gold statue is being raised from a sunken ship. What is the tension in the hoisting cable when the statue is at rest and a) completely immersed; b) out of the water?

SOLUTION a) When the statue is immersed, it experiences an upward buoyant force equal in magnitude to the weight of the water displaced (Fig. 14–10a). To find this force, we first find

(a)   (b)

14–10 (a) The completely immersed statue at rest. (b) Free-body diagram for the immersed statue.

the volume of the statue, using the density of gold from Table 14–1:

$$V = \frac{m}{\rho_{\text{gold}}} = \frac{15.0 \text{ kg}}{19.3 \times 10^3 \text{ kg/m}^3} = 7.77 \times 10^{-4} \text{ m}^3.$$

Using Table 14–1 again, we find the weight of this volume of seawater:

$$\begin{aligned} w_{\text{sw}} = m_{\text{sw}}g &= \rho_{\text{sw}}Vg \\ &= (1.03 \times 10^3 \text{ kg/m}^3)(7.77 \times 10^{-4} \text{ m}^3)(9.80 \text{ m/s}^2) \\ &= 7.84 \text{ N}. \end{aligned}$$

This equals the buoyant force B.

To find the cable tension T, we draw a free-body diagram for the statue (Figure 14–10b). Since the statue is at rest, the net external force acting on it is zero:

$$\Sigma F_y = B + T + (-mg) = 0,$$
$$T = mg - B = (15.0 \text{ kg})(9.80 \text{ m/s}^2) - 7.84 \text{ N}$$
$$= 147 \text{ N} - 7.84 \text{ N} = 139 \text{ N}.$$

The submerged statue seems to weigh 139 N, about 5% less than its actual weight of 147 N.

b) The density of air is about 1.2 kg/m³, so the buoyant force of air on the statue is

$$B = \rho_{\text{air}}Vg = (1.2 \text{ kg/m}^3)(7.77 \times 10^{-4} \text{ m}^3)(9.80 \text{ m/s}^2)$$
$$= 9.1 \times 10^{-3} \text{ N}.$$

This is only 62 parts per million of the statue's actual weight. This effect is not within the precision of our data, and we ignore it. Thus the tension in the cable with the statue in air is equal to the statue's weight, 147 N.

EXAMPLE 14–6

You place a container of seawater on a scale and note the reading on the scale. You now suspend the statue of Example 14–5 in the water (Fig. 14–11). How does the scale reading change?

SOLUTION Consider the water, the statue, and the container together as a system; the total weight of the system does not depend on whether or not the statue is immersed. The total supporting force, including the tension T and the upward force F of the scale on the container (equal to the scale reading), is the same in both cases. But we saw in Example 14–5 that T decreases by 7.84 N when the statue is immersed, so the scale reading F must *increase* by 7.84 N. An alternative viewpoint is that the water exerts an upward buoyant force of 7.84 N on the statue, so the statue must exert an equal downward force on the water, making the scale reading 7.84 N greater than the weight of water and container.

14–11 How does the scale reading change when the statue is immersed in the water?

14–5 SURFACE TENSION

A paper clip can rest atop a water surface even though its density is several times that of water. Some insects can walk on the surface of water; their feet make indentations in the surface but do not penetrate it. Such phenomena are examples of *surface tension:* the surface of the liquid behaves like a membrane under tension. The molecules of the liquid exert attractive forces on each other; there is zero net force on a molecule inside the volume of the liquid, but a surface molecule is drawn into the volume (Fig. 14–12). Thus the liquid tends to minimize its surface area, just as does a stretched membrane. Freely falling raindrops are spherical (*not* teardrop-shaped) because a sphere has a smaller surface area for a given volume than any other shape. Figure 14–13 is a beautiful example of the formation of a spherical droplet.

Figure 14–14 shows how we can make quantitative measurements of surface tension. A piece of wire is bent into a U shape, and a second piece of wire slides on the arms of the U. When the apparatus is dipped into a soap solution and removed, creating a liquid film, the film exerts a surface-tension force on the slider that quickly pulls it up toward the top of the inverted U (if the slider's weight w is not too great). When we pull the slider down, increasing the area of the film, molecules move from the interior of the liquid (which is many molecules thick, even in a thin film) into the surface layers. The surface layers do not simply stretch like a rubber sheet. Instead, more surface is created by molecules moving from the bulk liquid.

To hold the slider in equilibrium, we need a total downward force $F = w + T$. In equilibrium, F is also equal to the surface-tension force exerted by the soap film on the slider. Let l be the length of the wire slider. The film has both front and back surfaces, so the force F acts along a total length $2l$. The **surface tension** γ (Greek "gamma") in the film is defined as the ratio of the surface-tension force F to the length d along which the force acts:

$$\gamma = \frac{F}{d} \quad \text{(definition of surface tension).} \tag{14–10}$$

In this case, $d = 2l$ and

$$\gamma = \frac{F}{2l}.$$

Surface tension is a *force per unit length*. The SI unit is the newton per meter (N/m), but the cgs unit, the dyne per centimeter (dyn/cm), is more commonly used:

$$1 \text{ dyn/cm} = 10^{-3} \text{ N/m} = 1 \text{ mN/m}.$$

Table 14–2 shows some typical values of surface tension. The lowest values of γ occur in the liquified noble gases neon and helium, in which the attraction between atoms is very weak. (For the same reason, these elements do not form compounds.)

14–12 A molecule of a liquid is attracted by its fellow molecules. A molecule on the surface is attracted into the bulk liquid, which tends to reduce the surface area of the liquid.

14–14 Measuring the surface tension of a soap film (shown in purple). The horizontal slide wire is in equilibrium under the action of the upward surface force $2\gamma l$ and downward pull $w + T$.

14–13 The splash from a drop of water falling into a liquid produces a column called a Rayleigh jet. Surface tension pulls the top of the jet into a spherical droplet.

Water pressure p

R Fibers
Air pressure p_a

14–15 Surface tension makes it difficult to force water through small crevices. The required water pressure can be reduced by heating the water and adding soap, both of which lower the surface tension γ.

TABLE 14–2
EXPERIMENTAL VALUES OF SURFACE TENSION

LIQUID IN CONTACT WITH AIR	TEMPERATURE (°C)	SURFACE TENSION (mN/m, or dyn/cm)
Benzene	20	28.9
Carbon tetrachloride	20	26.8
Ethanol	20	22.3
Glycerin	20	63.1
Mercury	20	465.0
Olive oil	20	32.0
Soap solution	20	25.0
Water	0	75.6
Water	20	72.8
Water	60	66.2
Water	100	58.9
Oxygen	−193	15.7
Neon	−247	5.15
Helium	−269	0.12

Liquid between surfaces

R

O

Air fills interior

(a)

Surface tension force $(2\gamma)(2\pi R)$

Area $= \pi R^2$

R

(b)

14–16 (a) Cross section of a soap bubble showing the two surfaces, the thin layer of liquid between them, and the air inside the bubble. (b) Equilibrium of half of a soap bubble. The surface-tension force exerted by the other half is $(2\gamma)(2\pi R)$, and the force exerted by the air inside the bubble is the pressure p times the area πR^2.

The surface tension of a particular liquid usually decreases as temperature increases; the table shows this behavior for water. In Chapter 16 we will learn that temperature is related to the energy of molecular motion in a material. As temperature increases and the molecules of a liquid move faster, the interactions between molecules have less effect on their motion and the surface tension decreases.

To wash clothing thoroughly, water must be forced through the tiny spaces between the fibers (Fig. 14–15). This requires increasing the surface area of the water, which is difficult to do because of surface tension. The job is made easier by lowering the value of γ. Hence very hot water ($\gamma = 58.9$ mN/m at 100°C) is better for washing than tepid water ($\gamma = 72.8$ mN/m at 20°C), and soapy water ($\gamma = 25.0$ mN/m at 20°C) is better still.

PRESSURE INSIDE A BUBBLE

Surface tension causes a pressure difference between the inside and outside of a soap bubble or a liquid drop. A soap bubble consists of two spherical surface films with a thin layer of liquid between (Fig. 14–16a). Because of surface tension, the films tend to contract in an attempt to minimize their surface area. But as the bubble contracts, it compresses the inside air, eventually increasing the interior pressure to a level that prevents further contraction.

We can derive an expression for the excess pressure inside a bubble in terms of its radius R and the surface tension γ of the liquid. Assume first that there is no external pressure. Each half of the soap bubble is in equilibrium; the lower half is shown in Fig. 14–16b. The forces at the flat circular surface where this half joins the upper half are the upward force of surface tension and the downward force due to the pressure of air in the upper half. The circumference of the circle along which the surface tension acts is $2\pi R$. (We neglect the small difference between inner and outer radii.) The total surface-tension force for each surface (inner and outer) is $\gamma(2\pi R)$, for a total of $(2\gamma)(2\pi R)$. Air pressure pushes both downward and outward on the lower half of the bubble, but the *resultant* force due to air pressure is downward only; its magnitude is the pressure p times πR^2, the area of the circle where the two half-bubbles meet. For the sum of these forces to be zero, we must have

$$(2\gamma)(2\pi R) = p(\pi R^2),$$

$$p = \frac{4\gamma}{R}.$$

(14–11)

In general, the pressure outside the bubble is *not* zero. But Eq. (14–11) still gives the *difference* between inside and outside pressure. If the outside pressure is atmospheric pressure p_a, then

$$p - p_a = \frac{4\gamma}{R} \quad \text{(soap bubble).} \qquad (14\text{--}12)$$

A liquid drop has only *one* surface film. Hence the surface tension force is $\gamma(2\pi R)$, half that for a soap bubble. In equilibrium the difference between the pressure in the liquid and that in the outside air is also half that for a soap bubble:

$$p - p_a = \frac{2\gamma}{R} \quad \text{(liquid drop).} \qquad (14\text{--}13)$$

The smaller the radius of the bubble or drop, the greater the pressure difference. A large pressure is required to force water through small crevices, since the water must form droplets of small radius R (Fig. 14–15).

EXAMPLE 14-7

Calculate the excess pressure inside a drop of water at 20°C if the diameter is 2.00 mm, 20.0 μm, or 0.200 μm.

SOLUTION From Table 14–2, $\gamma = 72.8$ mN/m $= 72.8 \times 10^{-3}$ N/m. From Eq. (14–13), with a diameter $d = 2.00$ mm,

$$p - p_a = \frac{2\gamma}{R} = \frac{2(72.8 \times 10^{-3}\ \text{N/m})}{1.00 \times 10^{-3}\ \text{m}}$$

$$= 146\ \text{N/m}^2 = 146\ \text{Pa} = 0.00144\ \text{atm.}$$

For $d = 20.0\ \mu\text{m} = 20.0 \times 10^{-6}$ m, $p - p_a = 14{,}600$ Pa $= 0.144$ atm. If $d = 0.200\ \mu$m, $p - p_a = 1.46 \times 10^6$ Pa, or 14.4 atmospheres! The pressure inside such a minuscule droplet can be gigantic.

CAPILLARITY

When a gas-liquid interface meets a solid surface, such as the wall of a container (Fig. 14–17), the gas-liquid interface usually curves up or down near the solid surface. The angle θ at which it meets the surface is called the **contact angle.** When the molecules of the liquid are attracted to each other less strongly than to the solid, as with water and glass, the liquid "wets" or adheres to the solid surface. The gas-liquid interface curves up, and θ is less than 90°. The liquid is nonwetting when the attraction between liquid molecules is the stronger force, as with mercury and glass; the gas-liquid interface curves down, and θ is greater than 90°.

Surface tension causes elevation or depression of the liquid in a narrow tube (Fig. 14–18). This effect is called **capillarity.** When the contact angle is less than 90° (Fig. 14–18a), the total surface tension force along the line of contact with the tube wall acts upward; the liquid *rises* until it reaches an equilibrium height y at which the total surface tension force just balances the extra weight of the liquid in the tube. The curved liquid surface is called a *meniscus.* For a nonwetting liquid such as mercury (Figure 14–18b) the contact angle is greater than 90°. The meniscus curves down, and the surface is depressed, pulled *down* by the surface-tension forces.

Capillarity is responsible for the absorption of water by paper towels, the rise of melted wax in a candle wick, and many other everyday phenomena. Blood is pumped through the arteries and veins of your body, but capillarity is important in causing flow through the smallest blood vessels, which indeed are called capillaries.

Related to surface tension is the phenomenon of *negative pressure.* The stress in a liquid is ordinarily *compressive,* but in some circumstances, liquids can sustain *tensile* stresses. Consider a cylindrical tube, closed at one end, with a tight-fitting piston at the other. We fill the tube completely with a liquid that wets both the inner surface of the

14–17 When a gas-liquid interface meets a solid surface, the interface usually curves up or down near the solid surface.

14–18 Surface-tension forces on a liquid in a capillary tube. The liquid (a) rises if $\theta < 90°$ or (b) is depressed if $\theta > 90°$. The diameter of the tube is greatly exaggerated for clarity.

(a)   (b)

tube and the piston face; the molecules of liquid adhere to all the surfaces. If the surfaces are very clean and the liquid very pure, then when we pull the piston face, we observe a *tensile* stress and a slight *increase* in volume; we are *stretching* the liquid. Adhesive forces prevent it from pulling away from the walls of the container.

With water, tensile stresses as large as 300 atm have been observed in the laboratory. This situation is highly unstable; a liquid under tension tends to break up into many small droplets. In tall trees, however, negative pressures are a regular occurrence. Negative pressure is believed to be an important mechanism for transport of water and nutrients from the roots to the leaves in the small xylem tubes (diameter of the order of 0.1 mm) in the growing layers of the tree.

14–6 FLUID FLOW

We are now ready to consider *motion* of a fluid. Fluid flow can be extremely complex, as shown by the currents in river rapids or the swirling flames of a campfire. But some situations can be represented by relatively simple idealized models. An **ideal fluid** is a fluid that is *incompressible* (that is, its density cannot change) and has no internal friction (called *viscosity*). Liquids are approximately incompressible in most situations, and we may also treat a gas as incompressible if the pressure differences from one region to another are not too great. Internal friction in a fluid causes shear stresses when two adjacent layers of fluid move relative to each other, as when fluid flows inside a tube or around an obstacle. In some cases we can neglect these shear forces in comparison with forces arising from gravitation and pressure differences.

The path of an individual particle in a moving fluid is called a **flow line.** If the overall flow pattern does not change with time, the flow is called **steady flow.** In steady flow, every element passing through a given point follows the same flow line. In this case the "map" of the fluid velocities at various points in space remains constant, although the velocity of a particular particle may change in both magnitude and direction during its motion. A **streamline** is a curve whose tangent at any point is in the direction of the fluid velocity at that point. When the flow pattern changes with time, the streamlines do not coincide with the flow lines. We will consider only steady-flow situations, for which flow lines and streamlines are identical.

The flow lines passing through the edge of an imaginary element of area, such as the area A in Fig. 14–19, form a tube called a **flow tube.** From the definition of a flow line,

14–19 A flow tube bounded by flow lines. In steady flow, fluid cannot cross the walls of a flow tube.

in steady flow no fluid can cross the side walls of a flow tube; the fluids in different flow tubes cannot mix.

Figure 14–20 shows patterns of fluid flow from left to right around a number of obstacles and in a channel of varying cross section. The photographs were made by injecting dye into water flowing between two closely spaced glass plates. These patterns are typical of **laminar flow,** in which adjacent layers of fluid slide smoothly past each other and the flow is steady. (A *lamina* is a thin sheet.) At sufficiently high flow rates, or when boundary surfaces cause abrupt changes in velocity, the flow can become irregular and chaotic. This is called **turbulent flow.** In turbulent flow there is no steady-state pattern; the flow pattern continuously changes.

14–20 (a) (b) (c) Laminar flow around obstacles of different shapes. (d) Flow through a channel of varying cross-section area.

THE CONTINUITY EQUATION

The mass of a moving fluid doesn't change as it flows. This leads to an important quantitative relationship called the **continuity equation.** Consider a portion of a flow tube between two stationary cross sections with areas A_1 and A_2 (Fig. 14–21). The fluid speeds at these sections are v_1 and v_2, respectively. No fluid flows in or out across the sides of the tube because the fluid velocity is tangent to the wall at every point on the wall. During a small time interval dt the fluid at A_1 moves a distance $v_1\,dt$, so a cylinder of fluid with height $v_1\,dt$ and volume $dV_1 = A_1 v_1\,dt$ flows into the tube across A_1. During this same interval, a cylinder of volume $dV_2 = A_2 v_2\,dt$ flows out of the tube across A_2.

Let's first consider the case of an incompressible fluid so that the density ρ has the same value at all points. The mass dm_1 flowing into the tube across A_1 in time dt is $dm_1 = \rho A_1 v_1\,dt$. Similarly, the mass dm_2 that flows out across A_2 in the same time is $dm_2 = \rho A_2 v_2\,dt$. In steady flow the total mass in the tube is constant, so $dm_1 = dm_2$ and

$$\rho A_1 v_1 dt = \rho A_2 v_2 dt, \quad \text{or}$$

$$A_1 v_1 = A_2 v_2 \quad \text{(continuity equation, incompressible fluid).} \quad (14\text{–}14)$$

The product Av is the *volume flow rate* dV/dt, the rate at which volume crosses a section of the tube:

$$\frac{dV}{dt} = Av \quad \text{(volume flow rate).} \quad (14\text{–}15)$$

The *mass* flow rate is the mass flow per unit time through a cross section. This is equal to the density ρ times the volume flow rate dV/dt.

Equation (14–14) shows that the volume flow rate has the same value at all points along any flow tube. When the cross section of a flow tube decreases (Fig. 14–20d), the speed increases, and vice versa. The deep part of a river has larger cross section and slower current than the shallow part, but the volume flow rates are the same in both. This is the essence of the familiar maxim "Still waters run deep." The stream of water from a faucet narrows as it gains speed during its fall, but dV/dt is the same everywhere along

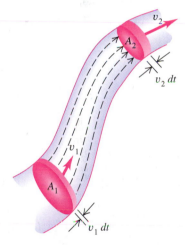

14–21 A flow tube with changing cross-section area. If the fluid is incompressible, the product Av has the same value at all points along the tube.

the stream. If a water pipe with 2-cm diameter is connected to a pipe with 1-cm diameter, the flow speed is four times as great in the 1-cm part as in the 2-cm part.

We can generalize Eq. (14–14) for the case in which the fluid is *not* incompressible. If ρ_1 and ρ_2 are the densities at sections 1 and 2, then

$$\rho_1 A_1 v_1 = \rho_2 A_2 v_2 \qquad \text{(continuity equation, compressible fluid).} \qquad (14\text{–}16)$$

We leave the details for an exercise. If the fluid is incompressible so that ρ_1 and ρ_2 are always equal, Eq. (14–16) reduces to Eq. (14–14).

14–7 BERNOULLI'S EQUATION

According to the continuity equation, the speed of fluid flow can vary along the paths of the fluid. The pressure can also vary; it depends on height as in the static situation (Section 14–3), and it also depends on the speed of flow. We can derive an important relationship called *Bernoulli's equation* that relates the pressure, flow speed, and height for flow of an ideal, incompressible fluid. Bernoulli's equation is an essential tool in analyzing plumbing systems, hydroelectric generating stations, and the flight of airplanes.

The dependence of pressure on speed follows from the continuity equation, Eq. (14–14). When an incompressible fluid flows along a flow tube with varying cross section, its speed *must* change, and so an element of fluid must have an acceleration. If the tube is horizontal, the force that causes this acceleration has to be applied by the surrounding fluid. This means that the pressure *must* be different in regions of different cross section; if it were the same everywhere, the net force on every fluid element would be zero. When a horizontal flow tube narrows and a fluid element speeds up, it must be moving toward a region of lower pressure in order to have a net forward force to accelerate it. If the elevation also changes, this causes an additional pressure difference.

To derive Bernoulli's equation, we apply the work-energy theorem to the fluid in a section of a flow tube. In Fig. 14–22 we consider the element of fluid that at some initial time lies between the two cross sections a and c. The speeds at the lower and upper ends are v_1 and v_2. In a small time interval dt the fluid that is initially at a moves to b, a distance $ds_1 = v_1 dt$, and the fluid that is initially at c moves to d, a distance $ds_2 = v_2 dt$. The cross-section areas at the two ends are A_1 and A_2, as shown. The fluid is incompressible; hence by the continuity equation, Eq. (14–14), the volume of fluid dV passing *any* cross section during time dt is the same. That is, $dV = A_1 ds_1 = A_2 ds_2$.

Let's compute the *work* done on this fluid element during dt. The pressures at the two ends are p_1 and p_2; the force on the cross section at a is $p_1 A_1$, and the force at c is $p_2 A_2$. The net work dW done on the element by the surrounding fluid during this displacement is therefore

$$dW = p_1 A_1 ds_1 - p_2 A_2 ds_2 = (p_1 - p_2)\, dV. \qquad (14\text{–}17)$$

The second term has a negative sign because the force at c opposes the displacement of the fluid.

The work dW is due to forces other than the conservative force of gravity, so it equals the change in the total mechanical energy (kinetic energy plus gravitational potential energy) associated with the fluid element. The mechanical energy for the fluid between sections b and c does not change. At the beginning of dt the fluid between a and b has volume $A_1 ds_1$, mass $\rho A_1 ds_1$, and kinetic energy $\frac{1}{2}\rho(A_1 ds_1)v_1^{2}$. At the end of dt the fluid between c and d has kinetic energy $\frac{1}{2}\rho(A_2 ds_2)v_2^{2}$. The net change in kinetic energy dK during time dt is

$$dK = \frac{1}{2}\rho\, dV(v_2^{2} - v_1^{2}). \qquad (14\text{–}18)$$

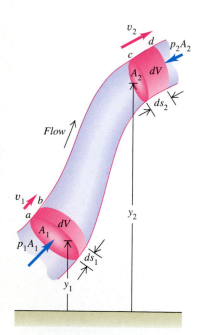

14–22 The net work done on a fluid element by the pressure of the surrounding fluid equals the change in the kinetic energy plus the change in the gravitational potential energy.

What about the change in gravitational potential energy? At the beginning of dt the potential energy for the mass between a and b is $dm\, gy_1 = \rho\, dV\, gy_1$. At the end of dt the potential energy for the mass between c and d is $dm\, gy_2 = \rho\, dV\, gy_2$. The net change in potential energy dU during dt is

$$dU = \rho\, dV\, g(y_2 - y_1). \tag{14–19}$$

Combining Eqs. (14–17), (14–18), and (14–19) in the energy equation $dW = dK + dU$, we obtain

$$(p_1 - p_2)\, dV = \frac{1}{2}\rho\, dV({v_2}^2 - {v_1}^2) + \rho\, dVg(y_2 - y_1),$$

$$p_1 - p_2 = \frac{1}{2}\rho({v_2}^2 - {v_1}^2) + \rho g(y_2 - y_1). \tag{14–20}$$

This is **Bernoulli's equation.** It states that the work done on a unit volume of fluid by the surrounding fluid is equal to the sum of the changes in kinetic and potential energies per unit volume that occur during the flow. We may also interpret Eq. (14–20) in terms of pressures. The first term on the right is the pressure difference associated with the change of speed of the fluid. The second term on the right is the additional pressure difference caused by the weight of the fluid and the difference in elevation of the two ends.

We can also express Eq. (14–20) in a more convenient form as

$$p_1 + \rho gy_1 + \frac{1}{2}\rho {v_1}^2 = p_2 + \rho gy_2 + \frac{1}{2}\rho {v_2}^2 \quad \text{(Bernoulli's equation).}$$
$$\tag{14–21}$$

The subscripts 1 and 2 refer to *any* two points along the flow tube, so we can also write

$$p + \rho gy + \frac{1}{2}\rho v^2 = \text{constant.} \tag{14–22}$$

Note that when the fluid is *not* moving (so $v_1 = v_2 = 0$), Eq. (14–21) reduces to the pressure relation we derived for a fluid at rest, Eq. (14–5).

CAUTION ▶ We stress again that Bernoulli's equation is valid *only* for incompressible, steady flow of a fluid with no viscosity. It's a simple equation that's easy to use; don't let this tempt you into using it in situations in which it doesn't apply! ◀

Problem–Solving Strategy

BERNOULLI'S EQUATION

Bernoulli's equation is derived from the work-energy relationship, so it isn't surprising that much of the problem-solving strategy suggested in Sections 7–2 and 7–3 is equally applicable here. In particular:

1. Always begin by identifying clearly the points 1 and 2 referred to in Bernoulli's equation.

2. Make lists of the known and unknown quantities in Eq. (14–21). The variables are p_1, p_2, v_1, v_2, y_1, and y_2, and the constants are ρ and g. What is given? What do you need to determine?

3. In some problems you will need to use the continuity equation, Eq. (14–14), to get a relation between the two speeds in terms of cross-section areas of pipes or containers. Or perhaps you will know both speeds and will need to determine one of the areas. You may also need to use Eq. (14–15) to find the volume flow rate.

4. As always, consistent units are essential. In SI units, pressure is in pascals, density in kilograms per cubic meter, and speed in meters per second. Also note that the pressures must be either all absolute pressures or all gauge pressures.

EXAMPLE 14-8

Water pressure in the home   Water enters a house through a pipe with an inside diameter of 2.0 cm at an absolute pressure of 4.0×10^5 Pa (about 4 atm). A 1.0-cm-diameter pipe leads to the second-floor bathroom 5.0 m above (Fig. 14–23). When the flow speed at the inlet pipe is 1.5 m/s, find the flow speed, pressure, and volume flow rate in the bathroom.

SOLUTION   Let points 1 and 2 be at the inlet pipe and at the bathroom, respectively. The speed v_2 at the bathroom is obtained from the continuity equation, Eq. (14–14):

$$v_2 = \frac{A_1}{A_2} v_1 = \frac{\pi (1.0 \text{ cm})^2}{\pi (0.50 \text{ cm})^2} (1.5 \text{ m/s}) = 6.0 \text{ m/s}.$$

We take $y_1 = 0$ (at the inlet) and $y_2 = 5.0$ m (at the bathroom). We are given p_1 and v_1, and can find p_2 from Bernoulli's equation:

$$p_2 = p_1 - \frac{1}{2} \rho (v_2{}^2 - v_1{}^2) - \rho g (y_2 - y_1)$$

$$= 4.0 \times 10^5 \text{ Pa} - \frac{1}{2} (1.0 \times 10^3 \text{ kg/m}^3)(36 \text{ m}^2/\text{s}^2 - 2.25 \text{ m}^2/\text{s}^2)$$

$$- (1.0 \times 10^3 \text{ kg/m}^3)(9.8 \text{ m/s}^2)(5.0 \text{ m})$$

$$= 4.0 \times 10^5 \text{ Pa} - 0.17 \times 10^5 \text{ Pa} - 0.49 \times 10^5 \text{ Pa}$$

$$= 3.3 \times 10^5 \text{ Pa} = 3.3 \text{ atm} = 48 \text{ psia}.$$

The volume flow rate is

$$\frac{dV}{dt} = A_2 v_2 = \pi (0.50 \times 10^{-2} \text{ m})^2 (6.0 \text{ m/s})$$

$$= 4.7 \times 10^{-4} \text{ m}^3/\text{s} = 0.47 \text{ L/s}.$$

Note that after the water is turned off, the term $\frac{1}{2} \rho (v_2{}^2 - v_1{}^2)$ in the equation for pressure vanishes, and the pressure rises to 3.5×10^5 Pa.

14–23 What is the water pressure in the second-story bathroom of this house?

EXAMPLE 14-9

Speed of efflux   Figure 14–24 shows a gasoline storage tank with cross-section area A_1, filled to a depth h. The space above the gasoline contains air at pressure p_0, and the gasoline flows out through a short pipe with area A_2. Derive expressions for the flow speed in the pipe and the volume flow rate.

SOLUTION   We can consider the entire volume of moving liquid as a single flow tube; v_1 and v_2 are the speeds at points 1 and 2, respectively, in Fig. 14–24. The pressure at point 2 is atmospheric pressure, p_a. Applying Bernoulli's equation to points 1 and 2 and taking $y = 0$ at the bottom of the tank, we find

$$p_0 + \frac{1}{2} \rho v_1{}^2 + \rho g h = p_a + \frac{1}{2} \rho v_2{}^2,$$

$$v_2{}^2 = v_1{}^2 + 2 \frac{p_0 - p_a}{\rho} + 2gh.$$

14–24 Calculating the speed of efflux for gasoline flowing out the bottom of a storage tank.

Because A_2 is much smaller than A_1, $v_1{}^2$ is very much smaller than $v_2{}^2$ and can be neglected. We then find

$$v_2{}^2 = 2\frac{p_0 - p_a}{\rho} + 2gh.$$

The speed v_2, sometimes called the *speed of efflux*, depends on both the pressure difference ($p_0 - p_a$) and the height h of the liquid level in the tank. If the top of the tank is vented to the atmosphere, there is no excess pressure: $p_0 = p_a$ and $p_0 - p_a = 0$. In that case,

$$v_2 = \sqrt{2gh}.$$

That is, the speed of efflux from an opening at a distance h below the top surface of the liquid is the *same* as the speed a body would acquire in falling freely through a height h. This result is called *Torricelli's theorem*. It is valid not only for an opening in the bottom of a container, but also for a hole in a side wall at a depth h below the surface. We find the volume flow rate from Eq. (14–15):

$$\frac{dV}{dt} = A_2\sqrt{2gh}.$$

EXAMPLE 14–10

The Venturi meter Figure 14–25 shows a *Venturi meter*, used to measure flow speed in a pipe. The narrow part of the pipe is called the *throat*. Derive an expression for the flow speed v_1 in terms of the cross-section areas A_1 and A_2 and the difference in height h of the liquid levels in the two vertical tubes.

SOLUTION We apply Bernoulli's equation to the wide (point 1) and narrow (point 2) parts of the pipe, with $y_1 = y_2$:

$$p_1 + \frac{1}{2}\rho v_1{}^2 = p_2 + \frac{1}{2}\rho v_2{}^2.$$

From the continuity equation, $v_2 = (A_1/A_2)v_1$. Substituting this and rearranging, we get

$$p_1 - p_2 = \frac{1}{2}\rho v_1{}^2\left(\frac{A_1{}^2}{A_2{}^2} - 1\right).$$

Because A_1 is greater than A_2, v_2 is greater than v_1 and the pressure p_2 in the throat is *less* than p_1. A net force to the right accelerates the fluid as it enters the throat, and a net force to the

14–25 The Venturi meter.

left slows it as it leaves. The pressure difference $p_1 - p_2$ is also equal to ρgh, where h is the difference in liquid level in the two tubes. Combining this with the above result and solving for v_1, we get

$$v_1 = \sqrt{\frac{2gh}{(A_1/A_2)^2 - 1}}.$$

EXAMPLE 14–11

Lift on an airplane wing Figure 14–26 shows flow lines around a cross section of an airplane wing. The flow lines crowd together above the wing, corresponding to increased flow speed and reduced pressure in this region, just as in the Venturi throat. The upward force on the underside of the wing is greater than the downward force on the top side; there is a net upward force, or *lift*. Lift is not simply due to the impulse of air striking the underside of the wing; in fact, it turns out that the reduced pressure on the upper wing surface makes the greatest contribution to the lift. (This highly simplified discussion ignores the formation of vortices; a more complete discussion would take these into account.)

We can also understand the lift force on the basis of momentum changes. Figure 14–26 shows that there is a net *downward* change in the vertical component of momentum of the air flowing past the wing, corresponding to the downward force the wing exerts on the air. The reaction force *on* the wing is *upward*, as we concluded above.

14–26 Flow lines around an airplane wing. The momentum of a parcel of air (relative to the wing) is $\vec{p}_i$ before encountering the wing and $\vec{p}_f$ afterwards.

14–8 TURBULENCE

When the speed of a flowing fluid exceeds a certain critical value, the flow is no longer laminar. The flow pattern becomes extremely irregular and complex, and it changes continuously with time; there is no steady-state pattern. This irregular, chaotic flow is called **turbulence.** The contrast between laminar and turbulent flow is shown in Fig. 14–27 for two familiar systems: a stream of water and rising smoke in air. Bernoulli's equation is *not* applicable to regions where there is turbulence, because the flow is not steady.

The transition from laminar to turbulent flow is often very sudden. A flow pattern that is stable at low speeds suddenly becomes unstable when a critical speed is reached. Irregularities in the flow pattern can be caused by roughness in the pipe wall, variations in density of the fluid, and many other factors. At small flow speeds these disturbances damp out; the flow pattern is *stable* and tends to maintain its laminar nature. But when the critical speed is reached, the flow pattern becomes unstable. The disturbances no longer damp out, but grow until they destroy the entire laminar-flow pattern. Normal blood flow in the human aorta is laminar, but a small disturbance such as a heart pathology can cause the flow to become turbulent. Turbulence makes noise, which is why listening to blood flow with a stethoscope is a useful diagnostic tool.

Turbulence poses some profound questions for theoretical physics. The motion of any particular particle in a fluid is presumably determined by Newton's laws. If we know the motion of the fluid at some initial time, shouldn't we be able to predict the motion at any later time? After all, when we launch a projectile, we can compute the entire trajectory if we know the initial position and velocity. How is it that a system that obeys well-defined physical laws can have unpredictable, chaotic behavior? There is as yet no simple answer to these questions. Indeed, the study of *chaos*—seemingly unpredictable behavior in systems that should be entirely predictable—is a very active field of research. A particularly significant result of chaos research has been the discovery of similar behavior in a broad range of systems, including the evolution of biological populations, the growth of crystals, the shapes of coastlines, and many others.

14–27 (a) Laminar flow of water. (b) Turbulent flow of water. (c) First laminar, then turbulent flow of smoke.

(a)   (b)   (c)

EXAMPLE 14–12

The curve ball Does a curve ball *really* curve? Yes, it certainly does, and the reason is turbulence. Figure 14–28a shows a ball moving through the air from left to right. To an observer moving with the center of the ball, the air stream appears to move from right to left, as shown by the flow lines in the figure. Because of the large speeds that are ordinarily involved (near 160 km/h, or 100 mi/h), there is a region of *turbulent* flow behind the ball.

Figure 14–28b shows a *spinning* ball with "top spin." Layers of air near the ball's surface are pulled around in the direction of spin by friction between the ball and air and by the air's internal friction (viscosity). The speed of air relative to the ball's surface becomes smaller at the top of the ball than at the bottom, and turbulence occurs farther forward on the top side than on the bottom. This asymmetry causes a pressure difference; the average pressure at the top of the ball is now greater than that at the bottom. The net force deflects the ball downward, as shown in Fig. 14–28c. This is why "top spin" is used in tennis to keep a

very fast serve in the court (Fig. 14–28d). In a baseball curve pitch, the ball spins about a nearly vertical axis, and the actual deflection is sideways. In that case, Fig. 14–28c is a *top* view of the situation. A curve ball thrown by a left-handed pitcher curves *toward* a right-handed batter, making it harder to hit (Fig. 14–28e).

A similar effect occurs with golf balls, which always have "back spin" from impact with the slanted face of the golf club. The resulting pressure difference between the top and bottom of the ball causes a lift force that keeps the ball in the air considerably longer than would be possible without spin. A well-hit drive appears from the tee to "float" or even curve *upward* during the initial portion of its flight. This is a real effect, not an illusion. The dimples on the ball play an essential role; the viscosity of air gives an undimpled ball a much shorter trajectory than a dimpled one with the same initial velocity and spin. Figure 14–29 shows the backspin of a golf ball just after it is struck by a club.

v_ball

(a)   (b)   (c)

(d)   (e)

14–28 Motion of the air from right to left relative to the ball corresponds to motion of a ball through still air from left to right. (a) A nonspinning ball has a symmetric region of turbulence behind it. (b) A spinning ball drags along layers of air near its surface. (c) The resulting asymmetric region of turbulence and the deflection of the air stream by the spinning ball. The force shown is exerted on the ball by the air stream; it pushes the ball in the direction of the tangential velocity of the front of the ball. The force can (d) push a tennis ball downward or (e) make a baseball curve.

14–29 Stroboscopic photograph of a golf ball being struck by a club. The picture was taken at 1000 flashes per second. The ball rotates about once in eight pictures, corresponding to an angular speed of 125 rev/s, or 7500 rpm.

*14–9 Viscosity

Viscosity is internal friction in a fluid. Viscous forces oppose the motion of one portion of a fluid relative to another. Viscosity is the reason it takes effort to paddle a canoe through calm water, but it is also the reason the paddle works. Viscous effects are important in the flow of fluids in pipes, the flow of blood, the lubrication of engine parts, and many other situations.

A viscous fluid tends to cling to a solid surface in contact with it. There is a thin *boundary layer* of fluid near the surface, in which the fluid is nearly at rest with respect to the surface. That's why dust particles can cling to a fan blade even when it is rotating rapidly and why you can't get all the dirt off your car by just squirting a hose at it.

The simplest example of viscous flow is motion of a fluid between two parallel plates (Fig. 14–30). The bottom plate is stationary, and the top plate moves with constant velocity $\vec{v}$. The fluid in contact with each surface has the same velocity as that surface. The flow speeds of intermediate layers of fluid increase uniformly from one surface to the other, as shown by the arrows, so the fluid layers slide smoothly over one another; the flow is laminar.

A portion of the fluid that has the shape *abcd* at some instant has the shape *abc′ d′* a moment later and becomes more and more distorted as the motion continues. That is, the fluid is in a state of continuously increasing *shear strain*. To maintain this motion, we have to apply a constant force F to the right on the upper plate to keep it moving and a force of equal magnitude toward the left on the lower plate to hold it stationary. If A is the surface area of each plate, the ratio F/A is the *shear stress* exerted on the fluid.

In Section 11–7 we defined shear strain as the ratio of the displacement dd' to the length l. In a solid, shear strain is proportional to shear stress. In a fluid the shear strain *increases continuously and without limit* as long as the stress is applied. The stress depends not on the shear strain, but on its *rate of change*. The rate of change of strain, also called the *strain rate,* equals the rate of change of dd' (the speed v of the moving surface) divided by l. That is,

$$\text{Rate of change of shear strain} = \text{strain rate} = \frac{v}{l}.$$

We define the **viscosity** of the fluid, denoted by η ("eta"), as the ratio of the shear stress, F/A, to the strain rate:

$$\eta = \frac{\text{Shear stress}}{\text{Strain rate}} = \frac{F/A}{v/l} \qquad \text{(definition of viscosity)}. \qquad (14\text{–}23)$$

Rearranging Eq. (14–23), we see that the force required for the motion in Fig. 14–30 is directly proportional to the speed:

$$F = \eta A \frac{v}{l}. \qquad (14\text{–}24)$$

Fluids that flow readily, such as water or gasoline, have smaller viscosities than "thick" liquids such as honey or motor oil. Viscosities of all fluids are strongly temper-

Lava is an example of a viscous fluid. The viscosity decreases with increasing temperature: the hotter the lava, the more easily it can flow.

14–30 Laminar flow of a viscous fluid. The plates above and below the fluid each have area A. Viscosity is the ratio of shear stress F/A to strain rate v/l.

ature dependent, increasing for gases and decreasing for liquids as the temperature increases. An important goal in the design of oils for engine lubrication is to *reduce* the temperature variation of viscosity as much as possible.

From Eq. (14–23) the unit of viscosity is that of force times distance, divided by area times speed. The SI unit is

$$1 \text{ N} \cdot \text{m}/[\text{m}^2 \cdot (\text{m/s})] = 1 \text{ N} \cdot \text{s/m}^2 = 1 \text{ Pa} \cdot \text{s}.$$

The corresponding cgs unit, 1 dyn · s/cm², is the only viscosity unit in common use; it is called a **poise,** in honor of the French scientist Jean Louis Marie Poiseuille (pronounced "pwa-*zoo*-yuh"):

$$1 \text{ poise} = 1 \text{ dyn} \cdot \text{s/cm}^2 = 10^{-1} \text{ N} \cdot \text{s/m}^2.$$

The centipoise and the micropoise are also used. The viscosity of water is 1.79 centipoise at 0°C and 0.28 centipoise at 100°C. Viscosities of lubricating oils are typically 1 to 10 poise, and the viscosity of air at 20°C is 181 micropoise.

For a *Newtonian fluid* the viscosity η is independent of the speed v, and from Eq. (14–24) the force F is directly proportional to the speed. Fluids that are suspensions or dispersions are often *non*-Newtonian in their viscous behavior. One example is blood, which is a suspension of corpuscles in a liquid. As the strain rate increases, the corpuscles deform and become preferentially oriented to facilitate flow, causing η to decrease. The fluids that lubricate human joints show similar behavior.

Figure 14–31 shows the flow speed profile for laminar flow of a viscous fluid in a long cylindrical pipe. The speed is greatest along the axis and zero at the pipe walls. The motion is like a lot of concentric tubes sliding relative to one another, with the central tube moving fastest and the outermost tube at rest. By applying Eq. (14–23) to a cylindrical fluid element, we could derive an equation describing the speed profile. We'll omit the details; the flow speed v at a distance r from the axis of a pipe with radius R is

$$v = \frac{p_1 - p_2}{4\eta L}(R^2 - r^2), \tag{14–25}$$

where p_1 and p_2 are the pressures at the two ends of a pipe with length L. The speed at any point is proportional to the change of pressure per unit length, $(p_2 - p_1)/L$ or dp/dx, called the *pressure gradient*. (The flow is always in the direction of decreasing pressure.) To find the total volume flow rate through the pipe, we consider a ring with inner radius r, outer radius $r + dr$, and cross-section area $dA = 2\pi r \, dr$. The volume flow rate through this element is $v\,dA$; the total volume flow rate is found by integrating from $r = 0$ to $r = R$. The result is

$$\frac{dV}{dt} = \frac{\pi}{8}\left(\frac{R^4}{\eta}\right)\left(\frac{p_1 - p_2}{L}\right) \quad \text{(Poiseuille's equation).} \tag{14–26}$$

This relation was first derived by Poiseuille and is called **Poiseuille's equation.** The volume flow rate is inversely proportional to viscosity, as we would expect. It is also proportional to the pressure gradient $(p_2 - p_1)/L$, and it varies as the *fourth power* of the radius R. If we double R, the flow rate increases by a factor of 16. This relation is important for the design of plumbing systems and hypodermic needles. Needle size is much more important than thumb pressure in determining the flow rate from the needle; doubling the needle diameter has the same effect as increasing the thumb force sixteenfold. Similarly, blood flow in arteries and veins can be controlled over a wide range by relatively small changes in diameter, an important temperature-control mechanism in warm-blooded animals. Relatively slight narrowing of arteries due to arteriosclerosis can result in elevated blood pressure and added strain on the heart muscle.

One more useful relation in viscous fluid flow is the expression for the force F exerted on a sphere of radius r moving with speed v through a fluid with viscosity η.

14–31 Velocity profile for a viscous fluid in a cylindrical pipe.

When the flow is laminar, the relationship is simple:

$$F = 6\pi\eta rv \qquad \text{(Stokes's law)}. \qquad (14\text{--}27)$$

We encountered this kind of velocity-proportional force in Section 5–4. Equation (14–27) is called **Stokes's law.**

EXAMPLE 14-13

Terminal speed in a viscous fluid Derive an expression for the terminal speed v_t of a sphere falling in a viscous fluid in terms of the sphere's radius r and density ρ and the fluid viscosity η, assuming that the flow is laminar so that Stokes's law is valid.

SOLUTION As in Example 5–19, the terminal speed is reached when the total force is zero, including the weight of the sphere, the viscous retarding force, and the buoyant force (Fig. 14–32). Let ρ and ρ' be the densities of the sphere and of the fluid, respectively. The weight of the sphere is then $\frac{4}{3}\pi r^3 \rho g$, and the buoyant force is $\frac{4}{3}\pi r^3 \rho' g$. At terminal speed,

$$\Sigma F_y = F_{\text{buoyancy}} + F_{\text{visc}} + (-mg)$$

$$= \frac{4}{3}\pi r^3 \rho' g + 6\pi\eta r v_t - \frac{4}{3}\pi r^3 \rho g = 0,$$

$$v_t = \frac{2}{9}\frac{r^2 g}{\eta}(\rho - \rho').$$

We can use this formula to *measure* the viscosity of a fluid. Or, if we know the viscosity, we can determine the radius of the sphere (or the approximate sizes of other small particles) by measuring the terminal speed. Robert Millikan used this method to determine the radii of very small electrically charged oil drops by observing their terminal speeds in air. He used these drops to measure the charge of an individual electron.

14-32 A sphere falling in a viscous fluid reaches a terminal speed when the sum of the forces acting on it is zero.

The terminal speed for a water droplet of radius 10^{-5} m in air is of the order of 1 cm/s; the Stokes's-law force is what keeps clouds in the air. At higher speeds the flow often becomes turbulent, and Stokes's law is no longer valid. In air the drag force at highway speeds is approximately proportional to v^2. We discussed the applications of this relation to skydiving in Section 5–4 and to air drag of a moving automobile in Section 6–6.

SUMMARY

KEY TERMS

fluid statics, 427

fluid dynamics, 427

density, 427

specific gravity, 427

pressure, 429

pascal, 429

atmospheric pressure, 429

Pascal's law, 430

gauge pressure, 431

absolute pressure, 431

mercury barometer, 432

buoyancy, 432

Archimedes' principle, 432

buoyant force, 433

surface tension, 435

contact angle, 437

capillarity, 437

- Density is mass per unit volume. If a mass m of homogeneous material has volume V, its density ρ is

$$\rho = \frac{m}{V}. \qquad (14\text{--}1)$$

Specific gravity is the ratio of the density of a material to the density of water.

- Pressure is normal force per unit area. Pascal's law states that pressure applied to the surface of an enclosed fluid is transmitted undiminished to every portion of the fluid. Absolute pressure is the total pressure in a fluid; gauge pressure is the difference between absolute pressure and atmospheric pressure. The SI unit of pressure is the pascal (Pa): 1 Pa = 1 N/m².

- If a fluid of uniform density ρ (an incompressible fluid) is at rest, the pressure difference between points 1 and 2 at elevations y_1 and y_2 is

$$p_2 - p_1 = -\rho g(y_2 - y_1). \qquad (14\text{--}5)$$

If the pressure at the surface of an incompressible fluid at rest is p_0, the pressure at a depth h is

$$p = p_0 + \rho g h. \qquad (14\text{--}6)$$

- Archimedes' principle states that when a body is immersed in a fluid, the fluid exerts an upward buoyant force on the body equal to the weight of the fluid the body displaces.
- The surface of a liquid behaves like a membrane under tension; the force per unit length across a line on the surface is called the surface tension, denoted by γ.
- An ideal fluid is incompressible and has no viscosity. A flow line is the path of a fluid particle; a streamline is a curve tangent at each point to the velocity vector at that point. A flow tube is a tube bounded at its sides by flow lines. In laminar flow, layers of fluid slide smoothly past each other. In turbulent flow there is great disorder and a constantly changing flow pattern.
- Conservation of mass in an incompressible fluid is expressed by the equation of continuity: for two cross sections A_1 and A_2 in a flow tube, the flow speeds v_1 and v_2 are related by

$$A_1 v_1 = A_2 v_2. \qquad (14\text{--}14)$$

The product Av is the *volume flow rate, dV/dt,* the rate at which volume crosses a section of the tube:

$$\frac{dV}{dt} = Av. \qquad (14\text{--}15)$$

- Bernoulli's equation relates the pressure p, flow speed v, and elevation y for steady flow in an ideal fluid. For any two points, denoted by subscripts 1 and 2,

$$p_1 + \rho g y_1 + \frac{1}{2}\rho v_1{}^2 = p_2 + \rho g y_2 + \frac{1}{2}\rho v_2{}^2. \qquad (14\text{--}21)$$

- The viscosity of a fluid characterizes its resistance to shear strain. In a Newtonian fluid the viscous force is proportional to strain rate. When such a fluid flows in a cylindrical pipe of inner radius R, and length L is the length of pipe, the total volume rate is given by Poiseuille's equation:

$$\frac{dV}{dt} = \frac{\pi}{8}\left(\frac{R^4}{\eta}\right)\left(\frac{p_1 - p_2}{L}\right). \qquad (14\text{--}26)$$

where p_1 and p_2 are the pressures at the two ends and η is the viscosity.

- A sphere of radius r moving with speed v through a fluid having viscosity η experiences a viscous resisting force F given by Stokes's law:

$$F = 6\pi\eta r v. \qquad (14\text{--}27)$$

DISCUSSION QUESTIONS

Q14–1 If the weight of a roomful of water is so great (see Example 14–1 in Section 14–2), why don't the first floors of houses built over basements or crawl spaces all collapse when flooded to the ceiling?

Q14–2 A rubber hose is attached to a funnel, and the free end is bent around to point upward. When water is poured into the funnel, it rises in the hose to the same level as in the funnel, even though the funnel has a lot more water in it than the hose. Why?

Q14–3 Comparing Examples 14–1 and 14–2 in Sections 14–2 and 14–3, it seems that 700 N of air is exerting a downward force of 2.0×10^6 N on the floor. How is this possible?

Q14–4 Equation (14–7) shows that an area ratio of 100 to 1 can give 100 times more output force than input force. Doesn't this violate conservation of energy? Explain.

Q14–5 You have probably noticed that the lower the tire pressure, the greater the contact area between the tire and the pavement. Why?

Q14–6 In hot-air ballooning a large balloon is filled with air heated by a gas burner at the bottom. Why must the air be heated? How does the balloonist control ascent and descent?

Q14–7 In describing the size of a large ship, one uses such expressions as "it displaces 20,000 tons." What does this mean? Can the weight of the ship be obtained from this information?

Q14–8 You drop a solid sphere of aluminum in a bucket of water that sits on the ground. The buoyant force equals the weight of water displaced; this is less than the weight of the sphere, so the sphere sinks to the bottom. If you take the bucket with you on an elevator that accelerates upward, the apparent weight of the water increases and the buoyant force on the sphere increases. Could the acceleration of the elevator be great enough to make the sphere pop up out of the water? Explain.

Q14–9 A rigid lighter-than-air dirigible filled with helium cannot continue to rise indefinitely. Why not? What determines the maximum height it can attain?

Q14–10 Air pressure decreases with increasing altitude. So why isn't air near the surface continuously drawn upward toward the lower-pressure regions above?

Q14–11 The purity of gold can be tested by weighing it in air and in water. How? Do you think you could get away with making a fake gold brick by gold-plating some cheaper material?

Q14–12 During the Great Mississippi Flood of 1993, the levees in St. Louis tended to rupture first at the bottom. Why?

Q14–13 A cargo ship travels from the Atlantic Ocean (salt water) to Lake Ontario (fresh water) via the St. Lawrence River. The ship rides several centimeters lower in the water in Lake Ontario than it did in the ocean. Explain why.

Q14–14 A submarine is more compressible than water. Why, then, can a submarine completely surrounded by water be only in unstable equilibrium?

Q14–15 An old question is "Which weighs more, a pound of feathers or a pound of lead?" If the weight in pounds is the gravitational force, will a pound of feathers balance a pound of lead on opposite pans of an equal-arm balance? Explain, taking into account buoyant forces.

Q14–16 Suppose the door of a room makes an airtight but frictionless fit in its frame. Do you think you could open the door if the air pressure on one side were standard atmospheric pressure and that on the other side differed from standard by 1%? Explain.

Q14–17 A balloon is less compressible than air. Why, then, is there a height at which a helium balloon is in stable equilibrium?

Q14–18 A piece of iron is glued to the top of a block of wood. When the block is placed in a bucket of water with the iron on top, the block floats. The block is now turned over so that the iron is submerged beneath the wood. Does the block float or sink? Does the water level in the bucket rise, drop, or stay the same? Explain your answers.

Q14–19 You take an empty glass jar and push it into a tank of water with the open mouth of the jar downward, so that the air inside the jar is trapped and cannot get out. If you push the jar deeper into the water, does the buoyant force on the jar stay the same? If not, does it increase or decrease? Explain your answer.

Q14–20 An ice cube floats in a glass of water. The ice cube contains many air bubbles. When the ice melts, will the water level in the glass rise, fall, or remain unchanged? Explain.

Q14–21 A small drop on the surface of a freshly waxed car is nearly spherical, but a larger drop is more flattened. Why?

Q14–22 Water drops easily roll off a duck's back. What can you conclude about the contact angle between water and the duck?

Q14–23 Can mercury in a clean glass tube with a clean glass piston exhibit the tensile stress of negative pressure? Explain.

Q14–24 You are told, "Bernoulli's equation tells us where there is higher fluid speed, there is lower fluid pressure, and vice versa." Is this statement always true, even for an idealized fluid? Explain.

Q14–25 As you walk past a cup of coffee sitting on a desk, the coffee has a velocity relative to you. Would you describe the coffee with fluid statics or fluid dynamics? Why?

Q14–26 If the velocity at each point in space in steady-state fluid flow is constant, how can a fluid particle accelerate?

Q14–27 Why do jet airplanes usually fly at altitudes above 30,000 ft, though it takes a lot of fuel to climb that high?

Q14–28 In a store-window vacuum cleaner display, a table-tennis ball is suspended in midair in a jet of air blown from the outlet hose of a tank-type vacuum cleaner. The ball bounces around a little but always moves back toward the center of the jet, even if the jet is tilted from the vertical. How does this behavior illustrate Bernoulli's equation?

Q14–29 How does a baseball pitcher give the ball the spin that makes it curve? Can he make it curve in either direction? Does it matter whether he is right-handed or left-handed?

Q14–30 What causes the sharp hammering sound that is sometimes heard from a water pipe when an open faucet is suddenly turned off?

Q14–31 A tornado consists of a rapidly whirling air vortex. Why is the pressure always much lower in the center than at the outside? How does this condition account for the destructive power of a tornado?

Q14–32 Airports at high elevations have longer runways for takeoffs and landings than do airports at sea level. One reason is that aircraft engines develop less power in the thin air well above sea level. What is another reason?

Q14–33 Why would a fire department add a viscosity-decreasing substance to its water for fighting fires in tall buildings?

Q14–34 When a smooth-flowing stream of water comes out of a faucet, it narrows as it falls; if it falls far enough it eventually breaks up into drops. Why does it narrow? Why does it break up?

Q14–35 Spreading margarine with a knife is an example of viscous flow. Is margarine a Newtonian fluid, or is it non-Newtonian? Explain.

EXERCISES

SECTION 14–2 DENSITY

14–1 On a part-time job, you are asked to bring a cylindrical iron rod of length 85.8 cm and diameter 2.85 cm from a storage room to a machinist. Will you need a cart? (To answer, calculate the weight of the rod.)

14–2 The moon has a radius of 1740 km and a mass of 7.35×10^{22} kg. What is its average density?

14–3 You purchase a rectangular piece of metal that has dimensions $5.0 \times 15.0 \times 30.0$ mm and a mass of 0.0158 kg. The seller tells you that the metal is gold. To check this, you compute the average density of the piece. What value do you get? Were you cheated?

14–4 A kidnapper demands a 40.0-kg cube of platinum as a ransom. What is the length of a side?

SECTION 14–3 PRESSURE IN A FLUID

14–5 A barrel contains a 0.120-m layer of oil floating on water that is 0.250 m deep. The density of the oil is 600 kg/m³ a) What is the gauge pressure at the oil-water interface? b) What is the gauge pressure at the bottom of the barrel?

14–6 An empty sport utility vehicle weighs 16.5 kN. Each of its tires has a gauge pressure of 205 kPa (29.7 lb/in.²). a) What is the total contact area of the four tires with the pavement? (Assume that the tire walls are flexible so that the pressure exerted by the tire on the pavement equals the air pressure inside the tire.) b) For the same tire pressure, what is that area when the vehicle is loaded with 9.1 kN of passengers and cargo?

14–7 You are designing a diving bell to withstand the pressure of seawater at a depth of 250 m. a) What is the gauge pressure at this depth? (Neglect changes in the density of the water with depth.) b) At this depth, what is the net force due to the water outside and the air inside the bell on a circular glass window 30.0 cm in diameter if the pressure inside the diving bell equals the pressure at the surface of the water? (Neglect the small variation of pressure over the surface of the window.)

14–8 What gauge pressure must a pump produce to pump water from the bottom of the Grand Canyon (elevation 730 m) to Indian Gardens (1370 m)? Express your results in pascals and in atmospheres.

14–9 The liquid in the open-tube manometer in Fig. 14–6a is mercury, $y_1 = 3.00$ cm, and $y_2 = 7.00$ cm. Atmospheric pressure is 980 millibars. a) What is the absolute pressure at the bottom of the U-shaped tube? b) What is the absolute pressure in the open tube at a depth of 4.00 cm below the free surface? c) What is the absolute pressure of the gas in the tank? d) What is the gauge pressure of the gas in pascals?

14–10 There is a maximum depth at which a diver can breathe through a snorkel tube (Fig. 14–33) because as the depth increases, so does the pressure difference, tending to collapse the diver's lungs. Since the snorkel connects the air in the lungs to the atmosphere at the surface, the pressure inside the lungs is atmospheric pressure. What is the external-internal pressure difference when the diver's lungs are at a depth of 6.1 m (about 20 ft)? Assume that the diver is in fresh water. (A scuba diver breathing from compressed air tanks can operate at greater depths than can a snorkeler, since the pressure of the air inside the scuba diver's lungs increases to match the external pressure of the water.)

14–11 An electrical short cuts off all power to a submersible when it is 30 m below the surface of the ocean. The crew must push out a hatch of area 0.75 m² and weight 300 N on the bottom to escape. If the pressure inside is 1.0 atm, what downward force must they exert on the hatch to open it?

14–12 You are assigned the design of a cylindrical pressurized water tank for a future colony on Mars, where the acceleration due to gravity is 3.71 m/s². The pressure at the surface of the water will be 130 kPa and the depth of the water will be 14.2 m.

P_a

6.1 m

FIGURE 14–33 Exercise 14–10.

The pressure of the air in the building outside the tank will be 93 kPa. Find the net downward force on the tank's flat bottom, of area 2.00 m², exerted by the water and air inside the tank and the air outside the tank.

14–13 A tapered, pressurized tank for a rocket contains 0.250 m³ of kerosene, which has a mass of 205 kg. The pressure at the top of the kerosene is 2.01×10^5 Pa. The kerosene exerts a force of 16.4 kN on the tank's bottom, which has an area of 0.0700 m². Find the depth of the kerosene.

14–14 The piston of a hydraulic automobile lift is 0.30 m in diameter. What gauge pressure, in pascals, is required to lift a car with a mass of 1200 kg? Also express this pressure in atmospheres.

SECTION 14–4 BUOYANCY

14–15 A slab of ice floats on a freshwater lake. What minimum volume must the slab have for a 45.0-kg woman to be able to stand on it without getting her feet wet?

14–16 An ore sample weighs 17.50 N in air. When the sample is suspended by a light cord and totally immersed in water, the tension in the cord is 11.20 N. Find the total volume and the density of the sample.

14–17 An object of average density ρ floats at the surface of a fluid of density ρ_{fluid}. a) How must the two densities be related? b) In view of the answer to part (a), how can steel ships float in water? c) In terms of ρ and ρ_{fluid}, what fraction of the object is submerged and what fraction is above the fluid? Check that your answers give the correct limiting behavior as $\rho \rightarrow \rho_{fluid}$ and as $\rho \rightarrow 0$. d) While on board your yacht, your cousin Throckmorton cuts a rectangular piece (dimensions $5.0 \times 4.0 \times 3.0$ cm) out of a life preserver and throws it into the ocean. The piece has a mass of 42 g. As it floats in the ocean, what percentage of its volume is above the surface?

14–18 A hollow plastic sphere is held below the surface of a freshwater lake by a cord anchored to the bottom of the lake.

The sphere has a volume of 0.650 m^3 and the tension in the cord is 900 N. a) Calculate the buoyant force exerted by the water on the sphere. b) What is the mass of the sphere? c) The cord breaks and the sphere rises to the surface. When the sphere comes to rest, what fraction of its volume will be submerged?

14–19 A cubical block of wood 10.0 cm on a side floats at the interface between oil and water with its lower surface 1.50 cm below the interface (Fig. 14–34). The density of the oil is 790 kg/m^3. a) What is the gauge pressure at the upper face of the block? b) What is the gauge pressure at the lower face of the block? c) What is the mass and density of the block?

FIGURE 14–34 Exercise 14–19.

14–20 A solid aluminum ingot weighs 89 N in air. a) What is its volume? b) The ingot is suspended from a rope and totally immersed in water. What is the tension in the rope (the *apparent* weight of the ingot in water)?

SECTION 14–5 SURFACE TENSION

14–21 Find the gauge pressure in pascals in a soap bubble 3.00 cm in diameter. The surface tension is 25.0×10^{-3} N/m.

14–22 Find the excess pressure at 20°C a) inside a large raindrop with a radius of 1.00 mm; b) inside a water drop with a radius of 0.0100 mm (typical of water droplets in fog).

14–23 How to Stand on Water. Estimate the upward surface-tension force that would be exerted on your feet if you attempted to stand on water. (You will need to measure your feet.) What maximum body mass could the water support this way?

14–24 Why Trees Don't Suck Air. The negative pressures that occur in a column of sap in a tall tree have been measured to be as great as −20 atm. These columns of sap are open at the top to the outside air, and water evaporates from the leaves. But if the pressure is negative, why doesn't air get sucked into the leaves? To answer this question, estimate the pressure difference required to force air into the interstices of the cell walls within the leaves (about 10^{-8} m across) and explain why outside air cannot in fact get into the tree. (Take the surface tension of the sap to be that of water at 20°C. The situation is the reverse of that in Fig. 14–15, with air displacing the sap in the interstices.)

14–25 The soap film in Fig. 14–14 is from the soap solution of Table 14–2. It is 22.0 cm wide and is at 20°C. The slide wire has a mass of 0.700 g. What is the magnitude of the downward pull T required to keep the slide wire in equilibrium?

SECTION 14–6 FLUID FLOW

14–26 Water is flowing in a pipe with varying cross-section area, and at all points the water completely fills the pipe. At

point 1 the cross-section area of the pipe is 0.070 m^2, and the magnitude of the fluid velocity is 3.50 m/s. a) What is the fluid speed at points in the pipe where the cross-section area is i) 0.105 m^2? ii) 0.047 m^2? b) Calculate the volume of water discharged from the open end of the pipe in 1.00 hour.

14–27 Water is flowing in a circular pipe of varying cross-section area, and at all points the water completely fills the pipe. a) At one point in the pipe the radius is 0.150 m. What is the speed of the water at this point if the volume flow rate in the pipe is 1.20 m^3/s? b) At a second point in the pipe the water speed is 3.80 m/s. What is the radius of the pipe at this point?

14–28 a) Derive Eq. (14–16). b) If the density increases by 1.50% from point 1 to point 2, what happens to the volume flow rate?

SECTION 14–7 BERNOULLI'S EQUATION

14–29 A sealed tank containing seawater to a height of 11.0 m also contains air above the water at a gauge pressure of 3.00 atm. Water flows out from the bottom through a small hole. Calculate the efflux speed of the water.

14–30 A small circular hole 6.00 mm in diameter is cut in the side of a large water tank, 14.0 m below the water level in the tank. The top of the tank is open to the air. Find a) the speed of efflux; b) the volume discharged per unit time.

14–31 What gauge pressure is required in the city mains for a stream from a fire hose connected to the mains to reach a vertical height of 15.0 m? (Assume that the mains have a much larger diameter than the fire hose.)

14–32 At one point in a pipeline the water's speed is 3.00 m/s and the gauge pressure is 5.00×10^4 Pa. Find the gauge pressure at a second point in the line, 11.0 m lower than the first, if the pipe diameter at the second point is twice that at the first.

14–33 Lift on an Airplane. Air streams horizontally past a small airplane's wings such that the speed is 70.0 m/s over the top surface and 60.0 m/s past the bottom surface. If the plane has a mass of 1340 kg and a wing area of 16.2 m^2, what is the net vertical force (including the effects of gravity) on the airplane? The density of the air is 1.20 kg/m^3.

14–34 A soft drink (mostly water) flows in a pipe at a beverage plant with a mass flow rate that would fill 220 0.355-L cans per minute. At point 2 in the pipe, the gauge pressure is 152 kPa and the cross-section area is 8.00 cm^2. At point 1, 1.35 m above point 2, the cross-section area is 2.00 cm^2. Find the a) mass flow rate; b) volume flow rate; c) flow speeds at points 1 and 2; d) gauge pressure at point 1.

14–35 Water discharges from a horizontal cylindrical pipe at the rate of 465 cm^3/s. At a point in the pipe where the radius is 2.05 cm the absolute pressure is 1.60×10^5 Pa. What is the pipe radius at a constriction if the pressure there is reduced to 1.20×10^5 Pa?

14–36 At a certain point in a horizontal pipeline the water's speed is 2.50 m/s, and the gauge pressure is 1.80×10^4 Pa. Find the gauge pressure at a second point in the line if the cross-section area at the second point is twice that at the first.

*SECTION 14–9 VISCOSITY

*14–37 Water at 20°C is flowing in a pipe of radius 10.0 cm. The viscosity of water at 20°C is 1.005 centipoise. If the water's speed in the center of the pipe is 2.50 m/s, what is the water's speed a) 5.0 cm from the center of the pipe (halfway between the center and the walls)? b) at the walls of the pipe?

*14–38 Water at 20°C flows through a pipe of radius 8.50 mm. The viscosity of water at 20°C is 1.005 centipoise. If the flow speed at the center is 0.200 m/s and the flow is laminar, find the pressure drop due to viscosity along a 3.00-m section of pipe.

*14–39 Water at 20°C is flowing in a horizontal pipe that is 15.0 m long; the flow is laminar, and the water completely fills the pipe. A pump maintains a gauge pressure of 1200 Pa at a large tank at one end of the pipe. The other end of the pipe is open to the air. The viscosity of water at 20°C is 1.005 centipoise. a) If the pipe has diameter 9.00 cm, what is the volume flow rate? b) What gauge pressure must the pump provide to achieve the same volume flow rate for a pipe with a diameter of 3.00 cm? c) For the pipe in part (a) and the same gauge pressure maintained by the pump, what does the volume flow rate become if the water is at a temperature of 60°C? (The viscosity of water at 60°C is 0.469 centipoise.)

*14–40 The South American insect *Rhodinus prolixus* sucks the blood of mammals. Its mouthparts are like a very fine hypodermic needle (which allows it to draw blood from its victims painlessly, and so without being noticed). The narrowest part of the "needle" is 0.20 mm long and is only 10 μm in diameter. a) What must be the gauge pressure in the insect's mouth cavity if it is to take a large meal of 0.25 cm^3 of blood in 15 minutes? Express your answer in Pa and in atm. (The viscosity of human blood in such a narrow tube is 1.0 centipoise. To get an approximate answer, apply Poiseuille's equation to the blood even though it is a non-Newtonian fluid.) b) Why is it a good approximation to ignore the dimensions of the rest of the insect's mouthparts?

*14–41 What speed must an aluminum sphere with radius 2.00 mm have in castor oil at 20°C for the viscous drag force to be one-fourth the weight of the sphere? (The viscosity of castor oil at this temperature is 9.86 poise.)

*14–42 Measuring Viscosity. A brass sphere with a mass of 0.35 g falls with a terminal speed of 5.0 cm/s in an unknown liquid. If the density of the liquid is 2900 kg/m^3, what is the viscosity of the liquid?

*14–43 Keeping all other variables constant, what happens to the volume flow rate in laminar flow if you double a) the pipe diameter? b) the viscosity? c) the pressure difference? d) the pressure gradient? e) the pipe length?

*14–44 For normal shots in basketball (not the desperation half-court-line ones) the force of air resistance is negligible. To show this, consider the ratio of the Stokes's law force to the weight of a 0.600-kg basketball. The basketball has radius 0.124 m and is moving through air of density 1.20 kg/m^3 at 5.00 m/s.

*14–45 A very narrow, high-intensity laser beam drills a cylindrical hole through the hull of a Federation spaceship; the hole is 0.180 m long and only 50.0 μm in radius. Air at 20°C with a viscosity of 181 micropoise begins to rush out in laminar flow from the 1.00-atm interior to the vacuum outside. a) What is the air speed along the axis of the cylinder, at the edge, and halfway between? b) How many days will it take for a cubic meter of air to flow through the hole? (Assume that the interior pressure remains 1.00 atm.) c) By what factor are the answers to parts (a) and (b) changed if the radius of the hole is twice as much and the flow remains laminar?

PROBLEMS

14–46 In a lecture demonstration, a professor pulls apart two hemispherical steel shells (diameter D) with ease using their attached handles. She then places them together, pumps out the air to an absolute pressure of p, and hands them to a bodybuilder in the back row to pull apart. a) If atmospheric pressure is p_0, how much force must the bodybuilder exert on each shell? b) Evaluate your answer for the case $p = 0.025$ atm, $D = 10.0$ cm.

14–47 The deepest point known in any of the earth's oceans is in the Marianas Trench, 10.92 km deep. a) Assuming water to be incompressible, what is the pressure at this depth? Use the density of seawater. b) The actual pressure is 1.16×10^8 Pa; your calculated value will be less because the density actually varies with depth. Using the compressibility of water and the actual pressure, find the density of the water at the bottom of the Marianas Trench. What is the percent change in the density of the water?

14–48 A swimming pool measures 5.0 m long × 4.0 m wide × 3.0 m deep. Compute the force exerted by the water against a) the bottom; b) either end. (*Hint:* Calculate the force on a thin horizontal strip at a depth h, and integrate this over the end of the pool.) Do not include the force due to air pressure.

14–49 The upper edge of a gate in a dam runs along the water surface. The gate is 2.00 m high and 4.00 m wide and is hinged along a horizontal line through its center (Fig. 14–35). Calculate the torque about the hinge arising from the force due to the water. (*Hint:* Use a procedure similar to that used in Problem 14–48; calculate the torque of a thin horizontal strip at a depth h and integrate this over the gate.)

FIGURE 14–35 Problem 14–49.

14–50 Force and Torque on a Dam. A dam has the shape of a rectangular solid. The side facing the lake has area A and height H. The surface of the freshwater lake behind the dam is at the top of the dam. a) Show that the net horizontal force exerted by the water on the dam equals $\frac{1}{2}\rho gHA$, that is, the average gauge pressure across the face of the dam times the area. (See Problem 14–48.) b) Show that the torque exerted by the water about an axis along the bottom of the dam is $\rho gH^2A/6$. c) How do the force and torque depend on the size of the lake?

14–51 An astronaut is standing at the north pole of a newly discovered spherically symmetric planet of radius R. In his hands he holds a container full of a liquid of mass m and volume V. At the surface of the liquid, the pressure is p_0; at a depth d below the surface, the pressure has a greater value p. From this information, determine the mass of the planet.

14–52 To find the density at a given point within a material, consider a small volume dV centered on that point. If the mass within that volume is dm, the density at the point is $\rho = dm/dV$. Consider a cylindrical rod of mass M, radius R, and length L with a density that is proportional to the square of the distance from one end, $\rho = Cx^2$. a) Show that $C = 3M/\pi R^2L^3$. b) Show that the average density, given by Eq. (14–1), is one-third of the density at the $x = L$ end.

14–53 The earth does not have a uniform density; it is most dense at its center and least dense at its surface. An approximation to its density is $\rho(r) = A - Br$, where $A = 12{,}700$ kg/m^3 and $B = 1.50 \times 10^{-3}$ kg/m^4. Use $R = 6.37 \times 10^6$ m for the radius of the earth approximated as a sphere. a) Geological evidence indicates that the densities are 13,100 kg/m^3 and 2,400 kg/m^3 at the earth's center and surface. What values does the linear approximation model give for the densities at these two locations? b) Imagine dividing the earth into concentric spherical shells. Each shell has radius r, thickness dr, volume $dV = 4\pi r^2\,dr$, and mass $dm = \rho(r)\,dV$. By integrating from $r = 0$ to $r = R$, show that the mass of the earth in this model is $M = \frac{4}{3}\pi R^3(A - \frac{3}{4}BR)$. c) Show that the given values of A and B give the correct mass of the earth to within 0.4%. d) We saw in Section 12–7 that a uniform spherical shell gives no contribution to g inside it. Show that $g(r) = \frac{4}{3}\pi Gr(A - \frac{3}{4}Br)$ inside the earth in this model. e) Verify that the expression of part (d) gives $g = 0$ at the center of the earth and $g = 9.85$ m/s^2 at the surface. f) Show that in this model g does *not* decrease uniformly with depth but rather has a maximum of $4\pi GA^2/9B = 10.01$ m/s^2 at $r = 2A/3B = 5640$ km.

14–54 In Example 12–9 (Section 12–7) we saw that inside a planet of uniform density (not a very realistic assumption for the earth) the acceleration due to gravity increases uniformly with distance from the center of the planet. That is, $g(r) = g_sr/R$, where g_s is the acceleration due to gravity at the surface, r is the distance from the center of the planet, and R is the radius of the planet. The interior of the planet can be treated approximately as an incompressible fluid of density ρ. a) Replace the height y in Eq. (14–4) with the radial coordinate r and integrate to find the pressure inside a uniform planet as a function of r. Let the pressure at the surface be zero. (This means ignoring the pressure of the planet's atmosphere.) b) Using this model, calculate the pres-

sure at the center of the earth. (Use a value of ρ equal to the average density of the earth, calculated from the mass and radius given in Appendix F.) c) Geologists estimate the pressure at the center of the earth to be approximately 4×10^{11} Pa. Does this agree with your calculation for the pressure at $r = 0$? What might account for any differences?

14–55 A U-shaped tube open to the air at both ends contains some mercury. A quantity of water is carefully poured into the left arm of the U-shaped tube until the vertical height of the water column is 15.0 cm (Fig. 14–36). a) What is the gauge pressure at the water-mercury interface? b) Calculate the vertical distance h from the top of the mercury in the right-hand arm of the tube to the top of the water in the left-hand arm.

FIGURE 14–36 Problem 14–55.

14–56 The Great Molasses Flood. On the afternoon of January 15, 1919, an unusually warm day in Boston, a 27.4-m-high (90 ft), 27.4-m-diameter cylindrical metal tank used for storing molasses ruptured. Molasses flooded into the streets in a 9-m-deep stream, killing pedestrians and horses and knocking down buildings. The molasses had a density of 1600 kg/m^3. If the tank was full before the accident, what was the total outward force the molasses exerted on its sides? (*Hint:* Consider the outward force on a circular ring of the tank wall of width dy and at a depth y below the surface. Integrate to find the total outward force. Assume that before the tank ruptured, the pressure at the surface of the molasses was equal to the air pressure outside the tank.)

14–57 An open barge has the dimensions shown in Fig. 14–37. If the barge is made out of 4.0-cm-thick steel plate on each of its four sides and its bottom, what mass of coal can the barge carry in fresh water without sinking? Is there enough room in the barge to hold this amount of coal? (The density of coal is about 1500 kg/m^3.)

FIGURE 14–37 Problem 14–57.

14–58 A hot-air balloon has a volume of 2200 m^3. The balloon fabric (the envelope) weighs 900 N. The basket with gear and full propane tanks weighs 1700 N. If the balloon can barely lift an additional 3200 N of passengers, breakfast, and champagne when the outside air density is 1.23 kg/m^3, what is the average density of the heated gases in the envelope?

14–59 Advertisements for a certain small car claim it floats in water. a) If the car's mass is 900 kg and its interior volume 3.0 m^3, what fraction of the car is immersed when it floats? The volume of steel and other materials may be neglected. b) Water gradually leaks in and displaces the air in the car. What fraction of the interior volume is filled with water when the car sinks?

14–60 A single ice cube of mass 9.70 g floats in a glass completely full of 420 cm^3 of water. Ignore the water's surface tension and its variation in density with temperature (as long as it remains a liquid). a) What volume of water does the ice cube displace? b) When the ice cube has completely melted, has any water overflowed? If so, how much? If not, explain why this is so. c) Suppose the water in the glass had been very salty water of density 1050 kg/m^3. What volume of salt water would the 9.70-g ice cube displace? d) Redo part (b) for the freshwater ice cube in the salty water.

14–61 A piece of wood is 0.600 m long, 0.250 m wide, and 0.080 m thick. Its density is 600 kg/m^3. What volume of lead must be fastened underneath to sink the wood in calm water so that its top is just even with the water level? What is the mass of this volume of lead?

14–62 A hydrometer consists of a spherical bulb and a cylindrical stem with a cross-section area of 0.400 cm^2 (Fig. 14–9a). The total volume of bulb and stem is 13.2 cm^3. When immersed in water, the hydrometer floats with 8.00 cm of the stem above the water surface. When immersed in an organic fluid, 3.20 cm of the stem is above the surface. Find the density of the organic fluid. (*Note:* This illustrates the precision of such a hydrometer. Relatively small density differences give rise to relatively large differences in hydrometer readings.)

14–63 The densities of air, helium, and hydrogen (at $p = 1.0$ atm and $T = 20°$C) are 1.20 kg/m^3, 0.166 kg/m^3, and 0.0899 kg/m^3, respectively. a) What is the volume in cubic meters displaced by a hydrogen-filled airship that has a total "lift" of 120 kN? (The "lift" is the amount by which the buoyant force exceeds the weight of the gas that fills the airship.) b) What would be the "lift" if helium were used instead of hydrogen? In view of your answer, why is helium used in modern airships like advertising blimps?

14–64 SHM of a Floating Object. An object with height h, mass M, and uniform cross-section area A floats upright in a liquid with density ρ. a) Calculate the vertical distance from the surface of the liquid to the bottom of the floating object at equilibrium. b) A downward force with magnitude F is applied to the top of the object. At the new equilibrium position, how much farther below the surface of the liquid is the bottom of the object than it was in part (a)? (Assume that some of the object remains above the surface of the liquid.) c) Your result in part (b) shows that if the force is suddenly removed, the object will oscillate up and down in simple harmonic motion. Calculate the period of this motion in terms of the density ρ of the liquid and the mass

M and cross-section area A of the object. Neglect the damping due to fluid friction (Section 13–8).

14–65 A 950-kg cylindrical can buoy floats vertically in salt water. The diameter of the buoy is 0.900 m. a) Calculate the additional distance the buoy will sink when a 70.0-kg man stands on top. (Use the expression derived in part (b) of Problem 14–64.) b) Calculate the period of the resulting vertical simple harmonic motion when the man dives off. (Use the expression derived in part (c) of Problem 14–64, and as in that problem neglect the damping due to fluid friction.)

14–66 In seawater a life preserver with a volume of 0.0400 m^3 will support a 75.0-kg person (average density 980 kg/m^3) with 20% of the person's volume above water when the life preserver is fully submerged. What is the density of the material composing the life preserver?

14–67 A block of balsa wood placed in one scale pan of an equal-arm balance is exactly balanced by a 0.0950-kg brass mass in the other scale pan. Find the true mass of the balsa wood if its density is 150 kg/m^3. Explain why it is accurate to neglect the buoyancy in air of the brass but *not* the buoyancy in air of the balsa wood.

14–68 Block A in Fig. 14–38 hangs by a cord from spring balance D and is submerged in a liquid C contained in beaker B. The mass of the beaker is 1.00 kg; the mass of the liquid is 1.80 kg. Balance D reads 3.50 kg and balance E reads 7.50 kg. The volume of block A is 3.80×10^{-3} m^3. a) What is the density of the liquid? b) What will each balance read if block A is pulled up out of the liquid?

FIGURE 14–38 Problem 14–68.

14–69 A hunk of aluminum is completely covered with a gold shell to form an ingot of weight 45.0 N. When you suspend the ingot from a spring balance and submerge the ingot in water, the balance reads 39.0 N. What is the weight of the gold in the shell?

14–70 Untethered helium balloons floating in a car that has all the windows rolled up and outside air vents closed move in the direction of the car's acceleration, but loose balloons filled with air move opposite. To show why, consider only the horizontal forces acting on the balloons. Let a be the magnitude of the car's forward acceleration. Consider a horizontal tube of air of cross-section area A that extends from the windshield, where $x = 0$ and $p = p_0$, back along the x-axis. Now consider a volume element of thickness dx in this tube. The pressure on its front surface is p and the pressure on its rear surface is $p + dp$. Assume the air has a constant density ρ. a) Apply Newton's second law to the volume element to show $dp = \rho a\, dx$. b) Integrate the result of part (a) to find the pressure at the front surface in terms of a and x. c) To show that considering ρ constant is reasonable, calculate the pressure difference in atm for a distance as long as 2.5 m and a large acceleration of 5.0 m/s². d) Show that the net horizontal force on a balloon of volume V is $\rho V a$. e) For negligible friction forces, show that the acceleration of the balloon (average density ρ_{bal}) is $(\rho/\rho_{bal})a$, so that the acceleration relative to the car is $a_{rel} = [(\rho/\rho_{bal}) - 1]a$. f) Use the expression for a_{rel} in part (e) to explain the movement of the balloons.

14–71 The weight of a king's solid crown is w. When suspended by a light rope and completely immersed in water, the tension in the rope (the crown's apparent weight) is fw. a) Prove that the crown's relative density (specific gravity) is $1/(1 - f)$. Discuss the meaning of the limits as f approaches 0 and 1. b) If the crown is solid gold and weighs 12.9 N in air, what is its apparent weight when completely immersed in water? c) Repeat part (b) if the crown is solid lead with a very thin gold plating, but still has a weight in air of 12.9 N.

14–72 A piece of steel has a weight w, an apparent weight (see Problem 14–71) w_{water} when completely immersed in water, and an apparent weight w_{fluid} when completely immersed in an unknown fluid. a) Prove that the fluid's density relative to water (specific gravity) is $(w - w_{fluid})/(w - w_{water})$. b) Is this result reasonable for the three cases of w_{fluid} greater than, equal to, or less than w_{water}? c) The apparent weight of the piece of steel in water of density 1000 kg/m³ is 87.2% of its weight. What percentage of its weight will its apparent weight be in formic acid (density 1220 kg/m³)?

14–73 You cast some metal of density ρ_m in a mold, but you are worried that there might be cavities within the casting. You measure the weight of the casting to be w, and the buoyant force when it is completely surrounded by water to be B. a) Show that $V_0 = B/(\rho_{water}g) - w/(\rho_m g)$ is the total volume of any enclosed cavities. b) If your metal is copper, the casting's weight is 156 N, and the buoyant force is 20 N, what is the total volume of any enclosed cavities in your casting? What fraction is this of the total volume of the casting?

14–74 A cubical block of wood 0.100 m on a side and with a density of 550 kg/m³ floats in a jar of water. Oil with a density of 750 kg/m³ is poured on the water until the top of the oil layer is 0.035 m below the top of the block. a) How deep is the oil layer? b) What is the gauge pressure at the block's lower face?

14–75 Dropping Anchor. An iron anchor with mass 35.0 kg and density 7860 kg/m³ lies on the deck of a small barge that has vertical sides and that is floating in a freshwater river. The area of the bottom of the barge is 8.00 m². The anchor is thrown overboard but is suspended above the bottom of the river by a rope; the mass and volume of the rope are small enough to ignore. After the anchor is overboard and the barge has finally stopped bobbing up and down, has the barge risen or sunk down in the water? By what vertical distance?

14–76 Assume that crude oil from a supertanker has density 750 kg/m³. The tanker runs aground on a sandbar. To refloat the tanker, its oil cargo is pumped out into steel barrels, each of which has a mass of 15.0 kg when empty and holds 0.120 m³ of oil. Neglect the volume occupied by the steel from which the barrel is made. a) If a salvage worker accidentally drops a filled, sealed barrel overboard, will it float or sink in the seawater? b) If the barrel floats, what fraction of its volume will be above the water surface? If it sinks, what minimum tension would have to be exerted by a rope to haul the barrel up from the ocean floor? c) Repeat parts (a) and (b) if the density of the oil is 910 kg/m³ and the mass of each empty barrel is 32.0 kg.

14–77 A cubical block of density ρ_B and with sides of length L floats in a liquid of greater density ρ_L. a) What fraction of the block's volume is above the surface of the liquid? b) The liquid is denser than water (density ρ_w) and does not mix with it. If water is poured on the surface of the liquid, how deep must the water layer be so that the water surface just rises to the top of the block? Express your answer in terms of L, ρ_B, ρ_L, and ρ_w. c) Find the depth of the water layer in part (b) if the liquid is mercury, the block is made of iron, and the side length is 10.0 cm.

14–78 A barge is in a rectangular lock on a freshwater river. The lock is 60.0 m long and 20.0 m wide, and the steel doors on each end are closed. With the barge floating in the lock, a 2.50×10^6 N load of scrap metal is put onto the barge. The metal has density 9000 kg/m³. a) When the load of scrap metal, initially on the bank, is placed onto the barge, what vertical distance does the water in the lock rise? b) The scrap metal is now pushed overboard into the water. Does the water level in the lock rise, fall, or remain the same? If it rises or falls, by what vertical distance does it change?

14–79 A U-shaped tube with a horizontal portion of length l (Fig. 14–39) contains a liquid. What is the difference in height between the liquid columns in the vertical arms a) if the tube has an acceleration a toward the right? b) if the tube is mounted on a horizontal turntable rotating with an angular speed ω with one of the vertical arms on the axis of rotation? c) Explain why the difference in height does not depend on the density of the liquid or on the cross-section area of the tube. Would it be the same if the vertical tubes did not have equal cross-section areas? Would it be the same if the horizontal portion were tapered from one end to the other? Explain.

FIGURE 14–39 Problem 14–79.

14–80 A cylindrical container of an incompressible liquid with density ρ rotates with constant angular speed ω about its axis of symmetry, which we take to be the y-axis (Fig. 14–40). a) Show that the pressure at a given height within the fluid increases in the radial direction (outward from the axis of rotation) according to $\partial p/\partial r = \rho \omega^2 r$. b) Integrate this partial differential equation to find the pressure as a function of distance from the axis of rotation along a horizontal line at $y = 0$. c) Combine the result of part (b) with Eq. (14–5) to show that the surface of the rotating liquid has a *parabolic* shape, that is, the height of the liquid is given by $h(r) = \omega^2 r^2/2g$. (This technique is used for making parabolic telescope mirrors; liquid glass is rotated and allowed to solidify while rotating.)

FIGURE 14–40 Problem 14–80.

14–81 An incompressible fluid with density ρ is in a horizontal test tube of inner cross-section area A. The test tube spins in a horizontal circle in an ultracentrifuge at an angular speed ω. Gravitational forces are negligible. Consider a volume element of the fluid of area A and thickness dr' a distance r' from the rotation axis. The pressure on its inner surface is p and on its outer surface is $p + dp$. a) Apply Newton's second law to the volume element to show $dp = \rho \omega^2 r' \, dr'$. b) If the surface of the fluid is at a radius r_0 where the pressure is p_0, show that the pressure p at a distance $r \geq r_0$ is $p = p_0 + \rho \omega^2 (r^2 - r_0^2)/2$. c) An object of volume V and density ρ_{ob} has its center of mass at a distance R_{cmob} from the axis. Show that the net horizontal force on the object is $\rho V \omega^2 R_{cm}$, where R_{cm} is the distance from the axis to the center of mass of the displaced fluid. d) Explain why the object will move inward if $\rho R_{cm} > \rho_{ob} R_{cmob}$ and outward if $\rho R_{cm} < \rho_{ob} R_{cmob}$. e) For small objects of uniform density, $R_{cm} = R_{cmob}$. What happens to a mixture of small objects of this kind with different densities in an ultracentrifuge?

14–82 What radius must a water drop have for the difference between inside and outside pressures to be 0.0250 atm? Assume $T = 20°C$.

14–83 A cubical block of wood 0.30 m on a side is weighted so that its center of gravity is at the point shown in Fig. 14–41a and it floats in water with one-half its volume submerged. If the block is "heeled" at an angle of 45.0° as in Fig. 14–41b, compute the net torque about a horizontal axis perpendicular to the block and passing through its geometrical center.

FIGURE 14–41 Problem 14–83.

14–84 Water stands at a depth H in a large open tank whose side walls are vertical (Fig. 14–42). A hole is made in one of the walls at a depth h below the water surface. a) At what distance R from the foot of the wall does the emerging stream strike the floor? b) How far above the bottom of the tank could a second hole be cut so that the stream emerging from it could have the same range as for the first hole?

FIGURE 14–42 Problem 14–84.

14–85 A cylindrical bucket, open at the top, is 25.0 cm high and 10.0 cm in diameter. A circular hole with cross-section area 1.50 cm² is cut in the center of the bottom of the bucket. Water flows into the bucket from a tube above it at the rate of 2.40×10^{-4} m³/s. How high will the water in the bucket rise?

14–86 Water flows steadily from an open tank as in Fig. 14–43. The elevation of point 1 is 10.0 m, and the elevation of points 2 and 3 is 2.00 m. The cross-section area at point 2 is 0.0480 m²; at point 3 it is 0.0160 m². The area of the tank is very large compared with the cross-section area of the pipe. Assuming that Bernoulli's equation applies, compute a) the discharge rate in cubic meters per second; b) the gauge pressure at point 2.

FIGURE 14–43 Problem 14–86.

14–87 Modern airplane design calls for a lift due to the net force of the moving air on the wing of about 2000 N per square meter of wing area. Assume that air flows past the wing of an aircraft with streamline flow. If the speed of flow past the lower wing surface is 120 m/s, what is the required speed over the upper surface to give a lift of 2000 N/m²? The density of the air is 1.20 kg/m³.

14–88 The radius of 1993's Hurricane Emily was about 350 km. The wind speed near the center ("eye") of the hurricane, whose radius was about 30 km, reached about 200 km/h. As air swirls in from the rim of the hurricane toward the eye, its angular momentum remains roughly constant. a) Estimate the wind speed at the rim of the hurricane. b) Estimate the pressure difference at the earth's surface between the eye and the rim. (*Hint:* See Table 14–1.). Where is the pressure greater? c) If the kinetic energy of the swirling air in the eye could be converted completely to gravitational potential energy, how high would the air go? d) In fact the air in the eye is lifted to heights of several kilometers. How can you reconcile this with your answer to part (c)?

14–89 Two very large open tanks A and F (Fig. 14–44) contain the same liquid. A horizontal pipe BCD, having a constriction at C and open to the air at D, leads out of the bottom of tank A, and a vertical pipe E opens into the constriction at C and dips into the liquid in tank F. Assume streamline flow and no viscosity. If the cross-section area at C is one half that at D and if D is a distance h_1 below the level of the liquid in A, to what height h_2 will liquid rise in pipe E? Express your answer in terms of h_1.

FIGURE 14–44 Problem 14–89.

14–90 The horizontal pipe shown in Fig. 14–45 has a cross-section area of 40.0 cm^2 at the wider portions and 10.0 cm^2 at the constriction. Water is flowing in the pipe, and the discharge from the pipe is 6.00×10^{-3} m^3/s (6.00 L/s). Find a) the flow speeds at the wide and the narrow portions; b) the pressure difference between these portions; c) the difference in height between the mercury columns in the U-shaped tube.

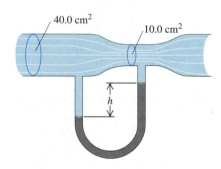

FIGURE 14–45 Problem 14–90.

14–91 Figure 14–27a shows a liquid flowing from a vertical pipe. Notice that the vertical stream of the liquid has a very definite shape as it flows from the pipe. To get the equation for this shape, assume that the liquid is in free fall once it leaves the pipe. Just as it leaves the pipe, the liquid has speed v_0 and the radius of the stream of liquid is r_0. a) Find an equation for the speed of the liquid as a function of the distance y it has fallen. Combining this with the equation of continuity, find an expression for the radius of the stream as a function of y. b) If water flows out of a vertical pipe at a speed of 1.20 m/s, how far below the outlet will the radius be one-half the original radius of the stream?

*14–92 a) With what speed is a solid brass sphere 2.50 mm in radius falling in a tank of glycerin at an instant when its acceleration is one half that of a freely falling body? The viscosity of the glycerin is 8.30 poise. b) What is the terminal speed of the sphere?

*14–93 **Speed of a Bubble in Liquid.** a) With what terminal speed does an air bubble 2.00 mm in diameter rise in a liquid with viscosity 1.50 poise and density 900 kg/m^3? (Assume that the density of the air is 1.20 kg/m^3 and that the bubble diameter remains constant.) b) What is the terminal speed of the same bubble in water at 20°C, which has viscosity 1.005 centipoise?

*14–94 Oil of viscosity 3.00 poise and density 860 kg/m^3 is to be pumped from one large open tank to another through a smooth, horizontal steel pipe 1.50 km in length and 0.110 m in diameter. The pipe discharges into the air. a) What gauge pressure, in pascals and in atmospheres, must the pump exert in order to maintain a flow of 0.0600 m^3/s? b) Explain why the power consumed by the pump is equal to the volume flow rate multiplied by the gauge pressure that the pump maintains. What is the numerical value of the power?

*14–95 The tank at the left in Fig. 14–46a has a very large cross-section area and is open to the atmosphere. The depth $y = 0.600$ m. The cross-section areas of the horizontal tubes leading out of the tank are 1.00 cm^2, 0.40 cm^2, and 0.20 cm^2, respectively. The liquid is ideal, having zero viscosity. a) What is the volume flow rate out of the tank? b) What is the speed in each portion of the horizontal tube? c) What is the height of the liquid in each of the five vertical side tubes? d) Suppose that the liquid in Fig. 14–46b has a viscosity of 0.0600 poise and a density of 800 kg/m^3 and that the depth of liquid in the large tank is such that the volume rate of flow is the same as that in part (a). The distance between the side tubes at c and d, and between those at e and f, is 0.200 m. The cross-section areas of the horizontal tubes are the same in both diagrams. What is the difference in level between the tops of the liquid columns in tubes c and d? e) In tubes e and f? f) What is the flow speed on the axis of each part of the horizontal tube?

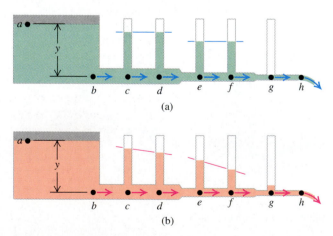

FIGURE 14–46 Problem 14–95.

CHALLENGE PROBLEMS

14–96 A rock with mass $m = 3.00$ kg is suspended from the roof of an elevator by a light cord. The rock is totally immersed in a bucket of water that sits on the floor of the elevator, but the rock doesn't touch the bottom or sides of the bucket. a) When the elevator is at rest, the tension in the cord is 21.0 N. Calculate the volume of the rock. b) Derive an expression for the tension in the cord when the elevator is accelerating *upward* with an acceleration of magnitude a. Calculate the tension when $a = 2.50$ m/s^2 upward. c) Derive an expression for the tension in the cord when the elevator is accelerating *downward* with an acceleration of magnitude a. Calculate the tension when $a = 2.50$ m/s^2 downward. d) What is the tension when the elevator is in free fall with a downward acceleration equal to g?

14–97 Suppose a piece of styrofoam, $\rho = 180$ kg/m^3, is held completely submerged in water (Fig. 14–47). a) What is the tension in the cord? Find this using Archimedes's principle. b) Use $p = p_0 + \rho gh$ to directly calculate the force exerted by the water on the two sloped sides and the bottom of the styrofoam; then show the vector sum of these forces is the buoyant force.

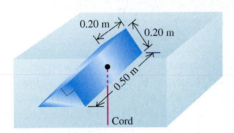

FIGURE 14–47 Challenge Problem 14–97.

14–98 A large tank with diameter D, open to the air, contains water to a height H. A small hole with diameter d ($d \ll D$) is made at the base of the tank. Neglecting any effects of viscosity, calculate the time it takes for the tank to drain completely.

14–99 A *siphon*, as depicted in Fig. 14–48, is a convenient device for removing liquids from containers. To establish the flow, the tube must be initially filled with fluid. Let the fluid have density ρ and let the atmospheric pressure be p_a. Assume that the cross-section area of the tube is the same at all points along it. a) If the lower end of the siphon is at a distance h below the surface of the liquid in the container, what is the speed of the fluid as it flows out the lower end of the siphon? (Assume that the container has a very large diameter, and neglect any effects of viscosity.) b) A curious feature of a siphon is that the fluid initially flows "uphill." What is the greatest height H that the high point of the tube can have if flow is still to occur?

FIGURE 14–48 Challenge Problem 14–99.

14–100 The following is quoted from a letter. *It is the practice of carpenters hereabouts, when laying out and leveling up the foundations of relatively long buildings, to use a garden hose filled with water, into the ends of the hose being thrust glass tubes 10 to 12 inches long. The theory is that water, seeking a common level, will be of the same height in both the tubes and thus effect a level. Now the question rises as to what happens if a bubble of air is left in the hose. Our greybeards contend the air will not affect the reading from one end to the other. Others say that it will cause important inaccuracies. Can you give a relatively simple answer to this question, together with an explanation?* Fig. 14–49 gives a rough sketch of the situation that caused the dispute.

FIGURE 14–49 Challenge Problem 14–100.

15

Temperature
and Heat

Perhaps the highest-temperature material you will ever see is the sun's outer atmosphere, or corona. At a temperature of about 2,000,000°C (3,600,000°F), the corona glows with a light that is literally unearthly. But because the corona is also very thin, its light is rather faint. You can only see the corona during a total solar eclipse when the sun's disk is covered by the moon, as shown here.

15–1 INTRODUCTION

Whether it's a sweltering summer day or a frozen midwinter night, your body needs to be kept at a nearly constant temperature. It has effective temperature-control mechanisms, but sometimes it needs help. On a hot day you wear less clothing to improve heat transfer from your body to the air and for better cooling by evaporation of perspiration. You probably drink cold beverages, possibly with ice in them, and sit near a fan or in an air-conditioned room. On a cold day you wear more clothes or stay indoors where it's warm. When you're outside, you keep active and drink hot liquids to stay warm. The concepts in this chapter will help you understand the basic physics of keeping warm or cool.

First, we need to define *temperature,* including temperature scales and ways to measure temperature. Next we discuss how the dimensions and volume of a body are affected by temperature changes. We'll encounter the concept of *heat,* which describes energy transfer caused by temperature differences, and we'll learn how to calculate the *rates* of such energy transfers.

Our emphasis in this chapter is on the concepts of temperature and heat as they relate to *macroscopic* objects such as cylinders of gas, ice cubes, and the human body. In Chapter 16 we'll look at these same concepts from a *microscopic* viewpoint in terms of the behavior of individual atoms and molecules. These two chapters lay the groundwork for the subject of **thermodynamics,** the study of energy transformations involving heat, mechanical work,

and other aspects of energy and how these transformations relate to the properties of matter. Thermodynamics forms an indispensable part of the foundation of physics, chemistry, and the life sciences, and its applications turn up in such places as car engines, refrigerators, biochemical processes, and the structure of stars. We'll explore the key ideas of thermodynamics in Chapters 17 and 18.

15–2 TEMPERATURE AND THERMAL EQUILIBRIUM

The concept of **temperature** is rooted in qualitative ideas of "hot" and "cold" based on our sense of touch. A body that feels hot usually has a higher temperature than a similar body that feels cold. That's pretty vague, and the senses can be deceived. But many properties of matter that we can *measure* depend on temperature. The length of a metal rod, steam pressure in a boiler, the ability of a wire to conduct an electric current, and the color of a very hot glowing object—all these depend on temperature.

Temperature is also related to the kinetic energies of the molecules of a material. In general this relationship is fairly complex, so it's not a good place to start in *defining* temperature. In Chapter 16 we will look at the relationship between temperature and the energy of molecular motion for an ideal gas. It is important to understand, however, that temperature and heat are inherently *macroscopic* concepts. They can and must be defined independently of any detailed molecular picture. In this section we'll develop a macroscopic definition of temperature.

To use temperature as a measure of hotness or coldness, we need to construct a temperature scale. To do this, we can use any measurable property of a system that varies with its "hotness" or "coldness." Figure 15–1a shows a familiar system that is used to measure temperature. When the system becomes hotter, the colored liquid (usually mercury or ethanol) expands and rises in the tube, and the value of L increases. Another simple system is a quantity of gas in a constant-volume container (Fig. 15–1b). The pressure p, measured by the gauge, increases or decreases as the gas becomes hotter or colder. A third example is the electrical resistance R of a conducting wire, which also varies when the wire becomes hotter or colder. Each of these properties gives us a number (L, p, or R) that varies with hotness and coldness, so each property can be used to make a **thermometer.**

To measure the temperature of a body, you place the thermometer in contact with the body. If you want to know the temperature of a cup of hot coffee, you stick the thermometer in the coffee; as the two interact, the thermometer becomes hotter and the coffee cools off a little. After the thermometer settles down to a steady value, you read the temperature. The system has reached an *equilibrium* condition, in which the interaction between the thermometer and the coffee causes no further change in the system. We call this a state of **thermal equilibrium.**

If two systems are separated by an insulating material or **insulator** such as wood, plastic foam, or fiberglass, they influence each other more slowly. Camping coolers are made with insulating materials to delay the ice and cold food inside from warming up and attaining thermal equilibrium with the hot summer air outside. An *ideal insulator* is a material that permits no interaction at all between the two systems. It prevents the systems from attaining thermal equilibrium if they aren't in thermal equilibrium at the start. An ideal insulator is just that, an idealization; real insulators, like those in camping coolers, aren't ideal, so the contents of the cooler will warm up eventually.

We can discover an important property of thermal equilibrium by considering three systems, A, B, and C, that initially are not in thermal equilibrium (Fig. 15–2). We surround them with an ideal insulating box so that they cannot interact with anything except each other. We separate systems A and B with an ideal insulating wall (the blue slab in Fig. 15–2a), but we let system C interact with both systems A and B. This interaction is shown in the figure by an orange slab representing a thermal **conductor,** a material that *permits* thermal interactions through it. We wait until thermal equilibrium is attained; then A and B are each in thermal equilibrium with C. But are they in thermal equilibrium *with each other?*

To find out, we separate system C from systems A and B with an ideal insulating wall (Fig. 15–2b), and then we replace the insulating wall between A and B with a *conducting* wall that lets A and B interact. What happens? Experiment shows that *nothing*

15–1 (a) A system whose temperature is specified by the value of length L. (b) A system whose temperature is given by the value of the pressure p.

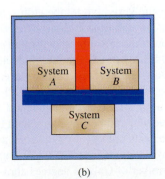

(a)   (b)

15–2 The zeroth law of thermodynamics. (a) If systems A and B are each in thermal equilibrium with system C, then (b) A and B are in thermal equilibrium with each other. Blue slabs represent insulating walls; orange slabs represent conducting walls.

happens; there are no additional changes to A or B. We conclude that **if C is initially in thermal equilibrium with both A and B, then A and B are also in thermal equilibrium with each other.** This result is called the **zeroth law of thermodynamics.** (The importance of this law was recognized only after the first, second, and third laws of thermodynamics had been named. Since it is fundamental to all of them, the name "zeroth" seemed appropriate.) It may seem trivial and obvious, but even so it needs to be verified by experiment.

Now suppose system C is a thermometer, such as the tube-and-liquid system of Fig. 15–1a. In Fig. 15–2a the thermometer C is in contact with both A and B. In thermal equilibrium, when the thermometer reading reaches a stable value, the thermometer measures the temperature of both A and B; hence A and B both have the *same* temperature. Experiment shows that thermal equilibrium isn't affected by adding or removing insulators, so the reading of thermometer C wouldn't change if it were in contact only with A or only with B. We conclude that **two systems are in thermal equilibrium if and only if they have the same temperature.** This is what makes a thermometer useful; a thermometer actually measures *its own* temperature, but when a thermometer is in thermal equilibrium with another body, the temperatures must be equal. When the temperatures of two systems are different, they *cannot* be in thermal equilibrium.

15–3 THERMOMETERS AND TEMPERATURE SCALES

To make the liquid-in-tube device shown in Fig. 15–1a into a useful thermometer, we need to mark a scale on the tube wall with numbers on it. These numbers are arbitrary, and historically many different schemes have been used. Suppose we label the thermometer's liquid level at the freezing temperature of pure water "zero" and the level at the boiling temperature "100," and divide the distance between these two points into 100 equal intervals called *degrees*. The result is the **Celsius temperature scale** (formerly called the *centigrade* scale in English-speaking countries). The Celsius temperature for a state colder than freezing water is a negative number. The Celsius scale is used, both in everyday life and in science and industry, almost everywhere in the world.

Another common type of thermometer uses a *bimetallic strip*, made by bonding strips of two different metals together (Fig. 15–3a). When the system gets hotter, one metal expands more than the other, so the composite strip bends when the temperature changes. This strip is usually formed into a spiral, with the outer end anchored to the thermometer case and the inner end attached to a pointer (Fig. 15–3b). The pointer rotates in response to temperature changes.

In a *resistance thermometer* the changing electrical resistance of a coil of fine wire, a carbon cylinder, or a germanium crystal is measured. Because resistance can be measured very precisely, resistance thermometers are usually more precise than most other types.

To measure very high temperatures, an *optical pyrometer* can be used. It measures the intensity of radiation emitted by a red-hot or white-hot substance. The instrument does not touch the hot substance, so the optical pyrometer can be used at temperatures that would destroy most other thermometers.

In the **Fahrenheit temperature scale,** still used in everyday life in the United States, the freezing temperature of water is 32°F (thirty-two degrees Fahrenheit) and the boiling temperature 212°F, both at standard atmospheric pressure. There are 180 degrees between freezing and boiling, compared to 100 on the Celsius scale, so one Fahrenheit degree represents only $\frac{100}{180}$, or $\frac{5}{9}$, as great a temperature change as one Celsius degree.

To convert temperatures from Celsius to Fahrenheit, note that a Celsius temperature T_C is the number of Celsius degrees above freezing; the number of Fahrenheit degrees above freezing is $\frac{9}{5}$ of this. But freezing on the Fahrenheit scale is at 32°F, so to obtain

15–3 (a) A bimetallic strip bends when heated because one metal (shown in red) expands more than the other. (b) A bimetallic strip, usually in the form of a spiral, may be used as a thermometer.

the actual Fahrenheit temperature T_F, multiply the Celsius value by $\frac{9}{5}$ and then add $32°$. Symbolically,

$$T_F = \frac{9}{5} T_C + 32° . \tag{15–1}$$

To convert Fahrenheit to Celsius, solve this equation for T_C:

$$T_C = \frac{5}{9} (T_F - 32°) . \tag{15–2}$$

In words, subtract $32°$ to get the number of Fahrenheit degrees above freezing, and then multiply by $\frac{5}{9}$ to obtain the number of Celsius degrees above freezing, that is, the Celsius temperature.

We don't recommend memorizing Eqs. (15–1) and (15–2). Instead, try to understand the reasoning that led to them so that you can derive them on the spot when you need them, checking your reasoning with the relation $100°C = 212°F$.

It is useful to distinguish between an actual temperature and a temperature *interval* (a difference or change in temperature). An actual temperature of 20° is stated as 20°C (twenty degrees Celsius), and a temperature *interval* of 10° is 10 C° (ten Celsius degrees). A beaker of water heated from 20°C to 30°C has a temperature change of 10 C°.

15–4 GAS THERMOMETERS AND THE KELVIN SCALE

When we calibrate two thermometers, such as a liquid-in-tube system and a resistance thermometer, so that they agree at 0°C and 100°C, they may not agree exactly at intermediate temperatures. Any temperature scale defined in this way always depends somewhat on the specific properties of the material used. Ideally, we would like to define a temperature scale that *doesn't* depend on the properties of a particular material. To establish a truly material-independent scale, we first need to develop some principles of thermodynamics. We'll return to this fundamental problem in Chapter 18. Here we'll discuss a thermometer that comes close to the ideal, the *gas thermometer*.

The principle of a gas thermometer is that the pressure of a gas at constant volume increases with temperature. A quantity of gas is placed in a constant-volume container (Fig. 15–4a), and its pressure is measured by one of the devices described in Section

(a)   (b)

15–4 (a) A constant-volume gas thermometer. (b) Graph of absolute pressure versus temperature for a constant-volume low-density gas thermometer. A straight-line extrapolation predicts that the pressure would become zero at −273.15°C if that temperature could be reached and the proportionality of pressure to temperature held exactly.

14–3. To calibrate a constant-volume gas thermometer, we measure the pressure at two temperatures, say 0°C and 100°C, plot these points on a graph, and draw a straight line between them (Fig. 15–4b). Then we can read from the graph the temperature corresponding to any other pressure.

By extrapolating this graph, we see that there is a hypothetical temperature, −273.15°C, at which the absolute pressure of the gas would become zero. We might expect that this temperature would be different for different gases, but it turns out to be the same for many different gases (at least in the limit of very low gas density). We can't actually observe this zero-pressure condition. Gases liquefy and solidify at very low temperatures, and the proportionality of pressure to temperature no longer holds.

We use this extrapolated zero-pressure temperature as the basis for a temperature scale with its zero at this temperature. This is the **Kelvin temperature scale,** named for the British physicist Lord Kelvin (1824–1907). The units are the same size as those on the Celsius scale, but the zero is shifted so that 0 K = −273.15°C and 273.15 K = 0°C; that is,

$$T_K = T_C + 273.15. \qquad (15\text{–}3)$$

This scale is shown in Fig. 15–4b. A common room temperature, 20°C (= 68°F), is 20 + 273.15, or about 293 K.

In SI nomenclature, "degree" is not used with the Kelvin scale; the temperature mentioned above is read "293 kelvins," not "degrees Kelvin." We capitalize Kelvin when it refers to the temperature scale; however, the *unit* of temperature is the *kelvin,* which is not capitalized (but is nonetheless abbreviated as a capital K).

EXAMPLE 15–1

Body temperature You place a small piece of melting ice in your mouth. Eventually, the water all converts from ice at $T_1 = 32.00°F$ to body temperature, $T_2 = 98.60°F$. Express these temperatures as °C and K, and find $\Delta T = T_2 - T_1$ in both cases.

SOLUTION First we find the Celsius temperatures. We know that $T_1 = 32.00°F = 0.00°C$, and 98.60°F is 98.60 − 32.00 = 66.60 F° above freezing; multiply this by (5 C°/9 F°) to find 37.00 C° above freezing, or $T_2 = 37.00°C$.

To get the Kelvin temperatures, we just add 273.15 to each Celsius temperature: $T_1 = 273.15$ K and $T_2 = 310.15$ K. "Normal" body temperature is 37.0°C, but if your doctor says that your temperature is 310 K, don't be alarmed.

The temperature *difference* $\Delta T = T_2 - T_1$ is 37.00 C° = 37.00 K. The Celsius and Kelvin scales have different zero points but the same size degrees. Therefore any temperature difference is the *same* on the Celsius and Kelvin scales but not the same on the Fahrenheit scale.

The Celsius scale has two fixed points, the normal freezing and boiling temperatures of water. But we can define the Kelvin scale using a gas thermometer with only a single reference temperature. We define the ratio of any two temperatures T_1 and T_2 on the Kelvin scale as the ratio of the corresponding gas-thermometer pressures p_1 and p_2:

$$\frac{T_2}{T_1} = \frac{p_2}{p_1} \qquad \text{(constant-volume gas thermometer, } T \text{ in kelvins).} \qquad (15\text{–}4)$$

The pressure p is directly proportional to the Kelvin temperature, as shown in Fig. 15–4b. To complete the definition of T, we need only specify the Kelvin temperature of a single specific state. For reasons of precision and reproducibility the state chosen is the *triple point* of water. This is the unique combination of temperature and pressure at which solid water (ice), liquid water, and water vapor can all coexist. This occurs at a temperature of 0.01°C and a water-vapor pressure of 610 Pa (about 0.006 atm). (This is the pressure of the *water;* it has nothing to do directly with the gas pressure in the *thermometer.*) The triple-point temperature T_{triple} of water is *defined* to have the value $T_{triple} = 273.16$ K, corresponding to 0.01°C. From Eq. (15–4), if p_{triple} is the pressure in a gas thermometer at temperature T_{triple} and p is the pressure at some other temperature T,

	K	C	F
Water boils	373	100°	212°
	100 K	100 C°	180 F°
Water freezes	273	0°	32°
CO_2 solidifies	195	−78°	−109°
Oxygen liquifies	90	−183°	−298°
Absolute zero	0	−273°	−460°

15–5 Relations among Kelvin, Celsius, and Fahrenheit temperature scales. Temperatures have been rounded off to the nearest degree.

then T is given on the Kelvin scale by

$$T = T_{triple} \frac{p}{p_{triple}} = (273.16 \text{ K}) \frac{p}{p_{triple}}. \qquad (15\text{–}5)$$

Low-pressure gas thermometers using various gases are found to agree very closely, but they are large, bulky, and very slow to come to thermal equilibrium. They are used principally to establish high-precision standards and to calibrate other thermometers.

EXAMPLE 15–2

Suppose a constant-volume gas thermometer has a pressure of 1.50×10^4 Pa at temperature T_{triple} and a pressure of 1.95×10^4 Pa at some unknown temperature T. What is T?

SOLUTION From Eq. (15–5) the Kelvin temperature is

$$T = (273.16 \text{ K}) \frac{1.95 \times 10^4 \text{ Pa}}{1.50 \times 10^4 \text{ Pa}} = 355 \text{ K } (= 82°\text{C}).$$

The relationships among the three temperature scales we have discussed are shown graphically in Fig. 15–5. The Kelvin scale is called an **absolute temperature scale,** and its zero point ($T = 0$ K $= −273.15°$C, the temperature at which $p = 0$ in Eq. (15–5)) is called **absolute zero.** At absolute zero a system of molecules (such as a quantity of a gas, a liquid, or a solid) has its *minimum* possible total energy (kinetic plus potential); because of quantum effects, however, it is *not* correct to say that all molecular motion ceases at absolute zero. To define more completely what we mean by absolute zero, we need to use the thermodynamic principles developed in the next several chapters. We will return to this concept in Chapter 18.

15–5 THERMAL EXPANSION

Most materials expand when their temperatures increase. The decks of bridges need special joints and supports to allow for expansion. A completely filled and tightly capped bottle of water cracks when it is heated, but you can loosen a metal jar lid by running hot water over it. These are all examples of *thermal expansion*.

LINEAR EXPANSION

Suppose a rod of material has a length L_0 at some initial temperature T_0. When the temperature changes by ΔT, the length changes by ΔL. Experiment shows that if ΔT is not too large (say, less than 100 C° or so), ΔL is *directly proportional* to ΔT. If two rods made of the same material have the same temperature change, but one is twice as long as the other, then the *change* in its length is also twice as great. Therefore ΔL must also be proportional to L_0. Introducing a proportionality constant α (which is different for different materials), we may express these relations in an equation:

$$\Delta L = \alpha L_0 \Delta T \qquad \text{(linear thermal expansion).} \qquad (15\text{–}6)$$

(b)

15–6 (a) We can visualize the forces between neighboring atoms in a solid by imagining them to be connected by springs that are easier to stretch than to compress. (b) A graph of potential energy versus distance between neighboring atoms shows that the forces are not symmetric. Compare this figure to Fig. 13–17b.

If a body has length L_0 at temperature T_0, then its length L at a temperature $T = T_0 + \Delta T$ is

$$L = L_0 + \Delta L = L_0 + \alpha L_0 \Delta T = L_0(1 + \alpha \Delta T). \qquad (15\text{–}7)$$

The constant α, which describes the thermal expansion properties of a particular material, is called the **coefficient of linear expansion.** The units of α are K^{-1} or $(C°)^{-1}$. (Remember that a temperature *interval* is the same in the Kelvin and Celsius scales.) For many materials, every linear dimension changes according to Eq. (15–6) or (15–7). Thus L could be the thickness of a rod, the side length of a square sheet, or the diameter of a hole. Some materials, such as wood or single crystals, expand differently in different directions. We won't consider this complication.

We can understand thermal expansion qualitatively on a molecular basis. Picture the interatomic forces in a solid as springs, as in Fig. 15–6 (we explored the analogy between spring forces and interatomic forces in Section 13–5). Each atom vibrates about its equilibrium position. When the temperature increases, the energy and amplitude of the vibration also increase. The interatomic spring forces are not symmetric about the equilibrium position; they usually behave like a spring that is easier to stretch than to compress. As a result, when the amplitude of vibration increases, the *average* distance between molecules also increases. As the atoms get farther apart, every dimension increases.

CAUTION ▶ If a solid object has a hole in it, what happens to the size of the hole when the temperature of the object increases? A common misconception is that if the object expands, the hole will shrink because material expands into the hole. But the truth of the matter is that if the object expands, the hole will expand too; as we stated above, *every* linear dimension of an object changes in the same way when the temperature changes. If you're not convinced, think of the atoms in Fig. 15–6a as outlining a cubical hole. When the object expands, the atoms move apart and the hole increases in size. ◀

The direct proportionality expressed by Eq. (15–6) is not exact; it is *approximately* correct only for sufficiently small temperature changes. For a given material, α varies somewhat with the initial temperature T_0 and the size of the temperature interval. We'll ignore this complication here, however. Average values of α for several materials are listed in Table 15–1. Within the precision of these values we don't need to worry whether T_0 is 0°C or 20°C or some other temperature. Note that typical values of α are very small; even for a temperature change of 100 C°, the fractional length change $\Delta L/L_0$ is only of the order of 1/1000 for the metals in the table.

VOLUME EXPANSION

Increasing temperature usually causes increases in *volume* for both solid and liquid materials. Just as with linear expansion, experiments show that if the temperature change ΔT is not too great (less than 100 C° or so), the increase in volume ΔV is approximately

TABLE 15–1

COEFFICIENTS OF LINEAR EXPANSION

MATERIAL	$\alpha\,[K^{-1}$ or $(C°)^{-1}]$
Aluminum	2.4×10^{-5}
Brass	2.0×10^{-5}
Copper	1.7×10^{-5}
Glass	$0.4\text{–}0.9 \times 10^{-5}$
Invar (nickel-iron alloy)	0.09×10^{-5}
Quartz (fused)	0.04×10^{-5}
Steel	1.2×10^{-5}

TABLE 15–2
COEFFICIENTS OF VOLUME EXPANSION

SOLIDS	$\beta[\text{K}^{-1}$ or $(\text{C}°)^{-1}]$	LIQUIDS	$\beta[\text{K}^{-1}$ or $(\text{C}°)^{-1}]$
Aluminum	7.2×10^{-5}	Ethanol	75×10^{-5}
Brass	6.0×10^{-5}	Carbon disulfide	115×10^{-5}
Copper	5.1×10^{-5}	Glycerin	49×10^{-5}
Glass	$1.2–2.7 \times 10^{-5}$	Mercury	18×10^{-5}
Invar	0.27×10^{-5}		
Quartz (fused)	0.12×10^{-5}		
Steel	3.6×10^{-5}		

proportional to both the temperature change ΔT and the initial volume V_0:

$$\Delta V = \beta V_0 \Delta T \quad \text{(volume thermal expansion)}. \qquad (15\text{–}8)$$

The constant β characterizes the volume expansion properties of a particular material; it is called the **coefficient of volume expansion.** The units of β are K^{-1} or $(\text{C}°)^{-1}$. As with linear expansion, β varies somewhat with temperature, and Eq. (15–8) is an approximate relation that is valid only for small temperature changes. For many substances, β decreases at low temperatures. Several values of β in the neighborhood of room temperature are listed in Table 15–2. Note that the values for liquids are generally much larger than those for solids.

For solid materials there is a simple relation between the volume expansion coefficient β and the linear expansion coefficient α. To derive this relation, we consider a cube of material with side length L and volume $V = L^3$. At the initial temperature the values are L_0 and V_0. When the temperature increases by dT, the side length increases by dL and the volume increases by an amount dV given by

$$dV = \frac{dV}{dL} dL = 3L^2 \, dL.$$

Now we replace L and V by the initial values L_0 and V_0. From Eq. (15–6), dL is

$$dL = \alpha L_0 \, dT;$$

since $V_0 = L_0^3$, this means that dV can also be expressed as

$$dV = 3L_0^2 \alpha L_0 \, dT = 3\alpha V_0 \, dT.$$

This is consistent with the infinitesimal form of Eq. (15–8), $dV = \beta V_0 \, dT$, only if

$$\beta = 3\alpha. \qquad (15\text{–}9)$$

You should check this relation for some of the materials listed in Tables 15–1 and 15–2.

When this SR-71 aircraft is sitting on the ground, its wing panels fit together so loosely that fuel leaks out of the wings onto the ground. But once it is in flight at over three times the speed of sound, air friction heats the panels so much that they expand to make a perfect fit.

Problem–Solving Strategy

THERMAL EXPANSION

1. Identify which quantities in Eq. (15–6) or (15–8) are known and which are unknown. Often you will be given two temperatures and will have to compute ΔT. Or you may be given an initial temperature T_0 and have to find a final temperature corresponding to a given length or volume change. In this case, find ΔT first; then the final temperature is $T_0 + \Delta T$.

2. Unit consistency is crucial, as always. L_0 and ΔL (or V_0 and ΔV) must have the same units, and if you use a value of α or β in K^{-1} or $(\text{C}°)^{-1}$, then ΔT must be in kelvins or Celsius degrees (C°). But you can use K and C° interchangeably.

3. Remember that the sizes of holes in a material expand with temperature just like any other linear dimension, and the volume of a hole (such as the volume of a container) expands in the same way as the corresponding solid shape.

EXAMPLE 15-3

Length change due to temperature change A surveyor uses a steel measuring tape that is exactly 50.000 m long at a temperature of 20°C. What is its length on a hot summer day when the temperature is 35°C?

SOLUTION We have $L_0 = 50.000$ m, $T_0 = 20°C$, $T = 35°C$, and $\Delta T = T - T_0 = 15$ C° = 15 K. From Eq. (15–6),

$$\Delta L = \alpha L_0 \, \Delta T = (1.2 \times 10^{-5} \text{ K}^{-1})(50 \text{ m})(15 \text{ K})$$
$$= 9.0 \times 10^{-3} \text{ m} = 9.0 \text{ mm},$$
$$L = L_0 + \Delta L = 50.000 \text{ m} + 0.009 \text{ m} = 50.009 \text{ m}.$$

Thus the length at 35°C is 50.009 m. Note that L_0 is given to five significant figures but that we need only two of them to compute ΔL.

EXAMPLE 15-4

In Example 15–3 the surveyor uses the measuring tape to measure a distance when the temperature is 35°C; the value that she reads off the tape is 35.794 m. What is the actual distance?

SOLUTION As we saw in Example 15–3, at 35°C the tape has expanded slightly; the distance between two successive meter

marks on the tape is a little more than a meter, in the ratio (50.009 m)/(50.000 m). Hence the true distance is greater than the value read off the tape by this same factor:

$$\frac{50.009 \text{ m}}{50.000 \text{ m}} (35.794 \text{ m}) = 35.800 \text{ m}.$$

EXAMPLE 15-5

Volume change due to temperature change A glass flask with volume 200 cm³ is filled to the brim with mercury at 20°C. How much mercury overflows when the temperature of the system is raised to 100°C? The coefficient of linear expansion of the glass is 0.40×10^{-5} K⁻¹.

SOLUTION Mercury overflows because β is much larger for mercury than for glass. From Table 15–2 the coefficient of volume expansion for mercury is $\beta_{\text{mercury}} = 18 \times 10^{-5}$ K⁻¹; from Eq. (15–9) the coefficient of volume expansion for the glass is

$$\beta_{\text{glass}} = 3\alpha_{\text{glass}} = 3(0.40 \times 10^{-5} \text{ K}^{-1}) = 1.2 \times 10^{-5} \text{ K}^{-1}.$$

The increase in volume of the glass flask is

$$\Delta V_{\text{glass}} = \beta_{\text{glass}} V_0 \Delta T$$
$$= (1.2 \times 10^{-5} \text{ K}^{-1})(200 \text{ cm}^3)(100°C - 20°C) = 0.19 \text{ cm}^3.$$

The increase in volume of the mercury is

$$\Delta V_{\text{mercury}} = \beta_{\text{mercury}} V_0 \Delta T$$
$$= (18 \times 10^{-5} \text{ K}^{-1})(200 \text{ cm}^3)(100°C - 20°C) = 2.9 \text{ cm}^3.$$

The volume of mercury that overflows is

$$\Delta V_{\text{mercury}} - \Delta V_{\text{glass}} = 2.9 \text{ cm}^3 - 0.19 \text{ cm}^3 = 2.7 \text{ cm}^3.$$

This is basically how a mercury-in-glass thermometer works, except that instead of letting the mercury overflow and run all over the place, the thermometer has it rise inside a sealed tube as T increases.

As Tables 15–1 and 15–2 show, glass has smaller coefficients of expansion α and β than do most metals. This is why you can use hot water to loosen a metal lid on a glass jar; the metal expands more than the glass does.

15–7 The volume of one gram of water in the temperature range from 0°C to 10°C. By 100°C the volume has increased to 1.034 cm³. If the coefficient of volume expansion were constant, the curve would be a straight line.

THERMAL EXPANSION OF WATER

Water, in the temperature range from 0°C to 4°C, *decreases* in volume with increasing temperature. In this range its coefficient of volume expansion is *negative*. Above 4°C, water expands when heated (Fig. 15–7). Hence water has its greatest density at 4°C. Water also expands when it freezes, which is why ice humps up in the middle of the compartments in an ice cube tray. By contrast, most materials contract when they freeze.

This anomalous behavior of water has an important effect on plant and animal life in lakes. A lake cools from the surface down; above 4°C, the cooled water at the surface flows to the bottom because of its greater density. But when the surface temperature drops below 4°C, the water near the surface is less dense than the warmer water below. Hence the downward flow ceases, and the water near the surface remains colder than that at the bottom. As the surface freezes, the ice floats because it is less dense than water.

The water at the bottom remains at 4°C until nearly the entire lake is frozen. If water behaved like most substances, contracting continuously on cooling and freezing, lakes would freeze from the bottom up. Circulation due to density differences would continuously carry warmer water to the surface for efficient cooling, and lakes would freeze solid much more easily. This would destroy all plant and animal life that cannot withstand freezing. If water did not have this special property, the evolution of life would have taken a very different course.

THERMAL STRESS

If we clamp the ends of a rod rigidly to prevent expansion or contraction and then change the temperature, tensile or compressive stresses called **thermal stresses** develop. The rod would like to expand or contract, but the clamps won't let it. The resulting stresses may become large enough to strain the rod irreversibly or even break it. Concrete highways and bridge decks usually have gaps between sections, filled with a flexible material or bridged by interlocking teeth (Fig. 15–8), to permit expansion and contraction of the concrete. Long steam pipes have expansion joints or U-shaped sections to prevent buckling or stretching with temperature changes. If one end of a steel bridge is rigidly fastened to its abutment, the other end usually rests on rollers.

To calculate the thermal stress in a clamped rod, we compute the amount the rod *would* expand (or contract) if not held and then find the stress needed to compress (or stretch) it back to its original length. Suppose that a rod with length L_0 and cross-section area A is held at constant length while the temperature is reduced (negative ΔT), causing a tensile stress. The fractional change in length if the rod were free to contract would be

$$\left(\frac{\Delta L}{L_0}\right)_{\text{thermal}} = \alpha\,\Delta T. \qquad (15\text{–}10)$$

Both ΔL and ΔT are negative. The tension must increase by an amount F that is just enough to produce an equal and opposite fractional change in length $(\Delta L/L_0)_{\text{tension}}$. From the definition of Young's modulus, Eq. (11–10),

$$Y = \frac{F/A}{\Delta L/L_0}, \qquad \text{so} \qquad \left(\frac{\Delta L}{L_0}\right)_{\text{tension}} = \frac{F}{AY}. \qquad (15\text{–}11)$$

If the length is to be constant, the *total* fractional change in length must be zero. From Eqs. (15–10) and (15–11), this means that

$$\left(\frac{\Delta L}{L_0}\right)_{\text{thermal}} + \left(\frac{\Delta L}{L_0}\right)_{\text{tension}} = \alpha\,\Delta T + \frac{F}{AY} = 0.$$

Solving for the tensile stress F/A required to keep the rod's length constant, we find

$$\frac{F}{A} = -Y\alpha\,\Delta T \qquad \text{(thermal stress)}. \qquad (15\text{–}12)$$

For a decrease in temperature, ΔT is negative, so F and F/A are positive; this means that a *tensile* force and stress are needed to maintain the length. If ΔT is positive, F and F/A are negative, and the required force and stress are *compressive*.

If there are temperature differences within a body, non-uniform expansion or contraction will result and thermal stresses can be induced. You can break a glass bowl by pouring very hot water into it; the thermal stress between the hot and cold parts of the bowl exceeds the breaking stress of the glass, causing cracks. The same phenomenon makes ice cubes crack when dropped into warm water. Heat-resistant glasses such as Pyrex™ have exceptionally low expansion coefficients and high strength.

15–8 The interlocking teeth of an expansion joint on a bridge. These joints are needed to accommodate changes in length that result from thermal expansion.

EXAMPLE 15-6

An aluminum cylinder 10 cm long, with a cross-section area of 20 cm^2, is to be used as a spacer between two steel walls. At 17.2°C it just slips in between the walls. When it warms to 22.3°C, calculate the stress in the cylinder and the total force it exerts on each wall, assuming that the walls are perfectly rigid and a constant distance apart.

SOLUTION Equation (15–12) relates the stress to the temperature change. From Table 11–1 we find $Y = 7.0 \times 10^{10}$ Pa, and from Table 15–1, $\alpha = 2.4 \times 10^{-5}$ K^{-1}. The temperature change is $\Delta T = 22.3°C - 17.2°C = 5.1$ C° = 5.1 K. The stress is F/A; from Eq. (15–12),

$$\frac{F}{A} = -Y\alpha \, \Delta T = -(0.70 \times 10^{11} \text{ Pa})(2.4 \times 10^{-5} \text{ K}^{-1})(5.1 \text{ K})$$

$$= -8.6 \times 10^6 \text{ Pa (or } -1200 \text{ lb/in.}^2).$$

The negative sign indicates that compressive rather than tensile stress is needed to keep the cylinder's length constant. This stress is independent of the length and cross-section area of the cylinder. The total force F is the cross-section area times the stress:

$$F = A\left(\frac{F}{A}\right) = (20 \times 10^{-4} \text{ m}^2)(-8.6 \times 10^6 \text{ Pa})$$

$$= -1.7 \times 10^4 \text{ N},$$

or nearly two tons. The negative sign indicates compression.

15–6 QUANTITY OF HEAT

When you put a cold spoon into a cup of hot coffee, the spoon warms up and the coffee cools down as they approach thermal equilibrium. The interaction that causes these temperature changes is fundamentally a transfer of *energy* from one substance to another. Energy transfer that takes place solely because of a temperature difference is called *heat flow* or *heat transfer,* and energy transferred in this way is called **heat.**

An understanding of the relation between heat and other forms of energy emerged gradually during the eighteenth and nineteenth centuries. Sir James Joule (1818–1889) studied how water can be warmed by vigorous stirring with a paddle wheel (Fig. 15–9a). The paddle wheel adds energy to the water by doing *work* on it, and Joule found that *the temperature rise is directly proportional to the amount of work done.* The same temperature change can also be caused by putting the water in contact with some hotter body (Fig. 15–9b); hence this interaction must also involve an energy exchange. We will explore the relation between heat and mechanical energy in greater detail in Chapters 17 and 18.

CAUTION ▶ It is absolutely essential for you to keep clearly in mind the distinction between *temperature* and *heat.* Temperature depends on the physical state of a material and is a quantitative description of its hotness or coldness. In physics the term "heat" always refers to energy in transit from one body or system to another because of a temperature difference, never to the amount of energy contained within a particular system. We can change the temperature of a body by adding heat to it or taking heat away, or by adding or subtracting energy in other ways such as mechanical work (Fig. 15–9a). If we cut a body in half, each half has the same temperature as the whole; but to raise the temperature of each half by a given interval, we add *half* as much heat as for the whole. ◀

(a)

(b)

15–9 The same temperature change of the same system may be accomplished by (a) doing work on it or (b) adding heat to it.

We can define a *unit* of quantity of heat based on temperature changes of some specific material. The **calorie** (abbreviated cal) is defined as *the amount of heat required to raise the temperature of one gram of water from* 14.5°C *to* 15.5°C. The kilocalorie (kcal), equal to 1000 cal, is also used; a food-value calorie is actually a kilocalorie. A corresponding unit of heat using Fahrenheit degrees and British units is the **British thermal unit,** or Btu. One Btu is the quantity of heat required to raise the temperature of one pound (weight) of water 1 F° from 63°F to 64°F.

Because heat is energy in transit, there must be a definite relation between these units and the familiar mechanical energy units such as the joule. Experiments similar in concept to Joule's have shown that

$$1 \text{ cal} = 4.186 \text{ J},$$

$$1 \text{ kcal} = 1000 \text{ cal} = 4186 \text{ J},$$

$$1 \text{ Btu} = 778 \text{ ft} \cdot \text{lb} = 252 \text{ cal} = 1055 \text{ J}.$$

The calorie is not a fundamental SI unit. The International Committee on Weights and Measures recommends using the joule as the basic unit of energy in all forms, including heat. We will follow that recommendation in this book.

SPECIFIC HEAT CAPACITY

We use the symbol Q for quantity of heat. When it is associated with an infinitesimal temperature change dT, we call it dQ. The quantity of heat Q required to increase the temperature of a mass m of a certain material from T_1 to T_2 is found to be approximately proportional to the temperature change $\Delta T = T_2 - T_1$. It is also proportional to the mass m of material. When you're heating water to make tea, you need twice as much heat for two cups as for one if the temperature interval is the same. The quantity of heat needed also depends on the nature of the material; raising the temperature of one kilogram of water by 1 C° requires 4190 J of heat, but only 910 J are needed to raise the temperature of one kilogram of aluminum by 1 C°.

Putting all these relationships together, we have

$$Q = mc\,\Delta T \qquad \text{(heat required for temperature change of mass } m\text{)}, \qquad (15\text{–}13)$$

where c is a quantity, different for different materials, called the **specific heat capacity** (or sometimes *specific heat*) of the material. For an infinitesimal temperature change dT and corresponding quantity of heat dQ,

$$dQ = mc\,dT, \qquad (15\text{–}14)$$

$$c = \frac{1}{m}\frac{dQ}{dT} \qquad \text{(specific heat capacity)}. \qquad (15\text{–}15)$$

In Eqs. (15–13), (15–14), and (15–15), Q (or dQ) and ΔT (or dT) can be either positive or negative. When they are positive, heat enters the body and its temperature increases; when they are negative, heat leaves the body and its temperature decreases.

CAUTION ▶ Remember that dQ does not represent a change in the amount of heat *contained* in a body; this is a meaningless concept. Heat is always energy *in transit* as a result of a temperature difference. There is no such thing as "the amount of heat in a body." The term *specific heat capacity* is unfortunate because it tends to suggest the erroneous idea that a body *contains* a certain amount of heat. ◀

The specific heat capacity of water is approximately

$$4190 \text{ J/kg} \cdot \text{K}, \qquad 1 \text{ cal/g} \cdot \text{C}°, \qquad \text{or } 1 \text{ Btu/lb} \cdot \text{F}°.$$

The specific heat capacity of a material always depends somewhat on the initial temperature and the temperature interval. Figure 15–10 shows this variation for water. In the problems and examples in this chapter we will usually ignore this small variation.

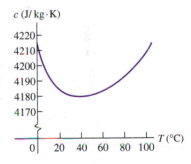

15–10 Specific heat capacity of water as a function of temperature. The value of c varies by less than 1% between 0°C and 100°C.

EXAMPLE 15–7

Feed a cold, starve a fever During a bout with the flu an 80-kg man ran a fever of 2.0 C° above normal, that is, a body temperature of 39.0°C (102.2°F) instead of the normal 37.0°C (98.6°F). Assuming that the human body is mostly water, how much heat is required to raise his temperature by that amount?

SOLUTION The temperature change is $\Delta T = 39.0°C - 37.0°C = 2.0$ C° = 2.0 K. From Eq. (15–13),

$$Q = mc\,\Delta T = (80 \text{ kg})(4190 \text{ J/kg} \cdot \text{K})(2.0 \text{ K}) = 6.7 \times 10^5 \text{ J}.$$

This corresponds to 160 kcal, or 160 food-value calories. In fact the specific heat capacity of the human body is more nearly equal to 3480 J/kg · K, about 83% that of water. The difference is due to the presence of protein, fat, and minerals, which have lower specific heat capacities. With this value of c, the required heat is 5.6×10^5 J = 133 kcal.

EXAMPLE 15–8

Overheating electronics You are designing an electronic circuit element made of 23 mg of silicon. The electric current through it adds energy at the rate of 7.4 mW = 7.4×10^{-3} J/s. If your design doesn't allow any heat transfer out of the element, at what rate does its temperature increase? The specific heat capacity of silicon is 705 J/kg·K.

SOLUTION In one second, $Q = (7.4 \times 10^{-3}$ J/s)(1 s) = 7.4×10^{-3} J. From Eq. (15–13), $Q = mc\,\Delta T$, the temperature change in one second is

$$\Delta T = \frac{Q}{mc} = \frac{7.4 \times 10^{-3}\ \text{J}}{(23 \times 10^{-6}\ \text{kg})(705\ \text{J/kg}\cdot\text{K})} = 0.46\ \text{K}.$$

Alternatively, we can divide both sides of Eq. (15–14) by dt and rearrange:

$$\frac{dT}{dt} = \frac{dQ/dt}{mc}$$

$$= \frac{7.4 \times 10^{-3}\ \text{J/s}}{(23 \times 10^{-6}\ \text{kg})(705\ \text{J/kg}\cdot\text{K})} = 0.46\ \text{K/s}.$$

At this rate of temperature rise (27 K every minute) the circuit element would soon self-destruct. Heat transfer is an important design consideration in electronic circuit elements (see Section 15–9).

MOLAR HEAT CAPACITY

Sometimes it's more convenient to describe a quantity of substance in terms of the number of *moles n* rather than the *mass m* of material. Recall from your study of chemistry that a mole of any pure substance always contains the same number of molecules. (We will discuss this point in more detail in Chapter 16.) The *molar mass* of any substance, denoted by M, is the mass per mole. (The quantity M is sometimes called *molecular weight,* but *molar mass* is preferable; the quantity depends on the mass of a molecule, not its weight.) For example, the molar mass of water is 18.0 g/mol = 18.0×10^{-3} kg/mol; one mole of water has a mass of 18.0 g = 0.0180 kg. The total mass m of material is equal to the mass per mole M times the number of moles n:

$$m = nM. \tag{15–16}$$

Replacing the mass m in Eq. (15–13) by the product nM, we find

$$Q = nMc\,\Delta T. \tag{15–17}$$

The product Mc is called the **molar heat capacity** (or *molar specific heat*) and is denoted by C (capitalized). With this notation we rewrite Eq. (15–17) as

$$Q = nC\,\Delta T \quad \text{(heat required for temperature change of } n \text{ moles).} \tag{15–18}$$

Comparing to Eq. (15–15), we can express the molar heat capacity C (heat per mole per temperature change) in terms of the specific heat capacity c (heat per mass per temperature change) and the molar mass M (mass per mole):

$$C = \frac{1}{n}\frac{dQ}{dT} = Mc \quad \text{(molar heat capacity).} \tag{15–19}$$

For example, the molar heat capacity of water is

$$C = Mc = (0.0180\ \text{kg/mol})(4190\ \text{J/kg}\cdot\text{K}) = 75.4\ \text{J/mol}\cdot\text{K}.$$

Values of specific and molar heat capacities for several substances are given in Table 15–3.

Precise measurements of heat capacities require great experimental skill. Usually, a measured quantity of energy is supplied by an electric current in a heater wire wound around the specimen. The temperature change ΔT is measured with a resistance thermometer or thermocouple embedded in the specimen. This sounds simple, but great care is needed to avoid or compensate for unwanted heat transfer between the sample and its surroundings. Measurements for solid materials are usually made at constant atmo-

TABLE 15–3
APPROXIMATE SPECIFIC AND MOLAR HEAT CAPACITIES (CONSTANT PRESSURE)

SUBSTANCE	SPECIFIC HEAT CAPACITY, c (J/kg · K)	M (kg/mol)	MOLAR HEAT CAPACITY, C (J/mol · K)
Aluminum	910	0.0270	24.6
Beryllium	1970	0.00901	17.7
Copper	390	0.0635	24.8
Ethanol	2428	0.0461	111.9
Ethylene glycol	2386	0.0620	148.0
Ice (near 0°C)	2100	0.0180	37.8
Iron	470	0.0559	26.3
Lead	130	0.207	26.9
Marble ($CaCO_3$)	879	0.100	87.9
Mercury	138	0.201	27.7
Salt (NaCl)	879	0.0585	51.4
Silver	234	0.108	25.3
Water (liquid)	4190	0.0180	75.4

Water has a much higher specific heat capacity than the glass or metals used to make cookware. This helps explain why it takes several minutes to boil water on a stove, even though the pot or kettle reaches a high temperature very quickly.

spheric pressure; the corresponding values are called *heat capacities at constant pressure,* denoted by c_p and C_p. For a gas it is usually easier to keep the substance in a container with constant *volume;* the corresponding values are called *heat capacities at constant volume,* denoted by c_V and C_V. For a given substance, C_V and C_p are different. If the system can expand while heat is added, there is additional energy exchange through the performance of *work* by the system on its surroundings. If the volume is constant, the system does no work. For gases the difference between C_p and C_V is substantial. We will study heat capacities of gases in detail in Section 17–8.

The last column of Table 15–3 shows something interesting. The molar heat capacities for most elemental solids are about the same, about 25 J/mol · K. This correlation, named the *rule of Dulong and Petit* (for its discoverers), forms the basis for a very important idea. The number of atoms in one mole is the same for all elemental substances. This means that on a *per atom* basis, about the same amount of heat is required to raise the temperature of each of these elements by a given amount, even though the *masses* of the atoms are very different. The heat required for a given temperature increase depends only on *how many* atoms the sample contains, not on the mass of an individual atom. We will see the reason why the rule of Dulong and Petit works so well when we study the molecular basis of heat capacities in greater detail in Chapter 16.

15–7 CALORIMETRY AND PHASE CHANGES

Calorimetry means "measuring heat." We have discussed energy transfer (heat) involved in temperature changes. Heat is also involved in *phase changes,* such as the melting of ice or boiling of water. Once we understand these additional heat relationships, we can analyze a variety of problems involving quantity of heat.

PHASE CHANGES

We use the term **phase** to describe a specific state of matter, such as a solid, liquid, or gas. The compound H_2O exists in the *solid phase* as ice, in the *liquid phase* as water, and in the *gaseous phase* as steam. A transition from one phase to another is called a **phase change** or *phase transition.* For any given pressure a phase change takes place at a definite temperature, usually accompanied by absorption or emission of heat and a change of volume and density.

A familiar example of a phase change is the melting of ice. When we add heat to ice at 0°C and normal atmospheric pressure, the temperature of the ice *does not* increase. Instead, some of it melts to form liquid water. If we add the heat slowly, to maintain the system very close to thermal equilibrium, the temperature remains at 0°C until all the ice is melted. The effect of adding heat to this system is not to raise its temperature but to change its *phase* from solid to liquid.

To change 1 kg of ice at 0°C to 1 kg of liquid water at 0°C and normal atmospheric pressure requires 3.34×10^5 J of heat. The heat required per unit mass is called the **heat of fusion** (or sometimes *latent heat of fusion*), denoted by L_f. For water at normal atmospheric pressure the heat of fusion is

$$L_f = 3.34 \times 10^5 \text{ J/kg} = 79.6 \text{ cal/g} = 143 \text{ Btu/lb.}$$

More generally, to melt a mass m of material that has a heat of fusion L_f requires a quantity of heat Q given by

$$Q = mL_f.$$

This process is *reversible*. To freeze liquid water to ice at 0°C, we have to *remove* heat; the magnitude is the same, but in this case, Q is negative because heat is removed rather than added. To cover both possibilities and to include other kinds of phase changes, we write

$$Q = \pm mL \qquad \text{(heat transfer in a phase change).} \qquad (15\text{–}20)$$

The plus sign (heat entering) is used when the material melts; the minus sign (heat leaving) is used when it freezes. The heat of fusion is different for different materials, and it also varies somewhat with pressure.

For any given material at any given pressure the freezing temperature is the same as the melting temperature. At this unique temperature the liquid and solid phases (liquid water and ice, for example) can coexist in a condition called **phase equilibrium.**

We can go through this whole story again for *boiling* or *evaporation,* a phase transition between liquid and gaseous phases. The corresponding heat (per unit mass) is called the **heat of vaporization** L_v. At normal atmospheric pressure the heat of vaporization L_v for water is

$$L_v = 2.256 \times 10^6 \text{ J/kg} = 539 \text{ cal/g} = 970 \text{ Btu/lb.}$$

That is, it takes 2.256×10^6 J to change 1 kg of liquid water at 100°C to 1 kg of water vapor at 100°C. By comparison, to raise the temperature of a kilogram of water from 0°C to 100°C requires $Q = mc\Delta T = (1.00 \text{ kg})(4190 \text{ J/kg} \cdot \text{C°})(100 \text{ C°}) = 4.19 \times 10^5$ J, less than one-fifth as much heat as is required for vaporization at 100°C. This agrees with everyday kitchen experience; a pot of water may reach boiling temperature in a few minutes, but it takes a much longer time to completely evaporate all the water away.

Like melting, boiling is a reversible transition. When heat is removed from a gas at the boiling temperature, the gas returns to the liquid phase, or *condenses,* giving up to its surroundings the same quantity of heat (heat of vaporization) that was needed to vaporize it. At a given pressure the boiling and condensation temperatures are always the same; at this temperature the liquid and gaseous phases can coexist in phase equilibrium.

Both L_v and the boiling temperature of a material depend on pressure. Water boils at a lower temperature (about 95°C) in Denver than in Pittsburgh because Denver is at higher elevation and the average atmospheric pressure is less. The heat of vaporization is somewhat greater at this lower pressure, about 2.27×10^6 J/kg.

Table 15–4 lists heats of fusion and vaporization for several materials and their melting and boiling temperatures at normal atmospheric pressure. Very few *elements* have melting temperatures in the vicinity of ordinary room temperatures; one of the few is the metal gallium, shown in Fig. 15–11.

15–11 The metal gallium, shown here melting in a person's hand, is one of the few elements that melt in the vicinity of room temperature. Its melting temperature is 29.8°C, and its heat of fusion is 8.04×10^4 J/kg.

TABLE 15–4
HEATS OF FUSION AND VAPORIZATION

SUBSTANCE	NORMAL MELTING POINT		HEAT OF FUSION, L_f	NORMAL BOILING POINT		HEAT OF VAPORIZATION, L_v
	K	°C	(J/kg)	K	°C	(J/kg)
Helium	*	*	*	4.216	−268.93	20.9×10^3
Hydrogen	13.84	−259.31	58.6×10^3	20.26	−252.89	452×10^3
Nitrogen	63.18	−209.97	25.5×10^3	77.34	−195.8	201×10^3
Oxygen	54.36	−218.79	13.8×10^3	90.18	−183.0	213×10^3
Ethanol	159	−114	104.2×10^3	351	78	854×10^3
Mercury	234	−39	11.8×10^3	630	357	272×10^3
Water	273.15	0.00	334×10^3	373.15	100.00	2256×10^3
Sulfur	392	119	38.1×10^3	717.75	444.60	326×10^3
Lead	600.5	327.3	24.5×10^3	2023	1750	871×10^3
Antimony	903.65	630.50	165×10^3	1713	1440	561×10^3
Silver	1233.95	960.80	88.3×10^3	2466	2193	2336×10^3
Gold	1336.15	1063.00	64.5×10^3	2933	2660	1578×10^3
Copper	1356	1083	134×10^3	1460	1187	5069×10^3

*A pressure in excess of 25 atmospheres is required to make helium solidify. At 1 atmosphere pressure, helium remains a liquid down to absolute zero.

Figure 15–12 shows how the temperature varies when we add heat continuously to a specimen of ice with an initial temperature below 0°C (point *a*). The temperature rises until we reach the melting point (point *b*). Then as more heat is added, the temperature remains constant until all the ice has melted (point *c*). Then the temperature starts to rise again until the boiling temperature is reached (point *d*). At that point the temperature again is constant until all the water is transformed into the vapor phase (point *e*). If the rate of heat input is constant, the slope of the line for the solid phase (ice) has a steeper slope than does the line for the liquid phase (water). Do you see why? (See Table 15–3.)

A substance can sometimes change directly from the solid to the gaseous phase. This process is called *sublimation,* and the solid is said to *sublime.* The corresponding heat is called the *heat of sublimation, L_s.* Liquid carbon dioxide cannot exist at a pressure lower than about 5×10^5 Pa (about 5 atm), and "dry ice" (solid carbon dioxide) sublimes at atmospheric pressure. Sublimation of water from frozen food causes freezer burn. The reverse process, a phase change from gas to solid, occurs when frost forms on cold bodies such as refrigerator cooling coils.

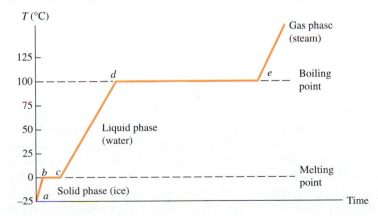

15–12 Graph of temperature versus time for a specimen of water initially in the solid phase (ice). Heat is added to the specimen at a constant rate. The temperature remains constant during each change of phase, provided that the pressure remains constant.

Very pure water can be cooled several degrees below the freezing temperature without freezing; the resulting unstable state is described as *supercooled*. When a small ice crystal is dropped in or the water is agitated, it crystallizes within a second or less. Supercooled water *vapor* condenses quickly into fog droplets when a disturbance, such as dust particles or ionizing radiation, is introduced. This principle is used in "seeding" clouds, which often contain supercooled water vapor, to cause condensation and rain.

A liquid can sometimes be *superheated* above its normal boiling temperature. Again, any small disturbance such as agitation or the passage of a charged particle through the material causes local boiling with bubble formation.

Steam heating systems for buildings use a boiling-condensing process to transfer heat from the furnace to the radiators. Each kilogram of water that is turned to steam in the boiler absorbs over 2×10^6 J (the heat of vaporization L_v of water) from the boiler and gives it up when it condenses in the radiators. Boiling-condensing processes are also used in refrigerators, air conditioners, and heat pumps. We will discuss these systems in Chapter 18.

The temperature-control mechanisms of many warm-blooded animals make use of heat of vaporization, removing heat from the body by using it to evaporate water from the tongue (panting) or from the skin (sweating). Evaporative cooling enables humans to maintain normal body temperature in hot, dry desert climates where the air temperature may reach 55°C (about 130°F). The skin temperature may be as much as 30°C cooler than the surrounding air. Under these conditions a normal person may perspire several liters per day. Unless this lost water is replaced, dehydration, heat stroke, and death result. Old-time desert rats (such as one of the authors) state that in the desert, any canteen that holds less than a gallon should be viewed as a toy!

The water may be warm and it may be a hot day, but these children will feel cold when they first step out of the swimming pool. That's because as water evaporates from their skin, it removes the heat of vaporization from their bodies. To stay warm, they will need to dry off immediately.

Evaporative cooling is also used to cool buildings in hot, dry climates and to condense and recirculate "used" steam in coal-fired or nuclear-powered electric-generating-plants. That's what is going on in the large, tapered concrete towers that you see at such plants.

Chemical reactions such as combustion are analogous to phase changes in that they involve definite quantities of heat. Complete combustion of one gram of gasoline produces about 46,000 J or about 11,000 cal, so the **heat of combustion** L_c of gasoline is

$$L_c = 46,000 \text{ J/g} = 4.6 \times 10^7 \text{ J/kg}.$$

Energy values of foods are defined similarly; the unit of food energy, although called a calorie, is a *kilocalorie,* equal to 1000 cal or 4186 J. When we say that a gram of peanut butter "contains 6 calories," we mean that 6 kcal of heat (6,000 cal or 25,000 J) is released when the carbon and hydrogen atoms in the peanut butter react with oxygen (with the help of enzymes) and are completely converted to CO_2 and H_2O. Not all of this energy is directly useful for mechanical work. We will study the *efficiency* of energy utilization in Chapter 18.

HEAT CALCULATIONS

Let's look at some examples of calorimetry calculations (calculations with heat). The basic principle is very simple: when heat flow occurs between two bodies that are isolated from their surroundings, the amount of heat lost by one body must equal the amount gained by the other. Heat is energy in transit, so this principle is really just conservation of energy. We take each quantity of heat *added to* a body as *positive* and each quantity *leaving* a body as *negative*. When several bodies interact, the *algebraic sum* of the quantities of heat transferred to all the bodies must be zero. Calorimetry, dealing entirely with one conserved quantity, is in many ways the simplest of all physical theories!

Problem–Solving Strategy

CALORIMETRY PROBLEMS

1. To avoid confusion about algebraic signs when calculating quantities of heat, use Eqs. (15–13) and (15–20) consistently for each body, noting that each Q is positive when heat enters a body and negative when it leaves. The algebraic sum of all the Q's must be zero.

2. Often you will need to find an unknown temperature. Represent it by an algebraic symbol such as T. Then if a body has an initial temperature of 20°C and an unknown final temperature T, the temperature change for the body is $\Delta T = T_{final} - T_{initial} = T - 20°C$ (not $20°C - T$).

3. In problems in which a phase change takes place, as when ice melts, you may not know in advance whether all the material undergoes a phase change or only part of it. You can always assume one or the other, and if the resulting calculation gives an absurd result (such as a final temperature that is higher or lower than all of the initial temperatures), you know that the initial assumption was wrong. Back up and try again!

EXAMPLE 15–9

A temperature change with no phase change A geologist working in the field drinks her morning coffee out of an aluminum cup. The cup has a mass of 0.120 kg and is initially at 20.0°C when she pours in 0.300 kg of coffee initially at 70.0°C. What is the final temperature after the coffee and the cup attain thermal equilibrium? (Assume that coffee has the same specific heat capacity as water and that there is no heat exchange with the surroundings.)

SOLUTION No phase changes occur in this situation, so we need use only Eq. (15–13). Call the final temperature T. By using Table 15–3, the (negative) heat gained by the coffee is

$$Q_{coffee} = m_{coffee} c_{water} \, \Delta T_{coffee}$$
$$= (0.300 \text{ kg})(4190 \text{ J/kg} \cdot \text{K})(T - 70.0°C).$$

The (positive) heat gained by the aluminum cup is

$$Q_{aluminum} = m_{aluminum} c_{aluminum} \, \Delta T_{aluminum}$$
$$= (0.120 \text{ kg})(910 \text{ J/kg} \cdot \text{K})(T - 20.0°C).$$

We equate the sum of these two quantities of heat to zero, obtaining an algebraic equation for T:

$$Q_{coffee} + Q_{aluminum} = 0, \quad \text{or}$$

$$(0.300 \text{ kg})(4190 \text{ J/kg} \cdot \text{K})(T - 70.0°C)$$
$$+ (0.120 \text{ kg})(910 \text{ J/kg} \cdot \text{K})(T - 20.0°C) = 0.$$

Solution of this equation gives $T = 66.0°C$. The final temperature is much closer to the initial temperature of the coffee than to that of the cup; water has a much larger specific heat capacity than aluminum, and we have over twice as much mass of water. We can also find the quantities of heat by substituting this value for T back into the original equations. We find that $Q_{coffee} = -5.0 \times 10^3$ J and $Q_{aluminum} = +5.0 \times 10^3$ J; Q_{coffee} is negative, which means that the coffee loses heat.

EXAMPLE 15–10

Changes in both temperature and phase A physics student wants to cool 0.25 kg of Diet Omni-Cola (mostly water), initially at 25°C, by adding ice initially at −20°C. How much ice should she add so that the final temperature will be 0°C with all the ice melted if the heat capacity of the container may be neglected?

SOLUTION The Omni-Cola loses heat, so the heat added to it is negative:

$$Q_{Omni} = m_{Omni} c_{water} \, \Delta T_{Omni}$$
$$= (0.25 \text{ kg})(4190 \text{ J/kg} \cdot \text{K})(0°C - 25°C)$$
$$= -26,000 \text{ J}.$$

From Table 15–3, the specific heat capacity of ice (not the same as for liquid water) is 2.1×10^3 J/kg · K. Let the mass of ice be m_{ice}; then the heat Q_1 needed to warm it from −20°C to 0°C is

$$Q_1 = m_{ice} c_{ice} \, \Delta T_{ice}$$
$$= m_{ice} (2.1 \times 10^3 \text{ J/kg} \cdot \text{K})(0°C - (-20°C))$$
$$= m_{ice} (4.2 \times 10^4 \text{ J/kg}).$$

From Eq. (15–20) the additional heat Q_2 needed to melt this mass of ice is the mass times the heat of fusion. Using Table 15–4, we find

$$Q_2 = m_{ice} L_f$$
$$= m_{ice} (3.34 \times 10^5 \text{ J/kg}).$$

The sum of these three quantities must equal zero:

$$Q_{Omni} + Q_1 + Q_2 = -26,000 \text{ J} + m_{ice} (42,000 \text{ J/kg})$$
$$+ m_{ice} (334,000 \text{ J/kg}) = 0.$$

Solving this for m_{ice}, we get $m_{ice} = 0.069$ kg = 69 g (three or four medium-size ice cubes).

EXAMPLE 15–11

What's cooking? A heavy copper pot of mass 2.0 kg (including the copper lid) is at a temperature of 150°C. You pour 0.10 kg of water at 25°C into the pot, then quickly close the lid of the pot so that no steam can escape. Find the final temperature of the pot and its contents, and determine the phase (liquid or gas) of the water. Assume that no heat is lost to the surroundings.

SOLUTION There are three conceivable outcomes in this situation. One, none of the water boils, and the final temperature is less than 100°C; two, a portion of the water boils, giving a mixture of water and steam at 100°C; or three, all the water boils, giving 0.10 kg of steam at a temperature of 100°C or greater.

The simplest case to calculate is the first possibility, so let's try that first. Let the common final temperature of the liquid water and the copper pot be T. Since we are assuming that no phase changes take place, the sum of the quantities of heat added to the two materials is

$$Q_{water} + Q_{copper} = m_{water}c_{water}(T - 25°C)$$
$$+ m_{copper}c_{copper}(T - 150°C)$$
$$= (0.10 \text{ kg})(4190 \text{ J/kg} \cdot \text{K})(T - 25°C)$$
$$+ (2.0 \text{ kg})(390 \text{ J/kg} \cdot \text{K})(T - 150°C)$$
$$= 0.$$

Solving this for T, we find $T = 106°C$. But this is above the boiling point of water, which contradicts our assumption that none of the water boils! So this assumption can't be correct; at least some of the water undergoes a phase change.

If we try the second possibility, in which the final temperature is 100°C, we have to find the fraction of water that changes to the gaseous phase. Let this fraction be x; the (positive)

amount of heat needed to vaporize this water is $(xm_{water}) L_v$. Setting the final temperature T equal to 100°C, we have

$$Q_{water} = m_{water}c_{water}(100°C - 25°C) + xm_{water}L_v$$
$$= (0.10 \text{ kg})(4190 \text{ J/kg} \cdot \text{K})(75 \text{ K})$$
$$+ x(0.10 \text{ kg})(2.256 \times 10^6 \text{ J/kg})$$
$$= 3.14 \times 10^4 \text{ J} + x(2.256 \times 10^5 \text{ J}),$$
$$Q_{copper} = m_{copper}c_{copper}(100°C - 150°C)$$
$$= (2.0 \text{ kg})(390 \text{ J/kg} \cdot \text{K})(-50 \text{ K}) = -3.90 \times 10^4 \text{ J}.$$

The requirement that the sum of all the quantities of heat be zero then gives

$$Q_{water} + Q_{copper} = 3.14 \times 10^4 \text{ J} + x(2.256 \times 10^5 \text{ J})$$
$$- 3.90 \times 10^4 \text{ J} = 0,$$
$$x = \frac{3.90 \times 10^4 \text{ J} - 3.14 \times 10^4 \text{ J}}{2.256 \times 10^5 \text{ J}} = 0.034.$$

This makes sense, and we conclude that the final temperature of the water and copper is 100°C. Of the original 0.10 kg of water, 0.034(0.10 kg) = 0.0034 kg = 3.4 g has been converted to steam at 100°C.

Had x turned out to be greater than 1, we would have again had a contradiction (the fraction of water that vaporized can't be greater than 1). In this case the third possibility would have been the correct description, all the water would have vaporized, and the final temperature would have been greater than 100°C. Can you show that this would have been the case if we had originally poured less than 15 g of 25°C water into the pot?

EXAMPLE 15–12

Combustion, temperature change, and phase change In a particular gasoline camp stove, 30% of the energy released in burning the fuel actually goes to heating the water in the pot on the stove. If we heat 1.00 L (1.00 kg) of water from 20°C to 100°C and boil 0.25 kg of it away, how much gasoline do we burn in the process?

SOLUTION The heat required to raise the temperature of the water from 20°C to 100°C is

$$Q_1 = mc \, \Delta T = (1.00 \text{ kg})(4190 \text{ J/kg} \cdot \text{K})(80 \text{ K})$$
$$= 3.35 \times 10^5 \text{ J}.$$

To boil 0.25 kg of water at 100°C requires

$$Q_2 = mL_v = (0.25 \text{ kg})(2.256 \times 10^6 \text{ J/kg}) = 5.64 \times 10^5 \text{ J}.$$

The total energy needed is the sum of these, or 8.99×10^5 J. This is only 0.30 of the total heat of combustion, so that energy is $(8.99 \times 10^5 \text{ J})/0.30 = 3.00 \times 10^6$ J. As we mentioned above, each gram of gasoline releases 46,000 J, so the mass of gasoline required is

$$\frac{3.00 \times 10^6 \text{ J}}{46,000 \text{ J/g}} = 65 \text{ g},$$

or a volume of about 0.09 L of gasoline.

15–8 MECHANISMS OF HEAT TRANSFER

We have talked about *conductors* and *insulators,* materials that permit or prevent heat transfer between bodies. Now let's look in more detail at *rates* of energy transfer. In the kitchen you use an aluminum pot for good heat transfer from the stove to whatever you're cooking, but your refrigerator is insulated with a material that *prevents* heat from flowing into the food inside the refrigerator. How do we describe the difference between these two materials?

The three mechanisms of heat transfer are conduction, convection, and radiation. *Conduction* occurs within a body or between two bodies in contact. *Convection* depends on motion of mass from one region of space to another. *Radiation* is heat transfer by electromagnetic radiation, such as sunshine, with no need for matter to be present in the space between bodies.

CONDUCTION

If you hold one end of a copper rod and place the other end in a flame, the end you are holding gets hotter and hotter, even though it is not in direct contact with the flame. Heat reaches the cooler end by **conduction** through the material. On the atomic level, the atoms in the hotter regions have more kinetic energy, on the average, than their cooler neighbors. They jostle their neighbors, giving them some of their energy. The neighbors jostle *their* neighbors, and so on through the material. The atoms themselves do not move from one region of material to another, but their energy does.

Most metals also use another, more effective mechanism to conduct heat. Within the metal, some electrons can leave their parent atoms and wander through the crystal lattice. These "free" electrons can rapidly carry energy from the hotter to the cooler regions of the metal, so metals are generally good conductors of heat. A metal rod at 20°C feels colder than a piece of wood at 20°C because heat can flow more easily from your hand into the metal. The presence of "free" electrons also causes most metals to be good electrical conductors.

Heat transfer occurs only between regions that are at different temperatures, and the direction of heat flow is always from higher to lower temperature. Figure 15–13 shows a rod of conducting material with cross-section area A and length L. The left end of the rod is kept at a temperature T_H and the right end at a lower temperature T_C, and heat flows from left to right. The sides of the rod are covered by an ideal insulator, so no heat transfer occurs at the sides.

When a quantity of heat dQ is transferred through the rod in a time dt, the rate of heat flow is dQ/dt. We call this rate the **heat current,** denoted by H. That is, $H = dQ/dt$. Experiments show that the heat current is proportional to the cross-section area A of the rod and to the temperature difference $(T_H - T_C)$ and is inversely proportional to the rod length L. Introducing a proportionality constant k called the **thermal conductivity** of the material, we have

$$H = \frac{dQ}{dt} = kA\frac{T_H - T_C}{L} \quad \text{(heat current in conduction).} \quad (15\text{–}21)$$

The quantity $(T_H - T_C)/L$ is the temperature difference *per unit length;* it is called the magnitude of the **temperature gradient.** The numerical value of k depends on the material of the rod. Materials with large k are good conductors of heat; materials with small k are poor conductors or insulators. Equation (15–21) also gives the heat current through a slab or through *any* homogeneous body with uniform cross section A perpendicular to the direction of flow; L is the length of the heat-flow path.

The units of heat current H are units of energy per time, or power; the SI unit of heat current is the watt (1 W = 1 J/s). We can find the units of k by solving Eq. (15–21) for k. We invite you to verify that the SI units of k are W/m · K. Some numerical values of k are given in Table 15–5.

The thermal conductivity of "dead" (that is, nonmoving) air is very small. A wool sweater keeps you warm because it traps air between the fibers. In fact, many insulating

15–13 Steady-state heat flow due to conduction in a uniform rod.

TABLE 15–5
THERMAL CONDUCTIVITIES

SUBSTANCE	k (W/m · K)
Metals	
Aluminum	205.0
Brass	109.0
Copper	385.0
Lead	34.7
Mercury	8.3
Silver	406.0
Steel	50.2
Various solids (representative values)	
Brick, insulating	0.15
Brick, red	0.6
Concrete	0.8
Cork	0.04
Felt	0.04
Fiberglass	0.04
Glass	0.8
Ice	1.6
Rock wool	0.04
Styrofoam	0.01
Wood	0.12–0.04
Gases	
Air	0.024
Argon	0.016
Helium	0.14
Hydrogen	0.14
Oxygen	0.023

15-14 This protective tile, developed for use in the space shuttle, has extraordinary thermal properties. The extremely small thermal conductivity and small heat capacity of the material make it possible to hold the tile by its edges, even though its temperature is high enough to emit the light for this photograph.

materials such as styrofoam and fiberglass are mostly dead air. Figure 15-14 shows a ceramic material with very unusual thermal properties, including very small conductivity.

If the temperature varies in a non-uniform way along the length of the conducting rod, we introduce a coordinate x along the length and generalize the temperature gradient to be dT/dx. The corresponding generalization of Eq. (15-21) is

$$H = \frac{dQ}{dt} = -kA\frac{dT}{dx}. \tag{15-22}$$

The negative sign shows that heat always flows in the direction of *decreasing* temperature.

For thermal insulation in buildings, engineers use the concept of **thermal resistance,** denoted by R. The thermal resistance R of a slab of material with area A is defined so that the heat current H through the slab is

$$H = \frac{A(T_H - T_C)}{R}, \tag{15-23}$$

where T_H and T_C are the temperatures on the two sides of the slab. Comparing this with Eq. (15-21), we see that R is given by

$$R = \frac{L}{k}, \tag{15-24}$$

where L is the thickness of the slab. The SI unit of R is $1 \text{ m}^2 \cdot \text{K/W}$. In the units used for commercial insulating materials in the United States, H is expressed in Btu/h, A is in ft^2, and $T_H - T_C$ in F°. (1 Btu/h = 0.293 W.) The units of R are then $\text{ft}^2 \cdot \text{F°} \cdot \text{h/Btu}$, though values of R are usually quoted without units; a 6-inch-thick layer of fiberglass has an R value of 19 (that is, $R = 19 \text{ ft}^2 \cdot \text{F°} \cdot \text{h/Btu}$), a two-inch-thick slab of polyurethane foam has a value of 12, and so on. Doubling the thickness doubles the R-value. Common practice in new construction in severe northern climates is to specify R values of around 30 for exterior walls and ceilings. When the insulating material is in layers, such as a plastered wall, fiberglass insulation, and wood exterior siding, the R values are additive. Do you see why? (See Problem 15-100.)

Electronics engineers concerned with cooling microprocessor chips use a different definition of thermal resistance. This subject is discussed in Section 15-9.

Problem-Solving Strategy

HEAT CONDUCTION

1. Identify the direction of heat flow in the problem (from hot to cold). In Eq. (15-21), L is always measured along this direction, and A is always an area perpendicular to this direction. Often when a box or other container has an irregular shape but uniform wall thickness, you can approximate it as a flat slab with the same thickness and total wall area.

2. In some problems the heat flows through two different materials in succession. The temperature at the interface between the two materials is then intermediate between T_H and T_C; represent it by a symbol such as T. The temperature differences for the two materials are then $(T_H - T)$ and $(T - T_C)$. In steady-state heat flow, the same heat has to pass through both materials in succession, so

the heat current H must be *the same* in both materials. It's like a series electric circuit (to be discussed in Chapter 25).

3. If there are two *parallel* heat flow paths, so that some heat flows through each, then the total H is the sum of the quantities H_1 and H_2 for the separate paths. An example is heat flow from inside to outside a house, both through the glass in a window and through the surrounding frame. In this case the temperature difference is the same for the two paths, but L, A, and k may be different for the two paths. This is like a parallel electric circuit.

4. As always, it is essential to use a consistent set of units. If you use a value of k expressed in W/m · K, don't use distances in cm, heat in calories, or T in °F!

EXAMPLE 15-13

Conduction through a picnic cooler A Styrofoam box used to keep drinks cold at a picnic has total wall area (including the lid) of 0.80 m² and wall thickness 2.0 cm. It is filled with ice, water, and cans of Omni-Cola at 0°C. What is the rate of heat flow into the box if the temperature of the outside wall is 30°C? How much ice melts in one day?

SOLUTION We assume that the total heat flow is approximately the same as it would be through a flat slab of area 0.80 m² and thickness 2.0 cm = 0.020 m (Fig. 15–15). We find k from Table 15–5. From Eq. (15–21) the heat current (rate of heat flow) is

$$H = kA \frac{T_H - T_C}{L} = (0.010 \text{ W/m} \cdot \text{K})(0.80 \text{ m}^2) \frac{30°C - 0°C}{0.020 \text{ m}}$$

$$= 12 \text{ W} = 12 \text{ J/s}.$$

The total heat flow Q in one day (86,400 s) is

$$Q = Ht = (12 \text{ J/s})(86,400 \text{ s}) = 1.04 \times 10^6 \text{ J}.$$

The heat of fusion of ice is 3.34×10^5 J/kg, so the quantity of ice melted by this quantity of heat is

$$m = \frac{Q}{L_f}$$

$$= \frac{1.04 \times 10^6 \text{ J}}{3.34 \times 10^5 \text{ J/kg}} = 3.1 \text{ kg}.$$

15–15 Conduction of heat. We can approximate heat flow through the walls of a picnic cooler by heat flow through a single flat slab of styrofoam.

EXAMPLE 15-14

A steel bar 10.0 cm long is welded end-to-end to a copper bar 20.0 cm long (Fig. 15–16). Both bars are insulated perfectly on their sides. Each bar has a square cross section, 2.00 cm on a side. The free end of the steel bar is maintained at 100°C by placing it in contact with steam, and the free end of the copper bar is maintained at 0°C by placing it in contact with ice. Find the temperature at the junction of the two bars and the total rate of heat flow.

15–16 Heat flow along two metal bars, one of steel and one of copper, connected end-to-end.

Rearranging and solving for T, we find

$$T = 20.7°C.$$

Even though the steel bar is shorter, the temperature drop across it is much greater than across the copper bar because steel is a much poorer conductor.

We can find the total heat current by substituting this value for T back into either of the above expressions:

$$H_{steel} = \frac{(50.2 \text{ W/m} \cdot \text{K})(0.0200 \text{ m})^2(100°C - 20.7°C)}{0.100 \text{ m}} = 15.9 \text{ W},$$

or

$$H_{copper} = \frac{(385 \text{ W/m} \cdot \text{K})(0.0200 \text{ m})^2(20.7°C)}{0.200 \text{ m}} = 15.9 \text{ W}.$$

SOLUTION As we discussed in the Problem-Solving Strategy, the heat currents in the two bars must be equal; this is the key to the solution. Let T be the unknown junction temperature; we use Eq. (15–21) for each bar and equate the two expressions:

$$H_{steel} = \frac{k_{steel}A(100°C - T)}{L_{steel}} = H_{copper} = \frac{k_{copper}A(T - 0°C)}{L_{copper}}.$$

The areas A are equal and may be divided out. Substituting $L_{steel} = 0.100$ m, $L_{copper} = 0.200$ m, and numerical values of k from Table 15–5, we find

$$\frac{(50.2 \text{ W/m} \cdot \text{K})(100°C - T)}{0.100 \text{ m}} = \frac{(385 \text{ W/m} \cdot \text{K})(T - 0°C)}{0.200 \text{ m}}.$$

EXAMPLE 15–15

In Example 15–14, suppose the two bars are separated. One end of each bar is maintained at 100°C, and the other end of each bar is maintained at 0°C (Fig. 15–17). What is the *total* rate of heat flow in the two bars?

SOLUTION In this case the bars are in parallel rather than in series. The total heat current is now the *sum* of the currents in the two bars, and for each bar, $T_H - T_C = 100°C - 0°C = 100$ K:

$$H = H_{steel} + H_{copper} = \frac{k_{steel}A(T_H - T_C)}{L_{steel}} + \frac{k_{copper}A(T_H - T_C)}{L_{copper}}$$

$$= \frac{(50.2 \text{ W/m} \cdot \text{K})(0.0200 \text{ m})^2(100 \text{ K})}{0.100 \text{ m}}$$

$$+ \frac{(385 \text{ W/m} \cdot \text{K})(0.0200 \text{ m})^2(100 \text{ K})}{0.200 \text{ m}}$$

$$= 20.1 \text{ W} + 77.0 \text{ W} = 97.1 \text{ W}.$$

The heat flow in the copper bar is much greater than that in the steel bar, even though it is longer, because the thermal conductivity of copper is much larger. The total heat flow is much greater than in Example 15–14, partly because the total cross

15–17 Heat flow along two metal bars, one of steel and one of copper, parallel to and separated from each other.

section for heat flow is greater and partly because the full 100-K temperature difference appears across each bar.

CONVECTION

Convection is transfer of heat by mass motion of a fluid from one region of space to another. Familiar examples include hot-air and hot-water home heating systems, the cooling system of an automobile engine, and the flow of blood in the body. If the fluid is circulated by a blower or pump, the process is called *forced convection;* if the flow is caused by differences in density due to thermal expansion, such as hot air rising, the process is called *natural convection* or *free convection.*

Free convection in the atmosphere plays a dominant role in determining the daily weather (Fig. 15–18), and convection in the oceans is an important global heat-transfer mechanism. On a smaller scale, soaring hawks and glider pilots make use of thermal updrafts from the warm earth. Sometimes the updrafts are strong enough to form a thunderstorm (Fig. 15–19). The most important mechanism for heat transfer within the human body (needed to maintain nearly constant temperature in various environments) is *forced* convection of blood, with the heart serving as the pump.

Convective heat transfer is a very complex process, and there is no simple equation to describe it. Here are a few experimental facts:

1. The heat current due to convection is directly proportional to the surface area. This is the reason for the large surface areas of radiators and cooling fins.
2. The viscosity of fluids slows natural convection near a stationary surface, giving a surface film that on a vertical surface typically has about the same insulating value as 1.3 cm of plywood (R value 0.7). Forced convection decreases the thickness of this film, increasing the rate of heat transfer. This is the reason for the "wind-chill factor"; you get cold faster in a cold wind than in still air with the same temperature.
3. The heat current due to convection is found to be approximately proportional to the $\frac{5}{4}$ power of the temperature difference between the surface and the main body of fluid.

A heating element in the tip of this submerged tube warms the surrounding water, producing a complex pattern of free convection.

RADIATION

Radiation is the transfer of heat by electromagnetic waves such as visible light, infrared, and ultraviolet radiation. Everyone has felt the warmth of the sun's radiation

(a)   (b)

and the intense heat from a charcoal grill or the glowing coals in a fireplace. Most of the heat from these very hot bodies reaches you not by conduction or convection in the intervening air but by *radiation*. This heat transfer would occur even if there were nothing but vacuum between you and the source of heat.

Every body, even at ordinary temperatures, emits energy in the form of electromagnetic radiation. At ordinary temperatures, say 20°C, nearly all the energy is carried by infrared waves with wavelengths much longer than those of visible light. As the temperature rises, the wavelengths shift to shorter values. At 800°C a body emits enough visible radiation to be self-luminous and appears "red-hot," although even at this temperature most of the energy is carried by infrared waves. At 3000°C, the temperature of an incandescent lamp filament, the radiation contains enough visible light so the body appears "white-hot."

The rate of energy radiation from a surface is proportional to the surface area A. The rate increases very rapidly with temperature, depending on the fourth power of the absolute (Kelvin) temperature. The rate also depends on the nature of the surface; this

15–18 Water has a higher specific heat capacity than does land. The heat of the sun has comparatively little effect on the temperature of the sea; by contrast, the land heats up during the day and cools off rapidly at night. Near the shore the temperature difference between land and sea gives rise to daytime sea breezes and nighttime land breezes.

15–19 Convection in a thunderstorm. Hot, moist air is less dense than cool, dry air and hence is lifted by buoyancy. The lifting can continue to altitudes of 12,000 to 18,000 m (40,000 to 60,000 ft). A typical thunderstorm 5 km in diameter may contain 5×10^8 kg of water; when this lifted moisture condenses to form raindrops or hailstones, about 10^{15} J of heat is released into the upper atmosphere (equal to the electrical energy used by a city of 100,000 people in a month). Air is pulled along with the falling rain and hail, creating strong downdrafts and violent surface winds. The downdrafts weaken as the energy supplied by the lifted moisture is used up.

dependence is described by a quantity e called the **emissivity.** This is a dimensionless number between 0 and 1, representing the ratio of the rate of radiation from a particular surface to the rate of radiation from an equal area of an ideal radiating surface at the same temperature. Emissivity also depends somewhat on temperature. Thus the heat current $H = dQ/dt$ due to radiation from a surface area A with emissivity e at absolute temperature T can be expressed as

$$H = Ae\sigma T^4 \qquad \text{(heat current in radiation)}, \qquad (15\text{--}25)$$

where σ is a fundamental physical constant called the **Stefan-Boltzmann constant.** This relation is called the **Stefan-Boltzmann law** in honor of its late nineteenth century discoverers. The current best numerical value of σ is

$$\sigma = 5.67051(19) \times 10^{-8} \text{ W/m}^2 \cdot \text{K}^4.$$

We invite you to check unit consistency in Eq. (15–25). Emissivity (e) is often larger for dark surfaces than for light ones. The emissivity of a smooth copper surface is about 0.3, but e for a dull black surface can be close to unity.

EXAMPLE 15–16

A thin square steel plate, 10 cm on a side, is heated in a black-smith's forge to a temperature of 800°C. If the emissivity is 0.60, what is the total rate of radiation of energy?

SOLUTION The total surface area, including both sides, is $2(0.10 \text{ m})^2 = 0.020 \text{ m}^2$. We must convert the temperature to the

Kelvin scale; 800°C = 1073 K. Then Eq. (15–25) gives

$$H = Ae\sigma T^4$$
$$= (0.020 \text{ m}^2)(0.60)(5.67 \times 10^{-8} \text{ W/m}^2 \cdot \text{K}^4)(1073 \text{ K})^4$$
$$= 900 \text{ W.}$$

While a body at absolute temperature T is radiating, its surroundings at temperature T_s are also radiating, and the body *absorbs* some of this radiation. If it is in thermal equilibrium with its surroundings, $T = T_s$ and the rates of radiation and absorption must be equal. For this to be true, the rate of absorption must be given in general by $H = Ae\sigma T_s^4$. Then the *net* rate of radiation from a body at temperature T with surroundings at temperature T_s is

$$H_{\text{net}} = Ae\sigma T^4 - Ae\sigma T_s^4 = Ae\sigma(T^4 - T_s^4). \qquad (15\text{--}26)$$

In this equation a positive value of H means a net heat flow *out of* the body. Equation (15–26) shows that for radiation, as for conduction and convection, the heat current depends on the temperature *difference* between two bodies.

EXAMPLE 15–17

Radiation from the human body If the total surface area of the human body is 1.20 m^2 and the surface temperature is 30°C = 303 K, find the total rate of radiation of energy from the body. If the surroundings are at a temperature of 20°C, what is the *net* rate of heat loss from the body by radiation? The emissivity of the body is very close to unity, irrespective of skin pigmentation.

SOLUTION The rate of radiation of energy per unit area is given by Eq. (15–25). Taking $e = 1$, we find

$$H = Ae\sigma T^4$$
$$= (1.20 \text{ m}^2)(1)(5.67 \times 10^{-8} \text{ W/m}^2 \cdot \text{K}^4)(303 \text{ K})^4$$
$$= 574 \text{ W.}$$

This loss is partly offset by *absorption* of radiation, which depends on the temperature of the surroundings. The *net* rate of radiative energy transfer is given by Eq. (15–26):

$$H = Ae\sigma(T^4 - T_s^4)$$
$$= (1.20 \text{ m}^2)(1)(5.67 \times 10^{-8} \text{ W/m}^2 \cdot \text{K}^4)[(303 \text{ K})^4 - (293 \text{ K})^4]$$
$$= 72 \text{ W.}$$

This is positive because the body is losing heat to its colder surroundings.

Heat transfer by radiation is important in some surprising places. A premature baby in an incubator can be cooled dangerously by radiation if the walls of the incubator happen to be cold, even when the *air* in the incubator is warm. Some incubators regulate the air temperature by measuring the baby's skin temperature.

A body that is a good absorber must also be a good emitter. An ideal radiator, with an emissivity of unity, is also an ideal absorber, absorbing *all* of the radiation that strikes it. Such an ideal surface is called an ideal black body or simply a **blackbody.** Conversely, an ideal *reflector,* which absorbs *no* radiation at all, is also a very ineffective radiator.

This is the reason for the silver coatings on vacuum ("Thermos") bottles, invented by Sir James Dewar (1842–1923). A vacuum bottle has double glass walls. The air is pumped out of the spaces between the walls; this eliminates nearly all heat transfer by conduction and convection. The silver coating on the walls reflects most of the radiation from the contents back into the container, and the wall itself is a very poor emitter. Thus a vacuum bottle can keep coffee or soup hot for several hours. The Dewar flask, used to store very cold liquefied gases, is exactly the same in principle.

15–9 INTEGRATED CIRCUITS

A Case Study in Heat Transfer

Integrated circuits (ic's) and VLSI (very large scale integration) chips are at the heart of most modern electronic devices, including computers, stereo systems, and fuel-injection systems in car engines (Fig. 15–20). The widespread use of these devices has posed new and interesting problems in heat transfer. Some of the electrical energy associated with the electric current in a chip is dissipated in the chip as heat; if a chip gets too hot as a result, the circuits become unreliable or are permanently destroyed. Keeping chips cool during operation is vitally important.

To introduce the problem, let's think about what happens when you turn on a light bulb. Electrical energy is dissipated as heat in the filament; the light bulb warms up until it reaches a temperature at which the rate of transfer of energy from the light bulb to its surroundings (by radiation as well as by conduction and convection to the surrounding air) just balances the rate of electrical energy input. A standard 60-W light bulb has a power input of 60 W and a surface area of about 120 cm². In equilibrium its power loss to its surroundings per unit surface area, or *power density,* is (60 W)/(120 cm²) = 0.5 W/cm².

15–20 (a) Modern ic chips are usually placed in plastic or ceramic packages for protection. The packages contain conducting paths that lead from the chip to outside pins, which connect the circuit to the overall circuit board in the final product. (b) This ceramic package was specifically designed to ensure r_{th} of less than 1 K/W.

(a)   (b)

The heating element on an electric range has a greater power loss relative to its surface area, so it has a larger power density of 2–3 W/cm^2.

Now consider the *much larger* power dissipation values that occur in microelectronics. Many thousands of electronic elements are packed tightly together on a silicon chip only a few millimeters on a side. Power densities for present-day VLSI chips range up to 40 W/cm^2. For comparison, the ceramic tiles that protect the space shuttle when it reenters the earth's atmosphere typically have to dissipate 100 W/cm^2 (see Fig. 15–14).

For an ic chip in a plastic package, the maximum safe temperature is about 100°C. A chip in a ceramic package can operate reliably up to about 120°C. To see whether the temperature of an operating chip is within limits, we use the same principle as for a light bulb: power input equals power output. The power output H (the rate of heat transfer out of the chip) is approximately proportional to the difference $T_{ic} - T_{amb}$ between the temperature T_{ic} of the chip and the ambient temperature T_{amb} (temperature of the surroundings). Using a proportionality constant r_{th} that depends on the shape and size of the ic, we express the rate of heat loss H as

$$H = \frac{T_{ic} - T_{amb}}{r_{th}}. \tag{15–27}$$

(Note that r_{th} is the same as R/A in Eq. (15–23).) When the ic reaches its final operating temperature, the rate of heat loss must equal the electrical power P dissipated in the device. Equating H and P and solving for T_{ic}, we find

$$T_{ic} = T_{amb} + r_{th}P. \tag{15–28}$$

Values of r_{th} for common ic packages in still air vary from 30 to 70 K/W (30 to 70 C°/W). For example, one watt of electrical power ($P = 1$ W) into an ic will raise the temperature of a 40-pin plastic package to about 62 C° = 62 K above the surrounding air temperature. For this unit, $r_{th} = 62$ K/W.

EXAMPLE 15–18

A 40-pin ceramic package has $r_{th} = 40$ K/W. If the maximum temperature the circuit can safely reach is 120°C, what is the highest power level at which the circuit can safely operate in an ambient temperature of 75°C?

This power level is adequate in many typical ic applications, but chips used in high-speed computing applications often require considerably higher power levels.

SOLUTION We use Eq. (15–27), replacing H by P:

$$P = \frac{T_{ic} - T_{amb}}{r_{th}}$$
$$= \frac{120°C - 75°C}{40 K/W} = 1.1 \text{ W}.$$

To remove heat from the chip more efficiently, we can force air past the circuit, improving the heat transfer by convection. Blowing air through the system at a rate of about 20 m^3/min can reduce the effective value of r_{th} for the chip and package by about 10 to 15 K/W. This is an improvement, but in many cases it is still not enough cooling for a high-performance ic.

One method of cooling that is currently being investigated is direct immersion of the ic package in a fluorocarbon fluid. These fluids are electrical insulators and are chemically inert, making them compatible with operation of electronic components. Their thermal transport properties are not very favorable because of small values of thermal conductivity and heat of vaporization. To improve heat transfer in these fluids, design engineers have tried forced convection, structural modification of packages, and boiling.

In forced convection, microscopic channels are cut into the back of the silicon base of the ic chip. Typical channel dimensions are 50 μm wide and 300 μm deep. Experiments with water (not suitable for cooling an actual ic) have achieved heat transfers as large as 790 W per square centimeter of surface area.

Adding fins to the basic cylindrical pin form of an ic package can increase the heated surface area of the package by a factor of 8 to 12. The larger surface area increases heat transfer by convection, conduction, and radiation (see Eqs. (15–23) and (15–25)), reducing the effective value of r_{th} by as much as a factor of 20.

Heat flow values of 45 W/cm^2 have been obtained by allowing the cooling liquid to boil so that heat transfer by convection involves both liquid and vapor at the same time. However, the sudden drop in temperature at the surface of the ic package when boiling begins results in thermal stress on the package. Further work in package design is underway to try to minimize this problem.

Chip packaging has become both an art and a science. Careful layout of electronic components on the circuit board can minimize the overall thermal resistance of the chip, allowing the heat generated to dissipate more readily. Much recent attention has been devoted to research and development of new packages, so there will be rapid changes over the next few years in the heat limitations of ic technology.

SUMMARY

- A thermometer measures temperature. Two bodies in thermal equilibrium must have the same temperature. A conducting material between two bodies permits interaction leading to thermal equilibrium; an insulating material prevents or impedes this interaction.

- The Celsius and Fahrenheit temperature scales are based on the freezing temperature ($0°C = 32°F$) and boiling temperature ($100°C = 212°F$) of water. They are related by

$$T_F = \frac{9}{5} T_C + 32°, \quad (15\text{–}1)$$

$$T_C = \frac{5}{9} (T_F - 32°), \quad (15\text{–}2)$$

and $1\,C° = \frac{9}{5}F°$.

- The Kelvin scale has its zero at the extrapolated zero-pressure temperature for a constant-volume gas thermometer, which is $-273.15°C$. Thus $0\,K = -273.15°C$, and

$$T_K = T_C + 273.15. \quad (15\text{–}3)$$

In the gas-thermometer scale, the ratio of two temperatures is defined to be equal to the ratio of the two corresponding gas-thermometer pressures:

$$\frac{T_2}{T_1} = \frac{p_2}{p_1}. \quad (15\text{–}4)$$

The triple-point temperature of water ($0.01°C$) is defined to be 273.16 K.

- Under a temperature change ΔT, any linear dimension L_0 of a solid body changes by an amount ΔL given approximately by

$$\Delta L = \alpha L_0 \,\Delta T, \quad (15\text{–}6)$$

where α is the coefficient of linear expansion. Under a temperature change ΔT, the change ΔV in the volume V_0 of any solid or liquid material is given

KEY TERMS

thermodynamics, 460
temperature, 460
thermometer, 461
thermal equilibrium, 461
insulator, 461
conductor, 461
zeroth law of thermodynamics, 462
Celsius temperature scale, 462
Fahrenheit temperature scale, 462
Kelvin temperature scale, 464
absolute temperature scale, 465
absolute zero, 465
coefficient of linear expansion, 466
coefficient of volume expansion, 467
thermal stress, 469
heat, 470
calorie, 470
British thermal unit, 470
specific heat capacity, 471
molar heat capacity, 472
phase, 473
phase change, 473
heat of fusion, 474
phase equilibrium, 474
heat of vaporization, 474

approximately by

$$\Delta V = \beta V_0 \, \Delta T, \qquad (15\text{–}8)$$

where β is the coefficient of volume expansion. For solids, $\beta = 3\alpha$.

- When a material is cooled or heated and held so that it cannot contract or expand, the tensile stress F/A is given by

$$\frac{F}{A} = -Y\alpha \, \Delta T. \qquad (15\text{–}12)$$

- Heat is energy in transit from one place to another as a result of a temperature difference. The quantity of heat Q required to raise the temperature of a mass m of material by a small amount ΔT is

$$Q = mc \, \Delta T, \qquad (15\text{–}13)$$

where c is the specific heat capacity of the material. When the quantity of material is given by the number of moles n, the corresponding relation is

$$Q = nC \, \Delta T, \qquad (15\text{–}18)$$

where $C = Mc$ is the molar heat capacity (M is the molar mass). The number of moles n and the mass m of material are related by $m = nM$.

- The molar heat capacities of many solid elements are approximately 25 J/mol $\cdot$ K; this is the rule of Dulong and Petit.

- To change a mass m of a material to a different phase at the same temperature (such as liquid to vapor or liquid to solid) requires the addition or subtraction of a quantity of heat Q given by

$$Q = \pm mL, \qquad (15\text{–}20)$$

where L is the heat of fusion, vaporization, or sublimation.

- When heat is added to a body, the corresponding Q is positive; when it is removed, Q is negative. The basic principle of calorimetry comes from conservation of energy. In an isolated system whose parts interact by heat exchange, the algebraic sum of the Q's for all parts of the system must be zero.

- The three mechanisms of heat transfer are conduction, convection, and radiation. Conduction is transfer of energy of molecular motion within materials without bulk motion of the materials. Convection involves mass motion from one region to another, and radiation is energy transfer through electromagnetic radiation.

- The heat current H for conduction depends on the area A through which the heat flows, the length L of the heat path, the temperature difference $(T_H - T_C)$, and the thermal conductivity k of the material:

$$H = \frac{dQ}{dt} = kA \frac{T_H - T_C}{L}. \qquad (15\text{–}21)$$

- Heat transfer by convection is a complex process, depending on surface area, orientation, and the temperature difference between a body and its surroundings.

- The heat current H due to radiation is given by

$$H = Ae\sigma T^4, \qquad (15\text{–}25)$$

where A is the surface area, e the emissivity of the surface (a pure number between 0 and 1), T the absolute temperature, and σ a fundamental constant called the Stefan-Boltzmann constant. When a body at temperature T is surrounded by material at temperature T_s, the net heat current from the body to its surroundings is

$$H_{\text{net}} = Ae\sigma T^4 - Ae\sigma T_s^{\,4} = Ae\sigma(T^4 - T_s^{\,4}). \qquad (15\text{–}26)$$

DISCUSSION QUESTIONS

Q15–1 A thermometer is laid out in direct sunlight. Does it measure the temperature of the air, or of the sun, or of something else? Explain.

Q15–2 What is the temperature of vacuum? Explain.

Q15–3 Does it make sense to say that one body is twice as hot as another? Why or why not?

Q15–4 Many automobile engines have cast-iron cylinders and aluminum pistons. What kinds of problems could occur if the engine gets too hot? (The coefficient of volume expansion of cast iron is approximately the same as that of steel.)

Q15–5 A piece of copper with a hole through it is placed in a bell jar and the air is pumped out. A student tells you that if the piece of copper warms up, the hole cannot expand because there are no longer any atoms in the hole to move apart. How would you explain the student's error to him?

Q15–6 Telescope mirrors must hold their curved shape precisely so that the images they produce will be in focus. Explain why this is a reason for not making the mirrors of large telescopes from ordinary glass, but from glasslike materials that have very low coefficients of thermal expansion.

Q15–7 Why do frozen water pipes burst? Would a mercury thermometer break if the temperature went below the freezing temperature of mercury? Why or why not?

Q15–8 Two bodies made of the same material have the same external dimensions and appearance, but one is solid and the other is hollow. When their temperature is increased, is the overall volume expansion the same or different? Why?

Q15–9 The inside of an oven is at a temperature of 200°C (392°F). You can put your hand in the oven without injury as long as you don't touch anything. But since the air inside the oven is also at 200°C, why isn't your hand burned just the same?

Q15–10 A newspaper article about the weather states that "the temperature of a body measures how much heat the body contains." Is this description correct? Why or why not?

Q15–11 To raise the temperature of an object, must you add heat to it? If you add heat to an object, must you raise its temperature? Explain.

Q15–12 A student asserted that a suitable unit for specific heat capacity is $1 \text{ m}^2/\text{s}^2 \cdot \text{C}°$. Is she correct? Why or why not?

Q15–13 Why do you think that the heat of vaporization for water is so much larger than the heat of fusion?

Q15–14 In some household air conditioners used in dry climates, air is cooled by blowing it through a water-soaked filter, evaporating some of the water. How does this cool the air? Would such a system work well in a high-humidity climate? Why or why not?

Q15–15 The units of specific heat capacity c are J/kg · K, but the units of heat of fusion L_f or heat of vaporization L_v are simply J/kg. Why don't the units of L_f and L_v include a factor of $(\text{K})^{-1}$ to account for a temperature change?

Q15–16 If you set an ice cube on your kitchen table, it will stay frozen longer if it is wrapped in a wet paper towel. Why?

Q15–17 Why does food cook faster in a pressure cooker than in an open pot of boiling water?

Q15–18 Desert travelers sometimes keep water in a canvas bag. Some water seeps through the bag and evaporates. How does this cool the water inside?

Q15–19 When you first step out of the shower, you feel cold. But as soon as you are dry you feel warmer, even though the room temperature does not change. Why?

Q15–20 The climate of regions adjacent to large bodies of water (like the Pacific or Atlantic coasts) is usually more moderate than that of regions far from large bodies of water (like the prairies). Why?

Q15–21 When water is placed in ice-cube trays in a freezer, why does the water not freeze all at once when the temperature has reached 0°C? In fact the water freezes first in a layer adjacent to the sides of the tray. Why?

Q15–22 Before giving you an injection, a physician swabs your arm with isopropyl alcohol at room temperature. Why does this make your arm feel cold? (*Hint:* The reason is *not* the fear of the injection! The boiling point of isopropyl alcohol is 82.4°C.)

Q15–23 A cold block of metal feels colder than a block of wood at the same temperature. Why? A *hot* block of metal feels hotter than a block of wood at the same temperature. Again, why? Is there any temperature at which the two blocks feel equally hot or cold? What temperature is this?

Q15–24 A person pours a cup of hot coffee, intending to drink it five minutes later. To keep the coffee as hot as possible, should she put cream in it now, or wait until just before she drinks it? Explain.

Q15–25 When a freshly baked apple pie has just been removed from the oven, the crust and filling are both at the same temperature. Yet if you sample the pie, the filling will burn your tongue but the crust will not. Why is there a difference? (*Hint:* The filling is moist while the crust is dry.)

Q15–26 Old-time kitchen lore suggests that things cook better (evenly and without burning) in heavy cast-iron pots. What desirable characteristics do such pots have?

Q15–27 Figure 15–18 shows that the land has a higher temperature than the ocean during the day, but a lower temperature at night. Explain why. (*Hint:* The specific heat capacity of soil is only 0.2–0.8 times as great as that of water.)

Q15–28 Why does moisture condense on the insides of windows on cold days? If it is cold enough outside, frost will form, but only when the outside temperature is far below freezing. Why is this so? Why are the bottoms of windows frostier than the tops? How do storm windows change the situation?

Q15–29 Why do glasses containing cold liquids sweat in warm weather? Glasses containing alcoholic drinks sometimes form frost on the outside. How can this happen?

Q15–30 It is well known that a potato bakes faster if a large nail is stuck through it. Why? Is an aluminum nail better than a steel one? Why or why not? (*Note:* Don't try this in a microwave oven!) There is also a gadget on the market to hasten roasting of meat, consisting of a hollow metal tube containing a wick and some water; this is claimed to be much better than a solid metal rod. How does this work?

Q15–31 Glider pilots in the Midwest know that thermal updrafts are likely to occur in the vicinity of freshly plowed fields. Why?

Q15–32 Some folks claim that ice cubes freeze faster if the trays are filled with hot water, because hot water cools off faster than cold water. What do you think?

Q15–33 We're lucky that the earth isn't in thermal equilibrium with the sun (which has a surface temperature of 5800 K). But why isn't it?

Q15–34 A microwave oven works by producing electromagnetic radiation that is absorbed by water molecules, raising the temperature of the water. This radiation can easily penetrate nonmetallic objects. Explain why a microwave oven can cook massive objects (such as potatoes and beef roasts) much faster than can a conventional oven. What disadvantages are associated with the reasons for this greater speed?

EXERCISES

SECTION 15–3 THERMOMETERS AND TEMPERATURE SCALES

15–1 Convert the following Celsius temperatures to Fahrenheit: a) $-62.8°C$, the lowest temperature ever recorded in North America (February 3, 1947, Snag, Yukon); b) $56.7°C$, the highest temperature ever recorded in the United States (July 10, 1913, Death Valley, California); c) $31.1°C$, the world's highest average annual temperature (Lugh Ferrandi, Somalia).

15–2 Find the Celsius temperatures corresponding to a) a winter night in Seattle ($41.0°F$); b) a hot summer day in Palm Springs ($107.0°F$); c) a cold winter day in northern Manitoba ($-18.0°F$).

15–3 One day in Saskatoon, Saskatchewan, the Fahrenheit and Celsius temperatures are the same. What are the temperatures and the season?

15–4 a) On January 22, 1943, the temperature in Spearfish, South Dakota, rose from $-4.0°F$ to $45.0°F$ in just two minutes. What was the temperature change in Celsius degrees? b) The temperature in Browning, Montana, was $44.0°F$ on January 23, 1916. The next day the temperature plummeted to $-56.0°F$. What was the temperature change in Celsius degrees?

15–5 a) You feel sick and are told that you have a temperature of $40.2°C$. What is your temperature in $°F$? Should you be concerned? b) The morning weather report in Sydney gives a current temperature of $12°C$. What is this temperature in $°F$?

15–6 A "blue norther" passes through Lubbock, Texas, one September afternoon and the temperature drops 11.8 C° in one hour. What is this temperature change in F°?

SECTION 15–4 GAS THERMOMETERS AND THE KELVIN SCALE

15–7 Convert the following record-setting temperatures to the Kelvin scale: a) the lowest temperature recorded in the 48 contiguous states ($-70.0°F$ at Rogers Pass, Montana, on January 20, 1954); b) Australia's highest temperature ($127.0°F$ at Cloncurry, Queensland, on January 16, 1889); c) the lowest temperature recorded in the northern hemisphere ($-90.0°F$ at Verkhoyansk, Siberia, in 1892).

15–8 Convert the following Kelvin temperatures to the Celsius and Fahrenheit scales: a) the midday temperature at the surface of the moon (400 K); b) the temperature at the tops of the clouds in the atmosphere of Saturn (95 K); c) the temperature at the center of the sun (1.55×10^7 K).

15–9 The normal boiling point of liquid neon is $-245.92°C$. What is this temperature on the Kelvin scale?

15–10 The ratio of the pressures of a gas at the melting point of platinum and its pressure at the triple point of water, when the gas is kept at constant volume, is found to be 7.476. What is the Celsius temperature of the melting point of platinum?

15–11 A gas thermometer registers an absolute pressure corresponding to 325 mm of mercury when in contact with water at the triple point. What pressure does it read when in contact with water at the normal boiling point?

15–12 Like the Kelvin scale, the *Rankine scale* is an absolute temperature scale: absolute zero is zero degrees Rankine ($0°R$). However, the units of this scale are the same size as those on the Fahrenheit scale rather than the Celsius scale. What is the numerical value of the triple-point temperature of water on the Rankine scale?

15–13 **A Constant-Volume Gas Thermometer.** An experimenter using a gas thermometer found the pressure at the triple point of water ($0.01°C$) to be 4.80×10^4 Pa and the pressure at the normal boiling point ($100°C$) to be 6.50×10^4 Pa. a) Assuming that the pressure varies linearly with temperature, use these two data points to find the Celsius temperature at which the gas pressure would be zero (that is, find the Celsius temperature of absolute zero). b) Does the gas in this thermometer obey Eq. (15–4) precisely? If that equation were precisely obeyed, and the pressure at $100°C$ were 6.50×10^4 Pa, what pressure would the experimenter have measured at $0.01°C$? (As we will learn in Section 16–2, Eq. (15–4) is accurate only for gases at very low density.)

SECTION 15–5 THERMAL EXPANSION

15–14 **Air Friction and Thermal Expansion.** The supersonic airliner Concorde is 62.1 m long when sitting on the ground on a typical ($15°C$) day. It is primarily made of aluminum. In flight at twice the speed of sound, friction with the air warms Concorde's skin and causes the aircraft to lengthen by 25 cm. (The passenger cabin is on rollers, and the airplane expands around the passengers.) What is the temperature of Concorde's skin in flight?

15–15 The Humber Bridge in England has the world's longest single span, 1410 m in length. Calculate the change in length of the steel deck of the span when the temperature increases from $-5.0°C$ to $18.0°C$.

15–16 **Ensuring a Tight Fit.** Aluminum rivets used in airplane construction are made slightly larger than the rivet holes and cooled by "dry ice" (solid CO_2) before being driven. If the diameter of a hole is 4.500 mm, what should be the diameter of a rivet at 23.0°C if its diameter is to equal that of the hole when the rivet is cooled to −78.0°C, the temperature of dry ice? Assume that the expansion coefficient remains constant at the value given in Table 15–1.

15–17 A U.S. one-cent coin has a diameter of 1.9000 cm at 20.0°C. The coin is made of a metal alloy (mostly zinc) for which the coefficient of linear expansion is 2.6×10^{-5} K^{-1}. What would its diameter be a) on a hot day in Death Valley (48.0°C); b) on a cold night in the mountains of Greenland (−53.0°C)?

15–18 The pendulum shaft of a clock is made of brass. What is the fractional change in length of the shaft when it is cooled from 19.50°C to 5.00°C?

15–19 A metal rod is 40.125 cm long at 20.0°C and 40.148 cm long at 45.0°C. Calculate the average coefficient of linear expansion of the rod for this temperature range.

15–20 A copper cylinder is initially at 20.0°C. At what temperature will its volume be 0.150% larger than it is at 20.0°C?

15–21 An underground tank with a capacity of 1700 L (1.70 m^3) is filled with ethanol that has an initial temperature of 19.0°C. After the ethanol has cooled off to the temperature of the tank and ground, which is 10.0°C, how much air space will there be above the ethanol in the tank? (Assume that the volume of the tank doesn't change.)

15–22 Determine the coefficient of volume expansion for water at a temperature of 9°C. (*Hint:* You will need to make measurements on Fig. 15–7.)

15–23 A glass flask whose volume is 1000.00 cm^3 at 0.0°C is completely filled with mercury at this temperature. When flask and mercury are warmed to 55.0°C, 8.95 cm^3 of mercury overflow. If the coefficient of volume expansion of mercury is 18.0×10^{-5} K^{-1}, compute the coefficient of volume expansion of the glass.

15–24 a) If an area measured on the surface of a solid body is A_0 at some initial temperature and then changes by ΔA when the temperature changes by ΔT, show that

$$\Delta A = (2\alpha)A_0\Delta T.$$

where α is the coefficient of linear expansion. b) A circular sheet of aluminum is 55.0 cm in diameter at 15.0°C. By how much does the area of one side of the sheet change when the temperature increases to 27.5°C?

15–25 A machinist bores a hole of diameter 1.350 cm in a steel plate at a temperature of 25°C. What is the cross-section area of the hole a) at 25°C; b) when the temperature of the plate is increased to 175°C? Assume that the coefficient of linear expansion remains constant over this temperature range. (*Hint:* See Exercise 15–24.)

15–26 A brass rod is 185 cm long and 1.60 cm in diameter. What force must be applied to each end of the rod to prevent it from contracting when it is cooled from 120°C to 10°C?

15–27 a) A wire that is 1.50 m long at 20°C is found to increase in length by 1.9 cm when warmed to 420°C. Compute its average coefficient of linear expansion for this temperature range. b) The wire is stretched just taut (zero tension) at 420°C.

Find the stress in the wire if it is cooled to 20°C without being allowed to contract. Young's modulus for the wire is 2.0×10^{11} Pa.

15–28 Steel train rails are laid in 12.0-m-long segments placed end-to-end. The rails are laid on a winter day when their temperature is −2.0°C. a) How much space must be left between adjacent rails if they are to just touch on a summer day when their temperature is 33.0°C? b) If the rails are originally laid in contact, what is the stress in them on a summer day when their temperature is 33.0°C?

SECTION 15–6 QUANTITY OF HEAT

15–29 Heat Loss During Breathing. In very cold weather a significant mechanism for heat loss by the human body is energy expended in warming the air taken into the lungs with each breath. a) On a cold winter day when the temperature is −20°C, what is the amount of heat needed to warm to body temperature (37°C) the 0.50 L of air exchanged with each breath? Assume that the specific heat capacity of air is 1020 J/kg · K and that 1.0 L of air has mass 1.3×10^{-3} kg. b) How much heat is lost per hour if the respiration rate is 20 breaths per minute?

15–30 While running, a 70-kg student generates thermal energy at a rate of 1200 W. To maintain a constant body temperature of 37°C, this energy must be removed by perspiration or other mechanisms. If these mechanisms failed and the heat could not flow out of the student's body, for what amount of time could a student run before irreversible body damage occurs? (Protein structures in the body are irreversibly damaged if body temperature rises to 44°C or above. The specific heat capacity of a typical human body is 3480 J/kg · K, slightly less than that of water. The difference is due to the presence of protein, fat, and minerals, which have lower specific heat capacities).

15–31 While painting the top of an antenna 225 m in height, a worker accidentally lets a 1.00-L water bottle fall from his lunchbox. The bottle lands in some bushes at ground level and does not break. If a quantity of heat equal to the magnitude of the change in mechanical energy of the water goes into the water, what is its increase in temperature?

15–32 A crate of fruit of mass 35.0 kg and specific heat capacity 3650 J/kg · K slides down a ramp inclined at 36.9° below the horizontal. The ramp is 8.00 m long. a) If the crate was at rest at the top of the incline and has a speed of 2.50 m/s at the bottom, how much work was done on the crate by friction? b) If an amount of heat equal to the magnitude of the work done by friction goes into the crate of fruit and the fruit reaches a uniform final temperature, what is its temperature change?

15–33 An engineer is working on a new engine design. One of the moving parts contains 1.60 kg of aluminum and 0.30 kg of iron and is designed to operate at 210°C. How much heat is required to raise its temperature from 20° to 210°C?

15–34 A nail being driven into a board increases in temperature. If we assume that 60% of the kinetic energy delivered by a 1.80-kg hammer with a speed of 7.80 m/s is transformed into heat that flows into the nail and does not flow out, what is the temperature increase of an 8.00-g aluminum nail after it is struck ten times?

15–35 An aluminum tea kettle with a mass of 1.50 kg and containing 1.80 kg of water is placed on a stove. If no heat is lost to the surroundings, how much heat must be added to raise the temperature from 20.0°C to 85.0°C?

15–36 In an effort to stay awake for an all-night study session, a student makes a cup of coffee by first placing a 200-W electric immersion heater in 0.320 kg of water. a) How much heat must be added to the water to raise its temperature from 20.0°C to 80.0°C? b) How much time is required? Assume that all of the heater power goes into heating the water.

15–37 A technician measures the specific heat capacity of an unidentified liquid by immersing an electrical resistor in it. Electrical energy is converted to heat transferred to the liquid for 120 s at a constant rate of 65.0 W. The mass of the liquid is 0.780 kg, and its temperature increases from 18.55°C to 22.54°C. a) Find the average specific heat capacity of the liquid in this temperature range. Assume that negligible heat is transferred to the container that holds the liquid and that no heat is lost to the surroundings. b) Suppose that in this experiment heat transfer from the liquid to the container or surroundings cannot be neglected. Is the result calculated in part (a) an *overestimate* or an *underestimate* of the average specific heat capacity? Explain.

SECTION 15–7 CALORIMETRY AND PHASE CHANGES

15–38 Before going in for his annual physical, a 70.0-kg man whose body temperature is 37.0°C consumes an entire 0.355-L can of a soft drink (mostly water) at 12.0°C. a) What will be his body temperature after equilibrium is attained? Ignore any heating by the man's metabolism. The specific heat capacity of the man's body is 3480 J/kg · K. b) Is the change in his body temperature great enough to be measured by a medical thermometer?

15–39 In the situation described in Exercise 15–38, the man's metabolism will eventually return the temperature of his body (and of the soft drink that he consumed) to 37.0°C. If his body releases energy at a rate of 7.00×10^3 kJ/day (the *basal metabolic rate,* or BMR), how long does this take? Assume that all of the released energy goes into raising the temperature.

15–40 An ice-cube tray of negligible mass contains 0.350 kg of water at 18.0°C. How much heat must be removed to cool the water to 0.0°C and freeze it? Express your answer in joules, calories, and Btu.

15–41 How much heat is required to convert 12.0 g of ice at −10.0°C to steam at 100.0°C? Express your answer in joules, calories, and Btu.

15–42 An open container has in it 0.550 kg of ice at −15.0°C. The mass of the container can be neglected. Heat is supplied to the container at the constant rate of 800 J/min for 500 min. a) After how many minutes does the ice *start* to melt? b) After how many minutes, from the time when the heating is first started, does the temperature start to rise above 0°C? c) Plot a curve showing the temperature as a function of the elapsed time.

15–43 The capacity of commercial air conditioners is sometimes expressed in "tons," the number of tons of ice (1 ton = 2000 lb) that can be frozen from water at 0°C in 24 h by the unit. Express the capacity of a 2-ton air conditioner in Btu/h and in watts.

15–44 Steam Burns vs. Water Burns. What is the amount of heat input to your skin when it receives the heat released a) by 25.0 g of steam initially at 100.0°C when it is cooled to skin temperature (34.0°C)? b) by 25.0 g of water initially at 100.0°C when it is cooled to 34.0°C? c) What does this tell you about the relative severity of steam and hot water burns?

15–45 What must be the initial speed of a lead bullet at a temperature of 25°C so that the heat developed when it is brought to rest will be just sufficient to melt it? Assume that all the initial mechanical energy of the bullet is converted to heat and that no heat flows from the bullet to its surroundings. (Typical rifles have muzzle speeds that exceed the speed of sound in air, which is 347 m/s at 25°C.)

15–46 Evaporation of sweat is an important mechanism for temperature control in some warm-blooded animals. a) What mass of water must evaporate from the skin of a 70.0-kg man to cool his body 1.00 C°? The heat of vaporization of water at body temperature (37°C) is 2.42×10^6 J/kg · K. The specific heat capacity of a typical human body is 3480 J/kg · K. (See Exercise 15–30.) b) What volume of water must the man drink to replenish the evaporated water? Compare to the volume of a soft-drink can (355 cm³).

15–47 "The Ship of the Desert." Camels require very little water because they are able to tolerate relatively large changes in their body temperature. While humans keep their body temperatures constant to within one or two Celsius degrees, a dehydrated camel permits its body temperature to drop to 34.0°C overnight and rise to 40.0°C during the day. To see how effective this mechanism is for saving water, calculate how many liters of water a 400-kg camel would have to drink if it instead attempted to keep its body temperature at a constant 34.0°C by evaporation of sweat during the day (12 hours) instead of rising to 40.0°C. (The specific heat capacity of a camel or other mammal is about the same as that of a typical human, 3480 J/kg · K. The heat of vaporization of water at 34°C is 2.42×10^6 J/kg.)

15–48 In a physics lab experiment a student immersed 200 one-cent coins (each having a mass of 3.00 g) in boiling water. After they reached thermal equilibrium, she fished them out and dropped them into 0.240 kg of water at 20.0°C in an insulated container of negligible mass. What was the final temperature of the coins? (One-cent coins are made of a metal alloy—mostly zinc—with a specific heat capacity of 390 J/kg · K.)

15–49 A copper pot with a mass of 0.500 kg contains 0.170 kg of water at a temperature of 20.0°C. A 0.250-kg block of iron at 85.0°C is dropped into the pot. Find the final temperature, assuming no heat loss to the surroundings.

15–50 A laboratory technician drops a 0.0850-kg sample of unknown material, at a temperature of 100.0°C, into a calorimeter. The calorimeter can, initially at 19.0°C, is made of 0.150 kg of copper and contains 0.200 kg of water. The final temperature of the calorimeter can is 26.1°C. Compute the specific heat capacity of the sample.

15–51 An insulated beaker with negligible mass contains 0.250 kg of water at a temperature of 75.0°C. How many kilograms of ice at a temperature of −20.0°C must be dropped in the water so that the final temperature of the system will be 30.0°C?

15–52 A glass vial containing a 16.0-g sample of an enzyme is cooled in an ice bath. The bath contains water and 0.120 kg of ice. The sample has specific heat capacity 2250 J/kg · K; the glass vial has mass 6.0 g and specific heat capacity 2800 J/kg · K. How much ice melts in cooling the enzyme sample from room temperature (19.5°C) to the temperature of the ice bath?

15–53 A 4.00-kg silver ingot is taken from a furnace where its temperature is 750°C and placed on a large block of ice at 0°C. Assuming that all the heat given up by the silver is used to melt the ice, how much ice is melted?

15–54 A copper calorimeter can with a mass of 0.100 kg contains 0.160 kg of water and 0.018 kg of ice in thermal equilibrium at atmospheric pressure. If 0.750 kg of lead at a temperature of 255°C is dropped into the calorimeter can, what is the final temperature? Assume that no heat is lost to the surroundings.

15–55 A vessel whose walls are thermally insulated contains 2.40 kg of water and 0.450 kg of ice, all at a temperature of 0.0°C. The outlet of a tube leading from a boiler in which water is boiling at atmospheric pressure is inserted into the water. How many grams of steam must condense inside the vessel (also at atmospheric pressure) to raise the temperature of the system to 28.0°C? Neglect the heat transferred to the container.

SECTION 15–8 MECHANISMS OF HEAT TRANSFER

15–56 Use Eq. (15–21) to show that the SI units of thermal conductivity are W/m · K.

15–57 Suppose that the rod in Fig. 15–13 is made of copper, is 45.0 cm long, and has a cross-section area of 1.25 cm². Let $T_H = 100.0$°C and $T_C = 0.0$°C. a) What is the final steady-state temperature gradient along the rod? b) What is the heat current in the rod in the final steady state? c) What is the final steady-state temperature at a point in the rod 12.0 cm from its left end?

15–58 One end of an insulated metal rod is maintained at 100°C, and the other end is maintained at 0°C by an ice-water mixture. The rod is 60.0 cm long and has a cross-section area of 1.25 cm². The heat conducted by the rod melts 8.50 g of ice in 10.0 min. Find the thermal conductivity k of the metal.

15–59 A carpenter builds an outer house wall with a layer of wood 3.0 cm thick on the outside and a layer of styrofoam insulation 2.2 cm thick as the inside wall surface. The wood has $k = 0.080$ W/m · K and the styrofoam has $k = 0.010$ W/m · K. The interior surface temperature is 19.0°C, and the exterior surface temperature is −10.0°C. a) What is the temperature at the plane where the wood meets the styrofoam? b) What is the rate of heat flow per square meter through this wall?

15–60 An electric kitchen range has total wall area of 1.40 m² and is insulated with a layer of fiberglass 4.0 cm thick. The inside surface of the fiberglass has a temperature of 175°C and its outside surface is at 35°C. The fiberglass has a thermal conductivity of 0.040 W/m · K. a) What is the heat current through the insulation, assuming it may be treated as a flat slab with an area of 1.40 m²? b) What electric-power input to the heating element is required to maintain this temperature?

15–61 The ceiling of a room has an area of 125 ft². The ceiling is insulated to an R value of 30 (in units of ft² · F° · h/Btu). The surface in the room is maintained at 69°F, and the surface in the attic has a temperature of 35°F. What is the heat flow through the ceiling into the attic in 5.0 h? Express your answer in Btu and in joules.

15–62 A long rod, insulated to prevent heat loss along its sides, is in perfect thermal contact with boiling water (at atmospheric pressure) at one end with an ice-water mixture at the other (Fig. 15–21). The rod consists of 1.00-m section of copper (one end in steam) joined end-to-end to a length L_2 of steel (one end in ice). Both sections of the rod have cross-section areas of 4.00 cm². The temperature of the copper-steel junction is 65.0°C after a steady state has been set up. a) How much heat per second flows from the steam bath to the ice-water mixture? b) What is the length L_2 of the steel section?

FIGURE 15–21 Exercise 15–62.

15–63 A pot with a steel bottom 8.50 mm thick rests on a hot stove. The area of the bottom of the pot is 0.150 m². The water inside the pot is at 100.0°C, and 0.390 kg are evaporated every 3.00 min. Find the temperature of the lower surface of the pot, which is in contact with the stove.

15–64 What is the rate of energy radiation per unit area of a blackbody at a temperature of a) 273 K? b) 2730 K?

15–65 What is the net rate of heat loss by radiation in Example 15–17 (Section 15–8) if the temperature of the surroundings is 5.0°C?

15–66 The emissivity of tungsten is 0.35. A tungsten sphere with a radius of 1.50 cm is suspended within a large evacuated enclosure whose walls are at 290 K. What power input is required to maintain the sphere at a temperature of 3000 K if heat conduction along the supports is neglected?

15–67 Size of a Light-Bulb Filament. The operating temperature of a tungsten filament in an incandescent light bulb is 2450 K, and its emissivity is 0.35. Find the surface area of the filament of a 150-W bulb if all the electrical energy consumed by the bulb is radiated by the filament as electromagnetic waves. (Only a fraction of the radiation appears as visible light.)

15–68 The Sizes of Stars. The hot, glowing surfaces of stars emit energy in the form of electromagnetic radiation. It is a good approximation to assume $e = 1$ for these surfaces. Find the radii of the following stars (assumed to be spherical): a) Rigel, the bright blue star in the constellation Orion, which radiates energy at a rate of 2.7×10^{32} W and has surface temperature 11,000 K; b) Procyon B (only visible using a telescope), which radiates energy at a rate of 2.1×10^{23} W and has surface temperature 10,000 K. c) Compare your answers to the radius of the earth, the radius of the sun, and the distance between the earth and the sun. (Rigel is an example of a *supergiant* star, and Procyon B is an example of a *white dwarf* star.)

SECTION 15–9 INTEGRATED CIRCUITS: A CASE STUDY IN HEAT TRANSFER

15–69 A night light has a 12-W bulb. The glass envelope of the bulb has surface area 20 cm^2. For this bulb, what is the power dissipation per unit surface area (by radiation, conduction, and convection to the surrounding air)?

15–70 The light bulb in Exercise 15–69 has $r_{th} = 5.0$ K/W. What is the operating temperature of the bulb surface if the ambient temperature is 18°C?

15–71 Maximum Temperature for an Operating Chip. One method for controlling the temperature of an operating VLSI chip is to decrease the ambient temperature by air conditioning or by immersing the chip in a cold liquid. A VLSI chip has a heat power output of 28 W and a value of $r_{th} = 4.5$ K/W. If the chip is to operate at temperatures up to 120°C, what is the maximum allowable ambient temperature?

15–72 The value of r_{th} for a computer chip running at 300 MHz is measured to determine the effectiveness of forced air cooling. Without the forced air, when the electrical power dissipation (by conversion to thermal energy) in the chip is 22 W, the chip rises to 88 C° above the ambient temperature. With the forced air system operating, the chip rises 58 C° above the ambient temperature for the same 22-W power dissipation. What is the value of r_{th} for the chip both a) without the forced air cooling and b) with the forced air cooling?

15–73 A certain computer chip has an average power consumption of 14.0 W when operating at a clock speed of 300 MHz. If the ambient air temperature is 20.0°C and the maximum allowable temperature of the chip is 100.0°C, what is the minimum value of r_{th} that will ensure safe operation of the chip?

PROBLEMS

15–74 Suppose that a steel hoop could be constructed to just fit around the earth's equator at a temperature of 20.0°C. What would be the thickness of space between the hoop and the earth if the temperature of the hoop were increased by 0.50 C°?

15–75 At an absolute temperature T_0, a cube has sides of length L_0 and has density ρ_0. The cube is made of a material with coefficient of volume expansion β. a) Show that if the temperature increases to $T_0 + \Delta T$, the density of the cube becomes, approximately,

$$\rho \approx \rho_0(1 - \beta\Delta T).$$

(*Hint:* Use the expression $(1 + x)^n \approx 1 + nx$, valid for $|x| \ll 1$.) Explain why this approximate result is only valid if ΔT is much less than $1/\beta$, and explain why you would expect this to be the case in most situations. b) A copper cube has sides of length 1.25 cm at 20.0°C. Find the change in its volume and density when its temperature is increased to 70.0°C.

15–76 At a temperature of 20.0°C the volume of a certain glass flask, up to a reference mark on the long stem of the flask, is exactly 100 cm^3. The flask is filled to this point with a liquid whose coefficient of volume expansion is 8.00×10^{-4} K^{-1}, with both flask and liquid at 20.0°C. The coefficient of volume expansion of the glass is 2.00×10^{-5} K^{-1}. The cross-section area of the stem is 50.0 mm^2 and can be considered constant. a) Explain why it is a good approximation to ignore the change in the cross-section area of the stem. b) How far will the liquid rise or fall in the stem when the temperature is raised to 50.0°C?

15–77 A metal rod that is 30.0 cm long expands by 0.0650 cm when its temperature is raised from 0°C to 100°C. A rod of a different metal and of the same length expands by 0.0350 cm for the same rise in temperature. A third rod, also 30.0 cm long, is made up of pieces of each of the above metals placed end-to-end and expands 0.0580 cm between 0°C and 100°C. Find the length of each portion of the composite bar.

15–78 On a cool (4°C) Saturday morning, a pilot fills the fuel tanks of her Pitts S-2C (a two-seat aerobatic airplane) to their full capacity of 106.0 L. Before flying on Sunday morning, when the temperature is again 4°C, she checks the fuel level and finds only 103.4 L of gasoline in the tanks. She realizes that it was hot on Saturday afternoon, and that thermal expansion of the gasoline caused the missing fuel to empty out of the tank's vent. a) What was the maximum temperature (in °C) reached by the fuel and tank on Saturday afternoon? The coefficient of volume expansion of gasoline is 9.5×10^{-4} K^{-1}, and the tank is made of aluminum. b) In order to have the maximum amount of fuel available for flight, when should the pilot have filled the fuel tanks?

15–79 a) Equation (15–12) gives the stress required to keep the length of a rod constant as its temperature changes. Show that if the length is permitted to change by an amount ΔL when its temperature changes by ΔT, the stress is equal to

$$\frac{F}{A} = Y\left(\frac{\Delta L}{L_0} - \alpha\Delta T\right),$$

where F is the tension on the rod, L_0 is the original length of the rod, A its cross-section area, α its coefficient of linear expansion, and Y its Young's modulus. b) A heavy brass bar has projections at its ends, as in Fig. 15–22. Two fine steel wires fastened between the projections are just taut (zero tension) when the whole system is at 20°C. What is the tensile stress in the steel wires when the temperature of the system is raised to 140°C? Make any simplifying assumptions you think are justified, but state what they are.

FIGURE 15–22 Problem 15–79.

15–80 A steel rod with a length of 0.350 m and an aluminum rod with a length of 0.250 m, both with the same diameter, are placed end-to-end between rigid supports with no initial stress in the rods. The temperature of the rods is now raised by 60.0 C°. What is the stress in each rod? (*Hint:* The length of the combined rod remains the same, but the lengths of the individual rods do not. See Problem 15–79.)

15–81 A steel ring with 2.5000-in. inside diameter at 20.0°C is to be warmed and slipped over a brass shaft with 2.5020-in. outside diameter at 20.0°C. a) To what temperature should the ring be warmed? b) If the ring and shaft together are cooled by some means such as liquid air, at what temperature will the ring just slip off the shaft?

15–82 Bulk Stress Due to a Temperature Increase. a) Prove that if an object under pressure is raised in temperature but not allowed to expand, the increase in pressure is

$$\Delta p = B\beta\Delta T,$$

where the bulk modulus B and the average coefficient of volume expansion β are both assumed positive and constant. b) What pressure is necessary to prevent a steel block from expanding when its temperature is increased from 20.0°C to 35.0°C?

15–83 A liquid is enclosed in a metal cylinder that is provided with a piston of the same metal. The system is originally at a pressure of 1.00 atm (1.013×10^5 Pa) and at a temperature of 30.0°C. The piston is forced down until the pressure on the liquid is increased by 50.0 atm, and then clamped in this position. Find the new temperature at which the pressure of the liquid is again 1.00 atm. Assume that the cylinder is sufficiently strong so that its volume is not altered by changes in pressure, but only by changes in temperature. Use the result derived in Problem 15–82. (*Hint:* See Section 11–6.)

Compressibility of liquid: $k = 8.50 \times 10^{-10}$ Pa^{-1}.

Coefficient of volume expansion of liquid: $\beta = 4.80 \times 10^{-4}$ K^{-1}.

Coefficient of volume expansion of metal: $\beta = 3.90 \times 10^{-5}$ K^{-1}.

15–84 A thirsty mechanic cools a 2.00-L bottle of a soft drink (mostly water) by pouring it into a large aluminum mug with a mass of 0.257 kg and adding 0.120 kg of ice initially at −15.0°C. If the soft drink and mug are initially at 20.0°C, what is the final temperature of the system, assuming no heat losses?

15–85 Spacecraft Reentry. A spacecraft made of aluminum circles the earth at a speed of 7700 m/s. a) Find the ratio of its kinetic energy to the energy required to raise its temperature from 0°C to 600°C. (The melting point of aluminum is 660°C. Assume a constant specific heat capacity of 910 J/kg · K.) b) Discuss the bearing of your answer on the problem of the reentry of a manned space vehicle into the earth's atmosphere.

15–86 A capstan is a rotating drum or cylinder over which a rope or cord slides in order to provide a great amplification of the rope's tension while keeping both ends free (Fig. 15–23). Since the added tension in the rope is due to friction, the capstan generates thermal energy. a) If the difference in tension between the two ends of the rope is 520 N and the capstan has a diameter of 10.0 cm and turns once in 0.900 s, find the rate at which thermal energy is being generated. Why does the number of turns not matter? b) If the capstan is made of iron and has a mass of 6.00 kg, at what rate does its temperature rise? Assume that the temperature in the capstan is uniform and that all the thermal energy generated flows into it.

FIGURE 15–23 Problem 15–86.

15–87 Debye's T^3 Law. At very low temperatures the molar heat capacity of rock salt varies with temperature according to Debye's T^3 law:

$$C = k\frac{T^3}{\Theta^3},$$

where $k = 1940$ J/mol · K and $\Theta = 281$ K. a) How much heat is required to raise the temperature of 1.50 mol of rock salt from 10.0 K to 40.0 K? (*Hint:* Use Eq. (15–18) in the form $dQ = nCdT$ and integrate.) b) What is the average molar heat capacity in this range? c) What is the true molar heat capacity at 40.0 K?

15–88 A person of mass 70.0 kg is sitting in the bathtub. The bathtub is 190 cm by 80 cm; before the person got in, the water was 10 cm deep. The water is at a temperature of 37.0°C. Suppose that the water were spontaneously to cool down to form ice at 0.0°C, and that all the energy released was used to launch the hapless bather vertically into the air. How high would the bather go? (As you will see in Chapter 18, this event is allowed by energy conservation but is prohibited by the second law of thermodynamics.)

15–89 Hot Air in a Physics Lecture. a) A typical student listening attentively to a physics lecture has a heat output of 100 W. How much heat energy does a class of 90 physics students release into a lecture hall over the course of a 50-min lecture? b) Assume that all the heat energy in part (a) is transferred to the 3200 m^3 of air in the room. The air has specific heat capacity 1020 J/kg · K and density 1.20 kg/m^3. If none of the heat escapes and the air conditioning system is off, how much will the temperature of the air in the room rise during the 50-min lecture? c) If the class is taking an exam, the heat output per student rises to 280 W. What is the temperature rise during 50 min in this case?

15–90 The molar heat capacity of a certain substance varies with temperature according to the empirical equation

$$C = 29.5 \text{ J/mol} \cdot \text{K} + (8.20 \times 10^{-3} \text{ J/mol} \cdot \text{K}^2)T.$$

How much heat is necessary to change the temperature of 3.00 mol of this substance from 27°C to 227°C? (*Hint:* Use Eq. (15–18) in the form $dQ = nC \, dT$ and integrate.)

15–91 a) An ice cube with a mass of 0.075 kg is taken from a freezer, where the cube's temperature was −10.0°C, and dropped into a glass of water at 0.0°C. If no heat is gained or lost from outside, how much water will freeze onto the cube? b) Is it possible to add so much ice to the glass that all of the water freezes? Explain.

15–92 **Hot Water Versus Steam Heating.** In a household hot-water heating system, water is delivered to the radiators at 70.0°C (158.0°F) and leaves at 28.0°C (82.4°F). The system is to be replaced by a steam system in which steam at atmospheric pressure condenses in the radiators and the condensed steam leaves the radiators at 35.0°C (95.0°F). How many kilograms of steam will supply the same heat as was supplied by 1.00 kg of hot water in the first system?

15–93 A copper calorimeter can with a mass of 0.446 kg contains 0.0950 kg of ice. The system is initially at 0.0°C. a) If 0.0350 kg of steam at 100.0°C and 1.00 atm pressure is added to the can, what is the final temperature of the calorimeter can and its contents? b) At the final temperature, how many kilograms are there of ice, how many of liquid water, and how many of steam?

15–94 In a container of negligible mass, 0.140 kg of ice initially at −15.0°C is added to 0.190 kg of water that has a temperature of 35.0°C. a) If no heat is lost to the surroundings, what is the final temperature of the system? b) At the final temperature, how many kilograms are there of ice and how many of liquid water?

15–95 In a container of negligible mass, 0.0400 kg of steam at 100°C and atmospheric pressure is added to 0.200 kg of water at 50.0°C. a) If no heat is lost to the surroundings, what is the final temperature of the system? b) At the final temperature, how many kilograms are there of steam and how many of liquid water?

15–96 A tube leads from a 0.150-kg calorimeter to a flask in which water is boiling under atmospheric pressure. The calorimeter has specific heat capacity 420 J/kg · K, and it originally contains 0.340 kg of water at 15.0°C. Steam is allowed to condense in the calorimeter at atmospheric pressure until the temperature of the calorimeter and contents reaches 71.0°C, at which point the total mass of calorimeter and contents is found to be 0.525 kg. Compute the heat of vaporization of water from these data.

15–97 In a container of negligible mass, 0.150 kg of ice at 0°C and 0.0950 kg of steam at 100°C are added to 0.200 kg of water that has a temperature of 50.0°C. a) If no heat is lost to the surroundings and the pressure in the container is a constant 1.00 atm, what is the final temperature of the system? b) At the final temperature, how many kilograms are there of ice, how many of liquid water, and how many of steam? c) Repeat parts (a) and

(b) if 0.350 kg of ice at 0°C and 0.012 kg of steam at 100°C are added to 0.200 kg of water that has a temperature of 40.0°C.

15–98 One experimental method of measuring an insulating material's thermal conductivity is to construct a box of the material and measure the power input to an electric heater inside the box that maintains the interior at a measured temperature above the outside surface. Suppose that in such an apparatus a power input of 180 W is required to keep the interior surface of the box 65.0 C° (about 120 F°) above the temperature of the outer surface. The total area of the box is 2.18 m², and the wall thickness is 3.9 cm. Find the thermal conductivity of the material in SI units.

15–99 **Effect of a Window in a Door.** A carpenter builds a solid wood door with dimensions 2.00 m × 0.95 m × 5.0 cm. Its thermal conductivity is $k = 0.120$ W/m · K. The air film on the inner and outer surfaces of the door have the same combined thermal resistance as an additional 1.8 cm thickness of solid wood. The inside air temperature is 20.0°C and the outside air temperature is −8.0°C. a) What is the rate of heat flow through the door? b) By what factor is the heat flow increased if a window 0.50 m on a side is inserted in the door? The glass is 0.45 cm thick and the glass has a thermal conductivity of 0.80 W/m · K. The air films on the two sides of the glass have a total thermal resistance that is the same as an additional 12.0 cm of glass.

15–100 A wood ceiling with thermal resistance R_1 is covered with a layer of insulation with thermal resistance R_2. Prove that the effective thermal resistance of the combination is $R = R_1 + R_2$.

15–101 Compute the ratio of the rate of heat loss through a single-pane window with area 0.15 m² to that for a double-pane window with the same area. The glass of a single pane is 4.2 mm thick, and the air space between the two panes of the double-pane window is 7.0 mm thick. The glass has thermal conductivity 0.80 W/m · K. The air films on the room and outdoor surfaces of either window have a combined thermal resistance of 0.15 m² · K/W.

15–102 Rods of copper, brass, and steel are welded together to form a Y-shaped figure. The cross-section area of each rod is 2.00 cm². The free end of the copper rod is maintained at 100.0°C and the free ends of the brass and steel rods at 0.0°C. Assume that there is no heat loss from the surfaces of the rods. The lengths of the rods are as follows: copper, 13.0 cm; brass, 18.0 cm; steel, 24.0 cm. a) What is the temperature of the junction point? b) What is the heat current in each of the three rods?

15–103 **Time Needed for a Lake to Freeze Over.** a) When the air temperature is below 0°C, the water at the surface of a lake freezes to form an ice sheet. Why doesn't freezing occur throughout the entire volume of the lake? b) Show that the thickness of the ice sheet formed on the surface of a lake is proportional to the square root of the time if the heat of fusion of the water freezing on the underside of the ice sheet is conducted through the sheet. c) Assuming that the upper surface of the ice sheet is at −10°C and that the bottom surface is at 0°C, calculate the time it will take to form an ice sheet 25 cm thick. d) If the lake in part (c) is uniformly 40 m deep, how long would it take to freeze all the water in the lake? Is this likely to occur?

15–104 A rod is initially at a uniform temperature of 0°C throughout. One end is kept at 0°C, and the other is brought into contact with a steam bath at 100°C. The surface of the rod is insulated so that heat can flow only lengthwise along the rod. The cross-section area of the rod is 2.50 cm^2, its length is 120 cm, its thermal conductivity is 380 W/m · K, its density is 1.00×10^4 kg/m^3, and its specific heat capacity is 520 J/kg · K. Consider a short cylindrical element of the rod 1.00 cm in length. a) If the temperature gradient at the cooler end of this element is 140 C°/m, how many joules of heat energy flow across this end per second? b) If the average temperature of the element is increasing at the rate of 0.250 C°/s, what is the temperature gradient at the other end of the element?

15–105 If the solar radiation energy incident per second on the frozen surface of a lake is 600 W/m^2 and 70% of this energy is absorbed by the ice, how much time will it take for a 2.50-cm-thick layer of ice to melt? The ice and the water beneath it are at a temperature of 0°C.

15–106 The rate at which radiant energy from the sun reaches the earth's upper atmosphere is about 1.50 kW/m^2. The distance from earth to sun is 1.50×10^{11} m, and the radius of the sun is 6.96×10^8 m. a) What is the rate of radiation of energy per unit area from the sun's surface? b) If the sun radiates as an ideal blackbody, what is the temperature of its surface?

15–107 A Thermos for Liquid Helium. A physicist uses a cylindrical metal can 0.250 m high and 0.090 m in diameter to store liquid helium at 4.22 K; at that temperature the heat of vaporization of helium is 2.09×10^4 J/kg. Completely surrounding the metal can are walls maintained at the temperature of liquid nitrogen, 77.3 K, with vacuum between the can and the surrounding walls. How much helium is lost per hour? The emissivity of the metal can is 0.200. The only heat transfer between the metal can and the surrounding walls is by radiation.

15–108 Thermal Expansion of an Ideal Gas. a) The pressure p, volume V, number of moles n, and Kelvin temperature T of an ideal gas are related by the equation $pV = nRT$, where R is a constant. Prove that the coefficient of volume expansion for an ideal gas is equal to the reciprocal of the Kelvin temperature if the expansion occurs at constant pressure. b) Compare the coefficients of volume expansion of copper and air at a temperature of 20°C. Assume that air may be treated as an ideal gas and that the pressure remains constant.

15–109 An engineer is developing an electric water heater to provide a continuous supply of hot water. One trial design is shown in Fig. 15–24. Water is flowing at the rate of 0.500 kg/min, the inlet thermometer registers 18.0°C, the voltmeter reads 120 V, and the ammeter reads 15.0 A (corresponding to a power input of (120 V)(15.0 A) = 1800 W). a) When a steady state is finally reached, what is the reading of the outlet thermometer? b) Why is it unnecessary to take into account the heat capacity mc of the apparatus itself?

FIGURE 15–24 Problem 15–109.

15–110 Food Intake of a Hamster. The energy output of an animal engaged in an activity is called the basal metabolic rate (BMR) and is a measure of the conversion of food energy into other forms of energy. A simple calorimeter to measure the BMR consists of an insulated box with a thermometer to measure the temperature of the air. The air has density 1.20 kg/m^3 and specific heat capacity 1020 J/kg · K. A 50.0-g hamster is placed in a calorimeter that contains 0.0500 m^3 of air at room temperature. a) When the hamster is running in a wheel, the temperature of the air in the calorimeter rises 1.60 C° per hour. How much heat does the running hamster generate in an hour? Assume that all this heat goes into the air in the calorimeter. Neglect the heat that goes into the walls of the box and into the thermometer, and assume that no heat is lost to the surroundings. b) Assuming that the hamster converts seed into heat with an efficiency of 10% and that hamster seed has a food energy value of 24 J/g, how many grams of seed must the hamster eat per hour to supply this energy?

CHALLENGE PROBLEMS

15–111 a) A spherical shell has inner and outer radii a and b, respectively, and the temperatures at the inner and outer surfaces are T_2 and T_1. The thermal conductivity of the material of which the shell is made is k. Derive an equation for the total heat current through the shell. b) Derive an equation for the temperature variation within the shell in part (a); that is, calculate T as a function of r, the distance from the center of the shell. c) A hollow cylinder has length L, inner radius a, and outer radius b, and the temperatures at the inner and outer surfaces are T_2 and T_1. (The cylinder could represent an insulated hot-water pipe, for example.) The thermal conductivity of the material of which the cylinder is made is k. Derive an equation for the total heat current through the walls of the cylinder. d) For the cylinder of part (c), derive an equation for the temperature variation inside the cylinder walls. e) For the spherical shell of part (a) and the hollow cylinder of part (c), show that the equation for the total heat current in each case reduces to Eq. (15–21) for linear heat flow when the shell or cylinder is very thin.

15–112 A steam pipe with a radius of 2.00 cm, carrying steam at 140°C, is surrounded by a cylindrical jacket with inner and outer radii 2.00 cm and 4.00 cm and made of a type of cork with thermal conductivity 4.00×10^{-2} W/m · K. This in turn is surrounded by a cylindrical jacket made of a brand of styrofoam with thermal conductivity 1.00×10^{-2} W/m · K and having inner and outer radii 4.00 cm and 6.00 cm (Fig. 15–25). The outer surface of the styrofoam is in contact with air at 15°C. Assume that this outer surface has a temperature of 15°C. a) What is the temperature at a radius of 4.00 cm, where the two insulating layers meet? b) What is the total rate of transfer of heat out of a 2.00-m length of pipe? (*Hint:* Use the expression derived in part (c) of Challenge Problem 15–111.)

FIGURE 15–25 Challenge Problem 15–112.

15–113 Suppose that both ends of the rod in Fig. 15–13 are kept at a temperature of 0°C, and that the initial temperature distribution along the rod is given by $T = (100°C) \sin \pi x/L$, where x is measured from the left end of the rod. Let the rod be copper, with length $L = 0.100$ m and cross-section area 1.00 cm². a) Show the initial temperature distribution in a diagram. b) What is the final temperature distribution after a very long time has elapsed? c) Sketch curves that you think would represent the temperature distribution at intermediate times. d) What is the initial temperature gradient at the ends of the rod? e) What is the initial heat current from the ends of the rod into the bodies making contact with its ends? f) What is the initial heat current at the center of the rod? Explain. What is the heat current at this point at any later time? g) What is the value of the *thermal diffusivity k/ρc* for copper, and in what unit is it expressed? (Here k is the thermal conductivity, $\rho = 8.9 \times 10^3$ kg/m³ is the density, and c is the specific heat capacity.) h) What is the initial time rate of change of temperature at the center of the rod? i) How much time would be required for the center of the rod to reach its final temperature if the temperature continued to decrease at this rate? (This time is called the *relaxation time* of the rod.) j) From the graphs in part (c), would you expect the magnitude of the rate of change of temperature at the midpoint to remain constant, increase, or decrease as a function of time? k) What is the initial rate of change of temperature at a point in the rod 2.5 cm from its left end?

15–114 Temperature Change in a Clock. A pendulum clock is designed to tick off one second on each side-to-side swing of the pendulum (two ticks per complete period). a) Will a pendulum clock gain time in hot weather and lose in cold, or the reverse? Explain your reasoning. b) A particular pendulum clock keeps correct time at 20.0°C. The pendulum shaft is of steel, and its mass may be neglected compared with that of the bob. What is the fractional change in length of the shaft when it is cooled to 10.0°C? c) How many seconds per day will the clock gain or lose at 10.0°C? d) How closely must the temperature be controlled if the clock is not to gain or lose more than 1.00 s a day? Does the answer depend on the period of the pendulum?

15–115 A solid cylindrical copper rod 0.200 m long has one end maintained at a temperature of 20.00 K. The other end is blackened and exposed to thermal radiation from surrounding walls at 500 K. The sides of the rod are insulated, so no energy is lost or gained except at the ends of the rod. When equilibrium is reached, what is the temperature of the blackened end? (*Hint:* Since copper is a very good conductor of heat at low temperature, with $k = 1670$ W/m · K at 20 K, the temperature of the blackened end is only slightly greater than 20.00 K.)

15–116 A Walk in the Sun. Consider a poor lost soul walking at 5 km/h on a hot day in the desert, wearing only a bathing suit. This person's skin temperature tends to rise due to four mechanisms: i) energy is generated by metabolic reactions in the body at a rate of 280 W, and almost all of this energy is converted to heat that flows to the skin; ii) heat is delivered to the skin by convection from the outside air at a rate equal to $k'A_{skin}(T_{air} - T_{skin})$, where k' is 54 J/h · C° · m², the exposed skin area A_{skin} is 1.5 m², the air temperature T_{air} is 47°C, and the skin temperature T_{skin} is 36°C; iii) the skin absorbs radiant energy from the sun at a rate of 1400 W/m²; iv) the skin absorbs radiant energy from the environment, which has temperature 47°C. a) Calculate the net rate (in watts) at which the person's skin is heated by all four of these mechanisms. Assume that the emissivity of the skin is $e = 1$ and that the skin temperature is initially 36°C. Which mechanism is the most important? b) At what rate (in L/h) must perspiration evaporate from this person's skin to maintain a constant skin temperature? (The heat of vaporization of water at 36°C is 2.42×10^6 J/kg.) c) Suppose instead the person is protected by light-colored clothing ($e \approx 0$) so that the exposed skin area is only 0.45 m². What rate of perspiration is required now? Discuss the usefulness of the traditional clothing worn by desert peoples.

Thermal Properties of Matter

Hot-air balloons use the principle that a gas tends to expand when its temperature is increased. When the gas expands so much that the average density of the balloon—including its envelope, gondola, and crew—is the same as that of the surrounding air, the balloon floats. The first successful flight of a human-carrying hot-air balloon was in 1783, more than a century before the first airplane.

16–1 INTRODUCTION

The kitchen is a great place to learn about how the properties of matter depend on temperature. When you boil water in a tea kettle, the increase in temperature produces steam that whistles out of the spout at high pressure. If you forget to poke holes in a potato before baking it, the high-pressure steam produced inside the potato can cause it to explode messily. Water vapor in the air can condense into droplets of liquid on the sides of a glass of ice water; if the glass is just out of the freezer, frost will form on the sides as water vapor changes to a solid.

All of these examples show the interrelations between the large-scale or *macroscopic* properties of a substance such as pressure, volume, temperature, and mass of substance. But we can also describe a substance using a *microscopic* perspective. This means investigating small-scale quantities such as the masses, speeds, kinetic energies, and momenta of the individual molecules that make up a substance.

The macroscopic and microscopic descriptions are intimately related. For example, the (microscopic) collision forces that occur when air molecules strike a solid surface (such as your skin) cause (macroscopic) atmospheric pressure. Standard atmospheric pressure is 1.01×10^5 Pa; to produce this pressure, 10^{32} molecules strike your skin every day with an average speed of over 1700 km/h (1000 mi/h)!

In this chapter we'll use both macroscopic and microscopic approaches to gain an understanding of the thermal properties of matter. One of the simplest kinds of matter to understand is the *ideal gas*. For such a gas, we'll be able to relate pressure, volume, temperature, and amount of substance to one another and to the speeds and masses of individual molecules. We will explore the molecular basis of heat capacities of both gases and solids, and we will take a look at the various *phases* of matter—gas, liquid, and solid—and the conditions under which each occurs.

16–2 EQUATIONS OF STATE

The conditions in which a particular material exists are described by physical quantities such as pressure, volume, temperature, and amount of substance. For example, a tank of oxygen in a welding outfit has a pressure gauge and a label stating its volume. We could add a thermometer and place the tank on a scale to determine its mass. These variables describe the *state* of the material and are called **state variables.**

The volume V of a substance is usually determined by its pressure p, temperature T, and amount of substance, described by the mass m or number of moles n. Ordinarily, we can't change one of these variables without causing a change in another. When the tank of oxygen gets hotter, the pressure increases. If the tank gets too hot, it explodes; this happens occasionally with overheated steam boilers.

In a few cases the relationship among p, V, T, and m (or n) is simple enough that we can express it as an equation called the **equation of state.** When it's too complicated for that, we can use graphs or numerical tables. Even then, the relation among the variables still exists; we call it an equation of state even when we don't know the actual equation.

Here's a simple (though approximate) equation of state for a solid material. The temperature coefficient of volume expansion β is the fractional volume change $\Delta V/V_0$ per unit temperature change, and the compressibility k is the negative of the fractional volume change $\Delta V/V_0$ per unit pressure change. If a certain amount of material has volume V_0 when the pressure is p_0 and the temperature is T_0, the volume V at slightly differing pressure p and temperature T is approximately

$$V = V_0[1 + \beta(T - T_0) - k(p - p_0)]. \qquad (16\text{–}1)$$

(There is a negative sign in front of the term $k(p - p_0)$ because an *increase* in pressure causes a *decrease* in the volume.) Equation (16–1) is called an *equation of state* for the material.

THE IDEAL GAS EQUATION

Another simple equation of state is the one for an *ideal gas*. Figure 16–1 shows an experimental setup to study the behavior of a gas. The cylinder has a movable piston to vary the volume, the temperature can be varied by heating, and we can pump any desired amount of any gas into the cylinder. We then measure the pressure, volume, temperature, and amount of gas. Note that *pressure* refers both to the force per unit area exerted by the cylinder on the gas and to the force per unit area exerted by the gas on the cylinder; by Newton's third law, these must be equal.

It is usually easiest to describe the amount of gas in terms of the number of moles n, rather than the mass. We did this when we defined molar heat capacity in Section 15–6; you may want to review that section. The **molar mass** M of a compound (sometimes called *molecular weight*) is the mass per mole, and the total mass m_{tot} of a given quantity of that compound is the number of moles n times the mass per mole M:

$$m_{tot} = nM \qquad \text{(total mass, number of moles, and molar mass).} \qquad (16\text{–}2)$$

We are calling the total mass m_{tot} because later in the chapter we will use m for the mass of one molecule.

Measurements of the behavior of various gases lead to several conclusions. First, the volume V is proportional to the number of moles n. If we double the number of moles, keeping pressure and temperature constant, the volume doubles.

Second, the volume varies *inversely* with the absolute pressure p. If we double the pressure while holding the temperature T and number of moles n constant, the gas compresses to one half of its initial volume. In other words, $pV = $ constant when n and T are constant.

Third, the pressure is proportional to the *absolute* temperature. If we double the absolute temperature, keeping the volume and number of moles constant, the pressure doubles. In other words, $p = (\text{constant})T$ when n and V are constant.

These three relationships can be combined neatly into a single equation, called the **ideal-gas equation:**

$$pV = nRT \qquad \text{(ideal-gas equation),} \qquad (16\text{–}3)$$

where R is a proportionality constant. An **ideal gas** is one for which Eq. (16–3) holds precisely for *all* pressures and temperatures. This is an idealized model; it works best at very low pressures and high temperatures, when the gas molecules are far apart and in rapid motion. It is reasonably good (within a few percent) at moderate pressures (such as a few atmospheres) and at temperatures well above those at which the gas liquefies.

We might expect that the constant R in the ideal-gas equation would have different values for different gases, but it turns out to have the same value for *all* gases, at least at

16–1 A hypothetical setup for studying the behavior of gases. The pressure p, volume V, temperature T, and number of moles of a gas can be controlled and measured.

8.4
State Variables and Ideal Gas Law

sufficiently high temperature and low pressure. It is called the **gas constant** (or *ideal-gas constant*). The numerical value of R depends on the units of p, V, and T. In SI units, in which the unit of p is Pa (1 Pa $= 1$ N/m^2) and the unit of V is m^3, the current best numerical value of R is

$$R = 8.314510(70) \text{ J/mol} \cdot \text{K}.$$

Note that the units of pressure times volume are the same as units of work or energy (for example, N/m^2 times m^3); that's why R has units of energy per mole per unit of absolute temperature. In chemical calculations, volumes are often expressed in liters (L) and pressures in atmospheres (atm). In this system, to four significant digits,

$$R = 0.08206 \frac{\text{L} \cdot \text{atm}}{\text{mol} \cdot \text{K}}.$$

We can express the ideal-gas equation, Eq. (16–3), in terms of the mass m_{tot} of gas, using $m_{tot} = nM$ from Eq. (16–2):

$$pV = \frac{m_{tot}}{M} RT. \tag{16–4}$$

From this we can get an expression for the density $\rho = m_{tot}/V$ of the gas:

$$\rho = \frac{pM}{RT}. \tag{16–5}$$

CAUTION ▶ When using Eq. (16–5), be certain that you distinguish between the Greek letter ρ ("rho") for density and the letter p for pressure. ◀

For a *constant mass* (or constant number of moles) of an ideal gas the product nR is constant, so the quantity pV/T is also constant. If the subscripts 1 and 2 refer to any two states of the same mass of a gas, then

$$\frac{p_1 V_1}{T_1} = \frac{p_2 V_2}{T_2} = \text{constant} \qquad \text{(ideal gas, constant mass).} \tag{16–6}$$

Notice that you don't need the value of R to use this equation.

The proportionality of pressure to absolute temperature is familiar; in fact, in Chapter 15 we *defined* a temperature scale in terms of pressure in a constant-volume gas thermometer. That may make it seem that the pressure-temperature relation in the ideal-gas equation, Eq. (16–3), is just a result of the way we define temperature. But the equation also tells us what happens when we change the volume or the amount of substance. Also, the gas-thermometer scale turns out to correspond closely to a temperature scale that we will define in Chapter 18 that doesn't depend on the properties of any particular material. For now, consider this equation as being based on this genuinely material-independent temperature scale, even though we haven't defined it yet.

Problem–Solving Strategy

IDEAL GASES

1. In some problems you will be concerned with only one state of the system; some of the quantities in Eq. (16–3) will be known, some unknown. Make a list of what you know and what you have to find. For example, $p = 1.0 \times 10^6$ Pa, $V = 4$ m^3, $T = ?$, $n = 2$ mol, or something comparable.

2. In other problems you will compare two different states of the same amount of gas. Decide which is state 1 and

which is state 2, and make a list of the quantities for each: p_1, p_2, V_1, V_2, T_1, T_2. If all but one of these quantities are known, you can use Eq. (16–6). Otherwise, you have to use Eq. (16–3). For example, if only p_1, V_1, and n are given, you can't use Eq. (16–6) to find T_1.

3. As always, be sure to use a consistent set of units. If you need the value of the gas constant R, choose its units and then convert the units of the other quantities accordingly.

You may have to convert atmospheres to pascals or liters to cubic meters (1 m³ = 10³ L = 10⁶ cm³). Sometimes the problem statement will make one system of units clearly more convenient than others. Decide on your system, and stick to it.

4. Don't forget that T must always be an *absolute* temperature. If you are given temperatures in °C, be sure to convert them to Kelvin temperatures by adding 273.15 (to three significant figures, 273). Likewise, p is always the absolute pressure, never the gauge pressure.

5. You may sometimes have to convert between mass m_{tot} and number of moles n. The relationship is $m_{tot} = Mn$, where M is the molar mass. Here's a tricky point: If you use Eq. (16–4), you *must* use the same mass units for m_{tot} and M. So if M is in grams per mole (the usual units for molar mass), then m_{tot} must also be in grams. If you want to use m_{tot} in kg, then you must convert M to kg/mol. For example, the molar mass of oxygen is 32 g/mol or 32×10^{-3} kg/mol. Be careful!

EXAMPLE 16–1

Volume of a gas at STP The condition called **standard temperature and pressure** (STP) for a gas is defined to be a temperature of 0°C = 273.15 K and a pressure of 1 atm $= 1.013 \times 10^5$ Pa. If you want to keep a mole of an ideal gas in your room at STP, how big a container do you need?

SOLUTION From Eq. (16–3), using R in J/mol · K,

$$V = \frac{nRT}{p} = \frac{(1 \text{ mol})(8.315 \text{ J/mol} \cdot \text{K})(273.15 \text{ K})}{1.013 \times 10^5 \text{ Pa}}$$

$$= 0.0224 \text{ m}^3 = 22.4 \text{ L}.$$

This is almost exactly the volume of three basketballs. A cube 0.282 m on a side would also do the job.

EXAMPLE 16–2

Compressing gas in an automobile engine In an automobile engine, a mixture of air and gasoline is compressed in the cylinders before being ignited. A typical engine has a compression ratio of 9.00 to 1; this means that the gas in the cylinders is compressed to 1/(9.00) of its original volume (Fig. 16–2). The initial pressure is 1.00 atm, and the initial temperature is 27°C. If the pressure after compression is 21.7 atm, find the temperature of the compressed gas.

SOLUTION Let state 1 be the uncompressed gas, and let state 2 be the fully compressed gas. Then $p_1 = 1.00$ atm, $p_2 = 21.7$ atm, and $V_1 = 9.00 V_2$. Converting temperature to the Kelvin scale by adding 273, $T_1 = 300$ K; the final temperature T_2 is unknown. The intake and exhaust valves at the top of the cylinder in Fig. 16–2 stay closed during the compression, so the number of moles of gas n is constant. Hence we can use Eq. (16–6):

$$\frac{p_1 V_1}{T_1} = \frac{p_2 V_2}{T_2}.$$

The temperature of the compressed gas is

$$T_2 = T_1 \frac{p_2 V_2}{p_1 V_1} = (300 \text{ K}) \frac{(21.7 \text{ atm}) V_2}{(1.00 \text{ atm})(9.00 V_2)}$$

$$= 723 \text{ K} = 450°\text{C}.$$

16–2 Cutaway of an automobile engine. While the air-gasoline mixture is being compressed prior to ignition, the intake and exhaust valves are both in the closed (up) position.

We didn't need to know the values of V_1 or V_2, only their ratio. Note that T_2 is the temperature of the air-gasoline mixture *before* the mixture is ignited; when burning starts, the temperature becomes higher still.

EXAMPLE 16–3

Mass of air in a scuba tank A typical tank used for scuba diving has a volume of 11.0 L (about 0.4 ft³) and a gauge pressure, when full, of 2.10×10^7 Pa (about 3000 psig). The "empty" tank contains 11.0 L of air at 21°C and 1 atm (1.013×10^5 Pa). When the tank is filled with hot air from a compressor, the temperature is 42°C and the gauge pressure is 2.10×10^7 Pa. What mass of air was added? (Air is a mixture of gases, about 78% nitrogen, 21% oxygen, and 1% miscellaneous; its average molar mass is 28.8 g/mol = 28.8×10^{-3} kg/mol.)

SOLUTION Let's first find the number of moles in the tank at the beginning (state 1) and at the end (state 2). We must remember to convert the temperatures to the Kelvin scale by adding 273 and to convert the pressure to absolute by adding 1.013×10^5 Pa. From Eq. (16–3), the number of moles n_1 in the "empty" tank is

$$n_1 = \frac{p_1 V_1}{RT_1} = \frac{(1.013 \times 10^5 \text{ Pa})(11.0 \times 10^{-3} \text{ m}^3)}{(8.315 \text{ J/mol} \cdot \text{K})(294 \text{ K})} = 0.46 \text{ mol}.$$

The volume of the metal tank is hardly affected by the increased pressure, so $V_1 = V_2$. The number of moles in the full tank is

$$n_2 = \frac{p_2 V_2}{RT_2} = \frac{(21.1 \times 10^6 \text{ Pa})(11.0 \times 10^{-3} \text{ m}^3)}{(8.315 \text{ J/mol} \cdot \text{K})(315 \text{ K})} = 88.6 \text{ mol}.$$

We added $n_2 - n_1 = 88.6$ mol $- 0.46$ mol $= 88.1$ mol to the tank. The added mass is not insubstantial: $m(n_2 - n_1)$ = (88.1 mol)$(28.8 \times 10^{-3}$ kg/mol) = 2.54 kg.

Could this problem have been solved in the same way as Example 16–2? The volume is constant, so $p/nT = R/V$ is constant and $p_1/n_1 T_1 = p_2/n_2 T_2$; this can be solved for n_2/n_1, the ratio of the final and initial number of moles. But we need the *difference* of these two numbers, not the ratio, so this equation by itself isn't enough to solve the problem.

EXAMPLE 16–4

Variation of atmospheric pressure with elevation Find the variation of atmospheric pressure with elevation in the earth's atmosphere, assuming that the temperature is 0°C at all elevations. Ignore the variation of g with elevation.

SOLUTION We begin with the pressure relation that we derived in Section 14–3, Eq. (14–4): $dp/dy = -\rho g$. The density ρ is given by Eq. (16–5): $\rho = pM/RT$. We substitute this into Eq. (14–4), separate variables, and integrate, letting p_1 be the pressure at elevation y_1 and p_2 be the pressure at y_2:

$$\frac{dp}{dy} = -\frac{pM}{RT} g,$$

$$\int_{p_1}^{p_2} \frac{dp}{p} = -\frac{Mg}{RT} \int_{y_1}^{y_2} dy,$$

$$\ln \frac{p_2}{p_1} = -\frac{Mg}{RT}(y_2 - y_1),$$

$$\frac{p_2}{p_1} = e^{-Mg(y_2 - y_1)/RT}.$$

Now let $y_1 = 0$ be at sea level, and let the pressure at that point be $p_0 = 1.013 \times 10^5$ Pa. Then our final expression for the pressure p at any height y is

$$p = p_0 e^{-Mgy/RT}.$$

At the summit of Mount Everest, where $y = 8863$ m,

$$\frac{Mgy}{RT} = \frac{(28.8 \times 10^{-3} \text{ kg/mol})(9.80 \text{ m/s}^2)(8863 \text{ m})}{(8.315 \text{ J/mol} \cdot \text{K})(273 \text{ K})} = 1.10,$$

$$p = (1.013 \times 10^5 \text{ Pa})e^{-1.10} = 0.337 \times 10^5 \text{ Pa}$$

$$= 0.33 \text{ atm}.$$

The assumption of constant temperature isn't realistic, and g decreases a little with increasing elevation (see Challenge Problem 16–76). Even so, this example shows why mountaineers need to carry oxygen on Mount Everest.

The ability of the human body to absorb oxygen from the atmosphere depends critically on atmospheric pressure. Absorption drops sharply when the pressure is less than about 0.65×10^5 Pa, corresponding to an elevation above sea level of about 4700 m (15,000 ft). There is no permanent human habitation on earth above 6000 m (about 20,000 ft), although survival for short periods of time is possible at higher elevations. Jet airplanes, which typically fly at altitudes of 8000 to 12,000 m, *must* have pressurized cabins for passenger comfort and health.

THE VAN DER WAALS EQUATION

The ideal-gas equation, Eq. (16–3), can be obtained from a simple molecular model that ignores the volumes of the molecules themselves and the attractive forces between them. We'll examine that model in Section 16–4. Meanwhile, we mention another equation of state, the **van der Waals equation,** that makes approximate corrections for these two omissions. This equation was developed by the 19th-century Dutch physicist J. D. van der Waals; the interaction between atoms that we discussed in Section 13–5 was named the *van der Waals interaction* after him. The van der Waals equation is

$$\left(p + \frac{an^2}{V^2}\right)(V - nb) = nRT. \tag{16–7}$$

The constants a and b are empirical constants, different for different gases. Roughly speaking, b represents the volume of a mole of molecules; the total volume of the molecules is then nb, and the net volume available for the molecules to move around in is $V - nb$. The constant a depends on the attractive intermolecular forces, which reduce the pressure of the gas for given values of n, V, and T by *pulling* the molecules together as they *push* on the walls of the container. The decrease in pressure is proportional to the number of molecules per unit volume in a layer near the wall (which are exerting the pressure on the wall) and is also proportional to the number per unit volume in the next layer beyond the wall (which are doing the attracting). Hence the decrease in pressure due to intermolecular forces is proportional to n^2/V^2.

When n/V is small (that is, when the gas is *dilute*), the average distance between molecules is large, the corrections in the van der Waals equation become insignificant, and Eq. (16–7) reduces to the ideal-gas equation. As an example, for carbon dioxide gas (CO_2) the constants in the van der Waals equation are $a = 0.364$ J $\cdot$ m^3/mol^2 and $b = 4.27 \times 10^{-5}$ m^3/mol. We found in Example 16–1 that one mole of an ideal gas at $T = 0°C = 273.15$ K and $p = 1$ atm $= 1.013 \times 10^5$ Pa occupies a volume $V = 0.0224$ m^3; according to Eq. (16–7), one mole of CO_2 occupying this volume at this temperature would be at a pressure 532 Pa less than 1 atm, a difference of only 0.5% from the ideal-gas value.

pV-DIAGRAMS

We could in principle represent the p-V-T relationship graphically as a *surface* in a three-dimensional space with coordinates p, V, and T. This representation sometimes helps in grasping the overall behavior of the substance, but ordinary two-dimensional graphs are usually more convenient. One of the most useful of these is a set of graphs of pressure as a function of volume, each for a particular constant temperature. Such a diagram is called a **pV-diagram.** Each curve, representing behavior at a specific temperature, is called an **isotherm,** or a *pV-isotherm*.

Figure 16–3 shows *pV*-isotherms for a constant amount of an ideal gas. The highest temperature is T_4; the lowest is T_1. This is a graphical representation of the ideal-gas equation of state. We can read off the volume V corresponding to any given pressure p and temperature T in the range shown.

Figure 16–4 shows a *pV*-diagram for a material that *does not* obey the ideal-gas equation. At temperatures below T_c the isotherms develop flat regions in which we can compress the material without an increase in pressure. Observation of the gas shows that it is *condensing* from the vapor (gas) to the liquid phase. The flat parts of the isotherms in the shaded area of Fig. 16–4 represent conditions of liquid-vapor *phase equilibrium*. As the volume decreases, more and more material goes from vapor to liquid, but the pressure does not change. (To keep the temperature constant during condensation, we have to remove the heat of vaporization, discussed in Section 15–7.)

When we compress such a gas at a constant temperature T_2 in Fig. 16–4, it is vapor until point a is reached. Then it begins to liquefy; as the volume decreases further, more material liquefies, and *both* the pressure and the temperature remain constant. At point b, all the material is in the liquid state. After this, any further compression results in a very rapid rise of pressure, because liquids are in general much less compressible than gases. At a lower constant temperature T_1, similar behavior occurs, but the condensation begins at lower pressure and greater volume than at the constant temperature T_2. At temperatures greater than T_c, *no* phase transition occurs as the material is compressed; at the highest temperatures, such as T_4, the curves resemble the ideal-gas curves of Fig. 16–3. We call T_c the *critical temperature* for this material. In Section 16–7 we'll discuss what happens to the phase of the gas above the critical temperature.

We will use *pV*-diagrams often in the next two chapters. We will show that the *area* under a *pV*-curve (whether or not it is an isotherm) represents the *work* done by the system during a volume change. This work, in turn, is directly related to heat transfer and changes in the *internal energy* of the system, which we'll get to in Chapter 17.

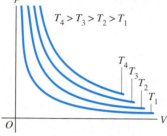

16–3 Isotherms, or constant-temperature curves, for a constant amount of an ideal gas. For each curve, the product $pV = nRT$ is constant, so p is proportional to $1/V$; the proportionality constant increases with increasing T.

16–4 A *pV*-diagram for a non-ideal gas, showing isotherms for temperatures above and below the critical temperature T_c. The liquid-vapor equilibrium region is shown as a green shaded area. At still lower temperatures the material might undergo phase transitions from liquid to solid or from gas to solid; these are not shown in this diagram.

16–3 MOLECULAR PROPERTIES OF MATTER

We have studied several properties of matter in bulk, including elasticity, density, surface tension, heat capacities, and equations of state, with only passing references to molecular structure. Now we want to look in more detail at the relation of bulk behavior to microscopic structure. We begin with a general discussion of the molecular structure of matter. Then in the next two sections we develop the kinetic-molecular model of an ideal gas, obtaining from this molecular model the equation of state and an expression for heat capacity.

All familiar matter is made up of **molecules.** For any specific chemical compound, all the molecules are identical. The smallest molecules contain one atom each and are of the order of 10^{-10} m in size; the largest contain many atoms and are at least 10,000 times that large. In gases the molecules move nearly independently; in liquids and solids they are held together by intermolecular forces that are electrical in nature, arising from interactions of the electrically charged particles that make up the molecules. Gravitational forces between molecules are negligible in comparison with electrical forces.

The interaction of two *point* electric charges is described by a force (repulsive for like charges, attractive for unlike charges) with a magnitude proportional to $1/r^2$, where r is the distance between the points. We will study this relationship, called *Coulomb's law,* in Chapter 22. Molecules are *not* point charges but complex structures containing both positive and negative charge, and their interactions are more complex. The force between molecules in a gas varies with the distance r between molecules somewhat as shown in Fig. 16–5, where a positive F corresponds to a repulsive force and a negative F to an attractive force. When molecules are far apart, the intermolecular forces are very small and usually attractive. As a gas is compressed and its molecules are brought closer together, the attractive forces increase. The intermolecular force becomes zero at an equilibrium spacing r_0, corresponding roughly to the spacing between molecules in the liquid and solid states. In liquids and solids, relatively large pressures are needed to compress the substance appreciably. This shows that at molecular distances slightly *less* than the equilibrium spacing, the forces become *repulsive* and relatively large.

Figure 16–5 also shows the potential energy as a function of r. This function has a *minimum* at r_0, where the force is zero. The two curves are related by $F(r) = -dU/dr$, as we showed in Section 7–5. Such a potential energy function is often called a **potential well.** A molecule at rest at a distance r_0 from a second molecule would need an additional energy $|U_0|$, the "depth" of the potential well, to "escape" to an indefinitely large value of r.

Molecules are always in motion; their kinetic energies usually increase with temperature. At very low temperatures the average kinetic energy of a molecule may be much *less* than the depth of the potential well. The molecules then condense into the liquid or solid phase with average intermolecular spacings of about r_0. But at higher temperatures the average kinetic energy becomes larger than the depth $|U_0|$ of the potential well. Molecules can then escape the intermolecular force and become free to move independently, as in the gaseous phase of matter.

In *solids,* molecules vibrate about more or less fixed points. In a crystalline solid these points are arranged in a recurring *crystal lattice.* Figure 16–6 shows the cubic crystal structure of sodium chloride (ordinary salt). A scanning tunneling microscope image of individual silicon atoms on the surface of a crystal is shown in Fig. 16–7.

The vibration of molecules in a solid about their equilibrium positions may be nearly simple harmonic if the potential well is approximately parabolic in shape at distances close to r_0. (We discussed this kind of simple harmonic motion in Section 13–5.) But if the potential-energy curve rises more gradually for $r > r_0$ than for $r < r_0$, as in Fig. 16–5, the average position shifts to larger r with increasing amplitude. As we pointed out in

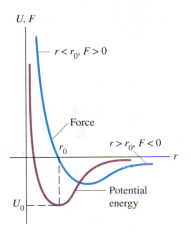

16–5 The force between two molecules (blue curve) is zero at a separation $r = r_0$, where the potential energy (dark red curve) is a minimum. The force is attractive when the separation is greater than r_0 and repulsive when the separation is less than r_0. (Compare this figure to Fig. 13–17.)

Chlorine
ions

Sodium
ions

16–6 Schematic representation of the cubic crystal structure of sodium chloride.

Section 15–5, this is the basis of thermal expansion.

In a *liquid,* the intermolecular distances are usually only slightly greater than in the solid phase of the same substance, but the molecules have much greater freedom of movement. Liquids show regularity of structure only in the immediate neighborhood of a few molecules. This is called *short-range order,* in contrast with the *long-range order* of a solid crystal.

The molecules of a *gas* are usually widely separated and so have only very small attractive forces. A gas molecule moves in a straight line until it collides with another molecule or with a wall of the container. In molecular terms, an *ideal gas* is a gas whose molecules exert *no* attractive forces on each other and therefore have no *potential* energy.

At low temperatures, most common substances are in the solid phase. As the temperature rises, a substance melts and then vaporizes. From a molecular point of view, these transitions are in the direction of increasing molecular kinetic energy. Thus temperature and molecular kinetic energy are closely related.

We have used the mole as a measure of quantity of substance. One **mole** of any pure chemical element or compound contains a definite number of molecules, the same number for all elements and compounds. The official SI definition is that **one mole is the amount of substance that contains as many elementary entities as there are atoms in 0.012 kilogram of carbon 12.** In our discussion, the "elementary entities" are molecules. (In a monatomic substance such as carbon or helium, each molecule is a single

16–7 A scanning tunneling microscope image of the surface of a silicon crystal. The area shown is only 9.0 nm $(9.0 \times 10^{-9}$ m) wide. Each blue "bead" is an individual silicon atom; you can clearly see how these atoms are arranged in a (nearly) perfect array of hexagons.

atom, but we'll still call it a molecule here.) Note that atoms of a given element may occur in any of several isotopes, which are chemically identical but have different atomic masses; "carbon 12" refers to a specific isotope of carbon.

The number of molecules in a mole is called **Avogadro's number** and denoted by N_A. The current best numerical value of N_A is

$$N_A = 6.0221367(36) \times 10^{23} \text{ molecules/mol} \qquad \text{(Avogadro's number)}.$$

The *molar mass* M of a compound is the mass of one mole. This is equal to the mass m of a single molecule multiplied by Avogadro's number.

$$M = N_A m \qquad \text{(molar mass, Avogadro's number, and mass of a molecule).} \qquad (16\text{–}8)$$

When the molecule consists of a single atom, the term *atomic mass* is often used instead of molar mass or molecular weight.

EXAMPLE 16-5

Find the mass of a single hydrogen atom and the mass of an oxygen molecule.

SOLUTION We use Eq. (16–8). From the periodic table of the elements (Appendix D) the mass per mole of atomic hydrogen (that is, the atomic mass) is 1.008 g/mol. Therefore the mass m_H of a single hydrogen atom is

$$m_H = \frac{1.008 \text{ g/mol}}{6.022 \times 10^{23} \text{ molecules/mol}} = 1.674 \times 10^{-24} \text{ g/molecule}.$$

From Appendix D, the atomic mass of oxygen is 16.0 g/mol, so the molar mass of oxygen, which has diatomic (two-atom)

molecules, is 32.0 g/mol. The mass of a single molecule of O_2 is

$$m_{O_2} = \frac{32.0 \text{ g/mol}}{6.022 \times 10^{23} \text{ molecules/mol}} = 53.1 \times 10^{-24} \text{ g/molecule}.$$

We note that the values in Appendix D are for the *average* atomic masses of a natural sample of each element. Such a sample may contain several different isotopes of the element, each with a different atomic mass. Natural samples of hydrogen and oxygen are almost entirely made up of just one isotope; this is not the case for all elements, however.

16–4 KINETIC-MOLECULAR MODEL OF AN IDEAL GAS

The goal of any molecular theory of matter is to understand the *macroscopic* properties of matter in terms of its atomic or molecular structure and behavior. Such theories are of tremendous practical importance; once we have this understanding, we can design materials to have specific desired properties. Such analysis has led to the development of high-strength steels, glasses with special optical properties, semiconductor materials for electronic devices, and countless other materials essential to contemporary technology.

In this section and the ones that follow, we will consider a simple molecular model of an ideal gas. This *kinetic-molecular model* represents the gas as a large number of particles bouncing around in a closed container. In this section we use the kinetic-molecular model to understand how the ideal-gas equation of state, Eq. (16–3), is related to Newton's laws. In the following section we'll use the kinetic-molecular model to predict the molar heat capacity of an ideal gas. We'll go on to elaborate the model to include "particles" that are not points but have a finite size. We will be able to see why polyatomic gases have larger molar heat capacities than monatomic gases.

The following discussion of the kinetic-molecular model has several steps, and you may need to go over it several times to grasp how it all goes together. Don't get discouraged!

Here are the assumptions of our model:

1. A container with volume V contains a very large number N of identical molecules, each with mass m.
2. The molecules behave as point particles; their size is small in comparison to the average distance between particles and to the dimensions of the container.

10 – 12:15

8.1
Characteristics of a Gas

16–8 Elastic collision of a molecule with an idealized container wall. The velocity component parallel to the wall does not change; the velocity component perpendicular to the wall reverses direction. The speed v does not change.

16–9 A molecule moving toward the wall with speed v_x collides with the area A during the time interval dt only if it is within a distance $v_x\,dt$ of the wall at the beginning of the interval. All such molecules are contained within a volume $Av_x\,dt$.

3. The molecules are in constant motion; they obey Newton's laws of motion. Each molecule collides occasionally with a wall of the container. These collisions are perfectly elastic.

4. The container walls are perfectly rigid and infinitely massive and do not move.

CAUTION ▶ Make sure you don't confuse N, the number of *molecules* in the gas, with n, the number of *moles*. ◀

During collisions the molecules exert *forces* on the walls of the container; this is the origin of the *pressure* that the gas exerts. In a typical collision (Fig. 16–8) the velocity component parallel to the wall is unchanged, and the component perpendicular to the wall reverses direction but does not change in magnitude.

Here is our program. First we determine the *number* of collisions per unit time for a certain wall area A. Then we find the total momentum change associated with these collisions and the force needed to cause this momentum change. Then we can determine the pressure, which is force per unit area, and compare the result to the ideal-gas equation. We'll find a direct connection between the temperature of the gas and the kinetic energy of the gas molecules.

To begin, let v_x be the *magnitude* of the x-component of velocity of a molecule of the gas. For now we assume that all molecules have the same v_x. This isn't right, but making this temporary assumption helps to clarify the basic ideas. We will show later that this assumption isn't really necessary.

As shown in Fig. 16–8, for each collision the x-component of velocity changes from $-v_x$ to $+v_x$. So the x-component of momentum changes from $-mv_x$ to $+mv_x$, and the *change* in the x-component of momentum is $mv_x - (-mv_x) = 2mv_x$.

If a molecule is going to collide with a given wall area A during a small time interval dt, then at the beginning of dt it must be within a distance $v_x\,dt$ from the wall (Fig. 16–9) and it must be headed toward the wall. So the number of molecules that collide with A during dt is equal to the number of molecules within a cylinder with base area A and length $v_x\,dt$ that have their x-velocity aimed toward the wall. The volume of such a cylinder is $Av_x\,dt$. Assuming that the number of molecules per unit volume (N/V) is uniform, the *number* of molecules in this cylinder is $(N/V)(Av_x\,dt)$. On the average, half of these molecules are moving toward the wall and half are moving away from it. So the number of collisions with A during dt is

$$\frac{1}{2}\left(\frac{N}{V}\right)(Av_x\,dt).$$

For the system of all molecules in the gas, the total momentum change dP_x during dt is the *number* of collisions multiplied by $2mv_x$:

$$dP_x = \frac{1}{2}\left(\frac{N}{V}\right)(Av_x\,dt)(2mv_x) = \frac{NAmv_x^2\,dt}{V}. \qquad (16\text{–}9)$$

(We are using capital P for total momentum and small p for pressure. Be careful!) The *rate* of change of momentum component P_x is

$$\frac{dP_x}{dt} = \frac{NAmv_x^2}{V}. \qquad (16\text{–}10)$$

According to Newton's second law, this rate of change of momentum equals the force exerted by the wall area A on the gas molecules. From Newton's *third* law this is equal and opposite to the force exerted *on* the wall *by* the molecules. Pressure p is the magnitude of the force exerted on the wall per unit area, and we obtain

$$p = \frac{F}{A} = \frac{Nmv_x^2}{V}. \qquad (16\text{–}11)$$

The pressure exerted by the gas depends on the number of molecules per volume (N/V), the mass m per molecule, and the speed of the molecules.

We mentioned that v_x is really *not* the same for all the molecules. But we could have sorted the molecules into groups having the same v_x within each group, then added up the resulting contributions to the pressure. The net effect of all this is just to replace v_x^2 in Eq. (16-11) by the *average* value of v_x^2, which we denote by $(v_x^2)_{av}$. Furthermore, $(v_x^2)_{av}$ is related simply to the *speeds* of the molecules. The speed v of any molecule is related to the velocity components v_x, v_y, and v_z by

$$v^2 = v_x^2 + v_y^2 + v_z^2.$$

We can average this relation over all molecules:

$$(v^2)_{av} = (v_x^2)_{av} + (v_y^2)_{av} + (v_z^2)_{av}.$$

But there is no real difference in our model between the x-, y-, and z-directions. (Molecular speeds are very fast in a typical gas, so the effects of gravity are negligibly small.) It follows that $(v_x^2)_{av}$, $(v_y^2)_{av}$, and $(v_z^2)_{av}$ must all be *equal*. Hence $(v^2)_{av}$ is equal to $3(v_x^2)_{av}$, and

$$(v_x^2)_{av} = \frac{1}{3}(v^2)_{av},$$

so Eq. (16-11) becomes

$$pV = \frac{1}{3}Nm(v^2)_{av} = \frac{2}{3}N\left[\frac{1}{2}m(v^2)_{av}\right]. \tag{16-12}$$

We notice that $\frac{1}{2}m(v^2)_{av}$ is the average translational kinetic energy of a single molecule. The product of this and the total number of molecules N equals the total random kinetic energy K_{tr} of translational motion of all the molecules. (The notation K_{tr} reminds us that this energy is associated with *translational* motion. There may be additional energies associated with rotational and vibrational motion of molecules.) The product pV equals two thirds of the total translational kinetic energy:

$$pV = \frac{2}{3}K_{tr}. \tag{16-13}$$

Now we compare this with the ideal-gas equation,

$$pV = nRT,$$

which is based on experimental studies of gas behavior. For the two equations to agree, we must have

$$K_{tr} = \frac{3}{2}nRT \qquad \text{(average translational kinetic energy of } n \text{ moles} \tag{16-14}$$
$$\text{of ideal gas).}$$

This remarkably simple result shows that K_{tr} is *directly proportional* to the absolute temperature T. We will use this important result several times in the following discussion.

The average translational kinetic energy of a single molecule is the total translational kinetic energy K_{tr} of all molecules divided by the number of molecules, N:

$$\frac{K_{tr}}{N} = \frac{1}{2}m(v^2)_{av} = \frac{3nRT}{2N}.$$

Also, the total number of molecules N is the number of moles n multiplied by Avogadro's number N_A, so

$$N = nN_A, \qquad \frac{n}{N} = \frac{1}{N_A},$$

and

$$\frac{K_{tr}}{N} = \frac{1}{2} m(v^2)_{av} = \frac{3}{2}\left(\frac{R}{N_A}\right)T. \qquad (16\text{--}15)$$

The ratio R/N_A occurs frequently in molecular theory. It is called the **Boltzmann constant**, k:

$$k = \frac{R}{N_A} = \frac{8.315 \text{ J/mol}\cdot\text{K}}{6.022\times 10^{23} \text{ molecules/mol}}$$

$$= 1.381\times 10^{-23} \text{ J/molecule}\cdot\text{K}.$$

In terms of k we can rewrite Eq. (16–15) as

$$\frac{1}{2} m(v^2)_{av} = \frac{3}{2} kT \qquad \text{(average translational kinetic energy of a gas molecule).} \qquad (16\text{--}16)$$

This shows that the average translational kinetic energy *per molecule* depends only on the temperature, not on the pressure, volume, or kind of molecule. We can obtain the average translational kinetic energy *per mole* by multiplying Eq. (16–16) by Avogadro's number and using the relation $M = N_A m$:

$$N_A \frac{1}{2} m(v^2)_{av} = \frac{1}{2} M(v^2)_{av} = \frac{3}{2} RT \qquad \text{(average translational kinetic energy per mole of gas).} \qquad (16\text{--}17)$$

The translational kinetic energy of a mole of ideal-gas molecules depends only on T.

Finally, it is sometimes convenient to rewrite the ideal-gas equation on a molecular basis. We use $N = N_A n$ and $R = N_A k$ to obtain the alternative form of the ideal-gas equation:

$$pV = NkT. \qquad (16\text{--}18)$$

This shows that we can think of the Boltzmann constant k as a gas constant on a "per-molecule" basis instead of the usual "per-mole" basis for R.

MOLECULAR SPEEDS

From Eqs. (16–16) and (16–17) we can obtain expressions for the square root of $(v^2)_{av}$, called the **root-mean-square speed** v_{rms}:

$$v_{rms} = \sqrt{(v^2)_{av}} = \sqrt{\frac{3kT}{m}} = \sqrt{\frac{3RT}{M}} \qquad \text{(root-mean-square speed of a gas molecule).}$$

$$(16\text{--}19)$$

It might seem more natural to characterize molecular speeds by their *average* value rather than by v_{rms}, but we see that v_{rms} evolves more directly from Eqs. (16–16) and (16–17). To compute the rms speed, we square each molecular speed, add, divide by the number of molecules, and take the square root; v_{rms} is the *root* of the *mean* of the *squares*. Example 16–7 illustrates this procedure.

When the sun goes down and the air temperature drops, the average speed of air molecules decreases as well.

Equations (16–16) and (16–19) show that at a given temperature T, gas molecules of different mass m have the same average kinetic energy but different root-mean-square speeds. On average, the nitrogen molecules ($M = 28$ g/mol) in the air around you are moving faster than are the oxygen molecules ($M = 32$ g/mol). Hydrogen molecules ($M = 2$ g/mol) are fastest of all; this is why there is hardly any hydrogen in the earth's atmosphere, despite its being the most common element in the universe. A sizable fraction of any H_2 molecules in the atmosphere would have speeds greater than the earth's "escape speed" of 1.12×10^4 m/s (calculated in Example 12–5, Section 12–4) and would escape into space. The heavier, slower-moving gases cannot escape so easily, which is why they dominate our atmosphere.

The assumption that individual molecules undergo perfectly elastic collisions with the container wall is actually a little too simple. More detailed investigation has shown that in most cases, molecules actually adhere to the wall for a short time and then leave again with speeds that are characteristic of the temperature *of the wall*. However, the gas and the wall are ordinarily in thermal equilibrium and have the same temperature. So there is no net energy transfer between gas and wall, and this discovery does not alter the validity of our conclusions.

Problem–Solving Strategy

KINETIC-MOLECULAR THEORY

As usual, using a consistent set of units is essential. The following are several places where caution is needed.

1. The usual units for molar mass M are grams per mole; the molar mass of oxygen (O_2) is 32 g/mol, for example. These units are often omitted in tables. In equations such as Eq. (16–19), when you use SI units you *must* express M in kilograms per mole by multiplying the table value by $(1 \text{ kg}/10^3 \text{ g})$. Thus in SI units, M for oxygen molecules is 32×10^{-3} kg/mol.

2. Are you working on a "per-molecule" basis or a "per-mole" basis? Remember that m is the mass of a single

molecule and M is the mass of a mole of molecules; N is the number of molecules, n is the number of moles; k is the gas constant per molecule, and R is the gas constant per mole. Although N, the number of molecules, is in one sense a dimensionless number, you can do a complete unit check if you think of N as having the unit "molecules"; then m has units "mass per molecule," and k has units "joules per molecule per kelvin."

3. Remember that T is always *absolute* (Kelvin) temperature.

EXAMPLE 16–6

Calculating molecular kinetic energy and v_{rms} a) What is the average translational kinetic energy of a molecule of an ideal gas at a temperature of 27°C? b) What is the total random translational kinetic energy of the molecules in one mole of this gas? c) What is the root-mean-square speed of oxygen molecules at this temperature?

SOLUTION a) To use Eq. (16–16), we first convert the temperature to the Kelvin scale: 27°C = 300 K. Then

$$\frac{1}{2}m(v^2)_{av} = \frac{3}{2}kT = \frac{3}{2}(1.38 \times 10^{-23} \text{ J/K})(300 \text{ K})$$

$$= 6.21 \times 10^{-21} \text{ J}.$$

This answer does not depend on the mass of the molecule.
b) The total translational kinetic energy K_{tr} is just the average translational kinetic energy of a molecule (which we just calculated) multiplied by the number of molecules, in this case Avogadro's number:

$$K_{tr} = N_A\left(\frac{1}{2}m(v^2)_{av}\right)$$

$$= (6.022 \times 10^{23} \text{ molecules})(6.21 \times 10^{-21} \text{ J/molecule})$$

$$= 3740 \text{ J}.$$

We can also get this result from Eq. (16–14):

$$K_{tr} = \frac{3}{2}nRT = \frac{3}{2}(1 \text{ mol})(8.315 \text{ J/mol} \cdot \text{K})(300 \text{ K})$$

$$= 3740 \text{ J}.$$

This is about the same kinetic energy as that of a sprinter in a 100-m dash.
c) From Example 16–5 (Section 16–3) the mass of an oxygen molecule is

$$m_{O_2} = (53.1 \times 10^{-24} \text{ g})(1 \text{ kg}/10^3 \text{ g}) = 5.31 \times 10^{-26} \text{ kg}.$$

From Eq. (16–19),

$$v_{rms} = \sqrt{\frac{3kT}{m}} = \sqrt{\frac{3(1.38 \times 10^{-23} \text{ J/K})(300 \text{ K})}{5.31 \times 10^{-26} \text{ kg}}}$$

$$= 484 \text{ m/s}.$$

This is 1740 km/h, or 1080 mi/h! Alternatively,

$$v_{rms} = \sqrt{\frac{3RT}{M}} = \sqrt{\frac{3(8.315 \text{ J/mol} \cdot \text{K})(300 \text{ K})}{32.0 \times 10^{-3} \text{ kg/mol}}}$$

$$= 484 \text{ m/s}.$$

Note again that when we use Eq. (16–19) with R in SI units, we have to express M in *kilograms* per mole, not grams per mole. In this example we use $M = 32.0 \times 10^{-3}$ kg/mol, *not* 32.0 g/mol.

EXAMPLE 16-7

Five gas molecules chosen at random are found to have speeds of 500, 600, 700, 800, and 900 m/s. Find the rms speed. Is it the same as the *average* speed?

SOLUTION The average value of v^2 for the five molecules is

$$(v^2)_{av} = \frac{\begin{bmatrix} (500\text{ m/s})^2 + (600\text{ m/s})^2 + (700\text{ m/s})^2 \\ +(800\text{ m/s})^2 + (900\text{ m/s})^2 \end{bmatrix}}{5}$$

$$= 5.10 \times 10^5 \text{ m}^2/\text{s}^2.$$

The square root of this is v_{rms}:

$$v_{rms} = 714 \text{ m/s}.$$

The *average* speed v_{av} is given by

$$v_{av} = \frac{500\text{ m/s} + 600\text{ m/s} + 700\text{ m/s} + 800\text{ m/s} + 900\text{ m/s}}{5}$$

$$= 700 \text{ m/s}.$$

We see that in general, v_{rms} and v_{av} are *not* the same. Roughly speaking, v_{rms} gives greater weight to the larger speeds than does v_{av}.

EXAMPLE 16-8

Volume of a gas at STP, revisited Find the number of molecules and the number of moles in one cubic meter of air at atmospheric pressure and 0°C.

SOLUTION From Eq. (16–18),

$$N = \frac{pV}{kT} = \frac{(1.013 \times 10^5 \text{ Pa})(1\text{ m}^3)}{(1.38 \times 10^{-23}\text{ J/K})(273\text{ K})}$$

$$= 2.69 \times 10^{25} \text{ molecules}.$$

The number of moles n is

$$n = \frac{N}{N_A} = \frac{2.69 \times 10^{25} \text{ molecules}}{6.022 \times 10^{23} \text{ molecules/mol}} = 44.7 \text{ mol}.$$

To check: The total volume is 1.00 m³, so the volume of one mole is

$$\frac{1.00 \text{ m}^3}{44.7 \text{ mol}} = 0.0224 \text{ m}^3/\text{mol} = 22.4 \text{ L/mol}.$$

This is the same result that we obtained in Example 16–1 (Section 16–2) using only macroscopic quantities.

COLLISIONS BETWEEN MOLECULES

We have ignored the possibility that two gas molecules might collide. If they are really points, they *never* collide. But consider a more realistic model in which the molecules are rigid spheres with radius r. How often do they collide with other molecules? How far do they travel, on average, between collisions? We can get approximate answers from the following rather primitive model.

Consider N spherical molecules with radius r in a volume V. Suppose only one molecule is moving. When it collides with another molecule, the distance between centers is $2r$. Suppose we draw a cylinder with radius $2r$, with its axis parallel to the velocity of the molecule (Fig. 16–10). The moving molecule collides with any other molecule whose center is inside this cylinder. In a short time dt a molecule with speed v travels a distance $v\,dt$; during this time it collides with any molecule that is in the cylindrical volume of radius $2r$ and length $v\,dt$. The volume of the cylinder is $4\pi r^2 v\,dt$. There are N/V molecules per unit volume, so the number dN with centers in this cylinder is

$$dN = 4\pi r^2 v\,dt\,N/V.$$

Thus the number of collisions *per unit time* is

$$\frac{dN}{dt} = \frac{4\pi r^2 v N}{V}.$$

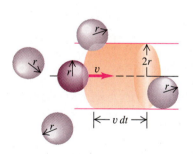

16–10 In a time dt a molecule with radius r will collide with any other molecule within a cylindrical volume of radius $2r$ and length $v\,dt$.

This result assumes that only one molecule is moving. The analysis is quite a bit more involved when all the molecules move at once. It turns out that in this case the collisions are more frequent, and the above equation has to be multiplied by a factor of $\sqrt{2}$:

$$\frac{dN}{dt} = \frac{4\pi\sqrt{2}r^2 v N}{V}.$$

The average time t_{mean} between collisions, called the *mean free time,* is the reciprocal of this expression:

$$t_{mean} = \frac{V}{4\pi\sqrt{2}r^2vN}. \tag{16-20}$$

The average distance traveled between collisions is called the **mean free path,** denoted by λ (the Greek letter "lambda"). In our simple model, this is just the molecule's speed v multiplied by t_{mean}:

$$\lambda = vt_{mean} = \frac{V}{4\pi\sqrt{2}r^2N} \quad \text{(mean free path of a gas molecule)}. \tag{16-21}$$

The mean free path is inversely proportional to the number of molecules per unit volume (N/V) and inversely proportional to the cross-section area πr^2 of a molecule; the more molecules there are and the larger the size of a molecule, the shorter the mean distance between collisions. Note that the mean free path *does not* depend on the speed of the molecule.

We can express Eq. (16–21) in terms of macroscopic properties of the gas, using the ideal-gas equation in the form of Eq. (16–18), $pV = NkT$. We find

$$\lambda = \frac{kT}{4\pi\sqrt{2}r^2p}. \tag{16-22}$$

If the temperature is increased at constant pressure, the gas expands, the average distance between molecules increases, and λ increases. If the pressure is increased at constant temperature, the gas compresses and λ decreases.

If you try to walk through a crowd, your mean free path—the distance you can travel on average without running into another person—depends on how large the people are and how closely they are spaced.

EXAMPLE 16-9

Calculating mean free path a) Estimate the mean free path of a molecule of air at 27°C and 1 atm. Model the molecules as spheres with radius $r = 2.0 \times 10^{-10}$ m. b) Estimate the mean free time of an oxygen molecule with $v = v_{rms}$.

SOLUTION a) From Eq. (16–22),

$$\lambda = \frac{kT}{4\pi\sqrt{2}r^2p}$$

$$= \frac{(1.38 \times 10^{-23}\ \text{J/K})(300\ \text{K})}{4\pi\sqrt{2}(2.0 \times 10^{-10}\ \text{m})^2(1.01 \times 10^5\ \text{Pa})}$$

$$= 5.8 \times 10^{-8}\ \text{m}.$$

The molecule doesn't get very far between collisions, but the distance is still several hundred times the radius of the molecule. To get a mean free path of one meter, the pressure must be about 5.8×10^{-8} atm. Pressures this low are found 100 km or so above

the earth's surface, at the outer fringe of our atmosphere.
b) We could use Eq. (16–20) to find the mean free time t_{mean}, but it's more convenient to use the basic relationship in Eq. (16–21) between t_{mean} and the mean free path: $\lambda = vt_{mean}$, so $t_{mean} = \lambda/v$. From Example 16–6, for oxygen at 27°C the root-mean-square speed is $v_{rms} = 484$ m/s, so the mean free time for a molecule with this speed is

$$t_{mean} = \frac{\lambda}{v} = \frac{5.8 \times 10^{-8}\ \text{m}}{484\ \text{m/s}} = 1.2 \times 10^{-10}\ \text{s}.$$

This molecule undergoes about 10^{10} collisions per second!
Note that the mean free path calculated in part (a) doesn't depend on the molecule's speed, but the mean free time does. Slower molecules have a longer average time interval t_{mean} between collisions than do fast ones, but the average *distance* λ between collisions is the same no matter what the molecule's speed.

16–5 HEAT CAPACITIES

When we introduced the concept of heat capacity in Section 15–6, we talked about ways to *measure* the specific or molar heat capacity of a particular material. Now we'll see how these numbers can be *predicted* on theoretical grounds. That's a significant step forward.

HEAT CAPACITIES OF GASES

The basis of our analysis is that heat is *energy* in transit. When we add heat to a substance, we are increasing its molecular energy. In this discussion we will keep the volume of the gas constant so that we don't have to worry about energy transfer through mechanical work. If we were to let the gas expand, it would do work by pushing on moving walls of its container, and this additional energy transfer would have to be included in our calculations. We'll return to this more general case in Chapter 17. For now, with the volume held constant, we are concerned with C_V, the molar heat capacity *at constant volume*.

In the simple kinetic-molecular model of Section 16–4 the molecular energy consists only of the translational kinetic energy K_{tr} of the pointlike molecules. This energy is directly proportional to the absolute temperature T, as shown by Eq. (16–14), $K_{tr} = \frac{3}{2}nRT$. When the temperature changes by a small amount dT, the corresponding change in kinetic energy is

$$dK_{tr} = \frac{3}{2}nR\,dT. \qquad (16-23)$$

From the definition of molar heat capacity at constant volume, C_V (Section 15–6), we also have

$$dQ = nC_V\,dT, \qquad (16-24)$$

where dQ is the heat input needed for a temperature change dT. Now if K_{tr} represents the total molecular energy, as we have assumed, then dQ and dK_{tr} must be *equal* (Fig. 16–11). Equating the expressions given by Eqs. (16–23) and (16–24), we get

$$nC_V\,dT = \frac{3}{2}nR\,dT,$$

$$C_V = \frac{3}{2}R \qquad \text{(ideal gas of point particles).} \qquad (16-25)$$

This surprisingly simple result says that the molar heat capacity (at constant volume) of *every* gas whose molecules can be represented as points is equal to $3R/2$.

To see whether this makes sense, let's first check the units. The gas constant *does* have units of energy per mole per kelvin, the correct units for a molar heat capacity. But more important is whether Eq. (16–25) agrees with *measured* values of molar heat capacities. In SI units, Eq. (16–25) gives

$$C_V = \frac{3}{2}(8.315 \text{ J/mol} \cdot \text{K}) = 12.47 \text{ J/mol} \cdot \text{K}.$$

For comparison, Table 16–1 gives measured values of C_V for several gases. We see that for *monatomic* gases our prediction is right on the money, but that it is way off for diatomic and polyatomic gases.

This comparison tells us that our point-molecule model is good enough for monatomic gases but that for diatomic and polyatomic molecules we need something more sophisticated. For example, we can picture a diatomic molecule as *two* point masses, like a little elastic dumbbell, with an interaction force between the atoms of the kind shown in Fig. 16–5. Such a molecule can have additional kinetic energy associated with *rotation* about axes through its center of mass. The atoms may also have a back-and-forth *vibrating* motion along the line joining them, with additional kinetic and potential energies. These possibilities are shown in Fig. 16–12.

When heat flows into a *monatomic* gas at constant volume, *all* of the added energy goes into an increase in random *translational* molecular kinetic energy. Equation

16–11 When an amount of heat dQ is added to (a) a constant volume of monatomic ideal gas molecules, (b) the total translational kinetic energy increases by $dK_{tr} = dQ$, and the temperature increases by $dT = dQ/nC_V$.

TABLE 16–1
MOLAR HEAT CAPACITIES OF GASES

TYPE OF GAS	GAS	C_V (J/mol · K)
Monatomic	He	12.47
	Ar	12.47
Diatomic	H_2	20.42
	N_2	20.76
	O_2	21.10
	CO	20.85
Polyatomic	CO_2	28.46
	SO_2	31.39
	H_2S	25.95

(16–23) shows that this gives rise to an increase in temperature. But when the temperature is increased by the same amount in a *diatomic* or *polyatomic* gas, additional heat is needed to supply the increased rotational and vibrational energies. Thus polyatomic gases have *larger* molar heat capacities than monatomic gases, as Table 16–1 shows.

But how do we know how much energy is associated with each additional kind of motion of a complex molecule, compared to the translational kinetic energy? The new principle that we need is called the principle of **equipartition of energy.** It can be derived from sophisticated statistical-mechanics considerations; that derivation is beyond our scope, and we will treat the principle as an axiom.

The principle of equipartition of energy states that each velocity component (either linear or angular) has, on average, an associated kinetic energy per molecule of $\frac{1}{2}kT$, or one-half the product of the Boltzmann constant and the absolute temperature. The number of velocity components needed to describe the motion of a molecule completely is called the number of **degrees of freedom.** For a monatomic gas, there are three degrees of freedom (for the velocity components v_x, v_y, and v_z); this gives a total average kinetic energy per molecule of $3(\frac{1}{2}kT)$, consistent with Eq. (16–16).

For a *diatomic* molecule there are two possible axes of rotation, perpendicular to each other and to the molecule's axis. (We don't include rotation about the molecule's own axis because in ordinary collisions there is no way for this rotational motion to change.) If we assign five degrees of freedom to a diatomic molecule, the average total kinetic energy per molecule is $\frac{5}{2}kT$ instead of $\frac{3}{2}kT$. The total kinetic energy of n moles is $K_{tot} = nN_A(\frac{5}{2}kT) = \frac{5}{2}n(kN_A)T = \frac{5}{2}nRT$, and the molar heat capacity (at constant volume) is

$$C_V = \frac{5}{2}R \quad \text{(diatomic gas, including rotation).} \quad (16\text{–}26)$$

In SI units,

$$C_V = \frac{5}{2}(8.315 \text{ J/mol} \cdot \text{K}) = 20.79 \text{ J/mol} \cdot \text{K}.$$

This agrees within a few percent with the measured values for diatomic gases given in Table 16–1.

Vibrational motion can also contribute to the heat capacities of gases. Molecular bonds are not rigid; they can stretch and bend, and the resulting vibrations lead to additional degrees of freedom and additional energies. For most diatomic gases, however, vibrational motion does *not* contribute appreciably to heat capacity. The reason for this is a little subtle and involves some concepts of quantum mechanics. Briefly, vibrational energy can change only in finite steps. If the energy change of the first step is much larger than the energy possessed by most molecules, then nearly all the molecules remain in the minimum-energy state of motion. In that case, changing the temperature

(a) Translational motion

(b) Rotational motion

(c) Vibrational motion

16–12 A diatomic molecule. Almost all the mass of each atom is in its tiny nucleus. (a) The center of mass has three independent velocity components. (b) The molecule has two independent axes of rotation through its center of mass. (c) The atoms and "spring" have additional kinetic and potential energies of vibration.

16-13 Experimental values of C_V for hydrogen gas (H_2). The temperature is plotted on a logarithmic scale. Appreciable rotational motion begins to occur above 50 K, and above 600 K the molecule begins to appreciably increase its vibrational motion.

does not change their average vibrational energy appreciably, and the vibrational degrees of freedom are said to be "frozen out." In more complex molecules the gaps between permitted energy levels are sometimes much smaller, and then vibration *does* contribute to heat capacity. The rotational energy of a molecule also changes by finite steps, but they are usually much smaller; the "freezing out" of rotational degrees of freedom occurs only in rare instances, such as for the hydrogen molecule below about 100 K.

In Table 16–1 the large values of C_V for some polyatomic molecules show the contributions of vibrational energy. In addition, a molecule with three or more atoms that are not in a straight line has three, not two, rotational degrees of freedom.

From this discussion we expect heat capacities to be temperature-dependent, generally increasing with increasing temperature. Figure 16–13 is a graph of the temperature dependence of C_V for hydrogen gas (H_2), showing the temperatures at which the rotational and vibrational energies begin to contribute to the heat capacity.

HEAT CAPACITIES OF SOLIDS

16-14 The forces between neighboring particles in a crystal may be visualized by imagining every particle as being connected to its neighbors by springs.

We can carry out a similar heat-capacity analysis for a crystalline solid. Consider a crystal consisting of N identical atoms (a *monatomic solid*). Each atom is bound to an equilibrium position by interatomic forces. The elasticity of solid materials shows us that these forces must permit stretching and bending of the bonds. We can think of a crystal as an array of atoms connected by little springs (Fig. 16–14). Each atom can *vibrate* about its equilibrium position.

Each atom has three degrees of freedom, corresponding to its three components of velocity. According to the equipartition principle, each atom has an average kinetic energy of $\frac{1}{2}kT$ for each degree of freedom. In addition, each atom has *potential* energy associated with the elastic deformation. For a simple harmonic oscillator (discussed in Chapter 13) it is not hard to show that the average kinetic energy of an atom is *equal* to its average potential energy. In our model of a crystal, each atom is essentially a three-dimensional harmonic oscillator; it can be shown that the equality of average kinetic and potential energies also holds here, provided that the "spring" forces obey Hooke's law.

Thus we expect each atom to have an average kinetic energy $\frac{3}{2}kT$ and an average potential energy $\frac{3}{2}kT$, or an average total energy $3kT$ per atom. If the crystal contains N atoms or n moles, its total energy is

$$E_{\text{tot}} = 3NkT = 3nRT. \tag{16-27}$$

16–15 Experimental values of C_V for lead, aluminum, silicon, and diamond. At high temperatures, C_V for each solid approaches about $3R$, in agreement with the rule of Dulong and Petit. At low temperatures, C_V is much less than $3R$.

From this we conclude that the molar heat capacity of a crystal should be

$$C_V = 3R \qquad \text{(ideal monatomic solid).} \qquad (16-28)$$

In SI units,

$$C_V = (3)(8.315 \text{ J/mol} \cdot \text{K}) = 24.9 \text{ J/mol} \cdot \text{K}.$$

This is the **rule of Dulong and Petit,** which we encountered as an *empirical* finding in Section 15–6: elemental solids all have molar heat capacities of about 25 J/mol · K. Now we have *derived* this rule from kinetic theory. The agreement is only approximate, to be sure, but considering the very simple nature of our model, it is quite significant.

At low temperatures, the heat capacities of most solids *decrease* with decreasing temperature (Fig. 16–15) for the same reason that vibrational degrees of freedom of molecules are frozen out at low temperatures. At very low temperatures the quantity kT is much *smaller* than the smallest energy step the vibrating atoms can take. Hence most of the atoms remain in their lowest energy states because the next higher energy level is out of reach. The average vibrational energy per atom is then *less* than $3kT$, and the heat capacity per molecule is *less* than $3k$. At higher temperatures when kT is *large* in comparison to the minimum energy step, the equipartition principle holds, and the total heat capacity is $3k$ per molecule or $3R$ per mole as the Dulong and Petit relation predicts. Quantitative understanding of the temperature variation of heat capacities was one of the triumphs of quantum mechanics during its initial development in the 1920s.

*16–6 MOLECULAR SPEEDS

As we mentioned in Section 16–4, the molecules in a gas don't all have the same speed. Figure 16–16 shows one experimental scheme for measuring the distribution of molecular speeds. A substance is vaporized in a hot oven; molecules of the vapor escape through an aperture in the oven wall and into a vacuum chamber. A series of slits blocks all molecules except those in a narrow beam, which is aimed at a pair of rotating disks. A molecule passing through the slit in the first disk is blocked by the second disk unless it arrives just as the slit in the second disk is lined up with the beam. The disks function as a speed selector that passes only molecules within a certain narrow speed range. This range can be varied by changing the disk rotation speed, and we can measure how many molecules lie within each of various speed ranges.

To describe the results of such measurements, we define a function $f(v)$ called a *distribution function*. If we observe a total of N molecules, the number dN having speeds in

16–16 A molecule with a speed v is passing through the first slit. When it reaches the second slit, the slits have rotated through the offset angle θ. If $v = \omega x/\theta$, the molecule passes through the second slit and reaches the detector.

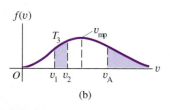

(a)

(b)

16–17 (a) Curves of the Maxwell-Boltzmann distribution function $f(v)$ for various temperatures. As the temperature increases, the curve becomes flatter, and its maximum shifts to higher speeds. (b) At temperature T_3 the fraction of molecules having speeds in the range v_1 to v_2 is shown by the shaded area under the T_3 curve. The fraction with speeds greater than v_A is shown by the area from v_A to infinity.

the range between v and $v + dv$ is given by

$$dN = Nf(v) \, dv. \tag{16–29}$$

We can also say that the *probability* that a randomly chosen molecule will have a speed in the interval v to $v + dv$ is $f(v) \, dv$. Hence $f(v)$ is the probability per unit speed *interval*; it is *not* equal to the probability that a molecule has speed exactly equal to v. Since a probability is a pure number, $f(v)$ has units of reciprocal speed (s/m).

Figure 16–17a shows distribution functions for several different temperatures. At each temperature the height of the curve for any value of v is proportional to the number of molecules with speeds near v. The peak of the curve represents the *most probable speed* v_{mp} for the corresponding temperature. As the temperature increases, the average molecular kinetic energy increases, and so the peak of $f(v)$ shifts to higher and higher speeds.

Figure 16–17b shows that the area under a curve between any two values of v represents the fraction of all the molecules having speeds in that range. Every molecule must have *some* value of v, so the integral of $f(v)$ over all v must be unity for any T.

If we know $f(v)$, we can calculate the most probable speed v_{mp}, the average speed v_{av}, and the rms speed v_{rms}. To find v_{mp}, we simply find the point where $df/dv = 0$; this gives the value of the speed where the curve has its peak. To find v_{av}, we take the number $Nf(v) \, dv$ having speeds in each interval dv, multiply each number by the corresponding speed v, add all these products (by integrating over all v from zero to infinity), and finally divide by N. That is,

$$v_{av} = \int_0^\infty v f(v) \, dv. \tag{16–30}$$

The rms speed is obtained similarly; the average of v^2 is given by

$$(v^2)_{av} = \int_0^\infty v^2 f(v) \, dv, \tag{16–31}$$

and v_{rms} is the square root of this.

The function $f(v)$ describing the actual distribution of molecular speeds is called the **Maxwell-Boltzmann distribution.** It can be derived from statistical-mechanics considerations, but that derivation is beyond our scope. Here is the result:

$$f(v) = 4\pi \left(\frac{m}{2\pi kT} \right)^{3/2} v^2 e^{-mv^2/2kT} \qquad \text{(Maxwell-Boltzmann distribution).}$$

$$\tag{16–32}$$

We can also express this function in terms of the translational kinetic energy of a molecule, which we denote by ϵ. That is, $\epsilon = \frac{1}{2}mv^2$. We invite you to verify that when this is substituted into Eq. (16–32), the result is

$$f(v) = \frac{8\pi}{m}\left(\frac{m}{2\pi kT}\right)^{3/2}\epsilon e^{-\epsilon/kT}. \qquad (16\text{–}33)$$

This form shows that the exponent in the Maxwell-Boltzmann distribution function is $-\epsilon/kT$ and that the shape of the curve is determined by the relative magnitude of ϵ and kT at any point. We leave it as an exercise to prove that the *peak* of each curve occurs where $\epsilon = kT$, corresponding to a most probable speed v_{mp} given by

8.2
Maxwell-Boltzmann Distribution—
Conceptual Analysis

8.3
Maxwell-Boltzmann Distribution—
Quantitative Analysis

$$v_{mp} = \sqrt{\frac{2kT}{m}}. \qquad (16\text{–}34)$$

To find the average speed, we substitute Eq. (16–32) into Eq. (16–30) and carry out the integration, making a change of variable $v^2 = x$ and then integrating by parts. We leave the details as an exercise; the result is

$$v_{av} = \sqrt{\frac{8kT}{\pi m}}. \qquad (16\text{–}35)$$

Finally, to find the rms speed, we substitute Eq. (16–32) into Eq. (16–31). Evaluating the resulting integral takes some mathematical acrobatics, but we can find it in a table of integrals. The result is

$$v_{rms} = \sqrt{\frac{3kT}{m}}. \qquad (16\text{–}36)$$

This result agrees with Eq. (16–19); it *must* agree if the Maxwell-Boltzmann distribution is to be consistent with the equipartition theorem and our other kinetic-theory calculations.

Table 16–2 shows the fraction of all the molecules in an ideal gas that have speeds *less than* various multiples of v_{rms}. These numbers were obtained by numerical integration; they are the same for all ideal gases.

The distribution of molecular speeds in liquids is similar, although not identical, to that for gases. We can understand the vapor pressure of a liquid and the phenomenon of boiling on this basis. Suppose a molecule must have a speed at least as great as v_A in Fig. 16–17b to escape from the surface of a liquid into the adjacent vapor. The number of such molecules, represented by the area under the "tail" of each curve (to the right of

TABLE 16–2

FRACTIONS OF MOLECULES IN AN IDEAL GAS WITH SPEEDS LESS THAN VARIOUS MULTIPLES OF v/v_{rms}

v/v_{rms}	FRACTION
0.20	0.011
0.40	0.077
0.60	0.218
0.80	0.411
1.00	0.608
1.20	0.771
1.40	0.882
1.60	0.947
1.80	0.979
2.00	0.993

v_A), increases rapidly with temperature. Thus the rate at which molecules can escape is strongly temperature-dependent. This process is balanced by another one in which molecules in the vapor phase collide inelastically with the surface and are trapped back into the liquid phase. The number of molecules suffering this fate per unit time is proportional to the pressure in the vapor phase. Phase equilibrium between liquid and vapor occurs when these two competing processes proceed at exactly the same rate. So if the molecular speed distributions are known for various temperatures, we can make a theoretical prediction of vapor pressure as a function of temperature. When liquid evaporates, it's the high-speed molecules that escape from the surface. The ones that are left have less energy on average; this gives us a molecular view of evaporative cooling.

Rates of chemical reactions are often strongly temperature-dependent, and the reason is contained in the Maxwell-Boltzmann distribution. When two reacting molecules collide, the reaction can occur only when the molecules are close enough for the electric-charge distributions of their electrons to interact strongly. This requires a minimum energy, called the *activation energy,* and thus a certain minimum molecular speed. Figure 16–17a shows that the number of molecules in the high-speed tail of the curve increases rapidly with temperature. Thus we expect the rate of any reaction that depends on an activation energy to increase rapidly with temperature. Similarly, many plant growth processes have strongly temperature-dependent rates, as can be seen by the rapid and diverse growth in tropical rain forests.

16–7 PHASES OF MATTER

We've talked a lot about ideal gases in the last few sections. An ideal gas is the simplest system to analyze from a molecular viewpoint because we ignore the interactions between molecules. But those interactions are the very thing that makes matter condense into the liquid and solid phases under some conditions. So it's not surprising that theoretical analysis of liquid and solid structure and behavior is a lot more complicated than that for gases. We won't try to go far here with a microscopic picture, but we can talk in general about phases of matter, phase equilibrium, and phase transitions.

In Section 15–7 we learned that each phase is stable only in certain ranges of temperature and pressure. A transition from one phase to another ordinarily takes place under conditions of **phase equilibrium** between the two phases, and for a given pressure this occurs at only one specific temperature. We can represent these conditions on a graph with axes p and T, called a **phase diagram;** Fig. 16–18 shows an example. Each point on the diagram represents a pair of values of p and T. Only a single phase can exist at each point, except for points on the solid lines, where two phases can coexist in phase equilibrium.

16–18 A typical pT phase diagram, showing regions of temperature and pressure at which the various phases exist and where phase changes occur.

TABLE 16–3
TRIPLE-POINT DATA

SUBSTANCE	TEMPERATURE (K)	PRESSURE (Pa)
Hydrogen	13.80	0.0704×10^5
Deuterium	18.63	0.171×10^5
Neon	24.56	0.432×10^5
Nitrogen	63.18	0.125×10^5
Oxygen	54.36	0.00152×10^5
Ammonia	195.40	0.0607×10^5
Carbon dioxide	216.55	5.17×10^5
Sulfur dioxide	197.68	0.00167×10^5
Water	273.16	0.00610×10^5

These lines separate the diagram into solid, liquid, and vapor regions. For example, the fusion curve separates the solid and liquid areas and represents possible conditions of solid-liquid phase equilibrium. Similarly, the vaporization curve separates the liquid and vapor areas, and the sublimation curve separates the solid and vapor areas. The three curves meet at the **triple point,** the only condition under which all three phases can coexist. In Section 15–4 we used the triple-point temperature of water to define the Kelvin temperature scale. Triple-point data for several substances are given in Table 16–3.

If we add heat to a substance at a constant pressure p_a, it goes through a series of states represented by the horizontal line (a) in Figure 16–18. The melting and boiling temperatures at this pressure are the temperatures at which the line intersects the fusion and vaporization curves, respectively. When the pressure is p_s, constant-pressure heating transforms a substance from solid directly to vapor. This process is called *sublimation;* the intersection of line (s) with the sublimation curve gives the temperature T_s at which it occurs for a pressure p_s. At any pressure less than the triple-point pressure, no liquid phase is possible. The triple-point pressure for carbon dioxide is 5.1 atm. At normal atmospheric pressure, solid carbon dioxide ("dry ice") undergoes sublimation; there is no liquid phase at this pressure.

Line (b) in Fig. 16–18 represents compression at a constant temperature T_b. The material passes from vapor to liquid and then to solid at the points where line (b) crosses the vaporization curve and fusion curve, respectively. Line (d) shows constant-temperature compression at a lower temperature T_d; the material passes from vapor to solid at the point where line (d) crosses the sublimation curve.

We saw in the pV-diagram of Fig. 16–4 that a liquid-vapor phase transition occurs only when the temperature and pressure are less than those at the point lying at the top of the green shaded area labeled "Liquid-vapor phase equilibrium region." This point corresponds to the end point at the top of the vaporization curve in Fig. 16–18. It is called the **critical point,** and the corresponding values of p and T are called the critical pressure and temperature, p_c and T_c. A gas at a pressure *above* the critical pressure does not separate into two phases when it is cooled at constant pressure (along a horizontal line above the critical point in Fig. 16–18). Instead, its properties change gradually and continuously from those we ordinarily associate with a gas (low density, large compressibility) to those of a liquid (high density, small compressibility) *without a phase transition*.

If this stretches credibility, think about liquid-phase transitions at successively higher points on the vaporization curve. As we approach the critical point, the *differences* in physical properties (such as density, bulk modulus, and viscosity) between the liquid and vapor phases become smaller and smaller. Exactly *at* the critical point they all become zero, and at this point the distinction between liquid and vapor disappears. The

The average temperature and pressure at the earth's surface are close to the triple point of water. That's why water can be found on earth in all three phases: vapor (in the atmosphere), liquid (in the oceans), and solid (like this iceberg).

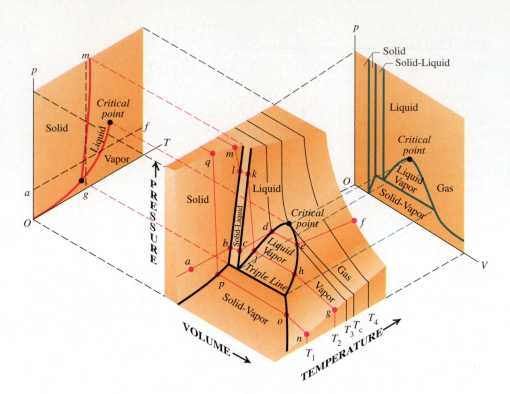

16–19 A *pVT*-surface for a substance that expands on melting. Projections of the boundaries on the surface on the *pT*- and *pV*-planes are also shown.

heat of vaporization also grows smaller and smaller as we approach the critical point, and it too becomes zero at the critical point.

For nearly all familiar materials the critical pressures are much greater than atmospheric pressure, so we don't observe this behavior in everyday life. For example, the critical point for water is at 647.4 K and 221.2×10^5 Pa (about 218 atm or 3210 psi). But high-pressure steam boilers in electric generating plants regularly run at pressures and temperatures well above the critical point.

Many substances can exist in more than one solid phase. A familiar example is carbon, which exists as noncrystalline soot and crystalline graphite and diamond. Water is another example; at least eight types of ice, differing in crystal structure and physical properties, have been observed at very high pressures.

We remarked in Section 16–2 that the equation of state of any material can be represented graphically as a surface in a three-dimensional space with coordinates *p*, *V*, and *T*. Such a surface is seldom useful in representing detailed quantitative information, but it can add to our general understanding of the behavior of materials at various temperatures and pressure. Figure 16–19 shows a typical *pVT*-surface. The light lines represent *pV*-isotherms; projecting them onto the *pV*-plane would give a diagram similar to Fig. 16–4. The *pV*-isotherms represent contour lines on the *pVT*-surface, just as contour lines on a topographic map represent the elevation (the third dimension) at each point. The projections of the edges of the surface onto the *pT*-plane gives the *pT*-phase diagram of Fig. 16–18.

Line *abcdef* represents constant-pressure heating, with melting along *bc* and vaporization along *de*. Note the volume changes that occur as *T* increases along this line. Line *ghjklm* corresponds to an isothermal (constant temperature) compression, with liquefaction along *hj* and solidification along *kl*. Between these, segments *gh* and *jk* represent isothermal compression with increase in pressure; the pressure increases are much greater in the liquid region *jk* and the solid region *lm* than in the vapor region *gh*. Finally, *nopq* represents isothermal solidification directly from the vapor phase; this is the process involved in growth of crystals directly from vapor, as in the formation of

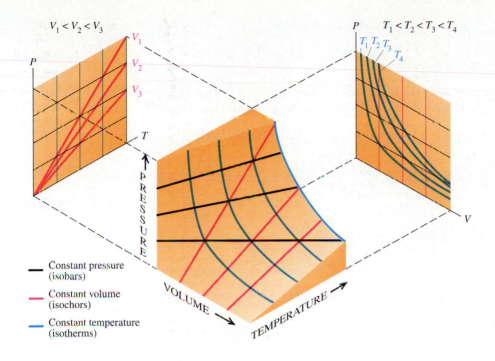

Constant pressure (isobars)

Constant volume (isochors)

Constant temperature (isotherms)

16–20 A pVT-surface for an ideal gas. At the left, each red line corresponds to a certain constant volume; at the right, each blue-green line corresponds to a certain constant temperature.

snowflakes or frost and in the fabrication of some solid-state electronic devices. These three lines on the pVT-surface are worth careful study.

For contrast, Fig. 16–20 shows the much simpler pVT-surface for a substance that obeys the ideal-gas equation of state under all conditions. The projections of the constant-temperature curves onto the pV-plane correspond to the curves of Fig. 16–3, and the projections of the constant-volume curves onto the pT-plane show the direct proportionality of pressure to absolute temperature.

SUMMARY

- The pressure p, volume V, and temperature T of a given quantity of a substance are called state variables. They are related by an equation of state. The ideal-gas equation of state is

$$pV = nRT, \qquad (16\text{–}3)$$

where n is the number of moles of gas and T is the absolute temperature. The constant R is the same for all gases for conditions in which this equation is applicable.

- A pV-diagram is a set of graphs, called isotherms, each showing pressure as a function of volume for a constant temperature.

- The molar mass M of a pure substance is the mass per mole. The total mass m_{tot} is related to the number of moles n by

$$m_{\text{tot}} = nM. \qquad (16\text{–}2)$$

Avogadro's number N_A is the number of molecules in a mole. The mass m of an individual molecule is related to M and N_A by

$$M = N_A m. \qquad (16\text{–}8)$$

- The average translational kinetic energy of the molecules of an ideal gas is directly

KEY TERMS

state variables, 499
equation of state, 499
molar mass, 500
ideal-gas equation, 500
ideal gas, 500
gas constant, 501
standard temperature and pressure, 502
van der Waals equation, 503
pV-diagram, 504
isotherm, 504
molecule, 505
potential well, 505
mole, 506
Avogadro's number, 507
Boltzmann constant, 510
root-mean-square speed, 510
mean free path, 513

proportional to absolute temperature:

$$K_{\text{tr}} = \frac{3}{2} nRT. \tag{16-14}$$

By using the Boltzmann constant, $k = R/N_{\text{A}}$, this can be expressed in terms of the average translational kinetic energy per molecule:

$$\frac{1}{2} m(v^2)_{\text{av}} = \frac{3}{2} kT. \tag{16-16}$$

The root-mean-square speed of molecules in an ideal gas is

$$v_{\text{rms}} = \sqrt{(v^2)_{\text{av}}} = \sqrt{\frac{3kT}{m}} = \sqrt{\frac{3RT}{M}}. \tag{16-19}$$

■ The mean free path λ of molecules in an ideal gas is

$$\lambda = vt_{\text{mean}} = \frac{V}{4\pi\sqrt{2}r^2N}. \tag{16-21}$$

■ The molar heat capacity C_V at constant volume for an ideal monatomic gas is

$$C_V = \frac{3}{2} R. \tag{16-25}$$

For an ideal diatomic gas, including rotational kinetic energy,

$$C_V = \frac{5}{2} R. \tag{16-26}$$

For an ideal monatomic solid,

$$C_V = 3R. \tag{16-28}$$

■ The speeds of molecules in an ideal gas are distributed according to the Maxwell-Boltzmann distribution:

$$f(v) = 4\pi \left(\frac{m}{2\pi kT} \right)^{3/2} v^2 e^{-mv^2/2kT}. \tag{16-32}$$

■ Ordinary matter exists in the solid, liquid, and gas phases. A phase diagram shows conditions under which two phases can coexist in phase equilibrium. All three phases can coexist at the triple point. The vaporization curve ends at the critical point, above which the distinction between liquid and gas phases disappears.

DISCUSSION QUESTIONS

Q16-1 Section 16-2 states that ordinarily, pressure, volume, and temperature cannot change individually without one affecting the others. Yet when a liquid evaporates, its volume changes, even though its pressure and temperature are constant. Is this inconsistent? Why or why not?

Q16-2 In the ideal-gas equation, could an equivalent Celsius temperature be used instead of the Kelvin one if an appropriate numerical value of the constant R is used? Why or why not?

Q16-3 On a chilly morning you can "see your breath." Can you really? What are you really seeing? Does this phenomenon depend on the temperature of the air, the humidity, or both? Explain.

Q16-4 When a car is driven some distance, the air pressure in the tires increases. Why? Should you let out some air to reduce the pressure? Why or why not?

Q16-5 The coolant in an automobile radiator is kept at a pressure higher than atmospheric pressure. Why is this desirable? The radiator cap will release coolant when the gauge pressure of the coolant reaches a certain value, typically 15 lb/in.^{2} or so. Why not just seal the system completely?

Q16-6 Unwrapped food placed in a freezer experiences dehydration, known as "freezer burn." Why?

Q16-7 "Freeze-drying" food involves the same process as "freezer burn," referred to in the previous question. For freeze-drying, the food is usually frozen first, then placed in a vacuum chamber and irradiated with infrared radiation. What is the purpose of the vacuum? The radiation? What advantages might freeze-drying have in comparison to ordinary drying?

Q16–8 A group of students drove from their university (near sea level) up into the mountains for a skiing weekend. Upon arriving at the slopes, they discovered that the bags of potato chips they had brought for snacks had all burst open. What caused this to happen?

Q16–9 An old or burned-out incandescent light bulb usually has a dark gray area on part of the inside surfaces. What is this? Why is it not a uniform deposit all over the inside of the bulb?

Q16–10 Helium is a mixture of two isotopes, one having molar mass 3 g/mol and the other 4 g/mol. In a sample of helium gas which atoms move faster, on the average? Explain why.

Q16–11 Which has more atoms, a kilogram of hydrogen or a kilogram of lead? Which has more mass? Explain.

Q16–12 In the kinetic-molecular model of an ideal gas, is the pressure of the gas due to collisions of molecules with the container walls, collisions of molecules with each other, or both? How does the kinetic-molecular model deal with collisions between molecules?

Q16–13 The proportion of various gases in the earth's atmosphere changes somewhat with altitude. Would you expect the proportion of oxygen at high altitude to be greater or less than at sea level compared to the proportion of nitrogen? Why?

Q16–14 Comment on the following statement: *When two gases are mixed, if they are to be in thermal equilibrium they must have the same average molecular speed.* Is the statement correct? Why or why not?

Q16–15 The kinetic-molecular model contains a hidden assumption about the temperature of the container walls. What is this assumption? What would happen if this assumption was not valid?

Q16–16 The temperature of an ideal gas is directly proportional to the average kinetic energy of its molecules. If a container of ideal gas is moving past you at 2000 m/s, is the temperature of the gas higher than if the container was at rest? Defend your answer.

Q16–17 You are sitting at the back of a chemistry lecture hall when your instructor opens a gas valve. You can *hear* the sound of the escaping gas almost immediately, but it takes several seconds before you can *smell* the gas. What is the reason for the "smell delay"?

Q16–18 In deriving the ideal-gas equation from the kinetic-molecular model, we ignored potential energy due to the earth's gravity. Is this omission justified? Why or why not?

Q16–19 The derivation of the ideal-gas equation included the assumption that the number of molecules is very large, so that we could compute the average force due to many collisions. However, the ideal-gas equation holds accurately only at low pressures, where the molecules are few and far between. Is this inconsistent? Why or why not?

Q16–20 A gas storage tank has a small leak. The pressure in the tank drops more quickly if the gas is hydrogen or helium than if it is oxygen. Why?

Q16–21 Consider two specimens of ideal gas at the same temperature. Specimen A has the same total mass as specimen B, but the molecules in specimen A have greater molar mass than they do in specimen B. In which specimen is the total kinetic energy of the gas greater? Does your answer depend on the molecular structure of the gases? Why or why not?

Q16–22 Some elements that form solid crystals have molar heat capacities *greater* than $3R$. What could account for this?

*Q16–23 In a gas that contains N molecules, would it be accurate to say that the number of molecules with speed v is equal to $f(v)$? Or would it be accurate to say that this number is given by $Nf(v)$? Explain your answers.

*Q16–24 Figure 16–16 shows a velocity selector. The two disks are a distance x apart, the slits in the two disks are at an angle θ to each other, and the disks rotate together with angular speed ω. Explain why a molecule that passes through the first slit of the selector will only be able to pass through the second slit if its speed is $v = \omega x/\theta$.

*Q16–25 Imagine a special air filter placed in a window of a house. The tiny holes in the filter only allow air molecules moving faster than a certain speed to exit the house, and only allow air molecules moving slower than that speed to enter the house from outside. What effect would this filter have on the temperature inside the house? (It turns out that the second law of thermodynamics—which we will discuss in Chapter 18—tells us that such a wonderful air filter would be quite impossible to make.)

Q16–26 A beaker of water at room temperature is placed in an enclosure, and the air pressure in the enclosure is slowly reduced. When the air pressure is reduced sufficiently, the water begins to boil. The temperature of the water does not rise when it boils; in fact, the temperature *drops* slightly. Explain these phenomena.

Q16–27 Ice is slippery to walk on, and especially slippery if you wear ice skates. What does this tell you about how the melting temperature of ice depends on pressure? Explain.

Q16–28 Hydrothermal vents are openings in the ocean floor that discharge very hot water. The water emerging from one such vent off the Oregon coast, 2400 m below the surface, has a temperature of 279°C. Despite its high temperature, the water doesn't boil. Why not?

Q16–29 The dark areas on the moon's surface are called *maria*, Latin for "seas," and were once thought to be bodies of water. In fact the maria are not "seas" at all, but plains of solidified lava. Given that there is no atmosphere on the moon, how can you explain the absence of liquid water on the moon's surface?

Q16–30 In addition to the normal cooking directions printed on the back of a box of rice, there are also "high altitude directions." The only difference is that the "high altitude directions" suggest increasing the cooking time and using a greater volume of boiling water in which to cook the rice. Why should the directions depend on the altitude in this way?

EXERCISES

SECTION 16–2 EQUATIONS OF STATE

16–1 A 20.0-L tank contains 0.225 kg of helium at 18.0°C. The molar mass of helium is 4.00 g/mol. a) How many moles of helium are in the tank? b) What is the pressure in the tank, in pascals and in atmospheres?

16–2 Helium gas with a volume of 2.60 L, under a pressure of 1.30 atm and at a temperature of 41.0°C, is warmed until both pressure and volume are doubled. a) What is the final temperature? b) How many grams of helium are there? The molar mass of helium is 4.00 g/mol.

16–3 A cylindrical tank has a tight-fitting piston that allows the volume of the tank to be changed. The tank originally contains 0.110 m^3 of air at a pressure of 3.40 atm. The piston is slowly pulled out until the volume of the gas is increased to 0.390 m^3. If the temperature remains constant, what is the final value of the pressure?

16–4 A 3.00-L tank contains air at 3.00 atm and 20.0°C. The tank is sealed and cooled until the pressure is 1.00 atm. a) What is the temperature then in degrees Celsius? Assume that the volume of the tank is constant. b) If the temperature is kept at the value found in part (a) and the gas is compressed, what is the volume when the pressure again becomes 3.00 atm?

16–5 A Jaguar XK8 convertible has an eight-cylinder engine. At the beginning of its compression stroke, one of the cylinders contains 499 cm^3 of air at atmospheric pressure (1.01×10^5 Pa) and a temperature of 27.0°C. At the end of the stroke the air has been compressed to a volume of 46.2 cm^3 and the gauge pressure has increased to 2.72×10^6 Pa. Compute the final temperature.

16–6 A welder using a tank of volume 0.0750 m^3 fills it with oxygen (molar mass = 32.0 g/mol) at a gauge pressure of 3.00×10^5 Pa and temperature of 37.0°C. The tank has a small leak, and in time some of the oxygen leaks out. On a day when the temperature is 22.0°C, the gauge pressure of the oxygen in the tank is 1.80×10^5 Pa. Find a) the initial mass of oxygen; b) the mass of oxygen that has leaked out.

16–7 A large cylindrical tank contains 0.750 m^3 of nitrogen gas at 27°C and 1.50×10^5 Pa (absolute pressure). The tank has a tight-fitting piston that allows the volume to be changed. What will be the pressure if the volume is decreased to 0.480 m^3 and the temperature is increased to 157°C?

16–8 A room with dimensions 7.00 m × 8.00 m × 2.50 m is filled with pure oxygen at 22.0°C and 1.00 atm. The molar mass of oxygen is 32.0 g/mol. a) How many moles of oxygen are required? b) What is the mass of this oxygen, in kilograms?

16–9 The gas inside a balloon will always have a pressure nearly equal to atmospheric pressure, since that is the pressure applied to the outside of the balloon. You fill a balloon with helium (a nearly ideal gas) to a volume of 0.600 L at a temperature of 19.0°C. What is the volume of the balloon if you cool it to the boiling point of liquid nitrogen (77.3 K)?

16–10 Deviations from the Ideal-Gas Equation. For carbon dioxide gas (CO_2), the constants in the van der Waals equation

are $a = 0.364$ J · m^3/mol^2 and $b = 4.27 \times 10^{-5}$ m^3/mol. a) If 1.00 mol of CO_2 gas at 350 K is confined to a volume of 400 cm^3, find the pressure of the gas using the ideal-gas equation and the van der Waals equation. b) Which equation gives a lower pressure? Why? What is the percentage difference of the van der Waals result from the ideal-gas equation result? c) The gas is kept at the same temperature as it expands to a volume of 4000 cm^3. Repeat the calculations of parts (a) and (b). d) Explain how your calculations show that the van der Waals equation is equivalent to the ideal-gas equation if n/V is small.

16–11 The total lung volume for a typical physics student is 6.00 L. A physics student fills her lungs with air at an absolute pressure of 1.00 atm. Then, holding her breath, she compresses her chest cavity, decreasing her lung volume to 5.70 L. What is then the pressure of the air in her lungs? Assume that the temperature of the air remains constant.

16–12 A diver observes a bubble of air rising from the bottom of a lake (where the absolute pressure is 3.50 atm) to the surface (where the pressure is 1.00 atm). The temperature at the bottom is 4.0°C, and the temperature at the surface is 23.0°C. a) What is the ratio of the volume of the bubble as it reaches the surface to its volume at the bottom? b) Would it be safe for the diver to hold his breath while ascending from the bottom of the lake to the surface? Why or why not?

16–13 A metal tank with volume 3.10 L will burst if the absolute pressure of the gas it contains exceeds 100 atm. a) If 11.0 mol of an ideal gas is put into the tank at a temperature of 23.0°C, to what temperature can the gas be warmed before the tank ruptures? Neglect the thermal expansion of the tank. b) Based on your answer to part (a), is it reasonable to neglect the thermal expansion of the tank? Explain.

16–14 With the assumption (not very realistic) that the air temperature is a uniform 0°C, what is the atmospheric pressure at an altitude of 5000 m? This is roughly the maximum altitude usually attained by aircraft without pressurized cabins.

16–15 At an altitude of 11,000 m (a typical cruising altitude for a jet airliner), the air temperature is −56.5°C and the air density is 0.364 kg/m^3. What is the pressure of the atmosphere at that altitude? (Note that the temperature at this altitude is not the same as at the surface of the earth, so that the calculation of Example 16–4 (Section 16–2) doesn't apply.)

SECTION 16–3 MOLECULAR PROPERTIES OF MATTER

16–16 Modern vacuum pumps make it easy to attain pressures of the order of 10^{-13} atm in the laboratory. At a pressure of 9.00×10^{-14} atm and an ordinary temperature (say $T = 300$ K), how many molecules are present in a volume of 1.00 cm^3?

16–17 The Lagoon Nebula (Fig. 16–21) is a cloud of hydrogen gas located 3900 light years from earth. The cloud is about 45 light years in diameter, and glows because of its high temperature of 7500 K. (The gas is raised to this temperature by the stars that lie within the nebula). The cloud is also very thin; there are only 80 molecules per cubic centimeter. a) Find the gas pressure (in atmospheres) in the Lagoon Nebula. Compare to the laboratory pressure referred to in Exercise 16–16. b) Science fiction films sometimes show starships being buffeted by turbulence as they fly through gas clouds such as the Lagoon Nebula. Does this seem realistic? Why or why not?

FIGURE 16–21 Exercise 16–17.

16–18 In a gas at standard conditions, what is the length of the side of a cube that contains a number of molecules equal to the population of the earth (about 6×10^9)?

16–19 How many moles are there in a 1.00-kg bottle of water? How many molecules? The molar mass of water is 18.0 g/mol.

16–20 Consider an ideal gas at 27°C and 1.00 atm pressure. Imagine the molecules to be, on average, uniformly spaced, with each molecule at the center of a small cube. a) What is the length of an edge of each small cube if adjacent cubes touch but don't overlap? b) How does this distance compare with the diameter of a molecule?

16–21 Consider 5.00 mol of liquid water. a) What volume is occupied by this amount of water? The molar mass of water is 18.0 g/mol. b) Imagine the molecules to be, on average, uniformly spaced, with each molecule at the center of a small cube. What is the length of an edge of each small cube if adjacent cubes touch but don't overlap? c) How does this distance compare with the diameter of a molecule?

SECTION 16–4 KINETIC-MOLECULAR MODEL OF AN IDEAL GAS

16–22 A flask contains a mixture of neon (Ne), krypton (Kr), and radon (Rn) gases. Compare a) the average kinetic energies of the three types of atoms; b) the root-mean-square speeds. (*Hint:* The periodic table in Appendix D shows the molar mass (in g/mol) of each element under the chemical symbol for that element.)

16–23 Gaseous Diffusion of Uranium. a) A process called *gaseous diffusion* is often used to separate isotopes of uranium, that is, atoms of the elements that have different masses, such as ^{235}U and ^{238}U. The only gaseous compound of uranium at ordinary temperatures is uranium hexafluoride, UF_6. Speculate on how $^{235}UF_6$ and $^{238}UF_6$ molecules might be separated by diffusion. b) The molar masses for $^{235}UF_6$ and $^{238}UF_6$ molecules are

0.349 kg/mol and 0.352 kg/mol, respectively. If uranium hexafluoride acts as an ideal gas, what is the ratio of the root-mean-square speed of $^{235}UF_6$ molecules to that of $^{238}UF_6$ molecules if the temperature is uniform?

16–24 The ideas of average and root-mean-square value can be applied to any distribution. A class of 150 students had the following scores on a 100-point quiz:

Score	Number of Students
10	11
20	12
30	24
40	15
50	19
60	10
70	12
80	20
90	17
100	10

a) Find the average score for the class. b) Find the root-mean-square score for the class.

16–25 a) Oxygen (O_2) has a molar mass of 32.0 g/mol. What is the average translational kinetic energy of an oxygen molecule at a temperature of 300 K? b) What is the average value of the square of its speed? c) What is the root-mean-square speed? d) What is the momentum of an oxygen molecule traveling at this speed? e) Suppose an oxygen molecule traveling at this speed bounces back and forth between opposite sides of a cubical vessel 0.10 m on a side. What is the average force it exerts on one of the walls of the container? (Assume that the molecule's velocity is perpendicular to the two sides that it strikes.) f) What is the average force per unit area? g) How many oxygen molecules traveling at this speed are necessary to produce an average pressure of 1 atm? h) Compute the number of oxygen molecules that are actually contained in a vessel of this size at 300 K and atmospheric pressure. i) Your answer for part (h) should be three times as large as the answer for part (g). Where does this discrepancy arise?

16–26 Calculate the mean free path of air molecules at a pressure of 3.50×10^{-13} atm and a temperature of 300 K. (This pressure is readily attainable in the laboratory; see Exercise 16–16.) As in Example 16–9, model the air molecules as spheres of radius 2.0×10^{-10} m.

16–27 At what temperature is the root-mean-square speed of nitrogen molecules equal to the root-mean-square speed of hydrogen molecules at 20.0°C? (*Hint:* The periodic table in Appendix D shows the molar mass (in g/mol) of each element under the chemical symbol for that element. The molar mass of H_2 is twice the molar mass of hydrogen atoms, and similarly for N_2.)

16–28 Smoke particles in the air typically have masses of the order of 10^{-16} kg. The Brownian motion (rapid, irregular movement) of these particles, resulting from collisions with air molecules, can be observed with a microscope. a) Find the root-mean-square speed of Brownian motion for a particle with a mass of 3.00×10^{-16} kg in air at 300 K. b) Would the root-mean-square speed be different if the particle were in hydrogen gas at the same temperature? Explain.

SECTION 16–5 HEAT CAPACITIES

16–29 a) Calculate the specific heat capacity at constant volume of water vapor, assuming the nonlinear triatomic molecule has three translational and three rotational degrees of freedom and that vibrational motion does not contribute. The molar mass of water is 18.0 g/mol. b) The actual specific heat capacity of water vapor at low pressures is about 2000 J/kg · K. Compare this with your calculation and comment on the actual role of vibrational motion.

16–30 a) The specific heat of ice at constant volume is 833 J/kg · K at −180°C, 1640 J/kg · K at −60°C, and 2060 J/kg · K at −5.0°C. Calculate C_V, the molar heat capacity at constant volume for ice, at each of these temperatures. The molar mass of H_2O is 18.0 g/mol. b) Why do you suppose the value of C_V increases with increasing temperature? c) Do the values that you calculate approach the value of $3R$ (given in the rule of Dulong and Petit) as the temperature increases? Speculate about why this should be so.

16–31 a) How much heat does it take to increase the temperature of 2.50 mol of a diatomic ideal gas by 30.0 K near room temperature if the gas is held at constant volume? b) What is the answer to the question in part (a) if the gas is monatomic rather than diatomic?

16–32 a) Compute the specific heat capacity at constant volume of nitrogen (N_2) gas, and compare with the specific heat capacity of liquid water. The molar mass of N_2 is 28.0 g/mol. b) You warm 1.00 kg of water at a constant volume of 1.00 L from 20.0°C to 30.0°C in a kettle. For the same amount of heat, how many kilograms of 20.0°C air would you be able to warm to 30.0°C? What volume (in liters) would this air occupy at 20.0°C and a pressure of 1.00 atm? Make the simplifying assumption that air is 100% N_2.

SECTION 16–6 MOLECULAR SPEEDS

*16–33 For a gas of nitrogen molecules (N_2), what must be the temperature if 94.7% of all the molecules have speeds less than

a) 1500 m/s; b) 1000 m/s; c) 500 m/s? Use Table 16–2. The molar mass of N_2 is 28.0 g/mol.

*16–34 Derive Eq. (16–33) from Eq. (16–32).

*16–35 Prove that $f(v)$ as given by Eq. (16–33) is maximum for $\epsilon = kT$. Use this result to obtain Eq. (16–34).

*16–36 For diatomic carbon dioxide gas (CO_2, molar mass = 44.0 g/mol) at $T = 300$ K, calculate a) the most probable speed v_{mp}; b) the average speed v_{av}; c) the root-mean-square speed v_{rms}.

SECTION 16–7 PHASES OF MATTER

16–37 Puffy cumulus clouds, which are made of water droplets, occur at lower altitudes in the atmosphere. Wispy cirrus clouds, which are made of ice crystals, occur only at higher altitudes. Find the altitude y (measured from sea level) above which only cirrus clouds can occur. On a typical day and at altitudes less than 11 km, the temperature at an altitude y is given by $T = T_0 - \alpha y$, where $T_0 = 15.0°C$ and $a = 6.0$ C°/1000 m.

16–38 Solid water (ice) is slowly warmed from a very low temperature. a) What minimum external pressure p_1 must be applied to the solid if a melting phase transition is to be observed? Describe the sequence of phase transitions that occur if the applied pressure p is such that $p < p_1$. b) Above a certain maximum pressure p_2, no boiling transition is observed. What is this pressure? Describe the sequence of phase transitions that occur if $p_1 < p < p_2$.

16–39 A physicist places a piece of ice at 0.00°C and a beaker of water at 0.00°C inside a glass box and closes the lid of the box. All the air is then removed from the box. If the ice, water, and beaker are all maintained at a temperature of 0.00°C, describe the final equilibrium state inside the box.

16–40 The atmosphere of the planet Mars is 95.3% carbon dioxide (CO_2) and about 0.03% water vapor. The atmospheric pressure is only 600 Pa, and the surface temperature varies from −30°C to −100°C. The polar ice caps contain both CO_2 ice and water ice. Could there be *liquid* CO_2 on the surface of Mars? Could there be liquid water? Why or why not?

PROBLEMS

16–41 a) Use Eq. (16–1) to estimate the change in the volume of a solid steel sphere of volume 11 L when the temperature and pressure increase from 21°C and 1.013×10^5 Pa to 42°C and 2.10×10^7 Pa. (*Hint:* Consult Chapters 11 and 15 to determine the values of β and k.) b) In Example 16–3 the change in volume of an 11-L steel scuba tank was neglected. Was this a good approximation? Explain.

16–42 At an absolute pressure of 2.00×10^{-13} atm, a partial vacuum easily obtained in laboratories (see Exercise 16–16), calculate the mass of nitrogen present in a volume of 3000 cm^3 if the temperature of the gas is 22.0°C. The molar mass of nitrogen (N_2) is 28.0 g/mol.

16–43 A cylinder 1.00 m tall with inside diameter 0.120 m is used to hold propane gas (molar mass 44.1 g/mol) for use in a barbecue. It is initially filled with gas until the gauge pressure is 1.30×10^6 Pa and the temperature is 22.0°C. The temperature of the gas remains constant as it is partially emptied out of the tank,

until the gauge pressure is 2.50×10^5 Pa. Calculate the mass of propane that has been used.

16–44 During a test dive in 1939 prior to being accepted by the U.S. Navy, the submarine *Squalus* sank at a point where the depth of water was 73.0 m. The temperature at the surface was 27.0°C and at the bottom it was 7.0°C. The density of seawater is 1030 kg/m^3. a) A diving bell was used to rescue 33 trapped crewmen from the *Squalus*. The diving bell was in the form of a circular cylinder 2.30 m high, open at the bottom and closed at the top. When the diving bell was lowered to the bottom of the sea, to what height did water rise within the diving bell? (*Hint:* You may ignore the relatively small variation in water pressure between the bottom of the bell and the surface of the water within the bell.) b) At what gauge pressure must compressed air have been supplied to the bell while on the bottom to expel all the water from it?

16–45 A glassblower makes a barometer using a tube 0.900 m long and with a cross-section area of 0.620 cm^2. Mercury stands in this tube to a height of 0.750 m. The room temperature is 20.0°C. A small amount of air is accidentally introduced into the evacuated space above the mercury, and the column drops to a height of 0.690 m. How many grams of air were introduced? The molar mass of air is 28.8 g/mol.

16–46 A hot-air balloon stays aloft because hot air at atmospheric pressure is less dense than cooler air at the same pressure; the calculation of the buoyant force is discussed in Chapter 14. If the volume of the balloon is 500 m³ and the surrounding air is at 15.0°C, what must be the temperature of the air in the balloon for it to lift a total load of 290 kg (in addition to the mass of the hot air)? The density of air at 15.0°C and atmospheric pressure is 1.23 kg/m³.

16–47 An automobile tire has a volume of 0.0150 m³ on a cold day when the temperature of the air in the tire is 5.0°C and atmospheric pressure is 1.02 atm. Under these conditions the gauge pressure is measured to be 1.70 atm (about 25 lb/in.²). After the car is driven on the highway for 30 min, the temperature of the air in the tires has risen to 45.0°C and the volume to 0.0159 m³. What then is the gauge pressure?

16–48 A flask with a volume of 1.50 L, provided with a stopcock, contains ethane gas (C_2H_6) at 300 K and atmospheric pressure (1.013×10^5 Pa). The molar mass of ethane is 30.1 g/mol. The system is warmed to a temperature of 380 K, with the stopcock open to the atmosphere. The stopcock is then closed, and the flask cooled to its original temperature. a) What is the final pressure of the ethane in the flask? b) How many grams of ethane remain in the flask?

16–49 A balloon whose volume is 750 m³ is to be filled with hydrogen at atmospheric pressure (1.01×10^5 Pa). a) If the hydrogen is stored in cylinders with a volume of 1.90 m³ at a gauge pressure of 1.20×10^6 Pa, how many cylinders are required? Assume that the temperature of the hydrogen remains constant. b) What is the total weight (in addition to the weight of the gas) that can be supported by the balloon if the gas in the balloon and the surrounding air are both at 15.0°C? The molar mass of hydrogen (H_2) is 2.02 g/mol. The density of air at 15.0°C and atmospheric pressure is 1.23 kg/m³. See Chapter 14 for a discussion of buoyancy. c) What weight could be supported if the balloon were filled with helium (with a molar mass of 4.00 g/mol) instead of hydrogen, again at 15.0°C?

16–50 A vertical cylindrical tank 0.900 m high has its top end closed by a tightly fitting frictionless piston of negligible weight. The air inside the cylinder is at an absolute pressure of 1.00 atm. The piston is depressed by pouring mercury on it slowly (Fig. 16–22). How far will the piston descend before mercury spills over the top of the cylinder? The temperature of the air is kept constant.

16–51 A large tank of water has a hose connected to it, as shown in Fig. 16–23. The tank is sealed at the top and has compressed air between the water surface and the top. When the water height h has the value 3.50 m, the absolute pressure p of the compressed air is 4.20×10^5 Pa. Assume that the air above the water expands at constant temperature, and take the atmospheric pressure to be 1.00×10^5 Pa. a) What is the speed with which water flows out of the hose when h = 3.50 m? b) As water flows out of

FIGURE 16–22 Problem 16–50.

FIGURE 16–23 Problem 16–51.

the tank, h decreases. Calculate the speed of flow for h = 3.00 m and for h = 2.00 m. c) At what value of h does the flow stop?

16–52 In one hour, an average person at rest consumes 14.5 L of oxygen at a pressure of 1.00 atm and a temperature of 20.0°C. a) Express this rate of oxygen consumption in terms of number of molecules per second. b) A person at rest inhales and exhales 0.5 L of air with each breath. The inhaled air is 21.0% oxygen, and the exhaled air is 16.3% oxygen. How many breaths per minute will satisfy this person's oxygen requirements? c) Repeat part (b) for a resting person at an elevation of 3000 m, where the pressure is 0.72 atm and the temperature is 0.0°C. Assume that the oxygen percentages and volume per inhalation are the same as stated in part (b). (To maintain its functions, the body still requires the same number of oxygen molecules per second as at sea level.) Explain why some people report "shortness of breath" at such elevations.

16–53 **How Many Atoms Are You?** Estimate the number of atoms in the body of a 50-kg physics student. Note that the human body is mostly water, which has molar mass 18.0 g/mol, and that each water molecule contains three atoms.

16–54 The size of an oxygen molecule is about 2.0×10^{-10} m. Make a rough estimate of the pressure at which the finite volume of the molecules should cause noticeable deviations from ideal-gas behavior at ordinary temperatures (T = 300 K).

16–55 **Successive Approximations and the van der Waals Equation.** In the ideal-gas equation, the number of moles per volume n/V is simply equal to p/RT. In the van der Waals equation, solving for n/V in terms of the pressure p and temperature

T is somewhat more involved. a) Show that the van der Waals equation can be written as

$$\frac{n}{V} = \left(\frac{p + an^2/V^2}{RT} \right)\left(1 - \frac{bn}{V}\right).$$

b) The van der Waals parameters for hydrogen sulfide gas (H_2S) are $a = 0.448$ J $\cdot$ m³/mol² and $b = 4.29 \times 10^{-5}$ m³/mol. Determine the number of moles per volume of H_2S gas at 127°C and an absolute pressure of 9.80×10^5 Pa as follows: i) Calculate a first approximation using the ideal-gas equation, $n/V = p/RT$. ii) Substitute this approximation for n/V into the right-hand side of the equation in part (a). The result is a new, improved approximation for n/V. iii) Substitute the new approximation for n/V into the right-hand side of the equation in (a). The result is a further-improved approximation for n/V. iv) Repeat step (iii) until successive approximations agree to the desired level of accuracy (in this case, to three significant figures). c) Compare your final result in part (b) to the result p/RT obtained using the ideal-gas equation. Which result gives a larger value of n/V? Why?

16-56 a) Compute the increase in gravitational potential energy for a nitrogen molecule (molar mass 28.0 g/mol) for an increase in elevation of 400 m near the earth's surface. b) At what temperature is this equal to the average kinetic energy of a nitrogen molecule? c) Is it possible that a nitrogen molecule near sea level where $T = 15.0°C$ could rise to an altitude of 400 m? Is it likely that it could do so without hitting any other molecules along the way? Explain.

16-57 The Lennard-Jones Potential. A commonly used potential-energy function for the interaction of two molecules (Fig. 16–5) is the Lennard-Jones 6-12 potential

$$U(r) = U_0\left[\left(\frac{R_0}{r}\right)^{12} - 2\left(\frac{R_0}{r}\right)^6\right],$$

where r is the distance between the centers of the molecules and U_0 and R_0 are positive constants. The corresponding force $F(r)$ is given in Eq. (13–26). a) Graph $U(r)$ and $F(r)$ versus r. b) Let r_1 be the value of r at which $U(r) = 0$, and let r_2 be the value of r at which $F(r) = 0$. Show the locations of r_1 and r_2 on your graphs of $U(r)$ and $F(r)$. Which of these values represents the equilibrium separation between the molecules? c) Find the values of r_1 and r_2 in terms of R_0, and find the ratio r_1/r_2. d) If the molecules are located a distance r_2 apart (as calculated in part (c)), how much work must be done to pull them apart so that $r \rightarrow \infty$?

16-58 a) What is the total random translational kinetic energy of 5.00 L of hydrogen gas, with pressure 1.01×10^5 Pa and temperature 300 K? b) If the tank containing the gas is moved with a speed of 30.0 m/s, by what percentage is the total kinetic energy of the gas increased? The molar mass of hydrogen (H_2) is 2.016 g/mol.

16-59 The speed of propagation of a sound wave in air at 27°C is about 350 m/s. Calculate, for comparison, a) v_{rms} for nitrogen molecules; b) the rms value of v_x at this temperature. The molar mass of nitrogen (N_2) is 28.0 g/mol.

16-60 The surface of the sun has a temperature of about 5800 K and consists largely of hydrogen atoms. a) Find the rms speed of

a hydrogen atom at this temperature. (The mass of a single hydrogen atom is 1.67×10^{-27} kg.) b) The escape speed for a particle to leave the gravitational influence of the sun is given by $(2GM/R)^{1/2}$, where M is the sun's mass, R its radius, and G the gravitational constant (Example 12–5 of Section 12–4). Use the data in Appendix F to calculate this escape speed. c) Can appreciable quantities of hydrogen escape from the sun? Can *any* hydrogen escape? Explain.

16-61 a) Show that a projectile with mass m can "escape" from the surface of a planet if it is launched vertically upward with a kinetic energy greater than mgR_p, where g is the acceleration due to gravity at the planet's surface and R_p is the planet's radius. Ignore air resistance. (See Problem 16–60.) b) If the planet in question is the earth, at what temperature does the average translational kinetic energy of a nitrogen molecule (molar mass = 28.0 g/mol) equal that required to escape? What about a hydrogen molecule (molar mass = 2.02 g/mol)? c) Repeat part (b) for the moon, for which $g = 1.63$ m/s² and $R_p = 1740$ km. d) While the earth and the moon have similar average surface temperatures, the moon has essentially no atmosphere. Use your results from parts (b) and (c) to explain why.

16-62 Planetary Atmospheres. a) The temperature near the top of Jupiter's multicolored cloud layer is about 140 K. The temperature at the top of the earth's troposphere, at an altitude of about 20 km, is about 220 K. Calculate the rms speed of hydrogen molecules in both these environments. Give your answers in m/s and as a fraction of the escape speed from the respective planet. (See Problem 16–60.) b) Hydrogen gas (H_2) is a rare element in the earth's atmosphere. In the atmosphere of Jupiter, by contrast, 89% of all molecules are H_2. Explain why, using your results from part (a). c) Suppose an astronomer claims to have discovered an oxygen (O_2) atmosphere on the asteroid Ceres. How likely is this? Ceres has a mass equal to 0.014 times the mass of the moon, a density of 2400 kg/m³, and a surface temperature of about 200 K.

16-63 a) For what mass of molecule or particle is v_{rms} equal to 1.00 mm/s at 300 K? b) If the particle is an ice crystal, how many molecules does it contain? The molar mass of water is 18.0 g/mol. c) Calculate the diameter of the particle if it is a spherical piece of ice. Would it be visible to the naked eye?

16-64 In describing the heat capacities of solids in Section 16–5, we stated that the potential energy $U = \frac{1}{2}kx^2$ of a harmonic oscillator averaged over one period of the motion is equal to the kinetic energy $K = \frac{1}{2}mv^2$ averaged over one period. Prove this result using Eqs. (13–13) and (13–15) for the position and velocity of a simple harmonic oscillator. For simplicity, assume that the initial position and velocity make the phase angle ϕ equal to zero. (*Hint:* Use the trigonometric identities $\cos^2(\theta) = [1 + \cos(2\theta)]/2$ and $\sin^2(\theta) = [1 - \cos(2\theta)]/2$. What is the average value of $\cos(2\omega t)$ averaged over one period?)

16-65 It is possible to make crystalline solids that are only one layer of atoms thick. Such "two-dimensional" crystals can be created by depositing atoms on a very flat surface. a) If the atoms in such a two-dimensional crystal can only move within the plane of the crystal, what will be its molar heat capacity near room temperature? Give your answer as a multiple of R and in J/mol · K. b) At very low temperatures, will the molar heat capacity of a two-dimensional crystal be greater, less, or equal to the result you found in part (a)? Explain why.

16–66 a) Calculate the total *rotational* kinetic energy of the molecules in 1.00 mol of a diatomic gas at 300 K. b) Calculate the moment of inertia of an oxygen molecule (O_2) for rotation about either the y- or z-axis shown in Fig. 16–12. Treat the molecule as two massive points (representing the oxygen atoms) separated by a distance of 1.21×10^{-10} m. The molar mass of oxygen *atoms* is 16.0 g/mol. c) Find the rms angular velocity of rotation of an oxygen molecule about either the y- or z-axis shown in Fig. 16–12. How does your answer compare to the angular velocity of a typical piece of rapidly rotating machinery (10,000 rev/min)?

16–67 For each polyatomic gas in Table 16–1, compute the value of the molar heat capacity at constant volume, C_V, on the assumption that there is no vibrational energy. Compare with the measured values in the table, and compute the fraction of the total heat capacity that is due to vibration for each of the three gases. (*Note:* CO_2 is linear; SO_2 and H_2S are not. Recall that a linear polyatomic molecule has two rotational degrees of freedom, and a nonlinear molecule has three.)

*16–68 a) Show that $\int_0^\infty f(v)\,dv = 1$, where $f(v)$ is the Maxwell–Boltzmann distribution of Eq. (16–32). b) In terms of the physical definition of $f(v)$, explain why the integral in part (a) *must* have this value.

*16–69 Calculate the integral in Eq. (16–31), $\int_0^\infty v^2 f(v)\,dv$, and compare this result to $(v^2)_{av}$ as given by Eq. (16–16). (*Hint:* You may use the tabulated integral

$$\int_0^\infty x^{2n} e^{-\alpha x^2}\,dx = \frac{1 \cdot 3 \cdot 5 \cdots (2n-1)}{2^{n+1}\alpha^n}\sqrt{\frac{\pi}{\alpha}},$$

where n is a positive integer and α is a positive constant.)

*16–70 Calculate the integral in Eq. (16–30), $\int_0^\infty v f(v)\,dv$, and compare this result to v_{av} as given by Eq. (16–35). (*Hint:* Make the change of variable $v^2 = x$ and use the tabulated integral

$$\int_0^\infty x^n e^{-\alpha x}\,dx = \frac{n!}{\alpha^{n+1}},$$

where n is a positive integer and α is a positive constant.)

*16–71 a) Explain why in a gas of N molecules, the number of molecules having speeds in the *finite* interval v to $v + \Delta v$ is $\Delta N = N \int_v^{v+\Delta v} f(v)\,dv$. b) If Δv is small, then $f(v)$ is approximately constant over the interval and $\Delta N \approx Nf(v)\,\Delta v$. For oxygen gas ($O_2$, molar mass = 32.0 g/mol) at $T = 300$ K, use this approximation to calculate the number of molecules with speeds within $\Delta v = 20$ m/s of v_{mp}. Express your answer as a multiple of N. c) Repeat part (b) for speeds within $\Delta v = 20$ m/s of $7v_{mp}$.

d) Repeat parts (b) and (c) for a temperature of 600 K. e) Repeat parts (b) and (c) for a temperature of 150 K. f) What do your results tell you about the shape of the distribution as a function of temperature? Do your conclusions agree with what is shown in Fig. 16–17?

16–72 The *vapor pressure* is the pressure of the vapor phase of a substance when it is in equilibrium with the solid or liquid phase of the substance. The *relative humidity* is the partial pressure of water vapor in the air divided by the vapor pressure of water at that same temperature, expressed as a percentage. The air is saturated when the humidity is 100%. a) The vapor pressure of water at 20.0°C is 2.34×10^3 Pa. If the air temperature is 20.0°C and the relative humidity is 60%, what is the partial pressure of water vapor in the atmosphere (that is, the pressure due to water vapor alone)? b) Under the conditions of part (a), what is the mass of water in 1.00 m^3 of air? (The molar mass of water is 18.0 g/mol. Assume that water vapor can be treated as an ideal gas.)

16–73 The Dew Point. The vapor pressure of water (Problem 16–72) decreases as the temperature decreases. If the amount of water vapor in the air is kept constant as the air is cooled, a temperature is reached, called the *dew point,* at which the partial pressure and vapor pressure coincide and the vapor is saturated. If the air is cooled further, vapor condenses to liquid until the partial pressure again equals the vapor pressure at that temperature. The temperature in a room is 30.0°C. A meteorologist cools a metal can by gradually adding cold water. When the can temperature reaches 16.0°C, water droplets form on its outside surface. What is the relative humidity of the 30.0°C air in the room? The table below lists the vapor pressure of water at various temperatures:

Temperature (°C)	Vapor Pressure (Pa)
10.0	1.23×10^3
12.0	1.40×10^3
14.0	1.60×10^3
16.0	1.81×10^3
18.0	2.06×10^3
20.0	2.34×10^3
22.0	2.65×10^3
24.0	2.99×10^3
26.0	3.36×10^3
28.0	3.78×10^3
30.0	4.25×10^3

16–74 Altitude at Which Clouds Form. On a spring day in the midwestern United States, the air temperature at the surface is 28.0°C. Puffy cumulus clouds form at an altitude where the air temperature equals the dew point (Problem 16–73). If the air temperature decreases with altitude at a rate of 0.6 C°/100 m, at approximately what height above the ground will clouds form if the relative humidity at the surface is a) 35% b) 80%? (*Hint:* Use the table in Problem 16–73.)

CHALLENGE PROBLEMS

16–75 Dark Nebulae and the Interstellar Medium. The dark area in Fig. 16–24 that appears devoid of stars is a *dark nebula*, a cold gas cloud in interstellar space that contains enough material to block out light from the stars behind it. A typical dark nebula is about 20 light years in diameter and contains about 50 hydrogen atoms per cubic centimeter (monatomic hydrogen, *not* H_2) at a temperature of about 20 K. (A light year is the distance light travels in vacuum in one year and is equal to

9.46×10^{15} m.) a) Estimate the mean free path for a hydrogen atom in a dark nebula. The radius of a hydrogen atom is 5.0×10^{-11} m. b) Estimate the rms speed of a hydrogen atom and the mean free time (the average time between collisions for a given atom). Based on this result, do you think that atomic collisions, such as those leading to H_2 molecule formation, are very important in determining the composition of the nebula? c) Estimate the pressure inside a dark nebula. d) Compare the rms speed of a hydrogen atom to the escape speed at the surface of the nebula (assumed spherical). If the space around the nebula were a vacuum, would such a cloud be stable or would it tend to evaporate? e) The stability of dark nebulae is explained by the presence of the *interstellar medium* (ISM), an even thinner gas that permeates space and into which the dark nebulae are embedded. Show that for dark nebulae to be in equilibrium with the ISM, the numbers of atoms per volume (N/V) and the temperatures (T) of dark nebulae and the ISM must be related by

$$\frac{(N/V)_{\text{nebula}}}{(N/V)_{\text{ISM}}} = \frac{T_{\text{ISM}}}{T_{\text{nebula}}} .$$

f) In the vicinity of the sun, the ISM contains about 1 hydrogen atom per 200 cm^3. Estimate the temperature of the ISM in the vicinity of the sun. Compare to the temperature of the sun's surface, about 5800 K. Would a spacecraft coasting through interstellar space burn up? Why or why not?

FIGURE 16–24 Challenge Problem 16–75.

16–76 In the *troposphere*, the part of the atmosphere that extends from the surface to an altitude of about 11 km, the temperature is not uniform but decreases with increasing elevation. a) Show that if the temperature variation is approximated by the linear relation

$$T = T_0 - \alpha y,$$

where T_0 is the temperature at the earth's surface and T is the temperature at height y, the pressure p at height y is given by

$$\ln\left(\frac{p}{p_0}\right) = \frac{Mg}{R\alpha} \ln\left(\frac{T_0 - \alpha y}{T_0}\right),$$

where p_0 is the pressure at the earth's surface and M is the molar mass for air. The coefficient α is called the lapse rate of temperature. It varies with atmospheric conditions, but an average value

is about 0.6 C°/100 m. b) Show that the above result reduces to the result of Example 16–4 (Section 16–2) in the limit that $\alpha \rightarrow 0$. c) With $\alpha = 0.6$ C°/100 m, calculate p for $y = 8863$ m and compare your answer to the result of Example 16–4. Take $T_0 = 288$ K and $p_0 = 1.00$ atm.

16–77 Van der Waals Equation and Critical Points. a) In pV-diagrams the slope $\partial p / \partial V$ along an isotherm is never positive. Explain why. b) Regions where $\partial p / \partial V = 0$ represent equilibrium between two phases; volume can change with no change in pressure, as when water boils at atmospheric pressure. We can use this to determine the temperature, pressure, and volume per mole at the critical point using the equation of state $p = p(V, T, n)$. If $T > T_c$, then $p(V)$ has no maximum along an isotherm, but if $T < T_c$, then $p(V)$ has a maximum. Explain how this leads to the following condition for determining the critical point:

$$\frac{\partial p}{\partial V} = 0 \quad \text{and} \quad \frac{\partial^2 p}{\partial V^2} = 0 \quad \text{at the critical point.}$$

c) Solve the van der Waals equation (Eq. 16–7) for p. That is, find $p(V, T, n)$. Find $\partial p/\partial V$ and $\partial^2 p/\partial V^2$. Set these equal to zero to obtain two equations for V, T, and n. d) Simultaneous solution of the two equations obtained in part (c) gives the temperature and volume per mole at the critical point, T_c and $(V/n)_c$. Find these constants in terms of a and b. (*Hint:* Divide one equation by the other to eliminate T.) e) Substitute these values into the equation of state to find p_c, the pressure at the critical point. f) Use the results from parts (d) and (e) to find the ratio $RT_c/p_c(V/n)_c$. This should not contain either a or b, and so should have the same value for all gases. g) Compute the ratio $RT_c/p_c(V/n)_c$ for the gases H_2, N_2, and H_2O using the critical point data given below.

Gas	T_c (K)	p_c (Pa)	$(V/n)_c$ (m³/mol)
H_2	33.3	13.0×10^5	65.0×10^{-6}
N_2	126.2	33.9×10^5	90.1×10^{-6}
H_2O	647.4	221.2×10^5	56.0×10^{-6}

h) Discuss how well the results of part (g) compare to the prediction of part (f) based on the van der Waals equation. What do you conclude about the accuracy of the van der Waals equation as a description of the behavior of gases near the critical point?

*16–78 In Example 16–7 (Section 16–4) we saw that $v_{\text{rms}} > v_{\text{av}}$. It is not difficult to show that this is *always* the case. (The only exception is when the particles have the same speed, in which case $v_{\text{rms}} = v_{\text{av}}$.) a) For two particles with speeds v_1 and v_2, show that $v_{\text{rms}} \geq v_{\text{av}}$, regardless of the numerical values of v_1 and v_2. Then show that $v_{\text{rms}} > v_{\text{av}}$ if $v_1 \neq v_2$. b) Suppose that for a collection of N particles you know that $v_{\text{rms}} > v_{\text{av}}$. Another particle, with speed u, is added to the collection of particles. If the new rms and average speeds are denoted as v'_{rms} and v'_{av}, show that

$$v'_{\text{rms}} = \sqrt{\frac{Nv_{\text{rms}}^2 + u^2}{N+1}} \quad \text{and} \quad v'_{\text{av}} = \frac{Nv_{\text{av}} + u}{N+1} .$$

c) Use the expressions in part (b) to show that $v'_{\text{rms}} > v'_{\text{av}}$, regardless of the numerical value of u. d) Explain why your results for (a) and (c) together show that $v_{\text{rms}} > v_{\text{av}}$ for any collection of particles if the particles do not all have the same speed.

The First Law of Thermodynamics

This steam locomotive operates using the first law of thermodynamics. When heat is generated by burning wood or coal in the locomotive's engine, part of the energy goes into increasing the temperature of the engine's water (which boils and produces steam). The remainder is used by the expanding steam to do work and propel the locomotive forward.

17-1 INTRODUCTION

Every time you drive a car, turn on an air conditioner, or use an electrical appliance, you reap the practical benefits of *thermodynamics,* the study of relationships involving heat, mechanical work, and other aspects of energy and energy transfer. If you've seen demonstrations that use liquid nitrogen, you may have wondered how gases are liquefied. One method is by first compressing the gas to very high pressure while keeping the temperature constant, then insulating it and allowing it to expand. The gas cools so much during the expansion that it liquefies. This is an example of a *thermodynamic process.*

The first law of thermodynamics, central to the understanding of such processes, is an extension of the principle of conservation of energy. It broadens this principle to include energy exchange by both heat transfer and mechanical work and introduces the concept of the *internal energy* of a system. Conservation of energy plays a vital role in every area of physical science, and the first law has extremely broad usefulness. To state energy relationships precisely, we need the concept of a *thermodynamic system,* and we discuss *heat* and *work* as two means of transferring energy into or out of such a system.

17-2 THERMODYNAMIC SYSTEMS

We have studied energy transfer through mechanical work (Chapter 6) and through heat transfer (Chapters 15 and 16).

Now we are ready to combine and generalize these principles. We will always talk about energy transfer to or from some specific *system.* The system might be a mechanical device, a biological organism, or a specified quantity of material such as the refrigerant in an air conditioner or steam expanding in a turbine. A **thermodynamic system** is a system that can interact (and exchange energy) with its surroundings, or environment, in at least two ways, one of which is heat transfer. A familiar example is a quantity of popcorn kernels in a pot with a lid. When the pot is placed on a stove, energy is added to the popcorn by conduction of heat; as the popcorn pops and expands, it does work as it exerts an upward force on the lid and moves it through a displacement (Fig. 17–1). The *state* of the popcorn changes in this process, since the volume, temperature, and pressure of the popcorn all change as it pops. A process such as this one, in which there are changes in the state of a thermodynamic system, is called a **thermodynamic process.**

In mechanics we used the concept of *system* on a regular basis in connection with free-body diagrams and conservation of energy and momentum. With *thermodynamic* systems, as with all others, it is essential to define clearly at the start exactly what is and is not included in the system. Only then can we describe unambiguously the energy transfers into and out of that system. For instance, in our popcorn example we defined the system to include the popcorn but not the pot, lid, or stove.

17–1 The popcorn in the pot is a thermodynamic system. In the thermodynamic process shown here, heat is added to the system, and the system does work on its surroundings to lift the lid of the pot.

Thermodynamics has its roots in many practical problems other than popping popcorn (Fig. 17–2). The engine in an automobile and the jet engines in an airplane use the heat of combustion of their fuel to perform mechanical work in propelling the vehicle. Muscle tissue in living organisms metabolizes chemical energy in food and performs mechanical work on the organism's surroundings. A steam engine or steam turbine uses the heat of combustion of coal or other fuel to perform mechanical work such as driving an electric generator or pulling a train.

SIGNS FOR HEAT AND WORK IN THERMODYNAMICS

We describe the energy relations in any thermodynamic process in terms of the quantity of heat Q added *to* the system and the work W done *by* the system. Both Q and W may be positive, negative, or zero (Fig. 17–3). A positive value of Q represents heat flow *into* the system, with a corresponding input of energy to it; negative Q represents heat flow *out of* the system. A positive value of W represents work done *by* the system against its surroundings, such as work done by an expanding gas, and hence corresponds to energy *leaving* the system. Negative W, such as work done during compression of a gas in which work is done *on the gas* by its surroundings, represents energy *entering* the system. We will use these conventions consistently in the examples in this chapter and the next.

CAUTION ▶ Note that our sign rule for work is *opposite* to the one we used in mechanics, in which we always spoke of the work done by the forces acting *on* a body. In thermodynamics it is usually more convenient to call W the work done *by* the system so that when a system expands, the pressure, volume change, and work are all positive. Take care to use the sign rules for work and heat consistently! ◀

17–3 WORK DONE DURING VOLUME CHANGES

17–2 (a) A jet engine uses the heat of combustion of its fuel to do work propelling the plane. (b) Humans and other biological organisms are more complicated systems than we can analyze fully in this book, but the same basic principles of thermodynamics apply to them.

A gas in a cylinder with a movable piston is a simple example of a thermodynamic system. Internal-combustion engines, steam engines, and compressors in refrigerators and air conditioners all use some version of such a system. In the next several sections we will use the gas-in-cylinder system to explore several kinds of processes involving energy transformations.

(a)

(b)

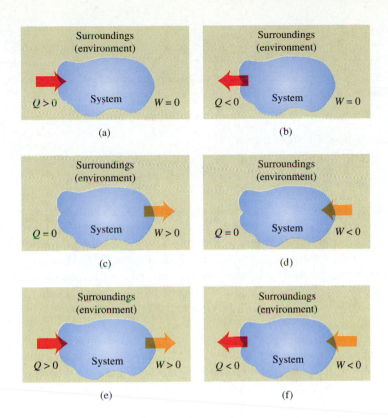

(a) (b)

(c) (d)

(e) (f)

17–3 A thermodynamic system may exchange energy with its surroundings (environment) by means of heat and work. (a) When heat is added *to* the system, Q is positive. (b) When heat is transferred *out* of the system, Q is negative. (c) When work is done *by* the system, W is positive. (d) When work is done *on* the system, W is negative. Energy transfer by both heat and work can occur simultaneously; in (e), heat is added to the system and work is done by the system, and in (f), heat is transferred out of the system and work is done on the system.

Motion of piston

(a)

Motion of piston

(b)

17–4 (a) When a molecule strikes a wall moving away from it, the molecule does work on the wall; the molecule's speed and kinetic energy decrease. The gas does positive work on the piston. (b) When a molecule strikes a wall moving toward it, the wall does work on the molecule; the molecule's speed and kinetic energy increase. The gas does negative work on the piston.

We'll use a microscopic viewpoint, based on the kinetic and potential energies of individual molecules in a material, to develop intuition about thermodynamic quantities. But it is important to understand that the central principles of thermodynamics can be treated in a completely *macroscopic* way, without reference to microscopic models. Indeed, part of the great power and generality of thermodynamics is that it does *not* depend on details of the structure of matter.

First we consider the *work* done by the system during a volume change. When a gas expands, it pushes outward on its boundary surfaces as they move outward. Hence an expanding gas always does positive work. The same thing is true of any solid or fluid material that expands under pressure, such as the popcorn in Fig. 17–1.

We can understand the work done by a gas in a volume change by considering the molecules that make up the gas. When one such molecule collides with a stationary surface, it exerts a momentary force on the wall but does no work because the wall does not move. But if the surface is moving, such as a piston in a gasoline engine, the molecule *does* do work on the surface during the collision. If the piston in Fig. 17–4a moves to the right, so that the volume of the gas increases, the molecules that strike the piston exert a force through a distance and do *positive* work on the piston. If the piston moves toward the left as in Fig. 17–4b, so the volume of the gas decreases, then positive work is done *on* the molecule during the collision. Hence the gas molecules do *negative* work on the piston.

Figure 17–5 shows a solid or fluid in a cylinder with a movable piston. Suppose that the cylinder has a cross-section area A and that the pressure exerted by the system at the piston face is p. The total force F exerted by the system on the piston is $F = pA$. When the piston moves out an infinitesimal distance dx, the work dW done by this force is

$$dW = F\,dx = pA\,dx.$$

dx

pA

A

17–5 The infinitesimal work done by the system during the small expansion dx is $dW = pA\,dx$.

8.5

Work Done By a Gas

When fuel is burned in one of the cylinders of an automotive engine, the heated fuel-air mixture expands and does an amount of work given by Eq. (17–2).

But

$$A \, dx = dV,$$

where dV is the infinitesimal change of volume of the system. Thus we can express the work done by the system in this infinitesimal volume change as

$$dW = p \, dV. \tag{17–1}$$

In a finite change of volume from V_1 to V_2,

$$W = \int_{V_1}^{V_2} p \, dV \qquad \text{(work done in a volume change).} \tag{17–2}$$

In general the pressure of the system may vary during the volume change. To evaluate the integral in Eq. (17–2), we have to know how the pressure varies as a function of volume. We can represent this relationship as a graph of p as a function of V (a pV-diagram, described at the end of Section 16–2). Figure 17–6a shows a simple example. In this figure, Eq. (17–2) is represented graphically as the *area* under the curve of p versus V between the limits V_1 and V_2. (In Section 6–4 we used a similar interpretation of the work done by a force F as the area under the curve of F versus x between the limits x_1 and x_2.)

According to the rule we stated in Section 17–2, work is *positive* when a system *expands*. In an expansion from state 1 to state 2 in Fig. 17–6a the area under the curve and the work are positive. A *compression* from 1 to 2 in Fig. 17–6b gives a *negative* area; when a system is compressed, its volume decreases and it does *negative* work on its surroundings (see also Fig. 17–4b).

If the pressure p remains constant while the volume changes from V_1 to V_2 (Fig. 17–6c), the work done by the system is

$$W = p(V_2 - V_1) \qquad \begin{array}{l} \text{(work done in a volume change at} \\ \text{constant pressure).} \end{array} \tag{17–3}$$

In any process in which the volume is *constant,* the system does no work because there is no displacement.

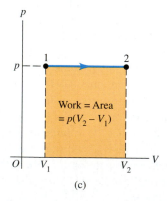

(a)   (b)   (c)

17–6 The work done equals the area under the curve on a pV-diagram. (a) In this change from state 1 to state 2 the volume increases and the work and area are positive. (b) In this change from state 1 to state 2 the volume decreases; the area is taken to be negative to agree with the sign of W. The two states are reversed from part (a). (c) The area of the rectangle gives the work done for a constant-pressure process. In this case, volume increases and $W > 0$.

EXAMPLE 17-1

Isothermal expansion of an ideal gas An ideal gas undergoes an *isothermal* (constant-temperature) *expansion* at temperature T, during which its volume changes from V_1 to V_2. How much work does the gas do?

SOLUTION From Eq. (17–2),

$$W = \int_{V_1}^{V_2} p\, dV.$$

From Eq. (16–3) the pressure p of n moles of ideal gas occupying volume V at absolute temperature T is

$$p = \frac{nRT}{V},$$

where R is the gas constant. We substitute this into the integral,

take the constants n, R, and T outside, and evaluate the integral:

$$W = nRT \int_{V_1}^{V_2} \frac{dV}{V} = nRT \ln \frac{V_2}{V_1} \quad \text{(ideal gas, isothermal process).}$$

In an expansion, $V_2 > V_1$ and W is positive. Also, when T is constant,

$$p_1 V_1 = p_2 V_2 \qquad \text{or} \qquad \frac{V_2}{V_1} = \frac{p_1}{p_2},$$

so the isothermal work may also be expressed as

$$W = nRT \ln \frac{p_1}{p_2} \quad \text{(ideal gas, isothermal process).}$$

In an isothermal expansion the volume increases and the pressure drops, so $p_1 > p_2$ and again the work is positive.

17–4 PATHS BETWEEN THERMODYNAMIC STATES

We've seen that if a thermodynamic process involves a change in volume, the system undergoing the process does work (either positive or negative) on its surroundings. Heat also flows into or out of the system during the process if there is a temperature difference between the system and its surroundings. Let's now examine how the work done by and heat added to the system during a thermodynamic process depend on the details of how the process takes place.

When a thermodynamic system changes from an initial state to a final state, it passes through a series of intermediate states. We call this series of states a **path.** There are always infinitely many different possibilities for these intermediate states. When they are all equilibrium states, the path can be plotted on a pV-diagram (Fig. 17–7a). Point 1 represents an initial state with pressure p_1 and volume V_1, and point 2 represents a final state with pressure p_2 and volume V_2. To pass from state 1 to state 2, we could keep the pressure constant at p_1 while the system expands to volume V_2 (point 3 in Fig. 17–7b), then reduce the pressure to p_2 (probably by decreasing the temperature) while keeping the volume constant at V_2 (to point 2 on the diagram). The work done by the system during this process is the area under the line $1 \rightarrow 3$; no work is done during the constant-volume process $3 \rightarrow 2$. Or the system might traverse the path $1 \rightarrow 4 \rightarrow 2$ (Fig. 17–7c); in that case the work is the area under the line $4 \rightarrow 2$, since no work is done during the constant-volume process $1 \rightarrow 4$. The smooth curve from 1 to 2 is another

17–7 (a) Three different paths between state 1 and state 2. (b)–(d) The work done by the system during a transition between two states depends on the path chosen.

(a)   (b)   (c)   (d)

possibility (Fig. 17–7d), and the work for this path is different from that for either of the other paths.

We conclude that *the work done by the system depends not only on the initial and final states, but also on the intermediate states, that is, on the path.* Furthermore, we can take the system through a series of states forming a closed loop, such as $1 \rightarrow 3 \rightarrow 2 \rightarrow 4 \rightarrow 1$. In this case the final state is the same as the initial state, but the total work done by the system is *not* zero. (In fact, it is represented on the graph by the area enclosed by the loop; can you prove that? See Exercise 17–6.) It follows that it doesn't make sense to talk about the amount of work *contained in* a system. In a particular state, a system may have definite values of the state coordinates p, V, and T, but it wouldn't make sense to say that it has a definite value of W.

Like work, the *heat* added to a thermodynamic system when it undergoes a change of state depends on the path from the initial state to the final state. Here's an example. Suppose we want to change the volume of a certain quantity of an ideal gas from 2.0 L to 5.0 L while keeping the temperature constant at $T = 300$ K. Figure 17–8 shows two different ways in which we can do this. In Fig. 17–8a the gas is contained in a cylinder with a piston, with an initial volume of 2.0 L. We let the gas expand slowly, supplying heat from the electric heater to keep the temperature at 300 K. After expanding in this slow, controlled, isothermal manner, the gas reaches its final volume of 5.0 L; it absorbs a definite amount of heat in the process.

Figure 17–8b shows a different process leading to the same final state. The container is surrounded by insulating walls and is divided by a thin, breakable partition into two compartments. The lower part has volume 2.0 L, and the upper part has volume 3.0 L. In the lower compartment we place the same amount of the same gas as in Fig. 17–8a, again at $T = 300$ K. The initial state is the same as before. Now we break the partition; the gas undergoes a rapid, uncontrolled expansion, with no heat passing through the insulating walls. The final volume is 5.0 L, the same as in Fig. 17–8a. The gas does no work during this expansion because it doesn't push against anything that moves. This uncontrolled expansion of a gas into vacuum is called a **free expansion;** we will discuss it further in Section 17–7.

Experiments have shown that when an ideal gas undergoes a free expansion, there is no temperature change. Therefore the final state of the gas is the same as in Fig. 17–8a. The intermediate states (pressures and volumes) during the transition from state 1 to state 2 are entirely different in the two cases; Figs. 17–8a and 17–8b represent *two different paths* connecting the *same states* 1 and 2. For the path in Fig. 17–8b, *no* heat is transferred into the system, and the system does no work. Like work, *heat depends not only on the initial and final states but also on the path.*

Because of this path dependence, it would not make sense to say that a system "contains" a certain quantity of heat. To see this, suppose we assign an arbitrary value to "the

17–8 (a) Slow, controlled isothermal expansion of a gas from an initial state 1 to a final state 2 with the same temperature but lower pressure. (b) Rapid, uncontrolled expansion of the same gas starting at the same state 1 and ending at the same state 2.

heat in a body" in some reference state. Then presumably the "heat in the body" in some other state would equal the heat in the reference state plus the heat added when the body goes to the second state. But that's ambiguous, as we have just seen; the heat added depends on the *path* we take from the reference state to the second state. We are forced to conclude that there is *no* consistent way to define "heat in a body"; it is not a useful concept.

While it doesn't make sense to talk about "work in a body" or "heat in a body," it *does* make sense to speak of the amount of *internal energy* in a body. This important concept is our next topic.

17-5 INTERNAL ENERGY AND THE FIRST LAW OF THERMODYNAMICS

Internal energy is one of the most important concepts in thermodynamics. In Section 7–4, when we discussed energy changes for a body sliding with friction, we stated that warming a body increased its internal energy and that cooling the body decreased its internal energy. But what *is* internal energy? We can look at it in various ways; let's start with one based on the ideas of mechanics. Matter consists of atoms and molecules, and these are made up of particles having kinetic and potential energies. We *tentatively* define the **internal energy** of a system as the sum of the kinetic energies of all of its constituent particles, plus the sum of all the potential energies of interaction among these particles.

Note that internal energy does *not* include potential energy arising from the interaction between the system and its surroundings. If the system is a glass of water, placing it on a high shelf increases the gravitational potential energy arising from the interaction between the glass and the earth. But this has no effect on the interaction between the molecules of the water, and so the internal energy of the water does not change.

We use the symbol U for internal energy. (We used this same symbol in our study of mechanics to represent potential energy. You may have to remind yourself occasionally that U has a different meaning in thermodynamics.) During a change of state of the system the internal energy may change from an initial value U_1 to a final value U_2. We denote the change in internal energy as $\Delta U = U_2 - U_1$.

We know that heat transfer is energy transfer. When we add a quantity of heat Q to a system and it does no work during the process, the internal energy increases by an amount equal to Q; that is, $\Delta U = Q$. When a system does work W by expanding against its surroundings and no heat is added during the process, energy leaves the system and the internal energy decreases. That is, when W is positive, ΔU is negative, and conversely. So $\Delta U = -W$. When *both* heat transfer and work occur, the *total* change in internal energy is

$$U_2 - U_1 = \Delta U = Q - W \quad \text{(first law of thermodynamics).} \quad (17\text{–}4)$$

We can rearrange this to the form

$$Q = \Delta U + W. \quad (17\text{–}5)$$

The message of Eq. (17–5) is that in general when heat Q is added to a system, some of this added energy remains within the system, changing its internal energy by an amount ΔU; the remainder leaves the system again as the system does work W against its surroundings. Because W and Q may be positive, negative, or zero, ΔU can be positive, negative, or zero for different processes (Fig. 17–9).

Equation (17–4) or (17–5) is the **first law of thermodynamics.** It is a generalization of the principle of conservation of energy to include energy transfer through heat as

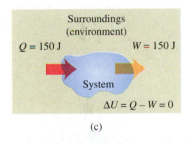

17-9 In a thermodynamic process, the internal energy of a system may increase, decrease, or stay the same. (a) If more heat is added to the system than the system does work, ΔU is positive and the internal energy increases. (b) If more heat flows out of the system than work is done on the system, ΔU is negative and the internal energy decreases. (c) If the heat added to the system equals the work done by the system, $\Delta U = 0$ and the internal energy is unchanged.

well as mechanical work. As you will see in later chapters, this principle can be extended to ever broader classes of phenomena by identifying additional forms of energy and energy transfer. In every situation in which it seems that the total energy in all known forms is not conserved, it has been possible to identify a new form of energy such that the total energy, including the new form, *is* conserved. There is energy associated with electric fields, with magnetic fields, and, according to the theory of relativity, even with mass itself.

At the beginning of this discussion we tentatively defined internal energy in terms of microscopic kinetic and potential energies. This has drawbacks, however. Actually *calculating* internal energy in this way for any real system would be hopelessly complicated. Furthermore, this definition isn't an *operational* one because it doesn't describe how to determine internal energy from physical quantities that we can measure directly.

So let's look at internal energy in another way. Starting over, we define the *change* in internal energy ΔU during any change of a system as the quantity given by Eq. (17–4), $\Delta U = Q - W$. This *is* an operational definition, because we can measure Q and W. It does not define U itself, only ΔU. This is not a shortcoming, because we can *define* the internal energy of a system to have a specified value in some reference state, and then use Eq. (17–4) to define the internal energy in any other state. This is analogous to our treatment of potential energy in Chapter 7, in which we arbitrarily defined the potential energy of a mechanical system to be zero at a certain position.

This new definition trades one difficulty for another. If we define ΔU by Eq. (17–4), then when the system goes from state 1 to state 2 by two different paths, how do we know that ΔU is the same for the two paths? We have already seen that Q and W are, in general, *not* the same for different paths. If ΔU, which equals $Q - W$, is also path-dependent, then ΔU is ambiguous. If so, the concept of internal energy of a system is subject to the same criticism as the erroneous concept of quantity of heat in a system, as we discussed at the end of Section 17–4.

The only way to answer this question is through *experiment*. For various materials we measure Q and W for various changes of state and various paths to learn whether ΔU is or is not path-dependent. The results of many such investigations are clear and unambiguous: While Q and W depend on the path, $\Delta U = Q - W$ *is independent of path. The change in internal energy of a system during any thermodynamic process depends only on the initial and final states, not on the path leading from one to the other.*

Experiment, then, is the ultimate justification for believing that a thermodynamic system in a specific state has a unique internal energy that depends only on that state. An equivalent statement is that the internal energy U of a system is a function of the state coordinates p, V, and T (actually, any two of these, since the three variables are related by the equation of state).

To say that the first law of thermodynamics, given by Eq. (17–4) or (17–5), represents conservation of energy for thermodynamic processes is correct, as far as it goes. But an important *additional* aspect of the first law is the fact that internal energy depends only on the state of a system. In changes of state, the change in internal energy is path-independent.

All this may seem a little abstract if you are satisfied to think of internal energy as microscopic mechanical energy. There's nothing wrong with that view, and we will make use of it at various times during our discussion. But in the interest of precise *operational* definitions, internal energy, like heat, can and must be defined in a way that is independent of the detailed microscopic structure of the material.

Two special cases of the first law of thermodynamics are worth mentioning. A process that eventually returns a system to its initial state is called a *cyclic* process. For such a process, the final state is the same as the initial state, and so the *total* internal

ActivPhysics

8.6

Heat, Internal Energy, and First Law of Thermodynamics

The internal energy of a cup of coffee just depends on its thermodynamic state—how much water and ground coffee it contains, and what its temperature is. It does not depend on the history of how the coffee was prepared—that is, the thermodynamic path that led to its current state.

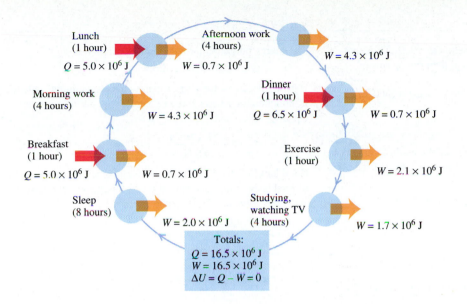

17–10 Every day, your body (a thermodynamic system) goes through a cyclic thermodynamic process like this one. Heat Q is added by metabolizing food, and your body does work W in breathing, walking, and other activities. If you return to the same state at the end of the day, $Q = W$ and the net change in your internal energy is zero.

energy change must be zero. Then

$$U_2 = U_1 \qquad \text{and} \qquad Q = W.$$

If a net quantity of work W is done by the system during this process, an equal amount of energy must have flowed into the system as heat Q. But there is no reason why either Q or W individually has to be zero (Fig. 17–10).

Another special case occurs in an *isolated* system, one that does no work on its surroundings and has no heat flow to or from its surroundings. For any process taking place in an isolated system,

$$W = Q = 0,$$

and therefore

$$U_2 - U_1 = \Delta U = 0.$$

In other words, *the internal energy of an isolated system is constant.*

Problem–Solving Strategy

THE FIRST LAW OF THERMODYNAMICS

1. The internal energy change ΔU in any thermodynamic process or series of processes is independent of the path, whether the substance is an ideal gas or not. This is of the utmost importance in the problems in this chapter and the next. Sometimes you will be given enough information about one path between given initial and final states to calculate ΔU for that path. Since ΔU is the same for every other path between the same two states, you can then relate the various energy quantities for other paths. For a cyclic process, $\Delta U = 0$.

2. As usual, consistent units are essential. If p is in Pa and V in m^3, then W is in joules. Otherwise, you may want to convert the pressure and volume units into Pa and m^3. If a

heat capacity is given in terms of calories, the simplest procedure is usually to convert it to joules. Be especially careful with moles. When you use $n = m_{tot}/M$ to convert between total mass and number of moles, remember that if m_{tot} is in kilograms, M must be in *kilograms* per mole. The usual units for M are *grams* per mole; be careful!

3. When a process consists of several distinct steps, it often helps to make a chart showing Q, W, and ΔU for each step. Put these quantities for each step on a different line, and arrange them so that the Q's, W's, and ΔU's form columns. Then you can apply the first law to each line; in addition, you can add each column and apply the first law to the sums. Do you see why?

EXAMPLE 17-2

Working off your dessert You propose to eat a 900-calorie hot fudge sundae (with whipped cream) and then run up several flights of stairs to work off the energy you have taken in. How high do you have to climb? Assume that your mass is 60 kg.

SOLUTION The system consists of you and the earth. Remember that one food-value calorie is 1 kcal = 1000 cal = 4190 J. The energy intake is

$$Q = 900 \text{ kcal} (4190 \text{ J/1 kcal}) = 3.77 \times 10^6 \text{ J}.$$

The energy output required to climb a height h is

$$W = mgh = (60 \text{ kg})(9.8 \text{ m/s}^2)h = (588 \text{ N})h.$$

If the final state of the system is the same as the initial state (that is, no fatter, no leaner), these two energy quantities must be equal: $Q = W$. Then

$$h = \frac{Q}{mg} = \frac{3.77 \times 10^6 \text{ J}}{588 \text{ N}} = 6410 \text{ m} \quad \text{(about 21,000 ft)}.$$

Good luck! We have assumed 100% efficiency in the conversion of food energy into mechanical work; this isn't very realistic. We'll talk more about efficiency later.

EXAMPLE 17-3

Figure 17–11 shows a pV-diagram for a *cyclic* process, one in which the initial and final states are the same. It starts at point a and proceeds counterclockwise in the pV-diagram to point b, then back to a, and the total work is $W = -500$ J. a) Why is the work negative? b) Find the change in internal energy and the heat added during this process.

SOLUTION a) The work done equals the area under the curve, with the area taken as positive for increasing volume and negative for decreasing volume. The area under the lower curve from a to b is positive, but it is smaller than the absolute value of the negative area under the upper curve from b back to a. Therefore the net area (the area enclosed by the path, shown with red stripes) and the work are negative. In other words, 500 more joules of work are done *on* the system than *by* the system.
b) For this and any other cyclic process (in which the beginning and end points are the same), $\Delta U = 0$, so $Q = W = -500$ J. That is, 500 joules of heat must come *out of* the system.

17-11 The net work done by the system in the process aba is -500 J. What would it have been if the process had proceeded clockwise in this pV-diagram?

EXAMPLE 17-4

A series of thermodynamic processes is shown in the pV-diagram of Fig. 17–12. In process ab, 150 J of heat are added to the system, and in process bd, 600 J of heat are added. Find a) the internal energy change in process ab; b) the internal energy change in process abd (shown in light blue); and c) the total heat added in process acd (shown in dark blue).

SOLUTION a) No volume change occurs during process ab, so $W_{ab} = 0$ and $\Delta U_{ab} = Q_{ab} = 150$ J.
b) Process bd occurs at constant pressure, so the work done by the system during this expansion is

$$W_{bd} = p(V_2 - V_1)$$
$$= (8.0 \times 10^4 \text{ Pa})(5.0 \times 10^{-3} \text{ m}^3 - 2.0 \times 10^{-3} \text{ m}^3)$$
$$= 240 \text{ J}.$$

The total work for process abd is

$$W_{abd} = W_{ab} + W_{bd} = 0 + 240 \text{ J} = 240 \text{ J},$$

17-12 A pV-diagram showing the various thermodynamic processes.

and the total heat is

$$Q_{abd} = Q_{ab} + Q_{bd} = 150 \text{ J} + 600 \text{ J} = 750 \text{ J}.$$

Applying Eq. (17–4) to *abd*, we find

$$\Delta U_{abd} = Q_{abd} - W_{abd} = 750 \text{ J} - 240 \text{ J} = 510 \text{ J}.$$

c) Because ΔU is independent of path, the internal energy change is the same for path *acd* as for path *abd*; that is,

$$\Delta U_{acd} = \Delta U_{abd} = 510 \text{ J}.$$

The total work for the path *acd* is

$$W_{acd} = W_{ac} + W_{cd} = p(V_2 - V_1) + 0$$
$$= (3.0 \times 10^4 \text{ Pa})(5.0 \times 10^{-3} \text{ m}^3 - 2.0 \times 10^{-3} \text{ m}^3)$$
$$= 90 \text{ J}.$$

Now we apply Eq. (17–5) to process *acd*:

$$Q_{acd} = \Delta U_{acd} + W_{acd} = 510 \text{ J} + 90 \text{ J} = 600 \text{ J}.$$

We see that although ΔU is the same (510 J) for *abd* and *acd*, W (240 J versus 90 J) and Q (750 J versus 600 J) have different values for the two processes.

Here is a tabulation of the various quantities:

STEP	Q	W	$\Delta U = Q - W$	STEP	Q	W	$\Delta U = Q - W$
ab	150 J	0 J	150 J	*ac*	?	90 J	?
bd	600 J	240 J	360 J	*cd*	?	0 J	?
abd	750 J	240 J	510 J	*acd*	600 J	90 J	510 J

EXAMPLE 17–5

Thermodynamics of boiling water One gram of water (1 cm^3) becomes 1671 cm^3 of steam when boiled at a constant pressure of 1 atm (1.013×10^5 Pa). The heat of vaporization at this pressure is $L_v = 2.256 \times 10^6$ J/kg. Compute a) the work done by the water when it vaporizes; b) its increase in internal energy.

SOLUTION a) For a constant-pressure process we may use Eq. (17–3) to compute the work done by the vaporizing water:

$$W = p(V_2 - V_1)$$
$$= (1.013 \times 10^5 \text{ Pa})(1671 \times 10^{-6} \text{ m}^3 - 1 \times 10^{-6} \text{ m}^3)$$
$$= 169 \text{ J}.$$

b) The heat added to the water is the heat of vaporization:

$$Q = mL_v = (10^{-3} \text{ kg})(2.256 \times 10^6 \text{ J/kg}) = 2256 \text{ J}.$$

From the first law of thermodynamics, Eq. (17–4), the change in internal energy is

$$\Delta U = Q - W = 2256 \text{ J} - 169 \text{ J} = 2087 \text{ J}.$$

To vaporize one gram of water, we have to add 2256 J of heat. Most (2087 J) of this added energy remains in the system as an increase in internal energy. The remaining 169 J leaves the system again as it does work against the surroundings while expanding from liquid to vapor. The increase in internal energy is associated mostly with the intermolecular forces that hold the molecules together in the liquid state. These forces are attractive, so the associated potential energies are greater after work has been done to pull the molecules apart, forming the vapor state. It's like increasing gravitational potential energy by pulling an elevator farther from the center of the earth.

INFINITESIMAL CHANGES OF STATE

In the preceding examples the initial and final states differ by a finite amount. Later we will consider *infinitesimal* changes of state in which a small amount of heat dQ is added to the system, the system does a small amount of work dW, and its internal energy changes by an amount dU. For such a process we state the first law in differential form as

$$dU = dQ - dW \qquad \text{(first law of thermodynamics, infinitesimal process).} \qquad (17\text{–}6)$$

For the systems we will discuss, the work dW is given by $dW = p\, dV$, so we can also state the first law as

$$dU = dQ - p\, dV. \qquad (17\text{–}7)$$

17–6 KINDS OF THERMODYNAMIC PROCESSES

In this section we describe four specific kinds of thermodynamic processes that occur often in practical situations. These can be summarized briefly as "no heat transfer" or *adiabatic*, "constant volume" or *isochoric*, "constant pressure" or *isobaric*, and "constant temperature" or *isothermal*. For some of these we can use a simplified form of the first law of thermodynamics.

ADIABATIC PROCESS

An **adiabatic process** (pronounced "ay-dee-ah-*bat*-ic") is defined as one with no heat transfer into or out of a system; $Q = 0$. We can prevent heat flow either by surrounding the system with thermally insulating material or by carrying out the process so quickly that there is not enough time for appreciable heat flow. From the first law we find that for every adiabatic process,

$$U_2 - U_1 = \Delta U = -W \qquad \text{(adiabatic process).} \qquad (17\text{--}8)$$

8.4
State Variables and Ideal Gas Law

When a system expands adiabatically, W is positive (the system does work on its surroundings), so ΔU is negative and the internal energy decreases. When a system is *compressed* adiabatically, W is negative (work is done on the system by its surroundings) and U increases. In many (but not all) systems an increase of internal energy is accompanied by a rise in temperature.

The compression stroke in an internal-combustion engine is an approximately adiabatic process. The temperature rises as the air-fuel mixture in the cylinder is compressed. The expansion of the burned fuel during the power stroke is also an approximately adiabatic expansion with a drop in temperature.

ISOCHORIC PROCESS

An **isochoric process** (pronounced "eye-so-*kor*-ic") is a *constant-volume* process. When the volume of a thermodynamic system is constant, it does no work on its surroundings. Then $W = 0$, and

$$U_2 - U_1 = \Delta U = Q \qquad \text{(isochoric process).} \qquad (17\text{--}9)$$

In an isochoric process, all the energy added as heat remains in the system as an increase in internal energy. Heating a gas in a closed constant-volume container is an example of an isochoric process. (Note that there are types of work that do not involve a volume change. For example, we can do work on a fluid by stirring it. In some literature, "isochoric" is used to mean that no work of any kind is done.)

ISOBARIC PROCESS

An **isobaric process** (pronounced "eye-so-*bear*-ic") is a *constant-pressure* process. In general, none of the three quantities ΔU, Q, and W is zero in an isobaric process, but calculating W is easy nonetheless. From Eq. (17–3),

$$W = p(V_2 - V_1) \qquad \text{(isobaric process).} \qquad (17\text{--}10)$$

Example 17–5 concerns an isobaric process, boiling water at constant pressure.

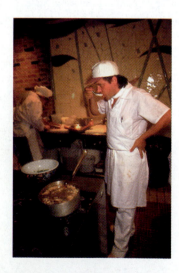

Most cooking involves isobaric processes. That's because the air pressure above a saucepan or frying pan, or inside a microwave oven, remains essentially constant while the food is being heated. (But see page 550 for an example of an *adiabatic* process in the kitchen.)

ISOTHERMAL PROCESS

An **isothermal process** is a *constant-temperature* process. For a process to be isothermal, any heat flow into or out of the system must occur slowly enough that thermal equilibrium is maintained. In general, none of the quantities ΔU, Q, or W is zero in an isothermal process.

In some special cases the internal energy of a system depends *only* on its temperature, not on its pressure or volume. The most familiar system having this special property is an ideal gas, as we'll discuss in the next section. For such systems, if the temperature is constant, the internal energy is also constant; $\Delta U = 0$ and $Q = W$. That is, any energy entering the system as heat Q must leave it again as work W done by the system. Example 17–1, involving an ideal gas, is an example of an isothermal process in which U is also constant. For most systems other than ideal gases the internal energy depends on pressure as well as temperature, so U may vary even when T is constant.

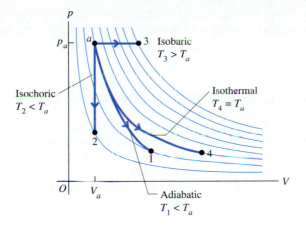

17–13 Four different processes for a constant amount of an ideal gas, all starting at state a. For the adiabatic process, $Q = 0$; for the isochoric process, $W = 0$; and for the isothermal process, $\Delta U = 0$. The temperature increases only during the isobaric expansion.

Figure 17–13 shows a pV-diagram for each of these four processes for a constant amount of an ideal gas. The path followed in an adiabatic process (a to 1) is called an **adiabat.** A vertical line (constant volume) is an **isochor,** a horizontal line (constant pressure) is an **isobar,** and a curve of constant temperature (shown as light blue lines in Fig. 17–13) is an **isotherm.**

17–7 INTERNAL ENERGY OF AN IDEAL GAS

We will now show that for an ideal gas, the internal energy U depends only on temperature, not on pressure or volume. Let's think again about the free-expansion experiment described in Section 17–4. A thermally insulated container with rigid walls is divided into two compartments by a partition (Fig. 17–14). One compartment has a quantity of an ideal gas, and the other is evacuated.

When the partition is removed or broken, the gas expands to fill both parts of the container. The gas does no work on its surroundings because the walls of the container don't move, and there is no heat flow through the insulation. So both Q and W are zero, and the internal energy U is constant. This is true of any substance, whether it is an ideal gas or not.

Does the *temperature* change during a free expansion? Suppose it *does* change, while the internal energy stays the same. In that case we have to conclude that the internal energy depends on both the temperature and the volume or on both the temperature and the pressure, but certainly not on the temperature alone. But if T is constant during a free expansion, for which we know that U is constant even though both p and V change, then we have to conclude that U depends only on T, not on p or V.

Many experiments have shown that when a low-density gas undergoes a free expansion, its temperature *does not* change. Such a gas is essentially an ideal gas. The conclusion is: **The internal energy of an ideal gas depends only on its temperature, not on its pressure or volume.** This property, in addition to the ideal-gas equation of state, is part of the ideal-gas model. Make sure that you understand that U depends only on T for an ideal gas, for we will make frequent use of this fact.

For non-ideal gases, some temperature change occurs during free expansions, even though the internal energy is constant. This shows that the internal energy cannot depend *only* on temperature; it must depend on pressure as well. From the microscopic viewpoint, in which internal energy U is the sum of the kinetic and potential energies for all the particles that make up the system, this is not surprising. Non-ideal gases usually have attractive intermolecular forces, and when molecules move farther apart, the associated potential energies increase. If the total internal energy is constant, the kinetic energies

17–14 The partition is broken (or removed) to start the free expansion of gas into the vacuum region.

must decrease. Temperature is directly related to molecular *kinetic* energy, and for such a gas a free expansion is usually accompanied by a *drop* in temperature.

17-8 HEAT CAPACITIES OF AN IDEAL GAS

We defined specific heat capacity and molar heat capacity in Section 15–6. We also remarked at the end of that section that the specific or molar heat capacity of a substance depends on the conditions under which the heat is added. It is usually easiest to measure the heat capacity of a gas in a closed container under constant-volume conditions. The corresponding heat capacity is the **molar heat capacity at constant volume,** denoted by C_V. Heat capacity measurements for solids and liquids are usually carried out in the atmosphere under constant atmospheric pressure, and we call the corresponding heat capacity the **molar heat capacity at constant pressure,** C_p. If neither p nor V is constant, we have an infinite number of possible heat capacities.

Let's consider C_V and C_p for an ideal gas. To measure C_V, we raise the temperature of an ideal gas in a rigid container with constant volume (neglecting its thermal expansion). To measure C_p, we let the gas expand just enough to keep the pressure constant as the temperature rises.

Why should these two molar heat capacities be different? The answer lies in the first law of thermodynamics. In a constant-volume temperature increase, the system does no work, and the change in internal energy ΔU equals the heat added Q. In a constant-pressure temperature increase, on the other hand, the volume *must* increase; otherwise, the pressure (given by the ideal-gas equation of state $p = nRT/V$) could not remain constant. As the material expands, it does an amount of work W. According to the first law,

$$Q = \Delta U + W. \tag{17-11}$$

For a given temperature increase, the internal energy change ΔU of an ideal gas has the same value no matter what the process (remember that the internal energy of an ideal gas depends only on temperature, not pressure or volume). Equation (17–11) then shows that the heat input for a constant-pressure process must be *greater* than that for a constant-volume process because additional energy must be supplied to account for the work W done during the expansion. So C_p is greater than C_V for an ideal gas. The pV-diagram in Fig. 17–15 shows this relationship. For air, C_p is 40% greater than C_V.

For a very few substances (one of which is water between 0°C and 4°C) the volume *decreases* during heating. In this case, W is negative, the heat input is *less* than in the constant-volume case, and C_p is *less* than C_V.

We can derive a simple relation between C_p and C_V for an ideal gas. First consider the constant-*volume* process. We place n moles of an ideal gas at temperature T in a constant-volume container. We place it in thermal contact with a hotter body; an infinitesimal quantity of heat dQ flows into the gas, and its temperature increases by an infinitesimal amount dT. By the definition of C_V, the molar heat capacity at constant volume,

$$dQ = nC_V \, dT. \tag{17-12}$$

The pressure increases during this process, but the gas does no work ($dW = 0$) because the volume is constant. The first law in differential form, Eq. (17–6), is $dQ = dU + dW$. Since $dW = 0$, $dQ = dU$ and Eq. (17–12) can also be written as

$$dU = nC_V \, dT. \tag{17-13}$$

Now consider a constant-*pressure* process with the same temperature change dT. We place the same gas in a cylinder with a piston that we can allow to move just enough to

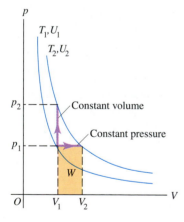

17–15 Raising the temperature of an ideal gas from T_1 to T_2 by a constant-volume or a constant-pressure process. For an ideal gas, U depends only on T, so ΔU is the same for both processes. In the constant-volume process, no work is done, so $Q = \Delta U$. But for the constant-pressure process, Q is greater, since it must include both ΔU and $W = p_1(V_2 - V_1)$. Thus $C_p > C_V$.

maintain constant pressure. Again we bring the system into contact with a hotter body. As heat flows into the gas, it expands at constant pressure and does work. By the definition of C_p, the molar heat capacity at constant pressure, the amount of heat dQ entering the gas is

$$dQ = nC_p \, dT. \tag{17–14}$$

The work dW done by the gas in this constant-pressure process is

$$dW = p \, dV.$$

8.7
Heat Capacity

8.8
Isochoric Process

8.9
Isobaric Process

8.10
Isothermal Process

We can also express dW in terms of the temperature change dT by using the ideal-gas equation of state, $pV = nRT$. Because p is constant, the change in V is proportional to the change in T:

$$dW = p \, dV = nR \, dT. \tag{17–15}$$

Now we substitute Eqs. (17–14) and (17–15) into the first law, $dQ = dU + dW$. We obtain

$$nC_p \, dT = dU + nR \, dT. \tag{17–16}$$

Now here comes the crux of the calculation. The internal energy change dU for the constant-pressure process is again given by Eq. (17–13), $dU = nC_V \, dT$, *even though now the volume is not constant*. Why is this so? Recall the discussion of Section 17–7; one of the special properties of an ideal gas is that its internal energy depends *only* on temperature. Thus the *change* in internal energy during any process must be determined only by the temperature change. If Eq. (17–13) is valid for an ideal gas for one particular kind of process, it must be valid for an ideal gas for *every* kind of process with the same dT. So we may replace dU in Eq. (17–16) by $nC_V \, dT$:

$$nC_p \, dT = nC_V \, dT + nR \, dT.$$

When we divide each term by the common factor $n \, dT$, we get

$$C_p = C_V + R \quad \text{(molar heat capacities of an ideal gas).} \tag{17–17}$$

As we predicted, the molar heat capacity of an ideal gas at constant pressure is *greater* than the molar heat capacity at constant volume; the difference is the gas constant R. (Of course, R must be expressed in the same units as C_p and C_V, such as J/mol · K.)

We have used the ideal-gas model to derive Eq. (17–17), but it turns out to be obeyed to within a few percent by many real gases at moderate pressures. Measured values of C_p and C_V are given in Table 17–1 for several real gases at low pressures; the difference in most cases is approximately $R = 8.315$ J/mol · K.

TABLE 17–1

MOLAR HEAT CAPACITIES OF GASES AT LOW PRESSURE

TYPE OF GAS	GAS	C_V (J/mol · K)	C_p (J/mol · K)	$C_p - C_V$ (J/mol · K)	$\gamma = C_p/C_V$
Monatomic	He	12.47	20.78	8.31	1.67
	Ar	12.47	20.78	8.31	1.67
Diatomic	H_2	20.42	28.74	8.32	1.41
	N_2	20.76	29.07	8.31	1.40
	O_2	20.85	29.17	8.31	1.40
	CO	20.85	29.16	8.31	1.40
Polyatomic	CO_2	28.46	36.94	8.48	1.30
	SO_2	31.39	40.37	8.98	1.29
	H_2S	25.95	34.60	8.65	1.33

The table also shows that the molar heat capacity of a gas is related to its molecular structure, as we discussed in Section 16–5. In fact, the first two columns of Table 17–1 are the same as Table 16–1.

The last column of Table 17–1 lists the values of the dimensionless **ratio of heat capacities,** C_p/C_V, denoted by the Greek letter γ ("gamma"):

$$\gamma = \frac{C_p}{C_V} \qquad \text{(ratio of heat capacities).} \tag{17–18}$$

(This is sometimes called the "ratio of specific heats.") For gases, C_p is always greater than C_V and γ is always greater than unity. This quantity plays an important role in *adiabatic* processes for an ideal gas, which we will study in the next section.

We can use our kinetic-theory discussion of the molar heat capacity of an ideal gas (Section 16–5) to predict values of γ. As an example, an ideal monatomic gas has $C_V = \frac{3}{2}R$. From Eq. (17–17),

$$C_p = C_V + R = \frac{3}{2}R + R = \frac{5}{2}R,$$

so

$$\gamma = \frac{C_p}{C_V} = \frac{\frac{5}{2}R}{\frac{3}{2}R} = \frac{5}{3} = 1.67.$$

As Table 17–1 shows, this agrees well with values of γ computed from measured heat capacities. For most diatomic gases near room temperature, $C_V = \frac{5}{2}R$, $C_p = C_V + R = \frac{7}{2}R$, and

$$\gamma = \frac{C_p}{C_V} = \frac{\frac{7}{2}R}{\frac{5}{2}R} = \frac{7}{5} = 1.40,$$

also in good agreement with measured values.

Here's a final reminder: For an ideal gas the internal energy change in *any* process is given by $\Delta U = nC_V \Delta T$, *whether the volume is constant or not.* This relation, which comes in handy in the following example, holds for other substances *only* when the volume is constant.

EXAMPLE 17–6

Cooling your room A typical dorm room or bedroom contains about 2500 moles of air. Find the change in the internal energy of this much air when it is cooled from 23.9°C to 11.6°C at a constant pressure of 1.00 atm. Treat the air as an ideal gas with $\gamma = 1.400$.

SOLUTION This is a constant-pressure process. Your first impulse may be to find C_p and then calculate Q from $Q = nC_p\Delta T$; find the volume change and find the work done by the gas from $W = p\,\Delta V$; then finally use the first law to find ΔU. This would be perfectly correct, but there's a much easier way. For an ideal gas the internal-energy change is $\Delta U = nC_V\Delta T$ for *every* process, *whether the volume is constant or not.* So all we have to do is find C_V and use this expression for ΔU. From Eqs.

(17–17) and (17–18),

$$\gamma = \frac{C_p}{C_V} = \frac{C_V + R}{C_V} = 1 + \frac{R}{C_V},$$

$$C_V = \frac{R}{\gamma - 1} = \frac{8.315 \text{ J/mol} \cdot \text{K}}{1.400 - 1} = 20.79 \text{ J/mol} \cdot \text{K}.$$

Then

$$\Delta U = nC_V\Delta T = (2500 \text{ mol})(20.79 \text{ J/mol} \cdot \text{K})(11.6 \text{ K} - 23.9 \text{ K})$$
$$= -6.39 \times 10^5 \text{ J}.$$

A room air conditioner must extract this much internal energy from the air in your room and transfer it to the air outside. We'll discuss how this is done in Chapter 18.

17–9 ADIABATIC PROCESSES FOR AN IDEAL GAS

An adiabatic process, defined in Section 17–6, is a process in which no heat transfer takes place between a system and its surroundings. Zero heat transfer is an idealization, but a process is approximately adiabatic if the system is well insulated or if the process takes place so quickly that there is not enough time for appreciable heat flow to occur.

In an adiabatic process, $Q = 0$, so from the first law, $\Delta U = -W$. An adiabatic process for an ideal gas is shown in the pV-diagram of Fig. 17–16. As the gas expands from volume V_a to V_b, it does positive work, so its internal energy decreases and its temperature drops. If point a, representing the initial state, lies on an isotherm at temperature $T + dT$, then point b for the final state is on a different isotherm at a lower temperature T. For an ideal gas an adiabatic curve (adiabat) at any point is always *steeper* than the isotherm passing through the same point. For an adiabatic *compression* from V_b to V_a the situation is reversed and the temperature rises.

The air in the output pipes of air compressors used in gasoline stations and in paint-spraying equipment and to fill scuba tanks is always warmer than the air entering the compressor; this is because the compression is rapid and hence approximately adiabatic. Adiabatic *cooling* occurs when you open a bottle of your favorite carbonated beverage. The gas just above the beverage surface expands rapidly in a nearly adiabatic process; the temperature of the gas drops so much that water vapor in the gas condenses, forming a miniature cloud.

CAUTION ▶ Keep in mind that when we talk about "adiabatic heating" and "adiabatic cooling," we really mean "raising the temperature" and "lowering the temperature," respectively. In an adiabatic process, the temperature change is due to work done by or on the system; there is *no* heat flow at all. ◀

We can derive a relation between volume and temperature changes for an infinitesimal adiabatic process in an ideal gas. Equation (17–13) gives the internal energy change dU for *any* process for an ideal gas, adiabatic or not, so we have $dU = nC_V dT$. Also, the work done by the gas during the process is given by $dW = p\,dV$. Then, since $dU = -dW$ for an adiabatic process, we have

$$nC_V\,dT = -p\,dV. \qquad (17\text{–}19)$$

To obtain a relation containing only the volume V and temperature T, we eliminate p using the ideal-gas equation in the form $p = nRT/V$. Substituting this into Eq. (17–19) and rearranging, we get

$$nC_V\,dT = -\frac{nRT}{V}\,dV,$$

$$\frac{dT}{T} + \frac{R}{C_V}\frac{dV}{V} = 0.$$

The coefficient R/C_V can be expressed in terms of $\gamma = C_p/C_V$. We have

$$\frac{R}{C_V} = \frac{C_p - C_V}{C_V} = \frac{C_p}{C_V} - 1 = \gamma - 1,$$

$$\frac{dT}{T} + (\gamma - 1)\frac{dV}{V} = 0. \qquad (17\text{–}20)$$

Because γ is always greater than unity for a gas, $(\gamma - 1)$ is always positive. This means that in Eq. (17–20), dV and dT always have opposite signs. An adiabatic *expansion* of an ideal gas ($dV > 0$) always occurs with a *drop* in temperature ($dT < 0$), and an adiabatic *compression* ($dV < 0$) always occurs with a *rise* in temperature ($dT > 0$); this confirms our earlier prediction.

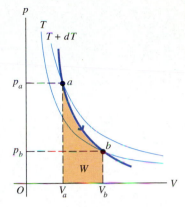

17–16 A pV-diagram of an adiabatic process for an ideal gas. As the gas expands from V_a to V_b, its temperature drops from $T + dT$ to T, corresponding to the decrease in internal energy due to the work W done by the gas (indicated by shaded area). For an ideal gas, when an isotherm and an adiabat pass through the same point on a pV-diagram, the adiabat is always steeper. (An adiabatic process is also shown in Fig. 17–13.)

8.11
Adiabatic Process

For finite changes in temperature and volume we integrate Eq. (17–20), obtaining

$$\ln T + (\gamma - 1)\ln V = \text{constant},$$

$$\ln T + \ln V^{\gamma-1} = \text{constant},$$

$$\ln\left(TV^{\gamma-1}\right) = \text{constant},$$

and finally,

$$TV^{\gamma-1} = \text{constant}. \qquad (17\text{–}21)$$

Thus for an initial state (T_1, V_1) and a final state (T_2, V_2),

$$T_1V_1^{\gamma-1} = T_2V_2^{\gamma-1} \qquad \text{(adiabatic process, ideal gas).} \qquad (17\text{–}22)$$

Because we have used the ideal-gas equation in our derivation of Eqs. (17–21) and (17–22), the T's must always be *absolute* (Kelvin) temperatures.

We can also convert Eq. (17–21) into a relation between pressure and volume by eliminating T, using the ideal-gas equation in the form $T = pV/nR$. Substituting this into Eq. (17–21), we find

$$\frac{pV}{nR}\,V^{\gamma-1} = \text{constant},$$

or, because n and R are constant,

$$pV^{\gamma} = \text{constant}. \qquad (17\text{–}23)$$

For an initial state (p_1, V_1) and a final state (p_2, V_2), Eq. (17–23) becomes

$$p_1V_1^{\gamma} = p_2V_2^{\gamma} \qquad \text{(adiabatic process, ideal gas).} \qquad (17\text{–}24)$$

We can also calculate the *work* done by an ideal gas during an adiabatic process. We know that $Q = 0$ and $W = -\Delta U$ for *any* adiabatic process. For an ideal gas, $\Delta U = nC_V(T_2 - T_1)$. If the number of moles n and the initial and final temperatures T_1 and T_2 are known, we have simply

$$W = nC_V(T_1 - T_2) \qquad \text{(adiabatic process, ideal gas).} \qquad (17\text{–}25)$$

We may also use $pV = nRT$ in this equation to obtain

$$W = \frac{C_V}{R}(p_1V_1 - p_2V_2) = \frac{1}{\gamma - 1}(p_1V_1 - p_2V_2) \qquad \begin{array}{l}\text{(adiabatic process,}\\ \text{ideal gas).}\end{array} \qquad (17\text{–}26)$$

(We used the result $C_V = R/(\gamma - 1)$ from Example 17–6.) If the process is an expansion, the temperature drops, T_1 is greater than T_2, p_1V_1 is greater than p_2V_2, and the work is *positive,* as we should expect. If the process is a compression, the work is negative.

Throughout this analysis of adiabatic processes we have used the ideal-gas equation of state, which is valid only for *equilibrium* states. Strictly speaking, our results are valid only for a process that is fast enough to prevent appreciable heat exchange with the surroundings (so that $Q = 0$ and the process is adiabatic), yet slow enough that the system does not depart very much from thermal and mechanical equilibrium. Even when these conditions are not strictly satisfied, though, Eqs. (17–22), (17–24), and (17–26) give useful approximate results.

Hot steam escapes from the top of this pressure cooker at high speed. Hence it has no time to exchange heat with its surroundings, and its expansion is nearly adiabatic. As the steam's volume increases, its temperature drops so much (see Eq. (17–22)) that it feels cool on this chef's hand.

EXAMPLE 17–7

Adiabatic compression in a diesel engine The compression ratio of a diesel engine is 15 to 1; this means that air in the cylinders is compressed to $\frac{1}{15}$ of its initial volume (Fig. 17–17). If the initial pressure is 1.01×10^5 Pa and the initial temperature is 27°C (300 K), find the final pressure and the temperature after compression. Air is mostly a mixture of diatomic oxygen and nitrogen; treat it as an ideal gas with $\gamma = 1.40$.

SOLUTION We have $p_1 = 1.01 \times 10^5$ Pa, $T_1 = 300$ K, and $V_1/V_2 = 15$. From Eq. (17–22),

$$T_2 = T_1\left(\frac{V_1}{V_2}\right)^{\gamma-1} = (300\text{ K})(15)^{0.40} = 886\text{ K} = 613°\text{C}.$$

From Eq. (17–24),

$$p_2 = p_1\left(\frac{V_1}{V_2}\right)^{\gamma} = (1.01 \times 10^5\text{ Pa})(15)^{1.40}$$

$$= 44.8 \times 10^5\text{ Pa} = 44\text{ atm}.$$

If the compression had been isothermal, the final pressure would have been 15 atm, but because the temperature also increases during an adiabatic compression, the final pressure is much greater. When fuel is injected into the cylinders near the end of the compression stroke, the high temperature of the air attained during compression causes the fuel to ignite spontaneously without the need for spark plugs.

17–17 Adiabatic compression of air in a cylinder of a diesel engine.

EXAMPLE 17-8

Work done in an adiabatic process In Example 17–7, how much work does the gas do during the compression if the initial volume of the cylinder is $1.00\text{ L} = 1.00 \times 10^{-3}\text{ m}^3$? Assume that C_V for air is 20.8 J/mol · K and $\gamma = 1.40$.

SOLUTION We may determine the number of moles using the ideal-gas equation $pV = nRT$ and then use Eq. (17–25), or we may use Eq. (17–26). With the first method we have

$$n = \frac{p_1 V_1}{RT_1} = \frac{(1.01 \times 10^5\text{ Pa})(1.00 \times 10^{-3}\text{ m}^3)}{(8.315\text{ J/mol} \cdot \text{K})(300\text{ K})}$$

$$= 0.0405\text{ mol},$$

and Eq. (17–25) gives

$$W = nC_V(T_1 - T_2)$$

$$= (0.0405\text{ mol})(20.8\text{ J/mol} \cdot \text{K})(300\text{ K} - 886\text{ K})$$

$$= -494\text{ J}.$$

With the second method,

$$W = \frac{1}{\gamma - 1}(p_1 V_1 - p_2 V_2)$$

$$= \frac{1}{1.40 - 1}\left[\begin{array}{l}(1.01 \times 10^5\text{ Pa})(1.00 \times 10^{-3}\text{ m}^3) \\ -(44.8 \times 10^5\text{ Pa})\left(\dfrac{1.00 \times 10^{-3}\text{ m}^3}{15}\right)\end{array}\right]$$

$$= -494\text{ J}.$$

The work is negative because the gas is compressed.

SUMMARY

- A thermodynamic system can exchange energy with its surroundings by heat transfer or by mechanical work and in some cases by other mechanisms. When a system at pressure p expands from volume V_1 to V_2, it does an amount of work W given by

$$W = \int_{V_1}^{V_2} p\, dV. \qquad (17\text{–}2)$$

If the pressure p is constant during the expansion,

$$W = p(V_2 - V_1) \qquad \text{(constant pressure only)}. \qquad (17\text{–}3)$$

- In any thermodynamic process, the heat added to the system and the work done by the system depend not only on the initial and final states but also on the path (the series of intermediate states through which the system passes).

KEY TERMS

thermodynamic system, 533
thermodynamic process, 533
path, 537
free expansion, 538
internal energy, 539
first law of thermodynamics, 539
adiabatic process, 544
isochoric process, 544
isobaric process, 544
isothermal process, 544
adiabat, 545
isochor, 545

■ The first law of thermodynamics states that when heat Q is added to a system while it does work W, the internal energy U changes by an amount

$$U_2 - U_1 = \Delta U = Q - W. \qquad (17\text{-}4)$$

In an infinitesimal process,

$$dU = dQ - dW. \qquad (17\text{-}6)$$

The internal energy of any thermodynamic system depends only on its state. The change in internal energy in any process depends only on the initial and final states, not on the path. The internal energy of an isolated system is constant.

■ Adiabatic process: No heat transfer into or out of a system; $Q = 0$.

■ Isochoric process: Constant volume; $W = 0$.

■ Isobaric process: Constant pressure; $W = p(V_2 - V_1)$.

■ Isothermal process: Constant temperature.

■ The internal energy of an ideal gas depends only on its temperature, not on its pressure or volume. For other substances the internal energy generally depends on both pressure and temperature.

■ The molar heat capacities C_V and C_p of an ideal gas are related by

$$C_p = C_V + R. \qquad (17\text{-}17)$$

The ratio of heat capacities, C_p/C_V, is denoted by γ:

$$\gamma = \frac{C_p}{C_V}. \qquad (17\text{-}18)$$

■ For an adiabatic process for an ideal gas the quantities $TV^{\gamma-1}$ and pV^γ are constant. For an initial state (p_1, V_1, T_1) and a final state (p_2, V_2, T_2),

$$T_1 V_1^{\gamma-1} = T_2 V_2^{\gamma-1}, \qquad (17\text{-}22)$$

$$p_1 V_1^\gamma = p_2 V_2^\gamma. \qquad (17\text{-}24)$$

The work done by an ideal gas during an adiabatic expansion is

$$W = nC_V(T_1 - T_2) = \frac{C_V}{R}(p_1 V_1 - p_2 V_2)$$

$$= \frac{1}{\gamma - 1}(p_1 V_1 - p_2 V_2). \qquad (17\text{-}25), (17\text{-}26)$$

DISCUSSION QUESTIONS

Q17-1 In Chapter 6 the relationship between work and changes in kinetic energy was derived from Newton's laws of motion. Would it be possible to derive the first law of thermodynamics, which relates work, heat, and changes in internal energy, from Newton's laws of motion? Why or why not?

Q17-2 It is not correct to say that a body contains a certain amount of heat, yet a body can transfer heat to another body. How can a body give away something it does not have in the first place?

Q17-3 If you are told the initial and final states of a system and the associated change in internal energy, can you determine whether the internal energy change was due to work or to heat transfer?

Q17-4 Discuss the application of the first law of thermodynamics to a mountaineer who eats food, gets warm and perspires a lot during a climb, and does a lot of mechanical work in raising herself to the summit. What about the descent? The mountaineer also gets warm during the descent. Is the source of this energy the same as during the ascent?

Q17-5 When ice melts at 0°C, its volume decreases. Is the internal energy change greater than, less than, or equal to the heat added? How can you tell?

Q17-6 Household refrigerators always have tubing on the outside, usually at the back or bottom. When the refrigerator is running, the tubing becomes quite hot. Where does the heat come from?

Q17-7 Example 17-5 (Section 17-5) describes a process that occurs at a constant temperature of 100°C. Why, then, can't we use in this example the expressions for W derived for an isothermal process in Example 17-1 (Section 17-3)?

Q17–8 Translated literally from the Greek, "isothermal" means "same heat." Give as many reasons as you can why such a literal translation is misleading.

Q17–9 Imagine a gas made up entirely of negatively charged electrons. Like charges repel, so the electrons exert repulsive forces on each other. Would you expect that the temperature of such a gas would rise, fall, or stay the same in a free expansion? Why?

Q17–10 There are a few materials that contract when their temperature is increased, such as water between 0°C and 4°C. Would you expect C_p for such materials to be greater or less than C_V? Why or why not?

Q17–11 In a constant-volume process, $dU = nC_V\,dT$. But in a constant-pressure process, it is *not* true that $dU = nC_p\,dT$. Why not?

Q17–12 When you blow on the back of your hand with your mouth wide open, your breath feels warm. But if you partially close your mouth to form an "o" and then blow on your hand, your breath feels cool. Why?

Q17–13 On a warm summer day a large cylinder of compressed gas (propane or butane) was used to supply several large gas burners at a cookout. After a while, frost formed on the outside of the tank. Why?

Q17–14 In the carburetor of an aircraft or automobile engine, air flows through a relatively small aperture, then expands. In cool, foggy weather, ice sometimes forms in this aperture even though the outside air temperature is above freezing. Why?

Q17–15 Air escaping from an air hose at a gas station always feels cold. Why?

Q17–16 When you use a hand pump to inflate the tires of your bicycle, the pump gets warm after a while. Why? What happens to the temperature of the air in the pump as you compress it? Why does this happen? When you raise the pump handle to draw outside air into the pump, what happens to the temperature of the air taken in? Again, why does this happen?

Q17–17 The prevailing winds on the Hawaiian island of Kauai blow from the northeast. The winds cool as they go up the slope of Mt. Waialeale (elevation 1523 m), causing water vapor to condense and rain to fall. There is much more precipitation at the summit than at the base of the mountain. In fact, Mt. Waialeale is the rainiest spot on earth, averaging 11.7 m of rainfall a year. But what makes the winds cool?

Q17–18 Applying the same considerations as in Question 17–17, explain why the island of Niihau, a few kilometers to the southwest of Kauai, is almost a desert, and farms there need to be irrigated.

Q17–19 When a gas is compressed adiabatically into the surrounding air, its temperature rises even though there is no heat input to the gas. Where does the energy come from to raise the temperature?

Q17–20 When a gas expands adiabatically, it does work on its surroundings. But if there is no heat input to the gas, where does the energy come from to do the work?

Q17–21 As described in Section 17–9, when you open a bottle of a carbonated beverage, a miniature cloud can form just above the surface of the beverage. This is more likely to happen if the bottle and its contents are cold before you open the bottle. Why? This effect does not occur with bottles of non-carbonated beverages such as fruit juice. Why not?

EXERCISES

SECTION 17–3 WORK DONE DURING VOLUME CHANGES
SECTION 17–4 PATHS BETWEEN THERMODYNAMIC STATES

17–1 Two moles of an ideal gas are heated at constant pressure from $T = 27°C$ to $107°C$. a) Draw a pV-diagram for this process. b) Calculate the work done by the gas.

17–2 Three moles of an ideal gas have an initial temperature of 127.0°C. While the temperature is kept constant, the volume is increased until the pressure drops to 40.0% of its original value. a) Draw a pV-diagram for this process. b) Calculate the work done by the gas.

17–3 A metal cylinder with rigid walls contains 2.50 mol of oxygen gas. The gas is cooled until the pressure increases to 30.0% of its original value. Neglect the thermal contraction of the cylinder. a) Draw a pV-diagram for this process. b) Calculate the work done by the gas.

17–4 A gas under a constant pressure of 1.50×10^5 Pa and with an initial volume of 0.0900 m^3 is cooled until its volume becomes 0.0600 m^3. a) Draw a pV-diagram for this process. b) Calculate the work done by the gas.

17–5 A gas undergoes two processes. In the first, the volume remains constant at 0.200 m^3 and the pressure increases from 2.00×10^5 Pa to 5.00×10^5 Pa. The second process is a compression to a volume of 0.120 m^3 at a constant pressure of 5.00×10^5 Pa. a) In a pV-diagram, show both processes. b) Find the total work done by the gas during both processes.

17–6 Work Done in a Cyclic Process. a) In Fig. 17–7a, consider the closed loop $1 \rightarrow 3 \rightarrow 2 \rightarrow 4 \rightarrow 1$. This is a *cyclic* process in which the initial and final states are the same. Find the total work done by the system in this cyclic process, and show that it is equal to the area enclosed by the loop. b) How is the work done for the process in part (a) related to the work done if the loop is traversed in the opposite direction, $1 \rightarrow 4 \rightarrow 2 \rightarrow 3 \rightarrow 1$? Explain.

SECTION 17–5 INTERNAL ENERGY AND THE FIRST LAW OF THERMODYNAMICS

17–7 In a certain chemical process, a lab technician supplies 254 J of heat to a system. At the same time, 73 J of work are done on the system by its surroundings. What is the increase in the internal energy of the system?

17–8 A gas in a cylinder expands from a volume of 0.110 m^3 to 0.320 m^3. Heat flows into the gas just rapidly enough to keep the pressure constant at 1.80×10^5 Pa during the expansion. The total heat added is 1.15×10^5 J. a) Find the work done by the gas. b) Find the change in internal energy of the gas. c) Does it matter whether or not the gas is ideal? Why or why not?

17–9 A gas in a cylinder is held at a constant pressure of 2.30×10^5 Pa and is cooled and compressed from 1.70 m^3 to 1.20 m^3. The internal energy of the gas decreases by 1.40×10^5 J. a) Find the work done by the gas. b) Find the absolute value $|Q|$ of the heat flow into or out of the gas, and state the direction of heat flow. c) Does it matter whether or not the gas is ideal? Why or why not?

17–10 A system is taken from state a to state b along the three paths shown in Fig. 17–18. a) Along which path is the work done by the system the greatest? The least? b) If $U_b > U_a$, along which path is the absolute value $|Q|$ of the heat transfer the greatest? For this path, is heat absorbed or liberated by the system?

FIGURE 17–18 Exercise 17–10.

17–11 Doughnuts: Breakfast of Champions! A typical doughnut contains 2.0 g of protein, 17.0 g of carbohydrates, and 7.0 g of fat. The average food-energy values of these substances are 4.0 kcal/g for protein and carbohydrates and 9.0 kcal/g for fat. a) During heavy exercise, an average person uses energy at a rate of 510 kcal/h. How long would you have to exercise to "work off" one doughnut? b) If the energy in the doughnut could somehow be converted into kinetic energy of your body as a whole, how fast could you move after eating the doughnut? Take your mass to be 60 kg, and express your answer in m/s and km/h.

17–12 A liquid is irregularly stirred in a well-insulated container and thereby undergoes a rise in temperature. Regard the liquid as the system. a) Has heat been transferred? How can you tell? b) Has work been done? How can you tell? Why is it important that the stirring is irregular? c) What is the sign of ΔU? How can you tell?

17–13 A system is taken around the cycle shown in Fig. 17–19, from state a to state b and then back to state a. The absolute value of the heat transfer during one cycle is 7200 J. a) Does the system absorb or liberate heat when it goes around the cycle in the direction shown in the figure? How can you tell? b) What is the work W done by the system in one cycle? c) If the system

FIGURE 17–19 Exercise 17–13.

goes around the cycle in a counterclockwise direction, does it absorb or liberate heat in one cycle? What is the magnitude of the heat absorbed or liberated in one counterclockwise cycle?

17–14 A thermodynamic system undergoes a cyclic process as shown in Fig. 17–20. The cycle consists of two closed loops, loop I and loop II. a) Over one complete cycle, does the system do positive or negative work? b) In each of loops I and II, is the net work done by the system positive or negative? c) Over one complete cycle, does heat flow into or out of the system? d) In each of loops I and II, does heat flow into or out of the system?

FIGURE 17–20 Exercise 17–14.

17–15 A student performs a combustion experiment by burning a mixture of fuel and oxygen in a constant-volume metal can surrounded by a water bath. During the experiment the temperature of the water is observed to rise. Regard the mixture of fuel and oxygen as the system. a) Has heat been transferred? How can you tell? b) Has work been done? How can you tell? c) What is the sign of ΔU? How can you tell?

17–16 Boiling Water at High Pressure. When water is boiled under a pressure of 2.00 atm, the heat of vaporization is 2.20×10^6 J/kg and the boiling point is 120°C. At this pressure, 1.00 kg of water has a volume of 1.00×10^{-3} m^3, and 1.00 kg of steam a volume of 0.824 m^3. a) Compute the work done when 1.00 kg of steam is formed at this temperature. b) Compute the increase in internal energy of the water.

SECTION **17–6 KINDS OF THERMODYNAMIC PROCESSES**

SECTION **17–7 INTERNAL ENERGY OF AN IDEAL GAS**

SECTION **17–8 HEAT CAPACITIES OF AN IDEAL GAS**

17–17 In an experiment to simulate conditions within an automobile engine, 645 J of heat is transferred to 0.185 mol of air contained within a cylinder of volume 40.0 cm^3. Initially the nitrogen is at a pressure of 3.00×10^6 Pa and a temperature of 780 K. a) If the volume of the cylinder is held fixed, what is the final temperature of the air? Assume that the air is essentially all nitrogen gas, and use the data in Table 17–1. Draw a pV-diagram for this process. b) Find the final temperature of the air if the volume of the cylinder is allowed to increase while the pressure remains constant. Draw a pV-diagram for this process.

17–18 A cylinder contains 0.0100 mol of helium at $T = 27.0$°C. a) How much heat is needed to raise the temperature to 67.0°C while keeping the volume constant? Draw a pV-diagram for this process. b) If instead the pressure of the helium is kept constant, how much heat is needed to raise the temperature from 27.0°C to 67.0°C? Draw a pV-diagram for this process. c) What accounts for the difference between your answers to parts (a) and (b)? In which case is more heat required? What becomes of the additional heat? d) If the gas is ideal, what is the change in its internal energy in part (a)? In part (b)? How do the two answers compare? Why?

17–19 The temperature of 0.150 mol of an ideal gas is held constant at 77.0°C while its volume is reduced to 25.0% of its initial volume. The initial pressure of the gas is 1.25 atm. a) Determine the work done by the gas. b) What is the change in its internal energy? c) Does the gas exchange heat with its surroundings? If so, how much? Does the gas absorb or liberate heat?

17–20 During an isothermal compression of an ideal gas, 335 J of heat must be removed from the gas to maintain constant temperature. How much work is done by the gas during the process?

17–21 Propane gas (C_3H_8) behaves like an ideal gas with $\gamma = 1.127$. Determine the molar heat capacity at constant volume and the molar heat capacity at constant pressure.

17–22 A cylinder contains 0.250 mol of carbon dioxide (CO_2) gas at a temperature of 27.0°C. The cylinder is provided with a frictionless piston, which maintains a constant pressure of 1.00 atm on the gas. The gas is heated until its temperature increases to 127.0°C. Assume that the CO_2 can be treated as an ideal gas. a) Draw a pV-diagram for this process. b) How much work is done by the gas in this process? c) On what is this work done? d) What is the change in internal energy of the gas? e) How much heat was supplied to the gas? f) How much work would have been done if the pressure had been 0.50 atm?

17–23 Gaseous ethane (C_2H_6) has $\gamma = 1.220$ and may be treated as an ideal gas. a) If 2.40 mol of ethane is to be heated from 20.0°C to 25.0°C at a constant pressure of 1.00 atm, how much heat is required? b) What will be the change in internal energy of the ethane?

PROBLEMS

17–30 A quantity of air is taken from state a to state b along a path that is a straight line in the pV-diagram (Fig. 17–21). a) In this process, does the temperature of the gas increase, decrease, or stay the same? Explain. b) If $V_a = 0.0700$ m^3, $V_b = 0.1100$ m^3, $p_a = 1.00 \times 10^5$ Pa, and $p_b = 1.40 \times 10^5$ Pa, what is the work W

SECTION 17–9 ADIABATIC PROCESSES FOR AN IDEAL GAS

17–24 A monatomic ideal gas that is initially at a pressure of 1.50×10^5 Pa and with a volume of 0.0800 m^3 is compressed adiabatically to a volume of 0.0400 m^3. a) What is the final pressure? b) How much work is done by the gas? c) What is the ratio of the final temperature of the gas to its initial temperature? Is the gas heated or cooled by this compression?

17–25 The engine of a Ferrari F355 F1 sports car takes in air at 20.0°C and 1.00 atm and compresses it adiabatically to 0.0900 times the original volume. The air may be treated as an ideal gas with $\gamma = 1.40$. a) Draw a pV-diagram for this process. b) Find the final temperature and pressure.

17–26 An ideal gas initially at 4.00 atm and 350 K is permitted to expand adiabatically to 1.50 times its initial volume. Find the final pressure and temperature if the gas is a) monatomic; b) diatomic with $C_V = \frac{5}{2}R$.

17–27 During an adiabatic expansion the temperature of 0.450 mol of argon (Ar) drops from 50.0°C to 10.0°C. The argon may be treated as an ideal gas. a) Draw a pV-diagram for this process. b) How much work does the gas do? c) Does heat flow into or out of the gas? If so, what is the direction and absolute value of this heat flow? d) What is the change in internal energy of the gas?

17–28 A cylinder contains 0.100 mol of an ideal monatomic gas. Initially the gas is at a pressure of 1.00×10^5 Pa and occupies a volume of 2.50×10^{-3} m^3. a) Find the initial temperature of the gas in kelvins. b) If the gas is allowed to expand to twice the initial volume, find the final temperature (in kelvins) and pressure of the gas if the expansion is i) isothermal; ii) isobaric; iii) adiabatic.

17–29 A quantity of sulfur dioxide gas (SO_2) occupies a volume of 5.00×10^{-3} m^3 at a pressure of 1.10×10^5 Pa. The gas expands adiabatically to a volume of 1.00×10^{-2} m^3, doing 285 J of work on its surroundings. Assume that the gas may be treated as ideal. a) Find the final pressure of the gas. (*Hint*: See Table 17–1.) b) How much work does the gas do on its surroundings? c) What is the ratio of the final temperature of the gas to its initial temperature?

done by the gas in this process? Assume that the gas may be treated as ideal.

17–31 When a system is taken from state a to state b in Fig. 17–22 along the path acb, 90.0 J of heat flows into the system and 60.0 J of work is done by the system. a) How much heat

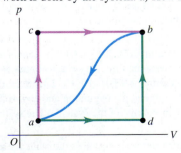

FIGURE 17–21 Problem 17–30.

FIGURE 17–22 Problem 17–31.

flows into the system along path *adb* if the work done by the system is 15.0 J? b) When the system is returned from *b* to *a* along the curved path, the absolute value of the work done by the system is 35.0 J. Does the system absorb or liberate heat? How much heat? c) If $U_a = 0$ and $U_d = 8.0$ J, find the heat absorbed in the processes *ad* and *db*.

17–32 A thermodynamic system is taken from state *a* to state *c* in Fig. 17–23 along either path *abc* or path *adc*. Along path *abc* the work *W* done by the system is 450 J. Along path *adc*, *W* is 120 J. The internal energies of each of the four states shown in the figure are $U_a = 150$ J, $U_b = 240$ J, $U_c = 680$ J, and $U_d = 330$ J. Calculate the heat flow *Q* for each of the four processes *ab*, *bc*, *ad*, and *dc*. In each process, does the system absorb or liberate heat?

17–33 Fig. 17–23 shows four states of a thermodynamic system, *a*, *b*, *c*, and *d*. The volume of the system is V_a for both state *a* and state *b*, and is V_c for both state *c* and state *d*. The pressure of the system is p_a for both state *a* and state *d*, and is p_c for both state *b* and state *c*. The internal energies of the four states are U_a, U_b, U_c, and U_d. For each of the processes *ab*, *bc*, *ad*, and *dc*, find a) the work done by the system during the process and b) the heat flow into the system during the process. c) The system can be taken from state *a* to state *c* along either path *abc* or path *adc*. Find the net heat flow into the system and the net work done by the system for each path. For which path is the net heat flow greater? For which path is the net work greater? d) A friend tells you that the amounts of heat flow should be the same for both path *abc* and path *adc*, since the initial state (*a*) and final state (*c*) of the system are the same for both paths. How would you respond?

FIGURE 17–23 Problems 17–32 and 17–33.

17–34 Nitrogen gas in an expandable container is cooled from 50.0°C to 10.0°C, with the pressure held constant at 3.00×10^5 Pa. The total heat liberated by the gas is 2.50×10^4 J. Assume that the gas may be treated as ideal. a) Find the number of moles of gas. b) Find the change in internal energy of the gas. c) Find the work done by the gas. d) How much heat would be liberated by the gas for the same temperature change if the volume were constant?

17–35 In a certain process, 2.15×10^5 J of heat is liberated by a system, and at the same time the system contracts under a constant external pressure of 9.50×10^5 Pa. The internal energy of the system is the same at the beginning and end of the process. Find the change in volume of the system. (The system is *not* an ideal gas.)

17–36 A cylinder with a frictionless movable piston, like that shown in Fig. 17–5, contains a quantity of helium gas. Initially the gas is at a pressure of 1.00×10^5 Pa, has a temperature of

300 K, and occupies a volume of 1.50 L. The gas then undergoes two processes. In the first, the gas is heated and the piston is allowed to move to keep the temperature equal to 300 K. This continues until the pressure reaches 2.50×10^4 Pa. In the second process, the gas is compressed at constant pressure until it returns to its original volume of 1.50 L. Assume that the gas may be treated as ideal. a) In a *pV*-diagram, show both processes. b) Find the volume of the gas at the end of the first process, and find the pressure and temperature at the end of the second process. c) Find the total work done by the gas during both processes. d) What would you have to do to the gas to return it to its original pressure and temperature?

17–37 A Thermodynamic Process in a Liquid. A chemical engineer is studying the properties of liquid methanol (CH_3OH). She uses a steel cylinder with a cross-section area of 0.0200 m^2 and containing 1.20×10^{-2} m^3 of methanol. The cylinder is equipped with a tightly fitting piston that supports a load of 3.00×10^4 N. The temperature of the system is increased from 20.0°C to 50.0°C. For methanol, the coefficient of volume expansion is 1.20×10^{-3} K^{-1}, the density is 791 kg/m^3, and the specific heat capacity at constant pressure is $c_p = 2.51 \times 10^3$ J/kg · K. Neglect the expansion of the steel cylinder. Find a) the increase in volume of the methanol; b) the mechanical work done by the methanol against the 3.00×10^4 N force; c) the amount of heat added to the methanol; d) the change in internal energy of the methanol. e) Based on your results, explain whether there is any substantial difference between the specific heat capacities c_p (at constant pressure) and c_V (at constant volume) for methanol under these conditions.

17–38 A Thermodynamic Process in a Solid. A cube of copper 2.00 cm on a side is suspended by a string. (The physical properties of copper are given in Tables 14–1, 15–2, and 15–3.) The cube is heated with a burner from 20.0°C to 90.0°C. The air surrounding the cube is at atmospheric pressure (1.01×10^5 Pa). Find a) the increase in volume of the cube; b) the mechanical work done by the cube to expand against the pressure of the surrounding air; c) the amount of heat added to the cube; d) the change in internal energy of the cube. e) Based on your results, explain whether there is any substantial difference between the specific heat capacities c_p (at constant pressure) and c_V (at constant volume) for copper under these conditions.

17–39 A Thermodynamic Process in an Insect. The African bombardier beetle *Stenaptinus insignis* can emit a jet of defensive spray from the moveable tip of its abdomen (Fig. 17–24). The beetle's body has reservoirs of two different chemicals; when the beetle is disturbed, these chemicals are combined in a reaction chamber, producing a compound that is warmed from 20°C to 100°C by the heat of reaction. The high pressure produced allows the compound to be sprayed out at speeds up to 19 m/s (68 km/h), scaring away predators of all kinds. (The beetle shown in the figure is tethered to a wire fastened to its back by wax. It is responding to having its left foreleg pinched by forceps. The beetle is 2 cm long.) Calculate the heat of reaction of the two chemicals (in J/kg). Assume that the specific heat capacity of the two chemicals and the spray is the same as that of water, 4.19×10^3 J/kg · K, and that the initial temperature of the chemicals is 20°C.

FIGURE 17–24 Problem 17–39.

17–40 A Compressed-Air Engine. You are designing an engine that runs on compressed air. Air enters the engine at a pressure of 1.60×10^6 Pa and leaves at a pressure of 2.80×10^5 Pa. What must be the temperature of the compressed air for there to be no possibility of frost forming in the exhaust ports of the engine? Assume that the expansion is adiabatic. (*Note:* Frost will form if the moist air is cooled below 0°C in the expansion.)

17–41 During certain seasons strong winds called "chinooks" blow from the west across the eastern slopes of the Rockies and downhill into Denver and adjoining areas. Although the mountains are cool, the wind in Denver is very hot; within a few minutes after the chinook wind arrives, the temperature can climb 20 C° ("chinook" is a Native American word meaning "snow eater"). Similar winds occur in the Alps (called "foehns") and in southern California (called "Santa Anas"). a) Explain why the temperature of the chinook wind rises as it descends the slopes. Why is it important that the wind be fast-moving? b) Suppose a strong wind is blowing toward Denver (elevation 1630 m) from Grays Peak (80 km west of Denver, at an elevation of 4350 m), where the air pressure is 5.60×10^4 Pa and the air temperature is −15.0°C. The temperature and pressure in Denver before the wind arrives are 2.0°C and 8.12×10^4 Pa. By how many Celsius degrees will the temperature in Denver rise when the chinook arrives?

17–42 A certain ideal gas has molar heat capacity at constant volume C_V. A sample of this gas initially occupies a volume V_0 at pressure p_0 and absolute temperature T_0. The gas expands isobarically to a volume $2V_0$, then expands further adiabatically to a final volume of $4V_0$. a) Draw a pV-diagram for this sequence of processes. b) Compute the total work done by the gas for this sequence of processes. c) Find the final temperature of the gas. d) Find the absolute value $|Q|$ of the total heat flow into or out of the gas for this sequence of processes, and state the direction of heat flow.

17–43 An air pump has a cylinder 0.250 m long with a movable piston. The pump is used to compress air from the atmosphere (at absolute pressure 1.01×10^5 Pa) into a very large tank at 4.20×10^5 Pa gauge pressure. (For air, $C_V = 20.8$ J/mol · K.) a) The piston begins the compression stroke at the open end of the cylinder. How far down the length of the cylinder has the piston moved when air first begins to flow from the cylinder into the tank? Assume that the compression is adiabatic. b) If the air is taken into the pump at 27.0°C, what is the temperature of the compressed air? c) How much work does the pump do in putting 20.0 mol of air into the tank?

17–44 Engine Turbochargers and Intercoolers. The power output of an automobile engine is directly proportional to the mass of air that can be forced into the volume of the engine's cylinders to react chemically with gasoline. Many cars have a *turbocharger* which compresses the air before it enters the engine, giving a greater mass of air per volume. This rapid, essentially adiabatic compression also heats the air. To compress it further, the air then passes through an *intercooler* in which the air exchanges heat with its surroundings at essentially constant pressure. The air is then drawn into the cylinders. In a typical installation, air is taken into the turbocharger at atmospheric pressure (1.01×10^5 Pa), density $\rho = 1.23$ kg/m^3, and temperature 15.0°C. It is compressed adiabatically to 1.45×10^5 Pa. In the intercooler, the air is cooled to the original temperature of 15.0°C at a constant pressure of 1.45×10^5 Pa. a) Draw a pV-diagram for this sequence of processes. b) If the volume of one of the engine's cylinders is 575 cm^3, what mass of air exiting from the intercooler will fill the cylinder at 1.45×10^5 Pa? Compared to the power output of an engine that takes in air at 1.01×10^5 Pa at 15.0°C, what percentage increase in power is obtained by using the turbocharger and intercooler? c) If the intercooler is not used, what mass of air exiting from the turbocharger will fill the cylinder at 1.45×10^5 Pa? Compared to the power output of an engine that takes in air at 1.01×10^5 Pa at 15.0°C, what percentage increase in power is obtained by using the turbocharger alone?

17–45 A monatomic ideal gas expands slowly to twice its original volume, doing 300 J of work in the process. Find the heat added to the gas and the change in internal energy of the gas if the process is a) isothermal; b) adiabatic; c) isobaric.

17–46 A cylinder with a piston contains 0.250 mol of oxygen at 2.40×10^5 Pa and 355 K. The oxygen may be treated as an ideal gas. The gas first expands isobarically to twice its original volume. It is then compressed isothermally back to its original volume, and finally it is cooled isochorically to its original pressure. a) Show the series of processes on a pV-diagram. b) Compute the temperature during the isothermal compression. c) Compute the maximum pressure. d) Compute the total work done by the piston on the gas during the series of processes.

17–47 Use the conditions and processes of Problem 17–46 to compute a) the work done by the gas, the heat added to it, and its internal-energy change during the initial expansion; b) the work done, the heat added, and the internal-energy change during the final cooling; c) the internal-energy change during the isothermal compression.

17–48 A cylinder with a piston contains 0.150 mol of nitrogen at 1.80×10^5 Pa and 300 K. The nitrogen may be treated as an ideal gas. The gas is first compressed isobarically to one-half its original volume. It then expands adiabatically back to its original volume, and finally it is heated isochorically to its original pressure. a) Show the series of processes on a pV-diagram. b) Compute the temperatures at the beginning and end of the adiabatic expansion. c) Compute the minimum pressure.

17–49 Use the conditions and processes of Problem 17–48 to compute a) the work done by the gas, the heat added to it, and its internal-energy change during the initial compression; b) the work done by the gas, the heat added to it, and its internal-energy change during the adiabatic expansion; c) the work done, the heat added, and the internal-energy change during the final heating.

17–50 Comparing Thermodynamic Processes. In a cylinder, 1.20 mol of an ideal monatomic gas initially at 3.60×10^5 Pa and 300 K expands until its volume triples. Compute the work done by the gas if the expansion is a) isothermal; b) adiabatic; c) iso-

baric. d) Show each process on a pV-diagram. In which case is the absolute value of the work done by the gas greatest? Least? e) In which case is the absolute value of the heat transfer greatest? Least? f) In which case is the absolute value of the change in internal energy of the gas greatest? Least?

17–51 A flexible balloon contains 0.350 mol of hydrogen sulfide gas (H_2S). Initially the H_2S has a volume of 7.00×10^3 cm^3 and a temperature of 27.0°C. The H_2S first expands isobarically until the volume doubles. Then it expands adiabatically until the temperature returns to its initial value. Assume that the H_2S can be treated as an ideal gas. a) Draw a diagram of the process in the pV-plane. b) What is the total heat supplied to the H_2S in the process? c) What is the total change in the internal energy of the H_2S? d) What is the total work done by the H_2S? e) What is the final volume?

CHALLENGE PROBLEMS

17–52 Oscillations of a Piston. A vertical cylinder of radius r contains a quantity of ideal gas, and is fitted with a piston of mass m that is free to move (Fig. 17–25). The piston and the walls of the cylinder are frictionless and are made of a perfect thermal insulator. The outside air pressure is p_0. In equilibrium, the piston sits a height h above the bottom of the cylinder. a) Find the absolute pressure of the gas trapped below the piston when in equilibrium. b) The piston is pulled up by a small distance and released. Find the net force acting on the piston when its base is a distance $h + y$ above the bottom of the cylinder, where y is much less than h. c) After the piston is displaced from equilibrium and released, it oscillates up and down. Find the frequency of these small oscillations. If the displacement is not small, are the oscillations simple harmonic? How can you tell?

17–53 The van der Waals equation of state, an approximate representation of the behavior of gases at high pressure, is given by Eq. (16–7):

$$\left(p + \frac{an^2}{V^2}\right)(V - nb) = nRT,$$

where a and b are constants having different values for different gases. (In the special case of $a = b = 0$, this is the ideal-gas equation.) a) Calculate the work done by a gas with this equation of state in an isothermal expansion from V_1 to V_2. Show that your answer agrees with the ideal-gas result found in Example 17–1 (Section 17–3) when you set $a = b = 0$. b) For ethane gas (C_2H_6), $a = 0.554$ J·m^3/mol^2 and $b = 6.38 \times 10^{-5}$ m^3/mol. Calculate the work W done by 1.80 mol of ethane when it expands from 2.00×10^{-3} m^3 to 4.00×10^{-3} m^3 at a constant temperature of 300 K. Do the calculation using i) the van der Waals equation of state and ii) the ideal-gas equation of state. c) How large is the difference between the two results for W in part (b)? For which equation of state is W larger? Use the interpretation of the terms a and b given in Section 16–2 to explain why this should be so. Are the differences between the two equations of state important in this case?

Open to outside air, pressure p_0

Piston, mass m

Ideal gas

h

FIGURE 17–25 Challenge Problem 17–52.

The Second Law of Thermodynamics

Could heat spontaneously flow out of these fried eggs, thus "unfrying" them and making them cooler, and flow back into the hot frying pan? Such a process would conserve energy and is allowed by the first law of thermodynamics. But such a process is in fact impossible because it would violate the *second* law of thermodynamics: heat cannot by itself flow from one body into a hotter body.

18–1 INTRODUCTION

Many thermodynamic processes proceed naturally in one direction but not the opposite. For example, heat always flows from a hot body to a cooler body, never the reverse. Heat flow from a cool body to a hot body would not violate the first law of thermodynamics; energy would be conserved. But it doesn't happen in nature. Why not? It is easy to convert mechanical energy completely into heat; this happens every time we use a car's brakes to stop it. Going in the reverse direction, there are plenty of devices that convert heat *partially* into mechanical energy. (An automobile engine is an example.) But even the cleverest would-be inventors have never succeeded in building a machine that converts heat *completely* into mechanical energy. Again, why not?

The answer to both of these questions has to do with the *directions* of thermodynamic processes and is called the *second law of thermodynamics.* This law places fundamental limitations on the efficiency of an engine or a power plant. It also places limitations on the minimum energy input needed to operate a refrigerator. So the second law is directly relevant for many important practical problems.

We can also state the second law in terms of the concept of *entropy,* a quantitative measure of the degree of disorder or randomness of a system. The idea of entropy helps explain why ink mixed with water never spontaneously unmixes and why a host of other seemingly possible processes are never observed to occur.

18–2 DIRECTIONS OF THERMODYNAMIC PROCESSES

Thermodynamic processes that occur in nature are all **irreversible processes.** These are processes that proceed spontaneously in one direction but not the other. The flow of heat from a hot body to a cooler body is irreversible, as is the free expansion of a gas discussed in Sections 17–4 and 17–7. Sliding a book across a table converts mechanical energy into heat by friction; this process is irreversible, for no one has ever observed the reverse process (in which a book initially at rest on the table would spontaneously start moving and the table and book would cool down). Our main topic for this chapter is the *second law of thermodynamics,* which determines the preferred direction for such processes.

Despite this preferred direction for every natural process, we can think of a class of idealized processes that *would* be reversible. A system that undergoes such an idealized **reversible process** is always very close to being in thermodynamic equilibrium within itself and with its surroundings. Any change of state that takes place can then be reversed (made to go the other way) by making only an infinitesimal change in the conditions of the system. For example, heat flow between two bodies whose temperatures differ only infinitesimally can be reversed by making only a very small change in one temperature or the other.

Reversible processes are thus **equilibrium processes,** with the system always in thermodynamic equilibrium. Of course, if a system were *truly* in thermodynamic equilibrium, no change of state would take place. Heat would not flow into or out of a system with truly uniform temperature throughout, and a system that is truly in mechanical equilibrium would not expand and do work against its surroundings. A reversible process is an idealization that can never be precisely attained in the real world. But by making the temperature gradients and the pressure differences in the substance very small, we can keep the system very close to equilibrium states and make the process nearly reversible. That's why we call a reversible process a *quasi-equilibrium process.*

By contrast, heat flow with finite temperature difference, free expansion of a gas, and conversion of work to heat by friction are all *irreversible* processes; no small change in conditions could make any of them go the other way. They are also all *nonequilibrium* processes, in that the system is not in thermodynamic equilibrium at any point until the end of the process.

There is a relationship between the direction of a process and the *disorder* or *randomness* of the resulting state. For example, imagine a tedious sorting job, such as alphabetizing a thousand book titles written on file cards. Throw the alphabetized stack of cards into the air. Do they come down in alphabetical order? No, their tendency is to come down in a random or disordered state. In the free expansion of a gas discussed in Sections 17–4 and 17–7, the air is more disordered after it has expanded into the entire box than when it was confined in one side, just as your clothes are more disordered when scattered all over your floor than when confined to your closet.

Similarly, macroscopic kinetic energy is energy associated with organized, coordinated motions of many molecules, but heat transfer involves changes in energy of random, disordered molecular motion. Therefore conversion of mechanical energy into heat involves an increase of randomness or disorder.

In the following sections we will introduce the second law of thermodynamics by considering two broad classes of devices: *heat engines,* which are partly successful in converting heat into work, and *refrigerators,* which are partly successful in transporting heat from cooler to hotter bodies.

18–3 HEAT ENGINES

The essence of our technological society is the ability to use sources of energy other than muscle power. Sometimes, mechanical energy is directly available; water power is an example. But most of our energy comes from the burning of fossil fuels (coal, oil, and gas) and from nuclear reactions. These supply energy that is transferred as *heat.* This is directly useful for heating buildings, for cooking, and for chemical processing, but to operate a machine or propel a vehicle, we need *mechanical* energy.

Thus it's important to know how to take heat from a source and convert as much of it as possible into mechanical energy or work. This is what happens in gasoline engines in automobiles, jet engines in airplanes, steam turbines in electric power plants, and many other systems. Closely related processes occur in the animal kingdom; food energy is "burned" (that is, carbohydrates combine with oxygen to yield water, carbon dioxide, and energy) and partly converted to mechanical energy as an animal's muscles do work on its surroundings.

Any device that transforms heat partly into work or mechanical energy is called a **heat engine.** Usually, a quantity of matter inside the engine undergoes inflow and outflow of heat, expansion and compression, and sometimes change of phase. We call this matter the **working substance** of the engine. In internal-combustion engines the working substance is a mixture of air and fuel; in a steam turbine it is water.

All motorized vehicles (except electric-powered vehicles) use heat engines for propulsion.

The simplest kind of engine to analyze is one in which the working substance undergoes a **cyclic process,** a sequence of processes that eventually leaves the substance in the same state in which it started. In a steam turbine the water is recycled and used over and over. Internal-combustion engines do not use the same air over and over, but we can still analyze them in terms of cyclic processes that approximate their actual operation.

8.12
Cyclic Process—Strategies
8.13
Cyclic Process—Problems

All heat engines *absorb* heat from a source at a relatively high temperature, perform some mechanical work, and *discard* or *reject* some heat at a lower temperature. As far as the engine is concerned, the discarded heat is wasted. In internal-combustion engines the waste heat is that discarded in the hot exhaust gases and the cooling system; in a steam turbine it is the heat that must flow out of the used steam to condense and recycle the water.

When a system is carried through a cyclic process, its initial and final internal energies are equal. For any cyclic process, the first law of thermodynamics requires that

$$U_2 - U_1 = 0 = Q - W, \quad \text{so} \quad Q = W.$$

That is, the net heat flowing into the engine in a cyclic process equals the net work done by the engine.

When we analyze heat engines, it helps to think of two bodies with which the working substance of the engine can interact. One of these, called the *hot reservoir,* represents the heat source; it can give the working substance large amounts of heat at a constant temperature T_H without appreciably changing its own temperature. The other body, called the *cold reservoir,* can absorb large amounts of discarded heat from the engine at a constant lower temperature T_C. In a steam-turbine system the flames and hot gases in the boiler are the hot reservoir, and the cold water and air used to condense and cool the used steam are the cold reservoir.

We denote the quantities of heat transferred from the hot and cold reservoirs as Q_H and Q_C, respectively. A quantity of heat Q is positive when heat is transferred *into* the working substance and is negative when heat leaves the working substance. Thus in a heat engine, Q_H is positive but Q_C is negative, representing heat *leaving* the working substance. This sign convention is consistent with the rules we stated in Section 17–2; we will continue to use those rules here. Frequently, it clarifies the relationships to state them in terms of the absolute values of the Q's and W's because absolute values are always positive. When we do this, our notation will show it explicitly.

We can represent the energy transformations in a heat engine by the *energy-flow diagram* of Fig. 18–1. The engine itself is represented by the circle. The amount of heat Q_H supplied to the engine by the hot reservoir is proportional to the width of the incoming "pipeline" at the top of the diagram. The width of the outgoing pipeline at the bottom is proportional to the magnitude $|Q_C|$ of the heat rejected in the exhaust. The branch line to the right represents the portion of the heat supplied that the engine converts to mechanical work, W.

When an engine repeats the same cycle over and over, Q_H and Q_C represent the quantities of heat absorbed and rejected by the engine *during one cycle;* Q_H is positive, and Q_C is negative. The *net* heat Q absorbed per cycle is

$$Q = Q_H + Q_C = |Q_H| - |Q_C|. \tag{18–1}$$

The useful output of the engine is the net work W done by the working substance. From the first law,

$$W = Q = Q_H + Q_C = |Q_H| - |Q_C|. \tag{18–2}$$

Ideally, we would like to convert *all* the heat Q_H into work; in that case we would have $Q_H = W$ and $Q_C = 0$. Experience shows that this is impossible; there is always some heat wasted, and Q_C *is never zero*. We define the **thermal efficiency** of an engine, denoted by

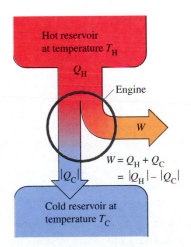

Hot reservoir at temperature T_H

Q_H

Engine

W

$W = Q_H + Q_C$
$= |Q_H| - |Q_C|$

$|Q_C|$

Cold reservoir at temperature T_C

18–1 Schematic energy-flow diagram for a heat engine.

e, as the quotient

$$e = \frac{W}{Q_H}. \tag{18–3}$$

The thermal efficiency e represents the fraction of Q_H that *is* converted to work. To put it another way, e is what you get divided by what you pay for. This is always less than unity, an all-too-familiar experience! In terms of the flow diagram of Fig. 18–1, the most efficient engine is the one for which the branch pipeline representing the work output is as wide as possible and the exhaust pipeline representing the heat thrown away is as narrow as possible.

When we substitute the two expressions for W given by Eq. (18–2) into Eq. (18–3), we get the following equivalent expressions for e:

$$e = \frac{W}{Q_H} = 1 + \frac{Q_C}{Q_H} = 1 - \left| \frac{Q_C}{Q_H} \right| \quad \text{(thermal efficiency of an engine).} \tag{18–4}$$

Note that e is a quotient of two energy quantities and thus is a pure number, without units. Of course, we must always express W, Q_H, and Q_C in the same units.

Problem–Solving Strategy

HEAT ENGINES

1. The suggestions in the Problem-Solving Strategy in Section 17–5 are equally useful throughout the present chapter, and we suggest that you re-read them.

2. Be very careful with the sign conventions for W and the various Q's. W is positive when the system expands and does work and is negative when the system is compressed. Each Q is positive if it represents heat entering the system and is negative if it represents heat leaving the system. When in doubt, use the first law when possible, to check consistency. When you know that a quantity is negative, such as Q_C in the above discussion, it sometimes helps to write it as $Q_C = - |Q_C|$.

3. Some problems deal with power rather than energy quantities. Power is work per unit time ($P = W/t$), and rate of heat transfer (heat current) H is heat transfer per unit time ($H = Q/t$). Sometimes it helps to ask, "What is W or Q in one second (or one hour)?"

EXAMPLE 18–1

A gasoline engine in a large truck takes in 10,000 J of heat and delivers 2000 J of mechanical work per cycle. The heat is obtained by burning gasoline with heat of combustion $L_c = 5.0 \times 10^4$ J/g. a) What is the thermal efficiency of this engine? b) How much heat is discarded in each cycle? c) How much gasoline is burned in each cycle? d) If the engine goes through 25 cycles per second, what is its power output in watts? In horsepower? e) How much gasoline is burned per second? Per hour?

SOLUTION We have $Q_H = 10,000$ J and $W = 2000$ J.
a) From Eq. (18–3) the thermal efficiency is

$$e = \frac{W}{Q_H} = \frac{2000 \text{ J}}{10,000 \text{ J}} = 0.20 = 20\%.$$

This is a fairly typical figure for cars and trucks if W includes only the work actually delivered to the wheels.

b) From Eq. (18–2), $W = Q_H + Q_C$ so

$$Q_C = W - Q_H = 2000 \text{ J} - 10,000 \text{ J}$$
$$= -8000 \text{ J}.$$

That is, 8000 J of heat leaves the engine during each cycle.
c) Let m be the mass of gasoline burned during each cycle. Then

$$Q_H = mL_c,$$

$$m = \frac{Q_H}{L_c} = \frac{10,000 \text{ J}}{5.0 \times 10^4 \text{ J/g}} = 0.20 \text{ g}.$$

d) The power P (rate of doing work) is the work per cycle multiplied by the number of cycles per second:

$$P = (2000 \text{ J/cycle})(25 \text{ cycles/s}) = 50,000 \text{ W} = 50 \text{ kW}$$

$$= (50,000 \text{ W}) \frac{1 \text{ hp}}{746 \text{ W}} = 67 \text{ hp}.$$

e) The mass of gasoline burned per second is the mass per cycle multiplied by the number of cycles per second:

$$(0.20 \text{ g/cycle})(25 \text{ cycles/s}) = 5.0 \text{ g/s}.$$

The mass burned per hour is

$$(5.0 \text{ g/s}) \frac{3600 \text{ s}}{1 \text{ h}} = 18,000 \text{ g/h} = 18 \text{ kg/h}.$$

The density of gasoline is about 0.70 g/cm^3, so this is about $25,700 \text{ cm}^3$, 25.7 L, or 6.8 gallons of gasoline per hour. If the truck is traveling at 55 mi/h (88 km/h), this represents fuel consumption of 8.1 miles/gallon (3.4 km/L).

18–4 INTERNAL-COMBUSTION ENGINES

The gasoline engine, used in automobiles and many other types of machinery, is a familiar example of a heat engine. Let's look at its thermal efficiency. Figure 18–2 shows the operation of one type of gasoline engine. First a mixture of air and gasoline vapor flows into a cylinder through an open intake valve while the piston descends, increasing the volume of the cylinder from a minimum of V (when the piston is all the way up) to a maximum of rV (when it is all the way down). The quantity r is called the **compression ratio;** for present-day automobile engines its value is typically 8–10. At the end of this *intake stroke,* the intake valve closes and the mixture is compressed, approximately adiabatically, to volume V during the *compression stroke.* The mixture is then ignited by the spark plug, and the heated gas expands, approximately adiabatically, back to volume rV, pushing on the piston and doing work; this is the *power stroke.* Finally, the exhaust valve opens, and the combustion products are pushed out (during the *exhaust stroke*), leaving the cylinder ready for the next intake stroke.

THE OTTO CYCLE

Figure 18–3 (page 564) is a *pV*-diagram for an idealized model of the thermodynamic processes in a gasoline engine. This model is called the **Otto cycle.** At point *a* the gasoline-air mixture has entered the cylinder. The mixture is compressed adiabatically to point *b* and is then ignited. Heat Q_H is added to the system by the burning gasoline along

18–2 Cycle of a four-stroke internal-combustion engine. (a) *Intake stroke:* Piston moves down, causing a partial vacuum in cylinder; gasoline-air mixture flows through open intake valve into cylinder. (b) *Compression stroke:* Intake valve closes, and mixture is compressed as piston moves up. (c) *Ignition:* Spark plug ignites mixture. (d) *Power stroke:* Hot burned mixture pushes piston down, doing work. (e) *Exhaust stroke:* Exhaust valve opens and piston moves up, pushing burned mixture out of cylinder. Engine is now ready for next intake stroke, and the cycle repeats.

18–3 The pV-diagram for the Otto cycle, an idealized model of the thermodynamic processes in a gasoline engine.

line bc, and the power stroke is the adiabatic expansion to d. The gas is cooled to the temperature of the outside air along line da; during this process, heat $|Q_C|$ is rejected. In practice, this gas leaves the engine as exhaust and does not enter the engine again. But since an equivalent amount of gasoline and air enters, we may consider the process to be cyclic.

We can calculate the efficiency of this idealized cycle. Processes bc and da are constant-volume, so the heats Q_H and Q_C are related simply to the temperatures:

$$Q_H = nC_V(T_c - T_b) > 0,$$

$$Q_C = nC_V(T_a - T_d) < 0.$$

The thermal efficiency is given by Eq. (18–4). Inserting the above expressions and cancelling out the common factor nC_V, we find

$$e = \frac{Q_H + Q_C}{Q_H} = \frac{T_c - T_b + T_a - T_d}{T_c - T_b}. \tag{18–5}$$

To simplify this further, we use the temperature-volume relation for adiabatic processes for an ideal gas, Eq. (17–22). For the two adiabatic processes ab and cd,

$$T_a(rV)^{\gamma-1} = T_bV^{\gamma-1} \qquad \text{and} \qquad T_d(rV)^{\gamma-1} = T_cV^{\gamma-1}.$$

We divide each of these equations by the common factor $V^{\gamma-1}$ and substitute the resulting expressions for T_b and T_c back into Eq. (18–5). The result is

$$e = \frac{T_dr^{\gamma-1} - T_ar^{\gamma-1} + T_a - T_d}{T_dr^{\gamma-1} - T_ar^{\gamma-1}} = \frac{(T_d - T_a)(r^{\gamma-1} - 1)}{(T_d - T_a)r^{\gamma-1}}.$$

Dividing out the common factor $(T_d - T_a)$, we get

$$e = 1 - \frac{1}{r^{\gamma-1}} \qquad \text{(thermal efficiency in Otto cycle).} \tag{18–6}$$

The thermal efficiency given by Eq. (18–6) is always less than unity, even for this idealized model. With $r = 8$ and $\gamma = 1.4$ (the value for air) the theoretical efficiency is $e = 0.56$, or 56%. The efficiency can be increased by increasing r. However, this also increases the temperature at the end of the adiabatic compression of the air-fuel mixture. If the temperature is too high, the mixture explodes spontaneously during compression instead of burning evenly after the spark plug ignites it. This is called *pre-ignition* or *detonation;* it causes a knocking sound and can damage the engine. The octane rating of a gasoline is a measure of its antiknock qualities. The maximum practical compression ratio for high-octane, or "premium," gasoline is about 10. Higher ratios can be used with more exotic fuels.

The Otto cycle, which we have just described, is a highly idealized model. It assumes that the mixture behaves as an ideal gas; it neglects friction, turbulence, loss of heat to cylinder walls, and many other effects that combine to reduce the efficiency of a real engine. Another source of inefficiency is incomplete combustion. A mixture of gasoline vapor with just enough air for complete combustion of the hydrocarbons to H_2O and CO_2 does not ignite readily. Reliable ignition requires a mixture that is "richer" in gasoline. The resulting incomplete combustion leads to CO and unburned hydrocarbons in the exhaust. The heat obtained from the gasoline is then less than the total heat of combustion; the difference is wasted, and the exhaust products contribute to air pollution. Efficiencies of real gasoline engines are typically around 35%.

THE DIESEL CYCLE

The Diesel engine is similar in operation to the gasoline engine. The most important difference is that there is no fuel in the cylinder at the beginning of the compression stroke.

A little before the beginning of the power stroke, the injectors start to inject fuel directly into the cylinder, just fast enough to keep the pressure approximately constant during the first part of the power stroke. Because of the high temperature developed during the adiabatic compression, the fuel ignites spontaneously as it is injected; no spark plugs are needed.

The idealized **Diesel cycle** is shown in Fig. 18–4. Starting at point a, air is compressed adiabatically to point b, heated at constant pressure to point c, expanded adiabatically to point d, and cooled at constant volume to point a. Because there is no fuel in the cylinder during most of the compression stroke, pre-ignition cannot occur, and the compression ratio r can be much higher than for a gasoline engine. This improves efficiency and ensures reliable ignition when the fuel is injected (because of the high temperature reached during the adiabatic compression). Values of r of 15–20 are typical; with these values and $\gamma = 1.4$, the theoretical efficiency of the idealized Diesel cycle is about 0.65 to 0.70. As with the Otto cycle, the efficiency of any actual engine is substantially less than this. Diesel engines are usually more efficient than gasoline engines. They are also heavier (per unit power output) and often harder to start. They need no carburetor or ignition system, but the fuel-injection system requires expensive high-precision machining.

18–4 The pV-diagram for the idealized Diesel cycle.

18–5 REFRIGERATORS

We can think of a **refrigerator** as a heat engine operating in reverse. A heat engine takes heat from a hot place and gives off heat to a colder place. A refrigerator does the opposite; it takes heat from a cold place (the inside of the refrigerator) and gives it off to a warmer place (usually the air in the room where the refrigerator is located). A heat engine has a net *output* of mechanical work; the refrigerator requires a net *input* of mechanical work. Using the sign conventions from Section 18–3, for a refrigerator Q_C is positive but both W and Q_H are negative; hence $|W| = -W$ and $|Q_H| = -Q_H$.

A flow diagram for a refrigerator is shown in Fig. 18–5. From the first law for a cyclic process,

$$Q_H + Q_C - W = 0, \qquad \text{or} \qquad -Q_H = Q_C - W,$$

or, because both Q_H and W are negative,

$$|Q_H| = Q_C + |W|. \tag{18–7}$$

Thus, as the diagram shows, the heat $|Q_H|$ leaving the working substance and given to the hot reservoir is always *greater* than the heat Q_C taken from the cold reservoir. Note that the absolute-value relation

$$|Q_H| = |Q_C| + |W| \tag{18–8}$$

is valid for both heat engines and refrigerators.

From an economic point of view, the best refrigeration cycle is one that removes the greatest amount of heat $|Q_C|$ from the inside of the refrigerator for the least expenditure of mechanical work, $|W|$. The relevant ratio is therefore $|Q_C|/|W|$; the larger this ratio, the better the refrigerator. We call this ratio the **coefficient of performance**, denoted by K. From Eq. (18–8), $|W| = |Q_H| - |Q_C|$, so

$$K = \frac{|Q_C|}{|W|} = \frac{|Q_C|}{|Q_H| - |Q_C|} \qquad \text{(coefficient of performance of a refrigerator).} \tag{18–9}$$

As always, we measure Q_H, Q_C, and W all in the same energy units; K is then a dimensionless number.

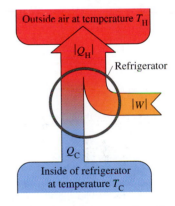

18–5 Schematic energy-flow diagram of a refrigerator.

18-6 Principle of the mechanical refrigeration cycle.

The principles of the common refrigeration cycle are shown schematically in Fig. 18-6. The fluid "circuit" contains a refrigerant fluid (the working substance). In the past this has usually been CCl_2F_2 or another member of the Freon family; because the release of such substances into the atmosphere depletes the ozone layer, alternative refrigerants are now in use. The left side of the circuit (including the cooling coils inside the refrigerator) is at low temperature and low pressure; the right side (including the condenser coils outside the refrigerator) is at high temperature and high pressure. Ordinarily, both sides contain liquid and vapor in phase equilibrium.

The compressor takes in fluid, compresses it adiabatically, and delivers it to the condenser coil at high pressure. The fluid temperature is then higher than that of the air surrounding the condenser, so the refrigerant gives off heat $|Q_H|$ and partially condenses to liquid. The fluid then expands adiabatically into the evaporator at a rate controlled by the expansion valve. As the fluid expands, it cools considerably, enough that the fluid in the evaporator coil is colder than its surroundings. It absorbs heat $|Q_C|$ from its surroundings, cooling them and partially vaporizing. The fluid then enters the compressor to begin another cycle. The compressor, usually driven by an electric motor, requires energy input and does work $|W|$ *on* the working substance during each cycle.

An air conditioner operates on exactly the same principle. In this case the refrigerator box becomes a room or an entire building. The evaporator coils are inside, the condenser is outside, and fans circulate air through these (Fig. 18-7). In large installations the condenser coils are often cooled by water. For air conditioners the quantities of greatest practical importance are the *rate* of heat removal (the heat current H from the region being cooled) and the *power* input $P = W/t$ to the compressor. If heat $|Q_C|$ is removed in time t, then $H = |Q_C|/t$. Then we can express the coefficient of performance as

$$K = \frac{|Q_C|}{|W|} = \frac{Ht}{Pt} = \frac{H}{P}.$$

Typical room air conditioners have heat removal rates H of 5000 to 10,000 Btu/h, or about 1500–3000 W, and require electric power input of about 600 to 1200 W. Typical coefficients of performance are about 2.5, with somewhat larger values for larger-capacity units. The actual values of K in operation depend on the inside and outside temperatures.

Unfortunately, K is often expressed commercially in mixed units, with H in Btu per hour and P in watts. In these units, H/P is called the **energy efficiency rating** (EER); the

18–7 An air conditioner works on the same principle as a refrigerator.

units, customarily omitted, are (Btu/h)/W. Because 1 W = 3.413 Btu/h, the EER is numerically 3.413 times as large as the dimensionless K. Room air conditioners typically have an EER of 7 to 10.

A variation on this theme is the **heat pump,** used to heat buildings by cooling the outside air. It functions like a refrigerator turned inside out. The evaporator coils are outside, where they take heat from cold air, and the condenser coils are inside, where they give off heat to the warmer air. With proper design, the heat $|Q_H|$ delivered to the inside per cycle can be considerably greater than the work $|W|$ required to get it there.

Work is *always* needed to transfer heat from a colder to a hotter body. Heat flows spontaneously from hotter to colder, and to reverse this flow requires the addition of work from the outside. Experience shows that it is impossible to make a refrigerator that transports heat from a colder body to a hotter body without the addition of work. If no work were needed, the coefficient of performance would be infinite. We call such a device a *workless refrigerator;* it is a mythical beast, like the unicorn and the free lunch.

18–6 THE SECOND LAW OF THERMODYNAMICS

Experimental evidence suggests strongly that it is *impossible* to build a heat engine that converts heat completely to work, that is, an engine with 100% thermal efficiency. This impossibility is the basis of one statement of the **second law of thermodynamics,** as follows:

> **It is impossible for any system to undergo a process in which it absorbs heat from a reservoir at a single temperature and converts the heat completely into mechanical work, with the system ending in the same state in which it began.**

We will call this the "engine" statement of the second law.

The basis of the second law of thermodynamics lies in the difference between the nature of internal energy and that of macroscopic mechanical energy. In a moving body the molecules have random motion, but superimposed on this is a coordinated motion of every molecule in the direction of the body's velocity. The kinetic energy associated with this *coordinated* macroscopic motion is what we call the kinetic energy of the moving body. The kinetic and potential energies associated with the *random* motion constitute the internal energy.

When a body sliding on a surface comes to rest as a result of friction, the organized motion of the body is converted to random motion of molecules in the body and in the surface. Since we cannot control the motions of individual molecules, we cannot convert this random motion completely back to organized motion. We can convert *part* of it, and this is what a heat engine does.

If the second law were *not* true, we could power an automobile or run a power plant by cooling the surrounding air. Neither of these impossibilities violates the *first* law of thermodynamics. The second law, therefore, is not a deduction from the first but stands by itself as a separate law of nature. The first law denies the possibility of creating or destroying energy; the second law limits the *availability* of energy and the ways in which it can be used and converted.

Our analysis of refrigerators in Section 18–5 forms the basis for an alternative statement of the second law of thermodynamics. Heat flows spontaneously from hotter to colder bodies, never the reverse. A refrigerator does take heat from a colder to a hotter body, but its operation requires an input of mechanical energy or work. Generalizing this observation, we state:

It is impossible for any process to have as its sole result the transfer of heat from a cooler to a hotter body.

We'll call this the "refrigerator" statement of the second law. It may not seem to be very closely related to the "engine" statement. In fact, though, the two statements are completely equivalent. For example, if we could build a workless refrigerator, violating the second or "refrigerator" statement of the second law, we could use it in conjunction with a heat engine, pumping the heat rejected by the engine back to the hot reservoir to be reused. This composite machine (Fig. 18–8a) would violate the "engine" statement of the second law because its net effect would be to take a net quantity of heat $Q_H - |Q_C|$ from the hot reservoir and convert it completely to work W.

Alternatively, if we could make an engine with 100% thermal efficiency, in violation of the first statement, we could run it using heat from the hot reservoir and use the work output to drive a refrigerator that pumps heat from the cold reservoir to the hot (Fig. 18–8b). This composite device would violate the "refrigerator" statement because its net effect would be to take heat Q_C from the cold reservoir and deliver it to the hot reservoir without requiring any input of work. Thus any device that violates one form of the second law can be used to make a device that violates the other form. If violations of the first form are impossible, so are violations of the second!

The conversion of work to heat, as in friction or viscous fluid flow, and heat flow from hot to cold across a finite temperature gradient, are *irreversible* processes. The "engine" and "refrigerator" statements of the second law state that these processes can be only partially reversed. We could cite other examples. Gases always seep sponta-

18–8 Energy-flow diagrams for equivalent forms of the second law. (a) A workless refrigerator (left), if it existed, could be used in combination with an ordinary heat engine (right) to form a composite device that functions as an engine with 100% efficiency, converting heat $Q_H - |Q_C|$ completely to work. (b) An engine with 100% efficiency (left), if it existed, would be used in combination with an ordinary refrigerator (right) to form a workless refrigerator, transferring heat Q_C from the cold reservoir to the hot with no net input of work. Since either of these is impossible, the other must be also.

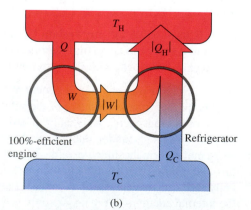

(a)   (b)

neously through an opening from a region of high pressure to a region of low pressure; gases and miscible liquids left by themselves always tend to mix, not to unmix. The second law of thermodynamics is an expression of the inherent one-way aspect of these and many other irreversible processes. Energy conversion is an essential aspect of all plant and animal life and of human technology, so the second law of thermodynamics is of the utmost fundamental importance in the world we live in.

18–7 THE CARNOT CYCLE

8.14
Carnot Cycle

According to the second law, no heat engine can have 100% efficiency. How great an efficiency *can* an engine have, given two heat reservoirs at temperatures T_H and T_C? This question was answered in 1824 by the French engineer Sadi Carnot (1796–1832), who developed a hypothetical, idealized heat engine that has the maximum possible efficiency consistent with the second law. The cycle of this engine is called the **Carnot cycle.**

To understand the rationale of the Carnot cycle, we return to a recurrent theme in this chapter: *reversibility* and its relation to directions of thermodynamic processes. Conversion of work to heat is an irreversible process; the purpose of a heat engine is a *partial* reversal of this process, the conversion of heat to work with as great an efficiency as possible. For maximum heat-engine efficiency, therefore, *we must avoid all irreversible processes.* This requirement turns out to be enough to determine the basic sequence of steps in the Carnot cycle, as we will show next.

Heat flow through a finite temperature drop is an irreversible process. Therefore, during heat transfer in the Carnot cycle there must be *no* finite temperature difference. When the engine takes heat from the hot reservoir at temperature T_H, the working substance of the engine must also be at T_H; otherwise, irreversible heat flow would occur. Similarly, when the engine discards heat to the cold reservoir at T_C, the engine itself must be at T_C. That is, every process that involves heat transfer must be *isothermal* at either T_H or T_C.

Conversely, in any process in which the temperature of the working substance of the engine is intermediate between T_H and T_C, there must be *no* heat transfer between the engine and either reservoir because such heat transfer could not be reversible. Therefore any process in which the temperature T of the working substance changes must be *adiabatic.* The bottom line is that every process in our idealized cycle must be either isothermal or adiabatic. In addition, thermal and mechanical equilibrium must be maintained at all times so that each process is completely reversible.

THE CARNOT CYCLE

The Carnot cycle consists of two reversible isothermal and two reversible adiabatic processes. Figure 18–9 (page 570) shows a Carnot cycle using as its working substance an ideal gas in a cylinder with a piston. It consists of the following steps:

1. The gas expands isothermally at temperature T_H, absorbing heat Q_H (*ab*).
2. It expands adiabatically until its temperature drops to T_C (*bc*).
3. It is compressed isothermally at T_C, rejecting heat $|Q_C|$ (*cd*).
4. It is compressed adiabatically back to its initial state at temperature T_H (*da*).

We can calculate the thermal efficiency e of a Carnot engine in the special case in which the working substance is an *ideal gas.* To carry out this calculation, we will first find the ratio Q_C/Q_H of the quantities of heat transferred in the two isothermal processes and then use Eq. (18–4) to find e.

For an ideal gas the internal energy U depends only on temperature and is thus constant in any isothermal process. For the isothermal expansion *ab*, $\Delta U_{ab} = 0$, and Q_H is

18–9 The Carnot cycle for an ideal gas. The light-blue lines in the pV-diagram are isotherms (curves of constant temperature); the dark-blue lines are adiabats (curves of zero heat flow).

equal to the work W_{ab} done by the gas during its isothermal expansion at temperature T_H. We calculated this work in Example 17–1 (Section 17–3); using that result, we have

$$Q_H = W_{ab} = nRT_H \ln \frac{V_b}{V_a}. \qquad (18\text{–}10)$$

Similarly,

$$Q_C = W_{cd} = nRT_C \ln \frac{V_d}{V_c} = -nRT_C \ln \frac{V_c}{V_d}. \qquad (18\text{–}11)$$

Because V_d is less than V_c, Q_C is negative ($Q_C = -|Q_C|$); heat flows out of the gas during the isothermal compression at temperature T_C.

The ratio of the two quantities of heat is thus

$$\frac{Q_C}{Q_H} = -\left(\frac{T_C}{T_H}\right)\frac{\ln(V_c/V_d)}{\ln(V_b/V_a)}. \qquad (18\text{–}12)$$

This can be simplified further by use of the temperature-volume relation for an adiabatic process, Eq. (17–22). We find for the two adiabatic processes:

$$T_H V_b^{\gamma-1} = T_C V_c^{\gamma-1} \qquad \text{and} \qquad T_H V_a^{\gamma-1} = T_C V_d^{\gamma-1}.$$

Dividing the first of these by the second, we find

$$\frac{V_b^{\gamma-1}}{V_a^{\gamma-1}} = \frac{V_c^{\gamma-1}}{V_d^{\gamma-1}} \qquad \text{and} \qquad \frac{V_b}{V_a} = \frac{V_c}{V_d}.$$

Thus the two logarithms in Eq. (18–12) are equal, and that equation reduces to

$$\frac{Q_C}{Q_H} = -\frac{T_C}{T_H}, \qquad \text{or} \qquad \frac{|Q_C|}{|Q_H|} = \frac{T_C}{T_H} \qquad \begin{array}{l}\text{(heat transfer in a} \\ \text{Carnot engine).}\end{array} \qquad (18\text{–}13)$$

The ratio of the heat rejected at T_C to the heat absorbed at T_H is just equal to the ratio T_C/T_H. Then from Eq. (18–4) the efficiency of the Carnot engine is

$$e_{\text{Carnot}} = 1 - \frac{T_C}{T_H} = \frac{T_H - T_C}{T_H} \quad \text{(efficiency of a Carnot engine).} \quad (18\text{–}14)$$

This simple result says that the efficiency of a Carnot engine depends only on the temperatures of the two heat reservoirs. The efficiency is large when the temperature *difference* is large, and it is very small when the temperatures are nearly equal. The efficiency can never be exactly unity unless $T_C = 0$; we'll see later that this, too, is impossible.

CAUTION ▶ In all calculations involving the Carnot cycle, you must make sure that you use *absolute* (Kelvin) temperatures only. That's because Eqs. (18–10) through (18–14) come from the ideal-gas equation $pV = nRT$, in which T is absolute temperature. ◀

EXAMPLE 18–2

A Carnot engine takes 2000 J of heat from a reservoir at 500 K, does some work, and discards some heat to a reservoir at 350 K. How much work does it do, how much heat is discarded, and what is the efficiency?

SOLUTION From Eq. (18–13) the heat Q_C discarded by the engine is

$$Q_C = -Q_H \frac{T_C}{T_H} = -(2000 \text{ J}) \frac{350 \text{ K}}{500 \text{ K}}$$

$$= -1400 \text{ J}.$$

Then from the first law, the work W done by the engine is

$$W = Q_H + Q_C = 2000 \text{ J} + (-1400 \text{ J})$$

$$= 600 \text{ J}.$$

From Eq. (18–14) the thermal efficiency is

$$e = 1 - \frac{T_C}{T_H} = 1 - \frac{350 \text{ K}}{500 \text{ K}} = 0.30 = 30\%.$$

Alternatively, from the basic definition of thermal efficiency,

$$e = \frac{W}{Q_H} = \frac{600 \text{ J}}{2000 \text{ J}} = 0.30 = 30\%.$$

EXAMPLE 18–3

Suppose 0.200 mol of an ideal diatomic gas ($\gamma = 1.40$) undergoes a Carnot cycle with temperatures 227°C and 27°C. The initial pressure is $p_a = 10.0 \times 10^5$ Pa, and during the isothermal expansion at the higher temperature the volume doubles. a) Find the pressure and volume at each of points a, b, c, and d in Fig. 18–9. b) Find Q, W, and ΔU for each step and for the entire cycle. c) Determine the efficiency directly from the results of part (b), and compare it with the result from Eq. (18–14).

SOLUTION a) We first remember to convert the Celsius temperatures to absolute temperatures by adding 273.15. The higher temperature is $T_H = (227 + 273.15)$ K = 500 K, and the lower temperature is $T_C = (27 + 273.15)$ K = 300 K. We then use the ideal-gas equation to find V_a:

$$V_a = \frac{nRT_H}{p_a} = \frac{(0.200 \text{ mol})(8.315 \text{ J/mol} \cdot \text{K})(500 \text{ K})}{10.0 \times 10^5 \text{ Pa}}$$

$$= 8.31 \times 10^{-4} \text{ m}^3.$$

The volume doubles during the isothermal expansion $a \to b$, so

$$V_b = 2V_a = 2(8.31 \times 10^{-4} \text{ m}^3)$$

$$= 16.6 \times 10^{-4} \text{ m}^3.$$

Also, during the isothermal expansion $a \to b$, $p_a V_a = p_b V_b$, so

$$p_b = \frac{p_a V_a}{V_b} = 5.00 \times 10^5 \text{ Pa}.$$

For the adiabatic expansion $b \to c$, $T_H V_b^{\gamma-1} = T_C V_c^{\gamma-1}$, so

$$V_c = V_b \left(\frac{T_H}{T_C}\right)^{1/(\gamma-1)} = (16.6 \times 10^{-4} \text{ m}^3) \left(\frac{500 \text{ K}}{300 \text{ K}}\right)^{2.5}$$

$$= 59.6 \times 10^{-4} \text{ m}^3.$$

Using the ideal-gas equation again for point c, we find

$$p_c = \frac{nRT_C}{V_c} = \frac{(0.200 \text{ mol})(8.315 \text{ J/mol} \cdot \text{K})(300 \text{ K})}{59.6 \times 10^{-4} \text{ m}^3}$$

$$= 0.837 \times 10^5 \text{ Pa}.$$

For the adiabatic compression $d \to a$, $T_C V_d^{\gamma-1} = T_H V_a^{\gamma-1}$, and

$$V_d = V_a \left(\frac{T_H}{T_C}\right)^{1/(\gamma-1)} = (8.31 \times 10^{-4} \text{ m}^3) \left(\frac{500 \text{ K}}{300 \text{ K}}\right)^{2.5}$$

$$= 29.8 \times 10^{-4} \text{ m}^3,$$

$$p_d = \frac{nRT_C}{V_d} = \frac{(0.200 \text{ mol})(8.315 \text{ J/mol} \cdot \text{K})(300 \text{ K})}{29.8 \times 10^{-4} \text{ m}^3}$$

$$= 1.67 \times 10^5 \text{ Pa}.$$

b) For the isothermal expansion $a \rightarrow b$, $\Delta U_{ab} = 0$. To find W_{ab} ($= Q_H$), we use Eq. (18–10):

$$W_{ab} = Q_H = nRT_H \ln \frac{V_b}{V_a}$$

$$= (0.200 \text{ mol})(8.315 \text{ J/mol} \cdot \text{K})(500 \text{ K})(\ln 2) = 576 \text{ J}.$$

For the adiabatic expansion $b \rightarrow c$, $Q_{bc} = 0$. From the first law of thermodynamics, $\Delta U_{bc} = Q_{bc} - W_{bc} = -W_{bc}$; hence the work W_{bc} done by the gas in this process equals the negative of the change in internal energy of the gas. From Eq. (17–13) we have $\Delta U = nC_V \Delta T$, where $\Delta T = T_C - T_H$ (final temperature minus initial temperature). Using $C_V = 20.8 \text{ J/mol} \cdot \text{K}$ for an ideal diatomic gas, we find

$$W_{bc} = -\Delta U_{bc} = -nC_V(T_C - T_H) = nC_V(T_H - T_C)$$

$$= (0.200 \text{ mol})(20.8 \text{ J/mol} \cdot \text{K})(500 \text{ K} - 300 \text{ K}) = 832 \text{ J}.$$

For the isothermal compression $c \rightarrow d$, $\Delta U_{cd} = 0$; Eq. (18–11) gives

$$W_{cd} = Q_C = nRT_C \ln \frac{V_d}{V_c}$$

$$= (0.200 \text{ mol})(8.315 \text{ J/mol} \cdot \text{K})(300 \text{ K})\left(\ln \frac{29.8 \times 10^{-4} \text{ m}^3}{59.6 \times 10^{-4} \text{ m}^3}\right)$$

$$= -346 \text{ J}.$$

For the adiabatic compression $d \rightarrow a$, $Q_{da} = 0$, and

$$W_{da} = -\Delta U_{da} = -nC_V(T_H - T_C) = nC_V(T_C - T_H)$$

$$= (0.200 \text{ mol})(20.8 \text{ J/mol} \cdot \text{K})(300 \text{ K} - 500 \text{ K})$$

$$= -832 \text{ J}.$$

We can tabulate the results as follows:

PROCESS	Q	W	ΔU
$a \rightarrow b$	576 J	576 J	0
$b \rightarrow c$	0	832 J	−832 J
$c \rightarrow d$	−346 J	−346 J	0
$d \rightarrow a$	0	−832 J	832 J
Total	230 J	230 J	0

Note that for the entire cycle, $Q = W$ and $\Delta U = 0$. Also note that the quantities of work in the two adiabatic processes are negatives of each other; it is easy to prove from the analysis leading to Eq. (18–13) that this must *always* be the case.

c) From the table, $Q_H = 576$ J and the total work is 230 J. Thus

$$e = \frac{W}{Q_H} = \frac{230 \text{ J}}{576 \text{ J}} = 0.40 = 40\%.$$

We can compare this with the result from Eq. (18–14):

$$e = \frac{T_H - T_C}{T_H} = \frac{500 \text{ K} - 300 \text{ K}}{500 \text{ K}} = 0.40 = 40\%.$$

THE CARNOT REFRIGERATOR

Because each step in the Carnot cycle is reversible, the *entire cycle* may be reversed, converting the engine into a refrigerator. The coefficient of performance of the Carnot refrigerator is obtained by combining the general definition of K, Eq. (18–9), with Eq. (18–13) for the Carnot cycle. We first rewrite Eq. (18–9) as

$$K = \frac{|Q_C|}{|Q_H| - |Q_C|} = \frac{|Q_C|/|Q_H|}{1 - |Q_C|/|Q_H|}.$$

Then we substitute Eq. (18–13), $|Q_C|/|Q_H| = T_C/T_H$, into this. The result is

$$K_{\text{Carnot}} = \frac{T_C}{T_H - T_C} \qquad \text{(coefficient of performance of a Carnot refrigerator).} \qquad (18\text{–}15)$$

When the temperature difference $T_H - T_C$ is small, K is much larger than unity; in this case a lot of heat can be "pumped" from the lower to the higher temperature with only a little expenditure of work. But the greater the temperature difference, the smaller the value of K and the more work is required to transfer a given quantity of heat.

EXAMPLE 18–4

If the cycle described in Example 18–3 is run backward as a refrigerator, what is its coefficient of performance?

SOLUTION From Eq. (18–9),

$$K = \frac{|Q_C|}{|W|} = \frac{346 \text{ J}}{230 \text{ J}} = 1.50.$$

Because the cycle is a Carnot cycle, we may also use

Eq. (18–15):

$$K = \frac{T_C}{T_H - T_C} = \frac{300 \text{ K}}{500 \text{ K} - 300 \text{ K}} = 1.50.$$

For a Carnot cycle, e and K depend only on the temperatures, as shown by Eqs. (18–14) and (18–15), and we don't need to calculate Q and W. For cycles containing irreversible processes, however, these two equations are not valid, and more detailed calculations are necessary.

THE CARNOT CYCLE AND THE SECOND LAW

We can prove that **no engine can be more efficient than a Carnot engine operating between the same two temperatures.** The key to the proof is the above observation that since each step in the Carnot cycle is reversible, the *entire cycle* may be reversed. Run backward, the engine becomes a refrigerator. Suppose we have an engine that is more efficient than a Carnot engine (Fig. 18–10). Let the Carnot engine, run backward as a refrigerator by negative work $-|W|$, take in heat Q_C from the cold reservoir and expel heat $|Q_H|$ to the hot reservoir. The superefficient engine expels heat $|Q_C|$, but to do this, it takes in a greater amount of heat $Q_H + \Delta$. Its work output is then $W + \Delta$, and the net effect of the two machines together is to take a quantity of heat Δ and convert it completely into work. This violates the engine statement of the second law. We could construct a similar argument that a superefficient engine could be used to violate the refrigerator statement of the second law. Note that we don't have to assume that the superefficient engine is reversible. In a similar way we can show that *no refrigerator can have a greater coefficient of performance than a Carnot refrigerator operating between the same two temperatures.*

Thus the statement that no engine can be more efficient than a Carnot engine is yet another equivalent statement of the second law of thermodynamics. It also follows directly that **all Carnot engines operating between the same two temperatures have the same efficiency, irrespective of the nature of the working substance.** Although we derived Eq. (18–14) for a Carnot engine using an ideal gas as its working substance, it is in fact valid for *any* Carnot engine, no matter what its working substance.

Equation (18–14), the expression for the efficiency of a Carnot engine, sets an upper limit to the efficiency of a real engine such as a steam turbine. To maximize this upper limit and the actual efficiency of the real engine, the designer must make the intake temperature T_H as high as possible and the exhaust temperature T_C as low as possible.

The exhaust temperature cannot be lower than the lowest temperature available for cooling the exhaust. For a steam turbine at an electric power plant, T_C may be the temperature of river or lake water; then we want the boiler temperature T_H to be as high as possible. The vapor pressures of all liquids increase rapidly with temperature, so we are limited by the mechanical strength of the boiler. At 500°C the vapor pressure of water is about 240×10^5 Pa (235 atm); this is about the maximum practical pressure in large present-day steam boilers.

The unavoidable exhaust heat loss in electric power plants creates a serious environmental problem. When a lake or river is used for cooling, the temperature of the body of water may be raised several degrees. This has a severely disruptive effect on the overall ecological balance, since relatively small temperature changes can have significant effects on metabolic rates in plants and animals. **Thermal pollution,** as this effect is called, is an inevitable consequence of the second law of thermodynamics, so careful planning is essential to minimize the ecological impact of new power plants.

To maximize efficiency, the temperatures inside a jet engine are made as high as possible. Exotic ceramic materials are used that can withstand temperatures in excess of 1000°C without melting or becoming soft.

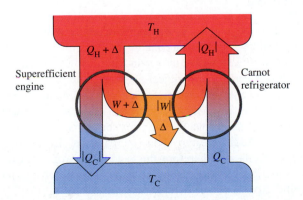

Superefficient engine

Carnot refrigerator

T_H

$Q_H + \Delta$

$|Q_H|$

$W + \Delta$ $|W|$

Δ

$|Q_C|$

Q_C

T_C

18–10 If there were a more efficient engine than a Carnot engine, it could be used in conjunction with a Carnot refrigerator to convert the heat Δ completely into work with no net heat transfer to the cold reservoir. By the second law of thermodynamics this cannot happen.

*18–8 THE KELVIN TEMPERATURE SCALE

In Chapter 15 we expressed the need for a temperature scale that doesn't depend on the properties of any particular material. We can now use the Carnot cycle to define such a scale. The thermal efficiency of a Carnot engine operating between two heat reservoirs at temperatures T_H and T_C is independent of the nature of the working substance and depends only on the temperatures. From Eq. (18–4), this thermal efficiency is

$$e = \frac{Q_H + Q_C}{Q_H} = 1 + \frac{Q_C}{Q_H}.$$

Therefore the ratio Q_C/Q_H is the same for *all* Carnot engines operating between two given temperatures T_H and T_C.

Kelvin proposed that we *define* the ratio of the temperatures, T_C/T_H, to be equal to the magnitude of the ratio Q_C/Q_H of the quantities of heat absorbed and rejected:

$$\frac{T_C}{T_H} = \frac{|Q_C|}{|Q_H|} = -\frac{Q_C}{Q_H} \qquad \text{(definition of Kelvin temperature).} \qquad (18–16)$$

Equation (18–16) looks identical to Eq. (18–13), but there is a subtle and crucial difference. The temperatures in Eq. (18–13) are based on an ideal-gas thermometer, as defined in Section 15–4, while Eq. (18–16) *defines* a temperature scale based on the Carnot cycle and the second law of thermodynamics and independent of the behavior of any particular substance. Thus the **Kelvin temperature scale** is truly *absolute*. To complete the definition of the Kelvin scale, we assign, as in Section 15–4, the arbitrary value of 273.16 K to the temperature of the triple point of water. When a substance is taken around a Carnot cycle, the ratio of the heats absorbed and rejected, $|Q_H|/|Q_C|$, is equal to the ratio of the temperatures of the reservoirs *as expressed on the gas-thermometer scale* defined in Section 15–4. Since the triple point of water is chosen to be 273.16 K in both scales, it follows that *the Kelvin and ideal-gas scales are identical.*

The zero point on the Kelvin scale is called **absolute zero.** Absolute zero can be interpreted on a molecular level; at absolute zero the system has its *minimum* possible total internal energy (kinetic plus potential). Because of quantum effects, however, it is *not* true that at $T = 0$, all molecular motion ceases. There are theoretical reasons for believing that absolute zero cannot be attained experimentally, although temperatures below 10^{-7} K have been achieved. The more closely we approach absolute zero, the more difficult it is to get closer. One statement of the *third law of thermodynamics* is that it is impossible to reach absolute zero in a finite number of thermodynamic steps.

*18–9 ENTROPY

The second law of thermodynamics, as we have stated it, is rather different in form from many familiar physical laws. It is not an equation or a quantitative relationship but rather a statement of *impossibility.* However, the second law *can* be stated as a quantitative relation with the concept of *entropy,* the subject of this section.

We have talked about several processes that proceed naturally in the direction of increasing disorder. Irreversible heat flow increases disorder because the molecules are initially sorted into hotter and cooler regions; this sorting is lost when the system comes to thermal equilibrium. Adding heat to a body increases its disorder because it increases average molecular speeds and therefore the randomness of molecular motion. Free expansion of a gas increases its disorder because the molecules have greater randomness of position after the expansion than before.

ENTROPY AND DISORDER

Entropy provides a *quantitative* measure of disorder. To introduce this concept, let's consider an infinitesimal isothermal expansion of an ideal gas. We add heat dQ and let the gas expand just enough to keep the temperature constant. Because the internal energy of an ideal gas depends only on its temperature, the internal energy is also constant; thus from the first law, the work dW done by the gas is equal to the heat dQ added. That is,

$$dQ = dW = p\, dV = \frac{nRT}{V}\, dV, \quad \text{so} \quad \frac{dV}{V} = \frac{dQ}{nRT}.$$

The gas is in a more disordered state after the expansion than before because the molecules are moving in a larger volume and have more randomness of position. Thus the fractional volume change dV/V is a measure of the increase in disorder, and the above equation shows that it is proportional to the quantity dQ/T. We introduce the symbol S for the entropy of the system, and we define the infinitesimal entropy change dS during an infinitesimal reversible process at absolute temperature T as

$$dS = \frac{dQ}{T} \quad \text{(infinitesimal reversible process).} \qquad (18\text{–}17)$$

If a total amount of heat Q is added during a reversible isothermal process at absolute temperature T, the total entropy change $\Delta S = S_2 - S_1$ is given by

$$\Delta S = S_2 - S_1 = \frac{Q}{T} \quad \text{(reversible isothermal process).} \qquad (18\text{–}18)$$

Entropy has units of energy divided by temperature; the SI unit of entropy is 1 J/K.

We can see how the quotient Q/T is related to the increase in disorder. Higher temperature means greater randomness of motion. If the substance is initially cold, with little molecular motion, adding heat Q causes a substantial fractional increase in molecular motion and randomness. But if the substance is already hot, the same quantity of heat adds relatively little to the greater molecular motion already present. So the quotient Q/T is an appropriate characterization of the increase in randomness or disorder when heat flows into a system.

EXAMPLE 18–5

One kilogram of ice at 0°C is melted and converted to water at 0°C. Compute its change in entropy, assuming that the melting is done reversibly. The heat of fusion of water is $L_f = 3.34 \times 10^5$ J/kg.

SOLUTION The temperature T is constant at 273 K. The heat needed to melt the ice is $Q = mL_f = 3.34 \times 10^5$ J. From Eq. (18–18) the increase in entropy of the system is

$$\Delta S = S_2 - S_1 = \frac{Q}{T} = \frac{3.34 \times 10^5 \text{ J}}{273 \text{ K}} = 1.22 \times 10^3 \text{ J/K.}$$

This increase corresponds to the increase in disorder when the water molecules go from the highly ordered state of a crystalline solid to the much more disordered state of a liquid (Fig. 18–11).

In any *isothermal* reversible process, the entropy change equals the heat transferred divided by the absolute temperature. When we refreeze the water, Q has the opposite sign, and the entropy change of the water is $\Delta S = -1.22 \times 10^3$ J/K. The water molecules rearrange themselves into a crystal to form ice, so disorder and entropy both decrease.

18–11 Water molecules are arranged in a regular, ordered way in an ice crystal. When the ice melts, the hydrogen bonds between molecules are broken, increasing the water's disorder and its entropy.

We can generalize the definition of entropy change to include *any* reversible process leading from one state to another, whether it is isothermal or not. We represent the process as a series of infinitesimal reversible steps. During a typical step, an infinitesimal quantity of heat dQ is added to the system at absolute temperature T. Then we sum (integrate) the quotients dQ/T for the entire process; that is,

$$\Delta S = \int_1^2 \frac{dQ}{T} \qquad \text{(entropy change in a reversible process).} \qquad (18-19)$$

The limits 1 and 2 refer to the initial and final states.

Because entropy is a measure of the disorder of a system in any specific state, it must depend only on the current state of the system, not on its past history. We will show later that this is indeed the case. When a system proceeds from an initial state with entropy S_1 to a final state with entropy S_2, the change in entropy $\Delta S = S_2 - S_1$ defined by Eq. (18-19) does not depend on the path leading from the initial to the final state but is the same for *all possible* processes leading from state 1 to state 2. Thus the entropy of a system must also have a definite value for any given state of the system. We recall that *internal energy,* introduced in Chapter 17, also has this property, although entropy and internal energy are very different quantities.

Since entropy is a function only of the state of a system, we can also compute entropy changes in *irreversible* (non-equilibrium) processes for which Eqs. (18-17) and (18-19) are not applicable. We simply invent a path connecting the given initial and final states that *does* consist entirely of reversible, equilibrium processes and compute the total entropy change for that path. It is not the actual path, but the entropy change must be the same as for the actual path.

As with internal energy, the above discussion does not tell us how to calculate entropy itself, but only the change in entropy in any given process. Just as with internal energy, we may arbitrarily assign a value to the entropy of a system in a specified reference state and then calculate the entropy of any other state with reference to this.

EXAMPLE 18-6

One kilogram of water at 0°C is heated to 100°C. Compute its change in entropy.

SOLUTION To use Eq. (18-19), we imagine that the temperature of the water is increased reversibly in a series of infinitesimal steps, in each of which the temperature is raised by an infinitesimal amount dT. From Eq. (15-14) the heat required to carry out each such step is $dQ = mc\,dT$. Substituting this into Eq. (18-19) and integrating, we find

$$\Delta S = S_2 - S_1 = \int_1^2 \frac{dQ}{T} = \int_{T_1}^{T_2} mc\,\frac{dT}{T} = mc \ln \frac{T_2}{T_1}$$

$$= (1.00 \text{ kg})(4190 \text{ J/kg} \cdot \text{K})\left(\ln \frac{373 \text{ K}}{273 \text{ K}}\right)$$

$$= 1.31 \times 10^3 \text{ J/K}.$$

In practice, this heating would be done irreversibly, perhaps by setting a pan of water on a stove with a finite temperature difference between the burner and pan. But the entropy change of the water is the same whether the process is reversible or irreversible.

In this calculation we assumed that the specific heat capacity c doesn't depend on temperature. That's a pretty good approximation, since c for water increases by only 1% between 0°C and 100°C.

EXAMPLE 18-7

A gas expands adiabatically and reversibly. What is its change in entropy?

SOLUTION In an adiabatic process, no heat enters or leaves the system. Hence $dQ = 0$, and there is *no* change in entropy in this reversible process: $\Delta S = 0$. Every *reversible* adiabatic process is

a constant-entropy process. (For this reason, reversible adiabatic processes are also called *isentropic* processes.) The increase in disorder resulting from the gas occupying a greater volume is exactly balanced by the decrease in disorder associated with the lowered temperature and reduced molecular speeds.

EXAMPLE 18-8

A thermally insulated box is divided by a partition into two compartments, each having volume V (Fig. 18–12). Initially, one compartment contains n moles of an ideal gas at temperature T, and the other compartment is evacuated. We then break the partition, and the gas expands to fill both compartments. What is the entropy change in this free-expansion process?

SOLUTION For this process, $Q = 0$, $W = 0$, $\Delta U = 0$, and therefore (because the system is an ideal gas) $\Delta T = 0$. We might think that the entropy change is zero because there is no heat exchange. But Eq. (18–19) can be used to calculate entropy changes for *reversible* processes only; this free expansion is *not* reversible, and there *is* an entropy change. The process is adiabatic, but it is not isentropic.

To calculate ΔS, we recall that the entropy change depends only on the initial and final states. We can devise a *reversible* process having the same endpoints, use Eq. (18–19) to calculate its entropy change, and thus determine the entropy change in the original process. An appropriate reversible process in this case is an isothermal expansion from V to $2V$ at temperature T. The gas does work W during this substitute expansion, so an equal amount of heat Q must be supplied to keep the internal energy constant. We found in Example 17–1 (Section 17–3) that the work done by n moles of ideal gas in an isothermal expansion from V_1 to V_2 is $W = nRT \ln (V_2/V_1)$. Using $V_1 = V$ and $V_2 = 2V$, we have

$$Q = W = nRT \ln \frac{2V}{V} = nRT \ln 2.$$

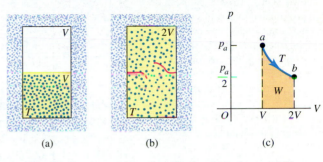

18–12 (a, b) Free expansion of an insulated ideal gas. (c) The free-expansion process doesn't pass through equilibrium states from a to b. However, the entropy change $S_b - S_a$ can be calculated by using the isothermal path shown or *any* reversible path from a to b.

Thus the entropy change is

$$\Delta S = \frac{Q}{T} = nR \ln 2,$$

and this is also the entropy change for the free expansion with the same initial and final states. For one mole,

$$\Delta S = (1 \text{ mol})(8.315 \text{ J/mol} \cdot \text{K})(\ln 2) = 5.76 \text{ J/K}.$$

EXAMPLE 18-9

Entropy and the Carnot cycle For the Carnot engine in Example 18–2 (Section 18–7), find the total entropy change in the engine during one cycle.

SOLUTION There is no entropy change during the adiabatic expansion or adiabatic compression. During the isothermal expansion at $T_H = 500$ K the engine takes in 2000 J of heat, and its entropy change is

$$\Delta S_H = \frac{Q_H}{T_H} = \frac{2000 \text{ J}}{500 \text{ K}} = 4.0 \text{ J/K}.$$

During the isothermal compression at $T_C = 350$ K the engine gives off 1400 J of heat, and its entropy change is

$$\Delta S_C = \frac{Q_C}{T_C} = \frac{-1400 \text{ J}}{350 \text{ K}} = -4.0 \text{ J/K}.$$

The total entropy change in the engine during one cycle is $\Delta S_H + \Delta S_C = 4.0 \text{ J/K} + (-4.0 \text{ J/K}) = 0$. The total entropy change of the two heat reservoirs is also zero, although each individual reservoir has an entropy change. This cycle contains no irreversible processes, and the total entropy change of the system and its surroundings is zero.

ENTROPY IN CYCLIC PROCESSES

Example 18–9 showed that the total entropy change for a cycle of a particular Carnot engine, which uses an ideal gas as its working substance, is zero. This result follows directly from Eq. (18–13), which we can rewrite as

$$\frac{Q_H}{T_H} + \frac{Q_C}{T_C} = 0. \qquad (18\text{--}20)$$

The quotient Q_H/T_H equals ΔS_H, the entropy change of the engine that occurs at $T = T_H$. Likewise, Q_C/T_C equals ΔS_C, the (negative) entropy change of the engine that occurs at $T = T_C$. Hence Eq. (18–20) says that $\Delta S_H + \Delta S_C = 0$, that is, there is zero net entropy change in one cycle.

18–13 (a) A reversible cyclic process for an ideal gas, shown as a red closed path on a pV-diagram. Several ideal-gas isotherms are shown in blue. (b) The path can be approximated by a series of long, thin Carnot cycles; one cycle is highlighted in gold. The total entropy change is zero for each Carnot cycle and for the actual cyclic process. (c) The entropy change $S_b - S_a$ is the same for all paths, such as paths 1 and 2, connecting points a and b.

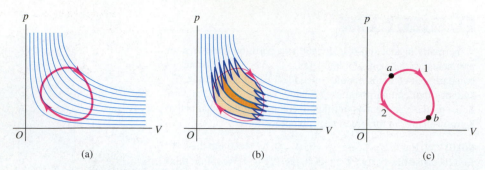

(a)   (b)   (c)

What about Carnot engines that use a different working substance? According to the second law, *any* Carnot engine operating between given temperatures T_H and T_C has the same efficiency $e = 1 - T_C/T_H$ (Eq. (18–14)). Combining this expression for e with Eq. (18–4), $e = 1 + Q_C/Q_H$, just reproduces Eq. (18–20). So Eq. (18–20) is valid for any Carnot engine working between these temperatures, whether its working substance is an ideal gas or not. We conclude that *the total entropy change in one cycle of any Carnot engine is zero.*

This result can be generalized to show that the total entropy change during *any* reversible cyclic process is zero. A reversible cyclic process appears on a pV-diagram as a closed path (Fig. 18–13a). We can approximate such as path as closely as we like by a sequence of isothermal and adiabatic processes forming parts of many long, thin Carnot cycles (Fig. 18–13b). The total entropy change for the full cycle is the sum of the entropy changes for each small Carnot cycle, each of which is zero. So **the total entropy change during *any* reversible cycle is zero:**

$$\int \frac{dQ}{T} = 0 \qquad \text{(reversible cyclic process).} \qquad (18–21)$$

It follows that when a system undergoes a reversible process leading from any state a to any other state b, *the entropy change is independent of the path* (Fig. 18–13c). If the entropy change for path 1 were different from the change for path 2, the system could be taken along path 1 and then backward along path 2 to the starting point, with a nonzero net change in entropy. This would violate the conclusion that the total entropy change in such a cyclic process must be zero. Because the entropy change in such processes is independent of path, we conclude that in any given state, the system has a definite value of entropy that depends only on the state, not on the processes that led to that state.

ENTROPY IN IRREVERSIBLE PROCESSES

In an idealized, reversible process involving only equilibrium states, the total entropy change of the system and its surroundings is zero. But all *irreversible* processes involve an increase in entropy. Unlike energy, *entropy is not a conserved quantity.* The entropy of an isolated system *can* change, but as we shall see, it can never decrease. The free expansion of a gas, described in Example 18–8, is an irreversible process in an isolated system in which there is an entropy increase.

EXAMPLE 18–10

Suppose 1.00 kg of water at 100°C is placed in thermal contact with 1.00 kg of water at 0°C. What is the total change in entropy? Assume that the specific heat capacity of water is constant at 4190 J/kg · K over this temperature range.

SOLUTION   This process involves irreversible heat flow because of the temperature differences. The final temperature is 50°C = 323 K. The first 4190 J of heat transferred cools the hot water to 99°C and warms the cold water from 0°C to 1°C. The

net change of entropy is approximately

$$\Delta S = \frac{-4190 \text{ J}}{373 \text{ K}} + \frac{4190 \text{ J}}{273 \text{ K}} = 4.1 \text{ J/K}.$$

Further increases in entropy occur as the system approaches thermal equilibrium at 50°C (323 K). The *total* increase in entropy can be calculated in the same way as in Example 18–6 by assuming that the process occurs reversibly. The entropy change of the hot water is

$$\Delta S_{\text{hot}} = mc \int_{T_1}^{T_2} \frac{dT}{T} = (1.00 \text{ kg})(4190 \text{ J/kg} \cdot \text{K}) \int_{373 \text{ K}}^{323 \text{ K}} \frac{dT}{T}$$

$$= (4190 \text{ J/K})\left(\ln \frac{323 \text{ K}}{373 \text{ K}}\right) = -603 \text{ J/K}.$$

The entropy change of the cold water is

$$\Delta S_{\text{cold}} = (4190 \text{ J/K})\left(\ln \frac{323 \text{ K}}{273 \text{ K}}\right) = +705 \text{ J/K}.$$

The *total* entropy change of the system is

$$\Delta S_{\text{total}} = \Delta S_{\text{hot}} + \Delta S_{\text{cold}} = (-603 \text{ J/K}) + 705 \text{ J/K} = +102 \text{ J/K}.$$

An irreversible heat flow in an isolated system is accompanied by an increase in entropy. We could have reached the same end state by simply mixing the two quantities of water. This, too, is an irreversible process; because the entropy depends only on the state of the system, the total entropy change would be the same, 102 J/K.

ENTROPY AND THE SECOND LAW

The results of Example 18–10 about the flow of heat from a higher to a lower temperature, or the mixing of substances at different temperatures, are characteristic of *all* natural (that is, irreversible) processes. When we include the entropy changes of all the systems taking part in the process, the increases in entropy are always greater than the decreases. In the special case of a *reversible* process, the increases and decreases are equal. Hence we can state the general principle: **When all systems taking part in a process are included, the entropy either remains constant or increases.** In other words, **no process is possible in which the total entropy decreases, when all systems taking part in the process are included.** This is an alternative statement of the second law of thermodynamics in terms of entropy. Thus it is equivalent to the "engine" and "refrigerator" statements discussed earlier. Figure 18–14 shows a specific example of this general principle.

The increase of entropy in every natural, irreversible process measures the increase of disorder or randomness in the universe associated with that process. Consider again the example of mixing hot and cold water. We *might* have used the hot and cold water as the high- and low-temperature reservoirs of a heat engine. While removing heat from the hot water and giving heat to the cold water, we could have obtained some mechanical work. But once the hot and cold water have been mixed and have come to a uniform temperature, this opportunity to convert heat to mechanical work is lost irretrievably. The lukewarm water will never *unmix* itself and separate into hotter and colder portions. No decrease in *energy* occurs when the hot and cold water are mixed. What has been lost is not *energy*, but *opportunity*, the opportunity to convert part of the heat from the hot water into mechanical work. Hence when entropy increases, energy becomes less *available*, and the universe becomes more random or "run down."

*18-10 MICROSCOPIC INTERPRETATION OF ENTROPY

We described in Section 17–5 how the internal energy of a system could be calculated, at least in principle, by adding up all the kinetic energies of its constituent particles and all the potential energies of interaction among the particles. This is called a *microscopic* calculation of the internal energy. We can also make a microscopic calculation of the entropy S of a system. Unlike energy, however, entropy is not something that belongs to each individual particle or pair of particles in the system. Rather, entropy is a measure of the disorder of the system as a whole. To see how to calculate entropy microscopically, we first have to introduce the idea of *macroscopic* and *microscopic* states.

Suppose you toss N identical coins on the floor, and half of them show heads and half show tails. This is a description of the large-scale or **macroscopic state** of the

18-14 The mixing of colored ink and water starts from a state of relative order (low entropy) in which each fluid is separate and distinct from the other. The final state after mixing is more disordered (has greater entropy) because particles of each fluid are randomly surrounded by particles of both fluids. Spontaneous unmixing of the ink and water, a process in which there would be a net decrease in entropy, is never observed.

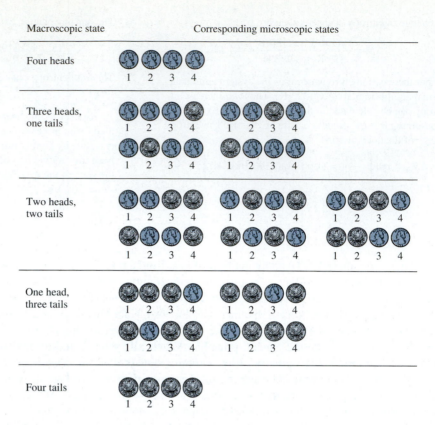

18–15 All possible microscopic states of four coins. There can be several possible microscopic states for each macroscopic state.

system of N coins. A description of the **microscopic state** of the system includes information about each individual coin: Coin 1 was heads, coin 2 was tails, coin 3 was tails, and so on. There can be many microscopic states that correspond to the same macroscopic description. For instance, with $N = 4$ coins there are six possible states in which half are heads and half are tails (Fig. 18–15). The number of microscopic states grows rapidly with increasing N; for $N = 100$ there are $2^{100} = 1.27 \times 10^{30}$ microscopic states, of which 1.01×10^{29} are half heads and half tails.

The least probable outcomes of the coin toss are the states that are either all heads or all tails. It is certainly possible that you could throw 100 heads in a row, but don't bet on it; the probability of doing this is only 1 in 1.27×10^{30}. The most probable outcome of tossing N coins is that half are heads and half are tails. The reason is that this *macroscopic* state has the greatest number of corresponding *microscopic* states, as shown in Fig. 18–15.

To make the connection to the concept of entropy, note that N coins that are all heads constitutes a completely ordered macroscopic state; the description "all heads" completely specifies the state of each one of the N coins. The same is true if the coins are all tails. But the macroscopic description "half heads, half tails" by itself tells you very little about the state (heads or tails) of each individual coin. We say that the system is *disordered* because we know so little about its microscopic state. Compared to the state "all heads" or "all tails," the state "half heads, half tails" has a much greater number of possible microscopic states, much greater disorder, and hence much greater entropy (which is a quantitative measure of disorder).

Now instead of N coins, consider a mole of an ideal gas containing Avogadro's number of molecules. The macroscopic state of this gas is given by its pressure p, volume V, and temperature T; a description of the microscopic state involves stating the position and velocity for each molecule in the gas. At a given pressure, volume, and temperature the gas may be in any one of an astronomically large number of microscopic states, depending on the positions and velocities of its 6.02×10^{23} molecules. If the gas undergoes a free expansion into a greater volume, the range of possible positions increases, as does the number of possible microscopic states. The system becomes more disordered, and the entropy increases as calculated in Example 18–8 (Section 18–9).

We can draw the following general conclusion: *For any system the most probable macroscopic state is the one with the greatest number of corresponding microscopic states, which is also the macroscopic state with the greatest disorder and the greatest entropy.*

Let w represent the number of possible microscopic states for a given macroscopic state. (For the four coins shown in Fig. 18–15 the state of four heads has $w = 1$, the state of three heads and one tails has $w = 4$, and so on.) Then the entropy S of a macroscopic state can be shown to be given by

$$S = k \ln w \quad \text{(microscopic expression for entropy)}, \qquad (18\text{–}22)$$

where $k = R/N_A$ is the Boltzmann constant (gas constant per molecule) introduced in Section 16–4. As Eq. (18–22) shows, increasing the number of possible microscopic states w increases the entropy S.

What matters in a thermodynamic process is not the absolute entropy S but the *difference* in entropy between the initial and final states. Hence an equally valid and useful definition would be $S = k \ln w + C$, where C is a constant, since C cancels in any calculation of an entropy difference between two states. But it's convenient to set this constant equal to zero and use Eq. (18–22). With this choice, since the smallest possible value of w is unity, the smallest possible value of S for any system is $k \ln 1 = 0$. Entropy can *never* be negative.

In practice, calculating w is a difficult task, so Eq. (18–22) is typically used only to calculate the absolute entropy S of certain special systems. But we can use this relation to calculate *differences* in entropy between one state and another. Consider a system that undergoes a thermodynamic process that takes it from macroscopic state 1, for which there are w_1 possible microscopic states, to macroscopic state 2, with w_2 associated microscopic states. The change in entropy in this process is

$$\Delta S = S_2 - S_1 = k \ln w_2 - k \ln w_1 = k \ln \frac{w_2}{w_1}. \qquad (18\text{–}23)$$

The *difference* in entropy between the two macroscopic states depends on the *ratio* of the numbers of possible microscopic states.

As the following example shows, using Eq. (18–23) to calculate a change in entropy from one macroscopic state to another gives the same results as considering a reversible process connecting those two states and using Eq. (18–19).

EXAMPLE 18–11

Use Eq. (18–23) to calculate the entropy change in the free expansion of n moles of gas at temperature T described in Example 18–8.

SOLUTION The situation is shown in Fig. 18–16. Let w_1 be the number of microscopic states when the gas occupies volume V. When the partition is broken, the velocities of the molecules are unaffected, since no work is done. But each molecule now has twice as much volume in which it can move and hence has twice the number of possible positions. The number of molecules is $N = nN_A$, so the number of possible microscopic states of the system as a whole increases by 2^N. Hence the number of microscopic states when the gas occupies volume $2V$ is $w_2 = 2^N w_1$.

Gas occupies volume V
number of microstates w_1

Gas occupies volume $2V$
number of microstates $w_2 = 2^N w_1$

18–16 In a free expansion of N molecules in which the volume doubles, the number of possible microscopic states increases by 2^N.

The change in entropy in this process is

$$\Delta S = k \ln \frac{w_2}{w_1} = k \ln \frac{2^N w_1}{w_1} = k \ln 2^N$$

$$= Nk \ln 2.$$

Since $N = nN_A$ and $k = R/N_A$, this becomes

$$\Delta S = (nN_A)(R/N_A) \ln 2 = nR \ln 2.$$

Thus we have found the same result as in Example 18–8, but without any reference to the thermodynamic path taken.

The relationship between entropy and the number of microscopic states gives us new insight into the entropy statement of the second law of thermodynamics, that the entropy of a closed system can never decrease. From Eq. (18–22) this means that a closed system can never spontaneously undergo a process that decreases the number of possible microscopic states.

An example of such a forbidden process would be if all of the air in your room spontaneously moved to one half of the room, leaving a vacuum in the other half. Such a "free compression" would be the reverse of the free expansion of Examples 18–8 and 18–11. This would decrease the number of possible microscopic states by a factor of 2^N. Strictly speaking, this process is not impossible! The probability of finding a given molecule in one half of the room is 1/2, so the probability of finding all of the molecules in one half of the room at once is $(1/2)^N$. (This is exactly the same as the probability of having a tossed coin come up heads N times in a row.) This probability is *not* zero. But lest you worry about suddenly finding yourself gasping for breath in the evacuated half of your room, consider that a typical room might hold 1000 moles of air, and so $N = 1000N_A = 6.02 \times 10^{26}$ molecules. The probability of all the molecules being in the same half of the room is therefore $(1/2)^{6.02 \times 10^{26}}$. Expressed as a decimal, this number has more than 10^{26} zeros to the right of the decimal point!

Because the probability of such a "free compression" taking place is so vanishingly small, it has almost certainly never occurred anywhere in the universe since the beginning of time. We conclude that for all practical purposes the second law of thermodynamics is never violated.

18–11 ENERGY RESOURCES

A Case Study in Thermodynamics

The laws of thermodynamics place very general limitations on conversion of energy from one form to another. In this time of increasing energy demand and diminishing resources, these matters are of the utmost practical importance. We conclude this chapter with a brief discussion of a few energy-conversion systems, present and proposed.

Over half of the electric power generated in the United States is obtained from coal-fired steam-turbine generating plants. Modern boilers can transfer about 80 to 90 percent of the heat of combustion of coal into steam. The maximum Carnot efficiency of a steam turbine is given by Eq. (18–14); T_H is the temperature of the boiler, and T_C is the temperature of the condenser. The highest practical value of T_H is about 800 K, and the condenser temperature is determined by the temperature of its surroundings, about 300 K. Hence $e_{Carnot} = 1 - (300 \text{ K})/(800 \text{ K}) = 0.63$, or 63%. In practice, the turbine efficiency is less than this ideal maximum, about 50%. The efficiency of large electrical generators in converting mechanical power to electrical is very high, typically 99%. Since efficiency is the ratio of the useful energy output to the energy input, the overall thermal efficiency of such a plant is the product of the individual efficiencies; this is roughly $(0.85)(0.50)(0.99)$, or about 42%.

EXAMPLE 18–12

The Tennessee Valley Authority's Paradise power plant has a generating capacity of about 1.3 gigawatt (1.3×10^9 W) of electric power. The plant burns 10,500 tons of coal per day; the heat of combustion of coal is about 6.2×10^6 J/kg. What is the overall thermal efficiency of the plant?

SOLUTION To calculate the thermal efficiency e, we need to know the heat input and energy output per time interval. One ton is 2000 lb, and one pound has a mass of 2.2 kg, so (10,500 tons)(2000 lb/ton)(2.2 kg/lb) = 4.62×10^7 kg of coal is used daily. Hence the heat supplied to the plant each day by burning coal is (6.2×10^6 J/kg)(4.62×10^7 kg) = 2.86×10^{14} J. The energy output in 1 d = 86,400 s is (1.3×10^9 J/s)(86,400 s) = 1.12×10^{14} J. Hence the overall efficiency is

$$e = \frac{W}{Q_H} = \frac{1.12 \times 10^{14} \text{ J}}{2.86 \times 10^{14} \text{ J}} = 0.39 = 39\%.$$

The efficiency of the steam turbine itself is the main limiting factor on the overall efficiency.

In a nuclear power plant, the heat to generate steam is supplied by a nuclear reaction, the fission of uranium, rather than the chemical reaction of burning coal. The steam turbines in nuclear power plants have the same theoretical efficiency limit as in coal-fired plants. At present it is not practical to run nuclear reactors at temperatures and pressures as high as those in coal boilers, so the net thermal efficiency of a nuclear plant is somewhat lower, typically 30%.

In 1990, 18% of the world's electric power supply came from nuclear power plants (20% in the United States). High construction costs, questions of public safety, and the problem of disposal of radioactive waste have slowed the development of additional nuclear power facilities. However, there is also increasing concern about the environmental impact of conventional coal-burning power plants. Coal smoke is responsible for serious and well-documented environmental hazards such as acid rain, and it contributes to global warming through the greenhouse effect. It is important to manage our development of energy resources so as to minimize the risks to human life and health, and nuclear power continues to receive attention as a future energy resource. We noted at the end of Section 18–7 that thermal pollution from both coal-fired and nuclear plants poses serious environmental problems.

Solar energy has promise for a variety of purposes. The power per unit surface area in the sun's radiation (before it passes through the earth's atmosphere) is about 1.4 kW/m². A maximum of about 1.0 kW/m² reaches the surface of the earth on a clear day, and the time average, over a 24-hour period, is about 0.2 kW/m². This radiation can be collected and used to heat water (Fig. 18–17a). A different solar-energy scheme is to use large banks of photovoltaic cells (Fig. 18–17b) for direct conversion of solar energy to electricity. A photovoltaic cell is not a heat engine in the usual sense and is not limited by the Carnot efficiency. There are other fundamental limitations on photocell efficiency, however; 25% is currently attainable, and 50% may be practicable with multilayer semiconductor photocells. Because of high capital costs, the price of photovoltaic-produced energy is about 25 cents per kilowatt-hour (compared with about 5 cents for conventional coal-fired steam-turbine plants). But such a system offers many

Absorber plate covered by glass

Tubes carry heated fluid

Insulation

(a)

(b)

18–17 (a) A passive solar collecting panel for home water heating. (b) An array of photovoltaic cells for direct conversion of sun power to electric power.

potential advantages, such as lack of noise, lack of moving parts, minimal maintenance requirements, and reduced pollution. Capital costs are likely to drop with improved technology, and photovoltaic systems continue to be studied and developed.

EXAMPLE 18–13

A solar water heater system is to be installed in a single-family dwelling to provide 300 L of 55°C hot water per day (normal use for a family of four). Estimate the required surface area of the collecting panels, assuming that water is supplied to the house at 15°C, that the average power per unit area from the sun is 130 W/m^2 (typical of much of the central United States), and that the collecting panels are 60% efficient.

SOLUTION The energy that must be supplied to heat 300 L of water (mass 300 kg) from 15°C to 55°C is

$$Q = mc\,\Delta T = (300 \text{ kg})(4190 \text{ J/kg} \cdot \text{C°})(55°\text{C} - 15°\text{C})$$

$$= 5.03 \times 10^7 \text{ J}.$$

The solar energy received by the collecting panels in one day is the average power P per unit area, multiplied by the time t (one day, or 86,400 s) and multiplied by the (unknown) panel area A. The heat Q delivered to the water is a fraction $e = 0.60$ of this, so

$$Q = ePtA,$$

$$A = \frac{Q}{ePt} = \frac{5.03 \times 10^7 \text{ J}}{(0.60)(130 \text{ W/m}^2)(86,400 \text{ s})} = 7.5 \text{ m}^2.$$

This area will fit handily onto the roof of the house.

EXAMPLE 18–14

The top of a small car (total area 3.0 m^2) is covered with photovoltaic cells that supply energy to an electric motor. It is a cloudless day, so the solar power received per unit area is 1.0 kW/m^2. If the efficiency of the photovoltaic cells is 25%, what is the power delivered to the motor?

SOLUTION The delivered power is the solar power per unit area,

multiplied by the collecting area and by the efficiency:

$$P = (1000 \text{ W/m}^2)(3.0 \text{ m}^2)(0.25)$$

$$= 750 \text{ W} = 1.0 \text{ hp}.$$

This is a very small amount of power for a car, so a solar-powered vehicle must be lightweight and have very low air and rolling resistance. The power is even less on an overcast day.

About 1% of the solar energy reaching the earth is transformed into wind energy. This can be gathered and converted by "forests" of wind turbines (Fig. 18–18). A typical turbine may have a blade diameter of 40 m; in an 8-m/s wind, such a turbine can extract about 400 kW from the air, of which about one-fourth can be converted to electric power. This is enough to provide the electric power needs of about 30 homes. About 7500 wind turbines (85% of the U.S. total) are located in California, where the state government encourages their use. As with photovoltaic systems, the capital costs of such systems are higher than for coal-fired generating plants of equal capacity, although some installations are producing energy at a cost of about 7 cents per kilowatt-hour.

Biomass is yet another mechanism for using solar energy. Plants absorb solar energy and store it; the energy can be utilized by burning or by fermentation and distillation to make alcohols. Ethanol made from fermented corn is used in the United States as a gasoline additive, although its cost is not quite competitive at present crude-oil price levels. An important feature of biomass materials is that they take as much carbon dioxide out of the atmosphere when they grow as is released when they burn; thus they don't contribute to the greenhouse effect.

An indirect scheme for collection and conversion of solar energy would use the temperature gradient in the ocean. In the Gulf of Mexico, for example, the water temperature near the surface is about 25°C, while at a depth of a few hundred meters it is about 5°C. The second law of thermodynamics forbids cooling the ocean and converting the extracted heat completely into work, but there is nothing to forbid running a heat engine between these two temperatures. The thermodynamic efficiency would be very low (less than 7%), but since two-thirds of the earth's surface is ocean, a vast supply of energy would be available.

18–18 An array of wind turbines to collect and convert wind and energy. The propellerlike blades turn electric generators, converting the kinetic energy of moving air into electrical energy.

The present level of activity in energy-conversion research is rather low in comparison to the 1970s, when the industrialized countries experienced a serious energy shortage. As we use up our available fossil-fuel resources, we will certainly have to become more active in developing alternative energy sources. The principles of thermodynamics will play a central role in this development.

SUMMARY

KEY TERMS

- A reversible process is one whose direction can be reversed by an infinitesimal change in the conditions of the process. A reversible process is also an equilibrium process.

- A heat engine takes heat Q_H from a source, converts part of it to work W, and discards the remainder $|Q_C|$ at a lower temperature. The thermal efficiency e of a heat engine is

$$e = \frac{W}{Q_H} = 1 + \frac{Q_C}{Q_H} = 1 - \left|\frac{Q_C}{Q_H}\right|. \qquad (18\text{-}4)$$

- A gasoline engine operating on the Otto cycle with compression ratio r has a theoretical maximum thermal efficiency e given by

$$e = 1 - \frac{1}{r^{\gamma-1}}. \qquad (18\text{-}6)$$

- A refrigerator takes heat Q_C from a colder place, has a work input $|W|$, and discards heat $|Q_H|$ at a warmer place. The coefficient of performance K is defined as

$$K = \frac{|Q_C|}{|W|} = \frac{|Q_C|}{|Q_H| - |Q_C|}. \qquad (18\text{-}9)$$

- The second law of thermodynamics describes the directionality of natural thermodynamic processes. It can be stated in several equivalent forms. The *engine* statement is that no cyclic process can convert heat completely into work. The *refrigerator* statement is that no cyclic process can transfer heat from a colder place to a hotter place with no input of mechanical work.

- The Carnot cycle operates between two heat reservoirs at temperatures T_H and T_C and uses only reversible processes. Its thermal efficiency is

$$e_{\text{Carnot}} = 1 - \frac{T_C}{T_H} = \frac{T_H - T_C}{T_H}. \qquad (18\text{-}14)$$

Two additional equivalent statements of the second law are that no engine operating between the same two temperatures can be more efficient than a Carnot engine and that all Carnot engines operating between the same two temperatures have the same efficiency.

- A Carnot engine run backward is a Carnot refrigerator. Its coefficient of performance is

$$K_{\text{Carnot}} = \frac{T_C}{T_H - T_C}. \qquad (18\text{-}15)$$

Another form of the second law is that no refrigerator operating between the same two temperatures can have a larger coefficient of performance than a Carnot refrigerator. All Carnot refrigerators operating between the same two temperatures have the same coefficient of performance.

KEY TERMS

irreversible process, 559
reversible process, 559
equilibrium process, 560
heat engine, 560
working substance, 560
cyclic process, 561
thermal efficiency, 561
compression ratio, 563
Otto cycle, 563
Diesel cycle, 565
refrigerator, 565
coefficient of performance, 565
energy efficiency rating, 566
heat pump, 567
second law of thermodynamics, 567
Carnot cycle, 569
thermal pollution, 573
Kelvin temperature scale, 574
absolute zero, 574
entropy, 575
macroscopic state, 579
microscopic state, 580

■ The Kelvin temperature scale is based on the efficiency of the Carnot cycle and is independent of the properties of any specific material. The zero point on the Kelvin scale is called absolute zero.

■ Entropy is a quantitative measure of the disorder of a system. The entropy change in any reversible process is

$$\Delta S = \int_{1}^{2} \frac{dQ}{T} \quad \text{(reversible process)}, \quad (18\text{–}19)$$

where T is absolute temperature. Entropy depends only on the state of the system, and the change in entropy between given initial and final states is the same for all processes leading from one to the other. This fact can be used to find the entropy change in an irreversible process.

■ The second law can be stated in terms of entropy S: The entropy of an isolated system may increase but can never decrease. When a system interacts with its surroundings, the total entropy change of system and surroundings can never decrease. When the interaction involves only reversible processes, the total entropy is constant and $\Delta S = 0$; when there is any irreversible process, the total entropy increases and $\Delta S > 0$.

■ When a system is in a particular macroscopic state, the particles that make up the system may be in any of w possible microscopic states. The greater the number w, the greater the entropy:

$$S = k \ln w. \quad (18\text{–}22)$$

DISCUSSION QUESTIONS

Q18–1 A pot is half-filled with water, and a lid is placed on the pot, forming a tight seal so that no water vapor can escape. The pot is heated on a stove, forming water vapor inside the pot. The heat is then turned off and the water vapor condenses back to liquid. Is this cycle reversible or irreversible? Why?

Q18–2 Give two examples of reversible processes and two examples of irreversible processes in purely mechanical systems, such as blocks sliding on planes, springs, pulleys, and strings. Explain what makes each process reversible or irreversible.

Q18–3 What irreversible processes occur in a gasoline engine? Why are they irreversible?

Q18–4 Suppose you try to cool the kitchen of your house by leaving the refrigerator door open. What happens? Why? Would the result be the same if you left open a picnic cooler full of ice? Explain the reason for any differences.

Q18–5 A member of the U.S. Congress proposed a scheme to produce energy as follows. Water molecules (H_2O) are to be broken apart to produce hydrogen and oxygen. The hydrogen is then burned (that is, combined with oxygen), releasing energy in the process. The only product of this combustion is water, so there is no pollution. In light of the second law of thermodynamics, what do you think of this energy-producing scheme?

Q18–6 Some critics of biological evolution claim that it violates the second law of thermodynamics, since evolution involves simple life forms developing into more complex, more highly ordered organisms. Explain why this is not a valid argument against evolution.

Q18–7 Why must a room air conditioner be placed in a window? Why can't it just be set on the floor and plugged in?

Q18–8 Is it a violation of the second law of thermodynamics to convert mechanical energy completely into heat? To convert heat completely into work? Explain your answers.

Q18–9 A growing plant creates a highly complex and organized structure out of simple materials such as air, water, and trace minerals. Does this violate the second law of thermodynamics? Why or why not? What is the plant's ultimate source of energy? Explain your reasoning.

Q18–10 Imagine a special air filter placed in a window of a house. The tiny holes in the filter only allow air molecules moving faster than a certain speed to exit the house, and only allows air molecules moving slower than that speed to enter the house from outside. Explain why such an air filter would cool the house, and why the second law of thermodynamics makes building such a filter an impossible task.

Q18–11 An electric motor has its shaft coupled to that of an electric generator. The motor drives the generator, and some current from the generator is used to run the motor. The excess current is used to light a home. What is wrong with this scheme?

Q18–12 When the sun shines on a glass-roofed greenhouse, the temperature becomes higher inside than outside. Does this phenomenon violate the second law of thermodynamics? Why or why not?

Q18–13 When a wet cloth is hung up in a hot wind in the desert, it is cooled by evaporation to a temperature that may be

20 C° or so below that of the air. Discuss this process in light of the second law of thermodynamics.

Q18–14 If no real engine can be as efficient as a Carnot engine operating between the same two temperatures, what is the point of developing and using Eq. (18–14)?

Q18–15 Suppose you want to increase the efficiency of a heat engine. Would it be better to increase T_H or to decrease T_C by an equal amount? Why?

Q18–16 What would be the efficiency of a Carnot engine operating with $T_H = T_C$? What would be the efficiency if $T_C = 0$ K and T_H were any temperature above 0 K? Interpret your answers.

Q18–17 Real heat engines, like the gasoline engine in a car, always have some friction between their moving parts, although lubricants keep the friction to a minimum. Would a heat engine with completely frictionless parts be 100% efficient? Why or why not? Does the answer depend on whether or not the engine runs on the Carnot cycle? Again, why or why not?

Q18–18 Does a refrigerator full of food consume more power if the room temperature is 20°C than if it is 15°C?

Or is the power consumption the same? Explain your reasoning.

Q18–19 Materials scientists are developing ceramics that are as strong as metals but can withstand much higher temperatures before melting. Explain the usefulness of such ceramics in engine design.

*Q18–20 Explain why each of the following processes are examples of increasing disorder or randomness: mixing hot and cold water; free expansion of a gas; irreversible heat flow; and developing heat by mechanical friction. Are entropy increases involved in all of these? Why or why not?

*Q18–21 Are the earth and sun in thermal equilibrium? Are there entropy changes associated with the transmission of energy from the sun to the earth? Does radiation differ from other modes of heat transfer with respect to entropy changes? Explain your reasoning.

*Q18–22 Discuss the entropy changes involved in the preparation and consumption of a hot-fudge sundae.

EXERCISES

SECTION 18–3 HEAT ENGINES

18–1 A diesel engine performs 2200 J of mechanical work and discards 4300 J of heat each cycle. a) How much heat must be supplied to the engine in each cycle? b) What is the thermal efficiency of the engine?

18–2 An aircraft engine takes in 9000 J of heat and discards 6400 J each cycle. a) What is the mechanical work output of the engine during one cycle? b) What is the thermal efficiency of the engine?

18–3 A Gasoline Engine. A gasoline engine takes in 16,100 J of heat and delivers 3700 J of work per cycle. The heat is obtained by burning gasoline with a heat of combustion of 4.60×10^4 J/g. a) What is the thermal efficiency? b) How much heat is discarded in each cycle? c) What mass of fuel is burned in each cycle? d) If the engine goes through 60.0 cycles per second, what is its power output in kilowatts? In horsepower?

18–4 A gasoline engine has a power output of 180 kW (about 241 hp). Its thermal efficiency is 28.0%. a) How much heat must be supplied to the engine per second? b) How much heat is discarded by the engine per second?

18–5 A certain nuclear-power plant has a mechanical-power output (used to drive an electric generator) of 330 MW. Its rate of heat input from the nuclear reactor is 1300 MW. a) What is the thermal efficiency of the system? b) At what rate is heat discarded by the system?

SECTION 18–4 INTERNAL-COMBUSTION ENGINES

18–6 What compression ratio r must an Otto cycle have to achieve an ideal efficiency of 65.0% if $\gamma = 1.40$?

18–7 For an Otto cycle with $\gamma = 1.40$ and $r = 9.50$, the temperature of the gasoline-air mixture when it enters the cylinder is 22.0°C (point a of Fig. 18–3). a) What is the temperature at the end of the compression stroke (point b)? b) The initial pressure of the gasoline-air mixture (point a) is 8.50×10^4 Pa, slightly

below atmospheric pressure. What is the pressure at the end of the compression stroke?

18–8 The Otto-cycle engine in a Mercedes-Benz SLK230 has a compression ratio of 8.8. a) What is the ideal efficiency of the engine? Use $\gamma = 1.40$. b) The engine in a Dodge Viper GT2 has a slightly higher compression ratio of 9.6. How much increase in the ideal efficiency results from this increase in the compression ratio?

SECTION 18–5 REFRIGERATORS

18–9 A refrigerator has a coefficient of performance of 2.10. Each cycle it absorbs 3.40×10^4 J of heat from the cold reservoir. a) How much mechanical energy is required each cycle to operate the refrigerator? b) During each cycle, how much heat is discarded to the high-temperature reservoir?

18–10 Liquid refrigerant at a pressure of 1.34×10^5 Pa leaves the expansion valve of a refrigerator at −23.0°C. It then flows through the evaporation coils inside the refrigerator and leaves as vapor at the same pressure and at −20.5°C, the same temperature as the inside of the refrigerator. The boiling point of the refrigerant at this pressure is −23.0°C, the heat of vaporization is 1.60×10^5 J/kg, and the specific heat capacity of the vapor at constant pressure is 485 J/kg · K. The coefficient of performance of the refrigerator is $K = 2.8$. If 8.00 kg of refrigerant flows through the refrigerator each hour, find the electric power that must be supplied to the refrigerator.

18–11 A window air conditioner unit absorbs 9.80×10^4 J of heat per minute from the room being cooled and in the same time period deposits 1.44×10^5 J of heat into the outside air. a) What is the power consumption of the unit in watts? b) What is the energy efficiency rating of the unit?

18–12 A freezer has a coefficient of performance of 2.40. The freezer is to convert 1.80 kg of water at 25.0°C to 1.80 kg of ice at −5.0°C in one hour. a) What amount of heat must be removed from the water at 25.0°C to convert it to ice at –5.0°C? b) How much electrical energy is consumed by the freezer during this hour? c) How much wasted heat is rejected to the room in which the freezer sits?

SECTION 18–7 THE CARNOT CYCLE

18–13 A Carnot engine whose high-temperature reservoir is at 620 K takes in 550 J of heat at this temperature in each cycle and gives up 335 J to the low-temperature reservoir. a) How much mechanical work does the engine perform during each cycle? b) What is the temperature of the low-temperature reservoir? c) What is the thermal efficiency of the cycle?

18–14 A Carnot engine is operated between two heat reservoirs at temperatures of 520 K and 300 K. a) If the engine receives 6.45 kJ of heat energy from the reservoir at 520 K in each cycle, how many joules per cycle does it reject to the reservoir at 300 K? b) How much mechanical work is performed by the engine during each cycle? c) What is the thermal efficiency of the engine?

18–15 An ice-making machine operates in a Carnot cycle. It takes heat from water at 0.0°C and rejects heat to a room at 24.0°C. Suppose that 85.0 kg of water at 0.0°C are converted to ice at 0.0°C. a) How much heat is rejected to the room? b) How much energy must be supplied to the device?

18–16 A Carnot refrigerator is operated between two heat reservoirs at temperatures of 320 K and 270 K. a) If in each cycle the refrigerator receives 415 J of heat energy from the reservoir at 270 K, how many joules of heat energy does it deliver to the reservoir at 320 K? b) If the refrigerator goes through 165 cycles each minute, what power input is required to operate the refrigerator? c) What is the coefficient of performance of the refrigerator?

18–17 A Carnot device extracts 5.00 kJ of heat from a body at –10.0°C. How much work is done if the device exhausts heat into the environment at a) 25.0°C; b) 0.0°C; c) –25.0°C? In each case, is the device acting as an engine or as a refrigerator?

18–18 An inventor claims to have developed an engine that in each cycle takes in 2.60×10^8 J of heat at a temperature of 400 K, does 42.0 kWh of mechanical work, and rejects heat at a temperature of 250 K. Would you advise investing money to put this engine on the market? Why or why not?

18–19 a) Show that the efficiency e of a Carnot engine and the coefficient of performance K of a Carnot refrigerator are related by $K = (1 - e)/e$. The engine and refrigerator operate between the same hot and cold reservoirs. b) What is K for the limiting values $e \to 1$ and $e \to 0$? Explain.

*SECTION 18–9 ENTROPY

*18–20 A sophomore with nothing better to do adds heat to 0.350 kg of ice at 0.0°C until it is all melted. a) What is the change in entropy of the water? b) The source of heat is a very massive body at a temperature of 25.0°C. What is the change in entropy of this body? c) What is the total change in entropy of the water and the heat source?

*18–21 Calculate the entropy change that occurs when 1.00 kg of water at 20.0°C is mixed with 2.00 kg of water at 80.0°C.

*18–22 Three moles of an ideal gas undergo a reversible isothermal compression at 20.0°C. During this compression, 1850 J of work is done on the gas. What is the change of entropy of the gas?

*18–23 What is the entropy change of 0.130 kg of helium gas at the normal boiling point of helium when it all condenses isothermally to 1.00 L of liquid helium? (*Hint:* See Table 15–4 in Section 15–7.)

*18–24 a) Calculate the change in entropy when 1.00 kg of water at 100°C is vaporized and converted to steam at 100°C. (See Table 15–4.) b) Compare your answer to the change in entropy when 1.00 kg of ice is melted at 0°C, calculated in Example 18–5 (Section 18–9). Is the entropy change greater for melting or for vaporization? Interpret your answer, using the idea that entropy is a measure of the randomness of a system.

*18–25 a) Calculate the change in entropy when 1.00 mol of water (molecular mass 18.0 g/mol) at 100°C evaporates to form water vapor at 100°C. b) Repeat the calculation of part (a) for 1.00 mol of liquid nitrogen, 1.00 mol of silver, and 1.00 mol of mercury when each is vaporized at its normal boiling point. (See Table 15–4 for the heats of vaporization, and Appendix D for the molar masses. Note that the nitrogen molecule is N_2.) c) Your results in parts (a) and (b) should be in relatively close agreement. (This is called the *rule of Drepez and Trouton.*) Explain why this should be so, using the idea that entropy is a measure of the randomness of a system.

*18–26 A block of copper with a mass of 3.50 kg, initially at 100.0°C, is dropped into 0.800 kg of water initially at 0.0°C. a) What is the final temperature of the system? b) What is the total change in entropy of the system?

*18–27 Two moles of an ideal gas undergo a reversible isothermal expansion from 0.0280 m³ to 0.0420 m³ at a temperature of 25.0°C. What is the change in entropy of the gas?

*SECTION 18–10 MICROSCOPIC INTERPRETATION OF ENTROPY

*18–28 A box is separated by a partition into two parts of equal volume. The left side of the box contains 500 molecules of nitrogen gas; the right side contains 100 molecules of oxygen gas. The two gases are at the same temperature. The partition is punctured, and equilibrium is eventually attained. Assume that the volume of the box is large enough for each gas to undergo a free expansion and not change temperature. a) On average, how many molecules of each type will there be in either half of the box? b) What is the change in entropy of the system when the partition is punctured? c) What is the probability that the molecules will be found in the same distribution as before the partition was punctured, that is, 500 nitrogen molecules in the left half and 100 oxygen molecules in the right half?

*18–29 Two moles of an ideal gas occupies a volume V. The gas expands isothermally and reversibly to a volume $3V$. a) Is the velocity distribution changed by the isothermal expansion? Explain. b) Use Eq. (18–23) to calculate the change in entropy of the gas. c) Use Eq. (18–18) to calculate the change in entropy of the gas. Compare this result to that obtained in part (b).

*18–30 You toss four identical coins on the floor. Each coin has an equal probability of showing heads or tails. a) What is the probability of all four coins showing heads? Of all four showing tails? b) What is the probability of three coins showing heads and one tails? What is the probability of three coins showing tails and one heads? c) What is the probability of two coins showing heads and two showing tails? d) What is the sum of the five probabilities calculated in parts (a) through (c)? Explain.

SECTION 18–11 ENERGY RESOURCES: A CASE STUDY IN THERMODYNAMICS

18–31 Solar Heating in Winter. A well-insulated house in Columbus, Ohio, has 150 m^2 of floor space and requires 1.50×10^{10} J of heat for the month of January. This heat is to be supplied by a solar collector with a collection efficiency of 60.0%. The average (night and day) solar energy input in Columbus in January is 65.7 W/m^2. What area of solar collector is required? Would the collector fit on the roof of the house?

18–32 a) A home owner in a cold climate has a coal-burning furnace that burns 4500 kg (about 5 tons) of coal during a winter. The coal used has a heat of combustion of 2.70×10^7 J/kg. If stack losses (the amount of heat energy lost up the chimney) are 20%, how many joules were actually used to heat the house? b) The home owner proposes to install a solar heating system, heating large tanks of water by solar radiation during the summer and using the stored energy for heating during the winter. Find the required dimensions of the storage tank to store a quantity of energy equal to that computed in part (a). Assume that the tank is a cube and that the water is raised to 49.0°C (120°F) in the summer and cooled to 27.0°C (81°F) in the winter.

18–33 The roof of a suburban house is equipped with 8.0 m^2 of passive solar collecting panels with collection efficiency 60%, used to heat water from 15.0°C to 55.0°C for household use. a) If the average solar-energy input is 150 W/m^2, what volume of water can be heated per hour? b) During an average day, a typical household uses about 75 L of 55.0°C hot water for each person. How many inhabitants can this water heating system accommodate?

18–34 A solar-power plant is to be built with a power output capacity of 850 MW. What land area must the solar-energy collectors occupy if they are a) photocells with 60% efficiency? b) mirrors that generate steam for a turbine-generator unit with overall efficiency of 30%? Take the average power in the sun's radiation to be 200 W/m^2 at the earth's surface. Express your answers in square kilometers and square miles.

18–35 A "solar house" has storage facilities for 4.00×10^9 J (about 4 million Btu). Compare the space requirements (in m^3) for this storage on the assumption a) that the energy is stored in water heated from a minimum temperature of 21.0°C (70°F) to a maximum of 49.0°C (120°F); b) that the energy is stored in Glauber salt heated in the same temperature range.

PROPERTIES OF GLAUBER SALT ($Na_2SO_4 \cdot 10H_2O$)

Specific heat capacity	
Solid	1930 J/kg · K
Liquid	2850 J/kg · K
Density	1600 kg/m^3
Melting point	32.0°C
Heat of fusion	2.42×10^5 J/kg

c) What is the advantage of using Glauber salt?

18–36 A coal-fired steam-turbine power plant has a mechanical-power output of 1100 MW and a thermal efficiency of 35.0%. a) At what rate must heat be supplied by burning coal? b) If West Virginia coal is used, which has a heat of combustion of 3.00×10^4 J/g, what mass of coal is burned per second? Per day? c) At what rate is heat discarded by the system? d) If the discarded heat is given to water in a river, and the water temperature is to rise by no more than 4.0 C°, what volume of water is needed per second? e) In part (d), if the river is of rectangular cross section with depth 5.0 m and width 100 m, what must be the speed of flow of the water?

18–37 Electric vs. Conventional Automobiles. a) Consider an energy-conversion process that involves two steps, such as using heat to boil water and using the steam to turn an electric generator. Each step has its own efficiency. Is the efficiency of the overall process equal to the sum of the two efficiencies, or the difference, or the product, or what? Explain your reasoning. b) Conventional automobiles are about 15% efficient; that is, only about 15% of the energy released by burning fuel is eventually converted to kinetic energy of the automobile. In an electric automobile with a battery-powered motor, the energy comes from a power plant that supplies the electricity to charge the batteries. Calculate the efficiency of an electric automobile using the following data: i) A typical electric power plant has an efficiency of 40%; ii) 10% of the energy supplied by the power plant is lost in transmission to the city; iii) 10% of the energy received by the city is lost in distribution within the city; iv) the efficiency of the battery-charging process is 80%; v) the efficiency of the electric motor is 90%. c) How does the overall efficiency compare to that of a conventional automobile?

PROBLEMS

18–38 A Carnot engine whose low-temperature reservoir is at −90.0°C has an efficiency of 40.0%. An engineer is assigned the problem of increasing this to 45.0%. a) By how many Celsius degrees must the temperature of the high-temperature reservoir be increased if the temperature of the low-temperature reservoir remains constant? b) By how many Celsius degrees must the temperature of the low-temperature reservoir be decreased if that of the high-temperature reservoir remains constant?

18–39 A heat engine takes 0.350 mol of a diatomic ideal gas around the cycle shown in the pV-diagram of Fig. 18–19. Process $1 \rightarrow 2$ is at constant volume, process $2 \rightarrow 3$ is adiabatic, and process $3 \rightarrow 1$ is at a constant pressure of 1.00 atm. The value of γ for this gas is 1.40. a) Find the pressure and volume at points 1, 2, and 3. b) Calculate Q, W, and ΔU for each of the three processes. c) Find the net work done by the gas in the cycle. d) Find the net heat flow into the engine in one cycle. e) What is the thermal efficiency of the engine? How does this compare to the efficiency of a Carnot-cycle engine operating between the same minimum and maximum temperatures T_1 and T_2?

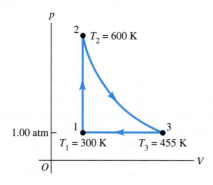

FIGURE 18–19 Problem 18–39.

18–40 An experimental power plant at the Natural Energy Laboratory of Hawaii generates electricity from the temperature gradient of the ocean. The surface and deep-water temperatures are 27°C and 6°C, respectively. a) What is the maximum theoretical efficiency of this power plant? b) If the power plant is to produce 210 kW of power, at what rate must heat be extracted from the warm water? At what rate must heat be absorbed by the cold water? Assume the maximum theoretical efficiency. c) The cold water that enters the plant leaves at a temperature of 10°C. What must be the flow rate of cold water through the system? Give your answer in kg/h and L/h.

18–41 What is the thermal efficiency of an engine that operates by taking n moles of diatomic ideal gas through the cycle $1 \rightarrow 2 \rightarrow 3 \rightarrow 4 \rightarrow 1$ shown in Fig. 18–20?

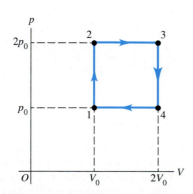

FIGURE 18–20 Problem 18–41.

18–42 A cylinder contains oxygen at a pressure of 2.00 atm. The volume is 4.00 L, and the temperature is 300 K. Assume that the oxygen may be treated as an ideal gas. The oxygen is carried through the following processes:

(i) Heated at constant pressure from the initial state (state 1) to state 2, which has $T = 450$ K.
(ii) Cooled at constant volume to 250 K (state 3).
(iii) Compressed at constant temperature to a volume of 4.00 L (state 4).
(iv) Heated at constant volume to 300 K, which takes the system back to state 1.

a) Show these four processes in a pV-diagram, giving the numerical values of p and V in each of the four states. b) Calculate Q and W for each of the four processes. c) Calculate the net work done by the oxygen. d) What is the efficiency of this device as a heat engine? How does this compare to the efficiency of a Carnot-cycle engine operating between the same minimum and maximum temperatures of 250 K and 450 K?

18–43 Thermodynamic Processes for a Refrigerator. A refrigerator operates on the cycle shown in Fig. 18–21. The compression ($d \rightarrow a$) and expansion ($b \rightarrow c$) steps are adiabatic. The temperature, pressure, and volume of the coolant in each of the four states a, b, c, and d are given in the table below.

State	T (°C)	P (kPa)	V (m³)	U (kJ)	Percentage That Is Liquid
a	80	2305	0.0682	1969	0
b	80	2305	0.00946	1171	100
c	5	363	0.2202	1005	54
d	5	363	0.4513	1657	5

a) In each cycle, how much heat is taken from inside the refrigerator into the coolant while the coolant is in the evaporator? b) In each cycle, how much heat is exhausted from the coolant into the air outside the refrigerator while the coolant is in the condenser? c) In each cycle, how much work is done by the motor that operates the compressor? d) Calculate the coefficient of performance of the refrigerator.

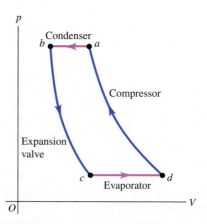

FIGURE 18–21 Problem 18–43.

18–44 A monatomic ideal gas is taken around the cycle shown in Fig. 18–22 in the direction shown in the figure. The path for process $c \to a$ is a straight line in the pV-diagram. a) Calculate Q, W, and ΔU for each process $a \to b$, $b \to c$, and $c \to a$. b) What are Q, W, and ΔU for one complete cycle? c) What is the efficiency of the cycle?

FIGURE 18–22 Problem 18–44.

18–45 A Stirling-cycle Engine. The *Stirling cycle* is similar to the Otto cycle, except that the compression and expansion of the gas is done at constant temperature, not adiabatically as in the Otto cycle. The Stirling cycle is used in *external* combustion engines, which means that the gas inside the cylinder is not used in the combustion process. Heat is supplied by burning fuel steadily outside the cylinder, instead of explosively inside the cylinder as in the Otto cycle. For this reason Stirling-cycle engines are quieter than Otto-cycle engines, since there are no intake and exhaust valves (a major source of engine noise). While small Stirling engines are used for a variety of purposes, Stirling engines for automobiles have not been successful because they are larger, heavier, and more expensive than conventional automobile engines. In the cycle, the working fluid goes through the following sequence of steps (Fig. 18–23):

(i) Compressed isothermally at temperature T_1 from the initial state a to state b, with a compression ratio r.
(ii) Heated at constant volume to state c at temperature T_2.
(iii) Expanded isothermally at T_2 to state d.
(iv) Cooled at constant volume back to the initial state a.

Assume that the working fluid is n moles of an ideal gas (for which C_V is independent of temperature). a) Calculate Q, W, and ΔU for each of the processes $a \to b$, $b \to c$, $c \to d$, and $d \to a$. b) In the Stirling cycle, the heat transfers in the processes $b \to c$ and $d \to a$ do not involve external heat sources, but rather use *regeneration*: the same substance that transfers heat to the gas inside the cylinder in the process $b \to c$ also absorbs heat back from the gas in the process $d \to a$. Hence the heat transfers $Q_{b \to c}$ and $Q_{d \to a}$ do not play a role in determining the efficiency of the engine. Explain this last statement by comparing the expressions for $Q_{b \to c}$ and $Q_{d \to a}$ calculated in part (a). c) Calculate the efficiency of a Stirling-cycle engine in terms of the temperatures T_1 and T_2. How does this compare to the efficiency of a Carnot-cycle engine operating between these same two temperatures? (Historically, the Stirling cycle was devised before the

Carnot cycle.) Does this result violate the second law of thermodynamics? Explain. Unfortunately, actual Stirling-cycle engines cannot achieve this efficiency, due to problems with the heat-transfer processes and pressure losses in the engine.

FIGURE 18–23 Problem 18–45.

18–46 A Carnot engine operates between two heat reservoirs at temperatures T_H and T_C. An inventor proposes to increase the efficiency by running one engine between T_H and an intermediate temperature T' and a second engine between T' and T_C using as input the heat expelled by the first engine. Compute the efficiency of this composite system and compare it to that of the original engine.

18–47 The maximum power that can be extracted by a wind turbine from an air stream is approximately

$$P = kd^2 v^3,$$

where d is the blade diameter, v is the wind speed, and the constant $k = 0.5 \ \text{W} \cdot \text{s}^3/\text{m}^5$. a) Explain the dependence of P on d and on v by considering a cylinder of air that passes over the turbine blades in time t (Fig. 18–24). This cylinder has diameter d, length $L = vt$, and density ρ. b) The Mod–5B wind turbine at Kahaku on the Hawaiian island of Oahu has a blade diameter of 97 m (slightly longer than a football field) and sits atop a 58-m tower. It can produce 3.2 MW of electric power. Assuming 25% efficiency, what wind speed is required to produce this amount of power? Give your answer in m/s and in km/h. c) Commercial wind turbines are commonly located in or downwind of mountain passes. Why?

FIGURE 18–24 Problem 18–47.

18–48 Fuel Economy and Automobile Performance. The Otto-cycle engine of a Volvo V70 has a compression ratio $r = 8.5$. The U.S. Environmental Protection Agency (EPA) fuel economy rating of this car is 25 miles per gallon on the highway (105 km/h, or 65 mi/h). Gasoline has a heat of combustion of

4.60×10^7 J/kg, and its density is 740 kg/m³. a) At 105 km/h, what is the rate of gasoline consumption in L/h? b) What is the theoretical efficiency of the engine? Use $\gamma = 1.40$. c) How much power is the engine producing at 105 km/h? Assume that the engine is operating at its theoretical efficiency, and give your answer in watts and in horsepower. By comparison, the engine of the Volvo V70 has a maximum power of 236 hp. d) Due to heat losses and friction in the drive train, the actual efficiency is approximately 15%. Repeat part (c) using this information. What fraction of the maximum possible power is used for highway driving?

18–49 Automotive Thermodynamics. A Volkswagen Passat has a six-cylinder Otto-cycle engine with compression ratio $r = 10.6$. The diameter of each cylinder, called the *bore* of the engine, is 82.5 mm. The distance that the piston moves during the compression in Fig. 18–2, called the *stroke* of the engine, is 86.4 mm. The initial pressure of the air-fuel mixture (at point *a* in Fig. 18–3) is 8.50×10^4 Pa, and the initial temperature is 300 K (the same as the outside air). Assume that 200 J of heat is added to each cylinder in each cycle by the burning gasoline, and that the gas has $C_V = 20.5$ J/mol · K and $\gamma = 1.40$. a) Calculate the total work done in one cycle in each cylinder of the engine, and the heat released when the gas is cooled to the temperature of the outside air. b) Calculate the volume of the air-fuel mixture at point *a* in the cycle. c) Calculate the pressure, volume, and temperature of the gas at points *b*, *c*, and *d* in the cycle. In a *pV*-diagram, show the numerical values of *p*, *V*, and *T* for each of the four states. d) Compare the efficiency of this engine with the efficiency of a Carnot-cycle engine operating between the same maximum and minimum temperatures.

*18–50 Recalculate the entropy change between points *a* and *b* of Fig. 18–12c if the reversible path is a) an isobaric expansion to 2V followed by an isochoric process; b) an isochoric cooling to $p_a/2$ followed by an isobaric expansion.

*18–51 A 0.0500-kg cube of ice at an initial temperature of −15.0°C is placed in 0.600 kg of water at $T = 45.0$°C in an insulated container of negligible mass. Calculate the entropy change of the system.

*18–52 a) For the Otto cycle shown in Fig. 18–3, calculate the entropy changes of the gas in each of the constant–volume processes $b \rightarrow c$ and $d \rightarrow a$ in terms of the temperatures T_a, T_b, T_c, and T_d and the number of moles *n* and the heat capacity C_V of the gas. b) What is the total entropy change in the engine during one cycle? (*Hint:* Use the relation between T_a and T_b and between T_d and T_c.) c) The processes $b \rightarrow c$ and $d \rightarrow a$ occur

irreversibly in a real Otto engine. Explain how can this be reconciled with your result in part (b).

*18–53 A *TS*-Diagram. a) Graph a Carnot cycle, plotting Kelvin temperature vertically and entropy horizontally. This is called a temperature-entropy diagram, or *TS*-diagram. b) Show that the area under any curve representing a reversible path in a temperature-entropy diagram represents the heat absorbed by the system. c) Derive from your diagram the expression for the thermal efficiency of a Carnot cycle. d) Draw a temperature-entropy diagram for the Stirling cycle, described in Problem 18–45. Use this diagram to relate the efficiency of the Carnot and Stirling cycles.

*18–54 A physics student immerses one end of a copper rod in boiling water at 100°C and the other end in an ice-water mixture at 0°C. The sides of the rod are insulated. After steady-state conditions have been achieved in the rod, 0.160 kg of ice melts in a certain time interval. For this time interval find a) the entropy change of the boiling water; b) the entropy change of the ice-water mixture; c) the entropy change of the copper rod; d) the total entropy change of the entire system.

*18–55 To heat one cup of water (250 cm³) to make coffee, you place an electric heating element in the cup. As the water temperature increases from 20°C to 65°C, the temperature of the heating element remains at a constant 120°C. Calculate the entropy change of a) the water; b) the heating element; c) the system of water and heating element. (Make the same assumption about the specific heat of water as in Example 18–10 (Section 18–9), and neglect the heat that flows into the ceramic coffee cup itself.) d) Is this process reversible or irreversible? Explain.

*18–56 An object of mass m_1, specific heat capacity c_1, and temperature T_1 is placed in contact with a second object of mass m_2, specific heat capacity c_2, and temperature $T_2 > T_1$. As a result, the temperature of the first object increases to T and the temperature of the second object decreases to T'. a) Show that the entropy increase of the system is

$$\Delta S = m_1 c_1 \ln \frac{T}{T_1} + m_2 c_2 \ln \frac{T'}{T_2},$$

and show that energy conservation requires that

$$m_1 c_1 (T - T_1) = m_2 c_2 (T_2 - T').$$

b) Show that the entropy change ΔS, considered as a function of *T*, is a *maximum* if $T = T'$, which is just the condition of thermodynamic equilibrium. c) Discuss the result of part (b) in terms of the idea of entropy as a measure of disorder.

CHALLENGE PROBLEM

18–57 Consider a diesel cycle that starts (at point *a* in Fig. 18–4) with air at temperature T_a. The air may be treated as an ideal gas. a) If the temperature at point *c* is T_c, derive an expres-

sion for the efficiency of the cycle in terms of the compression ratio *r*. b) What is the efficiency if $T_a = 300$ K, $T_c = 950$ K, $\gamma = 1.40$, and $r = 21.0$?

When an earthquake strikes, the news of the event travels through the body of the earth in the form of seismic waves. By studying such waves, geophysicists learn about the internal structure of the earth and where future earthquake activity is likely to occur.

19

Mechanical Waves

19–1 INTRODUCTION

When you're at the beach, enjoying the ocean surf, you're experiencing a wave motion. Ripples on a pond, musical sounds we can hear, other sounds we *can't* hear, the wiggles of a Slinky™ stretched out on the floor—all these are *wave* phenomena. Waves can occur whenever a system is disturbed from its equilibrium position and when the disturbance can travel or *propagate* from one region of the system to another. Sound, light, ocean waves, radio and television transmission, and earthquakes are all wave phenomena. Waves are important in all branches of physical and biological science; indeed, the wave concept is one of the most important unifying threads running through the entire fabric of the natural sciences.

This chapter and the next two are about mechanical waves, waves that travel within some material called a *medium*. We'll begin by deriving the basic equations for describing waves, including the important special case of *periodic* waves in which the pattern of the wave repeats itself as the wave propagates. Chapter 20 deals with what happens when two or more waves occupy the same space, giving rise to *interference,* and Chapter 21 is concerned with a particularly important type of mechanical wave called *sound.*

Not all waves are mechanical in nature. Another broad class is *electromagnetic* waves, including light, radio waves, infrared and ultraviolet radiation, x-rays, and gamma rays. *No* medium is needed for electromagnetic waves; they can travel through empty space. Yet another class of wave phenomena is the wavelike behavior of atomic and subatomic particles. This behavior forms part of the foundation of quantum mechanics, the basic theory that is used for the analysis of atomic and molecular structure. We will return to electromagnetic waves in later chapters. Meanwhile, we can learn the essential language of waves in the context of mechanical waves.

19–2 TYPES OF MECHANICAL WAVES

A **mechanical wave** is a disturbance that travels through some material or substance called the **medium** for the wave. As the wave travels through the medium, the particles that make up the medium undergo displacements of various kinds, depending on the nature of the wave.

Figure 19–1 shows three varieties of mechanical waves. In Fig. 19–1a the medium is a string or rope under tension. If we give the left end a small upward shake or wiggle, the wiggle travels along the length of the string. Successive sections of string go through the same motion that we gave to the end, but at successively later times. Because the displacements of the medium are perpendicular or *transverse* to the direction of travel of the wave along the medium, this is called a **transverse wave.**

In Fig. 19–1b the medium is a liquid or gas in a tube with a rigid wall at the right end and a movable piston at the left end. If we give the piston a single back-and-forth

19–1 (a) The hand moves the string up and then returns, producing a transverse wave. (b) The piston compresses the liquid or gas to the right, then returns, producing a longitudinal wave. (c) The board pushes to the right, then returns, producing a sum of transverse and longitudinal waves. In all three cases the solitary wave propagates to the right.

10.1

Properties of Mechanical Waves

motion, displacement and pressure fluctuations travel down the length of the medium. This time the motions of the particles of the medium are back and forth along the *same* direction that the wave travels, and we call this a **longitudinal wave.**

In Fig. 19–1c the medium is water in a channel, such as an irrigation ditch or canal. When we move the flat board at the left end forward and back once, a wave disturbance travels down the length of the channel. In this case the displacements of the water have *both* longitudinal and transverse components.

Each of these systems has an equilibrium state. For the stretched string it is the state in which the system is at rest, stretched out along a straight line. For the fluid in a tube it is a state in which the fluid is at rest with uniform pressure, and for the water in a trough it is a smooth, level water surface. In each case the wave motion is a disturbance from the equilibrium state that travels from one region of the medium to another. In each

19–2 The block of mass m is attached to a spring and undergoes simple harmonic motion, producing a sinusoidal wave that travels to the right on the string. In a real-life system a driving force would have to be applied to the mass m to replace the energy carried away by the wave.

case there are forces that tend to restore the system to its equilibrium position when it is displaced, just as the force of gravity tends to pull a pendulum toward its straight-down equilibrium position when it is displaced.

These examples have three things in common. First, in each case the disturbance travels or *propagates* with a definite speed through the medium. This speed is called the speed of propagation, or simply the **wave speed.** It is determined in each case by the mechanical properties of the medium. We will use the symbol v for wave speed. (The wave speed is *not* the same as the speed with which particles move when they are disturbed by the wave. We'll return to this point in Section 19–3.) Second, the medium itself does not travel through space; its individual particles undergo back-and-forth or up-and-down motions around their equilibrium positions. The overall pattern of the wave disturbance is what travels. Third, to set any of these systems into motion, we have to put in energy by doing mechanical work on the system. The wave motion transports this energy from one region of the medium to another. *Waves transport energy, but not matter, from one region to another.*

19–3 PERIODIC WAVES

The transverse wave on a stretched string in Fig. 19–1a is an example of a *wave pulse.* The hand shakes the string up and down just once, exerting a transverse force on it as it does so. The result is a single "wiggle," or pulse, that travels along the length of the string. The tension in the string restores its straight-line shape once the pulse has passed.

A more interesting situation develops when we give the free end of the string a repetitive, or *periodic,* motion. (You may want to review the discussion of periodic motion in Chapter 13 before going ahead.) Then each particle in the string will also undergo periodic motion as the wave propagates, and we have a **periodic wave.** In particular, suppose we move the string up and down in *simple harmonic motion* (SHM) with amplitude A, frequency f, angular frequency $\omega = 2\pi f$, and period $T = 1/f = 2\pi/\omega$. A possible experimental setup is shown in Fig. 19–2. As we will see, periodic waves with simple harmonic motion are particularly easy to analyze; we call them **sinusoidal waves.** It also turns out that *any* periodic wave can be represented as a combination of sinusoidal waves. So this particular kind of wave motion is worth special attention.

In Fig. 19–2 the wave that advances along the string is a *continuous succession* of transverse sinusoidal disturbances. Figure 19–3 shows the shape of a part of the string near the left end at time intervals of $\frac{1}{8}$ of a period, for a total time of one period. The wave shape advances steadily toward the right, as indicated by the short red arrow pointing to a particular wave crest. As the wave moves, any point on the string (the blue dot, for example) oscillates up and down about its equilibrium position with simple harmonic motion. *When a sinusoidal wave passes through a medium, every particle in the medium undergoes simple harmonic motion with the same frequency.*

CAUTION ▶ Be very careful to distinguish between the motion of the *transverse wave* along the string and the motion of a *particle* of the string. The wave moves with constant speed v *along* the length of the string, while the motion of the particle is simple harmonic and *transverse* (perpendicular) to the length of the string. ◀

For a periodic wave, the shape of the string at any instant is a repeating pattern. The length of one complete wave pattern is the distance from one crest to the next, or from one trough to the next, or from any point to the corresponding point on the next repetition of the wave shape. We call this distance the **wavelength** of the wave, denoted by λ (the Greek letter "lambda"). The wave pattern travels with constant speed v and advances a distance of one wavelength λ in a time interval of one period T. So the wave speed v is given by $v = \lambda/T$, or, because $f = 1/T$,

$$v = \lambda f \quad \text{(periodic wave).} \qquad (19\text{–}1)$$

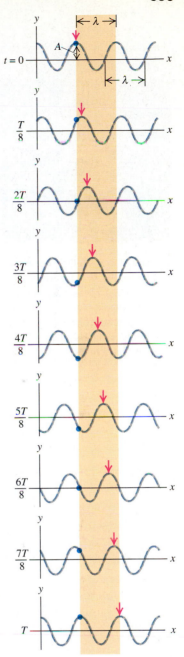

19–3 A sinusoidal transverse wave traveling to the right along a string. The shape of the string is shown at intervals of $\frac{1}{8}$ of a period; the vertical scale is exaggerated. All points on the string move up and down in SHM with amplitude A. The wave travels a distance of one wavelength λ in one period T. Any two particles in the string whose x-coordinates differ by λ, such as the blue dot at $x = \lambda$ and a particle at $x = 0$, oscillate in phase with each other.

The speed of propagation equals the product of wavelength and frequency. The frequency is a property of the *entire* periodic wave because all points on the string oscillate with the same frequency f.

In many important situations the wave speed v is determined entirely by the mechanical properties of the medium. In this case, increasing f causes λ to decrease so that the product $v = \lambda f$ remains the same, and waves of *all* frequencies propagate with the same wave speed. In this chapter we will consider *only* waves of this kind. (In later chapters we will study the propagation of light waves in matter for which the wave speed depends on frequency; this turns out to be the reason why prisms break white light into a spectrum and why raindrops create a rainbow.)

To understand the mechanics of a periodic *longitudinal* wave, we consider a long tube filled with a fluid, with a piston at the left end as in Fig. 19–1b. If we push the piston in, we compress the fluid near the piston, increasing the pressure in this region. This region then pushes against the neighboring region of fluid, and so on, and a wave pulse moves along the tube.

Now suppose we move the piston back and forth with simple harmonic motion along a line parallel to the axis of the tube (Fig. 19–4). This motion forms regions in the fluid where the pressure and density are greater or less than the equilibrium values. We call a region of increased pressure a *compression*. In the figure, compressions are represented by darkly shaded areas. A region of reduced pressure is an *expansion;* in the figure, expansions are represented by lightly shaded areas. The short red arrow shows the position of one particular compression; the compressions and expansions move to the right with constant speed v.

The motion of a single particle of the medium, such as the one shown by a blue dot in Fig. 19–4, is simple harmonic motion parallel to the direction of wave propagation. The wavelength is the distance from one compression to the next or from one expansion to the next. The fundamental equation $v = \lambda f$ holds for longitudinal waves as well as for transverse waves and indeed for *all* types of periodic waves. As for transverse waves, in this chapter we will consider only situations in which the speed of longitudinal waves does not depend on the frequency.

A series of drops falling vertically into water produces a periodic wave that spreads radially outward from the source. The wave crests and troughs are concentric circles. The distance along a radial line from one crest to the next, or from one trough to the next, is the wavelength.

EXAMPLE 19–1

Sound waves are longitudinal waves in air. The speed of sound depends on temperature; at 20°C it is 344 m/s (1130 ft/s). What is the wavelength of a sound wave in air at 20°C if the frequency is $f = 262$ Hz (the approximate frequency of middle C on a piano)?

SOLUTION From Eq. (19–1),

$$\lambda = \frac{v}{f} = \frac{344 \text{ m/s}}{262 \text{ s}^{-1}} = 1.31 \text{ m}.$$

The "high C" sung by coloratura sopranos (also designated C_6) is two octaves above middle C (designated C_4). Each octave corresponds to a factor of two in frequency, so the frequency of high C is four times that of middle C, $f = 4(262 \text{ Hz}) = 1048$ Hz. The speed of sound waves is unaffected by changes in frequency, so the wavelength is one-fourth as large, $\lambda = (1.31 \text{ m})/4 = 0.328$ m.

19–4 MATHEMATICAL DESCRIPTION OF A WAVE

Many characteristics of periodic waves can be described by using the concepts of wave speed, period, frequency, and wavelength. Often, though, we need a more detailed description of the positions and motions of individual particles of the medium at particular times during wave propagation. For this description we need the concept of a *wave function,* a function that describes the position of any particle in the medium at any time. We will concentrate on *sinusoidal* waves, in which each particle undergoes simple harmonic motion about its equilibrium position.

As a specific example, let's look at waves on a stretched string. If we ignore the sag of the string due to gravity, the equilibrium position of the string is along a straight line. We take this to be the x-axis of a coordinate system. Waves on a string are *transverse;* during wave motion a particle with equilibrium position x is displaced some distance y in the direction perpendicular to the x-axis. The value of y depends on which particle we are talking about (that is, y depends on x) and also on the time t when we look at it. Thus y is a *function* of x and t; $y = y(x, t)$. We call $y(x, t)$ the **wave function** that describes the wave. If we know this function for a particular wave motion, we can use it to find the displacement (from equilibrium) of any particle at any time. From this we can find the velocity and acceleration of any particle, the shape of the string, and anything else we want to know about the behavior of the string at any time.

WAVE FUNCTION FOR A SINUSOIDAL WAVE

Let's see how to determine the form of the wave function for a sinusoidal wave. Suppose a sinusoidal wave travels from left to right (the direction of increasing x) along the string, as in Fig. 19–3. Every particle of the string oscillates with simple harmonic motion with the same amplitude and frequency. But the oscillations of particles at different points on the string are *not* all in step with each other. The particle marked by the blue dot in Fig. 19–3 is at its maximum positive value of y at $t = 0$ and returns to $y = 0$ at $t = 2T/8$; these same events occur for a particle at the *center* of the colored band at $t = 4T/8$ and $t = 6T/8$, exactly one half-period later. For any two particles of the string, the motion of the particle on the right (in terms of the wave, the "downstream" particle) lags behind the motion of the particle on the left by an amount proportional to the distance between the particles.

Hence the cyclic motions of various points on the string are out of step with each other by various fractions of a cycle. We call these differences *phase differences,* and we say that the *phase* of the motion is different for different points. For example, if one point has its maximum positive displacement at the same time that another has its maximum negative displacement, the two are a half-cycle out of phase. (This is the case for the blue dot in Fig. 19–3 and a point at the center of the colored band.)

Suppose that the displacement of a particle at the left end of the string ($x = 0$), where the wave originates, is given by

$$y(x = 0, t) = A \sin \omega t = A \sin 2\pi f t. \tag{19-2}$$

That is, the particle oscillates in simple harmonic motion with amplitude A, frequency f, and angular frequency $\omega = 2\pi f$. The notation $y(x = 0, t)$ reminds us that the motion of this particle is a special case of the wave function $y(x, t)$ that describes the entire wave. At $t = 0$ the particle at $x = 0$ has zero displacement ($y = 0$) and is moving in the +y-direction (the value of y is increasing with time).

The wave disturbance travels from $x = 0$ to some point x to the right of the origin in an amount of time given by x/v, where v is the wave speed. So the motion of point x at time t is the same as the motion of point $x = 0$ at the earlier time $t - x/v$. Hence we can find the displacement of point x at time t by simply replacing t in Eq. (19–2) by $(t - x/v)$. When we do that, we find the following expression for the wave function:

$$y(x, t) = A \sin \omega \left(t - \frac{x}{v}\right) = A \sin 2\pi f \left(t - \frac{x}{v}\right) \tag{19-3}$$

(sinusoidal wave moving in +x-direction).

The displacement $y(x, t)$ is a function of both the location x of the point and the time t. We could make Eq. (19–3) more general by allowing for different values of the phase angle, as we did for simple harmonic motion in Section 13–3, but for now we omit this.

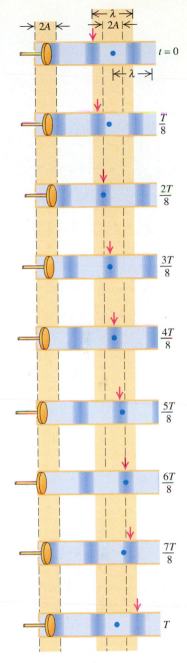

19–4 A sinusoidal longitudinal wave traveling to the right in a fluid. The disturbance of the fluid is shown at intervals of $\frac{1}{8}$ of a period. The piston and all points in the fluid move back and forth in SHM with amplitude A. The wave travels a distance of one wavelength λ in one period T. In equilibrium, the particle shown by the blue dot is $\frac{3}{2}$ wavelengths from the piston; its motion is one half-cycle out of phase with that of the piston.

We can rewrite the wave function given by Eq. (19–3) in several different but useful forms. We can express it in terms of the period $T = 1/f$ and the wavelength $\lambda = v/f$:

$$y(x,t) = A \sin 2\pi \left(\frac{t}{T} - \frac{x}{\lambda} \right)$$

(sinusoidal wave moving in $+x$-direction).   (19–4)

We get another convenient form of the wave function if we define a quantity k, called the **wave number:**

$$k = \frac{2\pi}{\lambda} \qquad \text{(wave number).}$$ (19–5)

Substituting $\lambda = 2\pi/k$ and $f = \omega/2\pi$ into the wavelength-frequency relation $v = \lambda f$ gives

$$\omega = vk \qquad \text{(periodic wave).}$$ (19–6)

We can then rewrite Eq. (19–4) as

$$y(x,t) = A \sin (\omega t - kx) \qquad \text{(sinusoidal wave moving in} \atop +x\text{-direction).}$$ (19–7)

Which of these various forms for the wave function $y(x, t)$ we use in any specific problem is a matter of convenience. Note that ω has units rad/s, so for unit consistency in Eqs. (19–6) and (19–7) the wave number k must have the units rad/m. (Some physicists define the wave number as $1/\lambda$ rather than $2\pi/\lambda$. When reading other texts, be sure to determine how this term is defined.)

The wave function $y(x, t)$ is graphed as a function of x for a specific time t in Fig. 19–5a. This graph gives the displacement y of a particle from its equilibrium position as a function of the coordinate x of the particle. If the wave is a transverse wave on a string, the graph in Fig. 19–5a represents the shape of the string at that instant, like a flash photograph of the string. In particular, at time $t = 0$,

$$y(x, t = 0) = A \sin (-kx) = -A \sin kx = -A \sin 2\pi \frac{x}{\lambda}.$$

A graph of the wave function versus time t for a specific coordinate x is shown in Fig. 19–5b. This graph gives the displacement y of the particle at that coordinate as a function of time. That is, it describes the motion of that particle. In particular, at the position $x = 0$,

$$y(x = 0, t) = A \sin \omega t = A \sin 2\pi \frac{t}{T}.$$

This is consistent with our original statement about the motion at $x = 0$, Eq. (19–2).

CAUTION▶ Make sure that you understand the difference between Figs. 19–5a and 19–5b. In particular, note that Fig. 19–5b is *not* a picture of the shape of the string; it is a graph of the position y of a particle at $x = 0$ as a function of time. ◀

We can modify Eqs. (19–3) through (19–7) to represent a wave traveling in the *negative* x-direction. In this case the displacement of point x at time t is the same as the

19–5 (a) The graph of $y(x, t)$ versus coordinate x for a specific time, in this case $t = 0$, describes the shape of the wave at that time. (b) The graph of $y(x, t)$ versus time t for a specific coordinate, in this case $x = 0$, describes the motion of a particle at that coordinate as a function of time. The vertical scale is exaggerated in both (a) and (b).

(a)   (b)

motion of point $x = 0$ at the *later* time $(t + x/v)$, so in Eq. (19–2) we replace t by $(t + x/v)$. For a wave traveling in the negative x-direction,

$$y(x, t) = A \sin 2\pi f \left(t + \frac{x}{v} \right) = A \sin 2\pi \left(\frac{t}{T} + \frac{x}{\lambda} \right) \quad (19–8)$$

$$= A \sin (\omega t + kx) \quad \text{(sinusoidal wave moving in } -x\text{-direction)}.$$

In the expression $y(x, t) = A \sin (\omega t \pm kx)$ for a wave traveling in the $-x$- or $+x$-direction, the quantity $(\omega t \pm kx)$ is called the **phase**. It plays the role of an angular quantity (always measured in radians) in Eq. (19–7) or (19–8), and its value for any values of x and t determines what part of the sinusoidal cycle is occurring at a particular point and time. For a positive crest (where $y = A$ and the sine function has the value 1), the phase could be $\pi/2$, $5\pi/2$, and so on; for a point of zero displacement it could be 0, π, 2π, and so on. The wave speed is the speed with which we have to move along with the wave to keep alongside a point of a given phase, such as a particular crest of a wave on a string. For a wave traveling in the $+x$-direction, that means $\omega t - kx = $ constant. Taking the derivative with respect to t, we find $\omega = k \, dx/dt$, or

$$\frac{dx}{dt} = \frac{\omega}{k}.$$

Comparing this with Eq. (19–6), we see that dx/dt is equal to the speed v of the wave. Because of this relationship, v is sometimes called the *phase velocity* of the wave. (*Phase speed* would be a better term.)

Problem–Solving Strategy

MECHANICAL WAVES

1. It helps to make a distinction between *kinematics* problems and *dynamics* problems. In kinematics problems we are concerned only with describing motion; the relevant quantities are wave speed, wavelength (or wave number), frequency (or angular frequency), amplitude, and the position, velocity, and acceleration of individual particles. In dynamics problems, concepts such as force and mass enter; the relation of wave speed to the mechanical properties of a system is an example. We'll get into these relations in the next few sections.

2. If f is given, you can find $T = 1/f$, and vice versa. If λ is given, you can find $k = 2\pi/\lambda$, and vice versa. If any two of the quantities v, λ, and f (or v, k, and ω) are known, you can find the third, using $v = \lambda f$ or $\omega = vk$. In some problems that's all you need. To determine the wave function completely, you need to know A and any two of v, λ, and f (or v, k, and ω). Once you have this information, you can use it in Eq. (19–3), (19–4), or (19–7) to get the specific wave function for the problem at hand. Once you have that, you can find the value of y at any point (value of x) and at any time by substituting into the wave function.

3. If the wave speed v is not given, you may be able to find it in one of two ways: Use the frequency-wavelength relation $v = \lambda f$, or use relations between v and the mechanical properties of the system, such as tension and mass per unit length for a string. We will develop these relations in the next three sections. Which method you use to find v will depend on what information you are given.

EXAMPLE 19-2

Wave on a clothesline Your cousin Throckmorton is playing with the clothesline. He unties one end, holds it taut, and wiggles the end up and down sinusoidally with frequency 2.00 Hz and amplitude 0.075 m. The wave speed is $v = 12.0$ m/s. At time $t = 0$ the end has zero displacement and is moving in the $+y$-direction. Assume that no wave bounces back from the far end to muddle up the pattern. a) Find the amplitude, angular frequency, period, wavelength, and wave number of the wave. b) Write a wave function describing the wave. c) Write equations for the displacement as a function of time of Throckmorton's end of the string and of a point 3.00 m from his end.

SOLUTION a) The amplitude A of the wave is just the amplitude of the motion of the end of the clothesline, $A = 0.075$ m. The angular frequency is

$$\omega = 2\pi f = (2\pi \text{ rad/cycle})(2.00 \text{ cycles/s}) = 4.00\pi \text{ rad/s}$$

$$= 12.6 \text{ rad/s.}$$

The period is $T = 1/f = 0.500$ s. We get the wavelength from Eq. (19–1):

$$\lambda = \frac{v}{f} = \frac{12.0 \text{ m/s}}{2.00 \text{ s}^{-1}} = 6.00 \text{ m.}$$

We find the wave number from Eq. (19–5) or (19–6):

$$k = \frac{2\pi}{\lambda} = \frac{2\pi \text{ rad}}{6.00 \text{ m}} = 1.05 \text{ rad/m,} \quad \text{or}$$

$$k = \frac{\omega}{v} = \frac{4.00\pi \text{ rad/s}}{12.0 \text{ m/s}} = 1.05 \text{ rad/m.}$$

b) We take the coordinate of Throckmorton's end of the string to be $x = 0$, and choose the direction in which the wave propagates along the string to be the positive x-direction. Then the wave function is given by Eq. (19–4):

$$y(x, t) = A \sin 2\pi \left(\frac{t}{T} - \frac{x}{\lambda} \right)$$

$$= (0.075 \text{ m}) \sin 2\pi \left(\frac{t}{0.500 \text{ s}} - \frac{x}{6.00 \text{ m}} \right)$$

$$= (0.075 \text{ m}) \sin [(12.6 \text{ rad/s})t - (1.05 \text{ rad/m})x].$$

We can also get this same equation from Eq. (19–7) by using the values of ω and k obtained above. The quantity $(12.6 \text{ rad/s})t - (1.05 \text{ rad/m})x$ is the phase of a point x on the string at time t.

c) With our choice of the positive x-direction the two points in question are at $x = 0$ and $x = +3.00$ m. For each point, we can obtain an expression for the displacement as a function of time by substituting these values of x into the wave function found in part (b):

$$y(x = 0, t) = (0.075 \text{ m}) \sin 2\pi \left(\frac{t}{0.500 \text{ s}} - \frac{0}{6.00 \text{ m}} \right)$$

$$= (0.075 \text{ m}) \sin (12.6 \text{ rad/s})t,$$

$$y(x = +3.00 \text{ m}, t) = (0.075 \text{ m}) \sin 2\pi \left(\frac{t}{0.500 \text{ s}} - \frac{3.00 \text{ m}}{6.00 \text{ m}} \right)$$

$$= (0.075 \text{ m}) \sin [(12.6 \text{ rad/s})t - \pi \text{ rad}].$$

The phases of these two points separated by one half-wavelength ($\lambda/2 = (6.00 \text{ m})/2 = 3.00$ m) differ by π radians. Both points oscillate in SHM with the same frequency and amplitude, but their oscillations are one half-cycle out of phase.

Using the above expression for $y(x = 0, t)$, can you show that the end of the string at $x = 0$ is moving in the positive direction at $t = 0$ as stated at the beginning of this example?

PARTICLE VELOCITY AND ACCELERATION IN A SINUSOIDAL WAVE

From the wave function we can get an expression for the transverse velocity of any *particle* in a transverse wave. We call this v_y to distinguish it from the wave propagation speed v. To find the transverse velocity v_y at a particular point x, we take the derivative of the wave function $y(x, t)$ with respect to t, keeping x constant. If the wave function is

$$y(x, t) = A \sin (\omega t - kx),$$

then

$$v_y(x, t) = \frac{\partial y(x, t)}{\partial t} = \omega A \cos (\omega t - kx). \tag{19–9}$$

The ∂ in this expression is a modified d, used to remind us that $y(x, t)$ is a function of *two* variables and that we are allowing only one (t) to vary. The other (x) is constant because we are looking at a particular point on the string. This derivative is called a *partial derivative*. If you haven't reached this point yet in your study of calculus, don't fret; it's a simple idea.

Equation (19–9) shows that the transverse velocity of a particle varies with time, as we expect for simple harmonic motion. The maximum particle speed is ωA; this can be greater than, less than, or equal to the wave speed v, depending on the amplitude and frequency of the wave.

The *acceleration* of any particle is the *second* partial derivative of $y(x, t)$ with respect to t:

$$a_y(x, t) = \frac{\partial^2 y(x, t)}{\partial t^2} = -\omega^2 A \sin (\omega t - kx) = -\omega^2 y(x, t). \tag{19–10}$$

The acceleration of a particle equals $-\omega^2$ times its displacement, which is the result we obtained in Section 13–3 for simple harmonic motion.

We can also compute partial derivatives of $y(x, t)$ with respect to x, holding t constant. This corresponds to studying the shape of the string at one instant of time, like a flash photo. The first derivative $\partial y(x, t)/\partial x$ is the *slope* of the string at any point. The second partial derivative with respect to x is the *curvature* of the string:

$$\frac{\partial^2 y(x, t)}{\partial x^2} = -k^2 A \sin(\omega t - kx) = -k^2 y(x, t). \qquad (19\text{–}11)$$

From Eqs. (19–10) and (19–11) and the relation $\omega = vk$ we see that

$$\frac{\partial^2 y(x, t)/\partial t^2}{\partial^2 y(x, t)/\partial x^2} = \frac{\omega^2}{k^2} = v^2,$$

$$\frac{\partial^2 y(x, t)}{\partial x^2} = \frac{1}{v^2}\frac{\partial^2 y(x, t)}{\partial t^2} \qquad \text{(wave equation).} \qquad (19\text{–}12)$$

The wave function $y = A \sin(\omega t + kx)$ also satisfies this relationship; we invite you to verify this.

Equation (19–12), called the **wave equation,** is one of the most important equations in all of physics. Whenever it occurs, we know that a disturbance can propagate as a wave along the x-axis with wave speed v. The disturbance need not be a sinusoidal wave; we'll see in the next section that *any* wave on a string obeys Eq. (19–12), whether the wave is periodic or not (see also Problems 19–36 and 19–39). In Chapter 33 we will find that electric and magnetic fields satisfy the wave equation; the wave speed will turn out to be the speed of light, which will lead us to the conclusion that light is an electromagnetic wave.

Figure 19–6a shows the velocity v_y and acceleration a_y, given by Eqs. (19–9) and (19–10), for several points on a string as a sinusoidal wave passes along it. Note that at points where the string has an upward curvature ($\partial^2 y/\partial x^2 > 0$), the acceleration of that point is positive ($a_y = \partial^2 y/\partial t^2 > 0$); this follows from the wave equation, Eq. (19–12). For the same reason the acceleration is negative ($a_y = \partial^2 y/\partial t^2 < 0$) at points where the string has a downward curvature ($\partial^2 y/\partial x^2 < 0$), and the acceleration is zero ($a_y = \partial^2 y/\partial t^2 = 0$) at points of inflection where the curvature is zero ($\partial^2 y/\partial x^2 = 0$). We

(a)

(b)

19–6 (a) Another view of the wave at $t = 0$ in Fig. 19–5a, showing the transverse velocity v_y and transverse acceleration a_y of several points on the string. The acceleration of each point is proportional to the string displacement at that point. (b) Light curve: wave at $t = 0$. Dark curve: the same wave at $t = 0.05T$. For each point, you can relate the magnitude and direction of the displacement during this time interval to v_y and a_y shown in (a).

emphasize again that v_y and a_y are the *transverse* velocity and acceleration of points on the string; these points move along the y-direction, not along the propagation direction of the wave. The transverse motions of several points on the string can be seen in Fig. 19–6b.

The concept of wave function is equally useful with *longitudinal* waves, and everything we have said about wave functions can be adapted to this case. The quantity y still measures the displacement of a particle of the medium from its equilibrium position; the difference is that for a longitudinal wave, this displacement is *parallel* to the x-axis instead of perpendicular to it. We'll discuss longitudinal waves in detail in Section 19–6.

19–5 SPEED OF A TRANSVERSE WAVE

10.2
Speed of Waves on a String

One of the key properties of any wave is the wave speed. Light waves in air have a much greater speed of propagation than do sound waves in air (3.00×10^8 m/s versus 344 m/s); that's why you see the flash from a bolt of lightning before you hear the clap of thunder. In this section we'll see what determines the speed of propagation of one particular kind of wave: transverse waves on a string. The speed of these waves is important to understand in its own right because it is an essential part of analyzing stringed musical instruments, as we'll discuss in Chapter 20. Furthermore, the speeds of many kinds of mechanical waves turn out to have the same basic mathematical expression, as does the speed of waves on a string.

The physical quantities that determine the speed of transverse waves on a string are the *tension* in the string and its *mass per unit length* (also called *linear mass density*). We might guess that increasing the tension should increase the restoring forces that tend to straighten the string when it is disturbed, thus increasing the wave speed. We might also guess that increasing the mass should make the motion more sluggish and decrease the speed. Both these guesses turn out to be right. We'll develop the exact relationship between wave speed, tension, and mass per unit length by two different methods. The first is simple in concept and considers a specific wave shape; the second is more general but also more formal. Choose whichever you like better.

WAVE SPEED ON A STRING: FIRST METHOD

We consider a perfectly flexible string (Fig. 19–7). In the equilibrium position the tension is F, and the linear mass density (mass per unit length) is μ. (When portions of the string are displaced from equilibrium, the mass per unit length decreases a little, and the

19–7 Propagation of a transverse wave on a string. (a) String in equilibrium; (b) part of the string in motion.

(a)

(b)

tension increases a little.) We ignore the weight of the string so that when the string is at rest in the equilibrium position, the string forms a perfectly straight line as in Fig. 19–7a.

Starting at time $t = 0$, we apply a constant transverse force F_y at the left end of the string. We might expect that the end would move with constant acceleration; that would happen if the force were applied to a *point* mass. But here the effect of the force F_y is to set successively more and more mass in motion. As shown in Fig. 19–7b, the wave travels with constant speed v, so the division point P between moving and nonmoving portions moves with the same constant speed v.

Figure 19–7b shows that all particles in the moving portion of the string move upward with constant *velocity* v_y, not constant acceleration. To see why this is so, we note that the *impulse* of the force F_y up to time t is $F_y t$. According to the impulse-momentum theorem (Section 8–2), the impulse is equal to the change in the total transverse component of momentum $(mv_y - 0)$ of the moving part of the string. Because the system started with *no* transverse momentum, this is equal to the total momentum at time t:

$$F_y t = mv_y.$$

The total momentum thus must increase proportionately with time. But since the division point P moves with constant speed, the length of string that is in motion and hence the total mass m in motion are also proportional to the time t that the force has been acting. So the *change* of momentum must be associated entirely with the increasing amount of mass in motion, not with an increasing velocity of an individual mass element. That is, mv_y changes because m, not v_y, changes.

At time t, the left end of the string has moved up a distance $v_y t$, and the boundary point P has advanced a distance vt. The total force at the left end of the string has components F and F_y. Why F? There is no motion in the direction along the length of the string, so there is no unbalanced horizontal force. Therefore F, the magnitude of the horizontal component, does not change when the string is displaced. In the displaced position the tension is $(F^2 + F_y^2)^{1/2}$ (greater than F), and the string stretches somewhat.

To derive an expression for the wave speed v, we again apply the impulse-momentum theorem to the portion of the string in motion at time t, that is, the portion to the left of P in Fig. 19–7b. The transverse *impulse* (transverse force times time) is equal to the change of transverse *momentum* of the moving portion (mass times transverse component of velocity). The impulse of the transverse force F_y in time t is $F_y t$. In Fig. 19–7b the right triangle whose vertex is at P, with sides $v_y t$ and vt, is similar to the right triangle whose vertex is at the position of the hand, with sides F_y and F. Hence

$$\frac{F_y}{F} = \frac{v_y t}{vt}, \qquad F_y = F\frac{v_y}{v},$$

and

$$\text{Transverse impulse} = F_y t = F\frac{v_y}{v}t.$$

The mass of the moving portion of the string is the product of the mass per unit length μ and the length vt, or μvt. The transverse momentum is the product of this mass and the transverse velocity v_y:

$$\text{Transverse momentum} = (\mu vt)v_y.$$

We note again that the momentum increases with time *not* because mass is moving faster, as was usually the case in Chapter 8, but because *more mass* is brought into motion. But the impulse of the force F_y is still equal to the total change in momentum of the system. Applying this relation, we obtain

$$F\frac{v_y}{v}t = \mu vt v_y.$$

Solving this for v, we find

$$v = \sqrt{\frac{F}{\mu}} \qquad \text{(speed of a transverse wave on a string).} \qquad (19\text{--}13)$$

Equation (19–13) confirms our prediction that the wave speed v should increase when the tension F increases but decrease when the mass per unit length μ increases.

Note that v_y does not appear in Eq. (19–13); thus the wave speed doesn't depend on v_y. Our calculation considered only a very special kind of pulse, but we can consider *any* shape of wave disturbance as a series of pulses with different values of v_y. So even though we derived Eq. (19–13) for a special case, it is valid for *any* transverse wave motion on a string, including the sinusoidal and other periodic waves we discussed in Section 19–3. Note that the wave speed doesn't depend on the amplitude or frequency of the wave, in accordance with our assumptions in Section 19–3.

WAVE SPEED ON A STRING: SECOND METHOD

Here is an alternative derivation of Eq. (19–13). If you aren't comfortable with partial derivatives, it can be omitted. We apply Newton's second law, $\Sigma \vec{F} = m\vec{a}$, to a small segment of string whose length in the equilibrium position is Δx (Fig. 19–8). The mass of the segment is $m = \mu\, \Delta x$; the forces at the ends are represented in terms of their x- and y-components. The x-components have equal magnitude F and add to zero because the motion is transverse and there is no component of acceleration in the x-direction. To obtain F_{1y} and F_{2y}, we note that the ratio F_{1y}/F is equal in magnitude to the *slope* of the string at point x and that F_{2y}/F is equal to the slope at point $x + \Delta x$. Taking proper account of signs, we find

$$\frac{F_{1y}}{F} = -\left(\frac{\partial y}{\partial x}\right)_x, \qquad \frac{F_{2y}}{F} = \left(\frac{\partial y}{\partial x}\right)_{x+\Delta x}. \qquad (19\text{--}14)$$

The notation reminds us that the derivatives are evaluated at points x and $x + \Delta x$, respectively. From Eq. (19–14) we find that the net y-component of force is

$$F_y = F_{1y} + F_{2y} = F\left[\left(\frac{\partial y}{\partial x}\right)_{x+\Delta x} - \left(\frac{\partial y}{\partial x}\right)_x\right]. \qquad (19\text{--}15)$$

We now equate F_y from Eq. (19–15) to the mass $\mu\, \Delta x$ times the y-component of acceleration $\partial^2 y/\partial t^2$. We obtain

$$F\left[\left(\frac{\partial y}{\partial x}\right)_{x+\Delta x} - \left(\frac{\partial y}{\partial x}\right)_x\right] = \mu\Delta x \frac{\partial^2 y}{\partial t^2}, \qquad (19\text{--}16)$$

These cables have a relatively large amount of mass per unit length (μ) and a relatively low tension (F). Hence, if the cables are disturbed—say, by a bird landing on them—transverse waves will travel along them quite slowly.

19–8 Free-body diagram for a segment of string whose length in its equilibrium position is Δx. The force at each end of the string is tangent to the string at the point of application; each force is represented in terms of its x- and y-components.

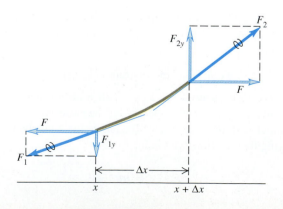

or, dividing by $F\,\Delta x$,

$$\frac{\left(\dfrac{\partial y}{\partial x}\right)_{x+\Delta x} - \left(\dfrac{\partial y}{\partial x}\right)_{x}}{\Delta x} = \frac{\mu}{F}\frac{\partial^{2}y}{\partial t^{2}}. \tag{19-17}$$

We now take the limit as $\Delta x \to 0$. In this limit, the left side of Eq. (19-17) becomes the derivative of $\partial y/\partial x$ with respect to x (at constant t), that is, the *second* (partial) derivative of y with respect to x:

$$\frac{\partial^{2}y}{\partial x^{2}} = \frac{\mu}{F}\frac{\partial^{2}y}{\partial t^{2}}. \tag{19-18}$$

Now, finally, comes the punch line of our story. Equation (19-18) has exactly the same form as the *wave equation*, Eq. (19-12), that we derived at the end of Section 19-4. That equation and Eq. (19-18) describe the very same wave motion, so they must be identical. Comparing the two equations, we see that for this to be so, we must have

$$v = \sqrt{\frac{F}{\mu}}, \tag{19-19}$$

which is the same expression as Eq. (19-13).

In going through this derivation, we didn't make any special assumptions about the shape of the wave. Since our derivation led us to rediscover Eq. (19-12), the wave equation, we conclude that the wave equation is valid for waves on a string that have *any* shape.

EXAMPLE 19-3

In Example 19-2 the linear mass density of the clothesline is 0.250 kg/m. How much tension does Throcky have to apply to produce the observed wave speed of 12.0 m/s?

SOLUTION We use Eq. (19-13); solving for F, we find

$$F = \mu v^{2} = (0.250\ \text{kg/m})(12.0\ \text{m/s})^{2} = 36.0\ \text{kg}\cdot\text{m/s}^{2}$$

$$= 36.0\ \text{N} = 8.09\ \text{lb}.$$

This much force is easily within Throcky's capability.

EXAMPLE 19-4

One end of a nylon rope is tied to a stationary support at the top of a vertical mine shaft 80.0 m deep (Fig. 19-9). The rope is stretched taut by a box of mineral samples with mass 20.0 kg attached at the lower end. The mass of the rope is 2.00 kg. The geologist at the bottom of the mine signals to his colleague at the top by jerking the rope sideways. a) What is the speed of a transverse wave on the rope? b) If a point on the rope is given a transverse simple harmonic motion with a frequency of 2.00 Hz, what is the wavelength of the wave?

SOLUTION a) Let's neglect the variation in tension between the bottom and top of the rope due to the rope's own weight. The tension F at the bottom of the rope equals the weight of the 20.0-kg load:

$$F = (20.0\ \text{kg})(9.80\ \text{m/s}^{2}) = 196\ \text{N}.$$

The mass per unit length is

$$\mu = \frac{m}{L} = \frac{2.00\ \text{kg}}{80.0\ \text{m}} = 0.0250\ \text{kg/m}.$$

$m_{\text{rope}} = 2.00\ \text{kg}$

80.0m

$m_{\text{samples}} = 20.0\ \text{kg}$

19-9 Sending signals by way of transverse waves on a vertical rope.

The wave speed is given by Eq. (19–13):

$$v = \sqrt{\frac{F}{\mu}} = \sqrt{\frac{196 \text{ N}}{0.0250 \text{ kg/m}}} = 88.5 \text{ m/s}.$$

b) From Eq. (19–1),

$$\lambda = \frac{v}{f} = \frac{88.5 \text{ m/s}}{2.00 \text{ s}^{-1}} = 44.3 \text{ m}.$$

If we account for the weight of the rope, the wave speed will increase and the wavelength will decrease as the wave travels up the rope because the tension increases. Can you verify that the wave speed at the top is 92.9 m/s?

19–6 SPEED OF A LONGITUDINAL WAVE

10.3

Speed of Sound in a Gas

Propagation speeds of longitudinal as well as transverse waves depend on the mechanical properties of the medium. We can derive relations for longitudinal waves that are analogous to Eq. (19–13) for transverse waves on a string. As in the wave-function discussion in Section 19–4, x is still the coordinate measured along the length of the wave medium. But for a longitudinal wave the displacement y is along that same direction rather than perpendicular to it as in a transverse wave.

Here is a derivation for the speed of longitudinal waves in a fluid in a pipe. This is a subject of some importance, since when the frequency of a longitudinal wave is within the range of human hearing, we call it **sound.** All musical wind instruments are fundamentally pipes in which a longitudinal wave (sound) propagates in a fluid (air). Human speech works on the same principle; sound waves propagate in your vocal tract, which is basically an air-filled pipe connected to the lungs at one end (your larynx) and to the outside air at the other end (your mouth). The steps in our derivation are completely parallel to the derivation of Eq. (19–13), and we invite you to compare the two developments.

Figure 19–10 shows a fluid (either liquid or gas) with density ρ in a pipe with cross-section area A. In the equilibrium state, the fluid is under a uniform pressure p. In Fig. 19–10a the fluid is at rest. At time $t = 0$ we start the piston at the left end moving toward the right with constant speed v_y. This initiates a wave motion that travels to the right along the length of the pipe, in which successive sections of fluid begin to move and become compressed at successively later times.

Figure 19–10b shows the fluid at time t. All portions of fluid to the left of point P are moving to the right with speed v_y, and all portions to the right of P are still at rest. The boundary between the moving and stationary portions travels to the right with a speed equal to the speed of propagation or wave speed v. At time t the piston has moved a distance $v_y t$, and the boundary has advanced a distance vt. As with a transverse disturbance in a string, we can compute the speed of propagation from the impulse-momentum theorem.

The quantity of fluid set in motion in time t is the amount that originally occupied a section of the cylinder with length vt, cross-section area A, and volume vtA. The mass of this fluid is ρvtA, and its longitudinal momentum (that is, momentum along the length of the pipe) is

$$\text{Longitudinal momentum} = (\rho vtA)v_y.$$

Equilibrium

(a)

$(p + \Delta p)A$

pA

Moving P At rest

(b)

19–10 Propagation of a longitudinal wave in a fluid confined in a tube. (a) Fluid in equilibrium. (b) Part of the fluid in motion. The net force on the moving fluid is to the left and is equal to $(p + \Delta p)A - pA = \Delta pA$.

Next we compute the increase of pressure, Δp, in the moving fluid. The original volume of the moving fluid, Avt, has decreased by an amount $Av_y t$. From the definition of the bulk modulus B, Eq. (11–13) in Section 11–6,

$$B = \frac{-\text{Pressure change}}{\text{Fractional volume change}} = \frac{-\Delta p}{-Av_y t/Avt},$$

$$\Delta p = B\frac{v_y}{v}.$$

The pressure in the moving fluid is $p + \Delta p$, and the force exerted on it by the piston is $(p + \Delta p)A$. The net force on the moving fluid (see Fig. 19–10b) is $\Delta p A$, and the longitudinal impulse is

$$\text{Longitudinal impulse} = \Delta p A t = B\frac{v_y}{v}At.$$

Because the fluid was at rest at time $t = 0$, the change in momentum up to time t is equal to the momentum at that time. Applying the impulse-momentum theorem, we find

$$B\frac{v_y}{v}At = \rho v t A v_y. \qquad (19\text{–}20)$$

When we solve this for v, we get

$$v = \sqrt{\frac{B}{\rho}} \qquad \text{(speed of a longitudinal wave in a fluid)}. \qquad (19\text{–}21)$$

Thus the speed of propagation of a longitudinal pulse in a fluid depends only on the bulk modulus B and the density ρ of the medium.

While we derived Eq. (19–21) for waves in a pipe, it also applies to longitudinal waves in a bulk fluid. Thus the speed of sound waves traveling in air or water is determined by this equation. We'll work out the details in the next section.

When a longitudinal wave propagates in a *solid* rod or bar, the situation is somewhat different. The rod expands sideways slightly when it is compressed longitudinally, while a fluid in a pipe with constant cross section cannot move sideways. Using the same kind of reasoning that led us to Eq. (19–21), we can show that the speed of a longitudinal pulse in the rod is given by

$$v = \sqrt{\frac{Y}{\rho}} \qquad \text{(speed of a longitudinal wave in a solid rod)}, \qquad (19\text{–}22)$$

where Y is Young's modulus, defined in Section 11–5.

CAUTION ▶ Equation (19–22) applies only to a rod or bar whose sides are free to bulge and shrink a little as the wave travels. It does not apply to longitudinal waves in a *bulk* solid, since in these materials, sideways motion in any element of material is prevented by the surrounding material. The speed of longitudinal waves in a bulk solid depends on the density, the bulk modulus, and the *shear* modulus; a full discussion is beyond the scope of this book. ◀

Note the similarity of form of Eqs. (19–13), (19–19), (19–21), and (19–22). In all these equations for wave speed, whether for transverse or longitudinal waves, the numerator is an elastic property describing the restoring force and the denominator is an inertial property of the medium. As with the derivation for a transverse wave on a string, Eqs. (19–21) and (19–22) are valid for sinusoidal and other periodic waves, not just for the special case discussed here.

Visualizing the relation between particle motion and wave motion is not as easy for longitudinal waves as for transverse waves on a string. Figure 19–11 will help you understand these motions. To use this figure, tape two index cards edge-to-edge with a gap of 1 mm or so between them, forming a thin slit. Place the cards over the figure with the slit horizontal at the top of the diagram, and move them downward with constant speed. The portions of the sine curves that are visible through the slit correspond to a row of particles in a medium in which a longitudinal sinusoidal wave is traveling. Each particle undergoes SHM about its equilibrium position, with delays or phase shifts that increase continuously along the slit. The regions of maximum compression and expansion move from left to right with constant speed. Moving the card upward simulates a wave traveling from right to left.

19–11 Diagram for illustrating longitudinal traveling waves.

Table 19–1 lists the speed of sound in several bulk materials. Sound waves travel slower in lead than in aluminum or steel because lead has a lower bulk modulus and shear modulus and a higher density.

TABLE 19–1

SPEED OF SOUND IN VARIOUS BULK MATERIALS

MATERIAL	SPEED OF SOUND (m/s)
Gases	
Air (20°C)	344
Helium (20°C)	999
Hydrogen (20°C)	1330
Liquids	
Liquid helium (4 K)	211
Mercury (20°C)	1451
Water (0°C)	1402
Water (20°C)	1482
Water (100°C)	1543
Solids	
Aluminum	6420
Lead	1960
Steel	5941

EXAMPLE 19–5

Wavelength of sonar waves A ship uses a sonar system to detect underwater objects (Fig. 19–12). The system emits underwater sound waves and measures the time interval for the reflected wave (echo) to return to the detector. Determine the speed of sound waves in water, and find the wavelength of a wave having a frequency of 262 Hz.

SOLUTION We use Eq. (19–21) to find the wave speed. From Table 11–2 we find that the compressibility of water, which is the reciprocal of the bulk modulus, is $k = 45.8 \times 10^{-11}$ Pa^{-1}. Thus $B = (1/45.8) \times 10^{11}$ Pa. The density of water is $\rho = 1.00 \times 10^3$ kg/m^3. We obtain

$$v = \sqrt{\frac{B}{\rho}} = \sqrt{\frac{(1/45.8) \times 10^{11} \text{ Pa}}{1.00 \times 10^3 \text{ kg/m}^3}}$$

$$= 1480 \text{ m/s}.$$

This value agrees well with the experimental value in Table 19–1; it is over four times the speed of sound in air at ordinary temperatures. The wavelength is given by

$$\lambda = \frac{v}{f} = \frac{1480 \text{ m/s}}{262 \text{ s}^{-1}} = 5.65 \text{ m}.$$

A wave of this frequency in air has a wavelength of 1.31 m, as we found in Example 19–1 (Section 19–3).

Dolphins emit high-frequency sound waves (typically 100,000 Hz) and use the echoes for guidance and for hunting. The corresponding wavelength in water is 1.48 cm. With this high-frequency "sonar" system they can sense objects that are roughly as small as the wavelength (but not much smaller). *Ultrasonic imaging* is a medical technique that uses exactly the

19–12 A sonar system uses underwater sound waves to detect and locate underwater objects.

same physical principle; sound waves of very high frequency and very short wavelength, called *ultrasound,* are scanned over the human body, and the "echoes" from interior organs are used to create an image. With ultrasound of frequency 5 MHz = 5×10^6 Hz, the wavelength in water (the primary constituent of the body) is 0.3 mm, and features as small as this can be discerned in the image. Ultrasound is used for the study of heart-valve action, detection of tumors, and prenatal examinations. Ultrasound is more sensitive than x rays in distinguishing various kinds of tissues and does not have the radiation hazards associated with x rays.

EXAMPLE 19–6

What is the speed of longitudinal sound waves in a lead rod?

SOLUTION This is the situation for which Eq. (19–22) is applicable. From Table 11–1, $Y = 1.6 \times 10^{10}$ Pa, and from Table 14–1, $\rho = 11.3 \times 10^3$ kg/m^3. We find

$$v = \sqrt{\frac{Y}{\rho}} = \sqrt{\frac{1.6 \times 10^{10} \text{ Pa}}{11.3 \times 10^3 \text{ kg/m}^3}} = 1.2 \times 10^3 \text{ m/s}.$$

This is more than three times the speed of sound in air. Note that our result is the speed with which a sound wave travels along a lead *rod*. You can see from Table 19–1 that sound waves in *bulk* lead travel faster still; the reason is that for lead the bulk modulus is greater than Young's modulus.

19–7 SOUND WAVES IN GASES

In the preceding section we derived Eq. (19–21), $v = \sqrt{B/\rho}$, for the speed of longitudinal waves in a fluid of bulk modulus B and density ρ. We can use this to find the speed of sound in an ideal gas.

The bulk modulus is defined in general as in Eq. (11–13); for infinitesimal pressure and volume changes, $B = -V \, dp/dV$. So we need to know how p varies with V for an ideal gas. If the temperature is constant, then according to the ideal gas law the product pV is constant, and we can use this to evaluate dp/dV. But when a gas is compressed *adiabatically* so that there is no heat flow, its temperature rises, and when it expands adiabatically, its temperature drops. In an adiabatic process for an ideal gas, Eq. (17–24) states that pV^γ is constant (recall that $\gamma = C_p/C_V$ is the dimensionless ratio of heat capacities), and we get a different result for B. When a wave travels through a gas, are the compressions and expansions adiabatic, or is there enough heat conduction between adjacent layers of gas to maintain a nearly constant temperature throughout?

Because thermal conductivities of gases are very small, it turns out that for ordinary sound frequencies, say 20 to 20,000 Hz, propagation of sound is very nearly adiabatic. Thus in Eq. (19–21) we use the **adiabatic bulk modulus** B_{ad}, derived from the assumption

$$pV^\gamma = \text{constant}. \tag{19–23}$$

We take the derivative of Eq. (19–23) with respect to V:

$$\frac{dp}{dV} V^\gamma + \gamma p V^{\gamma-1} = 0.$$

Dividing by $V^{\gamma-1}$ and rearranging, we find

$$B_{ad} = -V \frac{dp}{dV} = \gamma p. \tag{19–24}$$

For an *isothermal* process, $pV = \text{constant}$; we invite you to prove that the *isothermal* bulk modulus is

$$B_{iso} = p. \tag{19–25}$$

The adiabatic modulus is *larger* than the isothermal modulus by a factor γ.

Combining Eqs. (19–21) and (19–24), we find

$$v = \sqrt{\frac{\gamma p}{\rho}} \qquad \text{(speed of sound in an ideal gas).} \tag{19–26}$$

We can get a useful alternative form by using Eq. (16–5) for the density ρ of an ideal gas:

$$\rho = \frac{pM}{RT},$$

where R is the gas constant, M is the molar mass, and T is the absolute temperature. Combining this with Eq. (19–26), we find

$$v = \sqrt{\frac{\gamma RT}{M}} \qquad \text{(speed of sound in an ideal gas).} \qquad (19\text{–}27)$$

For any particular gas, γ, R, and M are constants, and the wave speed is proportional to the square root of the absolute temperature. Except for the numerical factor of 3 in one and γ in the other, this expression is identical to Eq. (16–19), which gives the root-mean-square speed of molecules in an ideal gas. This shows that sound speeds and molecular speeds are closely related; exploring that relationship in detail would be beyond our scope.

EXAMPLE 19–7

Compute the speed of sound waves in air at room temperature ($T = 20°C$).

SOLUTION From Example 16–3 (Section 16–2) the mean molar mass of air is 28.8×10^{-3} kg/mol. Also, $\gamma = 1.40$ for air, and $R = 8.315$ J/mol · K. At $T = 20°C = 293$ K we find

$$v = \sqrt{\frac{\gamma RT}{M}}$$

$$= \sqrt{\frac{(1.40)(8.315 \text{ J/mol} \cdot \text{K})(293 \text{ K})}{28.8 \times 10^{-3} \text{ kg/mol}}} = 344 \text{ m/s.}$$

This agrees with the measured speed of sound at this temperature to within 0.3 percent.

The ear is sensitive to a range of sound frequencies from about 20 Hz to about 20,000 Hz. From the relation $v = \lambda f$ the corresponding wavelength range at 20°C is from about 17 m, corresponding to a 20-Hz note, to about 1.7 cm, corresponding to 20,000 Hz.

Bats can hear much higher frequencies. Like dolphins, bats use high-frequency sound waves for navigation. A typical frequency is 100 kHz; the corresponding wavelength in air at 20°C is about 3.4 mm, small enough for them to detect the flying insects they eat.

In this discussion we have ignored the *molecular* nature of a gas and have treated it as a continuous medium. We know that a gas is actually composed of molecules in random motion, separated by distances that are large in comparison with their diameters. The vibrations that constitute a wave in a gas are superposed on the random thermal motion. At atmospheric pressure, a molecule travels an average distance (the mean free path) of about 10^{-7} m between collisions, while the displacement amplitude of a faint sound may be only 10^{-9} m. We can think of a gas with a sound wave passing through as being comparable to a swarm of bees; the swarm as a whole oscillates slightly while individual insects move about through the swarm, apparently at random.

19–8 ENERGY IN WAVE MOTION

Every wave motion has *energy* associated with it. The energy we receive from sunlight and the destructive effects of ocean surf and earthquakes bear this out. To produce any of the wave motions we have discussed in this chapter, we have to apply a force to a portion of the wave medium; the point where the force is applied moves, so we do *work* on the system. As the wave propagates, each portion of the medium exerts a force and does work on the adjoining portion. In this way a wave can transport energy from one region of space to another.

As an example of energy considerations in wave motion, let's look again at transverse waves on a string. How is energy transferred from one portion of string to another? Picture a wave traveling from left to right (the positive x-direction) on the string, and consider a particular point a on the string (Fig. 19–13a). The string to the left of a exerts a force on the string to the right of it, and vice versa. In Fig. 19–13b the string to the left of a has been removed, and the force it exerts at a is represented by the components F

19–13 (a) Point a on a string carrying a wave from left to right. (b) The components of the force exerted on the right part of the string by the left part at point a.

and F_y, as we did in Figs. 19–7 and 19–8. We note again that F_y/F is equal to the negative of the *slope* of the string at a, which is also given by $\partial y/\partial x$. Putting these together, we have

$$F_y(x, t) = -F \frac{\partial y(x, t)}{\partial x}. \tag{19–28}$$

We need the negative sign because F_y is negative when the slope is positive. We write the vertical force as $F_y(x, t)$ as a reminder that its value may be different at different points along the string and at different times.

When point a moves in the y-direction, the force F_y does *work* on this point and therefore transfers energy into the part of the string to the right of a. The corresponding power P (rate of doing work) at the point a is the transverse force $F_y(x, t)$ at a times the transverse velocity $v_y(x, t) = \partial y(x, t)/\partial t$ of that point:

$$P(x, t) = F_y(x, t)v_y(x, t) = -F \frac{\partial y(x, t)}{\partial x} \frac{\partial y(x, t)}{\partial t}. \tag{19–29}$$

This power is the *instantaneous* rate at which energy is transferred along the string. Its value depends on the position x on the string and on the time t. Note that energy is being transferred only at points where the string has a nonzero slope ($\partial y/\partial x$ is nonzero), so that there is a transverse component of the tension force, and where the string has a nonzero transverse velocity ($\partial y/\partial t$ is nonzero) so that the transverse force can do work.

Equation (19–29) is valid for *any* wave on a string, sinusoidal or not. For a sinusoidal wave with wave function given by Eq. (19–7), we have

$$y(x, t) = A \sin(\omega t - kx),$$

$$\frac{\partial y(x, t)}{\partial x} = -kA \cos(\omega t - kx),$$

$$\frac{\partial y(x, t)}{\partial t} = \omega A \cos(\omega t - kx),$$

$$P(x, t) = Fk\omega A^2 \cos^2(\omega t - kx). \tag{19–30}$$

By using the relations $\omega = vk$ and $v^2 = F/\mu$, we can also express Eq. (19–30) in the alternative form

$$P(x, t) = \sqrt{\mu F}\,\omega^2 A^2 \cos^2(\omega t - kx). \tag{19–31}$$

The $\cos^2$ function is never negative, so the instantaneous power in a sinusoidal wave is either positive (so that energy flows in the positive x-direction) or zero (at points where there is no energy transfer). Energy is never transferred in the direction opposite to the direction of wave propagation.

The maximum value of the instantaneous power $P(x, t)$ occurs when the $\cos^2$ function has the value unity:

$$P_{max} = \sqrt{\mu F}\,\omega^2 A^2. \tag{19–32}$$

To obtain the *average* power from Eq. (19–31), we note that the *average* value of the $\cos^2$ function, averaged over any whole number of cycles, is $\frac{1}{2}$. Hence the average power is

$$P_{av} = \frac{1}{2}\sqrt{\mu F}\,\omega^2 A^2 \qquad \text{(average power, sinusoidal wave on a string).} \tag{19–33}$$

The rate of energy transfer is proportional to the square of the amplitude and to the square of the frequency.

Power relationships analogous to Eq. (19–33) can be worked out for longitudinal waves. We won't go into detail; the results are most easily stated in terms of the average power *per unit cross-section area* in the wave motion. The average power per unit area is called the **intensity,** denoted by *I*. For fluids in a pipe it is given by

$$I = \frac{1}{2}\sqrt{\rho B}\,\omega^2 A^2, \qquad (19\text{–}34)$$

and for a solid rod it is given by

$$I = \frac{1}{2}\sqrt{\rho Y}\,\omega^2 A^2. \qquad (19\text{–}35)$$

Again the power is proportional to A^2 and to ω^2.

The proportionality of the rate of energy transfer to the square of the amplitude is a general result for waves of all types; the proportionality constant is different for different types of waves. However, the energy transfer rate is proportional to ω^2 only for mechanical waves. For electromagnetic waves, the intensity is independent of the value of ω.

EXAMPLE 19–8

a) In Examples 19–2 and 19–3, at what maximum rate does Throcky put energy into the rope? That is, what is his maximum instantaneous power? b) What is his average power? c) As Throcky tires, the amplitude decreases. What is the average power when the amplitude has dropped to 7.50 mm?

SOLUTION a) From Eq. (19–32),

$$P_{max} = \sqrt{\mu F}\,\omega^2 A^2$$
$$= \sqrt{(0.250 \text{ kg/m})(36.0 \text{ N})}\,(4.00\pi \text{ rad/s})^2 (0.075 \text{ m})^2$$
$$= 2.66 \text{ W}.$$

b) From Eqs. (19–32) and (19–33), $P_{av} = \frac{1}{2}P_{max}$, so

$$P_{av} = \frac{1}{2}(2.66 \text{ W}) = 1.33 \text{ W}.$$

c) This amplitude is $\frac{1}{10}$ of the value we used in parts (a) and (b). The average power is proportional to the *square* of the amplitude, so now the average power is

$$P_{av} = \frac{1}{10^2}(1.33 \text{ W}) = 0.0133 \text{ W} = 13.3 \text{ mW}.$$

EXAMPLE 19–9

Sound waves in the outer ear Sound waves entering the human ear first pass through the auditory canal before reaching the eardrum (Fig. 19–14). The auditory canal of a typical adult is 2.5 cm long and 7.0 mm in diameter. When you listen to ordinary conversation, the intensity of sound waves in the auditory canal is about 3.2×10^{-6} W/m^2. a) What is the average power delivered to the eardrum? b) Find the amplitude of these sound waves. Assume a frequency $f = 100$ Hz (typical of many of the sounds used in human speech).

SOLUTION a) The intensity *I* of the wave equals the average power divided by the cross-section area of the wave, so average power equals intensity *times* this area. The wave moves over the entire auditory canal of radius $r = (7.0 \text{ mm})/2 = 3.5$ mm, so the cross-section area is $\pi r^2 = \pi(3.5 \times 10^{-3} \text{ m})^2 = 3.8 \times 10^{-5}$ m^2. The average power is therefore

$$P_{av} = I\pi r^2 = (3.2 \times 10^{-6} \text{ W/m}^2)(3.8 \times 10^{-5} \text{ m}^2)$$
$$= 1.2 \times 10^{-10} \text{ W}.$$

This minuscule amount of power is absorbed by the eardrum, whose oscillations are transformed in the middle and inner ears into an electrical signal that is sent to the brain. A healthy ear can detect sound intensities as low as 10^{-12} W/m^2, in which case the average power delivered is only 3.8×10^{-17} W. Human hearing is sensitive indeed!

b) For waves propagating in a fluid in a pipe, the relationship between intensity *I* and amplitude *A* is given by Eq. (19–34), $I = \frac{1}{2}\sqrt{\rho B}\,\omega^2 A^2$. Solving this equation for *A*, we find

$$A = \frac{1}{\omega}\frac{\sqrt{2I}}{(\rho B)^{1/4}}.$$

The fluid is air, so we can use the results of Section 19–7 for sound waves in gases; the bulk modulus to be used is $B_{ad} = \gamma p$. For air, $\gamma = 1.40$, and normal atmospheric pressure is $p = 1.013 \times 10^5$ Pa, so $B_{ad} = (1.40)(1.013 \times 10^5 \text{ Pa}) = 1.42 \times 10^5$ Pa. We can find the density of the air using the ideal-gas law in the form of Eq. (16–5), $\rho = pM/RT$. Using room temperature (20°C,

19–14 The anatomy of the human ear.

or 293 K), we have

$$\rho = \frac{(1.013 \times 10^5 \text{ Pa})(28.8 \times 10^{-3} \text{ kg/mol})}{(8.315 \text{ J/mol} \cdot \text{K})(293 \text{ K})}$$

$$= 1.20 \text{ kg/m}^3.$$

Substituting into the above expression for the amplitude A, we get

$$A = \frac{1}{2\pi(100 \text{ s}^{-1})} \frac{\sqrt{2(3.2 \times 10^{-6} \text{ W/m}^2)}}{((1.20 \text{ kg/m}^3)(1.42 \times 10^5 \text{ Pa}))^{1/4}}$$

$$= 2.0 \times 10^{-7} \text{ m}.$$

This is a very tiny distance, yet this much displacement of air molecules is perfectly detectable by a person with normal hearing. In fact, a sound of this intensity would still be audible if we increased the frequency by a factor of 100 to $f = 10,000$ Hz. Our calculations show that the amplitude A is inversely proportional to the frequency, so the amplitude would be 0.01 as great, or 2.0×10^{-9} m. Your ear can detect the motion of air molecules over a distance that is only a few times larger than the size of the molecules themselves! We will explore the sensitivity of the ear further in Chapter 21.

SUMMARY

- A wave is any disturbance from an equilibrium condition that propagates from one region to another. A mechanical wave always travels within some material called the medium. In a periodic wave, the motion of each point of the medium is periodic. When the motion is sinusoidal, the wave is called a sinusoidal wave. The frequency f of a periodic wave is the number of repetitions per unit time, and the period T is the time for one cycle. The wavelength λ is the distance over which the wave pattern repeats. The speed of propagation v is the speed with which the wave disturbance travels. For any periodic wave, these quantities are related by

$$v = \lambda f. \tag{19–1}$$

- A wave function $y(x, t)$ describes the displacements of individual particles in the medium. For a sinusoidal wave moving in the $+x$-direction,

$$y(x,t) = A \sin \omega \left(t - \frac{x}{v} \right) = A \sin 2\pi f \left(t - \frac{x}{v} \right) \tag{19–3}$$

$$= A \sin 2\pi \left(\frac{t}{T} - \frac{x}{\lambda} \right), \tag{19–4}$$

or

$$y(x, t) = A \sin (\omega t - kx), \tag{19–7}$$

where the wave number k is defined as $k = 2\pi/\lambda$ and $\omega = vk$. In all three forms, A is the amplitude, the maximum displacement of a particle from its equilibrium position. If the wave is moving in the $-x$-direction, the minus sign in Eqs. (19–3), (19–4), and (19–7) is replaced by a plus sign.

■ The wave function must obey a partial differential equation called the wave equation,

$$\frac{\partial^2 y(x,t)}{\partial x^2} = \frac{1}{v^2}\frac{\partial^2 y(x,t)}{\partial t^2}. \qquad (19\text{--}12)$$

■ The speed of a transverse wave on a string with tension F and mass per unit length μ is

$$v = \sqrt{\frac{F}{\mu}} \qquad \text{(transverse wave on a string).} \qquad (19\text{--}13)$$

■ The speed of a longitudinal wave in a fluid with bulk modulus B and density ρ is

$$v = \sqrt{\frac{B}{\rho}} \qquad \text{(longitudinal wave in a fluid).} \qquad (19\text{--}21)$$

■ The speed of a longitudinal wave in a solid rod with Young's modulus Y and density ρ is

$$v = \sqrt{\frac{Y}{\rho}} \qquad \text{(longitudinal wave in a solid rod).} \qquad (19\text{--}22)$$

■ Sound propagation in gases is ordinarily an adiabatic process. The speed of sound in an ideal gas is

$$v = \sqrt{\frac{\gamma p}{\rho}} = \sqrt{\frac{\gamma RT}{M}} \qquad \text{(sound wave in an ideal gas).} \quad (19\text{--}26),(19\text{--}27)$$

■ Wave motion conveys energy from one region to another. For a sinusoidal wave on a stretched string, the average power is

$$P_{av} = \frac{1}{2}\sqrt{\mu F}\,\omega^2 A^2. \qquad (19\text{--}33)$$

For a longitudinal wave, the average power per unit cross-section area in the wave motion is called the intensity, I.

DISCUSSION QUESTIONS

Q19–1 A common pastime at sporting events is "doing the wave," in which the spectators in one section raise and lower their arms, signaling the spectators in the next section to do the same, and so on. In the language of this chapter, is this truly a wave? Why or why not?

Q19–2 Explain why it is necessary to include the factor 2π in Equation (19–4).

Q19–3 What kinds of energy are associated with waves on a stretched string? How could such energy be detected experimentally?

Q19–4 The amplitude of a wave decreases gradually as the wave travels down a long stretched string. What happens to the energy of the wave when this happens?

Q19–5 For the wave motions discussed in this chapter, does the speed of propagation depend on the amplitude? How can you tell?

Q19–6 The speed of ocean waves depends on the depth of the water; the deeper the water, the faster the wave travels. Use this to explain why ocean waves crest and "break" as the wave nears the shore.

Q19–7 Is it possible to have a longitudinal wave on a stretched string? Why or why not? Is it possible to have a transverse wave on a steel rod? Again, why or why not? If your answer is yes in either case, explain how you would create such a wave.

Q19–8 An echo is sound reflected from a distant object, such as a wall or a cliff. Explain how you can determine how far away the object is by timing the echo.

Q19–9 Why do you see lightning before you hear the thunder? A familiar rule of thumb is to start counting slowly, once per second, when you see the lightning; when you hear the thunder, divide the number you have reached by 3 to obtain your distance from the lightning in kilometers (or by 5 to obtain your distance in miles). Why does this work? Or does it?

Q19–10 For transverse waves on a string, is the wave speed the same as the maximum speed of any part of the string? Or is there a different relation between these two speeds? What about for longitudinal waves on a rod? Explain your reasoning.

Q19–11 Children make toy telephones by sticking each end of a long string through a hole in the bottom of a paper cup and knotting it so it will not pull out. When the spring is pulled taut, sound can be transmitted from one cup to the other. How does this work? Why is the transmitted sound louder than the sound traveling through air for the same distance?

Q19–12 The four strings on a violin have different thicknesses, but are all under approximately the same tension. Do waves travel faster on the thick strings or the thin strings? Why?

Q19–13 When a rock is thrown into a pond and the resulting ripples spread in ever-widening circles, the amplitude decreases with increasing distance from the center. Why?

Q19–14 When sound travels from air into water, does the frequency of the wave change? The speed? The wavelength? Explain your reasoning.

Q19–15 The hero of a western movie listens for an oncoming train by putting his ear to the track. Why does this method give early warning of the approach of a train?

Q19–16 In a transverse wave on a string, the motion of the string is perpendicular to the length of the string. How, then, is it possible for energy to move along the length of the string?

Q19–17 Justify the following statement: "A wave transfers energy fastest where the particles of the medium are moving fastest." In particular, explain how can this can be true for a transverse wave, in which the particles of the medium move in a direction perpendicular to that of wave propagation.

Q19–18 In speaker systems designed for high-fidelity music reproduction, the "tweeters" that reproduce the high frequencies are always much smaller than the "woofers" used for low frequencies. Explain why, using Eq. (19–34).

Q19–19 At very high frequencies, above 5×10^8 Hz or so, the propagation of sound in air is isothermal, not adiabatic. Explain why this should be so. (*Hint:* Very high frequencies mean very short wavelengths.)

EXERCISES

SECTION 19–3 PERIODIC WAVES

19–1 A fisherman notices that his boat is moving up and down periodically owing to waves on the surface of the water. It takes 2.5 s for the boat to travel from its highest point to its lowest, a total distance of 0.62 m. The fisherman sees that the wave crests are spaced 6.0 m apart. a) How fast are the waves traveling? b) What is the amplitude of each wave? c) If the total vertical distance traveled by the boat were 0.30 m, but the other data remained the same, how would the answers to parts (a) and (b) be affected? d) Would you expect that the motion of the boat would be purely vertical? Why or why not?

19–2 The sound waves from a loudspeaker spread out nearly uniformly in all directions when their wavelengths are large in comparison with the diameter of the speaker. When the wavelength is small in comparison with the diameter of the speaker, much of the sound energy is concentrated forward. For a speaker with a diameter of 12.0 cm, compute the frequency for which the wavelength of the sound waves in air ($v = 344$ m/s) is a) ten times the diameter of the speaker; b) equal to the diameter of the speaker; c) one-tenth the diameter of the speaker. (For a second factor in the optimal size of the speaker, see Question Q19–18.)

19–3 The speed of sound in air at 20°C is 344 m/s. a) What is the wavelength of a sound wave with a frequency of 784 Hz, corresponding to the note G_5 on a piano? b) What is the frequency of a sound wave with a wavelength of 0.0655 mm? (This frequency is too high for you to hear.)

19–4 The speed of radio waves in vacuum (equal to the speed of light) is 3.00×10^8 m/s. Find the wavelength for a) an AM radio station with frequency 540 kHz; b) an FM radio station with frequency 104.5 MHz.

19–5 Provided that the amplitude is sufficiently great, the human ear can respond to longitudinal waves over a range of frequencies from about 20.0 Hz to about 20,000 Hz. Compute the wavelengths corresponding to these frequencies a) for waves in air ($v = 344$ m/s); b) for waves in water ($v = 1480$ m/s).

SECTION 19–4 MATHEMATICAL DESCRIPTION OF A WAVE

19–6 A certain transverse wave is described by

$$y(x, t) = (6.50 \text{ mm}) \sin 2\pi \left(\frac{t}{0.0360 \text{ s}} - \frac{x}{28.0 \text{ cm}} \right)$$

Determine the wave's a) amplitude; b) wavelength; c) frequency; d) speed of propagation; e) direction of propagation.

19–7 Transverse waves on a string have wave speed 8.00 m/s, amplitude 0.0700 m, and wavelength 0.320 m. The waves travel in the $-x$-direction, and at $t = 0$ the $x = 0$ end of the string has zero displacement and is moving in the $+y$-direction. a) Find the frequency, period, and wave number of these waves. b) Write a wave function describing the wave. c) Find the transverse displacement of a particle at $x = 0.360$ m at time $t = 0.150$ s. d) How much time must elapse from the instant in part (c) until the particle at $x = 0.360$ m next has zero displacement? e) How much additional time must elapse from the instant in part (d) until the next time the particle has zero displacement?

19–8 Redraw Figs. 19–6a and 19–6b for the case in which the wave is propagating to the *left*.

19–9 Show that the following functions satisfy the wave equation, Eq. (19–12): a) $y(x, t) = A \sin(\omega t + kx)$; b) $y(x, t) = A \cos(\omega t + kx)$. c) In what directions are these waves traveling? How can you tell? d) For the wave of part (b), write the equations for the transverse velocity and transverse acceleration of a particle at point x.

19–10 a) For a wave described by $y(x, t) = A \sin (\omega t − kx)$, graph y, v_y, and a_y as functions of x for time $t = 0$. b) Consider the following points on the string: i) $x = 0$; ii) $x = \pi/4k$; iii) $x = \pi/2k$; iv) $x = 3\pi/4k$; v) $x = \pi/k$; vi) $x = 5\pi/4k$; vii) $x = 3\pi/2k$; and viii) $x = 7\pi/4k$. For a particle at each of these points at $t = 0$, describe in words whether the particle is moving and in what direction, and whether the particle is speeding up, slowing down, or instantaneously not accelerating.

19–11 A sinusoidal wave is propagating along a stretched string that lies along the x-axis. The displacement of the string as a function of time is graphed in Fig. 19–15 for particles at $x = 0$ and at $x = 0.0900$ m. a) What is the amplitude of the wave? b) What is the period of the wave? c) You are told that the two points $x = 0$ and $x = 0.0900$ m are within one wavelength of each other. If the wave is moving in the $+x$-direction, determine the wavelength and the wave speed. d) If instead the wave is moving in the $−x$-direction, determine the wavelength and the wave speed. e) Would it be possible to unambiguously determine the wavelength in parts (c) and (d) if you were not told that the two points were within one wavelength of each other? Why or why not?

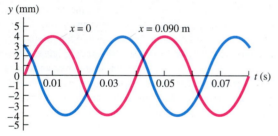

FIGURE 19–15 Exercise 19–11.

19–12 Speed of Propagation vs. Particle Speed. a) Show that Eq. (19–3) may be written as

$$y(x, t) = -A \sin \left[\frac{2\pi}{\lambda} (x - vt) \right].$$

b) Use $y(x, t)$ to find an expression for the transverse velocity v_y of a particle in the string on which the wave travels. c) Find the maximum speed of a particle of the string. Under what circumstances is this equal to the propagation speed v? Less than v? Greater than v?

19–13 A transverse wave on a string has amplitude 0.300 cm, wavelength 12.0 cm, and speed 6.00 cm/s. It is represented by $y(x, t)$ as given in Exercise 19–12. a) At time $t = 0$, compute y at 1.5-cm intervals of x (that is, at $x = 0$, $x = 1.5$ cm, $x = 3.0$ cm, and so on) from $x = 0$ to $x = 12.0$ cm. Graph the results. This is the shape of the string at time $t = 0$. b) Repeat the calculations for the same values of x at times $t = 0.400$ s and $t = 0.800$ s. Graph the shape of the string at these instants. In what direction is the wave traveling?

SECTION 19–5 SPEED OF A TRANSVERSE WAVE

19–14 With what tension must a rope of length 2.50 m and mass 0.120 kg be stretched for transverse waves of frequency 40.0 Hz to have a wavelength of 0.750 m?

19–15 One end of a horizontal rope is attached to a prong of an electrically driven tuning fork that vibrates at 120 Hz. The other end passes over a pulley and supports a 1.50-kg mass. The linear mass density of the rope is 0.0550 kg/m. a) What is the speed of a transverse wave on the rope? b) What is the wavelength? c) How would your answers to parts (a) and (b) be changed if the mass were increased to 3.00 kg?

19–16 A cowgirl ties one end of a 10.0-m-long rope to a fence post and pulls on the other end so the rope is stretched horizontally with a tension of 140 N. The mass of the rope is 0.800 kg. a) What is the speed of transverse waves on the rope? b) If the cowgirl moves the free end up and down with a frequency of 1.20 Hz, what is the wavelength of the transverse waves on the rope ? c) The cowgirl pulls harder on the rope so that the tension is doubled, to 280 N. With what frequency must she move the free end of the rope up and down to produce transverse waves of the same wavelength as in part (a)?

19–17 One end of a 14.0-m-long rubber tube, with a total mass of 0.800 kg, is fastened to a fixed support. A cord attached to the other end passes over a pulley and supports an object with a mass of 7.50 kg. The tube is struck a transverse blow at one end. Find the time required for the pulse to reach the other end.

19–18 If in Example 19–4 (Section 19–5) we do *not* neglect the weight of the rope, what is the wave speed a) at the bottom of the rope? b) at the middle of the rope? c) at the top of the rope?

SECTION 19–6 SPEED OF A LONGITUDINAL WAVE
SECTION 19–7 SOUND WAVES IN GASES

19–19 a) In a liquid with density 1300 kg/m^3, longitudinal waves with frequency 400 Hz are found to have wavelength 8.00 m. Calculate the bulk modulus of the liquid. b) A metal bar with a length of 1.50 m has density 6400 kg/m^3. Longitudinal sound waves take 3.90×10^{-4} s to travel from one end of the bar to the other. What is Young's modulus for this metal?

19–20 A major earthquake centered on Loma Prieta, California, near San Francisco, occurred at 5:04 P.M. local time on October 17, 1989 (in UTC, Coordinated Universal Time, 0h 4m 15s on October 18, 1989). The primary seismic waves (P waves) from such an earthquake are longitudinal waves that travel through the earth's crust. P waves were detected at Caracas, Venezuela, at 0h 13m 54s UTC; at Kevo, Finland, at 0h 15m 35s UTC; and at Vienna, Austria, at 0h 17m 02s UTC. The distances the P waves traveled from Loma Prieta were 6280 km to Caracas, 8690 km to Kevo, and 9650 km to Vienna. a) Use the arrival times to calculate the average speed of the P waves that traveled to these three cities. How can you account for any differences between the average speeds? b) The average density of the earth's crust is about 3.3 g/cm^3. Use this value to calculate the bulk modulus of the earth's crust along the path traveled by the P waves to each of the three cities. How do your answers compare to the bulk moduli in Table 11–1?

19–21 A submerged scuba diver hears the sound of a boat horn directly above her on the surface of the lake. At the same time, a friend on dry land 22.0 m from the boat also hears the horn (Fig. 19–16). The horn is 1.2 m above the surface of the water. What is the distance (labeled by "?" in Fig. 19–16) from the horn to the diver? Both air and water are at 20°C.

FIGURE 19–16 Exercise 19–21.

19–22 At a temperature of 27.0°C, what is the speed of longitudinal waves in a) hydrogen (molar mass 2.02 g/mol)? b) helium (molar mass 4.00 g/mol)? c) argon (molar mass 39.9 g/mol)? See Table 17–1 for values of γ. d) Compare your answers for parts (a), (b) and (c) with the speed in air at the same temperature.

19–23 A jet airliner is cruising at high altitude at a speed of 850 km/h (about 530 mi/h). This is equal to 0.85 times the speed of sound at that altitude (also called "Mach 0.85"). a) What is the air temperature at this altitude? b) With this data, can you determine the air pressure at this altitude? Explain.

19–24 Use the definition $B = -V\, dp/dV$ and the relation between p and V for an isothermal process to derive Eq. (19–25).

19–25 An 80.0-m-long brass rod is struck at one end. A person at the other end hears two sounds as a result of two longitudinal waves, one traveling in the metal rod and the other traveling in the air. What is the time interval between the two sounds? The speed of sound in air is 344 m/s; relevant information about brass can be found in Table 11–1 and Table 14–1.

19–26 What is the difference between the speed of longitudinal waves in air at 27.0°C and their speed at −13.0°C?

19–27 What must be the stress (F/A) in a stretched wire of a material whose Young's modulus is Y for the speed of longitudinal waves to equal 30 times the speed of transverse waves?

SECTION **19–8** **ENERGY IN WAVE MOTION**

19–28 A piano wire with mass 3.00 g and length 80.0 cm is stretched with a tension of 25.0 N. A wave with frequency 120.0 Hz and amplitude 1.6 mm travels along the wire. a) Calculate the average power carried by the wave. b) What happens to the average power if the wave amplitude is halved?

19–29 A longitudinal wave of frequency 220 Hz travels down a copper rod of radius 8.00 mm. The average power in the wave is 6.50 μW. (See Tables 11–1 and 14–1 for the necessary data about copper.) a) Find the wavelength of the wave. b) Find the amplitude A of the wave. c) Find the maximum longitudinal velocity of a particle in the rod.

19–30 Longitudinal Waves in Different Fluids. a) A longitudinal wave propagating in a water-filled pipe has intensity 3.00×10^{-6} W/m^2 and frequency 3400 Hz. Find the amplitude A and wavelength λ of the wave. Water has density 1000 kg/m^3 and bulk modulus 2.18×10^9 Pa. b) If the pipe is filled with air at pressure 1.00×10^5 Pa and density 1.20 kg/m^3, what will be the amplitude A and wavelength λ of a longitudinal wave with the same intensity and frequency as in part (a)? c) In which fluid is the amplitude larger, water or air? What is the ratio of the two amplitudes? Why is this ratio so different from one?

PROBLEMS

19–31 A transverse sine wave with an amplitude of 2.50 mm and a wavelength of 1.80 m travels from left to right along a long horizontal stretched string with a speed of 36.0 m/s. Take the origin at the left end of the undisturbed string. At time $t = 0$ the left end of the string is at the origin and is moving upward. a) What are the frequency, angular frequency, and wave number of the wave? b) What is the function $y(x, t)$ that describes the wave? c) What is $y(t)$ for a particle at the left end of the string? d) What is $y(t)$ for a particle 1.35 m to the right of the origin? e) What is the maximum magnitude of transverse velocity of any particle of the string? f) Find the transverse displacement and the transverse velocity of a particle 1.35 m to the right of the origin at time $t = 0.0625$ s.

19–32 A transverse wave on a rope is given by

$$y(x, t) = (0.750 \text{ cm}) \sin \pi[(250 \text{ s}^{-1})t + (0.400 \text{ cm}^{-1})x].$$

a) Find the amplitude, period, frequency, wavelength, and speed of propagation. b) Sketch the shape of the rope at the following values of t: 0, 0.0005 s, and 0.0010 s. c) Is the wave traveling in the +x- or −x-direction? d) The mass per unit length of the rope is 0.500 kg/m. Find the tension. e) Find the average power of this wave.

19–33 Three pieces of string, each of length L, are joined together end to end to make a combined string of length $3L$. The first piece of string has mass per unit length μ_1, the second piece has mass per unit length $\mu_2 = 4\mu_1$, and the third piece has mass per unit length $\mu_3 = \mu_1/4$. a) If the combined string is under tension F, how much time does it take a transverse wave to travel the entire length $3L$? Give your answer in terms of L, F, and μ_1. b) Does your answer to part (a) depend on the order in which the three pieces are joined together? Explain.

19–34 An ant of mass m is standing peacefully on top of a horizontal stretched rope. The rope has mass per unit length μ and is under tension F. Without warning, Throckmorton starts a sinusoidal transverse wave of wavelength λ propagating along the rope. The motion of the rope is in a vertical plane. What minimum wave amplitude will make the ant become momentarily weightless? Assume that m is so small that the presence of the ant has no effect on the propagation of the wave.

19–35 When a transverse sinusoidal wave is present on a string, the particles of the string undergo simple harmonic motion. This is the same motion as that of a mass m attached to an ideal spring of force constant k', for which the angular frequency of oscillation was found in Chapter 13 to be $\omega = \sqrt{k'/m}$. Consider a string with tension F and mass per unit length μ along which is propagating a sinusoidal wave with amplitude A and wavelength λ. a) Find the "force constant" k' of the restoring force that acts on a short segment of the string of length Δx (where $\Delta x \ll \lambda$). b) How does the "force constant" calculated in part (b) depend on F, μ, A, and λ? Explain the physical reasons why this should be so.

19–36 Motion of a Wave Pulse. The sinusoidal waves described in Section 19–4 can be produced on a string by continually oscillating one end of the string. If instead the end of the string is given a single shake, a *wave pulse* propagates down the string. A particular wave pulse is described by the function

$$y(x, t) = \frac{A^3}{A^2 + (x - vt)^2},$$

where $A = 1.00$ cm and $v = 20.0$ m/s. a) Sketch the pulse as a function of x at $t = 0$. How far along the string does the pulse extend? b) Sketch the pulse as a function of x at $t = 0.001$ s. c) At the point $x = 4.50$ cm, at what time t is the displacement maximum? At what two times is the displacement at $x = 4.50$ cm equal to half its maximum value? d) Show that $y(x, t)$ satisfies the wave equation, Eq. (19–12).

19–37 A Non-Sinusoidal Wave. The shape of a wave on a string at a specific instant is shown in Fig. 19–17. The wave is propagating to the right, in the +x-direction. a) Determine the direction of the transverse *velocity* of each of the six labeled points on the string. If the velocity is zero, state so. Explain your reasoning. b) Determine the direction of the transverse *acceleration* of each of the six labeled points on the string. Explain your reasoning. c) How would your answers be affected if the wave was propagating to the left, in the –x-direction?

FIGURE 19–17 Problem 19–37.

19–38 Locating Lightning by Radio. While doing her physics homework, a student in the Boston area is listening in her dorm room to the radio broadcast of a Red Sox–Yankees baseball game at Boston's Fenway Park. In the bottom of the fourth inning a thunderstorm moving from west to east makes its presence known in three ways: (1) The student sees a lightning flash (and hears the electromagnetic pulse on her radio receiver); (2) 3.00 s later, she hears the thunder over the radio. (3) 4.43 s after the lightning flash, the thunder rattles her window. By a previous careful measurement she knows that she is 1.12 km due north of the broadcast booth at the ballpark. The speed of sound is 344 m/s. Where did the lightning flash occur in relation to the ballpark?

19–39 Waves of Arbitrary Shape. a) Explain why *any* wave described by a function of the form $y(x, t) = f(t - x/v)$ moves in the +x-direction with speed v. b) Show that $y(x, t) = f(t - x/v)$ satisfies the wave equation, no matter what the functional form of f. To do this, write $y(x, t) = f(u)$, where $u = t - x/v$. Then, to take partial derivatives of $y(x, t)$, use the chain rule:

$$\frac{\partial y(x,t)}{\partial t} = \frac{df(u)}{du}\frac{\partial u}{\partial t} = \frac{df(u)}{du},$$

$$\frac{\partial y(x,t)}{\partial x} = \frac{df(u)}{du}\frac{\partial u}{\partial x} = \frac{df(u)}{du}\left(-\frac{1}{v}\right).$$

c) A wave pulse is described by the function $y(x, t) = De^{-(Bx - Ct)^2}$, where B, C, and D are all positive constants. What is the speed of this wave?

19–40 Equation (19–7) for a sinusoidal wave can be made more general by including a phase angle ϕ, where $0 \le \phi \le 2\pi$ (in radians) Then the wave function $y(x, t)$ becomes

$$y(x, t) = A \sin(\omega t - kx + \phi).$$

a) Sketch the wave as a function of x at $t = 0$ for $\phi = 0$, $\phi = \pi/4$, $\phi = \pi/2$, $\phi = 3\pi/4$, and $\phi = 3\pi/2$. b) Calculate the transverse velocity $v_y = \partial y/\partial t$. c) At $t = 0$, a particle on the string at $x = 0$ has displacement $y = A/\sqrt{2}$. Is this enough information to determine the value of ϕ? If you are told in addition that a particle at $x = 0$ is moving toward $y = 0$ at $t = 0$, what is the value of ϕ? d) Explain in general what you must know about the wave's behavior at a given instant to determine the value of ϕ.

19–41 a) Show that Eq. (19–33) can also be written as $P_{av} = \frac{1}{2}Fk\omega A^2$, where k is the wave number of the wave. b) If the tension F in the string is quadrupled while the amplitude A is kept the same, how must k and ω each change to keep the average power constant? (*Hint:* Recall Eq. (19–6).)

19–42 Instantaneous Power in a Wave. a) Graph $y(x, t)$ as given by Eq. (19–7) as a function of x for a given time t (say, $t = 0$). On the same axes, make a graph of the instantaneous power $P(x, t)$ as given by Eq. (19–31). b) Explain the connection between the slope of the graph of $y(x, t)$ versus x and the value of $P(x, t)$. In particular, explain what is happening at points where $P = 0$, where there is no instantaneous energy transfer. c) The quantity $P(x, t)$ always has the same sign. What does this imply about the direction of energy flow? d) Consider a wave moving in the –x-direction, for which $y(x, t) = A \sin(\omega t + kx)$. Calculate $P(x, t)$ for this wave, and make a graph of $y(x, t)$ and $P(x, t)$ as functions of x for a given time t (say, $t = 0$). What differences arise from reversing the direction of the wave?

19–43 A metal wire, with a density ρ and a Young's modulus Y, is stretched between rigid supports. At temperature T the speed of a transverse wave is found to be v_1. When the temperature is increased to $T + \Delta T$, the speed decreases to $v_2 < v_1$. Determine the coefficient of linear expansion of the wire.

CHALLENGE PROBLEMS

19–44 Longitudinal Waves on a Spring. A long spring such as a Slinky™ is often used to demonstrate longitudinal waves.

a) Show that if a spring that obeys Hooke's law has mass m, length L, and force constant k', the speed of longitudinal waves on the spring is $v = L\sqrt{k'/m}$. b) Evaluate v for a spring with $m = 0.250$ kg, $L = 2.00$ m, and $k' = 1.50$ N/m.

19–45 a) Show that for a wave on a string, the kinetic energy *per unit length of string* is

$$u_k(x, t) = \frac{1}{2}\mu v_y{}^2(x, t) = \frac{1}{2}\mu\left(\frac{\partial y(x, t)}{\partial t}\right)^2,$$

where μ is the mass per unit length. b) Calculate $u_k(x, t)$ for a sinusoidal wave given by Eq. (19–7). c) There is also elastic potential energy in the string, associated with the work required to deform and stretch the string. Consider a short segment of string at position x that has unstretched length Δx, as in Fig. 19–8. Neglecting the (small) curvature of the segment, its slope is $\partial y(x, t)/\partial x$. Assume that the displacement of the string from equilibrium is small, so that $\partial y/\partial x$ has a magnitude much less than unity. Show that the stretched length of the segment is approximately

$$\Delta x\left[1 + \frac{1}{2}\left(\frac{\partial y(x, t)}{\partial x}\right)^2\right].$$

(*Hint:* Use the relationship $\sqrt{1 + u} \approx 1 + \frac{1}{2}u$, valid for $|u| \ll 1$.) d) The potential energy stored in the segment equals the work done by the string tension F (which acts along the string) to stretch the segment from its unstretched length Δx to the length calculated in part (c). Calculate this work, and show that the potential energy *per unit length of string* is

$$u_p(x, t) = \frac{1}{2}F\left(\frac{\partial y(x, t)}{\partial x}\right)^2.$$

e) Calculate $u_p(x, t)$ for a sinusoidal wave given by Eq. (19–7). f) Show that $u_k(x, t) = u_p(x, t)$ for all x and t. g) Show $y(x, t)$, $u_k(x, t)$, and $u_p(x, t)$ as functions of x for $t = 0$ in one graph with all three functions on the same axes. Explain why u_k and u_p are maximum where y is zero, and vice versa. h) Show that the instantaneous power in the wave, given by Eq. (19–30), is equal to the total energy per unit length multiplied by the wave speed v. Explain why this result is reasonable.

19–46 A deep-sea diver is suspended beneath the surface of Loch Ness by a 100-m long cable that is attached to a boat on the surface (Fig. 19–18). The diver and his suit have a total mass of 120 kg and a volume of 0.0800 m³. The cable has a diameter of 2.00 cm and a linear mass density of $\mu = 1.10$ kg/m. The diver thinks he sees something moving in the murky depths and jerks the end of the cable back and forth to send transverse waves up the cable as a signal to his companions in the boat. a) What is

FIGURE 19–18 Challenge Problem 19–46.

the tension in the cable at its lower end, where it is attached to the diver? Do not forget to include the buoyant force that the water (density 1000 kg/m³) exerts on him. b) Calculate the tension in the cable a distance x above the diver. The buoyant force on the cable must be included in your calculation. c) The speed of transverse waves on the cable is given by $v = \sqrt{F/\mu}$ (Eq. 19–13) . The speed therefore varies along the cable, since the tension is not constant. (This expression neglects the damping force that the water exerts on the moving cable.) Integrate to find the time required for the first signal to reach the surface.

19–47 A uniform rope of length L and mass m is held at one end and whirled in a horizontal circle with angular velocity ω. Ignore the force of gravity on the rope. Find the time required for a transverse wave to travel from one end of the rope to the other.

19–48 For longitudinal waves in a solid rod, prove that the power per unit cross-section area in the wave motion (the *intensity*) is given by Eq. (19–35).

20

Wave Interference and Normal Modes

As these circular waves on the surface of a pond pass through each other, they combine in the simplest possible way: how much the water is disturbed at any point is simply the sum of the disturbances from each individual wave. This same principle of superposition applies to waves of many different kinds, including waves on a string and sound waves in a pipe.

20–1 INTRODUCTION

When a wave strikes the boundaries of its medium, all or part of the wave is *reflected*. When you yell at a building wall or a cliff face some distance away, the sound wave is reflected from the rigid surface, and an *echo* comes back. When you flip the end of a rope whose far end is tied to a rigid support, a pulse travels the length of the rope and is reflected back to you. In both cases the initial and reflected waves overlap in the same region of the medium. This overlapping of waves is called *interference*.

When there are *two* boundary points or surfaces, such as a guitar string that's tied down at both ends, we get repeated reflections. In such situations we find that sinusoidal waves can occur only for certain special frequencies, which are determined by the properties and dimensions of the medium. These special frequencies and their associated wave patterns are called *normal modes*. The pitches of most musical instruments are determined by the normal-mode frequencies of the vibrating part of the instrument. Normal-mode motion also explains why your singing voice sounds better in the shower and why a trained singer's amplified voice can break a crystal goblet if the correct note is sung.

In this chapter our focus will be on the interference of mechanical waves. But interference is also important in nonmechanical waves. It explains the colors seen in soap bubbles and is the basic principle behind holograms and the use of x rays to explore crystal structure. Interference and normal modes also appear in determining the energy levels of atoms. We'll explore these aspects of interference in later chapters.

20–2 BOUNDARY CONDITIONS FOR A STRING AND THE PRINCIPLE OF SUPERPOSITION

As a simple example of wave reflections and the role of the boundary of a wave medium, let's look again at transverse waves on a stretched string. What happens when a wave pulse or a sinusoidal wave arrives at the *end* of the string?

If the end is fastened to a rigid support, it is a *fixed* end that cannot move. The arriving wave exerts a force on the support; the reaction to this force, exerted *by* the support *on* the string, "kicks back" on the string and sets up a *reflected* pulse or wave traveling in the reverse direction. Figure 20–1 is a series of photographs showing the reflection of a pulse at the fixed end of a long coiled spring. The reflected pulse moves in the opposite direction from the initial, or *incident*, pulse, and its displacement is also opposite. This situation is illustrated for a wave pulse on a string in Fig. 20–2a.

The opposite situation from an end that is held stationary is a *free* end, one that is perfectly free to move in the direction perpendicular to the length of the string. For example, the string might be tied to a light ring that slides on a frictionless rod perpendicular to the rope, as in

20–1 A series of images of a wave pulse, equally spaced in time from top to bottom. The pulse starts at the right in the top image, travels to the left, and is reflected from the fixed end at the left.

(a)   (b)

20–2 Reflection of a wave pulse (a) at a fixed end of a string and (b) at a free end. Time increases from top to bottom in each figure.

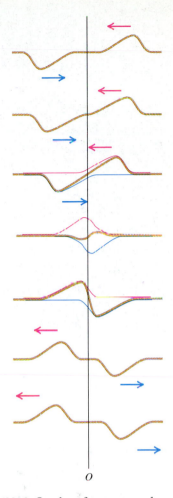

O

20–3 Overlap of two wave pulses traveling in opposite directions with one pulse inverted with respect to the other. Time increases from top to bottom.

Fig. 20–2b. The ring and rod maintain the tension but exert no transverse force. When a wave arrives at this free end, the ring slides along the rod. The ring reaches a maximum displacement, and both it and the string come momentarily to rest, as in the fourth drawing from the top in Fig. 20–2b. But the string is now stretched, giving increased tension, so the free end of the string is pulled back down, and again a reflected pulse is produced. As for a fixed end, the reflected pulse moves in the opposite direction from the initial pulse, but now the direction of the displacement is the same as for the initial pulse. The conditions at the end of the string, such as a rigid support or the complete absence of transverse force, are called **boundary conditions.**

The formation of the reflected pulse is similar to the overlap of two pulses traveling in opposite directions. Figure 20–3 shows two pulses with the same shape, one inverted with respect to the other, traveling in opposite directions. As the pulses overlap and pass each other, the total displacement of the string is the *algebraic sum* of the displacements at that point in the individual pulses. Because these two pulses have the same shape, the total displacement at point O in the middle of the figure is zero at all times. Thus the motion of the right half of the string would be the same if we cut the string at point O,

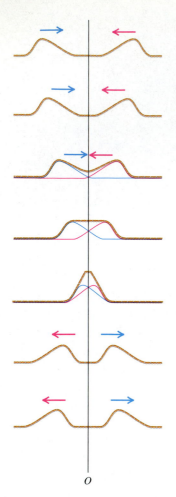

20-4 Overlap of two wave pulses traveling in opposite directions with no inversion of one pulse. Time increases from top to bottom.

threw away the left side, and held the end at *O* fixed. The two pulses on the right side then correspond to the incident and reflected pulses, combining so that the total displacement at *O* is *always* zero. For this to occur, the reflected pulse must be inverted relative to the incident pulse.

Figure 20–4 shows two pulses with the same shape, traveling in opposite directions but *not* inverted relative to each other. The displacement at point *O* in the middle of the figure is not zero, but the slope of the string at this point is always zero. According to Eq. (19–28), this corresponds to the absence of any transverse force at this point. In this case the motion of the right half of the string would be the same as if we cut the string at point *O* and anchored the end with a frictionless sliding ring (Fig. 20–2b) that maintains tension without exerting any transverse force. In other words, this situation corresponds to reflection of a pulse at a free end of a string at point *O*. In this case the reflected pulse is *not* inverted.

THE PRINCIPLE OF SUPERPOSITION

Combining the displacements of the separate pulses at each point to obtain the actual displacement is an example of the **principle of superposition:** when two waves overlap, the actual displacement of any point on the string at any time is obtained by adding the displacement the point would have if only the first wave were present and the displacement it would have if only the second wave were present. In other words, the wave function $y(x, t)$ that describes the resulting motion in this situation is obtained by *adding* the two wave functions for the two separate waves.

Mathematically, this additive property of wave functions follows from the form of the wave equation, Eq. (19–12) or (19–18), which every physically possible wave function must satisfy. Specifically, the wave equation is *linear;* that is, it contains the function $y(x, t)$ only to the first power (there are no terms involving $y(x, t)^2$, $y(x, t)^{1/2}$, etc.). As a result, if any two functions $y_1(x, t)$ and $y_2(x, t)$ satisfy the wave equation separately, their sum $y_1(x, t) + y_2(x, t)$ also satisfies it and is therefore a physically possible motion. Because this principle depends on the linearity of the wave equation and the corresponding linear-combination property of its solutions, it is also called the *principle of linear superposition.* For some physical systems, such as a medium that does not obey Hooke's law, the wave equation is *not* linear; this principle does not hold for such systems.

The principle of superposition is of central importance in all types of waves. When a friend talks to you while you are listening to music, you can distinguish the sound of speech and the sound of music from each other. This is precisely because the total sound wave reaching your ears is the algebraic sum of the wave produced by your friend's voice and the wave produced by the speakers of your stereo. If two sound waves did *not* combine in this simple linear way, the sound you would hear in this situation would be a hopeless jumble. Superposition also applies to electromagnetic waves (such as light) and many other types of waves.

20–3 STANDING WAVES ON A STRING

We have talked about the reflection of a wave *pulse* on a string when it arrives at a boundary point (either a fixed end or a free end). Now let's look at what happens when a *sinusoidal* wave is reflected by a fixed end of a string. We will again approach the problem by considering the superposition of two waves propagating through the string, one representing the original or incident wave and the other representing the wave reflected at the fixed end. The general term **interference** is used to describe the result of two or more waves passing through the same region at the same time.

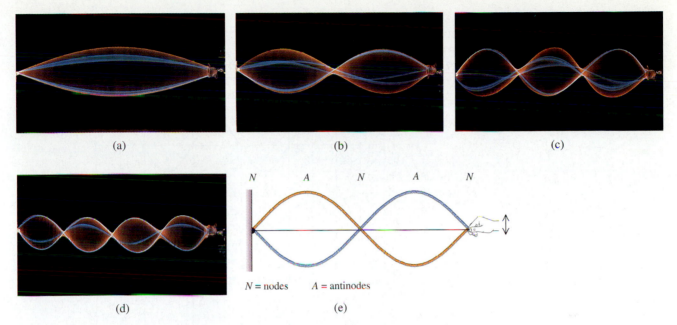

20–5 (a)–(d) Time exposures of standing waves in a stretched string. From (a) to (d) the frequency of oscillation of the right-hand end increases, and the wavelength of the standing wave decreases. (e) The extremes of the motion of the standing wave in (b), with nodes at the center and at the ends. The right-hand end of the string moves very little in comparison to the antinodes and so is essentially a node.

Figure 20–5 shows a string that is fixed at its left-hand end. Its right-hand end is moved up and down in simple harmonic motion to produce a wave that travels to the left; the wave reflected from the fixed end travels to the right. The resulting motion when the two waves combine no longer looks like two waves traveling in opposite directions. The string appears to be subdivided into a number of segments, as in the time-exposure photographs of Figs. 20–5a, 20–5b, 20–5c, and 20–5d. Figure 20–5e shows two instantaneous shapes of the string in Fig. 20–5b. Let's compare this behavior with the waves we studied in Chapter 19. In a wave that travels along the string, the amplitude is constant, and the wave pattern moves with a speed equal to the wave speed. Here, instead, the wave pattern remains in the same position along the string, and its amplitude fluctuates. There are particular points called **nodes** (labeled N in Fig. 20–5e) that never move at all. Midway between the nodes are points called **antinodes** (labeled A in Fig. 20–5e) where the amplitude of motion is greatest. Because the wave pattern doesn't appear to be moving in either direction along the string, it is called a **standing wave.** (To emphasize the difference, a wave that *does* move along the string is called a **traveling wave.**)

The principle of superposition explains how the incident and reflected wave combine to form a standing wave. In Fig. 20–6 the red curves show a wave traveling to the left. The blue curves show a wave traveling to the right with the same propagation speed, wavelength, and amplitude. The waves are shown at nine instants, one sixteenth of a period apart. At each point along the string, we add the displacements (the values of y) for the two separate waves; the result is the total wave on the string, shown in black.

At certain instants, such as $t = 8T/16$, the two wave patterns are exactly in phase with each other, and the shape of the string is a sine curve with twice the amplitude of either individual wave. At other instants, such as $t = 4T/16$, the two waves are exactly out of phase with each other, and the total wave at that instant is zero. The resultant displacement is *always* zero at those places marked N at the bottom of Fig. 20–6. These are the *nodes.* At a node the displacements of the two waves in red and blue are always equal

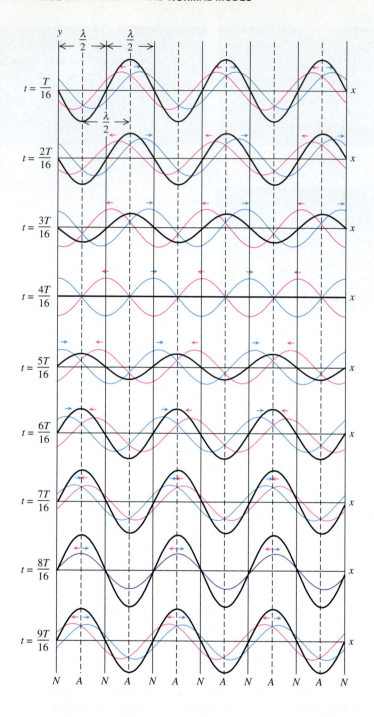

20–6 Formation of a standing wave. A wave traveling to the left (red curves) combines with a wave traveling to the right (blue curves) to form a standing wave (black curves). The horizontal x-axis in each part shows the equilibrium position of the string.

and opposite and cancel each other out. This cancellation is called **destructive interference.** Midway between the nodes are the points of *greatest* amplitude or *antinodes,* marked A. At the antinodes the displacements of the two waves in red and blue are always identical, giving a large resultant displacement; this phenomenon is called **constructive interference.** We can see from the figure that the distance between successive nodes or between successive antinodes is one half-wavelength, or $\lambda/2$.

We can derive a wave function for the standing wave of Fig. 20–6 by adding the wave functions $y_1(x, t)$ and $y_2(x, t)$ for two waves with equal amplitude, period, and wavelength traveling in opposite directions. Here $y_1(x, t)$ (the red curves in Fig. 20–6) represents an *incident* wave traveling to the left along the +x-axis, arriving at the point $x = 0$ and being reflected; $y_2(x, t)$ (the blue curves in Fig. 20–6) represents the *reflected* wave traveling to the right from $x = 0$. We noted in Section 20–2 that the wave reflected from a fixed end of a string is inverted, so we give a negative sign to one of the waves:

$$y_1(x, t) = A \sin(\omega t + kx) \qquad \text{(traveling to the left)},$$

$$y_2(x, t) = -A \sin(\omega t - kx) \qquad \text{(traveling to the right)}.$$

Note also that the change of sign corresponds to a shift in *phase* of 180° or π radians. At $x = 0$ the motion from the incident wave is $A \sin \omega t$, and the motion from the reflected wave is $-A \sin \omega t$, which we can also write as $A \sin(\omega t + \pi)$. The wave function for the standing wave is the sum of the individual wave functions:

$$y(x, t) = y_1(x, t) + y_2(x, t) = A[\sin(\omega t + kx) - \sin(\omega t - kx)].$$

We can rearrange this by using the identities for the sine of the sum and difference of two angles: $\sin(a \pm b) = \sin a \cos b \pm \cos a \sin b$. Using these and combining terms, we obtain the wave function for the standing wave:

$$y(x, t) = y_1(x, t) + y_2(x, t) = (2A \sin kx) \cos \omega t,$$

or

$$y(x, t) = (A_{SW} \sin kx) \cos \omega t \qquad \begin{array}{l}\text{(standing wave on a string,} \\ \text{fixed end at } x = 0).\end{array} \qquad (20\text{–}1)$$

The standing wave amplitude A_{SW} is twice the amplitude A of either of the original traveling waves:

$$A_{SW} = 2A.$$

Equation (20–1) has two factors: a function of x and a function of t. The factor $2A \sin kx$ shows that at each instant the shape of the string is a sine curve. But unlike a wave traveling along a string, the wave shape stays in the same position, oscillating up and down as described by the $\cos \omega t$ factor. This behavior is shown graphically by the black curves in Fig. 20–6. Each point in the string still undergoes simple harmonic motion, but all the points between any successive pair of nodes oscillate *in phase*. This is in contrast to the phase differences between oscillations of adjacent points that we see with a wave traveling in one direction.

We can use Eq. (20–1) to find the positions of the nodes; these are the points for which $\sin kx = 0$, so the displacement is *always* zero. This occurs when $kx = 0$, π, 2π, 3π, ..., or, using $k = 2\pi/\lambda$,

$$x = 0, \frac{\pi}{k}, \frac{2\pi}{k}, \frac{3\pi}{k}, \ldots$$

$$= 0, \frac{\lambda}{2}, \frac{2\lambda}{2}, \frac{3\lambda}{2}, \ldots \qquad \begin{array}{l}\text{(nodes of a standing wave on a} \\ \text{string, fixed end at } x = 0).\end{array} \qquad (20\text{–}2)$$

In particular, there is a node at $x = 0$, as there should be, since this point is a fixed end of the string.

A standing wave, unlike a traveling wave, *does not* transfer energy from one end to the other. The two waves that form it would individually carry equal amounts of power in opposite directions. There is a local flow of energy from each node to the adjacent antinodes and back, but the *average* rate of energy transfer is zero at every point. We invite you to evaluate the wave power given by Eq. (19–29) using the wave function of Eq. (20–1) and show that the average power is zero (see Challenge Problem 20–44).

EXAMPLE 20-1

A long string lies along the x-axis when in equilibrium. The end of the string at $x = 0$ is tied down. An incident sinusoidal wave (corresponding to the red curves in Fig. 20–6) travels along the string in the $-x$-direction at 84.0 m/s with an amplitude of 1.50 mm and a frequency of 120 Hz. This wave is reflected from the fixed end at $x = 0$, and the superposition of the incident traveling wave and the reflected traveling wave forms a standing wave. a) Find the equation giving the displacement of a point on the string as a function of position and time. b) Locate the points on the string that don't move at all. c) Find the amplitude, maximum transverse velocity, and maximum transverse acceleration at points of maximum oscillation.

SOLUTION a) Since there is a fixed end at $x = 0$, we may use Eq. (20–1) to describe this standing wave. The amplitude of the incident wave is $A = 1.50$ mm $= 1.50 \times 10^{-3}$ m; the reflected wave has the same amplitude, and the standing wave amplitude is $A_{sw} = 2A = 3.00 \times 10^{-3}$ m. The angular frequency ω and wave number k are

$$\omega = 2\pi f = (2\pi \text{ rad})(120 \text{ s}^{-1}) = 754 \text{ rad/s},$$

$$k = \frac{\omega}{v} = \frac{754 \text{ rad/s}}{84.0 \text{ m/s}} = 8.98 \text{ rad/m}.$$

Then Eq. (20–1) gives

$$y(x, t) = (A_{sw} \sin kx) \cos \omega t$$

$$= [(3.00 \times 10^{-3} \text{ m}) \sin (8.98 \text{ rad/m})x] \cos (754 \text{ rad/s})t.$$

b) The *nodes* are the points on the string that don't move. If we use our expression for $y(x, t)$ from part (a), these are the points at which $\sin (8.98 \text{ rad/m})x = 0$, so $(8.98 \text{ rad/m})x = 0, \pi, 2\pi, 3\pi, \ldots$, or

$$x = 0, \frac{\pi}{8.98 \text{ rad/m}}, \frac{2\pi}{8.98 \text{ rad/m}}, \frac{3\pi}{8.98 \text{ rad/m}}, \ldots$$

$$= 0, \ 0.350 \text{ m}, \ 0.700 \text{ m}, \ 1.050 \text{ m}, \ldots.$$

We could also find the positions of the nodes using Eq. (20–2). The wavelength is $\lambda = v/f = (84.0 \text{ m/s})/(120 \text{ s}^{-1}) = 0.700$ m, and Eq. (20–2) again gives $x = 0, 0.350$ m, 0.700 m, 1.050 m, ...

c) From the expression in part (a) for $y(x, t)$ we see that the maximum displacement from equilibrium is 3.00×10^{-3} m $= 3.00$ mm, which is just twice the amplitude of the incident wave. This maximum occurs at the antinodes, where $\sin (8.98 \text{ rad/m})x = \pm 1$, $(8.98 \text{ rad/m})x = \pi/2, 3\pi/2, 5\pi/2, \ldots$, and $x = 0.175$ m, 0.525 m, 0.875 m, ... Each antinode is midway between two nodes.

To find the transverse velocity (that is, velocity in the y-direction) of a particle on the string at any point x, we take the first partial derivative of $y(x, t)$ with respect to time:

$$v_y(x, t) = \frac{\partial y(x, t)}{\partial t} = [(3.00 \times 10^{-3} \text{ m}) \sin (8.98 \text{ rad/m})x]$$

$$\times [(-754 \text{ rad/s}) \sin (754 \text{ rad/s})t]$$

$$= [(-2.26 \text{ m/s}) \sin (8.98 \text{ rad/m})x] \sin (754 \text{ rad/s})t.$$

This is largest at an antinode, where $\sin (8.98 \text{ rad/m})x = \pm 1$ and the transverse velocity varies in value between 2.26 m/s and -2.26 m/s. As is always the case in simple harmonic motion, the maximum velocity occurs when the particle is passing through the equilibrium position ($y = 0$).

The transverse acceleration $a_y(x, t)$ is the first partial derivative of $v_y(x, t)$ with respect to time (that is, the *second* partial derivative of $y(x, t)$ with respect to time). We leave the calculation to you; the result is

$$a_y(x, t) = \frac{\partial v_y(x, t)}{\partial t} = \frac{\partial^2 y(x, t)}{\partial t^2}$$

$$= [(-1.71 \times 10^3 \text{ m/s}^2) \sin (8.98 \text{ rad/m})x] \cos (754 \text{ rad/s})t.$$

This is also largest at the antinodes; at these points the transverse acceleration varies in value between $+1.71 \times 10^3$ m/s^2 and -1.71×10^3 m/s^2. This maximum value is tremendous, 174 times the acceleration due to gravity!

EXAMPLE 20-2

Cousin Throckmorton is once again playing with the clothesline in Example 19–2 (Section 19–4). One end of the clothesline is attached to a vertical post; Throckmorton holds the other end loosely in his hand so that the speed of waves on the clothesline is a relatively slow 0.720 m/s. By experiment, Throckmorton finds several frequencies at which he can oscillate his end of the clothesline so that a light clothespin 45.0 cm from the post doesn't move. What are these frequencies?

SOLUTION There will be a standing wave on the clothesline formed by the superposition of the wave that Throcky produces and the wave reflected from the post. For the clothespin not to move, it must be at a node of the standing wave. From Eq. (20–2) the nodes are located at distances $\lambda/2, 2\lambda/2$,

$3\lambda/2, \ldots$ from the post; one of these must equal the clothespin distance $d = 45.0$ cm $= 0.450$ m. So $d = n\lambda/2$, where n could be any positive integer 1, 2, 3, ..., and so the wavelength is $\lambda = 2d/n$. Solving for the frequency, we get

$$f = \frac{v}{\lambda} = \frac{nv}{2d}$$

$$= n \frac{0.720 \text{ m/s}}{2(0.450 \text{ m})} = n(0.800 \text{ Hz})$$

$$= 0.800 \text{ Hz}, \ 1.60 \text{ Hz}, \ 2.40 \text{ Hz}, \ldots.$$

With $f = 0.800$ Hz the clothespin is at the first node beyond the post; with $f = 1.60$ Hz it is at the second node; with $f = 2.40$ Hz it is at the third node; and so on.

20–4 NORMAL MODES OF A STRING

10.4
Standing Waves on Strings
10.5
Tuning a Stringed Instrument:
Standing Waves
10.6
String Mass and Standing Waves

We have described standing waves on a string rigidly held at one end, as in Fig. 20–5. We made no assumptions about the length of the string or about what was happening at the other end. Let's now consider a string of a definite length L, rigidly held at *both* ends. Such strings are found in many musical instruments, including pianos, violins, and guitars. When a guitar string is plucked, a wave is produced in the string; this wave is reflected and re-reflected from the ends of the string, making a standing wave. This standing wave on the string in turn produces a sound wave in the air, with a frequency determined by the properties of the string. This is what makes stringed instruments so useful in making music.

To understand these properties of standing waves on a string fixed at both ends, let's first examine what happens when we set up a sinusoidal wave on such a string. The standing wave that results must have a node at *both* ends of the string. We saw in the preceding section that adjacent nodes are one half-wavelength ($\lambda/2$) apart, so the length of the string must be $\lambda/2$, or $2(\lambda/2)$, or $3(\lambda/2)$, or in general some integer number of half-wavelengths:

$$L = n\frac{\lambda}{2} \qquad (n = 1, 2, 3, \ldots) \qquad \text{(string fixed at both ends).} \qquad (20\text{–}3)$$

That is, if a string with length L is fixed at both ends, a standing wave can exist only if its wavelength satisfies Eq. (20–3).

Solving this equation for λ and labeling the possible values of λ as λ_n, we find

$$\lambda_n = \frac{2L}{n} \qquad (n = 1, 2, 3, \ldots) \qquad \text{(string fixed at both ends).} \qquad (20\text{–}4)$$

Waves can exist on the string if the wavelength is *not* equal to one of these values, but there cannot be a steady wave pattern with nodes and antinodes, and the total wave cannot be a standing wave. Equation (20–4) is illustrated by the standing waves shown in Figs. 20–5a, 20–5b, 20–5c, and 20–5d; these represent $n = 1, 2, 3,$ and 4, respectively.

Corresponding to the series of possible standing-wave wavelengths λ_n is a series of possible standing-wave frequencies f_n, each related to its corresponding wavelength by $f_n = v/\lambda_n$. The smallest frequency f_1 corresponds to the largest wavelength (the $n = 1$ case), $\lambda_1 = 2L$:

$$f_1 = \frac{v}{2L} \qquad \text{(string fixed at both ends).} \qquad (20\text{–}5)$$

This is called the **fundamental frequency.** The other standing-wave frequencies are $f_2 = 2v/2L, f_3 = 3v/2L$, and so on. These are all integer multiples of the fundamental frequency f_1, such as $2f_1, 3f_1, 4f_1$, and so on, and we can express *all* the frequencies as

$$f_n = n\frac{v}{2L} = nf_1 \qquad (n = 1, 2, 3, \ldots) \qquad \text{(string fixed at both ends).} \qquad (20\text{–}6)$$

These frequencies are called **harmonics,** and the series is called a **harmonic series.** Musicians sometimes call f_2, f_3, and so on **overtones;** f_2 is the second harmonic or the first overtone, f_3 is the third harmonic or the second overtone, and so on. The first harmonic is the same as the fundamental frequency.

For a string with fixed ends at $x = 0$ and $x = L$, the wave function $y(x, t)$ of the nth standing wave is given by Eq. (20–1) (which satisfies the condition that there is a node at $x = 0$), with $\omega = \omega_n = 2\pi f_n$ and $k = k_n = 2\pi/\lambda_n$:

$$y_n(x, t) = A_{sw} \cos \omega_n t \sin k_n x. \qquad (20\text{–}7)$$

We invite you to show that this wave function has nodes at both $x = 0$ and $x = L$, as it must.

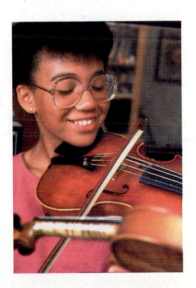

Each string of a violin naturally oscillates at one or more of its harmonic frequencies. These oscillations produce sound waves in the air with the same frequencies.

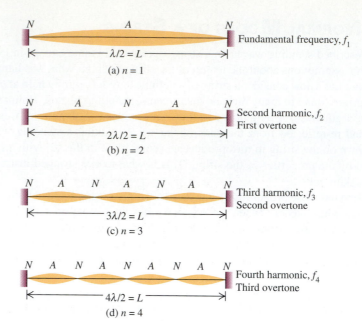

N A N Fundamental frequency, f_1

$\lambda/2 = L$

(a) $n = 1$

N A N A N Second harmonic, f_2
First overtone

$2\lambda/2 = L$

(b) $n = 2$

N A N A N A N Third harmonic, f_3
Second overtone

$3\lambda/2 = L$

(c) $n = 3$

N A N A N A N A N Fourth harmonic, f_4
Third overtone

$4\lambda/2 = L$

(d) $n = 4$

20–7 The first four normal modes of a string fixed at both ends. (Compare these to the photographs in Fig. 20–5.)

Astronomers have discovered that the sun oscillates in several different normal modes. In this theoretical model of one such mode, blue denotes places where solar material is moving outward and red denotes where it is moving inward.

A **normal mode** of an oscillating system is a motion in which all particles of the system move sinusoidally with the same frequency. For a system made up of a string of length L fixed at both ends, each of the wavelengths given by Eq. (20–4) corresponds to a possible normal mode pattern and frequency. There are infinitely many normal modes, each with its characteristic frequency and vibration pattern. Figure 20–7 shows the first four normal-mode patterns and their associated frequencies and wavelengths; these correspond to Eq. (20–7) with $n = 1$, 2, 3, and 4. By contrast, a harmonic oscillator, which has only one oscillating particle, has only one normal mode and one characteristic frequency. The string fixed at both ends has infinitely many normal modes because it is made up of a very large (effectively infinite) number of particles.

If we could displace a string so that its shape is the same as one of the normal-mode patterns and then release it, it would vibrate with the frequency of that mode. Such a vibrating string would displace the surrounding air with the same frequency, producing a traveling sinusoidal sound wave that your ears would perceive as a pure tone. But when a string is struck (as in a piano) or plucked (as is done to guitar strings), the shape of the displaced string is *not* as simple as one of the patterns in Fig. 20–7. The fundamental as well as many overtones are present in the resulting vibration. This motion is therefore a combination or *superposition* of many normal modes. Several simple-harmonic motions of different frequencies are present simultaneously, and the displacement of any point on the string is the sum (or superposition) of displacements associated with the individual modes. The sound produced by the vibrating string is likewise a superposition of traveling sinusoidal sound waves, which you perceive as a rich, complex tone with the fundamental frequency f_1. The standing wave on the string and the traveling sound wave in the air have similar **harmonic content** (the extent to which frequencies higher than the fundamental are present). The harmonic content depends on how the string is initially set into motion. If you pluck the strings of an acoustic guitar in the normal location over the sound hole, the sound that you hear has a different harmonic content than if you pluck the strings next to the fixed end on the guitar body.

It is possible to represent every possible motion of the string as some superposition of normal-mode motions. Finding this representation for a given vibration pattern is called *harmonic analysis*. The sum of sinusoidal functions that represents a complex wave is called a *Fourier series*. Figure 20–8 shows how a standing wave that is produced

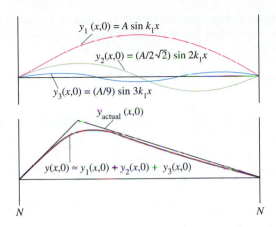

$y_1(x,0) = A \sin k_1 x$

$y_2(x,0) = (A/2\sqrt{2}) \sin 2k_1 x$

$y_3(x,0) = (A/9) \sin 3k_1 x$

$y_{actual}(x,0)$

$y(x,0) \approx y_1(x,0) + y_2(x,0) + y_3(x,0)$

N   N

20–8 The standing wave that results from plucking a guitar string into a triangular shape is well represented (except at the sharp maximum point) by the sum of only three sinusoidal functions. Including additional sinusoidal functions further improves the representation.

by plucking a guitar string of length L at a point $L/4$ from one end can be represented as a combination of sinusoidal functions.

As we have seen, the fundamental frequency of a vibrating string is $f_1 = v/2L$. The speed v of waves on the string is determined by Eq. (19–13), $v = \sqrt{F/\mu}$. Combining these, we find

ActivPhysics

10.10
Complex Waves: Fourier Analysis

$$f_1 = \frac{1}{2L}\sqrt{\frac{F}{\mu}} \qquad \text{(string fixed at both ends).} \qquad (20\text{–}8)$$

This is also the fundamental frequency of the sound wave created in the surrounding air by the vibrating string. Familiar musical instruments show how f_1 depends on the properties of the string. The inverse dependence of frequency on length L is illustrated by the long strings of the bass (low-frequency) section of the piano or the bass viol compared with the shorter strings on the treble section of the piano or the violin (Fig. 20–9). The pitch of a violin or guitar is usually varied by pressing a string against the fingerboard with the fingers to change the length L of the vibrating portion of the string. Increasing the tension F increases the wave speed v and thus increases the frequency (and the pitch). All string instruments are "tuned" to the correct frequencies by varying the tension; you tighten the string to raise the pitch. Finally, increasing the mass per unit length μ decreases the wave speed and thus the frequency. The lower notes on a steel guitar are

20–9 Comparison of ranges of a piano, a violin, a viola, a cello, and a string bass. In all cases, longer strings provide bass notes and shorter strings produce treble notes.

Bass

Cello

Viola

Violin

produced by thicker strings, and one reason for winding the bass strings of a piano with wire is to obtain the desired low frequency without resorting to a string that is inconveniently long.

Problem–Solving Strategy

STANDING WAVES

1. As with the problems in Chapter 19, it is useful to distinguish between the purely kinematic quantities, such as wave speed v, wavelength λ, and frequency f, and the dynamic quantities involving the properties of the medium, including F and μ (for transverse waves on a string) and B and ρ (for longitudinal waves in a pipe, described in the next section). You can compute the wave speed if you know either λ and f (or, equivalently, $k = 2\pi/\lambda$ and $\omega = 2\pi f$) or the properties of the medium. Try to determine whether the problem is only kinematic in nature or whether the properties of the medium are also involved.

2. In visualizing nodes and antinodes in standing waves, it is always helpful to draw diagrams. For a string you can draw the shape at one instant and label the nodes N and antinodes A. For longitudinal waves (discussed in the next section), it is not so easy to draw the shape, but you can still label the nodes and antinodes. The distance between two adjacent nodes or two adjacent antinodes is always $\lambda/2$, and the distance between a node and the adjacent antinode is always $\lambda/4$.

EXAMPLE 20-3

A giant bass viol In an effort to get your name in the *Guinness Book of World Records,* you set out to build a bass viol with strings that have a length of 5.00 m between fixed points. One string has a linear mass density of 40.0 g/m and a fundamental frequency of 20.0 Hz (the lowest frequency that the human ear can hear). a) Calculate the tension of this string. b) Calculate the frequency and wavelength on the string of the second harmonic. c) Calculate the frequency and wavelength on the string of the second overtone.

SOLUTION a) This problem involves the properties of the wave medium (the string). We are given the values of μ, L, and the fundamental frequency f_1, so we solve Eq. (20–8) for the string tension F:

$$F = 4\mu L^2 f_1^2 = 4(40.0 \times 10^{-3} \text{ kg/m})(5.00 \text{ m})^2 (20.0 \text{ s}^{-1})^2$$
$$= 1600 \text{ N} = 360 \text{ lb.}$$

(In a real bass viol the string tension is typically a few hundred newtons.)

b) From Eq. (20–6) the second harmonic frequency ($n = 2$) is

$$f_2 = 2f_1 = 2(20.0 \text{ Hz}) = 40.0 \text{ Hz.}$$

From Eq. (20–4) the wavelength on the string of the second harmonic is

$$\lambda_2 = \frac{2L}{2} = 5.00 \text{ m.}$$

c) The second overtone is the "second tone over" (above) the fundamental, that is, $n = 3$. Its frequency is

$$f_3 = 3f_1 = 3(20.0 \text{ Hz}) = 60.0 \text{ Hz.}$$

The wavelength on the string is

$$\lambda_3 = \frac{2L}{3} = 3.33 \text{ m.}$$

EXAMPLE 20-4

What are the frequency and wavelength of the sound waves that are produced in the air when the string in Example 20–3 is vibrating at its fundamental frequency?

SOLUTION When the string vibrates at a particular frequency, the surrounding air is forced to vibrate at the same frequency. So the frequency of the sound wave is the same as that of the standing wave on the string, $f_1 = 20.0$ Hz. But the relationship $\lambda = v/f$ shows that the *wavelength* of the sound wave will be different from the wavelength of the standing wave on the string if the two waves have different speeds. From Example 19–7 (Section 19–7) the speed of sound in air at 20°C is $v_{\text{sound}} = 344$ m/s; for the string on the giant bass viol, for

which $\lambda_1 = 2L = 2(5.00 \text{ m}) = 10.0$ m, the wave speed is $v_{\text{string}} = \lambda_1 f_1 = (10.0 \text{ m})(20.0 \text{ s}^{-1}) = 200$ m/s. The wavelength of the sound wave is

$$\lambda_{1 \text{ (sound)}} = \frac{v_{\text{sound}}}{f_1} = \frac{344 \text{ m/s}}{20.0 \text{ Hz}} = 17.2 \text{ m.}$$

The sound wave has a greater speed and longer wavelength than the standing wave on the string that produces the sound. In fact, for any normal mode on this string, the sound wave that is produced will have the same frequency as the wave on the string but will have a wavelength that is greater by a factor of $v_{\text{sound}}/v_{\text{string}} = (344 \text{ m/s})/(200 \text{ m/s}) = 1.72.$

20–10 Kundt's tube for determining the velocity of sound in a gas. The N's and A's are displacement nodes and antinodes. The blue shading represents the density of the gas at an instant when the gas pressure at the displacement nodes is a maximum or a minimum.

20–5 LONGITUDINAL STANDING WAVES AND NORMAL MODES

When longitudinal waves propagate in a fluid in a pipe with finite length, the waves are reflected from the ends in the same way that transverse waves on a string are reflected at its ends. The superposition of the waves traveling in opposite directions again forms a standing wave. Just as for transverse standing waves on a string, longitudinal standing waves (normal modes) in a pipe can be used to create sound waves in the surrounding air. This is the operating principle of the human voice as well as many musical instruments, including woodwinds, brasses, and pipe organs.

Transverse waves on a string, including standing waves, are usually described only in terms of the displacement of the string. But longitudinal waves in a fluid may be described either in terms of the displacement of the fluid or in terms of the pressure variation in the fluid. To avoid confusion, we'll use the terms **displacement node** and **displacement antinode** to refer to points where particles of the fluid have zero displacement and maximum displacement, respectively.

We can demonstrate longitudinal standing waves in a column of gas and measure the wave speed using an apparatus called Kundt's tube (Fig. 20–10). A horizontal glass tube a meter or so long is closed at one end and has a flexible diaphragm at the other end that can transmit vibrations. A nearby loudspeaker is driven by an audio oscillator and amplifier; this produces sound waves that force the diaphragm to vibrate sinusoidally with a frequency that we can vary. The sound waves within the tube are reflected at the other, closed end of the tube. We spread a small amount of light powder uniformly along the bottom of the tube. As we vary the frequency of the sound, we pass through frequencies at which the amplitude of the standing waves becomes large enough for the powder to be swept along the tube at those points where the gas is in motion. The powder therefore collects at the displacement nodes (where the gas is not moving). Adjacent nodes are separated by a distance equal to $\lambda/2$, and we can measure this distance. We read the frequency f from the oscillator dial, and we can then calculate the speed v of the waves from the relation $v = \lambda f$.

Figure 20–11 will help you to visualize longitudinal standing waves; it is analogous to Fig. 19–11 for longitudinal traveling waves. Again tape two index cards together edge-to-edge, with a gap of a millimeter or two forming a thin slit. Place the card over the diagram with the slit horizontal and move it vertically with constant velocity. The portions of the sine curves that appear in the slit correspond to the oscillations of the particles in a longitudinal standing wave. Each particle moves with longitudinal simple harmonic motion about its equilibrium position.

Figure 20–12 is an enlarged version of a portion of Fig. 20–11 centered on a displacement node. Note that particles on opposite sides of a displacement node vibrate in

20–11 Diagram for illustrating longitudinal standing waves. The straight vertical lines mark the locations of displacement nodes and pressure antinodes.

20–12 Particles on opposite sides of a displacement node (shown in green) vibrate in opposite phase, creating a pressure antinode. Particles on opposite sides of a displacement antinode vibrate in phase, creating a pressure node (a point of constant pressure).

Displacement node, pressure antinode

Displacement antinode, pressure node

opposite phase. When these particles approach each other, the gas between them is compressed and the pressure rises; when they recede from each other, there is an expansion and the pressure drops. Hence at a displacement *node* the gas undergoes the maximum amount of compression and expansion, and the variations in pressure and density above and below the average have their maximum value. By contrast, particles on opposite sides of a displacement *antinode* vibrate *in phase;* the distance between the particles is nearly constant, and there is *no* variation in pressure or density at a displacement antinode.

We use the term **pressure node** to describe a point in a longitudinal standing wave at which the pressure and density do not vary and the term **pressure antinode** to describe a point at which the variations in pressure and density are greatest. Using these terms, we can summarize our observations about longitudinal standing waves as follows: **a pressure node is always a displacement antinode, and a pressure antinode is always a displacement node.** Figure 20–10 depicts a longitudinal standing wave at an instant at which the pressure variations are greatest; the blue shading shows that the density and pressure of the gas have their maximum and minimum values at the displacement nodes (labeled *N*).

When reflection takes place at a *closed* end of a pipe (an end with a rigid barrier or plug), the displacement of the particles at this end must always be zero, analogous to a fixed end of a string. Thus a closed end of a pipe is a displacement node and a pressure antinode; the particles do not move, but the pressure variations are maximum. An *open* end of a pipe is a pressure node because it is open to the atmosphere, where the pressure is constant. Because of this, an open end is always a displacement *antinode,* in analogy to a free end of a string; the particles oscillate with maximum amplitude, but the pressure does not vary. (Strictly speaking, the pressure node actually occurs somewhat beyond an open end of a pipe. But if the diameter of the pipe is small in comparison to the wavelength, which is true for most musical instruments, this effect can safely be neglected.) Thus longitudinal waves in a column of fluid are reflected at the closed and open ends of a pipe in the same way that transverse waves in a string are reflected at fixed and free ends, respectively.

EXAMPLE 20–5

A directional loudspeaker aims a sound wave of frequency 200 Hz at a wall (Fig. 20–13). At what distances from the wall could you stand and hear no sound at all?

SOLUTION Your ear detects *pressure* variations in the air; increases or decreases in the pressure outside your eardrum

cause it to move slightly in or out, a motion that generates an electrical signal that is sent to the brain. (If you've ever had trouble getting your ears to "pop" on a drive up into the mountains or on an airline flight, you're familiar with just how sensitive your ears are to pressure changes.) Hence you will hear no sound if your ear is at a pressure node, which is a displacement

20–13 When a sound wave is directed at a wall, it interferes with the reflected wave to create a standing wave. The N's and A's are *displacement* nodes and antinodes.

antinode. The wall is a displacement node; the distance from a node to an adjacent antinode is $\lambda/4$, and the distance from one antinode to the next is $\lambda/2$. From Example 19–7 (Section 19–7),

the speed of sound waves in air at 20°C is 344 m/s; hence the wavelength of a 200-Hz sound wave in air is 1.72 m, and the distances d from the wall at which no sound will be heard are

$d = \lambda/4 = (1.72 \text{ m})/4$

   $= 0.43$ m   (first displacement antinode and pressure node),

$d = \lambda/4 + \lambda/2 = 3\lambda/4$

   $= 1.29$ m   (second displacement antinode and pressure node),

$d = 3\lambda/4 + \lambda/2 = 5\lambda/4$

   $= 2.15$ m   (third displacement antinode and pressure node),

and so on. If the loudspeaker is not highly directional, this effect is hard to notice because of multiple reflections of sound waves from the floor, ceiling, and other points on the walls.

ORGAN PIPES AND WIND INSTRUMENTS

The most important application of longitudinal standing waves is the production of musical tones by wind instruments. Organ pipes are one of the simplest examples. Air is supplied by a blower, at a gauge pressure typically of the order of 10^3 Pa (10^{-2} atm), to the bottom end of the pipe (Fig. 20–14). A stream of air emerges from the narrow opening at the edge of the horizontal surface and is directed against the top edge of the opening, which is called the *mouth* of the pipe. The column of air in the pipe is set into vibration, and there is a series of possible normal modes, just as with the stretched string. The mouth always acts as an open end; thus it is a pressure node and a displacement antinode. The other end of the pipe (at the top in Fig. 20–14) may be either open or closed.

In Fig. 20–15, both ends of the pipe are open, so both ends are pressure nodes and displacement antinodes. An organ pipe that is open at both ends is called an *open pipe*. The fundamental frequency f_1 corresponds to a standing-wave pattern with a displacement antinode at each end and a displacement node in the middle (Fig. 20–15a). The distance between adjacent antinodes is always equal to one half-wavelength, and in this case that is equal to the length L of the pipe; $\lambda/2 = L$. The corresponding frequency, obtained from the relation $f = v/\lambda$, is

$$f_1 = \frac{v}{2L} \quad \text{(open pipe).} \quad (20\text{--}9)$$

Figures 20–15b and 20–15c show the second and third harmonics (first and second overtones); their vibration patterns have two and three displacement nodes, respectively. For these, a half-wavelength is equal to $L/2$ and $L/3$, respectively, and the frequencies are twice and three times the fundamental, respectively. That is, $f_2 = 2f_1$ and $f_3 = 3f_1$. For *every* normal mode of an open pipe the length L must be an integer number of half-wavelengths, and the possible wavelengths λ_n are given by

$$L = n\frac{\lambda_n}{2} \quad \text{or} \quad \lambda_n = \frac{2L}{n} \quad (n = 1, 2, 3, \ldots) \quad \text{(open pipe).} \quad (20\text{--}10)$$

The corresponding frequencies f_n are given by $f_n = v/\lambda_n$, so all the normal-mode frequencies for a pipe that is open at both ends are given by

$$f_n = \frac{nv}{2L} \quad (n = 1, 2, 3, \ldots) \quad \text{(open pipe).} \quad (20\text{--}11)$$

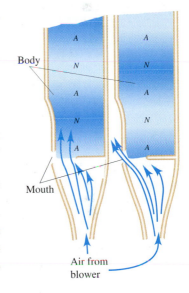

20–14 Cross sections of an organ pipe at two instants one half-period apart. Vibrations from the turbulent air flow set up standing waves in the pipe. The N's and A's are *displacement* nodes and antinodes; as the blue shading shows, these are points of maximum pressure variation and zero pressure variation, respectively.

20–15 A cross section of an open pipe showing the first three normal modes. The shading indicates the pressure variations. The red curves are graphs of the displacement along the pipe axis at two instants separated in time by one half-period. The N's and A's are the *displacement* nodes and antinodes; interchange these to show the *pressure* nodes and antinodes.

20–16 A cross section of a stopped pipe showing the first three normal modes as well as the *displacement* nodes and antinodes. Only odd harmonics are possible.

The value $n = 1$ corresponds to the fundamental frequency, $n = 2$ to the second harmonic (or first overtone), and so on. Alternatively, we can say

$$f_n = nf_1 \qquad (n = 1, 2, 3, \dots) \qquad \text{(open pipe)}, \qquad (20\text{–}12)$$

with f_1 given by Eq. (20–9).

Figure 20–16 shows a pipe that is open at the left end but closed at the right end. This is called a *stopped pipe*. The left (open) end is a displacement antinode (pressure node), but the right (closed) end is a displacement node (pressure antinode). The distance between a node and the adjacent antinode is always one quarter-wavelength. Figure 20–16a shows the lowest-frequency mode; the length of the pipe is a quarter-wavelength ($L = \lambda_1/4$). The fundamental frequency is $f_1 = v/\lambda_1$, or

$$f_1 = \frac{v}{4L} \qquad \text{(stopped pipe)}. \qquad (20\text{–}13)$$

This is one-half the fundamental frequency for an *open* pipe of the same length. In musical language, the *pitch* of a closed pipe is one octave lower (a factor of two in frequency) than that of an open pipe of the same length. Figure 20–16b shows the next mode, for which the length of the pipe is *three quarters* of a wavelength, corresponding to a frequency $3f_1$. For Fig. 20–16c, $L = 5\lambda/4$ and the frequency is $5f_1$. The possible wavelengths are given by

$$L = n\frac{\lambda_n}{4} \quad \text{or} \quad \lambda_n = \frac{4L}{n} \quad (n = 1, 3, 5, \dots) \quad \text{(stopped pipe)}. \qquad (20\text{–}14)$$

The normal-mode frequencies are given by $f_n = v/\lambda_n$, or

$$f_n = \frac{nv}{4L} \quad (n = 1, 3, 5, \ldots) \quad \text{(stopped pipe)}, \quad (20\text{–}15)$$

or

$$f_n = nf_1 \quad (n = 1, 3, 5, \ldots) \quad \text{(stopped pipe)}, \quad (20\text{–}16)$$

with f_1 given by Eq. (20–13). We see that the second, fourth, and all *even* harmonics are missing. In a pipe that is closed at one end, the fundamental frequency is $f_1 = v/4L$, and only the odd harmonics in the series ($3f_1$, $5f_1$, $\ldots$) are possible.

A final possibility is a pipe that is closed at *both* ends, with displacement nodes and pressure antinodes at both ends. This wouldn't be of much use as a musical instrument because there would be no way for the vibrations to get out of the pipe.

EXAMPLE 20–6

On a day when the speed of sound is 345 m/s, the fundamental frequency of a stopped organ pipe is 220 Hz. a) How long is this stopped pipe? b) The second *overtone* of this pipe has the same wavelength as the third *harmonic* of an open pipe. How long is the open pipe?

SOLUTION a) For a stopped pipe, $f_1 = v/4L$, so the length of the stopped pipe is

$$L_{\text{stopped}} = \frac{v}{4f_1} = \frac{345 \text{ m/s}}{4(220 \text{ s}^{-1})} = 0.392 \text{ m}.$$

b) The frequency of the first overtone of a stopped pipe is $f_3 = 3f_1$, and the frequency of the second overtone is $f_5 = 5f_1$:

$$f_5 = 5f_1 = 5(220 \text{ Hz}) = 1100 \text{ Hz}.$$

If the wavelengths are the same, the frequencies are the same, so the frequency of the third harmonic of the open pipe is also 1100 Hz. The third harmonic of an open pipe is at $3f_1 = 3(v/2L)$. If this equals 1100 Hz, then

$$1100 \text{ Hz} = 3\left(\frac{345 \text{ m/s}}{2L_{\text{open}}}\right) \quad \text{and} \quad L_{\text{open}} = 0.470 \text{ m}.$$

The stopped pipe is 0.392 m long and has a fundamental frequency of 220 Hz; the open pipe is longer, 0.470 m, but has a higher fundamental frequency of (1100 Hz)/3 = 367 Hz. If this seems like a contradiction, you should again compare Figs. 20–15a and 20–16a.

In an organ pipe in actual use, several modes are always present at once; the motion of the air is a superposition of these modes. This situation is analogous to a string that is struck or plucked, as in Fig. 20–8. Just as for a vibrating string, a complex standing wave in the pipe produces a traveling sound wave in the surrounding air with a harmonic content similar to that of the standing wave. A very narrow pipe produces a sound wave rich in higher harmonics, which we hear as a thin and "stringy" tone; a fatter pipe produces mostly the fundamental mode, heard as a softer, more flute-like tone. The harmonic content also depends on the shape of the pipe's mouth.

We have talked about organ pipes, but this discussion is also applicable to other wind instruments. The flute and the recorder are directly analogous. The most significant difference is that those instruments have holes along the pipe. Opening and closing the holes with the fingers changes the effective length L of the air column and thus changes the pitch. Any individual organ pipe, by comparison, can play only a single note. The flute and recorder behave as *open* pipes, while the clarinet acts as a *stopped* pipe (closed at the reed end, open at the bell).

Equations (20–11) and (20–15) show that the frequencies of any wind instrument are proportional to the speed of sound v in the air column inside the instrument. As Eq. (19–27) shows, v depends on temperature; it increases when temperature increases. Thus the pitch of all wind instruments rises with increasing temperature. An organ that has some of its pipes at one temperature and others at a different temperature is bound to sound out of tune.

Each pipe in a pipe organ has a different length and diameter, and emits sound waves with a different pitch and different harmonic content.

20–17 Two speakers driven by the same amplifier. The waves emitted by the speakers are in phase; they arrive at point P in phase because the two path lengths are the same. They arrive at point Q a half-cycle out of phase because the path lengths differ by $\lambda/2$.

20–6 INTERFERENCE OF WAVES

Wave phenomena that occur when two or more waves overlap in the same region of space are grouped under the heading *interference*. As we have seen, standing waves are a simple example of an interference effect: two waves traveling in opposite directions in a medium combine to produce a standing wave pattern with nodes and antinodes that do not move.

Figure 20–17 shows an example of another type of interference that involves waves that spread out in space. Two speakers, driven in phase by the same amplifier, emit identical sinusoidal sound waves with the same constant frequency. We place a microphone at point P in the figure, equidistant from the speakers. Wave crests emitted from the two speakers at the same time travel equal distances and arrive at point P at the same time; hence the waves arrive in phase, and there is constructive interference. The total wave amplitude at P is twice the amplitude from each individual wave, and we can measure this combined amplitude with the microphone.

Now let's move the microphone to point Q, where the distances from the two speakers to the microphone differ by a half-wavelength. Then the two waves arrive a half-cycle out of step, or *out of phase;* a positive crest from one speaker arrives at the same time as a negative crest from the other. Destructive interference takes place, and the amplitude measured by the microphone is much *smaller* than when only one speaker is present. If the amplitudes from the two speakers are equal, the two waves cancel each other out completely at point Q, and the total amplitude there is zero.

CAUTION ▶ Although this situation bears some resemblance to standing waves in a pipe, the total wave in Fig. 20–17 is a *traveling* wave, not a standing wave. To see why, recall that in a standing wave there is no net flow of energy in any direction. By contrast, in Fig. 20–17 there is an overall flow of energy from the speakers into the surrounding air; this is characteristic of a traveling wave. The interference between the waves from the two speakers simply causes the energy flow to be *channeled* into certain directions (for example, toward P) and away from other directions (for example, away from Q). You can see another difference between Fig. 20–17 and a standing wave by considering a point, such as Q, where destructive interference occurs. Such a point is *both* a displacement node *and* a pressure node because there is no wave at all at this point. Compare this to a standing wave, in which a pressure node is a displacement antinode and vice versa. ◀

20–18 Two speakers driven by the same amplifier, emitting waves in phase. Only the waves directed toward the microphone are shown, and they are separated for clarity. (a) Constructive interference occurs when the path difference is 0, λ, 2λ, 3λ, ... (b) Destructive interference occurs when the path difference is $\lambda/2$, $3\lambda/2$, $5\lambda/2$, ...; in this case little or no sound energy flows toward the microphone directly in front of the speakers. The energy is instead directed to the sides, where constructive interference occurs.

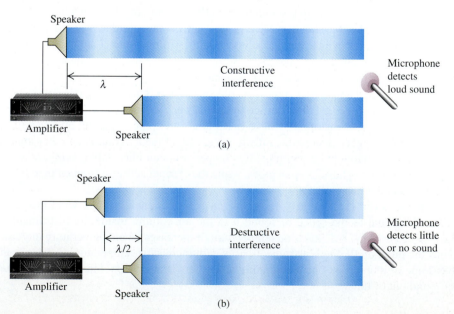

Constructive interference occurs wherever the distances traveled by the two waves differ by a whole number of wavelengths, $0, \lambda, 2\lambda, 3\lambda, \ldots$; in all these cases the waves arrive at the microphone in phase (Fig. 20–18a). If the distances from the two speakers to the microphone differ by any half-integer number of wavelengths, $\lambda/2, 3\lambda/2, 5\lambda/2, \ldots$, the waves arrive at the microphone out of phase and there will be destructive interference (Fig. 20–18b).

EXAMPLE 20-7

Two small loudspeakers, A and B (Fig. 20–19), are driven by the same amplifier and emit pure sinusoidal waves in phase. If the speed of sound is 350 m/s, a) for what frequencies does constructive interference occur at point P; b) for what frequencies does destructive interference occur at point P?

SOLUTION The nature of the interference at point P depends on the difference in path lengths from points A and B to point P; let's calculate this first. The distance from speaker A to point P is $[(2.00 \text{ m})^2 + (4.00 \text{ m})^2]^{1/2} = 4.47$ m, and the distance from speaker B to point P is $[(1.00 \text{ m})^2 + (4.00 \text{ m})^2]^{1/2} = 4.12$ m. The path difference is $d = 4.47 \text{ m} - 4.12 \text{ m} = 0.35$ m.

a) Constructive interference occurs when the path difference d is an integer number of wavelengths, $d = 0, \lambda, 2\lambda, \ldots$ or $d = 0, v/f, 2v/f, \ldots = nv/f$. So the possible frequencies are

$$f_n = \frac{nv}{d} = n\frac{350 \text{ n/s}}{0.35 \text{ m}} \qquad (n = 1, 2, 3, \ldots)$$

$$= 1000 \text{ Hz, } 2000 \text{ Hz, } 3000 \text{ Hz, } \ldots.$$

b) Destructive interference occurs when the path difference is a half-integer number of wavelengths, $d = \lambda/2, 3\lambda/2, 5\lambda/2, \ldots$ or $d = v/2f, 3v/2f, 5v/2f, \ldots.$ The possible frequencies are

$$f_n = \frac{nv}{2d} = n\frac{350 \text{ m/s}}{2(0.35 \text{ m})} \qquad (n = 1, 3, 5, \ldots)$$

$$= 500 \text{ Hz, } 1500 \text{ Hz, } 2500 \text{ Hz, } \ldots.$$

20–19 The distances from A to P and from B to P are hypotenuses of right triangles.

As we increase the frequency, the sound at point P alternates between large and small amplitudes; the maxima and minima occur at the frequencies we have found. It can be hard to notice this effect in an ordinary room because of multiple reflections from the walls, floor, and ceiling. Such an experiment is best done either outdoors or in an anechoic chamber, which has walls that absorb almost all sound and thereby eliminate reflections.

Experiments closely analogous to the one in Example 20–7, but using light, have provided both strong evidence for the wave nature of light and a means of measuring its wavelengths. We will discuss these experiments in detail in Chapter 37.

Interference effects are used to control noise from very loud sound sources such as gas-turbine power plants or jet engine test cells. The idea is to use additional sound sources that in some regions of space interfere destructively with the unwanted sound and cancel it out. Microphones in the controlled area feed signals back to the sound sources, which are continuously adjusted for optimum cancellation of noise in the controlled area.

20-7 RESONANCE

We have discussed several mechanical systems that have normal modes of oscillation. In each mode, every particle of the system oscillates with simple harmonic motion with the same frequency as the frequency of this mode. The systems we discussed in Sections 20–4 and 20–5 have an infinite series of normal modes, but the basic concept is closely

Speaker emits frequency f

Amplifier

(a)

(b)

20–20 (a) The air in an open pipe is forced to oscillate at the frequency f of the sound waves coming from the loudspeaker. (b) The graph of amplitude of oscillation of the air in the open pipe as a function of the driving frequency f. This graph is called the resonance curve of the open pipe. The peaks occur at the normal-mode frequencies f_1, $f_2 = 2f_1, f_3 = 3f_1, \ldots$

related to the simple harmonic oscillator, discussed in Chapter 13, which has only a single normal mode (that is, only one frequency at which it will oscillate after being disturbed).

Suppose we apply a periodically varying force to a system that can oscillate. The system is then forced to oscillate with a frequency equal to the frequency of the applied force (called the *driving frequency*). This motion is called a *forced oscillation*. We talked about forced oscillations of the harmonic oscillator in Section 13–9, and we suggest that you review that discussion. In particular, we described the phenomenon of mechanical **resonance.** A simple example of resonance is pushing Cousin Throckmorton on a swing. The swing is a pendulum; it has only a single normal mode, with a frequency determined by its length. If we push the swing periodically with this frequency, we can build up the amplitude of the motion. But if we push with a very different frequency, the swing hardly moves at all.

Resonance also occurs when a periodically varying force is applied to a system with many normal modes. An example is shown in Fig. 20–20a. An open organ pipe is placed next to a loudspeaker that is driven by an amplifier and emits pure sinusoidal sound waves of frequency f, which can be varied by adjusting the amplifier. The air in the pipe is forced to vibrate with the same frequency f as the *driving force* provided by the loudspeaker. In general the amplitude of this motion is relatively small, and the motion of the air inside the pipe will not be any of the normal-mode patterns shown in Fig. 20–15. But if the frequency f of the force is close to one of the normal-mode frequencies, the air in the pipe will move in the normal-mode pattern for that frequency, and the amplitude can become quite large. Figure 20–20b shows the amplitude of oscillation of the air in the pipe as a function of the driving frequency f. The shape of this graph is called the **resonance curve** of the pipe; it has peaks where f equals the normal-mode frequencies of the pipe. The detailed shape of the resonance curve depends on the geometry of the pipe.

If the frequency of the force is precisely *equal* to a normal-mode frequency, the system is in resonance, and the amplitude of the forced oscillation is maximum. If there were no friction or other energy-dissipating mechanism, a driving force at a normal-mode frequency would continue to add energy to the system, and the amplitude would increase indefinitely. In such an idealized case the peaks in the resonance curve of Fig. 20–20b would be infinitely high. But in any real system there is always some dissipation of energy, or damping, as we discussed in Section 13–9; the amplitude of oscillation in resonance may be large, but it cannot be infinite.

The "sound of the ocean" you hear when you put your ear next to a large seashell is due to resonance. The noise of the outside air moving past the seashell is a mixture of sound waves of almost all audible frequencies, which forces the air inside the seashell to oscillate. The seashell behaves like an organ pipe, with a set of normal-mode frequencies; hence the inside air oscillates most strongly at those frequencies, producing the seashell's characteristic sound. To hear a similar phenomenon, uncap a full bottle of your favorite beverage and blow across the open top. The noise is provided by your breath blowing across the top, and the "organ pipe" is the column of air inside the bottle above the surface of the liquid. If you take a drink and repeat the experiment, you will hear a lower tone because the "pipe" is longer and the normal-mode frequencies are lower.

Resonance also occurs when a stretched string is forced to oscillate. Suppose that one end of a stretched string is held fixed while the other is given a transverse sinusoidal motion with small amplitude, setting up standing waves. If the frequency of the driving mechanism is *not* equal to one of the normal-mode frequencies of the string, the amplitude at the antinodes is fairly small. However, if the frequency is equal to any one of the normal-mode frequencies, the string is in resonance, and the amplitude at the antinodes is very much larger than that at the driven end. The driven end is not precisely a node, but it lies much closer to a node than to an antinode when the string is in resonance. The photographs in Fig. 20–5 were made this way, with the left end of the string fixed and the right end oscillating vertically with small amplitude; large-amplitude standing waves resulted when the frequency of oscillation of the right end was equal to the fundamental frequency or to one of the first three overtones.

It is easy to demonstrate resonance with a piano. Push down the damper pedal (the right-hand pedal) so that the dampers are lifted and the strings are free to vibrate, and then sing a steady tone into the piano. When you stop singing, the piano seems to continue to sing the same note. The sound waves from your voice excite vibrations in the strings that have natural frequencies close to the frequencies (fundamental and harmonics) present in the note you sang.

A more spectacular example is a singer breaking a wine glass with her amplified voice. A good-quality wine glass has normal-mode frequencies that you can hear by tapping it. If the singer emits a loud note with a frequency corresponding exactly to one of these normal-mode frequencies, large-amplitude oscillations can build up and break the glass.

Resonance is a very important concept, not only in mechanical systems but in all areas of physics. In Chapter 32 we will see examples of resonance in electric circuits.

The frequency of the sound from this trumpet exactly matches one of the normal-mode frequencies of the goblet. The resonant vibrations of the goblet have such large amplitude that the goblet tears itself apart.

EXAMPLE 20–8

A stopped organ pipe is sounded near a guitar, causing one of the strings to vibrate with large amplitude. We vary the tension of the string until we find the maximum amplitude. The string is 80% as long as the stopped pipe. If both the pipe and the string vibrate at their fundamental frequency, calculate the ratio of the wave speed on the string to the speed of sound in air.

SOLUTION The large response of the string is an example of resonance; it occurs because the organ pipe and the guitar string have the same fundamental frequency. Letting the subscripts a and s stand for air and string, respectively, we have $f_{1a} = f_{1s}$.

We also know from Eq. (20–13) that $f_{1a} = v_a/4L_a$ and from Eq. (20–5) that $f_{1s} = v_s/2L_s$. Putting these together, we find

$$\frac{v_a}{4L_a} = \frac{v_s}{2L_s}.$$

Substituting $L_s = 0.80\,L_a$ and rearranging, we get

$$\frac{v_s}{v_a} = 0.40.$$

For example, if the speed of sound in air is 345 m/s, the wave speed on the string is $(0.40)(345\ \text{m/s}) = 138\ \text{m/s}$.

SUMMARY

- A wave that reaches a boundary of the medium in which it propagates is reflected. The principle of superposition states that the total wave displacement at any point where two or more waves overlap is the sum of the displacements of the individual waves.

- When a wave is reflected from a fixed or free end of a stretched string, the incident and reflected waves combine to form a standing wave containing nodes and anti-nodes. Adjacent nodes are spaced a distance $\lambda/2$ apart, as are adjacent antinodes. If the string has a fixed end at $x = 0$, the wave function of a standing wave with angular frequency $\omega = 2\pi f$, wave number $k = 2\pi/\lambda$, and amplitude A_{sw} is

$$y(x, t) = (A_{sw} \sin kx)\cos \omega t \quad \text{(standing wave on a string, fixed end at } x = 0). \tag{20-1}$$

- When both ends of a string with length L are held fixed, standing waves can occur only when L is an integer multiple of $\lambda/2$. The corresponding possible frequencies are

$$f_n = n\frac{v}{2L} = nf_1$$

$$(n = 1, 2, 3, \dots) \quad \text{(string fixed at both ends).} \tag{20-6}$$

Each frequency and its associated vibration pattern is called a normal mode. The lowest frequency f_1 is called the fundamental frequency. In terms of the mechanical properties F and μ of the string, the fundamental frequency is

$$f_1 = \frac{1}{2L}\sqrt{\frac{F}{\mu}} \quad \text{(string fixed at both ends).} \tag{20-8}$$

- For sound waves in a pipe or tube, a closed end is a displacement node and a pressure antinode; an open end is a displacement antinode and a pressure node. For a pipe open at both ends, the normal-mode frequencies are

$$f_n = \frac{nv}{2L} \quad (n = 1, 2, 3, \dots) \quad \text{(open pipe).} \tag{20-11}$$

For a pipe open at one end and closed at the other, the normal-mode frequencies are

$$f_n = \frac{nv}{4L} \quad (n = 1, 3, 5, \dots) \quad \text{(stopped pipe).} \tag{20-15}$$

- When two or more waves overlap in the same region of space, the resulting effects are called interference. The resulting amplitude can be either larger or smaller than the amplitude of each individual wave, depending on whether the waves are in phase or out of phase. When the waves are in phase, the result is called constructive interference; when they are one half-cycle out of phase, it is called destructive inter-ference.

- When a periodically varying force is applied to a system having normal modes, the system vibrates with the same frequency as that of the force; this is called a forced oscillation. If the force frequency is equal to or close to one of the normal-mode fre-quencies, the amplitude of the resulting forced oscillation can become very large; this phenomenon is called resonance.

DISCUSSION QUESTIONS

Q20–1 Energy can be transferred along a string by wave motion. However, in a standing wave on a string, no energy can ever be transferred past a node. Why not?

Q20–2 Can a standing wave be produced on a string by superposing two waves traveling in opposite directions with the same frequency but different amplitudes? Why or why not? Can a standing wave be produced by superposing two waves traveling in opposite directions with different frequencies but the same amplitude? Why or why not?

Q20–3 If you stretch a rubber band and pluck it, you hear a (somewhat) musical tone. How does the frequency of this tone change as you stretch the rubber band further? (Try it!) Does this agree with Eq. (20–8) for a string fixed at both ends? Explain.

Q20–4 A musical interval of an *octave* corresponds to a factor of two in frequency. By what factor must the tension in a guitar or violin string be increased to raise its pitch one octave? To raise it two octaves? Explain your reasoning.

Q20–5 By touching a string lightly at its center while bowing, a violinist can produce a note exactly one octave above the note to which the string is tuned, that is, a note with exactly twice the frequency. Why is this possible?

Q20–6 When an earthquake hits, tall buildings tend to oscillate from side to side. Is the top of the building a displacement node or antinode? What about the bottom of the building? Explain.

Q20–7 Does the pitch (or frequency) of an organ pipe increase or decrease with increasing temperature? Explain.

Q20–8 In most modern wind instruments the pitch is changed by using keys or valves to change the length of the vibrating air column. The bugle, however, has no valves or keys, yet it can play many notes. How is this possible? Are there restrictions on what notes a bugle can play?

Q20–9 Symphonic musicians always "warm up" their wind instruments by blowing them before a performance. What purpose does this serve?

Q20–10 When you inhale helium, your voice becomes high and squeaky. Why? (*Warning:* Inhaling too much helium can cause unconsciousness or death.)

Q20–11 The pitch of a kettledrum is determined by the frequencies of the normal modes of oscillation of the drumhead. How can a kettledrum be tuned?

Q20–12 As we discussed in Section 19–2, water waves are a combination of longitudinal and transverse waves. Defend the following statement: "When water waves hit a vertical wall, the wall is a node of the longitudinal displacement but an antinode of the transverse displacement."

Q20–13 A glass of water sits on a kitchen counter just above a dishwasher that is running. The surface of the water has a stationary pattern of concentric circular ripples. What causes this? Why is the pattern stationary? How does the speed of the dishwasher affect the number of concentric ripples in the water?

Q20–14 The rate at which exhaust gases flow out of a gasoline engine can be increased by using a "tuned" exhaust pipe of just the right length. (This improves the performance of the engine, since fuel can be forced into the engine at a faster rate.) How does this work? What determines the optimum length to use?

Q20–15 Some stringed instruments (such as the sitar of India) have two of the same string. When one string is plucked, the other one starts vibrating at the same frequency, even though it has not been touched. How is this possible?

Q20–16 Figure 12–13 shows the geometry of an *ellipse*. The points S and S' are called the *foci* of the ellipse (plural of *focus*); the sum of the distances SP and $S'P$ has the same value for any point P on the ellipse. If a room is built with curved walls in the shape of an ellipse, a person standing at one focus of the room can easily hear a person at the other focus, even if that person whispers and the overall dimensions of the room are quite large. Explain why, using the ideas of interference in space.

EXERCISES

SECTION 20–2 BOUNDARY CONDITIONS FOR A STRING AND THE PRINCIPLE OF SUPERPOSITION

20–1 A wave pulse on a string has the dimensions shown in Fig. 20–21 at $t = 0$. The wave speed is 5.0 m/s. a) If point O is a fixed end, draw the total wave on the string at $t = 1.0$ ms, 2.0 ms, 3.0 ms, 4.0 ms, 5.0 ms, 6.0 ms, and 7.0 ms. b) Repeat part (a) for the case in which point O is a free end.

20–2 A wave pulse on a string has the dimensions shown in Fig. 20–22 at $t = 0$. The wave speed is 40 cm/s. a) If point O is a fixed end, draw the total wave on the string at $t = 15$ ms, 20 ms, 25 ms, 30 ms, 35 ms, 40 ms, and 45 ms. b) Repeat part (a) for the case in which point O is a free end.

FIGURE 20–21 Exercise 20–1.

FIGURE 20–22 Exercise 20–2.

SECTION 20–3 STANDING WAVES ON A STRING

20–3 Standing waves on a wire are described by Eq. (20–1), with $A_{SW} = 2.50$ mm, $\omega = 942$ rad/s, and $k = 0.750\pi$ rad/m. The left-hand end of the wire is at $x = 0$. At what distances from the left-hand end are a) the nodes of the standing wave? b) the anti-nodes of the standing wave?

20–4 a) Prove by direct substitution that $y(x, t) = (A_{SW} \cos \omega t) \sin kx$ is a solution of the wave equation, Eq. (19–12), for $v = \omega/k$. b) Explain why the relation $v = \omega/k$ for *traveling* waves also applies to *standing* waves.

20–5 Adjacent antinodes of a standing wave on a string are 15.0 cm apart. A particle at an antinode oscillates in simple harmonic motion with amplitude 0.850 cm and period 0.0750 s. The string lies along the $+x$-axis and is fixed at $x = 0$. a) Find the displacement of a point on the string as a function of position and time. b) Find the speed of propagation of a transverse wave in the string. c) Find the amplitude at a point 3.0 cm to the right of an antinode.

20–6 Give the details of the derivation of Eq. (20–1) from $y_1(x, t) + y_2(x, t) = A[\sin(\omega t + kx) - \sin(\omega t - kx)]$.

20–7 Let $y_1(x, t) = A_1 \sin(\omega_1 t - k_1 x)$ and $y_2(x, t) = A_2 \sin(\omega_2 t - k_2 x)$ be two solutions to the wave equation, Eq. (19–12), for the same v. Show that $y(x, t) = y_1(x, t) + y_2(x, t)$ is also a solution to the wave equation.

SECTION 20–4 NORMAL MODES OF A STRING

20–8 A piano tuner stretches a steel piano wire with a tension of 800 N. The steel wire is 0.400 m long and has a mass of 3.00 g. a) What is the frequency of its fundamental mode of vibration? b) What is the number of the highest harmonic that could be heard by a person who is capable of hearing frequencies up to 10,000 Hz?

20–9 A wire of mass 40.0 g is stretched so that its ends are tied down at points 80.0 cm apart. The wire vibrates in its fundamental mode with frequency 60.0 Hz and with an amplitude at the antinodes of 0.300 cm. a) What is the speed of propagation of transverse waves in the wire? b) Compute the tension in the wire.

20–10 In deriving Eq. (20–2) for a string with a fixed end at $x = 0$, we found that nodes occur at positions x that satisfy $kx = n\pi$, where $n = 0, 1, 2, 3, \ldots$. Apply this condition to a string with fixed ends at $x = 0$ and $x = L$ to rederive Eq. (20–4) for the possible standing-wave wavelengths.

20–11 Playing a Cello. The portion of a cello string between the bridge and upper end of the finger board (that part of the string that is free to vibrate) is 60.0 cm long, and this length of the string has a mass of 2.00 g. The string sounds an A_4 note (440 Hz) when played. a) Where must the cellist put a finger (what distance x from bridge) to play a D_5 note (587 Hz)? (See Fig. 20–23.) For both the A_4 and D_5 notes the string vibrates in its fundamental mode. b) Without retuning, is it possible to play a G_4 note (392 Hz) on this string? Why or why not?

20–12 a) A horizontal string tied at both ends is vibrating in its fundamental mode. The standing wave has speed v, frequency f, amplitude A, and wavelength λ. Calculate the maximum trans-

FIGURE 20–23 Exercise 20–11.

verse velocity and maximum transverse acceleration of points located at i) $x = \lambda/2$, ii) $x = \lambda/4$, and iii) $x = \lambda/8$ from the left-hand end of the string. b) At each of the points in part (a), what is the amplitude of the motion? c) At each of the points in part (a), how much time does it take the string to go from its largest upward displacement to its largest downward displacement?

20–13 A 1.50-m-long rope is stretched between two supports with a tension that makes the speed of transverse waves 48.0 m/s. What are the wavelength and frequency of a) the fundamental? b) the second overtone? c) the fourth harmonic?

20–14 One of the 63.5-cm-long strings of an ordinary guitar is tuned to produce the note B_3 (frequency 245 Hz) when vibrating in its fundamental mode. a) Find the speed of transverse waves on this string. b) If the tension in this string is increased by 1.0%, what will be the new fundamental frequency of the string? c) If the speed of sound in the surrounding air is 344 m/s, find the frequency and wavelength of the sound wave produced in the air by the vibration of the B_3 string. How do these compare to the frequency and wavelength of the standing wave on the string?

SECTION 20–5 LONGITUDINAL STANDING WAVES AND NORMAL MODES

20–15 Standing sound waves are produced in a pipe that is 1.20 m long. For the fundamental and first two overtones, determine the locations along the pipe (measured from the left end) of the displacement nodes and the pressure nodes if a) the pipe is open at both ends; b) the pipe is closed at the left end and open at the right end.

20–16 Find the fundamental frequency and the frequency of the first three overtones of a pipe 45.0 cm long a) if the pipe is open at both ends; b) if the pipe is closed at one end. Use $v = 344$ m/s. c) For each of the above cases, what is the number of the highest harmonic that may be heard by a person who can hear frequencies from 20 Hz to 20,000 Hz?

20–17 The human vocal tract is a pipe that extends about 17 cm from the lips to the vocal folds (also called "vocal cords") near the middle of your throat. The vocal folds behave rather like the reed of a clarinet, and the vocal tract acts as a stopped pipe. Estimate the first three standing-wave frequencies of the vocal tract. Use $v = 344$ m/s. (The answers are only an estimate, since the position of lips and tongue affect the motion of air in the vocal tract.)

20–18 A certain pipe produces a fundamental frequency of 262 Hz in air. a) If the pipe is filled with helium at the same temperature, what fundamental frequency does it produce? (The molar mass of air is 28.8 g/mol, and the molar mass of helium is 4.00 g/mol.) b) Does your answer to part (a) depend on whether the pipe is open or stopped? Why or why not?

SECTION 20–6 INTERFERENCE OF WAVES

20–19 Figure 20–24 shows two rectangular wave pulses on a stretched string traveling toward each other. Each pulse is traveling with a speed of 1.00 mm/s and has the height and width shown in the figure. If the leading edges of the pulses are 8.00 mm apart at $t = 0$, sketch the shape of the string at $t = 4.00$ s, $t = 6.00$ s, and $t = 10.0$ s.

FIGURE 20–24 Exercise 20–19.

20–20 Two triangular wave pulses are traveling toward each other on a stretched string as shown in Fig. 20–25. Each pulse is identical to the other and travels at 2.00 cm/s. The leading edges of the pulses are 1.00 cm apart at $t = 0$. Sketch the shape of the string at $t = 0.250$ s, $t = 0.500$ s, $t = 0.750$ s, $t = 1.000$ s, and $t = 1.250$ s.

FIGURE 20–25 Exercise 20–20.

PROBLEMS

20–25 As shown in Exercise 20–4, a standing wave given by Eq. (20–1) satisfies the wave equation, Eq. (19–12). a) Show that a standing wave given by Eq. (20–1) also satisfies the equation

$$\frac{\partial^2 y(x, t)}{\partial t^2} = -\omega^2 y(x, t)$$

Interpret this equation in terms of what you know about simple harmonic motion. b) Does a traveling wave given by $y(x, t) =$

20–21 Two loudspeakers, A and B (Fig. 20–26), are driven by the same amplifier and emit sinusoidal waves in phase. Speaker B is 2.00 m to the right of speaker A. The speed of sound in the air is 344 m/s. Consider point Q along the extension of the line connecting the speakers, 1.00 m to the right of speaker B. Both speakers emit sound waves that travel directly from the speaker to point Q. a) What is the lowest frequency for which *constructive* interference occurs at point Q? b) What is the lowest frequency for which *destructive* interference occurs at point Q?

FIGURE 20–26 Exercises 20–21 and 20–22.

20–22 Two loudspeakers, A and B (Fig. 20–26), are driven by the same amplifier and emit sinusoidal waves in phase. Speaker B is 2.00 m to the right of speaker A. The frequency of the sound waves produced by the loudspeakers is 206 Hz, and their speed in the air is 344 m/s. Consider point P between the speakers and along the line connecting them, a distance x to the right of speaker A. Both speakers emit sound waves that travel directly from the speaker to point P. a) For what values of x will *destructive* interference occur at point P? b) For what values of x will *constructive* interference occur at point P? c) Interference effects like those in parts (a) and (b) are almost never a factor in listening to home stereo equipment. Why not?

SECTION 20–7 RESONANCE

20–23 You blow across the open mouth of an empty test tube and produce the fundamental standing wave of the air column inside the test tube. The speed of sound in air is 344 m/s and the test tube acts as a stopped pipe. a) If the length of the air column in the test tube is 14.0 cm, what is the frequency of this standing wave? b) What is the frequency of the fundamental standing wave in the air column if the test tube is half-filled with water?

20–24 The D note (588 Hz) from a trumpet causes a guitar string to vibrate in its second overtone with large amplitude. The vibrating portion of the guitar string has length 63.5 cm. What is the speed of transverse waves on the guitar string?

$A \sin (\omega t - kx)$ also satisfy the equation in part (a)? Interpret this result.

20–26 a) The red and blue waves in Fig. 20–3 combine so that the displacement of the string at O is always zero. To show this mathematically for a wave of arbitrary shape, consider a wave moving to the right along the string in Fig. 20–3 (shown in blue) that at time T is given by $y_1(x, T) = f(x)$, where f is some function of x. (The form of $f(x)$ determines the shape of the wave.) If the point O corresponds to $x = 0$, explain why at time T the wave moving to the left in Fig. 20–3 (shown in red) is given by the function $y_2(x, T) = -f(-x)$. b) Show that the total wave function $y(x, T) = y_1(x, T) + y_2(x, T)$ is zero at O, independent of the form of the function $f(x)$. c) The red and blue waves in Fig. 20–4 combine so that the slope of the string at O is always zero. To show this mathematically for a wave of arbitrary shape, again let the wave moving to the right in Fig. 20–4 (shown in blue) be given by $y_1(x, T) = f(x)$ at time T. Explain why the wave moving to the left (shown in red) at this same time T is given by $y_2(x, T) = f(-x)$. d) Show that the total wave function $y(x, T) = y_1(x, T) + y_2(x, T)$ has zero slope at O, independent of the form of the function $f(x)$, as long as $f(x)$ has a finite first derivative.

20–27 A string that lies along the $+x$-axis has a free end at $x = 0$. a) By using steps similar to those used to derive Eq. (20–1), show that an incident traveling wave $y_1(x, t) = A \sin (\omega t + kx)$ gives rise to a standing wave $y(x, t) = 2A \sin \omega t \cos kx$. b) Show that the standing wave has an antinode at its free end ($x = 0$). c) Find the maximum displacement, maximum speed, and maximum acceleration of the free end of the string.

20–28 A string with both ends held fixed is vibrating in its third harmonic. The waves have a speed of 192 m/s and a frequency of 240 Hz. The amplitude of the standing wave at an antinode is 0.400 cm. a) Calculate the amplitude at points on the string a distance of i) 40.0 cm; ii) 20.0 cm; and iii) 10.0 cm from the left-hand end of the string. b) At each point in part (a), how much time does it take the string to go from its largest upward displacement to its largest downward displacement? c) Calculate the maximum transverse velocity and the maximum transverse acceleration of the string at each of the points in part (a).

20–29 A wooden plank is placed over a pit which is 5.00 m wide. A physics student stands in the middle of the plank and begins to jump up and down such that she jumps upward from the plank two times each second. The plank oscillates with a large amplitude, with maximum amplitude at its center. a) What is the speed of transverse waves on the plank? b) At what rate does the student have to jump to produce large-amplitude oscillations if she is standing 1.25 m from the edge of the pit? (*Note:* The transverse standing waves of the plank have nodes at the two ends that rest on the ground on either side of the pit.)

20–30 A new musical instrument consists of a metal can with length L and diameter $L/10$. The top of the can is cut out, and a string is stretched across this open end of the can. a) The tension in the string is adjusted so that the fundamental frequency for longitudinal sound waves in the air column in the can equals the frequency of the third-harmonic standing wave on the string. What is the relationship between the speed v_t of transverse waves on the string and the speed v_a of sound waves in the air? b) What happens to the sound produced by the instrument if the tension in the string is doubled?

20–31 An organ pipe has two successive harmonics with frequencies 1372 and 1764 Hz. The speed of sound in air is 344 m/s. a) Is this an open or a stopped pipe? Explain. b) What two harmonics are these? c) What is the length of the pipe?

20–32 Longitudinal Standing Waves in a Solid. Longitudinal standing waves can be produced in a solid rod by holding it at some point between the fingers of one hand and stroking it with

the other hand. The rod oscillates with antinodes at both ends. a) Why are the ends antinodes and not nodes? b) The fundamental frequency can be obtained by stroking the rod while it is held at its center. Explain why this is the *only* place to hold the rod to obtain the fundamental. c) Calculate the fundamental frequency of a steel rod of length 1.50 m (see Table 19–1). d) What is the next possible standing-wave frequency of this rod? Where should the rod be held to excite a standing wave of this frequency?

20–33 a) Determine the first three normal-mode frequencies for a pipe of length L that is closed at *both* ends. Explain your reasoning. b) Use the results of part (a) to estimate the normal-mode frequencies of a shower stall. Use $v = 344$ m/s. Explain the connection between these frequencies and the observation that your singing voice probably sounds better in the shower, especially when you sing at certain frequencies.

20–34 One type of steel has a density of 7.8×10^3 kg/m^3 and a breaking stress of 7.0×10^8 N/m^2. A cylindrical guitar string is to be made of 4.00 g of this steel. a) What are the length and radius of the longest and thinnest string that can be placed under a tension of 900 N without breaking? b) What is the highest fundamental frequency that this string could have?

20–35 Combining Standing Waves. A guitar string of length L is plucked in such a way that the total wave produced is the sum of the fundamental and the second harmonic. That is, the standing wave is given by

$$y(x, t) = y_1(x, t) + y_2(x, t),$$

where

$$y_1(x, t) = C \cos \omega_1 t \sin k_1 x,$$
$$y_2(x, t) = C \cos \omega_2 t \sin k_2 x,$$

with $\omega_1 = vk_1$ and $\omega_2 = vk_2$. a) At what values of x are the nodes of y_1? b) At what values of x are the nodes of y_2? c) Graph the total wave at $t = 0$, $t = 1/8f_1$, $t = 1/4f_1$, $t = 3/8f_1$, and $t = 1/2f_1$. d) Does the sum of the two standing waves y_1 and y_2 produce a standing wave? Explain.

20–36 A massive aluminum sculpture is hung from a steel wire. The fundamental frequency for transverse standing waves on the wire is 200 Hz. The sculpture is then immersed in water so that one-third of its volume is submerged. a) What is the new fundamental frequency? b) Why is it a good approximation to treat the wire as being fixed at both ends?

20–37 A long tube contains air at a pressure of 1.00 atm and a temperature of 77.0°C. The tube is open at one end and closed at the other by a movable piston. A tuning fork near the open end is vibrating with a frequency of 500 Hz. Resonance is produced when the piston is at distances 18.0, 55.5, and 93.0 cm from the open end. a) From these measurements, what is the speed of sound in air at 77.0°C? b) From the result of part (a), what is the ratio γ of the molar heat capacities at constant pressure and constant volume for air at this temperature? (See Section 19–7. The molar mass of air is 28.8 g/mol.) c) These data show that a displacement antinode is slightly outside the open end of the tube. How far outside is it?

20–38 The auditory canal of the human ear (Fig. 19–14) extends about 2.5 cm from the outside ear to the eardrum. a) Explain why the human ear is especially sensitive to sounds at frequencies around 3500 Hz. Use $v = 344$ m/s. b) Would you expect the ear to be especially sensitive to frequencies around 7000 Hz? Around 10,500 Hz? Why or why not?

20–39 A cellist tunes the C-string of her instrument to a fundamental frequency of 65.4 Hz. The vibrating portion of the string is 0.600 m long and has a mass of 14.4 g. a) With what tension must it be stretched? b) What percent increase in tension is needed to increase the frequency from 65.4 Hz to 73.4 Hz, corresponding to a rise in pitch from C to D?

20–40 A platinum wire (density 21.4 g/cm^3) is 225 μm in diameter and 0.450 m long. One end of the wire is attached to the ceiling, while a 420-g mass is attached to the other end so that the wire hangs vertically under tension. If a vibrating tuning fork of just the right frequency is held next to the wire, the wire begins to vibrate as well. a) What tuning-fork frequencies will cause this to happen? You may assume that the bottom end of the wire (to which the mass is attached) is essentially stationary, and that the tension in the wire is essentially constant along its length. b) Justify the assumptions made in part (a).

20–41 The frequency of the note F_4 is 349 Hz. a) If an organ pipe is open at one end and closed at the other, what length must it have for its fundamental mode to produce this note at 20.0°C? b) At what air temperature will the frequency be 370 Hz, corresponding to a rise in pitch from F to F-sharp? (See Section 19–7. Ignore the change in length of the pipe due to the temperature change.)

20–42 A standing wave with a frequency of 1100 Hz in a column of methane (CH$_4$) at 20.0°C produces nodes that are 0.200 m apart. What is the ratio γ of the molar heat capacity at constant pressure to that at constant volume for methane? (See Section 19–7. The molar mass of methane is 16.0 g/mol.)

20–43 Two identical loudspeakers are located at points A and B, 2.00 m apart. The loudspeakers are driven by the same amplifier and produce sound waves with a frequency of 784 Hz. Take the speed of sound in air to be 344 m/s. A small microphone is moved out from point B along a line perpendicular to the line connecting A and B (line BC in Fig. 20–27). a) At what distances from B will there be *destructive* interference? b) At what distances from B will there be *constructive* interference? c) If the frequency is made low enough, there will be no positions along the line BC at which destructive interference occurs. How low must the frequency be for this to be the case?

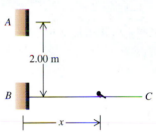

FIGURE 20–27 Problem 20–43.

CHALLENGE PROBLEMS

20–44 Instantaneous Power in a Standing Wave. From Eq. (19–29), the instantaneous rate at which a wave transmits energy along a string (instantaneous power) is

$$P(x, t) = -F \frac{\partial y(x, t)}{\partial x} \frac{\partial y(x, t)}{\partial t},$$

where F is the tension. a) Evaluate $P(x, t)$ for a standing wave of the form given by Eq. (20–1). b) Show that for all values of x, the *average* power P_{av} carried by the standing wave is zero. (Equation (19–33) does not apply here. Can you see why?) c) For a standing wave given by Eq. (20–1), graph $P(x, t)$ and the displacement $y(x, t)$ as functions of x for $t = 0$, $t = \pi/4\omega$, $t = \pi/2\omega$, and $t = 3\pi/4\omega$. (Positive $P(x, t)$ means energy is flowing in the $+x$-direction; negative $P(x, t)$ means the flow is in the $-x$-direction.) d) The *kinetic* energy per unit length of string is greatest where the string has the greatest transverse speed, and the *potential* energy per unit length of string is greatest where the string has the steepest slope (because there the string is stretched the most). (See Challenge Problem 19–45.) Using these ideas, discuss the flow of energy along the string.

20–45 The B-string of a guitar is made of steel (density 7800 kg/m^3), is 63.5 cm long, and has diameter 0.406 mm. The fundamental frequency is $f = 247.0$ Hz. a) Find the string tension. b) If the tension F is changed by a small amount ΔF, the frequency f changes by a small amount Δf. Show that

$$\frac{\Delta f}{f} = \frac{1}{2} \frac{\Delta F}{F}.$$

c) The string is tuned as in part (a) when its temperature is 18.5°C. Strenuous playing can make the temperature of the string rise, changing its vibration frequency. Find Δf if the temperature of the string rises to 29.5°C. The steel string has a Young's modulus of 2.00×10^{11} Pa and a coefficient of linear expansion of 1.20×10^{-5} (C°)$^{-1}$. Assume that the temperature of the body of the guitar remains constant. Will the pitch rise or fall?

21

Sound
and
Hearing

The human ear is an incredibly sensitive instrument, able to detect sounds so faint that they move the eardrum only a microscopic distance. But this sensitivity can be ruined by overly loud sounds. Excessive exposure to loud noises can destroy the hair cells in the inner ear, blocking the passage of signals along the auditory nerve from the ear to the brain. Hearing protection is essential for anyone working in a high-noise environment.

21–1 INTRODUCTION

Of all the mechanical waves that occur in nature, the most important in our everyday lives are longitudinal waves in a medium, usually air, called *sound* waves. The reason is that the human ear is tremendously sensitive and can detect sound waves even of very low intensity. Besides their use in spoken communication, our ears allow us to pick up a myriad of cues about our environment, from the welcome sound of a meal being prepared to the warning sound of an approaching car. The ability to hear an unseen nocturnal predator was essential to the survival of our ancestors, so it is no exaggeration to say that we humans owe our existence to our highly evolved sense of hearing.

In this chapter we will discuss several important properties of sound waves, including frequency, amplitude, and intensity. Up to this point we have described mechanical waves primarily in terms of displacement; however, a description of sound waves in terms of *pressure* fluctuations is often more appropriate, largely because the ear is primarily sensitive to changes in pressure. We'll study the relations among displacement, pressure fluctuation, and intensity and the connections between these quantities and human sound perception. We will find that interference of two sound waves differing slightly in frequency causes a phenomenon called *beats*. When a source of sound or a listener moves through the air, the listener may hear a different frequency than the one emitted by the source. This is the Doppler effect, another topic

of this chapter. We'll also look briefly at shock waves and "sonic booms."

21–2 SOUND WAVES

The most general definition of **sound** is that it is a longitudinal wave in a medium. Our main concern in this chapter is with sound waves in air, but sound can travel through any gas, liquid, or solid. You may be all too familiar with the propagation of sound through a solid if your neighbor's stereo speakers are right next to your wall.

The simplest sound waves are sinusoidal waves, which have definite frequency, amplitude, and wavelength. The human ear is sensitive to waves in the frequency range from about 20 to 20,000 Hz, called the **audible range,** but we also use the term *sound* for similar waves with frequencies above (**ultrasonic**) and below (**infrasonic**) the range of human hearing.

Sound waves usually travel out in all directions from the source of sound, with an amplitude that depends on the direction and distance from the source. We'll return to this point in the next section. For now, we concentrate on the idealized case of a sound wave that propagates in the positive x-direction only. As we discussed in Section 19–4, such a wave is described by a wave function $y(x, t)$, which gives the instantaneous displacement y of a particle in the medium at position x at time t. If the wave is sinusoidal, we can express it using Eq. (19–7):

$$y(x, t) = A \sin(\omega t - kx) \qquad \text{(sound wave propagating in the} \atop +x\text{-direction).}} \qquad (21\text{--}1)$$

Remember that in a longitudinal wave the displacements are *parallel* to the direction of travel of the wave, so distances x and y are measured parallel to each other, not perpendicular as in a transverse wave. The amplitude A is the maximum displacement of a particle in the medium from its equilibrium position.

Sound waves may also be described in terms of variations of *pressure* at various points. In a sinusoidal sound wave in air, the pressure fluctuates above and below atmospheric pressure p_a in a sinusoidal variation with the same frequency as the motions of the air particles. The human ear operates by sensing such pressure variations. A sound wave entering the ear canal exerts a fluctuating pressure on one side of the eardrum; the air on the other side of the eardrum, vented to the outside by the Eustachian tube, is at atmospheric pressure. The pressure difference on the two sides of the eardrum sets it into motion. Microphones and similar devices also usually sense pressure differences, not displacements, so it is very useful to develop a relation between these two descriptions.

Let $p(x, t)$ be the instantaneous pressure fluctuation in a sound wave at any point x at time t. That is, $p(x, t)$ is the amount by which the pressure *differs* from normal atmospheric pressure p_a. Think of $p(x, t)$ as the *gauge pressure* defined in Section 14–3; it can be either positive or negative. The *absolute* pressure at a point is then $p_a + p(x, t)$.

To see the connection between the pressure fluctuation $p(x, t)$ and the displacement $y(x, t)$ in a sound wave propagating in the $+x$-direction, consider an imaginary cylinder of air with cross-section area S and axis along the direction of propagation (Fig. 21–1). When no sound wave is present, the cylinder has length Δx and volume $V = S\Delta x$, as shown by the shaded volume in Fig. 21–1. When a wave is present, at time t the end of the cylinder that is initially at x is displaced by $y_1 = y(x, t)$, and the end that is initially at $x + \Delta x$ is displaced by $y_2 = y(x + \Delta x, t)$; this is shown by the red lines. If $y_2 > y_1$, as shown in Fig. 21–1, the cylinder's volume has increased, which causes a decrease in pressure. If $y_2 < y_1$, the cylinder's volume has decreased and the pressure has increased. If $y_2 = y_1$, the cylinder is simply shifted to the left or right; there is no volume change and no pressure fluctuation. The pressure fluctuation depends on the *difference* between the displacement at neighboring points in the medium.

Quantitatively, the change in volume ΔV of the cylinder is

$$\Delta V = S(y_2 - y_1) = S[y(x + \Delta x, t) - y(x, t)].$$

In the limit as $\Delta x \to 0$, the fractional change in volume dV/V is

$$\frac{dV}{V} = \lim_{\Delta x \to 0} \frac{y(x + \Delta x, t) - y(x, t)}{\Delta x} = \frac{\partial y(x, t)}{\partial x}. \qquad (21\text{--}2)$$

The fractional volume change is related to the pressure fluctuation by the bulk modulus B, which by definition [Eq. (11–13)] is $B = -p(x, t)/(dV/V)$. Solving for $p(x, t)$, we have

$$p(x, t) = -B\frac{\partial y(x, t)}{\partial x}. \qquad (21\text{--}3)$$

The negative sign arises because when $\partial y(x, t)/\partial x$ is positive, the displacement is greater at $x + \Delta x$ than at x, corresponding to an increase in volume and a *decrease* in pressure.

When we evaluate $\partial y(x, t)/\partial x$ for the sinusoidal wave of Eq. (21–1), we find

$$p(x, t) = BkA \cos(\omega t - kx). \qquad (21\text{--}4)$$

Figure 21–2 shows $y(x, t)$ and $p(x, t)$ for a sinusoidal sine wave at $t = 0$. While $y(x, t)$ and $p(x, t)$ describe the same wave, these two functions are one quarter-cycle out of phase; at a given time, the displacement is greatest where the pressure fluctuation is zero, and vice versa.

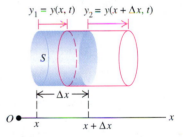

$y_1 = y(x, t)$ $y_2 = y(x + \Delta x, t)$

21–1 A cylindrical volume of gas with cross-section area S. The length in the undisplaced position is Δx. During wave propagation along the axis, the left end is displaced to the right a distance y_1 and the right end is displaced a different distance y_2. The resulting change in volume is $S(y_2 - y_1)$.

21–2 Pressure fluctuation $p(x, t)$ and displacement $y(x, t)$ in a sinusoidal traveling sound wave, shown as functions of x at a particular time t. The pressure fluctuation and displacement are a quarter-cycle out of phase; $p(x, t)$ is maximum where $y(x, t)$ is zero, and vice versa.

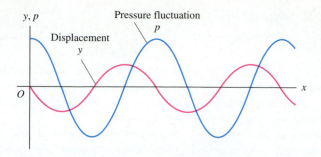

CAUTION ▶ Take care not to confuse the *traveling* sound wave described by Eqs. (21–1) and (21–2) with the *standing* sound waves in a pipe that we discussed in Section 20–5. In the traveling wave shown in Fig. 21–2, the patterns of the pressure fluctuation and the displacement move together in the $+x$-direction (to the right in Fig. 21–2); at any point x, particles oscillate in simple harmonic motion, and the pressure varies sinusoidally. By contrast, in a standing sound wave, the patterns of $p(x, t)$ and $y(x, t)$ do *not* move; there are points at which the displacement is always zero (the displacement nodes) and other points at which the pressure fluctuation is always zero (the pressure nodes). ◀

Equation (21–4) shows that the quantity BkA represents the maximum pressure fluctuation. We call this the **pressure amplitude,** denoted by p_{max}:

$$p_{max} = BkA \qquad \text{(sinusoidal sound wave).} \qquad (21\text{–}5)$$

The pressure amplitude is directly proportional to the displacement amplitude A, as we might expect, and it also depends on wavelength. Waves of shorter wavelength λ (larger wave number $k = 2\pi/\lambda$) have greater pressure variations for a given amplitude because the maxima and minima are squeezed closer together. A medium with a large value of bulk modulus B requires a relatively large pressure amplitude for a given displacement amplitude because large B means a less compressible medium, that is, greater pressure change required for a given volume change.

EXAMPLE 21–1

Amplitude of a sound wave in air In a sinusoidal sound wave of moderate loudness the maximum pressure variations are of the order of 3.0×10^{-2} Pa above and below atmospheric pressure p_a (nominally 1.013×10^5 Pa at sea level). Find the corresponding maximum displacement if the frequency is 1000 Hz and $v = 344$ m/s.

SOLUTION From Eq. (21–5) the maximum displacement is $A = p_{max}/Bk$. To determine k, we use Eq. (19–6), $\omega = vk$. We have $\omega = 2\pi f = (2\pi \text{ rad})(1000 \text{ Hz}) = 6283$ rad/s and

$$k = \frac{\omega}{v} = \frac{6283 \text{ rad/s}}{344 \text{ m/s}} = 18.3 \text{ rad/m.}$$

As described in Section 19–7, the appropriate value of B for

ordinary sound waves in air is the *adiabatic* bulk modulus. From Eq. (19–24), for air at normal atmospheric pressure,

$$B = \gamma p_a = (1.40)(1.013 \times 10^5 \text{ Pa}) = 1.42 \times 10^5 \text{ Pa.}$$

Then

$$A = \frac{p_{max}}{Bk} = \frac{3.0 \times 10^{-2} \text{ Pa}}{(1.42 \times 10^5 \text{ Pa})(18.3 \text{ rad/m})}$$

$$= 1.2 \times 10^{-8} \text{ m.}$$

This displacement amplitude is only about $\frac{1}{100}$ the size of a human cell. Remember that the ear actually senses pressure fluctuations; it detects these minuscule displacements only indirectly.

EXAMPLE 21–2

Amplitude of a sound wave in the inner ear When a sound wave enters the ear, it sets the eardrum into oscillation, which in turn causes oscillation of the three tiny bones in the middle ear called the *ossicles* (Fig. 21–3). This oscillation is finally transmitted to

the fluid-filled inner ear; the motion of the fluid disturbs hair cells within the inner ear, which transmit nerve impulses to the brain with the information that a sound is present. The moving part of the eardrum has an area of about 43 mm^2, and the area of

21–3 The anatomy of the human ear. The middle ear is the size of a small marble; the ossicles (hammer, anvil, and stirrup) are the smallest bones in the human body.

the stirrup (the smallest of the ossicles) where it connects to the inner ear is about 3.2 mm². For the sound in the previous example, determine a) the pressure amplitude and b) the displacement amplitude of the wave in the fluid of the inner ear.

SOLUTION a) We can safely neglect the mass of the ossicles (about 58 mg = 5.8×10^{-5} kg), so the force exerted by the ossicles on the fluid in the inner ear (mostly water) is the same as the force exerted on the eardrum and ossicles by the sound wave in air. Hence the pressure amplitude p_{max} is greater in the inner ear than in the outside air because the same force F is exerted on a smaller area S. Pressure equals force divided by area, so

$$p_{max\,(inner\,ear)} = \frac{F}{S_{inner\,ear}} = \frac{p_{max\,(air)}\,S_{eardrum}}{S_{inner\,ear}}$$

$$= \frac{(3.0 \times 10^{-2}\ \text{Pa})(43\ \text{mm}^2)}{3.2\ \text{mm}^2} = 0.40\ \text{Pa}.$$

The effect of the ossicles is to increase the pressure amplitude in the inner ear by a factor of (43 mm²)/(3.2 mm²) = 13. This amplification factor helps to give the human ear its great sensitivity.

b) To find the maximum displacement, we again use the relationship $A = p_{max}/Bk$ as in Example 21–1. The values of B, k, and p_{max} are different in the inner ear than in the air, however.

The fluid in the inner ear is mostly water, which has a much greater bulk modulus than air because water is much more difficult to compress. From Table 11–2 the compressibility of water (unfortunately also called k) equals 45.8×10^{-11} Pa^{-1}, so $B_{fluid} = 1/(45.8 \times 10^{-11}\ \text{Pa}^{-1}) = 2.18 \times 10^9$ Pa.

The wave in the inner ear has the same angular frequency ω as the wave in the air because the air, eardrum, ossicles, and inner ear fluid all oscillate together. But since the wave speed v is greater in water than in air, the wave number $k = \omega/v$ is less in the inner ear. From Table 19–1 the speed of sound in water is 1482 m/s at 20°C and increases somewhat with temperature; at human body temperature (37.0°C, or 98.6°F), v is about 1500 m/s. Using the value of ω from the previous example, we have

$$k_{inner\,ear} = \frac{\omega}{v_{inner\,ear}} = \frac{6283\ \text{rad/s}}{1500\ \text{m/s}} = 4.2\ \text{rad/m}.$$

When we put everything together, the maximum displacement of the fluid in the inner ear is

$$A_{inner\,ear} = \frac{p_{max\,(inner\,ear)}}{B_{fluid}\,k_{inner\,ear}} = \frac{0.40\ \text{Pa}}{(2.18 \times 10^9\ \text{Pa})(4.2\ \text{rad/m})}$$

$$= 4.4 \times 10^{-11}\ \text{m}.$$

This is an even smaller displacement amplitude than in the air. What really matters in the inner ear, however, is the *pressure* amplitude, since the pressure variations within the fluid cause the forces that set the hair cells into motion.

PERCEPTION OF SOUND WAVES

The physical characteristics of a sound wave are directly related to the perception of that sound by a listener. For a given frequency, the greater the pressure amplitude of a sinusoidal sound wave, the greater the perceived **loudness.** The relationship between pressure amplitude and loudness is not a simple one, and it varies from one person to another. One important factor is that the ear is not equally sensitive to all frequencies in the audible range. A sound at one frequency may seem louder than one of equal pressure amplitude at a different frequency. At 1000 Hz the minimum pressure amplitude that can be perceived with normal hearing is about 3×10^{-5} Pa; to produce the same loudness at 200 Hz or 15,000 Hz requires about 3×10^{-4} Pa. Perceived loudness also depends on the health of the ear. A loss of sensitivity at the high-frequency end usually happens naturally with age but can be further aggravated by excessive noise levels. Studies have shown that many young rock musicians have suffered permanent ear damage and have hearing that is typical of persons 65 years of age. Portable stereo headsets used at high volume pose similar threats to hearing. Be careful!

The frequency of a sound wave is the primary factor in determining the **pitch** of a sound, the quality that lets us classify the sound as "high" or "low." The higher the frequency of a sound (within the audible range), the higher the pitch that a listener will perceive. Pressure amplitude also plays a role in determining pitch. When a listener compares two sinusoidal sound waves with the same frequency but different pressure

21–4 (a) Graph of pressure fluctuation versus time from a clarinet with fundamental frequency $f_1 = 233$ Hz. (b) Harmonic content of the sound of the clarinet. (c) Graph of pressure fluctuation versus time from an alto recorder with fundamental frequency $f_1 = 523$ Hz. (d) Harmonic content of the sound of the alto recorder.

10.10

ActivPhysics

Complex Waves: Fourier Analysis

amplitudes, the one with the greater pressure amplitude is usually perceived as louder but also as slightly lower in pitch.

Musical sounds have wave functions that are more complicated than a simple sine function. The pressure fluctuation in the sound wave produced by a clarinet is shown in Fig. 21–4a. As we discussed in Section 20–5, the reason for the complexity is that the column of air in a wind instrument vibrates with the fundamental frequency and at many harmonics at the same time. The same occurs for a string that has been plucked, bowed, or struck. The sound wave produced in the surrounding air will have a similar amount of each harmonic, that is, a similar *harmonic content*. Figure 21–4b shows the harmonic content of the sound of a clarinet. Two tones produced by different instruments might have the same fundamental frequency (and thus the same pitch) but sound different because of the presence of different amounts of the various harmonics. The difference is called *tone color, quality,* or **timbre** and is often described in subjective terms such as reedy, golden, round, mellow, and tinny. A tone that is rich in harmonics, like the clarinet tone in Figs. 21–4a and 21–4b, usually sounds thin and "stringy" or "reedy," while a tone containing mostly a fundamental, like the alto recorder tone in Figs. 21–4c and 21–4d, is more mellow and flutelike. The same principle applies to the human voice, which is another example of a wind instrument; the vowels "a" and "e" sound different because of differences in harmonic content.

Another factor in determining tone quality is the behavior at the beginning (*attack*) and end (*decay*) of a tone. A piano tone begins with a thump and then dies away gradually. A harpsichord tone, in addition to having different harmonic content, begins much more quickly with a click, and the higher harmonics begin before the lower ones. When the key is released, the sound also dies away much more rapidly with a harpsichord than with a piano. Similar effects are present in other musical instruments. With wind and string instruments the player has considerable control over the attack and decay of the tone, and these characteristics help to define the unique characteristics of each instrument.

Unlike the tones made by musical instruments or the vowels in human speech, **noise** is a combination of *all* frequencies, not just frequencies that are integer multiples of a fundamental frequency. (An extreme case is "white noise," which contains equal amounts of all frequencies across the audible range.) Examples include the sound of the wind and the hissing sound you make in saying the consonant "s."

21–3 SOUND INTENSITY

Traveling sound waves, like all other traveling waves, transfer energy from one region of space to another. We define the **intensity** of a wave (denoted by I) to be *the time average rate at which energy is transported by the wave, per unit area,* across a surface perpendicular to the direction of propagation. That is, intensity I is average *power* per unit area.

In Section 6–5 we saw that power equals the product of force and velocity; see Eq. (6–18). So the power per unit area in a sound wave propagating in the x-direction equals the product of the pressure fluctuation $p(x, t)$ (force per unit area) and the *particle* velocity $v_y(x, t)$. For the sinusoidal wave described by Eq. (21–1), $p(x, t)$ is given by Eq. (21–4) and $v_y(x, t)$ is obtained by taking the time derivative of Eq. (21–1). We find

$$v_y(x, t) = \frac{\partial y(x, t)}{\partial t} = \omega A \cos(\omega t - kx),$$

$$p(x, t)v_y(x, t) = (BkA\cos(\omega t - kx))(\omega A\cos(\omega t - kx)) = B\omega k A^2\cos^2(\omega t - kx).$$

The intensity is, by definition, the average value of $p(x, t)v_y(x, t)$. For any value of x the average value of the function $\cos^2(\omega t - kx)$ over one period $T = 2\pi/\omega$ is 1/2, so

$$I = \frac{1}{2}B\omega k A^2. \tag{21–6}$$

By using the relations $\omega = vk$ and $v^2 = B/\rho$, we can transform Eq. (21–6) into the form

$$I = \frac{1}{2}\sqrt{\rho B}\,\omega^2 A^2 \qquad \text{(intensity of a sinusoidal sound wave),} \tag{21–7}$$

which we cited at the end of Section 19–8. This equation shows why in a stereo system a low-frequency woofer has to vibrate with much larger amplitude than a high-frequency tweeter to produce the same sound intensity.

It is usually more useful to express I in terms of the pressure amplitude p_{max}. Using Eq. (21–5) and the relation $\omega = vk$, we find

$$I = \frac{\omega p_{max}^2}{2Bk} = \frac{v p_{max}^2}{2B}. \tag{21–8}$$

By using the wave speed relation $v^2 = B/\rho$, we can also write Eq. (21–8) in the alternative forms

$$I = \frac{p_{max}^2}{2\rho v} = \frac{p_{max}^2}{2\sqrt{\rho B}} \qquad \text{(intensity of a sinusoidal sound wave).} \tag{21–9}$$

We invite you to verify these expressions (see Exercise 21–4). Comparison of Eqs. (21–7) and (21–9) shows that sinusoidal sound waves of the same intensity but different frequency have different displacement amplitudes A but the *same* pressure amplitude p_{max}. This is another reason why it is usually more convenient to describe a sound wave in terms of pressure fluctuations, not displacement.

The *total* average power carried across a surface by a sound wave equals the product of the intensity at the surface and the surface area if the intensity over the surface is uniform. The average total sound power emitted by a person speaking in an ordinary conversational tone is about 10^{-5} W, while a loud shout corresponds to about 3×10^{-2} W. If all the residents of New York City were to talk at the same time, the total sound power would be about 100 W, equivalent to the electric power requirement of a medium-sized light bulb. On the other hand, the power required to fill a large auditorium or stadium with loud sound is considerable (see Example 21–5 below.)

Problem–Solving Strategy

SOUND INTENSITY

Quite a few quantities are involved in characterizing the amplitude and intensity of a sound wave, and it's easy to get lost in the maze of relationships. It helps to put them in categories: The amplitude is described by A or p_{max}, and the frequency f can be determined from ω, k, or λ. These quantities are related through the wave speed v, which in turn is determined by the properties of the medium, B and ρ. Take a hard look at the problem at hand, identify which of these quantities are given and which you have to find; then start looking for relationships that take you where you want to go.

EXAMPLE 21-3

Intensity of a sound wave in air Find the intensity of the sound wave in Example 21–1, with $p_{max} = 3.0 \times 10^{-2}$ Pa, if the temperature is 20°C so that the density of air is $\rho = 1.20$ kg/m³ and the speed of sound is $v = 344$ m/s.

SOLUTION From Eq. (21–9),

$$I = \frac{p_{max}^2}{2\rho v} = \frac{(3.0 \times 10^{-2} \text{ Pa})^2}{2(1.20 \text{ kg/m}^3)(344 \text{ m/s})}$$

$$= 1.1 \times 10^{-6} \text{ J/(s} \cdot \text{m}^2) = 1.1 \times 10^{-6} \text{ W/m}^2.$$

A very loud sound wave at the threshold of pain has a pressure amplitude of about 30 Pa and an intensity of about 1 W/m². The pressure amplitude of the faintest sound wave that can be heard is about 3×10^{-5} Pa, and the corresponding intensity is about 10^{-12} W/m². We invite you to verify these correspondences.

EXAMPLE 21-4

Same intensity, different frequencies A 20-Hz sound wave has the same intensity as the 1000-Hz sound wave in Examples 21–1 and 21–3. What are the amplitude A and pressure amplitude p_{max} of the 20-Hz sound wave?

SOLUTION In Eq. (21–7), ρ and B depend on the medium, not the amplitude or frequency of the wave. For I to be constant, the product ωA must be constant. That is,

$$(20 \text{ Hz})A_{20} = (1000 \text{ Hz})(1.2 \times 10^{-8} \text{ m}),$$

$$A_{20} = 6.0 \times 10^{-7} \text{ m} = 0.60 \text{ } \mu\text{m}.$$

Do you understand why we didn't have to convert the frequencies to angular frequencies?

Since the intensity is the same for both frequencies, Eq. (21–9) shows that the *pressure* amplitude p_{max} must also be the same for both. Hence $p_{max} = 3.0 \times 10^{-2}$ Pa for $f = 20$ Hz. Note that using Eq. (21–5) and $k = \omega/v$, we get $p_{max} = BkA = (B/v)\omega A$; the bulk modulus B and wave speed v depend only on the medium, so we again conclude that the product ωA must have the same value for both frequencies.

EXAMPLE 21-5

"Play it loud!" For an outdoor concert we want the sound intensity at a distance of 20 m from the speaker array to be 1 W/m². Assuming that the sound waves have the same intensity in all directions, what acoustic power output is needed from the speaker array?

SOLUTION We make the assumptions that the speakers are near ground level and that the acoustic power is spread uniformly over a hemisphere 20 m in radius (that is, we assume that none

of the acoustic power is directed into the ground). The surface area of this hemisphere is equal to $(1/2)(4\pi)(20 \text{ m})^2$, or about 2500 m². The acoustic power needed is

$$(1 \text{ W/m}^2)(2500 \text{ m}^2) = 2500 \text{ W} = 2.5 \text{ kW}.$$

The electrical power input to the speaker would need to be considerably larger because the efficiency of such devices is not very high (typically a few percent for ordinary speakers, and up to 25% for horn-type speakers).

VARIATION OF INTENSITY WITH DISTANCE

If a source of sound can be considered as a point, the intensity at a distance r from the source is inversely proportional to r^2. This follows directly from energy conservation; if the power output of the source is P, then the average intensity I_1 through a sphere with radius r_1 and surface area $4\pi r_1^2$ is

$$I_1 = \frac{P}{4\pi r_1^2}.$$

The average intensity I_2 through a sphere with a different radius r_2 is given by a similar expression. If no energy is absorbed between the two spheres, the power P must be the same for both, and

$$4\pi r_1^2 I_1 = 4\pi r_2^2 I_2,$$

$$\frac{I_1}{I_2} = \frac{r_2^2}{r_1^2}. \tag{21–10}$$

TABLE 21-1

SOUND INTENSITY LEVELS FROM VARIOUS SOURCES (REPRESENTATIVE VALUES)

SOURCE OR DESCRIPTION OF SOUND	SOUND INTENSITY LEVEL, β (dB)	INTENSITY, I (W/m^2)
Threshold of pain	120	1
Riveter	95	3.2×10^{-3}
Elevated train	90	10^{-3}
Busy street traffic	70	10^{-5}
Ordinary conversation	65	3.2×10^{-6}
Quiet automobile	50	10^{-7}
Quiet radio in home	40	10^{-8}
Average whisper	20	10^{-10}
Rustle of leaves	10	10^{-11}
Threshold of hearing at 1000 Hz	0	10^{-12}

The intensity I at any distance r is therefore inversely proportional to r^2. This "inverse-square" relationship also holds for various other energy-flow situations with a point source, such as light emitted by a point source.

The inverse-square relationship between sound intensity and distance from the source does not apply indoors because sound energy can also reach a listener by reflection from the walls and ceiling. Indeed, part of the architect's job in designing an auditorium is to tailor these reflections so that the intensity is as nearly constant as possible over the auditorium.

THE DECIBEL SCALE

Because the ear is sensitive over such a broad range of intensities, a *logarithmic* intensity scale is usually used. The **sound intensity level** β of a sound wave is defined by the equation

$$\beta = (10 \text{ dB}) \log \frac{I}{I_0} \qquad \text{(definition of sound intensity level).} \qquad (21-11)$$

In this equation, I_0 is a reference intensity, chosen to be 10^{-12} W/m^2, approximately the threshold of human hearing at 1000 Hz. Recall that "log" means the logarithm to the base 10. Sound intensity levels are expressed in **decibels,** abbreviated dB. A decibel is $\frac{1}{10}$ of a *bel,* a unit named for Alexander Graham Bell (the inventor of the telephone). The bel is inconveniently large for most purposes, and the decibel is the usual unit of sound intensity level.

If the intensity of a sound wave equals I_0 or 10^{-12} W/m^2, its sound intensity level is 0 dB. An intensity of 1 W/m^2 corresponds to 120 dB. Table 21-1 gives the sound intensity levels in decibels of several familiar sounds. You can use Eq. (21-11) to check the value of sound intensity level β given for each intensity in the table.

Because the ear is not equally sensitive to all frequencies in the audible range, some sound-level meters weight the various frequencies unequally. One such scheme leads to the so-called dBA scale; this scale de-emphasizes the low and very high frequencies, where the ear is less sensitive than at midrange frequencies.

EXAMPLE 21-6

Temporary deafness A ten-minute exposure to 120-dB sound will typically shift your threshold of hearing at 1000 Hz from 0 dB up to 28 dB for a while. Ten years of exposure to 92-dB sound will cause a *permanent* shift up to 28 dB. What intensities correspond to 28 dB and 92 dB?

SOLUTION We rearrange Eq. (21-11) by dividing both sides by 10 dB and then using the relationship $10^{\log x} = x$:

$$I = I_0 10^{\beta/(10 \text{ dB})}.$$

When $\beta = 28$ dB,

$$I = (10^{-12} \text{ W/m}^2)10^{(28 \text{ dB}/10 \text{ dB})}$$
$$= (10^{-12} \text{ W/m}^2)10^{2.8} = 6.3 \times 10^{-10} \text{ W/m}^2.$$

Similarly, for $\beta = 92$ dB,

$$I = (10^{-12} \text{ W/m}^2)10^{(92 \text{ dB}/10 \text{ dB})} = 1.6 \times 10^{-3} \text{ W/m}^2.$$

If your answers are a factor of 10 too large, you may have entered 10×10^{-12} in your calculator instead of 1×10^{-12}. Be careful!

EXAMPLE 21-7

A bird sings in a meadow Consider an idealized model with a bird (treated as a point source) emitting constant sound power, with intensity inversely proportional to the square of the distance from the bird. By how many dB does the sound intensity level drop when you move twice as far away from the bird?

SOLUTION We label the two points 1 and 2 (Fig. 21–5), and we use Eq. (21–11) twice. The difference in sound intensity level, $\beta_2 - \beta_1$, is given by

$$\beta_2 - \beta_1 = (10 \text{ dB})\left(\log\frac{I_2}{I_0} - \log\frac{I_1}{I_0}\right)$$
$$= (10 \text{ dB})[(\log I_2 - \log I_0) - (\log I_1 - \log I_0)]$$
$$= (10 \text{ dB})\log\frac{I_2}{I_1}.$$

Now we use the reciprocal of Eq. (21–10); $I_2/I_1 = r_1^2/r_2^2$, so

$$\beta_2 - \beta_1 = (10 \text{ dB})\log\frac{r_1^2}{r_2^2} = (10 \text{ dB})\log\frac{r_1^2}{(2r_1)^2}$$
$$= (10 \text{ dB})\log\frac{1}{4} = -6.0 \text{ dB}.$$

A decrease in intensity of a factor of four corresponds to a 6-dB decrease in sound intensity level.

We invite you to prove that an intensity increase of a factor of two corresponds to a 3-dB increase in sound intensity level. This change is barely perceptible to the human ear. Most people usually interpret an increase of 8 to 10 dB in sound intensity level as a doubling of loudness.

21–5 When you double your distance from a point source of sound, by how much does the sound intensity level decrease?

Point source

P_2 P_1

D/V. HENNINGS.

21-4 BEATS

In Section 20–6 we talked about *interference* effects that occur when two different waves with the same frequency overlap in the same region of space. Now let's look at what happens when we have two waves with equal amplitude but slightly different frequencies. This occurs, for example, when two tuning forks with slightly different frequencies are sounded together or when two organ pipes that are supposed to have exactly the same frequency are slightly "out of tune."

Consider a particular point in space where the two waves overlap. The displacements of the individual waves at this point are plotted as functions of time in Fig. 21–6a. The total length of the time axis represents one second, and the frequencies are 16 Hz (blue graph) and 18 Hz (red graph). Applying the principle of superposition, we add the two displacements at each instant of time to find the total displacement at that time. The result is the graph of Fig. 21–6b. At certain times ($t = 0.25$ s and $t = 0.75$ s in Fig. 21–6) the two waves are in phase; their maxima coincide and their amplitudes add. But

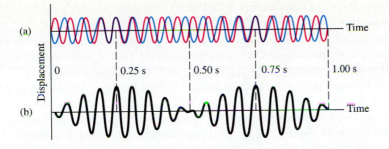

(a)

Displacement

0   0.25 s   0.50 s   0.75 s   1.00 s

Time

(b)   Time

21-6 Beats are fluctuations in amplitude produced by two sound waves of slightly different frequency, 16 Hz and 18 Hz. (a) Individual waves. (b) Resultant wave formed by superposition of the two waves. The beat frequency is 18 Hz − 16 Hz = 2 Hz.

because of their slightly different frequencies, the two waves cannot be in phase at all times. Indeed, at certain times (like $t = 0.50$ s in Fig. 21–6) the two waves are exactly *out* of phase. The two waves then cancel each other, and the total amplitude is zero.

The resultant wave in Fig. 21–6b looks like a single sinusoidal wave with a varying amplitude that goes from a maximum to zero and back. In this example the amplitude goes through two maxima and two minima in one second, so the frequency of this amplitude variation is 2 Hz. The amplitude variation causes variations of loudness called **beats,** and the frequency with which the loudness varies is called the **beat frequency.** In this example the beat frequency is the *difference* of the two frequencies. If the beat frequency is a few hertz, we hear it as a waver or pulsation in the tone.

We can prove that the beat frequency is *always* the difference of the two frequencies f_a and f_b. Suppose f_a is larger than f_b; the corresponding periods are T_a and T_b, with $T_a < T_b$. If the two waves start out in phase at time $t = 0$, they will again be in phase when the first wave has gone through exactly one more cycle than the second. This will happen at a value of t equal to T_{beat}, the *period* of the beat. Let n be the number of cycles of the first wave in time T_{beat}; then the number of cycles of the second wave in the same time is $(n - 1)$, and we have the relations

$$T_{\text{beat}} = nT_a \qquad \text{and} \qquad T_{\text{beat}} = (n - 1)T_b.$$

Eliminating n between these two equations, we find

$$T_{\text{beat}} = \frac{T_a T_b}{T_b - T_a}.$$

The reciprocal of the beat period is the beat *frequency,* $f_{\text{beat}} = 1/T_{\text{beat}}$, so

$$f_{\text{beat}} = \frac{T_b - T_a}{T_a T_b} = \frac{1}{T_a} - \frac{1}{T_b},$$

and finally

$$f_{\text{beat}} = f_a - f_b \qquad \text{(beat frequency)}. \qquad (21\text{–}12)$$

As claimed, the beat frequency is the difference of the two frequencies. In using Eq. (21–12), remember that f_a is the higher frequency.

An alternative way to derive Eq. (21–12) is to write functions to describe the curves in Fig. 21–6a and then add them. Suppose that at a certain position the two waves are given by $y_a(t) = A \sin 2\pi f_a t$ and $y_b(t) = -A \sin 2\pi f_b t$. We use the trigonometric identity

$$\sin a - \sin b = 2 \sin \tfrac{1}{2}(a - b) \cos \tfrac{1}{2}(a + b).$$

We can then express the total wave $y(t) = y_a(t) + y_b(t)$ as

$$y_a(t) + y_b(t) = [2A \sin \tfrac{1}{2}(2\pi)(f_a - f_b)t] \cos \tfrac{1}{2}(2\pi)(f_a + f_b)t.$$

ActivPhysics

10.7

Beats and Beat Frequency

If the two propellers on this airplane are not precisely synchronized, the pilots, passengers, and listeners on the ground will hear beats.

10.8
Doppler Effect: Conceptual Introduction

10.9
Doppler Effect: Problems

The amplitude factor (the quantity in brackets) varies slowly with frequency $\frac{1}{2}(f_a - f_b)$. The cosine factor varies with a frequency equal to the *average* frequency $\frac{1}{2}(f_a + f_b)$. The *square* of the amplitude factor, which is proportional to the intensity that the ear hears, goes through two maxima and two minima per cycle. So the beat frequency f_{beat} that is heard is twice the quantity $\frac{1}{2}(f_a - f_b)$, or just $f_a - f_b$, in agreement with Eq. (21–12).

Beats between two tones can be heard up to a beat frequency of about 6 or 7 Hz. Two piano strings or two organ pipes differing in frequency by 2 or 3 Hz sound wavery and "out of tune," although some organ stops contain two sets of pipes deliberately tuned to beat frequencies of about 1 to 2 Hz for a gently undulating effect. Listening for beats is an important technique in tuning all musical instruments.

At frequency differences greater than about 6 or 7 Hz, we no longer hear individual beats, and the sensation merges into one of *consonance* or *dissonance,* depending on the frequency ratio of the two tones. In some cases the ear perceives a tone called a *difference tone,* with a pitch equal to the beat frequency of the two tones. For example, if you listen to a whistle that produces sounds at 1800 Hz and 1900 Hz when blown, you will hear not only these tones but also a much lower 100-Hz tone.

The engines on multiengine propeller aircraft have to be synchronized so that the sounds don't cause annoying beats, which are heard as loud throbbing sounds. On some planes this is done electronically; on others the pilot does it by ear, just like tuning a piano.

21–5 THE DOPPLER EFFECT

You've probably noticed that when a car approaches you with its horn sounding, the pitch seems to drop as the car passes. This phenomenon, first described by the 19th-century Austrian scientist Christian Doppler, is called the **Doppler effect.** When a source of sound and a listener are in motion relative to each other, the frequency of the sound heard by the listener is not the same as the source frequency. A similar effect occurs for light and radio waves; we'll return to this later in this section.

To analyze the Doppler effect for sound, we'll work out a relation between the frequency shift and the velocities of source and listener relative to the medium (usually air) through which the sound waves propagate. To keep things simple, we consider only the special case in which the velocities of both source and listener lie along the line joining them. Let v_S and v_L be the velocity components along this line for the source and the listener, respectively, relative to the medium. We choose the positive direction for both v_S and v_L to be the direction from the listener L to the source S. The speed of sound relative to the medium, v, is always considered positive.

MOVING LISTENER

Let's think first about a listener L moving with velocity v_L toward a stationary source S (Fig. 21–7). The source emits a sound wave with frequency f_S and wavelength $\lambda = v/f_S$. The figure shows several wave crests, separated by equal distances λ. The wave crests approaching the moving listener have a speed of propagation *relative to the listener* of $(v + v_L)$. So the frequency f_L with which the crests arrive at the listener's position (that is, the frequency the listener hears) is

$$f_L = \frac{v + v_L}{\lambda} = \frac{v + v_L}{v/f_S},$$
(21–13)

or

$$f_L = \left(\frac{v + v_L}{v}\right)f_S = \left(1 + \frac{v_L}{v}\right)f_S \qquad \text{(moving listener, stationary source).}$$
(21–14)

21-7 A listener moving toward a stationary source hears a frequency that is higher than the source frequency because the relative speed of listener and wave is greater than the wave speed v.

So a listener moving toward a source ($v_L > 0$), as in Fig. 21-7, hears a higher frequency (higher pitch) than does a stationary listener. A listener moving away from the source ($v_L < 0$) hears a lower frequency (lower pitch).

MOVING SOURCE AND MOVING LISTENER

Now suppose the source is also moving, with velocity v_S (Fig. 21-8). The wave speed relative to the wave medium (air) is still v; it is determined by the properties of the medium and is not changed by the motion of the source. But the wavelength is no longer equal to v/f_S. Here's why. The time for emission of one cycle of the wave is the period $T = 1/f_S$. During this time, the wave travels a distance $vT = v/f_S$ and the source moves a distance $v_S T = v_S/f_S$. The wavelength is the distance between successive wave crests, and this is determined by the *relative* displacement of source and wave. As Fig. 21-8 shows, this is different in front of and behind the source. In the region to the right of the source in Fig. 21-8 (that is, in front of the source) the wavelength is

$$\lambda = \frac{v}{f_S} - \frac{v_S}{f_S} = \frac{v - v_S}{f_S} \qquad \text{(wavelength in front of a moving source). (21-15)}$$

In the region to the left of the source (that is, behind the source) it is

$$\lambda = \frac{v + v_S}{f_S} \qquad \text{(wavelength behind a moving source). (21-16)}$$

21-8 Wave crests emitted by a moving source are crowded together in front of the source (to the right of this source) and stretched out behind it (to the left of this source).

The Doppler effect explains why the siren on a fire engine or ambulance has a high pitch ($f_L > f_S$) when it is approaching you ($v_S < 0$) and a low pitch ($f_L < f_S$) when it is moving away ($v_S > 0$).

The waves in front of and behind the source are compressed and stretched out, respectively, by the motion of the source.

To find the frequency heard by the listener behind the source, we substitute Eq. (21–16) into the first form of Eq. (21–13):

$$f_L = \frac{v + v_L}{\lambda} = \frac{v + v_L}{(v + v_S)/f_S},$$

$$f_L = \frac{v + v_L}{v + v_S} f_S \qquad \text{(Doppler effect, moving source and moving listener).} \qquad (21–17)$$

This expresses the frequency f_L heard by the listener in terms of the frequency f_S of the source.

Equation (21–17) includes all possibilities for motion of source and listener (relative to the medium) along the line joining them. If the listener happens to be at rest in the medium, v_L is zero. When both source and listener are at rest or have the same velocity relative to the medium, then $v_L = v_S$ and $f_L = f_S$. Whenever the direction of the source or listener velocity is opposite to the direction from the listener toward the source (which we have defined as positive), the corresponding velocity to be used in Eq. (21–17) is negative.

Problem–Solving Strategy

DOPPLER EFFECT

1. Draw a sketch and establish a coordinate system. Define the positive direction to be the direction from listener to source, and make sure you know the signs of all the relevant velocities. A velocity in the direction from listener toward source is positive; a velocity in the opposite direction is negative. Also, the velocities must all be measured relative to the air in which the sound is traveling.

2. Use consistent notation to identify the various quantities: subscript S for source, L for listener.

3. When a wave is reflected from a surface, either stationary or moving, the analysis can be carried out in two steps. In

   the first, the surface plays the role of listener; the frequency with which the wave crests arrive at the surface is f_L. Then think of the surface as a new source, emitting waves with this same frequency f_L. Finally, determine what frequency is heard by a listener detecting this new wave.

4. Ask whether your final result makes sense. If the source and listener are moving toward each other, f_L is always greater than f_S; if they are moving apart, $f_L < f_S$. If the source and listener have no relative motion, $f_L = f_S$.

EXAMPLE 21-8

A police siren emits a sinusoidal wave with frequency $f_S = 300$ Hz. The speed of sound is 340 m/s. a) Find the wavelength of the waves if the siren is at rest in the air. b) If the siren is moving at 30 m/s (108 km/h, or 67 mi/h), find the wavelengths of the waves ahead of and behind the source.

21–9 Wavelengths ahead of and behind the police siren when the siren is moving through the air at 30 m/s.

SOLUTION a) When the source is at rest,

$$\lambda = \frac{v}{f_S} = \frac{340 \text{ m/s}}{300 \text{ Hz}} = 1.13 \text{ m}.$$

b) The situation is shown in Fig. 21–9. We use Eqs. (21–15) and (21–16). In front of the siren,

$$\lambda = \frac{v - v_S}{f_S} = \frac{340 \text{ m/s} - 30 \text{ m/s}}{300 \text{ Hz}} = 1.03 \text{ m}.$$

Behind the siren,

$$\lambda = \frac{v + v_S}{f_S} = \frac{340 \text{ m/s} + 30 \text{ m/s}}{300 \text{ Hz}} = 1.23 \text{ m}.$$

The wavelength is less in front of the siren and greater behind the siren.

EXAMPLE 21-9

If a listener L is at rest and the siren in Example 21-8 is moving away from L at 30 m/s (Fig. 21-10), what frequency does the listener hear?

SOLUTION We have $v_L = 0$ and $v_S = 30$ m/s. (The source velocity v_S is positive because the siren is moving in the same direction as the direction from listener to source.) From Eq. (21-17),

$$f_L = \frac{v}{v + v_S} f_S = \frac{340 \text{ m/s}}{340 \text{ m/s} + 30 \text{ m/s}} (300 \text{ Hz}) = 276 \text{ Hz}.$$

The source and listener are moving apart, so the frequency f_L heard by the listener is less than the frequency f_S emitted by the source.

Alternatively, we can use the result of Example 21-8 for the wavelength behind the source (which is where the listener in

21-10 The listener is at rest, and the siren moves away from the listener at 30 m/s.

Fig. 21-10 is located):

$$f_L = \frac{v}{\lambda} = \frac{340 \text{ m/s}}{1.23 \text{ m}} = 276 \text{ Hz}.$$

Even though the source is moving, the wave speed v relative to the stationary listener is unchanged.

EXAMPLE 21-10

If the siren is at rest and the listener is moving toward the left at 30 m/s (Fig. 21-11), what frequency does the listener hear?

SOLUTION Now $v_S = 0$. The positive direction (from listener to source) is still from left to right, so $v_L = -30$ m/s. Then Eq. (21-17) gives

$$f_L = \frac{v + v_L}{v} f_S = \frac{340 \text{ m/s} + (-30 \text{ m/s})}{340 \text{ m/s}} (300 \text{ Hz}) = 274 \text{ Hz}.$$

Again the frequency heard by the listener is less than the source frequency. Note that the *relative velocity* of source and listener is

21-11 The siren is at rest, and the listener moves away from it at 30 m/s.

the same as in the previous example, but the Doppler shift is different because the velocities relative to the *air* are different.

EXAMPLE 21-11

If the siren is moving away from the listener with a speed of 45 m/s relative to the air and the listener is moving toward the siren with a speed of 15 m/s relative to the air (Fig. 21-12), what frequency does the listener hear?

SOLUTION In this case, $v_L = 15$ m/s and $v_S = 45$ m/s. (Both velocities are positive because both velocity vectors point in the direction from listener to source.) From Eq. (21-17),

$$f_L = \frac{v + v_L}{v + v_S} f_S = \frac{340 \text{ m/s} + 15 \text{ m/s}}{340 \text{ m/s} + 45 \text{ m/s}} (300 \text{ Hz})$$

$$= 277 \text{ Hz}.$$

The frequency heard by the listener is again less than the source frequency, but the value is different than in the previous two examples, even though the source and listener move away from each other at 30 m/s in all three cases. The *sign* of the Doppler shift of frequency (that is, whether f_L is less than or greater than f_S) depends on how the source and listener are moving relative to each other; to determine the *value* of the Doppler shift of frequency, you must know the velocities of source and listener relative to the air.

21-12 The velocities relative to the air when the siren is moving away from the listener with a relative velocity of 30 m/s.

EXAMPLE 21–12

The police car with its 300-Hz siren is moving toward a warehouse at 30 m/s, intending to crash through the door. What frequency does the driver of the police car hear reflected from the warehouse?

SOLUTION In this situation there are *two* Doppler shifts, as shown in Fig. 21–13. In the first shift, the warehouse is the stationary "listener." The frequency of sound reaching the warehouse, which we call f_W, is greater than 300 Hz because the source is approaching. In the second shift, the warehouse acts as a source of sound with frequency f_W, and the listener is the driver of the police car; she hears a frequency greater than f_W because she is approaching the source.

To determine f_W, we use Eq. (21–17) with f_L replaced by f_W. The warehouse is at rest, so $v_L = 0$; the siren is moving in the negative direction from source to listener, so $v_S = -30$ m/s. Then

$$f_W = \frac{v}{v + v_S} f_S = \frac{340 \text{ m/s}}{340 \text{ m/s} + (-30 \text{ m/s})}(300 \text{ Hz}) = 329 \text{ Hz}.$$

To determine the frequency heard by the driver, we again use Eq. (21–17), but now with f_S replaced by f_W. The warehouse is now the source, so $v_S = 0$, and the velocity of the listener is $v_L = +30$ m/s; v_L is positive because it is in the direction from listener to source. Hence

$$f_L = \frac{v + v_L}{v} f_W = \frac{340 \text{ m/s} + 30 \text{ m/s}}{340 \text{ m/s}}(329 \text{ Hz}) = 358 \text{ Hz}.$$

21–13 (a) The police car with siren (source) moves at 30 m/s toward the stationary warehouse (listener). (b) The sound wave is reflected from the stationary warehouse (source) toward the police car (listener).

The reflected sound heard by the driver has an even higher frequency than the sound heard by a stationary listener in the warehouse.

DOPPLER EFFECT FOR ELECTROMAGNETIC WAVES

In the Doppler effect for sound, the velocities v_L and v_S are always measured relative to the *air* or whatever medium we are considering. There is also a Doppler effect for *electromagnetic* waves in empty space, such as light waves or radio waves. In this case there is no medium that we can use as a reference to measure velocities, and all that matters is the *relative* velocity of source and receiver. (By contrast, the Doppler effect for sound does not depend simply on this relative velocity, as discussed in Example 21–11.)

To derive the expression for the Doppler frequency shift for light, we have to use the special theory of relativity. We will discuss this in Chapter 39, but for now we quote the result without derivation. The wave speed is the speed of light, usually denoted by c, and it is the same for both source and receiver. In the frame of reference in which the receiver is at rest, the source is moving away from the receiver with velocity v. (If the source is *approaching* the receiver, v is negative.) The source frequency is again f_S. The frequency f_R measured by the receiver R (the frequency of arrival of the waves at the receiver) is then given by

$$f_R = \sqrt{\frac{c - v}{c + v}} f_S \qquad \text{(Doppler effect for light)}. \qquad (21\text{–}18)$$

When v is positive, the source is moving directly *away* from the receiver and f_R is always less than f_S; when v is negative, the source is moving directly *toward* the receiver and f_R is *greater* than f_S. The qualitative effect is the same as for sound, but the quantitative relationship is different.

A familiar application of the Doppler effect for radio waves is the radar device mounted on the side window of a police car to check other cars' speeds. The electromagnetic wave emitted by the device is reflected from a moving car, which acts as a

0

moving source, and the wave reflected back to the device is Doppler-shifted in frequency. The transmitted and reflected signals are combined to produce beats, and the speed can be computed from the frequency of the beats. Similar techniques ("Doppler radar") are used to measure wind velocities in the atmosphere.

The Doppler effect is also used to track satellites and other space vehicles. In Fig. 21–14 a satellite emits a radio signal with constant frequency f_S. As the satellite orbits past, it first approaches and then moves away from the receiver; the frequency f_R of the signal received on earth changes from a value greater than f_S to a value less than f_S as the satellite passes overhead.

The Doppler effect for electromagnetic waves, including visible light, is important in astronomy. Astronomers compare wavelengths of light from distant stars to those emitted by the same elements on earth. In a binary star system, in which two stars orbit about their common center of mass, the light is Doppler-shifted to higher frequencies when a star is moving toward an observer on earth and to lower frequencies when it's moving away. Measurements of the frequency shifts reveal information about the orbits and masses of the stars that comprise the binary system.

Light from most galaxies is shifted toward the longer-wavelength or red end of the visible spectrum, an effect called the *red shift*. This is often described as a Doppler shift resulting from motion of these galaxies away from us. But from the point of view of the general theory of relativity it is something much more fundamental, associated with the expansion of space itself. Distant galaxies have large red shifts because their light has been in transit for a long time and has shared in the expansion of all the space through which it moved. Extrapolating this expansion backward to over 10^{10} years ago leads to the "big bang" picture. From this point of view, the big bang was not an explosion in space but the initial rapid expansion of space itself.

21–14 Change of velocity component along the line of sight of a satellite passing a tracking station. The frequency received at the tracking station changes from high to low as the satellite passes overhead.

*21–6 SHOCK WAVES

You may have experienced "sonic booms" caused by an airplane flying over faster than the speed of sound. We can see qualitatively why this happens from Fig. 21–15. Let v_S denote the *speed* of the airplane relative to the air, so that it is always positive. The motion of the airplane through the air produces sound; if v_S is less than the speed of sound v, the waves in front of the airplane are crowded together with a wavelength given by Eq. (21–15):

$$\lambda = \frac{v - v_S}{f_S}.$$

As the speed v_S of the airplane approaches the speed of sound v, the wavelength approaches zero and the wave crests pile up on each other (Fig. 21–15a). The airplane must exert a large force to compress the air in front of it; by Newton's third law, the air exerts an equally large force back on the airplane. Hence there is a large increase in aerodynamic drag (air resistance) as the airplane approaches the speed of sound, a phenomenon known as the "sound barrier."

When v_S is greater in magnitude than v, the source of sound is **supersonic,** and Eqs. (21–15) and (21–17) for the Doppler effect no longer describe the sound wave in front of the source. Figure 21–15b shows a cross section of what happens. As the airplane moves, it displaces the surrounding air and produces sound. A series of wave crests is emitted from the nose of the airplane; each spreads out in a circle centered at the position of the airplane when it emitted the crest. After a time t the crest emitted from point S_1 has spread to a circle with radius vt, and the airplane has moved a greater distance $v_S t$ to position S_2. You can see that the circular crests interfere constructively at points along the green line that makes an angle α with the direction of the airplane velocity, leading

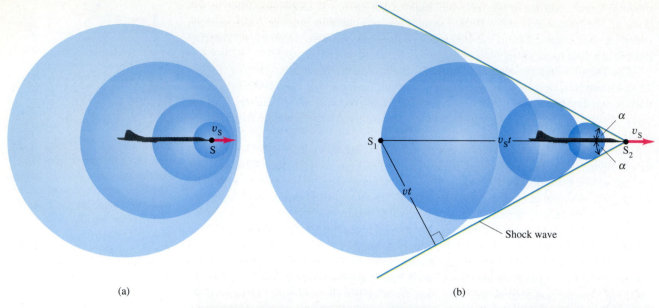

(a)   (b)

21–15 (a) As the speed of the source of sound S approaches the speed of sound, the wave crests begin to pile up in front of S. (b) A shock wave forms when the speed of the source is greater than the speed of sound. (c) Photograph of shock waves produced by a T-38 jet aircraft moving at 1.1 times the speed of sound. Separate shock waves are generated by the nose, wings, and tail. The angles of these waves vary because the air is accelerated and decelerated as it moves relative to the airplane, so the relative speed of the airplane and air is different at different points on the airplane.

to a very large-amplitude wave crest along this line. This large-amplitude crest is called a **shock wave** (Fig. 21–15c).

From the right triangle in Fig. 21–15b we can see that the angle α is given by

$$\sin \alpha = \frac{vt}{v_S t} = \frac{v}{v_S} \quad \text{(shock wave).} \tag{21–19}$$

In this relation, v_S is the *speed* of the source (the magnitude of its velocity) relative to the air and is always positive. The ratio v_S/v is called the **Mach number.** It is greater than unity for all supersonic speeds, and $\sin \alpha$ in Eq. (21–19) is the reciprocal of the Mach number. The first person to break the sound barrier was Capt. Chuck Yeager of the U. S. Air Force, flying the Bell X-1 at Mach 1.06 on October 14, 1947.

The actual situation is three-dimensional; the shock wave forms a *cone* around the direction of motion of the source. If the source (possibly a supersonic jet airplane or a rifle bullet) moves with constant velocity, the angle α is constant, and the shock-wave cone moves along with the source. It's the arrival of this shock wave that causes the sonic boom you hear after a supersonic airplane has passed by. The larger the airplane, the stronger the sonic boom; the shock wave produced at ground level by the Concorde supersonic airliner flying at 12,000 m (40,000 ft) causes a sudden jump in air pressure of about 20 Pa. In front of the shock-wave cone, there is no sound. Inside the cone a stationary listener hears the Doppler-shifted sound of the airplane moving away.

CAUTION ▶ We emphasize that a shock wave is produced *continuously* by any object that moves through the air at supersonic speed, not only at the instant that it "breaks the sound barrier." The sound waves that combine to form the shock wave, as in Fig. 21–15b, are created by the motion of the object itself, not by any sound source that the object may carry. The cracking noises of a bullet and of the tip of a circus whip are due to their supersonic motion. A supersonic jet airplane may have very loud engines, but these do not cause the shock wave. Indeed, a space shuttle makes a very loud sonic boom when coming in for a landing; its engines are out of fuel at this point, so it is a supersonic glider. ◀

Shock waves have applications outside of aviation. They are used to break up kidney stones and gallstones without invasive surgery, using a technique with the impressive name *extracorporeal shock-wave lithotripsy*. A shock wave produced outside the body is focused by a reflector or acoustic lens so that as much of it as possible converges on the stone. When the resulting stresses in the stone exceed its tensile strength, it breaks into small pieces and can be eliminated. This technique requires accurate determination of the location of the stone, which may be done using ultrasonic imaging techniques (see Example 19–5 in Section 19–6).

The first supersonic airplane, the Bell X-1, was shaped much like a .50-caliber bullet—which was known to be able to travel faster than sound.

EXAMPLE 21–13

Sonic boom of the Concorde The Concorde is flying at Mach 1.75 at an altitude of 8000 m, where the speed of sound is 320 m/s. How long after the plane passes directly overhead will you hear the sonic boom?

SOLUTION Figure 21–16 shows the situation just as the shock wave reaches you. From Eq. (21–19) the angle α is

$$\alpha = \arcsin \frac{1}{1.75} = 34.8°.$$

The speed of the plane is the speed of sound multiplied by the Mach number:

$$v_S = (1.75)(320 \text{ m/s}) = 560 \text{ m/s}.$$

From Fig. 21–16 we have

$$\tan \alpha = \frac{8000 \text{ m}}{v_S t},$$

$$t = \frac{8000 \text{ m}}{(560 \text{ m/s})(\tan 34.8°)} = 20.5 \text{ s}.$$

You hear the boom 20.5 s after the Concorde passes overhead, and at that time it has traveled (560 m/s)(20.5 s) = 11.5 km past the straight-overhead point.

In this calculation we assumed that the speed of sound is the same at all altitudes, so $\alpha = \arcsin v/v_S$ is a constant and the shock wave forms a perfect cone. In fact, the speed of sound decreases with increasing altitude. How would this affect the result?

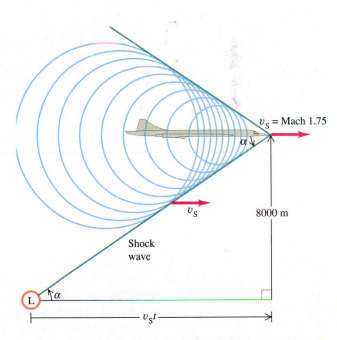

21–16 You hear a sonic boom when the shock wave reaches you at L (*not* just when the plane breaks the sound barrier). A listener to the right of L has not yet heard the sonic boom but will shortly; a listener to the left of L has already heard the sonic boom and now hears the Doppler-shifted sound of the airplane.

SUMMARY

KEY TERMS

sound, 646

audible range, 646

ultrasonic, 646

infrasonic, 646

pressure amplitude, 648

loudness, 649

pitch, 649

timbre, 650

noise, 650

intensity, 650

sound intensity level, 653

decibel, 653

beat, 655

beat frequency, 655

Doppler effect, 656

supersonic, 661

shock wave, 662

Mach number, 662

- Sound consists of longitudinal waves in a medium. A sinusoidal sound wave is characterized by its frequency f and wavelength λ (or angular frequency ω and wave number k) and by its amplitude A. The amplitude is also related to the pressure amplitude p_{max} by

$$p_{max} = BkA, \qquad (21\text{-}5)$$

where B is the bulk modulus of the wave medium.

- The loudness of a sound depends on its amplitude and frequency, while the pitch depends primarily on its frequency alone. The tone quality or timbre depends on the harmonic content and the attack and decay characteristics.

- The intensity I of a wave is the time average rate at which energy is transported by the wave, per unit area. For a sinusoidal wave of amplitude A and pressure amplitude p_{max},

$$I = \frac{1}{2}\sqrt{\rho B}\,\omega^2 A^2 = \frac{p_{max}^2}{2\rho v} = \frac{p_{max}^2}{2\sqrt{\rho B}}. \qquad (21\text{-}7),(21\text{-}9)$$

- The sound intensity level β of a sound wave of intensity I is defined as

$$\beta = (10\ \text{dB})\log\frac{I}{I_0}, \qquad (21\text{-}11)$$

where I_0 is a reference intensity defined to be $10^{-12}\ \text{W/m}^2$. Sound intensity levels are expressed in decibels (dB).

- Beats are heard when two tones with slightly different frequencies f_a and f_b are sounded together. With $f_a > f_b$, the beat frequency f_{beat} is

$$f_{beat} = f_a - f_b. \qquad (21\text{-}12)$$

- The Doppler effect for sound is the frequency shift that occurs when there is motion of a source of sound or a listener, or both, relative to the medium. The source and listener frequencies f_S and f_L and the source and listener velocities v_S and v_L relative to the medium (in which the speed of sound is v) are related by

$$f_L = \frac{v + v_L}{v + v_S}\,f_S. \qquad (21\text{-}17)$$

- A sound source moving with a speed v_S greater than the speed of sound v creates a shock wave. The wave front is a cone with angle α given by

$$\sin\alpha = \frac{v}{v_S}. \qquad (21\text{-}19)$$

DISCUSSION QUESTIONS

Q21-1 Lane dividers on highways sometimes have regularly spaced ridges or ripples. When the tires of a moving car roll along such a divider, a musical note is produced. Why? Explain how this phenomenon could be used to measure the car's speed.

Q21-2 The tone quality of an acoustic guitar is different when the strings are plucked near the bridge (the lower end of the strings) than when they are plucked near the sound hole (close to the center of the strings). Why?

Q21-3 Which has a more direct influence on the loudness of a sound wave: the *displacement* amplitude or the *pressure* amplitude? Explain your reasoning.

Q21-4 If the pressure amplitude of a sound wave is halved, by what factor does the intensity of the wave decrease? By what factor must the pressure amplitude of a sound wave be increased in order to increase the intensity by a factor of 16? Explain.

Q21-5 Defend the following statement: "*Adding* a certain number of decibels to the sound intensity level corresponds to *multiplying* the intensity by a corresponding factor."

Q21–6 A small fraction of the energy in a sound wave is absorbed by the air through which the sound passes. How does this modify the inverse-square relationship between intensity and distance from the source? Explain your reasoning.

Q21–7 According to Eq. (21–10), the intensity of sound at the back of a concert hall should be very much less than at the front of the hall, since the back row of seats is much farther from the stage. But in fact the intensity at the back is only slightly less than at the front. Why?

Q21–8 An organist in a cathedral plays a loud chord and then releases the keys. The sound persists for a few seconds and gradually dies away. Why does it persist? What happens to the sound energy when the sound dies away?

Q21–9 When you are shouting to someone a fair distance away, it is easier for her to hear you if the wind is blowing from you to her than if it is in the opposite direction. Explain the physical reason for this difference.

Q21–10 Two vibrating tuning forks have identical frequencies, but one is stationary and the other is mounted at the rim of a rotating platform. What does a listener hear? Explain.

Q21–11 A large church has part of the organ in the front of the church and part in the back. A person walking rapidly down the aisle while both segments are playing at once reports that the two segments sound out of tune. Why?

Q21–12 A sound source and a listener are both at rest on the earth, but a strong wind is blowing from the source toward the listener. Is there a Doppler effect? Why or why not?

Q21–13 Can you think of circumstances in which a Doppler effect would be observed for surface waves in water? For elastic waves propagating in a body of water deep below the surface?

If so, describe the circumstances and explain your reasoning. If not, explain why not.

*Q21–14 Does an aircraft make a sonic boom only at the instant its speed exceeds Mach 1? Explain your reasoning.

*Q21–15 If you are riding in a supersonic aircraft, what do you hear? Explain your reasoning. In particular, do you hear a continuous sonic boom? Why or why not?

*Q21–16 The speed of sound is not the same at all altitudes, but decreases with increasing altitude. How does this affect the calculation in Example 21–13 (Section 21–6)? Is the shape of the shock wave still a cone? Explain your reasoning.

*Q21–17 A jet airplane is flying at a constant altitude at a steady speed v_s greater than the speed of sound. Describe what is being heard by observers at points A, B, and C at the instant shown in Fig. 21–17, when the shock wave has just reached point B. Explain your reasoning.

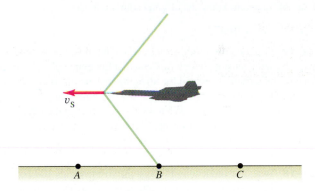

FIGURE 21–17 Question Q21–17.

EXERCISES

Unless indicated otherwise, assume the speed of sound in air to be $v = 344$ m/s.

SECTION 21–2 SOUND WAVES

21–1 Example 21–1 (Section 21–2) showed that for sound waves in air with frequency 1000 Hz, a displacement amplitude of 1.2×10^{-8} m produces a pressure amplitude of 3.0×10^{-2} Pa. a) What is the wavelength of these waves? b) For 1000-Hz waves in air, what displacement amplitude would be needed for the pressure amplitude to be at the pain threshold, which is 30 Pa? c) For what wavelength and frequency will waves with a displacement amplitude of 1.2×10^{-8} m produce a pressure amplitude of 1.5×10^{-3} Pa?

21–2 Example 21–1 (Section 21–2) showed that for sound waves in air with frequency 1000 Hz, a displacement amplitude of 1.2×10^{-8} m produces a pressure amplitude of 3.0×10^{-2} Pa. Water at 20°C has a bulk modulus of 2.2×10^9 Pa, and the speed of sound in water at this temperature is 1480 m/s. For 1000-Hz sound waves in 20°C water, what displacement amplitude is produced if the pressure amplitude is 3.0×10^{-2} Pa? Explain why your answer is much less than 1.2×10^{-8} Pa.

21–3 Consider a sound wave in air that has displacement amplitude 0.0200 mm. Calculate the pressure amplitude for frequencies of a) 150 Hz; b) 1500 Hz; c) 15,000 Hz. In each case compare the results to the pain threshold, which is 30 Pa.

SECTION 21–3 SOUND INTENSITY

21–4 Derive Eq. (21–9) from the equations that precede it.

21–5 A sound wave in air at 20°C has a frequency of 150 Hz and a displacement amplitude of 5.00×10^{-3} mm. For this sound wave calculate the a) pressure amplitude (in Pa); b) intensity (in W/m^2); c) sound intensity level (in decibels).

21–6 a) What is the sound intensity level in a car when the sound intensity is 0.500 μW/m^2? b) What is the sound intensity level in the air near a jackhammer when the pressure amplitude of the sound is 0.150 Pa and the temperature is 20.0°C?

21–7 For a person with normal hearing, the faintest sound that can be heard at a frequency of 400 Hz has a pressure amplitude of about 6.0×10^{-5} Pa. Calculate the corresponding intensity and sound intensity level at 20°C.

21–8 The intensity due to a number of independent sound sources is the sum of the individual intensities. a) When four quadruplets cry simultaneously, how many decibels greater is the sound intensity level than when a single one cries? b) To increase the sound intensity level again by the same number of decibels as in part (a), how many more crying babies are required?

21–9 A baby's mouth is 30 cm from her father's ear and 1.50 m from her mother's ear. What is the difference between the sound intensity levels heard by the father and by the mother?

21–10 A tornado warning siren on top of a tall pole radiates sound waves uniformly in all directions. At a distance of 15.0 m the intensity of the sound is 0.250 W/m². Neglect any effects from the reflection of the sound waves from the ground. a) At what distance from the siren is the intensity 0.010 W/m²? b) What is the total acoustical power output of the siren?

21–11 a) By what factor must the sound intensity be increased to increase the sound intensity level by 13.0 dB? b) Explain why you don't need to know the original sound intensity.

SECTION 21–4 BEATS

21–12 Two identical violin strings, when in tune and stretched with the same tension, have a fundamental frequency of 440.0 Hz. One of the strings is retuned by adjusting its tension. When this is done, 1.5 beats per second are heard when both strings are plucked simultaneously. a) What are the possible fundamental frequencies of the retuned string? b) By what fractional amount was the string tension changed if it was i) increased; ii) decreased?

21–13 Two sinusoidal sound waves with frequencies 108 Hz and 112 Hz arrive at your ear simultaneously. Each wave has an amplitude of 1.5×10^{-8} m when it reaches your ear. a) Describe in detail what you hear. b) What is the maximum amplitude of the total sound wave? What is the minimum amplitude?

SECTION 21–5 THE DOPPLER EFFECT

21–14 On the planet Arrakis a male ornithoid is flying toward his mate at 25.0 m/s while singing at a frequency of 1200 Hz. If the stationary female hears a tone of 1240 Hz, what is the speed of sound in the atmosphere of Arrakis?

21–15 In Example 21–12 (Section 21–5), suppose the police car is moving away from the warehouse at 20 m/s. What fre-

quency does the driver of the police car hear reflected from the warehouse?

21–16 Two train whistles, A and B, each have a frequency of 392 Hz. A is stationary and B is moving toward the right (away from A) at a speed of 35.0 m/s. A listener is between the two whistles and is moving toward the right with a speed of 15.0 m/s (Fig. 21–18). No wind is blowing. a) What is the frequency from A as heard by the listener? b) What is the frequency from B as heard by the listener? c) What is the beat frequency detected by the listener?

FIGURE 21–18 Exercise 21–16.

21–17 A railroad train is traveling at 25.0 m/s in still air. The frequency of the note emitted by the locomotive whistle is 400 Hz. What is the wavelength of the sound waves a) in front of the locomotive? b) behind the locomotive? What is the frequency of the sound heard by a stationary listener c) in front of the locomotive? d) behind the locomotive?

21–18 A swimming duck paddles the water with its feet once every 1.6 s, producing surface waves with this period. The duck is moving at constant speed in a pond where the speed of surface waves is 0.32 m/s, and the crests of the waves ahead of the duck are spaced 0.12 m apart. a) What is the duck's speed? b) How far apart are the crests behind the duck?

21–19 A railroad train is traveling at 30.0 m/s in still air. The frequency of the note emitted by the train whistle is 262 Hz. What frequency is heard by a passenger on a train moving in the opposite direction to the first at 18.0 m/s and a) approaching the first? b) receding from the first?

*SECTION 21–6 SHOCK WAVES

*21–20 A jet plane flies overhead at Mach 1.70 and at a constant altitude of 950 m. a) What is the angle α of the shock-wave cone? b) How much time after the plane passes directly overhead do you hear the sonic boom? Neglect the variation of the speed of sound with altitude.

PROBLEMS

21–21 a) Defend the following statement: "In a sinusoidal sound wave, the pressure variation given by Eq. (21–4) is greatest where the displacement given by Eq. (21–1) is zero." b) For a sinusoidal sound wave given by Eq. (21–1) with amplitude $A = 10.0\ \mu$m and wavelength $\lambda = 0.250$ m, graph the displacement y and pressure fluctuation p as functions of x at time $t = 0$. Show at least two wavelengths of the wave on your graphs. c) The displacement y in a *non*sinusoidal sound wave is shown in Fig. 21–19 as a function of x for $t = 0$. Draw a graph showing the pressure fluctuation p in this wave as a function of x at $t = 0$. This sound wave has the same 10.0-μm amplitude as the wave in part (b). Does it have the same pressure amplitude? Why or why not?

d) Is the statement in part (a) necessarily true if the sound wave is *not* sinusoidal? Explain your reasoning.

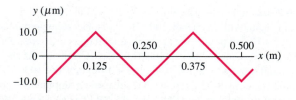

FIGURE 21–19 Problem 21–21.

21–22 Many airports have noise ordinances restricting the maximum sound intensity that an aircraft may produce when it takes off. At one California airport the maximum allowable sound intensity level is 98.5 dB as measured by a microphone at the end of the 1740-m-long runway. A certain airliner produces a sound intensity level of 100.0 dB on the ground when it flies over at an altitude of 100 m. On takeoff, this airliner rolls for 1200 m along the runway before leaving the ground, at which point it climbs at a 15° angle. Does this airliner violate the noise ordinance? Neglect any effects due to reflection of the sound waves from the ground.

21–23 The sound from a trumpet radiates uniformly in all directions in 20°C air. At a distance of 5.00 m from the trumpet the sound intensity level is 52.0 dB. The frequency is 587 Hz. a) What is the pressure amplitude at this distance? b) What is the displacement amplitude? c) At what distance is the sound intensity level 30.0 dB?

21–24 A sound wave entering a window of area A has a sound intensity level of β. a) How much acoustic power enters the window? b) Find the value of the acoustic power for the case of a 1.20-m² window and 55.0-dB sound.

21–25 A small sphere of radius R is arranged to pulsate so that its radius varies in simple harmonic motion between a minimum of $R - \Delta R$ and a maximum of $R + \Delta R$ with frequency f. This produces sound waves in the surrounding air of density ρ and bulk modulus B. a) Find the intensity of sound waves at the surface of the sphere. (The amplitude of oscillation of the sphere is the same as that of the air at the surface of the sphere.) b) Find the total acoustic power radiated by the sphere. c) At a distance $d \gg R$ from the center of the sphere, find the amplitude, pressure amplitude, and intensity of the sound wave.

21–26 A 2.00-MHz sound wave travels through a pregnant woman's abdomen and is reflected from the fetal heart wall of her unborn baby. The heart wall is moving toward the sound receiver as the heart beats. The reflected sound is then mixed with the transmitted sound, and 85 beats per second are detected. The speed of sound in body tissue is 1500 m/s. Calculate the speed of the fetal heart wall at the instant this measurement is made.

21–27 The sound source of a ship's sonar system operates at a frequency of 22.0 kHz. The speed of sound in water (assumed to be at a uniform 20°C) is 1482 m/s. a) What is the wavelength of the waves emitted by the source? b) What is the difference in frequency between the directly radiated waves and the waves reflected from a whale traveling directly toward the ship at 4.95 m/s? The ship is at rest in the water.

21–28 A police siren of frequency f_{siren} is attached to a vibrating platform. The platform and siren oscillate up and down in simple harmonic motion with amplitude A_p and frequency f_p. a) Find the maximum and minimum sound frequencies that you would hear at a position directly above the siren. b) At what point in the motion of the platform is the maximum frequency heard? The minimum frequency? Explain.

21–29 Horseshoe bats (genus *Rhinolophus*) emit sounds from their nostrils, then listen to the frequency of the sound reflected from their prey to determine the prey's speed. (The "horseshoe" that gives the bat its name is a depression around the nostrils that acts like a focusing mirror, so that the bat emits sound in a narrow beam like a flashlight.) A *Rhinolophus* flying at speed v_{bat} emits sound of frequency f_{bat}; the sound it hears reflected from an insect flying toward it has a higher frequency f_{refl}. a) Show that the speed of the insect is

$$v_{insect} = v \left[\frac{f_{refl}(v - v_{bat}) - f_{bat}(v + v_{bat})}{f_{refl}(v - v_{bat}) + f_{bat}(v + v_{bat})} \right].$$

where v is the speed of sound. b) If $f_{bat} = 80.7$ kHz, $f_{refl} = 83.5$ kHz, and $v_{bat} = 3.9$ m/s, calculate the speed of the insect.

21–30 A sound wave with frequency f_0 and wavelength λ_0 travels horizontally toward the right. It strikes and is reflected from a large, rigid, vertical plane surface, perpendicular to the direction of propagation of the wave and moving toward the left with a speed v_1. a) How many positive wave crests strike the surface in a time interval t? b) At the end of this time interval, how far to the left of the surface is the wave that was reflected at the beginning of the time interval? c) What is the wavelength of the reflected waves in terms of λ_0? d) What is the frequency of the reflected waves in terms of f_0? Is your result consistent with the assertion made in point 3 of the Problem-Solving Strategy in Section 21–5? e) A listener is at rest at the left of the moving surface. How many beats per second does she detect as a result of the combined effect of the incident and reflected waves?

21–31 a) Show that Eq. (21–18) can be written as

$$f_R = f_S \left(1 - \frac{v}{c}\right)^{1/2} \left(1 + \frac{v}{c}\right)^{-1/2}.$$

b) Use the binomial theorem to show that if $v \ll c$, this is approximately equal to

$$f_R = f_S \left(1 - \frac{v}{c}\right).$$

c) A pilotless reconnaissance aircraft emits a radio signal with a frequency of 243 MHz. It is flying directly toward a test engineer on the ground. The engineer detects beats between the received signal and a local signal also of frequency 243 MHz. The beat frequency is 46.0 Hz. What is the speed of the aircraft? (Radio waves travel at the speed of light, $c = 3.00 \times 10^8$ m/s.)

21–32 The gas cloud known as the Crab Nebula (Fig. 9–27) can be seen with even a small telescope. It is the remnant of a *supernova*, a cataclysmic explosion of a star. The explosion was seen on the earth on July 4, 1054 A.D. The streamers in Fig. 9–27 glow with the characteristic red color of heated hydrogen gas. In a laboratory on the earth, heated hydrogen produces red light with frequency 4.568×10^{14} Hz; the red light received from streamers in the Crab Nebula pointed toward the earth has frequency 4.586×10^{14} Hz. a) Estimate the speed with which the outer edges of the Crab Nebula are expanding. Assume that the speed of the center of the nebula relative to the earth is negligible. (You may use the formulas derived in Problem 21–31. The speed of light is 3.00×10^8 m/s.) b) Assuming that the expansion speed has been constant since the supernova explosion, estimate the diameter of the Crab Nebula. Give your answer in meters and in light years. c) The angular diameter of the Crab Nebula as seen from earth is about 5 arc minutes (1 arc minute = $\frac{1}{60}$ of a degree). Estimate the distance (in light years) to the Crab Nebula, and estimate the year in which the supernova explosion actually took place.

21–33 A radar installation used for monitoring thunderstorms emits radio waves at a frequency of 2800 MHz. A line of thunderstorms is approaching the installation at 42.0 km/h. a) In the frame of reference of the thunderstorms, is the frequency of the radio waves greater than or less than 2800 MHz? Why? By what amount does the frequency differ from 2800 MHz? (You may use the formulas derived in Problem 21–31. Radio waves travel at the speed of light, $c = 3.00 \times 10^8$ m/s.) b) Radio waves are reflected from the water drops in the thunderstorms, and the reflected waves are detected back at the installation. As measured by the receiver at the installation, is the frequency of these reflected waves greater than or less than 2800 MHz? Why? By what amount does the frequency of the reflected waves differ from 2800 MHz?

21–34 A woman stands at rest in front of a large, smooth wall. She holds a vibrating tuning fork of frequency f_0 directly in front of her (between her and the wall). a) She now runs toward the wall with a speed v_w. She detects beats due to the interference between the sound waves reaching her directly from the fork and those reaching her after being reflected from the wall. How many beats per second will she detect? (*Note:* If the beat frequency is too large, the woman may have to use some instrumentation other than her ears to detect and count the beats.) b) If she instead runs away from the wall, holding the tuning fork at her back so it is between her and the wall, how many beats per second will she detect?

*21–35 On a clear day you see a jet plane flying overhead. From the apparent size of the plane, you determine that it is flying at a constant altitude h. You hear the sonic boom a time T after the plane passes directly overhead. Show that if the speed of sound v is the same at all altitudes, the speed of the plane is

$$v_s = \frac{hv}{\sqrt{h^2 - v^2 T^2}}$$

(*Hint:* Trigonometric identities will be useful.)

CHALLENGE PROBLEMS

21–36 Figure 21–20 shows the pressure fluctuation p of a nonsinusoidal sound wave as a function of x for $t = 0$. The wave is traveling in the +x-direction. a) Graph the pressure fluctuation p as a function of t for $x = 0$. Show at least two cycles of oscillation. b) Graph the displacement y in this sound wave as a function of x at $t = 0$. At $x = 0$, the displacement at $t = 0$ is zero. Show at least two wavelengths of the wave. c) Graph the displacement y as a function of t for $x = 0$. Show at least two cycles of oscillation. d) Calculate the maximum velocity and the maximum acceleration of an element of the air through which this sound wave is traveling. e) Describe how the cone of a loudspeaker must move as a function of time to produce the sound wave in this problem.

21–37 Two loudspeakers, A and B, radiate sound uniformly in all directions in air at 20°C. The acoustic power output from A is 8.00×10^{-4} W, and from B it is 6.00×10^{-5} W. Both loudspeakers are vibrating in phase at a frequency of 172 Hz. a) Determine the difference in phase of the two signals at a point C along the line joining A and B, 3.00 m from B and 4.00 m from A (Fig. 21–21). b) Determine the intensity and sound intensity level at C from speaker A if speaker B is turned off and the intensity and sound intensity level at point C from speaker B if speaker A is turned off. c) With both speakers on, what are the intensity and sound intensity level at C?

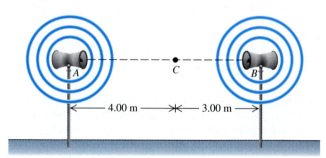

FIGURE 21–21 Challenge Problem 21–37.

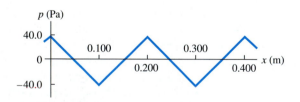

FIGURE 21–20 Challenge Problem 21–36.

The International System of Units

The Système International d'Unités, abbreviated SI, is the system developed by the General Conference on Weights and Measures and adopted by nearly all the industrial nations of the world. It is based on the mksa (meter-kilogram-second-ampere) system. The following material is adapted from B. N. Taylor, ed., National Bureau of Standards Technol. Spec. Pub. 330 (U.S. Govt. Printing Office, Washington, DC, 1991).

Quantity	Name of unit	Symbol	
	SI base units		
length	meter	m	
mass	kilogram	kg	
time	second	s	
electric current	ampere	A	
thermodynamic temperature	kelvin	K	
amount of substance	mole	mol	
luminous intensity	candela	cd	
	SI derived units		**Equivalent units**
area	square meter	m^2	
volume	cubic meter	m^3	
frequency	hertz	Hz	s^{-1}
mass density (density)	kilogram per cubic meter	kg/m^3	
speed, velocity	meter per second	m/s	
angular velocity	radian per second	rad/s	
acceleration	meter per second squared	m/s^2	
angular acceleration	radian per second squared	rad/s^2	
force	newton	N	$kg \cdot m/s^2$
pressure (mechanical stress)	pascal	Pa	N/m^2
kinematic viscosity	square meter per second	m^2/s	
dynamic viscosity	newton-second per square meter	$N \cdot s/m^2$	
work, energy, quantity of heat	joule	J	$N \cdot m$
power	watt	W	J/s
quantity of electricity	coulomb	C	$A \cdot s$
potential difference, electromotive force	volt	V	J/C, W/A
electric field strength	volt per meter	V/m	N/C
electric resistance	ohm	Ω	V/A
capacitance	farad	F	$A \cdot s/V$
magnetic flux	weber	Wb	$V \cdot s$
inductance	henry	H	$V \cdot s/A$
magnetic flux density	tesla	T	Wb/m^2
magnetic field strength	ampere per meter	A/m	
magnetomotive force	ampere	A	
luminous flux	lumen	lm	$cd \cdot sr$
luminance	candela per square meter	cd/m^2	
illuminance	lux	lx	lm/m^2
wave number	1 per meter	m^{-1}	
entropy	joule per kelvin	J/K	
specific heat capacity	joule per kilogram-kelvin	$J/kg \cdot K$	
thermal conductivity	watt per meter-kelvin	$W/m \cdot K$	

Quantity	Name of unit	Symbol	Equivalent units
radiant intensity	watt per steradian	W/sr	
activity (of a radioactive source)	becquerel	Bq	s^{-1}
radiation dose	gray	Gy	J/kg
radiation dose equivalent	sievert	Sv	J/kg
SI supplementary units			
plane angle	radian	rad	
solid angle	steradian	sr	

DEFINITIONS OF SI UNITS

meter (m)   The *meter* is the length equal to the distance traveled by light, in vacuum, in a time of 1/299,792,458 second.

kilogram (kg)   The *kilogram* is the unit of mass; it is equal to the mass of the international prototype of the kilogram. (The international prototype of the kilogram is a particular cylinder of platinum-iridium alloy that is preserved in a vault at Sèvres, France, by the International Bureau of Weights and Measures.)

second (s)   The *second* is the duration of 9,192,631,770 periods of the radiation corresponding to the transition between the two hyperfine levels of the ground state of the cesium-133 atom.

ampere (A)   The *ampere* is that constant current that, if maintained in two straight parallel conductors of infinite length, of negligible circular cross section, and placed 1 meter apart in vacuum, would produce between these conductors a force equal to 2×10^{-7} newton per meter of length.

kelvin (K)   The *kelvin,* unit of thermodynamic temperature, is the fraction 1/273.16 of the thermodynamic temperature of the triple point of water.

ohm (Ω)   The *ohm* is the electric resistance between two points of a conductor when a constant difference of potential of 1 volt, applied between these two points, produces in this conductor a current of 1 ampere, this conductor not being the source of any electromotive force.

coulomb (C)   The *coulomb* is the quantity of electricity transported in 1 second by a current of 1 ampere.

candela (cd)   The *candela* is the luminous intensity, in a given direction, of a source that emits monochromatic radiation of frequency 540×10^{12} hertz and that has a radiant intensity in that direction of 1/683 watt per steradian.

mole (mol)   The *mole* is the amount of substance of a system that contains as many elementary entities as there are carbon atoms in 0.012 kg of carbon 12. The elementary entities must be specified and may be atoms, molecules, ions, electrons, other particles, or specified groups of such particles.

newton (N)   The *newton* is that force that gives to a mass of 1 kilogram an acceleration of 1 meter per second per second.

joule (J)   The *joule* is the work done when the point of application of a constant force of 1 newton is displaced a distance of 1 meter in the direction of the force.

watt (W)   The *watt* is the power that gives rise to the production of energy at the rate of 1 joule per second.

volt (V)   The *volt* is the difference of electric potential between two points of a conducting wire carrying a constant current of 1 ampere, when the power dissipated between these points is equal to 1 watt.

weber (Wb)   The *weber* is the magnetic flux that, linking a circuit of one turn, produces in it an electromotive force of 1 volt as it is reduced to zero at a uniform rate in 1 second.

lumen (lm)   The *lumen* is the luminous flux emitted in a solid angle of 1 steradian by a uniform point source having an intensity of 1 candela.

farad (F)   The *farad* is the capacitance of a capacitor between the plates of which there appears a difference of potential of 1 volt when it is charged by a quantity of electricity equal to 1 coulomb.

henry (H)   The *henry* is the inductance of a closed circuit in which an electromotive force of 1 volt is produced when the electric current in the circuit varies uniformly at a rate of 1 ampere per second.

radian (rad)   The *radian* is the plane angle between two radii of a circle that cut off on the circumference an arc equal in length to the radius.

steradian (sr)   The *steradian* is the solid angle that, having its vertex in the center of a sphere, cuts off an area of the surface of the sphere equal to that of a square with sides of length equal to the radius of the sphere.

SI Prefixes   The names of multiples and submultiples of SI units may be formed by application of the prefixes listed in Appendix F.

Useful Mathematical Relations

ALGEBRA

$$a^{-x} = \frac{1}{a^x} \qquad a^{(x+y)} = a^x a^y \qquad a^{(x-y)} = \frac{a^x}{a^y}$$

Logarithms: If $\log a = x$, then $a = 10^x$. $\log a + \log b = \log (ab)$ $\log a - \log b = \log (a/b)$ $\log (a^n) = n \log a$
If $\ln a = x$, then $a = e^x$. $\ln a + \ln b = \ln (ab)$ $\ln a - \ln b = \ln (a/b)$ $\ln (a^n) = n \ln a$

Quadratic formula: If $ax^2 + bx + c = 0$, $x = \dfrac{-b \pm \sqrt{b^2 - 4ac}}{2a}$.

BINOMIAL THEOREM

$$(a+b)^n = a^n + na^{n-1}b + \frac{n(n-1)a^{n-2}b^2}{2!} + \frac{n(n-1)(n-2)a^{n-3}b^3}{3!} + \cdots$$

TRIGONOMETRY

In the right triangle ABC, $x^2 + y^2 = r^2$.

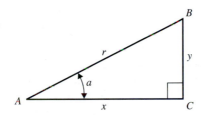

Definitions of the trigonometric functions: $\sin a = y/r$ $\cos a = x/r$ $\tan a = y/x$

Identities: $\sin^2 a + \cos^2 a = 1$

$\tan a = \dfrac{\sin a}{\cos a}$

$\sin 2a = 2 \sin a \cos a$

$\cos 2a = \cos^2 a - \sin^2 a = 2\cos^2 a - 1$
$\qquad\qquad\quad = 1 - 2\sin^2 a$

$\sin \tfrac{1}{2}a = \sqrt{\dfrac{1 - \cos a}{2}}$

$\cos \tfrac{1}{2}a = \sqrt{\dfrac{1 + \cos a}{2}}$

$\sin(-a) = -\sin a$

$\sin(a \pm b) = \sin a \cos b \pm \cos a \sin b$

$\cos(-a) = \cos a$

$\cos(a \pm b) = \cos a \cos b \mp \sin a \sin b$

$\sin(a \pm \pi/2) = \pm \cos a$

$\sin a + \sin b = 2 \sin \tfrac{1}{2}(a+b) \cos \tfrac{1}{2}(a-b)$

$\cos(a \pm \pi/2) = \mp \sin a$

$\cos a + \cos b = 2 \cos \tfrac{1}{2}(a+b) \cos \tfrac{1}{2}(a-b)$

GEOMETRY

Circumference of circle of radius r: $C = 2\pi r$
Area of circle of radius r: $A = \pi r^2$
Volume of sphere of radius r: $V = 4\pi r^3/3$
Surface area of sphere of radius r: $A = 4\pi r^2$
Volume of cylinder of radius r and height h: $V = \pi r^2 h$

CALCULUS

Derivatives:

$$\frac{d}{dx}x^n = nx^{n-1}$$

$$\frac{d}{dx}\sin ax = a\cos ax$$

$$\frac{d}{dx}\cos ax = -a\sin ax$$

$$\frac{d}{dx}e^{ax} = ae^{ax}$$

$$\frac{d}{dx}\ln ax = \frac{1}{x}$$

$$\int \frac{dx}{\sqrt{a^2 - x^2}} = \arcsin\frac{x}{a}$$

$$\int \frac{dx}{\sqrt{x^2 + a^2}} = \ln\left(x + \sqrt{x^2 + a^2}\right)$$

$$\int \frac{dx}{x^2 + a^2} = \frac{1}{a}\arctan\frac{x}{a}$$

$$\int \frac{dx}{(x^2 + a^2)^{3/2}} = \frac{1}{a^2}\frac{x}{\sqrt{x^2 + a^2}}$$

$$\int \frac{x\,dx}{(x^2 + a^2)^{3/2}} = -\frac{1}{\sqrt{x^2 + a^2}}$$

Power series (convergent for range of x shown):

$$(1 + x)^n = 1 + nx + \frac{n(n-1)x^2}{2!} + \frac{n(n-1)(n-2)}{3!}x^3 + \cdots \qquad (|x| < 1)$$

Integrals:

$$\int x^n\,dx = \frac{x^{n+1}}{n+1} \quad (n \neq -1)$$

$$\int \frac{dx}{x} = \ln x$$

$$\int \sin ax\,dx = -\frac{1}{a}\cos ax$$

$$\int \cos ax\,dx = \frac{1}{a}\sin ax$$

$$\int e^{ax}\,dx = \frac{1}{a}e^{ax}$$

$$\sin x = x - \frac{x^3}{3!} + \frac{x^5}{5!} - \frac{x^7}{7!} + \cdots \qquad (\text{all } x)$$

$$\cos x = 1 - \frac{x^2}{2!} + \frac{x^4}{4!} - \frac{x^6}{6!} + \cdots \qquad (\text{all } x)$$

$$\tan x = x + \frac{x^3}{3} + \frac{2x^2}{15} + \frac{17x^7}{315} + \cdots \qquad (|x| < \pi/2)$$

$$e^x = 1 + x + \frac{x^2}{2!} + \frac{x^3}{3!} + \cdots \qquad (\text{all } x)$$

$$\ln(1 + x) = x - \frac{x^2}{2} + \frac{x^3}{3} - \frac{x^4}{4} + \cdots \qquad (|x| < 1)$$

 APPENDIX C

The Greek Alphabet

Name	Capital	Lowercase	Name	Capital	Lowercase
Alpha	A	α	Nu	N	ν
Beta	B	β	Xi	Ξ	ξ
Gamma	Γ	γ	Omicron	O	o
Delta	Δ	δ	Pi	Π	π
Epsilon	E	ϵ	Rho	P	ρ
Zeta	Z	ζ	Sigma	Σ	σ
Eta	H	η	Tau	T	τ
Theta	Θ	θ	Upsilon	Υ	υ
Iota	I	ι	Phi	Φ	ϕ
Kappa	K	κ	Chi	X	χ
Lambda	Λ	λ	Psi	Ψ	ψ
Mu	M	μ	Omega	Ω	ω

Periodic Table of the Elements

Group	1	2	3	4	5	6	7	8	9	10	11	12	13	14	15	16	17	18
Period																		
1	1 **H** 1.008																	2 **He** 4.003
2	3 **Li** 6.941	4 **Be** 9.012											5 **B** 10.811	6 **C** 12.011	7 **N** 14.007	8 **O** 15.999	9 **F** 18.998	10 **Ne** 20.180
3	11 **Na** 22.990	12 **Mg** 24.305											13 **Al** 26.982	14 **Si** 28.086	15 **P** 30.974	16 **S** 32.066	17 **Cl** 35.453	18 **Ar** 39.948
4	19 **K** 39.098	20 **Ca** 40.078	21 **Sc** 44.956	22 **Ti** 47.867	23 **V** 50.942	24 **Cr** 51.996	25 **Mn** 54.938	26 **Fe** 55.845	27 **Co** 58.933	28 **Ni** 58.693	29 **Cu** 63.546	30 **Zn** 65.39	31 **Ga** 69.723	32 **Ge** 72.61	33 **As** 74.922	34 **Se** 78.96	35 **Br** 79.904	36 **Kr** 83.8
5	37 **Rb** 85.468	38 **Sr** 87.62	39 **Y** 88.906	40 **Zr** 91.224	41 **Nb** 92.906	42 **Mo** 95.94	43 **Tc** (98)	44 **Ru** 101.07	45 **Rh** 102.905	46 **Pd** 106.42	47 **Ag** 107.868	48 **Cd** 112.411	49 **In** 114.818	50 **Sn** 118.71	51 **Sb** 121.76	52 **Te** 127.6	53 **I** 126.904	54 **Xe** 131.29
6	55 **Cs** 132.905	56 **Ba** 137.327	71 **Lu** 174.967	72 **Hf** 178.49	73 **Ta** 180.948	74 **W** 183.84	75 **Re** 186.207	76 **Os** 190.23	77 **Ir** 192.217	78 **Pt** 195.078	79 **Au** 196.967	80 **Hg** 200.59	81 **Tl** 204.38	82 **Pb** 207.2	83 **Bi** 208.98	84 **Po** (210)	85 **At** (210)	86 **Rn** (222)
7	87 **Fr** (223)	88 **Ra** (226)	103 **Lr** (262)	104 **Rf** (261)	105 **Db** (262)	106 **Sg** (266)	107 **Bh** (264)	108 **Hs** (269)	109 **Mt** (268)	110 **Uun** (269)	111 **Uuu** (272)	112 **Uub** (277)	113 **Uut**	114 **Uuq** (285)	115 **Uup**	116 **Uuh** (289)	117 **Uus**	118 **Uuo** (293)

Lanthanoids

57 **La** 138.905	58 **Ce** 140.16	59 **Pr** 140.908	60 **Nd** 144.24	61 **Pm** (145)	62 **Sm** 150.36	63 **Eu** 151.964	64 **Gd** 157.25	65 **Tb** 158.925	66 **Dy** 162.5	67 **Ho** 164.930	68 **Er** 167.26	69 **Tm** 168.934	70 **Yb** 173.04

Actinoids

89 **Ac** (227)	90 **Th** (232)	91 **Pa** (231)	92 **U** (238)	93 **Np** (237)	94 **Pu** (244)	95 **Am** (243)	96 **Cm** (247)	97 **Bk** (247)	98 **Cf** (251)	99 **Es** (252)	100 **Fm** (257)	101 **Md** (258)	102 **No** (259)

For each element the average atomic mass of the mixture of isotopes occurring in nature is shown. For elements having no stable isotope, the approximate atomic mass of the longest-lived isotope is shown in parentheses. For elements that have been predicted but not yet detected, no atomic mass is given. All atomic masses are expressed in atomic mass units ($1\ u = 1.6605402(10) \times 10^{-27}$ kg), equivalent to grams per mole (g/mol).

Unit Conversion Factors

LENGTH
1 m = 100 cm = 1000 mm = 10^6 μm = 10^9 nm
1 km = 1000 m = 0.6214 mi
1 m = 3.281 ft = 39.37 in.
1 cm = 0.3937 in.
1 in. = 2.540 cm
1 ft = 30.48 cm
1 yd = 91.44 cm
1 mi = 5280 ft = 1.609 km
1 Å = 10^{-10} m = 10^{-8} cm = 10^{-1} nm
1 nautical mile = 6080 ft
1 light year = 9.461×10^{15} m

AREA
1 cm^2 = 0.155 in^2
1 m^2 = 10^4 cm^2 = 10.76 ft^2
1 $in.^2$ = 6.452 cm^2
1 ft^2 = 144 $in.^2$ = 0.0929 m^2

VOLUME
1 liter = 1000 cm^3 = 10^{-3} m^3 = 0.03531 ft^3 = 61.02 $in.^3$
1 ft^3 = 0.02832 m^3 = 28.32 liters = 7.477 gallons
1 gallon = 3.788 liters

TIME
1 min = 60 s
1 h = 3600 s
1 d = 86,400 s
1 y = 365.24 d = 3.156×10^7 s

ANGLE
1 rad = 57.30° = 180°/π
1° = 0.01745 rad = π/180 rad
1 revolution = 360° = 2π rad
1 rev/min (rpm) = 0.1047 rad/s

SPEED
1 m/s = 3.281 ft/s
1 ft/s = 0.3048 m/s
1 mi/min = 60 mi/h = 88 ft/s
1 km/h = 0.2778 m/s = 0.6214 mi/h
1 mi/h = 1.466 ft/s = 0.4470 m/s = 1.609 km/h
1 furlong/fortnight = 1.662×10^{-4} m/s

ACCELERATION
1 m/s^2 = 100 cm/s^2 = 3.281 ft/s^2
1 cm/s^2 = 0.01 m/s^2 = 0.03281 ft/s^2
1 ft/s^2 = 0.3048 m/s^2 = 30.48 cm/s^2
1 mi/h $\cdot$ s = 1.467 ft/s^2

MASS
1 kg = 10^3 g = 0.0685 slug
1 g = 6.85×10^{-5} slug
1 slug = 14.59 kg
1 u = 1.661×10^{-27} kg
1 kg has a weight of 2.205 lb when g = 9.80 m/s^2

FORCE
1 N = 10^5 dyn = 0.2248 lb
1 lb = 4.448 N = 4.448×10^5 dyn

PRESSURE
1 Pa = 1 N/m^2 = 1.450×10^{-4} $lb/in.^2$ = 0.209 lb/ft^2
1 bar = 10^5 Pa
1 $lb/in.^2$ = 6895 Pa
1 lb/ft^2 = 47.88 Pa
1 atm = 1.013×10^5 Pa = 1.013 bar
$\qquad$ = 14.7 $lb/in.^2$ = 2117 lb/ft^2
1 mm Hg = 1 torr = 133.3 Pa

ENERGY
1 J = 10^7 ergs = 0.239 cal
1 cal = 4.186 J (based on 15° calorie)
1 ft $\cdot$ lb = 1.356 J
1 Btu = 1055 J = 252 cal = 778 ft $\cdot$ lb
1 eV = 1.602×10^{-19} J
1 kWh = 3.600×10^6 J

MASS–ENERGY EQUIVALENCE
1 kg $\leftrightarrow$ 8.988×10^{16} J
1 u $\leftrightarrow$ 931.5 MeV
1 eV $\leftrightarrow$ 1.074×10^{-9} u

POWER
1 W = 1 J/s
1 hp = 746 W = 550 ft $\cdot$ lb/s
1 Btu/h = 0.293 W

Numerical Constants

FUNDAMENTAL PHYSICAL CONSTANTS*

Name	Symbol	Value
Speed of light	c	2.99792458×10^8 m/s
Magnitude of charge of electron	e	$1.60217733(49) \times 10^{-19}$ C
Gravitational constant	G	$6.67259(85) \times 10^{-11}$ N $\cdot$ m^2/kg^2
Planck's constant	h	$6.6260755(40) \times 10^{-34}$ J $\cdot$ s
Boltzmann constant	k	$1.380658(12) \times 10^{-23}$ J/K
Avogadro's number	N_A	$6.0221367(36) \times 10^{23}$ molecules/mol
Gas constant	R	$8.314510(70)$ J/mol $\cdot$ K
Mass of electron	m_e	$9.1093897(54) \times 10^{-31}$ kg
Mass of proton	m_p	$1.6726231(10) \times 10^{-27}$ kg
Mass of neutron	m_n	$1.6749286(10) \times 10^{-27}$ kg
Permeability of free space	μ_0	$4\pi \times 10^{-7}$ Wb/A $\cdot$ m
Permittivity of free space	$\epsilon_0 = 1/\mu_0 c^2$	$8.854187817 \ldots \times 10^{-12}$ C^2/N $\cdot$ m^2
	$1/4\pi\epsilon_0$	$8.987551787 \ldots \times 10^9$ N $\cdot$ m^2/C^2

OTHER USEFUL CONSTANTS*

Mechanical equivalent of heat		4.186 J/cal (15° calorie)
Standard atmospheric pressure	1 atm	1.01325×10^5 Pa
Absolute zero	0 K	-273.15°C
Electron volt	1 eV	$1.60217733(49) \times 10^{-19}$ J
Atomic mass unit	1 u	$1.6605402(10) \times 10^{-27}$ kg
Electron rest energy	$m_e c^2$	$0.51099906(15)$ MeV
Volume of ideal gas (0°C and 1 atm)		$22.41410(19)$ liter/mol
Acceleration due to gravity (standard)	g	9.80665 m/s^2

*Source: E. R. Cohen and B. R. Taylor, *Reviews of Modern Physics* **57**, 1121 (1987). Numbers in parentheses show the uncertainty in the final digits of the main number; for example, the number 1.6454(21) means 1.6454 ± 0.0021. Values shown without uncertainties are exact.

ASTRONOMICAL DATA[†]

Body	Mass (kg)	Radius (m)	Orbit radius (m)	Orbit period
Sun	1.99×10^{30}	6.96×10^8	—	—
Moon	7.35×10^{22}	1.74×10^6	3.84×10^8	27.3 d
Mercury	3.30×10^{23}	2.44×10^6	5.79×10^{10}	88.0 d
Venus	4.87×10^{24}	6.05×10^6	1.08×10^{11}	224.7 d
Earth	5.97×10^{24}	6.38×10^6	1.50×10^{11}	365.3 d
Mars	6.42×10^{23}	3.40×10^6	2.28×10^{11}	687.0 d
Jupiter	1.90×10^{27}	6.91×10^7	7.78×10^{11}	11.86 y
Saturn	5.69×10^{26}	6.03×10^7	1.43×10^{12}	29.42 y
Uranus	8.66×10^{25}	2.56×10^7	2.88×10^{12}	83.75 y
Neptune	1.03×10^{26}	2.48×10^7	4.50×10^{12}	163.7 y
Pluto	$1.5 \ \times 10^{22}$	1.15×10^6	5.92×10^{12}	248.0 y

[†]Source: **The Astronomical Almanac for 1995** (US Government Printing Office, Washington, DC, 1994), p. E88, and P. Kenneth Seidelmann, ed., **Explanatory Supplement to the Astronomical Almanac** (University Science Books, Mill Valley, CA, 1992), pp. 704–706. For each body, "radius" is its radius at its equator and "orbit radius" is its average distance from the sun (for the planets) or from the earth (for the moon).

PREFIXES FOR POWERS OF 10

Power of ten	Prefix	Abbreviation	Pronunciation
10^{-24}	yocto-	y	*yoc*-toe
10^{-21}	zepto-	z	*zep*-toe
10^{-18}	atto-	a	*at*-toe
10^{-15}	femto-	f	*fem*-toe
10^{-12}	pico-	p	*pee*-koe
10^{-9}	nano-	n	*nan*-oe
10^{-6}	micro-	μ	*my*-crow
10^{-3}	milli-	m	*mil*-i
10^{-2}	centi-	c	*cen*-ti
10^3	kilo-	k	*kil*-oe
10^6	mega-	M	*meg*-a
10^9	giga-	G	*jig*-a or *gig*-a
10^{12}	tera-	T	*ter*-a
10^{15}	peta-	P	*pet*-a
10^{18}	exa-	E	*ex*-a
10^{21}	zetta-	Z	*zet*-a
10^{24}	yotta-	Y	*yot*-a

Examples:

1 femtometer = 1 fm = 10^{-15} m

1 picosecond = 1 ps = 10^{-12} s

1 nanocoulomb = 1 nC = 10^{-9} C

1 microkelvin = 1 μK = 10^{-6} K

1 millivolt = 1 mV = 10^{-3} V

1 kilopascal = 1 kPa = 10^3 Pa

1 megawatt = 1 MW = 10^6 W

1 gigahertz = 1 GHz = 10^9 Hz

ANSWERS TO ODD-NUMBERED PROBLEMS

Chapter 1

1–1: 1.61 km
1–3: 3.34×10^3 ns
1–5: 5.36 L
1–7: a) 2330 km/h b) 648 m/s
1–9: 35.3 mi/gal
1–11: a) $1.1 \times 10^{-3}\%$ b) no
1–13: a) 0.1% b) 0.008% c) 0.03%
1–15: a) 2.8 ± 0.3 cm^3 b) 170 ± 20
1–17: Ten thousand
1–19: 10^6
1–21: 10^5 (with a full head of hair)
1–23: about 10^8
1–27: 9×10^8; about $3.50 per person
1–29: 7.8 m, 38° north of east
1–31: 104 m, 43° south of west
1–33: $A_x = 7.22$ m, $A_y = 9.58$ m; $B_x = 11.5$ m, $B_y = -9.64$ m; $C_x = -3.00$ m, $C_y = -5.20$ m
1–35: a) 11.1 m, 77.6° b) 11.1 m, 77.6° c) 28.5 m, 202.3° d) 28.5 m, 22.3°
1–37: 5.06 km, 20.2° north of west
1–39: a) 2.48 cm, 18.3° b) 4.10 cm, 83.7° c) 4.10 cm, 263.7°
1–41: $\vec{A} = (7.2\text{ m})\hat{\imath} + (9.6\text{ m})\hat{\jmath}$, $\vec{B} = (11.5\text{ m})\hat{\imath} - (9.6\text{ m})\hat{\jmath}$, $\vec{C} = (-3.0\text{ m})\hat{\imath} - (5.2\text{ m})\hat{\jmath}$
1–43: a) $A = 5.00$, $B = 5.39$ b) $-1.00\hat{\imath} + 5.00\hat{\jmath}$ c) 5.10, 101.3°
1–45: a) 14.00 b) 58.7°
1–47: a) out of the page b) into the page
1–49: $-23.0\hat{k}$; 23.0
1–51: a) 640 acres/mi^2 b) 43,560 ft^2/acre c) 3.26×10^5 gal/acre-ft
1–53: a) 7.04×10^{-10} s b) 5.11×10^{12} cycle/h c) 2.1×10^{26} d) 4.6×10^4 s
1–55: About 1 dentist per 1,000 population
1–57: b) $A_x = 3.03$ cm, $A_y = 8.10$ cm c) 8.65 cm, 69.5° measured in the sense from the $+x$-axis toward the $+y$-axis
1–59: 144 m, 41° south of west
1–61: b) 1.45 km
1–63: a) (87, 258) b) 137 pixels, 25° below straight left
1–67: b) 90°
1–69: a) $A = 5.39$, $B = 4.36$ b) $-5.00\hat{\imath} + 2.00\hat{\jmath} + 7.00\hat{k}$ c) 8.83; yes
1–73: a) 54.7° b) 35.3°
1–75: b) 72.2
1–77: 38.5 yd, 24.6° right of downfield
1–79: a) 76 ly b) 129°

Chapter 2

2–1: a) 197 m/s b) 169 m/s
2–3: 1 h 10 min
2–5: a) 14.0 m/s b) 11.4 m/s
2–7: a) 12.0 m/s b) i) 0 m/s ii) 15.0 m/s iii) 12.0 m/s c) 13.3 s
2–9: a) (in m/s^2) 0, 1.0, 2.0, 2.0, 3.0, 1.5, 1.5, 0; no; yes b) 2.5 m/s^2, 1.5 m/s^2, 0
2–11: a) near 5 s b) 30 s to 40 s c) 0 d) -1.7 m/s^2
2–13: about 10 m/s^2
2–17: a) 5.0 m/s b) 1.43 m/s^2
2–19: 1.70 m
2–21: a) 1.7 m/s^2 b) 12 s c) 240 m
2–25: a) 0, 6.3 m/s^2, -11.2 m/s^2 b) 100 m, 230 m, 320 m
2–27: a) 1.80×10^4 m/s b) 0.957 c) 6 h 11 min
2–29: b) 1 s, 3 s d) 2 s e) 3 s f) 1 s
2–31: b) $d/4$
2–33: a) 2.94 m/s b) 0.599 s
2–35: a) $t = \sqrt{2d/g}$ b) 0.190 s
2–37: a) 5.56 m/s, downward b) 9.80 m/s^2, downward c) 2.16 s

d) 16.1 m/s
2–39: a) 25.6 m/s b) 31.6 m c) 15.2 m/s
2–41: a) 249 m/s^2 b) 25.4 c) 101 m d) no
2–43: a) 7.5 m b) 180 m c) 20 m
2–45: a) $x(t) = (0.250\text{ m/s}^3)t^3 - (0.0100\text{ m/s}^4)t^4$, $v(t) = (0.750\text{ m/s}^3)t^2 - (0.0400\text{ m/s}^4)t^3$ b) 39.1 m/s
2–47: b) 0.627 s, 1.60 s c) negative at 0.627 s, positive at 1.60 s d) 1.11 s e) 2.45 m f) 2.00 s, 0 s
2–49: a) 82 km/h b) 31 km/h
2–51: a) 3.5 m/s^2 b) 0 c) 1.5 m/s^2
2–53: brake pedal
2–55: 4.6 m/s^2
2–57: a) 6.17 s b) 24.8 m c) $v_{\text{truck}} = 13.0$ m/s, $v_{\text{auto}} = 21.0$ m/s
2–59: a) 17 m/s b) 1.6 s
2–61: a) 15.9 s b) 393 m c) 29.5 m/s
2–63: a) -4.00 m/s b) 12.0 m/s
2–65: 28.6 m
2–67: a) no b) yes, 14.4 m/s; not physically attainable
2–69: a) 13.3 m b) 1.65 s
2–71: a) 7.59 m/s b) 5.14 m c) 1.60 s
2–73: a) 92.2 m b) 75.1 m/s
2–75: a) A b) 2.27 s, 5.73 s c) 1.00 s, 4.33 s d) 2.67 s
2–77: a) 9.55 s, 47.8 m b) 1.62 m/s d) 8.38 m/s e) no f) 3.69 m/s, 21.7 s, 80.0 m
2–79: a) 8.18 m/s b) i) 0.411 m ii) 1.15 km c) 9.80 m/s d) 4.90 m/s

Chapter 3

3–1: a) $(v_{\text{av}})_x = 1.4$ m/s, $(v_{\text{av}})_y = -1.3$ m/s b) 1.9 m/s, $-43°$
3–3: a) 7.1 cm/s, 45° b) 5.0 cm/s, 90°; 7.1 cm/s, 45°; 11 cm/s, 27°
3–5: b) $(a_{\text{av}})_x = -8.67$ m/s^2, $(a_{\text{av}})_y = -2.33$ m/s^2 c) 8.98 m/s^2, 195°
3–7: b) $\vec{v} = \alpha\hat{\imath} - 2\beta t\hat{\jmath}, \vec{a} = -2\beta\hat{\jmath}$ c) $v = 5.4$ m/s, $-63°$; $a = 2.4$ m/s^2, $-90°$ d) speeding up and turning right
3–9: a) 0.600 m b) 0.385 m c) $v_x = 1.10$ m/s, $v_y = -3.43$ m/s; $v = 3.60$ m/s, 72.2° below the horizontal
3–11: a) 1.08 s b) 6.18 m, 4.51 m; 11.5 m, 5.74 m; 16.8 m, 4.51 m c) 11.7 m/s, +24.8°; 10.6 m/s, 0°; 11.7 m/s, $-24.8°$ d) parallel: -4.11 m/s^2, 0, 4.11 m/s^2; perpendicular: 8.90 m/s^2, 9.80 m/s^2, 8.90 m/s^2
3–13: a) 1.4 km b) 8.5 km
3–15: a) 0.682 s, 2.99 s b) 24.0 m/s, 11.3 m/s; 24.0 m/s, -11.3 m/s c) 30.0 m/s, $-36.9°$
3–17: a) 1.5 m b) -0.89 m/s
3–19: a) 13.6 m b) 34.6 m/s c) 103 m
3–21: a) 0.034 m/s^2 = $0.0034g$ b) 1.4 h
3–23: a) 3.07 s b) 1.68 s
3–25: a) 3.50 m/s^2, upward b) 3.50 m/s^2, downward c) 12.6 s
3–27: b) no c) where the car is farthest from the center of the ellipse
3–29: a) 14 s b) 70 s
3–31: 0.36 m/s, 38° west of south
3–33: a) 4.7 m/s, 25° south of east b) 190 s c) 380 m
3–35: b) -7.1 m/s, -42 m/s c) 43 m/s, 9.5° west of south
3–37: a) $t = 0$ b) $t = 0$ c) $t = 0$, $x = 0$, $y = 15.0$ m; $t = 5.21$ s, $x = 6.25$ m, $y = 1.44$ m d) 6.41 m, 5.21 s
3–39: a) 1.0 m/s^2 b) 5.4° c) 2.3 m/s d) 31 s
3–41: 22 m/s
3–43: 274 m
3–45: a) 42.8 m/s b) 42.0 m
3–47: $4\sqrt{2}\, D$

3–49:   c) less than 45°

3–51:   b) 15°, 75°

3–53:   a) 17.8 m/s   b) in river, 28.4 m from the near bank

3–55:   a) 2.23 m   b) 3.84 m   c) 8.65 m/s   d) 3.09 m, 0.62 m

3–57:   c) $(18/25)gT^2$ d) $(3/2)gT^2$

3–61:   b) $v_x = R\omega(1 - \cos \omega t)$, $v_y = R\omega \sin \omega t$, $a_x = R\omega^2 \sin \omega t$, $a_y = R\omega^2 \cos \omega t$ c) $t = 0, 2\pi/\omega, 4\pi/\omega, ...$; $x = 0, 2\pi R, 4\pi R, ...$; $y = 0$; $a = R\omega^2$ in the +y-direction   d) no

3–63:   30 km

3–65:   a) 44.7 km/h, 26.6° west of south   b) 10.5° north of west

3–67:   a) 0.782 s   b) 7.67 m/s   c) 5.17 m/s   d) 1.04 m

3–69:   a) 80 m   b) 1.56×10^{-3} c) overall effect is to reduce the radius

3–71:   a) $\left(\dfrac{2v_0^2}{g}\right)[\tan(\theta + \phi) - \tan \theta]\dfrac{\cos^2(\theta + \phi)}{\cos \theta}$ b) $\dfrac{\pi}{4} - \dfrac{\theta}{2}$

3–73: $\Delta t = 0.5$ s: 9.589 m/s^2, 118.6°; $\Delta t = 0.1$ s: 9.983 m/s^2, 95.73°; $\Delta t = 0.05$ s: 9.996 m/s^2, 92.86°

3–75:   a) 1.5 km/h   b) 3.5 km/h

Chapter 4

4–1:   a) 0   b) 90°   c) 180°

4–3:   7.1 N horizontal, 7.1 N vertical

4–5:   494 N, 31.7°

4–7:   2.2 m/s^2

4–9:   16.0 kg

4–11:   a) 3.13 m, 3.13 m/s   b) 21.9 m, 6.25 m/s

4–13:   a) $\Sigma \vec{F} = 0$

4–15: 2.94×10^3 N

4–17:   a) 4.49 kg   b) 4.49 kg, 8.13 N

4–19:   825 N, blocks

4–23: 7.4×10^{-23} m/s^2

4–25:   5.83 m/s^2

4–27:   a) 25.0 N   b) 10.0 N

4–29:   1.47 m/s^2, upward

4–31: $F_x(t) = -6mBt$

4–33:   1840 N, 135°

4–35:   a) 17 N, 90° clockwise from +x-direction   b) 840 N

4–37: 2.36×10^3 N

4–39:   a) 4.4 m   b) 300 m/s   c) i) 2.7×10^4 N   ii) 9.0×10^3 N

4–41:   a) $4m\vec{a}$ b) $3m\vec{a}$ c) $2m\vec{a}$ d) $-m\vec{a}$ e) same magnitudes but opposite directions

4–43:   a) 2.93 m/s^2 b) 11.1 m/s^2

4–45:   b) 5.9×10^2 N   c) 2.2×10^3 N

4–47:   b) 3.53 m/s^2 c) 120 N   d) 93.3 N

4–49: $(k_1 t/m)\hat{\imath} + (k_2 t^4/4m)\hat{\jmath}$

4–51: $\vec{r} = \left(\dfrac{k_1}{2m}t^2 + \dfrac{k_2 k_3}{120 m^2}t^5\right)\hat{\imath} + \left(\dfrac{k_3}{6m}t^3\right)\hat{\jmath}$

         $\vec{v} = \left(\dfrac{k_1}{m}t + \dfrac{k_2 k_3}{24 m^2}t^4\right)\hat{\imath} + \left(\dfrac{k_3}{2m}t^2\right)\hat{\jmath}$

Chapter 5

5–1:   a) 25.0 N   b) 50.0 N

5–3:   a) 2.54×10^3 N   b) 1.01°

5–7:   a) 5.23×10^4 N   b) 3.36×10^4 N

5–9:   523 N

5–11:   a) 84.9 N   b) 60.0 N, 60.0 N

5–13:   a) $w \sin \alpha$ b) $2w \sin \alpha$ c) $w \cos \alpha$ for each block

5–15:   b) 2.96 m/s^2 c) 191 N; more than the bricks, less than the counterweight

5–17:   b) 2.50 m/s^2 c) 1.37 kg   d) $T < mg$

5–19:   a) 1.78 m/s^2, downward   b) 2.14 m/s^2, upward   c) yes

5–21:   a) in the direction down the slope   b) no deflection

5–25:   a) 22 N   b) 3.1 m

5–27:   a) 7.06 N   b) 8.14 N   c) 1.17 N, 2.25 N

5–29:   a) pulled down   b) 379 N

5–31:   11 times farther

5–33:   a) $\mu_k(m_A + m_B)g$ b) $\mu_k m_A g$

5–35:   a) 0.218 m/s   b) 11.7 N

5–37:   a) $\mu_k mg/(\cos \theta - \mu_k \sin \theta)$ b) $1/(\tan \theta) = \cot \theta$

5–39:   b) 8.75 N   c) 30.8 N   d) 1.54 m/s^2

5–41:   a) 0.44 kg/m   b) 42 m/s

5–43:   26.0 m/s

5–45:   a) 75°   b) no

5–47:   20.7°

5–49:   a) 1.49 rev/min   b) 0.918 rev/min

5–51:   a) 138 km/h   b) 3580 N

5–53:   2.43 m/s

5–55:   b) 10 m

5–57:   43 m/s

5–59:   14 m

5–61: $mg \tan \alpha$

5–63: $F[M + m(1 - x/L)]/(M + m)$

5–65:   a) 12.0 N   b) 15.0 N

5–67:   a) 16.9 N   b) 10.2 N

5–69: $\mu_s/(1 + \mu_s)$

5–71:   a) 66 N, northward   b) 59 N, southward

5–75: 3.09×10^3 N

5–77:   2.52 N

5–79:   a) 12.9 kg   b) 47.2 in left-hand cord, 101 N in right-hand cord

5–81: $a_1 = 2m_2 g/(4m_1 + m_2)$, $a_2 = m_2 g/(4m_1 + m_2)$

5–83:   1.46 m above the floor

5–85: g/μ_s

5–87:   b) 0.450

5–89:   a) $(F \cos \alpha)/(m_1 + m_2)$

5–91:   a) 21 m/s   b) 8.5 m/s

5–93:   a) move up   b) remains constant   c) remains constant   d) stop

5–95:   a) 6.00 m/s^2 b) 3.80 m/s^2 c) 7.36 m/s   d) 8.18 m/s   e) 7.78 m, 6.29 m/s, 1.38 m/s^2 f) 3.14 s

5–97:   a) 120 N   b) 3.79 m/s

5–99:   b) 0.28   c) no

5–101:   a) right   b) 117 m

5–103:   a) 81.1°   b) no   c) bead rides at bottom of loop ($\beta = 0$)

5–107: $T_{max} = 2\pi \sqrt{\left(\dfrac{h \tan \beta}{g}\right)\left(\dfrac{(\sin \beta + \mu_s \cos \beta)}{(\cos \beta - \mu_s \sin \beta)}\right)}$

         $T_{min} = 2\pi \sqrt{\left(\dfrac{h \tan \beta}{g}\right)\left(\dfrac{(\sin \beta - \mu_s \cos \beta)}{(\cos \beta + \mu_s \sin \beta)}\right)}$

5–109: $(M + m)g \tan \alpha$

5–111:   a) $F = \dfrac{\mu_k w}{\cos \theta - \mu_k \sin \theta}$ c) $\theta = \tan^{-1} \mu_k = 14.0°$

5–113:   a) $a_3 = g\dfrac{(-4m_1 m_2 + m_2 m_3 + m_3 m_1)}{(4m_1 m_2 + m_2 m_3 + m_3 m_1)}$ b) $a_B = -a_3$

         c) $a_1 = g\dfrac{(4m_1 m_2 - 3m_2 m_3 + m_3 m_1)}{(4m_1 m_2 + m_2 m_3 + m_3 m_1)}$

         d) $a_2 = g\dfrac{(4m_1 m_2 + m_2 m_3 - 3m_3 m_1)}{(4m_1 m_2 + m_2 m_3 + m_3 m_1)}$

         e) $T_A = \frac{1}{2}T_C$ f) $T_C = \dfrac{8gm_1 m_2 m_3}{(4m_1 m_2 + m_2 m_3 + m_3 m_1)}$

         g) $a_1 = a_2 = a_3 = a_B = 0$, $T_C = 2m_2 g$, $T_A = m_2 g$; yes

5–115: $\cos^2 \beta$

Chapter 6

6–1: a) 3.60 J   b) −0.900 J   c) 2.70 J
6–3: 300 J
6–5: a) 99.2 N   b) 387 J   c) −387 J   d) zero; zero   e) zero
6–7: 2.6×10^9 J
6–9: a) 4.3×10^3 J   b) 40 km/h
6–11: 74.2 J
6–13: a) $-(15/16)K_1$ b) no
6–15: 16.8 cm
6–17: a) −28.4 J   b) 15.3 m/s   c) no
6–19: a) 4.96 m/s   b) $a = 1.43$ m/s², $v = 4.96$ m/s
6–21: a) $v_0^2/2\mu_k g$ b) 51.3 m
6–23: a) 48.0 N, 64.0 N   b) 0.360 J, 0.640 J
6–25: a) 2.8 m/s   b) 3.5 m/s
6–27: a) 1.76   b) 0.67 m/s
6–29: a) 4.0 J   b) zero   c) −1.0 J   d) 3.0 J   e) −1.0 J
6–31: a) 2.83 m/s   b) 2.40 m/s
6–33: a) 5.65 cm   b) no; 0.57 J
6–35: 743 W = 0.995 hp
6–37: 0.23
6–39: 28
6–41: 8.1×10^6 N
6–43: b) 9   c) 108 W
6–45: b) 11 L   c) 0.10 L/km = 0.044 gal/mi
6–47: a) 358 N   b) 47.2 hp   c) 4.06 hp   d) 2.03%
6–49: a) 532 J   b) −315 J   c) zero   d) −203 J   e) 14.7 J
       f) 1.21 m/s
6–51: a) $1/\sin \alpha$ b) $W_{in} = W_{out}$
6–53: a) 2.59×10^{12} J   b) 4800 J
6–55: 2.59 m
6–57: a) 4.00 N   b) 32.0 N   c) −15.0 J
6–59: a) 0.15 N   b) 9.4 N   c) 0.44 J
6–61: a) 2.56 m/s   b) 5.28 N   c) 19.7 J
6–63: a) −910 J   b) 3.17×10^3 J
6–65: 1.0×10^5 N/m
6–67: 1.1 m from where the spring is released
6–69: a) 1.02×10^4 N/m, 8.16 m
6–71: a) 0.600 m   b) 1.50 m/s
6–73: 0.786
6–75: a) 1.10×10^5 J   b) 1.30×10^5 J   c) 3.99 kW
6–77: 3.6 h
6–79: 1.30×10^3 m³/s
6–81: a) 1.26×10^5 J   b) 1.46 W
6–83: a) 2.4 MW   b) 61 MW   c) 6.0 MW
6–85: a) 513 W   b) 355 W   c) 52.1 W
6–87: a) $\frac{1}{6}Mv^2$ b) 6.1 m/s   c) 3.9 m/s   d) $K_{ball} = 0.40$ J,
       $K_{spring} = 0.60$ J
6–89: a) 2.0×10^5 J   b) 2.8×10^5 J   c) 2.8×10^5 J
       d) 5 km/h

Chapter 7

7–1: 3.45×10^6 J
7–5: a) 24.0 m/s   b) 24.0 m/s   c) part (b)
7–9: 2.8 m/s
7–11: a) 880 J   b) −157 J   c) 471 J   d) 253 J   e) $a = 3.16$ m/s²,
       $v = 7.11$ m/s, $\Delta K = 253$ J
7–13: a) 80.0 J   b) 5.00 J
7–15: a) 6.32 cm   b) 12 cm
7–17: 1.72 m
7–19: a) 0.500 m/s   b) 0.500 m
7–21: a) 23.6 m/s   b) 69.2 m/s², upward
7–23: a) -1.2×10^2 J   b) $+1.2 \times 10^2$ J   c) zero   d) conservative
7–25: a) zero   b) −0.10 J   c) +0.10 J   d) conservative; $U(x) = \frac{1}{3}\alpha x^3$

7–27: a) −3.6 J   b) −3.6 J   c) −7.2 J   d) nonconservative
7–29: a) −59 J   b) −42 J   c) −59 J   d) nonconservative
7–31: $-6C_6/x^7$, attractive
7–33: $\vec{F} = -(2kx + k'y)\hat{\imath} - (k'x + 2ky)\hat{\jmath}$
7–35: a) $F(r) = (12a/r^{13}) - (6b/r^7)$ b) $(2a/b)^{1/6}$, stable   c) $b^2/4a$
       d) $a = 6.68 \times 10^{-138}$ J · m¹², $b = 6.41 \times 10^{-78}$ J · m⁶
7–37: a) 229 N   b) 823 J
7–39: 0.41
7–41: 15.5 m/s
7–43: 4.4 m/s
7–45: a) 919 N/m, 39.8 m   b) 17.0 m/s²
7–47: a) 7.00 m/s   b) 2.94 N
7–49: a) $mg(1 - h/d)$ b) 4.4×10^2 N   c) $\sqrt{2gh(1 - y/d)}$
7–51: 48.2°
7–53: a) 0.392   b) −0.832 J
7–55: a) $U(x) = (30.0 \text{ N/m})x^2 + (6.00 \text{ N/m}^2)x^3$ b) 7.85 m/s
7–57: 7.01 m/s
7–59: a) $m(g + a)^2/2gh$ b) $2gh/(g + a)$
7–61: 119 J
7–63: a) 3.87 m/s   b) 0.10 m
7–65: a) 3.1 m/s   b) 1.4 m   c) no
7–67: a) −50.6 J   b) −67.5 J   c) nonconservative
7–69: a) no   b) $x_0 = F/k$ d) no   e) $3F/k, -F/k$ f) $v_{max} = 2F/\sqrt{mk}$
       at $x = x_0 = F/k$
7–71: b) $v(x) = \left[\dfrac{2\alpha}{mx_0^2} \left(\left(\dfrac{x_0}{x}\right) - \left(\dfrac{x_0}{x}\right)^2 \right) \right]^{1/2}$

       c) $x = 2x_0$, $v = \sqrt{\alpha/2mx_0^2}$ d) zero

       e) $v(x) = \left[\dfrac{2\alpha}{mx_0^2} \left(\left(\dfrac{x_0}{x}\right) - \left(\dfrac{x_0}{x}\right)^2 - \dfrac{2}{9} \right) \right]^{1/2}$

       f) first case: x_0, ∞; second case: $3x_0/2, 3x_0$

Chapter 8

8–1: a) 1.20×10^5 kg · m/s   b) i) 60.0 m/s   ii) 26.8 m/s
8–3: b) baseball, 0.525   c) woman, 0.643
8–5: 0.256 kg · m/s, −y-direction
8–7: 562 N, no
8–9: a) 10.8 m/s to the right   b) 0.75 m/s to the left
8–11: a) $(18.8 \text{ kg} \cdot \text{m/s})\hat{\imath}$ b) $(-3.55 \times 10^{-3} \text{ kg} \cdot \text{m/s})\hat{\jmath}$
       c) $(7.50 \times 10^3 \text{ N})\hat{\imath}$
       d) $(13.0\hat{\imath} - 0.73\hat{\jmath})$ kg · m/s, $(89.3\hat{\imath} - 5.0\hat{\jmath})$ m/s
8–13: a) $At_2 + (B/3)t_2^3$ b) $(A/m)t_2 + (B/3m)t_2^3$
8–15: a) 5.68×10^{-2} m/s   b) 0.103 m/s
8–17: a) 4.66 m/s in the direction opposite the defender's original
       motion   b) −6580 J
8–19: a) 3.60 m/s   b) 8.64 J
8–21: a) $v_B = (m_A/m_B)v_A$
8–23: 525 N
8–25: a) $v_{A2} = 29.3$ m/s, $v_{B2} = 20.7$ m/s   b) 0.196 = 19.6%
8–27: a) 6.44 m/s, eastbound   b) 2.49 m/s   c) -2.81×10^5 J,
       -1.38×10^5 J; part (a)
8–29: 35.3 km/h, 19.3° west of south
8–31: 229 m/s
8–33: 0.150-kg glider: 3.2 m/s to the left; 0.300-kg glider: 0.20 m/s
       to the left
8–37: a) $9.00m$ b) 3.00×10^6 m/s
8–39: 7.42×10^8 m from the center of the sun; outside
8–41: a) 0.30 kg   b) $(2.0 \text{ kg} \cdot \text{m/s})\hat{\imath}$ c) $(6.7 \text{ m/s})\hat{\imath}$
8–43: a) 1.25 kg   b) $(1.5 \text{ m/s}^3)t\hat{\imath}$ c) $(5.6 \text{ N})\hat{\imath}$
8–45: a) 80.0 N   b) yes
8–47: 75.0 kg
8–49: a) 0.442   b) 800 m/s   c) 530 m/s
8–51: a) 7.2×10^{-66} b) 0.223

8–53: 3.21×10^{-25} kg
8–55: a) 6.32×10^{-22} kg $\cdot$ m/s b) 5.71×10^{-19} J
8–57: $(0.521\hat{\imath} + 7.81\hat{\jmath})$ m/s
8–59: 15
8–61: 30° east of north, 30° east of south
8–63: a) 5.00 m/s, east b) 5.71 m/s, east c) 3.78 m/s, east
8–65: a) Nash, 1.68 b) Packard, 0.933 c) F_P is larger, 0.933
       d) F_N is larger, 1.68
8–67: 23.2 cm
8–69: a) 25.8 m/s, 35.5° b) no
8–71: a) 0.222 b) −291 J c) 0.784 J
8–73: b) $M = m$ c) zero
8–75: b) $\frac{1}{2} M v_{cm}^{2}$
8–77: a) 3.56 m/s b) 5.22 m/s c) 4.67 m/s
8–79: A: 13.6 m/s; B: 6.34 m/s, 65.0°
8–81: a) $(L/2)\cos(\alpha/2)$, along axis from apex b) $(L/3)$, along
       bisector from bottom
       c) $L/\sqrt{8}$ along bisector d) $L/\sqrt{12}$ from each side
8–83: 0.400 m/s
8–85: 222 m/s, 1.01×10^3 m/s; $v_{Kr} = 1.5 v_{Ba}$
8–87: a) zero b) 1 d) 0.87 m f) 0.089 m
8–89: a) yes b) no; kinetic energy decreases by 4.8×10^3 J
8–91: a) $1.37 v_{ex}$ b) $1.18 v_{ex}$ c) $2.38 v_{ex}$ d) 2.94 km/s

8–93: a) $v = \begin{cases} (2400 \text{ m/s}) \ln\left[\dfrac{1}{1 - t/(120 \text{ s})}\right], & 0 \le t \le 90 \text{ s} \\ 3.33 \text{ km/s}, & t \ge 90 \text{ s} \end{cases}$

       b) $(20 \text{ m/s}^2)/[1 - t/(120 \text{ s})]$ for $0 \le t \le 90$ s
       c) 6000 N vs. 735 N on earth
8–95: b) 2L/3
8–97: a) $l^2 \lambda g/32$ b) $l^2 \lambda g/32$

Chapter 9

9–1: a) 0.600 rad = 34.4° b) 6.27 cm c) 1.05 m
9–3: a) 42 rad/s² b) 74 rad/s
9–5: a) $\omega(t) = (0.400 \text{ rad/s}) + (0.0360 \text{ rad/s}^3)t^2$ b) 0.400 rad/s
      c) $\omega = 1.30$ rad/s, $\omega_{av} = 0.700$ rad/s
9–7: a) $\alpha(t) = 2b - 6ct$ b) $b/3c$
9–9: a) −1.25 rev/s², 23.3 rev b) 2.67 s
9–11: a) 24.0 s b) 68.8 rev
9–13: 10.5 rad/s
9–15: 9.00 rev
9–17: a) 540 rad b) 12.3 s c) −8.17 rad/s²
9–19: a) 47.1 m/s b) 47.3 m/s
9–21: a) $a_{rad} = 18.0$ m/s² b) $v = 3.00$ m/s, $a_{rad} = 18.0$ m/s²
9–23: a) 0.180 m/s², 0, 0.180 m/s² b) 0.180 m/s², 0.377 m/s²,
       0.418 m/s²
       c) 0.180 m/s², 0.754 m/s², 0.775 m/s²
9–25: 10.7 cm; no
9–27: a) 0.831 m/s b) 109 m/s²
9–29: a) 2.29 b) 1.51 c) 15.7 m/s, 1.06×10^3 m/s² $= 108g$
9–31: $(M/12 + m/2)L^2$
9–33: a) 0.0640 kg $\cdot$ m² b) 0.0320 kg $\cdot$ m² c) 0.0320 kg $\cdot$ m²
9–35: 0.193 kg $\cdot$ m²
9–37: b) $K = \pi^2 I \omega^2/1800$
9–39: 0.600 kg $\cdot$ m²
9–41: 7.35×10^4 J
9–43: a) $K = 2\pi^2 I/T^2$ b) $dK/dt = (-4\pi^2 I/T^3)(dT/dt)$ c) 70 J
       b) −0.56 J/s
9–45: 823 J
9–47: an axis parallel to a diameter and $\left(2/\sqrt{15}\right)R$ from the center
       of the sphere

9–49: $\frac{1}{3} M(a^2 + b^2)$
9–51: $\frac{1}{3} M(L^2 - 3Lh + 3h^2)$
9–53: $\frac{1}{3} ML^2$
9–57: a) $(6.40 \text{ rad/s}^2)t - (1.50 \text{ rad/s}^3)t^2$
       b) $6.40 \text{ rad/s}^2 - (3.00 \text{ rad/s}^3)t$
       c) $\omega_{max} = 6.83$ rad/s at $t = 2.13$ s
9–59: a) 35.0 km/h = 9.72 m/s b) 8.51 J c) 652 rad/s
9–61: a) 211 rev/min b) 800 W
9–63: a) 1.70 m/s b) 84.8 rad/s
9–65: b) 2.00 m/s² d) 0.208 kg $\cdot$ m²
9–67: a) 2.14×10^{29} J b) 2.66×10^{33} J
9–69: a) −0.784 J b) 5.42 rad/s c) 5.42 m/s
       d) particle speed = 4.43 m/s
9–71: $\sqrt{(2gd)(m_B - \mu_k m_A)/(m_A + m_B + I/R^2)}$
9–73: $\sqrt{(g/R)(1 - \cos \beta)}$
9–75: a) 2.25×10^{-3} kg $\cdot$ m² b) 3.40 m/s c) 4.95 m/s
9–77: a) $(247/512)MR^2$ b) $(383/512)MR^2$
9–79: b) $\frac{1}{4}M(R_1^2 + R_2^2)$
9–81: a) $\frac{3}{5}MR^2$ b) larger
9–83: b) 5.97×10^{24} kg c) $0.334MR^2$
9–85: a) $s = r_0 \theta + \beta \theta^2/2$

       b) $\theta = \left(\dfrac{1}{\beta}\right)\left[\sqrt{r_0^2 + 2\beta vt} - r_0\right]$

       c) $\omega = \dfrac{v}{\sqrt{r_0^2 + 2\beta vt}}$, $\alpha = -\dfrac{\beta v^2}{(r_0^2 + 2\beta vt)^{3/2}}$; no
       d) $r_0 = 2.50$ cm, $\beta = 0.247$ μm/rad, 2.13×10^4 rev

Chapter 10

10–1: a) 40.0 N $\cdot$ m, out of the page b) 34.6 N $\cdot$ m, out of the page
       c) 20.0 N $\cdot$ m, out of the page d) 17.3 N $\cdot$ m, into the page
       e) zero f) zero
10–3: 2.50 N $\cdot$ m, counterclockwise
10–5: c) $(-1.05 \text{ N} \cdot \text{m})\hat{k}$
10–7: 1.2 m/s
10–9: a) $g(M + 3m)/(1 + 2m/M)$ b) less c) no effect
10–11: 0.482
10–13: a) 7.5 N in horizontal part, 18.2 N in hanging part
        b) 0.0160 kg $\cdot$ m²
10–15: a) 0.882 N b) 0.553 s c) 33.9 rad/s
10–17: a) 1/3 b) 2/7 c) 2/5 d) 5/13
10–19: 11.7 m
10–21: a) 0.309 rad/s b) 100 J c) 6.67 W
10–23: a) 0.38 N $\cdot$ m b) 160 rad c) 59 J d) 59 J
10–25: b) 65.6 N
10–27: a) 358 N $\cdot$ m b) 1.79×10^3 N c) 83.8 m/s
10–29: a) 115 kg $\cdot$ m²/s, into the page b) 125 kg $\cdot$ m²/s², out of
        the page
10–31: 4.71×10^{-6} kg $\cdot$ m²/s
10–33: b) 7.00 rad/s c) 1.03×10^{-2} J d) 1.03×10^{-2} J
10–35: 0.60 rev
10–37: a) 1.4 rad/s b) 1080 J before, 500 J after
10–41: a) 36.8 min b) 1.10×10^5 N $\cdot$ m
10–43: 5.4×10^{22} N $\cdot$ m
10–45: a) 0.955 kg $\cdot$ m² b) −0.0800 N $\cdot$ m c) 104 rev
10–47: b) 4500 W d) 2600 W
10–49: 0.675 s
10–51: a) $L/4$ from the end with the clay b) $(9g/8L) \sin \theta$
        c) $(3g/2L) \sin \theta$
10–53: 1200 N

10–55: a) 1.12 m/s^2 b) 14.0 N

10–57: $a = F/2m, f = F/2$

10–59: a) 0.957 m

10–61: a) $(27R - 17r)/10$ b) $(5R - 3r)/2$

10–63: a) 1.76 N b) 123 rad/s^2 c) 9.80 m/s^2
d) T would be unchanged, α and a would double

10–65: a) 9.34 m/s b) 18.7 m/s c) zero d) 5.60 m

10–67: 3.4 m

10–69: $g/3$

10–71: a) $6v/19L$ b) 3/19

10–73: a) 5.46 rad/s b) 3.17 cm c) 1.01×10^3 m/s

10–75: 3200 J

10–77: 0.30 rad/s, clockwise

10–81: 12.7 rad/s

10–85: a) $a = +\mu_k g$, $\alpha = -2\mu_k g/R$ b) $\omega_0^2 R^2/18\mu_k g$ c) $-M\omega_0^2 R^2/6$

10–87: a) $mv_1^2 r_1^2/r^3$ b) $(mv_1^2/2)[(r_1/r_2)^2 - 1]$ c) same

Chapter 11

11–1: 0.387 m from center of 1.00-kg sphere

11–5: 5450 N

11–7: 1000 N, 1.20 m from end where 400-N force is applied

11–11: a) 1920 N b) 1140 N

11–13: a) $T = 2.60\ w$; $F_{pivot} = 3.28\ w$ at 37.6° b) $T = 4.10\ w$; $F_{pivot} = 5.38\ w$ at 48.8°

11–15: 140 N by each hinge

11–17: $T_{left} = 270$ N, $T_{right} = 304$ N, $\theta = 40°$

11–19: a) 0.800 m b) clockwise c) 0.800 m, clockwise

11–21: 1.4 mm

11–23: 2.0×10^{11} Pa

11–25: a) upper: 3.1×10^{-3}; lower: 2.0×10^{-3} b) upper: 1.6 mm; lower: 0.98 mm

11–27: 9.1×10^6 N

11–29: a) 3.33×10^6 Pa b) 1.33×10^5 N

11–31: 4.8×10^9 Pa, 2.1×10^{-10} Pa^{-1}

11–33: 1/9

11–35: 3.41×10^7 Pa

11–37: 10.2 m/s^2

11–41: wing force: 7300 N upward; tail force: 600 N downward

11–43: a) 590 N b) neutral

11–47: 120 N to the right, 160 N upward

11–49: a) fore: 37 N; hind: 58 N b) fore: 53 N; hind: 55 N

11–51: a) $\tau = Fl \sin\phi$ b) 12.6 N $\cdot$ m with either method of calculation

11–53: a) $V = mg + w$, $H = T = (w + mg/4)\cot\theta$ b) 9.50×10^2 N c) 4.00°

11–55: a) 4.90 m b) 60 N

11–57: a) 300 N upward, 225 N $\cdot$ m counterclockwise b) $T = 132$ N; $F_{pivot} = 194$ N, 71.2° above horizontal

11–59: a) 60.0 N b) 53.1°

11–61: person above: 590 N; person below: 1370 N; above

11–63: a) 268 N b) 232 N c) 366 N

11–65: a) tips at 27°, slips at 31°; the bale tips before it slips b) tips at 27°, slips at 22°; the bale slips before it tips

11–67: a) $F_A = 80$ N, $F_B = 870$ N b) 1.92 m

11–69: a) 3.7 kN, 2.0 kN vertically upward

11–71: a) 5.4 mm b) 4.2 mm

11–73: a) 0.70 m from wire A b) 0.45 m from wire B

11–75: a) 1.63 m b) brass: 2.00×10^8 Pa; nickel: 4.00×10^8 Pa c) brass: 2.22×10^{-3}; nickel: 1.90×10^{-3}

11–77: a) 0.36 mm b) 0.045 mm c) 0.33 mm

11–79: a) $(F \cos^2\theta)/A$ b) $(F \sin 2\theta)/2A$ c) 0 d) 45°

11–81: a) 600 N b) 13.5 kN c) slide: $\mu_s w/(\sin\theta - \mu_s \cos\theta)$; tip: $w/[(1/9)\cos\theta + 2\sin\theta]$; 66°

11–83: the lesser of $h^2/L + L/2$ and L

11–85: $[(A^2 x/F) - k_0 V_0]/V_s$

11–87: a) 0.662 mm b) 2.20×10^{-2} J c) 8.33×10^{-3} J d) -3.04×10^{-2} J e) 3.04×10^{-2} J

Chapter 12

12–1: 2.18

12–3: F_{12}

12–5: a) 2.59×10^8 m

12–7: 2.4×10^{-3} N

12–9: a) 6.30×10^{20} N, toward sun b) 4.77×10^{20} N, 24.6° toward earth from sun c) 2.37×10^{20} N, toward sun

12–11: 2.1×10^{-9} m/s^2, down

12–13: 1.38×10^7 m

12–15: a) 0.37 m/s^2 b) 1700 kg/m^3

12–17: 5.98×10^{24} kg

12–19: 0.83 m/s; yes

12–21: a) 5.02×10^3 m/s b) 6.06×10^4 m/s

12–23: a) 175 min b) 3.71 m/s^2

12–25: 7.46×10^3 m/s

12–27: 8.3×10^4 m/s b) 1.3×10^6 s

12–29: b) 4.45×10^{12} m, 4.55×10^{12} m b) 248 y

12–33: a) $-GMm/\sqrt{a^2 + x^2}$ c) $GMmx/(a^2 + x^2)^{3/2}$ toward the center of the ring e) $-GMm/a$, zero

12–35: a) 53 N b) 52 N

12–37: a) 4.3×10^{37} kg $= 2.1 \times 10^7\ M_{sun}$ b) no c) 6.3×10^{10} m; yes

12–39: 1.39×10^{-9}

12–41: a) 2.20×10^{-10} N, 163° from +x-direction toward +y-direction b) $x = 0$, $y = 1.39$ m

12–43: a) 3.40 N, 0.61° b) 1.31×10^9 J

12–45: 3.83×10^8 m

12–47: a) 3.58×10^7 m

12–49: 1.8×10^2 m/s

12–51: $0.01R_E = 64$ km

12–53: 6.06×10^3 km/h

12–55: $\sqrt{2Gm_E h/(R_E^2 + hR_E)}$

12–57: a) 13.7 km/s b) 13.3 km/s c) 13.2 km/s

12–59: a) i) 2.8 y ii) 6.1 y b) 4.90×10^8 km c) 4.22×10^8 km

12–61: a) $GM^2/4R^2$ b) $\sqrt{GM/4R}$, $4\pi\sqrt{R^3/GM}$ c) $GM^2/4R$

12–63: 6.8×10^4 m/s

12–65: a) 7.91×10^3 s b) 1.53 c) 8.43×10^3 m/s, 5.51×10^3 m/s d) 2.41×10^3 m/s, 3.26×10^3 m/s; perigee

12–67: 3.22×10^9 J

12–69: 1.4×10^{14} m

12–71: $GMm/(x^2 + xL)$

12–73: b) 22 min c) 2.5×10^8 s $= 2900$ d $= 7.9$ y

12–75: a) 2.1 kN b) no

12–77: perpendicular: $GMm/(a\sqrt{L^2 + a^2})$; parallel: zero

Chapter 13

13–1: a) 4.54×10^{-3} s, 1.38×10^3 rad/s b) 1.14×10^{-3} s, 5.53×10^3 rad/s

13–3: 5.53×10^3 rad/s, 1.14×10^{-3} s

13–5: a) 0.167 s b) 37.7 rad/s c) 8.44×10^{-2} kg

13–7: a) 0.375 s b) 2.66 Hz c) 16.7 rad/s

13–9: a) $x = (3.0\ \text{mm})\cos[2\pi(440\ \text{Hz})t]$ b) 8.3 m/s, 2.3×10^4 m/s^3 c) jerk $= (6.3 \times 10^7$ m/s$^3)\sin[2\pi(440\ \text{Hz})t]$, 6.3×10^7 m/s^3

13–11: a) 0.383 m b) 1.02 rad $= 58.5°$ c) $x = (0.383\ \text{m})\cos[(12.2\ \text{rad/s})t + 1.02\ \text{rad}]$

13–13: 0.353 m

13–17:   a) $A/\sqrt{2}$, $\omega A/\sqrt{2}$ b) four times; $\pi/2\omega$ c) 3/4 kinetic, 1/4 potential

13–19:   a) 5.13 m/s^2, 0.961 m/s   b) –2.57 m/s^2, 0.833 m/s   c) 0.137 s   d) v, a can be found with energy approach; t cannot

13–21:   a) 2.84×10^{-2} J   b) 0.014 m   c) 0.615 m/s

13–23:   a) 5.31×10^3 N/m   b) 0.695 s

13–25:   3.97 cm

13–27:   a) 2.7×10^{-8} kg • m^2 b) 4.3×10^{-6} N • m/rad

13–29: 5.12×10^{-2} kg • m^2

13–33:   2.60 s

13–35:   2.00 m

13–37:   10.7 m/s^2

13–41:   0.58 s

13–43:   a) 0.393 Hz   b) 1.73 kg/s

13–45:   a) A b) magnitude = $bA/2m$, in $-x$-direction; slope is negative

     c) $a(0) = A\left(\dfrac{b^2}{2m^2} - \dfrac{k}{m}\right)$; if $b < \sqrt{2mk}$, $a(0) < 0$;

     if $b = \sqrt{2mk}$, $a(0) = 0$; if $b > \sqrt{2mk}$, $a(0) > 0$

13–47:   a) kg/s   b) i) $5.0F_{max}/k$ ii) $2.5F_{max}/k$

13–49:   a) 6.72×10^3 m/s^2 b) 3.02 kN   c) 18.3 m/s, 75.6 J   d) 17.6 kW

     e) 12.1 kN, 36.7 m/s, 302 J, 141 kW

13–51:   a) all unchanged   b) 1/4 as large   c) halved   d) $1/\sqrt{5}$ as great   e) U: unchanged; K: 1/5 as large

13–53:   a) 1.68 s   b) 0.090 m   c) 0.143

13–55: $\mu_s(m + M)g/k$

13–59:   a) 0.150 m/s   b) 0.112 m/s^2, downward   c) 0.700 s   d) 4.38 m

13–61:   a) 2.6 m/s   b) 0.21 m   c) 0.49 s

13–63:   0.933 s

13–65: $t_1 = T/4$

13–71:   d) $\sqrt{A/\pi^2 mR_0^3}$

13–73: $2\pi\sqrt{3M/2k}$

13–75:   0.88 m

13–77:   a) 3.97 m

13–79:   c) 0.430 m

13–81:   a) spring 1: 0.350 m; spring 2: 0.250 m   b) 0.702 s

13–83:   a) $\Delta T/T = -\frac{1}{2}(\Delta g/g)$ b) 9.7791 m/s^2

13–85:   a) $T/T_0 = \sqrt{(L^2 + 3y^2)/(L^2 + 2yL)}$ b) $y = 2L/3$

13–87:   a) $kl_0/(k - m\omega'^2)$

Chapter 14

14–1: $w = 41.8$ N; no

14–3: 7.02×10^3 kg/m^3; yes

14–5:   a) 706 Pa   b) 3160 Pa

14–7:   a) 2.52×10^6 Pa   b) 1.78×10^5 N

14–9:   a) 1.07×10^5 Pa   b) 1.03×10^5 Pa   c) 1.03×10^5 Pa   d) 5.33×10^3 Pa

14–11: 2.3×10^5 N

14–13:   4.14 m

14–15:   0.562 m^3

14–17:   a) $\rho < \rho_{fluid}$ c) submerged: ρ/ρ_{fluid}; above: $(\rho_{fluid} - \rho)/\rho_{fluid}$ d) 32%

14–19:   a) 116 Pa   b) 921 Pa   c) 0.822 kg, 822 kg/m^3

14–21:   6.67 Pa

14–23:   0.1 N, 0.01 kg

14–25: 4.14×10^{-3} N

14–27:   a) 17.0 m/s   b) 0.317 m

14–29:   28.4 m/s

14–31: 1.47×10^5 Pa

14–33:   500 N downward

14–35:   0.41 cm

14–37:   a) 1.88 m/s   b) 0

14–39:   a) 0.128 m^3/s   b) 9.72×10^4 Pa   c) 0.275 m^3/s

14–41:   5.96 mm/s

14–43:   a) 16 times greater   b) 1/2 as great   c) twice as great   d) twice as great   e) 1/2 as great

14–45:   a) 19.4 m/s, 0, 14.6 m/s   b) 152 d   c) in (a), 4; in (b), 1/16

14–47:   a) 1.10×10^8 Pa   b) 1080 kg/m^3, 5%

14–49: 2.61×10^4 N • m

14–51: $(p - p_0)VR^2/Gmd$

14–53:   a) 12,700 kg/m^3, 3140 kg/m^3

14–55:   a) 1470 Pa   b) 13.9 cm

14–57: 9.8×10^6 kg, yes

14–59:   a) 30%   b) 70%

14–61: 4.66×10^{-4} m^3, 5.27 kg

14–63:   a) 1.10×10^4 m^3 b) 112 kN

14–65:   a) 0.107 m   b) 2.42 s

14–67:   0.0958 kg

14–69:   33.5 N

14–71:   b) 12.2 N   c) 11.8 N

14–73:   b) 2.52×10^{-4} m^3, 0.124

14–75:   risen by 5.57×10^{-4} m

14–77:   a) $1 - \rho_B/\rho_L$ b) $(\rho_L - \rho_B)L/(\rho_L - \rho_W)$ c) 4.60 cm

14–79:   a) la/g b) $w^2l^2/2g$

14–83:   7.0 N • m

14–85:   13.1 cm

14–87:   133 m/s

14–89: $3h_1$

14–91:   a) $r = r_0(1 + 2gy/v_0^2)^{-1/4}$ b) 1.10 m

14–93:   a) 0.0130 m/s   b) 2.16 m/s

14–95:   a) 6.86×10^{-5} m^3/s   b) cd: 0.686 m/s; ef: 1.71 m/s; gh: 3.43 m/s

     c) c and d: 0.576 m; e and f: 0.450 m g and h: 0   d) 0.0264 m   e) 0.165 m

     f) cd: 1.37 m/s ef: 3.43 m/s gh: 6.86 m/s

14–97:   a) 80.4 N

14–99:   a) $\sqrt{2gh}$ b) $(p_a/\rho g) - h$

Chapter 15

15–1:   a) –81.0°F   b) 134.1°F   c) 88.0°F

15–3:   –40°, winter

15–5:   a) 104.4°F, yes   b) 54°F

15–7:   a) 216.5 K   b) 325.9 K   c) 205.4 K

15–9:   27.23 K

15–11:   444 mm Hg

15–13:   a) –282°C   b) no, 4.76×10^4 Pa

15–15:   0.39 m

15–17:   a) 1.9014 cm   b) 1.8964 cm

15–19: 2.3×10^{-5} (C°)$^{-1}$

15–21:   11 L

15–23: 1.7×10^{-5} (C°)$^{-1}$

15–25:   a) 1.431 cm^2 b) 1.437 cm^2

15–27:   a) 3.2×10^{-5} (C°)$^{-1}$ b) 2.5×10^9 Pa

15–29:   a) 38 J   b) 4.5×10^4 J

15–31:   0.526 C°

15–33: 3.03×10^5 J

15–35: 5.79×10^5 J

15–37:   a) 2.51×10^3 J/kg • K   b) overestimate

15–39:   8 min

15–41: 3.64×10^4 J = 8.69×10^3 cal = 34.5 Btu

15–43: 2.39×10^4 Btu/h = 7.01×10^3 W

15–45: 357 m/s

15–47: 3.45 L

15–49: 27.5°C

15–51: 0.0940 kg

15–53: 2.10 kg

15–55: 190 g

15–57: a) 222 K/m b) 10.7 W c) 73.3°C

15–59: a) −5.8°C b) 11 W/m^2

15–61: 7.1×10^2 Btu = 7.5×10^5 J

15–63: 105.5°C

15–65: 167 W

15–67: 2.1 cm^2

15–69: 6.0×10^3 W/m^2

15–71: −6°C

15–73: 5.71 K/W

15–75: b) 5.0×10^{-3} cm^3, −23 kg/m^3

15–77: 23.0 cm, 7.0 cm

15–79: b) 1.9×10^8 Pa

15–81: a) 87°C b) −80°C

15–83: 20.2°C

15–85: a) 54.3

15–87: a) 83.6 J b) 1.86 J/mol · K c) 5.60 J/mol · K

15–89: a) 2.7×10^7 J b) 6.89 K c) 19.3 K

15–91: a) 4.7×10^{-3} kg b) yes

15–93: a) 86.1°C b) no ice, 0.130 kg liquid water, no steam

15–95: a) 100°C b) 0.0214 kg steam, 0.219 kg liquid water

15–97: a) 100°C b) no ice, 0.419 kg liquid water, 0.0264 kg steam
c) 0°C, 0.154 kg ice, 0.408 kg liquid water, no steam

15–99: a) 94 W b) 1.3

15–101: 2.9

15–103: c) 6.0×10^5 s (about 170 h) d) 1.5×10^{10} s (about 500 y); no

15–105: 305 min

15–107: 5.82×10^{-3} kg

15–109: a) 69.6°C

15–111: a) $H = 4\pi kab\Delta T/(b - a)$ b) $T(r) = \dfrac{[T_1 b(r - a) + T_2 a(b - r)]}{r(b - a)}$

c) $H = 2\pi kL\Delta T/\ln(b/a)$ d) $T(r) = \dfrac{[T_1\ln(r/a) + T_2\ln(b/r)]}{\ln(b/a)}$

15–113: b) uniform 0°C d) 3.14×10^3 C°/m e) 121 W f) 0
g) 1.1×10^{-4} m^2/s
h) −11 C°/s i) 9.1 s j) decrease k) −7.7 C°/s

15–115: 20.42 K

Chapter 16

16–1: a) 56.2 mol b) 6.81×10^6 Pa = 67.2 atm

16–3: 0.959 atm

16–5: 503.0°C

16–7: 3.36×10^5 Pa

16–9: 0.159 L

16–11: 1.05 atm

16–13: a) 70.2°C b) yes

16–15: 2.28×10^4 Pa

16–17: a) 8.2×10^{-17} atm b) no

16–19: 55.6 mol, 3.35×10^{26} molecules

16–21: a) 9.00×10^{-5} m^3 b) 3.10×10^{-10} m c) about the same

16–23: b) 1.004

16–25: a) 6.21×10^{-21} J b) 2.34×10^5 m^2/s^2 c) 484 m/s
d) 2.57×10^{-23} kg · m/s e) 1.24×10^{-19} N
f) 1.24×10^{-17} Pa g) 8.15×10^{21} h) 2.45×10^{22}

16–27: 3800°C

16–29: a) 1390 J/kg · K

16–31: a) 1560 J b) 935 J

16–33: a) 987 K b) 438 K c) 110 K

16–37: 2500 m

16–39: solid and vapor, with no liquid

16–41: a) 6.9×10^{-3} L b) yes

16–43: 0.213 kg

16–45: 0.00123 g

16–47: 1.92 atm

16–49: a) 31 b) 8.41×10^3 N c) 7.80×10^3 N

16–51: a) 26.2 m/s b) 16.1 m/s, 5.44 m/s c) 1.74 m

16–53: 5.0×10^{27}

16–55: a) 303 mol/m^3 c) van der Waals

16–57: b) r_2 c) $r_1 = R_0/2^{1/6}$, $r_2 = R_0$, $r_1/r_2 = 2^{-1/6}$ d) U_0 e) $r \to \infty$

16–59: a) 517 m/s b) 299 m/s

16–61: b) 1.40×10^5 K, 1.01×10^4 K c) 6370 K, 459 K

16–63: a) 1.24×10^{-14} kg b) 4.16×10^{11} d) 2.95×10^{-6} m, no

16–65: a) $2R$ b) less

16–67: CO$_2$: 20.79 J/mol · K, 0.270; SO$_2$: 24.94 J/mol · K, 0.205;
H$_2$S, 24.94 J/mol · K, 0.039

16–71: b) $0.0420N$ c) $(2.94 \times 10^{-21})N$ d) $0.0297N$, $(2.08 \times 10^{-21})N$
e) $0.0595N$, $(4.15 \times 10^{-21})N$

16–73: 42.6%

16–75: a) 4.5×10^{11} m b) 703 m/s, 6.4×10^8 s (about 20 y), no
c) 1.4×10^{-14} Pa
d) about 650 m/s; evaporate f) 2×10^5 K; no

16–77: d) $T_c = 8a/27Rb$, $(V/n)_c = 3b$ e) $p_c = a/27b^2$ f) 8/3
g) 3.28, 3.44, 4.35

Chapter 17

17–1: b) 1330 J

17–3: b) 0

17–5: b) 0 for first process, -4.00×10^4 J for second process

17–7: 327 J

17–9: a) -1.15×10^5 J b) 2.55×10^5 J, out of gas c) no

17–11: a) 16.4 min b) 139 m/s = 501 km/h

17–13: a) absorbs b) 7200 J c) liberates 7200 J

17–15: a) yes b) no c) negative

17–17: a) 948 K b) 900 K

17–19: a) −605 J b) 0 c) yes, 605 J, liberates

17–21: $C_V = 65.47$ J/mol · K, $C_p = 73.78$ J/mol · K

17–23: a) 553 J b) 454 J

17–25: a) 495°C, 29.1 atm

17–27: b) 224 J c) $Q = 0$ d) −224 J

17–29: a) 4.50×10^4 Pa b) 345 J c) 0.818

17–31: a) 45.0 J b) liberates 65.0 J c) $Q_{ad} = 23.0$ J, $Q_{db} = 22.0$ J

17–33: a) $W_{ab} = 0$, $W_{bc} = p_c(V_c - V_a)$, $W_{ad} = p_a(V_c - V_a)$, $W_{dc} = 0$
b) $Q_{ab} = U_b - U_a$, $Q_{bc} = U_c - U_b + p_c(V_c - V_a)$,
$Q_{ad} = U_d - U_a + p_a(V_c - V_a)$, $Q_{dc} = U_c - U_d$
c) $Q_{abc} = U_c - U_a + p_c(V_c - V_a)$, $W_{abc} = p_c(V_d - V_a)$;
$Q_{adc} = U_c - U_a + p_a(V_c - V_a)$, $W_{adc} = p_a(V_d - V_a)$; abc, abc

17–35: -0.226 m^3

17–37: a) 4.32×10^{-4} m^3 b) 648 J c) 7.15×10^5 J d) 7.14×10^5 J
e) no substantial difference

17–39: 3.4×10^5 J/kg

17–41: b) 13.7 C°

17–43: a) 0.173 m b) 206°C c) 7.46×10^4 J

17–45: a) $Q = 300$ J, $\Delta U = 0$ b) $Q = 0$, $\Delta U = -300$ J c) $Q = 750$ J,
$\Delta U = 450$ J

17–47: a) $W = 738$ J, $Q = 2588$ J, $\Delta U = 1850$ J
b) $W = 0$, $Q = -1850$ J, $\Delta U = -1850$ J
c) 0

17–49: a) $W = -187$ J, $Q = -654$ J, $\Delta U = -467$ J
b) $W = 113$ J, $Q = 0$, $\Delta U = -113$ J
c) $W = 0$, $Q = 580$ J, $\Delta U = 580$ J

17–51: b) 3.60×10^3 J c) 0 d) 3.60×10^3 J e) 0.114 m^3

17–53: a) $W = nRT \ln[(V_2 - nb)/(V_1 - nb)] + an^2(1/V_2 - 1/V_1)$
b) i) 2.80×10^3 J ii) 3.11×10^3 J
c) 310 J; larger for ideal gas

Chapter 18

18–1:   a) 6500 J b) 0.34 = 34%

18–3:   a) 0.23 = 23% b) 12,400 J c) 0.350 g
d) 222 kW = 298 hp

18–5:   a) 0.25 = 25% b) 970 MW

18–7:   a) 453.2°C b) 1.99×10^6 Pa

18–9:   a) 1.62×10^4 J b) 5.02×10^4 J

18–11:   a) 767 W b) 7.27

18–13:   a) 215 J b) 378 K c) 0.391 = 39.1%

18–15:   a) 3.09×10^7 J b) 2.49×10^6 J

18–17:   a) −665 J, refrigerator b) −190 J, refrigerator
c) 285 J, engine

18–19:   b) $e \to 1$, $K = 0$; $e \to 0$, $K \to \infty$

18–21:   47.4 J/K

18–23:   −644 J/K

18–25:   a) 109 J/K b) N_2: 72.8 J/K; Ag: 103 J/K; Hg: 86.6 J/K

18–27:   6.74 J/K

18–29:   a) no b) 18.3 J/K c) 18.3 J/K

18–31:   142 m^2, yes

18–33:   a) 15 L b) 5

18–35:   a) 34.1 m^3 b) 8.02 m^3

18–37:   a) the product b) 23% c) better

18–39:   a) $p_1 = 1.00$ atm, $V_1 = 8.62 \times 10^{-3}$ m^3; $p_2 = 2.00$ atm,
$V_2 = 8.62 \times 10^{-3}$ m^3; $p_3 = 1.00$ atm, $V_3 = 1.31 \times 10^{-2}$ m^3
b) $1 \to 2$: $Q = 2183$ J, $W = 0$, $\Delta U = 2183$ J; $2 \to 3$: $Q = 0$,
$W = 1055$ J, $\Delta U = -1055$ J; $3 \to 1$: $Q = -1579$ J, $W = -451$ J,
$\Delta U = -1128$ J c) 604 J d) 604 J
e) $e = 27.7\%$, $e_{Carnot} = 50.0\%$

18–41:   10.5%

18–43:   a) 7.36×10^5 J b) 9.33×10^5 J c) 3.12×10^5 J d) 3.73
e) 3.71

18–45:   a) $a \to b$: $Q = -nRT_1 \ln r$, $W = -nRT_1 \ln r$, $\Delta U = 0$;
$b \to c$: $Q = nC_V(T_2 - T_1)$, $W = 0$, $\Delta U = nC_V(T_2 - T_1)$;
$c \to d$: $Q = nRT_2 \ln r$, $W = nRT_2 \ln r$, $\Delta U = 0$;
$d \to a$: $Q = -nC_V(T_2 - T_1)$, $W = 0$, $\Delta U = -nC_V(T_2 - T_1)$
c) $1 - T_1/T_2$, same, no

18–47:   b) 14 m/s = 50 km/h

18–49:   a) 122 J, 78 J b) 5.10×10^{-4} m^3
c) $p_b = 2.32 \times 10^6$ Pa, $V_b = 4.81 \times 10^{-5}$ m^3, $T_b = 771$ K;
$p_c = 4.00 \times 10^6$ Pa, $V_c = 4.81 \times 10^{-5}$ m^3, $T_c = 1333$ K;
$p_d = 1.47 \times 10^5$ Pa, $V_d = 5.10 \times 10^{-4}$ m^3, $T_d = 518$ K
d) $e_{Carnot} = 77.5\%$, $e = 61.1\%$

18–51:   10.5 J/K

18–55:   a) 150 J/K b) −120 J/K c) 30 J/K d) irreversible

18–57:   a) $e = 1 - \dfrac{1}{\gamma}\left[\dfrac{(T_c/T_a)^\gamma r^{\gamma(1-\gamma)} - 1}{(T_c/T_a) - r^{\gamma-1}}\right]$ b) 70.8%

Chapter 19

19–1:   a) 1.2 m/s b) 0.31 m c) a): unchanged; b): 0.15 m

19–3:   a) 0.439 m b) 5.25×10^6 Hz

19–5:   a) 17.2 m, 0.0172 m b) 74.0 m, 0.0740 m

19–7:   a) $f = 25.0$ Hz, $T = 0.0400$ s, $k = 19.6$ rad/m
b) $y(x, t) = (0.0700 \text{ m}) \sin 2\pi\left(\dfrac{t}{0.0400 \text{ s}} + \dfrac{x}{0.320 \text{ m}}\right)$
c) −0.0495 m d) 0.0050 s
e) 0.0200 s

19–9:   c) +x-direction for both
d) $v_y(x, t) = -\omega A \sin(\omega t + kx)$, $a_y(x, t) = -\omega^2 A \cos(\omega t + kx)$

19–11:   a) 4.0 mm b) 0.040 s c) 0.14 m, 3.6 m/s d) 0.24 m,
6.0 m/s e) no

19–13:   b) +x-direction

19–15:   a) 16.3 m/s b) 0.136 m c) both increase by $\sqrt{2}$

19–17:   0.390 s

19–19:   a) 1.33×10^{10} Pa b) 9.47×10^{10} Pa

19–21:   90.8 m

19–23:   a) −82°C b) no

19–25:   0.208 s

19–27: $Y/900$

19–29:   a) 16.0 m b) 3.29×10^{-8} m c) 4.55×10^{-5} m/s

19–31:   a) 20.0 Hz, 126 rad/s, 3.49 rad/m
b) $y(x, t) = (2.50 \text{ mm}) \sin [(126 \text{ rad/s})t - (3.49 \text{ rad/m})x]$
c) $y(t) = (2.50 \text{ mm}) \sin [(126 \text{ rad/s})t]$
d) $y(t) = (2.50 \text{ mm}) \sin [(126 \text{ rad/s})t - 3\pi/2]$
e) 0.314 m/s f) 0, −0.314 m/s

19–33:   a) $(7L/2)\sqrt{\mu_1/F}$ b) no

19–35:   a) $4\pi^2 F \Delta x/\lambda^2$

19–37:   a) 1, 0; 2, +; 3, −; 4, 0; 5, −; 6, +
b) 1, −; 2, +; 3, −; 4, +; 5, −; 6; 0
c) (a): answers would reverse sign; (b): no change

19–39:   c) C/B

19–41:   b) k decreases by factor of $2\sqrt{2}$; ω decreases by factor
of $\sqrt{2}$

19–43: $\alpha = (v_1^2 - v_2^2)\rho/Y\Delta T$

19–45:   b) $(1/2)\mu\omega^2 A^2 \cos^2(\omega t - kx)$ e) $(1/2)Fk^2A^2 \cos^2(\omega t - kx)$

19–47: $\pi/\omega\sqrt{2}$

Chapter 20

20–3:   a) $(1.33 \text{ m})n$, $n = 0, 1, 2, 3, \ldots$
b) $(1.33 \text{ m})(n + 1/2)$, $n = 0, 1, 2, 3, \ldots$

20–5:   a) $y(x, t) = (0.850 \text{ cm}) \sin [2\pi x/(0.300 \text{ m})] \cos [2\pi t/(0.0750 \text{ s})]$
b) 4.00 m/s c) 0.688 cm

20–9:   a) 96.0 m/s b) 461 N

20–11:   a) 45.0 cm b) no

20–13:   3.00 m, 16.0 Hz b) 1.00 m, 48.0 Hz c) 0.750 m, 64.0 Hz

20–15:   a) *fundamental:* displacement node at 0.60 m, pressure
nodes at 0 and 1.20 m; *first overtone:* displacement nodes
at 0.30 m and 0.90 m, pressure nodes at 0, 0.60 m, and 1.20
m; *second overtone:* displacement nodes at 0.20 m, 0.60 m,
and 1.00 m, pressure nodes at 0, 0.40 m, 0.80 m, and 1.20 m
b) *fundamental:* displacement node at 0, pressure node at
1.20 m; *first overtone:* displacement nodes at 0 and 0.80 m,
pressure nodes at 0.40 m and 1.20 m; *second overtone:*
displacement nodes at 0, 0.48 m, and 0.96 m, pressure nodes
at 0.24 m, 0.72 m, and 1.20 m

20–17:   506 Hz, 1520 Hz, 2530 Hz

20–21:   a) 172 Hz b) 86 Hz

20–23:   a) 614 Hz b) 1.23 kHz

20–25:   c) yes

20–27:   c) $2A$, $2A\omega$, $2A\omega^2$

20–29:   a) 20.0 m/s b) four times per second

20–31:   a) stopped b) $n = 7$, $n = 9$ c) 43.9 cm

20–33:   a) $v/2L$, v/L, $3v/2L$

20–35:   a) 0, L b) 0, $L/2$, L d) no

20–37:   a) 375 m/s b) 1.39 c) 0.8 cm

20–39:   a) 148 N b) 26%

20–41:   a) 0.246 m b) 56.3°C

20–43:   a) 0.026 m, 0.534 m, 1.27 m, 2.71 m, 9.01 m
b) 0.262 m, 0.861 m, 1.84 m, 4.34 m, ∞ c) 86 Hz

20–45:   a) 99.4 N c) −4.25 Hz, falls

Chapter 21

21–1: a) 0.344 m   b) 1.2×10^{-5} m   c) 6.9 m, 50 Hz

21–3: a) 7.78 Pa   b) 77.8 Pa   c) 778 Pa

21–5: a) 1.94 Pa   b) 4.58×10^{-3} W/m^2 c) 96.6 dB

21–7: 4.36×10^{-12} W/m^2, 6.40 dB

21–9: 14.0 dB

21–11: a) 20.0

21–13: a) 4 beats/s   b) 3.0×10^{-8} m, 0

21–15: 267 Hz

21–17: a) 0.798 m   b) 0.922 m   c) 431 Hz   d) 373 Hz

21–19: a) 302 Hz   b) 228 Hz

21–23: a) 0.0114 Pa   b) 7.51×10^{-9} m   c) 63.0 m

21–25: a) $I = 2\pi^2 \sqrt{\rho B} \, f^2 (\Delta R)^2$ b) $P = 8\pi^3 \sqrt{\rho B} \, f^2 R^2 (\Delta R)^2$

c) $A = (R/d)\Delta R$, $p_{max} = 2\pi \sqrt{\rho B} \, (Rf/d)\Delta R$,

$I = 2\pi^2 \sqrt{\rho B} \, (fR/d)^2 (\Delta R)^2$

21–27: a) 6.74 cm   b) 147 Hz

21–29: b) 2.0 m/s

21–31: c) 56.8 m/s

21–33: a) greater, 109 Hz   b) greater, 218 Hz

21–37: a) 180°   b) A: 3.98×10^{-6} W/m^2, 66.0 dB; B: 5.31×10^{-7} W/m^2, 57.2 dB
c) 1.60×10^{-6} W/m^2, 62.0 dB

PHOTO CREDITS

Cover Images   PhotoDisc, Inc.
FM   Hugh D. Young; John P. Surey.

Chapter 1 Opener: Jet Propulsion Laboratory/NASA Headquarters; 1-01a: John G. Ross/Photo Researchers, Inc.; 1-01b: NASA Headquarters; 1-03: ND-Viollet/Roger Viollet/ Liaison Agency, Inc.; U1-01: Jet Propulsion Laboratory/NASA Headquarters; U1-02: Addison Wesley Longman, Inc.

Chapter 2 Opener: NASA/Dryden Flight Research Center; 2-18: Richard Megna/Fundamental Photographs; U2-01: Tim Davis/Photo Researchers, Inc; U2-02: Michael Newman/ PhotoEdit; U2-03: The Military Picture Library/Corbis; U2-04: Wesley Hitt/Liaison Agency, Inc.

Chapter 3 Opener: Garmin International, Inc.; 3-14: Richard Megna/Fundamental Photographs; 3-16: Richard Megna/ Fundamental Photographs; U3-01: Stephen Dalton/Animals Animals/Earth Scenes; U3-02: Dean Abramson/Stock Boston; U3-03: Jack Fields/Photo Researchers, Inc.; U03-04: Eagles Aerobatic Flight Team.

Chapter 4 Opener: Jonathan Daniel/Allsport Photography (USA), Inc.; 4-113UL: Mark M. Lawrence/The Stock Market; 4-113UR, LL: John McDonough/Time Inc. Magazines/Sports Illustrated; U4-01: Wayne Eastep/Tony Stone Images; U4-02: Peter Beattie/Liaison Agency, Inc.; U4-03: James H. Robinson/ Animals Animals/Earth Scenes; U4-04: Steve Mason/ PhotoDisc, Inc.

Chapter 5 Opener: David Tipling/Tony Stone Images; 5-32a: NASA Headquarters; 5-32b: Helen Hansma, University of California at Santa Barbara; 5-32c: Frank Whitney/The Image Bank; 5-32d: Anglo-Australian Observatory, Photograph by David Malin; U5-01: NASA/Johnson Space Center; U5-02: David Madison/Tony Stone Images; U5-04: Didier Klein/Vandystadt/Allsport Photography (USA), Inc.

Chapter 6 Opener: Sami Sarkis/PhotoDisc, Inc.; U6-01: Dave Nagel/Liaison Agency, Inc.; U6-02: Michael Newman/ PhotoEdit; U6-03: Courtesy of Jeffrey H. Whitesell/Airliners of America.

Chapter 7 Opener: David Ball/Tony Stone Images; U7-01: Dennis O'Clair/Tony Stone Images; U7-02: Rudi Von Briel/PhotoEdit; U7-03: Paul A. Souders/Corbis; U7-04: Jeff Greenberg/Stock Boston.

Chapter 8 Opener: Addison Wesley Longman, Inc.; 8-04: Mike Powell/Allsport Photography (USA), Inc.; 8-21: Richard Megna/Fundamental Photographs; 8-23: John V.A.F. Neal/Photo Researchers, Inc.; 8-25: NASA Headquarters; 8-29: Anglo-Australian Observatory, Photograph by David Malin; U8-01: David Madison/Tony Stone Images; U8-02: Tony Freeman/PhotoEdit; U8-03: David Leah/Tony Stone Images.

Chapter 9 Opener: Stuart Dee/The Image Bank; 9-03b: Gorcher/Gorcher; 9-27: Smithsonian Astrophysical Observatory; U9-01: Blair Seitz/Photo Researchers, Inc.; U9-02: ©Jose Azel/Aurora/PNI; U9-03: NASA/Johnson Space Center.

Chapter 10 Opener: NASA/Johnson Space Center; 10-11: ©The Harold E. Edgerton 1992 Trust, courtesy of Palm Press, Inc.; U10-01: John A. Rizzo/PhotoDisc, Inc.; U10-02: The Goodyear Tire & Rubber Company; U10-03: Lori Adamski Peek/Tony Stone Images; U10-04: Gerard Lacz/Natural History Photographic Agency.

Chapter 11 Opener: Fred Maroon/Photo Researchers, Inc.; U11-01: NASA/Johnson Space Center; U11-02: E.B. Graphics/Liaison Agency, Inc.; U11-03: Paul Avis/Liaison Agency, Inc.

Chapter 12 Opener: Jerry Lodriguss/Jerry Lodriguss; 12-03L: NASA Headquarters; 12-03R: Jet Propulsion Laboratory/NASA Headquarters; 12-15b: Smithsonian Astro-

physical Observatory; 12-22: Space Telescope Science Institute; U12-01: George Hall/Corbis; U12-02: NASA/Goddard Space Flight Center; U12-03: Jet Propulsion Laboratory/NASA Headquarters; U12-04: NASA/Johnson Space Center.

Chapter 13 Opener: David Young-Wolff/PhotoEdit; 13-25L: AP/Wide World Photos; 13-25R: AP/Wide World Photos; U13-01: American Diagnostic Corporation; U13-02: Michael Newman/PhotoEdit.

Chapter 14 Opener: Dewitt Jones/Corbis; 14-04: Courtesy of Cenco; 14-13: Otto Rogge/The Stock Market; 14-20a–d: From Sears, Zemansky, and Young, *College Physics*, 7th ed., ©1991 by Addison-Wesley Publishing Company. Reprinted by permission of Addison Wesley Longman, Reading, Massachusetts; 14-27a–c: From Sears, Zemansky, and Young, *College Physics*, 7th ed., ©1991 by Addison-Wesley Publishing Company. Reprinted by permission of Addison Wesley Longman, Reading, Massachusetts; 14-29: ©The Harold E. Edgerton 1992 Trust, courtesy of Palm Press, Inc.; U14-01: Jack Vearey/Tony Stone Images; U14-02: Paul Chesley/Tony Stone Images.

Chapter 15 Opener: Akira Fujii; 15-04a: Courtesy of Cenco; 15-08: Marshall Henrichs/Marshall Henrichs; 15-11: Richard Megna/Fundamental Photographs; 15-14: Lockheed Martin Missiles & Space Company, Inc., Russ Underwood, photographer; 15-20a: John Raffo/Photo Researchers, Inc.; 15-20b: Texas Instruments Incorporated; U15-01: Jim Ross/NASA/ Dryden Flight Research Center; U15-02: Paul Seheult/Corbis; U15-03: Bill Bachmann/PhotoEdit; U15-04: Nancy Rogers/ Nancy Rogers.

Chapter 16 Opener: Gary Beck/Albuquerque Convention and Visitors Bureau; 16-07: Park Scientific Instruments; 16-21: Royal Observatory, Edinburgh/Anglo-Australian Observatory; 16-24: Anglo-Australian Observatory, Photograph by David Malin; U16-01: Jack Hollingsworth/PhotoDisc, Inc. U16-02: Keith Brofsky/PhotoDisc, Inc.; U16-03: PhotoDisc, Inc.

Chapter 17 Opener: Richard A. Cooke III/Tony Stone Images; 17-01: John P. Surey; 17-02a: Photri-Microstock, Inc.; 17-02b: AP/Wide World Photos; 17-24: Thomas Eisner and Daniel Aneshansley, Cornell University; U17-01: Stephen Frisch/Stock Boston; U17-02: John Kelly/Tony Stone Images; U17-03: Mark Adams Photography/Liaison Agency, Inc.; U17-04: Beaura Katherine Ringrose.

Chapter 18 Opener: Myrleen Ferguson/PhotoEdit; 18-14: Paul Silverman/Fundamental Photographs; 18-17b: Marvin Koner/Comstock; 18-18: Paul Gipe & Assoc.; U18-01: Mark Richards/PhotoEdit; U18-02: Mark Wagner/Tony Stone Images.

Chapter 19 Opener: U.S. Geological Survey, Denver; U19-01: Paul Morrell/Tony Stone Images; U19-02: R. Dolan/ Camerique/H. Armstrong Roberts

Chapter 20 Opener: Martin Dohrn/Science Photo Library/ Photo Researchers, Inc.; 20-01: Reproduced from *PSSC Physics*, 2nd ed. (1965), D.C. Heath & Company with Educational Development Center, Inc., Newton, Massachusetts; 20-05a–d: Richard Megna/Fundamental Photographs; 20-09M: Steinway & Sons; U20-02: Blair Seitz/Photo Researchers, Inc.; U20-03: National Optical Astronomy Observatories; U20-04: John Bacchus/Pearson Education Corporate Digital Archive; U20-05: Martin Bough/Fundamental Photographs.

Chapter 21 Opener: Brian Smith/Liaison Agency, Inc.; 21-15c: Leonard M. Weinstein/NASA/Langley Research Center; U21-01: Raytheon Aircraft Co.; U21-02: Dennis O'Clair/Tony Stone Images; U21-03: Robert A. Hoover/NASA/Dryden Flight Research Center.

INDEX